Proceedings Institution of Mechanical Engineers

International Conference

Vibrations in Rotating Machinery

13–15 September 1988
Heriot-Watt University
Edinburgh

Sponsored by
Institution of Mechanical Engineers

Co-sponsored by
Japan Society of Mechanical Engineers
Chinese Mechanical Engineering Society
Canadian Society for Mechanical Engineering
International Federation for the Theory of Machines and Mechanisms
Verein Deutscher Ingenieure
Koninklijk Instituut van Ingenieurs
Société des Ingénieurs et Scientifiques de France
Associazione Nationale di Meccanica
Vibration Institute

MechE Conference 1988–7

MEP Published for the Institution of Mechanical Engineers by
Mechanical Engineering Publications Limited

First published 1988

ISBN 0 85298 676 9

Printed by Waveney Print Services Ltd. Beccles, Suffolk.

Contents

IDENTIFICATION

PUMPS

GEARS

INSTABILITY

DYNAMIC ANALYSIS

LATE PAPERS

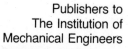

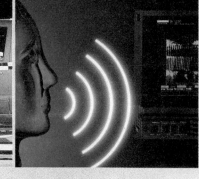

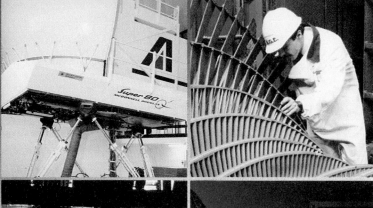

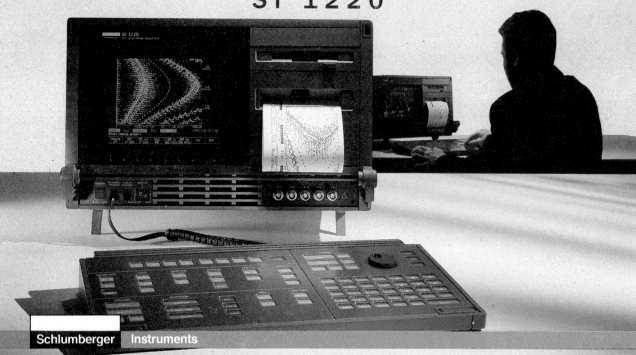

C318/88

Anti-earthquake considerations in rotordynamics

Y HORI, DrEng
Department of Mechanical Engineering, University of Tokyo, Tokyo, Japan

SYNOPSIS The influence of earthquakes on the safety of rotating machinery, such as sodium pumps in nuclear power plants, has been investigated in recent years, although the number of papers are still not very many inspite of the importance of the problem. Some of those papers will be reviewed.
Also, in connection with large turbo-generators, possible failure of a thrust bearing and possible earthquake-induced oil-whip of a generator rotor will be pointed out.
The importance of this kind of problems is and will be growing with the general trend of increase in the density of industrial facilities in many parts of the world.

1 INTRODUCTION

Earthquakes take place in many parts of the world, such as Japan, New Zealand, California (U. S.A.), Mexico, Chile, Italy, Greece, Yugoslavia, Rumania, Turkey, Central Asia (U.S.S.R.), Iran, China, India, etc.

They are sometimes so violent that many buildings are destroyed. Consequently, anti-earthquake designs of structures such as high-rise buildings, long bridges, etc. have been investigated extensively, especially since the advent of the computers. As the results of this, many high buildings, for example, are now constructed even in earthquake-ridden countries.

It seems that anti-earthquake designs of rotating machinery are equally important. For minimum social unrest after major earthquakes, electric power stations are, for example, certainly one of the most important facilities which we wish to survive such earthquakes. In this respect, anti-earthquake designs of rotating machinery such as turbo-generators, vertical multistage pumps, etc. in power plants are very important.

However, not very many papers have been published so far in this field. Survey of some of them with some additions of my own considerations [Sections 3 and 4.2] will be presented in this paper.

For reference, ground acceleration and some other data of three famous recorded earthquakes are given in Table 1. It should be noted that the acceleration on a building floor is often 2 or 3 times bigger than the values given in the table, depending on the vibration transimissibility of the building.

2 SEISMIC RESPONSE OF ROTORS IN LATERAL DIRECTION

Some recent papers on seismic responses of rotors in the lateral direction will be summarized. The topic will be confined to the problems of rotors supported by fluid film bearings.

2.1 Linear Responses

Lund [1] presented a mathematical formulation of a linear analysis of the response behaviour of a flexible rotor with flexible, damped supports. Effects of gyroscopic moment, internal damping, etc. were included. The supports were assumed to have linear characteristics which were allowed to be non-symmetric and frequency dependent. In this paper, he considered the response of a rotor to the foundation excitation (a shock pulse, a harmonic and a random excitation) and those by random mass unbalance excitation. As numerical examples, he calculated the transient response of a compressor rotor caused by a foundation shock pulse and also a random response (spectral density distribution of the amplitude) of a rotor to a random input (random vertical acceleration from the foundation of a uniform specral density over the frequency range 0-1000 Hz). Assuming Gaussian process, he calculated the probability that the response exceeds a certain given limit.

Measured Place, Year (Name of Earthquake)	Direction	Acceleraation m/s/s	Velocity m/s	Displacement m
El Centro, Calif., 1940 (Imperial Valley Earthq.)	N-S	3.42	0.342	0.181
	E-W	2.10	0.398	0.263
	U-D	2.06	-	-
Taft, Calif., 1952 (Kern County Earthq.)	N-S	1.77	0.0774	0.0226
	E-W	1.75	0.0934	0.0361
	U-D	1.03	-	-
Sendai, Japan, 1978 (Miyagi-Ken-Oki Earthq.)	N-S	2.58	0.355	0.140
	E-W	2.02	0.263	0.0759
	U-D	1.53	-	-

Table 1 Some Data of Earthquakes

Shimogo, Aida and Nakano [2] studied the seismic response of a flexible rotor supported by oil film bearings, using linealized oil film coefficents and three models of rotors: a rigid rotor, a flexible rotor with distributed mass and a flexible rotor with a lumped mass (Fig. 1). The gyroscopic effect was considered. The theory was verified by experiments on a test rig mounted on a sinusoidally vibrating table.

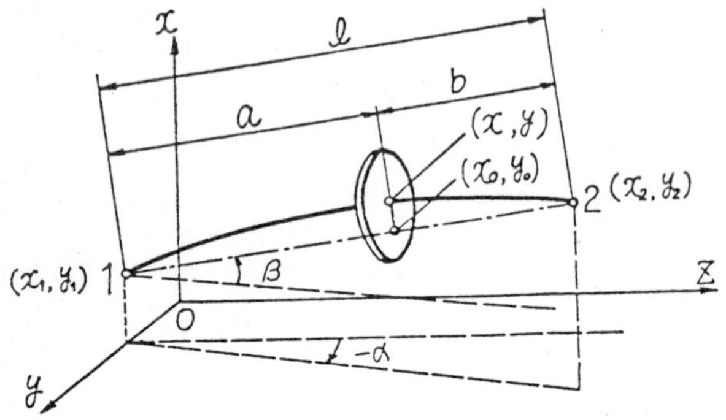

Fig 1 Analytical model of the flexible rotor with a lumped mass

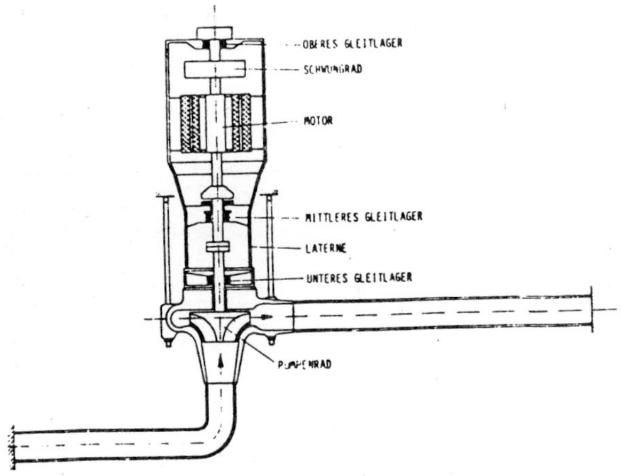

Fig 2a Sketch of the model

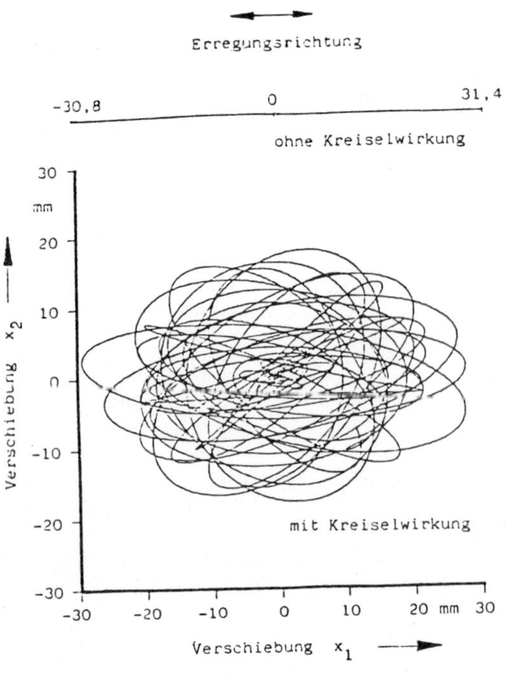

Fig 2b Locus of the upper end of the model for 15s

In calculations, an excitation (horizontal direction only) of a stationary Gaussian random process with a dominant frequency was given to the bearing pedestals. A power spectral density was assumed for the input acceleration (peak frequency = 2.62 Hz, r.m.s. value of acceleration = 196 gal) and those of the responses of the rotors (resonance frequencies of the flexible and rigid rotors = 15Hz & 22Hz) were calculated.

They found that the rotor accelerations and bearing loads in case of the flexible rotors were very much higher than those in case of the rigid rotor. On the Gaussian assumption, they obtained the probabilities that the bearing load exceeds the maximum allowable limit and that the rotor amplitude exceeds the gap between the rotor and the stator in their cases. They also made a case on the response of a multi-span rotor.

Klement, Merker and Schmied [3] discussed the influence of gyroscopic effect of a rotor on the vibrations of a vertical centrifugal pump during an earthquake. The numercal calculations using an artificial seismic wave showed that, besides the vibrations in the direction of the seismic wave, those in the transverse direction also result as shown in Fig. 2 if the gyroscopic effect is taken into consideration. However, the maxima of the displacement and the force remain almost the same if mechanical property of the support of the pump is isotropic. If it is strongly anisotropic (e.g. due to piping), the maxima will change and the neglect of the gyroscopic effect will give a wrong result.

2.2 Nonlinear Response (including contact problems)

In a paper on nonlinear dynamics of flexible multi-bearing rotors, Adams [4] pointed out the importance of the question whether or not the bearings in large power plant equipment are adequately designed to survive a major earthquake. He proposed a time-transient modal integration scheme to calculate the response of a full rotor-bearing-support structure in such cases.

Shimogo, Yoshida and Kazao [5] studied the seismic response of a vertical pump consisting of a water lifting pipe, an impeller at the bottom of the pipe and a driving shaft supported by synthetic rubber water bearings inside the pipe. Fig. 3 shows a simplified model of the pump. Linear characteristics of the bearing is assumed and the effect of interaction between the driving shaft and the water lifting pipe was included in the analysis. Assuming a stationary Gaussian

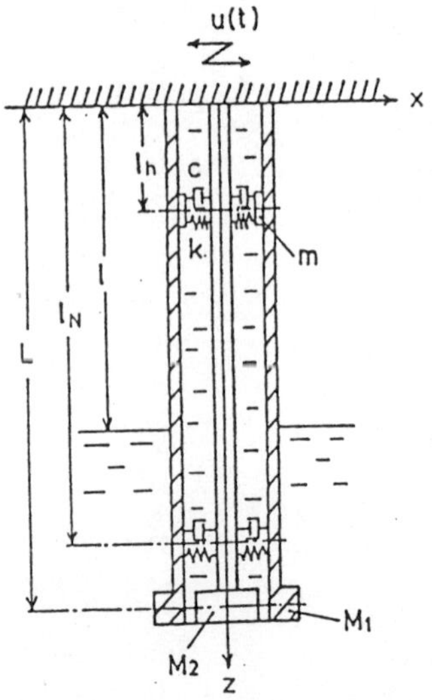

Fig 3 Simplified model of vertical shaft pump

horizontal input, they calculated the r.m.s. response of the shaft and the pipe and examined especially the effect of a snubber, provided at a certain height of the pipe, on the impact phenomena at the runner seal. From numerical studies, they found that the snubber reduces the bending stresses in the shaft and the pipe but increases the contact probability at the seal.

Kaga, Kikuchi, Matsushita, Takagi and Furudono [6] introduced a new "quasi-modal" method for the nonlinear analysis of seismic response of a rotor on the basis of modal synthesis and the substructure method.

They carried out vibration experiments and analysis on a vertical pump shown in Fig. 4, taking bearing nonlinearities into consideration. The pump consisted of a rotating shaft with an impeller, an inner casing and an outer casing and the shaft was supported by a ball bearing at the top and a water bearing at the bottom. The pump was excited by a sinusoidal and an earthquake wave and the time history responses were measured and calculated, taking nonlinearities such as contacts or collisions at the water bearing into consideration.

Calculations and experiments, which agree well, show that pulse-like peaks (collision) appear in bearing load curves corresponding to the seismic wave, especially when the rotational speed is low.

Matsushita, Takagi and Kikuchi [7] presented a general treatment of the seismic response of a rotor-bearing-casing system utilizing the modal method. They calculated the time history response to seismic excitations.

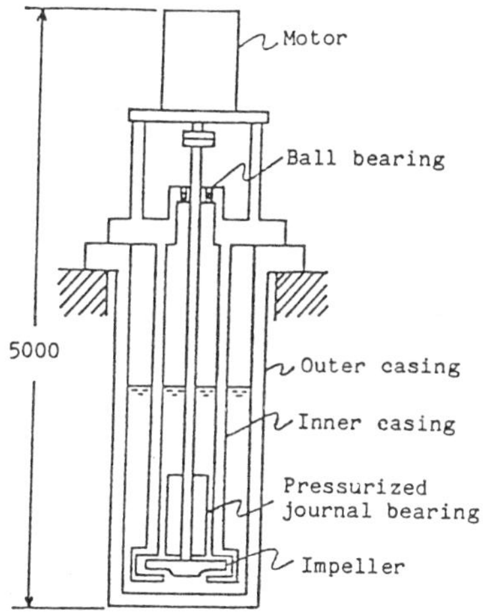

(a) Model pump

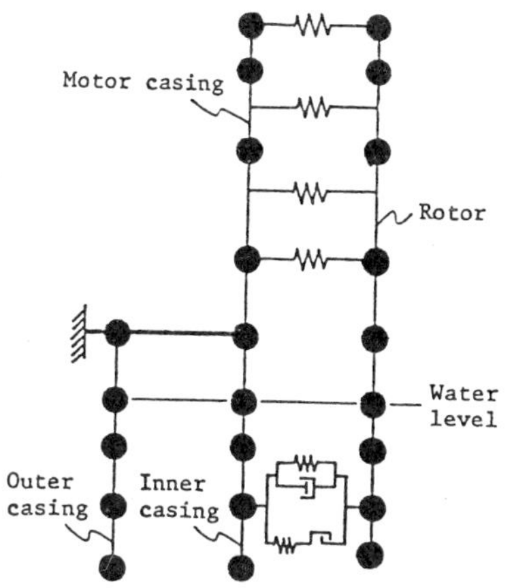

(b) Mathematical model

Fig 4 Model pump and mathematical model

They showed two numerical examples. The one concerns a horizontal, cylindrical, elastic rotor. Effects of the gyroscopic moment and the contact between the rotor and a deflection limiter are included. The other concerns a vertical, multistage pump shown in Fig. 5. Effects of the water bearings, the seal and the water virtual mass were included. While in case of small accelerations the rotor loci were almost on a straight line in the direction of the seismic excitation, they became two dimensional in case of large accelerations due to the contacts between the rotor and the seal ring .

Iwatsubo [8] calculated the response of a rotor in a vertical pump to seismic excitations. Then he considered the torsional vibrations of a driving shaft due to rubbing between an impeller and a wear ring caused by lateral vibrations of the shaft.

Choy, Padovan and Batur [9] investigated the effects of rub interaction between a rotor and its casing due to directional base excitations (e.g. seismic excitations) and a sudden increase of imbalance (e.g. blade loss). They carried out, using the modal method, a comprehensive analysis of the response of a complex rotor-bearing-blade-casing system under these extreme operating conditions. Component rub effects are included. Pattern recognition for signature analysis of rubs are also discussed.

It is interesting to notice in the numerical results that, in case of seismic excitation, the frequency of radial rub force is modulated by the seismic input frequency as shown in Fig. 6.

Subbiah and Rieger [10] extended Selvakumar and Sankar's time transfer matrix method to study the orbital behavior of non-linear rotor-bearing systems. The Houbolt time marching technique was adapted to model the system in the time domain.

They pointed out that the effect of pedestal random motions (= earthquake) may also be studied by their method.

2.3 Response of a Real Turbo-Generator Set to a Real Earthquake

Okabe, Aoki, Kashiwara and Matsushita [11]

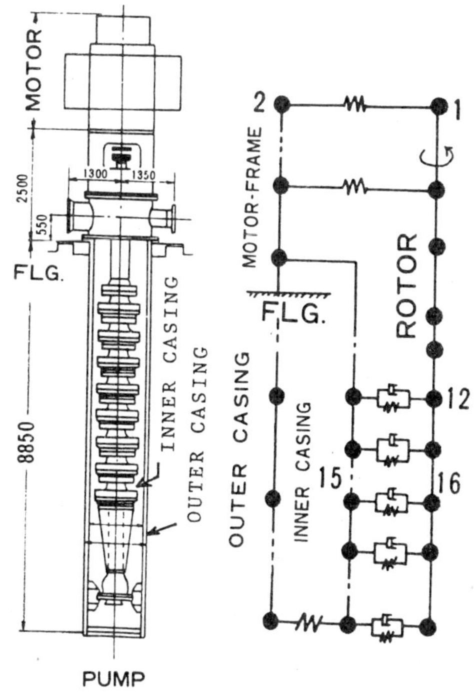

Fig 5 A multi-stage pump

calculated the seismic responses of a turbine-generator-bearing system of a real 784 MW set of a nuclear power plant to a seismic excitiation actually measured at the foundation during an earthquake. Peak acceleration was 21.8 gal.

The calculated results agreed well with the measured responses. Fig. 7 shows the comparison of measured and calculated shaft amplitudes at each bearings during the earthquake. Shaft amplitudes in normal operation are also given for reference.

2.4 An In-Situ Test of Earthquake Resistance

Chechenov [12] pointed out that, in case of complex structures like nuclear power plant equipment, it is important to check the dynamic characteristics of each part after everything has been assembled because it varies considerably during the assemblage of the equipment, pipeline connections, anti-earthquake fixing, special supports, etc.

He proposed a resonance method utilizing unbalance vibrators to obtain dynamics characteristics of units in an equipment in-situ quickly and economically.

3 SEISMIC RESPONSE OF ROTORS IN AXIAL DIRECTION

It should be noted that most rotating machines of a horizontal shaft supported by fluid film bearings are only very weakly protected against the axial disturbances. For example, all of a generator and several turbines of a large turbo-generator set is supported axially by only one tilting pad thrust bearing.

Let us consider an impulsive load on the tilting pad thrust bearing of a typical turbo-generator set under an earthquake excitation in the axial direction (Vibrational response will not be discussed here). Typical data of a nuclear set will be

total weight of rotating parts = 500 ton,
outer and inner diameter of
the thrust bearing = 80 cm and 66 cm,
total pad area = 1200 sqcm.

If we assume
earthquake acceleration = 0.3 g,
then
bearing load = 150 ton,
bearing pressure = 125 kgf/sqcm.

The transmissibity factor of a building on the machine floor is very often 2 or more, then these values become
bearing load = 300 ton,
bearing pressure = 250 kgf/sqcm.

These are tremendous values, although they are of short duration. The thrust bearing is usually designed for a much smaller load, because the thrust on the bearing is usually only 2 or 3 ton due to the pressure unbalance in turbines, etc.

It is recommended, therefore, to check the mechanical strength of pad supporting mechanisms and the transient tribological performance of pads of the bearing, if the set should be earthquake-proof.

4 EARTHQUAKE-INDUCED INSTABILITY OF ROTORS

Under certain conditions, an earthquake can trigger self-excited vibrations or whirling of a rotor.

4.1 Whirl due to Structural Damping

Self-excited whirl of rotors due to internal damping has been discussed by many investigators since Kimball [13]. It has been pointed out that two kinds of internal damping can be considered, namely, material damping and the so-called structural damping at shrink fits, joints, etc. and that in most cases the latter will be more responsible than the former for the occurence of self-excited whirl of rotors.

Tondl [A1] discussed the effect of material

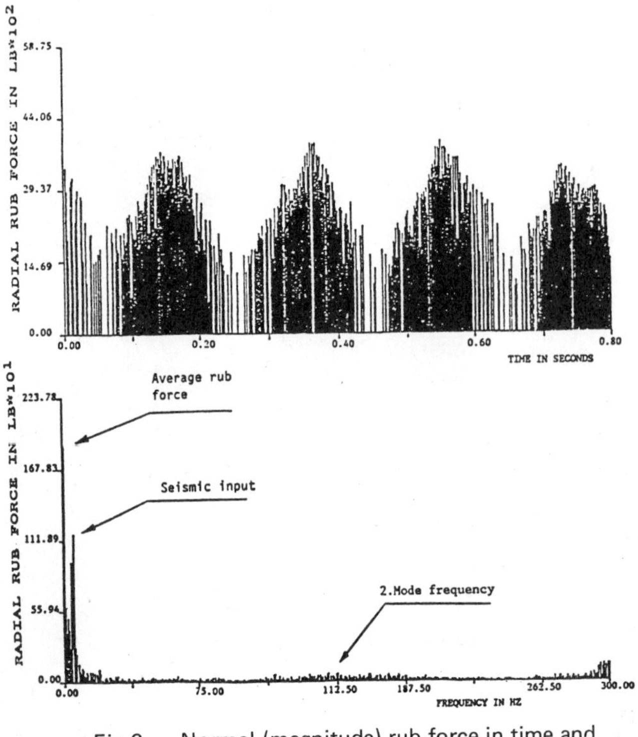

Fig 6 Normal (magnitude) rub force in time and frequency domain

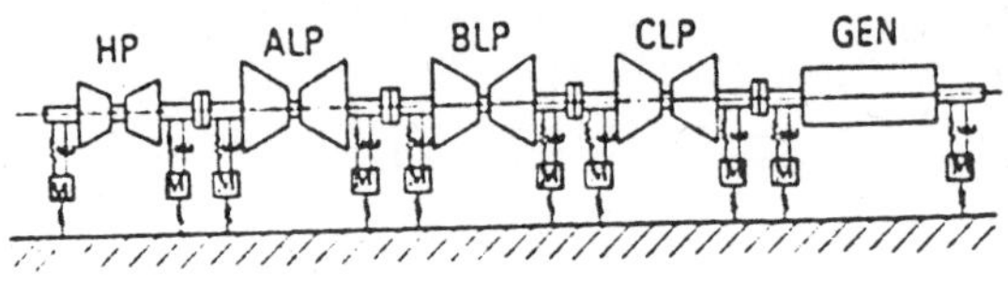

Fig 7a Turbine–generator set

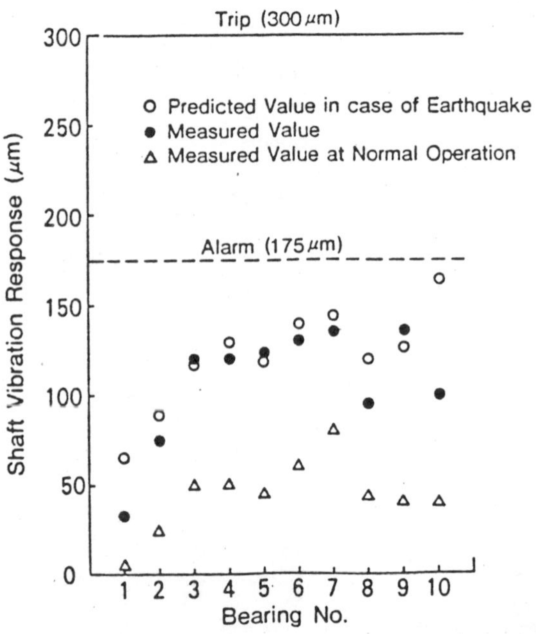

Fig 7b Shaft vibration response (maximum absolute displacement)

and structural damping on the stability of a rotating shaft in detail. In case of structural damping, he stated that as long as the shaft deformation is less than a certain limit and if no random deflection occurs which would cause the limit to be exceeded, dry friction will not assert itself. Conversely speaking, if a large enough deflection is given to a shaft, then the shaft can start unstable whirling.

Lund [14] discussed the destabilizing effect of friction in internal rotor joints with micro-slip. Several mechanism of friction, i.e. viscous damping, solid damping, dry friction and micro-slip friction, were analysed and compared.

He found, in case of micro-slip friction, that there is a certain critical amplitude below which the rotor is stable and above which unstable. This means that a disturbance in amplitude larger than the critical can cause an instability in a shaft.

These considerations suggest that a strong enough earthquake can cause an instability in a rotor with such structural damping, if it is running at speeds above the critical.

Nonami and Miyashita [15] studied experimentally the internal damping of various kind of rotors: one body rotors, shrink fit rotors, welded rotors, press fit rotors and bonded rotors. They found that the first three had less damping than the other and were more stable in running tests.

4.2 Oil Whip

Another kind of earthquake-induced instability, namely, earthquake-induced oil whip will be discussed in this section. This can occur in a rotor running smoothly at speeds higher than twice the critical. Large turbo-generators for fossil sets are typical examples of such rotors, because their critical speed is usually less than 1000rpm(e.g. 600rpm-900rpm) and they are operated at 3000 rpm or 3600 rpm. In such a case, a rotor which is running smoothly at the rated speed can be made to whip by a strong shock like that of an earthquake.

4.2.1 A Qualitative Explanation

In their early experimental studies on the stability limit of a rotor, Newkirk and Taylor [16] and Newkirk and Lewis [17] found that in some cases a slight shock was sufficient to trigger severe whipping in an otherwise stable rotor, but Pinkus [18] did not observe such a case. To understand the complicated effect of a shock on the stability of a rotor, it is helpful to distinguish the stability criterion for small vibrations of the rotor around its equilibrium point and the condition necessary for maintaining a violent oil whip [19].

It is well known that the oil whip (large amplitude whirling) can grow or continue to exist if the shaft speed is above twice the critical. But this is not a sufficient condition for the onset of the oil whip. A preliminary disturbance of a certain magnitude is neccesary for the oil whip to start.

On the other hand, it is known that the stability limit for small vibrations can be at any point below or above twice the critical speed, depending on the bearing pressure. The stability limit can be, at least theoretically, at any far point beyond twice the critical, if the bearing pressure is large enough.

Fig. 8 shows the case in which the stability limit for small vibrations (C) is far above twice the critical speed (B). Point (A) is the critical speed. In this case, the rotor becomes unstable when the shaft speed reaches (C) and then a violent whipping starts immediately after that as shown by an arrow. When the shaft speed is decreased, however, the rotor becomes stable again only after the shaft speed passes (B) as shown in the figure. It will be seen in such a case that the rotor can run smoothly at a point, say, (D), because it is still stable in the sense of small vibrations. But if a large enough shock is given to the rotor, then the rotor begins whipping immediately as shown by a broken line with an upward arrow.

In other words, a rotor with a large enough bearing pressure can run smoothly at speeds over twice the critical, unless a strong enough disturbance is given to the rotor. In Newkirk's cases, the shaft speeds were probably in the range between (B) and (C) and close enough to the stability limit (C) so that the slight shock was enough to cause whipping. In Pinkus' case, probably, either the shock was not strong enough or the speed was not in the range between (B) and (C).

To summarize, it can be said that a rotor running smoothly at a speed above twice the critical speed is stable only in the sense of metastability and can start whipping immediately if a strong enough shock is given to the rotor.

Most turbo-generator rotors seem to be running smoothly only in the sense of metastability at the rated speeds which are usually much higher than twice the critical. Such a design has been permitted only because we did not think of such a strong shock as can give a generato-rotor of 50 - 60 ton a sufficient effect. However, if we think of an earthquake, we realize that we must be a little more careful in this matter.

The ground acceleration during a strong earthquake can be 0.3 g or more and the acceleration on the generator floor can be 2 times these values or more. This means that the shock given to a generator-rotor by a strong earthquake can be comparable to the weight of the rotor. This seems to be big enough to cause whipping in a smoothly running rotor.

4.2.2 A Numerical Explanation

The discussion in the previous section will be confirmed numerically in the following. By numerical calculations, it is possible to determine qualitatively whether a certain rotor can maintain its stability in case of an earthquake of a certain strength or not.

For the sake of simplicity, we consider a symmetrical rotor composed of a mass and a massless shaft supported by oil film bearings as shown in Fig. 9. The governing equations will be as follows:

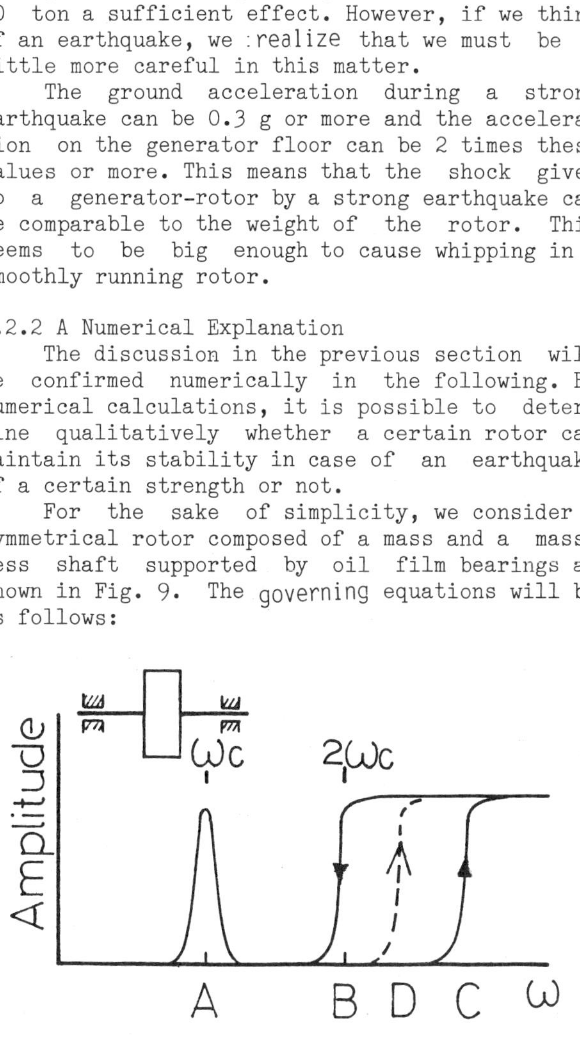

Fig 8 Effect of a shock on oil whip

$$m \ddot{x} + k (x - x_j) = m g - m \ddot{X}_b$$

$$m \ddot{y} + k (y - y_j) = - m \ddot{Y}_b$$

$$k (x - x_j) + 2 F_x = 0$$

$$k (y - y_j) + 2 F_y = 0$$

where $(\ddot{X}_b, \ddot{Y}_b)$ are the earthquake accelerations at the bearing pedestal, (x, y) and (x_j, y_j) are the co-ordinate of rotor and journal centre and (F_x, F_y) are the oil film force in x and y direction, respectively.

Starting from an equilibrium point of the journal center with a certain initial shock due to an earthquake (one cycle of a sinusoidal wave of a given acceleration), the locus of the journal centre was calculated step by step by the Runge-Kutta-Gill method. The Reynolds lubrication equation for a finite width bearing was integrated numerically at each step by means of a quick method utilizing the expansion of the pressure by orthogonal functions [20]. This method is 40 - 50 times faster than the usual finite difference method. The Reynolds boundary condition was used both at the outlet and inlet of the oil film.

Sample calculations were done for the following case.
slenderness ratio of the bearing:

$$L/D = 0.5$$

where L = bearing length, D = bearing diameter; bearing parameter:

$$Bp = 0.02$$

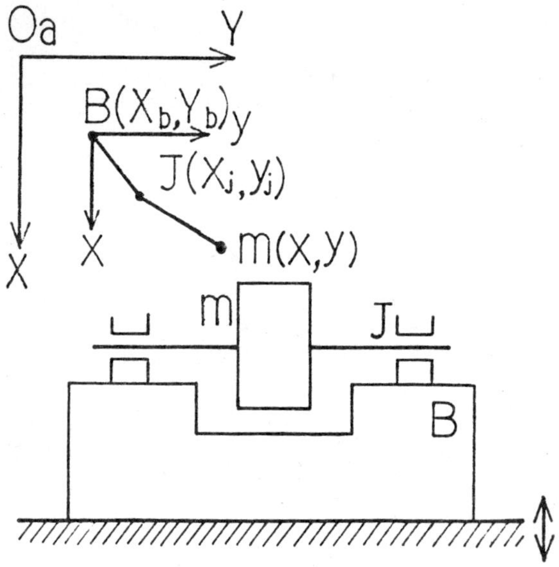

Fig 9 A model rotor

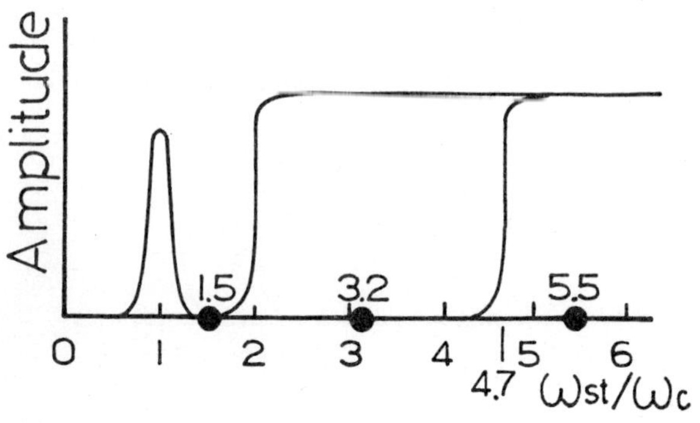

Fig 10 Shaft speed used in numerical calculations

stiffness parameter of the shaft:

$$(m g)/(c k) = 1.0$$

where Bp = [(ZL)/<3.14(W/g)>] sqrt(c/g) (r/c)**3, Z = viscosity, W = bearing load, c = clearance, r = D/2; normalized linear stability limit:

$$4.714$$

which is the ratio (the stability limit for small vibrations or the linear stasbility limit) / (the critical speed); and normalized shaft speed:

$$S = 1.5,\ 3.2,\ 5.5$$

where S = (rotating speed)/(critical speed), as shown in Fig. 10.

As a shock due to an earthquake at the bearing pedestal, we took one cycle of the sinusoidal waves in the horizontal direction of maximum accelerations of:

$$A = 0.05\ g,\ 0.1\ g,\ 0.3\ g$$

where g = acceleration of gravity. The circular frequency of the earthquake wave was assumed to be 0.8 times that of the rotor.

The results are shown in Fig. 11 and summarized in Table 2.

Since the non-dimensional stability limit for small vibrations in this case is S = 4.714, the case of S = 3.2 in Table 2 corresponds to that of (D) in Fig. 8. In this case, if the earthquake acceleration is small (A = 0.05 and 0.1), then the rotor remains stable. But if the acceleration is large (A = 0.3), then the rotor becomes unstable, as expected. By interpolation, it is possible to determine the exact allowable upper limit of the earthquake acceleration for stable running of the rotor. If we know the maximum acceleration expected, then it is possible to determine, by similar calculation, whether a particular machine is safe or not. Or, we can design a new machine which will not lose stability in case of an earthquake of certain expected strength.

In the case of S = 1.5, the rotor is stable for any value of acceleration as expected. In the case of S = 5.5 where the rotor should be unstable both in the sense of small vibrations and in that of large whirl, it is interesting to notice that the rotor seems to be circling along a small limit cycle when the initial disturbance is very small (A = 0.05). This may correspond to a limit cycle predicted by a nonlinear stability theory [21]; it is reported in the paper that if a nonlinearity of the second order of journal displacement and velocity in the oil film force

Table 2 Rotor Stability under an Earthquake

* S / A *	1.5	3.2	5.5
0.05 g	(a) stable	(b) stable	(c) limit cycle
0.1 g	(d) stable	(e) stable	(f) oil whip
0.3 g	(g) stable	(h) oil whip	(i) oil whip

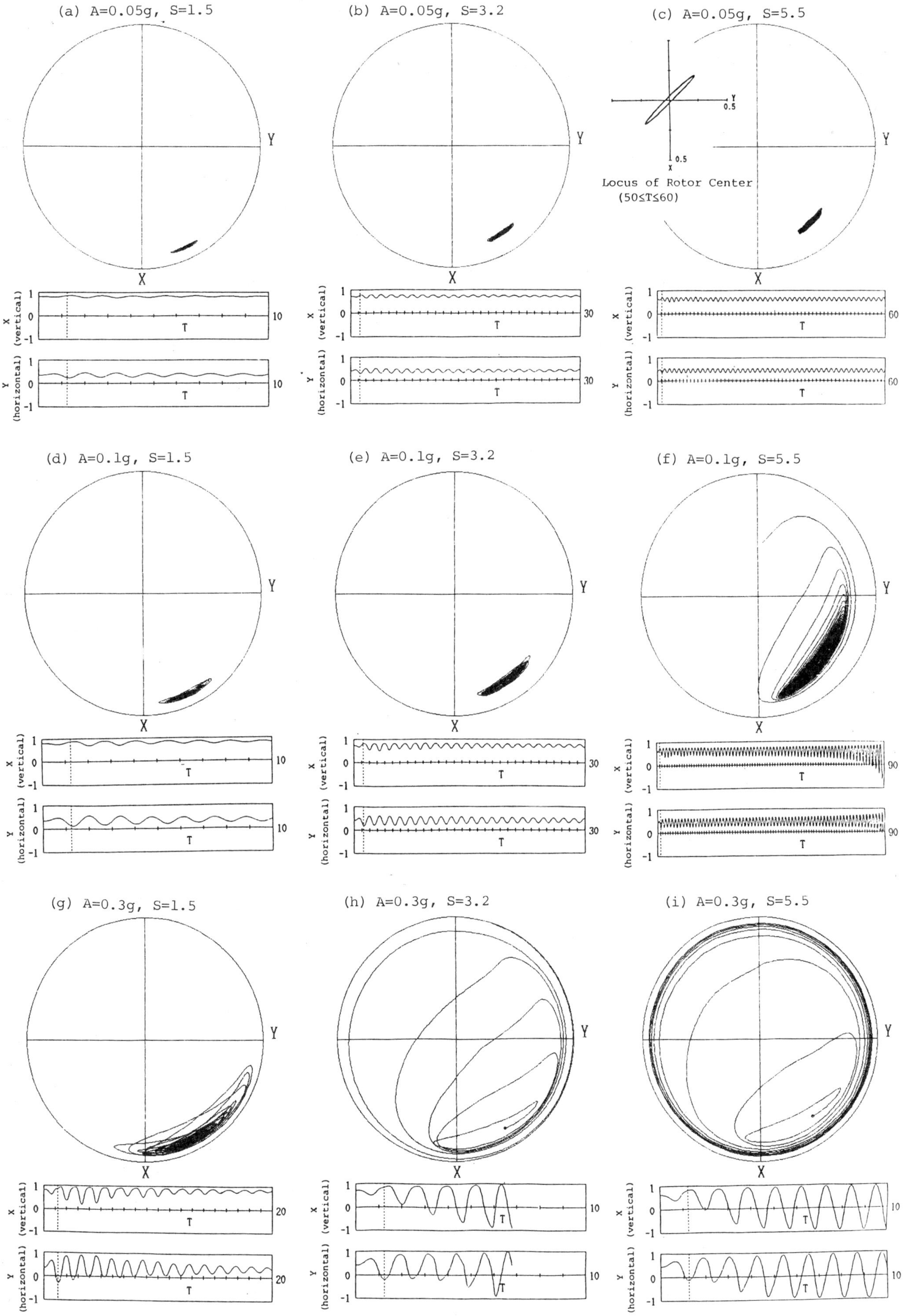

Fig 11 Journal loci (L/D = 0.5, $S = \omega/\omega_c$, $T = \omega_c t/2\pi$, ΔT = 1/50~1/500, linear stability limit; ω_{st}/ω_c = 4.714, disturbance; $\ddot{X}_b = 0$, $\ddot{Y}_b = A\sin(0.8\omega_c t)$ $[0 \leqslant t \leqslant 2\pi/0.8\omega_c]$, $\ddot{Y}_b = 0$ $[t \geqslant 2\pi/0.8\omega_c]$, quasi-Reynolds condition)

of the bearing is considered, then a small, stable limit cycle of the journal motion can exist in the unstable region predicted by the linear theory. It is said that the stability limit for the small limit cycle can be 15 - 20 % higher than the linear stability limit. In the present case, the percentage is 16.7 %.

In this paper, only the artificial sinusoidal shocks were considered. But it is possible to calculate the loci of journal and rotor centre during a real earthquake exactly in the same way, if the earthquake waves are given, and so to predict stability of the rotor under a real earthquake.

4.3 Simultaneous Effect of Structural Damping and Oil Film Force

If a rotor contains both the cause for structural damping instability and that for oil film instability (oil whip), then it is clear that they will cooperate, in parallel or series in time, to cause an whirling in the rotor.

5 CONCLUSION

Study on the response of rotating machinery to earthquakes has become increasingly important due to the increase in the density of industrial facilities in many countries. Some papers in this field were reviewed.

In connection with large turbo-generators, possible failure of thrust bearings and possible earthquake-induced self-excited whirl (friction whirl and oil whip) were pointed out. In case of earthquake-induced oil whip, it is possible to predict numerically whether a given rotor starts whipping at a given earthquake or not, if its acceleration data is available.

ACKNOWLEDGEMENT

The author thanks Dr. T. Kato, the University of Tokyo, for his great help in numerical calculations.

REFERENCES

Books

A1) TONDL, A.: "Some Problems in Rotor Dynamics", Chapman & Hall, 1965.

A2) GASCH, R. and PFUETZNER, H.: "Rotordynamik -- eine Einfuerung", Springer Verlag, 1975.

Papers

1) LUND,J.W.: "Response Characteristics of a Rotor with Flexible, Damped Supports", Dynamics of Rotors (IUTAM Symposium on Dyanamics of Rotors, August 1974, Lyngby, Denmark), Springer Verlag, 1975, pp 319-349.

2) SHIMOGO, T., AIDA, Y. and NAKANO, M.: "Seismic Response of a Flexible Rotor", Proceedings of IMechE Second International Conference on Vibrations in Rotating Machinery, September 1980, Cambridge, U.K., pp 321-326.

3) KLEMENT, H.-D., MERKER, H.-J. and SCHMIED, J.: "Einfluss der Kreiselwirkung bei Erdbebenrechnungen", VDI Berichte, No.496, 1983, pp 35-40.

4) ADAMS, M.L.: "Non-Linear Dynamics of Flexible Multi-Bearing Rotors", Journal of Sound and Vibration, Vol. 71, No.1, 1980, pp 129-144.

5) SHIMOGO, T., YOSHIDA, K. and KAZAO, Y.: "Seismic Response of Vertical Shaft Pump including Interaction between Shaft and Pipe", Proceedings of IFToMM International Conference on Rotordynamic Problems in Power Plants, September 1982, Rome, Italy, pp 439-448.

6) KAGA, M., KIKUCHI, K., MATSUSHITA, O., TAKAGI, M. and FURUDONO, M.: "Quasi-Modal Nonlinear Analysis for Seismic Response in Pump Rotors", Proceedings of IFToMM International Conference on Rotordynamics Problems in Power Plants, September 1982, Rome, Italy, pp 83-93

7) MATSUSHITA, O., TAKAGI, M. and KIKUCHI, K.: "Analysis of Rotor Vibrations Excited by Seimic Wave", Bulletin of JSME, Vol. 27, No. 224, February 1984, pp 278-288.

8) IWATSUBO, T.: "Evaluation of Cumulative Damage Failure of a Reactor Feed Vertical Pump under Seismic Excitation", Trans. of the 9th International Conference on Structural Mechanics in Reactor Technology (SMiRT), Lausanne, Switzerland, Auguat 1897, vol.K2 (Seismic Response Analysis of Nuclear Power Plant Systems), pp 1027-1032.

9) CHOY, F.K., PADOVAN, J. and BATUR, C.: "Rub Interactions of Flexible Casing Rotor Systems with Base Excitations", Rotating Machinery Dynamics, Proceedings of ASME Conference on Mechanical Vibration and Noise, September 1987, Boston, U.S.A., pp 477-484

10) SUBBIAH, R. and RIEGER, N.F.: "On the Transient Analysis of Rotor-Bearing Systems", Rotating Machinery Dynamics, Proceedings of ASME Conference on Mechanical Vibration and Noise, September 1987, Boston, U.S.A., pp 525-536.

11) OKABE, A., AOKI, S., KASHIWAHARA, K. and MATSUSHITA, O.: "Aseismic Vibration Estimation for Turbine-Generator Set", Proceedings of the KSME-JSME Vibration Conference, August 1987, Seoul, Korea, pp 379-387.

12) CHECHENOV, Kh.D. "Resonance Method of Testing the Eearthquake Resistance of Nuclear Power Plant Equipment", Energomashinostorenie, No. 8, 1986, pp 37-39.

13) KIMBALL, A.L.:"Internal Friction Theory of Shaft Whirling", General Electric Review, Vol. 27, No. 4, April 1924, pp244-251.

14) LUND, J.W.: "Destabilization of Rotors from Friction in Internal Joints with Micro-Slip", Proceedings of IFToMM International Conference on Rotordynamics, September 1986, Tokyo, Japan, pp 487-491.

15) NONAMI, K. and MIYASHITA, M.: "Internal Damping and Stability of Rotor-shaft Systems", Bulletin of JSME, Vol. 24, No. 187, January 1981, pp 200-207.

16) NEWKIRK, B.L. and TAYLOR, H.D.: "Shaft Whipping Due to Oil Action in Journal Bearings", General Electric Review, Vol. 28, No. 8, August 1925, pp 559-568.

17) NEWKIRK, B.L. and LEWIS, J.F.: "Oil-Film Whirl - An Investigation of Disturbances Due to Oil Films in Journal Bearings", Trans. ASME, Vol. 78, January 1956, pp 21-27.

18) PINKUS, O.: "Experimental Investigation of Resonant Whip", Trans. ASME, Vol. 78, July 1956, pp 975-983.

19) HORI, Y.: "A Theory of Oil Whip", Trans. of ASME, Series E, Journal of Applied Mechanics, June 1959, pp 189-198.

20) KATO, T. and HORI, Y.: "A Fast Method for Calculating Dynamic Coefficients of Finite Width Journal Bearings with Quasi-Reynolds Boundary Condition", (will be published in ASME Journal of Tribology.

21) MALIK, M. and HORI, Y.: "An Approximate Nonlinear Transient Analysis of Journal Bearing Response in Unstable Region of Linearized System", Proceeding of IFToMM International Conference on Rotordynamics, September 1986, Tokyo, Japan, pp 217-220

C257/88

Turbomachinery field balance techniques—three case histories

C JACKSON, BSME, FASME
Turbomachinery Consultant, Texas City, Texas, United States of America

Synopsis:Three case histories are shown for techniques in field balancing. The first is on a Methanol Syn Gas Steam Turbine, rated at 14,000 bhp at 11,000 rpm operating over the first two rigid mode criticals, i.e. cylindrical & pivotal. The second case is balancing an overhung impeller gas compressor rated at 9,000 bhp at 6,000 rpm operating immediately below it's first pivotal mode critical, using a coupling installed weight ring, without disassembly of the compressor. The third case describes a technique using the coupling assembly as a correction weight & determining the circumferential position in shaft mounting to counterbalance the rotor's residual unbalance.

1 CASE ONE-BALANCING PASS-OUT/CONDENSING TURBINE

This steam turbine operates on 600 psig(41.38 bar),700 deg.F.(371 C)steam with passout steam at 75 psig(5.2 bar) and condensing to 2-4 inches of mercury absolute(76 mm of Hg). The turbine drives two barrel type centrifugal compressors in a 1000 ton/day(991 tonnes/day) methanol plant. It has two trim balance disc, one at each end accessible without removing the turbine. This turbine had operated 13 years since an original field balance[2],in situ,using a two plane balance program, scope orbits[1] from proximity probes, and seismic bearing cap data from velocity sensors. Those records indicated that the H.P. inlet of the turbine was sensitive to calibration weights placed on the exhaust end balance plane. A single plane balance will correct both ends of this machine and the main thrust of the presentation will explain why. It should be noted that this turbine operates above two rigid mode resonances(criticals) and that the trim balance was deemed necessary to prevent an operating outage of about two weeks at a cost of $50,000 US dollars/day. Also, planned data gathering be taken in a manner allowing multi-plane techniques, if necessary.

The first Figure 1.1, shows the mode shapes of this turbine which are very useful in balancing many machines, as in many cases, it allows some guidance in modal placement of static and couple weights, though not necessary in this case. Below the mode diagram, the cross-section of the turbine exhaust end bearing/shaft/balance disc can be seen. A similar balance plane of drilled holes exists at the H.P. inlet end at a step in the shaft (used in 1972-2 plane).

Fig.1.2 shows the scope data of the shaft motion relative to the eddy current proximity probes attached to the bearing and observing a lissajous pattern from the dual probe mounted 90 degrees apart. This information depicting the "high spot" of the shaft relative to it's vibration sensor allows one to determine the correct quadrant of the balance plane to affect a positive correction, i.e. reduce the vibration. In reviewing the previous balance data, 13 years prior, 10 grams of mass placed at in the exhaust (coupling) end at a 5 inch(127 mm) radius had corrected over 1.5 mils(microns) peak-to-peak vibration at the steam inlet end. The sensitivity being 6.67 grams/mil (3.33 gm-mm/micron. Therefore; a straightforward balance shot in the exhaust end balance plane, without soloing the uncoupled turbine was considered; and would have been successful, but with 40% of the achieved results. The balance logic would have been correct but the rotor's sensitivity had increased almost 3:1 over 13 years.

2 CASE ONE- PROCEDURE

When a rotor is turning at slow speeds, the "heavy spot" of the residual imbalance is "in-phase" with "high spot" of the rotor as seen by proximity probes placed in cartesian coordinates observing this orbit(locus of high spots in rotation). These probes (sensors) only observe shaft outer surfaces in rotation (high spots). However, since the rotor's mass centre precesses forward in rotation and finally passes through it's first resonance to allow it to turn about it's true mass centre, the classic 180 degree(theoretical single mass-single degree of freedom) shift occurs and can be seen during the run-up. This places the "high spot" out-of-phase diametrically with the "heavy spot" and one would place a calibration weight at the "high spot". [see Figure 2.1] However, in this machine another critical speed is passed and the phase angles which were opposite at each end of the machine between the first and the second critical..are now back in phase again. The calibration weight will be placed 180 degrees opposite the "high spot".

After 13 years of continuous duty with some bruises along the way,e.g.trip-outs under full load; this rotor was slightly out of phase between the HP and LP end. Originally, this phase difference was negligible, but now a shift of about 50 degrees is indicated. The actual hole for the first calibration weight was taken from the uncoupled turbine run at a phase angle of 107 degrees.
This is nearest to Hole #4(Hole #4 is at 112.5 degrees) on the turbine exhaust end balance ring shown in Figure 2.2. Probably the most significant purpose of this paper is towards establishing a methodology in placing the first

calibration(trial) weight. It really should not be arbitrary as placing the weight on the existing "heavy spot" may (1)wreck the machine,(2) not allow a second data point,(3)encourage a security escort from the operating plant premises.

If the turbine rotor were operating at about 8,000 rpm,i.e. above the first critical, yet below the second; and, the phase angles are out-of-phase(say 170 degrees) it would seem fairly clear that the rotor is well into the pivotal (hour-glass-shaped) mode. The calibration weight would be placed at the actual phase angle measured,e.g.287 degrees, referring to Figs. 1.2,2.1,& 2.3-Exh.Vert.Z3,Run #1, which would have been hole #13(180 degrees from hole #4). However, since another critical(resonance) was passed, and the phase angles at each end of the turbine at 11,000 rpm were near in-phase,from Fig.2.3=287 vs 239, the calibration weight was placed 180 degrees opposite the phase angle indicated. On the keyphasor(once-per-rotation) triggered(blanked) oscilloscope (or) on the "Tracking Filter"(Vector Filter) display; the phase angle represents the angular number of degrees, against rotation, the "high spot" is from the vibration sensor(probe) being used(vertical,exhaust) at the precise moment in time that the rotor's timing mark(notch-in-shaft) passes under the keyphasor sensor(probe) which is mounted, in this case, to the exhaust end bearing shell. The logic of the orbit,time trace,and keyphasor "blanking" can be seen in Figure 2.4 for an example model,i.e. not the turbine in question-educational figure,relates to an example rotor as seen also in Fig. 2.1.

It should be noted that on the second run-up, the vibration level reduced significantly. The calibration weight used was two screws at 1/2"-13 NC x 1/2"(1.27 cm) length. These two screws were jammed, one-against-the-other, which is better than one screw at 1"(25.4 mm) from the ability to remain fixed in the hole. Given the opportunity, it is better to remove screws (which is done in the final step) rather than add screws since it is extremely difficult to "throw off a hole". The combined weight of the two screws is 14 grams, based on the experience from 13 years prior. Having no past experience with sensitivity (vibration response to an unbalance weight-mils/gm or gms/mil); using a force derived from the calibration weight,gms, and radial position,inches, and speed in rpm, a force equal to 1/10 the rotor weight can be determined. This should always give one a response yet not enough to be deleterious should the placement logic be in error. The choices in logic have 1 out of 2 results,i.e., one is generally 100 percent right or 100 percent wrong. The force in this choice was 438 lbs(199 kg) and for a rotor weighing 5,532.7 lbs(2515 kg.), it represents 8 percent of the rotor weight BUT 16 percent(0.16 g) of the exhaust end plane.

Most importantly, it should be noticed that both the HP and LP ends of the turbine were reduced, which means that the rotor wants to respond in a single plane fashion. There is nothing unique about the single plane solution used. It is conventional and found in many references. The solution is shown in Figure 2.5 as taken from Ref.[3]. A final correction weight of three holes with 14 grams in each was used. Since the solution called for weights to be added to Holes #2&3; weights at two 1/2"x 1/2"(1.27 cm x 1.27 cm) were removed from Holes #10 & #11 (Fig.2.2).

The final balance obtained was 0.21, 0.53.0.41,&0.25 mils p/p(5.33,13.46,10.41,& 6.35 micro-metres p/p). A further trim would have been fruitless and would have driven the HP end back up. This interestingly enough, was the end being corrected by weights added in the opposite (exhaust) end. The whole balance operation took less than one hour, after the exhaust end bearing had been fitted with a new spherical seat since a clearance is not desired, rather a 1-2 thou nip(interference). The bearings are pressure dam type with a diametrical clearance of 1 1/2 thou per inch of diameter(0.15 percent).Often, excessive clearance in a hydrodynamic bearing gives a synchronous clearance whirl and can be improperly blamed against a poor balance. These rotors have been sequentially balanced during assembly in no more than two stagings per balance to prevent introduction of coupled imbalances during assembly down the rotor shaft.

3 CASE THREE FINAL DISCUSSION & REVIEW

This steam turbine has three rows of blading before the passout valve and has a two stage double flow exhaust end blading for a total of seven rows of blading. It has a first rigid mode cylindrical mode at 3900 rpm, a second pivotal mode at 9200 rpm, a third "bending" mode at 15,650 rpm. The operating speed is normally at 11,000 rpm but has a 30 percent "turn down speed" to 8500 rpm. Therefore, it operates between the 2nd and 3rd criticals. It was commissioned in 1970 at 1.1 mils p/p vibration. This was the second,in situ, balance for better performance of a large methanol plant. One could expect,[N+2 Balance Planes], yet with some attention to responses, this rotor can be balanced with a single plane balance procedure at one of two API requested field balance planes. Specifically this balance plane overhangs the exhaust bearing centreline by 8" (towards the coupling). The coupling is a continuously lubricated flexible gear type with a 24"(61 cm) to allow adequate misalignment to the driven centrifugal compressors. The turbine has been hot aligned to the compressor using water stands & eight proximity probes per casing,i.e. two per bearing housing corner over a 24 hour time span under full temperature. The alignment is held to 1/2 thou per inch of coupling span(12 mils offset,305 microns). The gear tooth misalignment by the coupling manufacturer is 1/2 degree or 209 mils(5.3 mm).

The critical speed analysis of this rotor is performed through a CRITSPD program developed at the University of Virginia,USA. The ability of this rotor to be balanced in one plane caused some concern. In addition, the lack of response to a HP(governor end) calibration some 13 years before also brought about a need to do an UNBALANCED RESPONSE program wherein actual unbalances are placed in the rotor at different station so that one might develop a Rotor Deflection Diagram(not to be confused with a Mode (at Resonance) Diagram as shown in Fig. 1.1. This was in fact done in a different manner as the previous program was time consuming. Mr. Malcolm Leader, MS Graduate of UVA and Associate utilized the CRITSPD program as modified to show animated modes and simply interrupted the CRITSPD program after the 2 nd mode had iterated a solution and "single stepped" the program to 11,000 cpm(183.3 hz), changed the roots to affect a program solution. Fig. 3.1 is the result. Whilst a marginal error may be encouraged, this deflection diagram of this rotor is felt to show why the balance plane at the steam (HP) end is very near a node, and thus unresponsive. Further, it shows why the steam end will respond well to a weight placed at the exhaust end of this rotor.

A. It is good to develop a logic to assist the first weight placement. Polar plots are excellent, scope orbits help, synchronous tracking filter logic helps whether via displacement sensor or by velocity probes integrated to correct the 90 degree shifted phase angle or by simply plotting the shifted angles when using velocity(90 degrees) or accelerometers (180 degrees).

B. Mode diagrams are a big help in preparing, for critical machines.

C. Calculated weight amounts will give one a response and also not damage,e.g. 1/10 "g".

D. Layout of the balance planes, probes, rotation, will assist the ever present confusion.

E. Careful observance of all sensor responses can give one a accurate feedback of the overall conditions. One must be sensitive to phase angles stability. Often the rotor must be allowed to stabilize before good data can be obtained. In this case, data had to be taken quite quickly before the exhaust temperature rose. In solo, one is often on a constant enthalpy line rather than a constant entropy line for condensing turbines, particularly if the condenser is undersized to the maximum condenser steam flow. Actually, I have not found a machine allowing field balancing, to require more than a two plane solution. One must recognize that static and couple shots can be placed in a two plane program.

---

4 CASE TWO-BALANCING OVERHUNG COMPRESSOR WITHOUT DISASSEMBLY,IN SITU

This is a Case History of field balancing a 24 x 24 pipeline compressor being used in this plant as a cirulating gas compressor through a reactor. The compressor weighs 125,000 lbs.(56,818 kg) with 24"(60.96 cm) inlet and discharge nozzles. The compressor handles a 13 mol weight gas at a suction pressure of 700 psig(48.3 bar g), a discharge pressure of 750 psig(51.7 bar g) and over one million cu. ft./min.(28,320 cu. metres/min). The casing bore is over 6 ft.(1.8 m) and the casing inlet head weighs 8 tons(7273 kg). The rotor weighs 635 lbs(289 kg) with a single 30"(76.2 cm) impeller weighing 270 lbs.(123 kg). The compressor is driven by a 9000 bhp(rated) steam turbine through an 18"(45.7 cm) flexible gear tooth coupling with continuous lube spray. The impeller is overhung from two pressure dam hydrodynamic bearings @ 4 7/8"(12.38 cm) with 7 thou(0.1778 mm) clearance. The compressor operates at about 6,000 rpm which is generally on the ramp up towards the first rigid mode critical which is pivotal for this overhung machine (see Fig. 4.1).

The reason for field balancing this compressor, without disassembly, was economically simple but mechanically questionable as it had not been attempted before. The impeller was unbalanced from an accidental partial wash with methanol which removed part of a coating of iron carbonate, a waxy corrosion product, which coats the impellers fairly evenly over 6-8 years. The shutdown and overhaul would require 3-4 days, if everything went right, during which time methanol would have to be purchased on the open market. The plant can make syn gas without this compressor but it cannot make methanol without this compressor. The field balance procedure shown allowed this compressor to operate for three more years at savings over $300,000 US.

5 BALANCING RING FOR INSTALLATION ON THE COUPLING SLEEVE

Having operated this compressor for 14 years with baseline data via Bode', Polar, & Spectrum Plots; and after review of the mode shape and runup responses. It was decided to attempt a field balance using a weight ring (see Fig. 5.1) which would be installed on the coupling sleeve. There was no access to installing this balance ring on the compressor shaft at the coupling end. Further, since one proximity probe existed at the bearing nearest the impeller(wheel end) and two orthogonal proximity probes existed at the coupling end bearing, balance data should be logical and linear provided the coupling end probes were used in relationship to the weight placements in the balance ring(see Fig. 5.2). NOTE: If the location, plane of vibration measurement, along a rotor shaft is not separated by a "nodal point" from the weight placements, plane of the balance ring; then, some straightforward logic can be used.

6 CASE TWO - PROCEDURE

The balance ring was fabricated in the plant shops and pre balanced on the shop balancing machine. The compressor was isolated from the process and placed on nitrogen service. The initial data was recorded on all three proximity probes(eddy current type), a synchronous tracking filter(vector filter) was used to record the runup data, an oscilloscope (always used) recorded the orbits(lissajous photos), and an FM tape recorder was used to record all points for the record (or playback, if necessary). Since the proximity probes were installed in 1970, prior to the API 670 Standard, VIBRATION, AXIAL-POSITION, AND BEARING TEMPERATURE MONITORING SYSTEMS,1 st edition-1976,2 nd edition-1986,the probes will not split the top centreline in mounting. The actual positions are shown on the Balance Plane/Vector Plot Figure 6.1. The vertical coupling end probe is at 35 degrees anticlockwise from the top centreline of the bearing. The Keyphasor[Reg.TM] probe is 45 degrees anticlockwise from the same reference, and the horizontal coupling end probe is at 55 degrees clockwise from the same vertical reference(12:00 o'clock).

The original game plan was to balance the coupling end vibration which was operating under full load at 2.5 mils p/p(63.5 microns peak-to-peak) to a low amount. It was hoped that this would bring the full load impeller (wheel end) vibration of 3.5 mils p/p(88.9 microns p/p) to an acceptable amount, say 2 mils p/p(50 microns) to allow continued operation without damage.

The balance steps are shown in Figure 6.2. The calibration weights can be computed in ounces(gms) or in length, inches(cm)-both are shown. The original unloaded run(nitrogen),less than the previous loaded run(under process gas) values, is 1.96 mils p/p(49.8 microns) @ 49 degrees (phase angle). Since the phase angle(representing the rotor's "high spot"), below the first critical is more or less in phase with the "heavy spot", rotor's imbalance eccentricity, a calibration weight at about 180 degrees from that measured phase angle will be used(refer back to Fig. 2.1). The calibration weight of 0.61 ounces(17.3 gms) is used at a radius of 5 inches(12.7 cm). This is a 1" (25.4 mm) bolt with 1/2"- 13NC threads mounted in the balance with LOCTITE. The force is 195 lbs(88.5 kg) or 0.3 rotor "g's"(0.72 impeller "g's"). This is more than normal but considering the moments of the impeller-to-bearing versus the balance ring-to-bearing, it seemed more logical. The calibration weight was placed at 220 degrees, against rotation (CW), from the coupling end vertical probe. NOTE: In this balance procedure and in the balancing program used, the phase logic never changes. The vibration instrumentation measures the phase angle,(1) against rotation, from the sensor;(2) vector plotting and the weight placements are placed in the same manner.

The correction weights were three 1" long bolts, each placed at 170, 192.5, & 215 degree locations. The 22 1/2 degree hole locations were adjusted slightly in remeasuring the balance ring as it was clamped in a final assembly with thread locking compound on the mounting bolts and balancing screws. The bolt at 215 degrees was shortened in a "trim" balance to 3/4" (1.9 cm) length. The vector sum of these three weights was 1.36 ozs(38.5 gms) @ 182.5 degrees. The balance data in Fig. 6.2 table develops an Influence Coefficient Response Vector of 1.44 mils/oz @ 46.74 degrees(-313.26). This means that, in the future, one can take a vibration reading on this machine and affect a "one-shot" correction by dividing the vibration vector amount, for example here we had 1.96 mils @ 49 degrees, (by) the influence response vector,a=1.44 @ 46.74 degrees and determine an unbalance amount of 1.36 ounces @ 362.5 degrees. Thus, a correction weight of 1.36 ounces can be placed at a radius of 5" and an angle of 362.26-180=182.55 degrees [5]. It would be even better to calculate these values in oz-in or gm-in or gm-mm, in case the same radius is not always available.

Fig. 6.3 shows the orbit after the first balance and trim and reloading the compressor with process gas. The time traces of all three probes plus that of the steam turbine exhaust is shown in the same Fig. 6.3.

7 CASE TWO- DISCUSSION & CONCLUSION

This balance brought forth a problem in understanding by the author and balancer. This query held publication of this Case History for three years. The balance was very successful with all vibration below 1 mil actually in the 1/2 mil range. However, the balance runs on Nitrogen only showed the coupling end vibration to be corrected, even to the 1/4 mil levels. The impeller end correction does not manifest itself until the compressor is loaded with process gas, and then following a 100+ degree phase shift the vibration reduces below one mil,e.g. 3/4 mil p/p. In an effort to better understand this phenomenon, the data was checked over the next 1 1/2 years, only to find that the machine repeats itself quite well. A trim was made one year later as iron carbonate does recoat and it was predicted that the imbalance in the impeller(without physical verification) should tend to correct itself. The centre weight was removed one year later during some downtime in operations. The history of these other startups are recorded in Fig. 7.1. On the plus side, it is verified that this correction endured for the life or this compressor, which is being altered 3 years later in a process change and replaced with a 9000 hp induction generator.

---

8 CASE THREE-CPLGBAL-COMPENSATING THE ROTOR'S IMBALANCE WITH THE COUPLING'S IMBALANCE

This technique has been proven many times on real machinery and is demonstrated in the Vibration Institute's Rotor Dynamics & Balancing Course each May in Virginia. It was developed from an original paper by Mr. Anthony F. Winkler[6] following his practice at Dresser-Rand, Olean,NY. This will be excerpted from an original work[7] for The Vibration Institute. It was last used in clearing for shipment several gas turbine driven UK built injection pumps for service in Alaska. This incident will serve to explain the application need for the technique. In the case mentioned, each component had been balanced to within 1 mil p/p(25.4 microns),i.e. the turbine, the coupling, and the pump. HOWEVER, when coupled together the vibration was 2.5 mils p/p(63.5 microns) and rejected by the owner and his contractor. Using this procedure the couplings (dry type with hydraulic fitted hubs) were re-positioned in their fit (circumferential angle) to affect the matching of the coupling residual imbalance to correct the rotor's residual imbalance. Recognizing that an imbalance can be represented by a vector quantity of eccentric mass at an angle.

This procedure is limited to non-keyed couplings. Couplings which can be repositioned on a shaft or within defined hole templates, splines, etc. It works well with hydraulic fitted couplings and can work well with "marine type" integral flanges with coupling assemblies which are attached by indexed hole layouts provided in these integral flanges. Further, it is limited to a situation where the coupling and rotor have residual imbalances.

9 CASE THREE PROCEDURE

In this procedure, an original vibration is measured as 0.20 g's(0.25 ipsp) @323 degrees phase angle, using an accelerometer at 2880 rpm mounted on a bearing near the coupling(see Figure 9.1-polar plot). This is plotted as Vector ,Z1, and is recorded in the Worksheet (Fig. 9.2).

The coupling is then reversed 180 degrees from its current position. The machine is brought up to speed and the "reversed coupling" position is recorded,Z2, as 0.25 g's(0.32 ipsp,8.1 mm/sec peak) @ 240 degrees. This is plotted on the polar paper and recorded on the worksheet. A line is drawn connecting these to vector heads (arrows), and a mid-point,C, is determined in this line. A circle is drawn, inscribing the two vector heads, Z1 & Z2, with the midpoint,C, as the center. This circle represents the locus of all possible solutions created by rotating the coupling about the rotor shaft. The residual imbalance to be corrected, is represented by OC, called the C Vector. The amount of correction possible in the coupling is represented by CZ1=CZ2=0.15 g's (0.19 ipsp)(4.83 mm/sec).

The solution is rotating the coupling, from its current position, which caused Z2, by an angle equal to Z2-C-0, which will subtract the vector quantity of the coupling,CZ2, from the vector quantity,C, leaving a residual unbalance amount equal to the difference, in this case

0.02 g's(0.095 ipsp)(2.4 mm/sec peak). This shift angle is 257 degrees, against rotation, or -103 degrees, with rotation. A means of marking the coupling and the shaft must be established in the beginning. Further, the phase angle convention must be simple and consistent, in this case,again, it is from taken from the vibration sensor used to obtain the information,plus a Keyphasor probe(triggering).

10 CASE THREE- DISCUSSION & SUMMARY

Whilst I was working at Monsanto, it seemed that one could easily get the phase orientation confused and shift the coupling the right angle BUT the wrong direction. So, a program,CPLGBAL,(FIG.10.1) was written for the HP 41 Programmable Calculator, and is included in this paper. The instructions should be understood, straightaway, from the Worksheet example. Two examples are shown, despite the restrictions on artwork,to allow one to discern any differences with direction of rotation.

11 REFERENCES

1. Jackson, Charles,"Using the Orbit (Lissajous) to Balance Rotating Equipment,"ASME Paper #70-PET-30., (1970)
2. Jackson, Charles, "Two-Plane Field Balance Hybrid Systems-Vectors & Orbits, Trim Balance," Case History Series,Technology Interchange,Vibration Institute,Clarendon Hills,Il(1973)
3. Jackson, Charles, The Practical Vibration Primer, Gulf Publishing Co., Houston , Texas (1979).
4. Jackson, Charles, Trim Balancing a Syn-Gas Turbine in Situ, Vibrations, Sept. 1986,Vol.2, Number 2, Pgs. 13-18, The Vibration Institute, Clarendon Hills, Il.
5. Jackson, Charles,Part 4-Phase and Basic Balancing, Hydrocarbon Processing, Figure 11, Nov. 1975, Houston, Texas (repeated in Chapt. 4 of reference 3, above).
6. Winkler, Anthony F., High Speed Rotation Machine Unbalance, Coupling, or Rotor, The Vibration Institute, Houston, Texas, April, 1983.
7. Jackson, Chas.,Monsanto Distinguished Fellow, Repositioning A Coupling's Residual Unbalance to Correct a Rotor's Residual Unbalance-CPLGBAL,The Vibration Institute, Phila.,Pa.,1984.
8. Parkinson, A.G.,Darlow, M.S., & Smalley, A.J.,"A Theoretical Introduction to the Development of a Unified Aproach to Flexible Rotor Balancing",J.Sound Vibration.,68(4),pp 489-506 (1980).

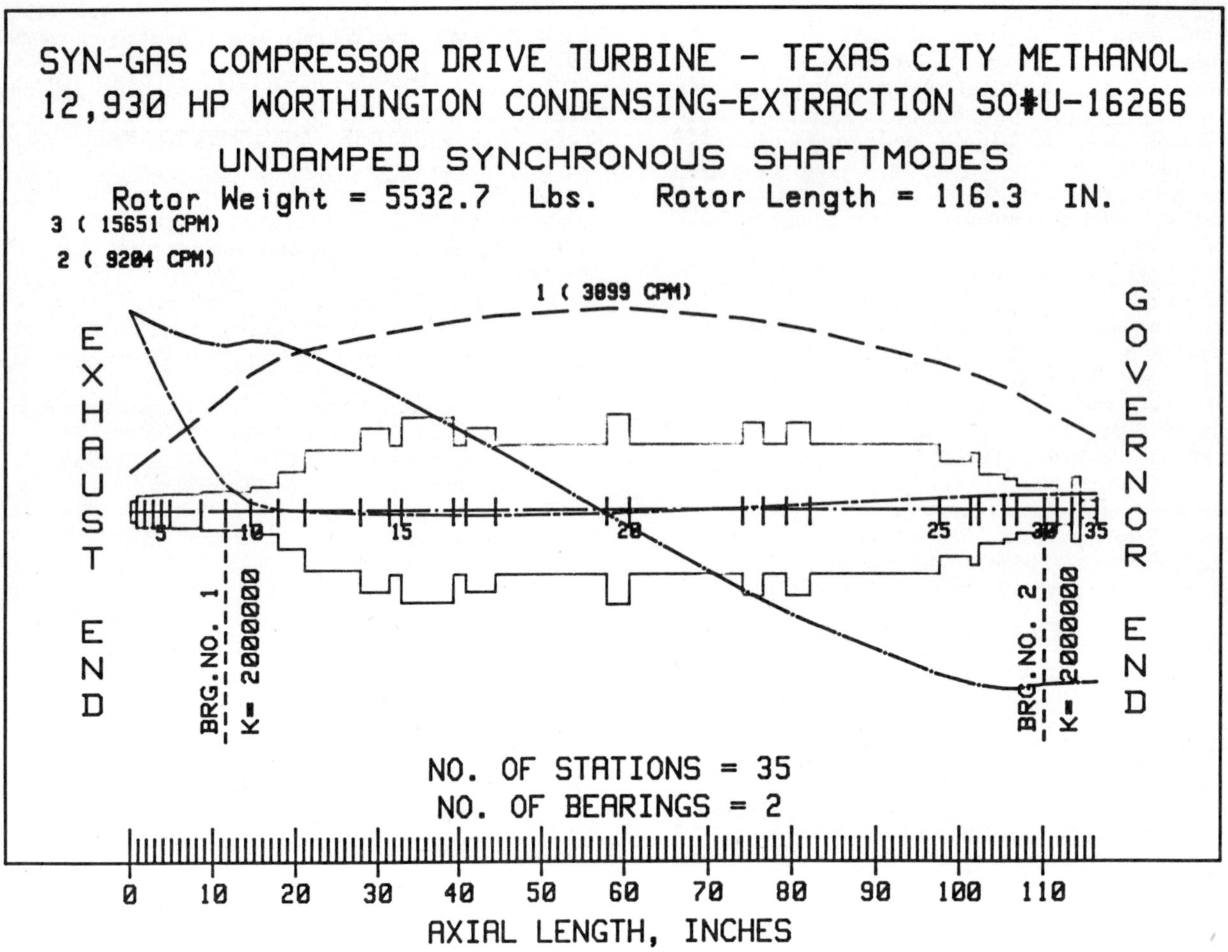

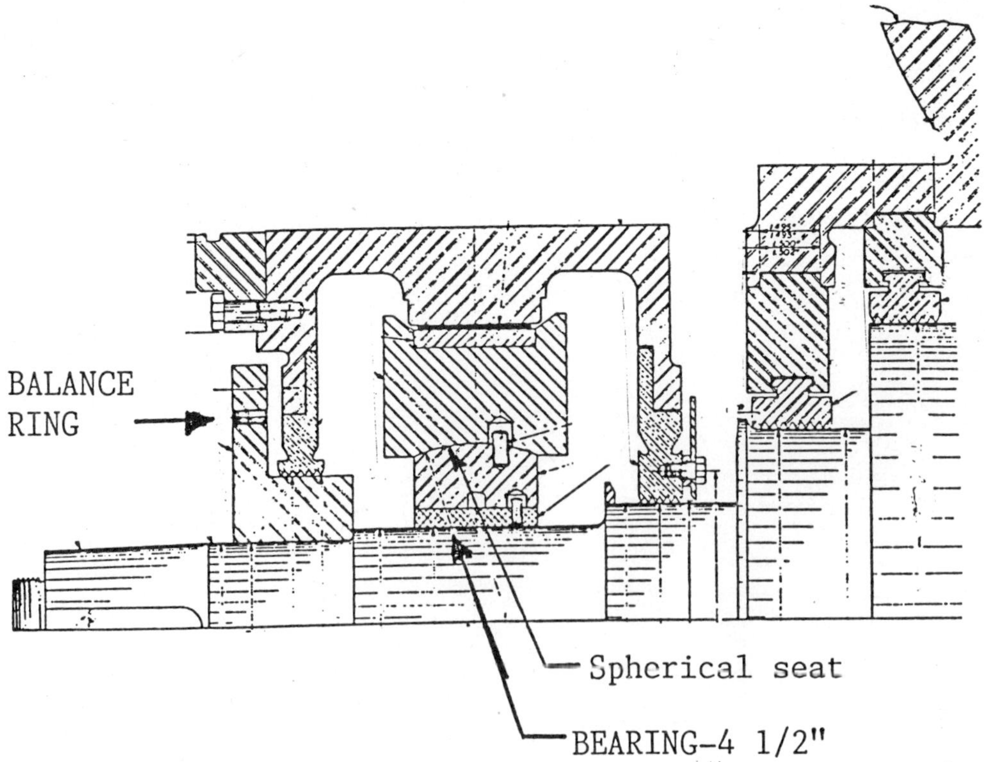

Fig 1.1 Rotor mode shapes and exhaust end balance ring details

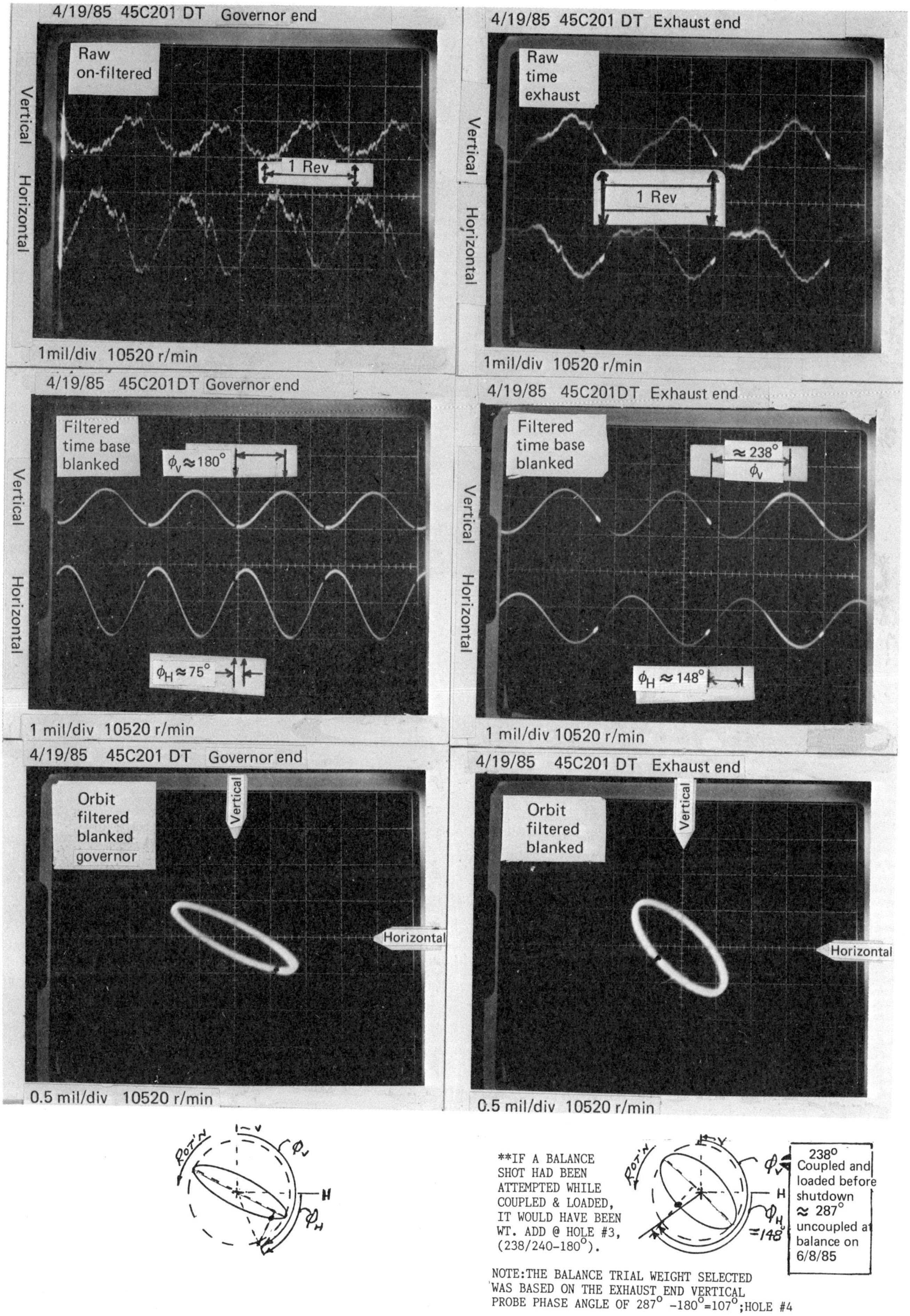

Fig 1.2 Scope traces taken two months prior to shutdown; governor (HP); exhaust (LP)

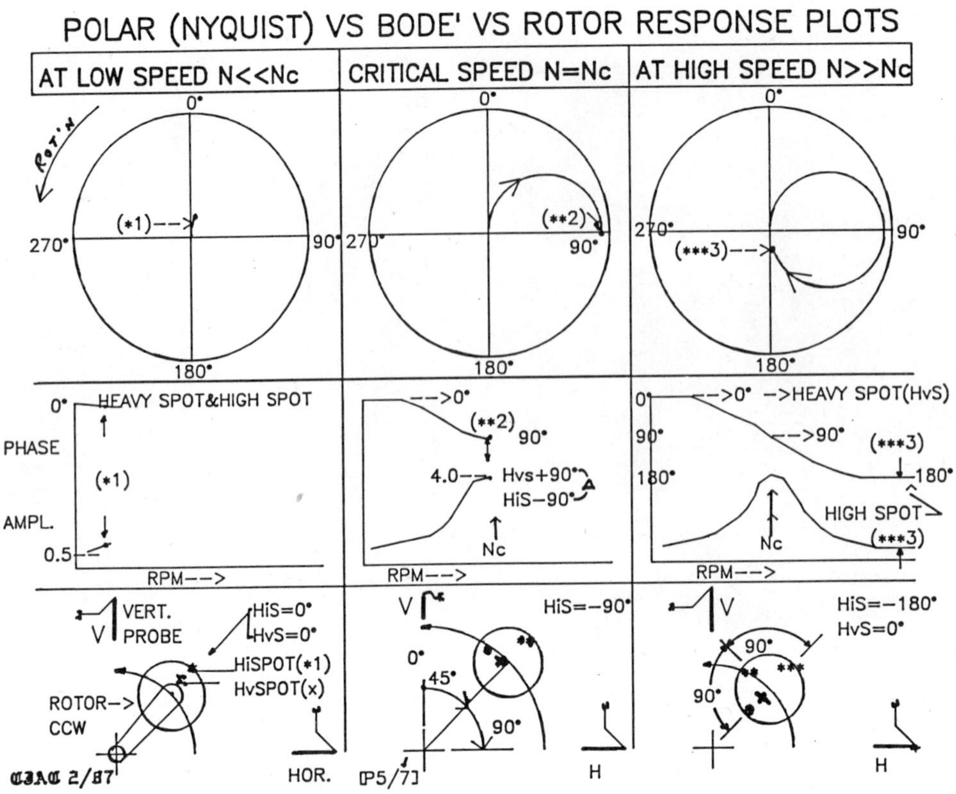

Fig 2.1 Logic of high spot phase angle changes going through a critical (rotor resonance)

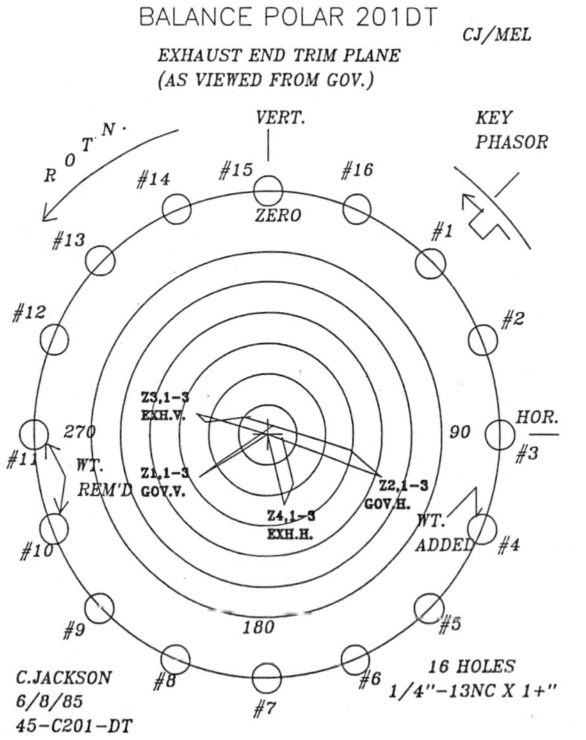

Fig 2.2 Polar layout combining the balance holes vectors and vibration sensors (probes)

SINGLE PLANE SOLUTION WITH RUNS

PROBE ID	ORIGINAL		TRIAL RUN		FINAL W/C.WT.	
	RUN#1,Z, ,1		RUN#2,Z, ,2		RUN#3,Z, ,3	
	AMP.	PHA.	AMP.	PHA.	AMP.	PHA.
GOV.V.Z1	1.32	239	0.85	243	0.21	37
GOV.H.Z2	2.15	109	1.45	101	0.53	303
EXH.V.Z3	1.27	287	1.05	280	0.41	317
EXH.H.Z4	1.21	165	0.90	151	0.25	100

XEQ "SGLBAL"
R.O.ANG.? 2.00
R.O.AMT.? 0.15
Z1 ANG? 287
Z1 AMT.? 1.27
Z2 ANG? 280
Z2 AMT? 1.05
T.W. ANG? 112.50
T.W. AMT.? 14GMS

C.W. AMT.= 66.42 GMS.
C.W. ANG.= 76.47 DEG.
T VEC.AMT.= 0.26
T VEC. ANG.= 136.32 DEG.
SENSITIVITY U/M= 53.58 GMS/MIL

CORRECTION USED= 40 GMS.

TRIM DATA:
ANGLE= 317
AMOUNT= 0.41
C.W. AMT.=22.8
C.W. ANG.=99.86 DEG.

NOTE: TRIM NOT USED*
*GOV.H CROSSED ZERO

CALCULATION–STATIC/COUPLE
SAME
SAME
Z1 ANG? 262DEG
Z1 AMT? 1.20 MILS
Z2 ANG? 263 DEG.
Z2 AMT? 0.90 MILS
SAME
SAME

C.W. AMT =57.52 GMS
C.W. ANG.= 108.63

U/M = 46.58 GMS/MIL

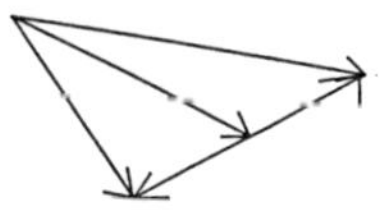

Fig 2.3 Balance data and solution

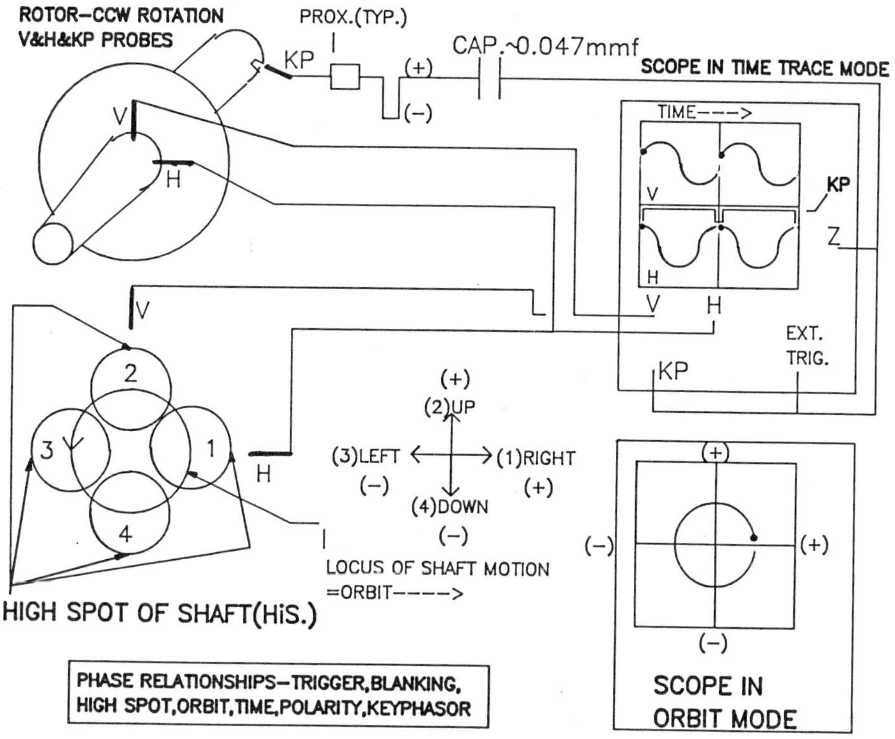

Fig 2.4 Orbit, motion, keyphase triggering and time trace phase logic

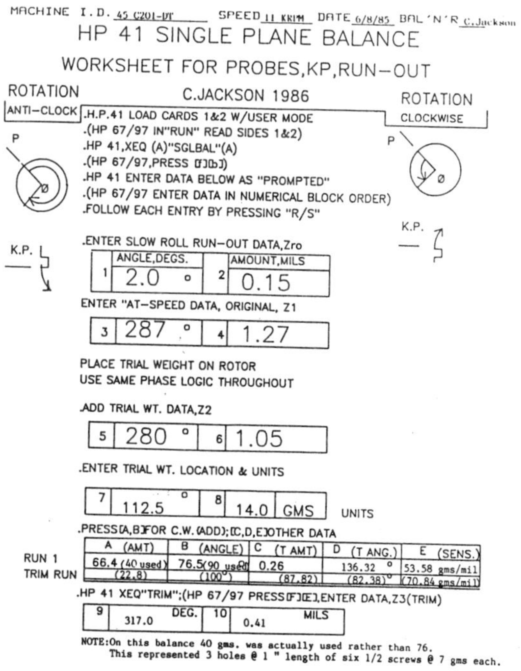

MACHINE I.D. 45 C201-DT SPEED 11 KRPM DATE 6/8/85 BAL'N'R C.Jackson

HP 41 SINGLE PLANE BALANCE

WORKSHEET FOR PROBES,KP,RUN–OUT

ROTATION ANTI–CLOCK — C.JACKSON 1986 — ROTATION CLOCKWISE

.H.P.41 LOAD CARDS 1&2 W/USER MODE
.(HP 67/97 IN"RUN" READ SIDES 1&2)
.HP 41,XEQ (A)"SGLBAL"(A)
.(HP 67/97,PRESS [f][b])
.HP 41 ENTER DATA BELOW AS "PROMPTED"
.(HP 67/97 ENTER DATA IN NUMERICAL BLOCK ORDER)
.FOLLOW EACH ENTRY BY PRESSING "R/S"

P — K.P. — Ø

.ENTER SLOW ROLL RUN–OUT DATA,Zro

	ANGLE,DEGS.		AMOUNT,MILS
1	2.0 °	2	0.15

ENTER "AT–SPEED DATA, ORIGINAL, Z1

3	287 °	4	1.27

PLACE TRIAL WEIGHT ON ROTOR
USE SAME PHASE LOGIC THROUGHOUT

.ADD TRIAL WT. DATA,Z2

5	280 °	6	1.05

.ENTER TRIAL WT. LOCATION & UNITS

7	112.5 °	8	14.0	GMS

UNITS

.PRESS[A,B]FOR C.W.(ADD);[C,D,E]OTHER DATA

	A (AMT)	B (ANGLE)	C (T AMT)	D (T ANG.)	E (SENS.)
RUN 1	66.4 (40 used)	76.5(90 used)	0.26	136.32 °	53.58 gms/mil
TRIM RUN	(22.8)	(100°)	(87.82)	(82.38)°	(70.84 gms/mil)

.HP 41 XEQ"TRIM";(HP 67/97 PRESS[F][E],ENTER DATA,Z3(TRIM)

9	317.0 DEG.	10	0.41 MILS

NOTE:On this balance 40 gms. was actually used rather than 76.
This represented 3 holes @ 1 " length of six 1/2 screws @ 7 gms each.

Fig 2.5 Balance worksheet/calculator

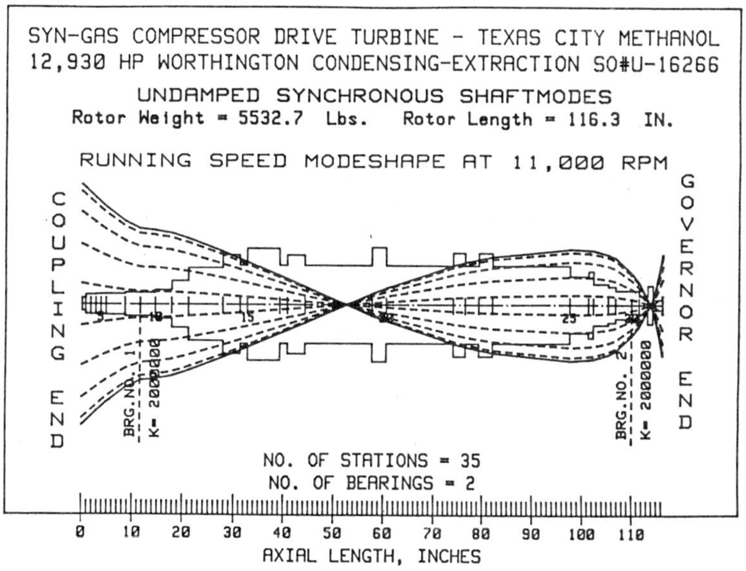

Fig 3.1 Deflection diagram (not mode) of rotor at operating speed

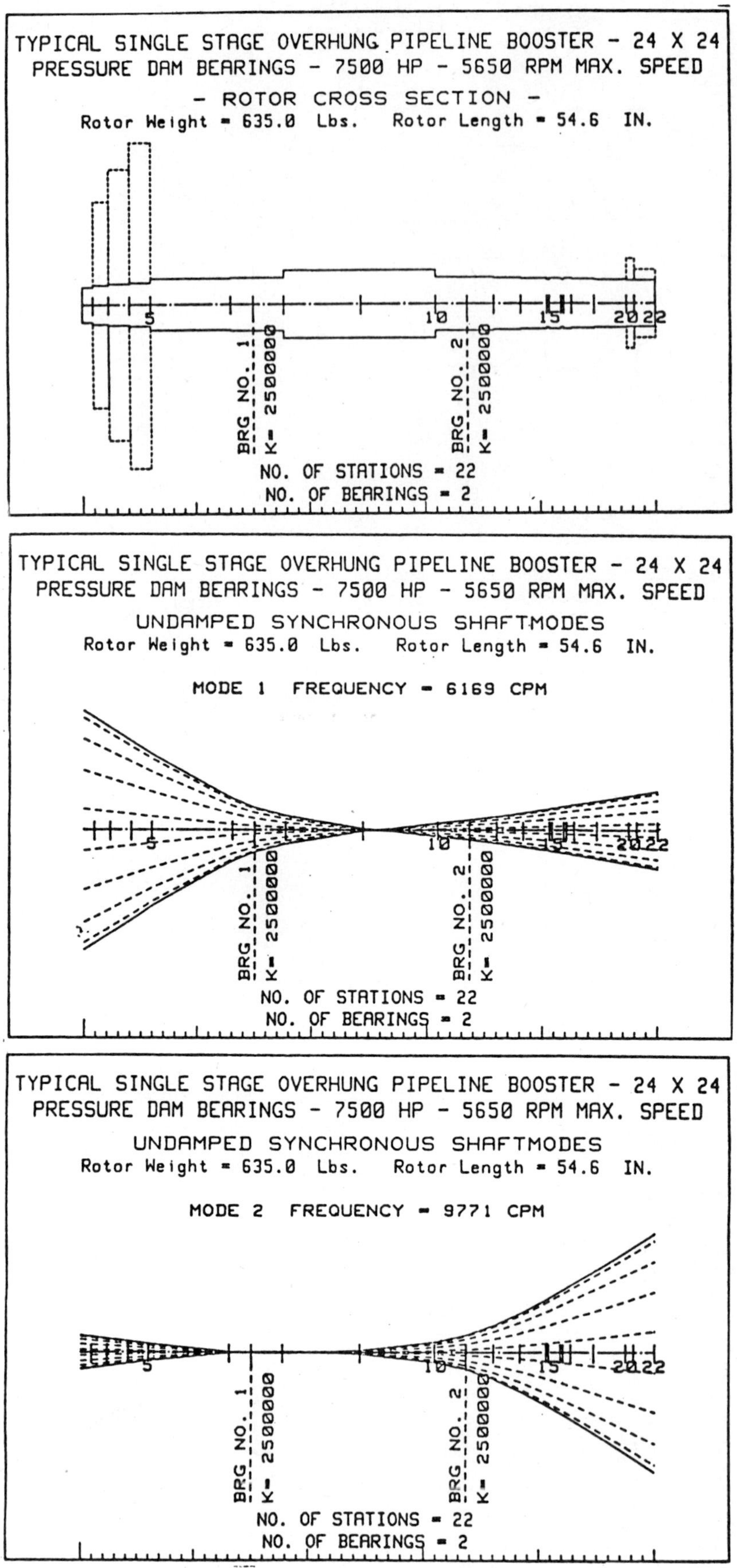

Fig 4.1 Mode diagrams and rotor model; first and second modes or pivotal

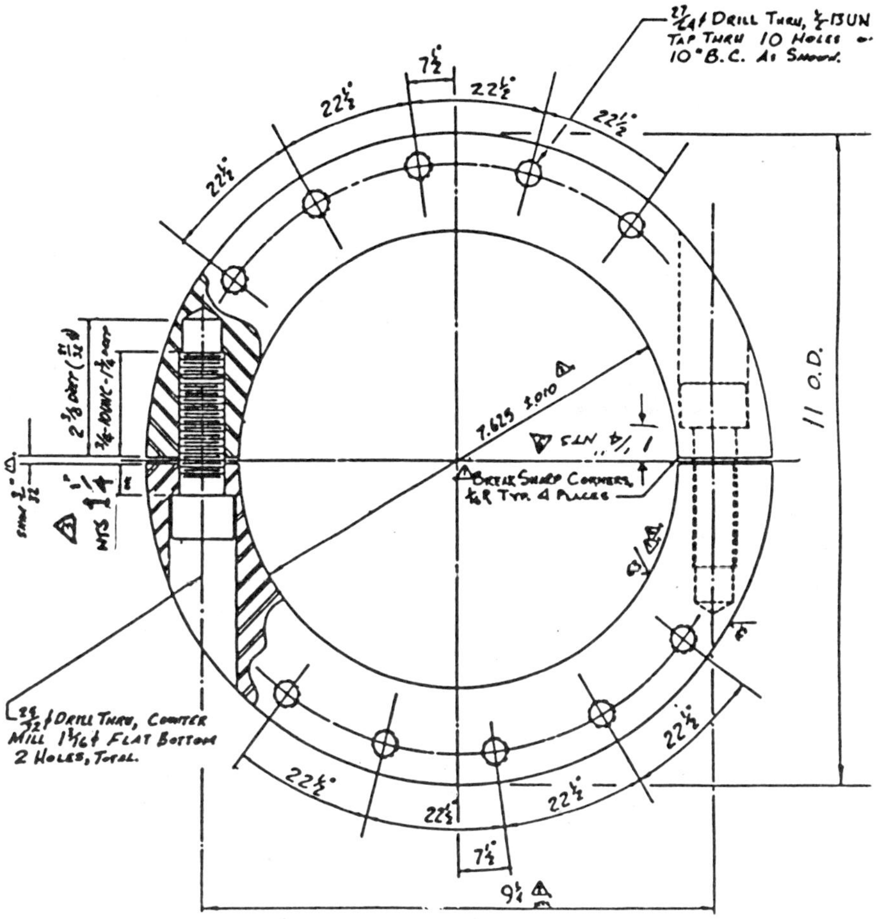

Fig 5.1 Balance ring assembly

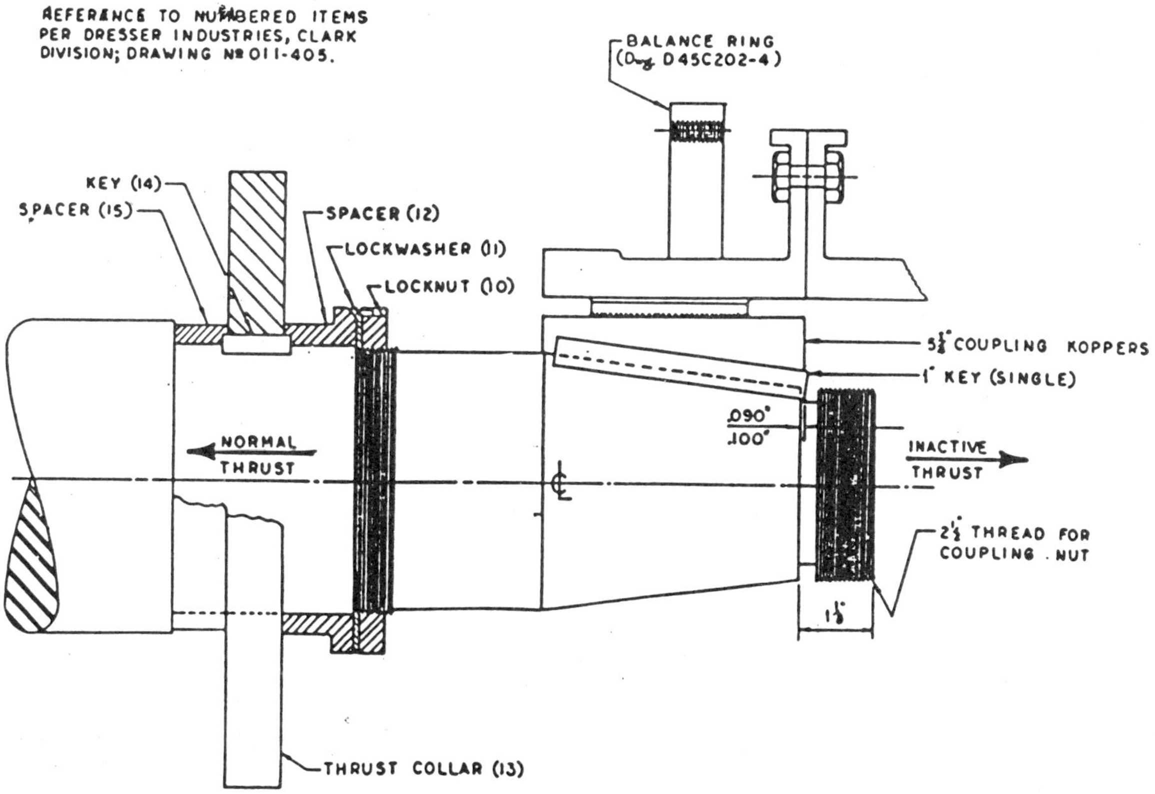

Fig 5.2 Balance ring/shaft detail

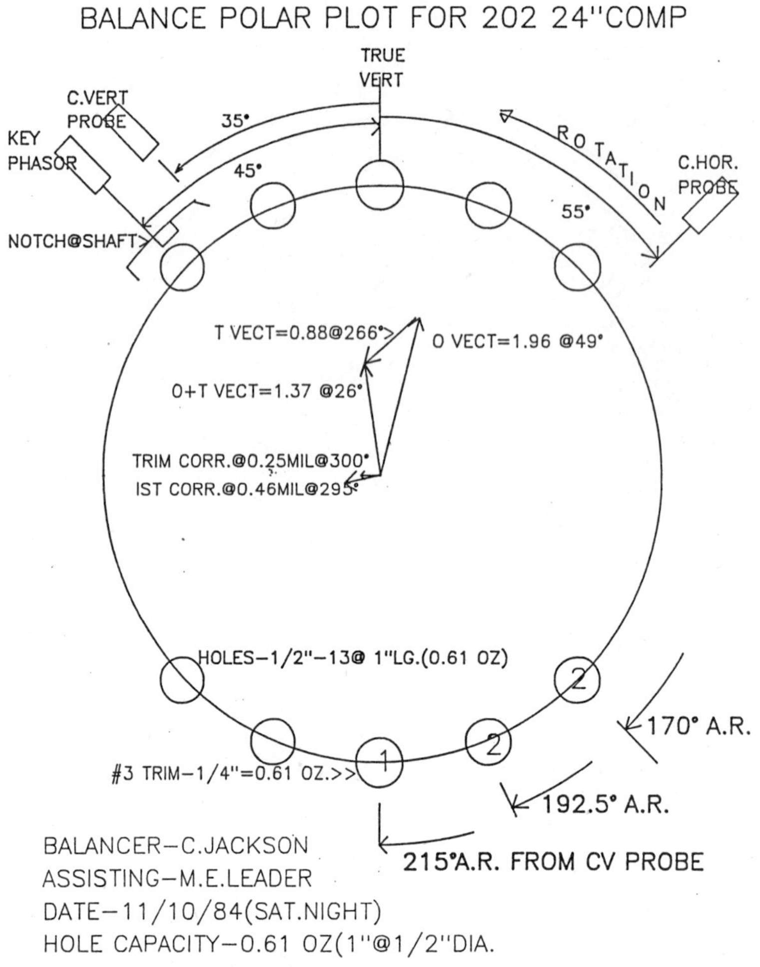

Fig 6.1 Balance plane, vectors, probe layout; W/angular locations

```
TABLE    REVIEW OF BALANCE  CALCULATIONS FOR SINGLE PLANE AS USED IN-SITU:
         REFERENCE:C.JACKSON,"THE PRACTICAL VIBRATION PRIMER", 1979 (OR)
                   C.JACKSON,FIG. 12,PART 4,HYDROCARBON PROCESSING, NOV. 1975
                                                 "SGLBAL"oz.    "SGLBAL"in.wt.
Orig. Vector      =0  =Z1=1.96 mils @ 49  deg.   Z1 ANG?=49.0      "Same"
Orig+Trial Vector=O+T=Z2=1.37 mils @ 26  deg.   Z1 AMT?=  1.96       "
Trial Weight Vect=    Ut=0.61 ozs. @ 220 deg.   Z2 ANG?= 26.0        "
Z1=a(Uu) ;a=influence coefficient;Uu=unbalance  Z2 AMT?=  1.37       "
Z2=a(Uu+Ut)                                     TW ANG?=220.00       "
Z2-Z1= a(Uu+Ut-Uu) = aUt                        TW AMT?=  0.61oz  1.00in.
                                                CW AMT =  1.36oz  2.23in.
Then; a= (Z2-Z1)/Ut                             CW ANG =182.55    "Same"
From Data a = [1.37@26 deg]-[1.96@49 deg.]/Ut   T  AMT =  0.88       "
a=1.37[cos 26+isin 26]-1.96[cos 49+isin 49]/Ut  T  ANG =266.45       "
a= -0.05-i0.88/Ut..(or) 0.88 @ -93.26 deg./Ut   U/M    =  0.69    1.14
a= (0.88@-93.26 deg.)/(0.61@220 deg.)           Mils/     /oz.     /inch
a= 1.44 @ -313.26 degs.                         XEQ "TRIM"
                                                ANG?   =295.00    "Same"
Unbalance = Uu= Z1/a                            AMT?   =  0.46       "
Uu = (1.96 @ 49 deg.) / (1.44@ -313.26 deg.)    CW AMT =  0.29oz  0.47in
Uu = 1.96/1.44 @ 49-(-313.26)=1.36@362.26 deg.  CW ANG = 57.48    "Same"
Corr. Wt.to balance=Ub=-Uu=1.36oz@362.26-180    T  AMT =  1.22       "
Corr. Wt.= 1.36 ozs. at 5"r. hole @182.55 deg.  T  ANG =171.48       "
                                                U/M    =  0.62    1.02
NOTE: THE HP-41 VERSION WITH TRIM FOR THE TRIAL Mils/     /oz.     /inch
WEIGHT IN INCHES & OUNCES IS SHOWN ON THE RIGHT.                CJAC 7/87
```

Fig 6.2 Table of balance calculations (a = influence coefficient response)

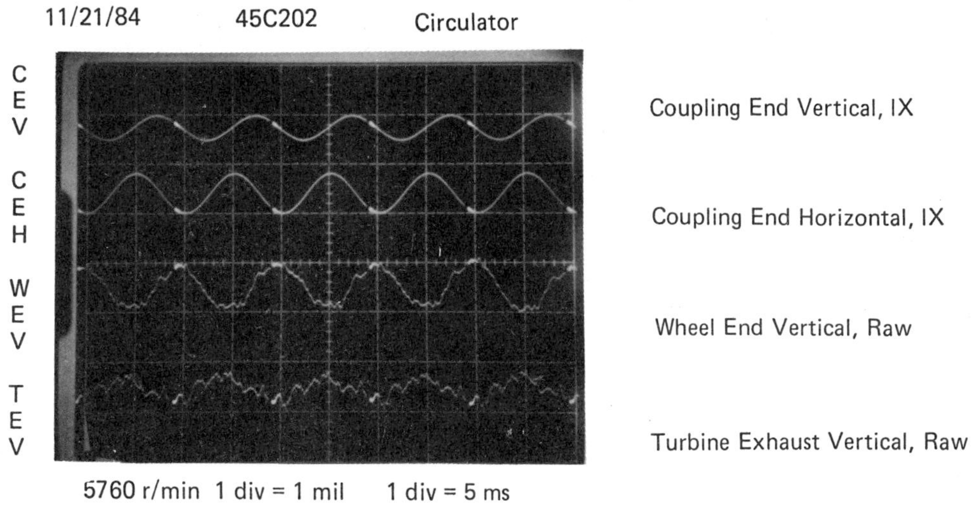

Final Scope Photos
45C202 - Second Startup
November 21, 1984 1 Div = 1 mil

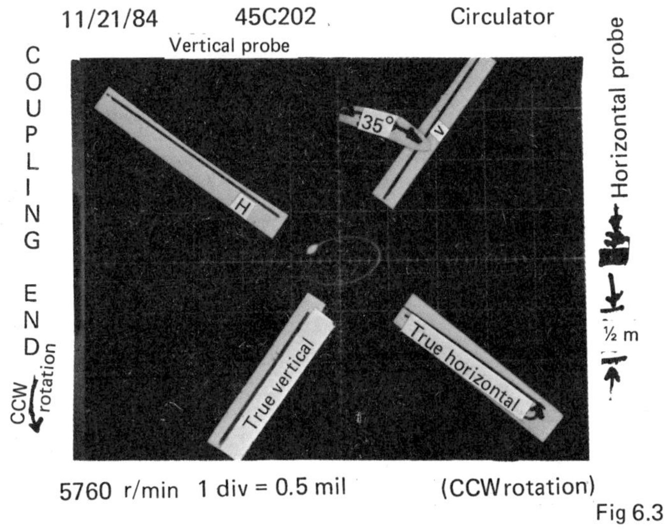

Fig 6.3 Final scope and time traces – balanced

Filtered Orbit of Coupling End Probes
11/21/84 1 Div = ½ mil

BALANCE DATA __ 24" Dresser Clark Pipeline Booster(Circulator) Compressor,Nov. 84
CJAC , 45C202, at 8000 hp, 6000 RPM, Balancing on Special Ring,Mt'd to Coupling Sleeve

Date	C.V.	Phase	C.H.	Phase	Wheel V.	Phase	Speed	Comments
11/10/84	1.96	49	1.76	320	2.75	185	6000	C.W. 1"@ 220°(1/2"-13)
	1.37	26	1.25	292	2.64	185	6000	3 @1"@182 (0.61 ounces
	0.46	295	0.84	206	2.58	186	6000	R=5";3.03 oz-in/hole
3/4 "bolt	0.25	300	0.66	172	2.41	204	5200	C.W.@215/220°N is less
[1"x1/2"-13NC=0.61 oz @ 5" R=3.03 oz-in/hole;;C.W.=8.34 oz-in;S=8.34/1.71mils=4.88 o-i/m [3/4" @ 215 deg.;1" @ 192.5 deg. ; 1" @ 170 deg.]								
After Start-up +10 days lock washer for thrust nut, fatigued,cutting thrust probes-SD,so								
11/20/81	2.17	74	1.45	352	3.31	219	5000	w/o ring(balance)
(1634hrs)	2.59	34	1.88	290	3.64	201	5200	3 screws+ring @ 68°
68°CCW	0.91	59	0.41	275	3.54	207	5300	Unloaded(N_2)M.E.L.Note
NOTE: Twelve Hours later after loading out the compressor with Syn Gas:								
11/21/84	0.49	303	0.83	227	0.88	203	5760	Loaded-Note Shifts
Shifts.....	-0.42	+116	+0.42	+ 48	-2.88	+ 4	+ 360	Second case below-L/UL
11/26/84	1.32	40	1.09	287	3.48	189	5425	Unloaded,N_2,at speed.
11/27/84	0.75	322	0.88	261	0.91	209	5720	Loaded
Shifts...	-0.57	+78	-0.21	+26	-2.57	-20	+ 295	Phases;2 w.r.+ 1 a.r.
1/8/85	1.42	302	1.38	233	0.80	231	5233	
4/1/85	0.36	320	2.15*	187*	0.67*	206*		(Instr.)Rev,d leads

Fig 7.1 History data after balance-compressor ran for three more years

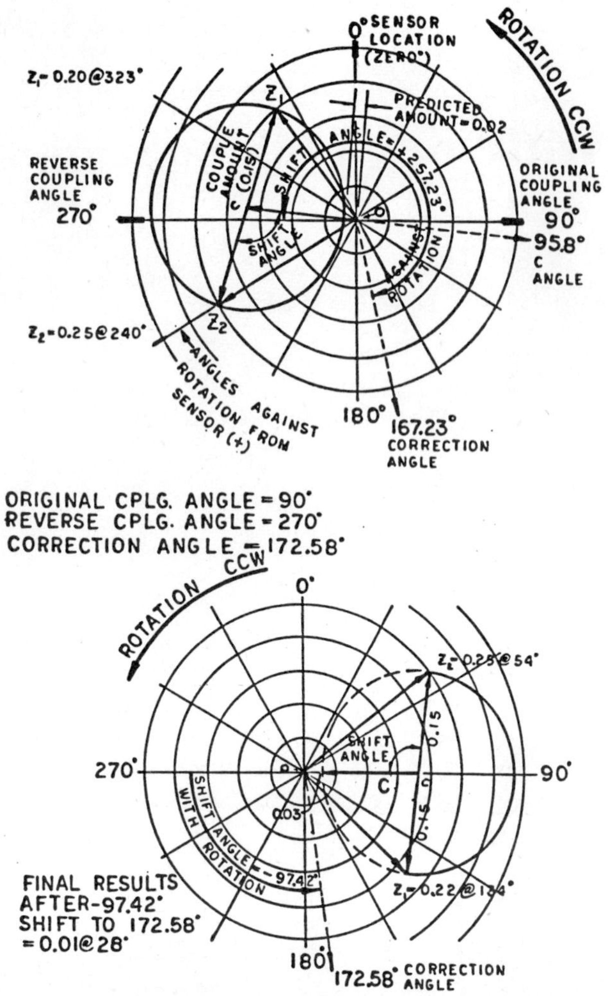

Fig 9.1 Polar plot (the two examples shown are balancing by repositioning coupling)

WORK/FILE SHEET

CPLGBAL –APPL. NOTE 84–01

Repositioning a coupling's residual unbalance to correct a rotor's residual unbalance

LIMITATION: Hydraulic or Marine type(Bolted) fitted coupling

Machine Indentification Monsanto Demo at Clear Lake, Texas Date 11/8/84

Coupling Type,Size,&Attachment Marine – Bolted Type

Operating Speed 2880 RPM;Key Phasor Probe @ 90°

Vibration sensor @ 0° ;Vibration sensor type Seismic Accel.

Rotor Rotation: (CCW) or CW (circle one)viewed from Coupling

•With rotor and coupling operating at speed, record Z1 phase angle & vibration amount (e.g. mils p/p). After DC coupling key phase probe on Bently DVF–2 to "manual" trigger from "auto". Record Z1 angle,amt.,(box 5&6 below)
•Stop rotor. Position K.P.notch–to–probe by turning in Rotation(onto notch),LED "on",DVF–2.
•Mark coupling and shaft(temp.);suggest ©probe.
•Record orig. coupling angle below..(box#1).
•Reverse coupling 180 deg.& record angle(box #2).
•Bring rotor to operating speed.Record Z2(box 3/4)

orig. cpl.		rev.cpl.		Z2 ang		Z2 amt		Z1 ang		Z1 amt	
1	90° r/s	2	270° r/s	3	240° r/s	4	0.25 r/s	5	323° r/s	6	0.2 r/s

•Plot values,polar,&/or use calc.(read sides 1–3)
•Calc."on",read mag.cards,"user",xeq A cplgbal A.
•Input data,boxes 1–6,on "prompt",key r/s,each.
When calc. completes solution,Press Keys A–F
If printer is used,"normal",headings & data PRT.

A	B	C	D	E	F
167.23°	257°	0.02	95.8°	0.17	0.15
Corr.∢	Shift ∢	Predict Amt.	C Vect.∢	C Vect. Amount	Couple Amount

C.Jackson
4/84

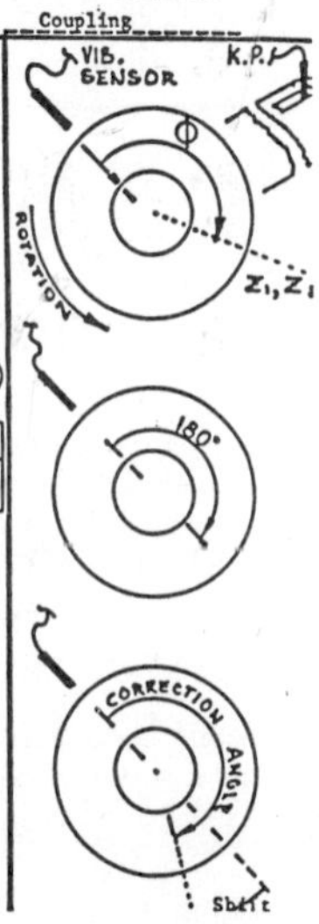

Fig 9.2 Worksheet for polar example (top is above)

```
PRP "CPLGBAL"

01♦LBL "CPLGBAL"
02 "CPLG.ANGLE=?"
03 PROMPT
04 STO 00
05 "REV.CPL.ANG=?"
06 PROMPT
07 STO 20
08 STO 21
09 "Z2 ANGLE=?"
10 PROMPT
11 STO 01
12 ENTER↑
13 "Z2 AMT.=?"
14 PROMPT
15 STO 02
16 P-R
17 STO 03
18 STO 13
19 X<>Y
20 STO 04
21 STO 14
22 "Z1 ANGLE=?"
23 PROMPT
24 STO 05
25 ENTER↑
26 "Z1 AMT.=?"
27 PROMPT
28 STO 06
29 P-R
30 STO 07
31 ST- 13
32 X<>Y
33 STO 08
34 ST- 14
35 RCL 14
36 ENTER↑
37 RCL 13
38 R-P
39 STO 16
40 X<>Y
41 STO 15
42 RCL 16
43 2.0
44 /
45 STO 17
46 STO 19

47 RCL 15
48 ENTER↑
49 RCL 17
50 P-R
51 STO 11
52 X<>Y
53 STO 12
54 RCL 03
55 ST- 11
56 RCL 04
57 ST- 12
58 RCL 12
59 ENTER↑
60 RCL 11
61 R-P
62 STO 10
63 ST- 19
64 X<>Y
65 STO 09
66 STO 18
67 RCL 15
68 ST- 18
69 RCL 18
70 ST+ 21
71 360
72 ENTER↑
73 RCL 21
74 X>Y?
75 XEQ 01
76 STO 21
77 RTN

78♦LBL 01
79 ENTER↑
80 360
81 -
82 RTN

83♦LBL A
84 "CORR. ANG"
85 AVIEW
86 PSE
87 RCL 21
88 VIEW X
89 RTN

90♦LBL B
91 "SHIFT ANG"
92 "SHIFT ANGLE"
93 AVIEW
94 PSE
95 RCL 18
96 VIEW X
97 RTN

98♦LBL C
99 "PRED. AMT"
100 AVIEW
101 PSE
102 RCL 19
103 VIEW X
104 RTN

105♦LBL D
106 "C ANGLE"
107 AVIEW
108 PSE
109 RCL 09
110 VIEW X
111 RTN

112♦LBL E
113 "C AMT."
114 AVIEW
115 PSE
116 RCL 10
117 VIEW X
118 RTN

119♦LBL F
120 "COUPLE AMT"
121 AVIEW
122 PSE
123 RCL 17
124 VIEW X
125 ADV
126 ADV
127 ADV
128 ADV
129 ADV
130 .END.
```

Fig 10.1 'CPLGBAL' programme

C273/88

Control strategies for use with magnetic bearings

C R BURROWS, BSc, PhD, CEng, FIMechE, FRSA and **M N SAHINKAYA**, BSc, MSc, PhD
School of Mechanical Engineering, University of Bath, Bath

SYNOPSIS This paper evaluates various strategies for applying a magnetic bearing to control the synchronous vibration of a flexible rotor. Both closed-loop and open-loop strategies are described. It is shown that an open-loop adaptive strategy is able to provide better vibration reduction than the closed-loop configuration. The control algorithms are evaluated using two experimental rigs. In one, a small rotor is supported on rolling element bearings, in the other a larger rotor with dimensions consistent with those found in process plant is supported on oil-film bearings.

NOTATION

j	$\sqrt{-1}$
p	number of control forces
$\underline{Q}$	controlled response vector in frequency domain
$\underline{Q}_0$	uncontrolled response vector in frequency domain
$\underline{R}$	partitioned form of receptence matrix
S	performance index
$\underline{U}$	control force vector in frequency domain
$\underline{\Lambda}$	diagonal weighting matrix
ω	rotational frequency
$()^{-1}$	inverse
$()^{T}$	transpose
$(\hat{})$	estimate
$()^{*}$	conjuqate

1 INTRODUCTION

The potential benefits to be gained by using electromagnetic forces to control rotor vibrations are well known [1,2]. These forces can be generated by electromagnetic devices either in open-loop or closed-loop configurations.

When used in closed-loop an electromagnetic bearing or actuator modifies the eigenvalues and eigenvectors of a rotor bearing system. If the device is controlled by feedback signals of the rotor displacement and velocity, measured only at the rotor section where the force is applied, then this is equivalent to introducing passive stiffness and damping elements acting on the rotor. If the electromagnetic forces are generated by measuring the displacement and velocity at a number of stations along a flexible rotor this allows system eigenvalues to be arbitrarily assigned but it may lead to system instability. When rotor instability may be induced by state feedback Burrows and Sahinkaya [3] have shown that an open-loop strategy can be used to minimise the system vibration.

Chen and Darlow [4] have described the use of an observer [5] to obtain the velocity and acceleration signals required to control the vibration of a rigid shaft. The acceleration feedback was used to cancel the out-of-balance force. In an earlier paper Ulbrich and Anton [6] discussed the use of an induction coil to measure rotor velocity but showed that stray magnetic flux can cause problems. Jayawant et. al. [7] have shown how a Hall effect probe can be used to measure acceleration.

In situations where shaft flexibility must be taken into account the number of states to be measured increases and this further favours the use of an observer instead of direct measurement. However, the presence of noise would probably require the use of a Kalman filter [8] rather than a simple Luenberger observer [5]. In the present state of development a more serious limitation is associated with rotor stability under closed-loop control as mentioned above, and it was this that led the authors to develop an open-loop strategy for multi-mass rotors [3]. Whether an open-loop or closed-loop configuration is adopted electromagnets provide a convenient method for applying the required control forces to the rotor.

Electromagnetic devices can be used either as bearings or force actuators. When used as a bearing they fulfil the dual role of supporting a rotor and providing a control force. If the machine configuration is such that passive bearings (oil film or rolling element) can be used to support the rotor, then a magnetic actuator can be introduced as a control element. This arrangement is studied later in the paper.

Two closed-loop strategies are assessed. In one, fixed gain control is used so that the magnetic actuator provides stiffness and damping effects. In the other, the gain in the position and velocity feedback paths is varied as a function of the rotor speed to study the effect of frequency dependent damping and stiffness, thus providing active control of vibration. In the latter case gain control is achieved by introducing a computer which can be placed within or outside the loop. Incorporation of a digital computer in the loop can lead to unwanted phenomenon due to the phase lag associated with hold circuits etc. [9].

The authors' open-loop strategy [3] is also evaluated. This consists of applying forces along two orthogonal axes at a given rotor section by using measurements of rotor displacement from a number of mass stations.

Two experimental rigs are described, one with a 15 mm diameter rotor supported on rolling element bearings and controlled by a magnetic actuator designed to minimise cost and complexity. The other rig has a 100 mm diameter shaft supported by plain oil film journal bearings and controlled by a high performance magnetic actuator. The rigs are used to examine open and closed strategies and to illustrate the implementation of active control of rotor vibration.

2 CLOSED-LOOP CONTROL

2.1 Description of Rig I

A simple magnetic bearing was designed by Lim [10] and applied to control a 1290 mm long flexible shaft of 15 mm diameter. supported on double row self-aligning ball bearings with a span of 1050 mm. Two silicon steel rigid disks of 100 mm diameter and 0.97 kg mass were located between the bearings. A 0.75 kW variable speed d.c. motor was used to drive the rotor through a 1:3.5 pulley arrangement and a pin chord type coupling. The bearings were pressed into a rectangular steel housing and this was positioned by four proof rings. These rings supported the static load and were fitted with strain gauges to measure the transmitted force. The shaft was modelled by 15 discrete mass stations as shown in Figure 1. The magnetic bearing is located at station 6 and the control loop is shown in Figure 2.

A microcomputer (6502 microprocessor) was used to change the feedback gains in the analogue feedback path as a function of the shaft speed. This configuration is suitable for implementing the adaptive control scheme discussed by Burrows, Sahinkaya and Turkay [11].

The force of attraction between the pole of a magnetic bearing and the journal is proportional to the square of the magnetisation current and inversely proportional to the square of the air gap. Schweitzer and Ulbrich [2] have shown that a premagnetisation current can be used to linearise the force gap relationship. To simplify the bearing design this approach was not adopted because it would have required additional coils on the stator rotor and larger magnets would be necessary to dissipate the heat produced by the current. This premagnetisation current would also produce a negative stiffness effect which would have to be compensated in the control system. Thus the arrangement used is not inherently unstable unlike that discussed later.

An analogue 'square root' circuit was designed to counteract the force-current square relation [10]. The current power amplifiers have negative current feedback to extend their bandwidth, reduce the nonlinearity, and make the gain less sensitive to temperature variations. Tests showed that the magnetic actuator operated linearly over 20% of the clearance space.

2.2 Experimental results

The magnetic bearing damping and stiffness characteristics for various values set by the microprocessor were estimated by implementing a frequency domain algorithm described by Burrows and Sahinkaya [12]. Details of the experiments are given by Lim [10]. The estimated stiffness and damping coefficients in the y direction are given in Figure 3. The nonlinear characteristics which are more pronounced than those reported by some workers are a consequence of the design simplifications mentioned above. The coupling terms between axes were found to be insignificant.

The system shown in Figure 1 was first simulated on a digital computer and values of stiffness and damping were determined to minimise the sum of squares of the displacement at the 15 mass stations as a function of rotor speed. The optimisation was subject to the constraints that the stiffness and damping coefficients are positive with maximum values limited by the magnetic controller. The results are shown in figure 4 and these were subsequently verified experimentally. The flat tops shown in both coefficients is a consequence of the physical limitations inherent in any device as seen in Figure 3. It is seen that simple on-off control of stiffness and damping is adequate to minimise the performance index. This observation is consistent with earlier work [11].

The effect of controlling stiffness and damping on the synchronous response is shown in Figure 5. The predicted response is compared with the experimental results to demonstrate the validity of the model. (For completion, Figure 5 also shows the theoretical response curve (c) obtained by implementing an open-loop strategy which is discussed later.)

Given that a good model can be derived for the system, then a simulation can be performed to determine suitable constant values of feedback gain over the operating range to achieve an 'acceptable' response. This is shown in Figure 6 which compares experimental and predicted values for displacement as a function of rotor speed. This approach provides less flexible control than the alternative adaptive open-loop or closed-loop strategies as discussed later. It has the merit of being simple and cheap to implement and it is an easy matter to ensure system stability. These advantages are outweighed by the inability to optimise the stiffness and damping as a function of ω, and cannot compensate for changes in system dynamics.

3 OPEN-LOOP CONTROL

3.1 Description of Rig II

The encouraging results produced by the preliminary work on a simple rotor led to the design of a rig which is more representative of the dimensions encountered in industrial plant. An overall view is shown in Figure 7. In addition to designing a more representative rotor the decision was made to introduce oil-film bearings so that subsequently it would be possible to study the ability of the magnetic controller to suppress oil whirl using the strategy developed by Sahinkaya and Burrows [13].

To obtain the forces required to control the rotor over an adequate bandwidth it was necessary to employ a more sophisticated electromagnetic force actuator than that used in the preliminary work. A prototype unit was designed and manufactured under the supervision of Professor G. Schweitzer of ETH. The nominal air gap between the force actuator stator and rotor was 1.2 mm, and 1 mm between the emergency bearing and rotor. Thus the pole faces were protected in the event of severe vibration levels being experienced.

The rotor consists of a 100 mm diameter mild steel shaft of nominal length 2358 mm supporting two 90 mm thick, 406 mm diameter overhung steel disks. The extended length of the 100 mm diameter section meant that the oil film bearings could be positioned at any point along the rotor. Drive was provided by a 25 kW d.c. variable speed motor through a flexible tooth belt and universal coupling.

The oil film bearings supporting the rotor were of the circumferential groove type. Each bearing land was 35 mm long with a measured diametral clearance of 0.29 mm. Tellus-T15 oil was supplied via a gravity feed system through a single hole on top of the bearing at which point the measured pressure was 0.4 bar. The complete rotor-drive assembly was mounted on a bedplate of approximate mass 2 tonnes.

Eddy current probes were used to measure vibration response at various positions along the rotor. The output of the transducers was sampled by a mainframe computer. This arrangement allowed easy development of software for computing the on-line control force. Sampling by ADC and DAC devices was controlled by a small slotted disk attached to the end of the rotor. Thus synchronisation of ADC, DAC and rotor was achieved.

3.2 Modelling and control

The distributed mass of the rotor is accounted for by discrete masses located at fourteen mass stations. Tests showed that this was adequate to predict the first two flexural modes of the system. The stiffness of the shaft connecting the mass stations is modelled by introducing lumped spring elements. The bearings are represented by the usual stiffness and damping coefficients. More details of the model are given by Clements [14].

The rotor response in the frequency domain is

$$\underline{Q}(j\omega) = \underline{Q}_o(j\omega) - \underline{R}(j\omega)\underline{U}(j\omega) \qquad (1)$$

where $\underline{Q}_o(j\omega)$ is the response due to the out of balance vector and $\underline{Q}(j\omega)$ is the controlled response. The matrix $\underline{R}(j\omega)$ of order 28xp contains the columns of the inverse impedance matrix corresponding to the chosen p control forces as shown by Burrows and Sahinkaya [3]. In this work p=2.

The parameters values in $\underline{R}(j\omega)$ associated with the bearing stiffness and damping coefficients, which vary with the eccentricity ratio, and hence with rotor speed, cannot be accurately predicted using theoretical expressions. This does not prevent the application of the control algorithm developed by Burrows and Sahinkaya [3] since this can also be used to include on-line estimation of the receptence matrix [14,15].

Equation (1) can be rewritten as

$$\underline{Q}_o(j\omega) = \underline{R}(j\omega)\underline{U}(j\omega) + \underline{Q}(j\omega) \qquad (2)$$

The object is to determine the control force $\underline{U}(j\omega)$ to minimise the displacement vector $\underline{Q}(j\omega)$. This is given by the weighted least square estimator [16] as

$$\underline{U} = (\underline{R}^{*T}\underline{\Lambda}\underline{R})^{-1}\underline{R}^{*T}\underline{\Lambda}\underline{Q}_o \qquad (3)$$

The weighting matrix $\underline{\Lambda}$ can be selected to give weights to the vibration amplitudes at critical sections of the rotor.

The absolute residuals are

$$\underline{Q} = \underline{Q}_o - \underline{R}\underline{U} \qquad (4)$$

The performance index minimised at each frequency ω is

$$S(\omega) = \underline{Q}^{*T}\underline{Q} \qquad (5)$$

In practice it is not feasible to measure the rotor displacements at all mass stations and this can affect the quality of control. However with careful selection of measurement sites it is possible to achieve a significant reduction in the synchronous response due to out of balance effects [15]. When only a limited number of measurements is taken, some elements in $\underline{Q}$ and $\underline{Q}_o$ are eliminated, thus eliminating the corresponding rows in $\underline{R}$ in Equations (1) to (4). This raises a question concerning the quality of control achievable from a limited number of measurements. This is discussed in more detail in reference [15].

3.3 Results

The reduction in synchronous response when the control force is generated from measurements at three shaft locations corresponding to displacements at the oil-film bearings and at the magnetic controller are shown in Figure (8). The response shown is that at the magnetic bearing and it is seen that the amplitude is reduced from 0.3 mm to approximately 0.02 mm at the critical speed. The corresponding forces generated by the magnetic controller are shown in Figure (9). The control signals to be fed to the magnetic controller were computed online by the control computer. The maximum force required to achieve the reduction was 250 Newtons. This compares with the maximum force that can be generated by the magnetic controller of 1 kN.

A wide range of other experiments was performed and details are given by Clements [14].

The benefits of applying the

open-loop control strategy to rig I described earlier was assessed by undertaking a computer simulation. The predicted results are shown as curve (c) in Figure (5) from which it is seen that this strategy is capable of providing better control than that attainable by optimising stiffness and damping coefficients. The forces required to obtain curve (c) in Figure 5 are shown in Figure 10. The stiffness and damping coefficients that would be required in a closed-loop configuration to produce these forces are shown in Figure 11. The corresponding stiffness is negative throughout the range and the damping coefficient is negative upto the first mode. These coefficient values could not be implemented because the system would be unstable.

Comparison of curves (a),(b) and (c) in Figure 5 shows that the open-loop strategy is capable of producing better control of the synchronous vibration than that achieved by closed-loop control whilst avoiding the possibility of rendering the system unstable. Open-loop control is generally regarded as being inferior to a closed-loop configuration. For example, in open-loop no account is taken of parameter variations, but this disadvantage is overcome by the strategy described earlier.

4 CONCLUSIONS

It has been shown that the open-loop control strategy developed by Burrows and Sahinkaya[3] can be used to overcome the inherent problems associated with the closed-loop control of flexible rotors. Unlike closed-loop control, which can cause a stable system to become unstable, the open-loop strategy is benign whilst significantly reducing the synchronous vibration.

The preliminary results presented suggest that open-loop control can attenuate vibration more than is possible using closed-loop control.

The algorithm adopted here overcomes the problems normally associated with open-loop control in that parameter changes are accounted for by estimating the system parameters before proceeding to compute the control forces.

ACKNOWLEDGEMENTS

The authors acknowledge the financial support provided by SERC. The assistance of Dr. Clements and Dr. Lim during various phases of this work is gratefully recorded. Paul Morton provided valuable advice during construction of the rig.

REFERENCES

(1) Habermann,H. and Liard, G., An active magnetic bearing system, Tribology Intl., 1980, 13, 85-88.

(2) Schweitzer, G. and Ulbrich, H., Magnetic bearings - A novel type of suspension, I.Mech.E. Conf. Vibrations in Rotating Machinery, 1980, C273/80, 151,156.

(3) Burrows, C.R. and Sahinkaya, M.N., Parameter estimation of multi-mode rotor bearing systems, Proc. Roy. Soc., Lond., 1982, A379, 367-387.

(4) Chen, H.M. and Darlow, M.A., Magnetic bearing with rotating force control, Trans. ASME, Jnl. of Tribology, 1988, 110, 100-105.

(5) Luenberger, D.G., An introduction to observers, IEEE Trans. on Auto. Cont., 1971, AC-16, No.6, 569-602.

(6) Ulbrich, H. and Anton, E., Theory and application of magnetic bearings with integrated displacements and velocity sensors, 3rd Intl. Conf. I.Mech.E., 1984, 543-551.

(7) Jayawant, B.V. et al., Development of 1-ton magnetically suspended vehicle using controlled D.C. electromagnets, Proc. IEE, 1976, 123, No.9, 941-948.

(8) Leondes, C.T., Control and dynamic systems, Academic Press, 12, 1976.

(9) Katz, P., Digital Control Using Microprocessors, Prentice Hall Int., 1981.

(10)Lim, T.M., The Development of a Simple Magnetic Bearing for Vibration Control, PhD thesis, 1987, University of Strathclyde.

(11)Burrows, C.R., Sahinkaya, M.N. and Turkay, O.S., An adaptive squeeze film bearing, Trans. ASME, Jnl. Lubr. Tech., 1983.

(12)Burrows, C.R. and Sahinkaya, M.N., Frequency domain estimation of linearised oil film coefficients, Jnl. Lubr. Tech., Trans. ASME, 1982, 104, 210-215.

(13)Sahinkaya, M.N. and Burrows, C.R., Control of stability and the synchronous vibration of a flexible rotor supported on oil film bearings, Trans ASME, Jnl. Dyn. Syst. Meas. & Control, 1985, 107, 139-144.

(14)Clements, S., On line vibration control of a flexible rotor bearing system, PhD thesis, 1987, University of Strathclyde.

(15) Burrows, C.R., Sahinkaya, M.N. and Clements, S., Electromagnetic control of oil film supported rotors using sparse measurements, Presented at the 1987 ASME Design Technology Conferences- 11th Biennial Conf. on Mechanical Vibrations and Noise, Boston, U.S.A., 1987.

(16) Johnston, J., Econometric Methods, NewYork:McGraw-Hill., 1972.

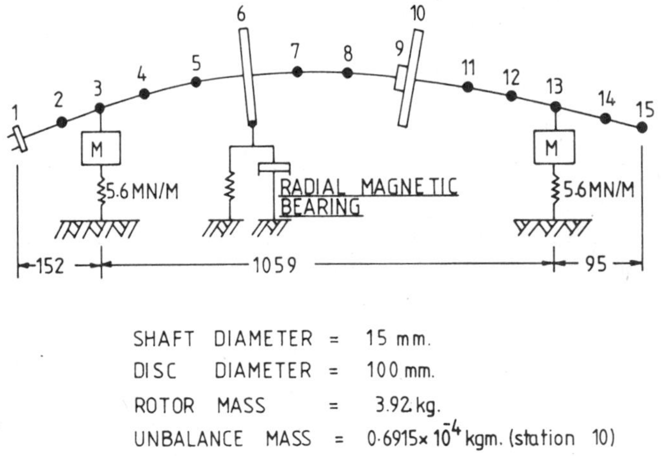

Fig 1 Mathematical model for experimental rig I

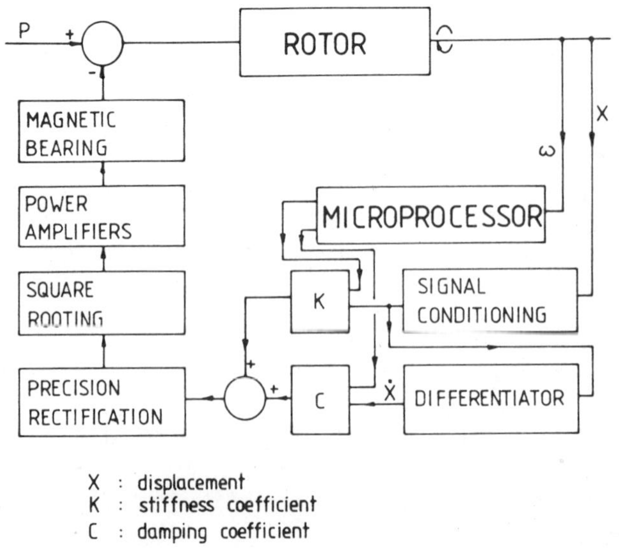

Fig 2 Magnetic bearing control loop (rig I)

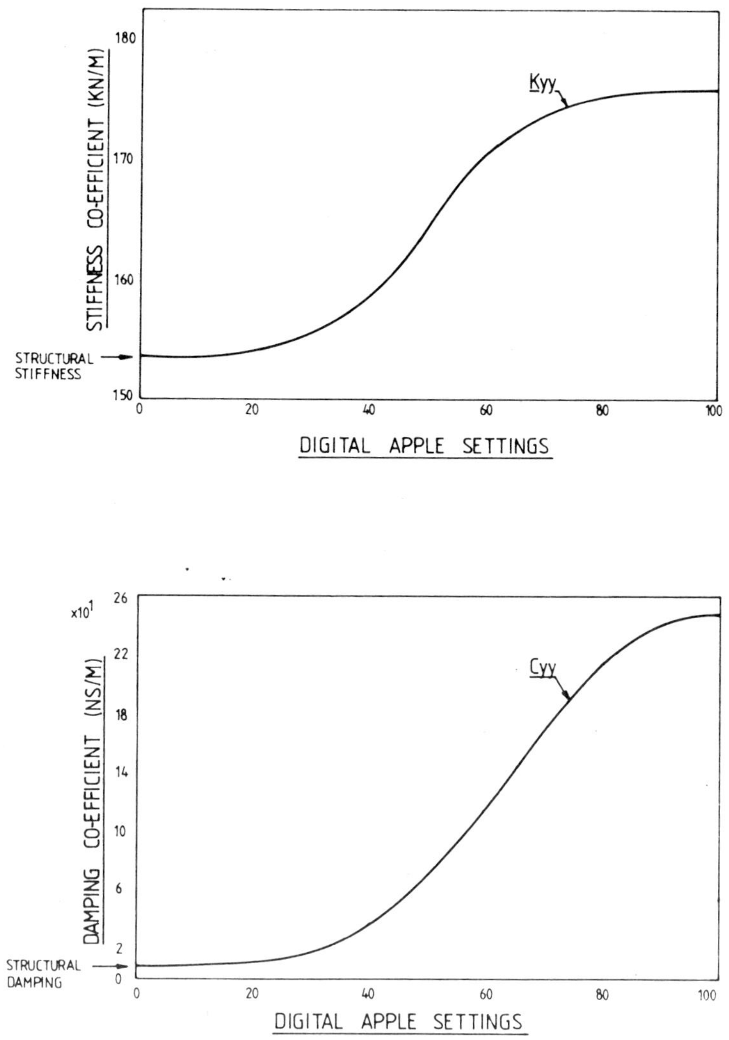

Fig 3 Estimation of stiffness and damping coefficients

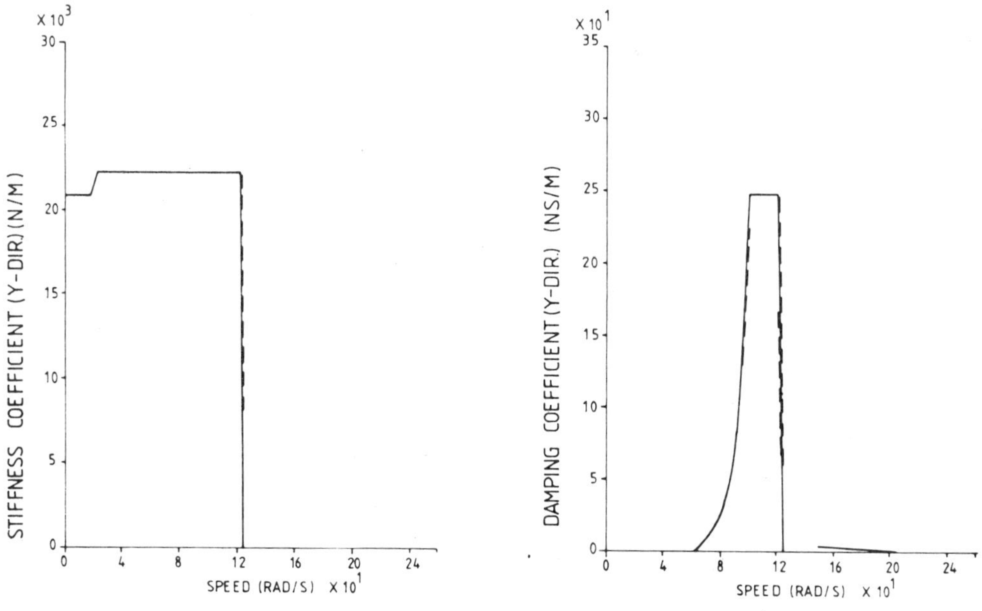

Fig 4 Stiffness and damping coefficients for optimum closed-loop vibration control

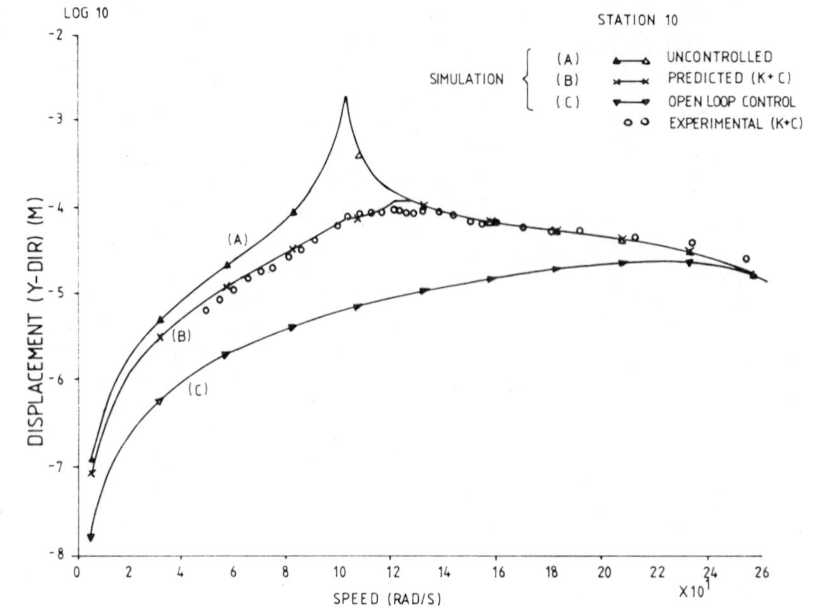

Fig 5 Comparison between measured and predicted optimum response

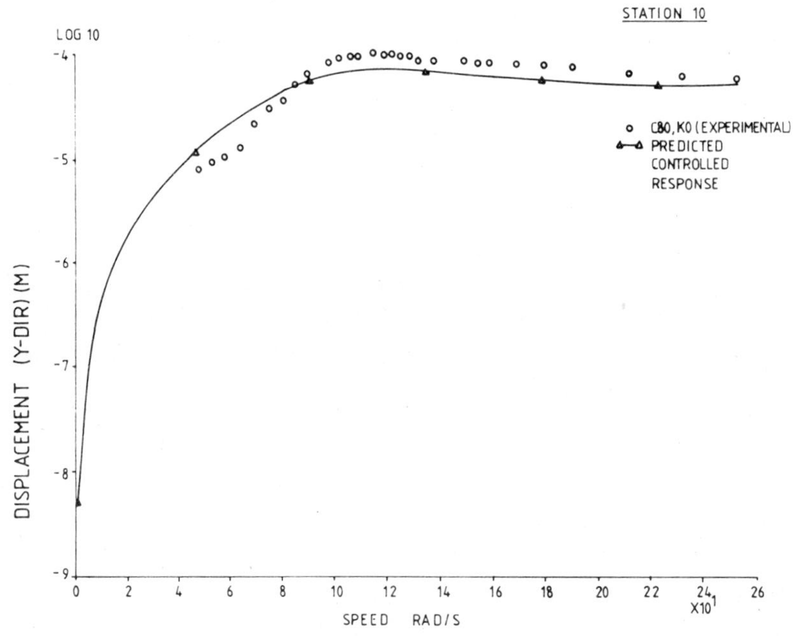

Fig 6 Controlled response with damping at 214 Ns/m

Fig 7 View of rig II

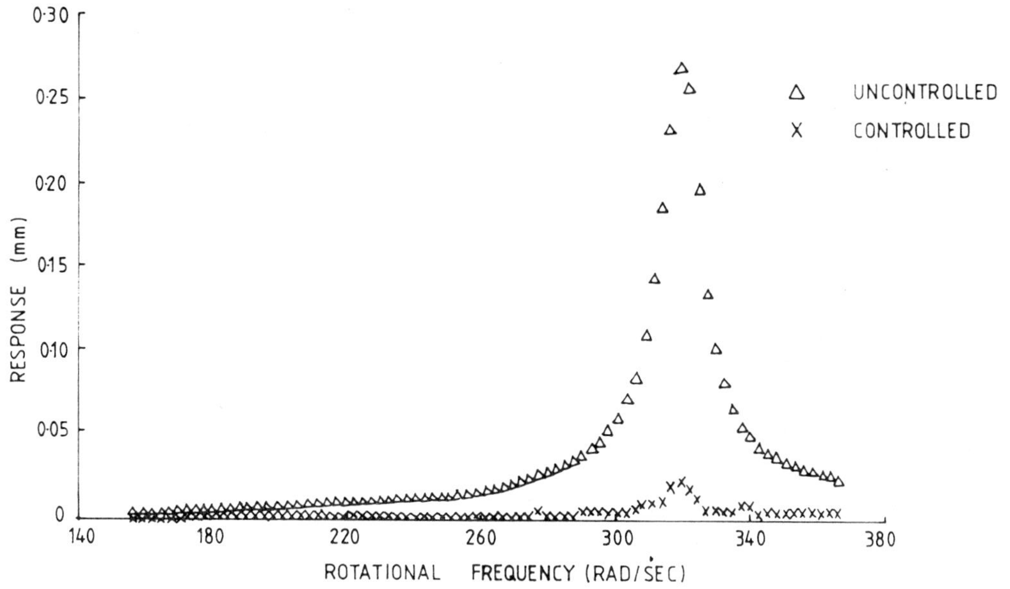

Fig 8 Controlled and uncontrolled response — mid-span

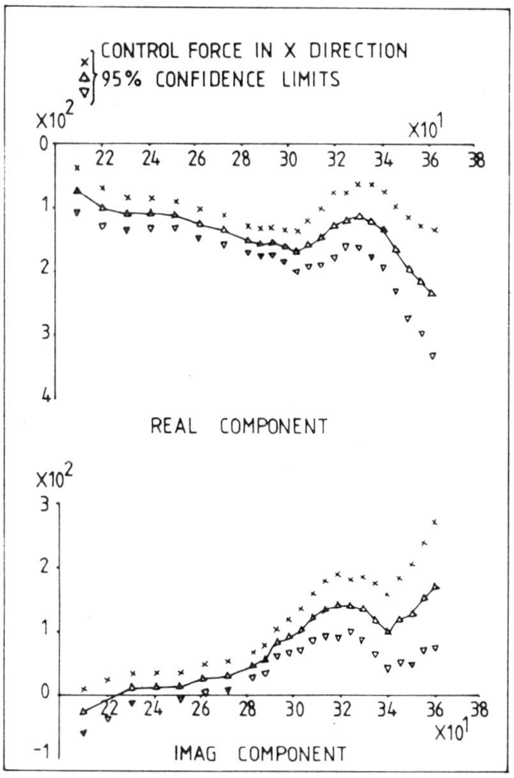

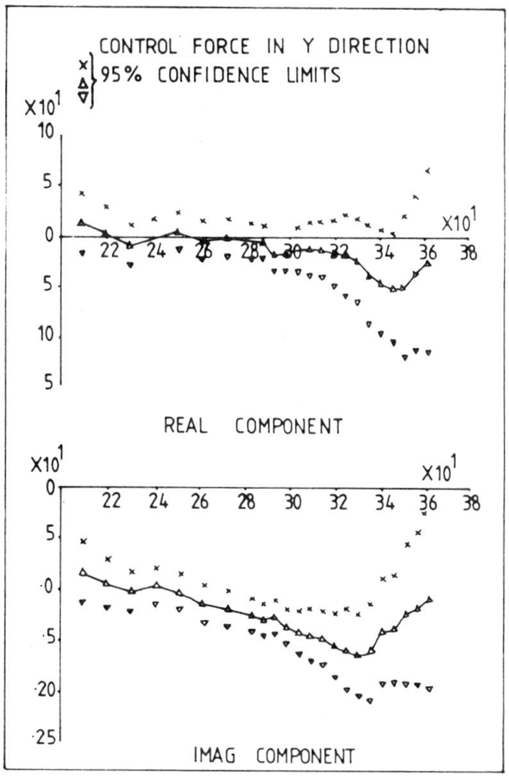

Fig 9 Control forces determined on-line

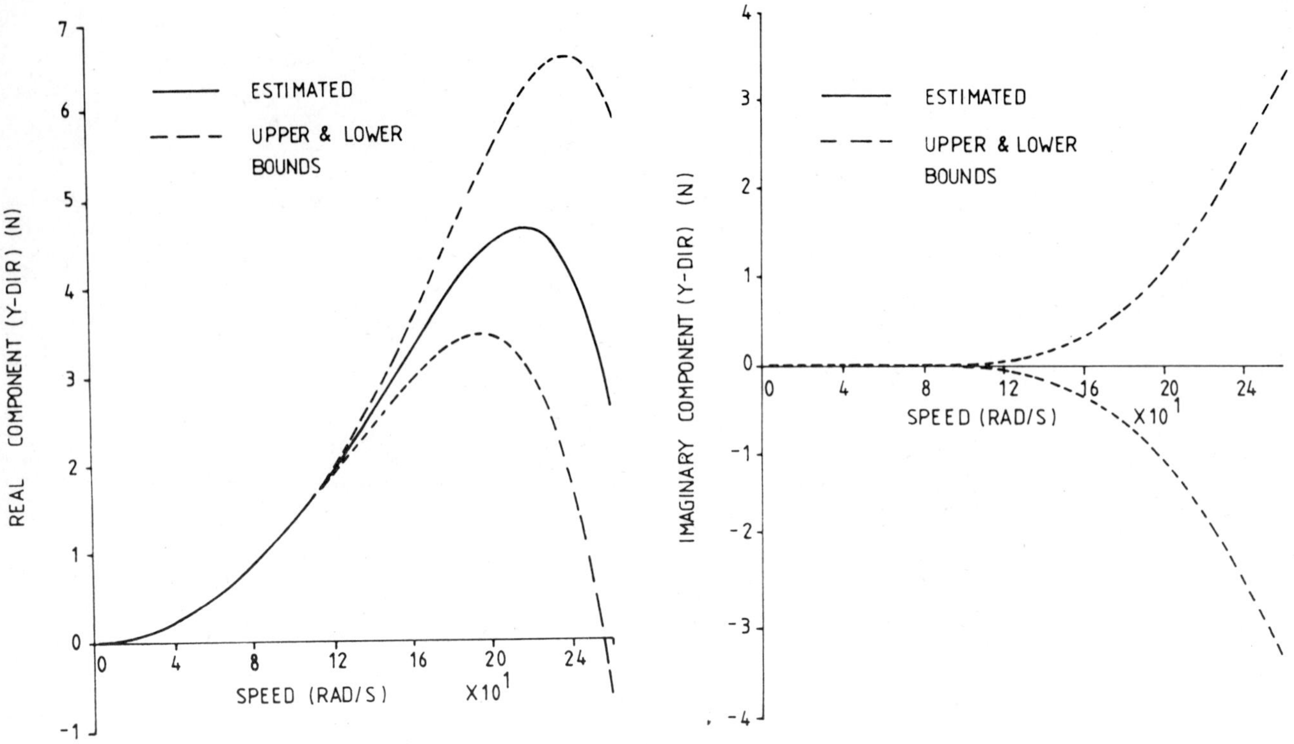

Fig 10 Estimated open-loop control forces for rig I

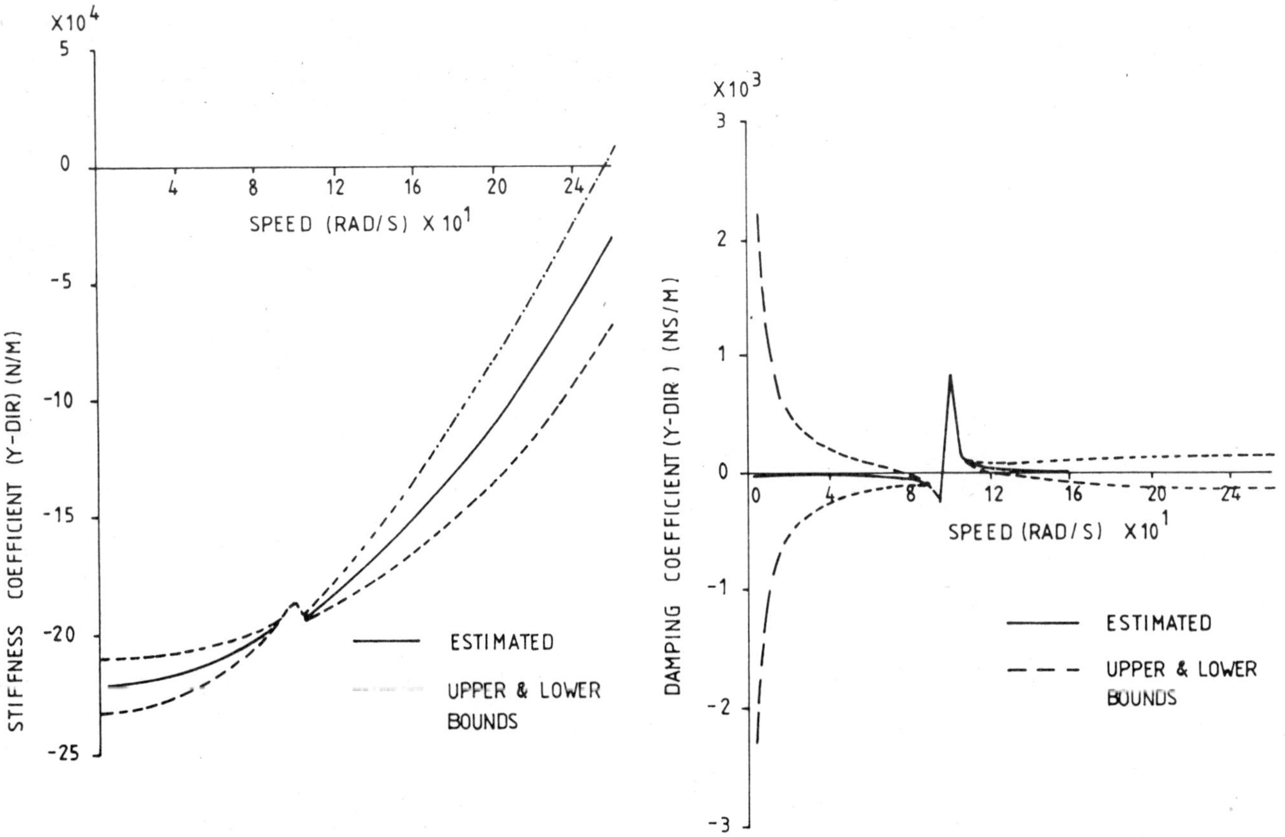

Fig 11 Stiffness and damping coefficients corresponding to open-loop control force (rig I)

C290/88

Active synchronous response control of rigid-body rotors

B G JOHNSON, DSc, MAIAA, MASME, R L HOCKNEY, MSc, PE, MIEEE and K M MISOVEC, MSc, MAIAA
Satcon Technology Corporation, Cambridge, Massachusetts, United States of America

SYNOPSIS This paper examines the control of "mass unbalance" in a rigid-body rotor suspended in magnetic bearings. The weaknesses of two conventional control approaches, simple lead-lag compensation and the use of tracking notch filters, are examined. An improved compensation scheme that combines the good stability performance of the lead-lag compensators with the good synchronous performance of the tracking notch filter is then presented. These controller algorithms are analytically compared using a dynamic model of the Combined Attitude, Reference, and Energy Storage (CARES) system as an example magnetically suspended rotor.

INTRODUCTION

The use of magnetic bearings to support rotating structures without contact has received increasing attention during the last ten years primarily because of improvements in magnetic materials and advances in low cost, reliable power and control electronics. Besides the advantages of frictionless support, long life, and high rotational speeds, magnetic bearings are also being used to achieve better rotordynamic behaviour than is possible with conventional bearings (1,2). One area where improved rotordynamics is possible with magnetic bearings is the control of synchronous vibrations caused by "unbalanced" rotors (3,4).

This paper will examine and compare three different approaches to active control of synchronous vibrations caused by mass unbalances. The goal of these approaches is to allow the rotor to spin about its true center of mass, therefore requiring no synchronous forces to be produced by the magnetic bearing. The first of these is the use of conventional lead-lag compensators that mimic the dynamic behaviour of conventional bearings modelled as a spring and damper in parallel. The second is the use notch filters, which eliminate the response to mass unbalances, in series with the lead-lag compensation (3,5,6). These tracking notch filters (TNF), however, introduce stability problems even for rigid-body rotors. The third approach is the recently developed tracking differential-notch filter (TDNF) which combines the good stability properties of the lead-lag compensator with the good synchronous response properties of the tracking notch filter approach.

In this paper, the control approaches will be examined for a specific rotor system, the Combined Attitude, Reference, and Energy Storage (CARES) system. This single system is designed to perform spacecraft attitude control, attitude determination, and energy storage functions that are currently performed separately by reaction or momentum wheels, inertial instruments, and batteries. A CARES system offers weight and life-cycle cost reductions compared to conventional single-function systems. Recent papers by Eisenhaure (7), Rockwell (8), and Boeing (9) provide analyses of the advantages of combined energy storage and attitude control system over conventional approaches.

The CARES system is based on magnetically suspended, high-energy-density flywheels used to store both energy and angular momentum. The laboratory module shown in Figure 1 consists of a central electromechanical hub connected to an outer flywheel. The electromechanical hub contains the large-angle magnetic suspension, motor/generator, and position sensors. A unique feature of this CARES design is the large-angle magnetic suspension, which allows limited-freedom gimballing of approximately 10 degrees. This gimballing freedom in the magnetic suspension and motor/generator allows the CARES system to be used as a control-moment-gyro for spacecraft attitude control. Detailed presentations of the various subsystems that comprise the CARES system can be found in four recent theses by Larkin (10), Downer (11), Foley (12), and Johnson (13).

MODELLING

For the purposes of this paper, the translational or radial dynamics will be modelled by considering the CARES rotor to be a rigid-body and assuming the magnetic bearings, because they are Lorentz-force actuators, behave as ideal current-controlled force sources. A general block-diagram of the magnetic bearing system is shown in Figure 2. For the translational dynamics of the CARES system, the outputs of the plant are the measured rotor radial position (z). This measured position is used as the input to the controller, which is a dynamic system implemented in either analog or digital electronics. The outputs of the controller are the control currents that drive the magnetic bearings. For the translational dynamics, the magnetic bearings produce radial forces that act on the hub of the rotor. Because the magnetic bearings used in the CARES system use Lorentz forces, the force-gap coupling typical of most magnetic bearings systems, shown by the dashed lines in Figure 2, is negligible and will be ignored.

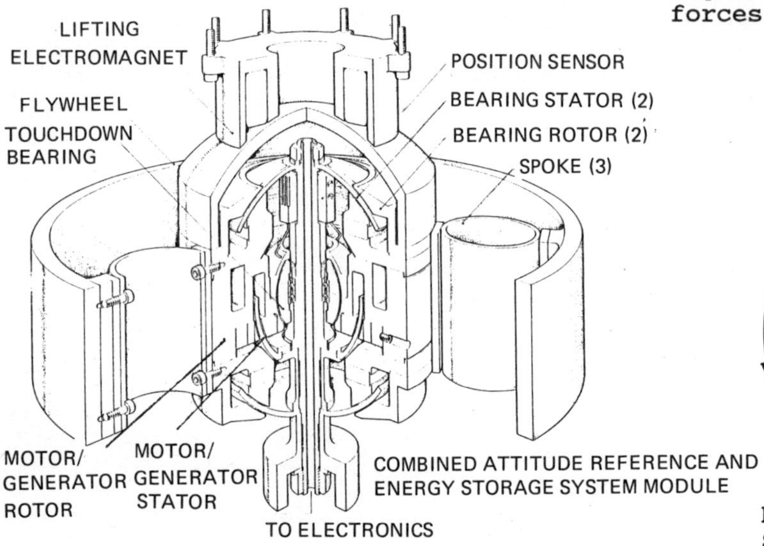

Fig 1 Cut-away view of CARES laboratory prototype

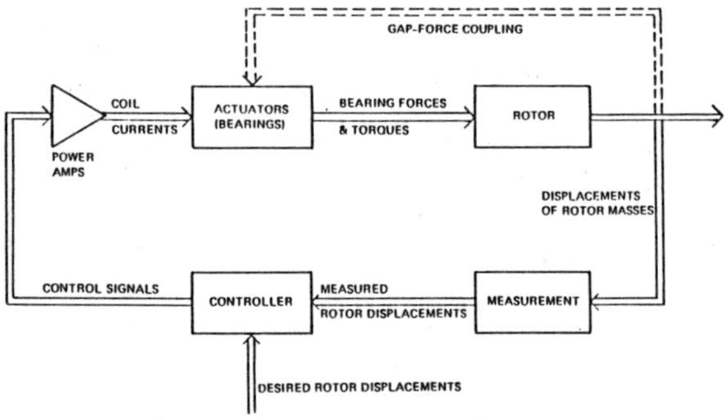

Fig 2 Magnetic bearing block diagram

The "mass unbalance" problem in this system arises because the measurement system, usually employing inductive or electo-optical position sensors, does not measure the position of the rotor center of mass. The position sensors actually measure the distance to a machined surface, which may be imperfect, giving a corrupted measurement of hub position. An exaggerated view of this is shown in Figure 3. In general, the measurement system will produce an output that consists of the center of mass position and components at harmonics of the rotational speed Ω. For machined surfaces, the dominant error mechanism is the misalignment between the geometrical center of the machined surface being measured (S_h) and the rotor center of mass (M_h).

Using complex notation, the position of the measurement center is given by

$$S_h = z = x + jy = z + \epsilon_s e^{j\Omega t} \quad (1)$$

where x and y are the radial positions of the measurement center, j is the square root of -1, and z is the complex position of the true center of mass. The effect of the measurement surface misalignment, therefore, is to add a sinusoidally varying disturbance signal to the measurement of the rotor center of mass position. This corrupted measurement will be used as a feedback signal for the magnetic bearing, producing synchronous forces in the magnetic bearing.

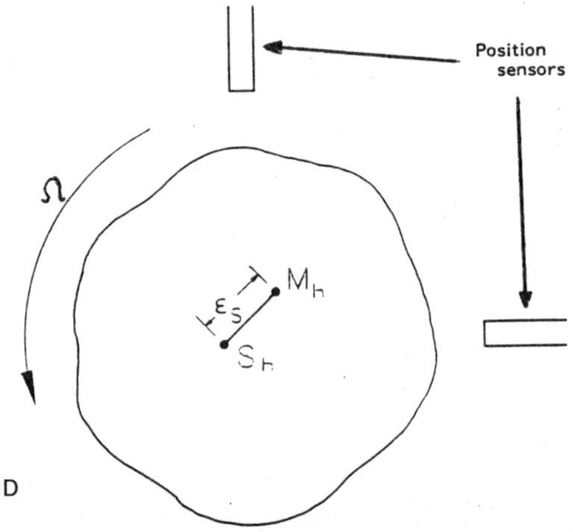

M_h -- Hub center of mass
S_h -- Center of measurement

Fig 3 Translational measurement geometry

LEAD-LAG COMPENSATION

Lead-lag compensation of the rotor position measurement is the underlying controller for the notch filter compensators considered in this paper. The block diagram of the lead-lag compensator is shown in Figure 4. Shown is the complex single input, single output (3,13) control loop which describes the dynamics of the two perpendicular, radial axes. Alternatively, this loop can be interpreted in a more conventional manner as a single axis of the two perpendicular and identical radial axes. The plant is shown as a simple rigid body, double integrator with the radial force as the input and rotor center of mass position as the output. The measured center of mass position is corrupted by the measurement error, as discussed above. The measurement error is an additive, synchronous error at the rotational frequency (Ω). The measured rotor position is subtracted from the commanded position and used as input to the lead-lag compensator. Note that the "mass unbalance" manifests itself as an additive output error.

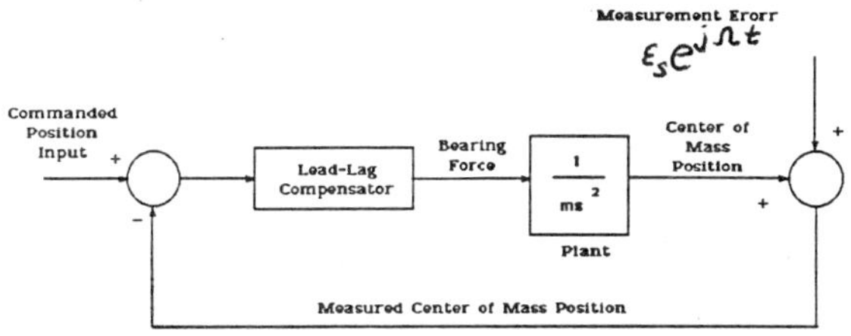

Fig 4 Lead–lag block diagram

The loop gain of the lead-lag compensator and plant is shown in Figure 5. The rigid-body cross-over frequency is chosen to be 300 rad/sec, yielding a moderate bandwith controller with good stability properties (13). Until after the cross-over frequency, the lead-lag and spring-damper are the same, yielding the same closed-loop dynamics inside the closed-loop bandwidth. Because of this, the lead-lag compensator is often used to mimic the behavior of a conventional spring-damper bearing.

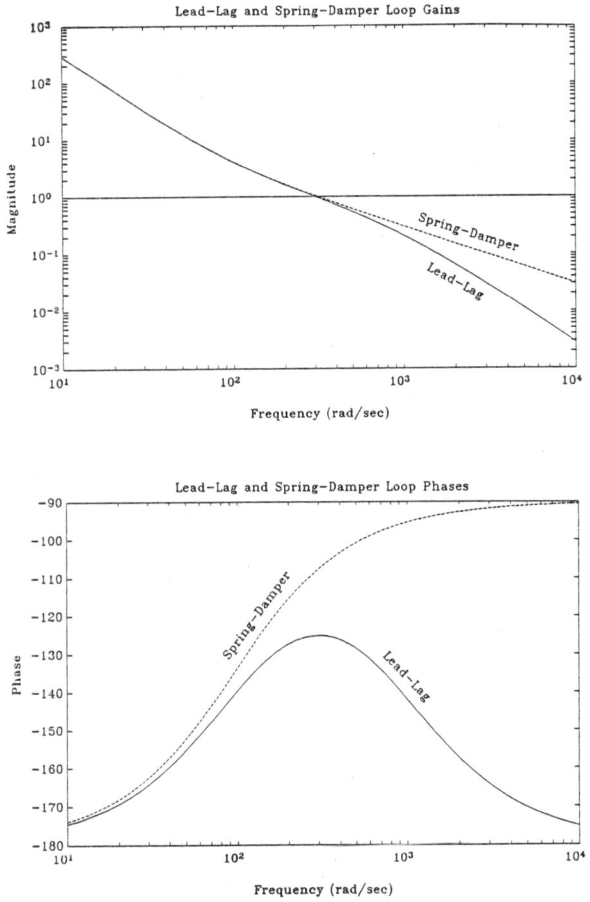

Fig 5 Lead–lag loop transfer function

In all the results presented in this report, one decade of lead will be centered about the cross-over frequency, as shown in Figure 5. This gives about 60 degrees of phase margin at cross-over resulting in well damped eigenvalues. Because of the good eigenvalue damping, the synchronous response of the system is well damped. This can be seen in solid line of Figure 6, a plot of the normalized center of mass amplitude versus rotational speed. The center of mass amplitude has been normalized by the measurement error distance (ϵ_s), which in conventional rotor systems corresponds to the mass unbalance distance.

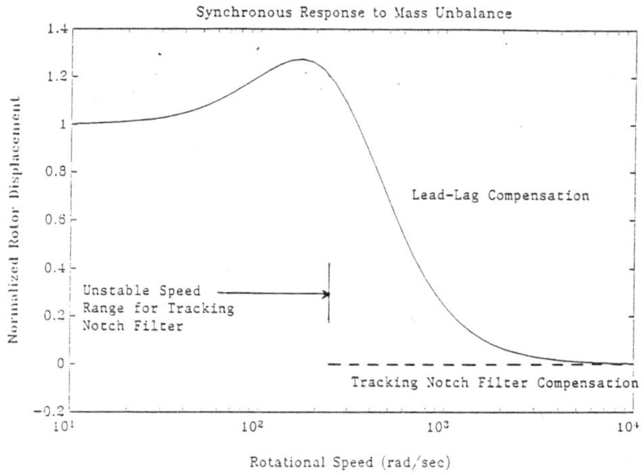

Fig 6 Synchronous response

At low, subcritical speeds, the system spins about the measurement center S_h with the center of mass whirling about S_h with an amplitude equal to the measurement error distance ϵ_s (see Figures 3 and 4). As the system goes through the critical speed, the center of mass position shows only slight peaking because of the good damping. At higher, supercritical speeds the system spins about the rotor center of mass.

LEAD-LAG COMPENSATOR WITH TRACKING NOTCH FILTER

This controller adds a notch filter at the synchronous frequency (rotational speed) into the feedback loop as shown in Figure 7. The notch frequency of this filter tracks the synchronous frequency (rotational speed), hence the name synchronous tracking filter. The tracking notch filter will eliminate all signals at the synchronous frequency from the control loop, including the measurement error ("mass unbalance"). Removing all signals at the synchronous frequency, however, can result in closed-loop instability, limiting the usefulness of this approach.

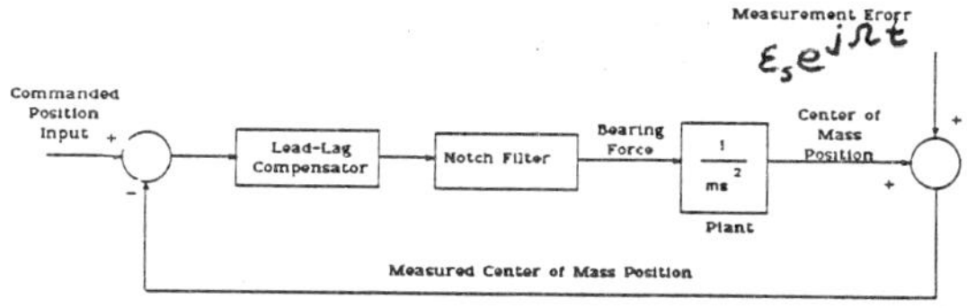

Fig 7 Tracking notch filter (TNF) block diagram

The notch used in this investigation is an ideal, two pole transfer function with infinite notch depth. The notch transfer function $N(s,\Omega)$ is a function of both frequency (Laplace variable s) and rotational speed (Ω). It is given by

$$N(s,\Omega) = \frac{s^2 + \Omega^2}{s^2 + \Omega s/Q + \Omega^2} \qquad (2)$$

The steepness of the notch is determined by the parameter Q. A larger Q makes a steeper notch[1]. In all the results presented in this paper, the notch Q has been set to 10.

Figure 8 is a Bode plot of the loop transfer function for a subcritical rotational speed of 100 rad/sec and a notch Q of 10. As can be seen, the Q of 10 leads to a relatively sharp notch. The infinite depth of the notch implies that at the synchronous frequency, the bearing system has no stiffness. Note that the phase of the loop transfer function is dramatically different than without the notch (Figure 5). The negative phase that the notch filter adds when the loop gain is greater than unity leads to instability in this case. At higher rotational speeds, speeds greater than the cross-over frequency, however, the phase loss caused by the notch filter occurs at frequencies past cross-over and does not cause instability.

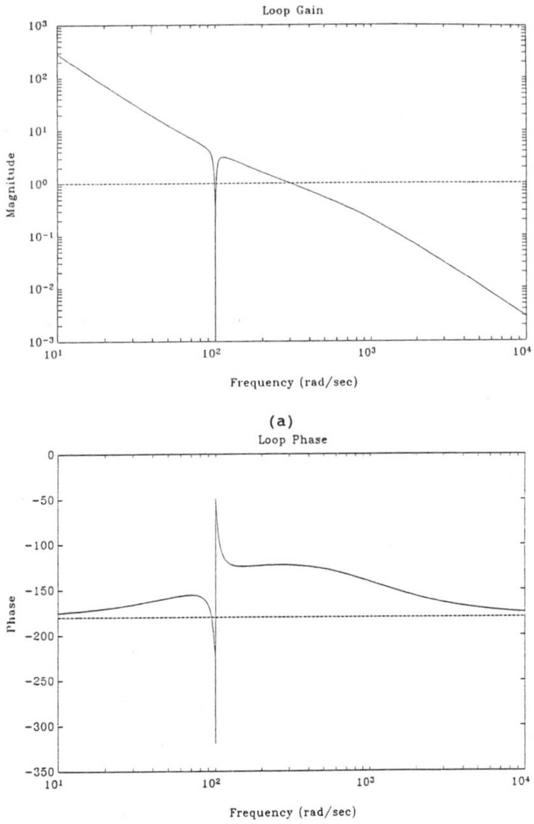

Fig 8 Tracking notch filter (TNF) loop transfer function

The stability results can be more easily seen in the dashed line of Figure 9. This is plot of the minimum closed-loop eigenvalue damping versus rotational speed for all three controllers. The solid line is the lead-lag compensator, the dashed line the tracking notch filter, and the dotted line the differential tracking notch filter to be presented in the next section. The tracking notch filter system is closed-loop unstable (negative damping) until a rotational speed of 250 rad/sec, approximately the cross-over frequency of the system.

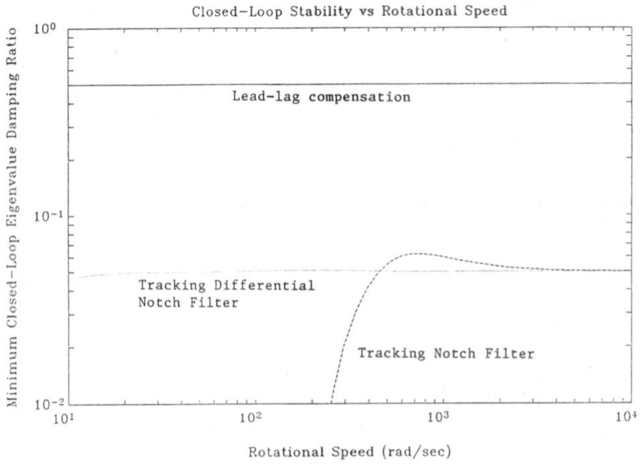

Fig 9 Stability versus rotational speed

The synchronous response of the tracking notch filter is shown in Figure 6 by the dashed line. In its stable speed range, the tracking notch filter perfectly rejects the "mass unbalance" and allows the rotor to spin about its center of mass. For comparison, the synchronous response of the lead-lag compensator by itself is also shown. The important result is that the tracking notch filter can only be used at supercritical rotational speeds. In this supercritical speed range the notch filter will perform better than the simple lead-lag compensator.

LEAD-LAG COMPENSATOR WITH TRACKING DIFFERENTIAL-NOTCH FILTER

As was shown in the last section, a tracking notch filter perfectly rejects the measurement error (mass unbalance) of a rigid-body rotor. It suffers, however, from a limited rotational speed range over which it is stable. The recently developed tracking differential-notch filter (TDNF) retains the good synchronous response performance of the conventional tracking notch filter but with vastly improved stability properties.

The stability problems of the tracking notch filter arise because the notch filter, while eliminating the synchronous measurement error, also eliminates any desirable control signals at the synchronous frequency. The differential notch filter alleviates this problem by re-inserting the desirable control signals into the feedback control loop. A block diagram of the system with the tracking differential-notch filter is shown in Figure 10. If the path through the plant model were eliminated, the system shown in Figure 10 would be the same as the tracking notch filter presented in the last section. Note, however, that the tracking notch filter has been implemented in a special way. Notch filtering of the center of mass position of the rotor achieved by subtracting a synchronous-tracking,

---

[1] The use of a more general finite depth notch filter has been investigated by Beatty (5).

bandpass-filtered version of the signal from itself. This results in a synchronous-tracking notch filter. The measurement error $\epsilon_s e^{j\Omega t}$ is therefore perfectly filtered from the feedback signal used to drive the compensator. The output of the compensator, which is assumed to be a simple lead-lag as before, is used produce the bearing force driving the plant.

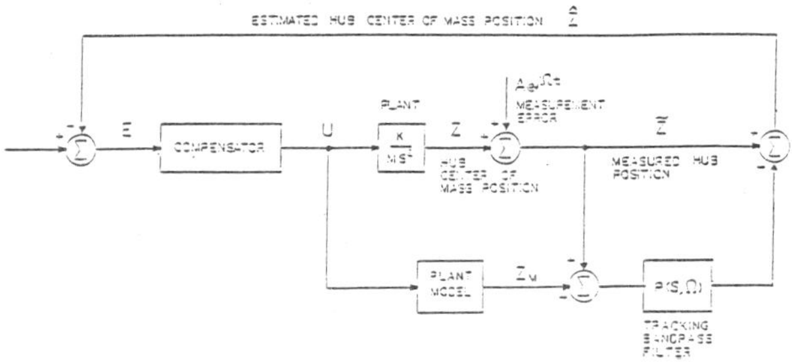

Fig 10 Tracking differential notch filter (TDNF) block diagram

The difference between the tracking differential-notch filter (TDNF) of Figure 10 and the tracking notch filter (TNF) of Figure 7 is the addition to the TDNF of a feedforward path utilizing a model of the plant. The purpose of this path is to re-insert the synchronous component of the plant signal that is lost because of the notch filtering of the plant output. This component is formed by routing the plant input signal through the plant model followed by the synchronous bandpass filter. This route provides an estimate of the desirable synchronous component that has been eliminated by the notch filtering of the actual plant output.

The estimated rotor center of mass position $\hat{z}(s)$ is given by

$$\hat{z}(s) = z(s) + (1 - P(s,\Omega))\cdot\epsilon_s e^{j\Omega t} - P(s,\Omega)\cdot(z(s) - z_m(s)) \qquad (3)$$

where

$\hat{z}(s)$ = Estimated rotor center of mass position
$z(s)$ = Rotor center of mass position
$P(s,\Omega)$ = Tracking bandpass filter
ϵ_s = Measurement error distance (mass unbalance distance)
Ω = Rotational speed
$z_m(s)$ = Output of plant model (modelled rotor center of mass position.

The bandpass filter $P(s,\Omega)$ is a function of the rotational speed Ω, with the passband tracking the rotational speed. The transfer function of this synchronous tracking bandpass filter is

$$P(s,\Omega) = \frac{\dfrac{\Omega_f s}{Q}}{s^2 + \dfrac{\Omega_f s}{Q} + \Omega_f^2} \qquad (4)$$

where as before, Q is a parameter of the filter that determines the steepness of the passband. The parameter Ω_f is the center frequency of the passband. It is nominally set equal to the actual rotational speed Ω. Note that at the synchronous frequency Ω, the bandpass filter has unity gain if the synchronous frequency Ω is equal to the passband center frequency Ω_f.

The rotor estimated center of mass position (Equation 3) can be rearranged to give

$$\hat{z}(s) = (1 - P(s,\Omega))\cdot z(s) + P(s,\Omega)\cdot z_m(s) \qquad (5)$$

where the bandpass filter $P(s,\Omega)$ is assumed to have unity gain at the synchronous frequency Ω. The important result here is that the estimated rotor center of mass position z(s) is not a function of the measurement error. In other words, this system perfectly rejects the measurement error (mass unbalance). Furthermore, if the plant model perfectly models the plant[2], the estimated rotor center of mass position is a perfect estimate of the actual rotor center of mass position.

An obvious choice for the plant model is the same as the plant, with a transfer function

$$G(s) \quad \frac{1}{ms_2} \qquad (6)$$

This plant model, however, does not work because the resulting system is uncontrollable through the input u(s). This happens because the input u(s) sees two identical, parallel paths. Only one of these can be stabilized by the input, as is shown in (3). A relatively simple solution to this problem exists. The output of the plant model is only used at the synchronous frequency Ω. At lower frequencies, the plant model need not be an accurate representation of the plant. Therefore the plant model can be made open-loop stable with

$$G_m(s) = \frac{\dfrac{1}{m}}{s^2 + 2\zeta\omega_o s + \omega_o^2} \qquad (7)$$

where

$G_m(s)$ = plant model
m = plant mass
ω_o = natural frequency of model
ζ = model damping

The natural frequency of the model (ω_o) is chosen to be below the range for which synchronous response control is important. The model damping (ζ) is

---

[2] A discussion of system performance when the plant model is inaccurate can be found in (3).

chosen to give a well damped model. For the examples in this section, ω_o is set to 5 rad/sec and ζ to 0.707.

The bode plot of the loop transfer function for this controller, with the loop broken at the compensator input, is shown in Figure 11. For this case, the rotational speed Ω is 100 rad/sec, the filter Q is 10 and the crossover frequency (ω_c) of the system, as determined by the lead-lag compensator, is 300 rad/sec. Since the rotational speed is less than the crossover frequency, this is a subcritical speed. The loop transfer function for this tracking differential notch filter (Figure 11) is essentially the same as for the simple lead-lag compensator (Figure 5). Importantly, the differential notch filter does not distort the loop gain as the simple notch filter does, as can be seen by comparing Figures 11 and 8. The tracking differential- notch filter remains stable at all speeds, as can be seen from the dotted line in Figure 9. The closed-loop damping ratios shown in Figure 9 indicate the speed ranges where the different control approaches are stable or unstable. Beyond determining the absolute stability of the control approaches, this plot is not useful in determining the relative stability since the notch filter approaches also feature transmission zeros that mitigate the effect of the lightly damped poles. In fact, the relative stability of the lead-lag and TDNF is nearly the same, as can be seen by the similar phase margins in Figures 5 and 11.

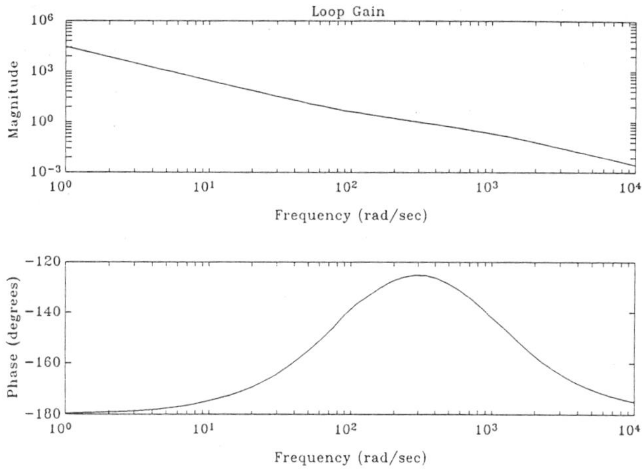

Fig 11 Tracking differential notch filter (TDNF) loop transfer function

The tracking differential-notch filter, however, also has the good synchronous response characteristics of the tracking notch filter. This can be seen in Figure 12, a plot of closed-loop transfer function from measurement error to rotor center of mass position for the same conditions as Figures 11. The important result is that the closed-loop gain at the synchronous frequency of 100 rad/sec is zero. This system, therefore, perfectly rejects the measurement error (mass unbalance). The TDNF, therefore, has "perfect" synchronous response. This contrasts with the synchronous performance of the simple lead-lag compensator as can be seen in Figure 6.

The important result of this section is that the differential notch filter nominally provides perfect rejection of the measurement error (mass unbalance) yet retains the good stability properties of the simple lead-lag compensator. Because the measurement error is rejected, the rotor spins about its center of mass at all speeds above the natural frequency of the plant model, which can be made arbitrarily small. Since the rotor spins about its center of mass, no synchronous forces are produced by the measurement error (mass unbalance). This means that the rotor spins as if it were perfectly balanced, for subcritical, critical and supercritical speed ranges.

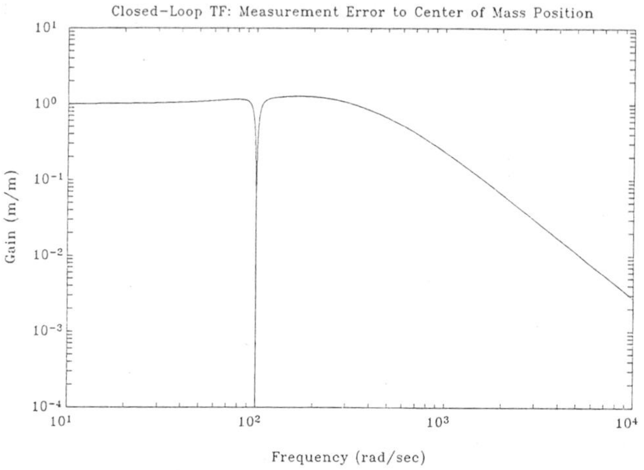

Fig 12 Tracking differential notch filter (TDNF) closed-loop gain

The tracking differential-notch filter (TDNF) developed in this section has been shown to provide the good stability of a simple lead-lag compensator combined with the good synchronous response performance gained by the addition of a notch filter. A relatively simple implementation of the TDNF has been developed comprised mainly of chip-level subcomponents and a low bandwidth digital section (3). In addition, the TDNF can be easily modified to allow adaptive control, through variation of the plant model used to augment the controller.

SUMMARY AND CONCLUSIONS

This paper has examined three different approaches to the control of mass-unbalance vibrations in magnetic bearing systems supporting rigid-body rotors. The goal of these systems is to allow the rotor to spin about its center of mass at all rotational speeds, thereby producing no synchronous forces in the magnetic bearings. The simplest approach, the use of lead-lag compensation, was shown to have good stability properties but produces synchronous forces. The other conventional approach, the use of tracking notch filters in the feedback path, was shown to have "perfect" synchronous response performance but stability only in limited speed ranges. An improved compensation scheme that combines the good stability performance of the lead-lag compensators with the good synchronous performance of the tracking notch filter was then developed. This tracking differential-notch filter is stable for all speeds and also produces "perfect" synchronous response to mass unbalance of the rigid-rotor.

REFERENCES

1) Schweitzer, G. Stabilization of Self-Excited Rotor Vibrations by an Active Damper. In Dynamics of Rotors, ed., FI Niordson, Springer-Verlag, New York: 1975.

2) Stanway, R. Active Vibration Control of a Flexible Rotor on Flexibly mounted Journal Bearings. ASME Journal of Dynamic Systems, Measurement and Control December 1981, v 103 p 383-388.

3) Johnson, BG, RL Hockney and KM Misovec. Synchronous Response Modelling and Control of an Annular Momentum Control Device. SatCon Technology Corporation Report R09-87, Cambridge Massachusetts: 1988.

4) Habermann, H and M Brunet. The Active Magnetic Bearing Enables Optimum Control of Machine Vibrations. ASME Gas Turbine Conference, No 85-GT-221, March 1985.

5) Beatty R. Synchronous Response Control of Rotors Using Tracking Notch Filters. S.M. Thesis, Massachusetts Institute of Technology, Cambridge Massachusetts: May 1988.

6) Weise, DA. Active Magnetic Bearings Provide Closed Loop Servo Control for Enhanced Dynamic Response. 27th IEEE Machine Tool Conference, Oct, 1985.

7) Eisenhaure, DB, Downer, JR, Bliamptis, T. and Hendrie, S. A Combined Attitude, Reference, and Energy Storage System for Satellite Applications. AIAA 22nd Aerospace Sciences Meeting, Reno, Nevada, January 9-12, 1984.

8) Rockwell International Corporation. Advanced Integrated Power and Attitude Control System (IPACS) Study. NASA Contractor Report 3912, 1985.

9) Boeing Aerospace Co. Study of Flywheel Energy Storage for Space Stations. Final Report NO.. D-180-27951-1, February 1984, Seattle, Washington.

10) Larkin, LL. Design and Optimization of a Motor/Generator for Use in Satellite Flywheel Energy Storage System. S.M. Thesis, Massachusetts Institute of Technology, Cambridge Massachusetts, September 1985.

11) Downer, JD. Design of Large Angle Magnetic Suspensions. ScD Thesis, Massachusetts Institute of Technology, Cambridge Massachusetts, June, 1986.

12) Foley, TF. Design of a Flexible Flywheel Support Structure for a Combined Energy Storage/Attitude Control System. S.M. Thesis, Massachusetts Institute of Technology, Cambridge Massachusetts, June 1986.

13) Johnson, BG. Active Control of a Flexible, Two-Mass Rotor: The Use of Complex Notation. ScD Thesis, Massachusetts Institute of Technology, Cambridge, Massachusetts, January 1987.

C303/88

The active magnetic bearing enables optimum control of machine vibrations

H ZLOTYKAMIEN
S2M-Société de Mécanique Magnétique, St Marcel, France

ABSTRACT

The active magnetic bearing is based on the use of forces created by a magnetic field to levitate the rotor without mechanical contact between the stationary and moving parts. A ferromagnetic ring fixed on the rotor 'floats' in the magnetic field generated by the electromagnets which are mounted as two sets of opposing pairs. The current is transmitted to the electromagnetic coils through amplifiers.

The four electromagnets control the rotor's position in response to the-signals transmitted from the sensors. The rotor is maintained in equilibrium under the control of the electromagnetic forces. Its position is determined by means of sensors which continuously monitor any displacements between rotor and stator through an electronic control system.

As in every control system, damping of the loop is provided by means of a phase advance command from one or more differenciating circuits of the position error signal.

The vibrations of the rotor and stator of a machine are generated by different forces:
- centrifugal forces due to the misalignment between the geometrical axis of the rotor (unbalance),
- reaction forces due to aerodynamical forces on the rotor and stator blades.

The active magnetic bearing allows the decrease and in many cases the fully cancelling of effects of these forces i.e. the vibrations of the machine. The inertial forces can be cancelled by shifting the axis of rotation of the rotor from the geometrical axis to the inertial axis (this system is usually called automatic balancing).

The reaction forces due to aerodynamical effects can be cancelled by the creation by the magnetic bearings of forces in opposition.
The vibrations are measured on the stator by accelerometers, and the signals drive magnetic bearings which generate forces having the same amplitude but in phase opposition. The improvement in vibrations amplitude usually ranges from 20 dB to 40 dB over a large band of frequencies.

1 INTRODUCTION

This paper is a continuation of the previous one titled: 'The active magnetic bearing enables optimum damping of flexible rotors.'

This time we will not focus on the damping capabilities of the magnetic bearing but on its vibrations control capabilities.

Nevertheless we will first recall:
- the magnetic bearings technology,

- the advantages of the magnetic bearings technology,

and then explain in details how a magnetic bearings suspension can control the machine vibrations.

2 MAGNETIC BEARINGS TECHNOLOGY

2.1 Principle of operation

As in any bearing, the active magnetic bearing is essentially composed of two parts: the rotor and the stator.

The rotor which may be internal or external, consists of a stack of ferromagnetic laminations without any slots or windings, force-fitted onto the shaft.

The stator, also made of stacked laminations, slotted and including windings, includes the electromagnet section of the bearing itself and the position sensor section.

The rotor is located in the magnetic field of the electromagnets which have to provide attractive forces which value must equal the load applied on the rotor.

The system is basically unstable so as soon as a small motion appears, the currents in the electromagnets must be adequately modified to maintain the equilibrium.

This is the reason why position sensors are required to provide information of actual rotor location.

For a complete rotor, where five degrees of freedom must be controlled, ten electromagnets are required.

It must be emphasized that, on the contrary of all other types of bearings, this bearing operates by attraction rather than repulsion.

Moreover, control is effective upon bearing activation and does not request any speed taking off.

2.2 Control loop and feedback system

The position sensor, which uses a high frequency carrier, provides a linear signal versus the rotor location.

This signal is compared with the reference set point (generally zero when a centered rotor is desired).

Then it is treated in a PID (Proportional, Integral, Demotive) network and amplified in power amplifiers which feed the electromagnets according to the error signal sign.

2.3 Radial bearing technology

The technology of a magnetic bearing is very close to that of the asynchronous induction motor.

The rotor is composed of ferromagnetic laminations without slots.

The thickness of these laminations varies from two mils to fourteen mils, according to the application, and are generally made of 3 percent Fe-Si non-oriented material, thereby allowing linear rotation speeds up to 200 m/s.

The stator also uses fourteen mils thick ferromagnetic laminations.

In most cases, the material used consists of non oriented 3 percent Fe-Si material. With this material, the specific load capacity of the magnetic bearing varies from eight to ten bars.

For a few applications where the load density of the machine is critical, Fe-Co laminations are used (51 percent Fe, 49 percent Co), enabling a higher specific load up to fourteen bars.

The airgap of the magnetic bearings varies from twelve mils to forty mils or more, according to the diameter of the bearing (50 mm/2" up to 1,000 mm/40").

These values, very high in respect to those encountered in other bearings, allow less accurate machining (generally a turning process is good enough).

The windings of the electromagnets, simpler than those of electric motors, use, however, the same technology and the same insulation classes.

2.4 Double axial thrust bearing

Axial control of the machines is generally performed by means of a double thrust bearing.

In the case of magnetic bearings, the stator of the bearing is of the annular type as well as the winding.

The laminating of the bearing is performed by sawing and filling the slots with ferromagnetic laminations.

The gap values are similar to those of the radial bearings.

The rotor generally consists of solid steel material allowing linear speeds up to 400 m/s. There is no necessity for the wheel to be laminated, because the bearing is not subject to any magnetization change during rotation, and therefore to any Eddy current losses.

The specific load is the same as for the radial bearings, with, in addition, the fact that all the surface is used in projection.

The axial sensor is not always located close to the thrust bearing. It is often at the shaft end as presently used, especially in machine tool applications.

2.5 Typical bearing arrangement

Fig 5 summarizes the magnetic bearings technology:

- the radial bearing rotor with the bearing and sensor laminations, and a thrust bearing flywheel,
- the stator of the radial bearing and associated sensor are split for assembly purposes,
- the axial bearing can be seen at the back, on the left. This axial bearing is in two sections with bean-shaped windings.

2.6 Auxiliary bearings

In order to protect the electromechanical parts in case of bearing overload or electronics failure, a set of auxiliary bearings is used for both radial and axial motions.

These bearings are usually ball bearings or carbon sleeves. They are located on the stator and therefore are standstill in normal operation.

The clearance between the inner race and the rotor is generally half the magnetic bearing clearance: eight mils to twenty mils.

This mechanical redundancy is not used in case of mains failure, the control electronics being protected by batteries.

2.7 Electronic cabinet

There are different standardized cabinets with various available power for the bearings.

The cabinet size is basically related to the power amplifier power which is determined by two factors:
- the load applied on the bearing which defines the maximum current to supply,

- the frequency of the disturbance which defines the minimum voltage required in all coils.

For very disturbed and heavy loaded machines like milling spindles, the standard cabinet produces twenty-five Amps. per power amplifier) with a water cooling system.

3 ADVANTAGES OF THE ACTIVE MAGNETIC BEARING TECHNOLOGY

There are two main areas of advantages:

3.1 Fluidless and contactless bearings

The first area of advantages is due to the fact that there is no contact between the rotor and the stator.
This feature brings several advantages and their related consequences such as:

NO WEAR	. eliminates mechanical maintenance . unlimited life
NO LUBRICATION	. eliminates oil, pumps, filters, tanks, piping,... . eliminates bearing oil seals . eliminates process fluid contamination
VERY HIGH PERIPHERAL SPEED:200 m/s	. very high rotation speed . large shaft diameters at the bearings
VERY LOW LOSSES	. negligible heat input to process fluid . very low power consumption
INSENSITIVITY TO HOSTILE ENVIRONMENTS	. can operate in environments such as: vacuum, steam, hydrocarbons and other hostile environments . elimination of seals needed to protect conventional bearings lube
WIDE OPERATING TEMPERATURE RANGE	. from cryogenic to elevated temperatures 20° K to 450°C

3.2 Controlled bearing

The second area of advantages is due to the fact that the magnetic bearing is permanently controlled with adaptive parameters.

This allows:

CONTROL OF ROTOR INSTANTANEOUS POSITION	. capability of rotor axial and radial position adjustment . automatic balancing (rotor rotation about its axis of inertia)eliminating vibrations due to unbalance . elimination of other vibrations by generation of opposite forces in the bearings
HIGH AND ADJUSTABLE STIFFNESS	. high rotor position accuracy
VERY HIGH AND ADJUSTABLE DAMPING	. passing through the critical speeds of the bearings and of the shaft
CONTINUOUS MONITORING OF OPERATING CONDITIONS WITHOUT ADDITIONAL EQUIPMENT	. speed . bearing loads . rotor position . rotor run out (unbalance and eccentricity)

4 VIBRATIONS CONTROL OF MACHINES DUE TO ACTIVE MAGNETIC BEARINGS

The vibrations of the rotor and stator of a machine are generated by different forces:

- centrifugal forces due to the misalignment between the geometrical axis and the inertial axis of the rotor (unbalance),
- reaction forces due to aerodynamic forces on the rotor and stator blades,

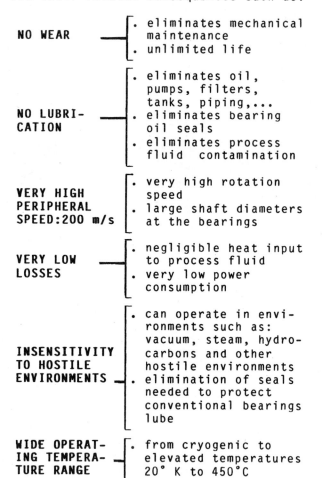

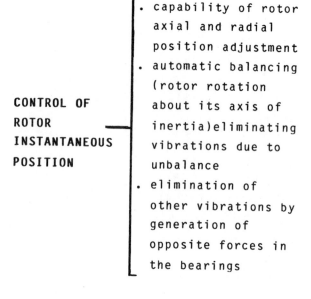

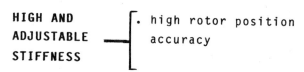

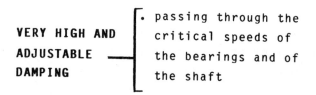

- forces created by the magnetic field of the stator of asynchronous electrical motor. These forces are located in the air gap of the motor.

4.1 Control of vibrations due to the unbalance

No matter what the machining quality of any rotating body, the inertial axis is never perfectly aligned with its geometrical axis.

When considering a rotor alone, without bearings (for instance in space or using very soft bearings) the rotation axis around which it will rotate, will automatically be its inertial axis. As its geometry is not perfect, there will be a run out relating to the remaining unbalance.

Putting that rotor on a ball or oil bearing system will automatically impose the rotation around the geometrical axis and, as a result, will create unbalance forces and vibrations when spinning up the shaft.

An AMB suspension totally solves the problem and enables a vibration-free functioning, whatever the unbalance value.

To do so, the control electronics is equipped with a special device called automatic balancing system. This device is simply a very narrow and selective filter located on the position sensor signal and acting at the rotation speed.

As a result the stiffness is zero on and only on the rotation frequency at which the unbalance problem is occurring.

As the bearings have no more stiffness at the speed of rotation, the rotor then rotates around its axis of inertia.

Of course, the bearing reacts with its regular stiffness for any frequency of perturbation other than the frequency of rotation.

This design allows out of balance operation without any additional load on the bearings that would result from the centrifugal forces and without vibration.

Naturally as in space conditions there is a small run out, but this shaft run out does not affect the machine operation.

4.2 Control of vibrations due to aerodynamic forces and to motor magnetic fields

4.2.1 Suspension behaviour with conventional bearings

When conventional bearing like ball bearings or hydrodynamic bearings are used in rotating machinery the only way to reduce vibration forces transmitted to ground is with the use of insulation pads.

Insulation pads can be represented by spring with high internal damping.

So the complete system, rotor, stator ground can be represented as per Fig 10.

A spring is used to provide a low stiffness of the connection to the ground and damping to avoid oscillation on the resonance frequency of this mass-spring system which is represented by the machine mass and the insulation pads.

The transmission ratio of the insulation pad is:

$$T = \frac{Fg}{Fs} = \frac{dp + \omega_0^2}{p^2 + dp + \omega_0^2}$$

The critical frequency of the pad suspension of the stator is:

$$fo = \frac{\omega_0}{2\pi} = \frac{1}{2\pi}\sqrt{\frac{K}{Ms}}$$

The value of this frequency is usually close to 20 Hz.

The damping is:

$$d \simeq \frac{a}{Ms} \simeq \sqrt{\frac{K}{Ms}}$$

to avoid too long oscillations on a critical frequency.

We can see three different bands of frequencies:

a) for frequencies lower than the critical frequency ($f \ll fo$) then T = 1.

All vibrations of the stator are transmitted to ground.

b) Close to and on the critical frequency (f = fo).

$$T \simeq \frac{dp + \frac{K}{Ms}}{dp} > 1$$

The vibrations transmitted to the ground are amplified.

c) For frequencies higher than fo ($f \gg fo$):

$$T = \frac{d}{p}$$

The forces transmitted decrease proportionally to the disturbance frequency.

We can conclude that the efficiency of the insulation pad is not too high, for stator frequencies up to several times the critical frequency.

4.2.2 Active magnetic bearings suspension behaviour

Now we shall see how an active magnetic bearings suspension with special feedback control loops using accelerometric sensors mounted on the stator allows use of insulation pads without internal damping and also enables compensation of stator vibrations.

Such a suspension and its control loop are depicted on Fig 11.

Three control loops are used:

a) the first one (ref. 9 on Fig 11), is the standard one giving good position control of the rotor relative to the stator.

The transfer function of the loop is a PID one. The position sensor is the induction sensor of the magnetic bearing. Its output is the rotor position related to the stator.

b) the second loop (ref. 10 on Fig 11), uses accelerometric sensor mounted on the stator. Its output is the acceleration of the stator. This signal goes through a highpass filter to eliminate zero drift error of the accelerometer.

After one integration, the signal becomes proportional to stator vibration speed and is used to damp the oscillations at the critical frequency of the insulation pad which is:

$$fo \simeq \frac{1}{2\pi}\sqrt{\frac{K}{Ms}}$$

No more internal mechanical damping of the insulation pad is needed and the transmission ratio of the pad becomes for high frequencies ($f \gg fo$):

$$T = \frac{K}{Ms} * \frac{1}{p^2}$$

The transmission ratio is decreasing with the square of the frequency. The insulation pad becomes very efficient for high frequencies, because the transmission by the internal damping of the insulation pad is not present and oscillations are damped by the active magnetic bearing with space-speed reference.

c) The third servoloop (ref. 11 on Fig 11) uses the space reference signal of the accelerometric sensor after a second integration.

This position signal in space goes through a very narrow band-pass filter and is introduced in the power amplifier with high gain in negative feedback.

At the specific frequency of the filter the stator is controlled by space reference with very high stiffness. Vibration forces at this frequency will remove displacement of the stator.

A narrow bandpass filter has to be used because, for stability reasons of the primary PID loop, high gain feedback is not possible in large bandwidth.

If there are several such specific perturbation frequencies, then several narrow bandpass filters may be used. We can now summarize the behaviour of the system equipped with AMB versus the frequency of rotation.

For all frequencies except the specific ones treated by the third control loop, the transmission ratio is:

$$T = \frac{Fg}{Fs} = \frac{\omega_o^2}{p^2 + 2\omega_o p + \omega_o^2}$$

and $$fo \simeq \frac{\omega_o}{2} \simeq \sqrt{\frac{K}{Ms}} \simeq 20 \text{ Hz}$$

a) For frequencies lower than the critical frequency ($f \ll fo$) as for conventional suspensions:

$$T = 1$$

All vibrations are transmitted to the ground.

b) Close to and around the critical frequency ($f \simeq fo$):

$$T = \frac{\omega_o}{2p} = \frac{1}{2}$$

There is no more amplification of the vibrations transmitted to ground as was the case with conventional bearings and pads, but a reduction of transmission by a factor of two is accomplished.

c) For frequencies higher than ($f \gg fo$):

$$T = \frac{\omega_o^2}{p^2}$$

The transmitted forces decrease proportionally to the square of the disturbance frequency (while it was decreasing proportionally to the frequency only with conventional bearings and pads).

In addition, for a limited number of well defined specific frequencies which are treated by the third control loop (ref. 11 on Fig 11), the transmission is no longer linked to the transfer function described above. This loop creates a space reference for the stator and corrects stator movement versus space at these frequencies.

The practical results achieved are usually an improvement of transmitted forces at these frequencies of greater than 40 dB.

4.3 Conclusion on vibration control

Thus, we have two types of conclusions. First, the active magnetic bearings enable the cancelling of the source of vibrations due to the unbalance, by letting the rotor rotate around its axis of inertia (which is called automatic balancing).

Second, due to special control loops, the active magnetic bearings enable the reduction of the transmission of the other vibrations (blade vibrations, vibrations due to motor magnetic field, etc) with a higher efficiency than conventional bearings and insulation pads.

In addition, at and only at a limited number of specific frequencies, the active magnetic bearings drastically reduce the vibrations due to its 'space reference' control loop.
These results are summarized on Fig 12:

5 CURRENT APPLICATIONS

The automatic balancing has been used for most of the active magnetic bearings of rotating machinery realized up to now by S2M. About 400 bearings of 40 types have been manufactured.

The smallest one realized was for a grinding spindle B1/3000 with a rotor weight of 1 kg at a speed of 180,000 rpm (Fig 13).

The biggest one was for a turbo-generator of 5 MW for blast furnace exhaust gas featuring a rotor weight of 8 tons at a speed of 3,000 rpm
In the middle, there is a wide range of compressors and steam turbines with rotor weights of some hundred kilograms working throughout the world (Fig 15).
Vibration control by accelerometic sensors is used for special applications where very low noise and vibrations of the machine frame are required.

6 CONCLUSION

The retrofit by exchange of bearings is not always possible, it is better to build a new well adapted machine.
But, generally speaking, active magnetic bearings can overcome most of the problems faced with conventional bearings.

REFERENCES

(1) S. EARNSHAW, "On the nature of the molecular forces", Trans. Cambridge Phil. Soc. 7, 97-112, 1842.

(2) M. TOURNIER and P. LAURENCEAU, "Suspension magnētique d'une maquette en soufflerie", La Recherche aēronautique 7-8, 1957.

(3) J. M. BEAMS, "Magnetic bearings", Society of automative engineers, automative engineering congress, Detroit, Michigan, Paper 810A, 13-17, January 1964.

(4) J. LYMAN, "Magnetic bearings - Pivot and Journal", paper given at 22nd ASLE Annual meeting in Toronto, May 1-4, 1967.

(5) Cambridge Thermionic Corporation, "A survey of magnetic bearings", 1972.

(6) H. HABERMANN, "Das aktive Magnetlager - ein neues Lagerungsprinzip", Haus der Technik,Vortragsveröffentlichungen Essen, 21.09.77.

(7) H. HABERMANN, "Le palier magnétique actif ACTIDYNE", AGARD Conference proceedings N°323, 1982.

(8) H. HABERMANN - M. BRUNET, "The active magnetic bearing enables optimum damping of flexible rotors", ASME International Gas Turbine Conference N° 84 - GT 117 - 1984.

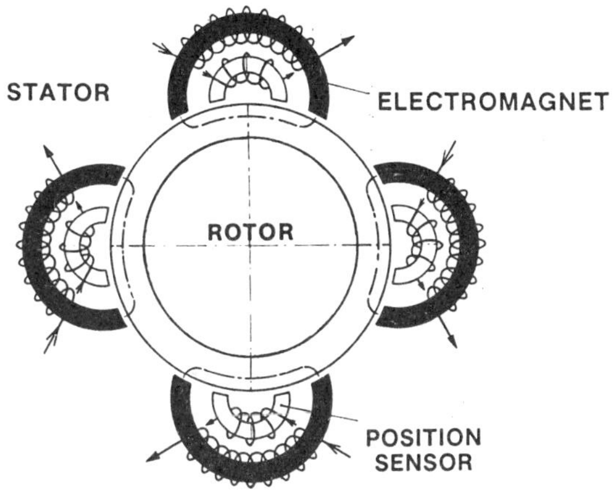

Fig 1 Bearing schematic diagram

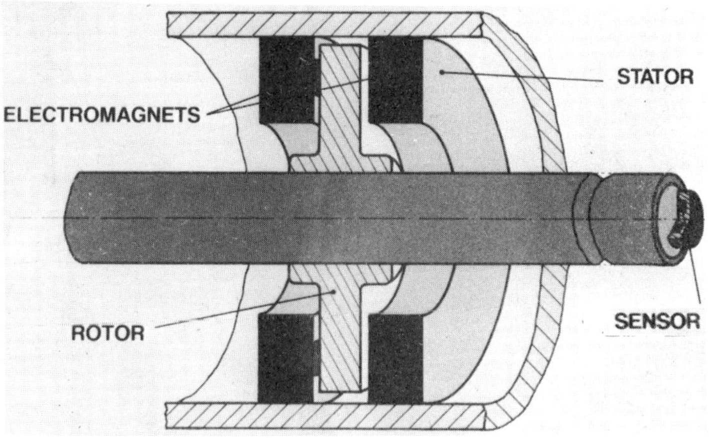

Fig 4 Double-axial thrust bearing

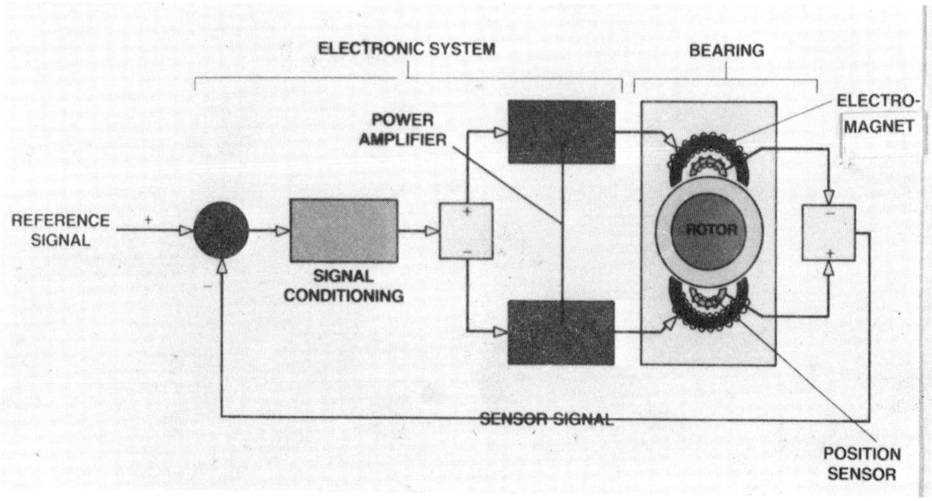

Fig 2 Control loop and schematic diagram

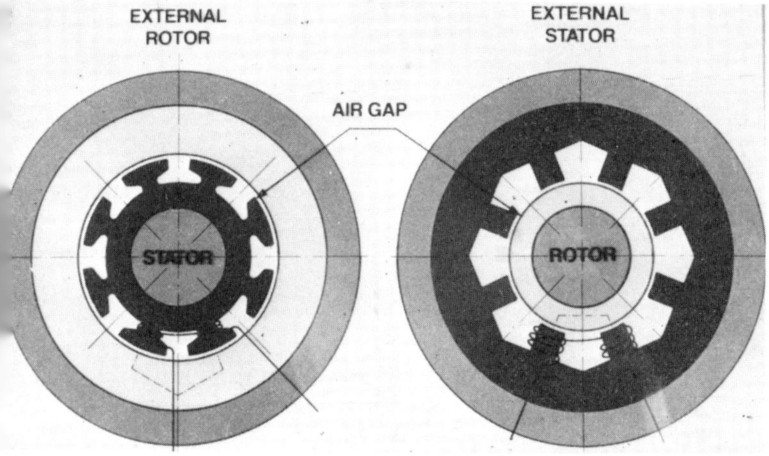

Fig 3 Radial bearing technology

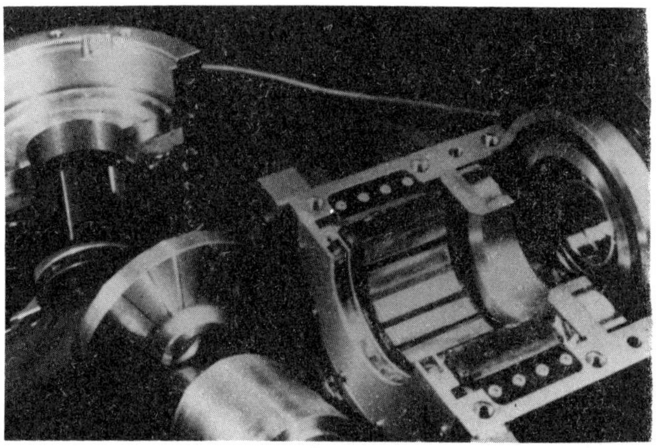

Fig 5 Typical bearing arrangement

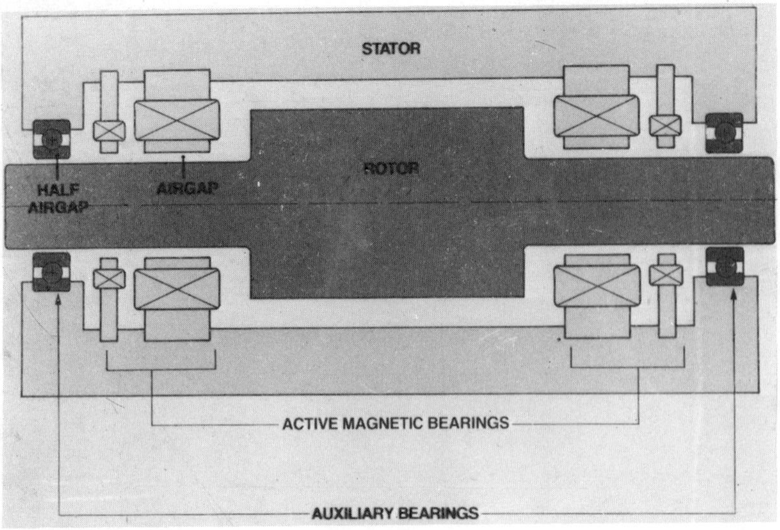

Fig 6 Auxiliary bearings

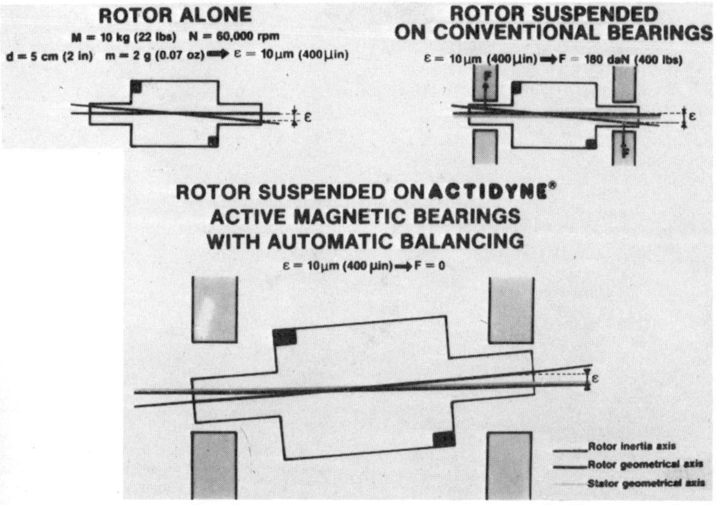

Fig 8 Automatic balancing with AMB

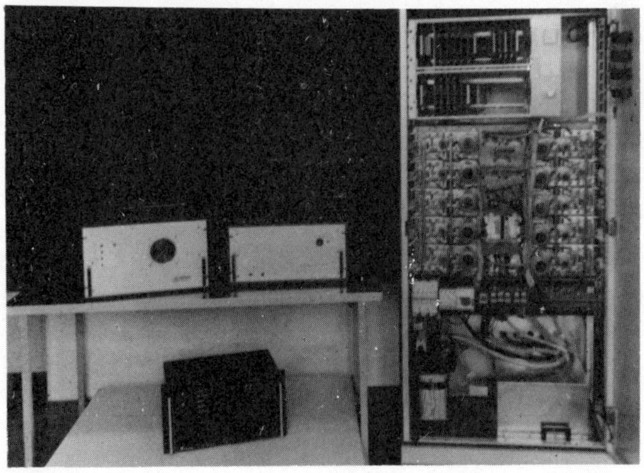

Fig 7 Electronic cabinets

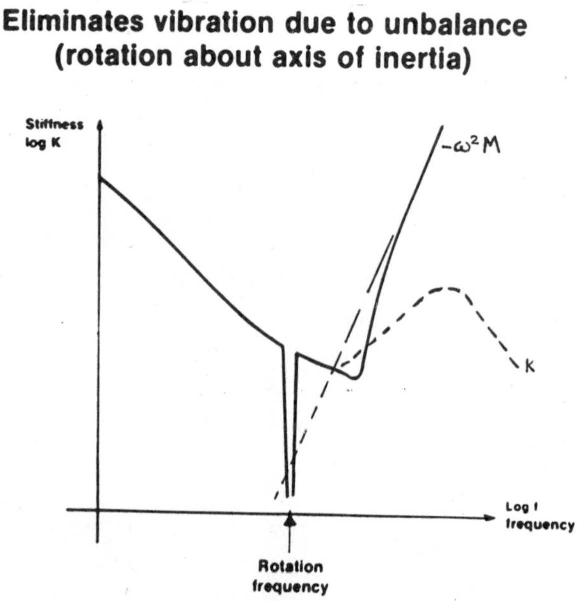

Fig 9 AMB stiffness characteristics with automatic balancing

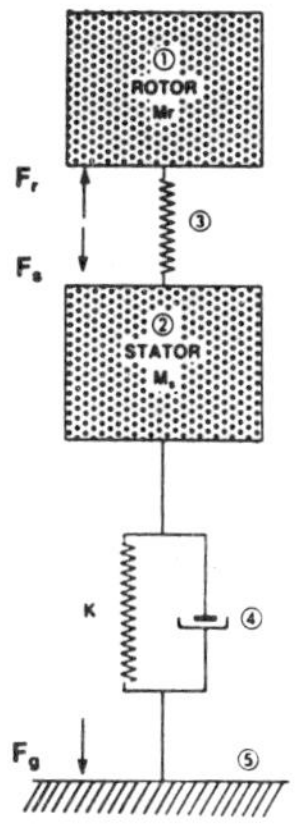

① Rotor
② Stator
③ Conventional bearing
④ Conventional insulation pad
⑤ Ground

F_r = Forces on Rotor
F_s = Forces on Stator
F_g = Forces on Ground

$$d = \frac{a}{M_s} \qquad f_o = \frac{\omega_o}{2\pi} = \frac{1}{2\pi}\sqrt{\frac{K}{M_s}} = 20\ \text{Hz}$$

$$T = \frac{F_g}{F_s} = \frac{dp + \omega_o^2}{p^2 + dp + \omega_o^2}$$

f	T	RESULTS ON VIBRATIONS TRANSMITTED TO THE GROUND
$\ll f_o$	= 1	ALL VIBRATIONS TRANSMITTED
$\sim f_o$	> 1	VIBRATIONS TRANSMITTED ARE AMPLIFIED
$\gg f_o$	< 1	VIBRATIONS TRANSMITTED ARE REDUCED PROPORTIONALLY TO THEIR FREQUENCY

Fig 10 Block diagram of a rotor suspended on conventional bearings and a stator suspended on conventional insulation pads

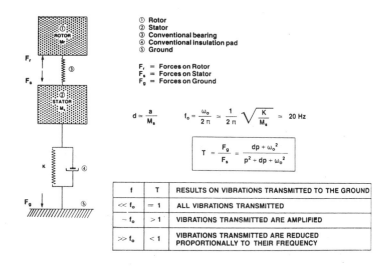

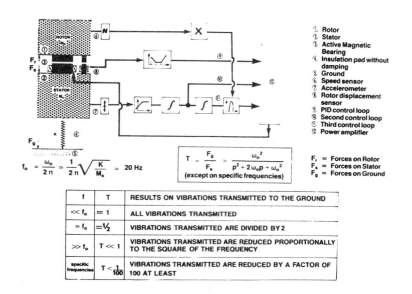

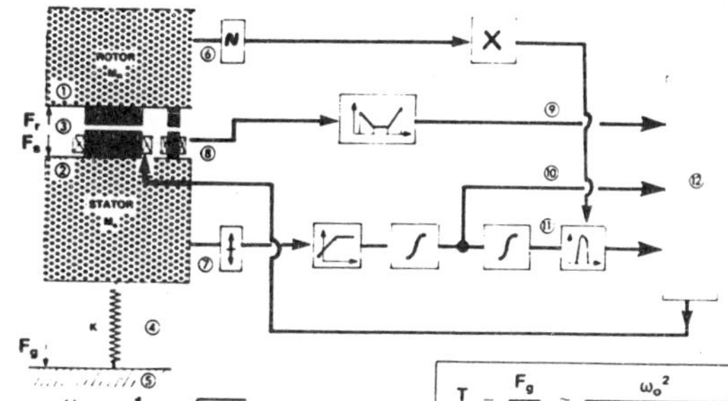

① Rotor
② Stator
③ Active Magnetic Bearing
④ Insulation pad without damping
⑤ Ground
⑥ Speed sensor
⑦ Accelerometer
⑧ Rotor displacement sensor
⑨ PID control loop
⑩ Second control loop
⑪ Third control loop
⑫ Power amplifier

$$f_o = \frac{\omega_o}{2\pi} = \frac{1}{2\pi}\sqrt{\frac{K}{M_s}} = 20\ \text{Hz}$$

$$T = \frac{F_g}{F_s} = \frac{\omega_o^2}{p^2 + 2\omega_o p + \omega_o^2}$$
(except on specific frequencies)

F_r = Forces on Rotor
F_s = Forces on Stator
F_g = Forces on Ground

f	T	RESULTS ON VIBRATIONS TRANSMITTED TO THE GROUND
$\ll f_o$	= 1	ALL VIBRATIONS TRANSMITTED
$\simeq f_o$	= ½	VIBRATIONS TRANSMITTED ARE DIVIDED BY 2
$\gg f_o$	$T \ll 1$	VIBRATIONS TRANSMITTED ARE REDUCED PROPORTIONALLY TO THE SQUARE OF THE FREQUENCY
specific frequencies	$T < \frac{1}{100}$	VIBRATIONS TRANSMITTED ARE REDUCED BY A FACTOR OF 100 AT LEAST

Fig 11 Block diagram of a rotor suspended on active magnetic bearings and a stator suspended on insulation pads without damping

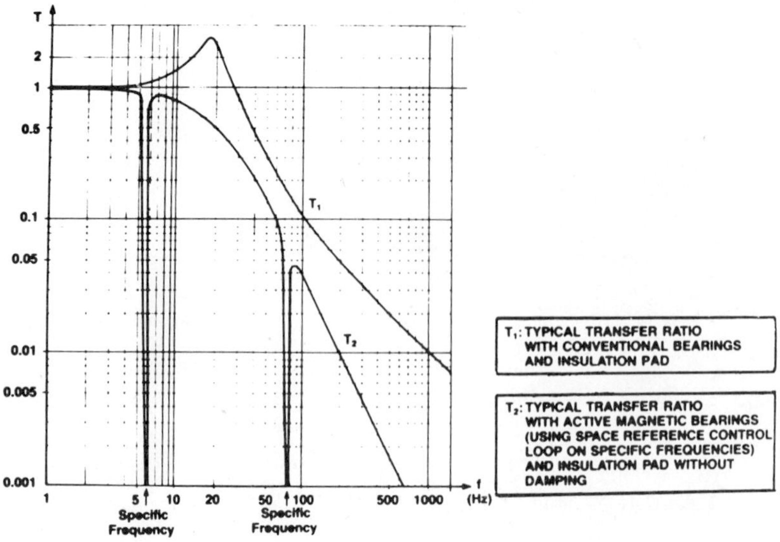

Fig 12 Comparison of transmission ratio between conventional bearings and active magnetic bearings

Fig 13 B1/3000 grinding electrospindle

Fig 14 Turbogenerator of 5 MW for blast furnace exhaust gas (rotor weight = 8 tons; speed = 3000 r/min)

Fig 15 The biggest magnetic bearing with a rotor diameter of 1250 mm was delivered to EdF (Electricité de France) for a test stand for the damping of critical frequencies of rotor of 300 MW turboalternators; the rotating force is 100 000 N (a bigger one will soon be realized)

C277/88

Magnetic bearing design optimization

J IMLACH, MS, **P E ALLAIRE**, PhD, MASME, MASLE, **R R HUMPHRIS**, DSc, MIEEE and **L E BARRETT**, PhD, MASME
Department of Mechanical and Aerospace Engineering, University of Virginia, Charlottesville, Virginia, United States of America

SYNOPSIS The design of magnetic bearings is concerned with three aspects: the magnetic component, the controls, and the rotor to which the bearing is to be applied. In particular, this paper is intended to be a design guide for the magnetic component. A general design optimization method is presented to enable the engineer to determine bearing parameters rapidly for particular applications. In this paper, the design method is applied only to a bearing which consists of four radial magnets. The method is based on the solution of four nonlinear equations. The parameters for which these equations are named are all inputs to the solution program. In addition to these, several other parameters are needed as program inputs. The program then solves a set of equations over a range of independent parameters.

NOMENCLATURE

$A_i = W_\ell L$ Area of minimum iron cross section.
$A_p = (W_\ell + 2t)L$ Projected area of pole face.
A_{cs} Coil surface area.
B_i Flux density.
c Average radial clearance.
D_o Outside diameter of bearing.
$G = Re\{G(j\omega)\}$ Real part of controller gain.
H_b Height of a bobbin.
H_ℓ Height of a magnet leg.
i_1 Steady state current in top coils.
i_2 " " " " bottom coils.
K_{eq} Effective bearing stiffness.
K_h Coil surface convection coefficient.
K_i Bearing current stiffness.
K_x Bearing position stiffness.
L Bearing axial length.
N Number of turns per coil.
PF Coil packing factor.
r_ℓ Radius of disk laminations.
r_m Mean radius of the coil.
r_w Radius of the bare wire.
t Thickness of a foot.
t_b Thickness of a bobbin.
t_w Thickness of wire insulation.
W Design load of the bearing.
W_b Width of a bobbin.
W_ℓ Width of a magnet leg.
θ_1 Angle subtended by a pole face.
θ_2 Half angle between opposite magnetic poles.
Θ Design temperature rise of the coil.
μ_0 Permeability of free space.
σ Conductivity of wire at design temperature.

1 INTRODUCTION

Magnetic bearings are the subject of much research today and are also being introduced into a wide range of industrial machinery. They are devices which have a number of advantages over conventional bearings, such as elimination of oil supply systems, control of vibrations, possible very long life, etc. The design of magnetic bearings is concerned with three aspects: the magnetic component, the controls and the rotor to which the bearing is to be applied. In particular, this paper is concerned with the design of the magnetic component. A design method is presented to enable the engineer to determine bearing parameters rapidly for particular applications.

The concept of magnetic suspension systems has been around for centuries but only recently has implementation proven feasible. Some of the early work on magnetic suspensions involving vertical centrifuges was carried out at the University of Virginia. Beams [1,2] described a high speed centrifuge rotor in a vacuum. The vertical position of the suspended mass was supported by the magnetic field of an axial solenoid. The horizontal position of the rotor was maintained by the symmetrically diverging field of the solenoid. This magnetic bearing system was extended to other devices such as a magnetically suspended molecular pump [3]. Interestingly, the molecular pump had an oil squeeze film damper combined with it to provide increased radial damping above that of the diverging field of the solenoid.

Magnetic suspensions continued to be developed for vertical devices. Applications to flywheels for potential use in space were reported in the literature [4]. In the 1970's, several commercial applications of magnetic bearings were reported by S2M of France [5]. These included a momentum wheel, image display unit, turbomolecular pump, electrogrinding spindle and a turbocompressor bearing. Some of these utilized radial bearings to support horizontal shaft devices. These applications continue to be reported in papers such as [6].

An extensive series of articles on various design aspects of radial magnetic bearings has been published by Schweitzer, et al. An early article [7] discussed the relation between control force and actuating current. It also discussed relations between the flux density, number of windings, air gaps and other bearing design parameters. Two characteristic parameters - "force current factor" and "force displacement factor" were derived. In this paper, these two parameters are called the current stiffness and position stiffness. Good agreement was obtained between theoretical calculations and experimental measurements on a specific bearing geometry [7,8].

The effects of centralized and decentralized control were discussed in [9]. It was found that decentralized control of each axis of the radial bearings decreased the complexity of the system without much loss in performance. More concepts related to the design of magnetic bearing components were presented by Schweitzer and Traxler [10]. These concepts included magnetic pole geometry, cooling capacity, optimum core width, and various losses. Traxler and Schweitzer [11] also reported on measured force characteristics for magnetic bearings. Recent work on a similar magnetic bearing was reported by Salm [12].

Nikolajsen, et al. [13] discussed the use of an electomagnetic damper on a transmission shaft. The geometry was a four pole design which differed from the Schweitzer work on an eight pole design [7,9]. Gondhalekar and Holmes [14] reported on a six pole design and the electronic drives required to operate it. Bradfield, et al. [15] presented the further application of the six pole bearing to a supercritical shaft. Hebbale and Taylor [16] discussed some nonlinear aspects of magnetic bearing designs, but primarily considered solid rotating magnetic components which are subject to large eddy current losses.

A theoretical model and the effects of control algorithms on an eight pole magnetic bearing were reported in Humphris, et al. [17]. Stiffness and damping coefficients were measured for the bearing. Rotor dynamic effects of the bearing were presented in several papers [18,19,20].

The design method described in this paper applies to a specific class of magnetic bearing. The general geometry of this bearing is shown in Figure 1. The bearing consists of four magnets each with two pole faces. The pole faces are located circumferentially around a laminated disk which is attached to the shaft. The area of each pole face is increased by adding tangential 'feet'. These feet also serve to hold the coils in place. The coils are wound either directly onto the magnets or onto thin walled bobbins. In either case the coils are assumed to be roughly rectangular in cross section (Figure 2).

The inputs to the solution method (i.e. the parameters which remain fixed throughout the solution) include: the axial length of the magnets; the size and type of wire used in the coils; the thickness of the bobbin material; the radius of the shaft laminations; the clearance between the laminations and the pole faces; and the coil packing factor. The basic results of the method include: the width and height of the magnet legs; the width and height of the bobbins; the foot thickness; the number of turns per coil; the steady state currents required in each coil; and the real part of the controller gain required to provide the desired stiffness. Once these values are determined, all remaining information about the bearing (inductance, voltage required, outside diameter, etc.) can easily be calculated. Figure 3 shows a block diagram of a typical control system which can be adjusted to attain the specified value of G, the controller gain.

2. MAJOR DESIGN PARAMETERS

Four of the most important quantities in the design of magnetic bearings are: 1) the bearing stiffness, 2) the bearing load capacity, 3) the maximum flux density in the magnets, and 4) the maximum bearing (coil) operating temperature. By determining governing equations for each of these parameters, the basis of a design method can be established.

While the bearing stiffness and load capacity can usually be determined by the specific design requirements, determining the best treatment of the flux density and operating temperature present a more difficult problem. For example, it may be desirable to minimize both the maximum flux density in the magnets and the temperature of the coils. Unfortunately this simple criteria results in bearings with very large outside diameters. This optimization is the area in which the designers must use their own judgement.

A linearized equation for the stiffness of magnetic bearings has been presented in the literature [17]. With a minor correction, it is used in this work. This relationship was derived via a perturbation of the magnet force equation about the centered rotor case. The correction to this relationship arises from recognizing that the position stiffness (K_x) of the bearing is independent of magnet orientation while the current stiffness (K_i) of the bearing is proportional to the cosine of the half angle between magnet pole faces. Thus the corrected linear stiffness equation is

$$K_{eq}=K_x+K_iG = -2\mu_o A_p\left(\frac{N}{c}\right)^2\left[\frac{(i_1{}^2+i_2{}^2)}{c} - (i_1+i_2)G\cos\phi\right] \qquad (1)$$

where ϕ can be shown to be $\frac{1}{2}\theta_1+\theta_2$ (see Figure 1). The actual stiffness of the bearings is a function of the rotor displacement. In order to determine the correctness of the linear assumption, the force predicted by the linear bearing equation (K_{eq} above) can been compared with the force predicted using the 'exact', or non-linear, bearing equations

$$F = \sqrt{2}\cos\phi\ \mu_o A_p N^2 [(\frac{i_1 - G\Delta x}{c - \Delta x})^2 - (\frac{i_2 + G\Delta x}{c + \Delta x})^2]$$

An example plot of both the linear and non-linear force equations for the bearings presented in section 5 is shown in Figure 4. This figure shows that the above theory predicts the magnetic bearing, in this particular case, acts as a stiffening spring for eccentricity ratios greater than about 0.1. This observation should be experimentally tested.

The basic load capacity equation has also been presented [17] for a bearing which is oriented with horizontal and vertical sets of magnet. By rotating the bearing 45 degrees the load can be supported by two magnet pairs (assuming the load is in the vertical direction). This change neccesitates a slight modification to the load equation. The resulting equation is

$$F = c_g \mu_o A_p (\frac{N}{c})^2 (i_1^2 - i_2^2) \qquad (2)$$

where c_g for this geometry is $\sqrt{2}\cos\phi$.

The flux density equation is easily derived from the conservation of flux requirement. This states that the maximum flux density in a magnetic circuit will occur at the point of minimum area (neglecting leakage). In general this will be in the magnet core. The flux generated by the magnets can be determined by dividing the magnetomotive force of the magnets by the reluctance of the flux path. For most practical radial bearings the total reluctance can be approximated by that of the gap between rotor and stator. Thus, equating the flux in the gap to the flux in the iron results in

$$\Phi = \mu_o N i_1 A_p/c = B_i A_i$$

Solving this equation for B_i yields

$$B_i = \mu_o N i_1 A_p/A_i c \qquad (3)$$

This equation represents the flux density in the stator when the bearing is operating in a centered position and flux leakage is neglected. In practice it is desirable to have this value be some fraction of the saturation flux density of the magnetic material. This is to insure that the bearing does not saturate under dynamic loads. The actual value will depend on the ratio of the maximum load (static and dynamic) to the operating load.

The final relation required is the heating equation for the bearing coils. This equation can quickly become very complex and cumbersome if put into a general form. It would ultimately rely on: the conduction/convection coefficient between the coils, bobbins and magnets; the conduction/convection coefficient between the coils and the medium surrounding the bearings; and the directionally dependent conduction coefficient within the coils. This last factor alone would depend on the wire used, the thickness and type of wire insulation, the potting material used in the coil and the coil packing factor.

To maintain some degree of generality in the design process several simplifying assumptions have been made in the derivation of the heat equation. The most important are, first, that the bobbin material is a perfect insulator. This implies that there will be no heat transfer between the coil and the magnets. The second assumption is that the coil itself is a perfect conductor. Thus, the surface temperature of the coil is assumed to represent the maximum coil temperature. Experimental and theoretical work is being conducted to ascertain the appropriateness of these assumptions and to obtain a better thermal model of the coils. At present it is expected that the designer make allowances for these assumptions when specifying the coil convection coefficient and maximum temperature.

The resulting heating equation is an equality between the heat generated by the coil resistance ($R = 2\pi r_m N/\pi r_w^2 \sigma$) and the heat removed by convection from the coil surface.

$$\theta K_n A_{cs} = 2 r_m i_1^2 N/\sigma r_w^2$$

or, solving for θ,

$$\theta = 2 r_m i_1^2 N/\sigma r_w^2 K_n A_{cs} \qquad (4)$$

The mean coil radius, r_m, is approximated by

$$r_m = \frac{1}{\pi}[W_\ell + D + 2(t_b + W_b)]$$

and the coil surface area, A_{cs}, is approximately

$$A_{cs} = 2(H_b - 2t_b)[D + W_\ell + 4W_b]$$

Four equations have been developed for the four major design parameters.

3. GEOMETRIC RELATIONS

Four equations, (1) through (4), have been derived but there are eight unknowns (W_ℓ, t, N, i_1, i_2, G, W_b, H_b) and only four equations. We have not, however, considered the geometric requirements of the bearing. It has been implicitly assumed that the flux leakage between pole faces is negligible. To reasonably meet this requirement it is recommended by magnet design [21] that the spacing between magnetically opposite pole faces be at least twice the pole foot thickness plus the gap thickness [2(t + c)]. The spacing between similar poles was chosen to be half this distance to reduce the risk of coupling between magnet pairs. The resulting configuration is shown in Figure 1 with the notation

$$\theta_2 = (t + c)/(r_d + c)$$

Several geometric aspects of the bearings can now be investigated. First, because of the choice of eight pole faces per bearing and the spacing requirements just described, we note from Figure 1 that $2\theta_1 + 3\theta_2 = \pi/2$ or, using the above relation,

$$\theta_1 = \frac{\pi}{4} - \frac{3}{2}\frac{(t + c)}{(r_d + c)}$$

Also, for practical bearings the width of a pole leg can be approximated by (Figure 2)

$$W_\ell = 2[(r_d + c)\sin(\frac{\theta_1}{2}) - t] \qquad (5)$$

Similarly, the maximum allowable bobbin width can be represented as (Figure 2)

$$W_b = (r_d + c + t)\tan[\frac{1}{2}(\theta_1 + \theta_2)] - \frac{1}{2}W_\ell \qquad (6)$$

An additional equation relating the height of the bobbin to the above parameters and the height of the pole leg can also be obtained from Figure 2.

$$H_b = -2(r_d+c) + [(r_d+c+H_\ell)^2 - W_b^2]^{1/2} + [(r_d+c)^2 - \frac{1}{4}W_\ell^2]^{1/2} \qquad (7)$$

Finally, by assuming the coil cross sectional area is equal to the open cross sectional area of the bobbin (Figure 2), another relation between unknowns can be obtained.

$$N\pi(r_w + t_{ins})^2/PF = (W_b - t_b)(H_b - 2t_b) \qquad (8)$$

There are now four more equations, bringing the total to eight. In equation (7) however, we have introduced an unknown, H_ℓ. At this point another equation, such as specifying the outer diameter of the bearing, could be introduced and the resulting set of coupled non-linear equations could be solved. Another approach, the one that is pursued, is to choose one of the unknown parameters and assign it a value. Then the effect of varying this parameter can be investigated. The parameter chosen for this study was t, the thickness of the foot.

4. SOLUTION PROCEDURE

For the solution procedure to be implemented several things must be assumed known. These are: B_i, K_{eq}, W, Θ, c, K_h, L, PF, r_ℓ, t_b and the wire properties. The first four of these parameters are recognized as those for which equations were derived in Section 2.

The maximum flux density will be determined by the material. It is recommended that this value should be chosen such that the magnet flux density will remain in the linear range (1.1 - 1.2 Tesla for standard magnet iron). The bearing stiffness should be chosen from the results of a dynamic analysis of the rotor to be controlled. The required load capacity of the bearing is also determined from an analysis of the rotor involved and should be sufficient to accommodate both the static and dynamic loads. The temperature rise of the bearing requires the most judgment on the part of the designer. Conditions that should be considered when choosing this value include: the ambient conditions in which the bearing will operate, the thermal ratings of the wire and bobbin material to be used, the effects of thermal loads on other portions of the system, etc.

The remaining parameters in the list of knowns are chosen in a similar manner. The bearing clearance must be determined by the specific circumstances and is usually the result of a compromise. The smaller the clearance the more tightly controlled the bearing must be. This requires larger voltages. The larger the clearance the larger the magnetomotive force must be for a given maximum loading. This requires a larger current or more turns of wire. The convection coefficient depends on the thermal properties and flow rate of the ambient fluid or medium. The bearing axial length will be determined in part by the space available for the bearing. The packing factor will be determined by the winding process (we have found values ranging from .65 to .75). The radius of the laminations occasionally must be determined by an iterative process. This is to insure that the radial width of the laminations is at least as great as the width of the magnet legs. The thickness of the bobbin depends somewhat on the material selected. Delrin material, .635 mm (.025 in) thick, has been used successfully.

When these parameters are determined, or initial guesses are made, a solution based on the equations developed in Sections 2 and 3 can be obtained. The solution procedure begins by assigning a value to the foot thickness, t. Once this value is assigned, the angular separation between all the pole faces (θ_2) and the angular dimension of the faces (θ_1) can be calculated. The width of the legs and the bobbins can then be calculated from equations (5) and (6).

The next step in the solution procedure is to solve the flux equation (3) for the product Ni_1. The weight equation (2) can then be solved explicitly for the product Ni_2, assuming a real solution exists. After these values are obtained, an iterative procedure is employed to obtain the values of N, i_1, and H_b. This is accomplished by: 1) assuming a value of H_b; 2) using this value and the product Ni_1 to solve equation (4) for i_1; 3) dividing Ni_1 by i_1 to obtain N; 4) solve equation (8) for H_b and compare it to the previous value; and 5) if H_b has not converged, correct its value (using Successive Over Relaxation) and continue with steps 2-5. A flowchart of this solution procedure is presented in Figure 5. The remaining parameters can be calculated directly from the other equations.

This procedure is carried out for several values of foot thickness, starting with some specified minimum thickness and continuing until no valid solution is found (the product Ni_2 becomes imaginary). A representative subset of results for each foot thickness can be presented for comparative purposes. Any interesting solutions can be examined in more detail. This procedure presents the designer with a range of possible designs meeting the input requirements.

These designs vary in outer diameter, steady state currents, number of turns, controller gain, etc. If a suitable design is encountered among these solutions, the process is terminated. If not the input parameters (usually θ and/or B) are adjusted and another solution is attempted. Parametric studies of the inputs can be easily conducted to establish general design guidelines. An example illustrates the design procedure.

5. AN EXAMPLE PROBLEM

A computer aided design software package based on the above method has been developed and used to design a set of magnetic bearings for a test rig. This rig is located in the ROMAC Laboratories at the University of Virginia.

The rotor for the test rig consists of a 12.7 mm (.5 inch) diameter shaft with three 2.55 kg (5.6 lb) disks supported between the bearings. In addition, the shaft has been fitted with two laminated iron disks which serve as the rotating portions of the magnetic bearings. These disks where sized to conform with a previously existing test rig. They are 58.4 mm (2.30 in.) in diameter and 25.4 mm (1.0 in) in axial length.

The load is shared equally between the bearings, thus the design load capacity of each bearing is 4.54 kg (10 lb). A bearing stiffness of 876 N/mm (5000 lb/in) was chosen. Also a maximum flux density of one third the saturation flux density, or .385 Tesla, was chosen to insure that the magnets did not saturate.

The design goal for this application was to achieve the smallest possible outside diameter practical while meeting the above criteria. A radial clearance of .64 mm (.025 in) was chosen. Also, the minimum practical thickness of the bobbin material was found to be .64 mm (.025 in). After several iterations of the other program inputs, an acceptable design was obtained. The copper wire used in this design is a #26 AWG with a 130°C rating. This resulted in a 75°C temperature rise over ambient for the coils. Other parameters for the design are given below, and one of the bearings is shown partially disassembled in figure 6, assembled and with coils attached in figure 7. The complete test rig is shown in figure 8.

GEOMETRIC PROPERTIES:

H_ℓ = 12.7 mm (.50 in)
W_ℓ = 11.4 mm (.45 in)
t = 3.18 mm (.125 in)
H_b = 11.3 mm (.445 in)
W_b = 5.9 mm (.234 in)
D_o = 114.3 mm (4.50 in)

ELECTRONIC PROPERTIES:

i_1 = .448 A
i_2 = .144 A
G = 8.2 A/MM (208 A/in)

6. CONCLUSIONS

Not much has been written in the literature about specific magnetic bearing design components and their optimization. While several authors have written about two or more design configurations, as discussed in the introduction of this paper, each author seems to choose a few reasonable configurations and analyse or construct them for testing.

This paper develops a method for modeling four major design parameters: stiffness, load capacity, maximum flux density and maximum temperature. Five geometric parameters are added to the design parameters. The specific equations presented in this paper are for a particular eight pole magnetic bearing geometry but the method of formulation is general and could be applied to many different magnetic bearing types. A method of solving these equations to obtain optimized designs is given and illustrated with an example.

The optimization presented in this paper is not a sophisticated one from an optimization expert's point of view. However, the modeling approach and successive over-relaxation scheme has been successful in obtaining designs which are optimized enough for several laboratory and industrial applications. From what the authors can find in the literature, this appears to be the first paper to document this type of systematic approach.

REFERENCES

1. Beams, J. W., "Magnetic-Suspension Ultracentrifuge Circuits," Electronics, March 1954.
2. Beams, J. W., Dixon, H. M., Robeson, A. and Snidow, "The Magnetically Suspended Equilibrium Centrifuge," Journal of Physical Chemistry, Vol. 59, 1955, pp. 915-922.
3. Williams, C. E. and Beams, J. W., "A Magnetically Suspended Molecular Pump," 1961 Transactions of the Eighth Vacuum Symposium and Second International Conference, Pergamon Press, 1962, pp. 295-299.
4. Sabnis, A. V., Dendy, J. B., and Schmitt, F. M., "Magnetically Suspended Large Momentum Wheel," Journal of Spacecraft, Vol. 12, 1975, pp. 420-427.
5. "Active Magnetic Bearing," Societe de Mechanique Magnetique, 1977.
6. Habermann, H. and Brunet, M., "The Active Magnetic Bearing Enables Optimum Control of Machine Vibrations, " ASME Gas Turbine Conference, Paper 85-GT-221.
7. Schweitzer, G. and Lange, R., " Characteristics of a Magnetic Rotor Bearing for Active Vibration Control," Institution of Mechanical Engineers, Conference on Vibrations in Rotating Machinery, September 1976, Paper C239/76, pp. 301-306.
8. Schweitzer, G. and Ulbrich, H., "Magnetic Bearings-A Novel Type of Suspension:

Institution of Mechanical Engineers, Conference on Vibrations in Rotating Machinery, September 1980, Paper C273/80, pp. 151-156.

9. Bleuler, H. and Schweitzer, G., "Dynamics of a Magnetically Suspended Rotor With Decentralized Control," First IASTED International Symposium on Applied Control and Identification, Copenhagen, June 28-July 1, 1983, pp. 17-22.

10. Schweitzer, G. and Traxler, A., "Design of Magnetic Bearings," International Symposium on Design and Synthesis, Tokyo, July 11-13, 1984.

11. Traxler, A. and Schweitzer, G., "Measurement of the Force Characteristics of a Contactless Electromagnetic Rotor Bearing," IMEKO International Measurement Confederation, Symposium on Measurement and Estimation, Brixen, Italy, May 8-12, 1984, pp. 187-191.

12. Salm, J. R., "Active Electromagnetic Suspension of an Elastic Rotor: Modeling, Control, and Experimental Results," Rotating Machinery Dynamics, ASME Design Technology Conference, Boston, Massachusetts, September 27-30, 1987, pp. 141-149.

13. Nikolajsen, J. N. and Holmes, R., "Investigation of an Electromagnetic Damper for Vibration Control of A Transmission Shaft," Proceedings of Institution of Mechanical Engineers, Vol. 193, 1979, pp. 331-336.

14. Gondhelkar, V. and Holmes, R., "Design of an Electromagnetic Bearing for the Vibration Control of a Flexible Transmission Shaft," Proceedings of Rotordynamic Instability Problems in High-Performance Turbomachinery, May 28-30, 1984, Texas A&M University, College Station, Texas.

15. Bradfield, C. D., Roberts, J. B., and Karunendiran, R., "Performance of an Electromagnetic Bearing for the Vibration Control of a Supercritical Shaft," Proceedings of Rotor Dynamics Instability Problems in High-Performance Turbomachinery, NASA 2443, Texas A&M University, College Station, Texas, June 2-4, 1986, pp. 431-460.

16. Hebbale, K. V., and Taylor, D. L., "Nonlinear Dynamics of Attractive Magnetic Bearings," Proceedings of Rotor Dynamics Instability Problems in High-Performance Turbomachinery, NASA 2443, Texas A&M University, College Station, Texas, June 2-4, 1986, pp. 397-418.

17. Humphris, R. R., Kelm, R. D., Lewis, D. W., and Allaire, P. E., "Effect of Control Algorithms on Magnetic Journal Bearing Properties," Journal of Engineering for Gas Turbines and Power, Trans. ASME, Vol. 108, October 1986, pp. 624-632.

18. Allaire, P. E., Humphris, R. R., and Barrett, L. E., "Critical Speeds and Unbalance Response of a Flexible Rotor in Magnetic Bearings," Proc. European Turbomachinery Symposium, October 27-28, 1986.

19. Allaire, P. E., Humphris, R. R., and Imlach, J., "Vibration Control of Flexible Rotors with Magnetic Bearing Supports," AFOSR/ARO Conference, Non-Linear Vibrations, Stability and Dynamics of Structures and Mechanisms, Virginia Tech, Blacksburg, VA, March 23-25, 1987.

20. Allaire, P. E., Humphris, R. R., Kasarda, M. E. F., and Koolman, M. I., "Magnetic Bearing/Damper Effects on Unbalance Response of Flexible Rotors," Proceedings AIAA Conference, Philadelphia, PA, August 10-14, 1987.

21. Rotors, H. C., Electromagnetic Devices, 1941, pp. 133-134, (John Wiley & Sons).

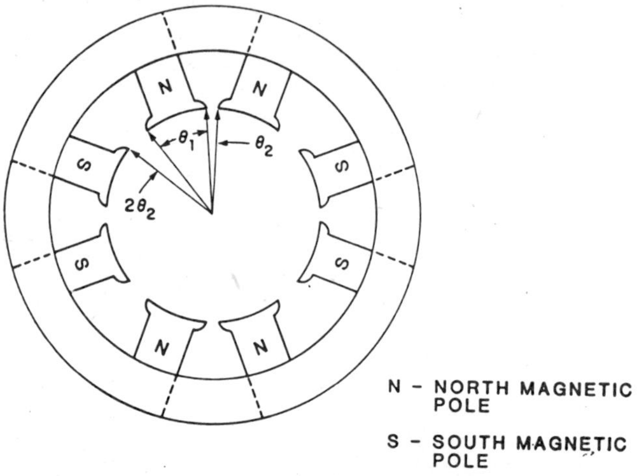

Fig 1 A generalized magnetic bearing assembly

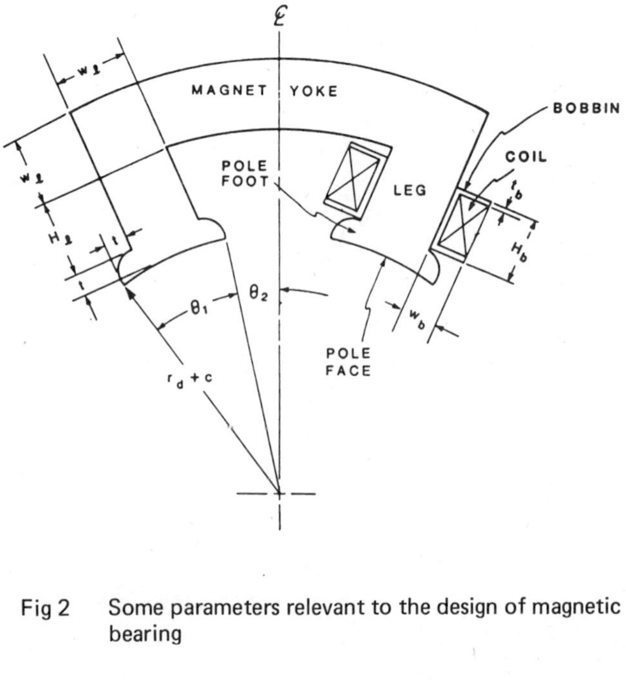

Fig 2 Some parameters relevant to the design of magnetic bearing

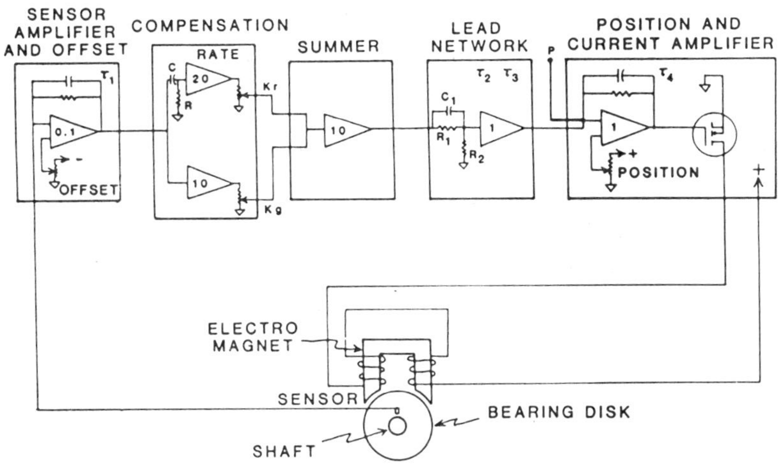

Fig 3 A representative controller block diagram

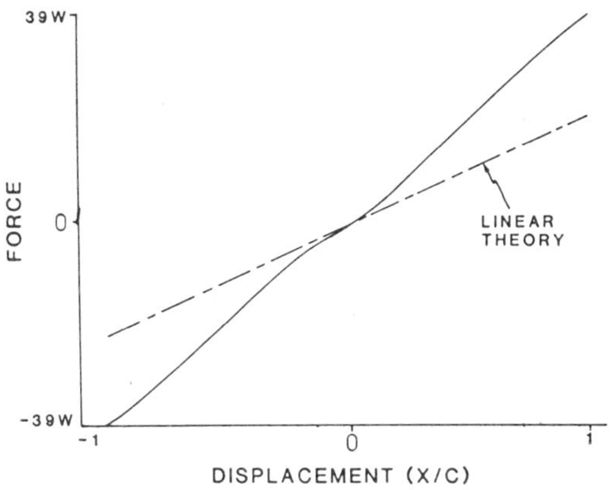

Fig 4 A comparison of the assumed linear stiffness theory to the non-linear theory for a specific case

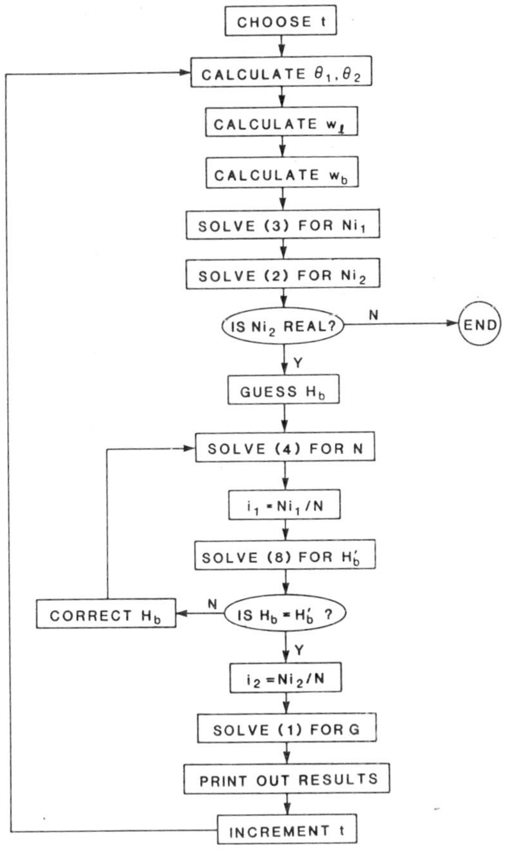

Fig 5 Flow chart of solution procedure

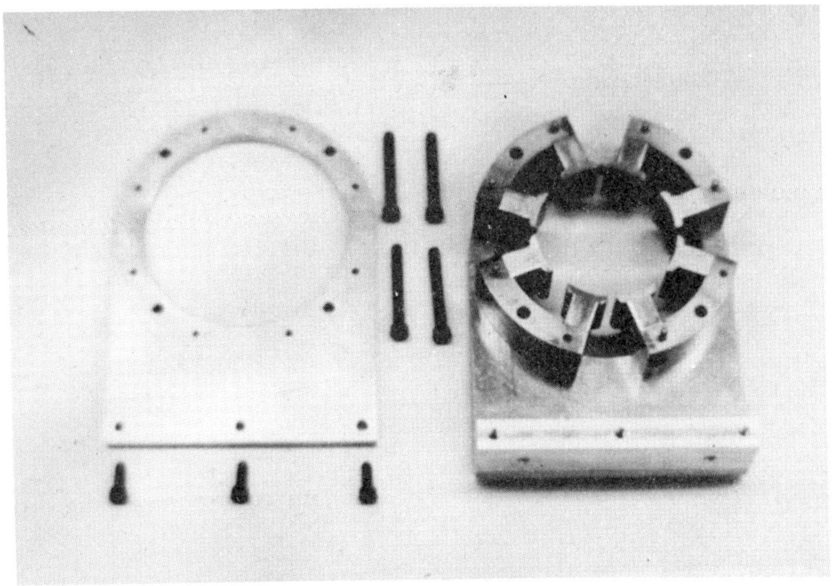

Fig 6 A disassembled view of a magnetic bearing designed for a test rig

Fig 7 Assembled bearing installed in test rig

Fig 8 Magnetic bearing test rig

C261/88

An active support system for rotors with oil-film bearings

S FÜRST, Dipl-Ing and **H ULBRICH**, DrDr
Department of Mechanical Engineering, Technical University of Munich, Munich, West Germany

SYNOPSIS The use of rotors in turbo-machinery, space applications, and ultra high vacuum technology always yields a number of problems that must be kept under control. The success of active vibration control depends mainly on modelling the mechanical system, finding problem-adapted control concepts, suitable actuators, as well as actuator and sensor locations. The attainable improvements of the dynamic behaviour of rotors with journal bearings by the use of active controlled supports are demonstrated by theory and experiments.

1 INTRODUCTION

The application of flexible rotors in modern turbines increases the dynamic sensitivity of this type of machinery. Dynamic problems can be caused by unbalance, support perturbations, instability due to internal damping, aerodynamic forces, and oil-film properties. The tuning of the system parameters (passive optimization) by itself often does not yield satisfying results. Further improvement may be achieved by using active control techniques. In numerous works the use of control forces, which act directly upon the rotor by magnetic bearings, is suggested (e.g. see Ref. /1-3/). Vibration control of an oil-film supported rotor by a magnetic bearing is developed and tested by BURROWS et al. /4/.

In the case of limited space at the rotor, an alternative approach may be the use of control forces acting via bearing housings. This method requires a flexible suspension of the bearings relative to the foundation. Controllers for such systems were designed by MOORE et al. /5/, NONAMI /6/, the authors /7/, and others. For a rotor with oil-film bearings, a state-feedback controller working with indirect control forces was introduced by STANWAY and O'REILLY /8/.

On the basis of the results of these studies, the present work focuses on the possibilities of vibration control of a flexible rotor with journal bearings by an active bearing suspension. The description of the rotor as a "hybrid multi - body system" (HMBS) allows for the consideration of additional effects which are mostly neglected in other investigations, such as gyroscopic effects, perturbation by aerodynamic and internal damping, and high order bending vibrations of the shaft. The goals of the vibration control may be manifold: reduction of resonance amplitudes, improvement of stability, etc..

2 NOTATION

$\mathbf{A}$	system matrix
$\mathbf{B}$	control matrix
$\mathbf{C}$	measurement matrix
c_{ij}	stiffness coefficients
d_{ij}	damping coefficients
F	force
f_e	number of "elastic degrees of freedom"
$\mathbf{h}(t)$	vector of external forces
i	control current
$\mathbf{K}$	feedback matrix
$\hat{\mathbf{K}}$	block diagonal matrix
k_i	force-current factor
k_x	force-displacement factor
$\mathbf{M}$	mass matrix
$\bar{\mathbf{M}}$	rectangular matrix of left hand eigenvectors
m	number of measurements
n	system order
$\mathbf{P}$	matrix of forces proportional to velocities
$\hat{\mathbf{P}}$	solution of the Matrix-Riccati-Equation
$\mathbf{Q}$	matrix of forces proportional to displacements
$\bar{\mathbf{Q}}$	weighting matrix
$\bar{\mathbf{R}}$	weighting matrix
q	cross-coupling coefficient
R	resonance factor (max. amplitude/ runner eccentricity)
r	number of actuators
$\mathbf{T}$	transformation matrix
t	time
$\mathbf{u}(t)$	control vector
$\mathbf{x}(t)$	displacement vector
$\bar{\mathbf{x}}(z)$	vector of the shape-functions
$\mathbf{x}_e(t)$	vector of the "elastic coordinates"
x, y	displacements
$\dot{x}, \dot{y}$	velocities
$\mathbf{y}(t)$	measurement vector
$\mathbf{z}(t)$	state vector
z	longitudinal coordinate of the rotor
Ω	rotational speed of the rotor
ω	eigenfrequency

Subscripts

a	actuator
D	design model
e	elastic body
M	modal controller
R	residual model
S	simulation model
s	sensor
Z	state controller

3 MATHEMATICAL MODEL

Under the assumption of small deviations from a given reference system, we arrive at the linear equations of motion in the usual form:

$$\mathbf{M}\ddot{\mathbf{x}}(t) + \mathbf{P}\dot{\mathbf{x}}(t) + \mathbf{Q}\mathbf{x}(t) = \mathbf{h}(t) \ . \qquad (1)$$

The description of the system as a HMBS leads to a subdivision of the generalized minimal coordinates. These are the coordinates of rigid bodies and the so-called "elastic coordinates". Coordinates of the latter typ result from the separation of the place- and time-dependent function, which are used to describe the elastic parts. The Ritz method has been applied:

$$x(z,t) = \bar{\mathbf{x}}(z)^T \mathbf{x}_e(t), \qquad \bar{\mathbf{x}},\mathbf{x}_e \in \mathbb{R}^{f_e} \ . \qquad (2)$$

$\bar{\mathbf{x}}(z)$ is the vector of the admissible shape-functions and $\mathbf{x}_e(t)$ the vector of corresponding time-functions, i.e. the "elastic coordinates". The eigenfunctions of the non-rotating "free-free" rotor are used as shape-functions for the running rotor. The determination of these functions is carried out in a preliminary stage by use of cubic spline functions. The system order is dependent upon the determination of the relevant frequency range. This in turn is given by the expected excitation frequencies. The journal bearings are described by a linear model

$$\begin{bmatrix} F_x \\ F_y \end{bmatrix} = \begin{bmatrix} c_{xx} & c_{xy} \\ c_{yx} & c_{yy} \end{bmatrix} \begin{bmatrix} x \\ y \end{bmatrix} + \begin{bmatrix} d_{xx} & d_{xy} \\ d_{yx} & d_{yy} \end{bmatrix} \begin{bmatrix} \dot{x} \\ \dot{y} \end{bmatrix}, \qquad (3)$$

with F_x, F_y the components of the reaction forces, c_{ij} the oil-film stiffness coefficients, d_{ij} the oil-film damping coefficients, x and y the displacements, and $\dot{x}$ and $\dot{y}$ the velocities of the journal /9/. The linear oil-film coefficients are functions of the working conditions, which are characterised by the Sommerfeld number. A comprehensive description of the derivation of the equation of motion may be found in /10/.

4 SPILLOVER EFFECTS

The starting point for the following control-concepts is the state equation of the HMBS

$$\dot{\mathbf{z}}(t) = \mathbf{A}\mathbf{z}(t) + \mathbf{B}\mathbf{u}(t), \qquad \mathbf{z} \in \mathbb{R}^n, \ \mathbf{u} \in \mathbb{R}^r$$
$$\text{and} \quad \mathbf{y}(t) = \mathbf{C}\mathbf{z}(t) \ , \qquad \mathbf{y} \in \mathbb{R}^m \ . \qquad (4)$$

A correct description of the elastic components demands an infinite number of shape-functions. In practice, only a finite number of eigen-movements can be described. In the following, this system of finite order n_S is called the simulation-model. Similarly, it is obvious that not all natural vibrations of the system can be extracted from measurements. The consequence of this incomplete system description are the spillover effects. They appear on two different levels. On the one hand, the measurements contain both the considered and disregarded eigen-functions, which influence the expected signals at the measurement locations (observation-spill-over). On the other hand, control forces may destabilize eigenfunctions of higher order (control-spillover) that have not been accounted for in the control design. Mostly, it is assumed that the internal material damping is large enough to prevent instabilities due to spillover effects.

Because this is not true in all cases, it is reasonable to already consider these effects in the control design stage. Other than by filtering the measurement signals, this may be done by consideration of the collocation condition

$$\mathbf{B} = \mathbf{C}^T \ , \qquad (5)$$

which requires that actuators and sensors are located in the same place /2/. For economic reasons, a different approach to minimize the spillover-effects is suggested here. Namely, it seems quite reasonable to use more sensors than actuators in order to receive as much information as possible about the system (i.e. m > r).

At first, the simulation model is split into a reduced-order design model (n_D) and a residual model (n_R):

$$\begin{bmatrix} \dot{\mathbf{z}}_D \\ \dot{\mathbf{z}}_R \end{bmatrix} = \begin{bmatrix} \mathbf{A}_D & \mathbf{0} \\ \mathbf{0} & \mathbf{A}_R \end{bmatrix} \begin{bmatrix} \mathbf{z}_D \\ \mathbf{z}_R \end{bmatrix} + \begin{bmatrix} \mathbf{B}_D \\ \mathbf{B}_R \end{bmatrix} \mathbf{u}$$
$$\text{and} \quad \mathbf{y} = \begin{bmatrix} \mathbf{C}_D & \mathbf{C}_R \end{bmatrix} \begin{bmatrix} \mathbf{z}_D \\ \mathbf{z}_R \end{bmatrix} \ . \qquad (6)$$

The design model includes all natural vibrations to be controlled. By definition, the design- and the residual-model are uncoupled through the use of the orthogonal eigenfunctions as shape-functions. The introduction of a controller $\mathbf{u} = \mathbf{K}\mathbf{y}$, which is designed for the reduced order model, causes the two subsystems of the simulation-model to be coupled by the measurement matrix **C**, the feedback matrix **K**, and the control matrix **B**:

$$\mathbf{A}_S = \left[\begin{array}{c|c} \mathbf{A}_D + \mathbf{B}_D\mathbf{K}\mathbf{C}_D & \mathbf{B}_D\mathbf{K}\mathbf{C}_R \\ \hline \mathbf{B}_R\mathbf{K}\mathbf{C}_D & \mathbf{A}_R + \mathbf{B}_R\mathbf{K}\mathbf{C}_R \end{array}\right] \ . \qquad (7)$$

If the controlled simulation-model is to have the required behaviour, the matrices $\mathbf{B}_R$ and $\mathbf{C}_R$ of the residual model must be zero matrices. This means that the neglected state variables are neither observable nor controllable. These requirements lead to a practical criterion for the choice of the actuator and sensor locations with regard to minimal spillover:

$$\sum_{i=n_D+1}^{n_S} |\bar{x}_i(z_a)| \rightarrow \min, \quad z_a\text{: position of actuators,}$$
$$(8)$$
$$\sum_{i=n_D+1}^{n_S} |\bar{x}_i(x_s)| \rightarrow \min, \quad z_s\text{: position of sensors.}$$

The demanded positions result from a minimum search for the sum of the absolute values of the high-order eigenfunctions (see equation (2)). In some simple cases, suitable positions may be found directly by an evaluation of the eigen-functions Fig. 1 . For HMBS with a very disproportionate mass concentration (e.g. rotor shaft with heavy runner), the high-order natural vibrations have almost no corresponding rigid body motions. Only the low-order relevant vibrations contain significant rigid body movements in addition to elastic deformations. Therefore, it is recommendable to place the actuators and sensors at the points of mass concentration, where, in practice, the high-order eigenfunctions have no or only small displacements.

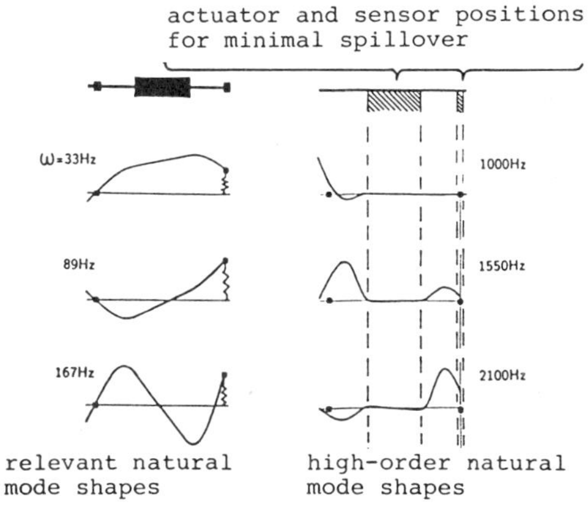

Fig 1 Minimization of spillover effects

5 DESIGN CONCEPTS FOR CONTROLLERS

Starting from a design model which is described by the time invariant and linear state equation

$$\dot{\mathbf{z}}_D = \mathbf{A}_D\mathbf{z}_D + \mathbf{B}_D\mathbf{u} \qquad (9)$$

and the measurement equation $\mathbf{y} = \mathbf{C}_D\mathbf{z}_D$, several different design concepts are tested for linear time invariant controllers of the form

$$\mathbf{u} = \mathbf{K}\mathbf{y}. \qquad (10)$$

All the following procedures, have in common that the controller design is made in the time domain.

5.1 Optimal state feedback

In this procedure the optimal feedback is defined by the minimization of the quadratic integral criterion

$$J = \int_0^\infty (\mathbf{z}_D^T\bar{\mathbf{Q}}_D\mathbf{z}_D + \mathbf{u}^T\bar{\mathbf{R}}\mathbf{u})\, dt \rightarrow \min. \qquad (11)$$

This method leads to an indirect pole placement. This means the closed control loop poles are fixed by the choice of the weighting matrices $\bar{\mathbf{Q}}_D$ and $\bar{\mathbf{R}}$. As mentioned earlier, it is impossible to design a state feedback for HMBS because the number of state variables is infinite. Yet , by introducing the described design model it is possible to use the theory of optimal feedback. In this case, the following conditions must be satisfied:

$$m = n_D \text{ und } \mathbf{z}_D = \mathbf{C}_D^{-1}\mathbf{y}. \qquad (12)$$

With this, one can calculate the feedback matrix for linear and time invariant systems from the solution $\hat{\mathbf{P}}$ of the Matrix-Riccati-Equation:

$$\mathbf{K}_Z = \bar{\mathbf{R}}^{-1}\mathbf{B}_D^T\hat{\mathbf{P}}\mathbf{C}_D^{-1} \quad . \qquad (13)$$

The main problem in the determination of a concrete controller is to choose suitable weighting matrices.

5.2 Modal controller

The aim of this method is to shift one or more eigenvalues of the design model. A detailed description of the procedure may be found in KORN, WILFERT /11/. Again assuming that the relevant system states are covered by the measurements, the feedback matrix of the modal controller results from

$$\mathbf{K}_M = (\bar{\mathbf{M}}\mathbf{B}_D)^{-1}\hat{\mathbf{K}}\bar{\mathbf{M}}\mathbf{C}_D^{-1} \qquad (14)$$

with $\hat{\mathbf{K}}$ being a matrix with diagonal block structure, representing the desired shift of the eigenvalues. $\bar{\mathbf{M}}$ is a rectangular matrix of the r left hand eigenvectors of the r eigenvalues which are to be shifted, and $\mathbf{B}_D$ the control matrix of the design model. Notice that only as many eigenvalues can be shifted as actuators exist. For instance, shifting one critical eigenvalue requires two actuators. A further method to calculate a modal controller uses the pseudo-inverse of the control matrix:

$$\mathbf{K}_M = \mathbf{B}_D^+\mathbf{T}\hat{\mathbf{K}}\mathbf{T}^{-1}\mathbf{C}_D^{-1}, \qquad (15)$$

with $\mathbf{B}_D^+$ as the pseudo inverse of $\mathbf{B}_D$ and $\mathbf{T}$ as the transformation matrix, which transforms the system matrix $\mathbf{A}_D$ into diagonal block structure. In contrast to the modal controller from equation (14), a theoretical limitation of the number of actuators or the number of eigenvalues to be shifted does not exist for this method. The pseudo inverse of $\mathbf{B}_D$ exists in any case, and consequently it is always possible to specify a feedback matrix $\mathbf{K}_M$. In practice, however, the results are only reasonable if the number of actuators is equal to the number of the eigenvalues which should be shifted.

5.3 Optimal velocity feedback

The local velocity feedback represents a particularly simple controller. According to the collocation condition, the actuators and sensors are placed at the same location, where only the velocity is measured and fed back. Thus one obtains pure damping, yet with the special advantage of the feedback gains being adaptable to changing working conditions (e.g. rotor speed). This seems to be especially important for rotors with journal bearings because of the strong speed dependence of the oil-film characteristics.

Some works already exist on the determination of the optimal velocity feedback (e.g. see /12/). All those works are based on eigenvalues being a function of the velocity feedback gain. The absolute or the relative stability reserve of the momentary critical natural mode is optimized according to the rotor speed. This leads to an adaptive support system for bearing houses with damping as a function of the rotor speed. For realizing this, one can imagine the use of an active system (e.g. electromagnetic actuators) for adaptation in addition to a basically passive damping (e.g. squeeze film damper).

6 TEST RIG

The test rig shown in Fig. 2 was designed and developed for the purpose of experimentally tes-

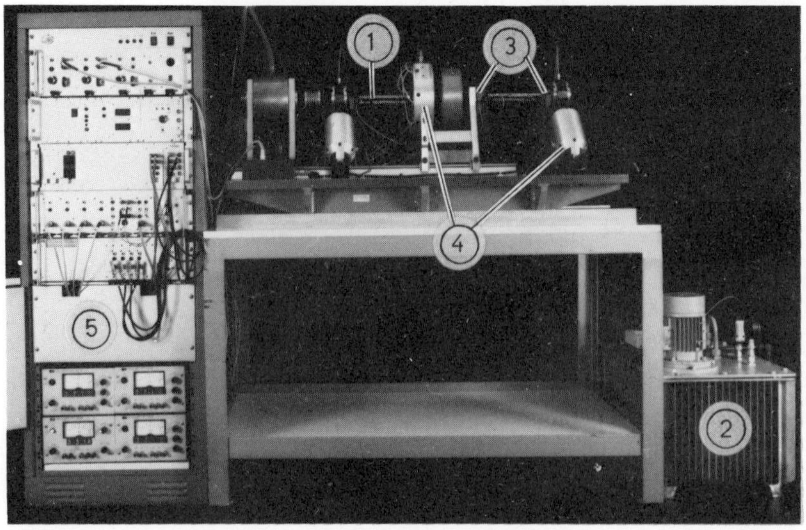

Fig 2 Full view of the test rig

ting controllers. Its components are as follows:

1 test rotor
2 oil supply system
3 system of sensors
4 actuators
5 control- and measurement-electronics

6.1 Test rotor

The rotor consists of a steel shaft (ø = 20 mm) and a runner (ø = 200 mm, m = 16.5 kg), which is frictionally mounted onto the shaft by clamping bushes. The journal bearings (four-lobe bearings, distance between bearings: 700 mm) are mounted via electromagnetic actuators on a foundation plate at 45° angles (Fig. 3). As a result of this configuration, the bearing stiffnesses are equal in magnitude both in vertical and in horizontal direction. Torque is transmitted onto the rotor shaft over an elastic coupling.

6.2 Sensors

Three systems of sensors are implemented for the analysis of vibrations. Two inductive displacement sensors are mounted perpendicularly in each of two moveable mountings. A third sensor system is integrated in a magnetic bearing. This allows for the simultaneous measurement of displacement and velocity without contact.

6.3 Actuators

The test rig is fitted with two different kinds of actuators. Forces can be applied onto the rotor directly via the magnetic bearing or indirectly through the electromagnetic actuators in the journal bearing supports. In both cases there is a linear dependence between the displacement x, the control current i and the force F:

$$F = k_i * i + k_x * x \quad (16)$$

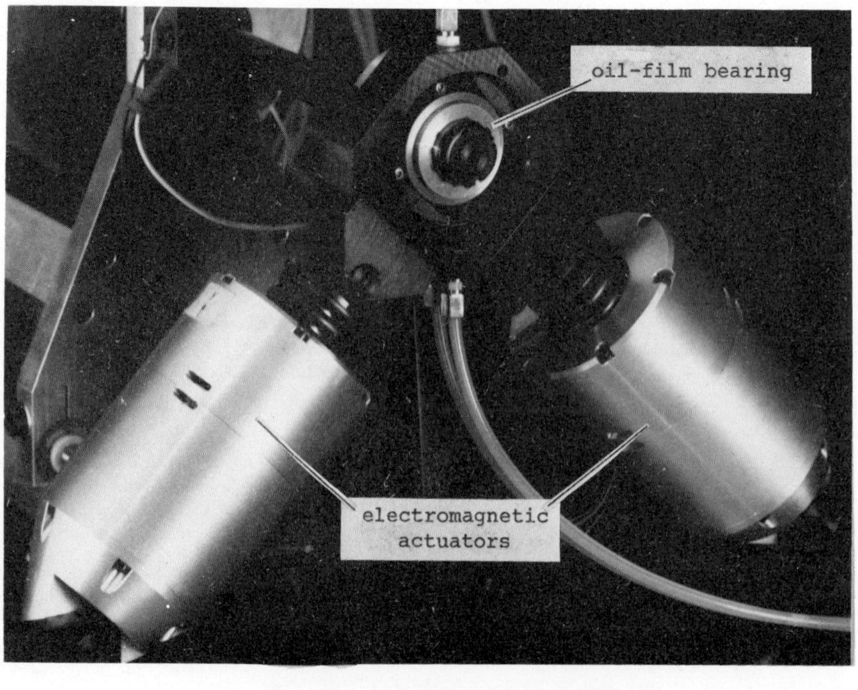

Fig 3 Active support bearing

A detailed description of these actuators may be found in /10/.

6.4 Control- and measurement-electronics

The test runs are monitored and controlled by a personal computer with various digital and analogueI/O-channels. A linear control circuit was developed, by which the feedback gain is adapted on-line to the momentary operational state. By this method various states of operation can be modulated (partial-/full-load, non-steady states, temperature fluctuations, etc.). Different types of excitations can also be simulated through the use of the control- and measurement systems. Among these are:

- predetermined bearing vibrations
- excitation by fluids (cross-coupling forces)
- impulse- and step-functions

Therefore, a good reproduction of existing excitations of the actual rotor system is possible.

7 NUMERICAL ANALYSIS

The applied analytical model of the test rig is shown in Fig. 4. Various computerized examinations have been conducted for this system model. In doing this, the following assumptions were made for deriving the equation of motion:

- the rotor is an elastic Raleigh-beam with variable cross-section
- visco-elastic material laws are valid for the rotor
- the rotor is rotationally symmetric
- the journal bearings may be described by a linear model

7.1 Uncontrolled system

At first various analyses of the passive system were conducted. This not only allowed for a verification of the applied model, but also made an estimation of additional influences possible.

In the course of this preliminary testing stage, the curve of the eigenvalues as a function of the rotor speed was determined. Also, the unbalance response and an impulse response were computed. Thus a general view is given of the steady and non-steady eigenbehaviour as well as the response to external excitation. Some of the results of these simulations will be briefly discussed.

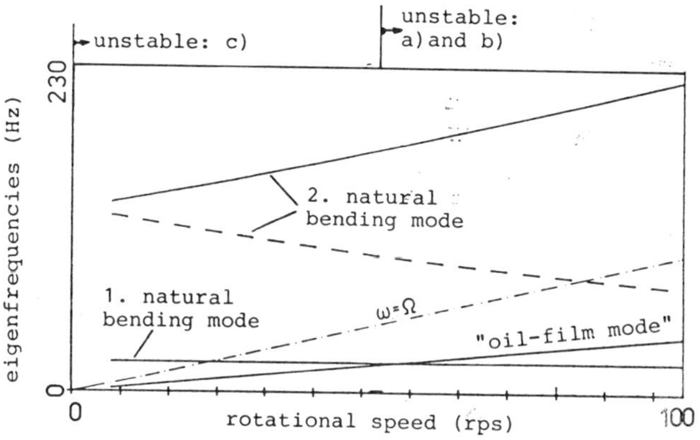

Fig 5 Calculated eigenfrequencies of the test rotor
(a) With internal damping
(b) Without internal damping
(c) With internal damping and cross-coupling forces

As can be seen from the curve of eigenfrequencies (Fig. 5), the 1. natural frequency in bending split up to one backward ($\omega < 0$) and one forward vibration mode ($\omega > 0$). Both these frequencies are independent of the rotor speed due to the almost symmetrical arrangement of the system. In contrast, two eigenfrequencies which result from the 2. natural frequency in bending are dependent upon the rotor speed due to gyroscopic effects. All these "elastic frequencies" practically cannot be influenced by the journal bearings because the stiffness of the oil film is much greater than that of the shaft. The oil film causes a further pair of eigenfrequencies that increase by a factor of about $\Omega/2$ with the rotor speed.

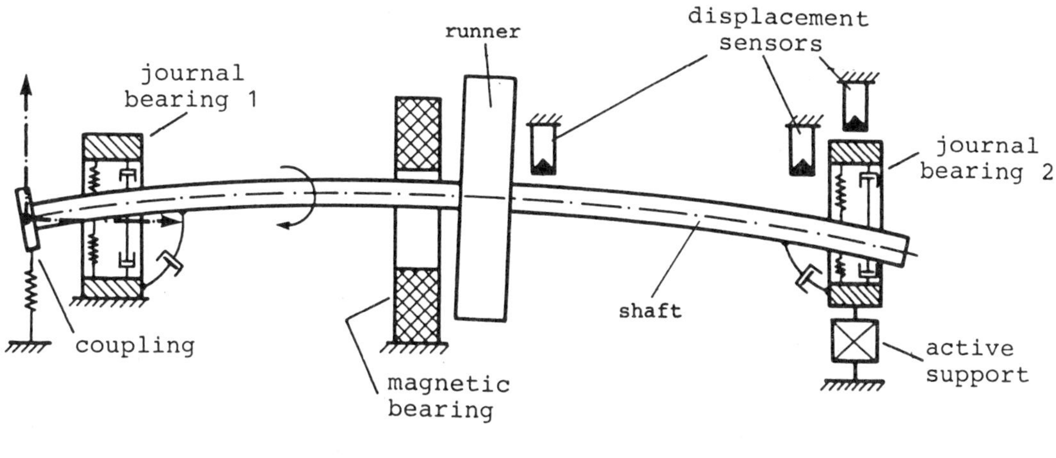

Fig 4 Analytical model of the test rig

Only one critical rotor speed due to an excitation through unbalance exists for the examined frequency range. This can be seen from the curve of the eigenfrequencies: There is exactly one point of intersection between the eigenfrequencies and the run-up line $\omega = \Omega$.

The additional boundaries to the regions of instability were gained by determining the real part of the eigenvalues. The 1. natural frequency in bending ($\omega > 0$) becomes unstable past $\Omega = 50$ Hz regardless of internal damping (self-excited vibration). In contrast, the presence of a cross-coupling force in the form of

$$\begin{bmatrix} F_x \\ F_y \end{bmatrix} = \begin{bmatrix} 0 & q \\ -q & 0 \end{bmatrix} \begin{bmatrix} x \\ y \end{bmatrix} \qquad (17)$$

results in instability across the whole range of rotor speed. The non-conservative terms lead to the transmission of the drive energy into the bending vibration of the rotor. The unbalance response of the system without cross-coupling forces shows a resonance factor $R > 25$ (R = max. amplitude/ runner excentricity), which indicates a very large strain on the system (Fig. 6).

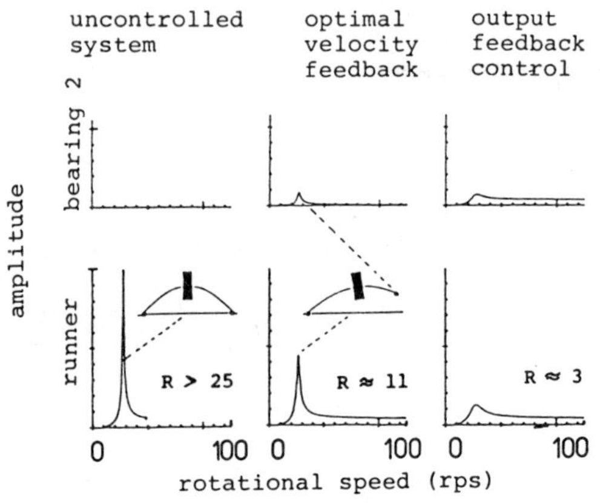

Fig 6 Calculated unbalance response

7.2 Controlled System

Three controllers will be presented here with one active bearing support (r = 2):

- velocity feedback controller (one measuring level, m = 2)
- output controller (two measuring levels, m = 8)
- state controller (five measuring levels, m = 20)

Velocity feedback controller

Because several papers on this topic already exist (e.g. see /12/,/13/), only a brief description will be given for the purpose of comparison. Through the use of uncoupled local velocity feedback the system damping can be improved to a point, where the rotor remains in the stable region throughout the whole speed range. The resonance factor is thereby reduced to $R \approx 11$ (Fig. 6).

Output controller

The following system characteristics were utilized in designing an output controller for practical applications:

- The 1. natural frequency in bending is of paramount importance for the vibrational behaviour of the rotor in the considered frequency range.
- A comparision between roller bearings and journal bearings shows insignificant differences in regard to their effect on the relevant mode of vibration.

Starting from a reduced model with roller bearings (only 1. natural frequency in bending considered; system order $n_D = 8$) an optimal state feedback controller was designed. This was in turn applied as an output controller in the simulation-model with journal bearings ($n_S = 20$). This controller is similar to a modal controller in that it influences mainly the 1. natural frequency in bending. The unbalance response shows that a better dynamic behaviour can be achieved by means of an output controller than through optimal bearing damping alone (Fig. 6).

Due to the fact that the journal bearing coefficients do not constitute part of the output controller design, this type of controller has no sensitivity towards fluctuations of these coefficients.

Furthermore, it should be noted that it is possible to stabilize a rotor in the presence of cross-coupling forces with an output controller. This an uncoupled velocity feedback cannot do.

State controller

Whereas this type of controller is impractical to implement because it requires an immense amount of measurement data, it can fundamentally demonstrate the range of possibilities given by an active bearing support. The system damping achieveable is only limited by the magnitude of attainable control forces. With a realistic choice of the feedback gain the system behaviour can be further improved (R = 2). Especially the damping of the 2. natural frequency in bending ($\omega > 0$) is affected. However, as long as this frequency is not excited by the unbalanced rotor, the state controller does not represent an improvement over the output controller for practical purposes.

8 EXPERIMENTAL RESULTS

The first measurements were conducted on the passive system in order to verify the system model. The results of the measured and the calculated eigenfrequencies in bending ($\Omega = 0$) correspond very well:

	calculated	measured
1. natural frequency	20.1 Hz	20.7 Hz
2. natural frequency	130 Hz	131 Hz

Additional measurements of test runs also verified the effectiveness of the active bearing support. This will be exemplified in three different cases. In each instance, an output controller with a combined displacement- and velocity-feedback is used. The measurements are limited to two levels (r = 2, m = 4).

8.1 Improvement of impulse response

Fig. 7 shows the impulse response of the uncontrolled and the controlled system. One can clearly recognize the greater system damping reflected by the faster decline of the disturbance.

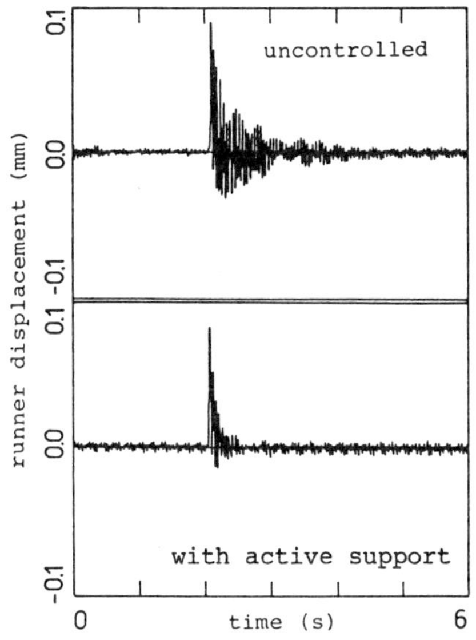

Fig 7 Impulse response of the rotor (rotor speed 40 r/s)

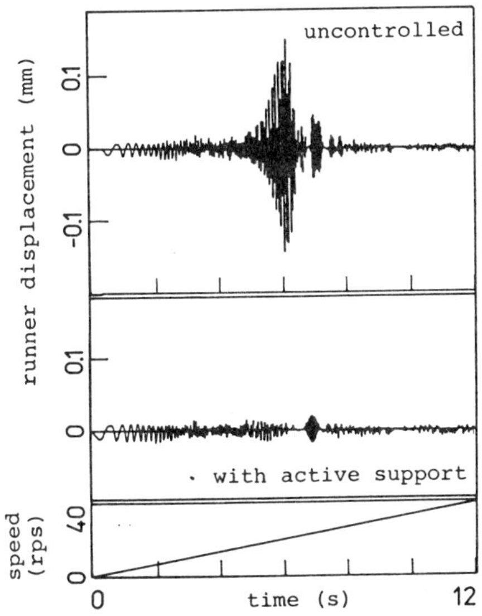

Fig 8 Run-up through the first critical speed

8.2 Reduction of amplitudes during unsteady run-up

The time-dependent course of the runner displacement is depicted for a rotor acceleration from 0 to 40 Hz (Fig. 8). Without active control, the amplitudes in the vicinity of the 1. critical speed (20.7 Hz) reach values of ca. 0.15 mm. The implementation of one active bearing support reduces the maximum displacement across the whole speed range to ca. 0.04 mm (73% reduction).

8.3 Stabilization of an unstable system

Experience has shown that the presence of cross-coupling forces can lead to an instability of the rotor. Such forces were simulated on the test rig by the magnetic bearing. The effect of instability is depicted in Fig. 9; the runner displacement rises exponentially. The application of control forces via the bearing housing clearly shows a stabilizing effect.

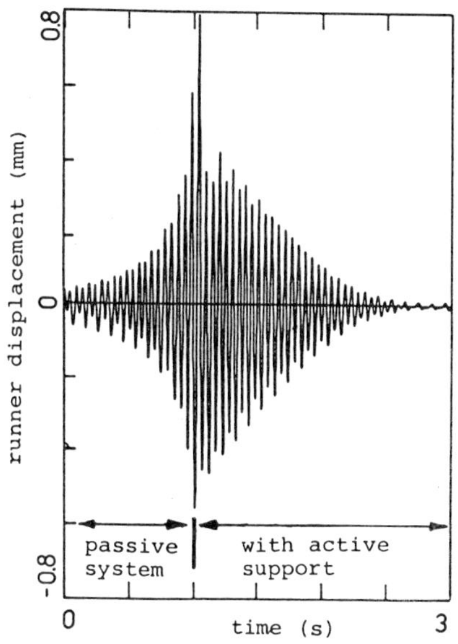

Fig 9 Control of an unstable system in the presence of cross-coupling forces

These experimental results demonstrate that it is fundamentally possible to avoid unstable regions through the use of an active bearing support.

9 CONCLUSIONS

The modelling of elastic rotor systems as "hybride multi body systems" is scientifically transparent and concurres with the demands set by control problems (low system order, simple model adaptation). Through the choice of suitable shape-functions one obtains a compact mathematical description, which takes various physical properties into account (internal damping, gyroscopic effects, etc.).

Numerical analyses and experimental results have shown that the dynamic behaviour of rotors on journal bearings can be decisively improved through active bearing supports. This applies to new designs as well as existing systems. The computer regulated controller introduced here makes an adaptation to vastly different operational states possible (adaptive control). The extent of attainable improvement is essentially determined by the available actuators and measurement data. Which type of control concept is to be chosen in each individual case depends on the objectives and also on economic reasons.

In view of technical implementation, future research must concentrate on developing new actuators, which can supply large control forces. Reliability and safety of such concepts also remain an important aspect.

REFERENCES

/1/ SAHINKAYA, M.N.; BURROWS, C.R.: Control of Stability and the Synchronous Vibration of a Flexible Rotor Supported on Oil-Film Bearings. Journal of Dynamic Systems, Measurement and Control, Vol. 107, 1985, pp. 139 - 144.

/2/ SALM J.; LARSONNEUR, R.: Modellbildung und aktive Dämpfung eines elastischen Rotors. Z. Angew. Math. u. Mech. 65 (1985) 4, S. 96 - 98.

/3/ REDMOND, I.; MC LEAN, R.F.; BURROWS, C.R.: Vibration Control of Flexible Rotors. ASME-Paper 85-DET-127, 1985.

/4/ BURROWS, C.; SAHINKAYA, M.; CLEMENTS, S.: Electromagnetic Control of Oil-Film Supported Rotors Using Sparse Measurements. 11 Biennial ASME Conference on Vibration and Noise, Session on Rotating Machinery Dynamics, Boston Massachusetts, Sept. 1987.

/5/ MOORE, J.W.; LEWIS, D.W.; HEINZMAN, J.: Feasibility of Activ Feedback Control of Rotordynamic Instability. Rotordynamic Instability Problems in High-Performance Turbomachinery, Proc. of a Workshop, Texas A & M University, College Station, Texas, 1980, pp. 467 - 478.

/6/ NONAMI, K.: Vibration Control of Rotor Shaft Systems by Active Control Bearings. ASME-Paper 85-DET-126, 1985.

/7/ ULBRICH, H.; FÜRST, S.: Modelling and Active Vibration Control of Flexible Rotors. The Theory of Machines and Mechanisms, Proc. of the 7th World Congress, Sept. 87, Sevilla, Spain, pp. 1739 - 1743.

/8/ STANWAY, R.; O'REILLY, J.: State-Variable Feedback Control of Rotor-Bearing Suspension Systems. Third Int. Conference on Vibrations in Rotating Machinery, Instn. Mechn. Engrs., 1984, Paper C274/84.

/9/ GLIENICKE, J.: Theoretische und experimentelle Ermittlung der Systemdämpfung gleitgelagerter Rotoren und ihre Erhöhung durch eine äußere Lagerdämpfung. Fortschr. Ber. VDI-Z., Reihe 8, Nr.34, 1972.

/10/ ULBRICH, H.: Dynamik und Regelung von Rotorsystemen. Fortschr.-Ber. VDI-Z, Reihe 11, Nr.86, 1986.

/11/ KORN, U.; WILFERT, H.: Mehrgrößenregelungen - Moderne Entwurfsprinzipien im Zeit- und Frequenzbereich. VEB-Verlag Technik, Berlin 1982.

/12/ SPRINGER, H.: Optimale Lagerdämpfung für hochflexible Rotoren. Z. Angew. Math. u. Mech. 65 (1985) 4, S. 105 - 107.

/13/ GUNTER, E.J.; BARRET, L.E.; ALLIARE, P.E.: Stabilization of Turbomachinery with Squeeze Film Dampers - Theorie and Application. I Mech E 1976, Paper C233/76.

C267/88

The use of fluid swirl to control the vibration behaviour of rotating machines

J A HART, BSc, PhD
Central Electricity Generating Board, Bedminster Down, Bristol
R D BROWN, BSc, MSc, CEng, MIMechE
Department of Mechanical Engineering, Heriot-Watt University, Edinburgh

SYNOPSIS A reduction in pre-swirl at entry to annular seals in turbo-machinery has been used by some manufacturers to improve rotor stability. A deliberate injection of negative swirl in an annular space has potential for reducing forced vibration near resonance. Experimental results on a 34kg mass rotor demonstrate that significant reductions in amplitude can be obtained at the synchronous critical speeds.

1 INTRODUCTION

During the past few decades the development of rotating machines e.g. compressors, pumps and gas turbines has led to increases in the energy density of the working fluid. Many of these machines are of a flexible design and even a small amount of the available energy when transformed into lateral whirling motion results in high levels of vibration. This vibration manifests itself as either an undue sensitivity to mass unbalance or an instability which occurs as speed or load crosses a stability boundary and a sub-synchronous frequency is observed in the displacement spectrum. BLACK et al (1) suggested that a deliberate negative pre-swirl at seal entry in centrifugal pumps would be useful in controlling vibration. Swirl brakes have been incorporated at labyrinth entry by a number of compressor manufacturers. KIRK (2) described a reverse flow injection in the interior of a labyrinth seal which was instrumental in suppressing a sub-synchronous whirl in the forward direction.

Experimental work on a small scale rig at Heriot-Watt to demonstrate sub-synchronous whirls under laboratory conditions (3) indicated that a retrograde flow in an annular space could significantly reduce unbalance response near a resonance. A simple theory demonstrated that the stiffness matrix was skew symmetric and hence the resulting forces were non-conservative.

The action of the reverse annular flow added substantial energy dissipation to the system. However unlike conventional dampers the damping action did not depend on rotor motion, it was primarily a function of rotor displacement.

A larger scale, fully instrumented experimental rig was constructed and measurements demonstrated that reductions of over 50 per cent could be obtained for synchronous response at speeds close to a critical for steady state conditions. Similar results were obtained when the 34 kg rotor was allowed to slow down from a speed well above the resonance.

Successful installation for industrial rotors could lead to dampers untilising the process fluid which efficiently reduce vibration when decelerating through a unbalance critical. A prior knowledge of the modal contributions to the overall response would be useful in deciding where to locate the device. A mid-span position would be most effective if the response was largely due to the first bending mode.

Many industrial rotors run for considerable periods at speeds well above the first critical. Depending on the type of service the rotor is subjected to unbalance modification due to various combinations of erosion, corrosion and possibly material deposited on the rotor. A coast down through a critical speed is therefore a potential hazard. If a damping device was fitted it could be left in a dormant state until required. The potential for reducing damage to labyrinth surfaces during a run-down and the possibility of reducing design clearances offers the prospect of improving operational efficiency significantly.

2 NOMENCLATURE

C	radial clearance
f	friction coefficient
h	local radial clearance
H	total head
L	length
R	radius
F_x, F_y	fluid forces
K_{xx}, K_{yy}	direct stiffness coefficients
K_{xy}, K_{yx}	cross stiffness coefficients

n	eccentricity ratio
u	local fluid circumferential velocity
$\bar{U}$	mean fluid circumferential velocity
θ	peripheral angle measured from minimum gap
ρ	density
δ	rotor displacement
τ_o	wall shear stress
ω	Angular velocity

3 A SIMPLE THEORY FOR ANNULAR FLOW

From the diagram in Figure 1 a circumferential flow is assumed to be maintained in the annular space between an eccentric rotor and a stator. If the mass flow is assumed constant then any change in local clearance modifies the local velocity. As the gap narrows there is an increase in velocity which leads to a local suction effect (Bernouilli); in addition the local shear stress increases significantly. Integrating these effects round the shaft section the restoring forces can be described as:-

$$\begin{bmatrix} F_x \\ F_y \end{bmatrix} = \rho \frac{\bar{U}^2 RL}{C} \begin{bmatrix} -\pi & -\pi f \\ \pi f & -\pi \end{bmatrix} \begin{bmatrix} x \\ y \end{bmatrix}$$

Details of this calculation are given in the Appendix.

It would appear that the direct terms are much bigger than the cross terms, the ratio being of the order 1/f, assuming that the friction effect extends right round the annulus. However the direct stiffness is in parallel with shaft stiffness and is only a small percentage of it. The cross-stiffness term is co-linear with an intrinsic damping and can increase damping effect when a rotor is lightly damped. A simple rotor model demonstrates that if this backward cross-coupling is too large a backward whirl can be initiated. On the other hand it is known that the increased damping can suppress a forward sub-synchronous whirl. It is clear from this elementary analysis that the value of the friction coefficient is of central importance. In practical situations a backward injection of flow into an annulus will experience grazing incidence conditions with the moving rotor surface. As far as the authors are aware there are no published measurements of friction data in these circumstances. It is clear that more elaborate models of the force system require experimental measurements of the friction coefficient.

4 EXPERIMENTAL ARRANGEMENT

The general arrangement of the test rotor is shown in Figure 2. The vertical shaft is mounted in self aligning ball bearings and was driven by a toothed belt drive from a 7.5 kW DC motor. The maximum speed was around 8000 r/min. The diameter of the rotor was 45 mm at the central section with three shrunk on discs, the two outer discs were made of brass and the central mass was steel 152 mm long and 149 mm diameter. The outer brass discs of 125 mm diameter were each located at the non-contact probe measurement planes. Tapped holes were machined in the end faces of the central mass to accept grubscrews for unbalance variation. The assembled rotor had a mass of 33.9 kg.

At the central plane of the assembled stator was a plenum chamber which discharged air through 8 nozzles into an annulus 0.508 mm deep surrounding the central mass of the rotor. The nozzles were formed in 6 layers about 25mm thick so that the nozzle throat was about 152 mm long and 0.5 mm wide. The total nozzle area was about 608mm^2. Because of flow limitations in the compressor the maximum pressure that could be maintained in the plenum chamber was about 48 kN/m^2. All tests were done with the nozzle discharge pointing backwards relative to shaft rotation. More details of the rig design and construction are described by Hart in (5).

Initial testing revealed that there was a structural resonance near a rotor critical speed. Four substantial beams were added externally to give additional suppport to the vertical stator. This modification was successful in raising the structural frequency so that it no longer interfered with measurements near the critical speed. Initially provision was made for recording orthogonal displacements at both brass disc stations. Pressure measurements were taken around the periphery of the plenum chamber to ensure that the pressure was distributed evenly. One pressure measurement point was placed just downsteam of a nozzle discharge. Other measurements were taken of flow and rotor speed. The displacement signals were fed to a two channel FFT analyser.

Subsequently the averaged spectra and time records were transmitted via an IEEE link to a small BBC computer and recorded on a floppy disc. These records could then be assessed, manipulated and presented as required. In this way it was possible to obtain the multi-spectra cascade plots which presented the overall results from several tests.

Initial tests were done by varying the plenum pressure in discrete steps at each speed selected throughout the speed range. Later tests were done by maintaining a constant pressure in the plenum and allowing the rotor to coast down from about 8000 r/min. In the main these tests showed similar reductions in vibration level when compared to the steady state method. However the absolute levels were lower as deceleration through the resonant peaks did not allow the full development of vibration response.

The tests were done at three different levels of unbalance added to the residually balanced rotor. Some tests were done with a deliberately roughened surface but these were inconclusive. It was thought that there were two major reasons, increased radial clearance and the pattern of the surface knurling used to provide the roughened surface. Due to the knurling pattern complete flow reversal would be possible

5 RESULTS

A large number of tests were carried out at three different levels of mass unbalance at the central mass. Initial testing was carried for a residual unbalance with no flow through the annular nozzles to provide datum levels. This data was later used to match with theoretical calculations of response. Two levels of increased unbalance 252 gm.mm and 404 gm.mm gave increased vibration amplitudes. At the highest unbalance the maximum displacement did not exceed 40% of the clearance. The majority of results presented here are concerned with the maximum unbalance i.e. 404 gm.mm or 11.9 gm.mm/kg. Results for the other levels of unbalance can be found in (4,5).

Figure 3 shows the effect of the backward circulation from the nozzle ring on the response with an upstream pressure of 41.4 kN/m^2 compared with no pressure induced circumferential flow. It is clear that the major axis of the response ellipse has been reduced to about 50 per cent of its former value at the two major peaks in the spectrum. These two peaks correspond to the two natural shaft frequencies.

The effect of plenum pressure is shown in more detail in figure 4. Synchronous orbits at two fixed speeds of 5040 r/min and 5400 r/min are given for pressure increments of 6.95 kN/m^2. In these figures the progressive reduction in orbit size as the pressure increases is clearly indicated. At both speeds the orbit shows a phase change which is related to the fact that the orbit variation is shown for a fixed speed. Consequently any change in damping or slight alterations in resonant frequency will lead to noticeable change in phase. A similar reduction in amplitude was also demonstrated by the rundown tests. In all the rundown experiments the shaft was run up to a speed of 7800 r/min before any air was allowed into the plenum chamber. The plenum pressure was then adjusted to the test value before the motor power was switched off. The vibration displacement from the proximity probes was recorded on a 4 channel F.M.tape recorder for subsequent analysis.

A cascade plot of vibration amplitude, frequency and rotor speed in figure 5 demonstrates the essential synchronous nature of the response for two different plenum pressures, 0.0 and 69.5 kN/m^2 as the rotor runs down . These results confirm the steady state results of figures 3 and 4. Figure 6 shows a plot of vibration amplitude against frequency, the third parameter being plenum pressure. This data was obtained using the "peak hold" feature of the FFT analyser thereby showing the envelope of the vibration amplitudes during rundown. A steady reduction in amplitude is shown as the pressure is increased.

6 THEORETICAL MODELLING

A numerical model of the rotor was obtained by using a normal mode method with 12 modes i.e. 2 rigid body and 4 bending modes for each lateral co-ordinate. The bending mode shapes were obtained from a lumped mass model of the rotor with 39 stations. This approach has been in use for over 15 years and is described in (6). The criterion used for modelling was the best fit that could be obtained for the first critical speed at 5040 r/min. Although it is possible to include additional stiffness terms to represent the effects of belt tension and casing structural resonances it was not considered necessary in this case. The only parameters that were altered to match experimental data are the bearing support stiffness, the overall shaft internal damping and the mass unbalance at the central disc. The comparison shown in figure (7) is for the residual unbalance case with no fluid cross-coupling effects. The support stiffness required is a typical value for a ball race mounted in a flexible structure. The overall proportional mass damping corresponds to a realistic damping ratio of 2 per cent. The required level of unbalance is 82 gm.mm or 2.42 gm.mm/kg. These values were then used in the model to investigate the effect of induced circumferential flow on response and stability. In order to assess the effect of negative cross-stiffness in a general way a non-dimensional value proportional to the modal stiffness is used. Figure 7 shows the 3% - 4% non-dimensional cross-stiffness is required to halve the vibration level. However the model is unstable at this level of cross-stiffness as is shown in figure 8. In the experimental results there was no sign of instability although there was evidence that the reduction in amplitude was levelling off as the cross-stiffness increased. A limit on pressure at the required flows meant that this regime could not be fully explored.

A comparison between the theoretical model and experimental results showed that the simple theory underestimates the magnitude of the cross stiffness terms. However the simple model does not include any increased damping arising from the impressed circumferential flow and also assumes that the friction effect is established over the whole circumference i.e. 360 degrees. As shown in the Appendix there is a considerable increase in cross-stiffness if the shear stress is assumed to act on half the rotor circumference.

7 DISCUSSION

The results obtained clearly demonstrate that significant reductions in vibration levels can be obtained with reverse circulation in an annular space. It is difficult to determine theoretically the amount of reduction without more definite experimental data on friction coefficients in these circumstances coupled with an examination of the velocity dependent effects. Details of the velocity distribution especially in the high shear region where the nozzle discharge encounters the moving rotor surface would be of interest.

The shape of the theoretical curve in Figure 8 does suggest another possible use for backward flow i.e. instability testing. If a flexible rotor is run at resonance then backward annular flow will initially reduce the amplitude. When the amplitude reduction has reached a minimum suggesting an approach to instability then it may be possible to measure or calculate the magnitude of the cross-stiffness. Extrapolation of the data will give an estimate of the cross-stiffness required to destabilise the rotor at running speed. This procedure could form a useful test which would give an indication of the sensitivity of instability to cross-stiffness before a full load and speed test is performed.

8 CONCLUSIONS

Although all the experimental data described above has been obtained under laboratory conditions there is sufficient evidence to suggest that full scale trials are worthwhile. All that is required in the initial stages is an annular collar that could be fitted to an exposed portion of shaft say near the coupling. It is assumed that there is a sufficient response at this location so that any additional forces can act efficiently in reducing vibration. If a moderate pressure of air, water or any other convenient fluid was admitted to the backward facing nozzles in the test collar then significant increases in damping could be obtained for lightly damped rotors. If the controlling valves were connected to a speed sensor then nozzle flow would only be required when the speed was close to a lightly damped resonance.

More sophisticated forms of control could be envisaged e.g. individual nozzle operation in a closed loop system. This would depend on the fact that cycle to cycle variation in the vibration pattern is normally small and total delays amounting to a complete period of rotation are acceptable. However it is doubtful whether the extra sophistication could be justified and the simpler system is attractive providing that the location of the critical speeds are known. The addition of a displacement transducer could be used if the critical speeds had not been determined. In any event a simple combination of speed and vibration transducers to control valve opening would be all that was necessary to minimise the vibration response when traversing critical speeds for both run-up and run-down conditions. This approach could be particularly useful as an extra safety measure for machines that experience unbalance changes during extended running at speeds above unbalance criticals.

REFERENCES

1. BLACK, H.F.; ALLAIRE, P.E. and BARRETT, L.E.: Inlet Flow Swirl in Short Turbulent Annular Seal Dynamics. 9th International Conference in Fluid Sealing, BHRA Fluids Engineering, Leeuwenhorst, The Netherlands, April 1981.

2. KIRK, R.G.: Labyrinth Seal Analysis for Centrifugal Compressor Design - Theory and Practice. International Conference on Rotordynamics, IFToMM, Tokyo, September 1986, pp 589-595

3. HART, J.A. and BROWN, R.D.: Laboratory Demonstration of Sub-Synchronous Vibration Induced by Fluid Friction. Journal of Mechanical Engineering Education, January 1987.

4. BROWN, R.D. and HART, J.A.: A Novel Form of Damper for Turbo-Machinery. Procedings of 1st European Turbomachinery Symposium, I.Mech.E, London October 1986, pp 55-63.

5. HART, J.A.: Experimental Investigation of Reverse Circumferential Flow and its Potential Application to the Vibration Reduction of Turbo-Machinery Rotors. Ph.D Thesis. Heriot-Watt University May 1987.

6. BLACK, H.F. and LOCH, N.E.; Computation of Lateral Vibration and Stability of Pump Rotors. I.Mech.E Conference on Computer-Aided Design of Pumps and Fans, Newcastle 1973, pp 71-79.

APPENDIX

The basic arrangement of a number of tangential nozzles provides a high circumferential flow acting against the forward rotation of the shaft. A simplified analysis assumes a constant circumferential flow in an annular channel at the periphery of a spinning rotor, figure 1. With a concentric rotor the fluid friction yields a pure torque due to the mean flow U in the radial clearance C. However if the rotor is perturbed laterally the nett friction acts at right angles to the displacement, thus providing an additional damping force acting against forward precession. There is also a direct negative stiffness due to the Bernouilli effect.

i) DIRECT STIFFNESS

With regard to Figure 1 a displacement in the positive x direction produces a normal pressure distribution on the rotor surface which can be obtained using Bernouilli. Resolving pressure forces in orthogonal directions yields

$$F_x = -RL \int_{-\pi}^{\pi} (H - \tfrac{1}{2}\rho u^2) \cos\theta d\theta$$

$$F_y = -RL \int_{-\pi}^{\pi} (H - \tfrac{1}{2}\rho u^2) \sin\theta d\theta$$

Since the local gap $h = C(1-n\cos\theta), n = \delta/C$ and letting $u = U C/h$ then by substitution and neglecting n^2 and higher powers,

$$F_x = RL\bar{U}^2 n\rho\pi \text{ and } F_y = 0$$

Therefore the direct stiffness coefficient

$$K_{xx} = -F_x/\delta = -RL\bar{U}^2 \rho\pi/C$$

ii) CROSS STIFFNESS

Again referring to Figure 1 for a displacement δ in the positive x direction the friction forces $F\chi$ and Fy are obtained by integrating the shear stress components round the rotor surface.

$$F_x = RL \int_{-\pi}^{\pi} \tau_o \sin\theta d\theta$$

$$F_y = -RL \int_{-\pi}^{\pi} \tau_o \cos\theta d\theta$$

now assuming that the shear stress $\tau = 1/2\rho\ u^2 f$ and proceeding as before

$$F_x = 0\ ,\ F_y = -RL\rho\bar{U}^2 f\ n\ \pi$$

and the cross stiffness coefficient

$$K_{yx} = -\ F_y\ /\ \delta = RL\rho\bar{U}^2\ f\ \pi/C$$

assuming that the shear stress acts thoughout the 360 degrees of the rotor circumference.

However if the active friction arc is assumed to be 180 degrees symmetrically located about the minimum gap then the force

$F_y = -RL\rho\bar{U}^2 f\ (\ 2\ +\ n\pi)$ and hence the cross stiffness coefficient increases to

$$K_{yx} = RL\rho\bar{U}^2\ f\ (\ 2/n\ +\ \pi)/C$$

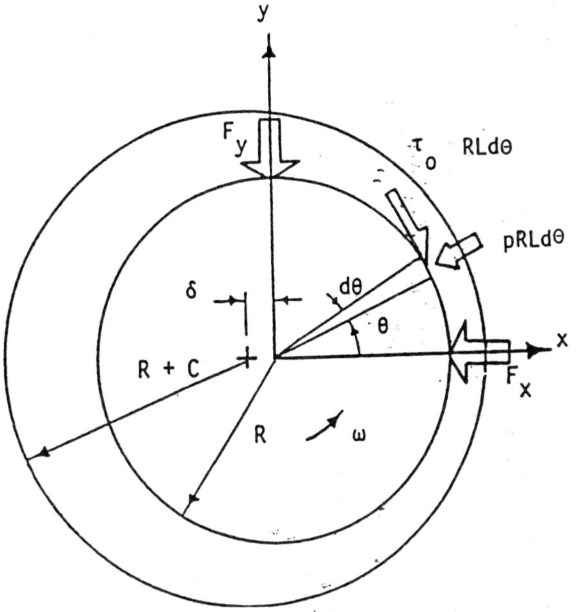

Fig 1 Geometry of annular gap and forces developed by circumferential flow

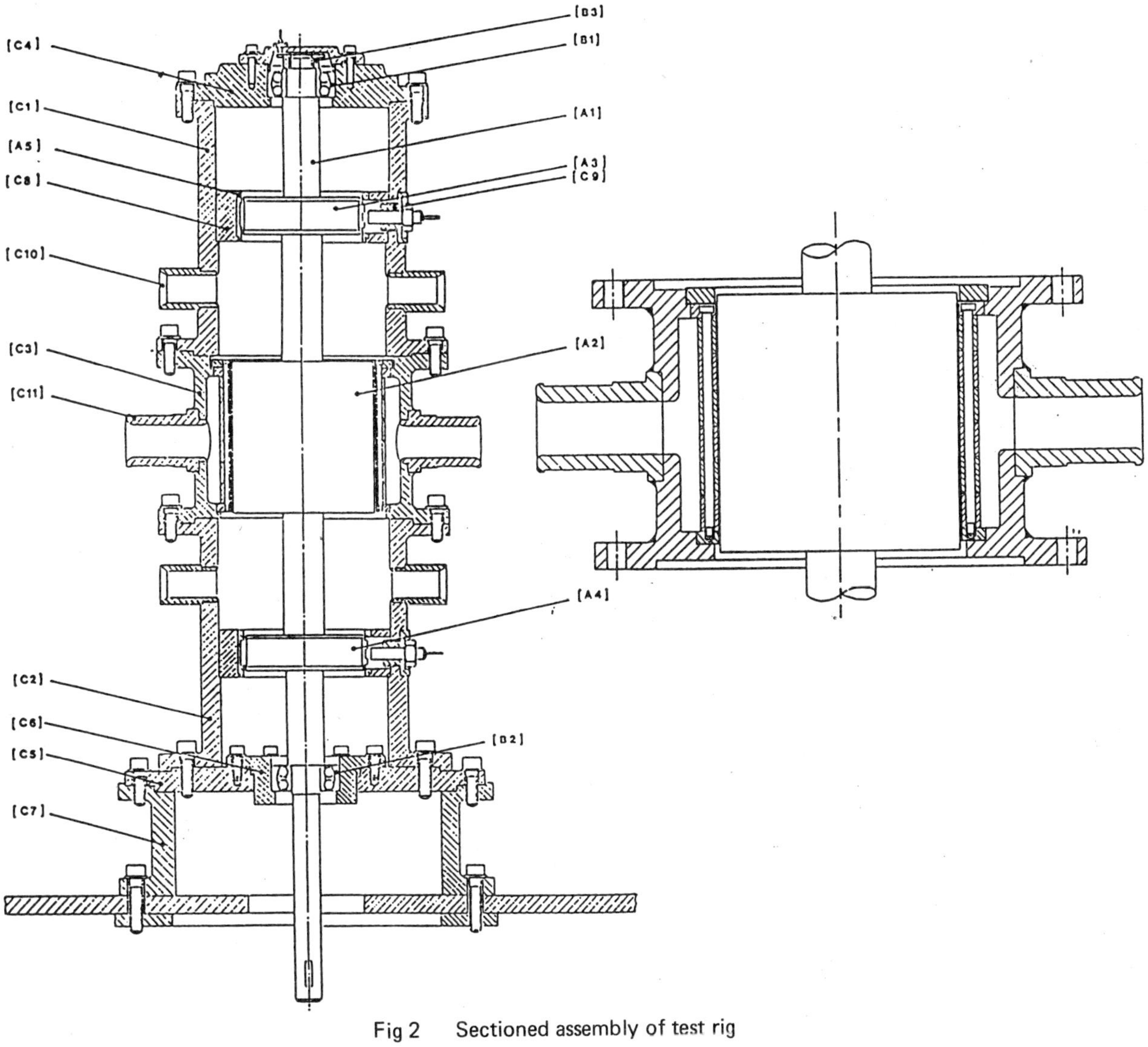

Fig 2 Sectioned assembly of test rig

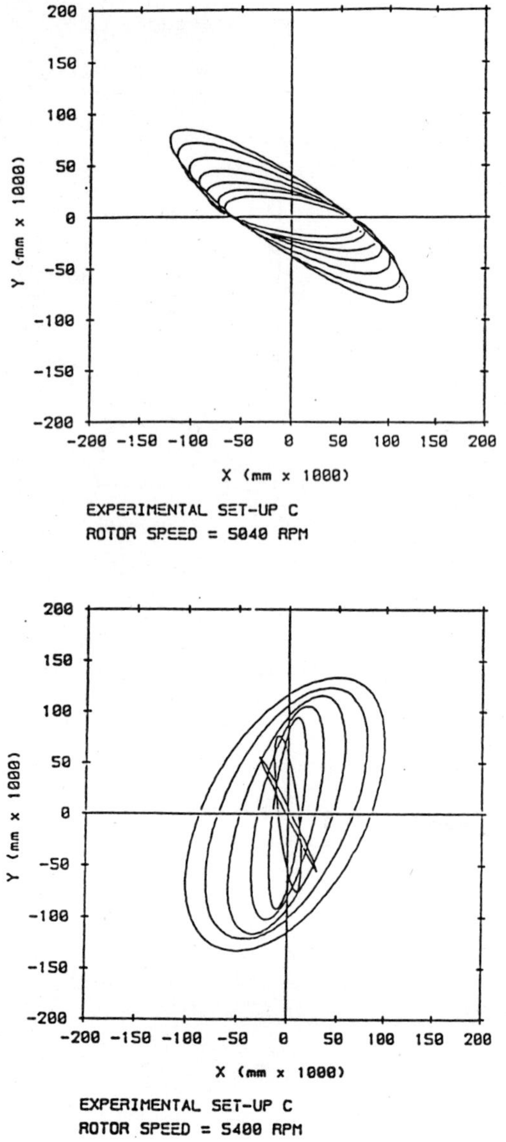

Fig 4 Effect of reverse flow on synchronous orbit (supply pressure increments = 6.9 kN/m^2)

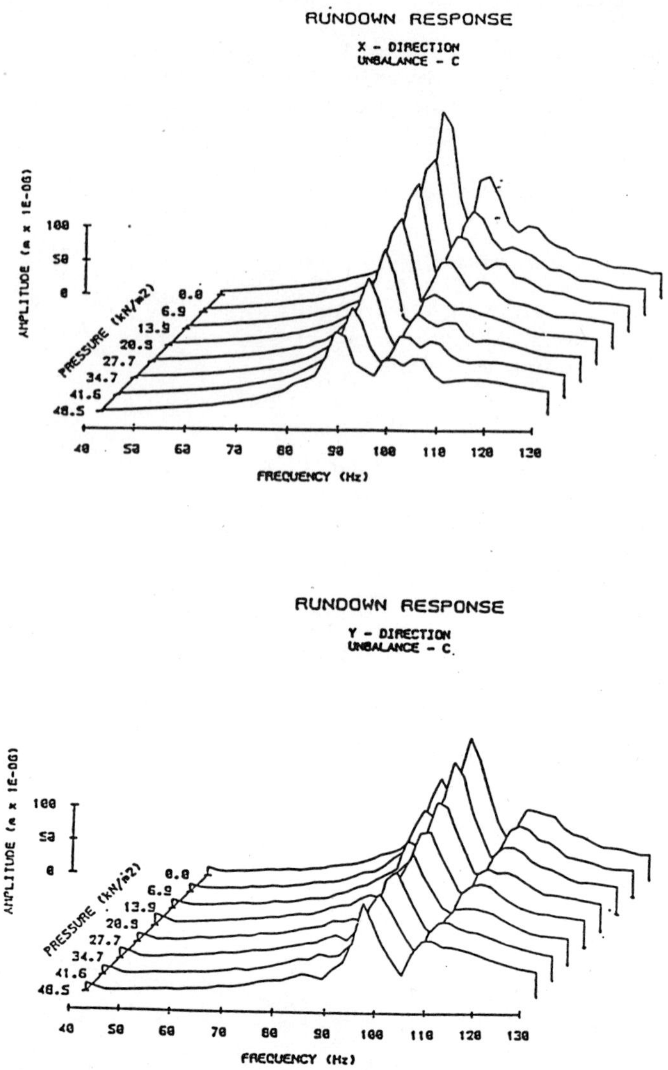

Fig 6 Effect of increasing supply pressure on 'peak-held' spectra of run-down response

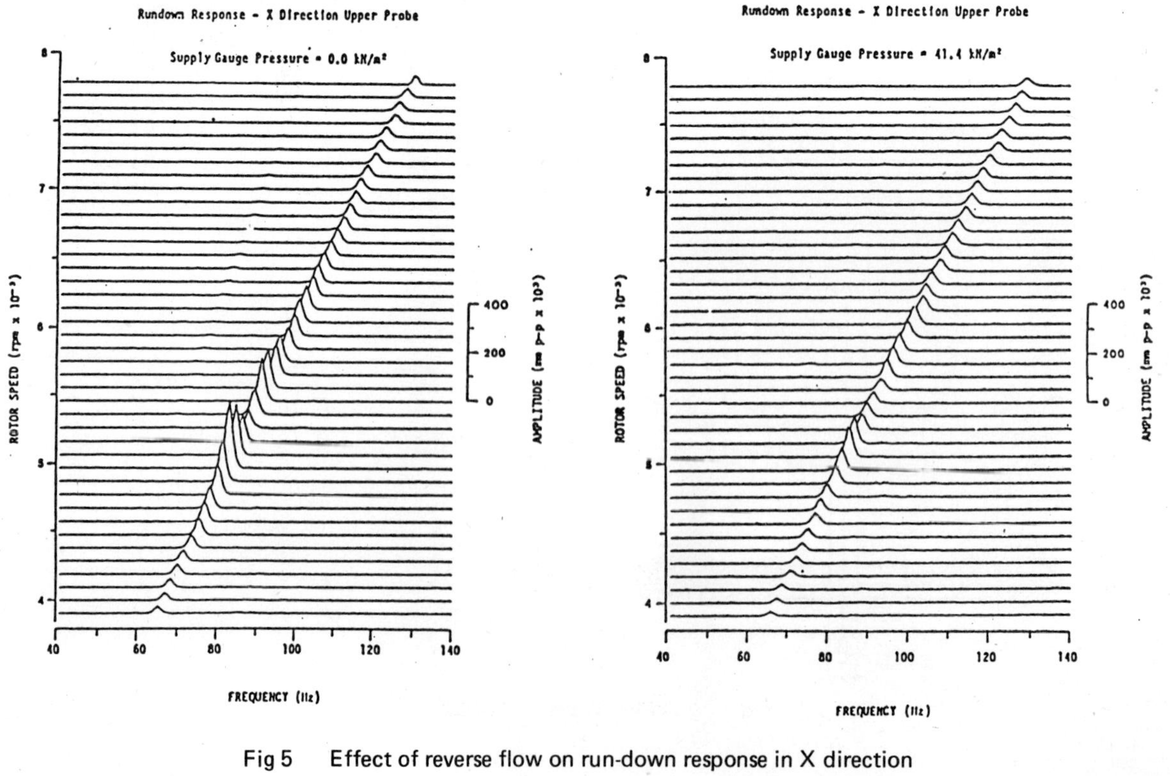

Fig 5 Effect of reverse flow on run-down response in X direction

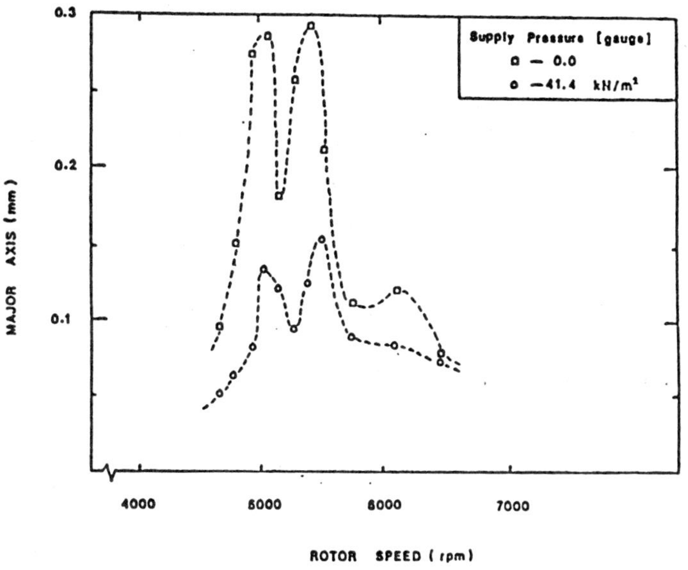

Fig 3 Effect of reverse flow on synchronous amplitude

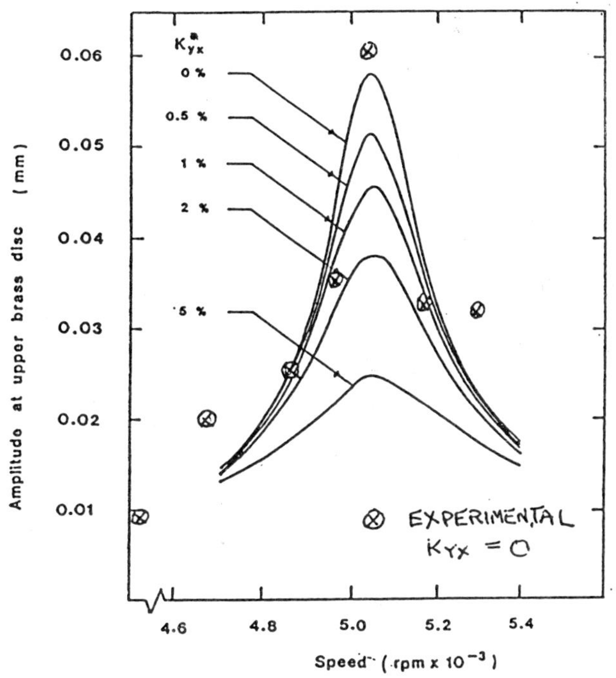

Fig 7 Effect of non-dimensional cross-coupled stiffness on response ($K^*_{XX} = 0$)

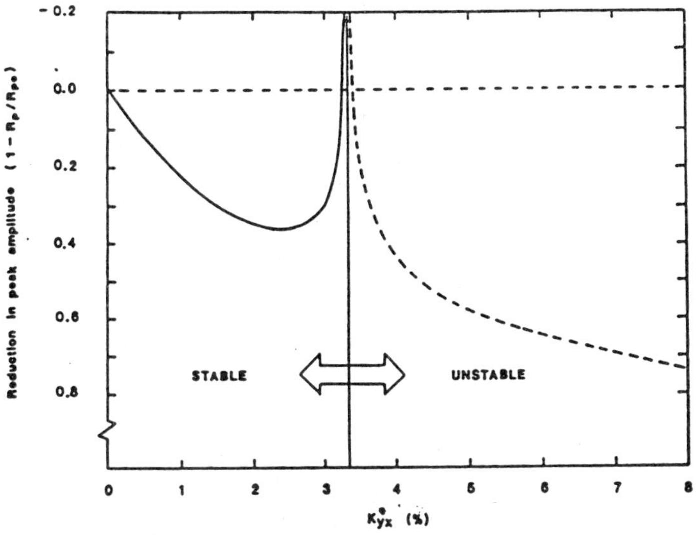

Fig 8 Relationship between reduction in peak amplitude and non-dimensional as a function of non-dimensional cross-coupled stiffness (DAMP = 21)

C320/88

Stabilization by cross stiffness control of electromagnetic damper for contained liquid rotor unstable vibration

O MATSUSHITA, DrEng, MJSME, **M TAKAGI**, DrEng, **M YONEYAMA, I SAITOH, A NAGATA** and **M AIZAWA**
Hitachi Limited, Tsuchiura-shi, Ibaraki-ken, Japan

SYNOPSIS It is a well-known phenomenon in rotordynamics that rotors, partially filled with liquid, encounter unstable vibrations within a certain rotational speed range. Similar self-excited vibrations appear in centrifuges, completely filled with heavy and light liquid mixtures inside the rotor. This instability is due to cross stiffness effect of $Kxy > 0$ and/or $Kyx < 0$ caused by wave vibrations of the border between heavy and light liquid laminations distributed by weight density difference.

An electromagnetic bearing control can produce particular damping effect for stabilization. A stabilizing control technique featured by an additional cross coupling network between X and Y directional channels is presented in this paper. This is equivalent to the damping effect owing to cross stiffness, i.e., $Kxy = -Kyx < 0$, made by the bearing control network. This optional control strategy, using the cross stiffness effect, is successfully achieved by combating unstable flow-induced forward rotor vibrations.

NOTATION

M	: mass matrix
Cg	: gyroscopic matrix
K	: shaft-bearing stiffness matrix
$Z = x + iy$	: rotor displacement in complex form
$Kd = Kxx = Kyy$	: direct stiffness of bearing
$Kc = Kxy = -Kyx$	: cross stiffness of bearing
$Cd = Cxx = Cyy$	: direct stiffness of bearing
$Cc = Cxy = -Cyx$	: cross stiffness of bearing
ζ	: damping ratio
Φ	: eigen mode shape
ω	: natural frequency
Ω	: rotational speed

1. INTRODUCTION

Rotors, partially containing liquid inside the rotating body, frequently encounter self-excited vibrations within a certain rotational speed range. Such unstable rotor vibrations, well-known in rotordynamics, are referred to as flow-induced self-excited vibrations due to the free surface wave at the border between the inside air and the outside liquid.[1]~[5]

Similar self-excited vibrations appear in large-scale and continuous flow-type ultra centrifuges, completely filled by liquids inside the rotating drum. Since the liquids form layers inside the drum according to the weight density difference, the laminar border of the liquids generates the same malfunctions as the instability due to the free surface.

In this study, stabilization for the liquid induced instability is achieved by employing an electromagnetic damper. The continuous-type centrifuge discussed here contains liquids in a drum vertically supported by passive bearings at the upper and lower ends. An additional electromagnetic damper is installed at the middle portion. The electromagnetic force directly acts upon the drum.

Ideas for electromagnetic damper stabilization are presented by Hendricks [6] and Simizu [7]. These studies are limited to theoretical approaches to control law with no experimental data about actual machine operation.

Our key for stabilization control is to consider rotor vibration features, i.e., rotor whirl motion and direction.

The instability due to contained liquids causes a forward whirl motion from X axis to Y axis, i.e., in same direction as rotor rotation. The X-directional vibration wave, thus, gains the phase lead by 90 degrees compared with the Y-directional one. In other words, the differential signal of the Y-directional vibration can be assumed from the X-directional vibration. It is found in the same manner that the X-directional differential signal is equal to the negative Y-directional vibration. These differential signals, which provide important damping effects, are simply obtained by crossing the respective X-and Y-directional displacement signals

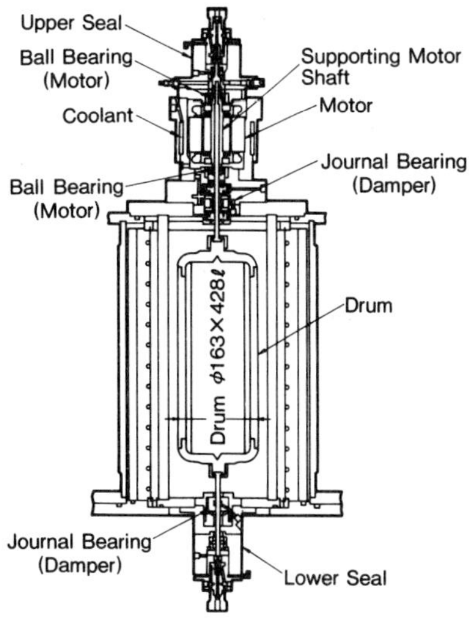

Fig 1 Conventional centrifuge

This cross damping effect can be realized by additional cross talk between two X and Y channels combined with a filter tuning only the unstable vibrations in the controller. In experiments, the damping effect owing to the cross was so great that the stabilization was successful.

2 CONVENTIONAL CENTRIFUGE

2. 1 Outline of conventional centrifuge

A cross-section of a conventional centrifuge is shown in Fig.1. The centrifuge has a drum suspended by upper and lower shafts. Both shafts are supported by journal bearings, that is, oil-film lubricated bearings. The outside of the journal bearings are encircled by thin laminar friction sheets which produce the frictional damping effect. The upper shaft is connected with a driving motor supported by ball bearings, placed in the upper casing. The specification of the machine is shown in Tab.1.

2. 2 Operation procedures

The liquid mixture to be separated by the centrifuge completely fills a cylindrical hollow space inside the rotating drum. The general operating procedure for the centrifuge is shown in Fig.2.

Advantages of the machine are [8] :
Owing to the new high frequency motor drive, the equipment can be operated just by connecting the power cord to the outlet.
Microcomputer-provided digital control ensures high accuracy in machine operation.
High performance in separation and refining with maximum rotor speed of 60,000 rpm and centrifugal force of 1785,000G in a wide range of application, etc.

2. 3 Cause of instability

Laminations of the liquid mixture are made according to weight density difference. The lamination of lighter liquid distributed inside behaves like air against the lamination of the heavier liquid distributed outside. Instability due to lamination border fluctuations in a liquid mixture completely filling a rotating drum induces self-excited rotor vibrations as well as would be expected in the case of a partially filled drum by liquid.

2. 4 Rotor vibration characteristis

This type of flow-induced unstable rotor vibration features forward whirl motions with varying natural frequency versus rotational speed. A natural frequency map of the centrifuge is drawn in Fig.3, where the horizontal axis is rotational speed and the vertical axis is natural frequencies in each mode. In Fig.3, the natural frequencies of the first and third mode shapes are almost flat because of no gyroscopic effect.

Tab.1 Tested Centrifuge Specifications

Item	Specifications
Max. rotor speed	35,000 rpm
Centrifugal acceleration	~90,000 G
Ultimate vacuum	1.3Pa (10×10^{-3}Torr)
Drum capacity	0.3 ℓ ~ 3.2 ℓ
Drum material	Titanium alloy

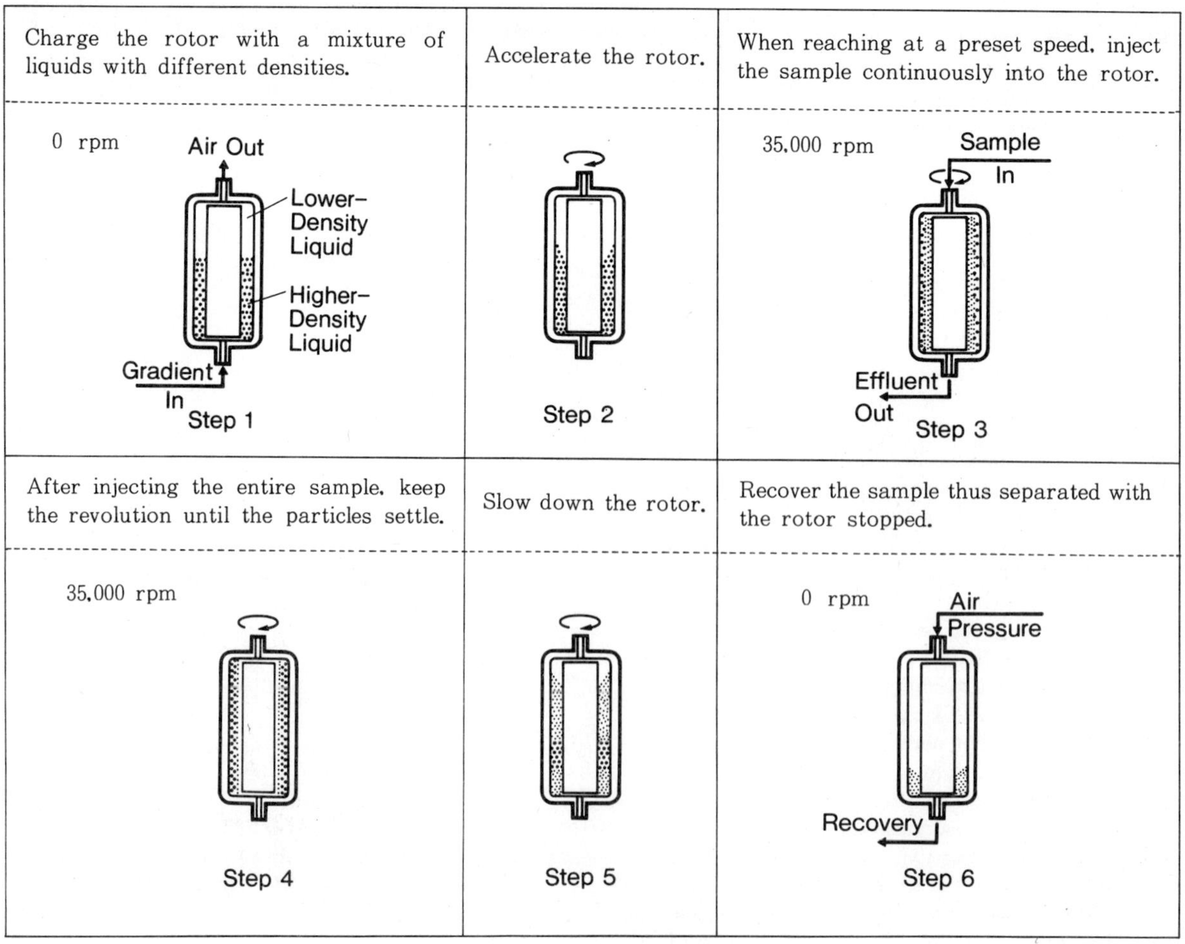

Charge the rotor with a mixture of liquids with different densities.
Accelerate the rotor.
When reaching at a preset speed, inject the sample continuously into the rotor.
0 rpm
Air Out
Lower-Density Liquid
Higher-Density Liquid
Gradient In
Step 1
Step 2
35,000 rpm
Sample In
Effluent Out
Step 3
After injecting the entire sample, keep the revolution until the particles settle.
Slow down the rotor.
Recover the sample thus separated with the rotor stopped.
35,000 rpm
Step 4
Step 5
0 rpm
Air Pressure
Recovery
Step 6

Fig 2 Operating procedures

The natural frequency of the second mode is separated from a natural frequency of non-rotating rotor into two natural frequencies as rotational speed increases, i.e., forward and backward natural frequencies. The former is noted by a solid line, the latter by a dotted line. This separation is caused by a gyroscopic effect of the conical mode shape.

2.5 Instability data

Rotational tests of this conventional centrifuge are done under various conditions of liquid capacity, and induced unstable vibrations are measured. The self-excited frequencies of these unstable vibrations which were met regardless of capacity difference are plotted in Fig.3, marked by stars ★.

It is clear in Fig.3 that the instability appears in the rotational speed range between 2,500 and 20,000 rpm. Unstable frequencies are not constant and are distributed on an increasing curve from about 43 Hz up to 90 Hz versus rotational speed.

Therefore, it is thought that unsable vibrations always correspond with forward rotor whirl.

This instability problem must be solved in order to develop a large capacity of centrifuge. A new damper element having a greater damping effect is required, compared with conventional journal bearing with laminar sheet friction damping effect. An active electromagnetic damper potentially having large damping effects is thus introduced into this machine.

3. EXPERIMENTAL EQUIPMENT

3.1 Electromagnetic damper-type centrifuge

The upper journal bearing, i.e., a passive damper, installed in the conventional centrifuge shown in Fig. 1 is removable. It is replaced by an active damper. The outline of a new centrifuge installing an electro magnetic damper experimentally used in this study is shown in Fig.4.

The motor, drum and shaft are the same as the conventional type. The electromagnetic bearing is located at the top of the drum to produce control force directly acting upon the drum.

Two displacement gap sensors are located beside the active bearing to detect the rotor position in X- and Y-directions. The detected displacement signals are input to the control network, and the controller determines the electrical current as output signals. The current activates the electromagnetic bearing to move the rotating drum.

The electromagnetic damper configuration is made of 8 poles like the usual motor, as shown in Fig.5. The coil is 44 turns per pole with 0.44 Ω resistance, and is made of 0.75mm diameter copper wire.

3.2 Tested liquids

A liquid mixture consisting of plain water (1.6 liter) and sucrose (1.6 liters, weight density 1.3) fills the drum. There is a core shaft with 6 blades in the drum. The ideal condition for the separation of the liquid is drawn in the picture of the drum section in Fig.4. The water is inside and the sucrose is outside. The boundary surface between the water and the sucrose is equivalent to the free surface of partially filled liquid in the drum.

4 CHARACTERISTIC OF ROTOR SYSTEM

4.1 Computational model

Rotor vibration analyses were done by the HIROT [9] rotordynamics computer program.

A computational model of the rotor system is

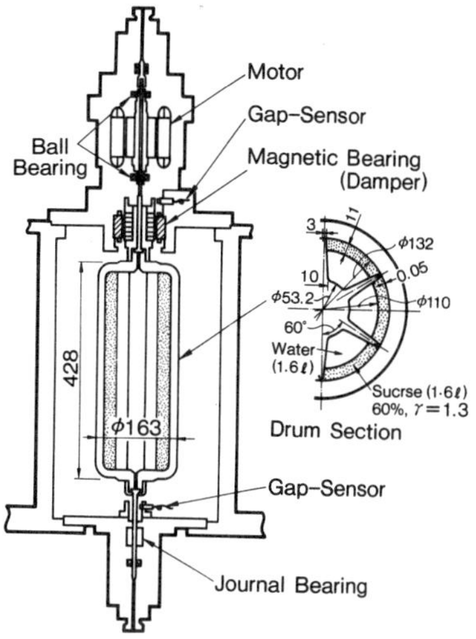

Fig 4 Centrifuge with electromagnetic damper

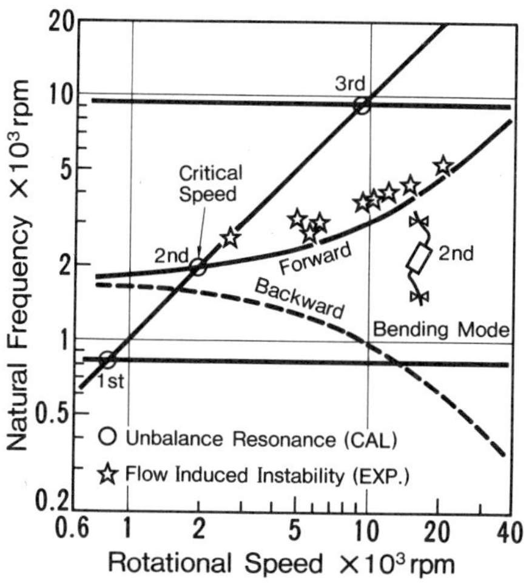

Fig 3 Eigenfrequency and unstable frequency

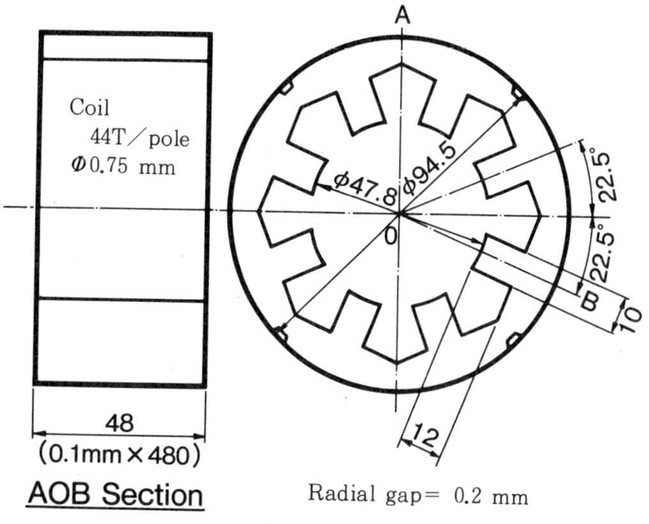

Fig 5 Configuration of electromagnet

divided into 48 stages by beam element as shown in Fig.6. The nodal points of No.5 and No.16 are ball bearings; of No.24 is the electromagnetic damper and of No.45 is the journal bearing.

4. 2 Eigensolutions

Corresponding eigen modes in the rotor free vibration are obtained as shown in Fig.7. with each natural frequency of non-rotating rotor. The shape of the second eigen mode is conical. which is most influenced by the gyroscopic effect.

The natural eigen frequency map is shown in Fig. 8. i.e.. natural frequencies versus rotational speed. The forward and backward natural frequencies are denoted by solid and dotted lines. respectively.

It must be noted that the second natural frequency of the conical mode is separated from 23.5 Hz at 0 rpm into one upward and one downward curve corresponding to the forward and backward whirl natural frequencies.

Intersections. marked ● by in Fig.8. between the rotational speed line and the forward natural frequency mean the critical speed of each mode. The first. second and third critical speeds are 700. 1,600 and 33,000 rpm. respectively.

5 INFLUENCE OF 4-COEFFICIENT

5. 1 4-Coefficient dynamics

In a general electromagnetic bearing-rotor-system. displacement sensors monitor the rotor position. This signal is compared with a reference. i.e.. a neutral position. The error signal is input to an electronical control network. which drives the power amplifier of the magnetic bearing.

The dynamic properties of electromagnetic bearings are determined by the controller. By simply using the same controller layouts for X and Y channels plus cross talk between both channels. axisymmetry of system description can easily be achieved. even with horizontal rotors or vertical rotors. This yields the expressions of the 4-coefficient system for the radial bearing characteristics shown in Fig. 9.

Magnitude adjustments and signs of the cross stiffness in the controller directly influence stability of whirl motions.

5. 2 Equation of Motion

The rotor system supported by electromagnetic bearings is described by combining the axisymmetrical equations of motion and the control network state equations. We assume here that active magnetic bearing force regulated by controllers can be represented by the 4 coefficients of bearing dynamics. as stated before. The equation of motion of the magnetically borne rotor can be written :

$$Mz+i\Omega Cg\ z+Kd\ z+\varepsilon Q\ (z.z)=0 \qquad (1)$$

where $\varepsilon Q=(\varepsilon Kd-i\varepsilon Kc)z+(\varepsilon Cd-i\varepsilon Cc)z$

and ε =small ($<<1$).

Hence. εQ is defined as a small deviation of the bearing reaction force depending upon the error signal from the reference signal introduced in the previous section. The reaction force is expressed by stiffnesses (εKd. εKc) and damping factors (εCd. εCc) small quantity ε stands for

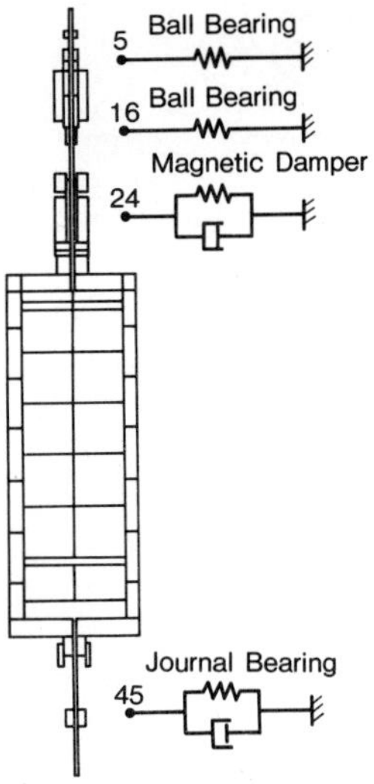

Fig 6 Computational model

4 Coefficients:

Active Magnetic Bearing Control

$$Q_x = K_d X + K_c Y + C_d\dot{X} + C_c\dot{Y}$$
$$Q_y = K_d Y - K_c X + C_d\dot{Y} - C_c\dot{X}$$
$$\text{i.e. } Q = (K_d - iK_c)Z + (C_d - iC_c)\dot{Z}$$
$$Q = Qx + iQy$$
$$= (Kd - iKc)\dot{Z} + (Cd - iCc)\dot{Z}$$

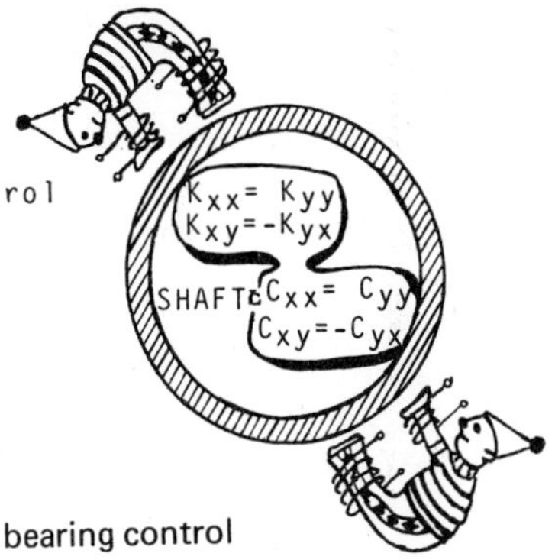

Fig 9 Active magnetic bearing control

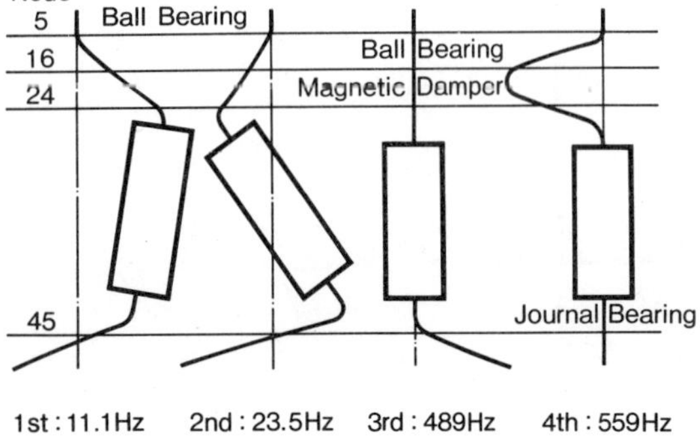

Fig 7 Eigenmode (0 r/min)

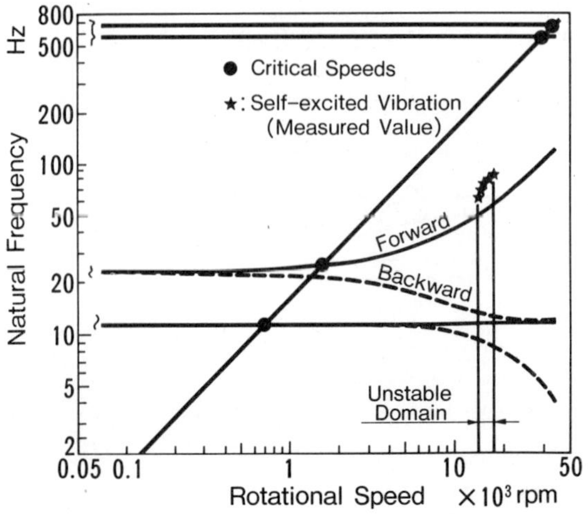

Fig 8 Eigenfrequency map

5. 3 Influence of the coefficients

The equation of motion (1) can be rewritten as state equation :

$$\begin{bmatrix} M & 0 \\ 0 & K \end{bmatrix}\begin{bmatrix} \ddot{z} \\ \dot{z} \end{bmatrix}+\begin{bmatrix} -i\Omega Cg & -K \\ K & 0 \end{bmatrix}\begin{bmatrix} \dot{z} \\ z \end{bmatrix}=-\begin{bmatrix} \varepsilon Q\ (\dot{z}.z) \\ 0 \end{bmatrix} \qquad (2)$$

The eigensolutions of this non conservative system are defined by asymptomatical concept;

$$z=\Phi\ a\ \exp\ (i\varphi)\ ,$$

that is,

$$\begin{bmatrix} \dot{Z} \\ Z \end{bmatrix}=\begin{bmatrix} Zv \\ Zd \end{bmatrix}=\begin{bmatrix} i\omega\Phi \\ \Phi \end{bmatrix} a\ \exp\ (i\varphi)=\Phi\ a\ \exp\ (i\varphi) \qquad (3)$$

where Φ = eigen mode, ω = natural frequency $da/dt=\varepsilon\alpha a$ and $d\varphi/dt=\omega+\varepsilon\beta$.

Introducing this into Eq. (1) and premultiplying with the transposed modal matrix finally enables an approximate eigensolution for the Φ mode using the asymptomatic method considering ε-influnce

$$\lambda=\varepsilon\ \alpha+i\ (\omega+\varepsilon\ \beta)=\frac{(A^*+\Phi^t\ \varepsilon\ Q)}{B^*}$$

$$=i\omega-\frac{i\omega\Phi^t\varepsilon\ Q\ (i\omega\Phi,\Phi)}{B^*} \qquad (4)$$

where $B^*=m\omega_n^2\ \{(1+(\frac{\omega}{\omega n})^2\}$ = positive

The influences of the 4 coefficients on the eigenvalues can now be expressed as follows :

$$\varepsilon\ \alpha=(\ \omega\ \ \varepsilon Cd+\omega\ \ \varepsilon Kc)\ /B^*$$

$$\varepsilon\ \beta=(\ \omega\ \ \varepsilon Kd+\omega\ \ \varepsilon Cc)\ /B^*$$

Therefore, natural frequency and modal damping vary with these 4 coefficients as shown in Fig.10.

As shown in the comment of Fig.10, the positive Kc due to liquid dynamics decreases forward damping ratio and the negative Kc, owing to the controller, increases forward damping ratio of the rotor system.

Therefore, in contrast to the oil-film reaction having positive Kc> 0 , the cross coupling coefficient Kc of the controller must be made negative for stabilization while suppressing the forward unstable whirl motion.

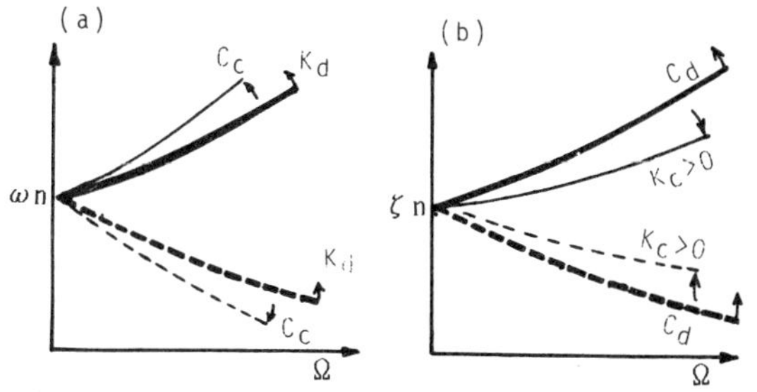

The bearing 4-coefficient moves natural frequency and damping ratio from thick lines to the thin lines according to the following rules.

Kd> 0 : Increase F. and B. natural frequency

Cd> 0 : Increase F. and B. damping ratio

Kc> 0 : Decrese F. damping ratio

: Increse B. damping ratio

Cc> 0 : Increses F. natural frequency

: Decrease B. natural frequency

Fig 10 Change due to cross-coupling

6 . CONTROL NETWORK

6. 1 PID controller

Conventional PID control law is fundamentally used for our experiment. The network of the PID controller for electromagnetic damper is shown in Fig. 11 and 12, according to complex variable form (z=x+iy) and an real X−Y channel form, respectively.

The X- and Y-directional control channels each have an independent PID control unit which is composed of a low pass filter, proportional (P−) operation, differential (D-) operation and integation (I−) operation.

The X- and Y-directional PID control networks are of the same specification and are independent of each other. Magnitudes of each operation vary with gain adjustment variables. These directional networks can produce directional stiffness Kd = Kxx = Kyy and damping factor Cd = Cxx = Cyy.

6. 2 Cross control units

A special stabilizing force is generated by a cross network between X and Y channels which is the origin of the cross stiffness of the bearing control. Part of the cross network is shown in the area bordered by the dotted line in Fig.11.

The corresponding X−Y form block diagram is given in Fig.12.

As shown in the actual whole network layout concerning X and Y channels in Fig.12, the main controller is based upon PID function. The cross network using a tuning filter is optional.

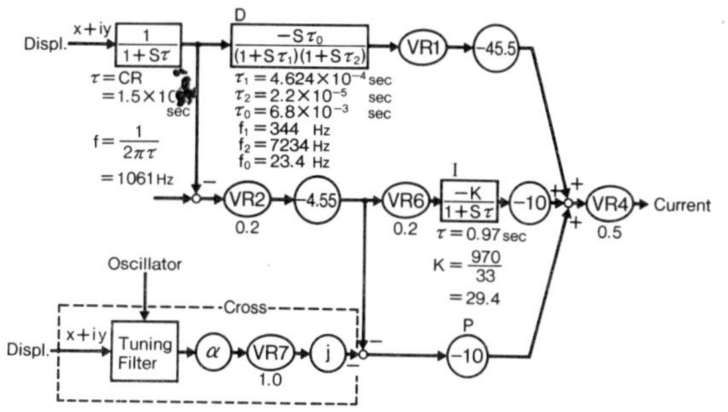

Fig 11 Controller network (z-form)

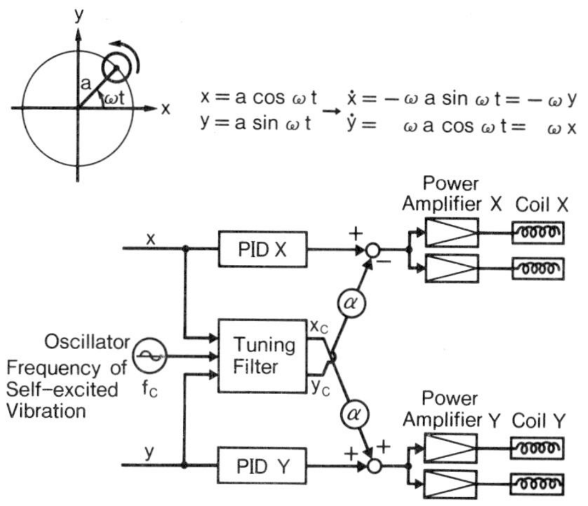

Fig 12 Controller network (x−y form)

A tuning filter in the option is necessary, because cross coupling influence must be limited so that it acts only upon unstable self-excited whirl vibration, not upon other vibration components. Then, only the vibration with the self-excited frequency component fc is filtered among all components included in the actual vibration signals. The tuning frequency is adjusted by an oscillator frequency so as to meet the induced self-excited frequency.

Then, the same frequency as the resultant unstable vibration frequency is oscillated and the output signal is input to a tuning filter. The tuned components in the X- and Y-directional vibrations are cross connected with the Y- and X-directional power amplifiers driving coils.

Regarding the option, the sign of the cross stiffness $Kc = Kxy = -Kyx$ must be negative for stabilization, because of the inputting damping effect into the system suffered to the forward whirl instability, stated in chapter 5.

Therefore, actual channels have cross connections from X-input to Y-power amplifier and from Y-input to X − power amplifier, respectively, as shown in Fig.12.

6.3 Transfer function of tuning filter

The Bode diagram concerning the transfer function of the tuning filter is shown in Fig.13. In this example, the tuning frequency is selected at 30 Hz with band width of 5 Hz.

The tuning filter phase changes from phase lead to lag. The frequency crossing 0 degree is 35 Hz, greater than the center frequncy 30 Hz by only 5 Hz. Only forward vibration waves pass. Even if there are backward vibration waves of the same frequency as the center frequency, they are filtered out.

6.4 Transfer function of whole control network

The Bode diagram of the whole network with the PID plus the cross coupling is shown in Fig.14. Three transfer function curves of each gain and phase, are measured which corresponds upon no cross, forward input ($X = \cos\omega t$ and $Y = \sin\omega t$) and backward input ($X=\cos\omega t$ and $Y=-\sin\omega t$).

If the tuning frequency in the cross network is set at 70 Hz, only vibration components with the 70 Hz frequency and forward direction can be passed, as shown in Fig.14.

Around the tuning frequency 70 Hz, the gain curves with the forward input fluctuates a little by ± 2 dB and the phase curve obtains phase lead up to about 90 degrees. According to this Bode diagram, a special damping effect for the forward rotor vibration can be expected by the cross coupling network with the tuning filter.

On the other hand, the Bode diagram for the backward input is almost uniform and idential to the no cross network case, because the tuning filter blocks backward input.

The Bode diagram of the forward input is different from the Bode diagram concerning no cross and the backward input. This is why the tuning filter is necessary.

The frequency of the flow-induced unstable vibration varies so greatly with the rotational speed, as shown in Fig.8, that the tuning frequency then is set manually by changing oscillator frequency.

7 RESULTS OF ROTATIONAL TEST

7.1 Water only operation

First of all, the rotational test with the rotor with only water in the drum is conducted. The test result is shown in Fig.15. In this case, unstable vibrations did not appear, though the rotor reached a rotational speed of 15,000 rpm. Rotor vibration was stable in the whole rotational speed range, because of no weight density difference in the water sample.

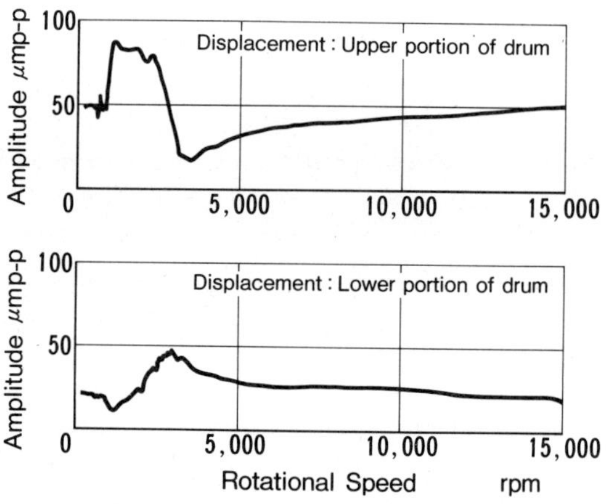

Fig 15 Vibration amplitude (water only)

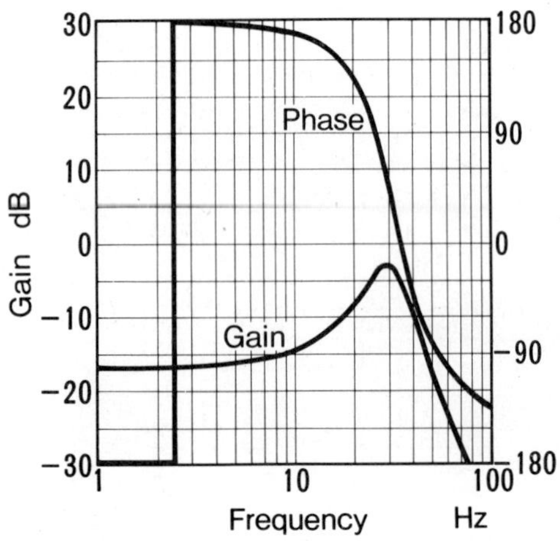

Fig 13 Tuning filter (fc = 30 Hz)

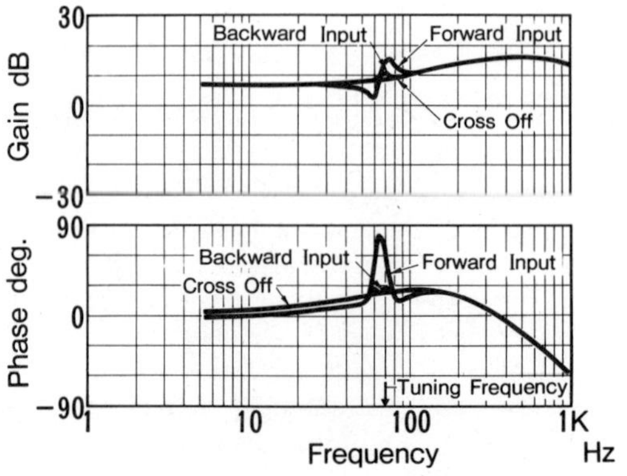

Fig 14 Controller bode diagram (PID plus cross network)

7. 2 Mixture liquid operation

Next, a liquid mixture with water (1.6 liters) and sucrose (1.6 liters) was loaded in the drum. The rotational test result is shown in Fig.16. In this rotation test, self-excited vibrations suddenly occured at the rotational speed 13,800 rpm. In the case of just the PID control, without the cross effect, the vibration amplitude grew and stabilization was impossible. In this case, unstable vibrations appeared in a rotational speed range of 13,800 rpm (230 rps) to 17,000 rpm (283 rps).

7. 3 Stabilization

However, by turning on the cross network, selecting the tuning frequency with the resultant unstable frequency and increasing magnitude of the cross function, the unstable vibrations disappeared. The stabilization was successfully achieved by the cross effect even in the unstable rotational domain of 13,800 rpm～17,000 rpm.

It was then discovered that the system was alway stable beyond this upper limit even without the cross function.

The self-excited vibration of this system was caused in the range from 13,800 rpm to 17,000 rpm. As shown by "★" in Fig. 8, the unstable vibration

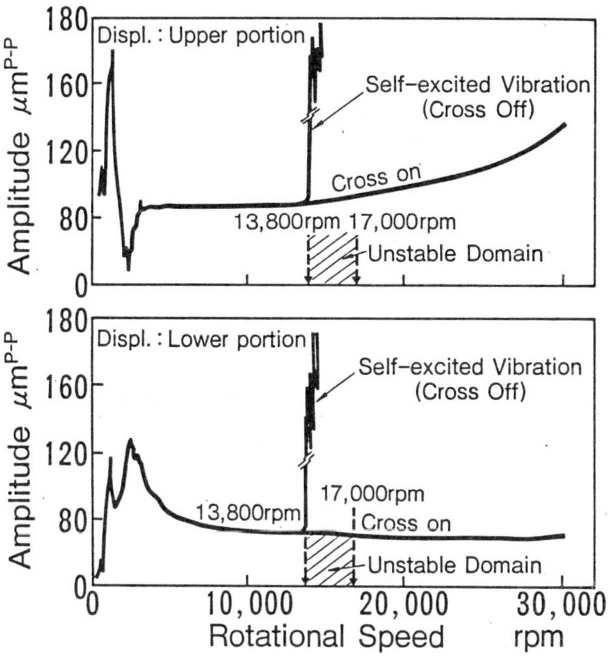

Fig 16 Vibration amplitude (mixture)

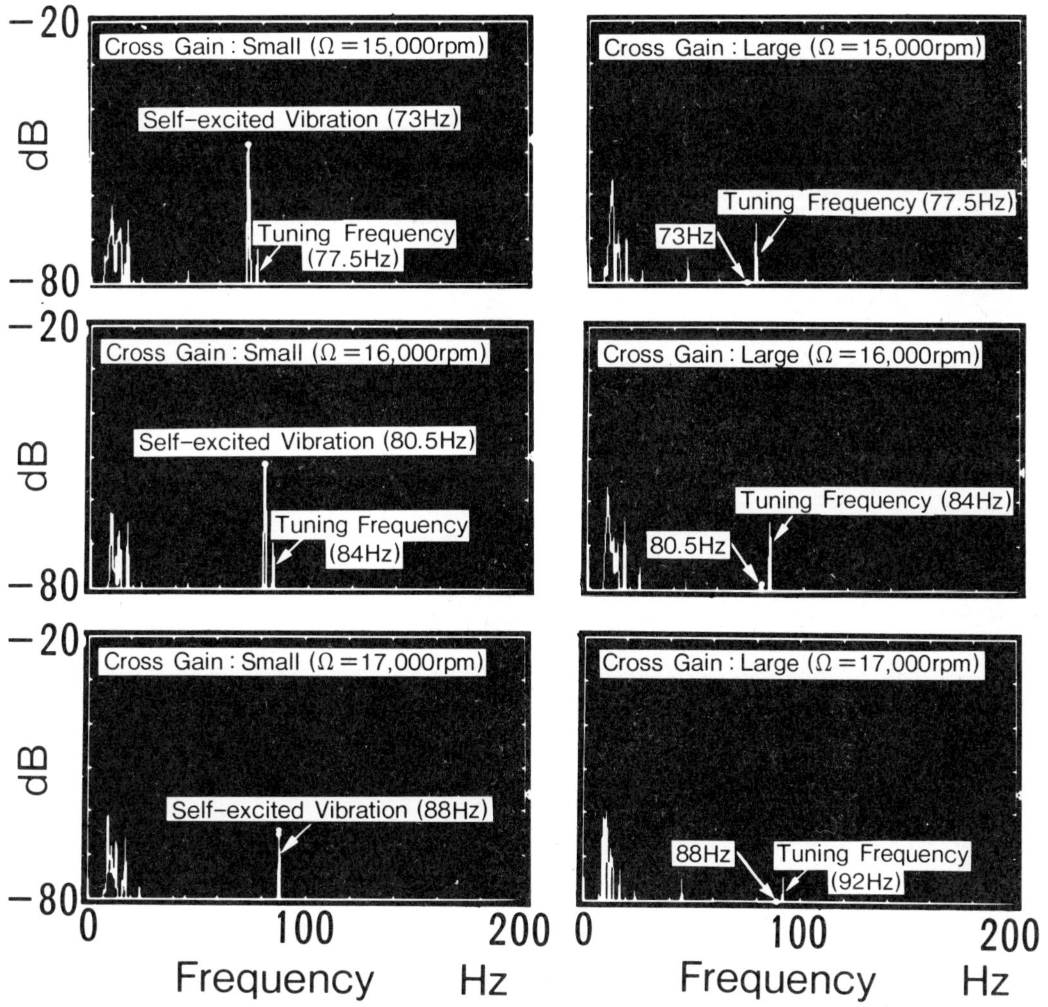

Fig 17 Frequency analysis before and after stabilization

frequencies change upward from 60 Hz to 100 Hz as the rotational speed increases. Tracking selection of the oscillator frequency was manual as it must always meet the increase of unstable frequency.

7. 4 Selection of tuning frequency

Phenomena measured during the unstable domain of rotational speed and stabilization are shown in Fig.17. Frequency analysis on the resulting unstable vibrations by FFT analyzer are done. The results of the vibration spectrum are shown concerning ratational speeds of 15,000 rpm, 16,000 rpm and 17,000 rpm. The left and right pictures correspond with small (instability) and large (stabilization) cross effects, respectively. For example, we can see unstable vibrations with the self-excited frequency of 73 Hz at a rotational speed of 15,000 rpm (250 rps), as shown in Fig. 17. (Cross Gain : Small)

In stabilization test to eliminate the unstable vibrations by using the cross stiffness by control network, the tuning filter frequency is adjusted by the oscillator with 77.5 Hz, closing to rotor unstable frequency of 73 Hz., and then the quantity of the cross network is incresed. Therefore, the instability is eliminated as shown in FFT analysis of Fig. 17 (Cross Gain : Large).

At another rotational speed, we changed the tuning frequency a little, depending upon the rotational speed.

Strictly speaking, effective stabilization was achieved by selecting a slightly higher tuning frequency than the resultant unstable frequency. This selection of higher oscillator frequency was better than the selection of the same frequency. For this reason it is thought that the peak of phase lead providing a large damping effect is located at a lower frequency than the tuning frequency, shown in Fig.14.

8. CONCLUSIONS

Cross stiffness of oil-film bearings in passive rotordynamics is a well-known devil's curse causing unstable rotor vibration, i.e., forward whirl motion, mostly due to the positive cross stiffness $Kc > 0$.

However, the cross stiffness control for electromagnetic bearing could be the angel in active rotordynamics, by suppressing unstable vibrations and eliminating system instability. Cross stiffness control is realized by the cross-talk between X and Y channel networks.

(1) Stabilization owing to cross stiffness control with the tuning filter is demonstrated by experiments with a centrifuge installed with an electromagnetic damper. The necessary technique for selecting the tuning frequency is to adjust it slightly higher than the resultant unstable frequency.

(2) Flow-induced unstable vibrations are caused within certain speed range, i.e., 13,800～17,000 rpm in this experiment. The unstable frequency increased as the rotational speed increased, because the corresponding mode was conical and was influenced by the gyroscopic effect.

(3) The instability did not occur in the case of the drum filled with only water.

9. REFERENCES

[1]Saitoh. S. et.al 2 ; Investingation into the vibration of a rotating hollow shaft partially filled with liquid (4th report) ;Trans.JSME(in Japanese), Vol.48, No.427,C (1982), p 321

[2]Yasuo, A. et.al 3 ; Vibrations of rotor containing liquid with partial free liguid surface ;Trans. Trans. JSME (in Japanese) ,Vol.51, No.46,C (1985], p.265

[3]Kaneko.S. et.al 1 ;Self-excited oscillation of a hollow rotating shaft partially filled with a liquid ; Trans.JSME(in Japanese), Vol.51,No.464,C (1985), p.765.

[4]Jinnouchi.Y. et.al 4 ; Self-excited vibrations of a cylindrical rotor partially filled with liquid ; (1 st report) Trans.JSME(in Japanese), Vol.51,No.467,C (1985), p.1463.

[5]OHTA H. et.al 3 ; Experiments on vibrations of a hollow rotor partially filled with liquid ; Trans.JSME(in Japanese), Vol.52,No.474,C (1986), p.474.

[6]HENDRICKS. S.L. et.al 1 ; Optimal control of a rotor partially filled with an inviscid incomoressible Fluid ; Trans.ASME, Vol.51,No.4, (1984), p.863.

[7]SHIMIZU H. at.al 2 ; Research on the self-exciting vibration of thrust − type magnetic bearing ; Trans.JSME(in Japanese), Vol.36,No.290,C (1970), p.1656.

[8]Hitachi Kouki Co.Ltd. ; Centrifuge Catalog and Its Mannual.

[9]MATUSHITA O. ; ICVPE.Xian June.19 (1986), p.817

C282/88

An analysis of combined squeeze-film and variable stiffness hydrostatic bearings, and their use in aircraft engine vibration control

M J GOODWIN, BSc, PhD, CEng, MIMechE and **M P ROACH**, BSc
Department of Mechanical and Computer Aided Engineering, North Staffordshire Polytechnic, Stafford, Staffordshire
J E T PENNY, BSc, PhD
Department of Mechanical and Production Engineering, Aston University, Birmingham

SYNOPSIS

This paper describes the performance of a new design of bearing proposed for use in aircraft engines. The design combines the advantages of variable stiffness hydrostatic bearings with the damping capacity of conventional squeeze-film bearings. When used to support aeroengine compressor shafts the bearings enable the system critical speeds to be tuned away from any instantaneous operating speed while the engine is running, and at the same time provide suitable damping to further attenuate the machine forced unbalance response, and to suppress instability. The new bearing is shown to be substantially superior to conventional squeeze-film bearings in machine vibration attentuation and isolation.

NOTATION

A	Effective area of hydrostatic pad, contained within land centre lines
B	Accumulator operating parameter
c	Bearing nominal radial clearance
C	Bearing damping coefficient
C	Dimensionless damping = $C/m\omega_{pp}$
e	Eccentricity of journal from bearing centre
f(x,y)	Function defined in appendix
F	Force
h	Bearing clearance
I	Capillary inertia coefficient
K	Inverse capillary flow resistance
m	Rotor mass
k	Stiffness
p	Lubricant pressure
Q	Lubricant flow rate
R	Bearing radius of curvature
s	Root of characteristic equation
t	Time
u	Rotor mass unbalance eccentricity
V	Bearing recess volume
W	Land width
x	Displacement in horizontal direction
X	Displacement amplitude generally
y	Displacement in vertical direction
z	Axial position on circumferential land
β	Ratio s/ω_{pp}
ϵ	Eccentricity ratio e/c
κ	Lubricant bulk modulus
μ	Lubricant dynamic viscosity
ϕ	Journal attitude angle
ψ	Angular position around bearing measured from eccentricity line
ω	Angular frequency

Subscripts

a	refers to accumulator or accumulator line
b	refers to bearing
c	refers to lubricant compressibility
d	refers to flow out over bearing lands
pp	refers to pin-pin natural frequency
R	refers to radial direction
s	refers to supply or supply line
s	refers to shaft stiffness
T	refers to tangential direction
v	refers to change in volume of bearing recess
o	refers to steady state condition.

1. INTRODUCTION

The method of suppressing machine resonance vibrations used by most aeroengine manufacturers is to mount the engine compressor shafts in one or more squeeze film bearings as discussed in references (1) to (4). Prior to the use of squeeze-film bearings, the engine compressor shafts were mounted in rolling element bearings which tend to behave like stiff springs and have little damping capacity. The squeeze film bearing, in contrast, includes an annular clearance space between the conventional rolling element bearing outer race and the main housing. The clearance space is filled with lubricant so that motion of the shaft towards the main housing results in a squeezing of the lubricant and generates a damping force. Separation of the squeeze film bearing surfaces is maintained either by the non-linear stiffness of partially cavitated squeeze films or by centralising springs in the case of 360° squeeze films.

Although the use of squeeze-films does considerably attenuate vibration levels, as compared with rigid supports, they do not always enable shaft supports to operate with optimum values of stiffness and damping. Other publications (5) (6) have shown that the value of support stiffness best able to attenuate vibrations can vary between zero and large values, depending upon the running speed of the machine and other design parameters.

An improvement in the operating characteristics of squeeze film bearings can only be obtained then, if a means is found of enabling the bearing to take on any required value of stiffness between zero and high values, whilst still enabling the bearing to support gravity loads and to operate with optimum damping.

Figure 1 shows a schematic diagram of an aeroengine compressor shaft mounted in combined variable stiffness hydrostatic and parallel squeeze-film bearings. In the proposed bearing design the rolling element bearing outer race is constrained not to rotate and is supported by a series of hydrostatic bearing pads spaced around its circumference; the circumferential lands of each of these hydrostatic pads are joined together to form two 360° oil squeeze films. A developed view of such a bearing surface is shown in Figure 2. The hydrostatic bearings are easily able to accommodate the bearing gravity loads and still maintain the shaft virtually concentric within the bearing clearance, and so no centralising springs are necessary. The hydrostatic bearings are inherently very stiff, but in the proposed design they are connected to gas bag type accumulators via variable flow resistance capillaries; when the capillaries are set to provide little flow resistance the accumulators can easily absorb instantaneous pressure pulsations in the bearing oil film by allowing lubricant to flow into them. When the accumulators are used in this way the bearing stiffness becomes very low as far as the application of dynamic loading is concerned. The steady load carrying capacity and the damping generated by the squeeze-films remain, of course, unaffected by the accumulator. This means that the variable stiffness hydrostatic bearing is able to support steady loads, for example gravity loading, whilst taking on almost any value of stiffness required as far as dynamic loads are concerned.

The paper describes an analysis of both the steady-state vibration characteristics and stability characteristics of a model aeroengine compressor shaft running in bearings of the design proposed.

2. THEORY

The theoretical analysis of the system described above included analyses of the hydrostatic bearing characteristics, the squeeze film bearing characteristics, the rotor and shaft system dynamic characteristics, and the investigation of the system stability characteristics.

Spatial limitations prevent the details of all of these aspects being described in detail in this paper, instead they are each discussed in general terms and references are provided for the reader who requires further information.

The analysis of the hydrostatic bearing characteristics was based on the unsteady flow equation for the lubricant within the bearing recess.

$$Q_s = Q_a + Q_d + Q_v + Q_c \qquad (1)$$

where Q_s is the flow of lubricant into the recess from the supply, Q_a is the flow out of the recess into the accummulator, Q_d is the flow of lubricant out of the recess over the bearing lands, Q_v the rate of increase of bearing recess volume as one bearing surface moves away from another, and Q_c the rate at which the recess lubricant volume decreases as a consequence of lubricant compressibility. These terms may be defined (7) as

$$Q_s = K_s\left\{(p_s - p_r) - I_s \frac{dQ_s}{dt}\right\} \qquad (2)$$

$$Q_a = K_a\left\{(p_r - p_a) - I_a \frac{dQ_a}{dt}\right\} \qquad (3)$$

$$Q_d = \frac{p_r . c^3}{12\mu W} . f(x,y) \qquad (4)$$

$$Q_v = A . \frac{dh}{dt} \qquad (5)$$

$$Q_c = \frac{V}{\kappa} . \frac{dp_r}{dt} \qquad (6)$$

where p_s, p_a, and p_r are the lubricant pressure at the supply, accumulator, and bearing recess respectively; K_s and K_a are inverse capillary flow resistances while c is the nominal bearing radial clearance, W the land width, and μ the lubricant dynamic viscosity. I_s and I_a are coefficients describing the inertia of the lubricant in the supply capillary and accumulator capillary respectively. The area contained within the land centre-line around each recess is defined as the area A while V is the recess volume and κ the lubricant bulk modulus. The term dh/dt is the rate of increase of the instantaneous bearing clearance at any particular recess and the function f(x,y) describes the bearing geometry and is defined in the appendix. The pressure in the bearing accumulator also changes according to

$$\frac{dp_a}{dt} = B . Q_a \qquad (7)$$

where B is a parameter describing the accumulator properties.

Equations (4) (5) and (6) may be substituted into equation (1) so that the resulting equations (1) (2) (3) and (7) may then be solved simultaneously to yield the flows Q_s and Q_a and the pressures P_r and P_a for any given displacement or velocity of the journal away from its equilibrium position. The changes in all of the bearing recess pressures P_r may be evaluated for given changes of instantaneous journal displacements and velocity about its equilibrium position, and so be used to evaluate the effective stiffness and damping of the hydrostatic bearing oil film. The hydrostatic bearing properties may thus be represented by eight linearised spring and damping coefficients as used by other researchers (8).

In previous investigations, (7) and (9), each of the bearing recesses was separated from the adjacent recesses by drain grooves sited between the axial lands. In the design investigated herein the bearing circumferential lands are all joined together to form a 360° squeeze film at each end of the bearing as shown in figure 2. The squeeze film bearing performance was analysed in isolation from the hydrostatic bearing characteristics using superposition. The squeeze film bearing analysis adopted was similar to that described in reference (10) where the squeeze-film lands are assumed to behave like short journal bearings with zero rotation. Reynolds' equation describing the lubricant pressure variation may then be written as

$$\frac{d}{dZ}\left(c^3 \frac{dp}{dz}\right) = 12\mu\,(e\,\dot{\emptyset}\,\sin\Psi + \dot{e}\,\mathrm{Cos}\,\Psi) \quad ..\ (8)$$

where p is the lubricant pressure at axial position z measured from the land centre line and c is the bearing clearance. The parameters e, $\emptyset$ and Ψ define the journal eccentricity, journal attitude angle, and angular position around the bearing respectively. Equation (8) may be integrated with respect to z, subject to the boundary conditions p = 0 at z = ± W/2, and with respect to Ψ to yield the net force acting on the journal in the radial and tangential directions. Since the hydrostatic bearing is capable of supporting gravity loads with negligible steady displacement of the journal from the concentric position, the resulting unbalance journal whirl orbits will be concentric with the bearing and will be circular; this means that in the resulting expressions for the fluid film forces, $\dot{e}$ can be set to zero, and $\dot{\emptyset}$ to ω. The resulting fluid film forces are then

$$F_R = 0 \quad \ldots\ldots\ldots\ldots\ldots\ldots\ldots\ldots\ldots\ldots \quad (9)$$

in the radial direction and

$$F_T = \frac{\mu R\, W^3\, \pi\, e\, \omega}{2\, c^3\, (1-\epsilon^2)^{3/2}} \quad \ldots\ldots\ldots\ldots\ldots\ldots \quad (10)$$

in the tangential direction. Since there is no fluid film force in the direction of the journal displacement the squeeze-film stiffness is zero; the instantaneous journal velocity in the tangential direction is of magnitude $e\omega$ so that the effective squeeze film damping is

$$C = \frac{F_T}{e\omega} = \frac{\pi\, R\, W^3\, \mu}{2\, c^3\, (1-\epsilon^2)^{3/2}} \quad \ldots\ldots\ldots\ldots\ldots\ldots \quad (11)$$

This value of damping coefficient is added to that developed for the hydrostatic bearing alone to form the net effective bearing damping coefficient.

In order to evaluate the steady state performance of the proposed bearing design the system represented in Figure 1 was modelled by a computer program which was based on the transfer matrix method described in reference (11). The various parameter values were chosen so that the system modelled the high speed compressor shaft of the General Electric TF34 turbofan engine. Part of the input data for this computer program comprised the effective bearing stiffness and damping coefficients evaluated as described above. The output from the computer program included shaft and rotor vibration amplitudes, and force transmitted through the bearings, at various running speeds.

In some instances it is possible for machine forced unbalance response to be satisfactory from a design standpoint, but for violent machine vibrations to still set in as a consequence of instability. To examine the implications of the proposed bearing design on system stability a theoretical analysis similar to that described in reference (12) was undertaken. This analysis makes use of equivalent bearing stiffness and damping coefficients which also allow for shaft flexibility; the method investigates the motion of the rotor modal mass when carried in such supports, the motion being of the form

$$x = X\, e^{st} \quad \ldots\ldots\ldots\ldots\ldots\ldots\ldots\ldots\ldots\ldots (12)$$

so that for stable motion (i.e. amplitude decreasing with time) the real part of s must be negative. The computer program outputs values of the real component of s for various bearing and rotor parameter values and for various running speeds.

3. RESULTS

The results of the investigation are shown in Figures 3 to 8. Figure 3 shows how the effective bearing stiffness is affected by changes in the flow resistance beween bearing recess and accumulators. Figures 4 and 5 show the variation of rotor response and force transmissibility with running speed for two values of bearing stiffness; these curves are plotted for a value of bearing damping coefficient C = 1.0 which was found to give rise to both low rotor vibration and low force transmissibility. Figures 6 and 7 show how the rotor response and force transmissibility of a system running in the proposed bearings compare with corresponding values for machines running in conventional squeeze-film bearings. For each of Figures 4 to 7 the unbalance was applied at the location of the rotor mass itself, and only the first and third modes of vibration were forced; however, similar results are also obtained when second mode forcing is allowed for. Finally, Figure 8 indicates the degree of system stability, showing the variation of the real component of s with running speed.

4. DISCUSSION

Figure 3 shows how effective changing the flow resistance between accumulators and bearing recesses is on the bearing stiffness. It can be seen that for high values of flow resistance, i.e. small values of K_a/K_s, the bearing stiffness is relatively high. This is because under these circumstances the accumulators are effectively shut off from the bearing and so are unable to influence its performance. This high value of bearing stiffness at low K_a/K_s values corresponds, for low running frequencies, to the bearing static load stiffness. In contrast, when the flow resistance between accumulators and bearing recesses is very small, i.e. high values of K_a/K_s, the accumulators are able to absorb pressure pulsations from the bearing recesses and so give rise to a value of stiffness which is less than five percent of the value in the absence of accumulators. The implication of this is that if the accumulators were connected to the bearing recesses only by a remotely controlled valve which was either open, to provide a negligible flow resistance, or closed, then the bearing stiffness can be 'switched' from a very high value to a negligible value whilst a machine is running.

Figure 4 shows how changing the bearing stiffness affects rotor vibration amplitude at various running speeds. It can be seen that when the bearing stiffness is equal to the shaft stiffness the machine first critical speed is well defined, at about 0.7 times the pin-pin critical speed with a peak rotor vibration amplitude of about 2.9 times the rotor unbalance eccentricity. When the stiffness is reduced to zero however the first critical speed is also shifted to zero and the peak level of rotor vibration is reduced to about 1.4 times the rotor unbalance eccentricity. It would also apparently be beneficial to run the machine up in speed from zero rev/min to about 1.2 times the pin-pin critical speed, and then switch to the higher bearing stiffness $k_b/K_s = 1$ before increasing speed further. This would result in the advantage of lowering rotor dimensionless vibration amplitudes, in the speed range 1.2 to 2.3 times the pin-pin natural frequency, from about 1.4 about 1.1 times the rotor unbalance eccentricity. For higher speeds still, both support stiffnesses give rise to rotor vibration amplitudes of about 1.0 times the rotor unbalance eccentricity.

Figure 5 is similar in form to Figure 4, but shows instead the variation of force transmissibility at the bearings with running speed. It can be seen that when the bearing stiffness is the same as the shaft stiffness, i.e. $k_b/k_s = 1.0$, the peak force transmissibility is about 3.1 at a running speed of about 0.7 times the system pin-pin critical speed. In contrast the zero stiffness support gives rise to a peak force transmissibility which is little greater than 1.0, and is generally much lower than the corresponding value for $k_b/k_s = 1.0$ in the running speed range from zero to about 1.2 times the pin-pin critical speed. For speeds greater than $\omega/\omega_{pp} = 1.2$ the higher stiffness support gives improved performance up to $\omega/\omega_{pp} = 2.0$ as compared with the zero stiffness support, and for yet higher speeds both supports give rise to very low support transmissibility.

In Figure 6 the performance of the proposed variable stiffness squeeze-film bearing is compared with that of conventional squeeze-film bearings. When compared with the conventional squeeze-film bearing which has centralising springs it can be seen that for speeds up to about 1.1 times the pin-pin critical speed the conventional bearing results in rotor vibration amplitudes which are up to about 30% greater than those for a system running in the proposed bearing design. At higher running speeds, both bearings result in similar rotor vibration amplitudes whose magnification tends to unity with increasing speed. When compared with conventional squeeze film bearings which do not have centralising springs, the conventional bearings result in peak rotor vibrations which are in excess of 300% of those of the proposed bearing.

Figure 7 shows the variation of force transmissibility with speed for the proposed bearing and for conventional squeeze-film bearings. The peak force transmissibility for the proposed bearing is about 1.1 whereas those values for the conventional designs are about 1.6, for the bearing with centralising springs, and about 3.2 for those designs without centralising springs. The proposed design offers a performance which is superior to those of conventional designs for running speeds up to about 2.0 times the pin-pin critical speed, and for higher speeds all three designs give rise to similar very low force transmissibility.

Figure 8 indicates how the stability of systems running in the proposed bearing design compares with those of systems running in conventional bearing designs. The figure shows the variation of the stability parameter β with running speed for an equivalent 'Jeffcott' rotor running in flexible bearings and in the absence of aerodynamic cross-coupling. A negative value of β indicates system stability. It can be seen that for bearings of moderate stiffness, where $k_b/k_s = 1.0$, the system is stable for all running speeds, although the degree of stability decreases with increasing speed. When running in the proposed (zero stiffness) bearings however, the system would appear to be only marginally stable at low running speeds where β is almost equal to zero, but for speed ratios in excess of about 0.3 the system is far more stable than when operated with moderately stiff bearings. In practice this potential design problem is circumvented by recognising that at very low frequencies the accumulators do not reduce bearing stiffness to zero (5) and that if the bearings are set to provide even a very low stiffness system stability is considerably improved; with this in mind the curve for a system whose bearing stiffness ratio $k_b/k_s = 0.2$ is also shown in Figure 8.

5. CONCLUSIONS

The use of accumulators to control the stiffness of the bearing type investigated allows the machine operator to 'switch' the bearing stiffness from a very high to a very low value by adjusting the lubricant flow resistance between bearing recesses and accumulators. This approach enables bearing stiffness to be tuned without adversely affecting the bearing damping which is provided by lubricant squeeze films.

For running speeds less than about 1.2 times the pin-pin critical speed the proposed bearing design gives rise to considerably lower rotor vibration levels and lower force transmissibility when operating with negligble effective stiffness. For higher speeds it is beneficial to 'switch' the accumulators out of operation and run with higher stiffness supports.

Conventional squeeze film bearings give rise to rotor vibration magnification and bearing force transmissibility values which are considerably higher, during operation at speeds up to 1.2 times the pin-pin critical speed, than those obtained when the proposed bearing design is used.

The proposed design of bearing can be utilised in such a manner as to maintain stable operation of the machine concerned.

6. ACKNOWLEDGEMENTS

The authors wish to express their gratitude to the United States Air Force for their support of the work described in this paper, conducted under contract grant no. AFOSR-84-0368-C and to the Programme Mananger Dr Anthony K Amos. The US government is authorised to reproduce and distribute reprints for governmental proposes notwithstanding any copyright notation thereon.

7. REFERENCES

(1) COOKSON, R. A. and KOSSA, S. S. (1980) The effectiveness of squeeze-film damper bearings supporting flexible rotors without a centralising spring. Int. J. Mech. Sci. Vol 22, p 313.

(2) CUNNINGHAM, R. E., FLEMING, D. P., and GUNTER, E. J. (1975) Design of a squeeze-film damper for a multi-mass flexible rotor. Trans. Am. Soc. Mech. Engrs., J. Eng. Ind., Nov. p.1383.

(3) HOLMES, R and HUMES, B (1978) An Investigation of vibration dampers in gas turbine engines. Proc 52nd. AGARD Symp, Cleveland, Oct.

(4) GUNTER, E. J., BARRETT, L. E., and ALLAIRE, P. E. (1976) Stabilisation of turbomachinery with squeeze film dampers - theory and applications. Conf. Vibration in rotating machinery. IMechE Spons. Cambridge.

(5) KRAMER, E. (1977) Computation of unbalance vibrations of turborotors. Trans. Am. Soc. Mech. Engrs. Paper No 77-DET-13.

(6) KIRK, R. G. and GUNTER, E. J. (1972). Effect of support flexibility and damping on the dynamic response of a single mass flexible rotor in elastic bearings. NASA report CR-2082.

(7) GOODWIN, M. J. (1981) Variable impedance bearings for large rotating machinery, PhD Thesis, the University of Aston in Birmingham.

(8) MORRISON, D. (1982) Influence of plain journal bearing on the whirling action of an elastic rotor. Proc Inst. Mech. Engrs. Vol 176 No 2 pp 542-553.

(9) GOODWIN, M. J., HOOKE, C. J., and PENNY, J. E. T. (1982) Controlling the dynamic characteristics of hydrostatic bearings using a pocket-connected accumulator, Proc. Int. Mech. Engrs, Vol 197 C. Dec.

(10) GUNTER, E. J., BARRETT, L. E., and ALLAIRE, P. E. (1976) Design of non-linear squeeze film dampers for aircraft engines. Trans. Am. Soc. Mech. Engrs., J. Lub. Tech. Paper No 76-Lub-25.

(11) RAO, J. S. (1983) Rotor dynamics. Published by Wiley Eastern Ltd.

(12) OGRODNIK, P. J., GOODWIN, M. J., and PENNY, J. E. T. The influence of design parameters on the occurence of oil whirl in rotor-bearing systems. Proc. Symp. Instability in Rotating Machines. NASA publicaton CP 276.

APPENDIX

The function f(x,y) in equation (4) is developed in reference (5) as

$$f(x,y) = [f(x_o, y_o) + f_{xo}.\ x_d + f_{yo}.y_d]$$

where

$$f(x_o\ y_o) = (2L + 4R\alpha) + \frac{1}{c}(12\ Ry_o \sin\alpha + 6L\ y_o \cos\alpha)$$

$$+ 6\ y_o^2\ L \cos^2\alpha + 6\ x_o^2\ L \sin^2\alpha)$$

$$+ \frac{1}{c^3}(4Ry_o^2\ [\sin\alpha - \frac{\sin^3\alpha}{3}]$$

$$+ 12\ R\ x\ o^2\ y_o\ \frac{\sin^3\alpha}{3}$$

$$+ 2Ly_o^3 \cos^3\alpha + 6Lx_o^2\ y_o \sin^2\alpha \cos\alpha)$$

and where f_{xo} and f_{yo} are the derivatives of $f(x_o\ y_o)$ with respect to x_o and y_o respectively. The quantities x_o and y_o are the steady displacements of the journal from the concentric position in the horizontal and vertical directions respectively, and x_d and y_d the dynamic displacement amplitudes away from the steady position in the horizontal and vertical directions respectively. The other parameters are the axial length of the bearing L, the bearing bore radius R, and half of the angle included by the circumferential land of the bearing recess, α.

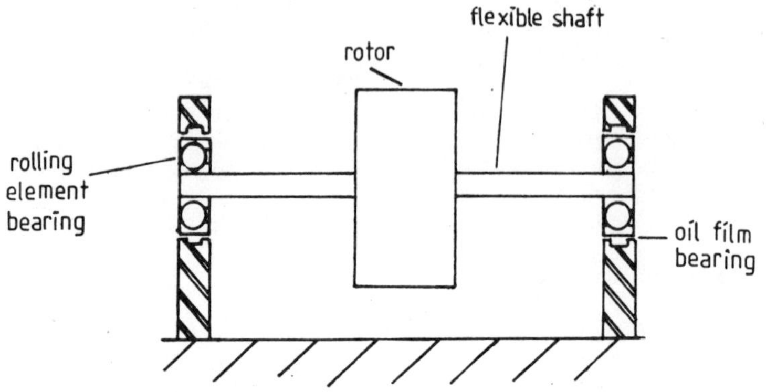

Fig 1 Schematic diagram of an aeroengine compressor shaft mounted in combined squeeze-film and variable stiffness hydrostatic bearings

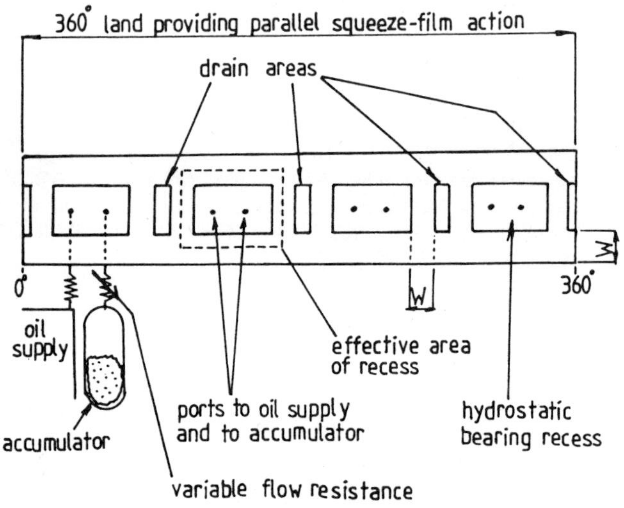

Fig 2 Developed view of bearing surface

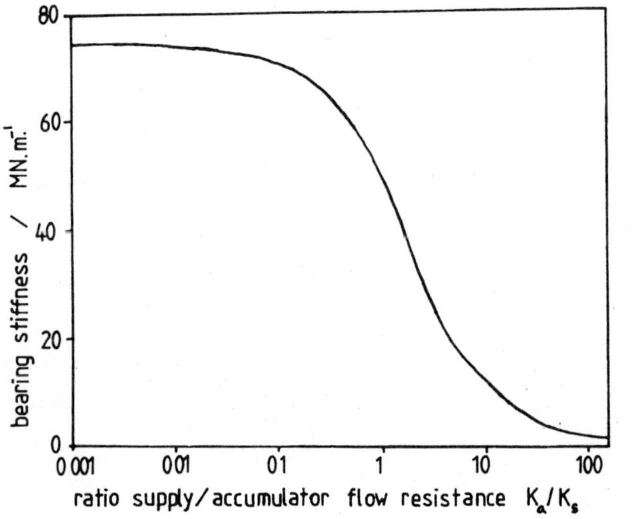

Fig 3 Variation of bearing stiffness with flow resistance in accumulator line

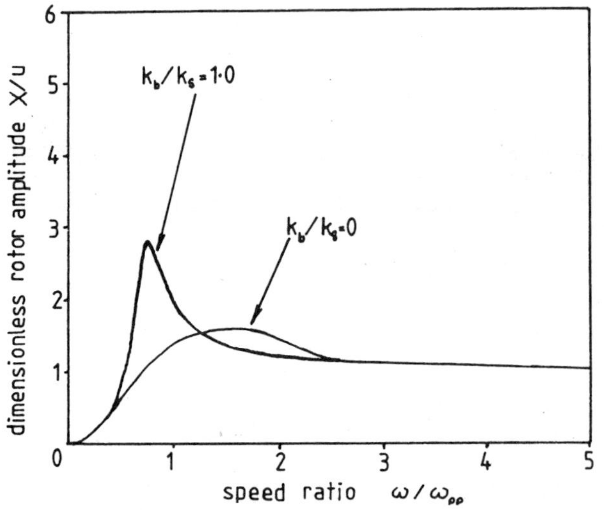

Fig 4 Variation of rotor vibration amplitude with running speed

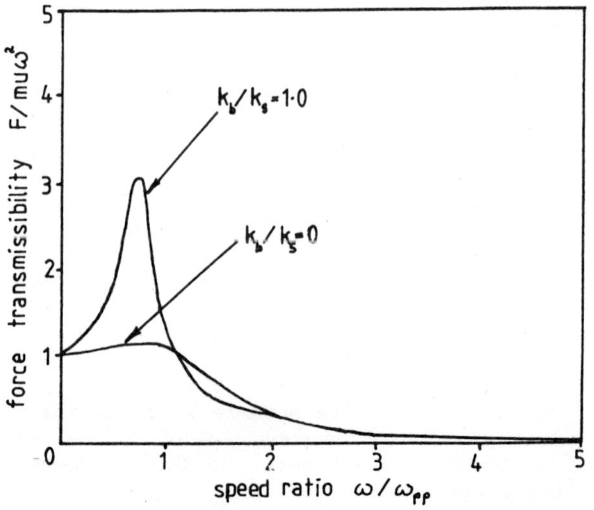

Fig 5 Variation of bearing force transmissibility with running speed

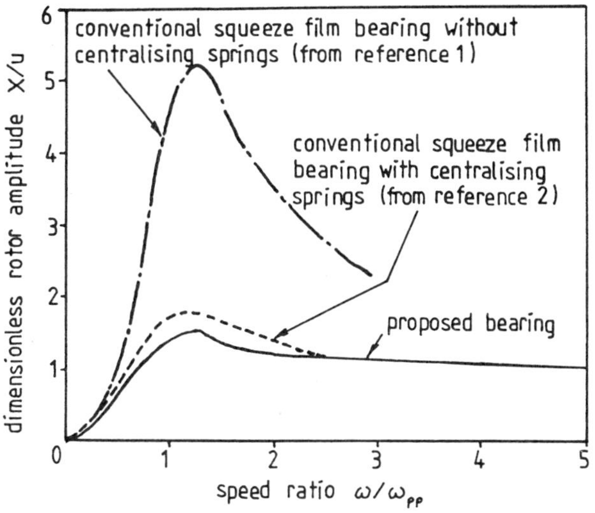

Fig 6 Comparison of rotor vibration amplitudes for system running in proposed bearing with those of a system utilising conventional squeeze-film bearings

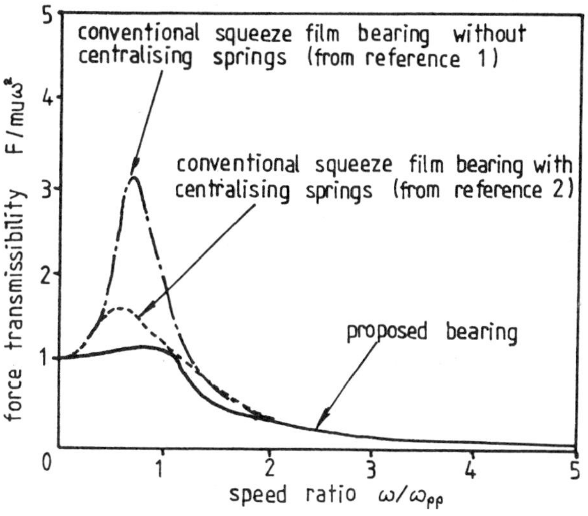

Fig 7 Comparison of bearing force transmissibility for system running in proposed bearing with that of a system utilising conventional squeeze-film bearings

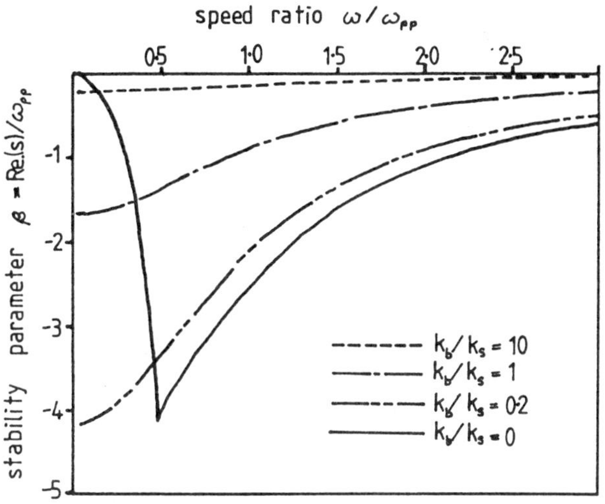

Fig 8 Variation of system stability parameter with speed and bearing stiffness

C278/88

Analysis and test results of two centrifugal compressors using active magnetic bearings

R G KIRK, BME, MME, PhDME, PE, MASME, MASLE
Department of Mechanical Engineering, Virginia Polytechnic Institute and State University, Blacksburg, Virginia, United States of America
J F HUSTAK, BME and K A SCHOENECK, BSME, MSME, BSIE, MASME
Dresser-Rand, Olean, New York, United States of America

SYNOPSIS

The design and shop test results are given for a high-speed eight-stage centrifugal compressor supported by active magnetic bearings. A brief summary of the basic operation of active magnetic bearings and the required rotor dynamics analysis are presented with specific attention given to design considerations for optimum rotor stability. The concerns for retrofits of magnetic bearings in existing machinery are discussed with supporting analysis of a four-stage centrifugal compressor. The current status of industrial machinery in North America using this new support system is presented and recommendations are given on design and analysis requirements for successful machinery operation of either retrofit or new design turbomachinery.

INTRODUCTION

A new technology in the form of active magnetic bearings (AMB) is being introduced into the marketplace for use on individual turbomachinery. The features of this technology when applied to turbocompressor design result in several economic, performance, and versatility improvements unavailable to the industry at the present time. Active magnetic bearings used in conjunction with dry gas seals and dry couplings now enable both the manufacturer and user to think in terms of oil-free centrifugal compressors.

Patent activity on passive, active, and combination magnetic bearing systems spans 150 years. The bulk of the initial investigations centered on permanent magnetic systems because they were easy to fabricate. It was later shown, however, that a passive magnetic suspension for three axes of displacement is unstable. This theory, first stated by Earnshaw (1) in 1842 is still valid today. The first totally active magnetic suspension system was described (2) and documented in a patent issued in 1957 but application to practical design conditions were not possible due to a lack of suitable electronic circuitry to switch the large DC currents required. In 1970, a totally active magnetic suspension system was developed for a communications satellite and in 1976, a new company was formed to further develop and commercially market active magnetic bearing systems internationally (3,4).

PRINCIPLE OF ACTIVE MAGNETIC BEARING OPERATION

The AMB is composed of two major mechanical parts consisting of the rotor and the stator. Both are made of ferromagnetic laminations. The rotor laminations are placed on the machine shaft at the selected journal location. The stator laminations are slotted and include windings to provide the magnetic levitation and position control. For each degree of freedom, two electromagnets are required since they operate by attraction only. Figure 1 shows the stator laminate construction of a radial bearing with the rotor laminate sleeve in the background.

Rotor position is monitored by sensors and this signal is compared to a nominal reference signal with a closed loop controller which supplies a command signal to the power amplifier. These amplifiers provide power to the electromagnets to resist rotor movement from the nominal position. The design of the control loop gives the option to select the effective bearing damping and stiffness. The details of this design procedure are not the subject of this paper but the values of stiffness and damping must be carefully selected to give the rotor system the desired optimum dynamic response and stability.

This design concept can be applied to both radial and axial thrust bearings to give total control of a rotating rotor system. The load capacity of the AMB using standard materials can be made equal to standard fluid-film bearing designs. The overload condition for the AMB is controlled by auxiliary bearings which must be designed to provide rotor constraint in the event of system power failure or momentary transient overload.

The many advantages and detailed design requirements for the AMB are discussed in greater detail by Haberman (3,4). The application of the AMB to industrial compressors with proper evaluation of the rotor dynamic response and stability is essential for success of this new technology. This latter concern will be addressed in the following discussion.

DEVELOPMENT CENTRIFUGAL COMPRESSOR WITH AMB

Figure 2-a is a view of an eight-stage, horizontally split, back-to-back centrifugal compressor equipped with magnetic radial and thrust bearings and gas seals on test at the authors' former company (1980). The eight-stage rotor housed inside the compressor, originally designed to run at 167 Hz (10 000 rpm) on hydrodynamic bearings, has since operated successfully for 750 hours at speeds up to 217 Hz (13 000 rpm) on magnetic bearings. The compressor is shown attached to two closed loops constructed for the purpose of operating the rotor in a pressurized environment over a wide range of pressures and flows from choke to surge.

One unique feature of this compressor was the installation of the thrust and journal bearings located on the free end of the rotor directly into the gas (nitrogen) pressurized environment, thereby eliminating the need for one shaft seal. To illustrate the concept of a nonlubricated centrifugal compressor, a gas seal was chosen as the main shaft seal on the coupling end of the rotor. Table 1 summarizes some of the important design features of the eight-stage back-to-back rotor while Fig. 2-b illustrates the appearance of the fully assembled test rotor.

Before power is applied to the bearings, the rotor is supported on two auxiliary ball bearings located in close proximity to the AMB. The clearance between the rotor and the inner race of the ball bearing is selected to prevent rotor contact with the AMB pole pieces or the internal seals of the compressor while the rotor is at rest or during an emergency shutdown. Typical shaft radial clearances in the AMB, internal seals, and auxiliary bearing inner race are 0.3 mm (0.012 in.), 0.254 mm (0.010 in.), and 0.15 mm (0.006 in.), respectively. When power is applied to the electronic controls, the electromagnets levitate the rotor in the magnetic field and rotation of the driving source, such as a motor or turbine, can be started. The sensors and control system regulate the strength and direction of the magnetic fields to maintain exact rotor position by continually adjusting to the changing forces on the rotor. Should both the main and redundant features of the AMB fail simultaneously, the auxiliary bearings and rotor system are designed to permit safe deceleration.

Magnetic bearing stiffness and damping properties are controlled by a PID (Proportional, Integral, Derivative) analog control loop that allows some flexibility for adjustment. The stiffness and damping characteristics are axi-symmetric, i.e., identical in both the horizontal and vertical directions. The resultant dynamic bearing stiffness, KD, is a complex number represented by the vectorial summation of a "real" stiffness component (K) and an "imaginary" stiffness component (cω).

$$KD = \sqrt{K^2 + (c\omega)^2} \qquad (1)$$

The phase relationship (α) between the "real" (K) and "imaginary (cw) components is given by

$$\alpha = \arctan\,(c\omega/K) \qquad (2)$$

where K = static real stiffness (N/m), ω = angular velocity of journal vibraton (sec^{-1}). Variation of the phase angle ranges from 0 to 90 degrees with typical values in the 30 to 40 degree range. Adjusting the amount of bearing damping is accomplished by changing the passive elements of the compensation circuit in the PID control loop. The amount of gain applied to the circuit through a potentiometer provides a proportional increase in the resultant dynamic stiffness, KD.

The undamped critical speed map shown in Fig. 3-a compares the standard fluid film bearing/oil seal design to the magnetic bearing/gas seal design for the 8-stage development compressor. The magnetic bearing design increased the first rigid bearing mode by reducing the bearing span but decreased the second, third, and fourth modes due to the additional weight of the ferromagnetic journal sleeves and the larger diameter thrust collar. Superimposed on this map are the stiffness and damping properties of the electromagnetic bearings as a function of rotor speed. For a design speed of 167 Hz (10 000 rpm), Fig. 3-a indicates the rotor would have to pass through three critical speeds and operate approximately 20 percent above the third critical and 40 percent below the fourth critical. Furthermore, since both the first and second criticals are rigid body modes, only a significant response at approximately 133 Hz (8000 rpm) would be expected as the rotor passed through its third (free-free mode) critical. Subsequent unbalance forced response calculations verified these expectations with acceptable operations of the compressor to 233 Hz (14 000 rpm) (see Fig. 3-b).

The damping required for optimum stability may be arrived at by plotting the systems' calculated damped natural frequencies vs. growth factors (5), (6), (7). For the first mode stiffness of 22.8 N/μm (130 000 lb/in), Fig. 4-a shows the movement of the eigenvalues as the bearing damping varies. Increased damping levels cause the first and second modes to become critically damped. The third mode increases in stability up to a point of 87.7 N-s/mm (501 lb-s/in.) but then decreases as damping is increased further. Since the first and second modes become critically damped, a second study was undertaken to determine if the third mode would go unstable at its corresponding stiffness of 40.1 N/μm (229 000 lb/in). The results of that analysis are presented in Fig. 4-b and indicate the third mode becomes critically damped as the damping is increased while the first mode damping would be at an optimum for 112.9 N-s/mm (645 lb-s/in.). The behavior of the first and third modes, resulting from the increase in bearing stiffness, is similar to results presented by Lund (5). These results indicate that a level of 70-87.5 N-s/mm (400-500 lb-s/in.) would be ideal for all modes up to and including the fourth.

The stiffness and damping characteristics for the magnetic bearing are given in Fig. 5. Active magnetic bearings have an important difference when compared to conventional fluid film bearings. Typical preloaded five-shoe tilt-pad bearings have characteristics generated predominantly by operating speed, with little influence from nonsynchronous excitations (8). The active magnetic bearing characteristics, shown in Fig. 5, are dependent on the frequency of excitation regardless of operating speed. For a given stiffness value selected to minimize unbalance forced response, the damping characteristics at subsynchronous excitation frequencies can be specified to assure optimum stability (3), (4).

The test program outlined for the compressor was directed toward confirming the analytical predictions for the dynamic behavior of the rotor and experimentally demonstrating the reliability of the complete system under typical operating conditions. The fully assembled compressor was installed on the test stand and operated at a maximum discharge pressure of 4.1 MPa (600 psig) with speeds up to 217 Hz (13 000 rpm). The results for a deceleration as recorded at the bearing probe locations are given in Figs. 6-a and 6-b. The test results indicate well-damped rotor responses at 67 Hz (4000 rpm) and 133 Hz (8000 rpm). In comparison, the synchronous response to unbalance shown in Fig. 3-b displays well-damped rotor response at 67 Hz (4000 rpm) and 133 Hz (8000 rpm). The amplitude of vibration is difficult to predict since the actual rotor unbalance configuration is made up of varying amounts of unbalance at different axial locations along the rotor. The predicted peak response frequencies are considered to be in good agreement with the test results.

DESIGN EVALUATION OF A FIELD RETROFIT

The economic advantages of gas seals and/or magnetic bearings have prompted interest in retrofit of existing units. For either retrofit or new machinery, attention must be given to placement of critical speeds for both main and backup bearings, response sensitivity, and overall stability considerations. The preliminary design study for a four-stage high-speed centrifugal compressor will illustrate in more detail the parameters that must be considered for total system dynamic analysis. The basic design parameters for this rotor are indicatd in the second column of Table 1.

The undamped critical speed map for the four-stage compressor is shown in Fig. 7-a. The magnetic bearing stiffness is positioned such that the compressor must pass through three critical speeds before reaching a maximum continuous operating speed of 241.7 Hz (14 500 rpm). Due to the rigid body nature of the second and third modes, the actual damped critical speeds will occur from the first and fourth modes. The frequencies at which these modes respond, shown in Fig. 7-b, are 71.7 Hz (4300 rpm) and 305 Hz (18 300 rpm). A plot of the systems calculated damped natural frequencies versus growth factor for a constant first mode stiffness of 15.1 N/μm (86 300 lb/in.) is shown in Fig. 8-a. The first forward mode typically goes unstable while the second and third modes become critically damped as the bearing damping is increased. For this compressor design, the first mode increases in stability as the damping increases up to 39.4 N-s/mm (225 lb-s/in.) but then decreases as the damping is increased further. The damping value initially supplied, 24.5 N-s/mm (140 lb-s/in.), should be increased by 61 percent based on the results of this analysis.

The optimum damping for stabiity was also calculated by an approximate method using the modal mass, rigid bearing critical frequency, and bearing stiffness (see Table 1). The equation from Ref. (9) can be written as follows (valid for high $\overline{K}$ ratios):

$$C_o = 2.893 \times 10^{-2} N_{cr} \left[M_m + \frac{4.98 \times 10^5 K_b}{N_{cr}^2} \right] \quad (3)$$

Example for four-stage first mode:

$$C_o = 2.893 \times 10^{-2} \times (90.39) \times \left(649 + (4.98 \times 10^5) \times (15.1)/(90.39)^2 \right) = 45.4 \text{ N-s/mm}$$

This quick calculation gives an answer 15 percent higher than the lengthy optimum damping method used in Fig. 8-a.

Figure 8-b shows a comparison of stability versus aerodynamic excitation between the conventional fluid film design and the magnetic bearing retrofit design. The increase in stability due to the magnetic bearing and dry gas seal design moves the log decrement from near zero to a value of 1.41.

Current Status of Active Magnetic Bearings

The application of magnetic bearings for industrial installations in North America has progressed to the point that both retrofit and new, original design applications have been initiated with successful operation on test and in the field. Table 2 shows the current status of new and retrofit turbomachinery in North America. The use of this new method of turbomachinery support and control has not been totally free of test stand problems. Machinery must withstand extremes of temperatures for various design applications which must be properly evaluated to assure adequate bearing materials selection. In addition, loading from rotor system balance variations due to initial build and transient excitation resulting from process upsets must be accounted for in the initial design considerations.

These potential problems make the design prediction capability and initial design studies for rotor dynamics analysis just as important for AMB as it has been for machinery supported on fluid-film bearings.

CONCLUSIONS AND RECOMMENDATIONS

The capability of an active magnetic bearing system to support a flexible turbocompressor rotor and simultaneously influence its vibrations has been successfully demonstrated. During 750 hours of accumulated operating time for the development compressor equipped with active magnetic bearings, the following observations were made:

1. The rotor behaved in a stable manner at all times when accelerating/decelerating through its first three critical speeds.
2. The rotor behaved in a stable manner while undergoing surge cycles at maximum discharge pressure.
3. The magnetic bearings were able to suppress rotor amplitude to avoid contact between rotational and stationary components through the first three modes up to a speed of 271 Hz (13 000 rpm). Speeds beyond this point were limited by impeller stress considerations.

The following recommendations can be made for the design and analysis of magnetic bearing suspension turbomachinery:

1. Bearing stiffness should be selected by evaluation of shaft stiffness ratio with typical placement at the beginning of the third mode ramp on the undamped critical speed map.
2. Bearing damping should be specified to give the optimum stability with consideration given to all modes below maximum operating speed.
3. Consideration must be given to the next mode above operating speed (typically the fourth mode) to avoid interference between operating speeds and system natural frequencies.
4. The machinery must be engineered such that the shaft critical speeds are at least 10 percent lower or higher than any continuous operating speed when the rotor system is assumed to be operating on the auxiliary bearings.

ACKNOWLEDGEMENTS

The authors wish to thank Mr. Howard Moses and Mr. David Weise of Magnetic Bearings Incorporated, Radford, VA for their assistance in the preparation of this paper and the permission to publish the information given in Table 2.

REFERENCES

(1) Earnshaw, S., "On the Nature of the Molecular Forces," Trans. Cambridge Phil. Soc. 7, 97-112, 1842.

(2) Tournier, M., and Laurenceau, P., "Suspension magnetique d'une maquette en soufflerie," La Recherche Aeronautique 7-8, 1957.

(3) Haberman, H., "The Active Magnetic Bearing Enables Optimum Damping of Flexible Rotors," ASME Paper 84-GT-117.

(4) Haberman, H., and Brunet, M., "The Active Magnetic Bearing Enables Optimum Control of Machine Vibrations," ASME Paper 85-GT-22, Presented at Gas Turbine Conf., Houston, TX, March 18-21, 1985.

(5) Lund, J. W., "Stability and Damped Critical Speeds of a Flexible Rotor in Fluid-Film Bearings," J. of Eng. for Industry, Trans. ASME, Series B. 96, 2, pp. 509-517, May (1974).

(6) Bansal, P., and Kirk, R. G., "Stability and Damped Critical Speeds of Rotor-Bearing Systems," J. of Eng. for Industry, Trans. ASME, Series B, 98, 1, pp. 108, February (1976).

(7) Kirk, R. G., "Stability and Damped Critical Speeds--How to Calculate and Interpret the Results," CAGI Technical Digest, 12, 2, pp. 1-14 (1980).

(8) Wilson, B. W., and Barrett, L. E., "The Effect of Eigenvalue-Dependent Tilt Pad Bearing Characteristics on the Stabiity of Rotor-Bearing Systems," Univ. of Virginia, Report No. UVA643092/MAES5/321, January 1985.

(9) Barrett, L. E. Gunter, E. J., and Allaire, P. E., "Optimum Bearing and Support Damping for Unbalance Response and Stability of Rotating Machinery," Trans. ASME, J. of Eng. for Power, pp. 1-6 (1978).

Table 1 Centrifugal Compressor Design Parameters and Nomenclature

Parameter, Nomenclature SI Units (US Units)	Eight-Stage		Four-Stage	
Operating Speed, N, Hz, (rpm)	217.7	(13 000)	241.7	(14 500)
Total Weight, W, N, (lb)	3678	(827)	1237	(278)
Bearing Span, mm, (in.)	1269	(49.97)	886	(34.88)
Shaft Length, mm, (in.)	1902	(74.90)	1253	(49.32)
Coupling End Overhang, mm, (in.)	304.8	(12)	184.9	(7.28)
Shaft Stiffness K_s, N/μm ,(lb/in.)	61.1	(3.49E5)	21.4	(1.22E5)
Bearing Stiffness @ MCOS, K_b, N/μm ,lb/in.	63.9	(3.65E5)	28	(1.60E5)
Stiffness Ratio, $\overline{K}$, Dim., (Dim.)	0.74	(0.74)	1.41	(1.41)
Mid-Span Diameter, mm, (in.)	123.9	(4.88)	76.2	(3)
Journal Diameter, mm, (in.)	187.5	(7.38)	93	(3.66)
First Rigid Bearing Critical, N_{cr}, Hz,(rpm)	92.52	(5551)	90.39	(5423)
First Peak Response Speed, FPS1, Hz, (rpm)	129.5	(7700)	71.67	(4300)
Second Peak Response Speed, FPS2, Hz, (rpm)	271.7	(16 300)	305	(18 300)
First Mode Modal Mass, M_m, N, (lb_m)	1775.	(399)	649	(146)
Bearing Stiffness @ N_1, K_{b1}, N/μm ,(lb/in.)	22.8	(1.30E5)	15.1	(8.63E4)
Optimum Damping (9), C_o, N-s/μm , (lb-s/in.)	9.18E-2	(524)	4.54E-2	(259)

Table 2 AMB Industrial Compressor Applications in North America

MACHINE	TYPE	SERVICE	DUTY	COMM	ROTOR WEIGHT (lb) N	THRUST LOAD (lb) N	SPEED (rpm) Hz	JOURNAL DIAMETER (in) cm	DRIVER RATING (HP) kw	OPERATING HOURS (*)
MTA-824BB	CENT COMP	DEVELOP-MENT	INTERMIT	1980	(850) 3778	(3150) 14 000	(13 000) 217	(7.5) 19.0	(5360) 3997	750
CDP-230	CENT COMP	PIPELINE	SEASONAL	1985	(3200) 14224	(12 000) 52 340	(5250) 87.5	(10.6) 26.9	(12 800) 9545	6800
CDP-416	CENT COMP	PIPELINE	SEASONAL	1986	(280) 1245	(3370) 14 980	(14 500) 242	(3.7) 9.4	(4250) 3095	6700
1B26	CENT COMP	PIPELINE	CONTINUOUS	1986	(780) 3467	(4050) 18 000	(11 000) 183	(6.5) 16.5	(5500) 4101	4800
IB26	CENT COMP	PIPELINE	CONTINUOUS	1987	(780) 3467	(4050) 18 000	(11 000) 183	(6.5) 16.5	(5500) 4101	300
CBF-842	CENT COMP	REFINERY	CONTINUOUS	1987	(1420) 6312	(4590) 20 403	(10 250) 171	(6.0) 15.2	(4500) 3356	750
5P2	CENT COMP	PIPELINE	SEASONAL	NEW	(1540) 6845	(5500) 24 448	(7140) 119	(6.0) 15.2	(16 600) 12 379	-

*AMB Cabinet operating hours as of December 15, 1987

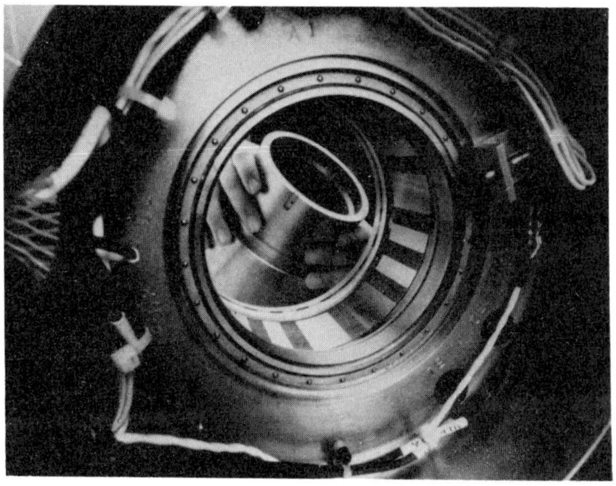

Fig 1 Active magnetic bearing stator with rotor sleeve in background

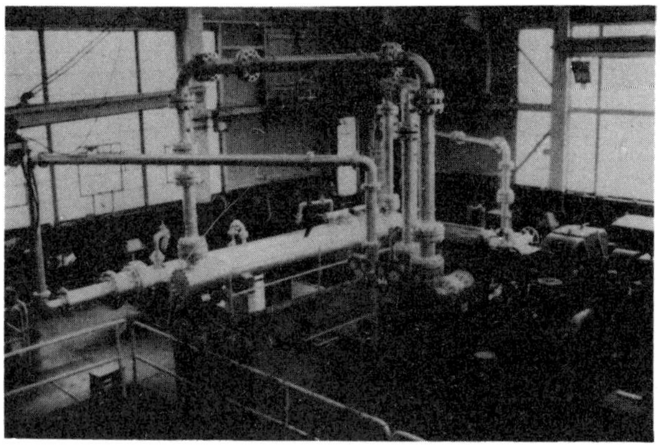

Fig 2a Eight-stage development compressor test set-up

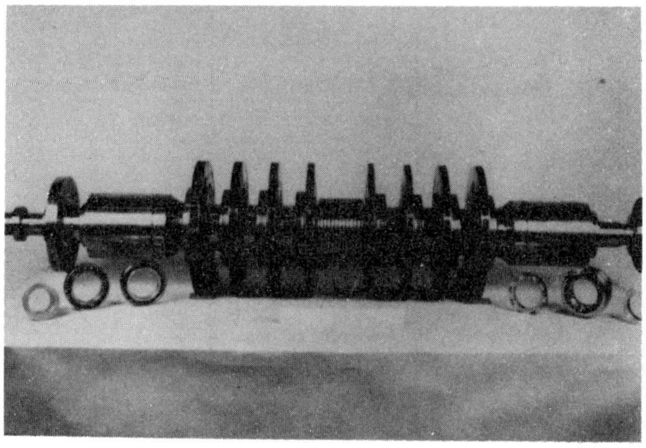

Fig 2b Eight-stage development compressor rotor

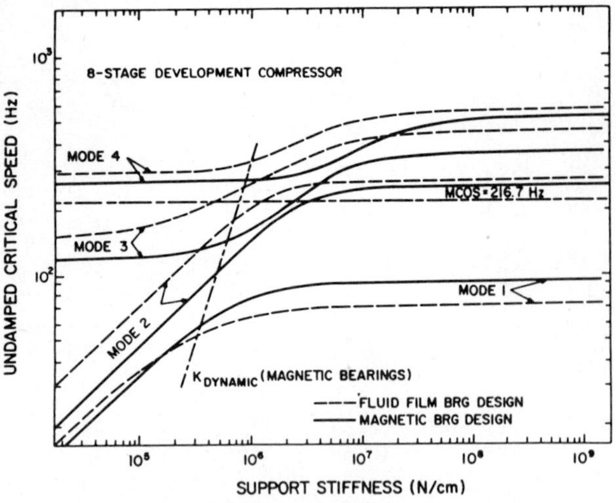

Fig 3a Eight-stage development compressor undamped critical speed map

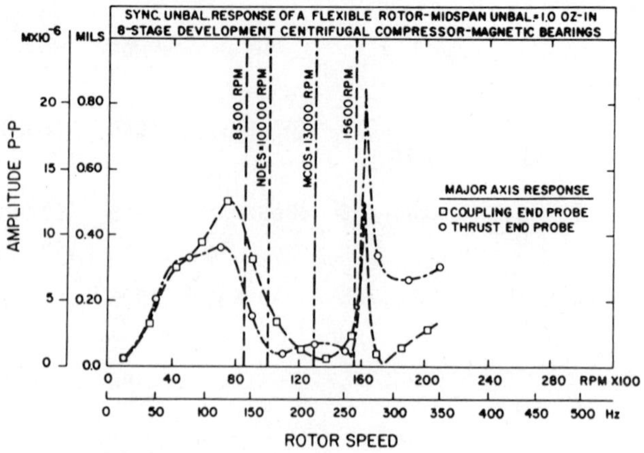

Fig 3b Eight-stage development compressor synchronous response to midspan unbalance

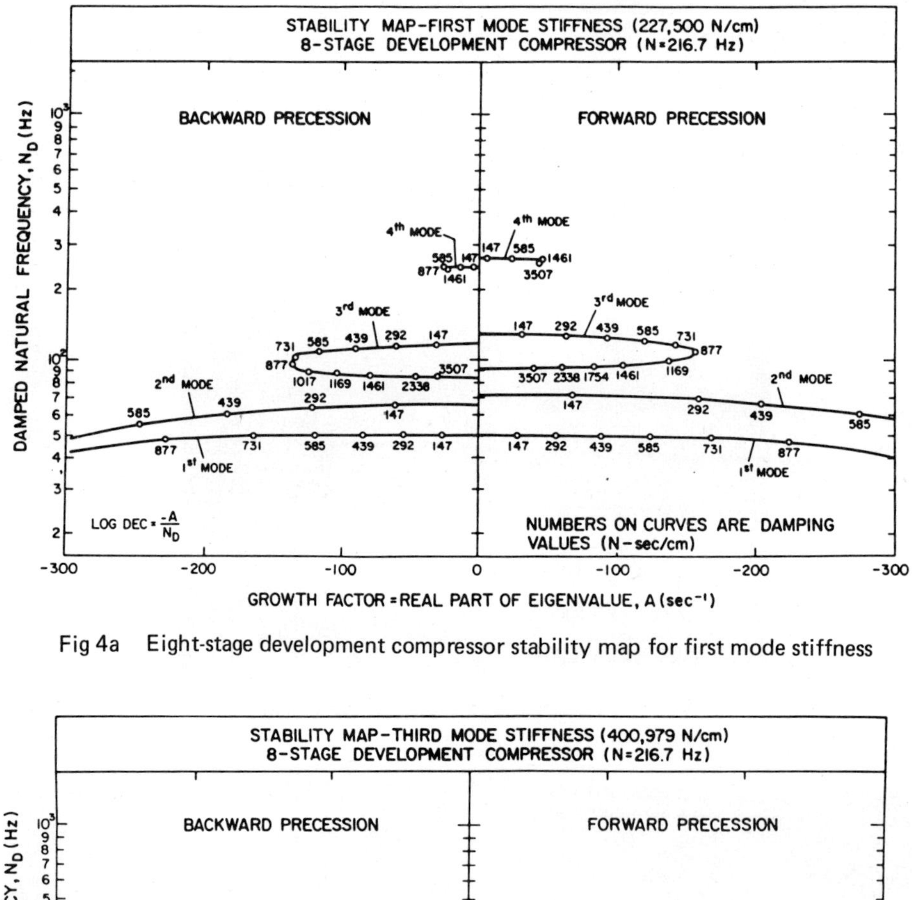

Fig 4a Eight-stage development compressor stability map for first mode stiffness

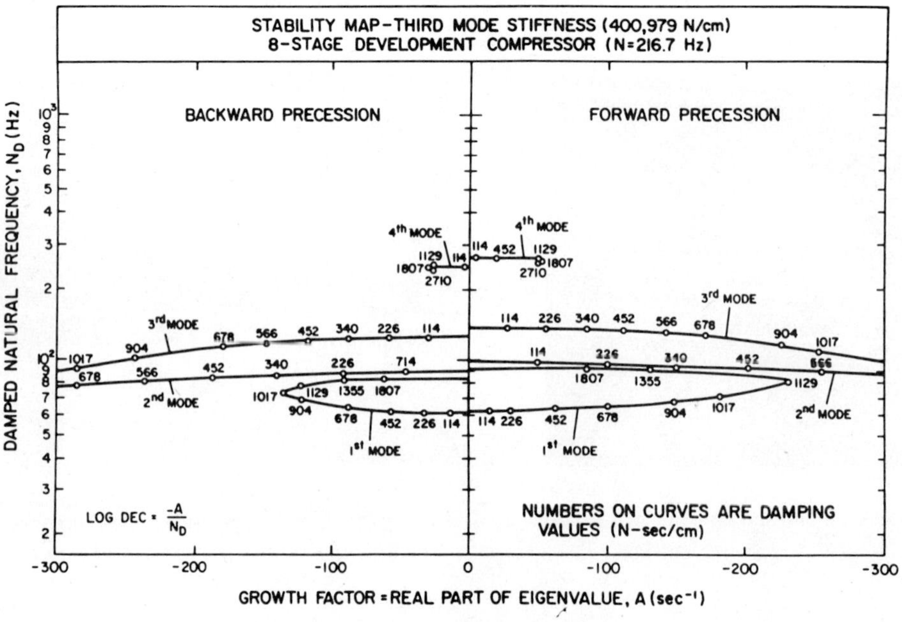

Fig 4b Eight-stage development compressor stability map for third mode stiffness

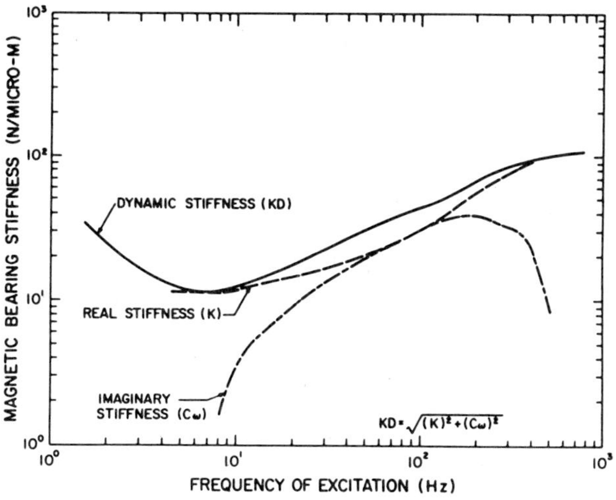

Fig 5 Eight-stage development compressor magnetic bearing characteristics

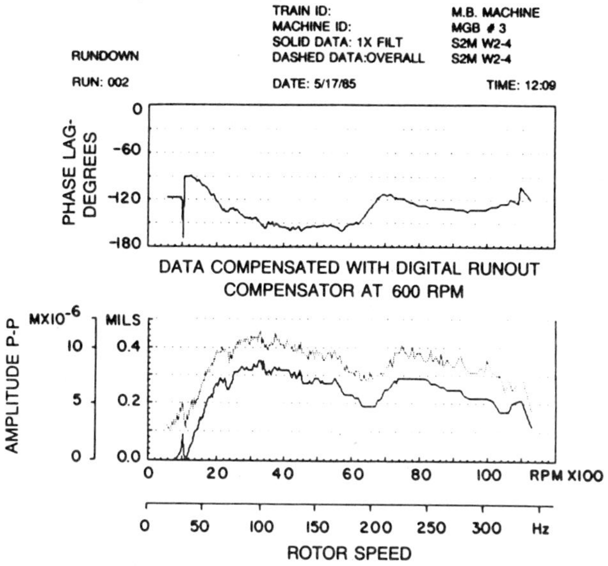

Fig 6a Eight-stage development compressor coupling end response from test results

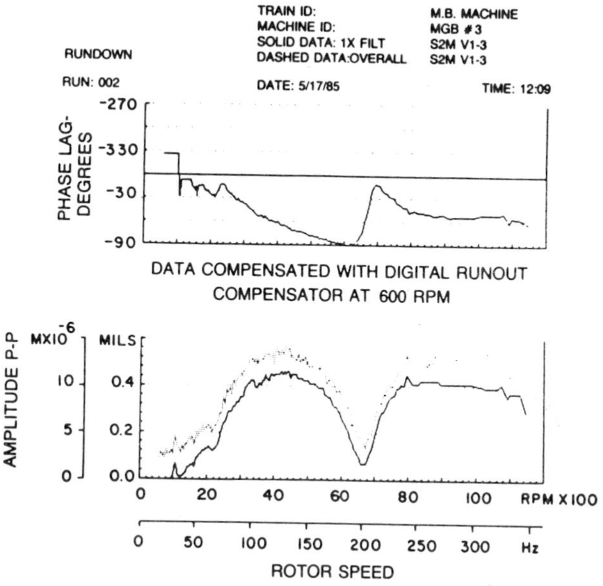

Fig 6b Eight-stage development compressor thrust end response from test results

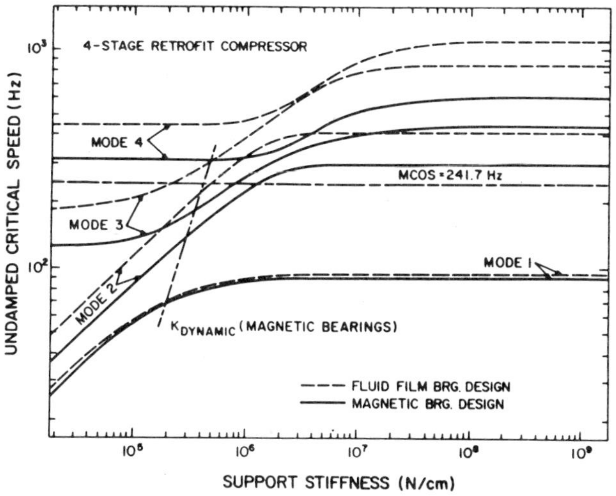

Fig 7a Four-stage retrofit compressor undamped critical speed map

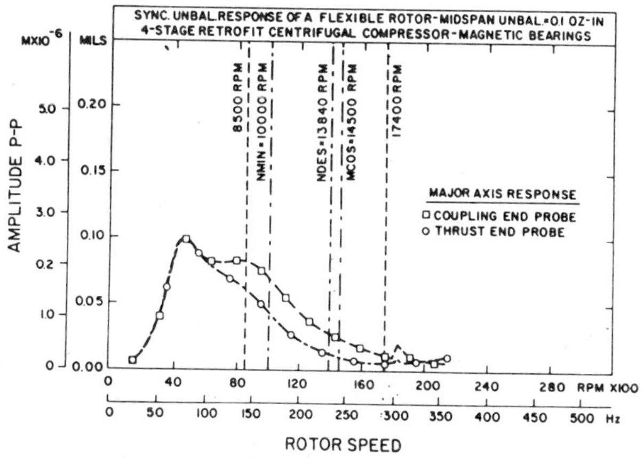

Fig 7b Four-stage retrofit compressor synchronous response to mid-span unbalance

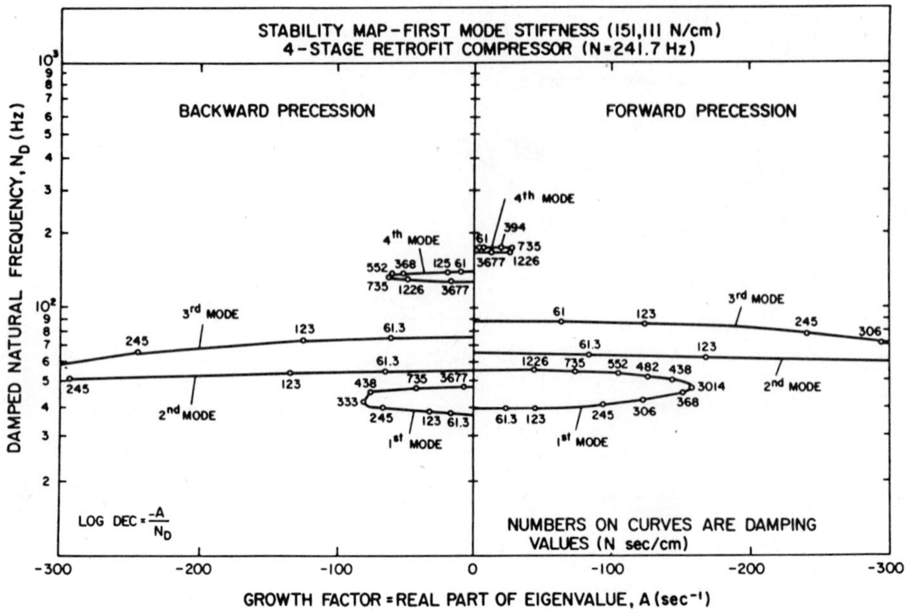

Fig 8a Four-stage retrofit compressor stability map for first mode stiffness

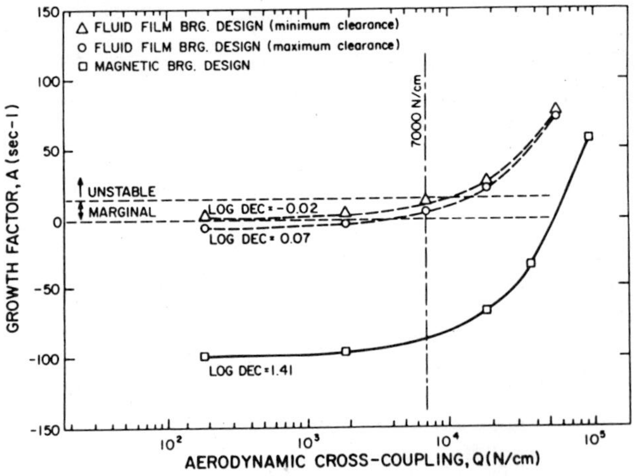

Fig 8b Four-stage retrofit compressor stability analysis results

C287/88

Active electromagnetic suspension and vibration control of an elastic rotor with a signal processor

H BLEULER, PhD
Institute of Mechanics, Swiss Federal Institute of Technology, Zürich, Switzerland
J SALM, PhD
Motoren Turbinen Union, Friedrichshafen, West Germany

Abstract: Active electromagnetic bearings are finding an increasing number of applications in various fields. Such bearings support a rotor without contact and can simultaneously be used to improve the dynamic properties of a flexible rotor.

In the experiments described in this paper, an elastic rotor borne by active magnetic bearings successfully passed four bending critical speeds. A rotational speed of 10000 rpm has been surpassed. Critical speeds are passed with vibrations remaining very small.

The theoretical part of this paper has been presented by the co-author in /1/, /2/ and extensively in /3/. The experimental part shows clearly the advantage a fast digital signal processor as opposed to a general purpose microprocessor. The control (fixed gain) is described in detail. Some aspects considered essential for the successful performance at high rotational speed (e.g. minimal time lag) are discussed.

1. Introduction, Goal of this Paper

One of the often mentioned properties of active electromagnetic bearings deals with the limiting of vibrations. This is of special interest for the passing of critical speeds of flexible rotors. There is a number of papers dealing with experiments on active electromagnetic bearings as dampers for otherwise conventionally borne rotors, e.g. /4/, /5/ or /6/ and others.

However, only very little experimental work is reported on the damping of critical speed vibrations of rotors supported by magnetic bearings only, e.g. /7/ or /8/. In /7/, analog control and cross-coupling of the radial directions are employed to reduce critical speed vibrations. In /8/, only one critical speed is passed.

Rotors supported completely without contact are finding increasing industrial applications in the most various fields. The present paper reports on experiments of the operation of a highly elastic rotor at several critical speeds. The control is digital. At the same time, publications /1/ /2/ and /3/ by the co-author are complemented by new experimental results, since much higher rotational speeds are reached with the new control.

Active electromagnetic bearings require a closed-loop control with the basic task of supporting the rotor. The basic elements of a radial bearing are shown in Fig. 1.

In the approach of the present paper, the additional task of **active damping** of the first few vibrational modes of an elastic rotor is achieved. The bearing control (a fixed gain control) has been designed in such a way, that the rotor theoretically can pass all critical speeds up to a high limit depending mainly on the frequency performance of the controller.

The section "Modelling and Control Layout" presents an abstract of the theoretical background based on /1/, /2/ and /3/.

The practical layout of the controller is then discussed. It is demonstrated, why the use of a digital signal processor (DSP) brings a crucial advantage for this control task. A DSP is a powerful microprocessor designed specially for signal processing purposes. DSPs are widely used in information processing and communications (digital filters, CD-players etc.) and they seem well suited for control applications in electro-mechanical (mechatronics) systems as well.

Experimental results and the conclusion close the paper.

2. Modelling and Control Layout

2.1 System Equations

A flexible rotor is considered, described by the **equation of motion** and the **output equation** in the following general matrix notation:

$$M \ddot{q}(t) + (D+G)\dot{q}(t) + K q(t) = B f(t) \qquad (1)$$

$$y(t) = \begin{bmatrix} C_s & 0 \\ 0 & C_v \end{bmatrix} \begin{bmatrix} q(t) \\ \dot{q}(t) \end{bmatrix}$$

These equations can be a low-order subset of equations obtained from a finite element model.

The vectors are:

q(t) n elements; mechanical degrees of freedom. The dimension n must be at least equal to the number of rigid modes plus the number of elastic modes to be modelled .

f(t) p elements; external actuator forces exerted by the p locally discrete bearings.

y(t) 2 s elements; output variables. The sensor signals are part of this vector. The structure of equations (1) contains the assumption of neglectable sensor dynamics. It is usually possible to provide a sensor system meeting this assumption with sufficient accuracy.

Velocity signals, although not measured directly in our real system, can be reconstructed from the position signals. By including them in the output vector, control design can be based on physical parameters like stiffness and damping. The derivation of the velocity signal is treated in section 3.

The matrices of (1) and their definiteness properties are:

- M nxn mass matrix $M = M^T > 0$
- D nxn damping matrix $D = D \geq 0$
- G nxn gyro matrix $G = -G^T$
- K nxn stiffness matrix $K = K^T \geq 0$
- B nxp influence matrix of external forces on the degrees of freedom.
- C_s sxn output (or measurement) matrix for positions.
- C_v sxn output (or measurement) matrix for velocities.

2.2 Control Design

The task of the controller can now be formulated as follows: Computing from the output vector y(t) the forces f(t) necessary to obtain a desired rotor behaviour.

As shown in /1/, /2/ and /3/, stability can be achieved by the most simple and obvious method, **direct fixed-gain output feedback**, with the constant feedback coefficient matrices G_s (stiffness) and G_v (damping):

$$f(t) = -[\,G_s \; G_v\,]\; y(t) \qquad (2)$$

if certain conditions stated below are met. To formulate these conditions, the total closed loop system (including unmodelled modes) is represented by combining (1) and (2):

$$M \ddot{q} + (D_{tot} + G)\; q + K_{tot}\, q = 0 \qquad (3)$$

where

$$K_{tot} = K + B\, G_s C_s$$

and

$$D_{tot} = D + B\, G_v\, C_v$$

A mechanical system as described by equation (3) is asymptotically stable in the sense of Liapunov if three conditions are met:

1) the mass matrix M is symmetric and positive definite.

2) the complete stiffness matrix K_{tot} (i.e. including the control) is symmetric and positive definite.

3) the symmetric part of the rate-dependent matrix D_{tot} is positive definite (or semi definite and contains pervasive damping).

In order to meet these conditions, feedback matrices Gs and Gv must be appropriately designed and the observability and controllability requirements familiar in control theory must be met. These requirements are commonly satisfied for practical sensor-actuator locations, except for some special cases, e.g. when all sensors (actuators) are placed at the nodes of a bending mode. Obviously, such a mode will not be observable (controllable) in this special case. Higher order bending modes are usually damped well enough, so that even such a special case will cause no problems in practice. Apart from such special cases, we may therefore take observability and controllabilty for granted.

In case of sensor-bearing collocation, the equivalence $B^T = C_s = C_v$ is satisfied. The stability conditions can then be met simply by selecting **positive definite** feedback matrices Gs and Gv (or semi-definite with additional conditions, see /1/, /2/ and /3/ for details).

3. Realizing the Control with the Digital Signal Processor

3.1 Why Digital Control ?

After some years of practical experience with analog and digital magnetic bearing control, the often mentionned tremendous advantages of digital control can be fully appreciated.

These advantages include the great **flexibility** at every level of an experiment, from the hardware layout to experimenting with different controllers to measurement and to the implementation of additional features in industrial equipement.

As an example for the simplifying of hardware layout, we do not use a constant current source and differential coils for linearizing the magnetic actuator characteristics, this task being performed by the controller software. Any additional function like sensor calibration, soft lifting of a rotor at start-up, position control, safety control, adaptive feedback etc. is performed by modifying only the controller program and not the hardware.

3.2 Design of the Feedback Matrices: Selecting Stiffness and Damping

A further simplification of control design may be applied for most magnetic bearing systems, especially so if collocation of sensors and bearings is realized. It is the **decoupling** /9/ of the bearings from each other. This decentralization results in diagonal feedback matrices Gs and Gv and greatly reduces the complexity of the controller thus decomposing into local **single channel** controllers.

The present rotor-bearing system features sensor-actuator collocation. Therefore, decentralized single-channel control is applied for each bearing and each radial direction. The stability conditions stated in section 2 are now automatically reached by the most simple requirement of selecting positive scalars as stiffness and damping coefficients.

The next question is how to select **stiffness** and **damping** of the bearings. The first three points are concerned with the upper limit:

1) A large stiffness tends to "pull" the nodes inside the bearings, rendering the control of bending vibrations difficult if not impossible.

2) Saturation of the actuators will restrict the range of linear operation

3) The system frequencies are increased by high bearing stiffness. The sampling rate of a digital controller must remain safely above at least the double of the highest frequency to be controlled.

4) The magnetic force increases with the inverse of the square of the air gap. Linearizing then yields a negative bearing stiffness coefficient k_s /10/. This coefficient must be compensated by the controller. For practical reasons, the minimum positive bearing stiffness which can be superposed to this compensation should be of the order of magnitude of k_s /9/.

In the next step, the **damping** value for each bearing can be chosen according to quite simple and classical methods. (near critical damping for the lowest frequency modes). Real poles (over critical damping) are usually better avoided as they result in very high damping coefficients susceptible to saturate the actuators and act as noise amplifiers. As an upper limit, the arguments 1) 2) and 3) above are applicable not only for stiffness, but for damping as well.

3.3 Digital Control Algorithm

The basic hardware elements of the digital control loop are shown in figure 2.

The core of the control algorithm consists of multiplying the digitalized position- and velocity signals with the feedback coefficients. This is simple when both signals are measured. In our case, as in most practical applications, only position sensors are used. Therefore, selection of stiffness and damping is only the first part of the control design. The second part, the choice of an appropriate derivation algorithm, is described later.

The rotor position can be made independent of static load force through addition of an integrating feedback path. Up to the saturation-force of the magnetic bearing, a quasi-infinite static stiffness is thus feasable.

A **lower limit** of the sampling frequency is given by the rule of thumb, that it should be roughly equal to five to ten times the highest system frequency to be controlled actively. The question then is, up to which mode can and should be influenced actively?

Inner damping will be sufficient to avoid high frequency vibrations, as long as these higher modes are not excited. The upper frequency limit depends not so critically on the sampling time, which can be made short enough with modern microprocessors. The most crucial factor here seems to be the time lag. It can cause excitation of higher modes in the following manner:

Time Lag

The conversion time of the analog-digital and digital-analog converters plus a part of the computation time all contribute to a time-lag between input and output signal. The actuator system (power amplifier & bearing coil) acts as a low-pass filter with corresponding phase lags which reinforces the effect of the time lag.

Damping always means **phase lead** of force versus position input signals. The time-lag translates into a phase lag increasing proportionally to frequency. As soon as the over-all phase changes from positive to negative, the controller excites vibrations rather than damping them.

If the time lag is large, special care must be taken not to excite vibrations in higher modes. In our earlier experiments, a carefully designed analog filter of 2nd order was necessary. The DSP made this filter obsolete thanks to smaller time lag.

Obtaining a Velocity Signal:

If, as in our case, the velocities are to be derived from displacement signals, some additional calculations are required. The first choice is between two basically different approaches, namely reconstructing the velocities from a Luenberger observer (i.e. a dynamic model of the rotor system, extensively treated in control theory) or using some discrete time derivation algorithm.

Observer:

It has been shown to give very good results for rotors allowing modelling as rigid bodies /9/ . In the case of strong elastic behaviour however, this would lead to relatively high order models. Even powerful processors are not yet able to do all the necessary calculations in the short sampling periods necessary. Our work up to now indicates, that observers based on reasonably reduced models seem to be parameter sensitive and perform poorly. Further work is continuing in this area. Transputers could possibly help solving this problem in the near future.

Derivating velocity from position:

The most direct method, using the difference sequence of the input signals, already gives good results for many applications. The velocity signal at a time kT_S (k being an integer) is approximated by

$$y = \{ y[kT_S] - y[(k-1)T_S] \} / T_S \qquad (4)$$

The resulting controller is of first order as it can be realized with a single time-lag element per channel.

In the case of rigid-bodies and restricting oneself to single channel systems, it is even possible to demonstrate an equivalence of such an algorithm and the reduced-order observer.

The simple formula (4) illustrates why the sampling period can also be selected too short: The rate dependent damping force must be calculated for the lowest as well as the highest eigenfrequencies to be controlled in the closed loop system. If the sampling period is very short and the mode frequency low, the difference of two nearly equal values is computed in formula (4) and multiplied by the large value $1/T_S$. The input signals $y(kT_S)$ are converted to digital values of limited precision, therefore small values of T_S quickly lead to intolerable noise in the system.

Higher order differentiating algorithms can be of advantage. The second order expression (parabola approximation)

$$y = \{3\, y[kT_S] - 4\, y[(k-1)T_S] + y[(k-2)T_S]\} / 2T_S$$

has been used successfully for the experiments described below. From the **signal-processing** point of view, this control represents a finite-impulse-response algorithm (FIR). A Luenberger observer as mentioned above would lead to an infinite-impulse-response algorithm (IIR). Classical algorithms, as found e.g. in "Applied Analysis" by C. Lanczos /12/ could also be applied.

4. Experimental Results & Conclusion

4.1 The experimental Rotor System

A photograph of the rotor (length 1 m , mass ca. 7 kg, shaft thickness 20 mm) is shown in figure 3, while figure 4 shows the complete rotor-bearing assembly. The bearing parameters are: Force-current factor of 60 N/A, due to the relatively large air-gap of 1mm, negative bearing-stiffness coeff. ks of 120 N/mm, pre-magnetizing current of 1 A.

The free-free **bending modes** of the rotor are shown in fig. 5 along with the corresponding eigenfrequencies.

Figure 6 shows how the **eigenfrequencies** are influenced by the bearing stiffness. For the following experiment, the **bearing stiffness** has been set at 50 N/mm and the damping at 500 Ns/m. Higher values are avoided for this elastic rotor, as the bending nodes would be drawn too close to the bearings. An advantage of these low values is the fact that only very small vibrational forces are transmitted to the foundation.

4.2 The Control System:

A single **digital signal processor** (DSP, model TMS 32020) is used for all four channels. It features a 32-bit accumulator and an integer multiplier with 16-bit operands performing an operation in the processor cycle time of 200 nsec.

The core of the control algorithm is programmed in assembler, while the more extensive service program running on a host personal computer is written in Pascal.

The analog-to-digital **converters** transform the sensor signals to 12-bit numbers in a conversion time of 20 μsec.

The **sampling frequency** for satisfactory operation can be selected on-line in the range from 2.5 kHz to 12 kHz. For the experiments described here, the sampling frequency of 2.7 kHz was sufficient.

The **delay time** between input and output of one channel, determined by the A/D and D/A conversion time and the calculation time for one channel is about 40 microseconds. It is this low value, which is responsible for the good performance of the controller.

In the earlier implementation, with two general purpose microprocessors, the sampling period was roughly the same (400 microseconds) than for the DSP. The delay time was however much longer (250 microseconds). At the eigenfrequency of 514 Hz the phase deterioration thus was already more than 45 degrees. To avoid exciting this and higher bending modes, a carefully designed analog filter was therefore required. This filter is not necessary for the much faster DSP, owing to the shorter time lag. This is believed to be the main reason for the substantial improvement of the control performance.

4.3 Mechanical Characteristics:

A good **damping** is achieved over a wide frequency range, well above the first four **critical speeds**. Therefore a rotational speed of 10000 rpm was reached with the control outlined above. This speed is near the capacity of the motor, the controller seems to allow still faster rotation.

As a measurement result, figure 7 shows the spectra of the position signal measured at a distance of 70 mm inward from one of the two radial bearings. It is a position were there are no bending nodes of the few lowest frequency modes. There is one measurement for every 10 Hz of rotational speed, from 10 Hz to 170 Hz (10200 rpm). The peak amplitude was reached at the first two critical speeds, at about 17 and 28 Hz. This peak amplitude is about 0.15 mm or 15% of the air gap in the bearing.

At higher rotational speeds, the rotor spins about the **principal axis** of the discs seen in fig. 3. The spinning discs tend to stabilize the rotation axis. This and the good damping characteristics explain why the rotor operates at the 3rd and 4th critical speed (at 47 Hz resp. 140 Hz) with vibration amplitudes below 0.1 mm.

The vibration amplitude in the bearings remains practically constant for the whole speed range above 30 Hz. This vibration is due to the unbalance of the rotor, i.e. misalignment of mass-geometrical and magnetical axis which in our case is more than 0.1 mm in the bearings. The rather low bearing stiffness of 50 N/mm explains, why the unbalance-vibration is not reduced to smaller amplitudes.

5. Conclusion

Magnetic bearings seem to be well suited for high rotational speeds not only for rigid rotors, but especially also for highly flexible rotors. **Very good damping characteristics** have been obtained for such a flexible rotor. It has easily been operated at four bending critical speeds and up to 10000 rpm, at the limit of motor capacity.

The rotor is allowed to move freely within the bearing clearance of in the order of tenths of a milimeter. This is by far enough to allow for the usual manufacturing tolerances like non-alignment of geometrical and inertial axis. The rotor used for this experiment has never been balanced. It even has a quite noticeable bent of its axis.

The rotation takes place about the principal axis. The geometric eccentricity will only generate a small bearing force proportional to the bearing stiffness. This small residual vibrational force transmitted to the foundation could even be eliminated completely by some appropriate feature of the controller /11/.

The digital **signal processor** system seems very well suited for this control task. Its high speed makes it possible to realize a **short time lag**. This is of crucial importance for the good damping characteristics.

Note that the complete control-layout procedure never uses the actual rotor model. The only parameters flowing directly into the control layout are the bearing parameters stiffness k_s and force-current factor k_i . A more accurate model description is needed only for simulation purposes.

It seems that the control for a wide range of practical magnetic bearing systems may be designed along the simple outline given in this paper.

References

/1/ J. Salm, G. Schweitzer: Modelling and Control of a Flexible Rotor with Magnetic Bearings. 3rd Conf. on Vibr. in Rot. Mach., IMechE, York, 1984

/2/ J. Salm: Active Electromagnetic Suspension of an Elastic Rotor: Modelling, Control and Experimental Results. 11th biennal conf. on Mech. Vibr. and Noise, ASME, Boston 1987

/3/ J. Salm: Eine aktive Magnetische Lagerung eines elastischen Rotors als Beispiel ordnungsreduzierter Regelung grosser elastischer Systeme. Dissertation Nr. 8465, Swiss Federal Inst. of Technology, 1987

/4/ C.D. Bradfield, J.B. Roberts, R. Karunendiran: Performance of an Electromagnetic Bearing for the Vibration Control of a Supercritical Shaft. 4th workshop on rotordynamic instability problems, Texas University, 1986

/5/ R. Nikolajsen, R. Holmes, V. Gondhalekar: Investigation of an Electromagnetic Damper for Vibration Control of a Transmission Shaft. Inst. Mech. Eng., vol. 193, Nr. 31, 1979

/6/ H. Ulbrich: Dynamik und Regelung von Rotorsystemen. VDI-Z Fortschritt Berichte, Reihe 11, Nr. 86, 1986

/7/ O. Matsushita, M. Takagi, N., Tsumaki, M. Yoneyama, T. Sugaya, H. Bleuler: Flexible Rotor Vibration Analysis Combined with Active Magnetic Bearing Control. Int. conf. on Rotordynamics, JSME and IFToMM, Tokyo, 1986

/8/ P.E. Allaire, R.R. Humphris, R.D. Kelm: Dynamics of a Flexible Rotor in Magnetic Bearings. 4th workshop on rotordynamic instability problems, Texas University, 1986

/9/ H. Bleuler: Decentralized Control of Magnetic Rotor Bearing Systems. Dissertation Nr. 7573, Swiss Federal Institute of Technology, 1984

/10/ G. Schweitzer: Magnetic Bearings. CISM lecture course on rotordynamics, Udine, 1985

/11/ W.D. Pietruszka, N. Wagner: Aktive Beeinflussung des Schwingungsverhaltens eines magnetisch gelagerten Rotors. VDI Fortschrittsberichte 456, 1982

/12/ C. Lanczos: "Applied Analysis", Pitman & Sons Ltd., London 1957

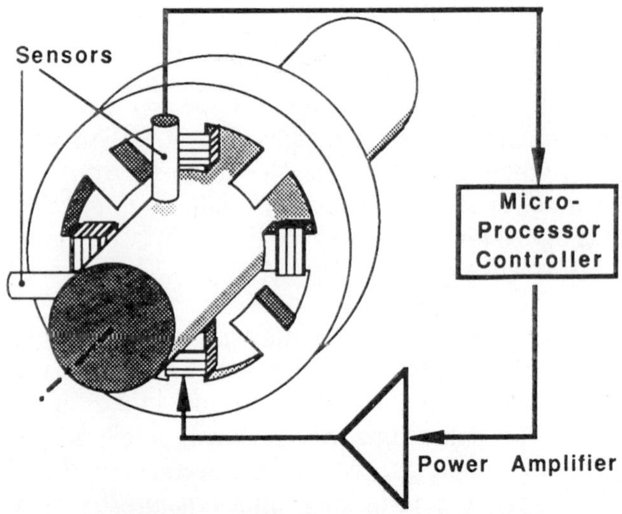

Fig 1 Basic elements of a contact-free electromagnetic bearing control loop. The radial bearing shown supports without contact the rotor in two degrees of freedom using attractive forces only; it has four coils and two control channels, one for each radial direction

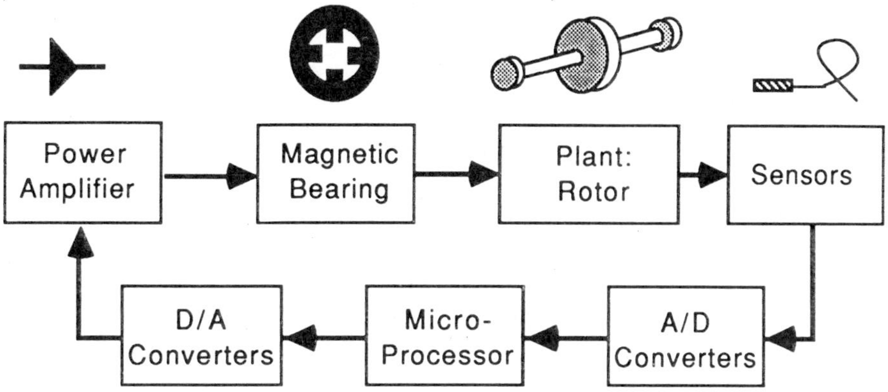

Fig 2 Hardware elements of a magnetic bearing control loop with digital control

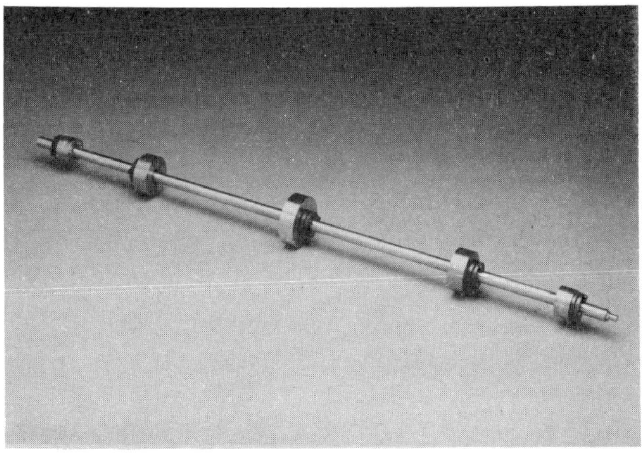

Fig 3 The rotor; length 1 m, mass ≈ 7 kg, shaft thickness 20 mm; the magnetic bearings are placed at both ends of the shaft

Fig 4 The rotor-bearing system. The clearance in the magnetic bearings is about 1 mm; the power amplifiers are seen below the table, the monitor screen for the microprocessor controller is also visible

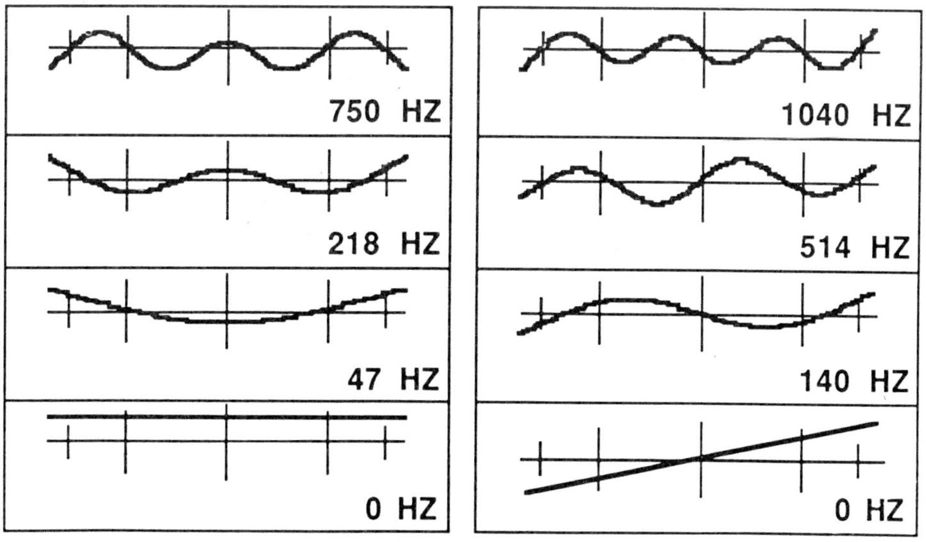

Fig 5 Free–free bending modes of the rotor along with the corresponding eigenfrequencies

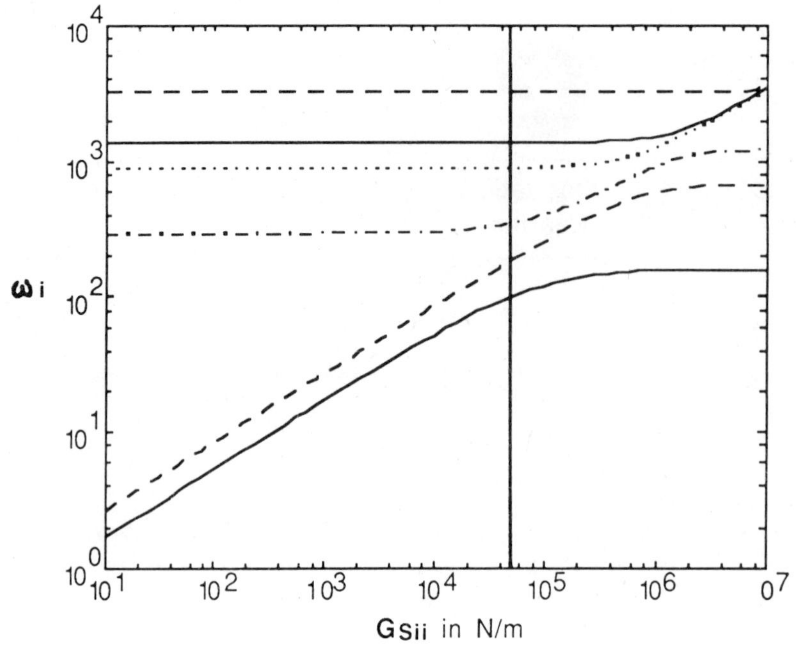

Fig 6 Influence of the bearing stiffness on the eigenfrequencies; for the experiment, the bearing stiffness has been set at 50 N/mm and the damping at 500 Ns/m

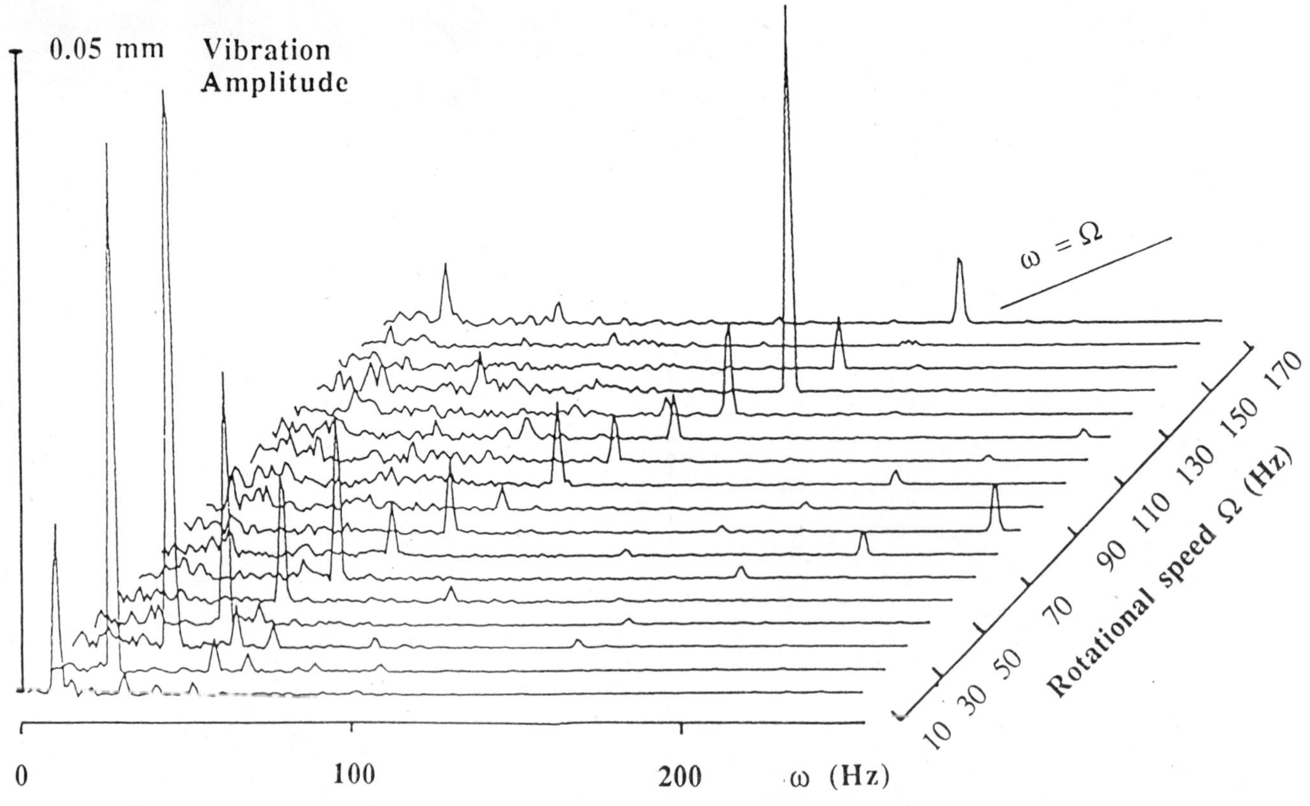

Fig 7 Measurement of the frequency spectrum of the vibration (position signal) at 7 cm from the bearing on the far side of the motor.
There is one measurement for every 10 Hz of rotational speed; the critical speeds are clearly seen on the $\omega = \Omega$ axis at 16 Hz, 28 Hz, 50 Hz and 140 Hz. The curves do not show actual peak amplitudes, as they are a spectrum averaged over 10 measurements; the actual peak amplitudes are however only slightly above the values seen in this representation

C313/88

Microprocessor-controlled automatic balancing systems

H KALISZER, DSc, MSc, Dipl-Ing, CEng, MIMechE, FIProdE, MSME, **A HOLNICKI**, MSc, Dipl-Ing, MSIMP and **M ORLOWSKI**, MSIMP
Department of Mechanical Engineering, University of Birmingham, Birmingham

SYNOPSIS

The paper discusses the types of grinding operations for which an adaptively controlled fully automatic balancing system is economically advantageous. It analyses the basic requirements of such a system in terms of reliability and accuracy.

The paper then describes the design of a fully automatic balancing system controlled by a microcomputer which monitors the vibration level of the grinding machine.

The developed and tested balancing system may be widely used for various types of rotating machinery where quick and automatic correction of rotating masses is essential.

1. INTRODUCTION

Vibration generated during grinding has a considerable effect upon the size and surface geometry formed on the ground components. To minimise the effect of vibration modern grinding machines are massively built with considerable rigidity and specially designed bearings for the wheel assembly. In many cases the machines are mounted on heavy foundations.

In spite of all these precautions, it is nearly impossible to obtain a satisfactory surface texture of the workpiece if the machine spindle together with the grinding wheel is not adequately balanced.

The grinding wheel rotating system usually consists of the following components: the wheel with flanges, the wheel spindle assembly mounted in precision bearings and comprising pulleys and driving belts. For simplicity the grinding machine system shown in Fig 1 represents one degree of freedom. Such simplification is fully adequate to analyse the effect of wheel unbalance in the relative motion between the wheel and work.

Although grinding wheels are manufactured within internationally defined tolerance limits and carefully inspected, they all have an accepted amount of unbalance. In addition the mounting of the wheels in flanges also produces an unbalance. Due to all these factors the resulting unbalance of the wheel assembly may be quite considerable.

As follows from previous investigations [1,2] a total amount of unbalance of a rotating wheel assembly in medium size cylindrical grinders may reach up to 15000 g-mm. (Fig. 2). Such unbalanced wheels are responsible for generating centrifugal forces which are proportional to the square of their rotational speed. This may lead to extensive vibrations and to a periodic loss of contact between the wheel and work. As a result waviness will appear on the work periphery. The number of waves will depend upon the ratio of the wheel and work rotational speeds, whereas the amplitude of such waviness will depend upon the wheel unbalance and wheel rotational speed.

Grinding which is usually the final finishing operation has a dominant effect upon the functional behaviour, reliability and wear of many components which form an essential part of most modern machines.

To reduce or eliminate the detrimental effect of wheel-spindle vibration, the system must be balanced. Although the various parts of the assembly can be balanced once only during the production stages by the machine tool manufacturer, the grinding wheel, by the nature of its work and wear requires periodic balancing.

In order to eliminate the unwanted effect of unbalance, a suitable balancing procedure is required to compensate for the changing mass distribution in the wheel.

2. METHODS OF BALANCING

There are two fundamental methods of balancing [1],

(a) Gravitational (non-rotating) method which is used to provide the information

for the location of the correction while the wheel and balancing arbor are supported on a balancing stand.

(b) Centrifugal (rotating) methods which are used to measure and locate the correction required while the wheel is revolving in its own bearings.

At the present time various kinds of gravitational balancing equipment are known and widely used.

The main disadvantages of this type of equipment are:

(a) The length of time required to complete the balancing operation (up to 50 mins on a Roller Stand)

(b) Low balancing accuracy and therefore substantial residual unbalance.

Many attempts have been made to design balancing devices which enable the wheel to be balanced quickly and precisely.

These types of balancing devices are based on the centrifugal method, where balancing is achieved without removing the wheel from the grinding machine [3].

These can be divided into two main groups:

(a) A group in which the information about the amount and position is determined while the wheel is rotating, but the fixing of the correction mass is determined manually when the wheel is stationary.

(b) A group in which the information of the unbalance and positioning of the correction mass is determined while the wheel is revolving on its spindle at the normal rotational speed.

The purpose of this paper is to describe and analyse an original fully automatic computer controlled balancing device designed and built in this Department [2,7].

All balancing devices operating in the centrifugal mode (type b) must be equipped with special mechanical, electrical or hydraulic systems to operate the correction masses in a predetermined manner.

There are several ways to achieve the movement of masses as shown in Fig. 3. [8]. Methods (a) and (c) use two compensating masses. During balancing operations, the masses are displaced until the resulting balancing moment from both masses is equal and opposite to the existing unbalance. Methods (b) and (d) use a single compensating mass. In method (b) the correction procedure is achieved in a polar mode. While in method (d) the movement of the compensating mass follows a spiral locus which is rather difficult to control.

The balancing system described in this paper is based on the principle shown on Fig. 3bdue to its simplicity and ease of controlling the movement of the single compensating mass.

3. THE BALANCING SYSTEM

3.1 Basic requirements of an effective balancing system.

An efficient balancing system should incorporate the following features:

1. A fully automatic balancing procedure, independent of the action of the operator, with an option for semi-automatic operation (manual control).

2. Adaptability of the balancing system to fit, with minimum design modifications, a wide range of grinding machines and with different wheel sizes.

3. The balancing unit should be compact and self-contained allowing to be mounted or removed with ease from the wheel spindle assembly.

4. The balancing capacity should be sufficient to compensate for the maximum unbalance that may be present on the wheel assembly.

5. Control unit should be universal, easy attachable and adjustable to any grinding machine.

3.2 Description of the balancing system

The balancing system developed in the department [2,3,6,7,8] consists of three basic units: balancing head, vibration sensor and control system.

1. The balancing head contains the compensating (balancing) mass which can move circumferentially or radially by means of two DC motors.

   A perspective view of the balancing head is shown on Fig. 5 [7].

   As can be seen, the balancing head contains the driving gears, the compensating mass, the pinion gears and the DC motors. The hollow shaft of the unit accommodates the slip-rings which pass the control signals from the microcomputer unit to the motors.

   The amplitude gear (1) has an eccentric groove to allow the compensating mass to slide radially. The phase gear (2) has two radial grooves so that the mass moves circumferentially when both gears rotate simultaneously.

   As can be seen from Fig 5 the correction mass is sandwiched between the two spur gears 1 and 2. If the two gears rotate at the same speed and the same direction relative to the casing, the angular position (phase angle) of the mass changes but the magnitude of correction will remain constant. If however the amplitude gear rotates and the phase angle remains stationary the correction mass will move radially. As a result the magnitude of correction will change but the phase angle will remain constant.

By considering all possible sources of unbalance in the grinding wheel assembly [2] and complying with BS ISO/DIS, 6103 standard [4] it was found that the maximum required balancing capacity for medium size grinders should be around 20000 g-mm.

To optimise the size and weight of the compensating mass of any required capacity a special computer program has been developed [7].

Besides the balancing head described above, a complete system consists of the following additional elements.(Fig.4)

2. A vibration velocity sensor mounted on a grinding wheel head for the measurement of revolution vibratory forces due to the unbalance of the wheel.

3. An input signal interface unit that contains low pass filter of the vibration signal, the signal amplifier and A/D converter, as well as the interfacing circuits of the control signals. There are also two DIP switches mounted on the interface that are used: one for the setting of the unbalance threshold and another for adjusting the filter cut-off frequency.

4. An output signal interface that contains two bi-directional DC motor drive circuits and drivers for process status indicators.

5. An operator board with a system status indicator and switch that activates the balancing procedure.

6. An attachable service board for initial adjustment and calibration to suit particular machine requirements, maintenance and troubleshooting of the system.

7. A Z-80 based mircrocomputer control unit that calculates the unbalance level and controls in the close loop the balancing action.

In the operational mode, the signal from the wheelhead vibration after being picked-up by the vibration sensor is filtered to select the relevant signal component of the frequency which corresponds with the grinding wheel rotations speed. The filtered signal is gained, digitized and its average amplitude is computed. The amplitude is then used as an indication of the wheel unbalance. When the unbalance exceeds an acceptable limit an indication lamp on the operator board signals the need to start the balancing action.

4. COMPUTER AIDED BALANCING PROCEDURE

As mentioned previously the correction procedure is achieved by applying a selected locus method (Fig. 3b). Such a method is also best suited for an automatic control. It works by correcting the unbalance in two separate actions i.e. varying the angular and radial positions of the compensating mass.

In case of repeated balancing of a previously well balanced wheel (for example after the wheel dressing operation) the compensating mass is moved from the actual to the final position without passing through its optimum central position.

To achieve the above, an optimising algorithm (Fig. 6) known as "hill-climbing" method has been adopted. The microcomputer collects a sample of the grinding machine vibration and computes the corresponding unbalance level. The compensating mass is then moved one step in the radial direction and a new unbalance value is computed. If it is smaller than that of the preceding sample, it is an indication that the balancing operation is carried out in the right direction and therefore compensating mass is shifted in the same direction until a further sample is examined. If the opposite situation takes place, the direction of operation of the mass is reversed.

If as a result of change in direction of the movement of the compensating mass the unbalance magnitude is not reduced any further, a similar action of a step motion of the angular (phase) drive commences.

When as a result of the balancing action, the resulting unbalance is reduced to a smaller value, the magnitude of the consequentive steps controlling the motion of the compensating mass is automatically readjusted to a smaller value. In this way it is possible to achieve a more precise movement of the compensating mass.

Finally, when the magnitude of the signal amplitude reaches a value below the predetermined unbalance limit the required state of balance is fully achieved. Such conditions occur when the position of the compensating mass is equivalent to a wheel assembly unbalance of approximately 150 g-mm, which corresponds to an unbalance mass of approximately 0.5g on the wheel periphery. This is the minimum system response which corresponds to the threshold of the sensitivity of the vibration sensor.

The balancing procedure is fully controlled by a specially developed software for a Z-80 based dedicated microcomputer module.

Such balancing procedure is always initiated whenever the computer check, shows that the amplitude of computed vibration is above the selected upper level of the wheel unbalance. For the system discussed in this paper the upper limit corresponds to a wheel unbalance of 300 g-mm.

It will be noted from Fig. 2 that the limit of human perception corresponds to an amplitude of vibration which will result from a wheel unbalance of 450 g-mm. According to DIN such limits of human perception is a good indication of smoothness of running.

In addition it can also be seen from Fig. 2 that the selected upper level of initiating the balancing procedure and equal to 300 g-mm is much below the required technological limit to achieve a surface texture and waviness for a good quality grinding finish.

For each step of the balancing operation the computer collects a sample of the wheelhead vibration signal and computes the corresponding amplitude of such a signal, to determine the magnitude of wheel unbalance component. The total time of data acquisition and computation is approximately 8 seconds.

5. ADJUSTMENT AND CALIBRATION OF THE BALANCING SYSTEM

In the case of mounting the balancing system on a new machine which is in a state of rest it is necessary to carry out a full adjustment procedure by connecting the Service Board (Fig. 7).

Firstly, by using one of the two DIP switches described earlier it is necessary to adjust the filler cut-off frequency in accordance with the rotational speed of the grinding wheel assembly.

In the next step it is necessary to control manually the radial movement of the compensating mass until the board indicator light has shown that the mass has reached its central position. In this position the centre of gravity of the mass and the geometrical centre of the balancing system will coincide.

To calibrate the system it is necessary to move manually the compensating mass into a position which corresponds to an acceptable minimum unbalance. This can be achieved by the switches on the Service Board. The manually set level of unbalance is then considered as a threshold for subsequent automatic action. To check the acceptable level of unbalance magnitude, any known measuring system can be used. The threshold setting will be indicated on the five bit binary display unit on the service board by pressing the "read" switch on the service board. To fix such desired balancing conditions for the wheel spindle assembly it is necessary to set the number on the DIP switch according to the displayed number.

The disconnection of the service board after calibration procedure automatically switches on the automatic mode of the balancing system and activates the Operation Board which contains a Balancing Status indicating lamp and a "start" switch.

The balancing mode of the Operating Board is shown in Fig. 8.

6. CONCLUSION

The fully automatic version of the balancing action is carried out while the grinding machine is in normal use. The Computer Operated Balancing System constantly monitors the unbalance value and gives the information when the balancing process is to be started.

The balancing head is compact, self-contained and can be easily mounted on or removed from the grinding wheel assembly.

By considering all possible sources of unbalance in the grinding wheel assembly and complying with BS and ISO/DIS 6103 standards it was found that the maximum required capacity for the Jones-Shipman grinder should be 20000 g-mm. This maximum capacity takes into account not only the maximum unbalance of the grinding wheel but also any additional unbalance in the rotating wheel spindle assembly. The selected capacity of 20000 g-mm is also appropriate for other medium size grinders allowing therefore for more flexibility.

By considering the computation time of around 8 seconds, the balancing of a newly mounted wheel may take up to 1.5 - 2 minutes, while the average balancing time for the wheel in use would normally not exceed 30 seconds.

The unit may be used independently or as a module of a multimicroprocessor control system adapted for the grinding process and described in detail elsewhere. [9]

ACKNOWLEDGEMENTS

The authors wish to thank Sigma Company for the financial support and for taking over the manufacture and distribution of the balancer.

Thanks are also due to Mr Abdulwahab and Mr Bird for their significant contribution in designing the Balancing Head.

REFERENCES

1. H. Kaliszer, Accuracy of balancing grinding wheels by using gravitational and centrifugal methods. Proc. 4th Int. MTDR Conf. 1963.

2. G. Trmal and H. Kaliszer, Adaptively controlled fully automatic balancing system. Proc. 17th Int MTDR Conf. 1976.

3. H. Kaliszer, Automatic balancing of grinding wheels, IMechE., Conf. Cambridge, 1980.

4. British Standards Inst. Draft Int.l Standard ISO/DIS 6103, Bonded Abrasive Products, Balancing of Grinding Wheels. General on Unbalance Tolerances Doc. 79/71092DC, 1979.

5. British Standard Inst. BS3851:1976 Glossary: Terms used in the mechanical balancing of rotating machinery.

6. S. Spiewak, H. Kaliszer and M. Kuchta, Computer controlled automatic balancing. SME Int. Grinding Conf. 1984 USA.

7. H. Kaliszer and M. Abdulwahab, Adaptively controlled fully automatic balancing system. JSME Tokyo 1986.

8. S. Hayes and H. Kaliszer, A new method for centrifugal balancing of rotors revolving in their own bearings. Proc. Int MTDR Conf 1964.

9. S. Spiewak and H. Kaliszer, Multimicroprocessor control for plunge grinding. Proc. Int. MTDR Conf 1983.

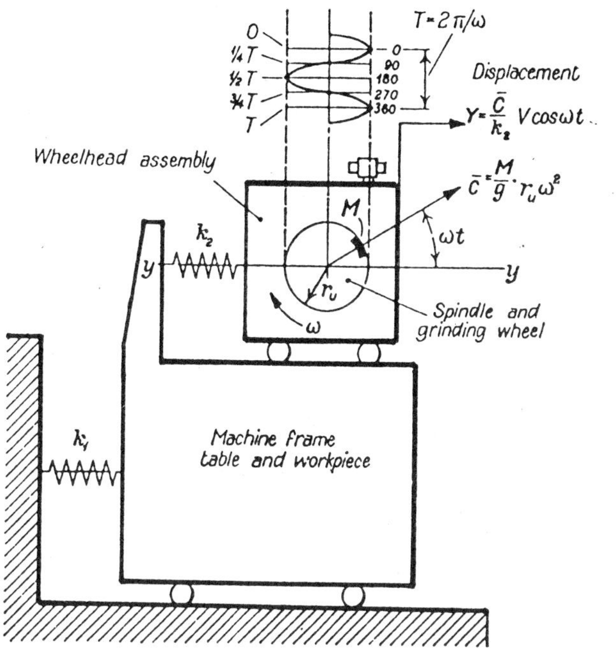

Fig 1 Dynamic system of a grinding machine

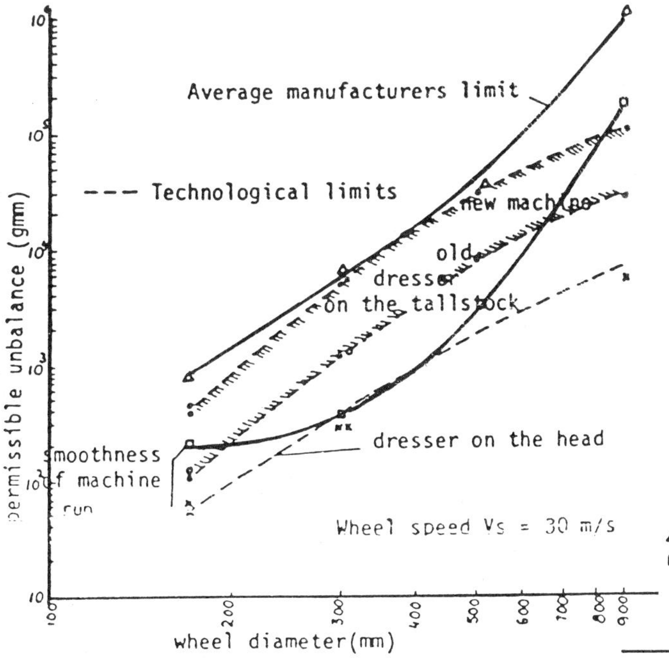

Fig 2 Permissible wheel unbalance as a function of a wheel diameter

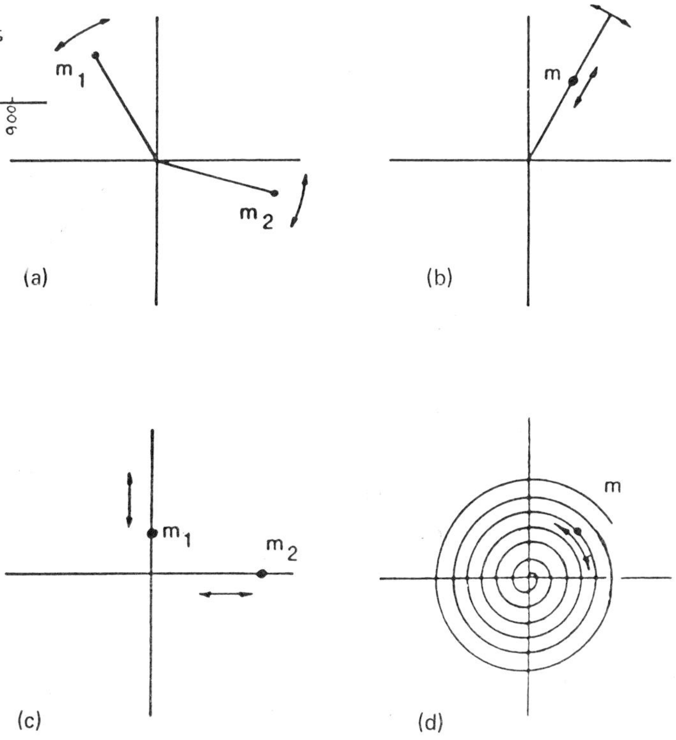

Fig 3 Common balancing methods

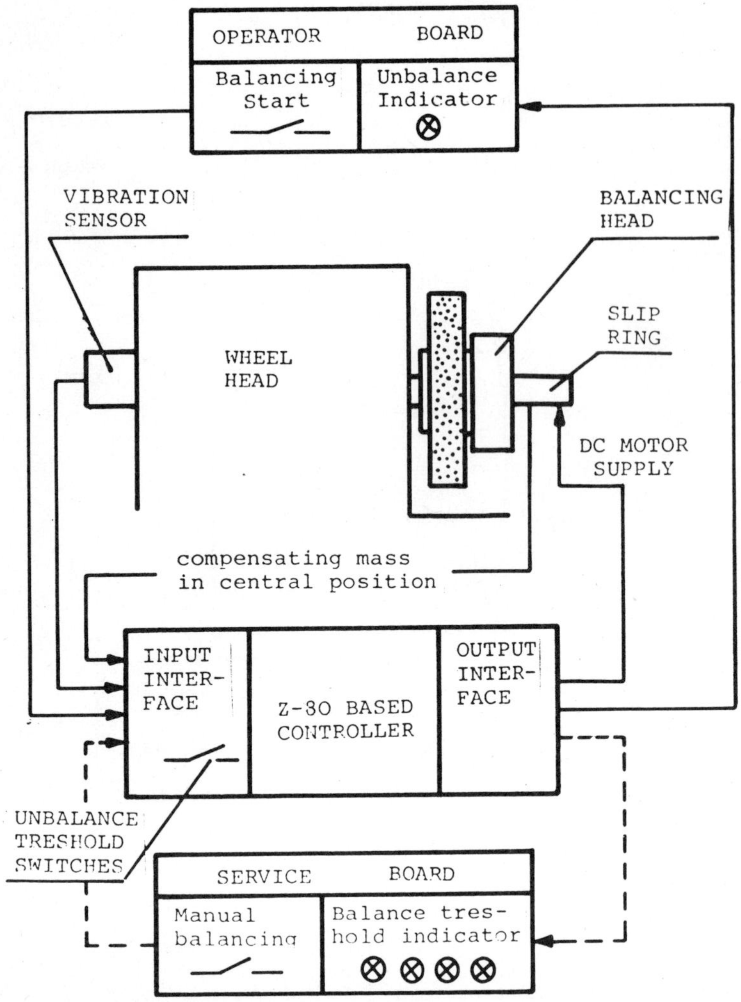

Fig 4 Block diagram of microcomputer-controlled balancing system

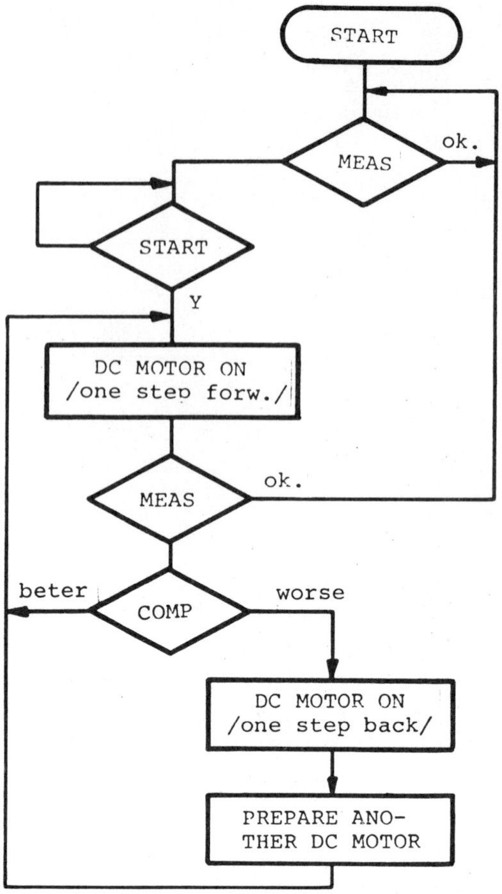

Fig 6 Algorithm showing the full automatic cycle

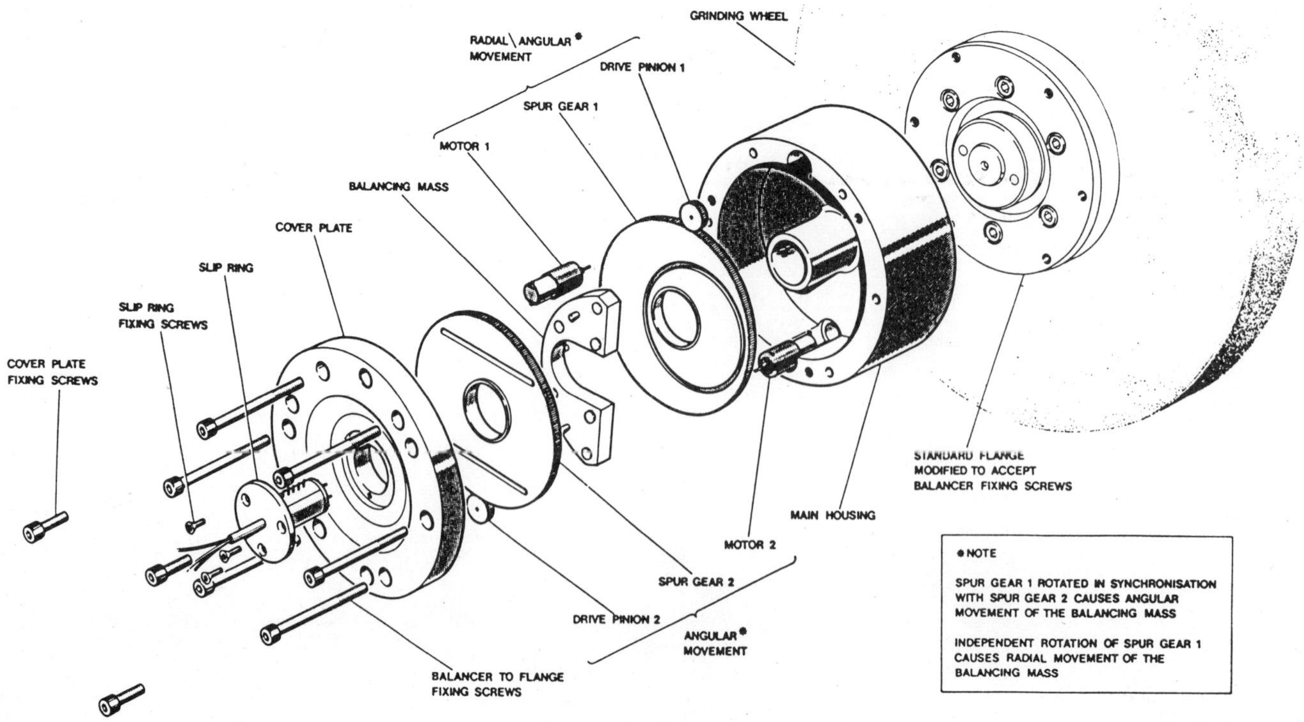

Fig 5 Balancing head

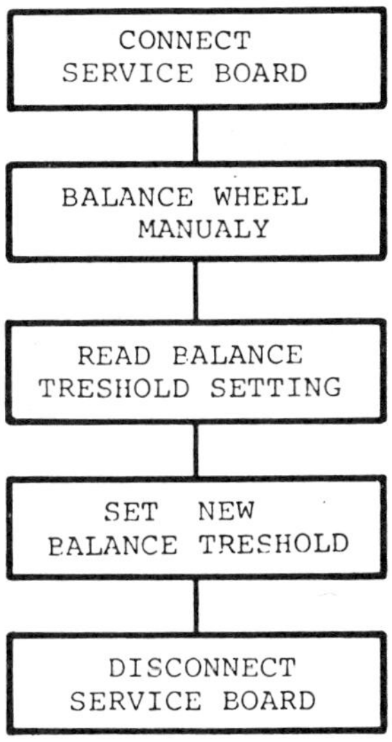

Fig 7 Algorithm showing calibration procedure

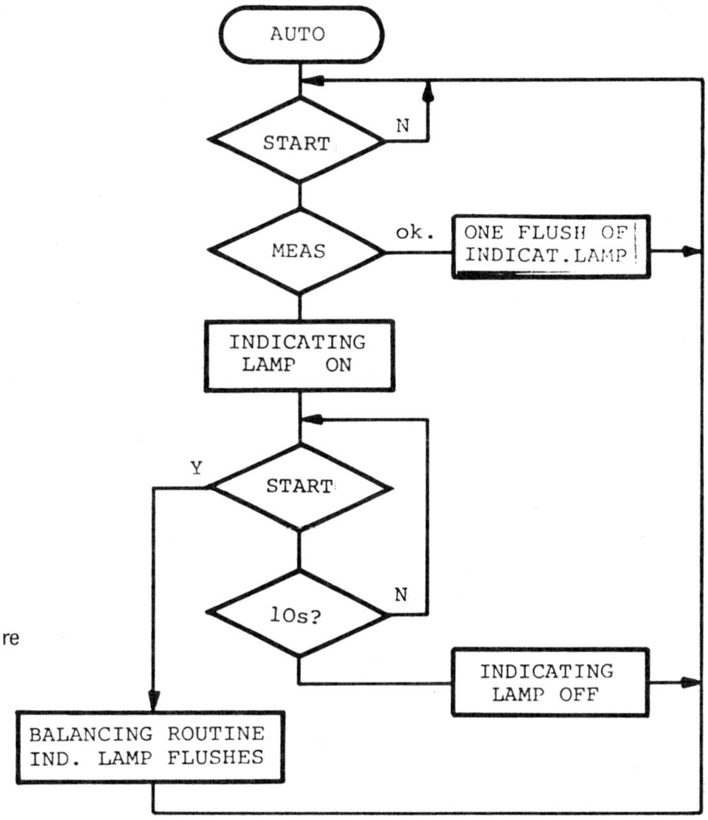

Fig 8 Algorithm showing the sequence of operations

C312/88

Structural loads due to surge in an axial compressor

N T BIRCH, BSc, PhD, **J B BROWNELL**, BTech, **A M CARGILL**, MA, PhD, **M R LAWSON**, BA, MSc, **R J PARKER**, BSc, ARCS and **K G TILLEN**
Rolls-Royce plc, Derby

SYNOPSIS This paper describes a model for predicting the large impulsive loads that occur during surge in an axial flow compressor. The model combines a computational fluid dynamics method for the unsteady aerodynamic blade - shock wave interaction with a finite element calculation of the resulting structural response. It is validated by means of a shock tube experiment which involves laser methods of flow visualisation and measurements of blade deflection using both strain gauges and Moiré photography. The results from this experiment are shown to be in good agreement with the computations, confirming the validity of the technique.

1 INTRODUCTION

In the axial compressor of a modern high pressure ratio aero-engine surge is a very violent event, which can cause large structural loads. If these loads are not allowed for at the design stage they can cause excessive blade deflections resulting in the contact of rotor and stator blades, and movement of the casing relative to the rotor resulting in blade rubs. These cause, at best, deterioration in performance and engine life and, at worst, can lead to mechanical failure. This paper describes progress in the prediction of the blade deflections during surge.

The physics of surge has been described in the present context by Mazzawy (1). He shows that surge can result from the stalling of the flow in the high pressure regions of the compressor. This stall spreads from its initial site around the whole of the annulus of the compressor blocking the flow so that it can no longer support the pressure rise of the compressor. The upshot is that a shock wave is formed which propagates forward to the engine inlet and quickly reduces the flow rate through the machine to slightly less than zero. This rapid reduction in flow is followed by a longer blow down phase in which the pressure is reduced to its zero flow value. Thereafter, the compressor begins to re-pressurise itself again. This occurs in two parts - an initial rapid part where the flow rate is re-established but the pressure rise remains low, and a slower part where the compressor becomes fully re-pressurised. If the flow then stalls, the whole process can then begin again.

It is during the initial phase that the large loads are produced. The asymmetry of the flow can cause large lateral forces on the casing of the compressor and the passage of the shock can result in large impulsive loads on the blades which give rise to large deflections. Existing prediction methods for these loads are mostly proprietary and not described in the open literature. The procedures described by Mazzawy (1) and Rudy (2) are believed to be typical, however. Mazzawy idealises the loading on the blade as being like that due to a shock wave passing through a constriction. He obtains the initial strength of the shock wave from engine measurements and then works out the over-pressure at the blades. Rudy uses this method to obtain the peak value of the loading on the blades and combines it with calculations by Favrat and Suter (3) of cascade-shock wave interactions, to estimate the pressure time history and hence the total impulse delivered to the blade. The time history of the blade deflection is then calculated by modal decomposition.

While these methods are undoubtedly useful, they probably contain too many simplifications and result in over-conservative designs. The modern numerical methods that are described below are intended to improve on them by properly accounting for physical features of the interaction. The deflection calculation is done in two parts - the calculation of the aerodynamic load due to the shock wave, and the deflection under the applied loads. This ignores aerodynamic forces set up by the motion of the blade, but these are believed to have little impact on the peak deflection produced.

The aerodynamic interaction is calculated by a fully viscous time-marching method which correctly accounts for the complex processes of wave reflection and diffraction that are present in reality but ignored in the simple models. The structural calculation is performed using a finite-element time integration method which calculates the deflection of the blades in real time. Compared with the more conventional modal approach used by Rudy, it is superior in its provision for including plastic deformations of the blade and, despite the small time increments used, is not much more costly than an approach involving the computation of lots of modes.

Before a procedure such as this can be used for design, it must first be validated. Ideally this should be done on an engine but that is impossible here due to the cost of carrying out tests that could result in structural damage and the unavailability of instrumentation that can withstand the high pressures and temperatures while still having a fast enough response for transient measurements. To validate the present model a shock tube has been used, in which a shock is fired at a cascade. The pressures in the blade passage are measured with fast response piezo-electric gauges to compare with the aerodynamic calculation and the blade deflections are measured using high speed Moiré photography and strain gauges. While these measurements are the best that can be done, for the reasons outlined above, it should not be forgotten that their purpose is prediction method validation and not simulation of the engine. In particular, the flow in the engine will be different because there will almost certainly not be pure shock waves as assumed here (an effect nearly impossible to either measure or model!) and because the presence of a mean flow through the cascade in the engine (as opposed to the initially stagnant flow of the shock tube) will result in different flow patterns. These differences can of course be calculated by the methods described below.

2 AERODYNAMIC CALCULATION

In the previous section the surge phenomenon was described and idealised to the situation of a normal shock wave propagating through a cascade. In this section a calculation of this idealisation is described using a Computational Fluid Dynamics (CFD) method. The calculation was performed to simulate, as precisely as possible, the shock tube experiment described below. The method used was the two-dimensional, finite-volume, Navier-Stokes scheme of Norton et al. (4). This scheme solves for the compressible, viscous flow through a cascade of aerofoil sections using an implicit time-marching algorithm. Usually, this method is used to calculate the steady flow through the cascade, however here it was adapted to calculate the unsteady flow where a shock wave propagates backwards through the cascade.

In its normal steady state mode of operation, calculations are performed by specifying the inlet flow angle (α), total pressure (P), and total temperature (T), along with the static pressure (p) at the cascade outlet. The remaining flow variables on the inlet and outlet boundaries are obtained by extrapolation along one-dimensional characteristics, outwards from the flow domain. In the present situation, however, stagnant flow conditions exist throughout the flow domain initially, and at later times, flow behind the propagating shock wave is directed into the domain across what is normally the cascade outlet. To simulate this situation the calculation procedure was adapted as follows :-

a) Initially, stagnant flow conditions are imposed throughout the flow domain at given total pressure and temperature, as shown in figure 1.

b) The flow conditions behind the shock (Mach number, total temperature and flow angle) are imposed at what would normally be the outlet boundary. These conditions are obtained from standard shock tables (e.g. Houghton and Brock (5)). The method was also modified to make this boundary behave as an inlet boundary, rather than an outlet, since flow is now directed into the domain here.

c) The flowfield is integrated forward in time using the method described by Norton et al (4) with appropriate choice of time-step (δt) and smoothing coefficients to ensure both numerical stability and temporal and spatial accuracy.

d) The calculation is terminated when the pressure wave reaches the inlet boundary. Further time integration cannot be performed as the boundary condition here is invalid once the flow conditions first start to change. For this reason, the flow domain is extended a distance of 2 axial chords away from the blade leading edges, to enable quasi-steady flow conditions to be established in the cascade, prior to the pressure wave reaching the flow domain boundary.

Flow calculations were performed for the cascade shock tube experimental conditions described below. The calculations and experiment were performed in parallel and, for this reason, the conditions imposed in the calculation do not agree precisely with those in the experiment. The flow conditions used are given in Table 1. A calculation grid of around 3000 points was used using the complex grid system employed by Norton et al. (4), as shown in Figure 1. A time-step of 0.005μs was used for a total duration of 100μs, the total calculation thus requiring 20 000 time-steps. Results in the form of blade passage pressure time histories are compared with the experiment in section 5, below.

3 SURGE SIMULATION USING A SHOCK TUBE - EXPERIMENTAL STUDY

The experimental simulation of surge in an engine was performed using a shock tube and a cascade of stator blades. This arrangement allowed a backward travelling shock wave to impinge on the cascade of blades and produce the transient aerodynamic loading which excited the vibration of the blades. Optical techniques were used to monitor the transient deflection of the blades and to visualize the shock wave as it traversed the test section.

3.1 <u>The Shock Tube</u>

The shock tube and cascade of blades are shown schematically in Figure 5. The shock tube had three main sections: the driver section, the driven section (containing the cascade) and a dump tank. Each section was separated by a thin diaphragm. The driven section was filled to a predetermined pressure using nitrogen gas. The driver section was then filled with helium gas until the pressure caused the aluminium diaphragm to rupture. This rupture and the rapid expansion of the gas in the driver section into the driven section launched a shock wave down the driven section. The shock wave was followed by a slug of gas at constant pressure and velocity which took several milliseconds to pass through the test section. When the shock reached the end of the driven section it burst through a thin membrane and discharged the pressure into the evacuated dump tank. The Mach number of the shock and the pressure change through the shock were controlled by varying the following parameters (6).

a) The initial pressure in the driven section.

b) The thickness and material of the driver section diaphragm and hence the pressure at which it bursts.

c) The gas constants of the gases in the driven and driver sections (these may be mixtures of gases).

3.2 <u>Typical test conditions</u>

The initial driven section:-

pressure 690kPa: Nitrogen at 288K

Driver Conditions - the initial fill:-

pressure 100kPa: Nitrogen at 288K

Driver Conditions - to diaphragm burst:-

pressure 4000kPa: Helium

Shock velocity at working section 550m/s (Mach 1.6)

Pressure change across shock 1300kPa

3.3 <u>Instrumentation</u>

The shock strength was measured with two fast-response Kistler piezo-electric pressure transducers mounted at two different axial stations up-stream of the cascade. Their spacing was measured accurately so that shock velocity could be calculated from the relative time of arrival of the shock at each probe. Pressure at the blade tips was measured using an array of miniature piezo-electric transducers manufactured in house (Figure 5). Data from these transducers was recorded using Analogic Data 6000 transient event recorders. Each recorder had two channels capable of recording 8192 samples at 100ns time intervals. The data was stored on the 5 1/4" floppy discs. It was later transferred to local micro computers for further analysis and then to the IBM mainframe for comparison with theoretical predictions. The pressure data from the array of blades was used to compare with the computer flow model and to provide pressure-field data for the deformation model.

3.4 Optical Instrumentation

Two optical techniques were employed, a shadowgraph (7) technique enabled visualization of the shock front as it travelled through the test section and a projected grid pattern with high-speed photography was used to measure the movement of the blade tips. Both techniques used an Oxford Lasers, copper vapour laser as a light source and a drum camera for recording the images.

The copper vapour laser produces a train of very short, high power light pulses so that rapid transient events can be recorded with ease. Laser output parameters during these experiments were typically:-

pulse repetition frequency	20 kHz
pulse energy -	0.5 mJ
pulse duration -	20 ns

The drum camera, normally used in streak photography, was used to capture 30-40 images of the shock tube test section. No shuttering of the drum camera was necessary and no other rotating parts required other than the drum which rotated at 30 000 rpm. The laser pulses were short enough to freeze both object and image motion without any discernible blurring. Optical access to the cascade section was provided by an acrylic window placed over the cascade blade tips.

The cascade was illuminated by a collimated beam of light from the laser delivered to the working area by a fibre-optic cable. The illumination was incident upon the cascade from a direction perpendicular to the shock tube axis. The cascade was viewed from the same direction as the illumination using a large aperture beam splitter.

The passage of the shock-front through the cascade test section creates the shadow/caustic pair that identifies the shock front in the photographs. The actual position of the shock front is at the outer border of the shadow i.e. the downstream edge. The speed of the shock meant that it was not possible to run the laser at a high enough pulse repetition frequency to capture a satisfactory number of shock positions in one test. Recording of the shock positions was accomplished, over a number of tests, by advancing and retarding the laser pulses relative to the shock's arrival. The technique produced clear photographs of the shock which could be compared with output from the theoretical flow model. In order to quantify the transient deformation of the cascade blades during and after the passage of the shock front, a projected grid pattern was used. The grid projected from a square wave amplitude grating, was focused onto the blade tips and acted as a static scale against which the blade movement was referenced. The grid pitch was 0.17 mm and the laser pulse separation was 70 µs. The first few frames on each record occurred before the shock arrival and gave the static reference position of the blade tips. Subsequent frames showed the blades response.

Analysis of the fringe patterns shows the movement of the blades at several points along the blade tip over a period of approximately 2 ms. This could be compared with the predicted response of the blades.

4 STRUCTURAL CALCULATION

The purpose of the structural calculation is the estimation of the blade deflections that result from the loads determined by the aerodynamaic calculation. The program used for the structural calculation was DYNA3D (8). This is a nonlinear finite element structural dynamics program. Explicit integration using the using the central difference method is used to solve the dynamic equations and the program uses 8 node brick elements, with one point integration to model solids.

The blade used here was a mid-stage stator from a modern high pressure ratio compressor and was discretised as 3742 brick elements, using four elements through the thickness of the blade as shown in Fig 6. The measured frequency of the first vibration mode of the blade was 3.4KHz and that of the second mode was 4.8KHz. In the shock tube experiment, the free shock velocity due to the overpressure was approximately 550m/s and resulted in a transit time of approximately 40 µs. This meant that the transit time for the shock was a significant part of the modal period.

The calculation was driven by either the pressures calculated by the CFD method only, or by the CFD pressures for the shock phase and the measured pressures from the shock tube for the period after the shock has passed the blade. For a two-dimensional section of the blade, calculated pressure were available for the first 60 µs of the event, at stations corresponding to the element centroids across the chord and at intervals of 4 µs. Measured pressures were available for the eight stations in the blade passage at intervals of 0.5 µs. In using these pressures, the following assumptions were made.

(a) The pressures calculated for the mid-height section held for the whole of the blade span.

(b) The pressures could be interpolated between the 4 µsec intervals of the CFD output.

5 RESULTS

5.1 Aerodynamic Calculation

In this section, the results of the aerodynamic CFD calculation described above are compared with the shock tube experimental pressure data and high-speed photography. For the purposes of comparison, the initial time, t=0, is taken to be the time when the shock wave first reaches the blade trailing edge position.

Figure 2a,b,c shows a comparison of the calculated surface pressure distribution and photographs of the experimental shock position at time t=20 µs, 40 µs, and 55 µs, respectively.

At t=20 µs (Fig 2a), the shock lies near mid-chord and is still nearly plane except for a slight lead and lag on the pressure and suction surfaces. Also evident from the photographs is the shock diffraction wave, which radiates in a circular pattern from the blade trailing edge. The calculation also shows the shock position further advanced on the pressure surface compared to the suction surface. Also, the calculated shock has, at this stage, become smeared out over about eight grid intervals. This is an inevitable consequence of the finite-volume discretisation errors on the calculation grid employed and the artificial smoothing, which is necessary for numerical stability.

At t=40 µs (Fig 3b), the shock on the pressure surface has almost reached the blade leading edge position, whilst lying some 15-20%

axial chord back on the suction side. The calculated pressure distribution shows a marked increase in blade loading at this point in time, and further smearing of the shock - particularly on the suction surface.

At t=55 μs (Fig 2c), the pressure side leg of the shock has left the blade passage and curves around the blade leading edge. The suction side leg lies at the leading edge and the entire shock structure is rather complex. Also evident from the photography is the shedding of a vortex from the leading edge. This vortex was not predicted in the calculation and this is attributed to the smearing of the calculated shock wave as it passes through the passage. The calculated total loading is now reduced and the blade surface pressures only change slowly from those shown in Fig 3c at later times.

Figure 3a,b shows a comparison between the calculated and measured static pressure histories at gauges positioned near the leading and trailing edges. Each plot shows the history from a pressure side and a suction side gauge. Agreement at the trailing edge (Fig 3a) is extremely good. The shock wave reaches the leading edge gauges (Fig 3b) some 30 μs later. Here, although the initial pressure rise is quite well predicted, the experiment indicates that the cross-passage pressure gradient subsequently changes sign - a feature not reproduced by the calculation. This is attributed to smoothing effects in the calculation which will not capture the fine detail of the complex, but weaker, wave interactions in the post-shock flow.

Overall, the calculation appears to give quite close agreement with the experiment, at least whilst the shock is within the passage. This suggests the calculated initial blade loading and impulse can be well estimated from the calculated blade surface pressure histories. At later times, however, the calculation tends towards a steady-state situation, unlike the experiment, where the blades are excited into vibration.

Finally, Figure 4 shows the calculated total blade loading history along with the loading normally experienced by the blade during hot day take-off conditions. The loading reaches a peak nearly ten times the engine conditions after about 40 μs. This corresponds to the point in time when the pressure surface leg of the shock wave just reaches the blade leading edge. After 40 μs, the loading drops to a constant level, which is due to the quasi-steady flow conditions behind the shock.

To conclude, it must be mentioned that further extensive calculations have been performed to simulate the real engine situation, in which a steady-state flow is established prior to the passing of the pressure wave. The presence of a flow (rather than stagnant conditions) leads to some notable differences in the flow pattern.In particular, it leads to a reduction in the shock speed relative to the blades, and a larger lag between the suction and pressure surface shock positions. This leads to generally higher impulses and is accentuated in blades with higher stagger angles, such as rotor blades. The purpose of the experiment here, however, is to provide data to calibrate the calculation method, which can then be applied in real engine situations.

5.2 <u>STRUCTURAL CALCULATION</u>

In this section the results of structural dynamics calculations are compared with measured data obtained from the shock tube test and it is shown how frequency, earthing constraints and pressure history can affect the results. The shock tube experiment displayed qualitatively all the factors involved in the engine case. The resulting leading and trailing edge displacements are shown in Figure 7.

Four analyses were run in all.

In Case 1 the calculated aerodynamic loads were applied to a finite element model consisting of only the aerofoil and fillet which was earthed at the aerofoil platform intersection. The model was loaded with the CFD calculated loads with the assumption that the calculated loads for the final timestep remained constant. The results of the calculation, showing the deflections of the leading edge, midchord and trailing edge tip positions are shown in Figure 8a.

When the results are compared with the measured results the major differences are

a) The calculated blade frequency is higher.

b) The measured mean displacement is higher than the calculated.

c) The ratio between the trailing edge and leading edge displacements is higher in the calculation.

In both measured and calculated results, modes other than the fundamental mode are excited by the shock impulse. In the former case the fundamental mode frequency is lower than that obtained from the frequency analysis of an individually clamped blade. It was noted that the steady state measured pressures (ie. after 100 μs were lower than the final state calculated by CFD.

In Case 2, the same structural model was analysed but this time using the measured flow pressures instead after the final CFD timestep. The results are shown in Figure 8b. This shows a decrease in twist due to the lower steady state pressures and a slight change of phase between the two major modes.

In Cases 3 and 4, the effect of including more of the blade in the model and changing the earthing constraints was analysed using the same pressure loads as case 2. In Case 3, the blade platform was included in the model and the blade earthed at the intersection of the platform base and the platform feet. In Case 4, the blade was earthed at the base of the platform feet.

The results of Case 3 and Case 4 can be seen in figures 8c and 8d respectively. The effect of increasing the scope of the model and changing the earthing constraints is to lower the blade frequencies. In Case 3, the effect was mainly on the fundamental mode. In Case 4, there was a further lowering of the first mode frequency though it was still higher than that measured in the shock tube.

In order to understand the frequency difference, a modal analysis was carried out of the cascade assembly. The mounting was found to be not completely rigid as consistent results for the blade frequencies could not be obtained for different amplitudes. For a rigidly clamped blade, a 4 to 1 trailing edge to leading edge deflection ratio was measured for the fundamental mode with the fundamental frequency corresponded to case 4, while for the cascade mounting this ratio was lower and varied with amplitude.

To investigate the effect of the change of aerodynamic loading on the vibration amplitude, the calculation was broken down. It was found that in Case 1, 75% of the maximum kinetic energy was produced during the shock phase of the blade loading.

6 CONCLUSIONS

1. The CFD method was successfully adapted to enable calculations of the propagating shock wave through a cascade to be performed. Agreement between calculated and measured static pressure histories was generally good, at least during the initial impulse phase.

2. The above structural calculations together with the measurements show that the amplitude of the blade vibration after a high pressure surge is largely determined by the shock pulse. The absolute value of the displacement is dependent also on the following flow and the fixity of the blade mounting. To produce absolute aggreement requires a more complete understanding of the blade mounting constraints.

7 REFERENCES

(1) MAZZAWY,R.S. Surge Induced Loads in Gas Turbines. Trans. ASME Jnl Eng for Power, 1980, 102, 162-168.

(2) RUDY,M.D. Transient Blade Response due to Surge Induces Structural Loads. SAE paper 821438, Society of Automotive Engineers, Aerospace Congress and Exposition, Anaheim, CA, 1982.

(3) FAVRAT,D. & SUTER,P. Interaction of the Rotor Blade Shock Waves in Supersonic Compressors with Upstream Stator Vanes. Trans. ASME Jnl Eng for Power, 1978, 100, 140-147.

(4) NORTON,R.J.G.,THOMPKINS,W.T., & HAIMES,R. Implicit Finite Difference Schemes with Non-Simply Connected Grids - a Novel Approach, AIAA paper 84-0003,1984.

(5) HOUGHTON,E.L. & BROCK,A.E. Tables for the Compressible Flow of Dry Air, Edward Arnold, London,1970.

(6) GAYDON, A.G., HURLE, I.R. The Shock Tube in High-Temperature Chemical Physics. Chapman & Hall, London. 1963.

(7) MERZKIRCH, W. Flow Visualisation. Academic Press, 1974.

(8) HALQUIST,J.O. DYNA3D USERS MANUAL Rev 2 Lawrence Livermore National Laboratory, November 1982, rev April 1984 and March 1986.

ACKNOWLEDGEMENT

The authors wish to express their gratitude to Rolls Royce plc. for their permission to publish the above work.

Parameter	Experiment	Aerodynamic calculation
Pre-shock stagnant flow pressure (P)	690KPa	690KPa
Pre-shock stagnant flow temperature (T)	288K	300K
Post-shock static pressure (P)	1990KPa	2000KPa
Post-shock total pressure (P)	–	3077KPa
Post-shock total temperature (T)	–	476K

Table 1 Shock tube flow conditions

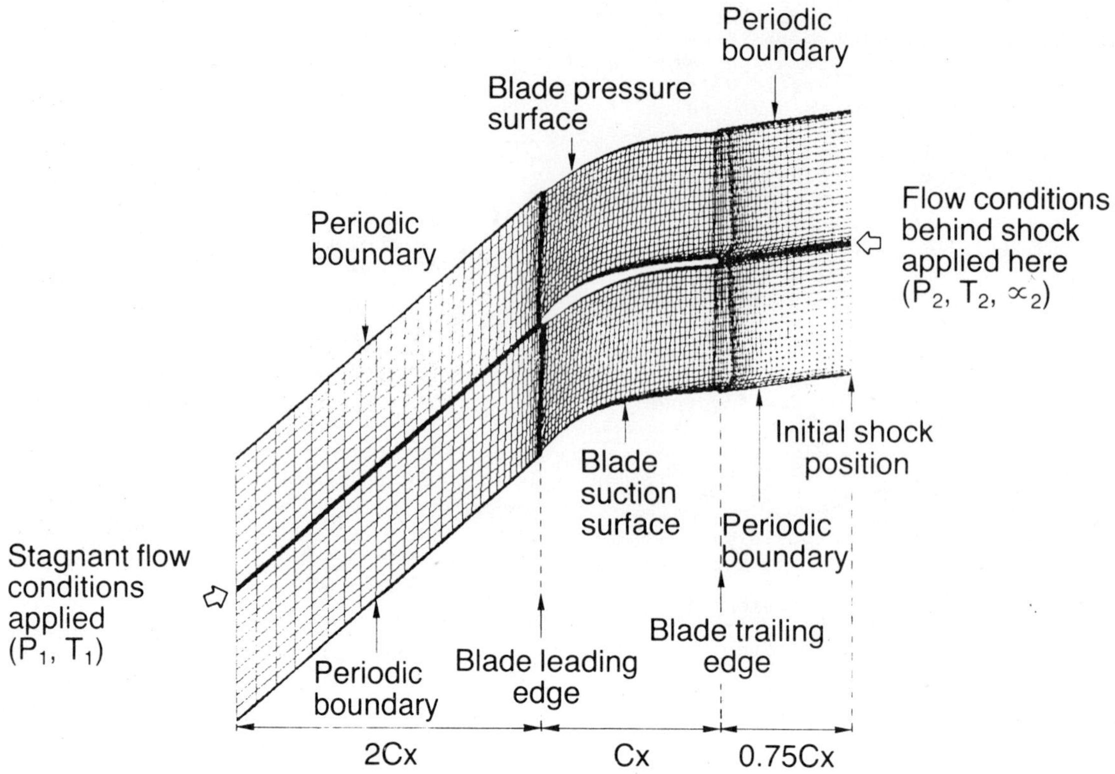

Fig 1 Aerodynamic calculation grid

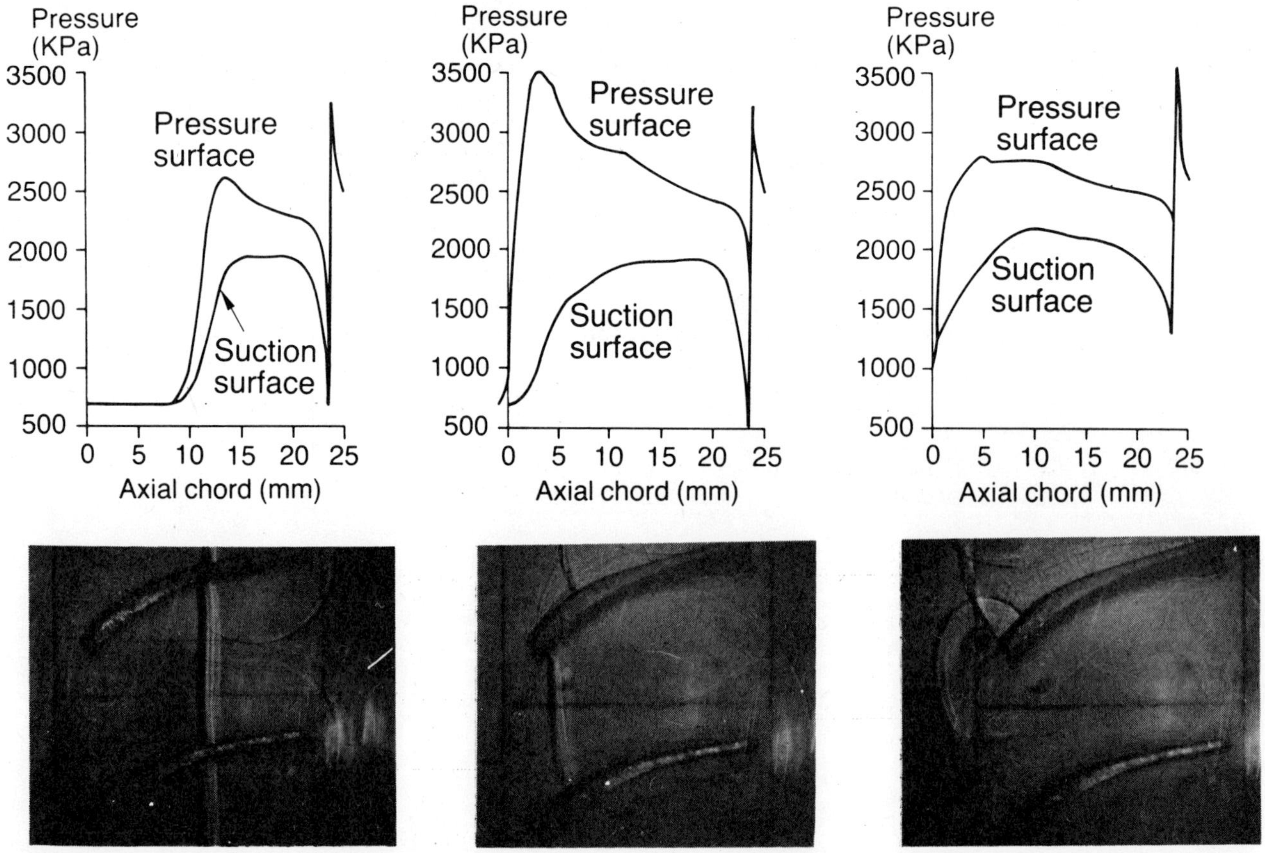

(a) t = 20 μs

(b) t = 40 μs

(c) t = 55 μs

Fig 2 Calculated surface pressure and flow visualization

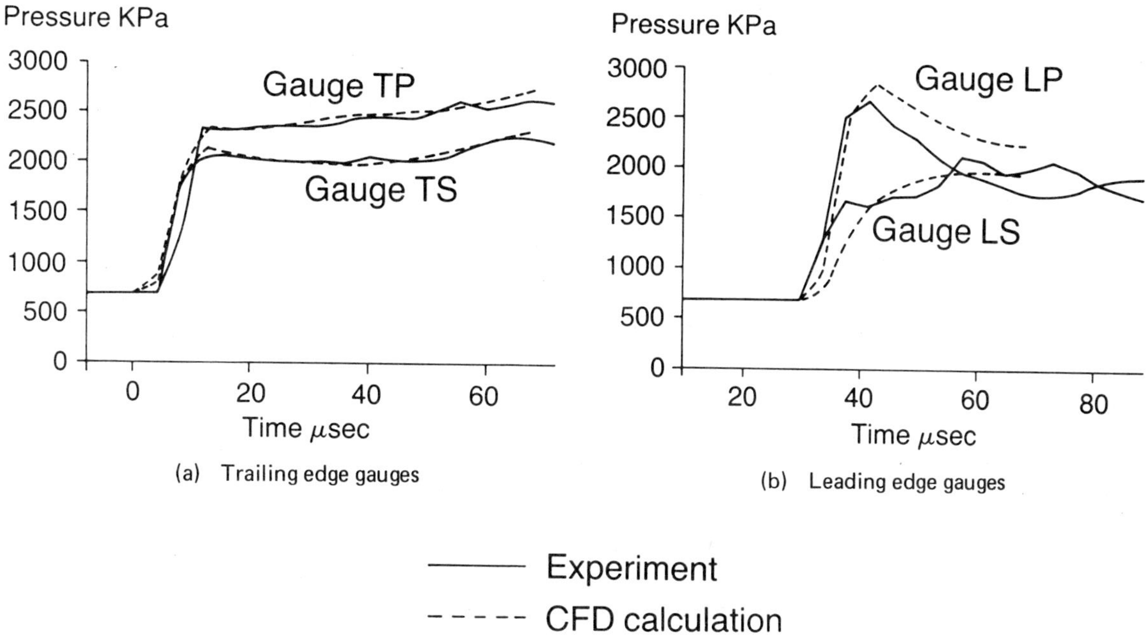

Fig 3 Calculated and experimental pressure histories

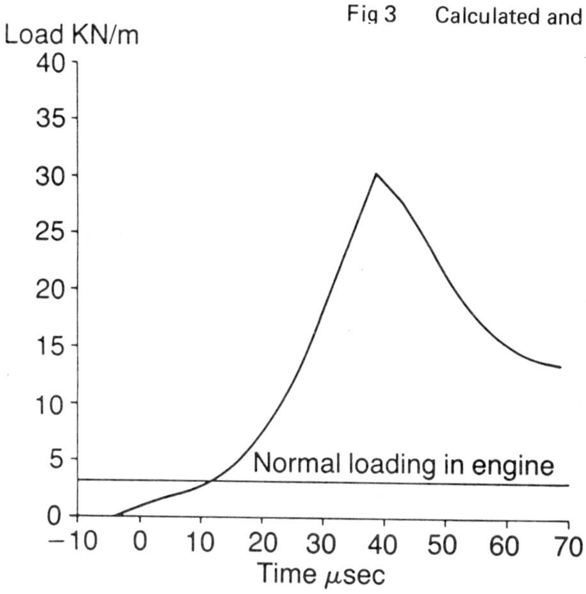

Fig 4 Calculated blade loading history

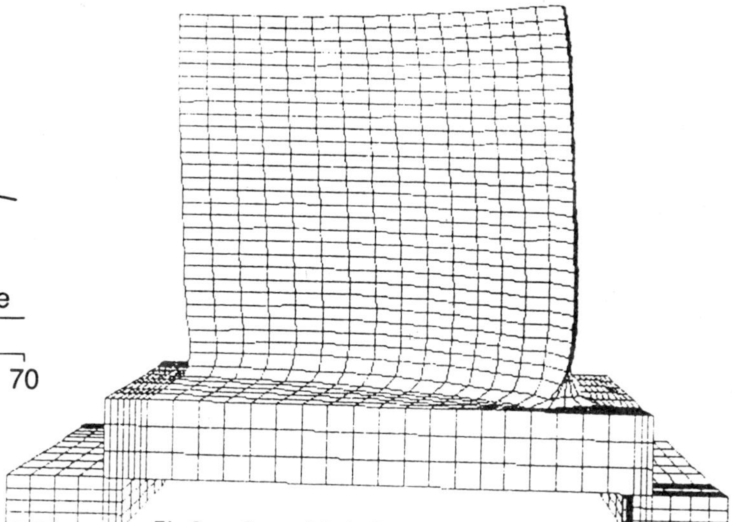

Fig 6 Stator blade finite element model

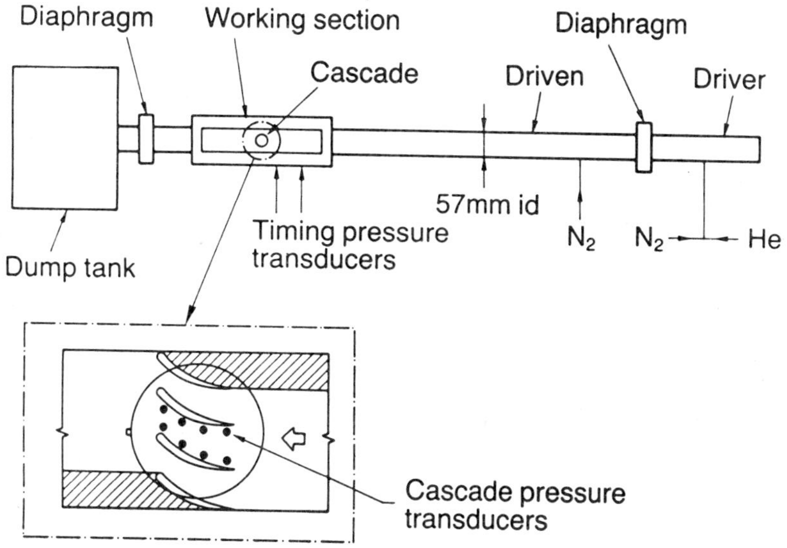

Fig 5 Diagram of shock tube experiment

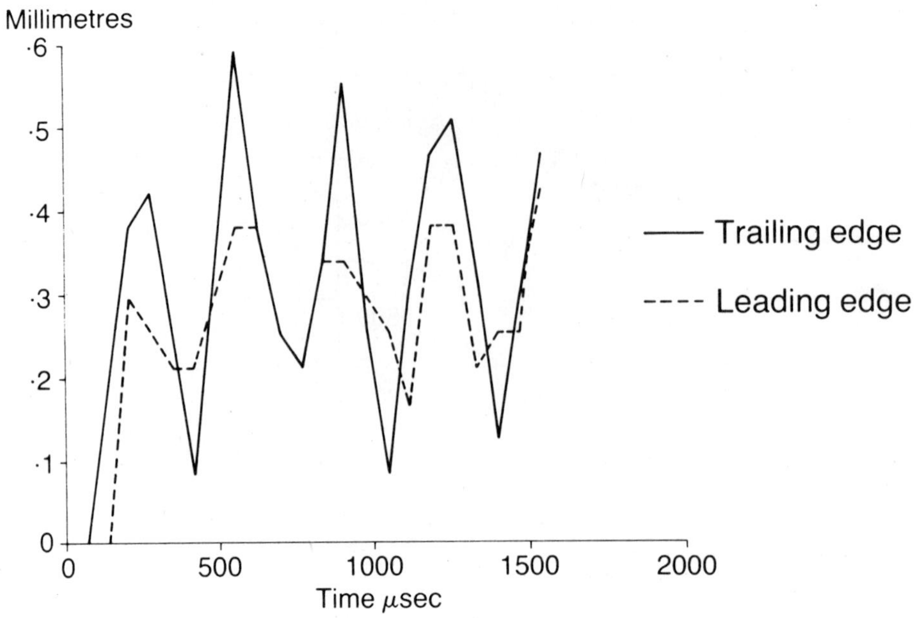

Fig 7 Measured displacements from shock tube experiments

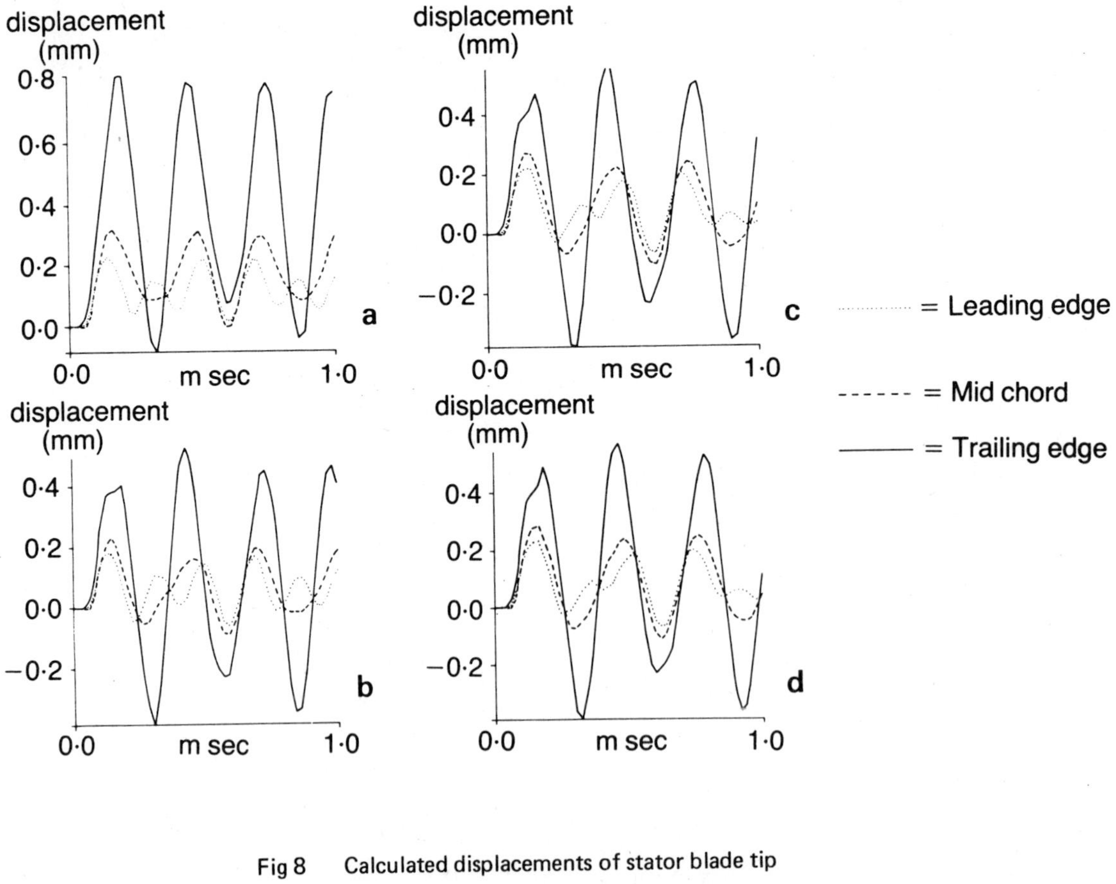

Fig 8 Calculated displacements of stator blade tip

C246/88

The influence of damping and angular acceleration on the vibratory response of rotating beam-like blades passing through resonance

H IRRETIER, Dr-Ing, VDI
Institute of Mechanics, University of Kassel, Kassel, West Germany

SYNOPSIS The response of beam-like turbine blades running through nozzle excitation resonances is calculated from a numerical model. The influence of damping and angular acceleration on the response maxima and the resonance shift are considered for two examples of blades.

NOTATION

A	blade cross section
B	width of rectangular blade cross section
e	Euler's number (e = 2.71828)
E	Young's modulus
f	frequency
f_D, $\dot{f}_D$	rotational frequency of the blade and its derivative
f^o_ℓ, f_ℓ	ℓ-th natural frequency of non-rotating and rotating blade
$f^r_{\ell i}$	resonance frequency at ℓ/i resonance
g_j	time-dependent functions of excitation
G	shear modulus
H	height of rectangular blade cross section
i	integer multiple of fundamental harmonic of excitation
$i_{\alpha\beta}$, $\tilde{i}_{\alpha\beta}$	radii of gyration referred to x_α and $\tilde{x}_\alpha$
L	blade length
m_{jj}	torsional and bending moments of the blade
$\overline{M}_{jj}$	Fourier-spectra of m_{jj}
m_1	exciting torsional moment
M_1	distribution of exciting torsional moment along the blade axis
n	number of modes considered in the modal transformation
p_α	exciting bending forces
$\underline{p}$	vector of exciting forces and moment
P_α	distribution of the exciting bending forces along the blade axis
r_ℓ	ℓ-th modal exciting force
R	disc radius
s	nozzle width
t	time
T_ℓ	ℓ-th modal coordinate
$T^r_{\ell i}$	time constant of damped natural motion after ℓ/i resonance
$V^r_{\ell i}$	maximum of response factor for ℓ/i resonance
w_α	bending displacement of the blade neutral axis
$W_{\alpha\ell}$	modal bending displacement of the blade
$\overline{W}_\alpha$	Fourier-spectra of w_α
$\underline{W}_\ell$	ℓ-th mode shape vector of the blade
x_j, $\tilde{x}_\alpha$	cartesian coordinates at the blade root and local
z	number of nozzles
α	stagger angle at the blade root
$\alpha_{\ell i}$	dimensionless angular acceleration at ℓ/i resonance
$\Gamma_{\alpha\ell}$	modal angles of deformed blade cross section
$\Delta f^r_{\ell i}$	resonance frequency shift at ℓ/i resonance
ζ_ℓ	ℓ-th modal damping ratio
φ_1	torsional angle of the blade
$\Phi_{1\ell}$	modal torsional angle of the blade
$\overline{\Phi}_1$	Fourier-Spectrum of φ_1
ψ	twisting angle
μ	rate of admission in the nozzle
ν	Poisson's ratio
ρ	density of material
ω^o_ℓ, ω_ℓ	ℓ-th circular natural frequency of the non-rotating and rotating blade
$\omega^r_{\ell i}$	resonance circular frequency at ℓ/i resonance
Ω	circular frequency
Ω_D, $\dot{\Omega}_D$	angular velocity and angular acceleration of the blade

Subscripts

j	index (j = 1, 2, 3)
ℓ	modal index (ℓ = 1, n)
i	index of harmonic of excitation (i = 1, ∞)
ℓi	index for resonance between ℓ-th natural frequency and i-th harmonic of excitation
α	index (α = 2, 3)
$\underline{\ \ }$	matrix or vector

Superscripts

r	resonance or response maximum
T	transposed matrix
$\cdot$	derivative with respect to time

1 INTRODUCTION

In recent times, the problem of calculating the forced vibrations of turbine blades became a subject of increasing importance in the field of turbomachinery. Papers were presented concerning various aspects of this problem, e.g.excitation mechanisms and forces /1/, /2/, /3/, stationary responses /4/, /5/ and fatigue analysis and life estimation /6/, /7/ and /8/. All these references are based on the assumption of stationary, time-independent excitation spectra i.e. constant rotational speed of the blade. The influence of time-varying excitation due to a change of the rotational speed during a run-up or a run-down was included in the analysis first in /9/, later in /10/, and recently in /11/ by the author. Some aspects concerning this question - to some extent similar to earlier considerations of the problem of systems with one degree of freedom - are also included in /12/ and /13/.

The present paper is a continuation of the work started by the author in /9/, /10/ and /11/. Two types of beam-like blades - both subjected to nozzle excitation - are considered. The transient response during a run-up is calculated and frequency response spectra are discussed with a particular view on the influence of damping and angular acceleration on the blade response.

2 THE MECHANICAL AND NUMERICAL MODEL OF THE BLADE

The blade model applied here is a one-dimensional beam described by an extended beam-theory including all important effects of a rotating turbine blade. It is shown in Fig. 1 and was described in all details in /9/, /10/ and /11/. The blade model can perform bending-bending vibrations in the x_2, x_3-plane and torsional vibrations around the x_1-axis, excited by time-dependent and distributed forces $p_2 = p_2(x_1,t)$ and $p_3 = p_3(x_1,t)$ and the moment $m_1 = m_1(x_1,t)$, respectively. It is assumed that these excitation functions are of the form

$$p_2 = P_2(x_1)\cdot g_2(t) \;,\; p_3 = P_3(x_1)\cdot g_3(t) \qquad (1a)$$

and

$$m_1 = M_1(x_1)\cdot g_1(t) \qquad (1b)$$

Thus, P_2, P_3 and M_1 stand for any distribution of exciting force or moment along the blade axis and g_1, g_2 and g_3 are functions which describe the time-dependence of the excitation e.g. periodic with constant frequency for stationary nozzle excitation or periodic with increasing or decreasing frequency for non-stationary nozzle excitation due to a blade run-up or run-down, respectively.

The differential equations of motion of the blade include shear deformation and rotary inertia, twisting, tapering and the centrifugal force field due to rotation. The boundary conditions include a linear-elastic clamping at the blade root. Again, details on both are given in /9/, /10/ and /11/.

To calculate the response of the blade due to the above mentioned excitations, a modal transformation technique is used. This yields n independent equations of motion

$$\ddot{T}_\ell + 2\zeta_\ell\omega_\ell\dot{T}_\ell + \omega_\ell^2 T_\ell = r_\ell \;;\; \ell = 1,n \qquad (2)$$

where $T_\ell = T_\ell(t)$ are the modal coordinates of the problem, ζ_ℓ the damping ratios on the basis of an assumed viscous Rayleigh-damping, ω_ℓ the circular natural frequencies of the blade, r_ℓ the modal excitation forces and n the number of modes considered in the modal transformation. More details concerning the derivation of equation (2) are presented in /9/, /10/ and /11/, too. The modal exciting forces r_ℓ are related to the exciting forces in the equations (1) by

$$r_\ell = \int_0^L \underline{W}_\ell^T \underline{p}\, dx_1 \;;\; \ell = 1,n \qquad (3)$$

where for the bending-bending vibrations $\underline{p}$ is given by

$$\underline{p} = [p_2 \;\; 0 \;\; p_3 \;\; 0]^T \qquad (4)$$

and

$$\underline{W}_\ell^T = [W_2 \;\; \Gamma_2 \;\; W_3 \;\; \Gamma_3]_\ell \qquad (5)$$

is the ℓ-th mode shape vector consisting for each mode of the deflections $W_2(x_1)$ and $W_3(x_1)$ of the axis x_1 and the angles $\Gamma_2(x_1)$ and $\Gamma_3(x_1)$ of the deformed cross sections. For the torsional vibrations, the corresponding equations for $\underline{p}$ and $\underline{W}_\ell^T$ are

$$\underline{p} = [m_1] \qquad (6)$$

and

$$\underline{W}_\ell^T = [\Phi_1]_\ell \;, \qquad (7)$$

where Φ_1 is the torsional angle of the ℓ-th torsional blade mode.

3 THE EXCITATION OF THE BLADE

The main source of forced vibrations of turbine blades are periodic loads in circumferential direction which originate either from partial admission, from wakes behind stator blades, or from both. All these types of excitation are summarized here with the notion 'nozzle', the significant mark of which is that during one revolution the blade passes a number z of nozzles as shown in Fig. 2. In each of the nozzles the blade is subjected to the fluid force within the μ-th part of the nozzle width s while the $(1-\mu)$-th part of s is free from load (Fig. 3). Thus, the excitation spectrum of the blade generally consists of the frequencies izf_D where i is any integer number from one to infinity and f_D is the rotational frequency of the blade. In some cases, however, only the odd integers i are part of the spectrum e.g. for a rectangular form of the force in the nozzle (cp. chapter 5 and figures 6 and 9).

4 SOLUTION

The first step in the solution procedure is the determination of the natural frequencies and the mode shapes of the blade including the stiffening effect of rotation. Both are calculated by a numerical integration of the differential equations and a shooting algorithm to fulfill the boundary conditions. Details of this method are presented in /14/.

The calculation of the forced vibrations of the blade starts with the computation of the modal exciting forces r_ℓ according to equation (3). First, referring to the space-time separation of the exciting functions p_2, p_3 and m_1 according to the equations (1), the values of the space-integrals in equation (3) are calculated numerically where any distribution of the exciting forces P_2,

P_3 and M_1 along the blade can be considered /9/, /10/. Second, the time-dependent parts g_1, g_2 and g_3 of the excitation functions are generated by subprograms based on the nozzle excitations shown in Fig. 2 and 3. Constant and, in particular, linear in time increasing or decreasing rotational frequencies can be considered. On principle, any arbitrary load distribution within each nozzle (Fig. 3) can be assumed. Third, the modal equations of motion (2) are integrated numerically by a Runge-Kutta-Fehlberg method for a chosen number n of modes. For the case of increasing or decreasing rotational frequencies, the corresponding change of the natural frequencies is taken into consideration during the integration procedure /9/, /10/, /11/.

After the calculation of the modal coordinates T_ℓ, the physical blade response is determined from the modal expansion. The blade bending deflections w_2 and w_3 and the torsional deflection φ_1 as well as the bending moments m_{22} and m_{33} and the torsional moment m_{11} can be considered as functions of time. Additionally, a Fourier-transformation can be performed within finite rectangular time windows during the considered time of blade operation to calculate the corresponding frequency spectra $\bar{W}_2$, $\bar{W}_3$, $\bar{\Phi}_1$, $\bar{M}_{22}$, $\bar{M}_{33}$ and $\bar{M}_{11}$ /9/, /10/, /11/.

5 NUMERICAL RESULTS

Two types of blades are considered in the following studies of transient forced vibrations of blades subjected to nozzle excitation. The first one is the simple rectangular beam of Fig. 4 which was already investigated partly in /9/, /10/ and /11/. It is excited by partial admission, described by two opposite sectors of 90° with constant load and two adjoining sectors of again 90° which are unloaded ($z = 2$; $\mu = 0{,}5$). The second blade considered here is a low pressure steam turbine blade shown in Fig. 5 and analysed in /15/ and /16/. This blade is excited by nozzle excitation of $z = 30$ nozzles, each with a constant load within $\mu = 0{,}5$ of its width.

In both cases, the blades are subjected only to a bending excitation p_3 while p_2 and m_1 are zero. P_3 is distributed along the blades as indicated in Fig. 4. Likewise in both cases, the blades run up from standstill to their speeds or operation following a linear increasing rotational frequency

$$f_D = \dot{f}_D \cdot t \quad , \tag{8}$$

where $\dot{f}_D$ denotes the linear increase of the rotational frequency related to the angular acceleration $\dot{\Omega}_D$ by $\dot{f}_D = \dot{\Omega}_D/2\pi$.

The aim of the following numerical studies is to consider the influences of both the angular acceleration $\dot{\Omega}_D$ as well as the modal damping ratios ζ_ℓ on the vibratory response of the two blades during these run-ups.

5.1 Rectangular Blade Model

The Campbell-diagram of the rectangular blade model shown in Fig. 4 is presented in Fig. 6. The first three natural frequencies of tangential bending vibrations are plotted besides the harmonics of excitation which - in this case of z=2 nozzles each with constant load in $\mu = 0{,}5$ of its width which yields a rectangular, periodic excitation function - consist of $2f_D$, $6f_D$, $10f_D$ etc. Running up from standstill into a range of rotational frequency above $f_D = 50$ Hz, the blade passes various resonances of which those indicated by circles in Fig. 6 are considered in the following response spectra in which the amplitudes $\tilde{M}_{22}$ of the bending moment m_{22}, related to the blade data by $\tilde{M}_{22} = L\,M_{22}/EAi_{22}^2$ are plotted versus frequency $f = \Omega/2\pi$ for 0.5 s time windows during the run-up.

First, only the transient response due to the resonance between the first natural frequenby f_1 and the fundamental harmonic of excitation $zf_D = 2f_D$ is discussed in Fig. 7 for a rotational frequency increase of $\dot{f}_D = 2$ Hz/s and two different damping ratios of $\zeta_1 = 0.001$ and $\zeta_1 = 0.05$, respectively. The plots show that for any time of operation the transient response of the blade consists of the natural frequency f_1 of the blade, which increases parabolically due to the increasing rotational speed, and the linearly increasing frequency $2f_D$ of the excitation.

For the small damping ratio of $\zeta_1 = 0.001$ (Fig. 7a), the contribution of the natural frequency is most significant shortly after the passage through the resonance and then decreases exponentially according to $e^{-\zeta_1\omega_1 t}$ due to the viscous damping. For the considered damping ratio of $\zeta_1 = 0.001$, the time constant $T^r_{11} = 1/\zeta_1\omega^r_{11}$ of the amplitude decrease, calculated with the resonance frequency $f^r_{11} = \omega^r_{11}/2\pi$ Hz for the coincidence between the first natural frequency f_1 and the first harmonic of excitation $2f_D$, amounts $T^r_{11} = 5.86$ s. The corresponding amplitude decrease immediately after the response maximum V^r_{11} is indicated in Fig. 7a and shows a decrease on 0.22 of V^r_{11} which is less than $1/e \approx 0.37$ of V^r_{11} within one time constant T^r_{11} due to the increasing natural frequency.
Due to the transient character of the response, its maximum V^r_{11} does not occur at coincidence between the first natural frequency f_1 and the first excitation frequency $2f_D$, i.e. not at the resonance frequency f^r_{11}, but is shifted to a higher frequency. This frequency shift Δf^r_{11} can be calculated for a system with one degree of freedom and time-independent natural frequency by a formula presented in /17/. This can be modified here for each resonance between one natural frequency $f_\ell = \omega_\ell/2\pi$ and one of the harmonics of excitation $iz\dot{f}_D$ into

$$\Delta f^r_{\ell i} = \alpha_{\ell i}\frac{2.17\sqrt{|\alpha_{\ell i}|}}{|\alpha_{\ell i}|+0.08\ \zeta_\ell+0.56\ \zeta_\ell\sqrt{|\alpha_{\ell i}|}} \tag{8}$$

where $\alpha_{\ell i}$ is a dimensionless angular acceleration of the run-up or run-down given by

$$\alpha_{\ell i} = \frac{iz\dot{\Omega}_D}{\omega_\ell^2} \tag{9}$$

For the case considered here,this value is $\alpha_{11} = 0.000864$ which yields for $\zeta_1 = 0.001$ a resonance frequency shift of $\Delta f^r_{11} = 0.063\ f^r_{11}$ which is indicated in Fig. 7a, too.

Besides the resonance frequency shift, the transient response has a maximum response factor

$V^r_{\ell i}$ which is less than in the case of steady-state vibrations. It can be calculated approximately by the formula

$$V^r_{\ell i} \approx \frac{1}{2\zeta_\ell + 0.5\sqrt{|\alpha_{\ell i}|}} \qquad (10)$$

presented in /18/. For the case discussed here, the maximum response factor amounts $V^r_{11} = 59.9$. The corresponding value is indicated in Fig. 7a on the basis of the response factor $V_{11} = 1$ for zero exciting frequency. Fig. 7a shows that the calculated response spectrum follows the response factor curve quite well up to the maximum response. Then - as discussed before - it follows the natural frequency curve according to the exponential amplitude decrease as the part of the entire response on the exciting frequency line is rather small in this case of very low damping in comparison to that part on the natural frequency curve.

For the second case of the relatively high damping ratio $\zeta_2 = 0.05$, the natural motion of the blade is smoothed out rapidly during all times of operations. Consequently, the response spectrum consists preponderantly of the response factor curve along the exciting frequency line as shown in Fig. 7b. Due to the high damping value and the relatively small angular acceleration, the shift of the maximum response from the resonance frequency f^r_{11} is rather small and the maximum response factor $V^r_{11} = 8.72$ according to equation (10) is near to the value of the steady-state excitation.

The results in Fig. 7 show that all important effects of a transient vibration due to passage through resonance are described sufficiently by the numerical model used in this paper. After these fundamental studies, a realistic, multiple frequency nozzle excitation of the rectangular blade model (Fig. 4) is considered next for different values of damping for the first natural mode - while that one for the second mode remains constant - and different angular accelerations. Again, the number of nozzles is assumed to be z = 2. The response spectra during the run-up are shown in Fig. 8, now for both the first and second mode of the considered blade. The three diagrams in Fig. 8 show transient resonance responses shortly after each intersection of one of the natural frequencies f_ℓ and one of the harmonics of excitation izf_D. For each mode, the response maximum decreases with increasing number i of the harmonic of rotational speed due to the decreasing Fourier-components of excitation.

In all cases considered in Fig. 8, the response of the second mode is much less than that one of the first mode for the same harmonic of excitation, even for the high damping of the first mode. The reason for that is the distribution of the exciting force along the blade shown in Fig. 4 which excites predominantly the first mode and less all higher ones.

The comparison of the response spectra plotted in Fig. 8a and 8b - the difference between both is the damping ratio of $\zeta_1 = 0.005$ and $\zeta_1 = 0.01$ - shows that the response due to resonance between the first natural frequency f_1 and the first harmonic excitation $2f_D$ (i=1) is reduced considerably by the increase of the modal damping ratio ζ_ℓ. However, the decrease is not as much as in the case of steady-state excitation because of the influence of angular acceleration. This tendency is well-known from /19/ and /20/ and also visible from the influence of $\alpha_{\ell i}$ on $V^r_{\ell i}$ in equation (10). For the maximum responses due to resonances between the first natural frequency f_1 and the higher harmonics of excitation $6f_D$, $10f_D$ etc. (i = 3,5 etc.), the influence of damping on the response maxima decreases more and more because the angular acceleration parameter $\alpha_{\ell i}$ in equation (9) increases linearly with the number i of the harmonic of excitation and, consequently, its influence increases in comparison to the influence of damping in the equation (10) for the maximum response factor.

In Fig. 8c a response spectrum is considered for a case of half angular acceleration in comparison to the plots in Fig. 8a and 8b. Again, as in Fig. 8a, the damping ratios $\zeta_1 = 0.005$ and $\zeta_2 = 0.001$. Besides the slower increase of the rotational speed, Fig. 8c shows stronger responses at resonances due to the same reason. In particular, this effect is visible for the first mode but also quite clearly from the resonance point between the second natural frequency f_2 and the second harmonic of excitation $6f_D$ (i=3).

5.2 Low Pressure Steam Turbine Blade

After discussing the blade model with rectangular cross section, now the low pressure blade shown in Fig. 5 is considered during a run-up from standstill subjected to nozzle excitation. Only its fundamental mode is taken into account. The Campbell-diagram is shown in Fig. 9 for the corresponding bending-bending natural frequency which increases from the value $f^o_1 = 104$ Hz for zero rotation only slightly in the considered range of rotational frequency. The harmonics of excitation for the number of z = 30 nozzles are also plotted in Fig. 9 and show that during a run-up from standstill up to a rotational frequency of $f_D = 4$ Hz, i.e. $n_D = 240$ rpm, all harmonics of excitation yield resonance with the first natural frequency of the blade.

Corresponding response spectra which show again a normalized bending moment $\tilde{M}_{22}$ at the blade root, are presented in Fig. 10 for various modal damping ratios and angular accelerations. All plots in Fig. 10 show the expected transient responses shortly after the coincidence between each harmonic of excitation and the natural frequency. Again, the influence of damping and angular acceleration is visible as discussed before for the blade model with rectangular cross section. The maximum responses decrease with increasing damping, however, as shown with equation (10) less than in the stationary case $\alpha_{\ell i} = 0$ for constant rotational frequency. The higher the harmonic i of excitation is, the higher is the value α_ℓ of equation (9) for a given angular acceleration $\dot{\Omega}_D$ and a given natural frequency ω_ℓ. Thus, the character of non-stationarity increases and, consequently, the influence of damping decreases for resonances with increasing harmonics of excitation.

6 CONCLUSIONS

Two types of beam-like turbine blades subjected to nozzle excitation were considered during various operations during a run-up from standstill with linearly increasing rotational frequency. It was proved - to some extent known from systems with one degree of freedom subjected to transient harmonic excitation - that the resonance responses are function of the modal damping ratio and of a modal frequency acceleration which depends linearly on the order of harmonic of excitation. Corresponding formulae were presented and their results were compared with those results from the numerical simulation of the run-ups and show a satisfactory agreement.

7 REFERENCES

(1) RAO, J.S.; RAO, V.V.R.; SESHADRI, V. Non-steady forces in turbomachine stage. 3rd Int. Conf. on Vibration in Rotating Machinery, York, 1984, 243 - 254

(2) RIEGER, N. Flow path excitation mechanisms for turbomachine blades. 7th IFToMM World Congress, Sevilla, 1987, Proc. 'Rotordynamics Session'

(3) RAO, V.V.R.; RAO, J.S. Effect of downwash on the non-steady forces in a turbomachine stage. 11th ASME Conf. on Mechanical Vibrations and Noise, Boston, 1987, Proc. 'Bladed Disk Assemblies', 9 - 20

(4) JADVANI, H.M.; RAO, J.S. Forced vibrations on rotating pretwisted blades. IFToMM Int. Conf. on Rotordynamics Problems in Power Plants, Rom, 1982, 259 - 266

(5) RAO, J.S.; GUPTA, K.; VYAS, N.S. Analytical and experimental investigations of vibratory stresses of a rotating steam turbine blade under NPF excitation, IFToMM Int. Conf. on Rotordynamics. Tokyo, 1986, 289 - 300

(6) STEELE, J.M.; LAM, T.C.T.; RIEGER, N.F. Turbine blade life prediction computer program. EPRI Workshop on Steam Turbine Blade Reliability WS 81 - 248, 1982, 1 - 14

(7) RIEGER, N.F. Blade fatigue. 6th IFToMM World Congress, Delhi, 1983, Proc. 'Rotordynamics Session', 66 - 75

(8) RAO, J.S.; VYAS, N.S. On life estimation of turbine blading. 7th IFToMM World Congress, Sevilla, 1987, Proc. 'Rotordynamics Session'

(9) IRRETIER, H. Numerical analysis of transient responses in blade dynamics. 3rd Int. Conf. on Vibration in Rotating Machinery, York, 1984, 255 - 267

(10) IRRETIER, H. Computer simulation of the run-up of a turbine blade subjected to partial admission. 10th ASME Conf. on Mechanical Vibration and Noise, 85-DET-128, Cincinnati, 1985, 1 - 11

(11) IRRETIER, H. Transient vibrations of turbine blades due to passage through partial admission and nozzle excitation resonances. IFToMM Int. Conf. on Rotordynamics, Tokyo, 1986, 301 - 306

(12) VYAS, N.S. Vibratory Stress Analysis and fatigue life estimation of turbine blades. PhD Thesis, 1986, Dept. of Mechanical Engineering, IIT Delhi, Delhi

(13) VYAS, N.S.; GUPTA, K.; RAO, J.S. Transient response of turbine blades. 7th IFTomm World Congress, Sevilla, 1987, 689 - 696

(14) IRRETIER, H.; MAHRENHOLTZ, O. Eigenfrequencies and mode shapes of a free-standing, twisted, tapered and rotating blade with respect to an elastically supported root. 8th ASME Conf. on Mechanical Vibration and Noise, 81-DET-125, Hartford, 1981, 1 - 9

(15) NUSCHKE, H. Berechnungen und meßtechnische Analysen freier und erzwungener Schwingungen stabförmiger Turbinenschaufeln. MSc Thesis, 1986, Dept. of Mechanical Engineering, Institute of Mechanics, University of Kassel, Kassel

(16) HOHLRIEDER, M.; IRRETIER, H. Experimentelle und numerische Ermittlung der Eigenschwingungsgrößen einer Dampfturbinenschaufel. Report, 1987, Dept. of Mechanical Engineering, Institute of Mechanics, University of Kassel, Kassel

(17) KAZ, A.M. Erzwungene Schwingungen beim Resonanzdurchgang. Inst. Mekh. Akad. Nauk UDSSR. Ing. Spornik III/2, 1947, 100 - 125

(18) MARKERT, R. An- und Auslaufvorgänge von periodisch erregten Systemen. GAMM-Conference, Stuttgart, 1987

(19) GOLOSKOKOW, E.G.; FILIPPOW, A.P. Instationäre Schwingungen mechanischer Systeme. Akademie-Verlag, Berlin, 1971

(20) DITTRICH, G.; SOMMER, J. Das instationäre Verhalten des linearen Schwingers mit einem Freiheitsgrad. VDI-Zeitschrift 118, 1976, 477 - 482

(21) MARKERT, R.; PFÜTZNER, H. An- und Auslaufvorgänge einfacher Schwinger. Forschung im Ingenieur-Wesen 47, 1981, 117 - 125

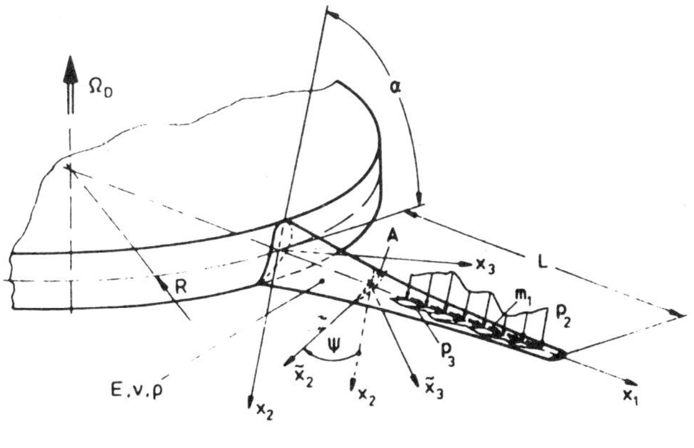

Fig 1 Model of a rotating turbine blade subjected to external exciting forces and moment

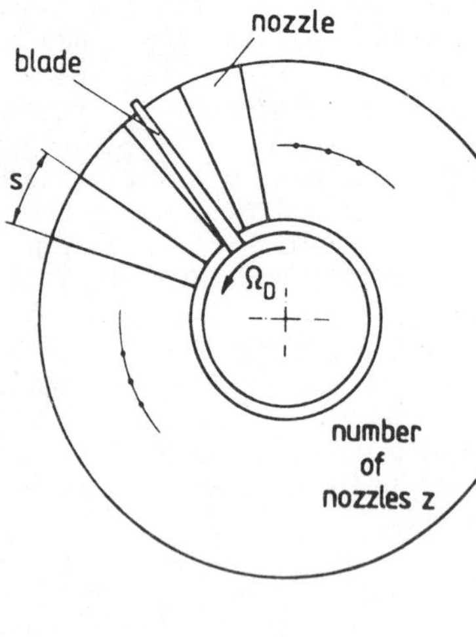

Fig 2 Nozzle excitation of a rotating turbine blade

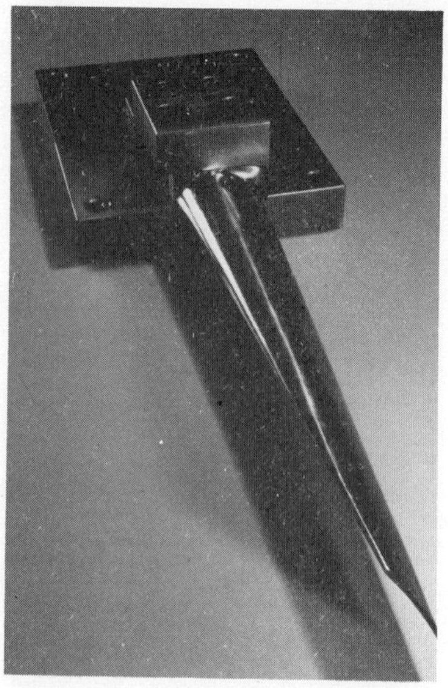

Fig 5 Low pressure steam turbine blade

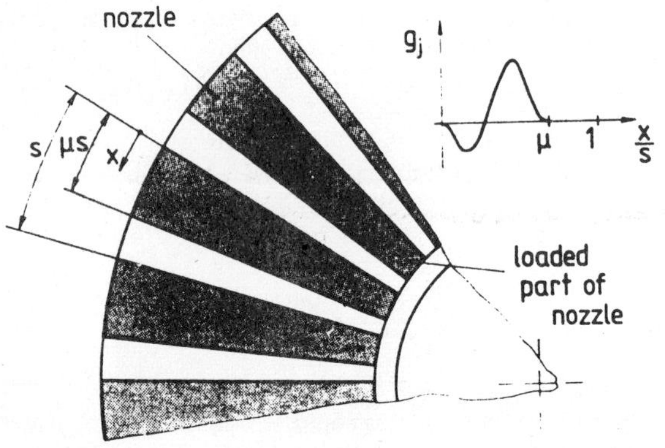

Fig 3 Distribution of exciting forces in one nozzle

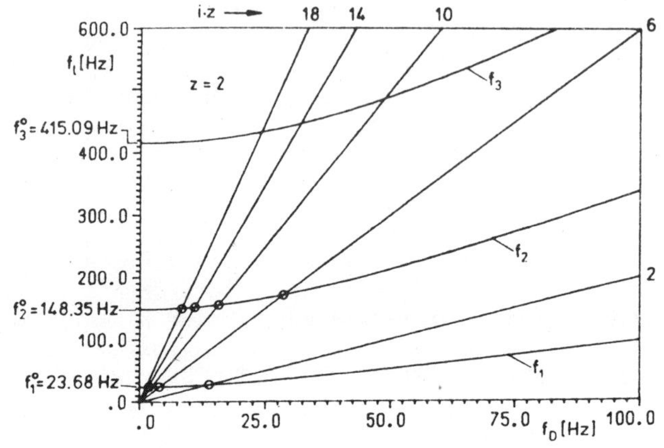

Fig 6 Campbell diagram of the rectangular blade model under partial admission excitation

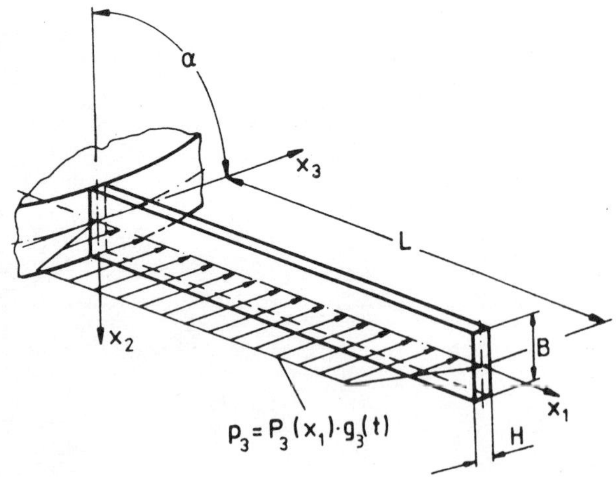

Fig 4 Rectangular blade model and exciting force

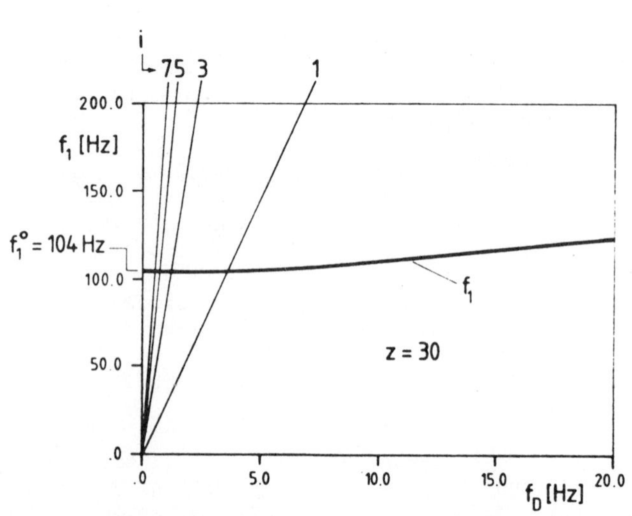

Fig 9 Campbell diagram for the first mode of the low pressure steam turbine blade under nozzle excitation

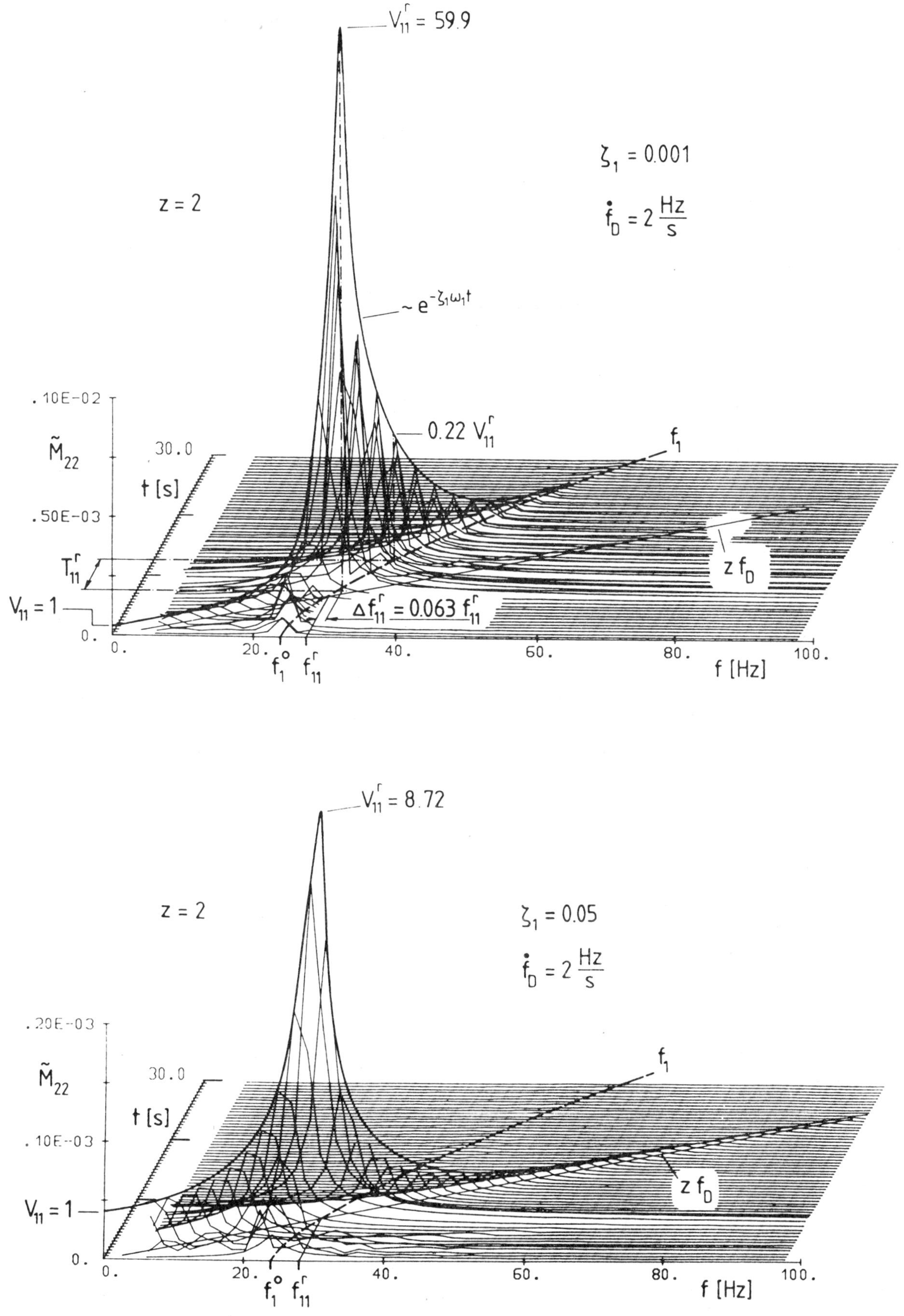

Fig 7 Transient response spectra for the first mode of the rectangular blade model during run-up under partial admission excitation

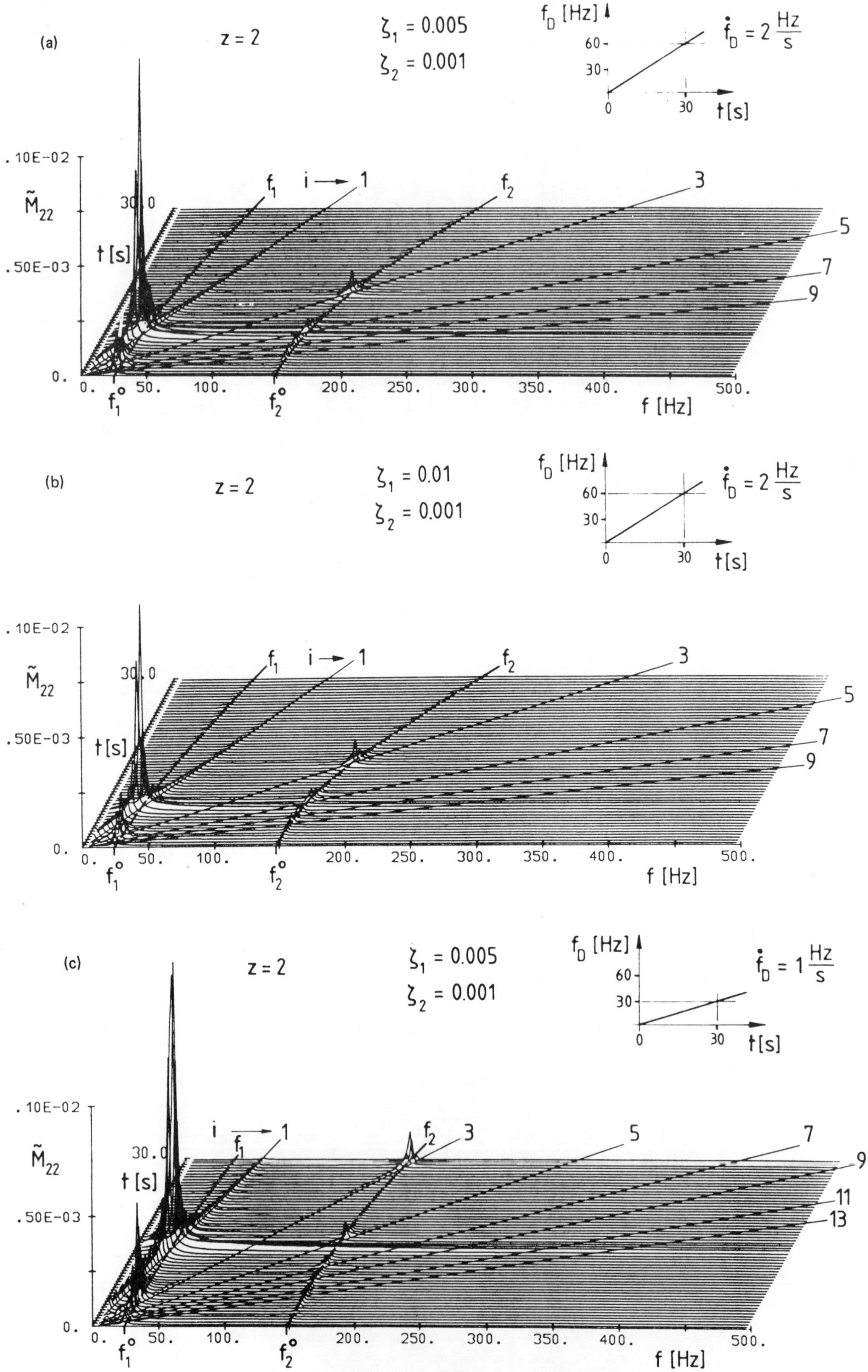

Fig 8 Transient response spectra of the rectangular blade model during run-up under partial admission excitation for various modal damping ratios and angular accelerations

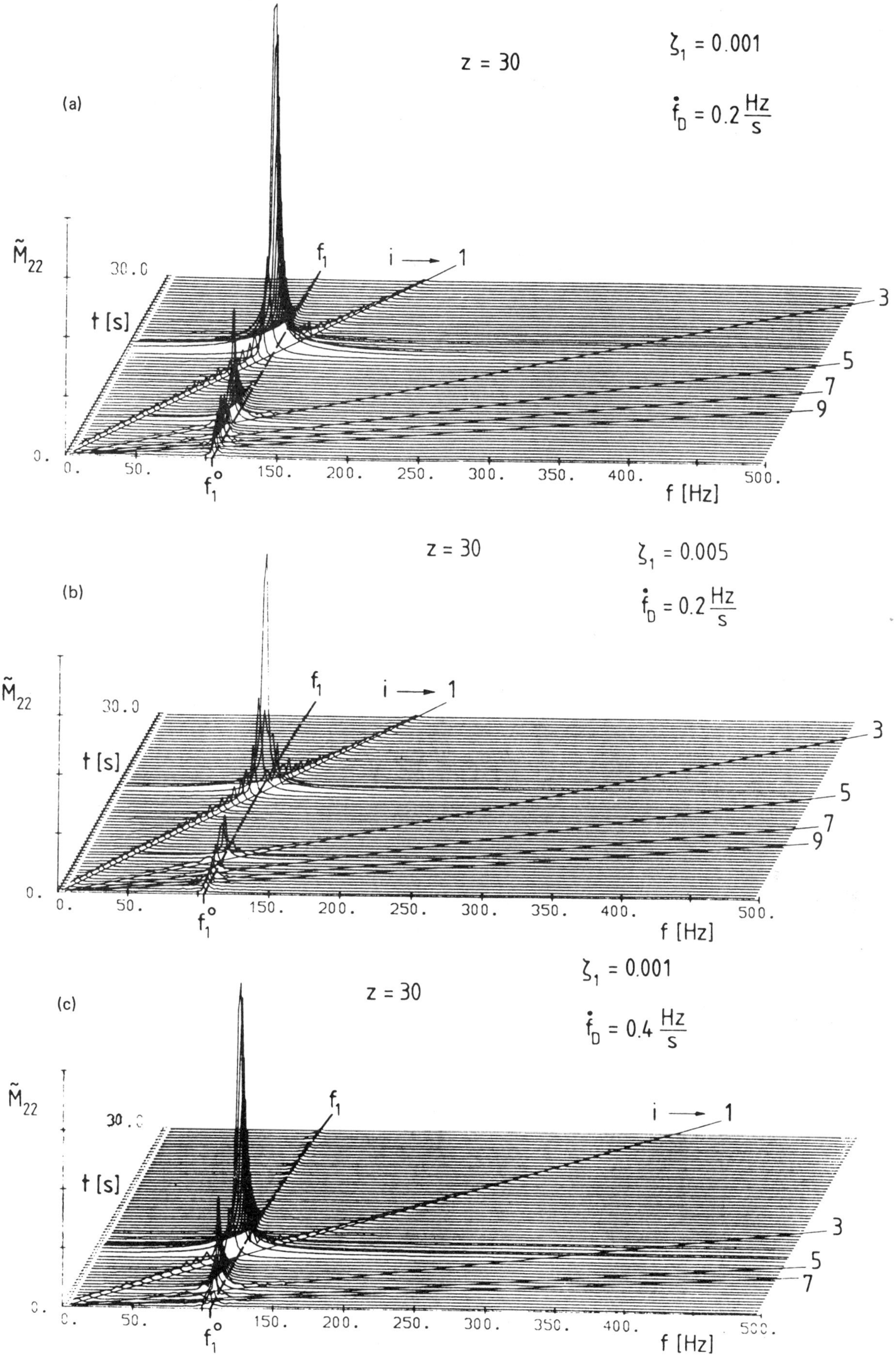

Fig 10 Transient response spectra for the first mode of the low pressure steam turbine blade during run-up under nozzle excitation for various modal damping ratios and angular accelerations

C321/88

Gasdynamic damping properties of steam turbine blades

N F RIEGER, PhD
Stress Technology Incorporated, Rochester, New York, United States of America

SYNOPSIS

A theory for determining the gasdynamic damping values of turbine blades is presented, which has special application to steam turbine blades. This theory is based on the pumping work which is done on the main gas stream by such blades during their vibration. This theory is suitable for blades of any airfoil cross-section under non-separated flow conditions, during any blade mode of vibration. Knowledge of the power output of a typical blade is needed, together with details of the blade mode shape. Details of the theory are given, together with a simple application to a practical case.

NOTATION

A,B	Integral Angles
C	Velocity Damping
D	(Superscript) Dynamic Component
F	Blade Force
$\underline{F}$	Force Vector
L	Length of Blade
N	Speed, rpm
S	(Superscript) Steady Component
$\underline{T}$	Torque Vector
W	Work Done
$\underline{i}, \underline{j}, \underline{k}$	Unit Vectors
m	Modal Response Harmonic
o	Excitation Harmonic
p	Pressure
$\underline{p}$	Pressure Vector
$\underline{q}$	Lift Force Vector
r	Radius
s	Blade
t	Time
u	Amplitude of Vibration
$\dot{u}$	Velocity of Vibration
$\underline{z}$	Radial Coordinate
α	Constant
γ	Frequency Rad/Sec
λ	Phase Angle Between Force and Displacement
$\underline{\omega}$	Disk Rotational Speed Vector
ϕ	Phase Angle From Time Datum
θ	Angle of Rotation
$\dot{\theta}$	Angular Velocity

INTRODUCTION

This paper presents a general theory for the prediction of gasdynamic damping of turbine blades. The contribution of gasdynamic damping to the suppression of blade resonant vibrations is widely recognized, though few attempts appear to have been made to evaluate its magnitude in practice. To some extent this may be due to difficulties which surround the development of a satisfactory predictive theory for damping in turbine blades. These difficulties arise from the obscure and variable structural assembly conditions with which the blades are secured to the disk, and they are also due to the lack of solid test data to guide the application of conventional damping theories to structural components within the turbine environment. Gasdynamic damping is often seen by turbine engineers as a minor contributor to the small overall damping values which blade systems possess. It is argued that such minor damping is not worth considering, as its value falls within the error band for structural damping. Specific values for gasdynamic damping are, however, difficult to obtain to test this idea.

More generally, it appears that the mechanism of gasdynamic damping itself is not widely understood, and discussions of this mechanism are not readily located within the subject literature. For these reasons the theory presented here is comprehensive in nature, to address the need for a discussion of the mechanics of gasdynamic damping, and to provide a procedure for obtaining numerical values of gasdynamic damping coefficients for particular stage geometry and power level details. The theory presented is suitable for the computer analysis of general blade geometries. It is shown that the gasdynamic damping mechanism is viscous and linear in nature for the relatively small vibration amplitudes associated with turbine blades.

PREVIOUS WORK

There appears to have been little previous work published in the open literature on the gasdynamic damping properties of steam turbine blades. Legendre [1], [2], presented two surveys of aerodynamic damping between 1967 and 1972. These surveys were directed toward compressor blades, with some reference to turbine blades. Hammons [3] included gasdynamic damping in his computer analyses of decay of torsional oscillations in turbine-generator systems, but gave no details of the theory upon which his coefficients were based. Subsequent work by Rieger [4] showed close correlation between Hammon's results and the theory described herein.

An extensive literature exists worldwide on the gasdynamic properties of compressor blades, which in recent years has focused on damping and dynamic response of long fan stage blades. This literature is primarily concerned with the influence of all primary damping sources, including the likely damping contributions of special devices, to the prediction of the instability threshold for such blades (rotating stall, flutter). It is less concerned with the identification of damping coefficients under the stable, forced vibration conditions addressed in this paper, although some gasdynamic damping data for selected airfoil operating conditions has been presented: see Legendre [1], [2], Fleeter [5], and others. The results obtained by these authors are in general accordance with the numerical values for gasdynamic damping presented in this paper. Further work on gasdynamic damping is needed, to compare predicted damping values with test data from steam turbine blades.

THEORY

Consider the elemental section ds.dz of the airfoil surface shown in figure 1 which is situated at radius z from the axis of rotation. The net lift force $\underline{q}$ which acts on this surface during operation arises from the variation of the pressure $\underline{p}$ around this surface, and is given by:

$$\underline{q} = \oint \underline{p}\ ds.dz$$

If the gas pressure $\underline{p}$ contains both a steady component $\underline{p}^S$ and a fluctuating component $\underline{p}^D \sin\gamma t$ of frequency γ, the net lift force per unit length can be written as:

$$\underline{q} = \oint \{\underline{p}^S + \underline{p}^D \sin\gamma t\}\ ds.dz$$

$$= \underline{q}^S + \underline{q}^D \sin\gamma t$$

where

$$\underline{q}^S = \underline{q}_T^S + \underline{q}_A^S = q_T^S \underline{i} + q_A^S \underline{j}$$

$$\underline{q}^D = \underline{q}_T^D + \underline{q}_A^D = q_T^D \underline{i} + q_A^D \underline{j}$$

and $\underline{i}$ and $\underline{j}$ are unit vectors aligned as shown in figure 2.

More generally, the blade is responding to a forcing spectrum which contains m harmonic components. The net gas force per unit length of blade is then:

$$\underline{q} = \underline{q}^S + \sum \underline{q}_m^D \sin(\gamma_m t + \phi_m)$$

and ϕ_m is the phase angle of the nth harmonic referred to some arbitrary time datum.

The blade responds to this forcing in each of its n modes of vibration.

The blade amplitude at any time t at radius z may be written:

$$\underline{u} = \underline{u}^S + \sum_{m=1} \sum_{n=1} \underline{u}_{mn}^D \sin(\gamma_m t + \phi_m - \lambda_{mn})$$

where λ_{mn} is the phase angle between the mth forcing harmonic and the nth modal amplitude.

The blade velocity $\dot{\underline{u}}$ at radius z is then:

$$\dot{\underline{u}} = \dot{\underline{u}}^S + \sum_{m=1} \sum_{n=1} \gamma_m u_{mn}^D \cos(\gamma_m t + \phi_m - \lambda_{mn})$$

$$= \underline{\omega}^S z + \sum_{m=1} \sum_{n=1} \gamma_m u_{mn}^D \cos(\gamma_m t + \phi_m - \lambda_{mn})$$

where $\underline{\omega}^S$ is the steady angular velocity of rotation of the rotor.

The torque $\underline{T}$ about the axis of rotation due to the total lift force $\underline{F}$ is given by:

$$\underline{T} = \underline{F} \times \underline{z} = (F_T\underline{i} + F_A\underline{j}) \times (r\ \underline{k})$$

$$= T_A\underline{i} - T_T\underline{j}$$

The total force due to the gas load $\underline{q}$ on the blade is:

$$\underline{F} = \int_{r_o}^{r_o+L} \underline{q}\, dz = \int_{r_o}^{r_o+L} \{q^S(z) + \sum_{m=1} q_m^D (z) \cdot \sin(\gamma_m t + \emptyset_m)\}\, dz$$

The work done per cycle due to the total torque $\underline{T}$ oscillating through the total angle θ is given by:

$$W = \oint \underline{T} \,.\, d\underline{\theta} = \oint T \,.\, \frac{d\theta}{dt}\, dt = \oint \underline{T} \,.\, \dot{\theta}\, dt$$

$$= \oint (\underline{F} \times \underline{z}) \cdot (\dot{\underline{u}} \times \frac{1}{z})\, dt = \oint \underline{F} \cdot \dot{\underline{u}}\, dt$$

Substituting gives:

$$W = \frac{1}{\gamma_m} \int_0^{2\pi} \left[\int_{r_o}^{r_o+L} \{\underline{q}^S(z) + \sum_{m=1} \underline{q}_m^D (z) \cdot \sin(\gamma_m t + \emptyset_m)\}\, dz \right] .$$

$$\left[\dot{u}^S (z) + \sum_{m=1} \sum_{n=1} \gamma_m\, \underline{u}_m(z) \cdot \cos(\gamma_m t + \emptyset - \lambda_{mn})\right] d(\gamma_m t)$$

$$= W^S(z) + W^D(z)$$

where $W^S(z)$ is the steady component, and $W^D(z)$ is the harmonic component of the total work. Considering $W^S(z)$, and writing $\underline{q}^S(z) = q_t^S(z)\underline{i}$ for the work-producing component, and $\dot{\underline{u}}^S(z) = \omega^S z\underline{i}$ for the steady velocity gives, upon expansion:

$$W^S(z) = \frac{1}{\gamma_m} \int_0^{2\pi} \left[\int_{r_o}^{r_o+L} q_t^S (z)\underline{i}\, dz \right] \cdot \left[\omega^S z\underline{i}\right] \cdot d(\gamma_m t)$$

For the simplest case $q_t^S(z) = q_t^S$, i.e., constant along the blade length. Integrating gives:

$$W^S(z) = \frac{2\pi}{\gamma_m} q_t^S\, L\, \omega^S\, (r_o + \frac{L}{2})$$

The period of integration $\gamma_m t$ is arbitrary and so may be extended to $\omega^S t$, the duration of one revolution of the shaft, i.e., one work-cycle. In this manner, the steady-state work expression becomes:

$$W^S(z) = 2\pi\, (q_t^S L)(r_o + \frac{L}{2})$$

which is the product of the steam force in the direction of motion times the distance moved in the direction of motion during one revolution, measured at the point of application of this force, as would be expected from elementary considerations.

The expression for the harmonic work term $W^D(z)$ is:

$$W^D(z) = \frac{1}{\gamma_m} \int_0^{2\pi} \int_{r_o}^{r_o+L} \sum_{m=1} \underline{q}_m^D (z) \cdot \sin(\gamma_m t + \emptyset_m) \cdot$$

$$\sum_{m=1} \sum_{n=1} \gamma_m \underline{u}(z)_{mn} \cos(\gamma_m t + \emptyset_m - \lambda_{mn})\, dz.d(\gamma_m t)$$

Each of the remaining products in the work integral involves terms which are constant in time, and which therefore represent zero harmonic work. The terms $q_m^D(z)$ and $u(z)_{mn}$ respectively describe the unit vector force and the vector velocity at location z along the blade. Expanding both terms into their components, and taking the scalar product gives the following work integrals for the tangential and axial directions:

$$W_T^D(z) = \frac{1}{\gamma_m} \int_0^{2\pi} \int_{r_o}^{r_o+L} \sum_{m=1} q_{T,m}^D(z) \cdot \sin(\gamma_m t + \emptyset m) .$$

$$\sum_{m=1} \sum_{n=1} \gamma_m\, u_{T,mn}(z) \cdot \cos(\gamma_m t + \emptyset_m - \lambda_{mn}) dz\, d(\gamma_m t)$$

$$W_A^D(z) = \frac{1}{\gamma_m} \int_0^{2\pi} \int_{r_o}^{r_o+L} \sum q_{A,m}^D(z) \cdot \sin(\gamma_m t + \emptyset_m)$$

$$\sum_{m=1} \sum_{n=1} \gamma_m u_{A,mn}(z) \cos(\gamma_m t + \phi_m - \lambda_{mn})\, dz\, d(\gamma_m t)$$

Recalling that:

$$\int_0^{2\pi} \sin A \cos (A - B)\, dA = \pi \sin B$$

allows the above expressions for tangential and axial harmonic work to be reduced to:

$$W_T^D(z) = \pi \int_{r_o}^{r_o+L} \sum_{m=1} q_{T,m}^D(z) \sum_{m=1} \sum_{n=1} u_{T,mn}(z)\, dz \,.\, \sin \lambda_{mn}$$

$$W_A^D(z) = \pi \int_{r_o}^{r_o+L} \sum_{m=1} q_{A,m}^D(z) \sum_{m=1} \sum_{n=1} u_{A,mn}(z)\, dz \,.\, \sin \lambda_{mn}$$

These expressions indicate that gasdynamic damping in a given mode depends upon the phase angle λ_{mn} between the nth gas harmonic forcing component and the response of the blade in the nth mode. For all practical cases the gasdynamic damping is evidently negligible unless the phase angle $\lambda_{mn} = \pi/2$, i.e., blade resonance exists between the forcing harmonic m and the vibration mode n.

The magnitude of the resonant gasdynamic damping may therefore be determined mode by mode for individual resonant modes, once the radial distributions of the tangential and axial forcing q_T^D (z) and q_A^D (z) are known, together with the associated mode shapes u_T (z) and u_A (z).

With the above concepts included the general work integrals for each resonant mode n become:

$$W_{T,mn}^D = \pi \int_{r_o}^{r_o+L} q_{T,m}^D(z)\, u_{T,mn}(z)\, dz$$

$$W_{A,mn}^D = \pi \int_{r_o}^{r_o+L} q_{A,m}^D(z)\, u_{A,mn}(z)\, dz$$

To establish the practical meaning of these expressions, consider the simple case of the uniform cantilever blade shown in figure 1 which rotates at ω^S rad/sec while vibrating in its lowest mode at γ_1 rad/sec. Assuming that the first mode of vibration in the tangential direction has the form:

$$u_{T,1}(z) = u_{T,1}^0 \{1 - \cos \frac{\pi z}{2L}\}$$

where u_T^0 is the vibration amplitude of the tip of the cantilever. In this instance the blade is encastre' at its attachment to the disk, and so,

$$u_{A1}(0) = 0.$$

For convenience the gasdynamic force per unit length along the blade is assumed to be constant, i.e., q_T^D (z) = q_T^D = Const. The work integral is therefore obtained between 0 and L, as the cantilever base has no harmonic motion. Then

$$W_{T,1}^D = \pi \int_0^L q_T^D u_{T,1}^0 \{1 - \cos \frac{\pi z}{2L}\}\, dz$$

$$= 0.3634\, \pi\, q_T^D\, L\, u_{T,1}^0$$

This expression indicates that the work product of the total excitation harmonic force $q_T^D L$ and the blade tip vibration amplitude $u_{T,1}^0$ must be modified by the factor 0.3634, to allow for the effect of the mode shape in assessing the damping work of this case.

MAGNITUDE OF HARMONIC GASDYNAMIC FORCE

The magnitude of the gasdynamic damping force which develops in response to blade oscillatory motions may be determined from the stage power (or torque vs. speed) curve. A typical stage power curve is shown in figure 3, in which the constant power characteristic shown is described by the relation:

$$\alpha P = TN,$$

where P is the output horsepower, T is the stage driving torque in lb.in., and N is the shaft speed in rpm. With these units the constant α has the value 63025.

Under conditions which involve small fluctuations of speed and torque but with constant power output, the differential relationship between these quantities is:

$$\alpha dP = NdT + TdN$$

For constant P, dP = 0 and so:

$$0 = NdT + TdN,$$

i.e.,

$$\frac{dT}{dN} = -\frac{T}{N} = -\frac{\alpha P}{N^2},$$

thus

$$dT = -\frac{T}{N}\, dN = -\, 9.55\, \frac{T}{N}\, d\dot{\theta},$$

$$= -\, C_S^{\theta}\, d\dot{\theta},$$

where

$$C_S^{\theta} = 9.55\ T/N$$

The fluctuating blade torque is governed by oscillations in the gasdynamic force along the length of the blade. For convenience this force may be referred to the blade mid-height by writing

$$dT = (r_o + \frac{1}{2} L)\ dF = -C_s^{\theta}\ d\dot{\theta}$$

$$= -\, C_S^{\theta}\, \dot{u}_T^D\, (L)\ /\ (r_o + L),$$

i.e.,

$$dF = -\, C_S^u\, \dot{u}_T^D(L) = -\, C_S^u\, \gamma\, u_T^D(L),$$

where

$$C_S^u = C_S^{\theta}\ /\ (r_o + \frac{1}{2} L)\ (r_o + L) = 9.55\ \frac{T}{N}\ \frac{1}{(r_o + \frac{1}{2} L)\ (r_o + L)}$$

Recalling the expression for the gasdynamic-work and substituting for dF therein gives:

$$W_{T,1}^D = -\, 0.3634\ \pi\ C_S^u\ \gamma\ \{u_T^D\ (L)\}^2$$

where C_S^u is given by the formula stated above, and $\{u_T^D(L)\}$ is the blade tip amplitude. This shows that the gasdynamic work on the blade is negative, i.e., dissipative, and that the form of the work expression is the same as the well-known form for viscous dissipation, i.e., $W = -(\text{const.})\ \pi\ C_D\ \gamma\ u^2$. A similar expression can be obtained for the axial gasdynamic work $W_{A,1}^D$, once the modal equation $u(z)_{A,1}$ is specified, and the distribution of the gas force along the length of the blade is known. The damping work for higher modes, and for coupled vibrations in both the axial and tangential directions, likewise depends mode shapes, and on the lengthwise distribution.

NUMERICAL EXAMPLE

In order to compare the magnitude of the gasdynamic damping with material damping consider a turbine blade idealized as a uniform cantilever beam 6 inches long, 1 inch wide, and 0.125 inches thick. Let there be 90 such blades equally spaced around the rim of a 30 inch diameter disk. This blade row delivers 4600 HP at 3600 rpm. The blade material is AISI 403 12% Cr stainless steel.

The fundamental natural frequency using Young's modulus $E = 30 \times 10^6$ psi and density $\rho = 0.283\ \text{lb/in}^3$ is given by:

$$f = \frac{1}{2\pi} \cdot \left(\frac{1.8751}{L}\right)^2 \left(\frac{EIg}{wA}\right)^{1/2}$$

$$= \frac{1}{2\pi} \left(\frac{1.8751}{6}\right)^2 \left[\frac{30\text{x}10^6\text{x}0.125^2\text{x}386.4}{0.283 \times 12}\right]^{1/2}$$

= 114 Hz at zero rpm.

At 3600 rpm this frequency rises to 166 Hz. This latter value is used to calculate the gasdynamic damping.

As the horsepower H = TN/63025, with $r_o = 15$ in., L = 6 in., and N = 3600 rpm, the steady torque T is:

$$T = (63025\text{x}4600)/(3600\text{x}90) = 894.8\ \text{lb. in.}$$

For a fiber stress of 5 ksi at the cantilever support, the corresponding tip displacement becomes:

$$\{U_T^D\} = \sigma_{b,\ max} L^2/3Ec$$

$$= (2)(5000)(6)^2/3(30 \times 10^6)(0.125)$$

$$= 0.032\ \text{in.}$$

The gasdynamic damping expression obtained previously gives:

$$E_g = 0.3634\pi \cdot \frac{9.55T\gamma/N}{(r_o+L/2)(r_o+L)} \cdot \{U_T^D\}^2$$

$$= 0.3634\pi \cdot \frac{9.55\ (894.8)}{(15+3)(15+6)} \cdot \frac{2\pi\ (116)}{3600} \cdot (0.032)^2$$

$$= 7.66 \times 10^{-3}\ \text{lb. in.}$$

The elastic strain energy U_b due to bending of the blade is given by:

$$U_b = 1/2 \int_V \sigma_i\ \varepsilon_i\ dV$$

where i = 1, . . . 6 for both the stress and strain tensors. Rejecting the transverse stress components gives the following approximate expression for U_b:

$$U_b = 1/2 \int_V (\sigma^2/E)\,dV$$

$$= 1/2 \int_V F^2x^2/EI\ .\ dV$$

$$= F^2L^3/6EI$$

$$= \sigma^2_{b,\ max}\ V/18E$$

Substituting the above numerical values gives:

$$U_b = \frac{(6x1x0.125)\ (5000)^2}{18 \times 30 \times 10^6}$$

$$= 3.47 \times 10^{-2}\ \text{in.lb.}$$

The strain energy U_c due to steady centrifugal loading of the blade can be obtained from the relation:

$$U_c = 1/2 \int_{r_o}^{r_o+L} (\rho^2\ A\Omega^4/E)\ [(r_o+L)^2 - x^2]\ dx$$

$$= (\rho^2 A\Omega^4/120E)r_o^5[(r_o+L/r_o)^2\ \{8(r_o+L/r_o)^3$$
$$- 15(r_o+L/r_o)^2 + 10\} - 3]$$

where $\rho = w/g$ is mass density. Substituting gives:

$$U_c = (0.283/386.4)^2\ \frac{(2\pi\cdot 60)^4}{8}$$

$$\times \frac{(1x0.125)}{(30x10^6)}\ [101347.2]$$

$$= 0.560\ \text{lb. in.}$$

The influence of the steady centrifugal load on the gasdynamic damping may be included by writing:

$$\delta = Eg/2\ (U_b+U_c)$$

$$= (7.66x10^{-3})/2(3.47x10^{-2}+0.567)$$

$$= 6.44 \times 10^{-3},\ \text{or } 0.644\%.$$

The relationship between the gasdynamic damping and the material damping is also of interest for this example. The material damping energy for the blade is given by Lazan's law [6], i.e.,

$$E_m = \int_V J\ \sigma_a^{\ n}\ dV$$

where J and n are material constants obtained by testing, and σ_a is alternating stress. Again representing the blade as a uniform cantilever and substituting gives:

$$E_m = 2 \int_o^L \int_o^{h/2} J\ (Fxy/I)^n\ b\ dy\ dx$$

$$= JV\ \sigma^n_{b,\ max}/(n+1)^2$$

where V = bhL, the volume of the blade. For AISI 403 stainless steel reference [6] gives $J = 2 \times 10^{-10}$, n = 2.1 at 900°F. Substituting for E_m:

$$E_m = (2x10^{-10})(6x1x0.125)(5000)^{2.1}/(3.1)^2$$

$$= 9.15 \times 10^{-4}\ \text{lb. in.}$$

The ratio of the gasdynamic and damping energies is:

$$Eg/E_m = 8.38$$

For convenience the log. dec. for the blade with material damping may be written as:

$$\delta = (Eg+E_m)/2(U_b+U_C)$$

$$= 7.21 \times 10^{-3}, \text{ or } 0.72\%$$

It appears that gasdynamic damping exceeds the material damping at a vibratory stress of 5 ksi because the blade thickness is small in relation to the corresponding vibration amplitude. However, experience with long, thick blades has also shown that material damping may be greater than gasdynamic damping in certain instances. The variation of log. dec. and the gasdynamic and material damping energies in the above example are shown in Figure 4 for different levels of dynamic stress at N = 3600 rpm.

COMMENTS

The expression obtained for gasdynamic damping is of the same form as velocity damping, and this allows the gasdynamic damping effect to be readily included into a solution for the blade forced vibrations. However, it is observed that the gasdynamic damping coefficient C_S^u is based on the stage torque T. When the stage power output, i.e., the output per blade, is reduced, the blade damping will decrease correspondingly.

The studies in this paper have been directed toward the lowest mode of a typical cantilever blade. Numerical results were obtained for an idealized blade in its fundamental mode, to demonstrate the principles involved. The theory applies equally well to actual blades, and to higher modes of such blades. It is now common practice to model such blades numerically in great detail. The gasdynamic damping formulation given herein may either be incorporated directly into a forced response code, e.g., BLADE [7], or computed separately using numerical values from displacement mode shapes, suitably calibrated. Finally, it is worth emphasizing that the beneficial effects derived from gasdynamic damping are reduced at frequencies away from resonance, because the phase angle λ_{mn} decreases the damping effect where resonance does not exist.

CONCLUSION

A theory of gasdynamic damping for steam turbine blades has been presented, and a formula for evaluating such damping is given for a simple case. Proven test data for gasdynamic damping are unavailable, and so experimental verification of the theory remains as a future task. Numerical results obtained with this formula are consistent with practical experience.

ACKNOWLEDGEMENT

Details of the numerical example were calculated by Mr. Rajinder Singh, Senior Project Engineer, STI, to whom the author extends his grateful thanks for this assistance.

REFERENCES

1) Legendre, R., 'Aerodynamic Damping of Compressor Blades and Turbine Turbomachinery Proceedings of Colloquium on Turbomachinery , Cambridge (U.K.), July 1967. Published by I.Mech.E., London, pp. 111-116.

2) Legendre, R., Aerodynamic Damping of Two-Mode Vibrations, Colloquium on Flow Research on Blading. Elseveir Publishing Company, 1970, pp. 197-207.

3) Hammons, T. J., 'Accumulative Fatigue Life Expenditure of Turbine Generator Shafts Following Worst-Case System Disturbances,' Transactions IEEE Winter Annual Meeting, Preprint - February 1983.

4) Rieger, N. F., Vibrations of Rotating Machinery - Part 2: Blading and Torsional Vibrations, The Vibration Institute, Clarendon Hills, IL, Chapter 7, p. 130.

5) Fleeter, S., Proceedings, AGARD Conference on High Speed Flow Through Turbomachine Blading.

6) Lazan, B. J., Damping Materials and Members in Structural Mechanics, Pergammon Press, Incorporated, New York, NY, 1968.

7) BLADE. Steam Turbine Blade Life Prediction Code (Research Version 0.0), User's Manual. Electric Power Research Institute, Palo Alto, CA, April 14, 1987.

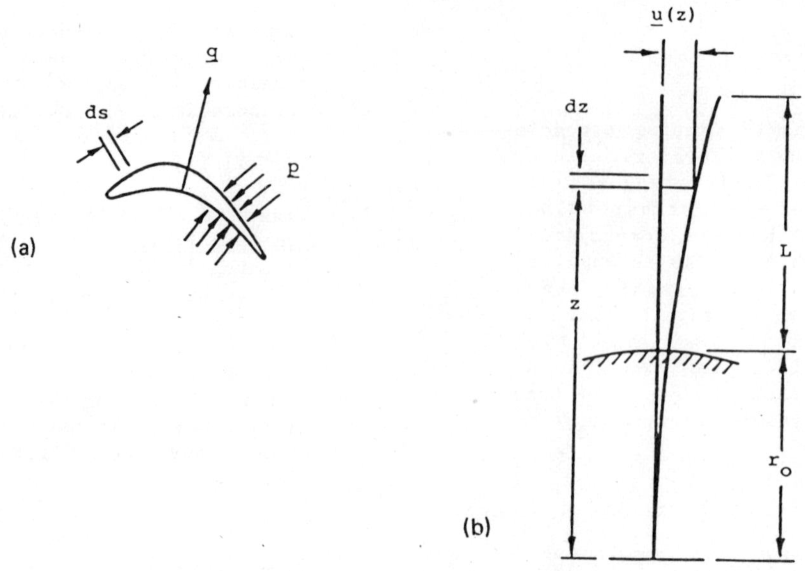

Fig 1 Blade loading and geometry details
(a) Details of gas loading on airfoil at radius z
(b) Blade geometry on disc

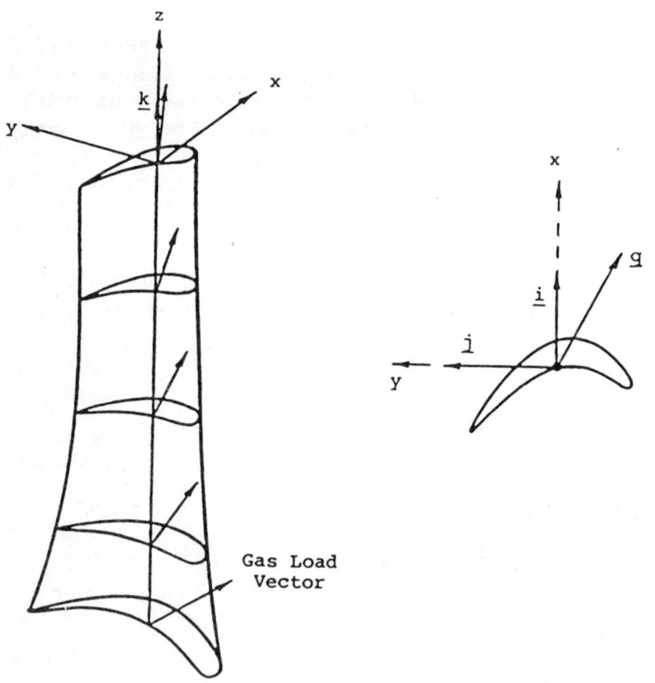

Fig 2 Blade coordinates and load vectors

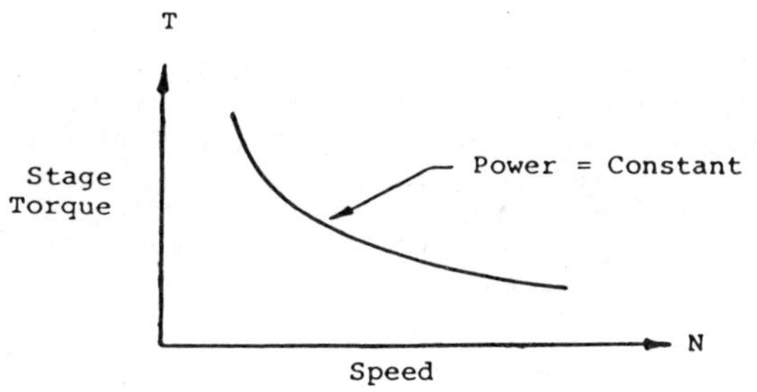

Fig 3 Stage power curve

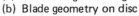

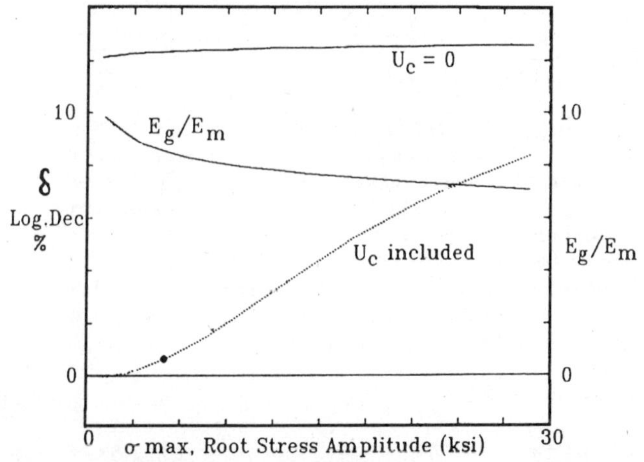

Fig 4 Total damping for rotating cantilever beam example

(10) EWINS, D. J., 'Estimation of Resonant Peak Amplitudes,' Journal of Sound and Vibration, Vol. 43, No. 4, 1975.

(11) DYE, RCF and HENRY, T. A., 'Vibration Amplitudes of Compressor Blades Resulting from Scatter in Natural Frequencies,' J. Eng for Power, ASME 91, 182-188, 1969.

(12) KAZA, K. R. V. and KIELB, R. E., 'Effects of Mistuning on Bending - Torsion Flutter and Response of a Cascade in Incompressible Flow,' DOE/NASA/1028-29, NASA TM-81674.

(13) KAZA, K. R. V. and KIELB, R. E., 'Coupled Bending-Torsion Flutter of a Mistuned Cascade with Nonuniform Blades,' NASA Tech. Memo 82813.

(14) KIELB, R. E. and KAZA, K. R. V., 'Aeroelastic Characteristics of a Cascade of Mistuned Blades in Subsonic and Supersonic Flows,' ASME Paper No. 81-DET-122.

(15) STANGE, W. A. and MACBAIN, J. C., 'An Investigation of Dual Mode Phenomena in a Mistuned Bladed Disk,' ASME Paper No. 81-DET-133.

(16) SINGH, M. P., 'Turbine Blade Dynamics: A Probabilistic Approach,' ASME Vibrations of Blades and Bladed Disk Assemblies, Book No. H00335.

(17) JOHNSON, D. C. and BISHOP, R.E.D., 'The Modes of Vibration of a Certain System Having a Number of Equal Frequencies,' Journal of Applied Mechanics, Sept. 1956, pp. 379-384.

APPENDIX

<u>Derivation of Equation of Motion</u>

The basic model used is based on one analyzed by Johnson and Bishop (17). In the present analysis, damping has been introduced between the large masses representing the disk and the small masses representing the blades.

Equations of motion have been derived using Lagrange's Method.

The kinetic energy of the system

$$2T = M\dot{q}_0^2 + m_1\dot{q}_1^2 + \cdots + m_n\dot{q}_n^2$$

The potential energy of the system

$$2V = k_1(q_0 - q_1)^2 + k_2(q_0 - q_2)^2 + \cdots + k_n(q_0 - q_n)^2$$

The dissipative energy due to damping

$$2D = C_1(\dot{q}_0 - \dot{q}_1)^2 + \cdots + C_n(\dot{q}_0 - \dot{q}_n)^2$$

The Lagrangian equation is expressed as:

$$\frac{d}{dt}\left(\frac{\partial T}{\partial \dot{q}_r}\right) + \frac{\partial D}{\partial \dot{q}_r} + \frac{\partial V}{\partial q_r} = 0$$

Equations of motion

$$M\ddot{q}_0 + \left(\sum_{r=1}^{n} C_r\right)\dot{q}_0 + \left(\sum_{r=1}^{n} k_r\right)q_0 - \sum_{r=1}^{n} C_r\dot{q}_r - \sum_{r=1}^{n} k_r q_r = 0$$

$$m_1\ddot{q}_1 + C_1\dot{q}_1 + k_1q_1 - C_1\dot{q}_0 - k_1q_0 = 0$$
$$\vdots$$
$$m_n\ddot{q}_n + C_n\dot{q}_n + k_nq_n - C_n\dot{q}_0 - k_nq_0 = 0$$

The determinant for the n mass system

$$\Delta = \begin{vmatrix} \left(\sum^{n} k_r - M\omega^2 + i\omega\sum^{n} C_r\right) & -(k_1 + i\omega C_1) & \cdots & -(k_n + i\omega C_n) \\ -(k_1 + i\omega C_1) & (k_1 - m_1\omega^2 + i\omega C_1 & \cdots & 0 \\ -(k_2 + i\omega C_2) & 0 & \cdots & 0 \\ \vdots & \vdots & & \vdots \\ -(k_n + i\omega C_n) & 0 & \cdots & (k_n - m_n\omega^2 + i\omega C_n) \end{vmatrix}$$

After reducing the expression

$$\Delta = \left\{\prod^{n}(k_r - m_r\omega^2 + i\omega C_r)\right\} \times \left[\left(\sum^{n} k_r - M\omega^2 + i\omega\sum^{n} C_r\right) - \sum^{n}\frac{(k_r + i\omega C_r)^2}{k_r - m_r\omega^2 + i\omega C_r}\right]$$

$$\Delta_{oo} = \prod^{n} (k_r - m_r\omega^2 + i\omega C_r)$$

$$\Delta_{lo} = \frac{-(k_1 + i\omega C_1)\prod^{n} (k_r - m_r\omega^2 + i\omega C_r)}{k_1 - m_1\omega^2 + i\omega C_1}$$

$$\Delta_{lp} = \frac{(k_1 + i\omega C_1)(k_p + i\omega C_p)\prod^{n} (k_r - m_r\omega^2 + i\omega C_r)}{(k_1 - m_1\omega^2 + i\omega C_1)(k_p - m_p\omega^2 + i\omega C_p)}$$

and

$$\Delta_{ll} = \frac{\left(\sum^{n} k_r - M\omega^2 + i\omega \sum^{n} C_r\right)\left\{\prod^{n} (k_r - m_r\omega^2 + i\omega C_r)\right\}}{k_1 - m_1\omega^2 + i\omega C_1} + \frac{\prod^{n} (k_r - m_r\omega^2 + i\omega C_r)}{k_1 - m_1\omega^2 + i\omega C_1} \times \left[\sum^{n} \frac{(k_r + i\omega C_r)^2}{k_r - m_r\omega^2 + i\omega C_r} - \frac{(k_1 + i\omega C_1)^2}{k_1 - m_1\omega^2 + i\omega C_1}\right]$$

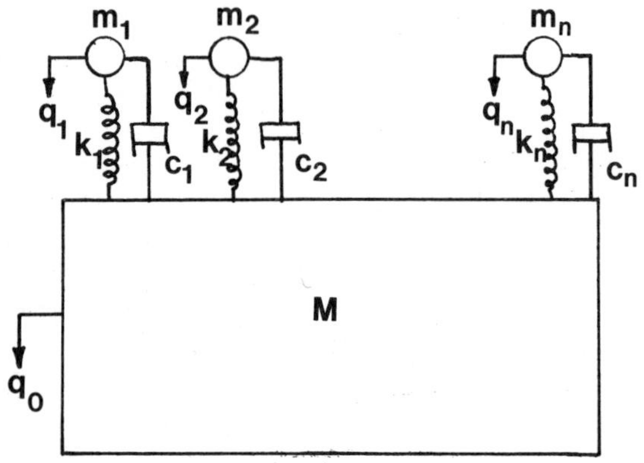

Fig 1 Basic model

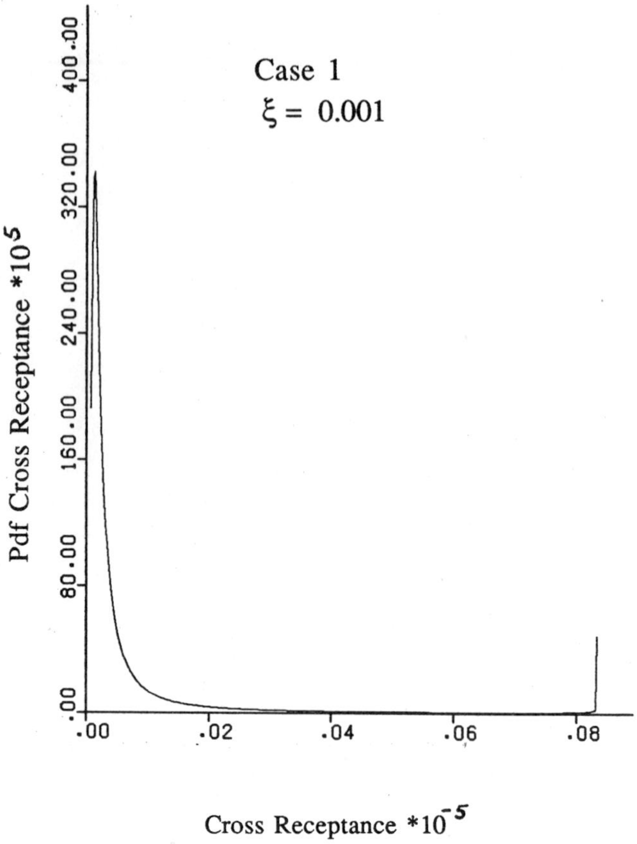

Fig 2 Pdf cross receptance

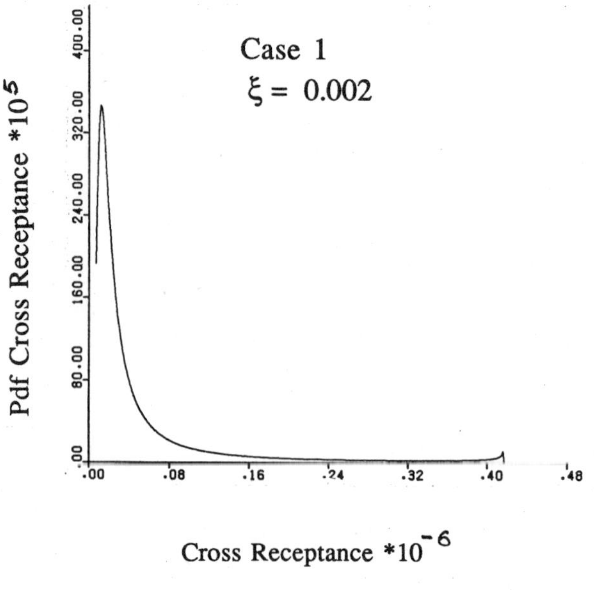

Fig 3 Pdf cross receptance

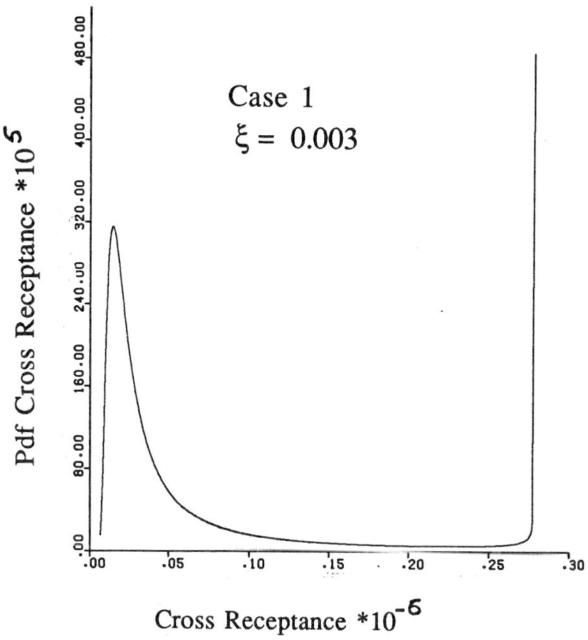

Fig 4 Pdf cross receptance

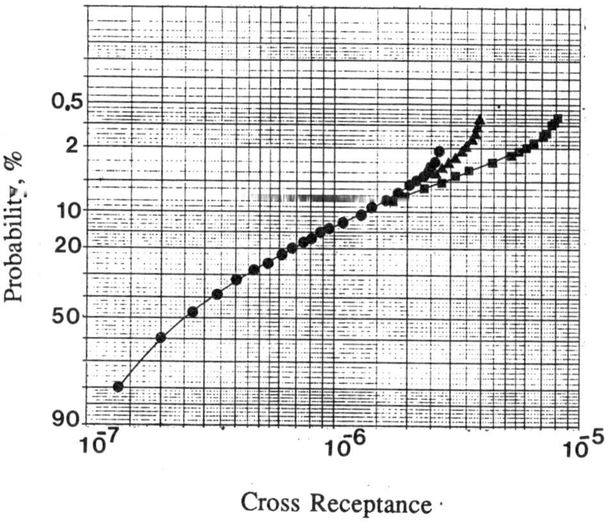

Fig 5 Cumulative distribution (case 1)

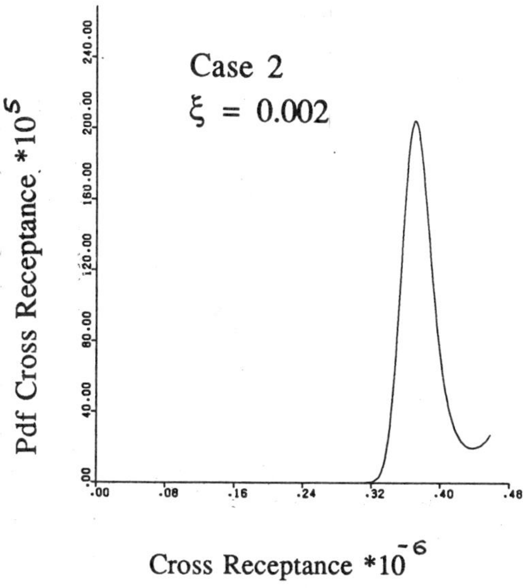

Fig 6 Pdf cross receptance

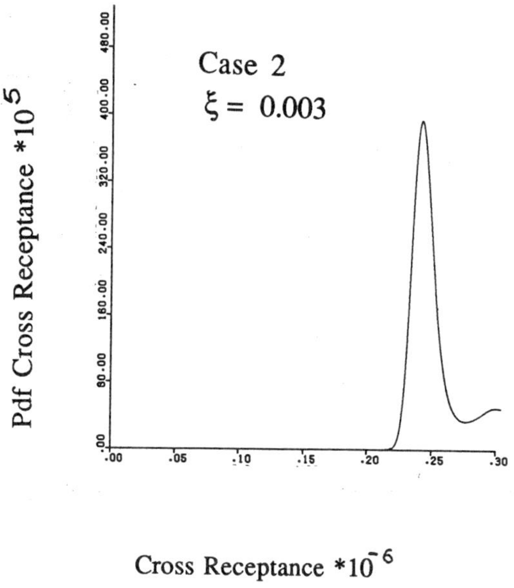

Fig 7 Pdf cross receptance

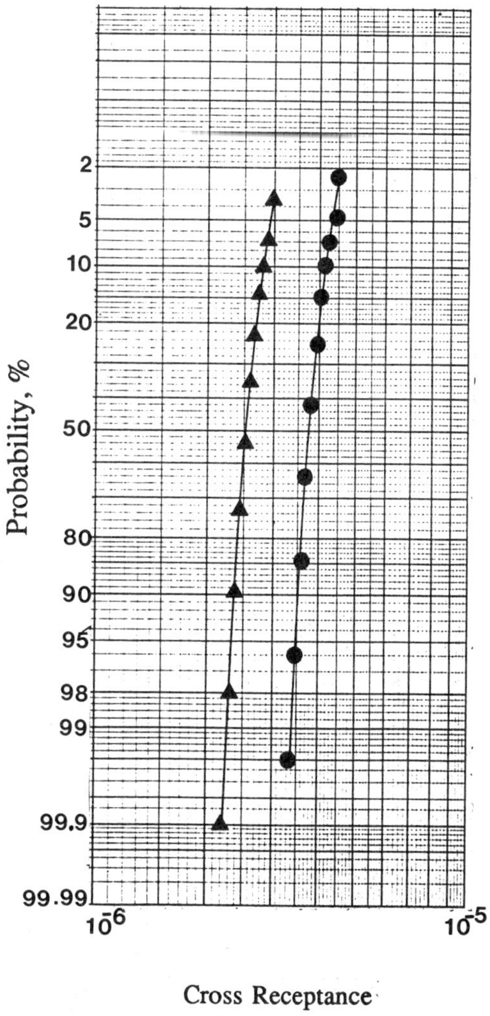

Fig 8 Cumulative distribution (case 2)

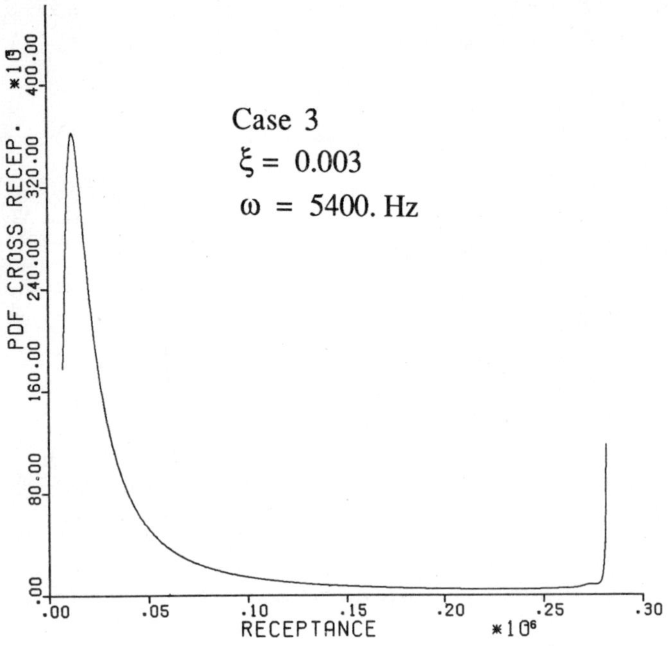

Fig 9 Pdf total receptance

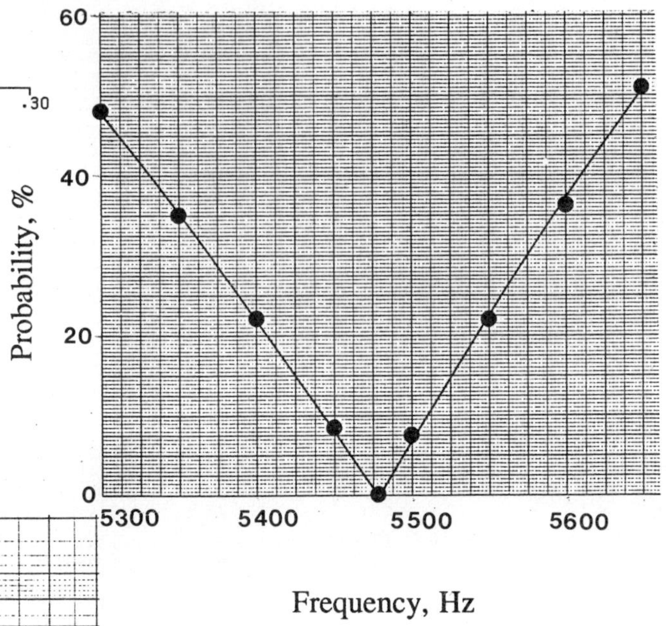

Fig 11 Probability of total receptance to be higher than the receptance at design condition

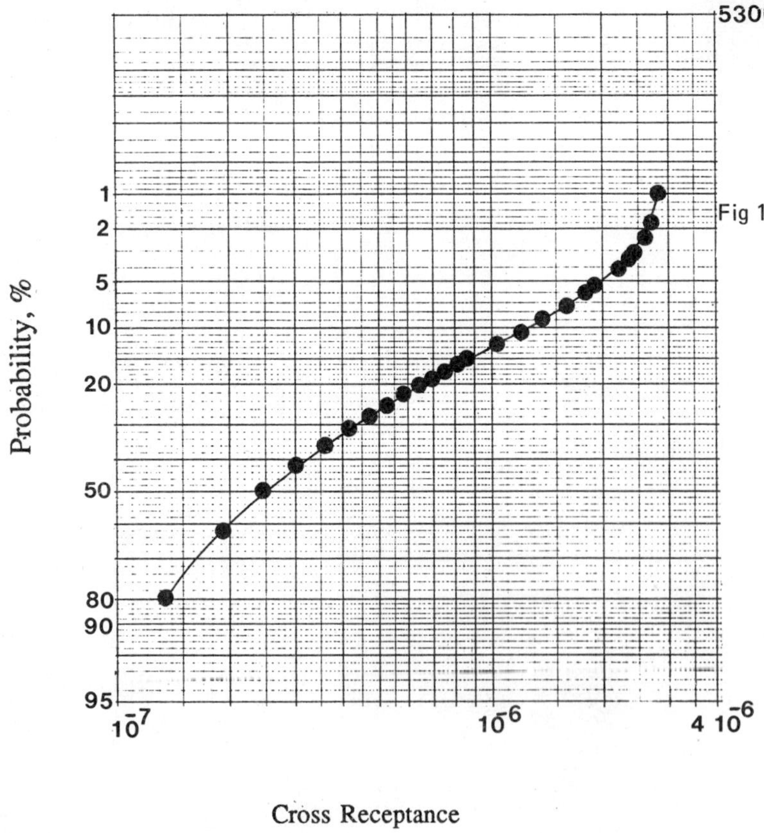

Fig 10 Cumulative distribution (case 3)

C266/88

Free vibration of the mis tuned bladed disc

R RZADKOWSKI, PhD
Institute of Fluid Flow Machinery, Polish Academy of Science, Gdansk, Poland

SYNOPSIS The problem of calculating the natural frequencies of a rotating bladed disk is solved by the Ritz Method. A consistent second-order thick plate theory is used for the disc. The blade is treated by a beam theory and includes shear deformation, rotary inertia effects, blade pretwist and centrifugal effects. An agreement between predicted and experimentally-determined frequencies of the system is shown to be good.

NOTATION

A area of the cross-section of the blade

$A_{\Omega B}$ work done by the centrifugal forces of the blade

$A_{d\Omega}$ work done by the centrifugal forces of the disc

A_{kB} work done by the Coriolis force of the blade

A_{dc} additional strain energy of the disc

$a_{ij}, b_{ij}, d_{ij}, e_{ij}$ parameters of the aproximated functions

a,b,d,e vectors of the parameters $a_{ij}, b_{ij}, d_{ij}, e_{ij}$

h thickness of the disc

$i_{\alpha\beta}$ maximal radius of gyration of the blade

$k_{\alpha\beta}$ shear factor

k nodal diameter

K stiffnes matrix

L blade length

M mas matrix

N number of blades

q generalize coordinate vector

p natural freqencie

R disc outer radius.

r_o disc inner radius

u_i coordinates of the displacements of the volume element of the blade

u_i^o coordinates of the displacements of the neutral axis of the blade

u_{id} coordinates of the displacements of the volume element of the disc

ω_i cross section rotation angles of the blade

ϵ_{ij} permutation symbol

χ_α deplananation function of the blade ?

φ warping function of the cross section of the blade

ϑ_o angle of the blade pretwist

γ_α angle of the shear forces

SUBSCRIPTS

i,j index(1,2,3)

α,β,μ,ν index(1,2)

SUPERSCRIPTS

' derivative with respect to x_3

˙ derivative with respect to t

1 INTRODUCTION

The structural dynamic aspects of a mistuned bladed disc have been studied by many researches e.g [1,2,4,5,7,11,13]. They have used discrete and continuous models to calculate natural frequencies and a dynamical response of the system. The conclusions of such studies have sometimes been inconsistent. For example it is desirable to identify in advance the blades that are likely to experience the largest vibration amplitudes.Some workers have observed that such blades are most likely to be those having extreme detune[1,5], while others disagree [6]with it. In order to solve this problem Afolabi [2] divided the spectrum of a mistuned assembly into three classes and tried to resolve some contradictions in the forced response characteristic of mistuned assemblies previously reported by earlier workers.In this papar we would like to give another answer to this problem and suggest the best distribution of the detuned blades around the disc from the stress point of view. The influence of the number of blades on the natural frequencies and modal shapes of the tuned system will be considered as well.

2 BASIC ASSUMPTIONS

A consistent second-order thick plate theory is used for the disc. This thick plate theory is due to Blocki and the general theory is reported in reference [3]. In this application the effect of prestress is also included. It may arise from centrifugal loads in the disc and centrifugal loads-from the blades. The disc is assumed to be rotationally symmetric, the middle surface is assumed to be a plane of symmetry. The blades are treated by a beam theory developed by Janecki[8]. It includes shear deformation, rotary inertia effects, blade pretwist,blade twist, centrifugal effects. The blades are assumed to be rigidly attached to the disc outer boundary.

3 THE BLADE MODEL

The blade is modelled by a twisted and pretwisted beam with the variable cross-section area. The blades are made from homogenous and isotropic material and are subject to Hooke's Law. The centre of the mass and the centre of bending $x_{s\alpha}$ of the blade cross-section are not coincident. The bending-twist vibration is due to: twist of the blade, the asymmetrical cross-section, shear and centrifugal forces and the deformation of the cross-section in the flexural direction. The blade is located at the right-handed coordinates x_{1i} with the origin at the centre of mass at the cross-section of the blade root and it rotates with an angular velocity Ω (Fig 1). The axis x_{12} is normal to the axis of rotation and has the same direction as the radius of the disc. The axes x_{0i} are the main central axes of inertia at any cross section of the blade. The angle of the blade pre-twist is denoted by ϑ_o. The displacement of a point in an arbitrary cross-section of x_{1i} is denoted respectively:

$$u_\alpha = u^\circ_\alpha - \epsilon_{\alpha\beta}\, \bar{x}_{1\beta}\, \omega_3,$$
(1)
$$u_3 = u^\circ_3 + \epsilon_{\alpha\beta}\, \omega_\alpha\, x_{1\beta} + \epsilon_{\alpha\beta}\, \gamma_\alpha \chi_\beta + \omega'_3\, \varphi\ ,$$

where $\bar{x}_{1\beta} = x_{1\beta} - x_{s\beta}$. (2)

From the linear theory the componets of the strain tensor are:

$$\varepsilon_{11} = \varepsilon_{12} = \varepsilon_{22} = 0,$$

$$\varepsilon_{\alpha 3} = 0.5\left(\frac{\partial\varphi}{\partial \bar{x}_{1\alpha}} - \epsilon_{\alpha\beta}\bar{x}_{1\beta}\right)\omega'_3 +$$

$$0.5\epsilon_{\mu\nu}\left(\delta_{\mu\alpha} - \frac{\partial\chi_\mu}{\partial \bar{x}_{1\alpha}}\right)\gamma_\nu \qquad (3)$$

$$\varepsilon_{33} = (u^\circ_3)' + \epsilon_{\alpha\beta}\, x_{1\beta}\, \omega'_\alpha + \vartheta'_o\, \rho^2\, \omega'_3 + \varphi\, \omega''_3 +$$
$$+\epsilon_{\alpha\beta}\, \gamma'_\alpha\, \chi_\beta,$$

$$\rho^2 = \bar{x}_{o2}\frac{\partial\varphi}{\partial \bar{x}_{o1}} - \bar{x}_{o1}\frac{\partial\varphi}{\partial \bar{x}_{o2}} \qquad (4)$$

From the Hooke's law for the isotropic material we obtain

$$\tau_{\alpha 3} = 2G\varepsilon_{\alpha 3}\ , \qquad \tau_{33} = E\varepsilon_{33}. \qquad (5)$$

The strain energy of the blade is

$$\mathcal{V} = 0.5\int_o^L\int_A \varepsilon_{i3}\tau_{i3}\, dA dl \qquad (6)$$

It is convenient from the numerical point of view to express the strain energy in terms of generalized forces Q_k, M_k ,B.

$$Q_\alpha = GA\ \epsilon_{\mu\nu}\ k_{\alpha\mu}\ \gamma_\nu\ ,$$

$$Q_3 = \int_A \tau_{33} dA,$$

$$M_\beta = \in_{\alpha\beta} \int_A x_{1\beta}\, \tau_{33}\, dA \quad ,$$

$$M_3 = M_\vartheta + \vartheta'_o B_\rho \qquad (7)$$

$$M_\vartheta = GI_\vartheta\, \omega'_3 \quad ,$$

$$B_\rho = \int_A \rho^2\, \tau_{33}\, dA \; .$$

$$B = \int \varphi\, \tau_{33}\, dA$$

$$\text{where} \quad k_{\alpha\mu} = \int_A (\delta_{\alpha\beta} - \frac{\partial \chi_\alpha}{\partial x_{1\beta}})(\delta_{\mu\beta} - \frac{\partial \chi_\mu}{\partial x_{1\beta}}) dA.$$

Substituting eqs (3)(5) for (7) results in:

$$Q_3 = E(Au^{o\,\prime}_3 + \vartheta'_o\, I_\rho\, \omega'_3) \; ,$$

$$M_1 = E\, (I_1 \omega'_1 - I_{12}\, \omega'_2 + \vartheta'_o\, I_{1\rho}\, \omega'_3 + I_{1\varphi}\, \omega''_3) \; ,$$

$$M_2 = -E\, (I_{12}\omega'_1 - I_{2\varphi}\omega'_2 + \vartheta'_o\, I_{2\varphi}\omega'_3 + I_{2\varphi}\omega''_3),$$

$$B = E\, (I_{1\varphi}\omega'_1 - I_{2\varphi}\, \omega'_2 + \vartheta'_o\, I_{\rho\varphi}\, \omega'_3 + I_\varphi\, \omega''_3), \qquad (8)$$

$$B_\rho = E\, (I_\rho\, u^{o\,\prime}_3 + I_{1\rho}\omega'_1 - I_{2\rho}\omega' + \vartheta'_o\, I_{\rho\rho}\, \omega'_3 + I_{\rho\varphi}\, \omega''_3),$$

where

$$I_\alpha = \int_A |\in_{\alpha\beta}| x^2_\beta\, dA \quad , \qquad I_{12} = \int_A x_1 x_2 dA \quad ,$$

$$I_\rho = \int_A \rho^2\, dA \quad , \qquad (9)$$

$$I_{\alpha\nu} = \int_A |\in_{\alpha\beta}| x_\beta\, \nu\, dA \quad , \; I_{\rho\nu} = \int_A \rho^2\, \nu\, dA,$$

$$I_\nu = \int (\frac{\partial \varphi}{\partial x_{1\mu}} - \in_{\mu\beta}\, \bar{x}_{1\beta})^2 dA \; . \qquad (10)$$

From the transformation formula between coordinates (O_o, x_{oi}), (O_1, x_{1i}) i.e. the global and global non-inertial system of reference leads to the following:

$$x_1 = R x_o \quad , \qquad x_1 = \mathrm{col}(x_{1i}),$$

$$\mathbf{R} = \begin{bmatrix} \cos\vartheta_o & -\sin\vartheta_o & 0 \\ \sin\vartheta_o & \cos\vartheta_o & 0 \\ 0 & 0 & 1 \end{bmatrix} . \qquad (11)$$

The strain energy is a function of M_{o1}, M_{o2}, M_3 Q_{o1}, Q_{o2}, Q_3, B.

In order to determine the kinetic energy of the rotating blades the displacements are considered in the stationary reference axes X_i:

$$U_1 = u_1,$$

$$U_2 = (R + x_{13} + u_3)\cos\Omega t + (u_2 + x_{11})\sin\Omega t, \qquad (12)$$

$$U_3 = (R + x_{13} + u_3)\sin\Omega t - (u_2 + x_{11})\sin\Omega t,$$

The kinetic energy of the rotating beam is:

$$T = 0.5\rho \iint_{OA}^{L} \dot{U}^2_i dA\, dx_{13}. \qquad (13)$$

Substituting eqs (12) for (13) :

$$T = T_o + A_\Omega + A_k. \qquad (14)$$

From eqs(8),(11) the kinetic energy of the beam is a function of parametrs M_{o1}, M_{o2}, M_3, Q_1, Q_2, Q_3, B.

4 THE DISC MODEL

The thick plate theory is due to Blocki[3]. The basic assumptions of this theory are: the deflection of the mid-surface of the plate is small in comparison with the plate thickness; normals to the mid-surface of the plate before deformation remain straight but not necessarily normal to the mid-surface after deformation . The deformation of the plate can be described by the five parameters u_{id} $(i=r,\theta,1)$, φ_{kd} $(k=r,\theta)$. Then the displacements of the plate at the coordinates (r,θ,x_1) can be expressed in the form:

$$u_{d\alpha} = u_{d\alpha} + x_1 \varphi_{d\alpha} + \frac{x_1^3}{6} \chi_{d\alpha}, (\alpha = r,\theta), \qquad (15)$$

$$u_{d1} = u_{d1},$$

where

$$\chi_{dr} = -\frac{8}{h^2}(1 - 0.5\,\nu)(\frac{\partial u_{d1}}{\partial r} + \varphi_{dr}) \qquad (16)$$

$$\chi_{d\theta} = -\frac{8}{h^2}(1 - 0.5\,\nu)(\frac{\partial u_{d1}}{r\partial\theta} + \varphi_{d\theta}) \qquad (17)$$

Using the linear theory, the stress tensor can be calculated.

The strain energy of the disc is given by:

$$\mathcal{E}_d = 0.5 \int_{r_o}^{R} \int_0^{2\pi} \int_{-h/2}^{h/2} \left[\frac{E}{1-\nu^2} (\varepsilon^2_r + 2\nu\varepsilon_r\varepsilon_\theta + \varepsilon^2_\theta) + \frac{5}{6} Gh\varepsilon^2_{r1} + \frac{5}{6} Gh\varepsilon^2_{\theta 1} \right] r d\theta dr dl. \qquad (18)$$

In order to take into account the

influence of the rotation of the disc on the vibrational characteristic of the bladed disc the displacement of the elementary part of the disc has to be investigated in the stationary reference axes x_i up to terms with the second order

$$U_{d1} = u_{d1} \qquad (19)$$

$$U_{dr} = (u_{dr} + r - \xi_d - u^2_{d\theta}/2r)\sin\Omega t - u_{d\theta}\cos\Omega t$$

$$U_{d\theta} = (u_{dr} + r - \xi_d - u^2_{d\theta}/2r)\cos\Omega t + u_{d\theta}\sin\Omega t.$$

where ξ_d the change of the coordinates of the elementary volume of the disc due to bending is:

$$\xi_d = 0.5\int_{r_o}^{r}\left(\frac{\partial u_{d1}}{\partial r}\right)^2 dr,$$

where

$u^2_{d\theta}/2r$ is the difference between the radial displacements with reference to different means of measurment.

The kinetic energy of the rotating disc is given by:

$$T_d = 0.5\rho\int_r^R\int_o^{2\pi}\int_{-h/2}^{h/2} \dot{u}^2_{di} r \, dr \, d\theta \, dx_1 \qquad (20)$$

Using eqs(19),(20) and integrating (20) with respect to x_1 gives:

$$T_d = T_{do} + A_{d\Omega} + A_{dk} \qquad (21)$$

The additional strain energy in bending due to initial in-plane stress τ_{ro}, $\tau_{\theta o}$ caused by the static components of centrifugal force is

$$A_{dc} = 0.5\int_{r_o}^{R}\int_0^{2\pi}\left[N_r\left(\frac{\partial u_{d1}}{\partial r}\right)^2 + N_\theta\left(\frac{\partial u_{d1}}{r\partial\theta}\right)^2\right] r\,dr\,d\phi \qquad (22)$$

where

$$N_r = \int_{-h/2}^{h/2}\tau_{ro}\,dx_1 \quad , \quad N_\theta = \int_{-h/2}^{h/2}\tau_{\theta o}\,dx_1.$$

The stresses τ_{ro}, $\tau_{\theta o}$ in radial and angular directions are calculated by means of a spline interpolation technique [10].

5 COUPLING THE SUBSTRUCTURES

Let us assume that the disc and the blades are one system and that the angular distance between the blades is $d\theta$. Assuming the rigid fixed blades in the rim the transitional equations for the displacements and angles are:

$$u_{1cn} = (u_{d1} - \varphi_{dr}\,x_{13}) + u_{1n},$$

$$u_{2cn} = -u_{d\theta}\left(\frac{R+x_{13}}{R}\right) + u_{2n},$$

$$u_{3cn} = u_{dr} + u_{3n}$$

$$\omega_{1cn} = \frac{u_{d\theta}}{R} - \frac{\partial u_{dr}}{r\partial\theta} + \omega_{1n},$$

$$\omega_{2cn} = -\varphi_{dr} + \omega_{2n},$$

$$\omega_{3cn} = -\varphi_{d\theta} + \omega_{3n}, \qquad n=1,\dots,N.$$

6.EIGENVALUE PROBLEM

The vibration analysis of the bladed disc can be carried out using variational methods. The functional of the problem contains the sum of the energies of all the elements of the system.

$$\delta\mathcal{P} = \delta\int_{t_2}^{t_1}\left[(\mathcal{E}_d - T_d - A_{d\Omega} + A_{dc} + \sum_{i=1}^{N}(\mathcal{E}_i - T_i - A_{i\Omega})\right]dt = 0 \qquad (24)$$

The equations governing the vibration of the bladed disc are obtained from the functional (24). The approximate solutions of these equations are to be found using the Ritz Method.The parameters describing vibration of the disc u_{id}, φ_{dk} were approximated by

$$l_d = \sum_{j=1}^{C}\sum_{k=1}^{K}\varphi_{ldj}(r)(a_{jk}\sin k\theta + b_{jk}\cos k\theta)\sin pt = \qquad (25)$$

$$= \sum_{k=1}^{K} A_{ld}(a \sin k\theta + b \cos k\theta)\sin pt.$$

where $\varphi_{ldj}(r)$ are the eigenfunctions of the cantilever beam. For example:

$$u_{d1} = \sum_{j=1}^{C}\sum_{k=1}^{K}\varphi_{ud1j}(r)(a_{jk}\sin k\theta + b_{jk}\cos k\theta)\sin pt, \qquad (26)$$

$$\varphi_{ud1j} = \Big[\cosh c_j (r-r_o)/(R-r_o) - \cos c_j (r-r_o)/(R-r_o) \Big] - \alpha_j \Big[\sinh c_j (r-r_o) - \sin c_j (r-r_o)/(R-r_o) \Big] ,$$

$$\cosh c_k \cos c_k + 1 = 0 ,$$

$$\alpha_j = (\cos c_j + \cosh c_j)/(\sin c_j + \sinh c_j) .$$

The parameters describing the vibration of the blades $M_{01}, M_{02}, M_3, Q_{01}, Q_{02}, Q_3, B$ are aproximated by:

$$l_{bn} = \sum_{j=1}^{C} \sum_{k=1}^{K} \varphi_{lbnj}(r) \Big[d_{jk} \sin(2\pi kn/N) + e_{jk} \cos(2\pi kn/N) \Big] \sin pt \qquad (27)$$

For example

$$M_{01n} = \sum_{j=1}^{C} \sum_{k=1}^{K} \varphi_{M1nj}(r) \Big[d_{jk} \sin(2\pi kn/N) + e_{jk} \cos(2\pi kn/N) \Big] \sin pt$$

where

$$\varphi_{M11j} = \Big[\cosh c_j \xi - \cos c_j \xi \Big] - \alpha_j \Big[\sinh c_j \xi - \sin c_j \xi \Big] , \quad \xi = x_{13}/L.$$

Putting eqs (25), (27) into (24) we obtain

$$\delta \mathcal{P} = \delta \int_{t_1}^{t_2} \Big[q_d^T (E_d + A_{dc} - A_{d\Omega}) q_d + (q_d^T T_d q_d) + q_B^T (E_B - A_{B\Omega}) q_B + q_B^T T_B q_B \Big] dt \qquad (28)$$

From Lagrange's equation we find that

$$(M - p^2 K)\, q = 0$$

7.VALIDATION OF THE PROGRAM

Jager [9] reports on the frequency measurement on a bladed disc with a simple geometric shape. The blade has a rectangular cross section. The calculations of this case have also been reported in [12]. The results of the various authors are given in the Table 1

Table 1. Measured and calculated natural frequencies of a bladed disc

Mode no	Measured Ref[9]	Calcul. Ref[9]	Calcul. Ref[12]	Calcul. This report
1	164	154	144	146
2	430	450	448	448
3	985	1005	993	1014
4	1930	2040	2011	2012
1	237	230	225	232
2	490	515	510	511
3	1215	1270	1295	1304
4	2050	2145	2143	2151
1	280	276	275	286
2	585	599	592	536
3	1500	1600	1580	1670
4	2200	2275	2332	2304

8.TUNED BLADED DISCS

In many works on calculating natural frequencies of a tuned bladed disk,the displacements of the middle plane are described as:

$$u_{d1} = u_{d1c} \cos k\theta + u_{d1s} \sin k\theta \qquad (29)$$

or

$$u_{d1} = u_{d1c} \cos k\theta . \qquad (30)$$

The wave number can attain the values 0,1,...,int(N/2). The analysis proceeds by fixing the value of k and performing all calculations of interest for that particular value of k. The process is repeated for all values of k which are of interest. In this paper the displacements of the middle plane of the disc are described by eq.(25) and forces and moments of the blades by eq (27).So the coupled vibrations of the disc and blades through the nodal diameters are taken into account. Putting eq.(25,27) to (24) we obtained

$$\sum_{n=1}^{N} G(r,\theta) \sin(k_1 2\Pi n/N) \sin(k_2 2\Pi n/N) \qquad (31)$$

$$\sum_{n=1}^{N} G(r,\theta) \sin(k_1 2\Pi n/N) \cos(k_2 2\Pi n/N), \qquad (32)$$

$$\sum_{n=1}^{N} G(r,\theta) \cos(k_1 2\Pi n/N) \sin(k_2 2\Pi n/N),$$

$$\sum_{n=1}^{N} G(r,\theta) \cos(k_1 2\Pi n/N) \cos(k_2 2\Pi n/N), \qquad (33)$$

where $G(r,\theta)$ are matrices of the geometrical characteristic of the blades and the disc.

The values of the trigonometric sum are known:

$$\sum_{n=1}^{N} \sin(k_1 2\Pi n/N)\sin(k_2 2\Pi n/N) = \begin{cases} N/2 \ , |k_1-k_2| = Nc \text{ and } k_1+k_2 \neq Nc \\ -N/2, |k_1 - k_2| \neq Nc \text{ and } k_1 + k_2 = Nc \\ 0 \ , k_1, k_2 \end{cases} \quad (34)$$

$$\sum_{n=1}^{N} \cos(k_1 2\Pi n/N)\cos(k_2 2\Pi n/N) = \begin{cases} N \ , |k_1-k_2| = Nc_1 \text{ and } k_1 + k_2 = Nc_2 \\ N/2 \ , |k_1-k_2| \neq Nc \text{ and } k_1+k_2 = Nc \\ \text{and } |k_1 - k_2| = Nc \text{ and } k_1 + k_2 \neq Nc \\ 0 \ , k_1, k_2. \end{cases} \quad (35)$$

$$\sum_{n=1}^{N} \sin(k_1 2\Pi n/N)\cos(k_2 2\Pi n/N) = \sum_{n=1}^{N} \cos(k_1 2\Pi n/N)\sin(k_2 2\Pi n/N) = 0 \quad (36)$$

From the trigonometric sum we can calculate that for a disc with three blades the following nodal diameters are conjugated (1,2),(1,4),(1,5),.. . Whereas for the disc with six blades are the following ones (1,5),(1,7),(2,8),....So in calculation for a small number of blades it is important to consider conjuction of the vibration through the nodal diameters. This conjunction causes the curvature of the nodal lines [11] but doesn't cause the splitting of the natural frequencies . The conclusion follows that the conjuction of vibration through the nodal diameters exerts influence on the mode shape of the system . Let us now consider the dynamical behaviour of the tuned bladed disk when the number of blades is increased. The calculations were made for the bladed disc Tab 2. The numerical results can be seen at Fig 2. From that figure it can be observed that the frequencies of a tuned bladed disc are divided into two groups

Table 2

Physical Properties of Bladed Disc Assembly

Disc	r_o=2.6cm ,R =30cm, h=2cm
Blade	L=20cm,width a=1cm,thickness b=2
E=2.1	10^6kG/cm^2,ν=0.3,ρ=7.98kGs^2/cm^4

One group is associated with the vibration of the cantilever blades(BB bending-bending frequencies) . The second group is associated with vibration of the disc. When the number of the blades increases the natural freqencies from the first group increase whereas in the second group the frequencies decrease. In the second group frequencies decrease from natural freqencies of the disc (r_o=2.6cm,R=30cm) BTm to the natural frequencies of the disc(r_o=2.6cm R=R+L=20+30cm)BTw. One more interesting property of the system should be noticed. When the number of blades changes the position of the nodal diameters changes as well. It has not been found out , what kind of relationship that is. Taking into account the unknown functions of the blades and the disc in the form(eqs(25,27))we can divide all natural frequencies into two groups. The first group comprises the single frequencies resulting from the vibration of the system for nodal diameters k=0, the second group includes the doubled frequencies connected with the vibration of the system for k greater then zero(these frequencies split when the detuning appears).

9. MISTUNED BLADED DISC

Let us consider the bladed disc with only one blade detuned (tab 3). Mistuning of the one blade causes both the splitting of the double natural frequencies and the curvature of the nodal lines. From the mathematical point of view splitting depends on the elements of the stifness and the mass matrix multiplied by trigonomertrical sum eqs(32) .

Table 3

Physical Properties of Bladed Disc Assembly

Disc	r =2.6 cm,R=30cm ,h=2cm		
			detuned
blades L=5 cm, axb		1 x2	1.1 x2.2
		0.75x1.5	0.825x1.6
		0.5 x1	0.555x1.1
number of blades N=60			

These being not equal to zero for a mistuned system.In order to find the worst position of one detuned blade around the disc from the stress point of view the position of the detuned blade has been changed and the natural frequencies as well as modal shapes have been calculated .It has been found out that changing of the position of one detuned blade doesn't change the value of the splitting of the doubled natural freqencies and the magnitude of the maximal curvature of the nodal lines. As a result of the position changing of the detuned blade the position of the nodal diameter has changed too. It has been observed that knowing the positon of the nodal line leads to finding the position of the detuned blade and vice versa. From the stress point of view the detuned blade has the maximal amplitude of vibration. Let us now consider bladed disc with two detuned blades (Tab 3.). One of the detuned blade is placed at the arbitrary position for example 8 (see Fig 3). The position of the other one is continuously changed around the the disc. Fig 4a shows the doubled natural frequencies of the modal shape with one nodal diameter, when the positon of the second detuned blade has been changed. Fig 4b shows the magnitude of the splitting of double frequencies. From Fig 4. it can be seen that the maximal splitting of the natural frequencies occurs when the positons of detuned blades are (8,7),(8,9),(8,22),(8,52). Does the maximal splitting correspond to the maximal value of the curvature of the nodal diameters? Fig 5 shows that the maximal curvature is equivalent to maximal splitting only when positions of the detuned blades are (8,7),(8,9). In these positons the detuned blades experience the maximal stress .As far as 3 and 4 and more detuned blades are concerned it is very difficult to say in advance which of the blade in the mistuned bladed disc experience the maximal stress. For the certain distribution of the blade natural frequencies around the disc the most detuned blade shows the maximal stress. In case of another distribution of the same mistuned blades the maximal stress can be observed at another blade which hasn't any maximal detuning. Everything depends on the distribution of all the mistuned blades around the disc. In order to examine the effects of various shapes of mistune distribution a series of numerical tests have been conducted[11]. It can be calculated that the maximal stress occurs in the most detuned blade when all detuned blades are close together. The best distribution of detuned blades around the disc is when the natural frequencies of a set of detuned blades are likely to conform to a n-periodic function. It is interesting to observe that when the blades of the mistuned disc have the rectangular cross sectional area, the position of the nodal diameters of the mistuned assemblies can be foreseen in the first modal shapes. For example if the position of the nodal lines with only one blade detuned at the position 8 (Fig 6) as well as the position of the nodal lines with only one blade detuned at the position 16 (Fig 6) are known, and treated nodal lines as vector that should be summed up, a position of the nodal line for a bladed disc with the mistuned blades at the positions 8 and 16 can be found. At the Figure 6 above the schema of the disc the values of natural frequencies are placed. On the schema of the disc the position of the nodal lines is shown. The dashed line shows in enlargment the maximal value of the curvature. Below the schema of the disc there are two sets of coordinates. In one set the form of curvature of the nodal line is shown. Picture (7,8)shows the change of the nodal line of the mistuned bladed disc (Tab 4) for modal shapes coresponding to

5 nodal diameters as well as one nodal circle (Fig 8) and one nodal diameter and one nodal circle (Fig 7).

Tabel 4

Physical Properties of Bladed Disk Assembly

Disc	$r = 2.6$ cm, $R = 30$cm, $h = 2$cm
detuned blades L=5cm	
1(1.1x2),2(0.9x2),3(0.6x2)cm^2	

From that picture it is seen how complicated the modal shapes of mistuned bladed disc which blades with only rectangular cross - sectional area can be.

10 CONLUSIONS

The Ritz method was used for calculating natural frequencies of mistuned bladed discs. The blade and disc models can be used for calculating the dynamic properties of the complex shaped bladed disc. From numerical results we come to the following conclusions: For the mistuning bladed disc with one detuned blade the detuned blade experiences the largest vibration amplitude in mode shapes. For the mistuning system with two detuning blades the detuned blade experinces the largest vibration amplitudes, when the detuned blades are close together. For the mistuning bladed disc with arbitrary distribution of the blades it is very difficult to predict in advance which of the blades experince the largest vibration amplitude,with the exception of the mostly detuned blades being close together. The best distribution of the detuned blades from the stress point of view is when the the natural frequencies of a set of blades are likely to conform to the n-periodic function.

11. REFERENCES

[1] Afolabi,D. The Frequency Respnse of Mistuned Bladed Disc Assemblies, Vibration of Blades and Bladed Disc Assemblies, Cincinnati, Ohio, September 10-13, 1985.

[2] Afolabi,D. The Eigenvalue Spectrum of Mistuned Bladed Disc, Vibration of Blades and Bladed Disc Assemblies, Cincinnati, Ohio, September 10-13, 1985.

[3] Blocki,J., Strength problem of the segmenal shells of revolution with ribs, loaded by surface forces and the temperature field, PhD. Thesis, IMP PAN Gdansk 1983. (in Polish).

[4] Ewins,D.J, "Vibration charcteristics of Bladed Disc Assemblies" , J Mech Eng 1973, I Mech Eng 15, 3, 165-186.

[5] Ewins,D.J, and Han,Z.S. Resonant Vibration Levels of Mistuned Bladed Disc, ASME J Vib Acous Stress and Reliability in Design, vol 106,p211-217,1984.

[6] El-Bayomy,L.E. and Srinivasean,A.V. Influence of Mistuning on Rotor Blade Vibration, AIAA Journal, vol 13,p460-464, 1975.

[7] Irretier,H., Schmidt,K.J.,Mistuned Bladed Discs, Dynamical Behaviour and Computation , Vibration Conference Italy 1983.

[8] Janecki,S., Dynamic of cantilever blades in the last stage of the steam turbines. Zeszyty Naukowe, IMP PAN 105/ /1010/80. (in Polish).

[9] Jager,B.,"Eigenfrequenzen einer Scheibe mit verwundenen Schaufeln" ,Zeitschrift fur Flugwissenschaft ,Vol.10, 1962, pp 439-446.

[10] Rzadkowski,R.,Receptance coefficients of rotating Damped Thin disc of Variable thickness:spline interpolation technique Calculation" ,Vibration of Blades and Bladed Disc Assemblies, Cincinnati,Ohio,September 10-13 ,1985

[11] Rzadkowski,R.,Vibration of Mistuned Bladed Disc, PhD Thesis, IMP PAN Gdansk 1987 (in Polish).

[12] Wildheim, S.J.,"Natural Frequencies of Rotating Bladed Discs Using Clamped-Free Blade Modes ".ASME Paper 81-Det-124.

[13] Vorobiev,J.S.,"Vibration of Elements of Turbomachinery"Kiev,Naukova Dumka,1978(in Russian).

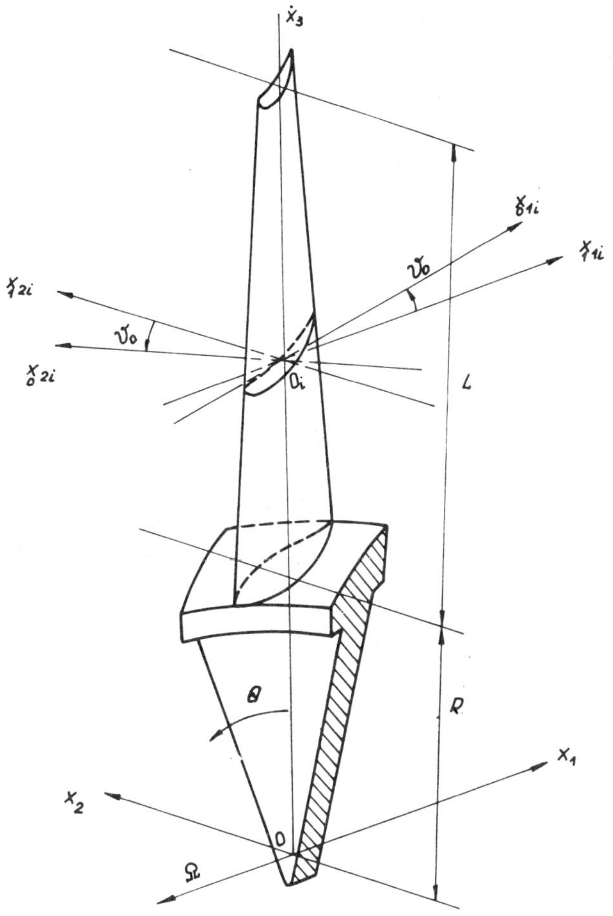

Fig 1 Theoretical model

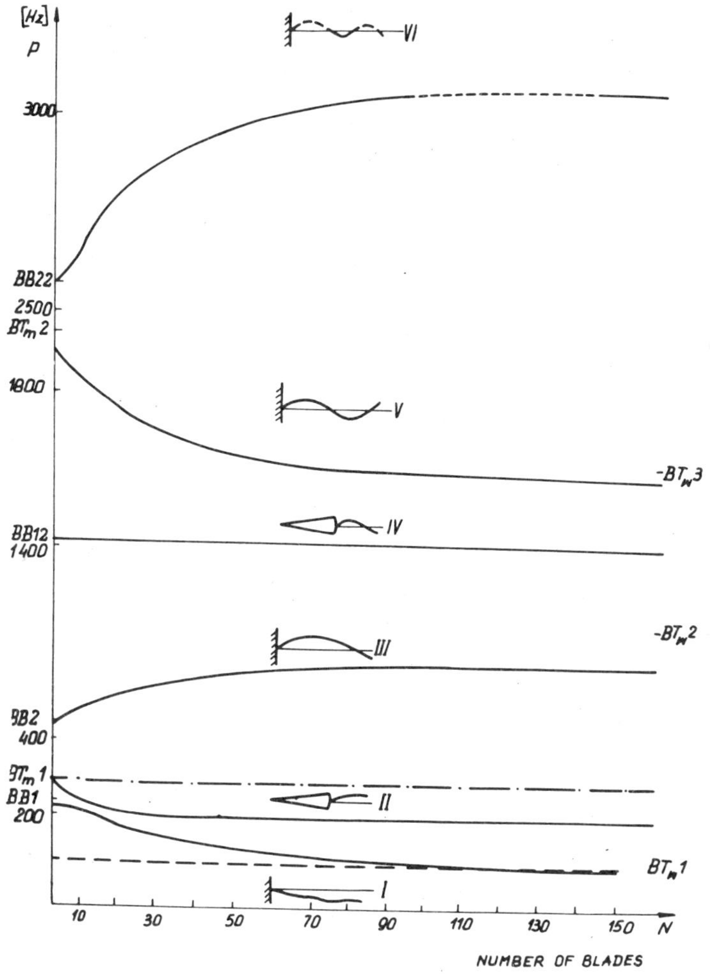

Fig 2 Change of natural frequencies for various number of blades;
— — — natural frequencies of the disc, R = 50 cm;
— · — natural frequencies of the disc, R = 30 cm

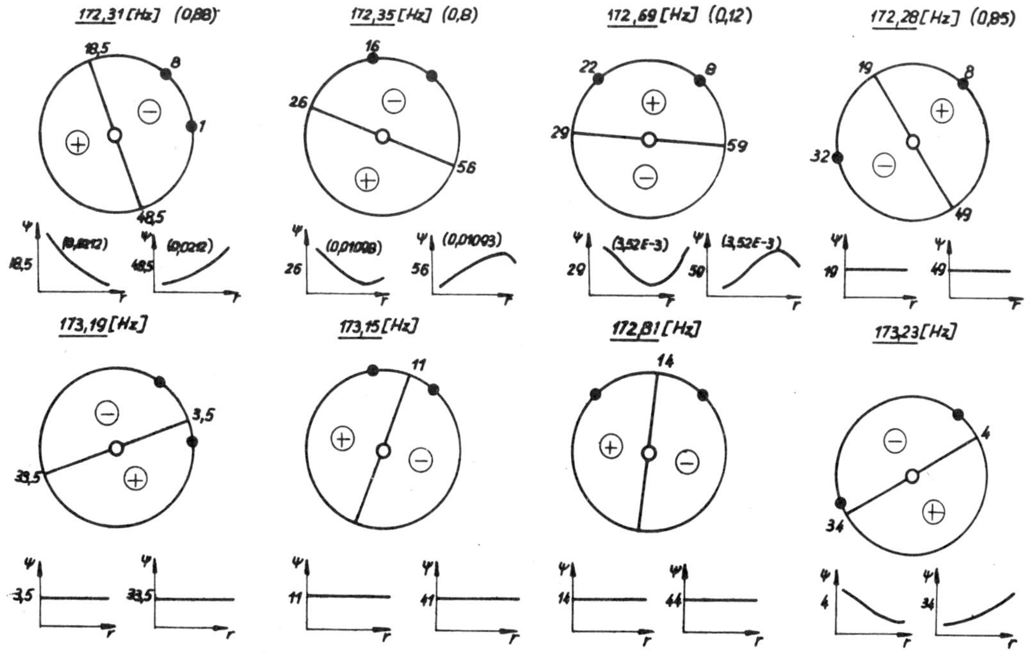

Fig 3 Natural frequencies and curvature of nodal lines for modal shapes corresponding to one nodal diameter of the disc with 60 blades, two of them being detuned and changing its position around the disc

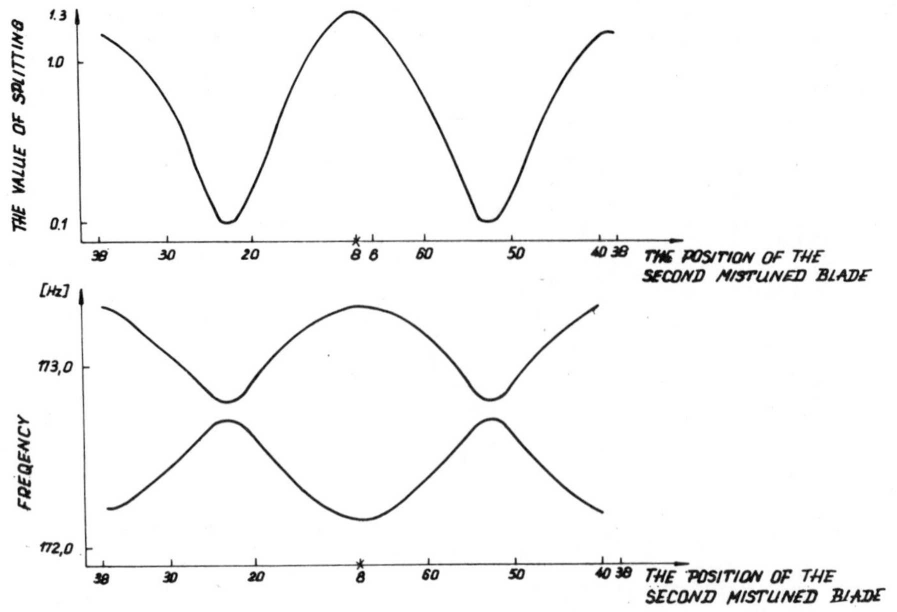

Fig 4 The value of splitting of double natural frequencies for modal shapes corresponding to one nodal diameter of the disc with 60 blades, two of them being detuned

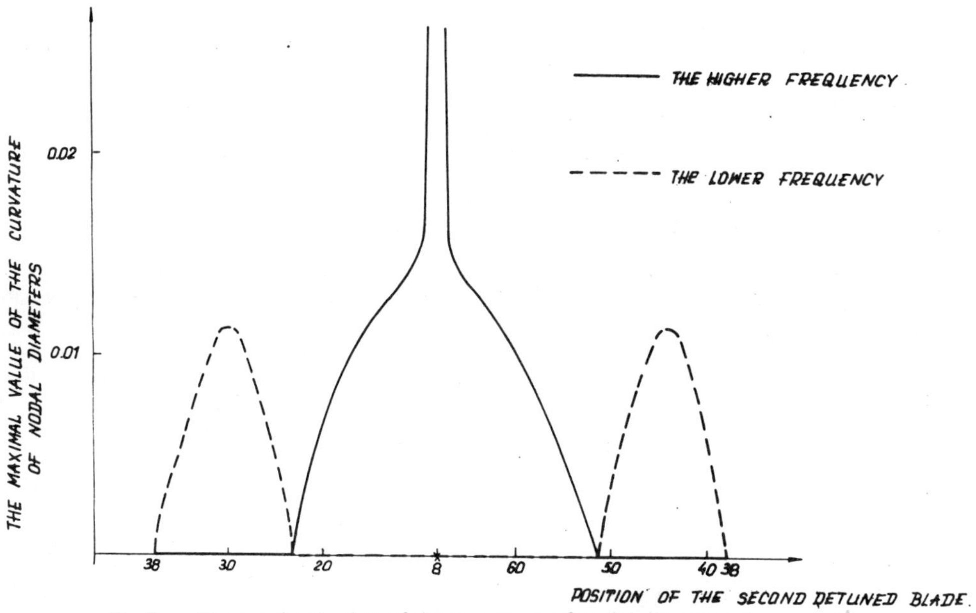

Fig 5 The maximal value of the curvature of nodal diameters for modal shapes corresponding to one nodal diameter of the disc with 60 blades, two of them being detuned

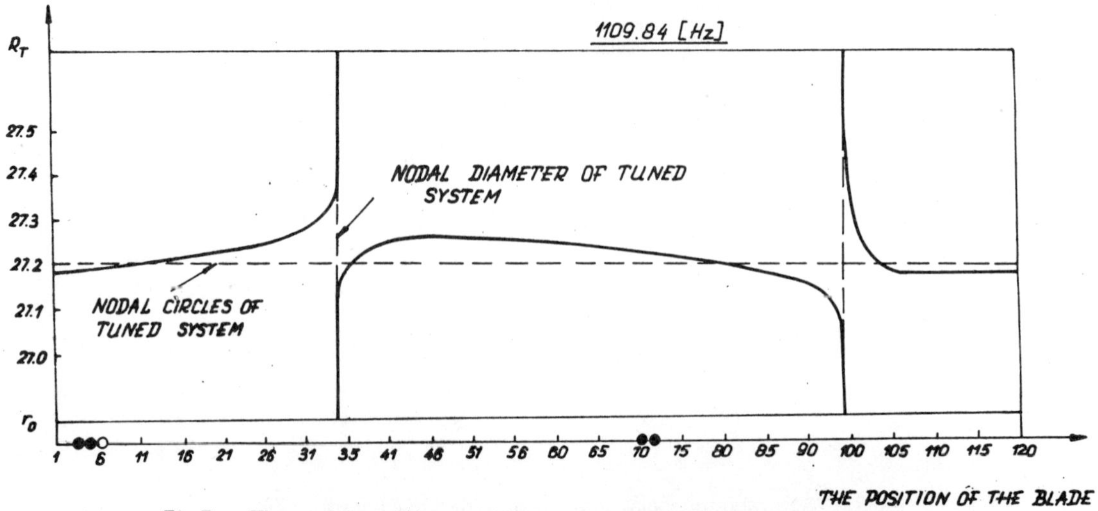

Fig 7 The position of the nodal lines of the disc with 120 blades, five of them being detuned for modal shape corresponding to one nodal diameter and one nodal circle

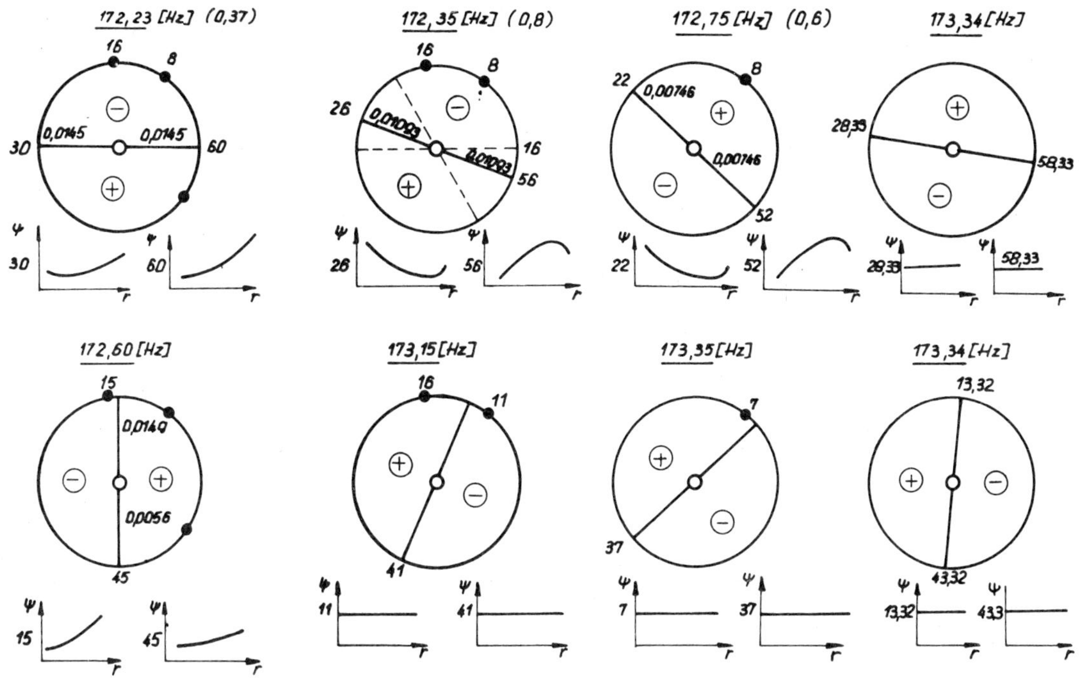

Fig 6 The positions of the nodal lines for modal shapes corresponding to one nodal diameter of the disc with 60 blades, with different number of detuned blades

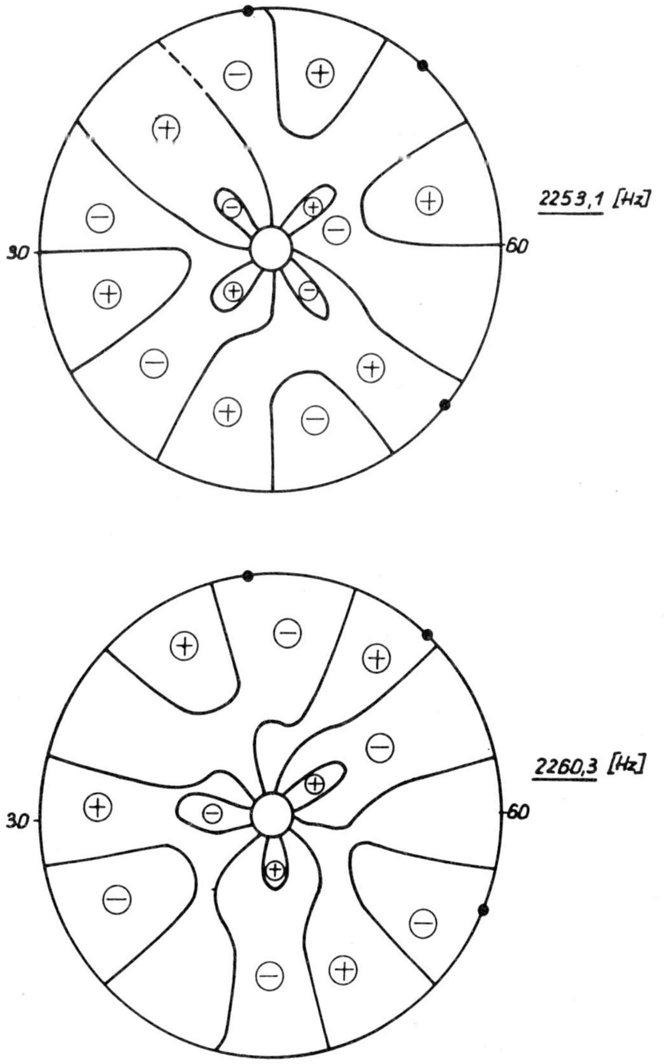

Fig 8 The position of the nodal lines of the disc with 60 blades, with three detuned blades for modal shape corresponding to five nodal diameters and one nodal circle

C319/88

The dynamics of n-bladed propellers with arbitrarily positioned hinges under the influence of aerodynamic forces

M PERSON, Dr-Ing
Technical University of Berlin, West Berlin

SYNOPSIS The equations of motion of n-bladed propellers with arbitrarily positioned hinges including mass forces, spring forces, gravity forces and aerodynamic forces are derived from the equations of an one-bladed propeller, by superposition. Different types of propellers are compared for time variances at the equations due to the aerodynamic forces. The stability (Floquet) of an one-bladed see-saw propeller with free yaw movement, of an one-bladed see-saw propeller with controlled yaw movement and of a three-bladed propeller with free yaw movement is analysed. The results illustrate the need to consider the interaction of the motions of nacelle or hub and blade.

1 INTRODUCTION

The subjects of our research are the dynamics of wind turbines with hinges to connect hub and blades. The coupling of blade dynamics and rotor dynamics and the coupling of a rotating structure (blade, propeller) and a non rotating structure (nacelle) lead to time variances at the equations of motion which may cause severe dynamic troubles.

Investigating a system of propeller and engine block Crandall and Dugundji /1,2/ showed that there may be stability problems neglecting aerodynamic forces. Chen /3/ proved that such stability problems may remain even if aerodynamic forces are included when he analysed a two-bladed propeller on non symmetrical supports.

The author analysed the equations of motion of n-bladed propellers including mass forces, spring forces and gravity forces but neglecting aerodynamic forces /4/. It was obvious that one-bladed and two-bladed propellers may have more dynamic problems than propellers with three or more blades which - due to their 'disc characters' - show no time variances at the parts of the equations of motion that belong to the nacelle.

The numerical analysis of an one-bladed propeller with free yaw movement displayed severe stability problems due to the interaction of yaw and pitch motions of the nacelle and the relative motion of the blade.

In this paper the aerodynamic forces will be added to the equations of motion, the time variances due to the aerodynamic forces will be presented and the stability behaviour of the above mentioned one-bladed propeller, of an one-bladed propeller with controlled yaw movement and of a three-bladed propeller with free yaw movement will be analysed under the influence of the aerodynamic forces.

2 MECHANICAL MODEL

The mechanical model (Fig. 1) consists of:

(a) hub (six degrees of freedom),
(b) n rigid blades,
(c) n hinges (one degree of freedom each) of arbitrary angle and position to connect hub and blade.

In spite of the rotational symmetry of the hub three different moments of inertia of the hub are allowed. Thus the complete nacelle instead of the hub can be studied if wanted. With the Coriolis forces, of course, only the moments of inertia of the hub are allowed.

3 THE EQUATIONS OF MOTION

The equations of motion include

(a) mass forces,
(b) spring forces,
(c) gravity forces,
(d) aerodynamic forces,
(e) a structural damping.

The equations of motion neglecting aerodynamic forces and structural damping were presented in /4/. To add the aerodynamic forces the following assumptions were made:

(a) Strip theory
That means that there is no interference between successive blade elements along the blade and that the forces acting on a blade element are solely due to the lift and drag characteristics of the local profile of the blade element.

(b) Quasistatic aerodynamics
That means that there is no time delay between changes of the flow along the blade element and changes of the resulting forces.

(c) Arbitrary angles of attack α
Stall controlled wind turbines have large angles of attack up to 90° during start up and above the rated wind velocity. Therefore arbitrary angles of attack have to be allowed.

(d) Because of the large angles of attack the drag forces are treated in the same way as the lift forces.

(e) The lift (c_L) and drag (c_D) coefficients depend on the degrees of freedom $\Delta\underline{q}$ of the system and on their velocities $\Delta\dot{\underline{q}}$:

$$c_L(\alpha) = c_L(\alpha(\Delta\underline{q}, \Delta\dot{\underline{q}}))$$

$$c_D(\alpha) = c_D(\alpha(\Delta\underline{q}, \Delta\dot{\underline{q}}))$$

They are taken from experimental data respectively from catalogues of profiles.

(f) The effects of inertia of the air are neglected.

The velocities and the resulting forces acting on a blade element are shown in Fig. 2.

After integration over the length of the blade and linearization for arbitrary reference positions q_{i0} of the angular degrees of freedom the aerodynamic forces thus distribute to the parts of the homogeneous equations of motion proportional to $\Delta\dot{\underline{q}}$ ($\underline{D}_{ae}(t)$) and proportional to $\Delta\underline{q}$ ($\underline{K}_{ae}(t)$) and to the inhomogeneous part ($\underline{k}_{ae}(t)$).

Therefore the equations of motion of the one-bladed propeller including aerodynamic forces result in

$$\underline{M}(t)\Delta\ddot{\underline{q}} + [\Omega\cdot\underline{G}(t) + \underline{D}_{ae}(t)]\Delta\dot{\underline{q}} + [\underline{K}_F + \Omega^2\cdot\underline{K}_{11}(t) + \dot{\Omega}\cdot\underline{K}_{21}(t) + g\cdot\underline{K}_{31}(t) + \underline{K}_{ae}(t)]\Delta\underline{q} =$$

$$= -\underline{K}_F\underline{q}_0 - \Omega^2\cdot\underline{k}_{10}(t) - \dot{\Omega}\cdot\underline{k}_{20}(t) - g\cdot\underline{k}_{30}(t) + \underline{k}_{ae}(t)$$

$$\underline{M}(t)\Delta\ddot{\underline{q}} + \underline{D}(t)\Delta\dot{\underline{q}} + \underline{K}(t)\Delta\underline{q} = \underline{k}(t) \qquad (1)$$

The equations (1) are time variant depending on the angular position $\varphi(t)$ of the propeller. At constant rotational speed Ω ($\dot{\Omega} = 0$) the equations (1) become periodically time variant:

$$\underline{M}(t+T) = \underline{M}(t)$$
$$\underline{D}(t+T) = \underline{D}(t)$$
$$\underline{K}(t+T) = \underline{K}(t)$$

where T is the time of one period respectively of one rotation of the propeller.

4 THE EQUATIONS OF MOTION OF THE N-BLADED PROPELLER

The equations of motion of multi-bladed propellers can be derived by superposition of the different blades. How that has to be done with mass forces, spring forces and gravity forces was presented in /4/.

With aerodynamic forces two different cases have to be distinguished:

(a) a non constant wind profile due to the roughness of the ground $v_w = v_w(h)$

(b) an idealised constant wind profile $v_w = \text{const.}$

With non constant wind profile each blade has to be calculated by itself for every new angular position of the propeller and then added to the equations of motion. Therefore the parts d_{ij}^w of the blades belonging to the elements d_{ij} of the 6x6 sub-matrix of the hub amount to:

$$d_{ij}^w = \sum_{k=1}^{n} (d_{ij}^w)_k$$

With idealised constant wind profile the aerodynamic forces at different blades differ only due to the fact that the angular position of each blade is marked by a constant angle ψ_k:

$$\varphi_k(t) = \varphi(t) + \psi_k$$

with

$$\psi_k = \frac{2(k-1)}{n}\pi \quad ; \quad k = 1(1)n$$

If one splits d_{ij}^w into constant and time variant part:

$$d_{ij}^{w} = d_{ij}^{wc} + d_{ij}^{wt}$$

$$= d_{ij}^{wc} + \left(\sum_{m=1}^{4} c_m^{wt} f_m^{wt}\right)_{ij}$$

with constant wind profile the constant terms and the coefficients (c_m^{wt}) of the time variant functions (f_m^{wt}) deriving from the different blades will be identical. The only difference between the time variant functions of the different blades will be the constant angles ψ_k in their arguments. Therefore the part of the elements d_{ij} of the 6x6 sub-matrices of the hub that derives from the n propeller blades amounts to

$$\sum_{k=1}^{n} (d_{ij}^{W})_k = n \cdot d_{ij}^{wc} + \left(\sum_{m=1}^{4} c_m^{wt} \sum_{k=1}^{n} f_{m_k}^{wt}\right)_{ij}$$

To calculate the second sum in this equation the rules for adding the trigonometrical functions can be used for the aerodynamic forces as has been done with mass forces, spring forces and gravity forces in /4/.

At the two-bladed see-saw propeller, moreover, it has to be considered that a motion of one blade with γ_s causes a motion of the other blade with $-\gamma_s$.

Thus with restriction to an idealised constant wind profile we get the qualitative structure of the equations of motion of different types of wind turbines due to the aerodynamic forces.

5 TIME VARIANCES DUE TO THE AERODYNAMIC FORCES AT DIFFERENT TYPES OF PROPELLERS (FIG. 3)

To get an idea of the relevant time variances due to the aerodynamic forces apart from the constant wind profile two more restrictions were applied:

(a) There are no side winds. That means that the axis of the propeller has the direction of the wind.

(b) The reference positions of the angular degrees of freedom of the hub (α_1, β_1, γ_1) are zero.

Neglecting a stochastic wind both restrictions seem to be justified if the propeller runs at normal conditions. The reference positions of the hinges depend on the velocity of the wind and on the rotational speed of the propeller. Therefore they have to remain arbitrary.

Different to mass forces, spring forces and gravity forces the aerodynamic forces are not the same with an one-bladed see-saw propeller as with a two-bladed see-saw propeller. An one-bladed see-saw propeller has always a rotating unbalance due to the aerodynamic forces. Moreover with a two-bladed see-saw propeller at constant wind profile it does not make sense to have another than a pure flap hinge ($\beta_g = 0^o$). Any other angle of the hinge would induce non symmetrical aerodynamic forces if the blades move out of their reference position. Therefore a pure flap hinge was assumed with the two-bladed see-saw propeller and - with respect to the constant wind profile - consequently the reference position of the degree of freedom of the hinge was taken to be zero.

Fig. 3(b) shows the different time variances due to the aerodynamic forces at the different types of propellers. As with mass forces, spring forces and gravity forces (Fig. 3(a), taken from /4/) the 'disc character' of the three-bladed propeller shows up clearly, because the time variances at the sub-matrices of the hub are missing. Consequently the three-bladed propeller without hinges has no time variances at all.

The two-bladed propellers have time variances according to twice the frequency of the propeller rotation at the sub-matrices of the hub apart from the elements belonging to thrust and torque, which have no time variances. At the two-bladed see-saw propeller the equations belonging to thrust and torque are not coupled to the other equations.

The one-bladed see-saw propeller has the same time variances as the two-bladed propellers apart from the equations belonging to thrust and torque. There we find time variances according to once the frequency of the propeller rotation because of its rotating unbalance due to the aerodynamic forces.

The non homogeneous parts of the equations due to the aerodynamic forces have no time variances apart from the one-bladed see-saw propeller.

6 STABILITY ANALYSIS VIA THE THEOREM OF FLOQUET

To analyse the stability behaviour of different types of propellers basically, the theorem of Floquet was used. It leads to an analysis of the Floquet transition matrix $\underline{Q}$ (T) which is the special fundamental matrix of the periodical system at the end of one periode of time respectively after one rotation of the propeller:

$$\underline{z}(T) = \underline{Q}(T)\, \underline{z}_0$$

where T is the time of one periode and $\underline{z}$ is the column of state variables. The eigenvalues of $\underline{Q}$ (T) give the information whether the system has stable or unstable solutions respectively:

$$\|\underline{z}(T)\| \leq \|\underline{z}_0\| \quad \text{or} \quad \|\underline{z}(T)\| \geq \|\underline{z}_0\|$$

7 STABILITY ANALYSIS OF AN EXPERIMENTAL ONE-BLADED PROPELLER WITH FREE YAW MOVEMENT, D = 6 M

In /4/ the author analysed the stability behaviour of an experimental one-bladed see-saw propeller (Fig. 4) neglecting the aerodynamic forces.

It is a propeller with free yaw movement. It has got a diameter of D = 6 m. The hinge is a flap hinge that is inclined for 22.5° within the rotor plane so that a flap motion of the blade induces a motion around the pitch axis of the blade to decrease the effective air forces. It was designed for a tip speed ratio of $\lambda_A = 12$ at a wind of v = 8 m/s.

The nacelle was included in the mechanical model of this propeller. Very important for the stability behaviour of this propeller are the very soft mounting of the nacelle and the very low moments of inertia of the non rotating parts (nacelle) compared to the axial moments of inertia of the propeller.

The very soft mounting of the nacelle shows up by comparing the rated rotational speed Ω_T at v = 8 m/s to the eigenfrequency ω_β (pitch motion of the nacelle):

$$\Omega_T / \omega_\beta = 23.3$$

The ratio of the maximum moment of inertia of the nacelle Θ_N according to its center of mass to the axial moment of inertia of the propeller Θ_P amounts to:

$$\Theta_N / \Theta_P = 0.032$$

Neglecting the aerodynamic forces there were stability problems due to mass forces, spring forces and gravity forces. The limit rotational speed of stable running decreased with increasing inclination of the angle of the hinge. The region of instability reached from this limit rotational speed up to the maximum rotational speed. These stability problems were due to the interaction of the yaw (α_1) and pitch (β_1) motion of the nacelle and the relative motion of the blade (γ_S).

For the stability analysis including the aerodynamic forces therefore only α_1, β_1 and γ_S were taken into account. Because the hinge has no spring the propeller would behave like a wheather vane and could not start rotating. Therefore the degree of freedom of the hinge was blocked up to a rotational speed of Ω = 20 rad/s and then released.

The analysis was carried out according to a non constant wind profile with a velocity of v = 9 m/s at a height of 12 m. For the blade a Wortmann profile was used (FX 63-137).

The results of this stability analysis are presented in Fig. 5. Fig. 5(a) shows the regions of stable running of the propeller according to the angle of the hinge within the rotor plane ($\beta_g = 0°$ for pure flap hinge) under the influence of aerodynamic forces. Very clearly there are two regions of instability of which the second one increases with increasing angle of the hinge. Above an angle of about 33° the propeller remains unstable up to the maximum rotational speed. Fig. 5(b) shows the size of these instabilities displayed by the eigenvalues μ of the Floquet transition matrix. With $|\mu| > 1$ there are unstable solutions, with $|\mu| < 1$ there are stable solutions of the homogeneous equations of motion. Obviously the second region of instability is devided into three seperate parts. To analyse their origin the system with $\beta_g = 22.5°$ was examined more closely (Fig. 6).

From the stability analysis neglecting aerodynamic forces /4/ we know that there are stability problems caused by mass forces, spring forces and gravity forces due to the interaction of yaw and pitch of the nacelle and the relative motion of the blade. Therefore this kind of instability can arise only when the degree of freedom of the hinge is released. This happens at a rotational speed of Ω = 20 rad/s and leads to a sudden and strong increase of instability. This kind of instability (III, Fig. 6) reaches up to the maximum rotational speed. The first region and the beginning of the second region (I and II, Fig. 6) are caused by the aerodynamic forces only. This kind of instability reaches up to about Ω = 33 rad/s. At this rotational speed there is a distinct decrease of instability. Above Ω = 33 rad/s the aerodynamic forces have a damping influence. But this damping influence does not completely overcome the instability due to mass forces, spring forces and gravity forces. The propeller does not run stable again until it has reached a tip speed ratio clearly beyond the one it was designed for ($\lambda_A = 12$).

To get an physical understanding of the instabilities due to the aerodynamic forces (I and II of Fig. 6) and to get an idea of their practical consequences it is necessary to find out what they are caused by. This can be explained examining the equations of motion. If the degrees of freedom of the nacelle are blocked there remains the equation of motion of the blade:

$$\Theta_H \Delta\ddot{\gamma}_S + D_{ae}(3,3)\,\Delta\dot{\gamma}_S + \left(K_{ae}(3,3) + \Omega^2 \Theta_S\right)\Delta\gamma_S = 0 \qquad (2)$$

With Θ_H and Θ_S being the according moments of inertia depending on the angle β_g of the hinge and $D_{ae}(3,3)$ and $K_{ae}(3,3)$ deriving from the aerodynamic forces. The later can be split into:

$$D_{ae}(3,3) = c_L' D_{c_L'}(3,3) + D_{ae_{red}}(3,3)$$

$$K_{ae}(3,3) = c_L' K_{c_L'}(3,3) + K_{ae_{red}}(3,3)$$

with c_L' being the derivative of the lift coefficient

$$c_L' = \frac{\partial}{\partial \alpha}(c_L)$$

Introducing this into equation (2) we get:

$$\Theta_H \Delta\ddot{\gamma}_S + \left[c_L' D_{c_L'}(3,3) + D_{ae_{red}}(3,3)\right]\Delta\dot{\gamma}_S + \left[c_L' K_{c_L'}(3,3) + K_{ae_{red}}(3,3) + \Omega^2 \Theta_S\right]\Delta\gamma_S = 0$$

$D_{c_L'}(3,3)$ and $K_{c_L'}(3,3)$ have a positive value. Therefore a negative c_L' will cause a destabilisation of the system.

With fixed pitch axis of the blade the angle of attack α at a section of the blade will vary from about 90° while starting to about 6 - 7° at rated speed. The diagram of the lift coefficient depending on the angle of attack (Fig. 8) shows two main regions of strongly negative c_L' between $\alpha \approx 90°$ and $\alpha \approx 67°$ (I, Fig.7) and between $\alpha \approx 27°$ and $\alpha \approx 13°$ (II, Fig.7). These regions correspond to the unstable regions due to the aerodynamic forces displayed in Fig. 6.

In reality this unstable behaviour was noticed in wind-tunnel measurements at the Institute for Air and Space Research at the Technical University of Berlin /6/ but it did not damage the system. The system ran into a limit cycle with a low amplitude level.

While thus the destabilising effects of the aerodynamic forces do not seem to be dangerous to a wind turbine their damping influence at higher rotational speeds is not sufficient to completely eliminate the instabilities due to mass forces, spring forces and gravity forces. That in the end caused the destruction of this wind turbine.

8 STABILITY ANALYSIS OF AN ONE-BLADED PROPELLER WITH CONTROLLED YAW MOVEMENT, D = 48 M

The stability problems of the above analysed propeller were due to the free yaw movement, the very soft mounting of the nacelle and to the very small moments of inertia of the nacelle compared to the axial moments of inertia of the propeller. Therefore another one-bladed propeller with rather stiff mounting and controlled yaw movement was analysed. Because of the controlled yaw movement the stiffnesses of the yaw motion (α_1) and of the pitch motion (β_1) of the nacelle are the according stiffnesses of its tower. The ratios of rated rotational speed to the eigenfrequencies of these motions show the comparably stiff mounting:

$$\Omega_T / \omega_\alpha = 0.32$$

$$\Omega_T / \omega_\beta = 0.13$$

The ratio of maximum moment of inertia of the nacelle according to its center of mass to the axial moment of inertia of the propeller is about 17 times as much as with the above analysed propeller:

$$\Theta_N / \Theta_P = 0.75$$

It is a propeller of 48 m diameter and again designed for a tip speed ratio of $\lambda_A = 12$. For the blade the Wortmann profile (FX 63-137) was used again.

Without aerodynamic forces this propeller behaves stable for the whole range of the rotational speed. The stability analysis including aerodynamic forces is presented in Fig. 8. Up to a rotational speed of $\Omega = 1.5$ rad/s the blade hinge was blocked and the analysis shows no unstable behaviour. Above $\Omega = 1.5$ rad/s there is an unstable region which increases with increasing angle of the hinge. Again this region corresponds to the strongly negative c_L' with angles of attack just above the maximum c_L. Comparing Fig. 8(a) and Fig. 8(b) which displayes the size of the instabilities again it can be recognized that the region of aerodynamic instability is followed by a small region of very small instabilities which are probably due to numerical inaccuracy and that the damping influence of the aerodynamic forces at higher rotational speeds is hardly noticeable at this stiff system.

As the stiffness belonging to the degrees of freedom of the nacelle at this wind turbine is mostly caused by the stiffness of its steel tower a structural damping can be introduced into the equations of motion. Therefore a small structural damping proportional to the stiffness matrix was added to the equations of motion:

$$\underline{D}_S \cdot \Delta\dot{\underline{q}} = k_S(D) \cdot \underline{K}_F \cdot \Delta\dot{\underline{q}}$$

D represents the damping ratio.

The results of the analysis including structural damping are presented in Fig. 8(a) and in Fig. 9. It is obvious that the structural damping does not affect the instabilities caused by the aerodynamic forces. The region of instability is the same with D = 0.2 % and D = 1 % (Fig. 8(a)). Compared to the analysis without structural damping the region of very small instabilities has vanished. Outside of the region of aerodynamic instability the damping influence of the structural damping is obvious (Fig. 9).

9 STABILITY ANALYSIS OF A THREE-BLADED PROPELLER WITH FREE YAW MOVEMENT, D = 7 M

Fig. 3 showed the 'disc character' of tree-bladed propellers because the time variances at the sub-matrices of the hub were missing. As the one-bladed see-saw propeller with free yaw movement showed severe stability problems due to the interaction of mass forces, spring forces and gravity forces of blade and nacelle it was interesting to analyse a three-bladed propeller with free yaw movement.

It is a propeller of 7 m diameter. The hinges are pure flap hinges ($\beta_g = 0$). It was designed for a tip speed ratio of $\lambda_A = 8.3$. Again the nacelle was included in the mechanical model and for the analysis including aerodynamic forces only α_1 (yaw), β_1 (pitch) and the relative degrees of freedom of the hinges γ_5 , γ_6 and γ_7 were considered. With this propeller the ratio of maximum moment of inertia of the nacelle according to its center of mass to the axial moment of inertia of the propeller amounts to:

$$\Theta_N / \Theta_P = 5.1$$

For the analysis including aerodynamic forces another blade profile (Göttingen 797) was used than with the two one-bladed propellers. It shows an even stronger decline of the lift coefficient c_L for angles of attack in the region of the flow separation above the maximum c_L.

The stability analysis is presented in Fig. 10. With this propeller the angles of the hinges were varied from pure flap hinges ($\beta_g = 0^\circ$, $\alpha_g = 0^\circ$) to pure lag hinges ($\beta_g = 0$, $\alpha_g = 90$) so that only the propeller with pure flap hinges can be compared to the one-bladed propeller. There are no stability problems caused by mass forces, spring forces and gravity forces at the propeller with pure flap hinges. The region of aerodynamic instability is smaller but stronger according to the stronger decline of c_L. The rated tip speed ratio $\lambda_A = 8.3$ is well beyond this region. Outside of this region there is a small but noticeable damping due to the aerodynamic forces.

As the aerodynamic instability does not lead to any damage of the system because of the limit cycle with small amplitudes - as was explained before - this wind turbine should run without any dynamic problems. And so it does in reality.

10 CONCLUSIONS

The equations of motion of the n-bladed propeller with arbitrarily positioned hinges including aerodynamic forces were obtained from the equations of motion of the one-bladed propeller by superposition.

A comparison of the periodical time variances that occur at the equations of motion of different types of propellers due to the aerodynamic forces showed that propellers with less than three blades have time variances according to the frequency and to twice the frequency of the rotational speed within the sub-matrices of the hub.

Analysing the stability behaviour of an experimental one-bladed see-saw propeller with free yaw movement it was proved that the damping effects of the aerodynamic forces do not completely eliminate the stability problems due to mass forces, spring forces and gravity forces at the interaction of blade and nacelle motions.

The aerodynamic forces themselves cause another region of instability but this does not lead to damage of the system in reality.

The analyses of an one-bladed see-saw propeller with controlled yaw movement and of a three-bladed propeller with free yaw movement showed a far better stability behaviour of these two propellers.

ACKNOLEDGEMENT

This research was sponsored by the Deutsche Forschungsgemeinschaft (DFG).

REFERENCES

1 Crandall, S.H. and Dugundji, J. Forced backward whirling of aircraft propeller - engine systems. IMechE 2nd Conference, Vibrations in Rotating Machinery, Cambridge, September 1980 (Mechanical Engineering Publications, London).

2 Crandall, S.H. and Dugundji, J. Resonant whirling of aircraft propeller - engine systems. J. Appl. Mech., 1981, 48.

3 Chen, S.Y. Stability of two-bladed aeroelastic rotors on flexible supports. JAHS, January 1983, pp 34-41

4 Person, M. The equations of motion of n-bladed propellers with arbitrarily positioned hinges and their application to an experimental one-bladed wind turbine. Proc Instn Mech Engrs Vol 199 No A4, pp 237-244.

5 Johnson, W. Helicopter theory. Princton University Press, 1980.

6 Harders, H. Windkanalversuche zum statischen und dynamischen Verhalten von Windturbinen. Unveröffentl. Diplomarbeit, TU Berlin, ILR, 1980.

7 Person, M. Zur Dynamik von Windturbinen mit Gelenkflügeln. VDI Fortschritt-Berichte, Reihe 11, Nr. 104 (1988).

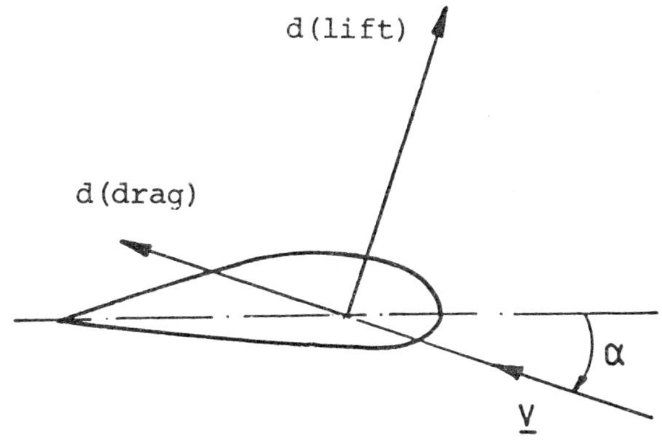

Fig 2 Velocity $\underline{v}$ and resulting forces acting on a blade element

Fig 1 Mechanical model

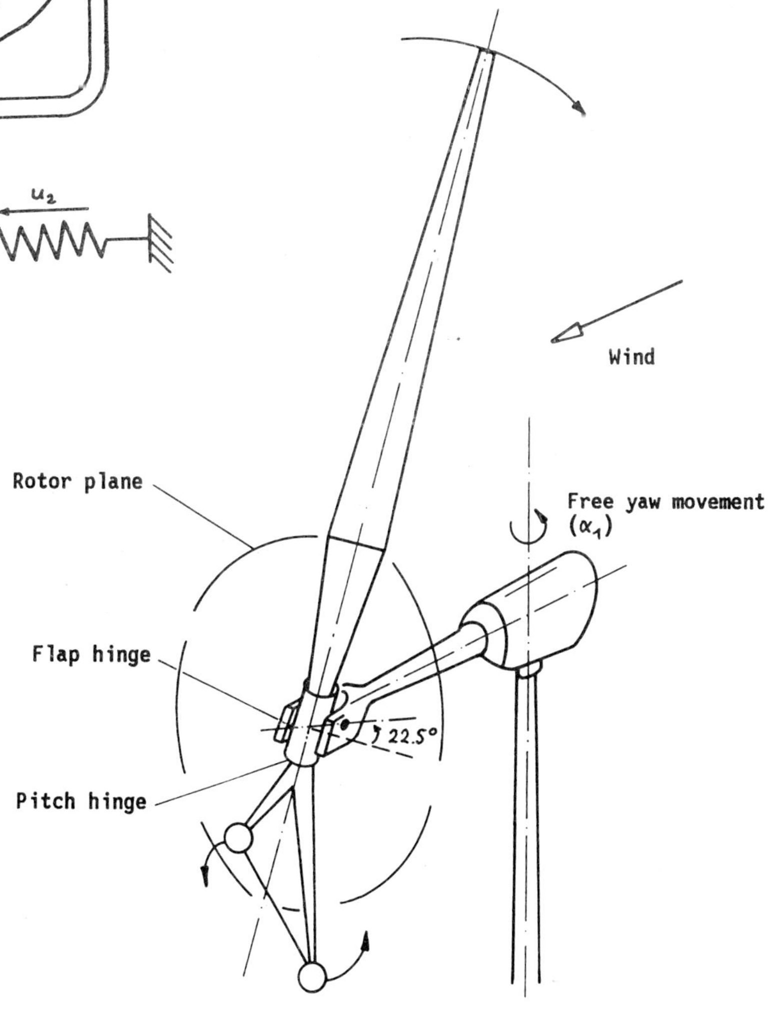

Fig 4 One-bladed see-saw propeller with free yaw movement

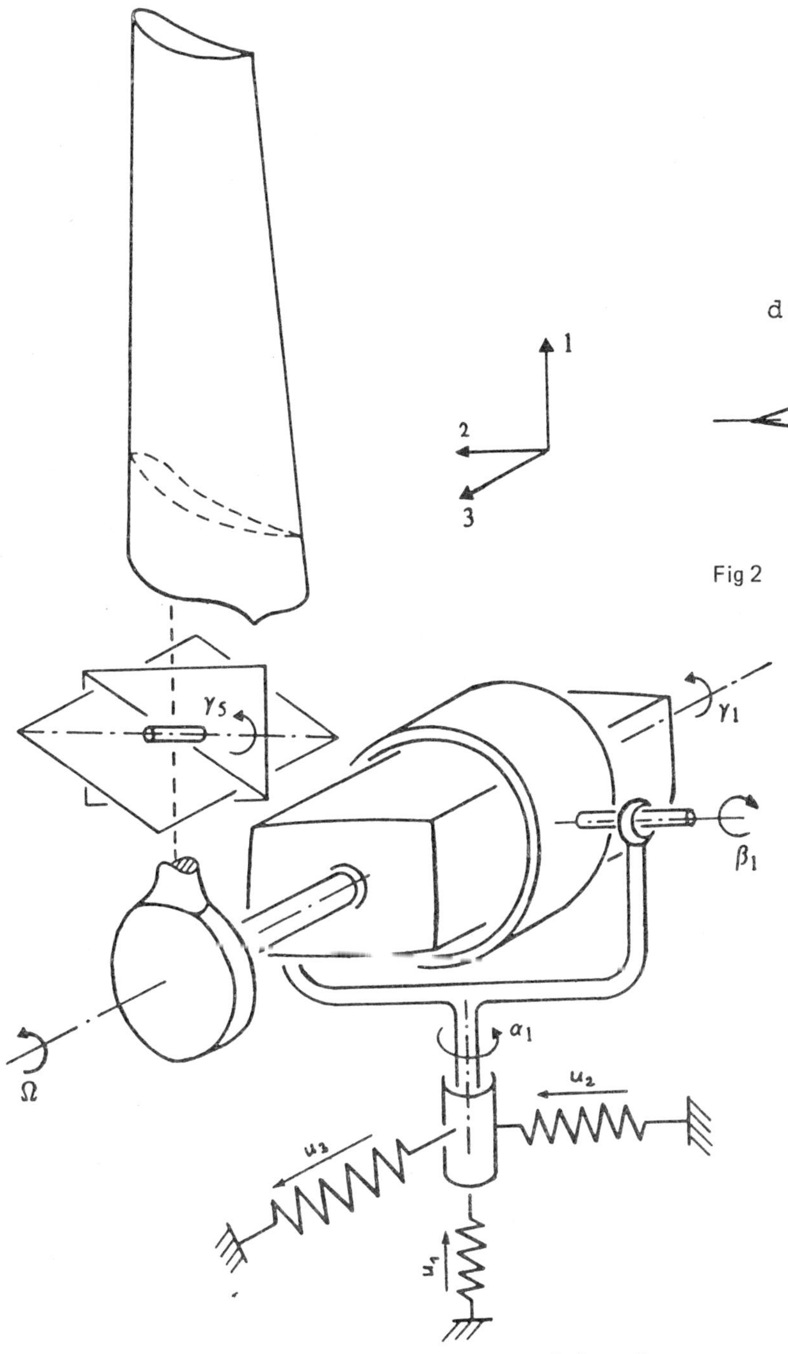

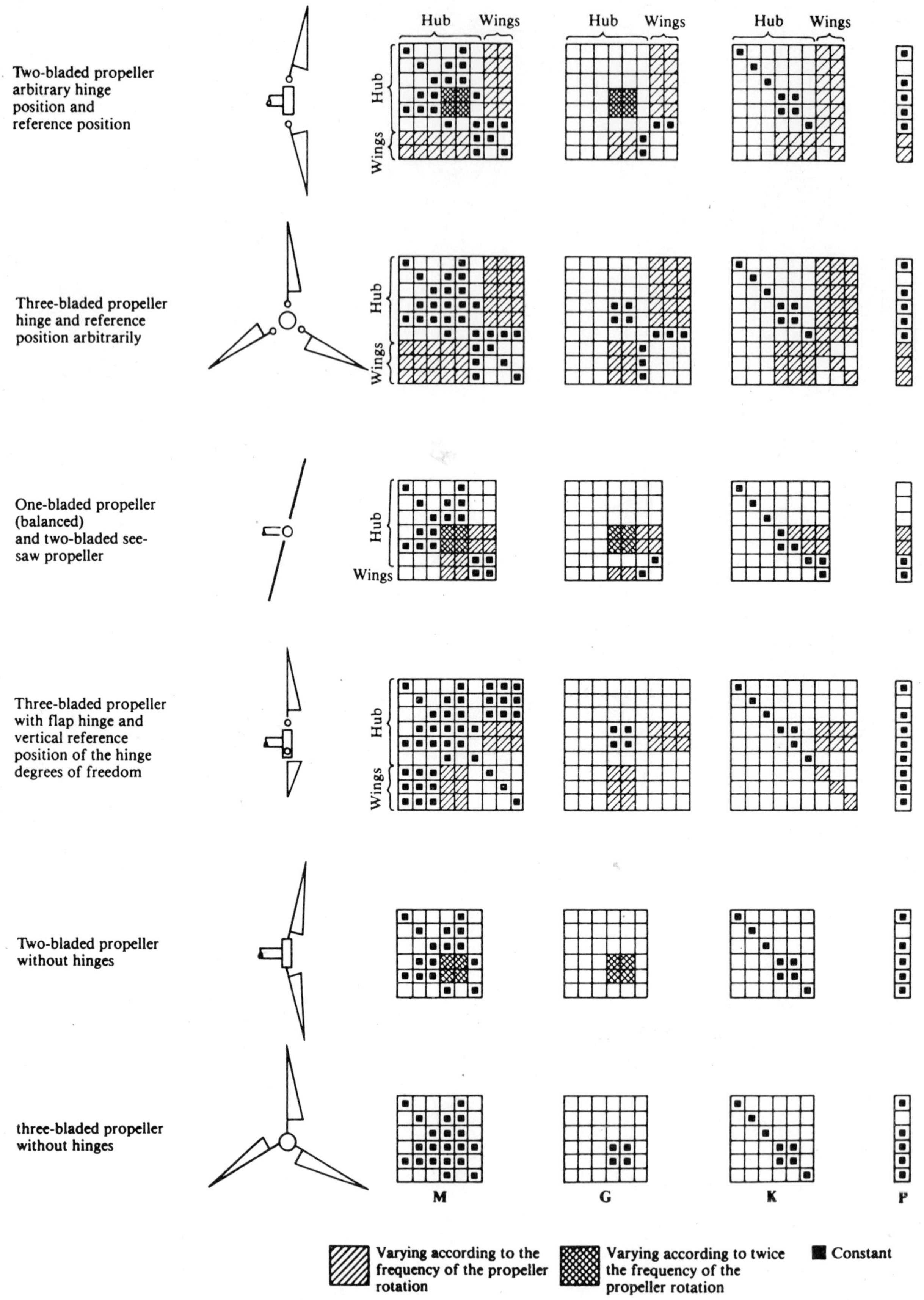

Fig 3a Time variances at the equations of motion of different types of propeller (included: mass forces, gravity forces, spring forces; not included: aerodynamic forces) [from 4]

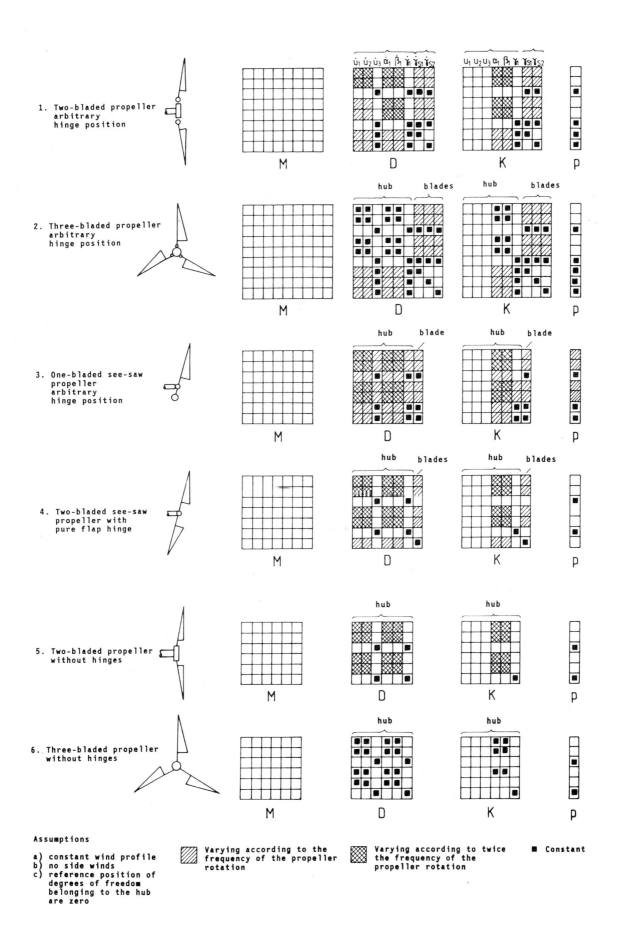

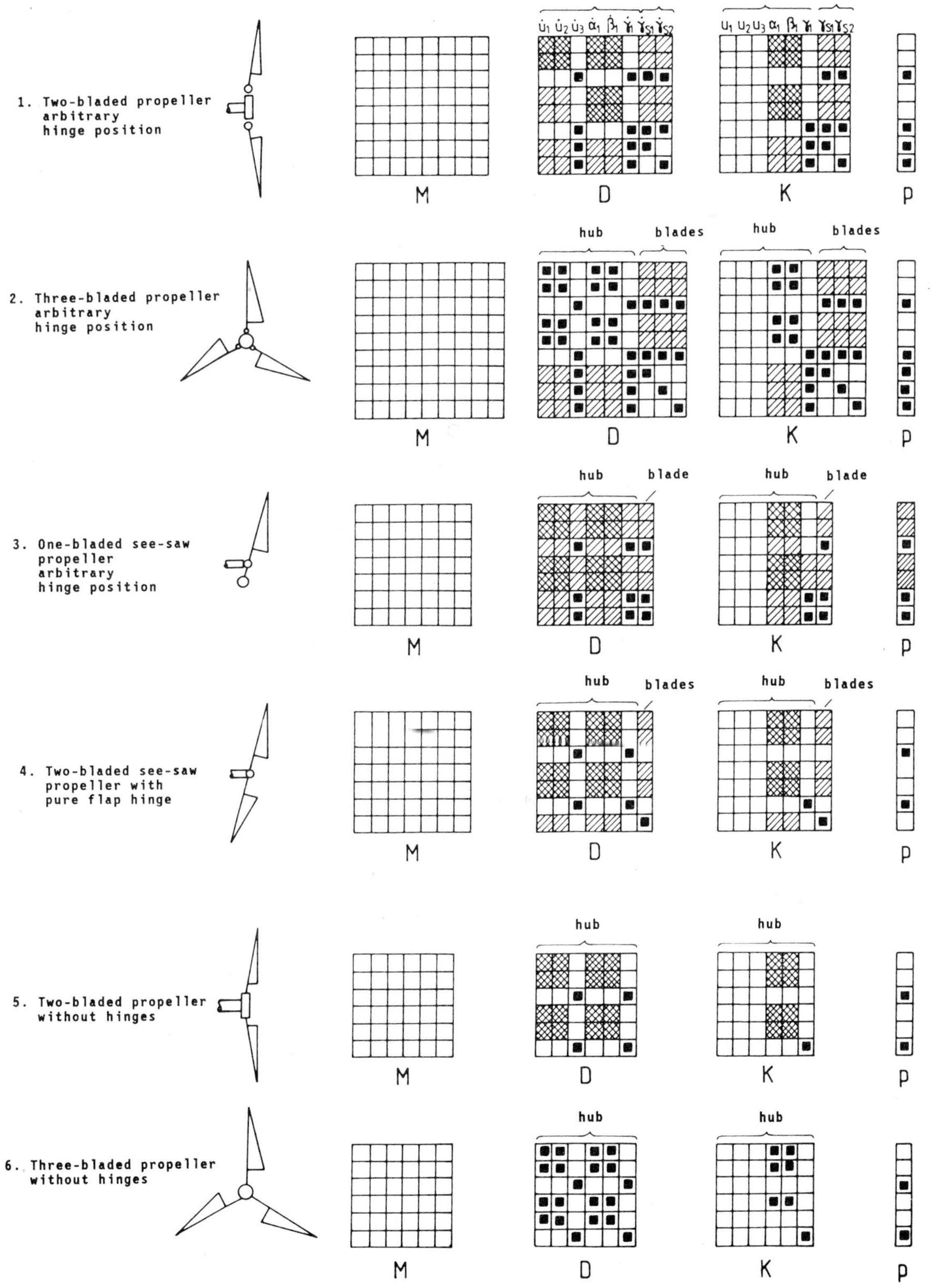

Assumptions

a) constant wind profile
b) no side winds
c) reference position of degrees of freedom belonging to the hub are zero

Varying according to the frequency of the propeller rotation

Varying according to twice the frequency of the propeller rotation

■ Constant

Fig 3b Time variances at the equations of motion of different types of propeller due to the aerodynamic forces

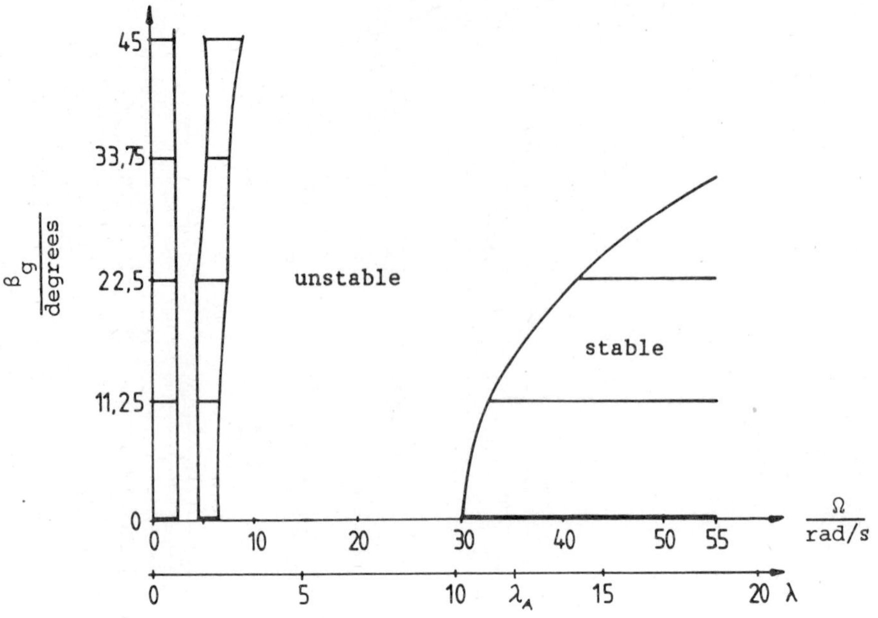

(a)

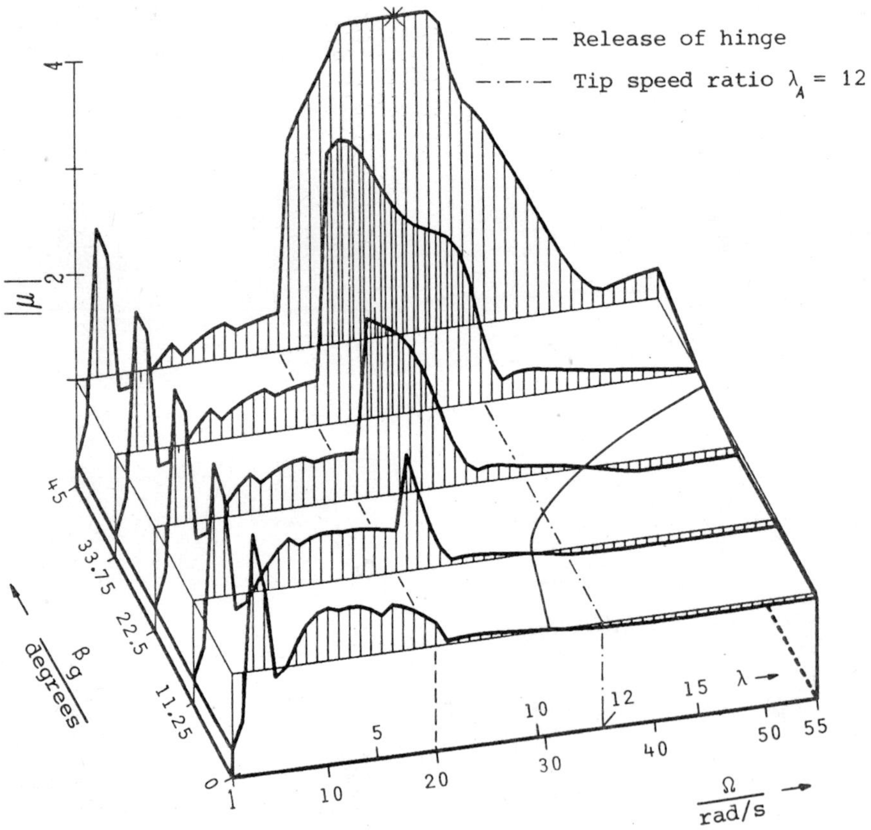

(b)

Fig 5 One-bladed see-saw propeller with free yaw movement
(a) Limit rotational speed of stable running including aerodynamic forces
(b) Limit rotational speed of stable running, size of instability (including aerodynamic forces)

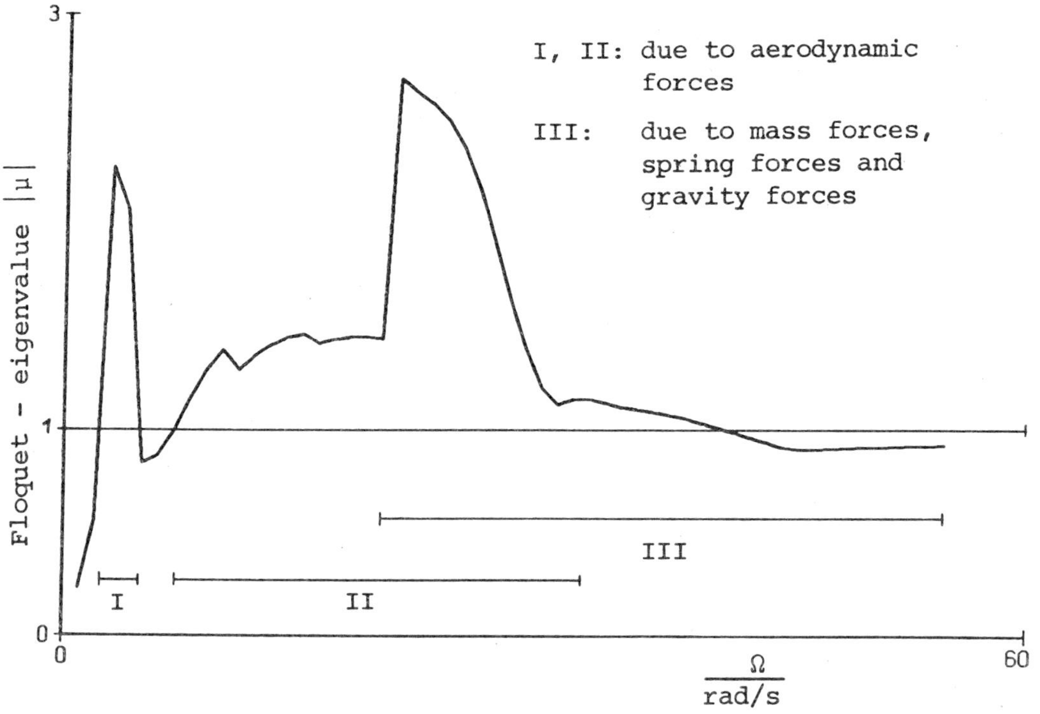

Fig 6 One-bladed see-saw propeller with free yaw movement (limit rotational speed of stable running at $\beta_g = 22.5°$)

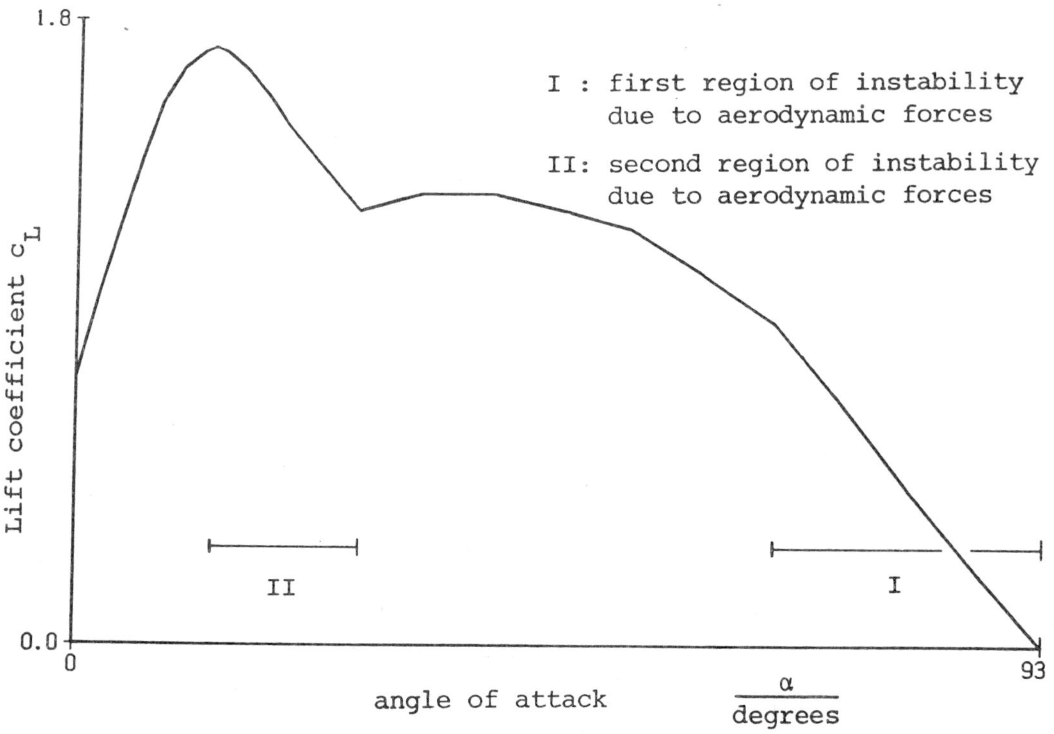

Fig 7 Lift coefficient c_L depending on angle of attack α (profile FX 63-137, Re = 360 000)

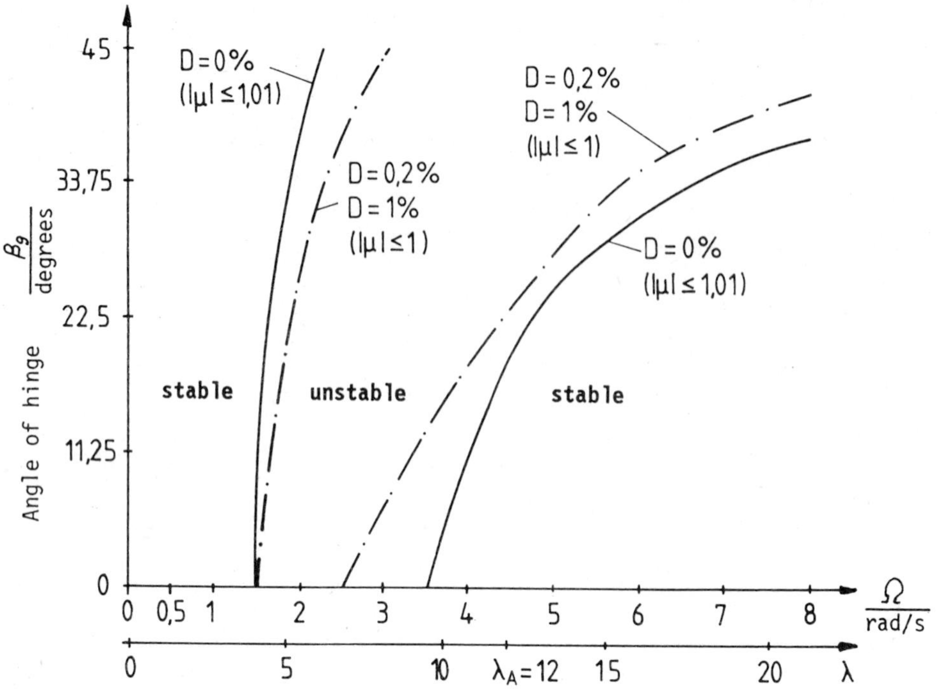

(a)

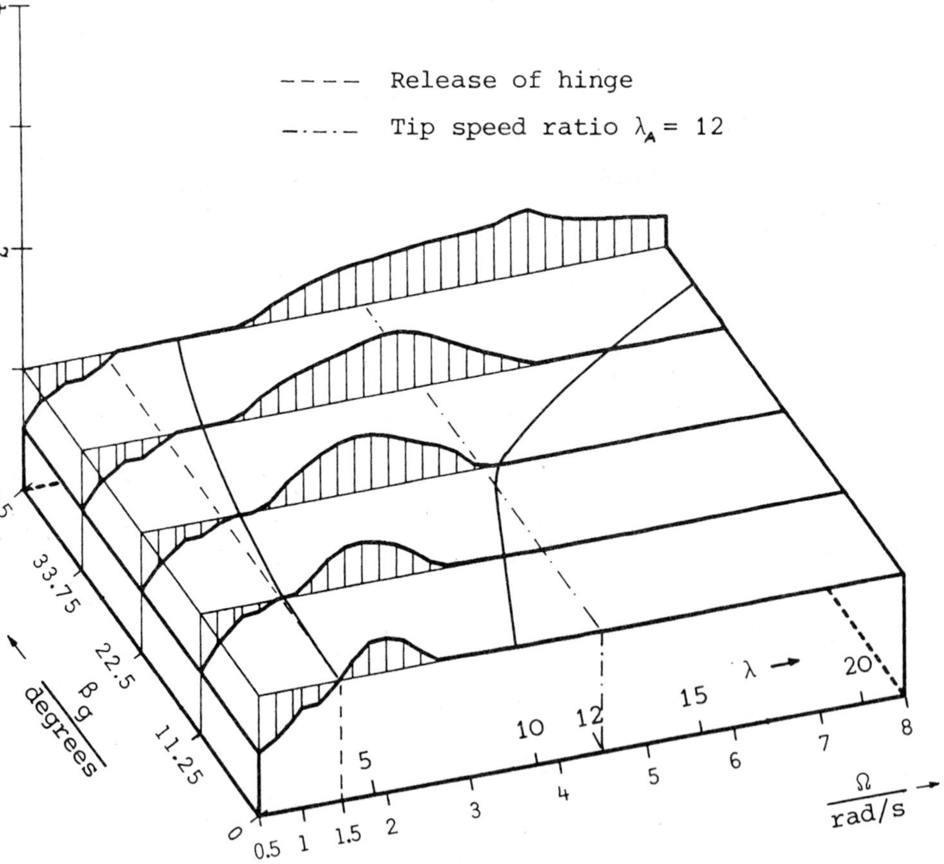

(b)

Fig 8 One-bladed see-saw propeller with controlled yaw movement
(a) Limit rotational speed of stable running including aerodynamic forces and a structural damping
(b) Limit rotational speed of stable running, size of instability, D = 0 per cent

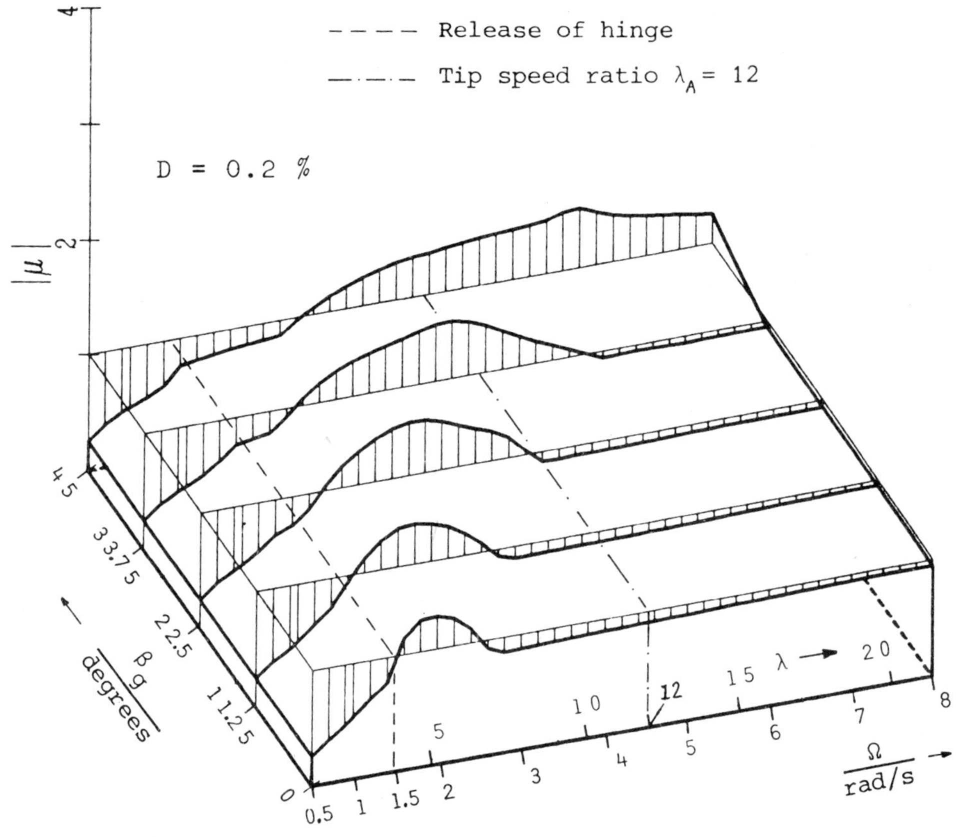

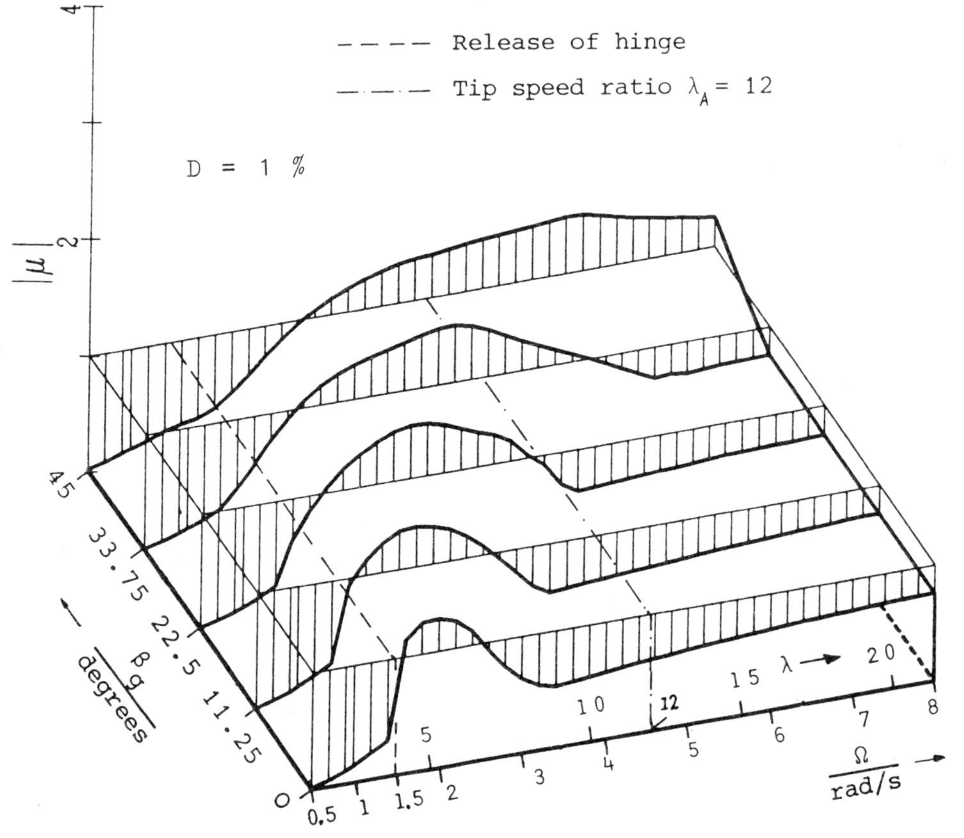

Fig 9 One-bladed see-saw propeller with controlled yaw movement (limit rotational speed of stable running including aerodynamic forces and a structural damping, size of instability)

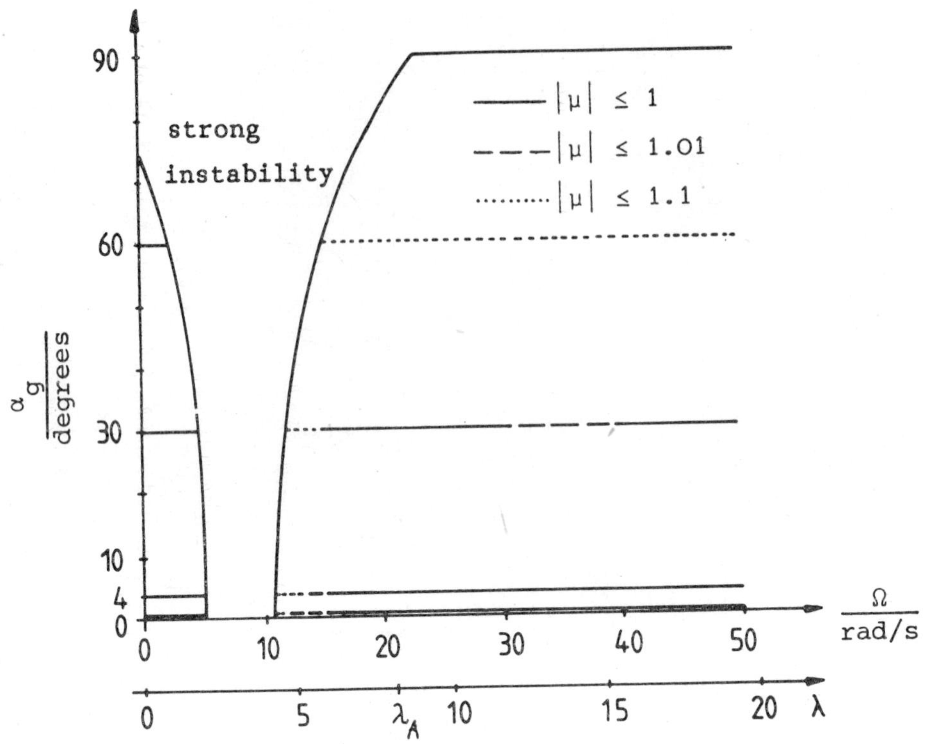

(a)

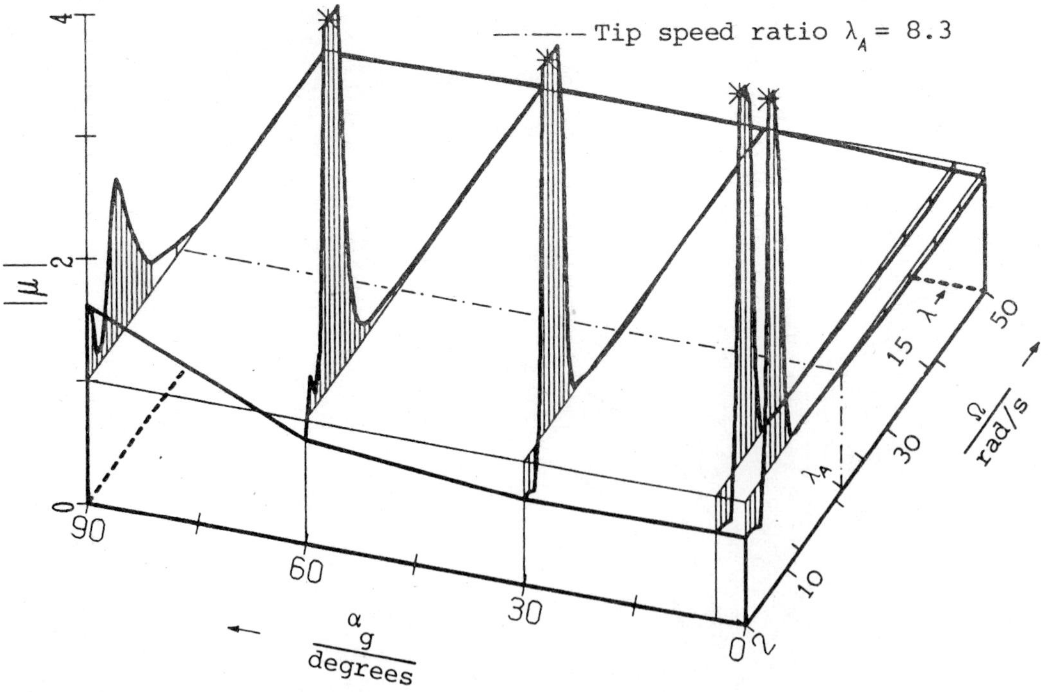

(b)

Fig 10 Three-bladed propeller with free yaw movement
(a) Limit rotational speed of stable running including aerodynamic forces
(b) Limit rotational speed of stable running including aerodynamic forces, size of instability

C310/88

Interpretation of strain gauge results for gas turbine engines with the aid of finite element analysis

R ELLIOTT, BSc and **A K McBRIDE**, BSc, CEng, MIMechE
Rolls-Royce plc, Derby

SYNOPSIS The fundamental objectives when assessing the dynamic characteristics of a turbine rotor or rotor assembly are to ensure that the component is:- (i) free from potentially hazardous resonances, and (ii) those resonances which cannot be eliminated from the engines normal running range do not impose unacceptable restrictions on operating limits or component lives.

In order to make an assessment of the fatigue life in service, accurate estimates are required of both the alternating stress levels and distribution within individual components at resonant conditions. The data can then be used in conjunction with steady stress levels and a knowledge of the materials fatigue properties to predict a life for the component and its possible failure position(s).

The most commonly used experimental method among aero engine companies is to strain gauge an assembly in an engine. This provides only limited direct measurement of the peak stresses/strains in hostile environments such as gas turbines, where surface temperatures can exceed 1,000 and high centrifugal forces can exist.

When dealing with bladed disc assemblies the situation is further complicated by the conflicting requirements of obtaining sufficient information on the distribution of stress and the need to cover an adequate sample size within the constraints of the telemetry system available.

Experience has shown that the most practical solution to this problem is to supplement the limited experimental data with comprehensive theoretical analyses using detailed finite element models, validated by laboratory measurements where possible.

The paper describes how this technique can be applied to real problems and is illustrated by reference to actual data and analysis.

1. INTRODUCTION

An important aspect of the mechanical design of a gas turbine engine is to ensure that the compressor and turbine rotating assemblies have satisfactory dynamic characteristics; failure to achieve an acceptable design could have serious implications, either from a safety aspect or from commercial considerations. In addition, in the civil aero engine field, stringent legal requirements have to be met to allow the engine to be certificated by the relevant government aviation authorities, prior to passenger - carrying flight.

Various techniques are available for the dynamic analysis of a rotor stage, ranging from pure theoretical analysis to direct measurement of stresses in the engine via straingauges. This paper concentrates on a method used for the first, high pressure, turbine stage of a high thrust version of a typical gas turbine engine. The HP turbine is probably the most difficult area of the engine for straingauge measurement. Figure 1 shows a schematic of the engine with the HP turbine stage highlighted; the inset shows the assembly in detail.

2. BASIC VIBRATION CHARACTERISTICS OF A BLADED TURBINE DISC

In describing the vibration characteristics of a turbine disc, the only deflections which need to be considered are those in the axial direction and the torsion about a radial line - the stiffnesses of disc's in the radial and circumferential directions are very large. Since the disc is circular and symmetrical there must be at least two points on a diameter (called nodes) where the axial deflections are zero. The line joining the nodes along a diameter is known as a nodal diameter. In a similar fashion, there is a radial variation in axial displacement along the non-nodal diameters; circles joining these nodes are known as nodal circles. Hence, it is possible to describe the displaced shape of the disc in terms of the number of nodal diameters and circles. Figure 2 shows a disc vibrating in a three nodal diameter (3D), two nodal circle (2nd family) mode of vibration. The natural frequency of each mode increases with the number of nodal diameters and circles.

The dynamic characteristics of the disc will be modified by the addition of a set of cantilevered blades. As far as the disc is concerned, this will have the effect of lowering its diametral frequencies.

Given a rigid clamp at their roots, the blades would resonate at their own natural frequencies. The effect of the disc is to vary the clamp support at each blade root. For low diametral modes, the flexibility of the rim support is such that it will interact with the blade flexibility and the frequency of the assembly will tend towards that of the disc alone. For high diametral modes, the rim support is seen as a rigid clamp and hence the assembly frequencies tend toward that of the blade alone.

The dynamic characteristics of the assembly will be further modified if, as in recent practice, the blades are interlocked at their tips. A typical interlock arrangement is shown in figure 3. The cross-interlock dimension is made an interference fit so that in order to accommodate the full ring of shrouds, each blade has to twist, thereby restraining each blade tip. This has the effect of increasing the first fundamental frequency of the blade alone. In terms of the behaviour of the bladed disc, for low diametral modes, the assembly frequencies will again tend towards the natural frequencies of the disc alone. For high diametral modes, the first family frequencies will tend towards the first natural frequency of the blade in its fixed-fixed configuration. This is shown diagrammatically in figure 4.

Figure 4a shows frequency plotted against the number of nodal diameters. Figure 4b shows the frequencies plotted in Campbell diagram form against engine speed. The straight lines emanating from the origin ('spokes') show typical sources of excitation.

3. STRAINGAUGE TESTING

3.1 The Need for Straingauge Testing

As figure 4b shows, the assembly frequencies change with increasing speed. Temperature effects cause a drop in frequency whilst engine rotation causes an increase due to centrifugal stiffening effects. These factors make the frequencies difficult to calculate accurately; a knowledge of the complex temperature distributions in the blades and disc is required. Even if the thermal distribution were to be known exactly, stresses are calculated in terms of ratios rather than absolute values - the level of aerodynamic forcing in the engine is not known and forced response methods are difficult to apply for a complex aerofoil shape.

Similar limitations apply to rig testing, particularly in terms of the temperature distribution. Therefore, absolute values of amplitude can only be measured with any certainty via straingauges in the actual engine.

3.2 Results from Straingauge Testing

A typical straingauge survey of a high pressure turbine rotor would involve measurements for a variety of engine manoeuvres, in order to simulate service operation in both typical and extreme conditions. The results yield, basically, three important pieces of information:-

i) which sources of excitation are prominent in the engine.

ii) which modes are being excited by i) and where the resonances occur, in terms of engine speed and frequency of vibration.

iii) a measure of the amplitude of vibration.

This information is limited in terms of both quality and quantity by several factors, some of which are outlined below.

i) currently available telemetry systems limit the number of gauges available for any one test manoeuvre, so that only one or two positions on a relatively small sample of blades can be covered.

ii) the hostile environment can lead to a high gauge mortality rate, which reduces still further the quantity of data that can be gathered.

iii) the uniaxial straingauges that are typically used for such applications can measure the average strain in one direction only, which is insufficient to describe the state of stress at that point. On simple structures it may be possible to align the gauge with the expected principal axes direction. For rotors with complex geometry which often have anisotropic material properties this strain cannot readily be converted into a stress.

iv) no indication can be obtained of the variation in stress levels throughout the rotor and it is fairly certain that the gauge will not be in the position of maximum stress for every mode of vibration.

These last two handicaps can be overcome by supplementing the available data with comprehensive theoretical analysis using detailed three dimensional finite element models. Where possible these are validated by laboratory measurements.

4. FINITE ELEMENT ANALYSIS

4.1 Use of F.E. Analysis

The use of finite element techniques for the solution of static and dynamic problems is now widespread in many branches of engineering and the aero engine business relies heavily on this type of analysis.

Complex geometrical shapes can be modelled by breaking the component down into an assembly of individual (finite) elements which are small enough to describe approximately the displacement/ stress field. It is then possible using a knowledge of the material properties and loadings/constraints on the system to determine the stress distribution throughout the structure.

The same model can also be used for dynamic problems such as the calculation of natural frequencies in undamped free vibration, i.e. the eigenvalue problem ref (1).

The dynamic characteristics of such a system are defined by equation (4.1) where M is the mass matrix, K the stiffness matrix and u the displacement response.

$$M\ddot{u} + Ku = 0 \qquad (4.1)$$

A solution for equation (4.1) can be written in the form

$$u = Y_j e^{iwjt} \qquad (4.2)$$

which can be substituted into the equation of motion (4.1) yielding the eigenvalue problem

$$Ky_j = \lambda_j My_j \text{ and } \lambda_j = W_j^2 \qquad (4.3)$$

where λ_j is the j th eigenvalue, Y_j is the corresponding eigenvector and W_j is the j th natural frequency.

The eigenvectors define the modeshape at each frequency and modal stresses can be recovered by treating them as physical displacements.

The magnitudes of these stresses have no significance as they are related only to an arbitrary normalisation of the eigenvectors, but the distribution is truly representative.

4.2 Advantages of Finite Element Techniques

One of the main advantages of using finite element analysis in place of experimental techniques is flexibility, in that any engine condition can be simulated quite accurately by suitable loading and constraint combinations. Loads such as gas pressures, centrifugal forces and thermal distributions, which all affect the stiffness of a component (and hence its frequency), are catered for by performing a geometrically non linear stress analysis prior to the vibration analysis. The stiffness matrix from the first analysis is then used in place of the unloaded stiffness matrix.

Other advantages are the ability to obtain complete stress/strain/energy distributions for the entire model (both internal and external), and that analysis can be carried out before hardware becomes available.

4.3 Limitations

Only structures which behave in a linear elastic fashion can be reasonably simulated, and for models which make use of cyclic symmetry, such as bladed disc assemblies it is assumed that they are perfectly tuned.

Boundary conditions such as at mating surfaces can also be difficult to simulate realistically at times.

5. VALIDATION USING OTHER METHODS

Comparisons between previous analyses and available experimental data have shown the finite element method to be accurate in terms of frequency, modeshape and stress level, providing the model is sufficiently detailed and geometrically accurate.

Various experimental techniques are available for comparison and some or all would normally be used to increase confidence in the results obtained from a particular model.

i) Frequency checks can be made for the lowest modes via spectral analysis or analysis of straingauge output under laboratory conditions.

ii) Modeshapes can usually be compared with holograms of components vibrating in particular modes as these show lines of constant displacement.

iii) Stress distributions can be checked by applying several straingauges or rosettes to the component and determining the relative magnitudes. Another technique used is SPATE, Stress Pattern Analysis by Thermal Emission where the stress distribution can be inferred by looking at the tiny fluctuations in surface temperature of the component which are proportional to the difference of the principal stresses.

6. ESTIMATING ENGINE STRESS LEVELS

As described in section 3. it is not usually possible to determine absolute peak stresses from straingauge tests alone and so the three-dimensional finite element model is used to supplement the data. The gauge results show which frequencies are being excited at particular speeds and indicate the magnitude of the strain in the gauge direction at that particular location.

By comparing these results with a Campbell diagram generated from theoretical results it is possible to identify the modes of interest and then the correct theoretical modal stress distribution can be scaled using the measured component strain.

7. PROCEDURE

Outlined below are the typical steps followed when determining the peak vibratory stresses in a high pressure turbine blade (or similar component) at realistic engine conditions.

i) Prior to test, identify one or two convenient positions on the blade aerofoil to attach the straingauges where a response should be seen in as many modes as possible but is away from any large stress gradients.

ii) During the test, record amplitude, frequency and speed variations with time during manoeuvres covering the required speed range. Other important engine parameters such as turbine gas temperature and pressure may also be noted.

iii) Post test, the recorded data is processed through a data reduction system which allows plots of the strain and frequency variation with speed or time to be produced (see fig 8). Examination of this data allows the maximum responses seen at each resonance to be determined. Because of the large variations in blade temperature with speed (typically 600°C) the readings may require correction for gauge factor differences to give an accurate result.

iv) Estimate for each mode that has been identified the calculated surface strain in the gauge direction, averaged over the gauge area, from the finite element model results. A scaling factor can now be calculated as:-

$\varepsilon_m / \varepsilon_{FE}$

Where ε_m is the measured strain and ε_{FE} that obtained from the model.

When this is applied to the stress distribution obtained from the finite element analysis it allows the principal stress magnitudes to be estimated at any position on the component. The level can also be compared with fatigue strength data making due allowance for the presence of steady stresses in the blade (also obtainable from the model) and lifing estimates made.

This step automatically allows for any anisotropy in the material as this is built into the model and also for the principal axes not being aligned normal and parallel to the gauge direction.

v) The more usual approach to scaling modal stresses is to use the blade tip amplitude but this is not easily measured on a shrouded blade in an engine environment.

8. EXAMPLE

To illustrate the use of the procedure outlined above a particular example has been chosen, which demonstrates the value of this technique.

8.1 Background

A new high pressure turbine blade was designed for development testing of an increased thrust variant of an engine. The blade was made from a cast nimonic material, directionally solidified to give enhanced creep resistance at high temperatures. It also featured an advanced internal cooling design which necessitates intricate cast passages within the blade, resulting in complex geometry and thermal distributions.

Satisfactory dynamic characteristics were to be ensured by use of an interlocking outer shroud ring and this was confirmed at the design stage by the usual 3D F.E. analysis with the blade restrained at the shroud interlock. Fig 5 shows the model used.

Initial engine straingauge testing produced responses which were indicative of a bladed disc assembly with tight interlocks (as was the design intent). Later endurance testing of the engine produced cracking in the blade aerofoil at several locations which metallurgical examinations revealed to be due to high cycle fatigue (high vibratory stress).

This was unexpected in view of the initial favourable results from the analysis and straingauge test but further inspections revealed that the interlocks had become loose, due to wear and deformation of the aerofoil. The cause of the deformation was felt to be increased shroud temperatures due to changes in the combustion chamber subsequent to the original straingauge testing.

The lack of interlock would cause the assembly to behave quite differently from the original intention, with the lower frequency cantilever type modes dropping into the engine running range. (compare figs 6a and 6b)

The problem was then to identify which were the damaging modes of vibration now present and to assess what could be done to eliminate the cracking.

8.2 Investigations

A repeat engine straingauge test was carried out using as far as possible components from the engine that had produced aerofoil cracking, in order to simulate as closely as possible the conditions experienced. The turbine blades had also come from the previous test.

The results of the test showed five significant responses in what could easily be identified as blade cantilever modes. There was no sign of any bladed disc interaction which gave confidence that a finite element analysis using a single blade model would be sufficient.

Stress and strain distributions were calculated for these problem modes (Fig 7 shows those for 1st flap mode) and the peak stress positions and magnitudes compared with the cracking that had occurred. The finite element results had been scaled previously using the method outlined in section 7. Table 1 shows the results obtained which have been normalised to the fatigue strength of the material.

It can be seen from this table that large vibratory stresses were present in the vicinity of the crack locations. Examination of the straingauge raw data (see fig 8) also shows that some of the modes were responding in overlapping speed ranges which would produce even higher total stresses.

A similar, retrospective, analysis was carried out using the original model constraints and the initial straingauge test results. This indicated a peak stress in the region of two thirds the value seen in cantilever modes and with only one significant response seen in the normal speed range.

Table 1 - Estimated Engine Peak Aerofoil Stresses*

Normalised to Fatigue strength

MODE	STRESS AT S/G POSITION	PEAK AEROFOIL STRESS
CANTILEVER		
1	0.45	0.96
2	0.16	0.50
3	0.2	70.51
4	0.30	0.71
8	0.40	0.66
RESTRICTED		
1	0.26	0.70

*No allowance has been made for scatter or steady stress.

based on max response obtained in each mode

It was concluded that if the cantilever modes could be eliminated, the assembly would perform acceptably. Therefore, use of the technique led to the conclusion that effort should be concentrated on keeping the interlocks tight, rather than modifying the blade geometry to alter the dynamic characteristics.

9. CONCLUSIONS

When setting up a straingauge test, care should be exercised in the positioning of the gauges. Finite element analysis can be used to ensure that the gauges are both responsive to all probable modes of vibration yet not in areas of high stress gradient.

Similar care should be taken in the interpretation of the engine results. Allowance should be made for the following effects:

.Strain averaging over the finite area of the gauge

.Blade anisotropic material properties

.Differences between gauge calibration temperature and the surface temperature at which the resonance occurs.

.Conversion of measured strain to maximum principal stress at a position.

Interpretation of the results is most easily accomplished with the aid of a finite element analysis of the component. Results of the analysis

should be validated by any suitable experimental means.

For high diametral modes, where the assembly frequencies tend toward the frequency of the blade alone, a finite element analysis of a single blade, rather than the bladed disc assembly, will often suffice.

10. REFERENCES

(1)VARIOUS,A finite element primer, 1986, ppl 73-180

Department of Trade and Industry, National

Engineering Laboratory.

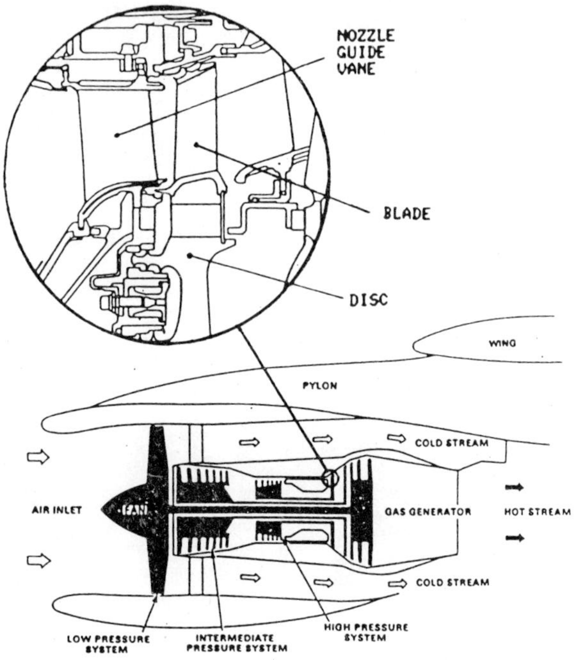

Fig 1 Schematic representation of engine showing HP turbine

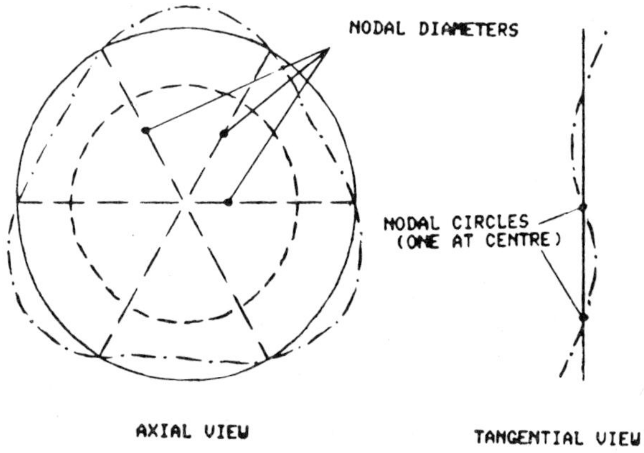

Fig 2 Displaced shape for three-dimensional/second family mode of vibration

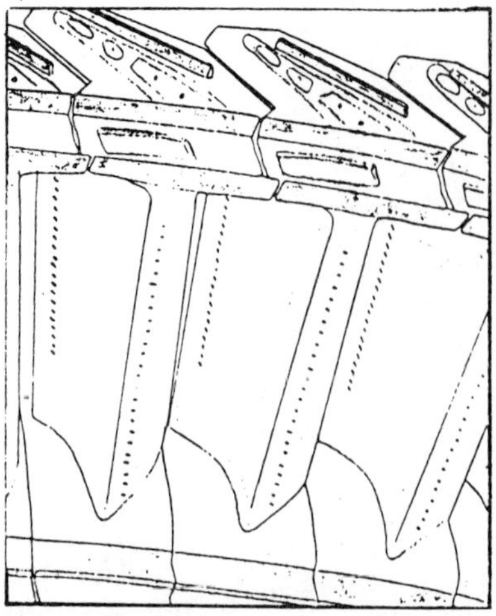

Fig 3 Typical interlock arrangement

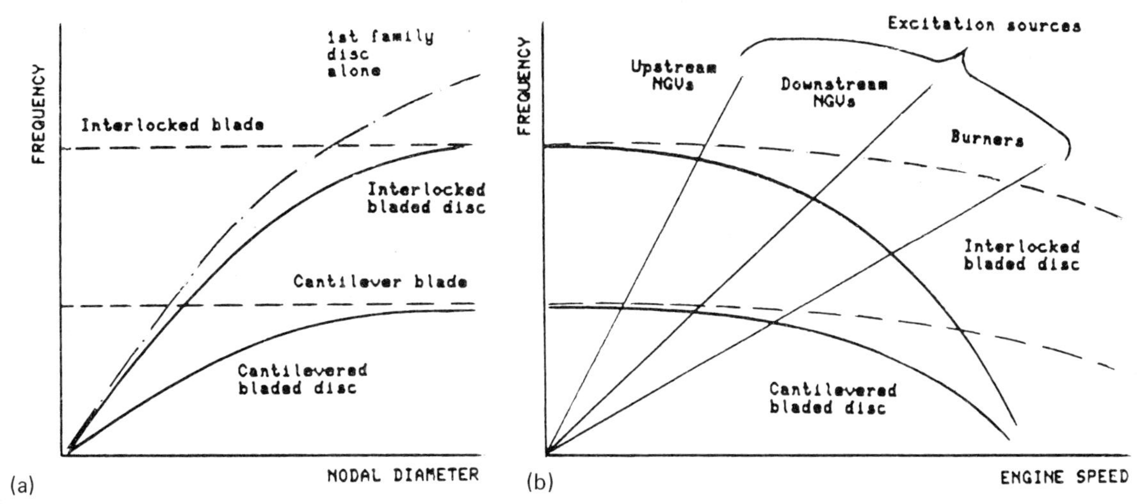

Fig 4 Bladed disc characteristics

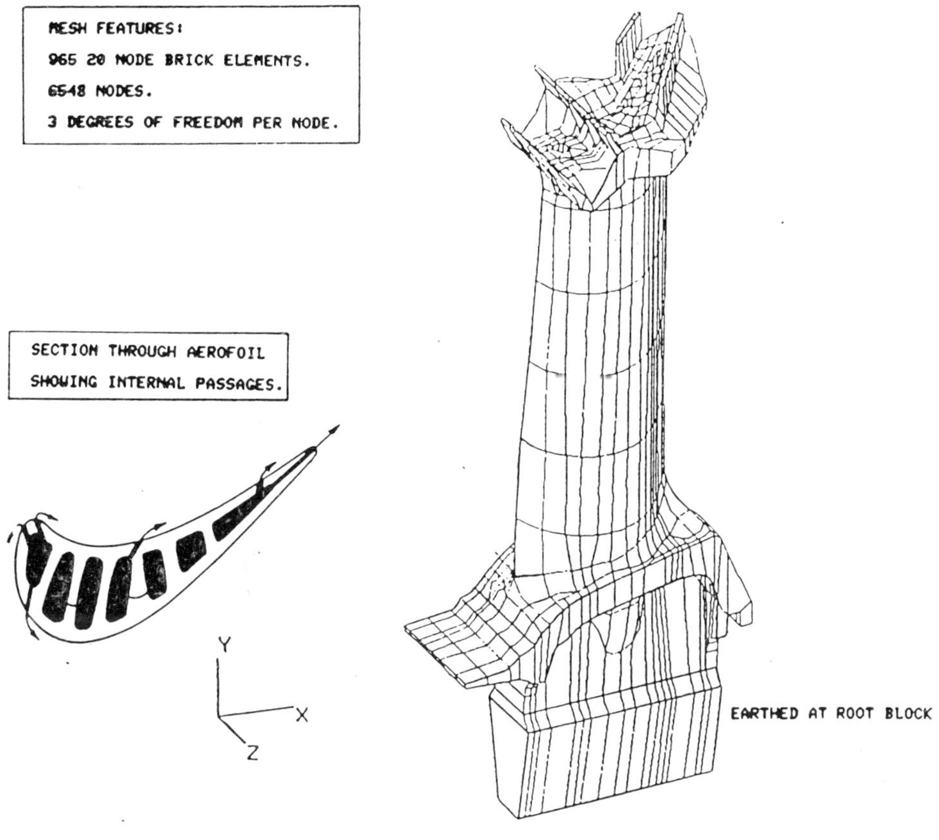

Fig 5 Three-dimensional finite element model of HP turbine blade

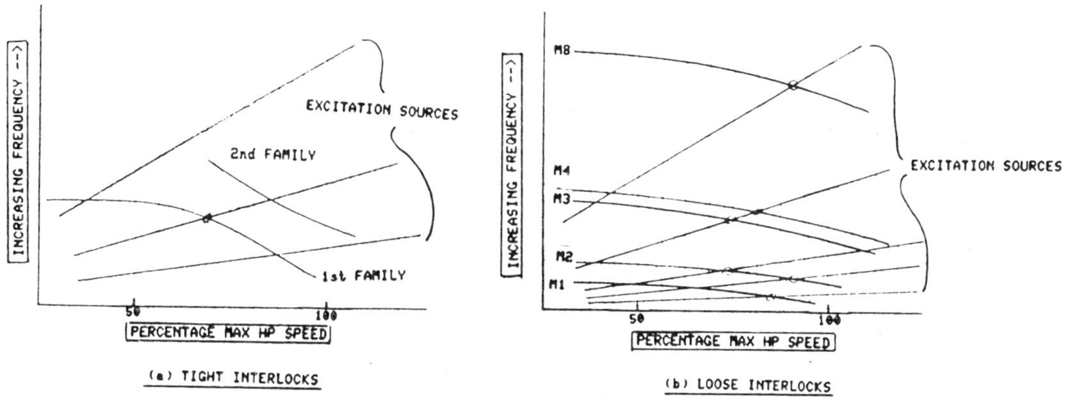

Fig 6 Theoretical Campbell diagrams for HP turbine blade

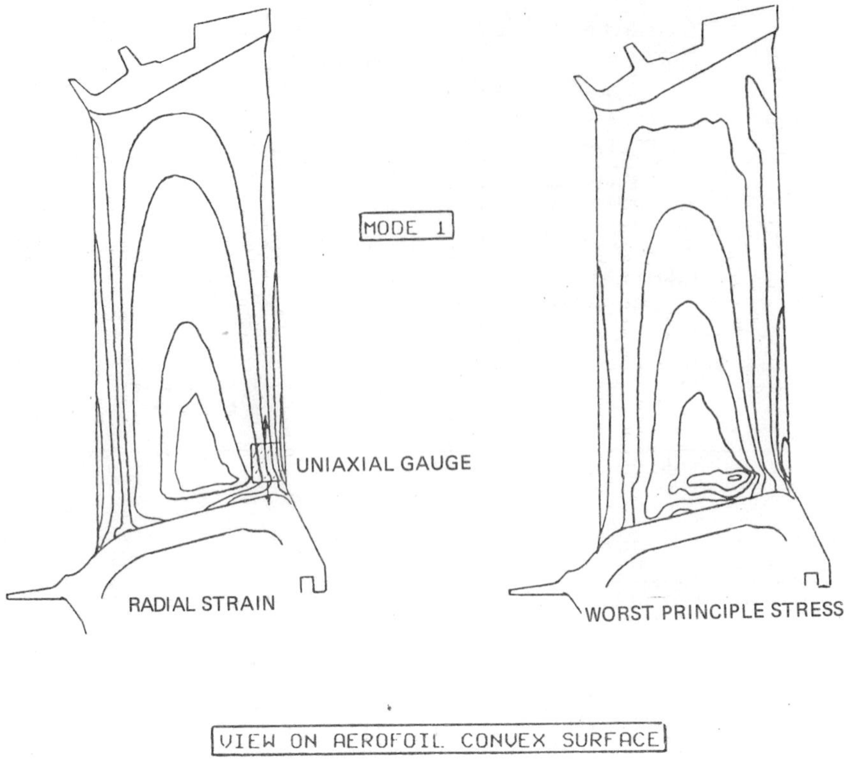

Fig 7 Contour plots of radial strain and worst principal stress

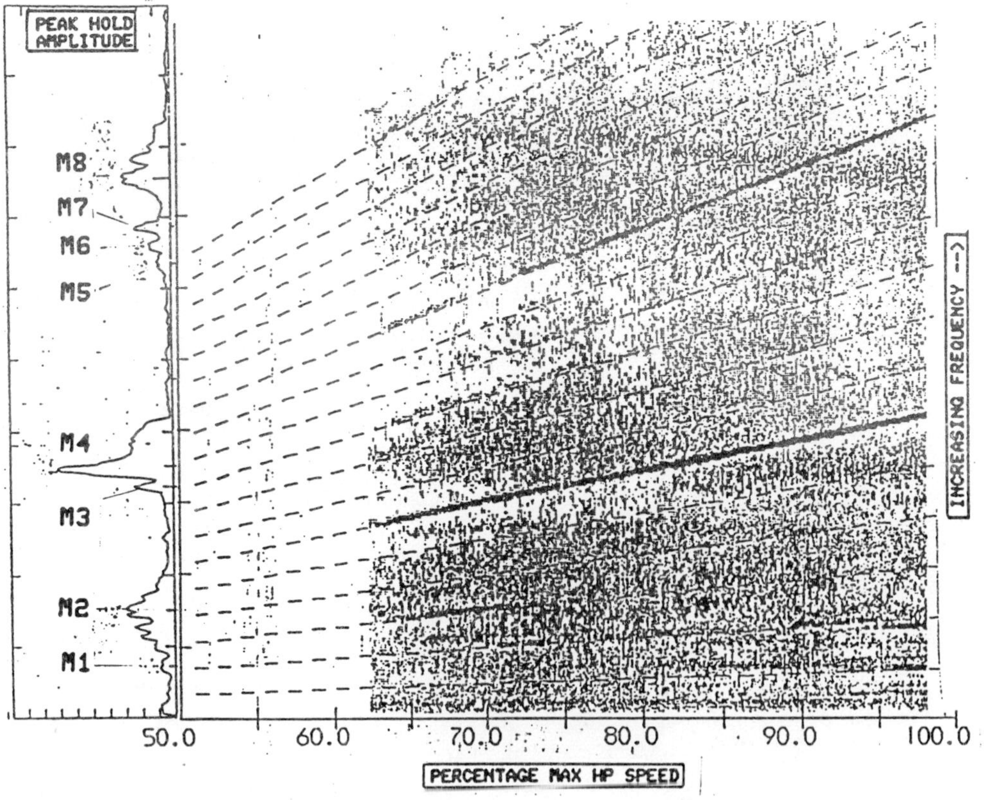

Fig 8 Engine strain gauge results showing cantilever modes

C264/88

On critical speeds of rotating discs

W SEEMANN, Dipl-Ing and **J WAUER**, Dr-Ing
Institut für Technische Mechanik, University of Karlsruhe, Karlsruhe, West Germany

SYNOPSIS Disc-shaped structures are important distributed parameter components in rotating machinery. To Complete classical investigations in this field, different dynamic influences are taken more accurately into consideration here than in the past, e.g. the pre-deformation by centrifugal forces, gyroscopic coupling and the inclination of disc relative to the rotational axis. Starting from the governing non-linear boundary value problem derived by Hamilton's principle, the stationary pre-deformation of a constant speed disc rotor is calculated first. After that, small vibrations around the basic configuration are analysed so that the speed-dependent eigenfrequencies and the corresponding critical speeds can be specified.

NOTATION

R	outer radius of disc
h	thickness of disc
r, ϕ, z	body-fixed cylindrical coordinates
t	time
u, v, w	displacements of a general point of middle surface
$U(r), V(r), W(r)$	radial distributions of displacement
U_k, V_k, W_k	shape functions
f_r, f_ϕ, f_z	body forces
E	Young's modulus
T	kinetic energy
V	elastic potential
δW	virtual work of non-potential forces
$\vec{v}$	absolute velocity of a general point of middle surface
L, M, N	differential operators
n	circumferential node number
ρ	mass density
ν	Poisson's ratio
α	angle of inclination of the disc relative to the rotational axis
$\vec{\varepsilon}$	strain tensor
ζ	thickness co-ordinate
Ω (Λ)	rotational speed (in non-dimensional form)
ω (λ)	eigenfrequency (eigenvalue)

1 INTRODUCTION

Starting from the classical Jeffcoat rotor, considerations about unbalance-excited vibrations were increasingly extended to distributed parameter rotor systems during the last decades. Slender shaft rotors have been mainly studied, not only in the form of more or less complex models, e.g. (1), (2), but also as practical fluid-film bearing rotor systems in turbomachinery, e.g. (3), (4). Untii now the vibration behaviour of rotating discs has not been investigated in detail. As a matter of fact, the classical book (5) contains Grammel's classical results, however, in actual research work on this topic, e.g. (6), (7), different simplifications were introduced so that the governing equations of motion are not so complicated and some conclusions are questionable.

The objective of this contribution is to obtain a more detailed knowledge and understanding of the dynamics of such rotating discs taking the different additional influences more accurately into consideration than in other papers. That means that not only the gyroscopic coupling should be regarded but also the pre-deformation caused by centrifugal forces of a constant speed rotor. Finally the inclination of the disc relative to the rotational axis should be included. Consequently, a non-linear theory of elasticity is needed and the in-plane and bending vibrations of the rotating disc are coupled.

The major point of interest is the calculation of the speed-dependent natural frequencies and the associated critical speeds. For this purpose, the governing non-linear boundary value problem is firstly derived by Hamilton's principle. Secondly, the pre-deformations during stationary operation with a constant rotational speed are calculated where a small inclination is assumed. Finally the usual method of small oscillations leads to a linearized boundary value problem describing free and forced vibrations around the stationary pre-deformation.

This paper gives some additional results to those listed in reference (8), where the stability of pure in-plane vibrations was mainly investigated.

2 ROTOR MODEL AND NON-LINEAR EQUATIONS OF MOTION

The object of the investigation, refer to Fig. 1, is a circular solid disc of radius R and mass density ρ rotating around a central axis with a constant angular velocity Ω. The disc plane should be inclined against the rotational axis described by a small constant angle $\alpha \ll 1$ between the normal of the disc and the axis of revolution. For mathematical simplification, the thickness h of the centrally clamped disc should be constant and only the two cases of a clamped or a stress-free periphery should be regarded.

The theory of thin discs on the basis of the usual Kirchhoff-Love hypothesis should be used.

The material should be elastic (Young's modulus E and Poisson's ratio ν), damping effects should be omitted. It can be stated that the extension to a viscoelastic material in the sense of Voigt-Kelvin is possible without fundamental difficulties (9).

Space and time-dependent radial and torsional vibrations should be discussed in the plane of the disc and also out-of-plane oscillations, all coupled very strongly. Assuming the above, the displacements of a general disc point can be reduced (in non-linear form) to the deflections of the corresponding middle surface point. The displacements $u(r,\phi,t)$, $v(r,\phi,t)$ and $w(r,\phi,t)$ (radial and circumferential co-ordinates r, ϕ and time t) should be measured in a body-fixed, rotating reference frame of the unit base vectors $\vec{e}_r$, $\vec{e}_\phi$, $\vec{e}_z$ and composed of a basic deformation u_0, v_0 and w_0 owing to centrifugal inertia and superimposed small vibrations Δu, Δv and Δw excited by the periodic body forces

$$
\begin{aligned}
f_r(r,\phi,t) &= \sum_{j=1}^{\infty} F_{rj}(r,\phi)\cos(j\Omega t+\alpha_j)\\
f_\phi(r,\phi,t) &= \sum_{j=1}^{\infty} F_{\phi j}(r,\phi)\cos(j\Omega t+\beta_j) \qquad (1)\\
f_z(r,\phi,t) &= \sum_{j=1}^{\infty} F_{zj}(r,\phi)\cos(j\Omega t+\gamma_j)
\end{aligned}
$$

caused by weight, unbalances or reactions of vibrating blades if the disc is, for instance, a component of a turbine. A second non-rotating reference frame of the unit base vectors $\vec{e}_X$, $\vec{e}_Y$, $\vec{e}_Z$ should exist where the origin O is located at the disc centre and the axis Oxyz fixed in the disc are derived from the inertial axis OXYZ by a rotation Ωt about the rotational axis OZ.

The governing non-linear boundary value problem should be generated here by means of Hamilton's principle.

$$
\delta\int_{t_1}^{t_2}(T-V)\,dt+\int_{t_1}^{t_2}\delta W\,dt=0 \qquad (2)
$$

where T is the kinetic energy, V the elastic potential and δW the virtuel work of the non-potential forces for the rotating disc.

For the kinetic energy one obtains

$$
T=\frac{1}{2}\rho h\int_0^{2\pi}\int_0^{R}\vec{v}^{\,2}\,r\,dr\,d\phi \qquad (3)
$$

(if rotational inertia on bending is neglected), where $\vec{v}$ is the absolute velocity of a middle surface point corresponding with a general disc point:

$$
\begin{aligned}
\vec{v} = {}&\{(-\dot{u}\cos\alpha\sin\phi-(r+u)\Omega\cos\phi-\dot{v}\cos\alpha\cos\phi+\\
&+v\Omega\sin\phi+\dot{w}\sin\alpha)\sin\Omega t+(\dot{u}\cos\phi-(r+u)\Omega\cos\alpha\cdot\\
&\cdot\sin\phi-\dot{v}\sin\phi-v\Omega\cos\alpha\cos\phi+w\Omega\sin\alpha)\cos\Omega t\}\vec{e}_X+\\
&+\{(\dot{u}\cos\phi-(r+u)\Omega\cos\alpha\sin\phi-\dot{v}\sin\phi-v\Omega\cos\alpha\cdot\\
&\cdot\cos\phi+w\Omega\sin\alpha)\sin\Omega t+(\dot{u}\cos\alpha\sin\phi+(r+u)\Omega\cdot \qquad (4)\\
&\cdot\cos\phi+\dot{v}\cos\alpha\cos\phi-v\Omega\sin\phi-\dot{w}\sin\alpha)\cos\Omega t\}\vec{e}_Y+\\
&+(\dot{u}\sin\alpha\sin\phi+\dot{v}\sin\alpha\cos\phi+\dot{w}\cos\alpha)\vec{e}_Z
\end{aligned}
$$

$$
\dot{} = \frac{\partial}{\partial t}
$$

The potential V of an isotropic, homogenous thin disc, where a plane stress state can be assumed, reads

$$
\begin{aligned}
V=\frac{E}{2(1-\nu^2)}\int_{-h/2}^{+h/2}\int_0^{2\pi}\int_0^{R}\{\varepsilon_{rr}^2+\varepsilon_{\phi\phi}^2+2\nu\varepsilon_{rr}\cdot\\
\cdot\varepsilon_{\phi\phi}+2(1-\nu)\varepsilon_{r\phi}^2\}\,r\,dr\,d\phi\,dz
\end{aligned}
\qquad (5)
$$

where the governing elements of Green's strain tensor $\vec{\varepsilon}$ can be calculated in a sufficient second order approximation (denoted by a superior parenthesized figure)

$$
\begin{aligned}
\varepsilon_{rr}^{(2)} = {}& u_{,r}+\frac{1}{2}(u_{,r}^2+v_{,r}^2+w_{,r}^2)+\zeta(-w_{,rr}+u_{,rr}w_{,r})+\\
&+\frac{\zeta}{r}v_{,rr}w_{,\phi}+\frac{1}{2}\zeta^2\{w_{,rr}^2+\frac{1}{r^2}(w_{,r\phi}-\frac{1}{r}w_{,\phi})^2\}\\
\varepsilon_{\phi\phi}^{(2)} = {}& \frac{1}{r}\{u+v_{,\phi}+\frac{1}{2r}\{(u_{,\phi}-v)^2+(u+v_{,\phi})^2+w_{,\phi}^2\}+\\
&+\zeta(-w_{,r}+u_{,r}w_{,r})+\frac{\zeta}{r}\{w_{,\phi\phi}+w_{,r}(u_{,\phi\phi}-\\
&-v_{,\phi})+v_{,r}w_{,\phi}-w_{,r}(u+v_{,\phi})\}+\frac{\zeta}{r^2}\{w_{,\phi}\cdot\\
&\cdot(u_{,\phi}+v_{,\phi\phi})+w_{,\phi}(u_{,\phi}-v)\}+\frac{\zeta^2}{2r}\{(w_{,r}+ \qquad (6)\\
&+\frac{1}{r}w_{,\phi\phi})^2+(w_{,r\phi}+\frac{1}{r}w_{,\phi})^2\}\}\\
\varepsilon_{r\phi}^{(2)} = {}& \frac{1}{2}\{v_{,r}+\frac{1}{r}(u_{,\phi}-v)+\frac{1}{r}u_{,r}(u_{,\phi}-v)+\frac{1}{r}v_{,r}(u+\\
&+v_{,\phi})+\frac{1}{r}w_{,r}w_{,\phi}+\frac{\zeta}{r}\{-2w_{,r\phi}+2w_{,r}(u_{,r\phi}-v_{,r})\}+\\
&+\frac{2\zeta}{r^2}\{w_{,\phi}-w_{,r}(u_{,\phi}-v)+w_{,\phi}(u_{,r}+v_{,r\phi})\}-\\
&-\frac{2\zeta}{r^3}w_{,\phi}(u+v_{,\phi})+\frac{\zeta^2}{r}(w_{,r\phi}-\frac{1}{r}w_{,\phi})\{w_{,rr}+\\
&+\frac{1}{r}(w_{,r}+\frac{1}{r}w_{,\phi\phi})\}\} \qquad {}_{,r}=\frac{\partial}{\partial r} \qquad {}_{,\phi}=\frac{\partial}{\partial \phi}
\end{aligned}
$$

With the assumptions of the classical Kirchhoff-Love hypothesis they can really be expressed in terms of the displacements of middle surface (ζ denotes the distance between the general disc point and the associated middle surface one).

Introduced into equation (5) the elastic potential V can be determined in a third order approximation $V^{(3)}$, where the integration over the thickness co-ordinate z has to be carried out.

Finally, on the postulated assumptions, the virtual work becomes

$$
\delta W = h\int_0^{2\pi}\int_0^{R}(f_r\delta u+f_\phi\delta v+f_z\delta w)\,r\,dr\,d\phi \qquad (7)
$$

Now, Hamilton's principle (2) can be evaluated. Carrying out the required variations leads to the non-linear field equations

$$
\begin{aligned}
L_1(u,v,w) &= 0\\
L_2(u,v,w) &= 0 \qquad (8.1)\\
L_3(u,v,w) &= 0
\end{aligned}
$$

and the corresponding boundary conditions

$$
\begin{aligned}
L_1^j\{u(j,\phi,t),v(j,\phi,t),w(j,\phi,t)\} &= 0\\
L_2^j\{u(j,\phi,t),v(j,\phi,t),w(j,\phi,t)\} &= 0
\end{aligned}
\qquad (8.2)
$$

$$L^j_{31}\{u(j,\phi,t),v(j,\phi,t),w(j,\phi,t)\} = 0$$
$$L^j_{32}\{u(j,\phi,t),v(j,\phi,t),w(j,\phi,t)\} = 0 \quad j=0,R$$

both are given (in a second order approximation in the sense of Kármán (10)) in appendix 1.

3 STATIONARY PRE-DEFORMATION

Without the influence of the time-dependent excitations f_r, f_ϕ, f_z (1) a time-independent deformation

$$\begin{aligned} u &= u_0(r,\phi) \\ v &= v_0(r,\phi) \\ w &= w_0(r,\phi) \end{aligned} \tag{9}$$

results during the stationary operation with a constant rotational speed Ω.

Setting the time derivatives within the equation (8) equal to zero yields a non-linear boundary value problem free of time to calculate the pre-deformations u_0, v_0, w_0 (9).

In a linear approximation (usually sufficient) it reads

$$\begin{aligned} M_1(u_0,v_0,w_0) &= 0 \\ M_2(u_0,v_0,w_0) &= 0 \\ M_3(u_0,v_0,w_0) &= 0 \end{aligned} \tag{10.1}$$

$$u_0(0,\phi) = v_0(0,\phi) = w_0(0,\phi) = w_{0,r}(0,\phi) = 0$$
$$u_0(R,\phi) \text{ or } ru_{0,r}+\nu(u_0+v_{0,\phi})|_{(r=R,\phi)} = 0$$
$$v_0(R,\phi) \text{ or } v_{0,r}+\frac{1}{r}(u_{0,\phi}-v_0)|_{(r=R,\phi)} = 0$$
$$w_0(R,\phi) \text{ or } w_{0,rrr}+(2-\nu)\{\frac{1}{r}w_{0,r}+\frac{1}{r^2}w_{0,\phi\phi}\}_{,r}|_{(r=R,\phi)} = 0 \tag{10.2}$$
$$w_{0,r}(R,\phi) \text{ or } w_{0,rr}+\nu(\frac{1}{r}w_{0,r}+\frac{1}{r^2}\cdot w_{0,\phi\phi})|_{(r=R,\phi)} = 0$$

where the field equations (10.1) are specified in appendix 2.

It can be seen that despite this simplification an exact solution seems to be impossible. To obtain an approximate solution by formulae, the wellknown theory of perturbation should be applied. Carrying this out, we assume that the deformations u_0, v_0 and w_0 can be represented by an expansion having the form

$$\begin{aligned} u_0 &= u_0^{(0)}(r) + \alpha u_0^{(1)}(r,\phi) + O(\alpha^2) \\ v_0 &= 0 + \alpha v_0^{(1)}(r,\phi) + O(\alpha^2) \\ w_0 &= 0 + \alpha w_0^{(1)}(r,\phi) + O(\alpha^2) \end{aligned} \tag{11}$$

where the small angle $\alpha<<1$ as a perturbation parameter is performed. Since for $\alpha=0$ a pure radial deformation $u_0^{(0)}$ results, the components $v_0^{(0)}$, $w_0^{(0)}$ vanish identically. Moreover, this quantity is only a function of r and not of ϕ because the field of centrifugal inertia is rotationally symmetric.

Substituting the relation (11) into the linear boundary value problem (10), a linear ordinary differential equation

$$\frac{E}{1-\nu^2}\{u^{(0)}_{0,rr}+\frac{1}{r}u^{(0)}_{0,r}-\frac{1}{r^2}u_0^{(0)}\}+\rho\Omega^2\{r+u_0^{(0)}\}=0 \tag{12.1}$$

results for the remaining pre-deformation $u_0^{(0)}(r)$ with associated boundary conditions

$$u_0^{(0)}(0) = 0$$
$$u_0^{(0)}(R) \text{ or } Ru^{(0)}_{0,r}(R)+\nu u_0^{(0)}(R) = 0 \tag{12.2}$$

and a boundary value problem

$$\frac{E}{1-\nu^2}\{u^{(1)}_{0,rr}+\frac{1}{r}u^{(1)}_{0,r}+\frac{1-\nu}{2r^2}u^{(1)}_{0,\phi\phi}-\frac{1}{r^2}u_0^{(1)}\}+ + \rho\Omega^2u_0^{(1)} = 0$$

$$\frac{E}{1-\nu^2}\{\frac{1}{r}u^{(1)}_{0,r\phi}+\frac{1}{r^2}u^{(1)}_{0,\phi}\}+\rho\Omega^2\frac{1}{1+\nu}v_0^{(1)} = 0 \tag{13.1}$$

$$\frac{Eh^2}{12(1-\nu^2)}\{w^{(1)}_{0,rrrr}+\frac{2}{r}w^{(1)}_{0,rrr}-\frac{1}{r^2}w^{(1)}_{0,rr}+ + \frac{1}{r^3}w^{(1)}_{0,r}+\frac{1}{r^4}w^{(1)}_{0,\phi\phi\phi\phi}+\frac{2(1-\nu)}{r^4}w^{(1)}_{0,\phi\phi}+ + \frac{2}{r^2}w^{(1)}_{0,rr\phi\phi}-\frac{2}{r^3}w^{(1)}_{0,\phi\phi}\}-\rho\Omega^2\{r+u_0^{(0)}\}\sin\phi = 0$$

unchanged boundary conditions (10.2), where the variables are marked by a superior parenthesized figure 1 (13.2)

for the corrections of first order $u_0^{(1)}$, $v_0^{(1)}$, $w_0^{(1)}$. The corrections of higher order should be neglected here.

All homogenous parts of differential equation within the boundary value problems (12) and (13) can be reduced to Bessel-differential equations so that, adding a special particular solution, the complete solution of the boundary value problems can be expressed in formulae, e.g. (11). Because it is a longer calculation, only the solution of (12) should be constructed in detail.

By the transformation

$$y = \frac{r}{R}\{\Lambda^2(1-\nu^2)\}^{1/2} \tag{14}$$

where a speed parameter in a non-dimensional form

$$\Lambda = \Omega R(\frac{\rho}{E})^{1/2} \tag{15}$$

is defined, equation (12.1) changes into a Bessel-differential equation

$$u^{(0)}_{0,yy}+\frac{1}{y}u^{(0)}_{0,y}+(1-\frac{1}{y^2})u_0^{(0)} = -\frac{y}{\{\Lambda^2(1-\nu^2)\}^{1/2}} \tag{16}$$

with the general solution

$$u_0^{(0)}(y) = C_1J_1(y)+C_2I_1(y)-\frac{y}{\{\Lambda^2(1-\nu^2)\}^{1/2}} \tag{17}$$

J_k and I_k denote the well-known Bessel-functions of the first and second kind of order k. Fitting the solution (17) to the boundary conditions (12.2), where the argument y is replaced by r according to relation (14), yields the completely calculated solution

$$u_0^{(0)}(r) = \frac{J_1\{\{\Lambda^2(1-\nu^2)\}^{1/2}\frac{r}{R}\}}{J_1\{\{\Lambda^2(1-\nu^2)\}^{1/2}\}}R - r \tag{18.1}$$

for the case of a clamped periphery and

$$u_0^{(0)}(r) = \frac{(1+\nu)J_1\{\{\Lambda^2(1-\nu^2)\}^{1/2}\frac{r}{R}\}}{Q}R - r$$

$$Q = \{\Lambda^2(1-\nu^2)^{1/2}\}J_0\{\{\Lambda^2(1-\nu^2)\}^{1/2}\}- -(1-\nu)J_1\{\{\Lambda^2(1-\nu^2)\}^{1/2}\} \tag{18.2}$$

for a stress-free outer edge.

Considering the boundary value problem (13) solutions of the form

$$u_0^{(1)} = u_0^{(1)}(r),\ v_0^{(1)} = 0,\ w_0^{(1)} = w_0^{(1)}(r,\phi) \qquad (19)$$

can be sought so that it can be simplified as

$$\frac{E}{(1-\nu^2)}\{u^{(1)}_{0,rr} + \frac{1}{r}u^{(1)}_{0,r} - \frac{1}{r^2}u^{(1)}_0\} + \rho\Omega^2 u_0^{(1)} = 0$$

$$\frac{Eh^2}{12(1-\nu^2)}\{w^{(1)}_{0,rrrr} + \frac{2}{r}w^{(1)}_{0,rrr} - \frac{1}{r^2}w^{(1)}_{0,rr} + \frac{1}{r^3}w^{(1)}_{0,r} + \frac{1}{r^4}w^{(1)}_{0,\phi\phi\phi\phi} + \frac{2(1-\nu)}{r^4}w^{(1)}_{0,\phi\phi} + \frac{2}{r^2}w^{(1)}_{0,rr\phi\phi} - \frac{2}{r^3}w^{(1)}_{0,\phi\phi}\} - \rho\Omega^2\{r+u_0^{(0)}\}\sin\phi = 0 \qquad (20.1)$$

boundary conditions (13.2) for $u_0^{(1)}$ and $w_0^{(1)}$ (20.2)

While the solution to $u_0^{(1)}$ is trivial,

$$u_0^{(1)} \equiv 0 \qquad (21)$$

the remaining boundary value problem within (20) describing the deformation quantity $w_0^{(1)}$ has to be solved in two steps. First we assume a solution

$$w_0^{(1)}(r,\phi) = W_0^{(1)}(r)\sin\phi \qquad (22)$$

so that the corresponding partial differential equation within (20.1) changes into an ordinary one. In a second step, the solution $W_0^{(1)}(r)$ to this inhomogenous equation of deformation together with adjusting to the associated boundary conditions within (20.2) has to be constructed. The result of such a calculation should not be given here but now it becomes evident that an approximate solution for the pre-deformations of a rotating disc according to equation (11) can be determined.

A discussion shows that the basic configuration will be unstable if the characteristic denominators of the governing expressions, refer to equation (18), for instance, go to zero. Fig. 2 and 3 represent the corresponding stability diagram for the planar deformation u_0, where, in the case of a clamped outer surface, an additional restriction $(u_0+r)_{,r} \geq 0$ has to be satisfied. The non-dimensional safe speed Λ_c is plotted as a function of the material property ν. For the bending deformation w_0, a similar stability diagram can be expected.

4 CRITICAL SPEEDS

The usual method of small vibrations (based on the above approximately calculated pre-deformations u_0, v_0 and w_0)

$$u(r,\phi,t) = u_0(r) + \Delta u(r,\phi,t)$$
$$v(r,\phi,t) = 0 + \Delta v(r,\phi,t) \qquad (23)$$
$$w(r,\phi,t) = w_0(r,\phi) + \Delta w(r,\phi,t)$$

leads from equation (10) to a linearized boundary value problem describing free and forced vibrations Δu, Δv and Δw around the stationary pre-deformation u_0, v_0 and w_0. In the following , a range of operation of the rotating disc should be considered, where the rotational speed Ω is sufficiently smaller than the critical value Ω_c. Subsequently, the pre-deformations u_0, v_0, w_0 are small, where the order of magnitudes of the components $u_0^{(0)}$, $v_0^{(0)}$, $w_0^{(0)}$ and $u_0^{(1)}$, $v_0^{(1)}$, $w_0^{(1)}$ are different once more. Hence, taking only linear terms of pre-deformation $u_0^{(0)}$, $v_0^{(0)}$, $w_0^{(0)}$ into consideration is allowed (the latter two vanish identically) and the equations of motion are

$$L_1^*(\Delta u,\Delta v) = 0$$
$$L_2^*(\Delta u,\Delta v) = 0 \qquad (24.1)$$
$$L_3^*(\Delta w) = 0$$

$$L_1^{*j}\{\Delta u(j,\phi,t),\Delta v(j,\phi,t)\} = 0$$
$$L_2^{*j}\{\Delta u(j,\phi,t),\Delta v(j,\phi,t)\} = 0 \qquad (24.2)$$
$$L_{31}^{*j}\{\Delta w(j,\phi,t)\} = 0$$
$$L_{32}^{j}\{\Delta w(j,\phi,t)\} = 0 \qquad j = 0,R$$

described in detail in appendix 3. It can be seen that on the postulated assumption of a small angle α, the disc inclination does not influence the vibration behaviour. So the initially coupled boundary value problem decomposes in a first one for radial and torsional vibrations and a second one for bending oscillations.

The associated literature shows that many researchers analysed the bending problem of rotating discs, e.g. (12), (13). However, a comparison between the different basic relations illustrates that mostly a simplified theory of elasticity was used and, therefore, a shortened boundary value problem results. On the other hand, such simplifications discussing the transverse vibrations of rotating discs are not significant so that the results within a technical consideration differ neither qualitatively nor quantitatively. That is the reason why in the following the interest is solely focused on the coupled in-plane vibrations.

Here, only the natural frequencies and the resulting critical speeds should be discussed so that the governing boundary value problem within equation (24) in its homogenous form may be considered. Let us assume solutions of product form

$$\Delta u(r,\phi,t) = U(r)\ e^{i(\omega t \pm n\phi)}$$
$$\Delta v(r,\phi,t) = V(r)\ e^{i(\omega t \pm n\phi)} \qquad (25)$$

which reduce the above mentioned boundary value problem to the corresponding eigenvalue problem

$$M(\vec{Y}) + \lambda N_1(\vec{Y}) + \lambda^2 N_2(\vec{Y}) = 0 \quad \vec{Y}=(U,V)^T$$
$$M^j\{\vec{Y}(j)\} = 0 \qquad j=0,R \qquad (26)$$

where the differential matrix operators are defined in appendix 4 and the relation

$$\lambda = R(\frac{\rho}{E})^{1/2}\omega \qquad (27)$$

combines the unknown natural frequency ω and the eigenvalue λ.

Without pre-deformation the eigenvalue problem (27) is identically the same which Burdess et al. (7) recently discussed and it is equivalent to another one on vibrations of rotating cylinders which Kürktschiev (14) considered twelve years earlier. In this way, the results calculated here can be verified. But unfortunately, the procedure presented in both references (7) and (14) can not be extended straightforwardly to preloaded discs because now the differential equations are no longer Bessel-differential equations. By a modified approach, a solution of the eigenvalue problem can be constructed as a power series expansion. This solution is shown in reference (8). However, some lengthy algebra is needed so that another method is preferred here.

The approximate free vibration characteristics should be analysed by the Galerkin technique. The procedure is established for single differential equations, refer to (15), for instance, but it can be extended in a straightforward manner to eigenvalue problems of vector form (26). An expansion into a series

$$\vec{Y} = \sum_{k=1}^{N} \vec{Y}_k(r)\, a_k \qquad (28)$$

where the shape functions $\vec{Y}_k(r)$ have to satisfy all boundary conditions (26.2), is used. Then the Galerkin equations can be formulated as

$$\int_0^R \vec{Y}_\ell^T \{M(\sum_k \vec{Y}_k a_k) + \lambda N_1(\sum_k \vec{Y}_k a_k) + \lambda^2 N_2(\sum_k \vec{Y}_k a_k)\} \cdot$$
$$\cdot\, dr = 0 \qquad \ell \text{ fixed, } \ell = 1(1)N \qquad (29)$$

Suitable shape functions Y_k are the eigenfunctions of non-rotating discs (so that both the pre-deformation and the gyroscopic coupling vanish identically) with the same boundary conditions as in the actual case, discussed in detail in reference (16), for example.

A numerical evaluation leads to approximate results of the speed-dependent eigenfrequencies of a rotating disc. If the frequency of the different harmonic components of a periodic excitation here f_r, f_ϕ (refer to equation (1)), coincides with one of the natural frequencies of the considered disc rotor, then resonance will follow. The critical speeds defined in this way can be best understood from Campbell's diagram. In this diagram the excitation frequencies Ω_e^j induced by the rotation with the speed Ω and the eigenfrequencies are plotted as functions of the rotational speed. The intersection points of the different speed-dependent excitation frequencies

$$\Omega_e^j(\Omega) = j\Omega$$
$$\Lambda_e^j = \Omega_e^j R(\tfrac{\rho}{E})^{1/2} \qquad j = 1\ (1)\ \ldots \qquad (30)$$

and eigenfrequencies $\omega_n(\Omega)$ determine the critical rotational speeds to be found.

Actually , a first result for the case of the simplest boundary conditions of a clamped outer edge can be presented, refer to Fig. 4. The eigenfrequencies without taking pre-deformation into consideration (denoted by dashed lines) are just the same as those found by other authors, refer to (14),for instance. The results without neglecting pre-deformation (denoted by full lines) demonstrate that the added influence of pre-deformation significantly changes the initial results. It has to be stated that for a centrally clamped disc, eigenfunctions with a node number n=1 do not exist. As an additional effect, the rotating disc can be shown to become dynamically unstable when it is run at a high speed. With regard to pre-deformation, this is discussed in more detail in reference (8).

It may be supposed that also for a stress-free outer surface, taking the pre-deformation into regard, modifies the corresponding results significantly. The conclusion of reference (7) that a rotating disc becomes unstable by flutter belongs to such questionable results, for instance.

5 CONCLUSION

Extensions of simple concentrated parameter rotor models to distributed parameter systems are important objects of various research work in dynamics of machines. Flexible disc rotors have been studied here.

The governing equations of motion of a thin rotating disc have been generated by an analytic principle of mechanics. It has been shown that a non-linear formulation is necessary if not only the stationary pre-deformations but also the superimposed vibrations should be described sufficiently.

These pre-deformations caused by centrifugal inertia of a constant speed rotor namely influence the natural frequencies and, therefore, the associated critical speeds significantly. Neglecting them can lead to a both qualitatively and quantitatively wrong description of the free and forced vibrations of rotating discs.

On the other hand, the inclination of the disc -if the governing angle is small- modifies only the stationary pre-deformations. Concerning the vibration behaviour inclination is an effect of higher order.

REFERENCES

(1) KELKEL, K. Auswuchten elastischer Rotoren in isotrop federnder Lagerung. Dr.-Ing. Thesis, 1978, University of Karlsruhe (Hochschulverlag Stuttgart).

(2) WAUER, J. A General Linear Approach to Symmetric Distributed Parameter Rotor Systems. Proc. 6th IFToMM-Conf., 1983, New Dehli, Vol. II, 1313-1317 (Wiley Eastern Lim., New Dehli/Bangalore/Bombay/Calcutta).

(3) LUND, J.W. Stability and Damped Critical Speeds of a Flexible Rotor in Fluid-Film Bearings. ASME J. Eng. for Industry, 1974, 96, 509-517.

(4) MURPHY, B.T. and VANCE, J.M. An Improved Method for Calculating Critical Speeds and Rotor Dynamic Stability of Turbomachinery. ASME J. of Eng. for Power, 1983, 105, 591--595.

(5) BIEZENO, C.B. and GRAMMEL, R. Technische Dynamik, Vol. II, 2nd Ed., 1953, 1-105 (Springer, Berlin/Göttingen/Heidelberg).

(6) DOBY, R. On the Elastic Stability of Coriolis-coupled Oscillations of a Rotating Disc. J. Franklin Inst., 1969, 288, 203--212.

(7) BURDESS, J.S., WREN, T. and FAWCETT, J.N. Plane Stress Vibrations in Rotating Discs. Proc. Instn Mech. Engrs, Part C, 1987, 201 (C1), 37-44

(8) SEEMANN, W. and WAUER, J. Vibrations of High Speed Disk Rotors. Proc. 2nd. Int. Symp. on Transport Phenomena, Dynamics, and Design of Rotating Machinery, 1988, Honolulu, to appear.

(9) MAASS,M. Dynamische Spannungskonzentrationsprobleme bei allseits berandeten, gelochten Scheiben. Dr.-Ing. Thesis, 1986, University of Karlsruhe

(10) WAUER, J. Die dünne, elastische Kreiszylinderschale unter axialem Druck - verschiedene Approximationen im Vergleich. Z. angew. Math. Mech., 1986, 66, T104-T106

(11) MAGRAB, E.B. Vibrations of elastic structural members, 1979, 73-92 and 215-272 (Sijthoff&Noordhoff, Alphen aan den Rijn/ Germantown).

(12) SOUTHWELL, R.V. On the Free Transverse Vibrations of a Uniform Circular Disc Clamped at Its Centre. Proc. Royal Society of London, 1922, 101, 133-153.

(13) MOTE JR.,C.D. Free Vibration of Initially Stressed Circular Disks, ASME J. Eng. for Industry, 1965, 87, 258-264.

(14) KÜRKTSCHIEV, R. Stabilität der Schwingungen eines rotierenden Zylinders aus linear-elastischem Material. Ing.-Arch., 1975, 44, 1-7.

(15) LEIPHOLZ, H. Die direkte Methode der Variationsrechnung und Eigenwertprobleme der Technik, 1975 (Braun, Karlsruhe).

(16) ZIMMERMANN, P. Erzwungene und freie ungedämpfte Schwingungen kreisförmig begrenzter Scheiben, Ing.-Arch., 1971, 40, 377-401.

APPENDIX 1

The left side of the field equations (8.1) are defined by

$$L_1(u,v,w)=\frac{Eh}{2(1-\nu^2)}\{\{2ru_{,r}+r(3u_{,r}^2+v_{,r}^2+w_{,r}^2)+\nu(u+v_{,\phi})(\tfrac{1}{r}(u+v_{,\phi})+2(1+u_{,r}))+(u_{,\phi}-v)(1-\nu)v_{,r}+\tfrac{1}{r}\cdot(u_{,\phi}-v))+\tfrac{\nu}{r}w_{,\phi}^2\}_{,r}+\{(u_{,\phi}-v)(\tfrac{2}{r}(u_{,r}+\tfrac{1}{r}(u+v_{,\phi}))+\tfrac{1-\nu}{r})+(1-\nu)(u_{,r}v_{,r}+\tfrac{1}{r}(w_{,r}w_{,\phi}+v_{,r}(u+v_{,\phi})))\}_{,\phi}+(1-\nu)v_{,r\phi}-\tfrac{1}{r}(u+v_{,\phi})\{2(1+\nu u_{,r})+3(u+v_{,\phi})\}-\tfrac{1}{r^2}\cdot\{(u_{,\phi}-v)^2+w_{,\phi}^2\}-2\nu u_{,r}-\nu(u_{,r}^2+w_{,r}^2)-v_{,r}^2+\tfrac{1-\nu}{r}v_{,r}\cdot(u_{,\phi}-v)\}+\rho hr\{-\ddot{u}+2\Omega(\cos\alpha\dot{v}-\sin\alpha\cos\phi\dot{w})+\Omega^2\cdot((\cos^2\phi+\cos^2\alpha\sin^2\phi)(r+u)+\sin\phi(\sin^2\alpha\cos\phi v-\sin\alpha\cos\alpha w))\}+rhf_r$$

$$L_2(u,v,w)=\frac{Eh}{2(1-\nu^2)}\{\{rv_{,r}(1-\nu+2u_{,r})+(u+v_{,\phi})(2v_{,r}+\tfrac{1-\nu}{r}(u_{,\phi}-v))+(1-\nu)(u_{,r}(u_{,\phi}-v)+w_{,r}w_{,\phi})\}_{,r}+\{\tfrac{1}{r}(u+v_{,\phi})(2+\tfrac{3}{r}(u+v_{,\phi})+2\nu u_{,r})+\tfrac{1}{r}(u_{,\phi}-v)((1-\nu)\cdot v_{,r}+\tfrac{1}{r}(u_{,\phi}-v))+\tfrac{1}{r^2}w_{,\phi}^2+2\nu u_{,r}+\nu(u_{,r}^2+w_{,r}^2)+v_{,r}^2\}_{,\phi}+(1-\nu)u_{,r\phi}+\tfrac{1}{r}(u_{,\phi}-v)\{\tfrac{2}{r}(u+v_{,\phi})(2u_{,r}+\tfrac{2}{r}(u+v_{,\phi})+1-\nu\}+(1-\nu)v_{,r}(u_{,r}+\tfrac{1}{r}(u+v_{,\phi}))+\tfrac{1-\nu}{r}w_{,r}w_{,\phi}\}+\rho rh\cdot\{-\ddot{v}-2\Omega(\cos\alpha\dot{u}-\sin\alpha\sin\phi\dot{w})+\Omega^2(\sin^2\alpha\sin\phi\cos\phi\cdot(r+u)-\sin\alpha\cos\alpha\cos\phi w+(\sin^2\phi+\cos^2\alpha\cos^2\phi)v\}+rhf_\phi$$

$$L_3(u,v,w)=\frac{Eh}{1-\nu^2}\{\{ru_{,r}w_{,r}+\nu w_{,r}(u+v_{,\phi})+\tfrac{1-\nu}{2}w_{,\phi}\cdot(v_{,r}+\tfrac{1}{r}(u_{,\phi}-v))\}_{,r}+\{\tfrac{1}{r^2}w_{,\phi}(u+v_{,\phi})+\tfrac{\nu}{r}u_{,r}w_{,\phi}+\tfrac{1-\nu}{2}\cdot w_{,r}(v_{,r}+\tfrac{1}{r}(u_{,\phi}-v))\}_{,\phi}\}-\frac{Eh^3r}{12(1-\nu^2)}\nabla^2_{r\phi}\nabla^2_{r\phi}w+\rho hr\{-\ddot{w}+2\Omega\sin\alpha(\cos\phi\dot{u}-\sin\phi\dot{v})-\Omega^2(\sin\alpha\cos\alpha(\sin\phi\cdot(r+u)+\cos\phi v)-\sin^2\alpha w\}+rhf_z$$

The expressions of boundary conditions (8.2) read

$$L_1^0\{u(0,\phi,t),v(0,\phi,t),w(0,\phi,t)\} = u(0,\phi,t)$$

$$L_2^0\{u(0,\phi,t),v(0,\phi,t),w(0,\phi,t)\} = v(0,\phi,t)$$

$$L_{31}^0\{u(0,\phi,t),v(0,\phi,t),w(0,\phi,t)\} = w(0,\phi,t)$$

$$L_{32}^0\{u(0,\phi,t),v(0,\phi,t),w(0,\phi,t)\} = w_{,r}(0,\phi,t)$$

$$L_1^R\{u(R,\phi,t),v(R,\phi,t),w(R,\phi,t)\} = u(R,\phi,t) \text{ or}$$

$$ru_{,r}+\nu(u+v_{,\phi})+ru_{,r}^2+\nu(u+v_{,\phi})u_{,r}+\tfrac{1-\nu}{2}\{v_{,r}+\tfrac{1}{r}(u_{,\phi}-v)\}(u_{,\phi}-v)\big|_{(r=R,\phi,t)}$$

$$L_2^R\{u(R,\phi,t),v(R,\phi,t),w(R,\phi,t)\} = v(R,\phi,t) \text{ or}$$

$$\tfrac{1-\nu}{2}r\{v_{,r}+\tfrac{1}{r}(u_{,\phi}-v)\}+ru_{,r}v_{,r}+\nu(u+v_{,\phi})v_{,r}+\tfrac{1-\nu}{2}\{v_{,r}+\tfrac{1}{r}(u_{,\phi}-v)\}(v_{,\phi}+u)\big|_{(r=R,\phi,t)}$$

$$L_{31}^R\{u(R,\phi,t),v(R,\phi,t),w(R,\phi,t)\} = w(R,\phi,t) \text{ or}$$

$$h^2\{w_{,rrr}+(2-\nu)(\tfrac{1}{r}w_{,r}+\tfrac{1}{r^2}w_{,\phi})_{,r}\}-\{ru_{,r}w_{,r}+\nu(u+v_{,\phi})w_{,r}+\tfrac{1-\nu}{2}(w_{,\phi}(v_{,r}+\tfrac{1}{r}(u_{,\phi}-v))\}\big|_{(r=R,\phi,t)}$$

$$L_{32}^R\{u(R,\phi,t),v(R,\phi,t),w(R,\phi,t)\} = w_{,r}(R,\phi,t) \text{ or}$$

$$w_{,rr}+\nu(\tfrac{1}{r}w_{,r}+\tfrac{1}{r^2}w_{,\phi\phi})\big|_{(r=R,\phi,t)}$$

APPENDIX 2

The left side of the differential equations (10.1) are given by

$$M_1(u_0,v_0,w_0)=\frac{E}{1-\nu^2}\{ru_{0,rr}+u_{0,r}-\tfrac{1}{r}u_0+\tfrac{1-\nu}{2r}u_{0,\phi\phi}+\tfrac{1+\nu}{2}v_{0,r\phi}-v_{0,\phi}\tfrac{3-\nu}{2r}\}+\rho r\Omega^2\{\sin^2\alpha\sin\phi\cos\phi v_0-\sin\alpha\cos\alpha\sin\phi w_0+(\cos^2\phi+\cos^2\alpha\sin^2\phi)(r+u_0)\}$$

$$M_2(u_0,v_0,w_0)=\frac{E}{1-\nu^2}\{\tfrac{1-\nu}{2r}v_{0,rr}+\tfrac{1-\nu}{2}v_{0,r}-\tfrac{1-\nu}{2r}v_0+\tfrac{1}{r}v_{0,\phi\phi}+\tfrac{1+\nu}{2}u_{0,r\phi}+\tfrac{3-\nu}{2r}u_{0,\phi}\}+\rho r\Omega^2\{\sin^2\alpha\sin\phi\cdot\cos\phi(r+u_0)-\sin\alpha\cos\alpha\cos\phi w_0+(\sin^2\phi+\cos^2\alpha\cos^2\phi)v_0\}$$

$$M_3(u_0,v_0,w_0)=\frac{Eh^2}{12(1-\nu^2)}\nabla^2_{r\phi}\nabla^2_{r\phi}w_0+\rho r\Omega^2\{\sin\alpha\cos\alpha\cdot\sin\phi(r+u_0)-\sin\alpha\cos\alpha\cos\phi v_0-\sin^2\alpha w_0\}$$

APPENDIX 3

The left side of the differential equations (24.1) has the form

$$L_1^*(\Delta u,\Delta v) = \frac{E}{1-\nu^2}(A_{rr}\Delta u_{,rr}+A_{\phi\phi}\Delta u_{,\phi\phi}+A_r\Delta u_{,r}+B_{r\phi}\Delta v_{,r\phi}+B_\phi\Delta v_{,\phi})+\rho r(-\Delta\ddot{u}+2\Omega\Delta\dot{v}+\Omega^2\Delta u)+rf_r$$

$$L_2^*(\Delta u,\Delta v) = \frac{E}{1-\nu^2}(B_{rr}\Delta v_{,rr}+B_{\phi\phi}\Delta v_{,\phi\phi}+B_r\Delta v_{,r}+$$
$$+A_{r\phi}\Delta u_{,r\phi}+A_\phi\Delta u_{,\phi})+\rho r(-\Delta\ddot{v}-2\Omega\Delta\dot{u}+$$
$$+\Omega^2\Delta v)+rf_\phi$$

$$L_3^*(\Delta w) = \frac{Eh^2 r}{12(1-\nu^2)}\nabla^2_{r\phi}\nabla^2_{r\phi}\Delta w-\frac{E}{1-\nu^2}(C_{rr}\Delta w_{,rr}+$$
$$+C_r\Delta w_{,r}+C_\phi\Delta w_{,\phi})+\rho r\Delta\ddot{w}+rf_z$$

$$A_{rr} = r(1+3u^{(0)}_{0,r})+\nu u^{(0)}_0$$
$$A_{\phi\phi} = \frac{1}{r^2}u^{(0)}_0+\frac{1}{2r}(1+2u^{(0)}_{0,r})-\frac{\nu}{2r}$$
$$A_r = 1+3u^{(0)}_{0,r}+3ru^{(0)}_{0,rr}+\nu u^{(0)}_{0,r}$$
$$A = -\frac{1}{r}-\frac{3+\nu}{r^2}u^{(0)}_0+\nu u^{(0)}_{0,rr}$$
$$A_{r\phi} = \frac{1+\nu}{2}(1+u^{(0)}_{0,r}+\frac{1}{r}u^{(0)}_0)$$
$$A_\phi = \frac{3-\nu}{2r}+\frac{7+\nu}{2r^2}u^{(0)}_0+\frac{3+\nu}{2r}u^{(0)}_{0,r}+\frac{1-\nu}{2}u^{(0)}_{0,rr}$$
$$B_{rr} = \frac{1-\nu}{2}r+u^{(0)}_0+ru^{(0)}_{0,r}$$
$$B_{\phi\phi} = \frac{1}{r}+\frac{3}{r^2}u^{(0)}_0+\frac{\nu}{r}u^{(0)}_{0,r}$$
$$B_r = \frac{1-\nu}{2}+2u^{(0)}_{0,r}+ru^{(0)}_{0,rr}$$
$$B = -\frac{1-\nu}{2r}-\frac{1+\nu}{2r^2}u^{(0)}_0-\frac{3-\nu}{2r}u^{(0)}_{0,r}-\frac{1-\nu}{2}u^{(0)}_{0,rr}$$
$$B_{r\phi} = \frac{1+\nu}{2}(1+u^{(0)}_{0,r}+\frac{1}{r}u^{(0)}_0)$$
$$B_\phi = -\frac{3-\nu}{2r}-\frac{4+\nu}{r^2}u^{(0)}_0-\frac{1}{r}u^{(0)}_{0,r}+\nu u^{(0)}_{0,rr}$$

$$C_{rr} = ru^{(0)}_{0,r}+\nu u^{(0)}_0$$
$$C_r = ru^{(0)}_{0,rr}+(1-\nu)u^{(0)}_{0,r}$$
$$C_\phi = \frac{1}{r}(\nu u^{(0)}_{0,r}+\frac{1}{r}u^{(0)}_0)$$

The corresponding expressions of the boundary conditions (24.2) are

$$L_1^{*0}\{\Delta u(0,\phi,t),\Delta v(0,\phi,t)\} = \Delta u(0,\phi,t)$$
$$L_2^{*0}\{\Delta u(0,\phi,t),\Delta v(0,\phi,t)\} = \Delta v(0,\phi,t)$$
$$L_{31}^{*0}\{\Delta w(0,\phi,t)\} = \Delta w(0,\phi,t)$$
$$L_{32}^{*0}\{\Delta w(0,\phi,t)\} = \Delta w_{,r}(0,\phi,t)$$
$$L_1^{*R}\{\Delta u(R,\phi,t),\Delta v(R,\phi,t)\} = \Delta u(R,\phi,t) \quad \text{or}$$
$$A_r^*\Delta u_{,r}+\nu A^*(\Delta u+\Delta v_{,\phi})|_{(r=R,\phi,t)}$$
$$L_2^{*R}\{\Delta u(R,\phi,t),\Delta v(R,\phi,t)\} = \Delta v(R,\phi,t) \quad \text{or}$$
$$B_r^*\Delta v_{,r}+B^*(\Delta v-\Delta u_{,\phi})|_{(r=R,\phi,t)}$$
$$L_{31}^{*R}\{\Delta w(R,\phi,t)\} = \Delta w(R,\phi,t) \quad \text{or} \quad h^2\{\Delta w_{,rrr}+$$
$$+(2-\nu)\{\frac{1}{r}\Delta w_{,r}+\frac{1}{r^2}\Delta w_{,\phi\phi}\}_{,r}\}-C_r^*\Delta w_{,r}|_{(r=R,\phi,t)}$$
$$L_{32}^{*R}\{\Delta w(R,\phi,t)\} = \Delta w_{,r}(R,\phi,t) \quad \text{or}$$
$$\Delta w_{,rr}+\nu(\frac{1}{r}\Delta w_{,r}+\frac{1}{r^2}\Delta w_{,\phi\phi})|_{(r=R,\phi,t)}$$

$$A_r^* = 1+u^{(0)}_0+2u^{(0)}_{0,r}$$
$$B_r^* = u^{(0)}_0+ru^{(0)}_{0,r}+\frac{1-\nu}{2}r$$
$$A^* = B^* = 1+\frac{1}{r}u^{(0)}_0+u^{(0)}_{0,r}$$
$$C_r^* = ru^{(0)}_{0,r}+\nu u^{(0)}_0$$

APPENDIX 4

The differential operators of equation (26) are defined by

$$M = \begin{bmatrix} m_{11} & m_{12} \\ m_{21} & m_{22} \end{bmatrix} \quad N_1 = \begin{bmatrix} 0 & n^1_{12} \\ n^1_{21} & 0 \end{bmatrix} \quad N_2 = \begin{bmatrix} n^2_{11} & 0 \\ 0 & n^2_{22} \end{bmatrix}$$

$$M^0 = \begin{bmatrix} 1 & 0 \\ 0 & 1 \end{bmatrix} \quad M^R = \begin{bmatrix} m^R_{11} & m^R_{12} \\ m^R_{21} & m^R_{22} \end{bmatrix}$$

$$m_{11} = A_{rr}(.)_{,rr}+A_r(.)_{,r}+A-n^2A_{\phi\phi}+r(1-\nu^2)\Lambda^2$$
$$m_{12} = \pm inB_\phi \qquad m_{21} = \pm inA_\phi$$
$$m_{22} = B_{rr}(.)_{,rr}+B_r(.)_{,r}+B-n^2B_{\phi\phi}+r(1-\nu^2)\Lambda^2$$
$$n^1_{12} = -n^1_{21} = 2ir(1-\nu^2)\Lambda$$
$$n^2_{11} = n^2_{22} = r(1-\nu^2)$$
$$m^R_{11} = 1 \quad \text{or} \quad A_r^*(.)_{,r}+\nu A^*$$
$$m^R_{12} = 0 \quad \text{or} \quad \nu(\pm in)A^*$$
$$m^R_{21} = 0 \quad \text{or} \quad -(\pm in)B^*$$
$$m^R_{22} = 1 \quad \text{or} \quad B_r^*(.)_{,r}+B^*$$

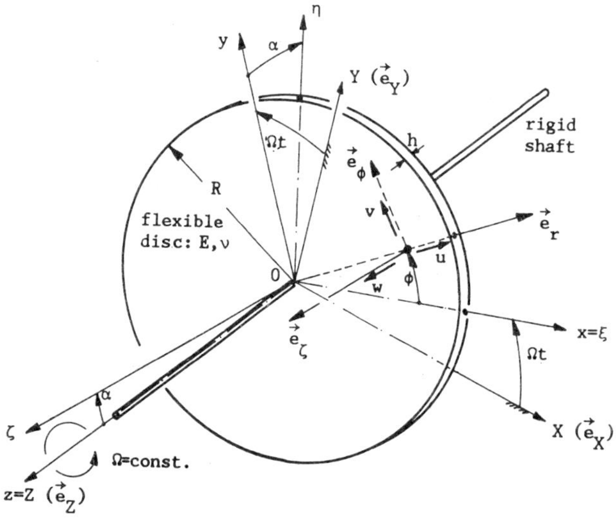

Fig 1 Model of a disc-rotor

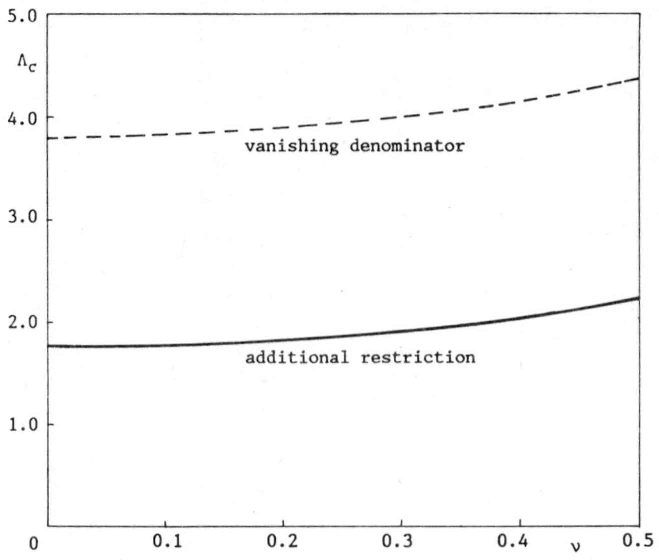

Fig 2 Stationary stability diagram, clamped outer surface

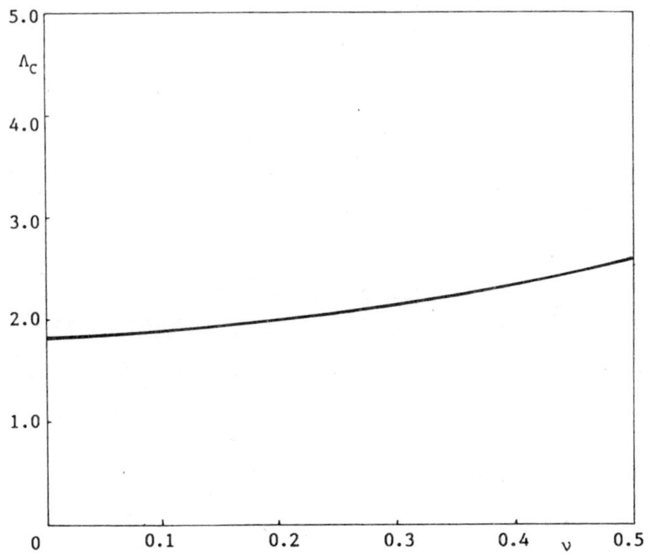

Fig 3 Stationary stability diagram, stress-free periphery

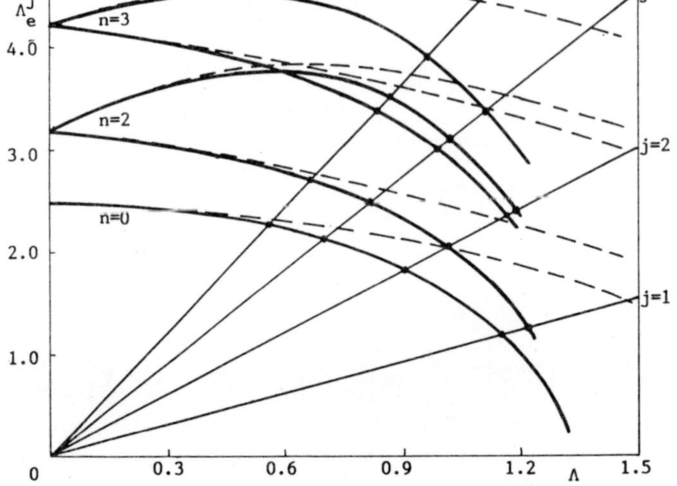

Fig 4 Campbell diagram of a disc-rotor with a clamped outer edge

C272/88

Strategies for the vibration analysis of a large turbocharger

J B TURNBULL, BSc and **G M CHAPMAN**, BSc, PhD, CEng, MIMechE
Department of Mechanical Engineering, Loughborough University of Technology, Loughborough, Leicestershire

1.0 ABSTRACT

A vibration analysis technique is shown, with particular reference to a turbocharger. A turbocharger rotor is tested experimentally using modal analysis and Electronic Speckle Interferometry to find the vibration characteristics. A relatively simple computer model has been generated using a commercial software package and some response predictions are presented. Vibration reducing strategies based on a satisfactory mathematical model have been indicated. Advantages and limitations for each phase are shown, as is the importance of the results gained from such an analysis.

2.0 INTRODUCTION

Turbochargers are already designed to their mechanical limits to meet today's exacting requirements with the consequence that vibration is a serious problem, catastrophic failure is additionally likely to occur as turbochargers run at very high speeds. Vibration of turbocharger components is easily stimulated owing to the very high frequency content of the excitation stimulus which is exacerbated by the large speed range covered by the turbocharger.

Three types of vibration occur in turbochargers : out of balance vibration, self excited vibration and gas excited vibration.

Out of balance vibrations may only be reduced by improving the manufacturing and assembly processes or by changing the support conditions. The vibrations are readily predictable and can be designed out of the system.[1]

Self excited vibrations, such as shaft whirl, have been researched extensively. The onset of these vibrations is a function of rotational speed and bearing geometry. Again these vibrations may be fairly well predicted from theory.[1]

Gas excited vibrations are less common than the previous two excitation sources but are less predictable. This type of vibration may manifest itself at both the impeller and the turbine end.

The most common gas excited vibration is that caused by pressure fluctuations as turbine or impeller blading moves past a stationary object. In the case where this object is a bladed nozzle, then the nozzle blades are themselves also excited. Reduction or elimination of this type of vibration is sometimes possible by using a Campbell diagram, (see Figure 1), and by knowing the operating speed range and natural frequencies of the rotor. Then by changing either a natural frequency or the number of blades of the affected part, a solution may be found.

There are three major drawbacks to this system. Sometimes it is impossible to change a parameter for solution if the operating speed range is large. Also a seemingly innocuous stimulus may be transmitted to some other part of the turbocharger with catastrophic results. Lastly if the turbocharger runs outside it's nominal operating range due to a malfunction of it's speed govenor, often it's engine, catastrophic failure can occur.

A second type of gas excited vibration is that experienced when the turbocharger is operating close to the surge line. This type is called rotating stall and involves low-pressure pockets revolving around the impeller at sub-synchronous speeds. This is not normally a problem unless the turbocharger is running outside it's normal speed range.

A third form of gas excitation is the vibration caused by engine gas pulsations. Exhaust gas can be input to the turbocharger at a constant rate but often the exhaust from several cylinders is pulsed to conserve kinetic energy. This pulsing, together with the interaction with inlet porting combines to produce a complicated waveform with a large frequency content. The excitation waveform may therefore excite a number of components in the turbocharger and is particularly difficult to design out of the system.

2.2 Project Initiation

The response of a turbocharger to gas excitation is not well understood. Work on individual turbine blades has been progressing since the 1960s with the latest techniques now being employed,[2], but little work has been done on the rotor as a whole.

As part of the SERC's design initiative, a project at Loughborough University was set up to investigate ways of reducing vibrations at the design stage.

Industrial support was given by a local manufacturer of large turbochargers. A rotor was supplied for practical evaluation together with the supporting design data.

2.3 Analytical Approach

The analysis technique adopted was as follows. Using a real rotor, the natural frequencies are found using various experimental techniques. A computer model is generated and modified to have the same vibration characteristics as the real rotor. Stimuli are imitated mathematically, and using the computer model, response tests are performed on the turbocharger. Large displacement areas are highlighted and vibration reducing strategies are used to alter the computer model to reduce the vibration response.

Complete validation is achieved for a stationary, room temperature rotor. The effects of centrifugal force can be easily added into the PAFEC program. The effects of temperature are much harder to quantify. High temperatures may change the Youngs Modulus, the geometry and the material damping of the rotor. These parameters may be estimated or experimentally determined for future inclusion in the model.

Once a good model has been completed and response data obtained, a second validation exercise will be performed. This will involve testing a running turbocharger using the signature ratio modal analysis technique. Individual peaks may be compared with the model predictions to give a general picture of the accuracy.

3.0 ANALYSIS

3.1 Experimental Work

The primary technique for determining the vibration characteristics of a rotor was modal analysis. Modal analysis is carried out on a Fast Fourier Analyser using an excitor, force and displacement transducers to monitor response characteristics. The Analyser gives the natural frequencies and by noting the polarity of the imaginary component of the transfer function, the mode shape may be plotted,[3], (see Figure 2).

For the rotor, excitation is achieved using either the hammer impact technique or by an electromagnetic excitor driven by white noise. The hammer impact technique is relatively quick and easy and gives reliable results. The excitor is needed when using a zoom band on the analyser in order to ensure energy is input into the system over a narrow frequency range.

The main advantage of modal analysis is that it can be used on extremely complex structures to yield complex mode shapes at specific natural frequencies. Two results obtained for a turbine disc are shown in Figure 3.

One major limitation to this technique is the time required to obtain satisfactory results particularly when using an excitor.

A second technique that can be used is Electronic Speckle Interferometry, (ESPI). This is a laser interferometry technique and is a developement from holography, (see Figure 4).

The object is excited at each natural frequency whilst viewed by an ESPI rig. Light and dark fringes are produced showing the mode shape and as a digital frequency generator is used, it is possible to accurately find the natural frequencies.

The main advantage of ESPI is that once set up, it is very quick and visual; the mode shapes may be easily seen.

ESPI does have many limitations. Only those surfaces that can be clearly seen may be investigated. Due to electronic noise and resolution problems, only the most dominant modes can be seen. Complicated modes are similarly difficult to determine. It is relatively difficult to get good quality hard copy of these results. Lastly, as the component is driven at a single frequency, the noise produced can be extremely uncomfortable and can cause disturbance to other personnel in the vicinity.

It is important to stress that more than one experimental technique should be used. Each has it's limitations but this may be countered, to a certain extent, by comparison between tests.

A turbocharger turbine disc was tested using both modal analysis and ESPI.The results are shown in Figure 5. Modal test (1) was performed with the rotor suspended by soft elastic. Modal test (6) was performed with the rotor held at the bearings in a test rig. These situations may be referred to as 'free-free' and 'in situ' respectively. Photographs of these test positions are shown in Figure 6. The stiffness of the bearings was simulated by assuming an overall direct stiffness and calculating the material properties required of a rubber mount. The bearings are not accurately modelled, therefore, so any results which involve shaft bending must be viewed with caution.

The force input for the free-free analysis was restricted to hammer impact whereas the 'in situ' testing was achieved with a combination of the two. The number of measurement points depended largely on the complexity of the mode shape. Figure 3 shows some complex modes and also the number of measurement points; each '+','-', or '.' indicates one such point.

The only difference between the ESPI tests was the power of the laser.ESPI (1) used a 5 mW laser and ESPI (2) used a 35 mW laser. For the ESPI results, the rotor was held 'in situ'. Several different excitation points were tested although the quantity and quality of mode shapes did not appear to be sensitive to this variation.

The nomenclature used in Figure 5 is as follows.

D Diametrical nodal line, number indicates quantity.

C Circular nodal line, number indicates quantity.

AV Axial vibration.

AR Axial rigid body, bearing vibration.

+ Indicates superimposed local vibration or blade effects

L,M and H These refer to the relative response levels of the vibration. They do not compare between tests.

The results show that comparison between tests is important. The mode shape of the frequency at 7400 Hz could not be identified by modal analysis, yet was clearly seen to be a 5D mode when using ESPI.

When testing the disc using ESPI, a frequency was found at 1579 Hz with a 2D mode shape. No such frequency was found using modal analysis. Both techniques had shown a 2D at 3135 Hz. By further investigation it became apparent that the driver excitor interface was imperfect and so was producing a spurious image at exactly half frequency. By reducing the amplitude of the shaker force, the needle interface should improve with the spurious image disappearing. This image remained, however, until a very low amplitude input was selected. Without performing a modal analysis test, this phenomenon is most easily discovered by the fact that exactly the same mode shape occurs at exactly half frequency. It must be emphasised that the use of a sole vibration analysis technique is unlikely to yield a complete solution.

The major modes, those with high response levels, have been determined, although further investigation would be necessary to classify the mode shapes of all the frequencies shown in the table.

3.2 Mathematical Modelling of the Rotor

The finite element method of mathematical modelling is extremely applicable to computers and many software packages are available. Loughborough University has access to PAFEC, which was investigated as to it's suitability. The primary project requirement was that the package had to be able to perform a dynamic analysis. Secondly it was important that the predicted frequencies should be accurate. Thirdly the package had to be able to do response calculations and fourthly that it should be flexible.

PAFEC has the ability for dynamic and response calculations, is flexible and work within the department has shown the element algorithms to be reliable.

The accuracy of the finite element method is heavily dependent on the number of pieces or finite elements used. Other influential factors are the method of the formation of the element mesh and the computational algorithm used. The limiting factor of the accuracy is dependent on the computing resources available. Finding the natural frequencies involves inversion of the system matrix, whose order is equal to the number of degrees of freedom with the consequence that for a large system the computer time required becomes very large indeed.

A computer model obtained for the turbocharger under investigation is shown in Figure 7. The bearings were represented using discrete direct equivelent stiffnesses. Bearing damping will be added later. The model shown uses 104 3-dimensional 20 noded brick elements. It should be noted that such elements are necessary for the analysis but extremely time consuming in their calculation. The model uses the maximum allowable computer time on the university system, a Prime C. Further work is to be undertaken using a supercomputer at Manchester, a Cyber 205 as the natural frequencies obtained initially are not very accurate. The results for the whole rotor are shown in Figure 8. The only comparable results to the modal analysis to date are the turbine disc 2D and 3D modes. The frequency of the 2D (modal analysis) is 3140 Hz, whilst the frequency of the 2D (finite element) is 5540 Hz. The 3D results are 4700 Hz (modal analysis) and 9290 Hz (finite element). Improved finite element modelling and further modal analysis is being undertaken for a broader and closer comparison.

3.3 Response Due to Gas Excitation

The two most important gas stimuli are the blade pass excitations and the engine gas pulsations. The blade pass excitation may be modelled by a sine wave but the engine gas pulsation is a complex waveform.[4] An approximation of it, generated by PAFEC, is shown in Figure 9.

The next stage of the analysis is to excite the mathematical model of the turbocharger by the computer generated stimulus and to observe the response. Previous calculated knowledge relating to natural frequency mode shapes will have indicated expected locations for excessive vibration response. This will highlight the areas of interest for the extended mathematical analysis for the transient response to the time history input excitation.

3.4 Vibration Reduction

The computer model will then be altered to reduce the vibration levels. Factors that will be investigated are bearing position, bearing characteristics, bearing seperation, component geometry, component material and component location. Other strategies that can be tried include testing vibration absorbers and friction dampers.

4.0 Analysis Results

Several different strategies are required to reduce vibration. Certain strategies will be effective against certain stimuli. As experimentation continues trends emerge, making it possible to compile a guide to the turbocharger designer which will help in designing vibration resistant turbochargers.

5.0 Conclusion

The use of the given analysis is of potentially great use to the designer. Much of turbocharger design is one of trial and error; the prototype is manufactured and run at constant speeds over the speed range with a suitable engine. It is difficult to measure the component vibration amplitudes: all that can be done is to measure the casing vibration amplitudes. This could mean that potentially catastrophic vibration is occuring locally.

There are other limitations inherent with this type of design and test approach. If the turbocharger runs outside the nominal speed range then the natural frequencies could move into an excitation band. Also, the turbocharger could stimulate a frequency by a transient speed fluctuation.

By understanding the vibration characteristics before actually building the rotor, a much more reliable turbocharger may be manufactured.

6.0 References

1. RAO,J.S. Rotor Dynamics, 1983. (Wiley Eastern).

2. CHAPMAN,G.M. and WANG,X. 'Interpretation of Experimental and Theoretical Data for Prediction of Mode Shapes of Vibrating Turbocharger Blades. ASME International Vibration Conference, Boston, 1987.

3. EWINS,D.J. Modal Testing, Theory and Practice,1984. (John Wiley and Sons).

4. CONNER et al. 'Excitation of Turbine Blade Vibrations in Large Turbochargers' The Third International Conference on Turbocharging and Turbochargers.1986.

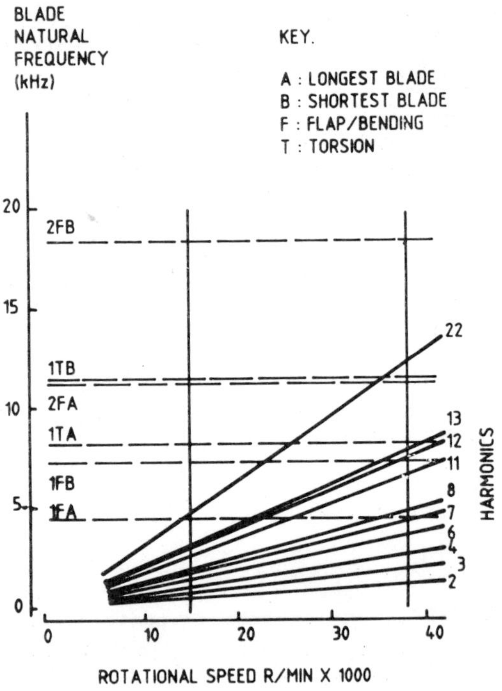

Fig 1 Campbell diagram for typical turbine blading

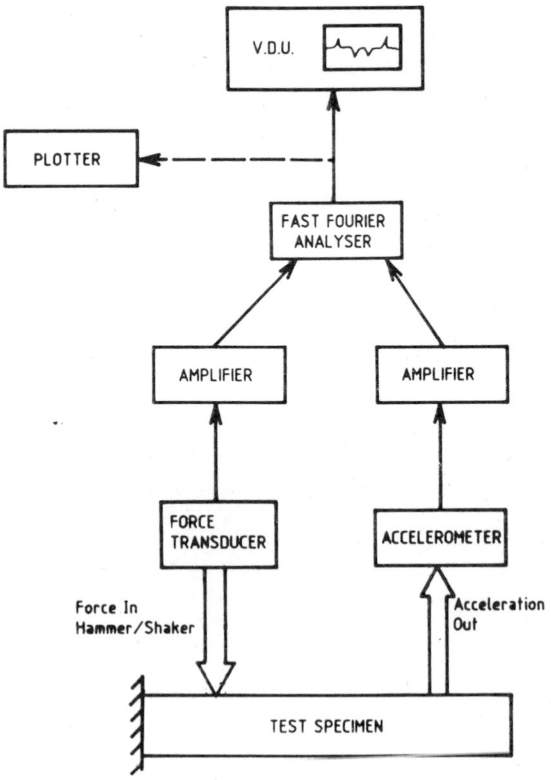

Fig 2 Test set-up for experimental modal analysis

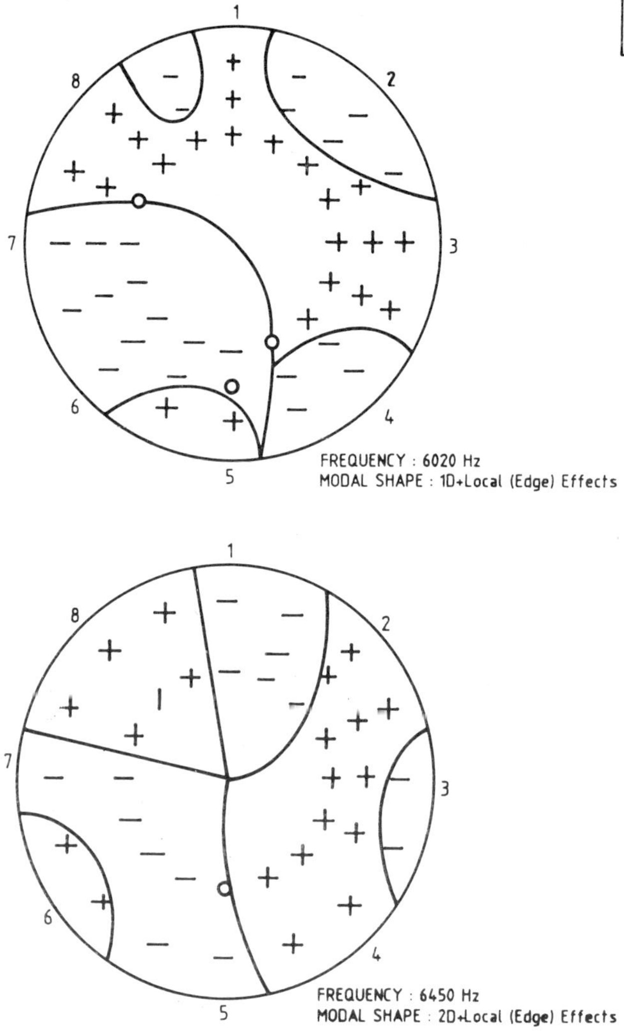

Fig 3 Typical complicated mode shapes obtained using modal analysis

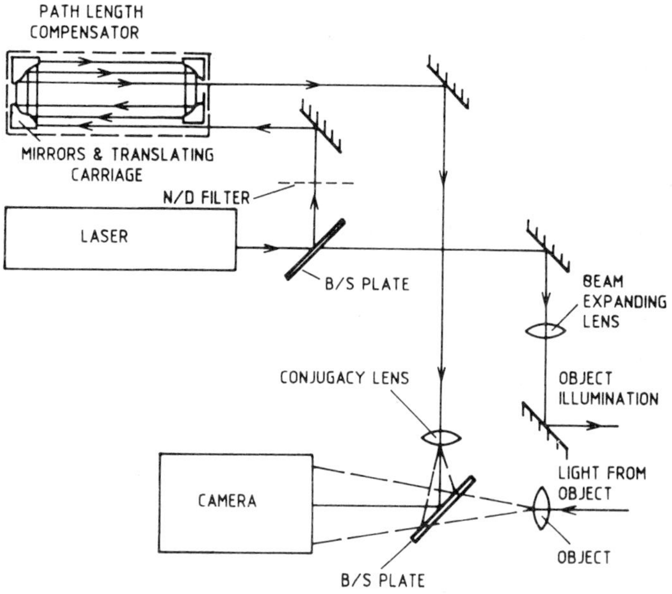

Fig 4a Principles of electronic speckle interferometry

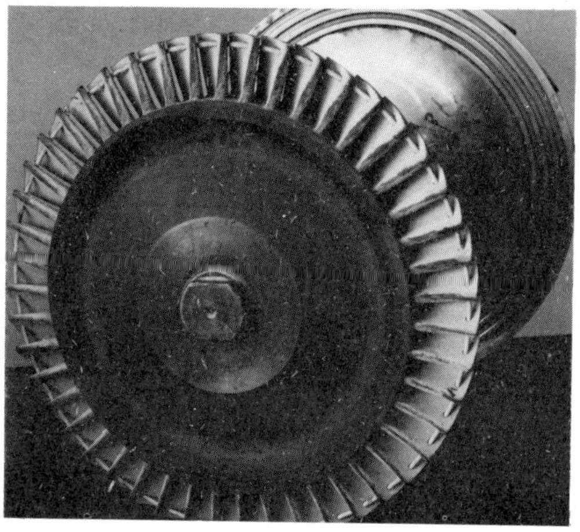

(1)

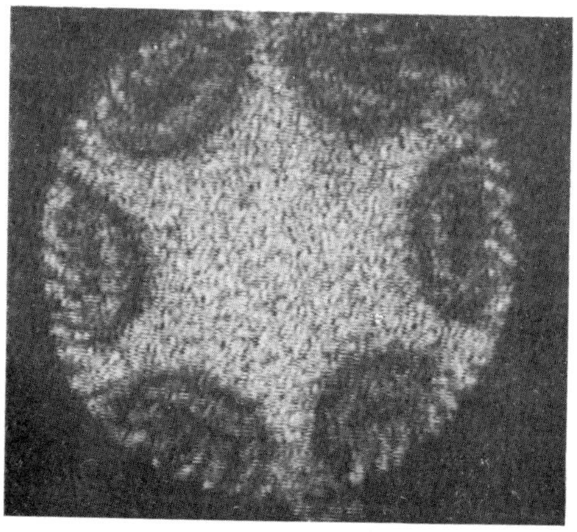

(2)

Fig 4b Rotor turbine disc (1) and typical ESPI result (2)

FREQUENCY Hz	FREE-FREE MODAL (1)	'IN - SITU' MODAL (6)	'IN - SITU' ESPI (1)	'IN - SITU' ESPI (2)
	425 L 1D	150 L AR		
1000	1025 L 1D			
			1567 / 1579 2D	1567 / 1579 2D
	1825 L 1D			
2000				
	2275 M C	2275 L AV		
3000				
	3125 / 3175 H 2D	3125 / 3175 H 2D	3135 / 3157 2D	3133 / 3157 2D
	3500 L 1D			
4000				
	4175 M C	4175 L 1C		
	4650 / 4725 H 3D	4650 / 4725 M 3D	4646 / 4727 3D	4646 / 4717 3D
5000				
	5500 M 1D+	5412 L AV/C		
	5675 M 1D	5650 L 1D		
6000		6020 L 1D+	6055 3D	6208 4D
	6200 H 4D			
	6425 H U	6450 L 2D+		
	6725 H 2D	6750 L 2D	6763 2D	
7000				
	7400 H4D/5D	7230 L 2D+	7252 2D/3D	7433 5D
	7650 M3D/4D		7446 5D	
8000				
9000				

Fig 5 Experimental turbine disc frequencies

Fig 6a View of the rotor; in situ

Fig 6b View of the rotor; freely supported

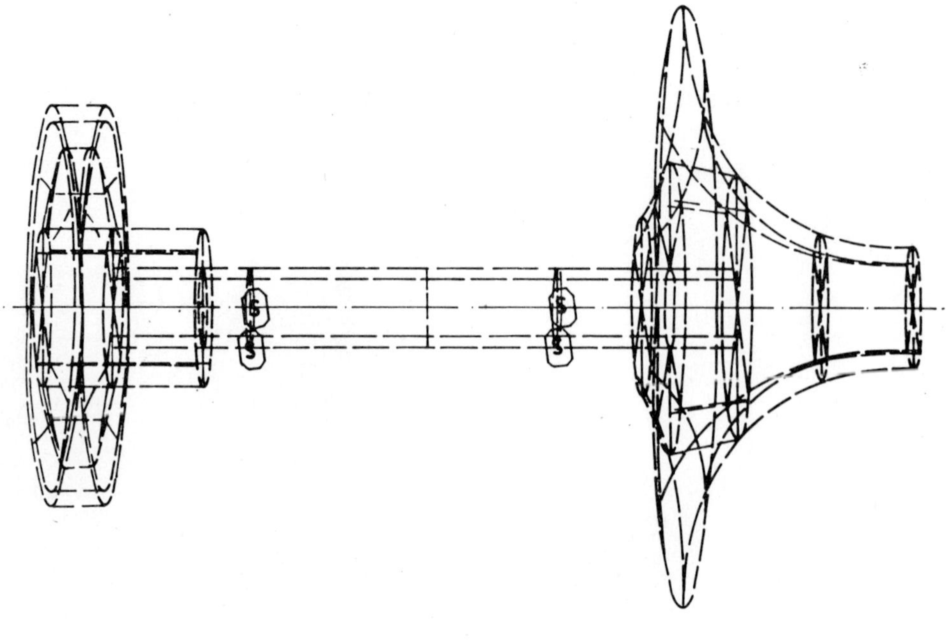

Fig 7 Finite element rotor model

MODES		FREQUENCY
DESCRIPTION	SYMBOL	
RIGID BODY CONICAL	RC	108
RIGID BODY TRANSLATIONAL	RT	202
1 st. TORSIONAL	1T	464
1 st BENDING	1B	606
2 nd BENDING	2B	1400
UNKNOWN	-	2920
3rd BENDING	3B	3820
2 nd TURBINE DIAMETER	2DT	5540
0 TURBINE CIRCLES	OCT	7450
COMBINATION	1DI/1DT/4B	7690
2 nd IMPELLER DIAMETER	2DI	8090
1 st RADIAL	1R	8220
3 rd TURBINE DIAMETER	3DT	9290

Fig 8 Prediction of resonance frequencies using FE model of in situ rotor

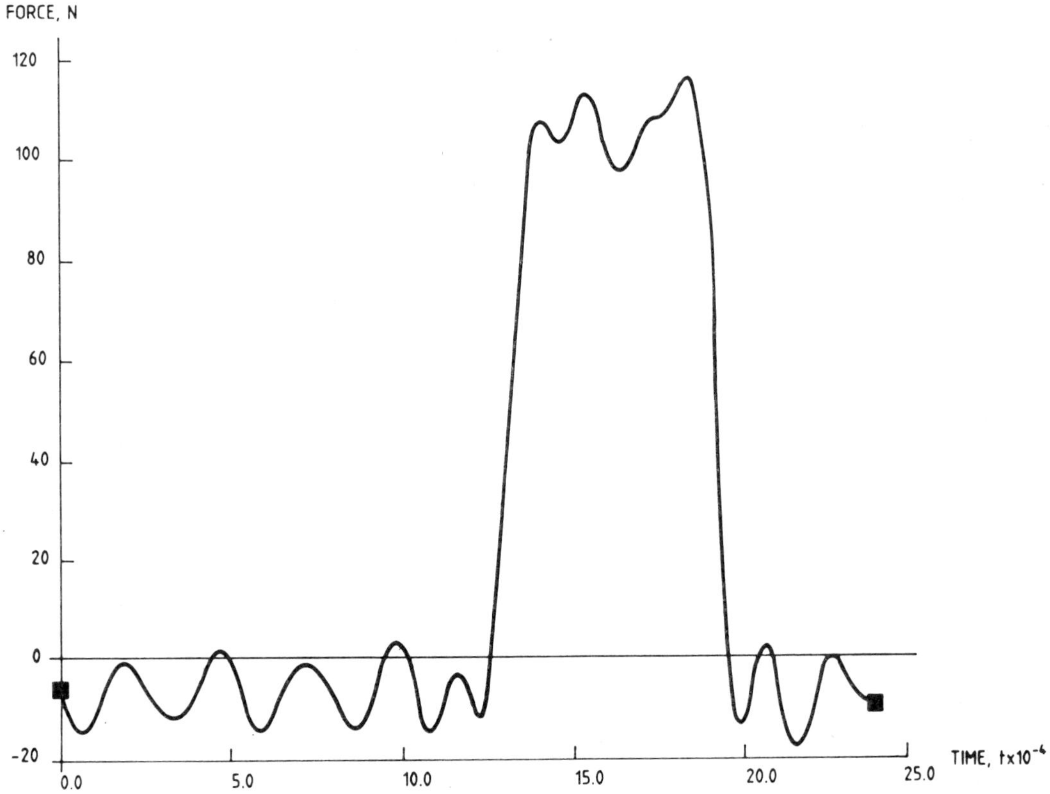

Fig 9 Model of gas pulse input force

C270/88

Identification of turbomachinery foundations from run-down records—a preliminary experimental study

T K TEE, BEng, **R STANWAY**, MSc, DPhil, CEng, MIMechE, MIEE and **J E MOTTERSHEAD**, BSc, PhD, CEng, MIMechE
Department of Mechanical Engineering, University of Liverpool, Liverpool
A W LEES, BSc, PhD, CPhys, MInstP
Central Electricity Generating Board, Bedminster Down, Bristol

SYNOPSIS In this paper the authors describe the theory and application of an experimental technique for determining the mass, damping and stiffness characteristics associated with turbomachinery steel foundations. The technique requires a record of bearing pedestal displacements during run-down and these are used to compute the forces transmitted through the bearing oil-films to the foundations. The force/displacement data is then processed using a recently-developed frequency-domain filtering algorithm to estimate the unknown foundation parameters. Preliminary experimental results from a portal frame rig are presented to demonstrate the feasibility of the proposed approach and to indicate aspects of the work which require further investigation.

NOTATION

$\underline{B}^{-1}(\omega)$	matrix of frequency response functions
$\underline{C}, c$	damping matrix and coefficient
$\underline{H}(\omega)$	$\frac{\partial}{\partial \hat{\underline{x}}} [\underline{B}(\hat{\underline{x}},\omega)\ \underline{z}(\omega)]^T$
J	cost functional
j	$\sqrt{-1}$
$\underline{K}, k$	stiffness matrix and coefficient
$\underline{M}, m$	mass matrix and coefficient
n	number of degrees of freedom
$\underline{P}$	error covariance matrix
P_{ij}	element of $\underline{P}$
$\underline{Q}$	weighting matrix
$\underline{q}(\omega)$	forcing vector
$T_1(\omega), T_2(\omega)$	scalar transfer functions
t	time
$\underline{u}(\omega)$	response vector
$\underline{x}$	state vector containing unknown parameters
$\underline{z}(\omega)$	vector of observations
$\underline{\varepsilon}$	equation error vector
ω	radian frequency
Ω	frequency interval
$\underline{\xi}$	measurement error vector
$(\dot{\ })$	denotes differentiation with respect to time
$(\ )^{-1}$	denotes matrix inverse
$(\ )^T$	denotes transpose
$(\hat{\ })$	denotes estimate

1 INTRODUCTION

In the power generation industries it is of great importance to be able to predict future service problems by interpreting vibration records from turbomachinery. In this respect, measurements of lateral vibration at the bearings, taken during the run-down regime, are especially valuable: during this regime the forces generated by rotor unbalance sweep through the entire frequency range and are liable to excite the most significant modes of vibration of the rotor, oil-film bearings and foundations.

If abnormal behaviour is to be detected through correct interpretation of the run-down records then mathematical models, characterising the "normal" behaviour of the various machine elements, must be available. Over the past twenty years industrial and academic research has been carried out to improve modelling techniques in this area and notable contributions have been made in the derivation of realistic models of large rotors (1) and oil-film bearings (2). However, much work remains to be done to develop realistic models of the steel foundations upon which modern power generating plant is supported (3).

Experience with the modelling of rotors and oil-film bearings has shown that theoretical descriptions, based upon construction data, invariably need to be refined using experimental data to account for observed phenomena. Preliminary investigations into the modelling of steel foundations seem to indicate that a similar approach will be required. Within the Central Electricity Generating Board it has been found that foundations built to identical specifications are capable of displaying quite distinct vibration signatures (3). Obviously, a model based solely upon construction data could not possibly predict the signatures of individual foundations.

In what follows here, the authors describe progress made towards the development of more realistic mathematical models of turbomachinery steel foundations. The overall approach involves two main steps:

i) a decomposition due to Lees (4) which allows the forces transmitted to the foundations during run-down to be computed;
ii) the application of a recently-developed parameter estimation (5) algorithm to experimental data.

As we shall show, in principle, steps i) and ii) enable a finite-element model of steel foundations to be improved to the point where it adequately characterises the vibrational behaviour of an individual foundation. The philosophy and theoretical development of the decomposition, step i), is described in reference (4), also to be presented at this conference. In the present paper, we focus attention on step ii), the signal processing aspects of identifying turbomachinery foundations and present some preliminary experimental results.

2 STATEMENT OF PROBLEM

It is well established (for example see Ewins (6)) that the linearised equations of motion for a vibrating structure with n degrees of freedom can be written

$$\underline{M}\,\underline{\ddot{u}}(t) + \underline{C}\,\underline{\dot{u}}(t) + \underline{K}\,\underline{u}(t) = \underline{q}(t) \qquad (1)$$

where $\underline{M}$, $\underline{C}$ and $\underline{K}$ are, respectively mass, damping and stiffness arrays of dimension (n x n) and $\underline{q}(t)$ and $\underline{u}(t)$ are (n x 1) vectors of applied forces and the corresponding displacement responses of the structure.

Such a model can be used to characterise the lateral vibrational behaviour of a rotor-bearing-foundation system. However many degrees of freedom are likely to be involved (for example, 24, for a relatively simple turbomachinery model (7)) and possibly many hundreds of parameters to characterise the behaviour of the various machine elements. The problem that we address here is that of estimating those parameters which are unknown, from experimental records of the applied forces $\underline{q}(t)$ and displacement responses $\underline{u}(t)$.

Perhaps the most obvious approach to parameter estimation is to use frequency response data to estimate modal parameters, i.e. damping factors and undamped natural frequencies (6). As we shall show, the approach adopted here is to work with the frequency domain model corresponding to equation (1), i.e.

$$(-\omega^2\underline{M} + j\omega\underline{C} + \underline{K})\underline{u}(\omega) = \underline{q}(\omega) \qquad (2)$$

but to estimate the physical parameters in arrays $\underline{M}$, $\underline{C}$ and $\underline{K}$ rather than their modal counterparts. The frequency response data for estimation is to be provided by the run-down signature taken from the rotating machine. The result due to Lees (4) enables us to isolate the contribution of the foundations from the dynamics of the overall machine. In this way we can establish an equivalent network of masses, dampers and springs to characterise the influence of the foundations on the run-down signature.

3 DECOMPOSITION OF ROTOR-BEARING MODEL

Although a full account of the decomposition of the rotor-bearing foundation model appears in a companion paper (4), it is worthwhile to summarise the main result here. In the decomposition it is assumed that there is prior knowledge of

i) the state of unbalance of the rotor;
ii) the eigenvalues and free-free mode shapes of the rotor;
iii) the linearised dynamics of the oil-film support bearings.

Such knowledge would typically come from tests such as those described in references (1) and (2). Given this information, Lees shows that from experimental measurements of the displacement responses of the bearing pedestals during run-down, the corresponding steady-state forces transmitted to the foundations, through the oil-film bearings, can be computed. These forces, each as a function of frequency, constitute the entries in the vector $\underline{q}(\omega)$ in equation (2). The measured displacements, which are used to compute the forces, provide the entries in the vector $\underline{u}(\omega)$ in the same equation. Thus an input/output record of the foundation dynamics is available for subsequent processing.

4 SIGNAL PROCESSING ASPECTS

From the arguments in Section 3, the identification of turbomachinery foundations from run-down records can be posed as a problem of estimating the elements of $\underline{M}$, $\underline{C}$ and $\underline{K}$ in equation (2) from the input/output records $\underline{q}(\omega)$ and $\underline{u}(\omega)$.

If, as may appear logical, $\underline{q}(\omega)$ is treated as the input vector and $\underline{u}(\omega)$ as the output vector then this leads to frequency response functions which are non-linear in the unknown parameters. To illustrate this point, consider the scalar transfer function:

$$T_1(\omega) = \frac{u(\omega)}{q(\omega)} = \frac{1}{-\omega^2 m + j\omega c + k}$$

where clearly $T_1(\omega)$ is non-linear in the parameters, m, c and k. Formulating parameter estimation in this way (the so-called "output-error" approach) is perfectly feasible but for large scale systems can lead to excessive computational requirements (8).

To reduce the computational requirements, an "equation-error" formulation can be adopted. Here the response vector $\underline{u}(\omega)$ is treated as input and the forcing vector $\underline{q}(\omega)$ is regarded as the output. In the scalar case, this is simply equivalent to inverting the transfer function $T_1(\omega)$ to obtain

$$T_2(\omega) = \frac{q(\omega)}{u(\omega)} = -\omega^2 m + j\omega c + k$$

such that $T_2(\omega)$ is linear in m, c and k. This result carries over to the general vector case which is described in reference (5). A general derivation giving full mathematical detail of filter formulations for both output error and equation error approaches has been given by Mottershead (9). In the following section a brief description of the formulation of an equation error filter is presented.

5 FREQUENCY-DOMAIN FILTERING

We begin by assuming that a vector of displacement responses is available and that, in general, the records contained in this vector will be contaminated with noise such that

$$\underline{z}(\omega) = \underline{u}(\omega) + \underline{\xi} \quad (3)$$

where $\underline{z}(\omega)$ is a complex vector containing measured displacements and $\underline{\xi}$ is a vector of zero-mean noise terms. Combining equations (2) and (3)

$$\underline{q}(\omega) = \underline{B}(\omega)\,[\underline{z}(\omega) - \underline{\xi}] \quad (4)$$

where $\underline{B}^{-1}(\omega)$ is the matrix of frequency response functions.

In the time-domain filtering approach to parameter estimation (for example, see (10)), the vector to be estimated contains both the unknown parameters and physical state variables. The computational burden is dramatically reduced by adopting the frequency-domain formulation, where, if we assume the unknown parameters are independent of frequency, we can write

$$\frac{d\underline{x}}{d\omega} = \underline{0} \quad (5)$$

where $\underline{x}$ is a vector containing the unknown parameters but which does not contain the physical state variables (displacements and velocities) which are inherent in the time-domain formulation (10).

Further simplification is achieved by adopting the equation-error formulation. To summarise, parameter estimation is approached by minimising the cost functional

$$J = \int_0^{\Omega} \underline{\varepsilon}^T \underline{Q}\, \underline{\varepsilon}\, d\omega \quad (6)$$

where $\underline{Q}$ is a pre-assigned weighting matrix and $\underline{\varepsilon}$ is the equation error vector given by

$$\underline{\varepsilon} = \underline{q}(\omega) - \underline{H}^T\underline{\hat{x}} \quad (7)$$

where $\underline{\hat{x}}$ denotes an estimate of the state vector containing the unknown parameters and the matrix $\underline{H}$ is given by

$$\underline{H}(\omega) = \frac{\partial}{\partial \underline{\hat{x}}}\,[\underline{B}(\underline{\hat{x}},\omega)\underline{z}(\omega)]^T \quad (8)$$

It is shown in reference (5) that the real part of the equations which minimise the functional J can be written in the form

$$\frac{d\underline{x}}{d\Omega} = 2\underline{P}(\Omega)\mathrm{Re}[\underline{H}(\Omega)]\underline{Q}\ \mathrm{Re}[\underline{q}(\Omega) - \underline{B}(\underline{\hat{x}},\Omega)\underline{z}(\Omega)]$$

and

$$\frac{d\underline{P}}{d\Omega} = -2\underline{P}(\Omega)\mathrm{Re}[\underline{H}(\Omega)]\underline{Q}\ \mathrm{Re}[\underline{H}^T(\Omega)]\underline{P}(\Omega) \quad (9)$$

and that two further equations of identical structure can be written to account for the imaginary component of the equation error.

The estimate, $\underline{\hat{x}}$, of the unknown parameters is obtained through the numerical integration of the real (and imaginary) parts of equations (9) over the frequency interval $0 \leq \omega \leq \Omega$. Integration of the equations in $\underline{P}$ over the same interval provides the appropriate weightings together with an indication of the confidence which can be placed in the estimates $\underline{\hat{x}}$. Because of the linear (equation error) formulation, the equations in $\underline{P}$ do not involve $\underline{\hat{x}}$ and thus can be computed independently of the equations in $\underline{\hat{x}}$.

6 EXPERIMENTAL STUDY

6.1 Experimental Facility

To examine the application of the approach outlined in the previous sections, data was obtained from a portal frame rig, Fig. 1. This rig was designed to display the most significant features (lightly-damped and closely-spaced modes) associated with steel foundations for turbomachinery.

The applied force was provided by an electro-magnetic shaker, suspended from above by wire ropes and acting at station 1, Fig. 1. The response of the portal frame was measured using accelerometers mounted at stations 1, 2, 3 and 4. It was anticipated that the data gathered with such an arrangement of actuator and sensors would enable the estimation of mass, stiffness and damping parameters in a model involving four degrees of freedom.

6.2 Parameter Estimation with a Pseudo-Random Forcing Signal

In the first series of experiments the portal frame was excited using a pseudo-random signal with a bandwidth of 200 Hz, capable of exciting the first three modes of vibration. Data were gathered from a load cell connected between the shaker and frame ($q(t)$) and from each of the four accelerometers ($\underline{z}(t)$). Each of the four input/output records was Fourier transformed using a "B and K" dual channel analyser to provide the appropriate frequency response functions $q(\omega)/z_i(\omega)$, $i = 1,2,3,4$ between the shaker and each of the four accelerometers.

To accommodate force/acceleration records, rather than force/displacement records as implied by equations (3) to (9), the matrix $\underline{B}(\omega)$ of frequency-response functions was re-defined such that

$$\underline{B}(\omega) = \underline{M} + \frac{\underline{C}}{j\omega} - \frac{\underline{K}}{\omega^2}$$

Starting values, $\underline{\hat{x}}(0)$, for parameter estimation were obtained from a finite element model of the portal frame. The starting values for the matrix $\underline{P}$ in equation (9) were taken as

$$P_{ij}(0) = 3.0 \ (i=j)$$

$$P_{ij}(0) = 1.0 \ (i\neq j)$$

which has been found, by trial and error, to provide rapid convergence without inducing numerical instability.

Subject to these initial conditions, equations (9), together with their imaginary counterparts, were solved numerically using a fourth-order Runge-Kutta-Merson algorithm. The frequency response data in each $q(\omega)/z_i(\omega)$ was supplied to the algorithm, one frequency at each sequential step, until the whole range of frequencies from 0 to 200 Hz had been covered.

The signal processing was arranged so that firstly estimates of the mass elements were adjusted, keeping the damping and stiffness terms constant. Then the damping and stiffness terms were treated in a similar way. Finally estimates were improved by using the zoom facility on the FFT analyser and applying the filter to each resonant peak in turn.

The effectiveness of the parameter estimation can be judged by reference to Figs. 2 and 3. Figure 2 shows the predictions of two of the four frequency response functions $z_1(\omega)/q(\omega)$ (i.e. displacement at station 1 against applied force) and $z_3(\omega)/q(\omega)$ (displacement at station 3 against applied force) using the starting values for parameter estimation obtained from the finite-element model. Figure 3 shows the predictions of $z_1(\omega)/q(\omega)$ and $z_3(\omega)/q(\omega)$ using the final values obtained after applying the parameter estimation algorithm to the measured data from the four frequency-response functions. The frequency-response functions shown in Figs. 2 and 3 are typical of those obtained from a large number of similar experiments.

6.3 The Application of a Swept-Sine Forcing Signal

Before applying the filtering algorithm, equations (9) to run-down data from a large rotating machine, it remained to establish that the type of forcing signal which will then be encountered - a swept-sine function with amplitude proportional to the square of applied frequency -will produce similar results to those obtained with a pseudo-random signal. As a preliminary step, the portal frame rig was excited by a swept-sine function with constant amplitude.

Figure 4 shows the results of obtaining frequency response functions $z_1(\omega)/q(\omega)$ and $z_3(\omega)/q(\omega)$ using a swept-sine function and the same pseudo-random signal as that used to obtain the results in Section 6.2. The parameter estimation algorithm was then applied to the swept-sine data to obtain estimates of the mass, damping and stiffness parameters in the four degrees-of-freedom model. The predictions of $z_1(\omega)/q(\omega)$ and $z_3(\omega)/q(\omega)$ obtained using the swept-sine forcing input are displayed in Fig. 5.

6.4 Discussion of Results

From inspection of Fig. 2, it is apparent that the finite-element model, used to start the parameter estimation is incapable of predicting the frequency responses obtained by FFT analysis which are taken as references. Specifically the resonant peaks predicted by the finite-element model are some 10 Hz higher than the FFT estimates.

Application of parameter estimation, Fig. 3 produces predictions which accurately reflect the higher modes although there is still some room for improvement at lower frequencies (i.e. below 40 Hz). Note that, in general, both amplitude and phase are accurately predicted.

Figure 4 shows that using a swept-sine input signal, it is possible to produce, by FFT analysis, frequency-response estimates which are almost identical to those obtained from random signal forces. However the swept-sine results are noticeably noisier than those produced using a random signal input, especially the low frequency portion of Fig. 4(a).

It appears from these preliminary results that the noise does have an effect on parameter estimation. The results of applying parameter estimation to swept-sine frequency responses, Fig. 5, are noticeable inferior to those obtained using random forcing, Fig. 3. In particular the parameter estimation algorithm overestimates the damping levels and this problem is currently receiving attention. It is well known that identification techniques based on the equation error will provide estimates which are asymptotically biased if noise is present on the measured data. In order to overcome the bias problem an algorithm (11) has recently been developed using an instrumental variable technique (12). Application of the new algorithm to the portal frame data is imminent.

7 CONCLUSIONS

The authors' ultimate objective is to develop a technique for obtaining mathematical models of turbomachinery foundations which can account for observed behaviour. It has been established in a companion paper (4) that the forces transmitted to the foundation during run-down can be computed. In the present paper, it has been shown how this information might be employed to estimate mass, damping and stiffness parameters associated with the foundations.

In a preliminary study, it has been demonstrated that experimental records of applied force together with acceleration response can be processed using a frequency-domain filtering algorithm to estimate the unknown parameters associated with a steel portal frame rig. Starting with a finite-element model, which is incapable of predicting observed frequency-response functions, and exciting the portal frame with a pseudo-random excitation signal, the application of the filtering algorithm to experimental data provides parameter estimates which provide accurate predictions of observed behaviour.

The results were extended by considering the application of a swept-sine excitation signal. It was shown that frequency-response functions, almost identical to those obtained with a pseudo-random forcing signal, could be obtained by FFT analysis of the swept-sine input and responses. The discrepancies, although they may appear almost insignificant, do require further investigation since the results indicate that the quality of parameter estimates from the swept-sine frequency-responses is inferior to those obtained from pseudo-random forcing. In particular it is intended to apply an instrumental variable algorithm to account for the noise.

Before applying the technique to data from a large rotating machine, the influence of employing a forcing signal with a frequency-squared amplitude characteristic will be investigated under controlled conditions on the portal frame rig.

ACKNOWLEDGEMENTS

T. K. Tee is supported by an SERC (CASE) award in collaboration with the CEGB.

REFERENCES

(1) MORTON, P. G. On the dynamics of large turbo-generator rotors. Proceedings of the Institution of Mechanical Engineers, 1965-66, 180, 295.

(2) MORTON, P. G. The derivation of bearing characteristics by means of transient excitation applied directly to a rotating shaft. G.E.C. Journal of Science and Technology, 1975, 42, 37-47.

(3) LEES, A. W. and SIMPSON, I. C. The dynamics of turbo-alternator foundations. Conference paper C6/83, Institution of Mechanical Engineers.

(4) LEES, A. W. The least squares method applied for investigating rotor/foundation parameters. Conference Paper C366/065, Institution of Mechanical Engineers.

(5) MOTTERSHEAD, J. E., LEES, A. W. and STANWAY, R. A linear, frequency domain filter for parameter identification of vibrating structures. ASME Journal of Vibrations, Acoustics, Stress and Reliability in Design, 198, 109, 262-269.

(6) EWINS, D. J. Modal testing: Theory and practice, 1984, Research Studies Press, Letchworth.

(7) FIROOZIAN, R. and STANWAY, R. Modelling and control of turbomachinery vibrations, ASME Journal of Vibrations, Acoustics, Stress and Reliability in Design, to appear.

(8) MOTTERSHEAD, J. E. and STANWAY, R. Identification of structural vibration parameters using a frequency-domain filter. Journal of Sound and Vibration, 1986, 109, 495-506.

(9) MOTTERSHEAD, J. E. A unified theory of recursive, frequency domain filters with application to system identification of vibrating structures. ASME Journal of Vibrations, Acoustics, Stress and Reliability in Design, in press.

(10) STANWAY, R., FIROOZIAN, R. and MOTTERSHEAD, J. E. Estimation of the linearised damping coefficients of a squeeze-film vibration isolator. Proceedings of the Institution of Mechanical Engineers, 1987, 201, 181-191.

(11) MOTTERSHEAD, J. E. and FOSTER, C. D. An instrumental variable method for the estimation of mass, stiffness and damping parameters from measured frequency response functions, Mechanical Systems and Signal Processing, submitted.

(12) KENDALL, M. G. and STUART, A. The advanced theory of statistics, Vol. 2, Griffin, London, 1961.

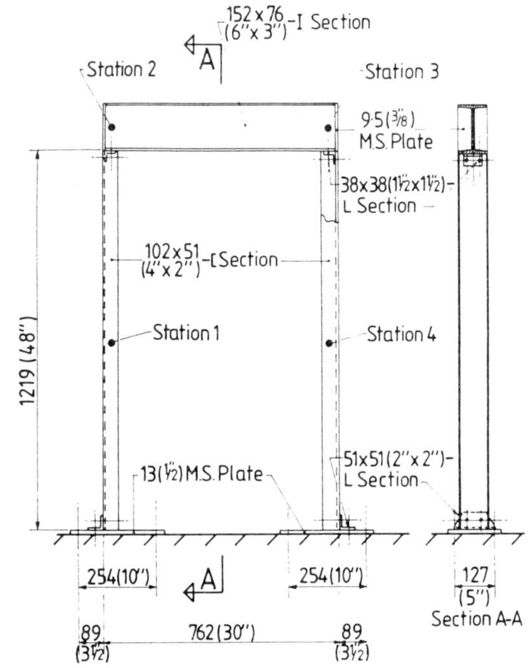

Fig 1 Portal frame rig

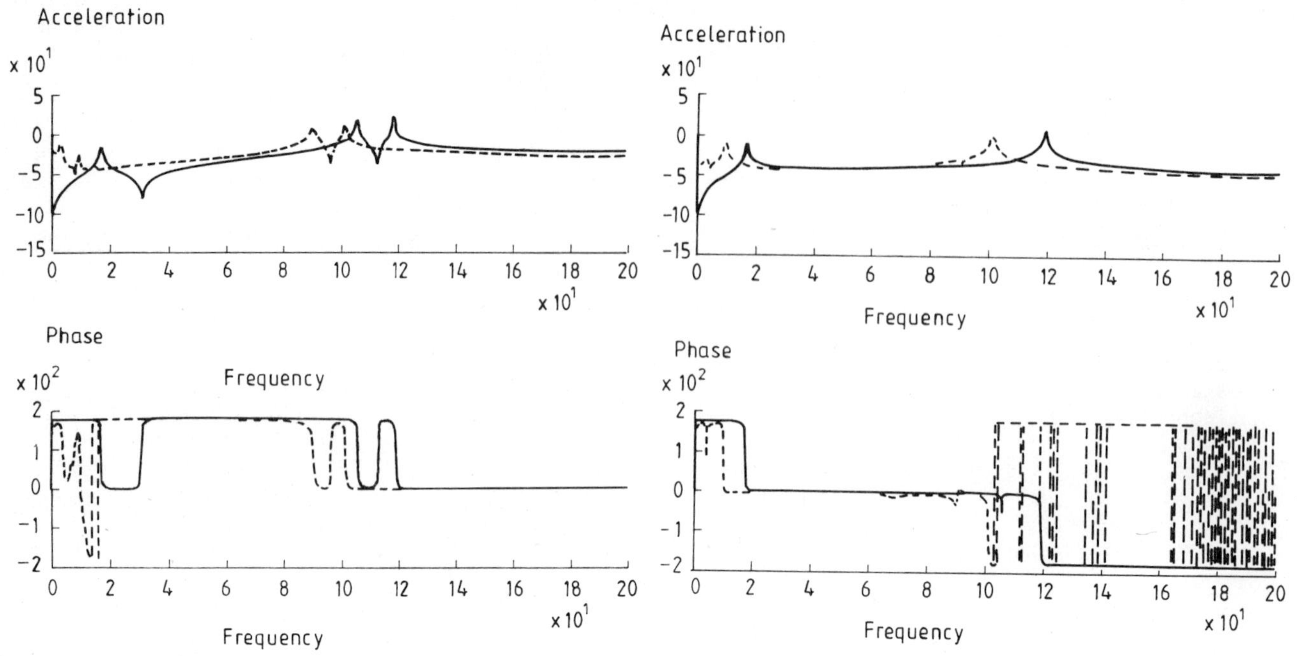

Fig 2 Frequency response functions prior to parameter estimation
(a) acceleration at station one/force at station one
(b) acceleration at station three/force at station one
– – – – – FFT analysis using random input
———— Predicted from finite-element model

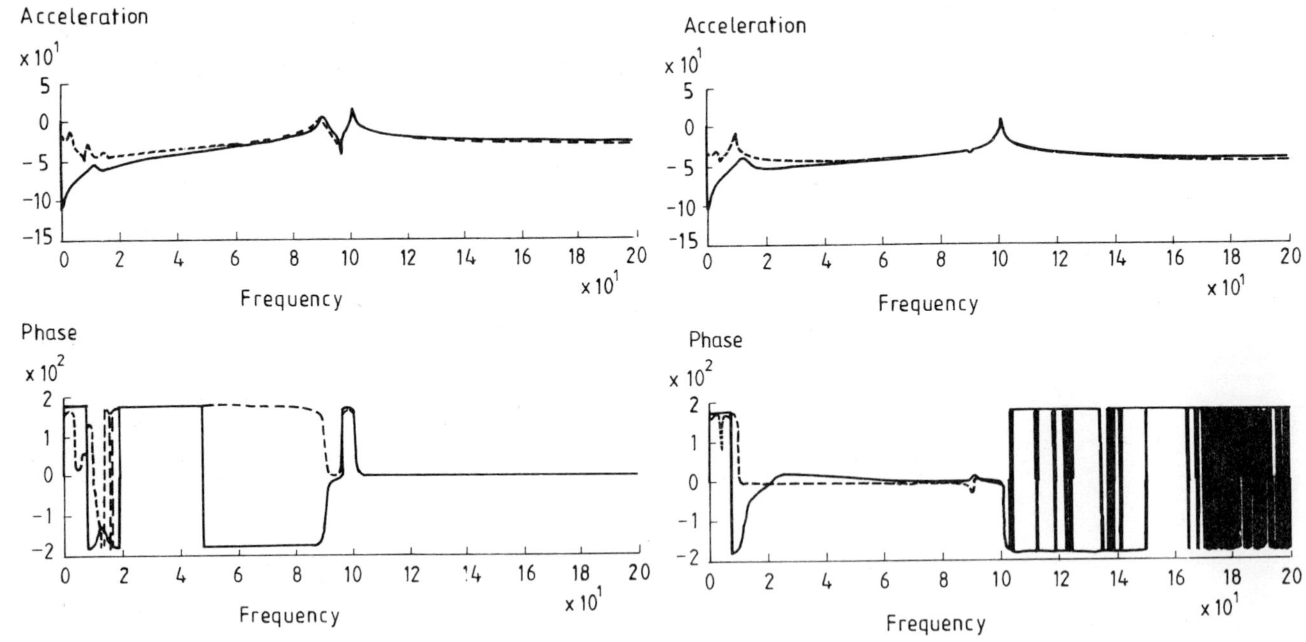

Fig 3 Frequency response functions following parameter estimation
(a) acceleration at station one/force at station one
(b) acceleration at station three/force at station one
– – – – – FFT analysis using random input
———— Predicted by model after frequency domain filtering

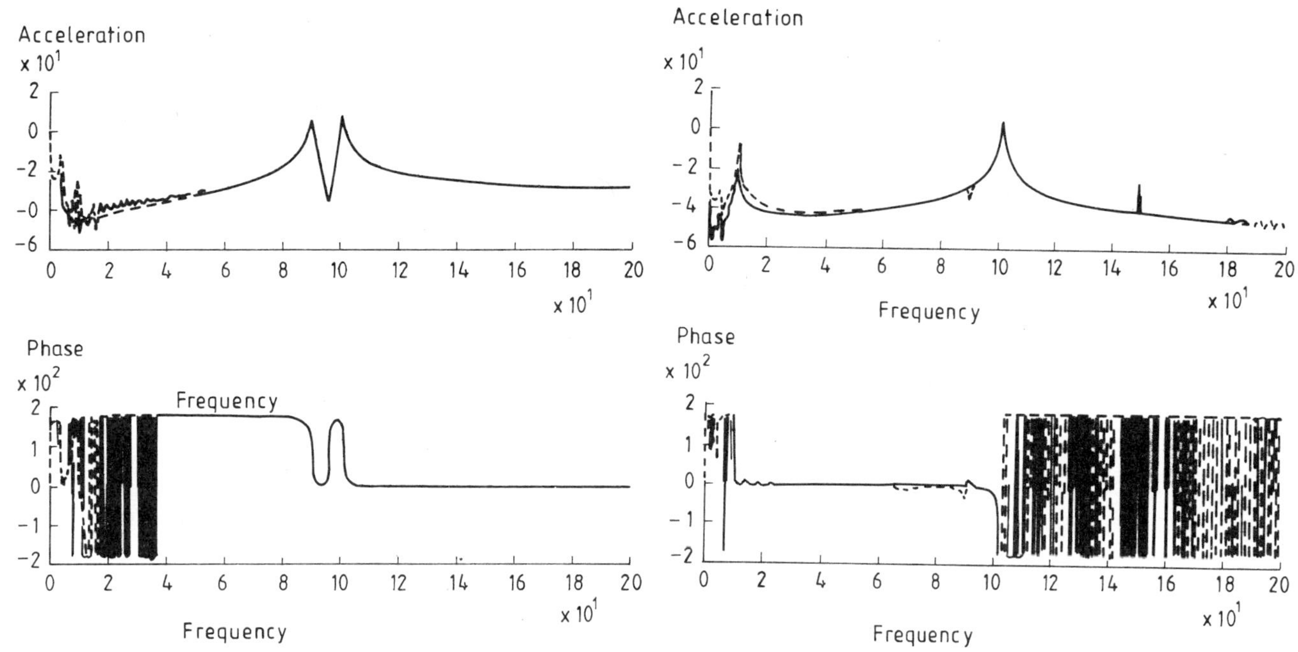

Fig 4 Comparison of random and swept-sine inputs for generating frequency response functions
(a) acceleration at station one/force at station one
(b) acceleration at station three/force at station one
– – – – – FFT analysis with random input
———— FFT analysis with swept-sine input

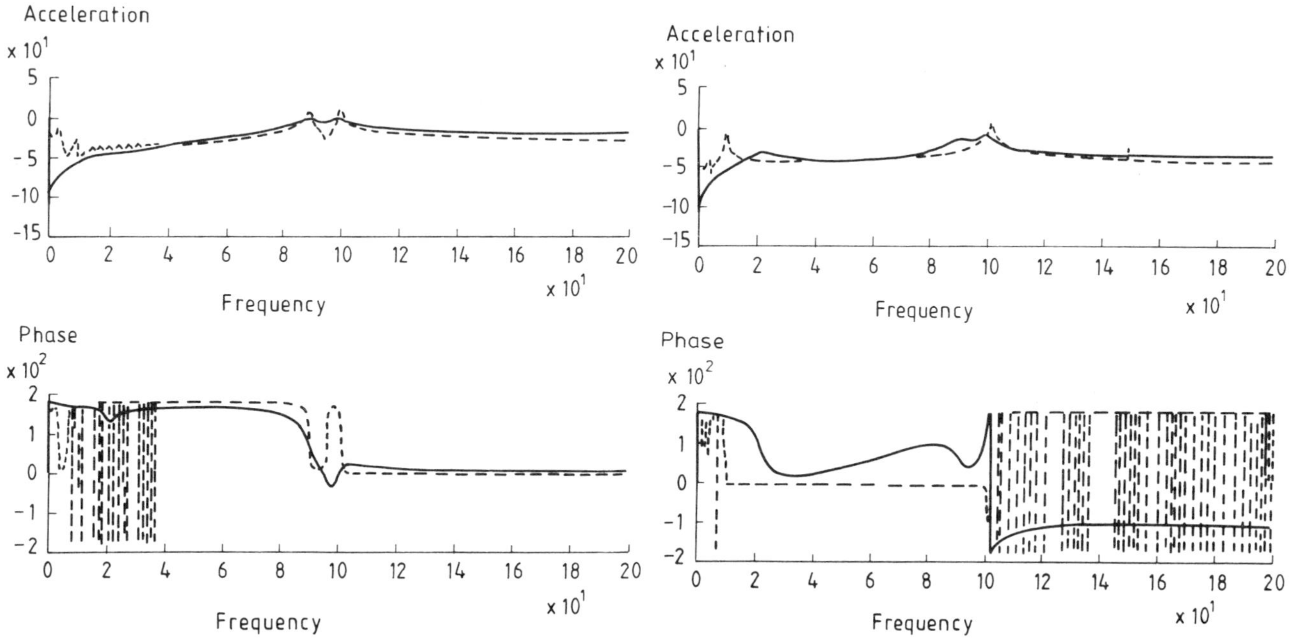

Fig 5 Parameter estimation using swept-sine input. Frequency response functions
(a) acceleration at station one/force at station one
(b) acceleration at station three/force at station one
– – – – – FFT analysis with swept-sine input
———— Predicted by model after frequency domain filtering of swept-sine data

The least squares method applied to identify rotor/foundation parameters

A W LEES, BSc, PhD, MInstP, MINucE
Central Electricity Generating Board, Bedminster Down, Bristol

SYNOPSIS

A method is presented to use measured vibrational data to directly construct a numerical model of the coupled rotor/foundation system. The method avoids costly test programmes on machine foundation structures yet leads to many of the benefits of such a series of measurements. This is achieved by considering the shaft and foundation as two separate structures linked by a known relationship at the bearings. Bearing pedestal motion is measured and rotor motion is eliminated to yield the forcing function on the structure in terms of the rotor's modal properties; hence the behaviour of the supporting structure can be inferred.

1 THE PROBLEM

During the decade 1960-70 there was a tremendous increase in the size of turbo alternators in use in British Power stations, the typical sizes, going from 60 to 500 MW with the introduction of 660 MW units in the early 1970's. Within this same period the use of low tuned steel foundations began to replace the more traditional massive concrete blocks. These two developments together accentuated the need for a more rigourous understanding of the mechanics of these machines. A variety of techniques were developed to assist design and fault diagnosis of these machines over several years and the properties of rotors mounted on stiff supports is now reasonably well understood, but the accurate treatment of the influence of foundation properties has obstinately remained as an unresolved area.

To date the models, even in their limited scope, have proved invaluable in the resolution of a wide variety of rotor dynamic problems and there is good reason to believe this value will be accentuated as their numerical accuracy in improved and scope of applicability is widened. In addition to this application to date, accurate models are now required to integrate with computerised monitoring schemes to help detect and diagnose faults. The commercial values involved can be put onto context by observing that a single 660 MW unit can be worth up £250000 per day, whilst on the other hand an undetected fault could be catastrophic with costs running into tens of millions of pounds.

Recognising the scale of these problems, a number of attempts have been made both in the UK and overseas (1-3) to offer some treatment of the foundations's influence on a machines behaviour. Several authors have attempted to construct finite element models of supporting structure and compare the models with measured data on the real structure. Some of this work is reviewed in Reference 1. Whilst on some of these cases there was some substantial agreement between theoretical and experimental results, there were also some very significant areas of difference apparent. When applied to the calculation of the vibrational performance of flexibly mounted machines, these models have not proved adequate for quantitative predictions, although they have been of great value in enhancing our understanding. The situation is further exacerbated by observing that foundation structures built to the same drawings can exhibit vibrational behaviour which, though showing generally comparable features, also show significant differences. The origin of such differences is likely to be ascribed to such causes as difference in machining tolerances, bolt tightness and so on indicating a lack of sufficient data to adequately model the overall system. In any event, these observations suggest a need for using measured foundation properties as input into the calculation of a machines overall dynamic behaviour. The problem in this is the acquisition of adequate data to use.

The performance of vibration tests on turbine foundations can be an extremely time demanding exercise, often requiring exclusive access to a machine during its overhaul. A dilemma arises that the machines for which an accurate model is required are precisely those where there is greatest pressure to return from overhaul period in the shortest possible time. There would appear to be two ways forward: either one can exert forces to a structure whilst the machine is on load, or one may seek ways of using the pattern of a machines overall behaviour under known circumstances to extract the required model properties. The second of these options is examined in this paper.

2. METHOD

In principle the physical basis of the approach presented here is clear enough. It is supposed that a good numerical model of the rotor train of the machine in question exists, as does knowledge of its state of unbalance. In subsequent work it is hoped that the ability will be developed to extract information for a machine on which the state of unbalance is unknown. However for the present work the objectives are more limited. Using a direct FE model of the rotor leads to a complicated sequence of matrix operations. The weakness of this point of view is that it leads to the inversion of a large matrix at each frequency step of the calculation, and a far more convenient approach is developed here by discussing the dynamic properties of the rotor in terms of its free-free eigenvalues and mode shapes.

In the method proposed here, it is assumed that the bearing behaviour is well understood (although it may prove possible to relax this constraint in later work). Given some unbalance distribution, the motion of the rotor is well defined, apart from the influence of the bearing forces. Hence a relationship may be used to eliminate the (unknown) rotor motion and so express the bearing force in terms of the (measured) pedestal vibration levels.

A satisfactory approach may be derived using the modal properties of the rotor system. Provided that the free natural frequencies and corresponding mode shapes of the rotor are known, from standard techniques the response $\{y\}$ of the rotor at point x, due to the application of continuous force F (x') at a frequency ω, is given by

$$\{y(x)\} = \int_0^L G(\omega,x,x')\{F(x')\}\, dx' \qquad (1)$$

where $G(\omega,x,x')$ is the dynamic flexibility or Green's function of the rotor, and x' is a dummy variable. Furthermore, standard analysis shows that the Green's function may be written as

$$G(\omega,x,x') = \sum_n \frac{\{\Psi n\ (x)\}\{\overset{*}{\Psi n}\ (x')\}^T}{\omega^2{}_n - \omega^2}$$

ω_n and Ψ_n being the free-free natural frequencies and mode shapes and where the summation is over all modes of the system, (in this case the rotor). As mentioned earlier, the forces acting on the rotor in a real machine, may be divided into two summation is over all modes of the system, (in this case the rotor). As mentioned earlier, the forces acting on the rotor in categories: a distributed unbalance force S(x), and the set of bearing reaction forces P acting only at a discrete set of locations. The shaft motion at some point x may now be written using equation 2 as:

$$\{y(x)\} = \int_0^L G(\omega,x,x')\{S(x')\}dx' + \sum G(\omega,x,x_r)\{P(x_r,\omega)\} \qquad (3)$$

which can be further simplified by observing that over any finite frequency range, the effect of a distributed unbalance may be adequately represented by a discrete set of unbalances at some number of unbalance planes. Hence 3 can be rewritten as:

$$\{y\ (x)\} = \sum_b G\ (\omega,x,x_b)\{S'(x_b)\} + \sum_r G\ (\omega,x,x_r)\{P(x_r,\omega)\} \qquad (4)$$

this gives the motion of an arbitrary point along the shaft, so consider the point in question to be $x = x_m$, the location of the mth bearing.

$$\{y\ (x_m)\} = \sum_b G(\omega,x_m,x_b)\ S'\ (x_b) + \sum_m G\ (\omega,x_m,x_r)\{P(x_r,\omega)\} \qquad (5a)$$

which may be rewritten, using an abbreviated notation, as:

$$\{y_m\} = G_{mb}\ (\omega)\ \{S'_b\} + G_{mr}\ (\omega)\ \{P_r\ (\omega)\} \qquad (5b)$$

where

$$G_{ab}\ (\omega) = G(\omega,x_a,x_b) \qquad (5c)$$

The shaft motion at the bearing location x_m can be related to the bearing force P_m through a knowledge of the bearing properties. This relationship can be written in the form:

$$\{P_m(\omega)\} = [B_m\ (\omega)]\ \{Z_m\ (\omega) - y_m\ (\omega)\} \qquad (6)$$

[B] is a 2 x 2 complex matrix giving the properties of the mth bearing. Since only sinusoidal time variations are under consideration, damping terms are incorporated into [B] by the use of complex numbers. Hence in equation 6, m is a specific bearing number and no summation over different bearings is implied. $\{Z_m\ (\omega)\}$ is the measured pedestal vibration at bearing m. Using equation (6) to eliminate y from (5b) yields.

$$\{Z_m\ (\omega)\} - {}_m[B\ (\omega)]^{-1}\ \{P_m\ (\omega)\} = G_{mb}\ (\omega)\ \{S'_b\} + G_{mr}\ (\omega)\ \{P_r\ (\omega)\} \qquad (7)$$

which may be written in matrix form as:

$$[C]\ \{P\} = \{T\} \qquad (8)$$

where $\{P\}$ is the vector of unknown forces

$\{T\}$ is a vector whose m^{th} component is given by:

$$T_m(\omega) = Z_m(\omega) - G_{mb}(\omega)\{S_b'(\omega)\} \quad (9)$$

and C is a square matrix whose general element is given by:

$$C_{mn}(\omega) = G_{mn}(\omega) + [B_m^{-1}(\omega)]\,\delta_{mn} \quad (10)$$

where δ_{mn} is the kronecker delta function.

3. A SIMPLE EXAMPLE

As a test on the method of separating rotor and support, consider a machine consisting of a uniform beam supported on two bearings. The model is highly simplified and is not intended to represent any real machine, but the dimensions used are representative of those on items of auxiliary power plant such as feed pumps.

The model used as a test case is shown in Figure 1. The rotor is 2.29 m long with a diameter of 0.1 m . The rotor has a total mass of 150 kg and the first free-free bending mode is 89.1 Hertz. In this simple test the rotor was idealised with only two beam finite elements. Whilst this would be totally inadequate to represent the dynamics of a real machine, it is adequate to test the consistency of the methods outlined in this paper.

The rotor is supported on bearings whose stiffness are $8.85.10^6$ and $1.77.10^6$ N/m respectively and thee stiffnesses are constant with rotational speed. The stiffness of both bearing supports is $8.85.10^6$ N/m. These two bearing locations carry masses of 90 and 135 Kg respectively. This model machine has its first two natural frequencies at 39 and 90 hertz respectively. The response of the bearing pedestal 1 is shown in Figure 2. The plots shown are for the original model; those derived show small discrepancies which are thought to be a consequence of the approximate data used, whilst the bearing loads are plotted in Figure 3. Owing to the parameters chosen for the model the load at the two bearings differed very little over the frequency range under consideration. The forces derived using the methods outlined above are in good agreement with the exact answers, the discrepancy being typically of order 2%. It is thought that the discrepancies are due to the limited accuracy of the natural frequency and mode shape data used (four figure accuracy). Subsequent testing should be carried out using full machine accuracy numbers involved.

4. THE SUPPORTING STRUCTURE

Using the method outlined in section 2 the force exerted on the structure may be inferred. Since the response of the supporting structure is measured (at the bearing pedestals), the dynamic properties of the foundation can be calculated. However, since coherent forces are applied to the bearings, conventional modal analysis techniques are not suitable. Instead a direct least square method is used to identify structural parameters. The equation of motion, for excitation at ω_ℓ may be written

$$\{\tilde{P}(\omega_\ell)\} = [K]\{\tilde{X}(\omega_\ell)\} + i\,\omega_\ell [C]\{\tilde{X}(\omega_\ell)\} - \omega_\ell^2 [M]\{\tilde{X}(\omega_\ell)\} \quad (11)$$

where the Matrices [K] [C] and [M] are all real but have yet to be determined. P may now be regarded as known. Given a vector of measurements at frequency ω_ℓ, $\{x(\omega_\ell)\}$, a vector of residues may be formed at each frequency.

$$\{\tilde{R}(\omega_\ell)\} = \{\tilde{P}(\omega_\ell)\} - [K]\{x(\omega_\ell)\} + \omega_\ell^2 [M]\{\tilde{x}(\omega_\ell)\} = \omega_\ell [C]\{\tilde{x}(\omega_\ell)\} \quad (12)$$

Then at each measurement frequency ω_ℓ, a deviation is defined as

$$\sigma(\omega_\ell) = \{\tilde{R}^{\star}(\omega_\ell)\}^T [\Lambda(\omega_\ell)]\{\tilde{R}(\omega_\ell)\} \quad (13)$$

from which the total deviation is given by summation over the m frequency measurements

$$\sigma_T = \sum_{\ell=1}^{m} \sigma(\omega_\ell) \quad (14)$$

The weighting matrix $[\Lambda(\omega_\ell)]$ in equ. 13 is an expression of the degree of confidence in each measurement of the components of $\{\tilde{R}(\omega)\}$. If these measurements are considered to be statistically independent, then $[\Lambda(\omega_\ell)]$ is a diagonal matrix. In the present work it is assumed that there is equal confidence in all measurements, ie $[\Lambda(\omega\ )] = [1]$ for all ω_ℓ.

Having established an expression for σ, the analysis proceeds by minimising σ_T with respect to each of the unknown elements of the matrices [K], [M] and [C], giving the set of equations.

$$\frac{\partial\sigma_T}{\partial K_{ij}} = \frac{\partial\sigma_T}{\partial M_{ij}} = \frac{\partial\sigma_T}{\partial C_{ij}} = 0 \quad (15)$$

giving 1 equation for each of the undetermined terms.

From (15) it is clear that $\sigma(\omega_\ell)$ is a scalar quantity which is quadratic in the unknown elements of matrices K, M and C.

The essential, and rather unusual, step in the method proposed here, is to express the total deviation in a form in which the matrix elements form a vector. Thus equation 10 may be written in the form

$$\{\tilde{R}\ (\omega_\ell)\} = \{\tilde{P}\ (\omega_\ell)\} - [W\ (\omega_\ell)]\ \{V\} \qquad (16)$$

In this equation the matrix $[W\ (\omega_\ell)]$ is made up of components of the vector $\{x\ (\omega_\ell)\}$, and $\{V\}$ is made up of the unknown elements of K, M and C. The ordering of components of $\{V\}$ is arbitrary, but the choice of ordering dictates the form of $[W\ (\omega_\ell)]$ and $\{V\}$ for a particular choice of ordering.

Differentiating σ with respect to $\{V\}$ using (13) gives

$$[\alpha(\omega_\ell)]\ \{V\} = [W(\omega_\ell)]^{*T}\ \{\tilde{P}(\omega_\ell)\} + [\tilde{W}(\omega_\ell)]^{T}\ \{P(\omega_\ell)\}^{*T} \qquad (17)$$

where

$$\alpha\ (\omega_\ell) = [W(\omega_\ell)]^{*T}[W(\omega_\ell)] \qquad (18)$$

The right hand side of equation 17 may be considered as a 'force' term by writing.

$$\{q\ (\omega_\ell)\} = [W\ (\omega_\ell)]^{*T}\ \{P\ (\omega_\ell)\} + [W\ (\omega_\ell)]^{T}\{P\ (\omega_\ell)\}^{*T} \qquad (19)$$

Since $\{V\}$ is independent of frequency, the overall equation to be solved is obtained by summation over all frequencies to give

$$[A]\ \{V\} = \{Q\} \qquad (20)$$

where

$$[A] = \sum_{\ell=1}^{m}\ [\alpha\ (\omega_\ell)] \qquad (21)$$

and

$$\{Q\} = \sum_{\ell=1}^{m}\ \{q\ (\omega_\ell)\} \qquad (22)$$

A simple example is given in the next section.

5 RESULTS

A simple test was performed with the following 2 dof model. The system was represented by the matrices

$$[M] = \begin{bmatrix} 1 & 0.2 \\ 0.2 & 1 \end{bmatrix} \text{Kg.},\quad [K] = \begin{bmatrix} 200000 & 0 \\ 0 & 100000 \end{bmatrix} \text{N/m}$$

having resonant frequencies at 49.4 and 73.98 Hz respectively. In the early tests a uniform weighting function was used and a constant unit force was applied at degree of freedom 1. The frequency range considered was 0 to 100 Hz, the upper limit being 50% above the higher resonance.

In the example used in the present work displacement data was generated numerically in a separate routine. Gaussian noise was then added in some cases.

Figure 4 shows the resulting model parameters.

Figure 5 shows the response of the system with the addition of a damping matrix of the form

$$D = \begin{bmatrix} 0.005 & 0 \\ 0 & 0.005 \end{bmatrix} \text{N.sec/rad.m}^{-1}$$

As will be seen from figure 2, good agreement is obtained between the input data and the response of the derived model.

Figure 6 shows how the various matrix terms extracted from the input data converge to their final values. The figures shown the values of the matrix elements when the summations have been carried out for frequencies zero up to the plotted value. These curves refer to an undamped model with an rms noise level of 10^{-4} mm. Note that the curves show the various matrix elements as zero over some parts of the frequency range. Within these parts of the range, the estimated matrices did not yield real positive values for the natural frequencies of the system and so it was possible to reject these estimates without further information.

6. DISCUSSION

A complete method of determining the dynamic properties of a machine mounted on flexible supports has been proposed in this paper and separate examples have been quoted to show how the two main strands of the argument work separately, namely the determination of force levels and the identification of structural matrices. Further work within the Board has made preliminary assessments of the behaviour of real machines with encouraging results, but this work has yet to be refined and reported. Further work has to be undertaken to determine accuracy of the results, prior to applying the method to a complete turbo alternator. The work presented in this paper forms a part of the overall project aimed at providing adequate modelling of the machine as a whole. A simple least squares approach has been adopted here with a view to reducing to a minimum the mathematical complexities involved. In practice it is likely that some form of sequential filter (4) will be used which allows a more realistic view to be taken of the weightings to be given to different methods. These techniques have been examined with regard to the purely structural aspects of the problems in Ref 5 which is being presented at this conference. At each stage in the analysis a number of points are being studied further, but the main aim of the present paper has been to discuss the overall strategy of the approach.

The most obvious criticism of the technique is the assumption that bearing properties are known. Although this is clearly not true in any accurate sense, at any given speed one has a much better idea of bearing behaviour than that of the supporting structure in many instances. Eventually it may be possible to treat some bearing parameters, perhaps loading , as quantities to be identified, but this and similar developments must await further studies. In the same way, it has been assumed that the state of unbalance is known. This can be achieved by using balancing runs to provide basic data.

7 CONCLUSIONS

A method has been presented to use operational vibration to help provide a mathematical model of a machine support structure. Component parts of this procedure have been checked against simple models with encouraging results. Whilst further work is required it is believed that the approach outlined here offers significant advantages over any known alternatives.

8 REFERENCES

1 Lees A W & Simpson IC. The dynamics of turbo-alternator foundations. Paper C5/83 presented to I.Mech.E. meeting, London, February 1983.

2 Stecco S S & Pinzauti M. On the influence of casing stiffness in surbomachinery vibration analysis. Paper C271/80 Second Int. Conf. on 'Vibrations in Routine Machinery'. I.Mech.E, Sept 1980, Churchill College, Cambridge, UK.

3 Jacker M. Vibration analysis of large rotor-bearing-foundation-systems using a modal condensation for the reduction of unknowns. Paper 279/80. Second Int. conf. on 'Vibrations in Rotating Machinery'. I.Mech.E., September 1980, Churchill.

4 Mottershead J E, Lees A W & Stanway R. A linear frequency domain filter for the paramter identification of vibrating structures. Trans ASME J.Vibration, Accoustics, Stress & Reliability in Design 87, 109, 262-269.

5 Tee T K, Mottershead J E, Stanway R and Lees A W. Identification of Turbogenerator foundations from run-downs records - a preliminary experimental study. Paper C366/031 also presented at this conference.

9 ACKNOWLEDGEMENT

This work is published by permission of the Central Electricity Generating Board

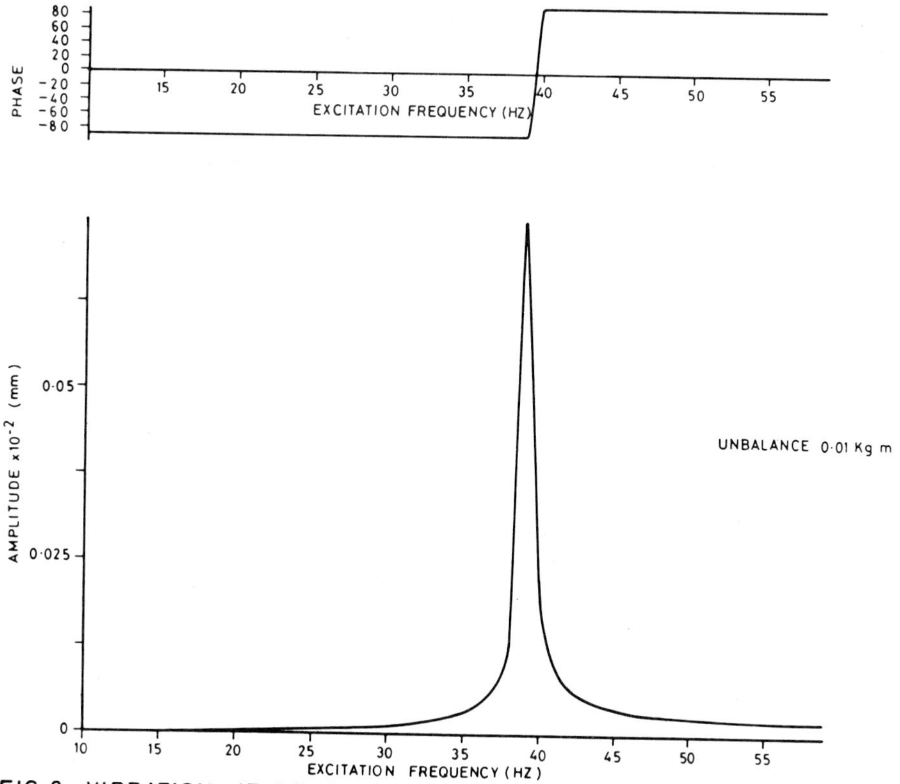

FIG. 2 VIBRATION AT BEARING 1

Fig 2 Variation at bearing 1

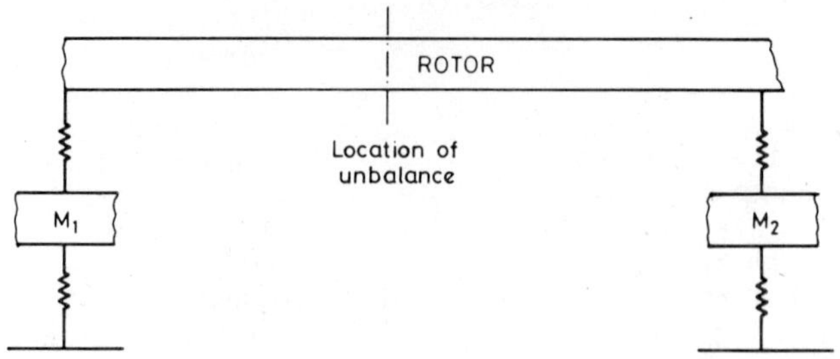

Fig 1 Simple test model

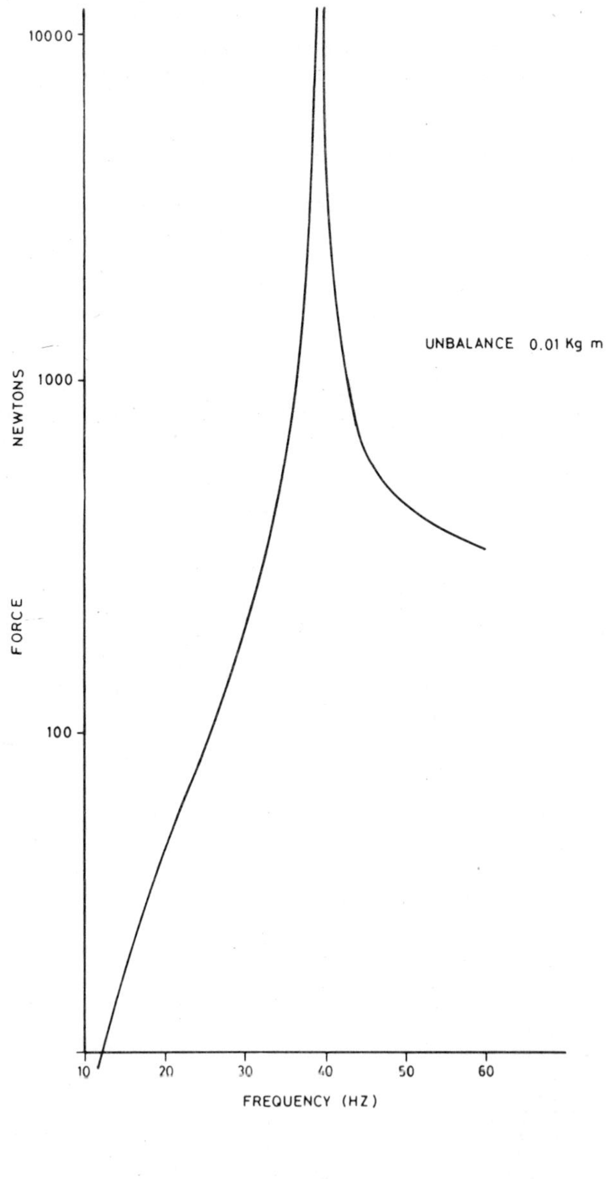

Fig 3 Bearing 1 force variation

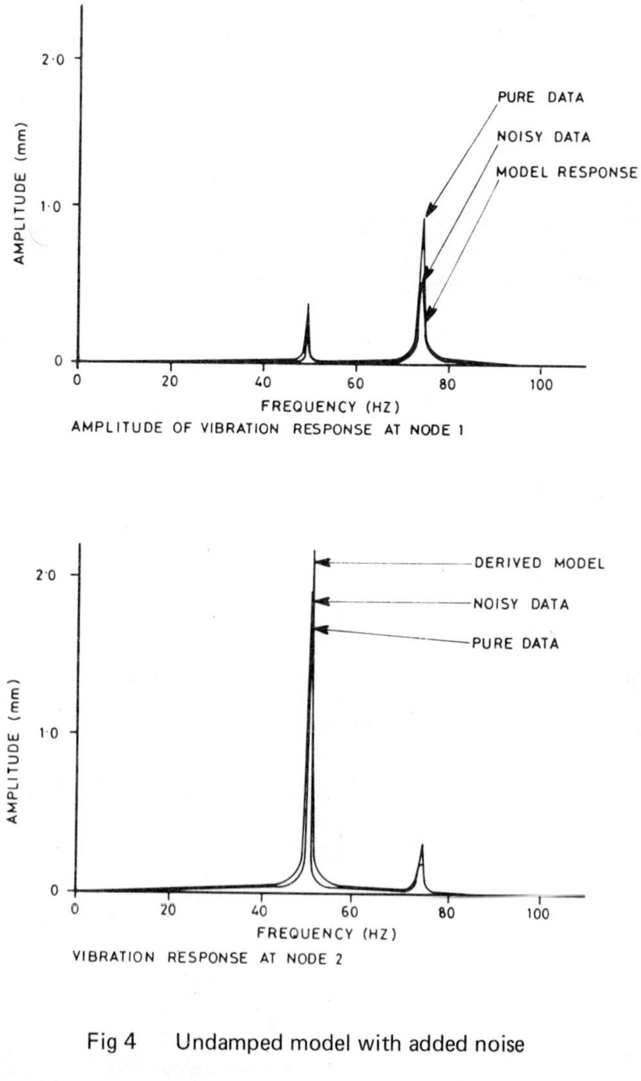

Fig 4 Undamped model with added noise

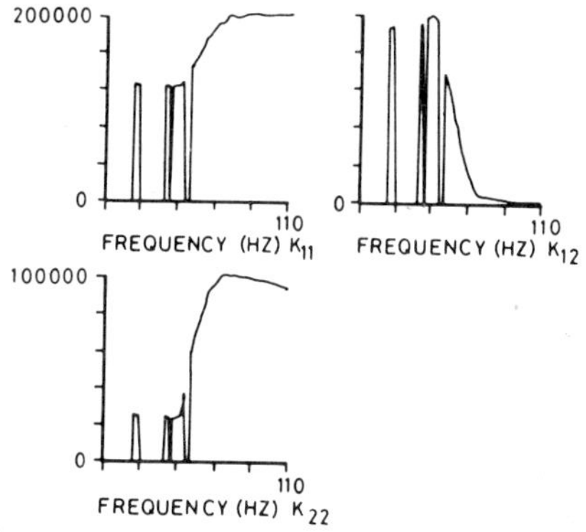

PREDICTIONS OF STIFFNESS VALUES

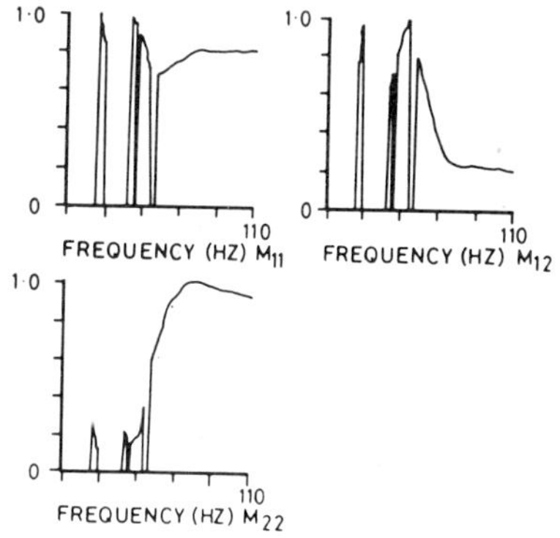

PREDICTIONS OF MASS MATRIX

Fig 6 Development of model matrices

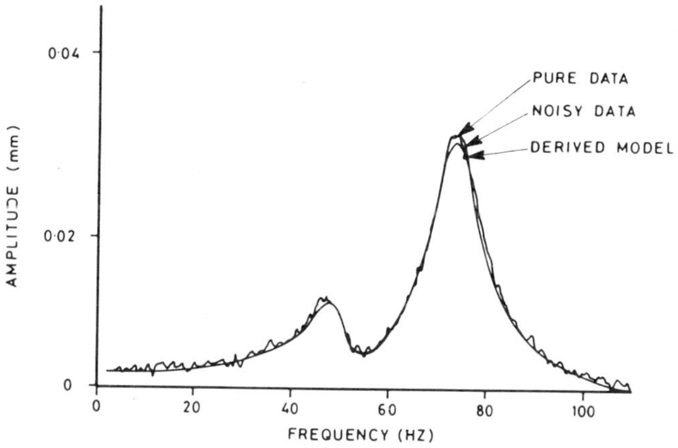

VIBRATION RESPONSE AT NODE 1

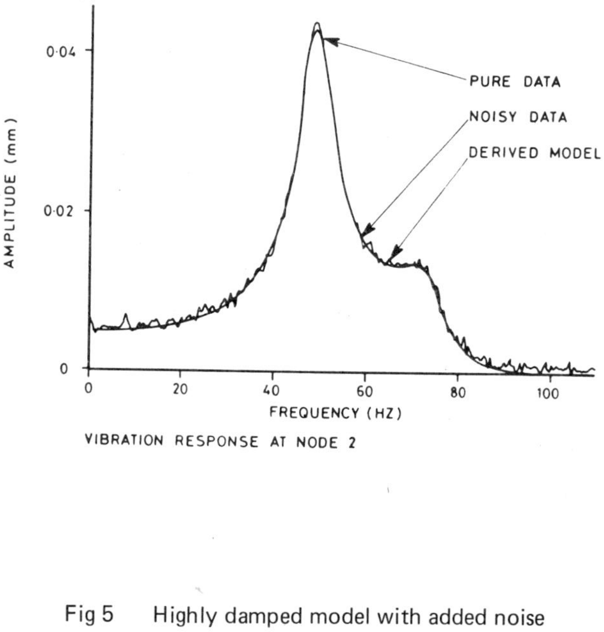

VIBRATION RESPONSE AT NODE 2

Fig 5 Highly damped model with added noise

C300/88

A method to identify the foundation modal parameters through measurements of the rotor vibrations

G DIANA and F CHELI
Dipartimento di Meccanica, Politecnico di Milano, Milan, Italy
A VANIA
Dipartimento di Meccanica Strutturale, Universita degli Studi di Pavia, Pavia, Italy

SYNOPSIS

This paper is aimed at illustrating an approach to computing the modal parameters of the rotor machinery supporting structure by measuring the shaft and support vibrations during run-ups or run-downs. The modal parameters can be evaluated by minimizing some functions derived from the experimental frequency response and the analytical one obtained through a mathematical model of the system composed of the shaft, the bearing and the foundation.

Two mathematical methods for evaluating the modal parameters of the supporting structure are described. Furthermore, the results obtained applying the proposed methodology to the experimental and the analytical frequency responses are illustrated.

NOTATION

- [A] modal matrix of the natural modes Φ_i of the foundation;
- [B] mechanical impedance matrix of the foundation;
- [G] "generalized impedance matrix" of the foundation;
- [K] ($[\bar{K}]$) stiffness matrix in $\underline{X}$ ($\underline{Z}$) coordinates;
- [M] ($[\bar{M}]$) mass matrix in $\underline{X}$ ($\underline{Z}$) coordinates;
- [R] ($[\bar{R}]$) damping matrix in $\underline{X}$ ($\underline{Z}$) coordinates;
- [m] mass matrix of the foundation in $\underline{q}$ coordinates;
- [r] damping matrix of the foundation in $\underline{q}$ coordinates;
- [k] stiffness matrix of the foundation in $\underline{q}$ coordinates;
- $\underline{F}_r$ generalized forces on the rotor;
- $\underline{F}_c$ generalized forces at the connecting nodes;
- $\underline{p}$ foundation modal parameters (m_i, r_i e k_i);
- $\underline{q}$ principal coordinates of the foundation;
- $\underline{X}_r$ rotor displacements;
- $\underline{X}_c$ connecting nodes displacements;
- $\underline{X}_{rc}$ relative rotor-bearing displacements;

Note: the experimental quantities are superscripted with the symbol ^{o}

1 INTRODUCTION

The interaction between the shaft and the supporting structure can considerably modify the rotor dynamic behaviour. To take into account the effects due to the foundation, the methodology usually accepted is to describe the supporting structure through its mechanical impedance ([2], [3], [4], [11], [12], [13]), which can be evaluated by means of the foundation modal parameters [6].

This paper is aimed at illustrating a methodology to identify the foundation modal parameters by minimizing some objective functions derived from the shaft experimental frequency response and the one evaluated by means of a mathematical model of the system composed of the shaft, the bearings and the foundation. The supporting structure modal parameters are the only unknown quantities of the problem.

This paper deals with the following issues:

- description of the motion equations of the system composed of the shaft, the bearings and the supporting structure;
- description of the mathematical methods suggested for evaluating the structure modal parameters;
- description of the objective functions used in the mathematical methods;
- illustration of the results obtained with some numerical applications;
- conclusions.

2 THE EQUATIONS OF MOTION

Let us consider separately the substructures composed of the shaft and the bearings and the one composed of the foundation, as illustrated in Fig 1. The motion equations of the first substructure can be expressed in the following form ([1], [2], and [3]):

$$[M]\ddot{\underline{X}} + [R]\dot{\underline{X}} + [K]\underline{X} = \underline{F} \qquad (2.1)$$

where $\underline{X}$ is a vector that contains the displacements $\underline{X}_r$ of the rotor and the displacements $\underline{X}_c$ of the supporting structure at the bearings:

$$\underline{X}^T = \underline{X}^T_r, \underline{X}^T_c \qquad (2.2)$$

Furthermore, the vector $\underline{F}$ contains the forces $\underline{F}_r$ applied to the rotor and the forces $\underline{F}_c$ transmitted by the foundation to the rotor at the connecting points $\underline{X}_c$:

$$\underline{F}^T = \left\{ \underline{F}^T_r, \underline{F}^T_c \right\} \qquad (2.3)$$

From the equation (2.1), taking into account of (2.2), we obtain the following two systems of equations:

$$[M_{rr}]\ddot{\underline{X}}_r + [R_{rr}]\dot{\underline{X}}_r + [R_{rc}]\dot{\underline{X}}_c + [K_{rr}]\underline{X}_r + [K_{rc}]\underline{X}_c = \underline{F}_r$$

$$[R_{cr}]\dot{\underline{X}}_r + [R_{cc}]\dot{\underline{X}}_c + [K_{cr}]\underline{X}_r + [K_{cc}]\underline{X}_c = \underline{F}_c \qquad (2.4)$$

being:

$$[M] = \begin{bmatrix} [M_{rr}] & [0] \\ [0] & [0] \end{bmatrix}$$

$$[R] = \begin{bmatrix} [R_{rr}] & [R_{rc}] \\ [R_{cr}] & [R_{cc}] \end{bmatrix} \qquad (2.4a)$$

$$[K] = \begin{bmatrix} [K_{rr}] & [K_{rc}] \\ [K_{cr}] & [K_{cc}] \end{bmatrix}$$

The forces $\underline{F}_c$ transmitted at the connecting points can be expressed in terms of the relative displacements $\underline{X}_{rc}$ between the rotor and the supports as:

$$\underline{F}_c = [\, R_o \,]\, \dot{\underline{X}}_{rc} + [\, K_o \,]\, \underline{X}_{rc} \qquad (2.5)$$

where $[R_o]$ and $[K_o]$ are the equivalent damping and stiffness matrices calculated by linearizing the oil film forces ([7], [8], [9]).

The supporting structure motion can be described by means of modal techniques considering, as independent variables, the generalized coordinates $\underline{q}$. The displacements $\underline{X}_c$ can be expressed by the relation:

$$\underline{X}_c = [\, A \,]\, \underline{q} \qquad (2.6)$$

where $[A]$ is a matrix the columns of which are the normal modes Φ_i of vibration evaluated at the connecting points. Therefore, the motion equations of the supporting structure can be written as:

$$[m]\, \ddot{\underline{q}} + [r]\, \dot{\underline{q}} + [k]\, \underline{q} = -[A]^T\, \underline{F}_c \qquad (2.7)$$

If the vibration modes are uncoupled, $[m]$, $[r]$ and $[k]$ are diagonal matrices. $[m]$, $[r]$ and $[k]$ contain the foundation modal parameters $\underline{p}$, that is the generalized mass m_i, the damping factor r_i and the generalized stiffness k_i corresponding to the i-th natural frequency of the foundation. Let us consider the independent variables $\underline{Z}$:

$$\underline{Z}^T = \left\{ \underline{X}^T{}_r ,\, \underline{q}^T \right\} \qquad (2.8)$$

with this position and considering the equations (2.6) and (2.7), the eq. (2.4) becomes:

$$[\, \bar{M} \,]\, \ddot{\underline{Z}} + [\, \bar{R} \,]\, \dot{\underline{Z}} + [\, \bar{K} \,]\, \underline{Z} = \bar{\underline{F}} \qquad (2.9)$$

where:

$$[\bar{M}] = \begin{bmatrix} [M_{rr}] & [0] \\ [0] & [m] \end{bmatrix}$$

$$[\bar{R}] = \begin{bmatrix} [R_{rr}] & [R_{rc}]\,[A] \\ [A]^T\,[R_{cr}] & [r] + [A]^T\,[R_{cc}]\,[A] \end{bmatrix} \qquad (2.10)$$

$$[\bar{K}] = \begin{bmatrix} [K_{rr}] & [K_{rc}]\,[A] \\ [A]^T\,[K_{cr}] & [k] + [A]^T\,[K_{cc}]\,[A] \end{bmatrix}$$

$$\bar{\underline{F}}^T = \left\{ \underline{F}^T{}_r ,\, \underline{0}^T \right\}$$

In the case of harmonic forces:

$$\bar{\underline{F}}_r = \underline{F}_{ro}\, e^{i\Omega t} \qquad (2.11)$$

the equation (2.9) becomes:

$$[H_{rr}]\, \underline{X}_{ro} + [H_{rc}]\, \underline{q}_o = \underline{F}_{ro}$$
$$[H_{cr}]\, \underline{X}_{ro} + ([H_{cc}] + [G(\Omega,\underline{p})])\underline{q}_o = \underline{0} \qquad (2.12)$$

where:

$$[H_{rr}] = (-\Omega^2 [M_{rr}] + i\Omega [R_{rr}] + [K_{rr}])$$
$$[H_{rc}] = (i\Omega [R_{rc}]\,[A] + [K_{rc}]\,[A])$$
$$[H_{cr}] = (i\Omega [A]^T [R_{cr}] + [A]^T [K_{cr}]) \qquad (2.13)$$
$$[H_{cc}] = (i\Omega [A]^T [R_{cc}]\,[A] + [A]^T [K_{cc}]\,[A])$$

The matrix $[G(\Omega, \underline{p})]$ depends on both the frequency Ω of the forces $\underline{F}_r$ and the modal parameters $\underline{p}$; this matrix can be called the "generalized mechanical impedance matrix" of the supporting structure. The matrix $[G]$ can be expressed in the form:

$$[G(\Omega, \underline{p})] = (-\Omega^2 [m] + i\Omega [r] + [k]) \qquad (2.14)$$

The mechanical impedance matrix $[B(\Omega, \underline{p})]$, (see [1]), can be defined from the relation:

$$[B(\Omega, \underline{p})]\, \underline{X}_{co} = -\underline{F}_{co} \qquad (2.15)$$

If the number of vibration modes Φ_i is equal to the number of degrees of freedom $\underline{X}_c$ associated with the connecting nodes of the model, $[A]$ is a square matrix and matrix $[B]$ can be obtained, from the equation (2.7), through the relation:

$$[B(\Omega, \underline{p})] = ([A]^T)^{-1}\, [G(\Omega, \underline{p})]\, [A]^{-1} \qquad (2.16)$$

The paper is aimed at illustrating the methodology, previously summarized, that allows us to calculate the foundation modal parameters. Introducing these parameters into a mathematical model it is possible to evaluate the dynamic behaviour of the system composed of the shaft, the bearings and the supporting structure. Introducing the values of the modal parameters into the matrices $[\bar{M}]$, $[\bar{R}]$ and $[\bar{K}]$ of the equation (2.9) it is possible to calculate the frequency response of the global system, in the time domain, when any type of external forces are applied. Furthermore, substituting the modal parameters in the equations (2.12) for calculating the matrix $[G]$ it is possible to evaluate the response of the system to an harmonic excitation.

3. IDENTIFICATION OF THE FOUNDATION MODAL PARAMETERS

The methodology suggested for the identification of the foundation modal parameters needs the knowledge of the rotor experimental frequency response measured during a run-up or a run-down, that is the displacements $\underline{X}^o{}_{ro}$ of the rotor and $\underline{X}^o{}_{co}$ of its supports at any angular velocity Ω of the machine. The frequency response must be compared with the one obtained by means of the mathematical model previously described, in which the oil film equivalent damping and stiffness matrices are obtained by linearizing the oil film forces in the neighbourhood of the static equilibrium position ([7], [8] and [9]). Furthermore, the components Φ_i of the vibration modes of the foundation, evaluated at the connecting points, are assumed to be known. The vibration modes can be calculated by means of a finite element model of the supporting structure or considering as modal shapes the ones corresponding to:

- uncoupled supports, schematized with a one degree of freedom model;
- rigid motions;
- the lowest vibration modes of a free-free unrestrained beam.

With these assumptions, the only unknown quantities of the mathematical model are the supporting structure modal parameters, that is the natural frequencies ω_i, the dimensionless damping factors $(r/r_c)_i$ and the modal masses m_i. The modal parameters $\underline{p}$ are calculated using numerical methods that minimize some objective functions derived from the rotor experimental frequency response and the analytical one.

4 THE OBJECTIVE FUNCTIONS

Some different objective functions for evaluating the modal parameters have been tested. These functions have been illustrated in [1] together with an analysis of the applicability limits in relation to the mechanical system characteristics. In this paper only two objective functions have been utilized.

4.1 Method I

The forces that the two substructures shown in Fig 1 exchange at the k-th support can be calculated either as a function of the characteristics of the rotor and the oil film or as a function of the support structure modal parameters. The forces $\underline{F}^u_{co}$ and $\underline{F}^1_{co}$ must satisfy the following relation:

$$\underline{F}^u_{co}(\Omega,\underline{F}_{ro},\underline{X}_{co}) = \underline{F}^1_{co}(\Omega,\underline{p},\underline{X}_{co}) \qquad (4.1)$$

where:

$$\underline{F}^u_{co}=(-[H_{cr}(\Omega)][H_{rr}(\Omega)]^{-1}[H_{rc}(\Omega)][A]^{-1})\underline{X}_{co}+$$
$$+([H_{cr}(\Omega)][H_{rr}(\Omega)]^{-1})\underline{F}_{ro}+([H_{cc}(\Omega)][A]^{-1})\underline{X}_{co} \qquad (4.2)$$

$$\underline{F}^1_{co} = -([G(\Omega,\underline{p})]\,[A]^{-1})X_{co}$$

The frequency response $\underline{X}^o_{co}$, corresponding to the forces $\underline{F}^o_{ro}$ applied to the rotor, can be determined experimentally. Therefore it is possible to calculate $\underline{F}^u_{co}(\Omega,\underline{F}^o_{ro},\underline{X}^o_{co})$ and $\underline{F}^1_{co}(\Omega,\underline{p},\underline{X}^o_{co})$, as function of the unknown quantities $\underline{p}$, using an iterative method. The objective function that must be minimized is:

$$f_I = \sum_{\Omega}\sum_{K}(F^u_{cok}(\Omega,\underline{F}^o_{ro},\underline{X}^o_{co}) - F^1_{cok}(\Omega,\underline{p},\underline{X}^o_{co}))^2 \qquad (4.3)$$

where $\underline{F}^u_{cok}$ and $\underline{F}^1_{cok}$ are the forces calculated at the k-th support of the shaft at the rotor angular velocity Ω.

Because the matrix [A] must be inverted, the number of vibration modes of the supported structure must be equal to the number of degrees of freedom associated to the connecting nodes of the model.

4.2 Method II

Introducing for the displacements $\underline{X}_{rc}$ the measured values into the equation (2.5) it is possible to calculate the forces $\underline{F}_c$: these forces, in the case of an harmonic excitation, are:

$$\underline{F}_{co} = (i\Omega[R_o] + [K_o])\,\underline{X}^o_{rco} \qquad (4.4)$$

The motion of the supporting structure, in principal coordinates $\underline{q}$, can be expressed as:

$$(-\Omega^2[\,m\,] + i\Omega[\,r\,] + [\,k\,])q_o =$$
$$= -\,[A]^T(i\Omega[R_o] + [K_o])\,\underline{X}^o_{rco} \qquad (4.5)$$

where the Lagrange's component can be evaluated using the experimental frequency response $\underline{X}^o_{rco}$.

Assigning some values to $\underline{p}$, for the first iteration, it is possible to obtain $\underline{q}_o$ from the equation (4.5) and, using the eq.(2.6), the displacements $\underline{X}_{co}$.

The objective function is the difference between the measured displacements $X^o_{cok}(\Omega)$ and the computed one $X_{cok}(\Omega,\underline{X}^o_{rco},\underline{p})$, both evaluated at the k-th connecting point:

$$f_{II} = \sum_{\Omega}\sum_{K}(X^o_{cok}(\Omega) - X_{cok}(\Omega,\underline{X}^o_{rco},\underline{p}))^2 \qquad (4.6)$$

This method has no restriction about the number of the foundation modes to take into account and does not need the knowledge of the forces $\underline{F}_{ro}$ applied to the rotor.

5 NUMERICAL APPLICATIONS

The mathematical techniques previously described have been checked using an analytical model (par.5.2) and subsequently have been used to evaluate the foundation modal parameters of a laboratory model ([10]), in order to point out the validity of the methodology in a real application (Par.5.3).

5.1 Experimental set-up

The model is a modification of the rotor kit, kindly gifted by Bently Nevada Corporation; it is composed of a motor driven rotor, supported by two bearings connected to a rigid base plate. Lubricated cylindrical bearings are mounted and the oil is supplied by an electric motor driven pump. A disk is fixed by two setscrews at about the middle of the rotor. This disk has a series of axial threaded holes placed at the same distance from its center. These holes can receive balancing/unbalancing masses.

Relative vibrations between the rotor and the bearings have been measured by a couple of proximitors placed, in vertical and horizontal directions, at each support and at about the middle of the rotor (Fig 2). Supports absolute vibrations have been measured by accelerometers placed on the bearing cases in vertical and horizontal directions. The model rotational speed is evaluated by a proximitor placed, in the horizontal direction, in front of a radial notch on the rotor. The signal from this probe is also used as phase reference for all the vibration signals.

A finite element model of the rotor and the positions of the probes are shown in Fig 3. The signals from all the probes have been acquired by a PDP 11/34 Digital computer equipped with an A/D converter and suitable acquisition and evaluation software, that allows us to compute a synchronous analysis of the signals, determining their first three synchronous harmonic components.

5.2 An analytical test

The methodology proposed for evaluating the modal parameters of the shaft supporting structure needs a mathematical model of the mechanical system composed of the rotor, the bearings and the foundation. The rotor has been modelled with finite element techniques as shown in Fig 3, whereas the matrices containing the oil film stiffness and damping coefficients have been determined by integrating the Reynolds equation. The analytical curve (dotted line) of the vertical frequency response of the rotor evaluated at the bearing n.1 (see Fig 3), with a known unbalance force and considering a rigid support structure, is shown in Fig 4. This curve shows a rotor resonance speed at 3600 rpm and fits very well the corresponding experimental frequency response curve (solid line), measured with the base plate rigidly constrained to the foundation block. This test shows that the mathematical model well reproduces the dynamic behaviour of the system and can be used for the subsequent analysis.

In order to test the mathematical methods previously

discussed, the modal parameters of an elastic foundation have been calculated using the analytical frequency response of the rotor, evaluated by means of the mathematical model described before, in which the effects due to the foundation have been taken into account through a finite element model.

The vertical frequency response of the rotor, evaluated at the bearing n.1, with the same unbalance force as in the case of Fig 4, is shown in Fig 5. The effects due to the elastic supporting structure considerably modify the dynamic behaviour of the rotor in comparison with the case of the rigid foundation. Table 1 contains the values of the foundation modal parameters obtained by the finite element model of the foundation together with the ones calculated applying the method II and considering the vibration modes of the supporting structure finite element model.

In order to execute a further applicability test of the proposed methodology, the foundation modal parameters, considering as vibration modes of the foundation the uncoupled modes of the one degree of freedom models for every support, have been evaluated. The vertical frequency response curve of the rotor, calculated at the bearing n.1 by means of the finite element model and the one obtained using the supporting structure modal parameters determined with the method II, are shown in Fig 6. Similar results have been obtained using the method I. This preliminary test confirms the validity of the methodology. Besides, even if using the uncoupled modal shapes, not corresponding to the real ones, it is possible to reproduce the behaviour of the system in a very good way (see Fig 6). This fact shows that it is possible to evaluate the modal parameters without the knowledge of the effective shapes of the different vibration modes of the foundation.

5.3 An experimental test

After having analyzed the dynamic behaviour of the laboratory model, with the base plate rigidly constrained to the foundation block, some elastic elements have been mounted between the plate and the foundation (see Fig 2). In this condition the rotor and support vibrations, with a known unbalance, were measured.

Using the experimental data, the supporting structure modal parameters have been calculated, considering rigid and uncoupled modal shapes. Applying the method II and evaluating the contact forces $\underline{F}_c$ through equation (4.4), the support X_{cok} displacements, as a function of the foundation modal parameters, are calculated (eq. 4.6). The experimental curve of the frequency response of the bearing 2, in the vertical direction, is shown in Fig 7 with the analytical one, computed with the modal parameters optimized by the method. This curve fits the experimental response in a satisfactory way.

Using the modal parameters so calculated, the frequency response of the global system composed of the shaft, the bearings and the foundation has been evaluated. The comparison between the analytical and the experimental frequency response, evaluated at the support 2, is illustrated in Fig 8. Furthermore, the comparison between the analytical and the experimental frequency response of the rotor, calculated at bearing 2, is illustrated in Fig 9.

The flexible foundation, through its natural frequencies, considerably modifies the rotor dynamic behaviour (see Fig 4 and Fig 7); nevertheless the method allows us to reproduce the experimental frequency response.

6. CONCLUSIONS

The objective functions proposed in this paper are able to identify the foundation modal parameters, as illustrated at par.5.2.

The proposed methods do not need the knowledge of the effective vibration modes of the foundation. Besides the method II can be used without knowing the forces applied to the rotor.

The results obtained applying this methodology on a laboratory model are very satisfactory. However the methods should be tested in a full scale rotor with a greater number of supports.

REFERENCES

(1) CHELI,F. DIANA,G. VANIA,A. Identificazione dei parametri modali delle fondazioni di macchine rotanti. L'Energia Elettrica, n.6, 1987.

(2) DIANA, G. P.A.L.L.A.: a Package to Analyze the Dynamic Behaviour of a Rotor-Supporting Structure System. CISM Rotordynamics Conf., Udine, 1986.

(3) CURAMI,A. PIZZIGONI,B. Un programma di calcolo automatico per l'analisi statica di una linea d'alberi. L'Energia Elettrica, n.12, 1981.

(4) DIANA,G. MASSA,E. et al. A Forced Method to Calculate the Oil Film Instability Threshold of a Rotor-Foundation System. IFToOM, Rome, Italy, 1982.

(5) DIANA,G. BACHSCHMID,N. Influenza della struttura portante sulle velocita' critiche flessionali degli alberi. L'Energia Elettrica, n.9, 1978.

(6) CHELI,F. DIANA,G. CURAMI,A. VANIA,A. On the Modal Analysis to Define Mechanical Impedance of a Foundation. ASME, Cincinnati, 1985.

(7) FRIGERI,C. GASPARETTO,M. VACCA,M. Cuscinetto lubrificato in regime laminare e turbolento - parte I - Analisi statica. L'Energia Elettrica, vol.LVII, 1980.

(8) BIRAGHI,B. FALCO,M. PASCOLO,P. SOLARI,A. Cuscinetto lubrificato in regime laminare e turbolento - parte II - Analisi dinamica. L'Energia Elettrica, vol.LVII, 1980.

(9) FALCO,M. MACCHI,M. Vallarino,G. Cuscinetto lubrificato in regime laminare e turbolento - parte III - Lubrificazione mista idrostatica-idrodinamica. L'Energia Elettrica, vol.LVII, 1980.

(10) DIANA,G. CHELI,F. MANENTI,A. PETRONE,F. Non Linear Effects in Lubricated Bearings. 4th Int. Conf. on Vibration in Rotating Machinery, Edinburgh, 1988.

(11) LUND,J.W. Response Characteristics of a Rotor with Flexible Damped Supports. Dynamics of Rotors, Symposium Lyngby/Denmark, Spring Verlag, 1975.

(12) KRAMER,E. Computation of Vibrations of the Coupled System Machine-Foundation. Proc. Vibrations in Rotating Machinery, Cambridge, 1976.

(13) RIEGER,N.F. THOMAS,C.B. WALTER,W.W. Dynamic Stiffness Matrix Approach for Rotor-Bearing System Analysis. Proc. Vibrations in Rotating Machinery, Cambridge, 1976.

Table 1 Comparison between assigned modal parameter of the foundation and the calculated ones

	Mode	Assigned Modal Parameters
Natural Frequency [Hz]	1	10.50
	2	35.33
	3	23.87
	4	47.75
Dimensionless damping factor (r/r_c)	1	.020
	2	.020
	3	.020
	4	.020
Modal mass	1	.42
	2	2.28
	3	1.53
	4	1.79

	Mode	Calculated Modal Parameters
Natural Frequency [Hz]	1	10.66
	2	35.33
	3	23.87
	4	47.75
Dimensionless damping factor (r/r_c)	1	.018
	2	.020
	3	.020
	4	.020
Modal mass	1	43
	2	2.26
	3	1.52
	4	1.77

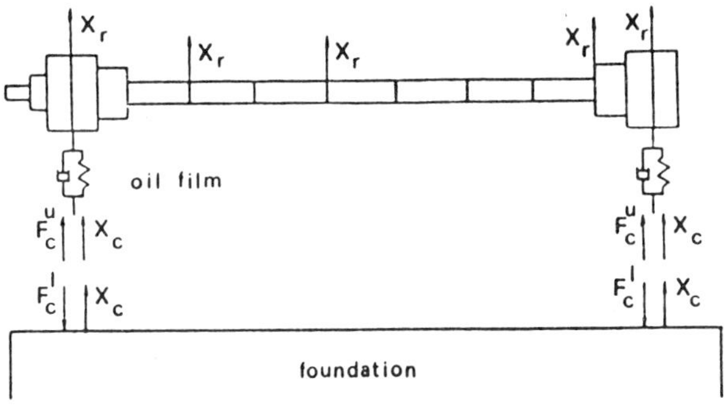

Fig 1 Key elements of rotor–foundation system

Fig 2 Laboratory model

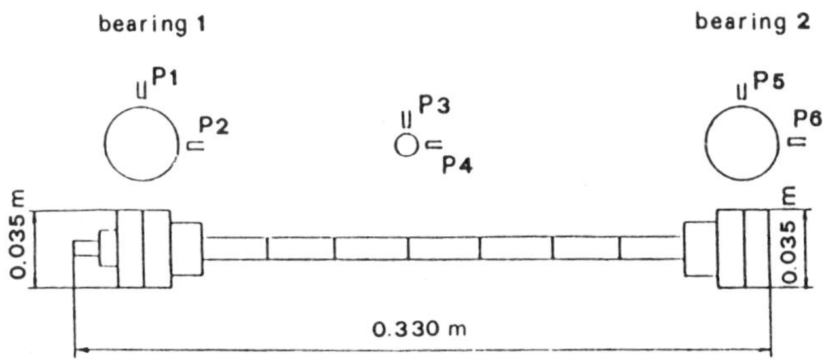

Fig 3 Finite element model of the rotor and positions of the probes

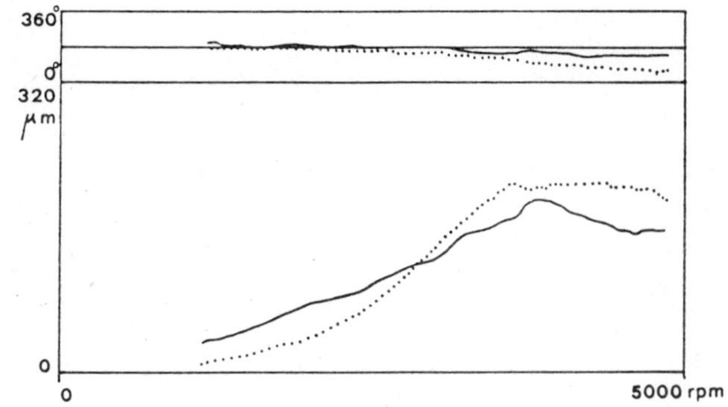

Fig 4 Rigid foundation; vertical frequency response of the rotor evaluated at the bearing 1 (see Figure 3) with a known unbalance (solid line = experimental response, dotted line = analytical response)

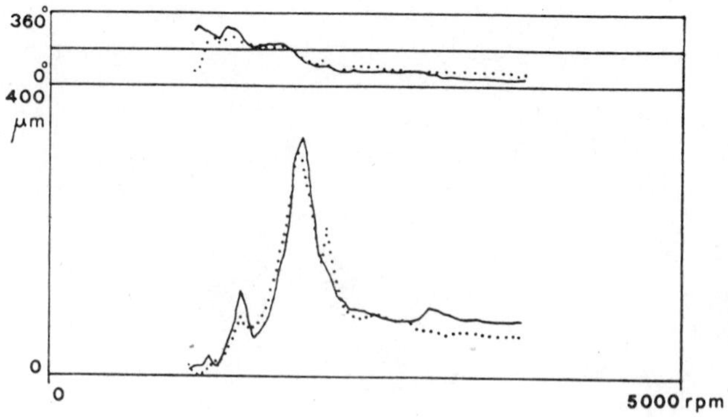

Fig 7 Flexible foundation; vertical frequency response of the support evaluated at the bearing 2 (see Figure 3) (solid line = X^{o}_{co2} (Ω) response, dotted line = fitting curve X_{co2} (Ω, X^{o}_{rco}, p), see equation (4.6))

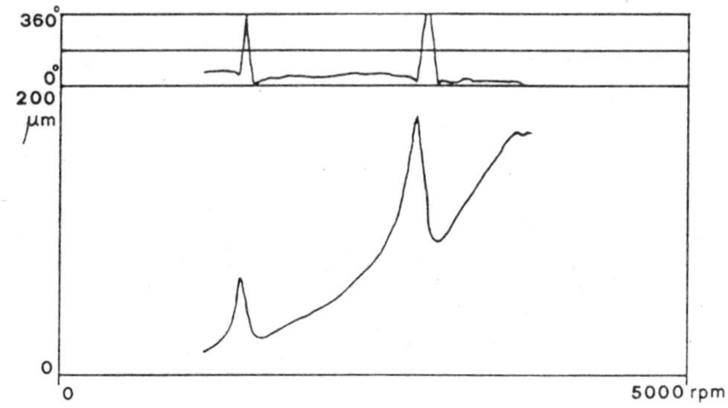

Fig 5 Flexible foundation; analytical frequency response of the rotor evaluated at the bearing 1 (see Figure 3) with a known unbalance in vertical direction

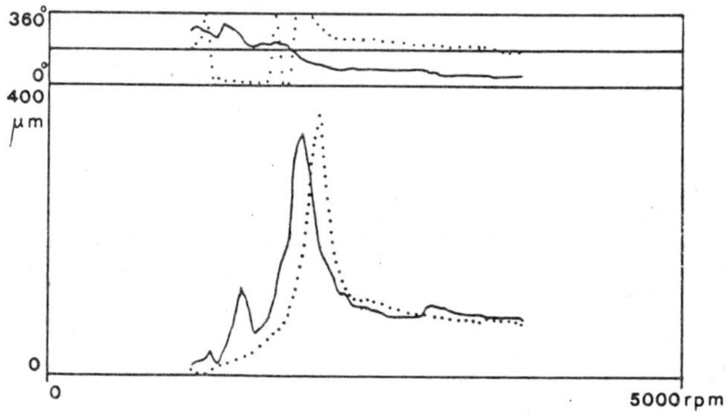

Fig 8 Flexible foundation; vertical frequency response of the support evaluated at the bearing 2 (see Figure 3) with a known unbalance (solid line = experimental response, dotted line = fitting curve)

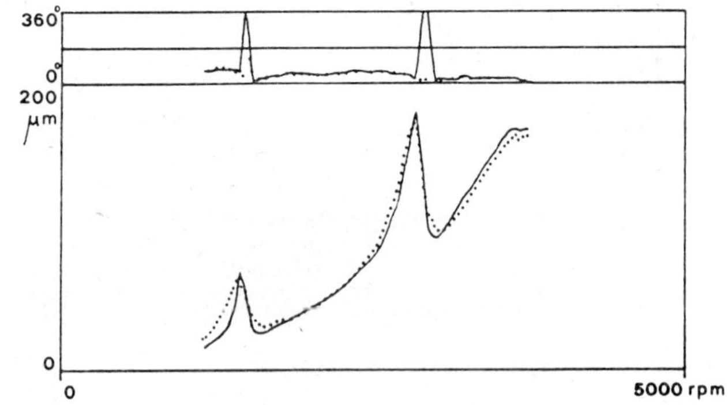

Fig 6 Flexible foundation; vertical frequency response of the rotor evaluated at the bearing 1 (see Figure 3) with a known unbalance (solid line = analytical response, dotted line = fitting curve)

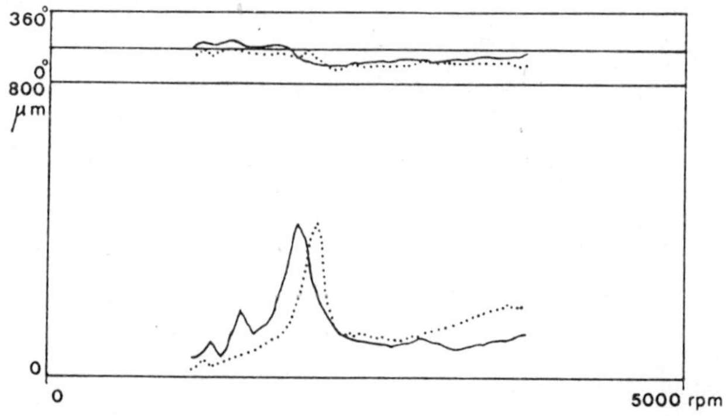

Fig 9 Flexible foundation; vertical frequency response of the rotor evaluated at the bearing 2 (see Figure 3) with a known unbalance (solid line = experimental response, dotted line = fitting curve)

C258/88

Impact excitation tests of a rotor supported in ball-bearings

J TONNESEN, MSME and **J W LUND**, PhD
Department of Machine Elements, Technical University of Denmark, Lyngby, Denmark

1 INTRODUCTION

An experimental program has been carried out on impact excitation of a rotor supported in ball bearings. The rotor, while running, is excited at several locations by an impact hammer, and the response is measured at the bearings by sensors for displacement, force and acceleration. The impact method has been used in previous investigations [1,2,3], but primarily for the purpose of determining the dynamic characteristics of the bearings. An FFT analyzer is employed to obtain the frequency dependent transfer functions, and it is shown how they may be used to derive the influence coefficients for balancing the rotor. Subsequent tests have demonstrated the practical validity of the method even though the balancing is based on non-synchronous coefficients. The balancing is performed using the influence coefficient method [4,5].

The results give information about the natural frequencies, the associated modeshapes and the modal whirl direction that may serve diagnostic purposes. Some problems connected with the interference from the rotational frequency are discussed briefly.

2 DESCRIPTION OF TEST RIG

The test rotor is shown schematically in Fig.1. The rotor consists of a shaft with a basic diameter of 35 mm, an overall length of 1.085 m and a bearing span of 0.785 m. The rotor configuration is almost symmetric with most of its mass concentrated in three discs mounted on the shaft between the bearings, further details may be found in [6].

The rotor system mass is 65.3 kg and the reactions at the bearings are 314 N at bearing no.1 and 327 N at bearing no.2. The maximum speed of the test rig is 200 Hz and the critical speeds of the rotor system have been measured to be:

1st critical horizontally:	36.0 Hz
1st critical vertically:	38.7 Hz
2nd critical horizontally:	120.0 Hz
2nd critical vertically:	133.5 Hz

2.1 Instrumentation

The rotor is excited by an impact generator consisting of a spring loaded piston in a pressurized airbearing and a release mechanism. The piston impacts the rotor through a piezo-electric force transducer and a relatively soft and curved cap, the selected cap material being a compromise between the required frequency response span and the ability to withstand the sliding speeds of the rotor without being worn flat too quickly. The impact generator is calibrated by impacting a reference mass to which a calibrated accelerometer is attached. In addition the rotor may be excited by a handheld impact hammer.

The response of the rotor system is measured at and near the bearing pedestals. At the bearing pedestals are installed piezoelectric transducers measuring the transmitted force and acceleration. The rotor displacement is measured by eddy current type non-contacting displacement transducers placed at impact planes PL1 and PL3. All the transducers sense the response in the horizontal direction. An FFT analyser is employed to obtain the frequency dependent transfer functions.

The rotational speed of the shaft is measured by a photoelectric tachometer probe, which is activated by markings painted on the shaft. The same markings provide also the reference phase information.

3 ANALYSIS

A harmonic signal, a, with frequency ω can be expressed as:

$$a = a_c\cos\omega t - a_s\sin\omega t = \mathrm{Re}\{(a_c + ia_s)e^{i\omega t}\} \quad (1)$$

By adopting the conventional complex notation, eq. (1) may be written formally as:

$$a = a_c + ia_s \quad (2)$$

with exp($i\omega t$) and the real part operator implied.

The rotor is assumed to behave linearly such that if it is excited by harmonic forces f_x and f_y in orthogonal directions, the resulting response becomes:

$$a = H_{ax}f_x + H_{ay}f_y \quad (3)$$

where the H's are transfer functions. They are complex and depend on frequency.

In the rotor there is a fixed reference coordinate system with a 1-axis and a 2-axis, located by means of the markings painted on the rotor. Thus, any mass unbalance can be specified by its two components U_1 and U_2 and the corresponding mass unbalance forces at the rotational frequency Ω become:

$$f_x = \Omega^2(U_1\cos\Omega t - U_2\sin\Omega t) \sim \Omega^2 U$$
$$f_y = \Omega^2(U_1\sin\Omega t + U_2\cos\Omega t) \sim -i\Omega^2 U \quad (4)$$

where:

$$U = U_1 + iU_2 \quad (5)$$

Upon substitution into eq.(3) it is found that:

$$a = H_{au}U \quad (6)$$

where:

$$H_{au} = \Omega^2(H_{ax} - iH_{ay}) \quad (7)$$

H_{au} is the transfer function (influence coefficient) to be used in balancing the rotor.

The transfer functions H_{ax} and H_{ay} are determined by impacting the rotor at the balancing planes and by performing a Fourier transform analysis of the response. They depend on speed because of the influence from gyroscopic moments and changing bearing characteristics. Hence, in eq.(7) H_{ax} and H_{ay} should be properly evaluated at the desired balancing speed and at synchronous frequency (ω=Ω). At that frequency, however, the response has a strong component, a_0, caused by the existing unbalance in the rotor, and that component must be subtracted out of the measured response, a, in order to get the signal due solely to the impact excitation:

$$a - a_0 = H_{ax}f_x \qquad (f_y=0) \quad (8)$$

such that H_{ax} should be computed as:

$$H_{ax} = (H_{ax})_{meas} - (H_{ax})_0 \quad (9)$$

Subsequent tests demonstrated the workability of the idea, but is was not found possible to supply enough impact energy to the rotor to get the difference between the two measured H_{ax}-values in eq.(9) significantly above the measurement tolerances. The scheme was therefore abandoned and, although in principle not correct, H_{ax} and H_{ay} are instead measured at non-synchronous frequencies as discussed later.

In conventional form the transfer function can be expressed as:

$$H_{ax} = \sum_n \frac{x_n(a)x_n(f)}{N_n(\omega_n^2-\omega^2+i2\omega\omega_n\beta_n)} \quad (10)$$

where ω_n is the n'th natural frequency, β_n is the modal damping factor, N_n is the associated modal mass, and $x_n(a)$ and $x_n(f)$ are the modal amplitudes at the measurement location and at the force location, respectively. The equation assumes a system where the damping is either small or else distributed, and where the gyroscopic effects are not important (also, the stiffness matrix must be symmetric which excludes rotors supported in journal bearings).

At a resonant frequency (ω=ω_n) with small damping, the summation is dominated by a single term such that H_{ax} becomes proportional to the modeshape. Examples are shown in Figs. 3 and 4 where the shaft has been impacted at nine places along the rotor while the response is measured by the force transducers at the bearings. Only the horizontal modeshapes are obtained because the transducers are in the horizontal plane.

If the response at any location is measured both in the x and the y-direction (the y-axis comes after the x-axis is the direction of rotation), then the response may be represented by a forward whirl and a backward whirl component:

$$a_f = \tfrac{1}{2}(x+iy) \qquad a_b = \tfrac{1}{2}(x-iy) \quad (11)$$

where x and y in analogy with eq.(2) can be written as:

$$x = x_c + ix_s \qquad y = y_c + iy_s \quad (12)$$

From this it is found that:

$$|a_f|^2 - |a_b|^2 = x_s y_c - x_c y_s \quad (13)$$

If X and Y are the Fourier transforms of x and y, then the cross spectral density is computed as:

$$S_{xy} = \lim\{\tfrac{2}{T}\bar{X}Y\} \quad (14)$$

where $\bar{X}$ is the complex conjugate of X, and T is the time length of the recorded signal.

At any given frequency, the imaginary part of the cross spectrum is:

$$\mathrm{Im}\{S_{xy}\} \sim -(x_s y_c - x_c y_s) \quad (15)$$

By comparing with eq.(13) it is seen that the response is in forward whirl when the imaginary part of S_{xy} is negative.

With amplitude measurements taken at plane no.3 (Fig.1), the imaginary part of the cross spectrum is shown in Fig.5. The first and second critical speeds in the horizontal plane are found to be backward whirl modes while the two critical speeds in the vertical plane are forward whirl modes.

This technique is primarily of usefulness in identifying modes and resonances for diagnostic purposes in rotor systems more complex than the present simple test rotor.

4 EXPERIMENTAL PROCEDURE

One of the purposes of the test program is to evaluate the accuracy of the calculated influence coefficients found from the impact excitation, so the procedure adopted for these tests consists of impact exciting the rotor at the same positions where the balance correction weights are installed. A comparison between the calculated influence coefficients and those found from the conventional method of placing trial weights gives a direct check on the accuracy of the method.

TABLE 1

TRANSDUCER TYPE: ACCELEROMETER, PIEZO-ELECTRIC																	
		BEARING #1								BEARING #2							
		IMPACT INFLUENCE COEFFICIENTS						TRIAL WEIGHT INFLUENCE COEFFICIENTS		IMPACT INFLUENCE COEFFICIENTS						TRIAL WEIGHT INFLUENCE COEFFICIENTS	
ROTOR SPEED WHEN IMPACTING; Hz.		0		13.5		87				0		13.5		87			
RESPONSE FREQ.Hz.	IMPACT OR T.W. POSITION	AMPL. μV/N	PHASE DEG.	AMPL. μV/N	PHASE DEG.	AMPL. μV/N	PHASE DEG.	AMPL. μV/N	PHASE DEG.	AMPL. μV/N	PHASE DEG.	AMPL. μV/N	PHASE DEG.	AMPL. μV/N	PHASE DEG.	AMPL. μV/N	PHASE DEG.
111.5	PL1	163.	175.2	170.	-175.3	150.	-178.9	143.	178.6	184.	2.4	167.	2.2	161.	2.5	156.	-1.1
	PL2	143.	178.7	148.	183.5	119.	182.4	117.	179.4	211.	2.1	213.	3.5	188.	-0.6	181.	-0.8
	PL3	161.	1.9	136.	6.4	129.	-0.9	129.	-0.5	189.	176.2	181.	177.2	160.	179.1	151.	179.9
	PL4	145.	-2.7	123.	8.3	125.	-2.0	120.	0	233.	180.4	214.	-174.2	217.	-179.4	202.	179.2
115.	PL1	227.	183.3	230.	181.6	215.	180.3	214.	178.8	280.	-1.7	273.	0.7	241.	0.3	247.	-2.0
	PL2	206.	181.6	230.	178.9	184.	-177.0	186.	179.6	327.	-1.5	313.	4.9	267.	-0.8	278.	-1.2
	PL3	234.	-2.3	210.	2.6	196.	-1.7	195.	-0.3	302.	-179.5	282.	177.6	243.	178.6	241.	178.9
	PL4	203.	0.2	216.	6.4	184.	-5.5	188.	-0.5	353.	180.6	303.	-176.9	312	-177.7	304.	178.9
117.	PL1	326.	179.0	325	181.2	284.	179.8	320.	176.8	390.	-3.7	379.	7.9	336.	-5.5	394.	-5.3
	PL2	309.	177.9	319.	180.6	288.	179.9	291.	177.7	451.	-3.3	434.	0.6	415.	-1.3	425.	-3.1
	PL3	316.	-3.5	292.	1.1	271.	-4.1	287.	-3.3	427.	177.3	381.	-179.0	353.	170.3	380.	176.8
	PL4	306.	-0.4	259.	4.1	283.	-2.3	291.	-3.5	463.	177.1	417.	-177.8	448.	-178.6	455.	175.2

The impacting or balance weight planes are numbered PL1 to PL4 as shown in Fig.1. The measurements for impacting are taken with the rotor at standstill and running at 13.5 Hz and 87 Hz, these speeds being well below an associated critical speed.

The response measurements are taken at frequencies as close as possible to the critical speed frequencies and several points are chosen, so if the rotor can not be operated to the highest selected response frequency, the next lower is used and so on. For the first critical speed are selected 30.5, 32.5 and 34.5 Hz and for the second critical speed are selected 111.5, 115 and 117 Hz. Fig.2 shows some typical response curves from which the data are obtained at the selected frequencies. Whereas the measurements with the force and displacement transducers are taken at all the six response frequencies, the signals from the accelerometers are too weak at the three lowest response frequencies to be of any use.

The measurements for the trial weight runs are taken at the six response frequencies just mentioned. However, prior to the actual test program, the rotor is balanced as well as possible. This is done only to ensure a sufficiently low level of vibration to allow installing a sufficiently large trial weight and still being able to pass the first critical speed.

The rotor is now balanced, using first the influence coefficients from the trial weights and afterwards for comparison, the calculated impact influence coefficients, and in both cases is used the rotor's original unbalance condition. The rotor is first balanced at 32.5 Hz using balancing planes nos. 2 and 3, and final balancing is done at 32.5 Hz and 115 Hz using balancing planes nos. 1 and 4 and the calculated influence coefficients from both the cases where the rotor is run at 13.5 Hz and 87 Hz.

In order to find the modeshapes of the rotor at its natural frequencies the rotor is excited at standstill with the handheld impact hammer. Some typical results are shown in Figs. 3 and 4 where the modeshapes are presented as measured by the force transducers at the bearing housings.

4.1 Experimental results

The results for the calculated influence coefficients are summarized in tables nos. 1 to 3, where they are compared directly to the influence coefficients obtained from the conventional trial weight method. In the calculations the influence coefficients are evaluated on the basis of two test series, where in the first series the impact excitation direction is in the horizontal and vertical planes, and in the second series where the direction is displaced 45 degrees. It is the average of the two test series that are shown in the tables. The average deviation between the two series is approximately 6 per cent on the amplitudes and 3 degrees on the phase.

5 DISCUSSION OF RESULTS

It is seen from tables 1 to 3 that the impact influence coefficients generally speaking are speed dependent as far as the amplitudes are concerned, the trend being a lowering as the speed of the rotor increases. This effect is most likely due to two not easily separable causes: the rolling element bearings' variable stiffness and damping, and the influence of gyroscopic moments. However, if a comparison is made between those influence coefficients where the rotor speed is closest to the response frequency it is seen that the impact influence coefficients amplitudes are still greater than those determined from the trial weight method, the average deviation being about 15 per cent at the lower frequencies and approximately 5 per cent at the higher frequencies. Based on

TABLE 2

TRANSDUCER TYPE: FORCE, PIEZO-ELECTRIC

		BEARING #1								BEARING #2							
		IMPACT INFLUENCE COEFFICIENTS						TRIAL WEIGHT INFLUENCE COEFFICIENTS		IMPACT INFLUENCE COEFFICIENTS						TRIAL WEIGHT INFLUENCE COEFFICIENTS	
ROTOR SPEED WHEN IMPACTING; Hz.		0		13.5		87				0		13.5		87			
RESPONSE FREQ.Hz.	IMPACT OR T.W. POSITION	AMPL. μV/N	PHASE DEG.	AMPL. μV/N	PHASE DEG.	AMPL. μV/N	PHASE DEG.	AMPL. μV/N	PHASE DEG.	AMPL. μV/N	PHASE DEG.	AMPL. μV/N	PHASE DEG.	AMPL. μV/N	PHASE DEG.	AMPL. μV/N	PHASE DEG:
30.5	PL1	384.	-2.7	306.	-3.7			247.	-0.8	419.	2.6	306.	-3.1			258.	7.2
	PL2	396.	-0.3	324.	3.3			280.	-0.4	656.	5.1	529.	1.9			449.	6.6
	PL3	278.	-2.5	204.	-4.4			196.	-2.9	924.	3.8	672.	2.1			693.	3.0
	PL4	186.	-9.9	127.	-12.2			107.	1.8	839.	5.5	587.	4.1			595.	4.1
32.5	PL1	474.	0.6	407.	6.8			359.	5.1	635.	4.7	560.	18.5			506.	9.5
	PL2	532.	3.0	465.	11.6			459.	6.5	988.	6.8	904.	12.2			839.	9.9
	PL3	405.	4.6	322.	17.8			367.	8.7	1372.	4.7	1090.	17.2			1093.	7.9
	PL4	274.	0.8	262.	28.9			231.	10.5	1241.	0.5	897.	16.4			875.	7.1
34.5	PL1	720.	-0.3	586.	12.3			609.	2.9	1248.	0.9	1038.	13.0			1051.	5.2
	PL2	967.	1.1	805.	11.3			879.	5.1	1959.	3.7	1676.	15.1			1768.	6.5
	PL3	870.	4.4	677.	9.3			777.	5.0	2285.	2.5	1817.	10.8			2035.	5.4
	PL4	553.	0.2	411.	10.9			498.	5.7	1767.	-3.0	1319.	13.0			1500.	5.4
111.5	PL1	400.	-4.6	402.	3.1	365.	1.5	341.	-0.9	745.	179.5	701.	-176.5	663.	-179.4	632.	179.8
	PL2	342.	-3.3	331.	0.3	292.	-0.2	276.	-0.4	841.	-179.0	858.	-176.9	758.	-177.2	739.	180.5
	PL3	362.	179.9	337.	-173.5	317.	-178.5	301.	179.8	783.	-0.3	706.	4.2	669.	3.1	626.	0.8
	PL4	327.	179.9	319.	-167.9	320.	-173.3	278.	179.8	978.	1.6	973.	5.1	867.	-1.7	846.	0.5
115.	PL1	522.	0.1	512.	3.6	489.	-0.6	471.	-1.0	1055.	-178.9	1059.	-174.9	908.	-179.4	934.	179.5
	PL2	486.	-2.7	504.	0.7	443.	1.7	414.	-0.5	1227.	180.0	1165.	-176.1	1105.	-177.9	1063.	180.2
	PL3	509.	177.4	484.	-175.9	433.	-176.5	429.	179.8	1182.	0.3	980.	8.0	976.	2.0	938.	0.4
	PL4	475.	-178.5	470.	-168.6	454.	-178.2	410.	179.5	1375.	0.3	1218.	7.7	1188.	1.9	1179.	0.2
117.	PL1	685.	-3.5	645.	5.7	637.	-3.1	671.	-3.0	1426.	178.3	1384.	-174.0	1198.	176.3	1420.	176.4
	PL2	660.	-1.1	625.	1.2	570.	0.5	617.	-1.9	1650.	178.7	1486.	-175.6	1438.	-178.3	1553.	178.5
	PL3	662.	178.7	668.	-174.2	595.	178.3	609.	177.5	1544.	-1.9	1560.	3.6	1274.	1.2	1383.	-2.2
	PL4	649.	178.9	628.	-173.4	616.	178.8	609.	176.5	1743.	-1.4	1641.	10.8	1731.	-3.9	1681.	-2.9

TABLE 3

TRANSDUCER TYPE: DISPLACEMENT, EDDY CURRENT

		BEARING #1								BEARING #2							
		IMPACT INFLUENCE COEFFICIENTS						TRIAL WEIGHT INFLUENCE COEFFICIENTS		IMPACT INFLUENCE COEFFICIENTS						TRIAL WEIGHT INFLUENCE COEFFICIENTS	
ROTOR SPEED WHEN IMPACTING; Hz.		0		13.5		87				0		13.5		87			
RESPONSE FREQ.Hz.	IMPACT OR T.W. POSITION	AMPL. μV/N	PHASE DEG.	AMPL. μV/N	PHASE DEG.	AMPL. μV/N	PHASE DEG.	AMPL. μV/N	PHASE DEG.	AMPL. μV/N	PHASE DEG.	AMPL. μV/N	PHASE DEG.	AMPL. μV/N	PHASE DEG.	AMPL. μV/N	PHASE DEG.
30.5	PL1	2839.	2.3	2318.	7.8			2095.	5.0	3874.	2.1	3299.	1.6			2732.	8.2
	PL2	4242.	2.5	3545.	10.5			3303.	8.3	6113.	2.2	5372.	6.4			4583.	7.2
	PL3	3955.	3.0	3052.	14.0			2961.	6.5	6541.	1.5	5281.	14.6			5089.	4.8
	PL4	2540.	1.9	1795.	14.5			1860.	11.2	4363.	2.0	2843.	8.3			3184.	8.3
32.5	PL1	4000.	4.0	3932.	14.1			3512.	7.0	5899.	3.6	5204.	13.5			5006.	6.3
	PL2	6029.	5.1	5482.	14.5			5451.	7.0	9068.	4.7	8473.	12.6			8020.	7.4
	PL3	5770.	4.4	5794.	19.0			5189.	7.4	9480.	4.3	8023.	16.5			8701.	7.3
	PL4	3845.	3.8	3179.	22.9			3355.	9.4	6316.	3.3	5038.	14.6			5714.	8.2
34.5	PL1	7345.	2.1	6141.	14.3			6350.	4.0	11020.	0.8	9292.	14.4			9513.	8.8
	PL2	11513.	3.7	9733.	15.4			10606.	5.1	17350.	3.6	14226.	14.7			16304.	5.4
	PL3	10839.	3.2	8923.	12.2			10210.	4.5	17702.	3.5	14539.	10.1			16560.	4.0
	PL4	7091.	1.4	5536.	13.4			6675.	5.3	11613.	2.0	9070.	14.8			10955.	5.9
111.5	PL1	927.	0.2	814.	8.7	790.	1.3	775.	-0.2	1147.	179.5	1002.	-173.0	1016.	-176.1	993.	179.7
	PL2	910.	1.8	791.	14.6	727.	3.3	715.	-0.6	1371.	179.6	1448.	-175.9	1168.	-175.5	1176.	-179.7
	PL3	1164.	179.9	1119.	-173.5	1016.	-178.2	978.	180.8	948.	-0.9	779.	14.4	695.	2.6	677.	0.1
	PL4	1041.	-179.8	986.	-177.4	897.	-178.2	897.	180.1	938.	-0.5	840.	9.5	755.	2.7	750.	0.3
115.	PL1	1296.	0.2	1295.	5.3	1182.	0.7	1147.	0.6	1512.	-178.3	1598.	-173.8	1429.	-179.3	1373.	179.4
	PL2	1369.	0.8	1296.	1.9	1111.	8.7	1088.	-0.5	1809.	-179.7	1835.	-178.2	1593.	-176.2	1555.	180.2
	PL3	1599.	179.6	1391.	-177.3	1348.	-179.7	1310.	180.2	1370.	-1.1	1323.	1.4	1129.	6.1	1074.	-0.1
	PL4	1447.	-179.0	1310.	-170.0	1247.	-177.6	1258.	179.7	1344.	1.0	1317.	6.6	1207.	3.5	1158.	-0.1
117.	PL1	1746.	-1.2	1760.	2.6	1696.	1.5	1695.	-3.6	2005.	178.1	1967.	-169.4	1842.	-176.6	1982.	177.0
	PL2	1843.	-1.5	1800.	3.1	1604.	4.6	1666.	-1.8	2335.	179.0	2222.	-177.6	2315.	-177.9	2164.	178.8
	PL3	2043.	178.3	1830.	-175.0	1880.	177.2	1846.	178.7	1872.	-1.8	1683.	7.5	1732.	0.0	1658.	1.4
	PL4	1843.	178.9	1699.	-171.0	1825.	179.9	1832.	177.1	1844.	-1.8	1738.	8.2	1617.	5.0	1804.	-2.8

this it may be expected that balancing of the rotor close to the first critical speed will not be as good as close to the second critical speed.

Using an FFT-analyser resolution of 125 mHz it was found that the phase deviation was small and in most cases within the measurement tolerances which is typically 1 degree. The associated coherence function was typically between 1 and 0.95. However, two other factors should theoretically have caused a phase delay. The first one is the error introduced by the friction between the impact head and the rotating shaft [6]. The phase delay would depend on the friction angle. The second factor is the "smearing-out" of the impact force. The impulse is typically 0.8 msec long and at the rotational speeds, with periods of 74.1 msec and 11.5 msec, the impact head touches the rotor surface over arcs of 4 and 25 degrees, respectively. The test results were inconclusive as to establishing a consistent phase discrepancy.

From Fig.2 it is seen that the transducer sensitivity varies greatly from the displacement probe to the accelerometer in particular near the first critical speed and also near the second critical speed. The sensitivity of the force probe is between the two other probes and is not as frequency dependant as the displacement and acceleration probes. Thus it appears advantageous to use force probes whenever it is possible but the present tests confirmed also that the widely used probes for displacement and acceleration will produce acceptable results. If displacement probes are used some care must be exercised in selecting the running speed of the rotor such that the higher harmonics of the rotor frequency do not coincide with the selected response frequencies. The higher harmonic components originate from the shaft's out-of-roundness and as can be seen from Fig.2 these components can actually be larger than the response one tries to measure. Based on this, it should now be evident why a rotor speed of 13.5 Hz was chosen such that the response frequencies at 111.5 Hz to 117 Hz were well away from, in this case, the 8th and 9th harmonics, marked 8f and 9f on Fig.2.

From Figs. 3 and 4 it is seen that the balancing planes nos. 1-4 are all suitable for correcting unbalance caused by the first and second modes and all 4 planes are used. However, in the experiments it was observed, that when balancing for the second mode, the balancing planes nos. 2 and 3 would not give satisfactory results, whereas planes nos. 1 and 4 always produced good results. Assuming that the third modeshape at 228 Hz has some influence already at the second mode, it can be seen from inspection of Figs. 3 and 4 that planes nos. 2 and 3 are very close to a nodal point for the third mode, thus causing the poor response. This theory was further substantiated by including the balancing planes of the center disc of the rotor, because the midplanes have a good response to the third modeshape. The authors are aware of similar cases and it is thus suggested that if measurements of the modeshapes are available that these are used for diagnostic purposes.

6 CONCLUSION

The investigation has demonstrated that the proposed method of calculating the influence coefficients for balancing purposes by impacting a rotor at a speed below a critical speed is of some practical value and should be considered as a supplementary method to the convential methods: trial weight and modal balance methods.

As shown, a wide range of measurement transducers can be employed and the measured values agree sufficiently well with those directly obtained influence coefficients.

The principal shortcomings of the method are that access is required to impact the rotor where the balancing planes are positioned, although interpolation techniques may be tried, and that the rotor has a fine machined surface at the points of impact. However, because of the established close agreement in the present experiments, the approach merits confidence and could be developed into a practical method of balancing.

Finally, the mode shapes may also be used for diagnostic purposes if a flexible rotor appears difficult to balance.

REFERENCES

1) MORTON, P.G., "The Derivation of Bearing Characteristics by Means of Transient Excitation Applied Directly to a Rotating Shaft", Dynamics of Rotors, IUTAM Symposium 1974, Lyngby, Denmark, Springer Verlag 1975, pp.350-379
2) GLIENICKE, J., "Experimental Investigation of the Stiffness and Damping Coefficients of Turbine Bearings", Proced. of the Institution of Mechanical Engineers, Vol.181, Part 3B, 1966-67. pp.116-129.
3) NORDMANN, R., and SCHOLHORN, K., "Identification of Stiffness and Damping Coefficients of Journal Bearings by Means of the Impact Methods", Proced. Second International Conf. on Vibrations in Rotating Machinery, Cambridge, England, The Institution of Mechanical Engineers 1980, pp.223-230.
4) LUND, J.W. and TONNESEN, J., "Analysis and Experiments on Multiplane Balancing of a Flexible Rotor", Journal of Engineering for Industry, Trans. ASME, Series B, Vol.94, No.1, 1972, pp.148-158.
5) GOODMAN, T.P., "A Least-Squares Method for Computing Balance Corrections", Journal of Engineering for Industry, Trans. ASME, Series B, Vol.86, No.3, Aug. 1964, pp.273-279.
6) TONNESEN, J. and LUND, J.W., "Impact Excitation Tests to Determine the Influence Coefficients for Balancing Lightly Damped Rotor", Contribution at ASME INT. GASTURBINE and Aero-engine Congress, Amsterdam, Holland, June 1988.

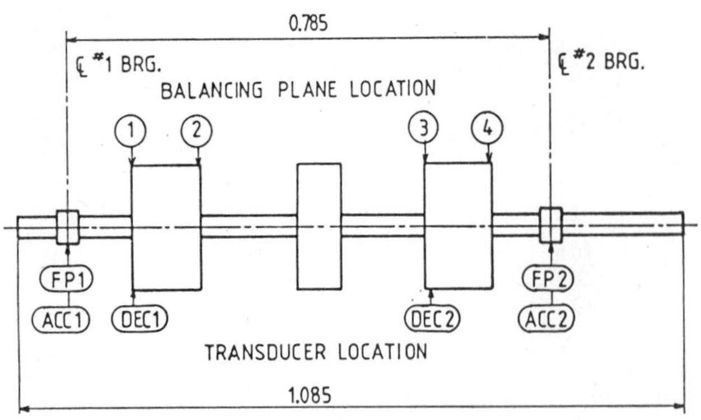

Fig 1 Schematic of test rotor

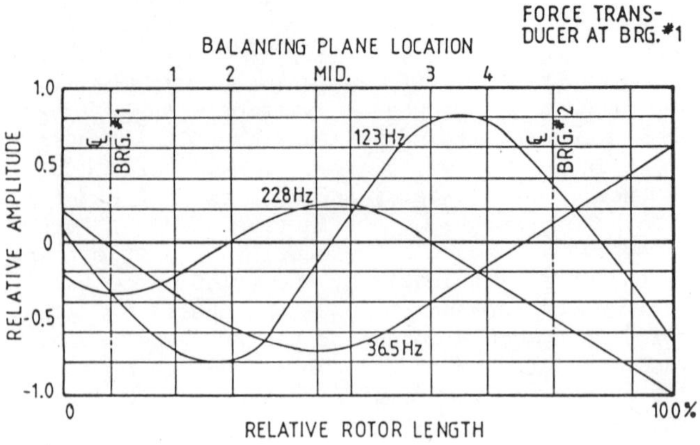

Fig 3 Horizontal modeshapes

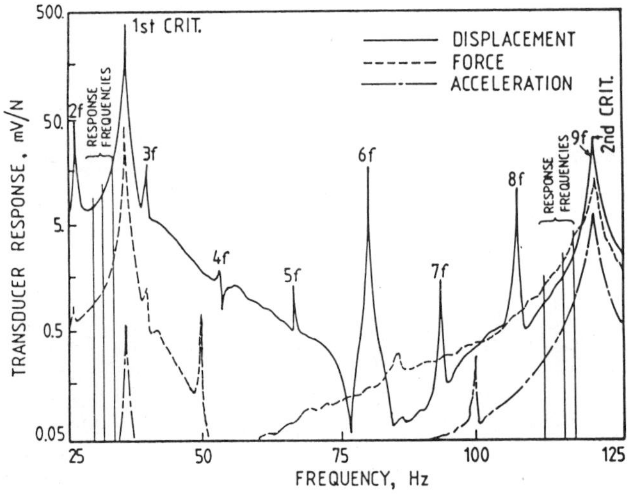

Fig 2 Measured transfer functions

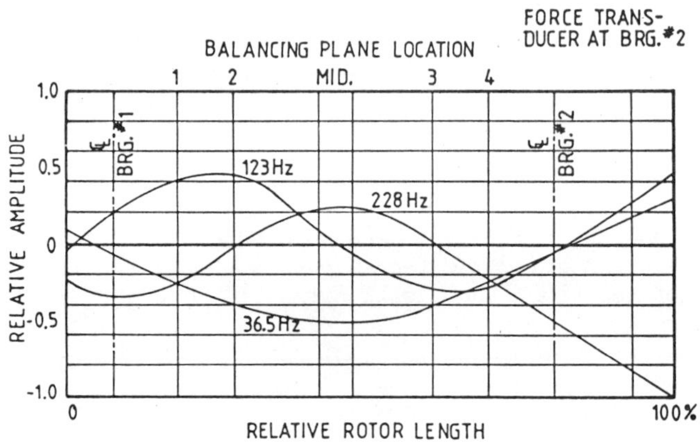

Fig 4 Horizontal modeshapes

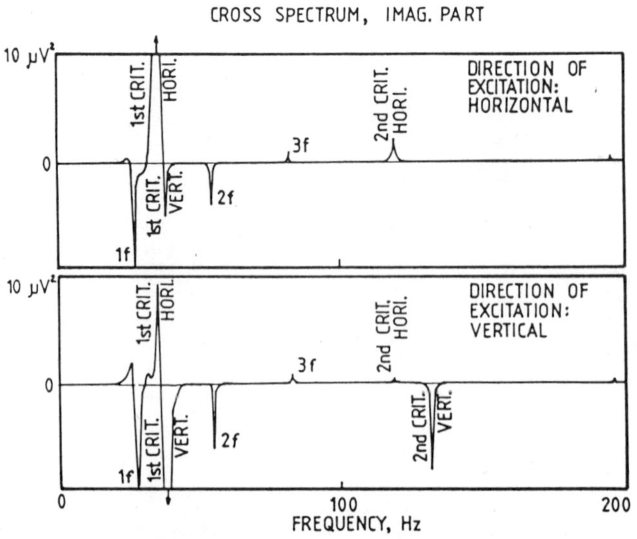

Fig 5 Identification of whirl directions

C251/88

A procedure for identifying non-linearity in rigid rotors supported in hydrodynamic and ball/roller bearing systems

L GARIBALDI, Dr-Ing
Department of Mechanical Engineering, Politecnico di Torino, Torino, Italy
G R TOMLINSON, MSc, PhD
Department of Mechanical Engineering, Heriot Watt University, Edinburgh

SYNOPSIS A procedure for identifying the non-linearity in a rigid rotor supported in either hydrodynamic or ball/roller bearings is shown. The method used is based upon a numerical Hilbert transform which relates the real and the imaginary part of an analytical signal. The distortions introduced by the HT in the frequency response characteristics are then related with non-linear behaviour of the support considered.

NOTATION

M	mass
C	viscous damping coefficient
K	stiffness
β	non-linear damping coefficient
G	non-linear stiffness coefficient
ω	rotational speed
ε	eccentricity
$\lvert\bar{Z}\rvert$	amplitude modulus
φ	phase angle

1 INTRODUCTION

The basis of this work is the application of the Hilbert transform in the rotor dynamic field. The Hilbert transform has been successfully applied in the domain of modal testing for detecting and identifying structural non-linearity [1]. Basically, the transform detects the presence of non-causality which is introduced into the frequency response function as a result of the non-linearity. The procedure requires the amplitude and phase or the real and imaginary parts of the frequency response functions. The Hilbert transform provides the prediction of the real part from the imaginary part (or vice versa) and any distortions between the measured data and the Hilbert transform are used as indicator of non-linearity [1].

As it is the first application in this sense the simple Jeffcott-rotor is used as the mathematical model in simulations to study non-linearity associated with hydrodynamic and ball/roller bearing systems. Thus we consider a two degrees of freedom rigid-rotor modelled as a point mass in the XY plane of the bearing with gyroscopic effects being ignored. Application of a numerical Hilbert transform to the frequency response characteristics shows that non-linearity associated with hydrodynamic and ball/roller bearing systems can be readily detected.

2 THE EQUATION OF MOTION IN POLAR COORDINATES

The most general equation for a non-linear rotor, using the Z notation, may be written as [4,8]

$$M\ddot{\bar{Z}} + (C_n + C_r)\dot{\bar{Z}} + K\bar{Z} - i\omega C_r\bar{Z} + [\beta_n(|\dot{\bar{Z}}|) + \beta_r(|\dot{\bar{Z}}|)]\dot{\bar{Z}} + G(|\bar{Z}|)\bar{Z} - i\omega\beta_r(|\dot{\bar{Z}}|)\bar{Z} = Mg + M\varepsilon\omega^2 e^{i\omega t} \quad (1)$$

where the subscripts n and r indicate the non-rotating or the rotating components.
Considering the unbalance response for a system in which the non-linear effects are a function of the displacement only, and neglecting the Mg term we obtain,

$$[1 - \omega^2\frac{M}{K} + \frac{G(|\bar{Z}_o|)}{K} + i\omega\frac{C_n}{K}]\bar{Z}_o = \frac{M}{K}\varepsilon\omega^2 \quad (2)$$

In equation (2) the rotating damping is considered to offer no contribution and the $G(|\bar{Z}_o|)$ function is assumed to be of the form [5],

$$G(|\bar{Z}_o|) = \mu K^*|\bar{Z}_o|^2 \quad (3)$$

which represents a cubic non-linear stiffness.

Because of the fact that the orbit is circular the solution $\bar{Z} = \bar{Z}_o e^{i\omega t}$ is the exact solution and not merely a first approximation as is the case in the analysis of systems governed by the classical Duffin equation subject to harmonic excitation. The equations for the amplitude $\bar{Z}_o$ and the phase φ as a function of ω are,

$$|\bar{Z}_o|^2\{[1 - \omega^2\frac{M}{K} + \frac{\mu K^*|\bar{Z}_o|^2}{K}]^2 + (\frac{\omega C_n}{K})^2\} = (\frac{M\varepsilon}{K})^2\omega^4 \quad (4)$$

$$tg\varphi = \{\omega C_n\,[K - \omega^2 M + \mu K^*|\bar{Z}_o|^2]^{-1}\} \quad (5)$$

Assuming $\mu^* = \mu K^*/K$,

$$\mu^{*2}|\bar{Z}_o|^6 + 2\mu^*|\bar{Z}_o|^4[(1 - \omega^2\frac{M}{K})] +$$

$$|\bar{Z}_o|^2[(\frac{\omega C_n}{K})^2 + (1 - \omega^2\frac{M}{K})^2] - (\frac{M\varepsilon}{K})^2\omega^4 = 0 \quad (6)$$

$$\varphi = tg^{-1}\{\omega C_n/[K - \omega^2 M + \mu^*|\bar{Z}_o|^2]\} \quad (7)$$

Solving equation (6) for $|\bar{Z}_o|^2$ it is possible to obtain the frequency response function of the system for different values of the dimensionless exciting eccentricity parameter $(\mu\varepsilon)^{1/2}$ and the non rotating damping C_n.

Once the frequency response functions are obtained it is possible to apply the Hilbert transform procedure which indicates differences between the data and the transform if non-linearity is significant. Figures 1 and 2 are examples of what happens when equations (6) and (7) are used to simulate the frequency response function of a system with cubic stiffness effects. The figures show the Nyquist plots of the response amplitude $|\bar{Z}_o|$ for the same value of Cn. Figure 1 represents a low level of excitation (i.e. small unbalance) and Figure 2 a level of excitation which is a factor of three greater than that used to generate Figure 1. In Figure 1 the Hilbert transform overlays the data whereas in Figure 2 there is a considerable difference between the transform and the data, this distortion being found to typically represent a hardening cubic stiffness function [10].

3 THE ROTOR WITH BALL/ROLLER BEARINGS

Utilizing the same approach as before, two models have been realized using a different function for $G(|\bar{Z}_o|)$ in equation (3) to simulate a ball and a roller bearing as reported in references [5] & [6],

$$G(|\bar{Z}_o|) = K^*\sqrt{(|\bar{Z}_o|)} \quad \text{for ball bearings and,} \quad (8)$$

$$G(|\bar{Z}_o|) = K^*(|\bar{Z}_o|)^{1/9} \quad \text{for roller bearings} \quad (9)$$

The modulus and phase of the response amplitude $|\bar{Z}_o|$ was computed numerically as a function of the rotational speed ω. From these the real and the imaginary parts were used to compute the Hilbert transforms. Figures 3a, b show the effect of increasing the excitation level (eccentricity level) for the case of the ball bearings. As before we see that the increased eccentricity results in a distortion between the data and the transform. However in the case of the roller bearings shown in Figures 4a, b the non-linearity is not strongly evident, even at the higher excitation level. This shows that roller bearings are less prone to non-linear behaviour (for the same excitation level) than ball bearings.

4 THE ROTOR WITH HYDRODYNAMIC BEARINGS

The general equation of motion with different coefficients in the X, Y directions and cross coupling terms typical of oil film bearings can be written in the form:

$$M\ddot{X} + C_{xx}\dot{X} + C_{xy}\dot{Y} + K_{xx}X + K_{xy}Y = M\varepsilon\omega^2\cos(\omega t) \quad (10)$$

$$M\ddot{Y} + C_{yy}\dot{Y} + C_{yx}\dot{X} + K_{yy}Y + K_{yx}X = M\varepsilon\omega^2\sin(\omega t) \quad (11)$$

where the gravitational effects are again neglected.

Equations (10) and (11) can be expressed as,

$$D^4M^2 + D^3M(C_{xx} + C_{yy}) + D^2(MK_{xx}MK_{yy} + C_{xx}C_{yy} - C_{xy}C_{yx}) + D(C_{xx}K_{yy} + C_{yy}K_{xx} - C_{xy}K_{yx} - C_{yx}K_{xy}) + (K_{xx}K_{yy} - K_{xy}K_{yx})Xe^{i\omega t} =$$

$$M\varepsilon\omega^2[(M + \frac{C_{xy}}{\omega})D^2 + (C_{yy} + \frac{K_{xy}}{\omega})D + K_{yy}]e^{i\omega t} \quad (12)$$

where D = d/dt. The frequency response function corresponding to equation (12) in Bode and Nyquist form are shown in Figures 5 and 6, together with their Hilbert transforms. The presence of two peaks arises from the different stiffnesses with respect to the X,Y directions, and indicates the non null value of cross-coupling terms. Figure 7 shows the Nyquist plot response of the same system with two negative cross-coupling terms. In all these cases the HT indicates causal, linear behaviour.

5 EFFECTS OF DIFFERENT DAMPING ASSUMPTIONS

It is usual to assume that the damping follows a low of the form,

$$C = C_o(a_o + a_1\omega + a_2\omega^2 + \ldots) \quad (13)$$

$$\text{i.e.} \quad C = \Sigma\, C_o a_n \omega^n \quad (14)$$

if equation (14) is used in equation (12) to represent the damping terms, the response is still a linear function of amplitude and the Hilbert transform will overlay the original data.

However, if we assume the damping law to be a quadratic form, arising from the variation in the eccentricity the equations of motion can be written as,

$$\begin{vmatrix} M & 0 \\ 0 & M \end{vmatrix} \begin{Bmatrix} \ddot{X} \\ \ddot{Y} \end{Bmatrix} + \begin{vmatrix} K_{xx} & K_{xy} \\ K_{yx} & K_{yy} \end{vmatrix} \begin{Bmatrix} X \\ Y \end{Bmatrix} + \begin{vmatrix} C_{xx} & C_{xy} \\ C_{yx} & C_{yy} \end{vmatrix} \begin{Bmatrix} \dot{X}|\dot{X}| \\ \dot{Y}|\dot{Y}| \end{Bmatrix} = M\varepsilon\omega^2 \begin{Bmatrix} \cos(\omega t) \\ \sin(\omega t) \end{Bmatrix} \quad (15)$$

To solve this equation a Runge-Kutta fourth order integration method was used to predict the time response at defined rotational speeds.
At each speed the ratio between the response and the excitation and the phase between these two signal was computed. Initially the system parameters were chosen to provide a linear response in order to have a comparison with analitical data and to achieve the required precision for the time integration steps to provide minimal computer time consumption.

Once this was achieved, the square law damping system was simulated using different stiffnesses in the X, Y directions and assuming a negative value for the damping cross terms.

The data obtained from this simulation are shown in Figures 8 and 9 using the Nyquist plot format. As expected, Figure 8 demonstrates that the system is linear. When the non-linear damping is introduced the Hilbert transform shows a considerable distortion from the data, thus detecting non-linearity.

6 CONCLUSIONS

Preliminary work using a numerical Hilbert transform which relates the reai and the imaginary parts of causal systems has shown that non-linearity , arising from stiffness and damping of a rigid rotor supported in ball/roller and hydrodynamic bearings, can be detected from the frequency response function. The results presented have been based on numerical simulations and analytical calculations. However, it is considered that if the response amplitude and phase with respect to the unbalance is obtained from a run-up or a coast-down signature then it should be possible to detect non-linearity associated with the bearing system. It is recognised that non-linearity may also arise from the bearing pedestals/supporting structure and this will also be detected by the Hilbert transform, causing difficulty in identifying the exact cause of the non-linearity. Therefore, further work is required to ascertain the usefulness of such a procedure for rotating systems before any definite conclusions can be drawn as to its usefulness as a diagnostic tool.

REFERENCES

[1] TOMLINSON G.R. and AHMED I. Hilbert transform procedures for detecting and quantifying non-linearity in modal testing. MECCANICA 1987.

[2] SIMON M and TOMLINSON G.R. Applications of the Hilbert transform in the modal analysis of linear and non-linear systems. Jnl. of Sound and Vib. 1984, 90 (2) pp.275-282.

[3] HAUOI A. Transformees de Hilbert et applica tiones aux systemes non-lineaires. These de Docteur d'Ingenieur, ISMCM St. Ouen Paris 1983.

[4] SAITO S. Calculation of non-linear inbalan ce response of horizontal Jeffcott rotor supported by ball bearigs with radial clea rance. J. of Acoustic, Stress & Reilabylity in Des., Oct 1985, Vol 107, pp. 416-420.

[5] PALMGREN A. Les roulements, II Ed, SKF 1967.

[6] GARGIULO E.P. A simple way to Estimate bearing stiffnesses. Machine design, july 1980, pp. 107-110.

[7] WHILE M.F. Rolling Element Bearing Vibration Transfer Characteristics: Effects of stiffnesses. J. of Appl. Mech., Sept 1979, Vol 46, pp 677-684.

[8] GENTA G., REPACI A. Circular whirling and unbalance response of non-linear rotors". XI Biennal Conf. on Mechanical Vibration and Noise, Boston, sept.1987.

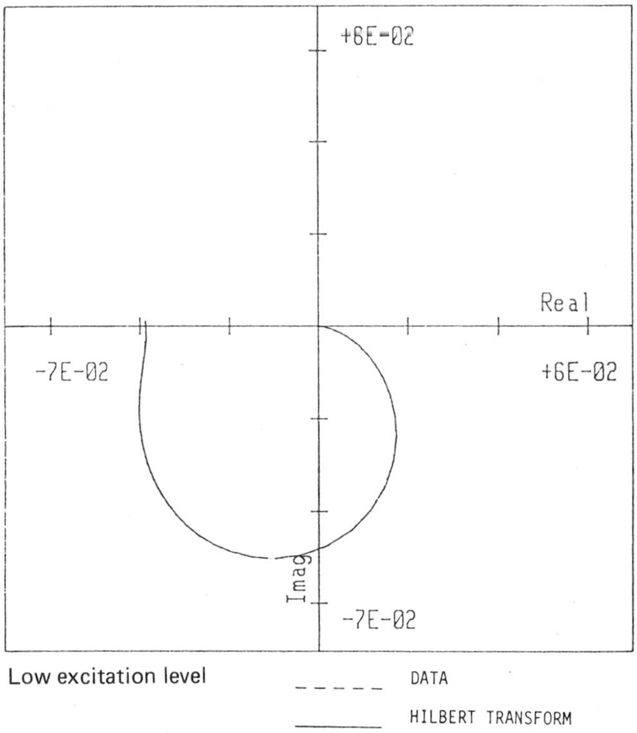

Fig 1 Nyquist plot of a simple rotor in bearings with non-linear stiffness characteristics as defined by equations (6) and (7). Since the Hilbert transform overlays the frequency response data, a linear response is indicated

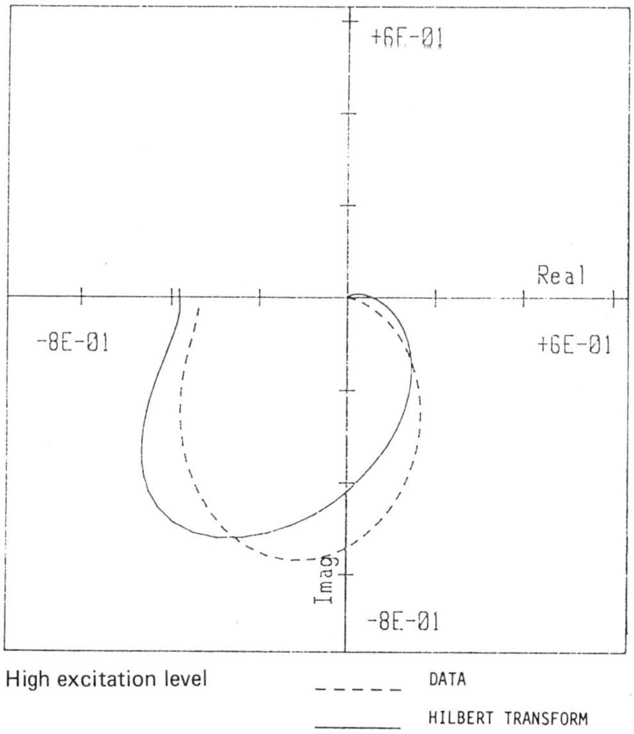

Fig 2 Effect of increasing the excitation level (that is increasing imbalance) on the response shown in Fig 1; the distortion between the Hilbert transform and the frequency response data indicates a non-linear system

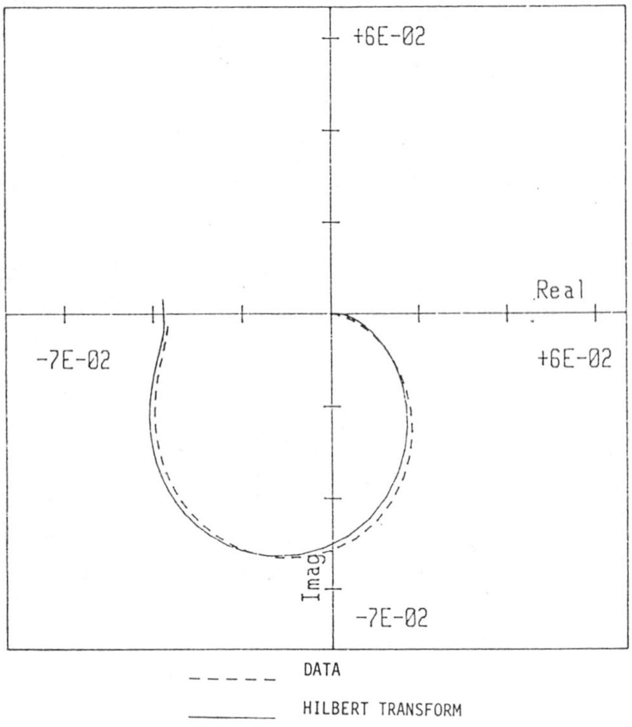

Fig 3a Nyquist plot of a rigid rotor in ball-bearings subject to a low excitation level; the small difference between the data and the Hilbert transform indicate the presence of non-linearity

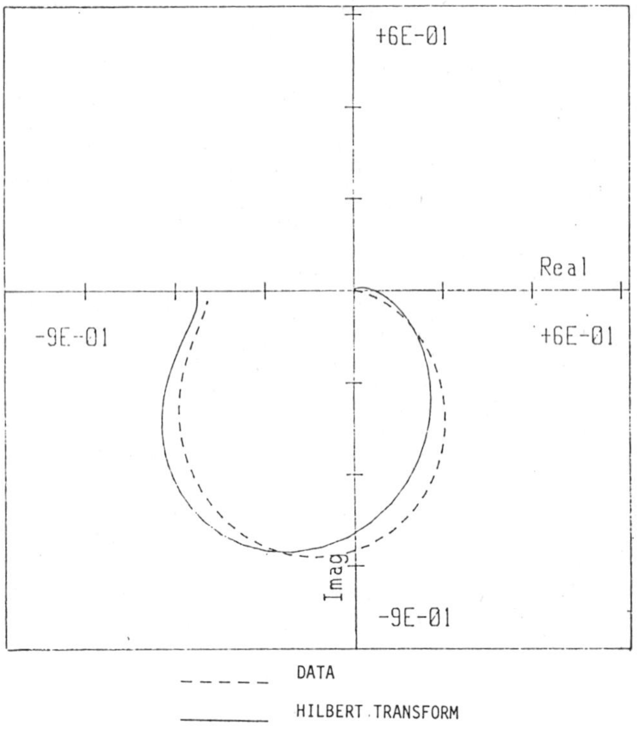

Fig 3b Increasing the excitation level (eccentricity) for the rotor in ball-bearings indicates an increasing non-linear characteristic

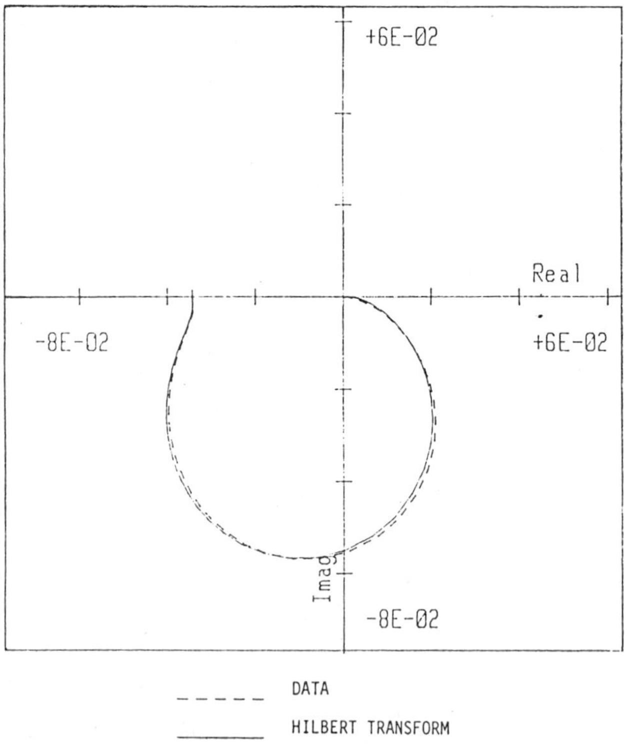

Fig 4a Effect of replacing the ball-bearings with roller-bearings in the system used to generate the response of Fig 3a; the non-linearity effects are significantly reduced

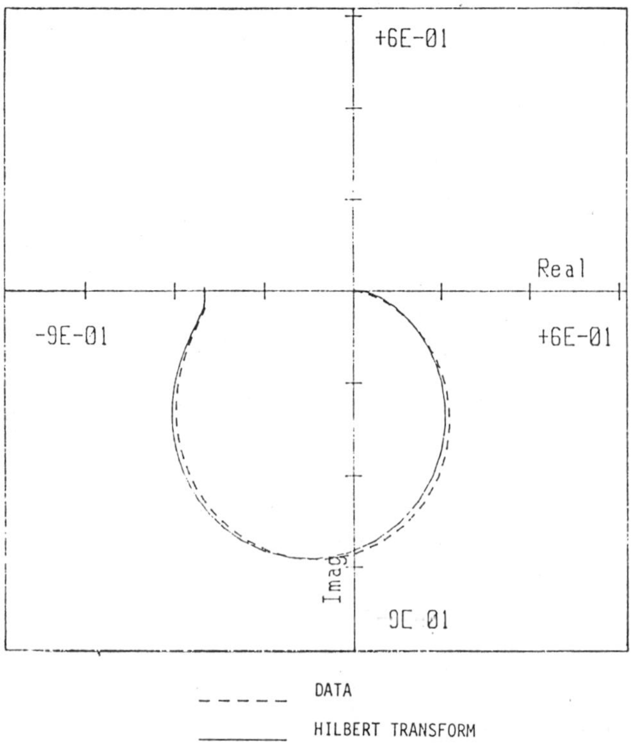

Fig 4b Increasing the excitation level to that used in Fig 3b; the small deviation between the Hilbert transform and the data is not as severe as that shown in Fig 3b

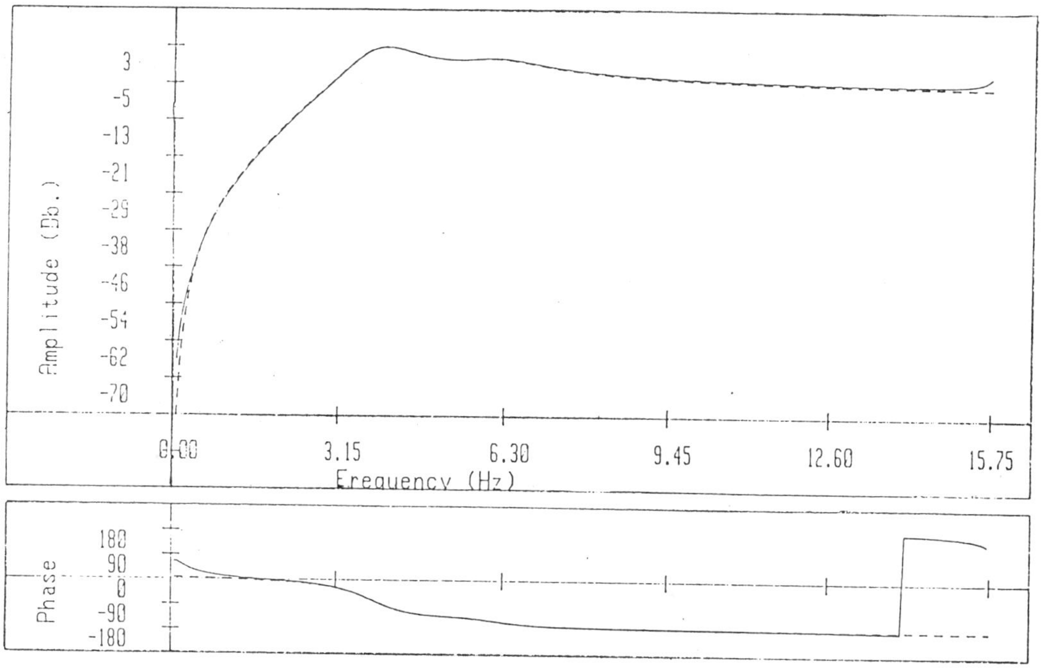

DATA - - - - -

HILBERT TRANSFORM ———

Fig 5 Rigid rotor in hydrodynamic bearings; bode plot of the frequency response from equation (12) and its Hilbert transform; system parameters are:
$K_{xx} = 16$, $K_{yy} = 36$; $C_{xx} = C_{yy} = 1$, $\epsilon = 0.5$
$K_{xy} = K_{yx} = 0$; $C_{xy} = 1$, $C_{yx} = -1$

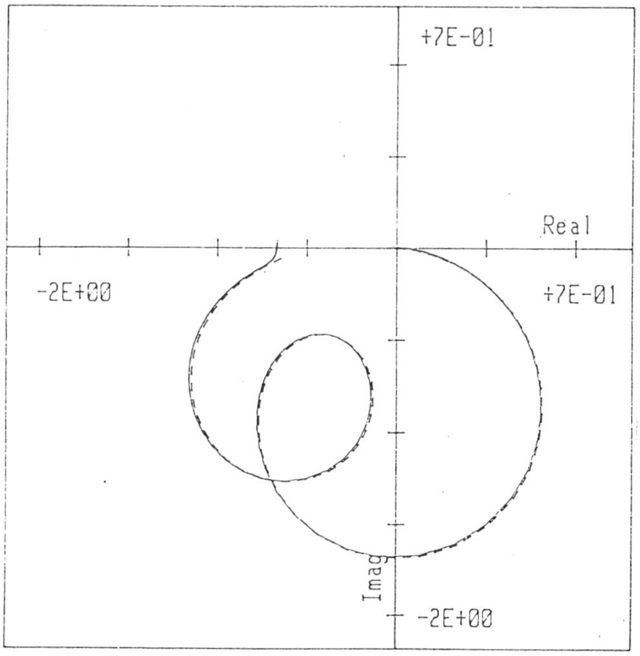

DATA - - - - -

HILBERT TRANSFORM ———

Fig 6 Nyquist plot of data shown in Fig 5; the similarity between the Hilbert transform and the data indicates linearity

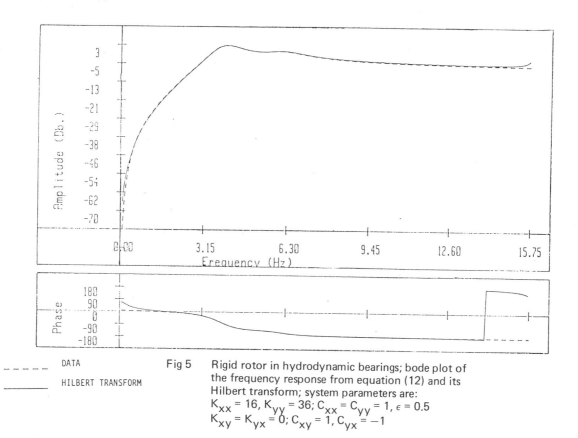

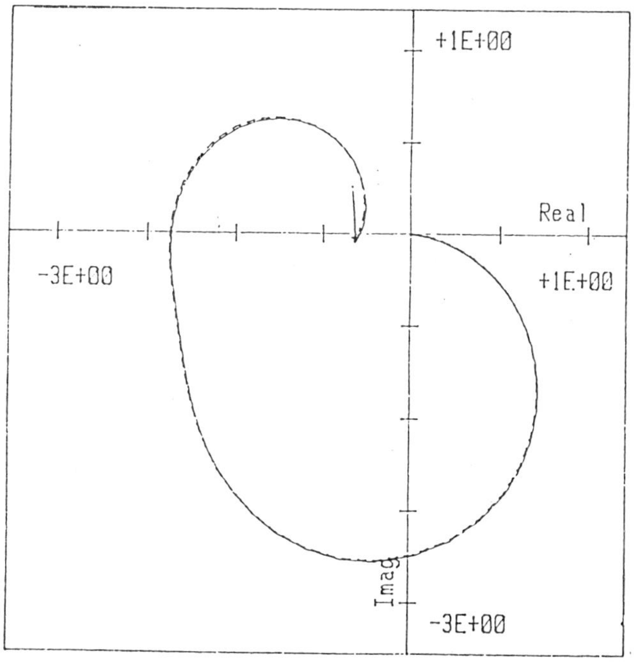

DATA - - - - -

HILBERT TRANSFORM ———

Fig 7 Effect of making the cross-coupling damping terms both negative. Although the frequency response characteristics have changed, the system is still linear as indicated by the overlay of the data and its Hilbert transform

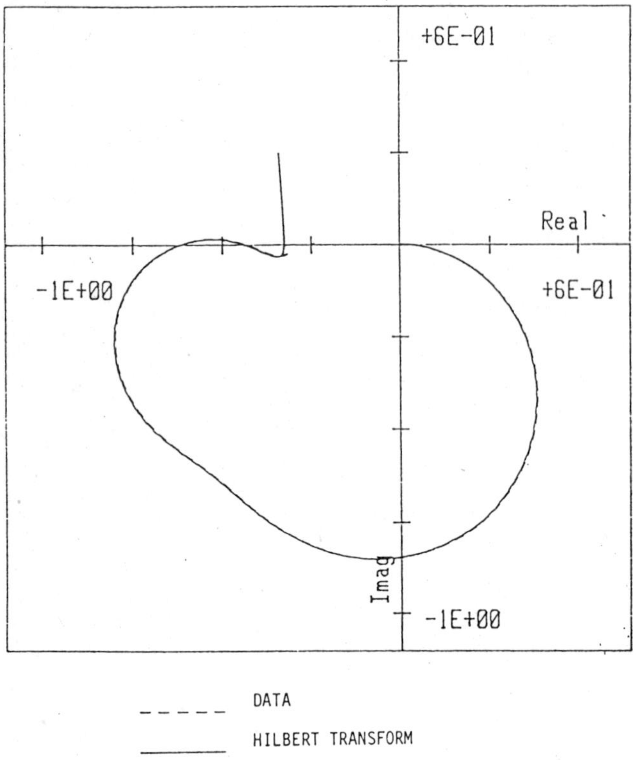

Fig 8 Nyquist plot of the dynamic response and Hilbert transform of a rigid rotor supported in hydrodynamic bearings governed by equation (15) with system parameters:
$M = 1$, $K_{xx} = 16$, $K_{yy} = 36$, $K_{xy} = K_{yx} = 4$
$C_{xx} = C_{yy} = 1.5$, $C_{xy} = C_{yx} = -0.5$

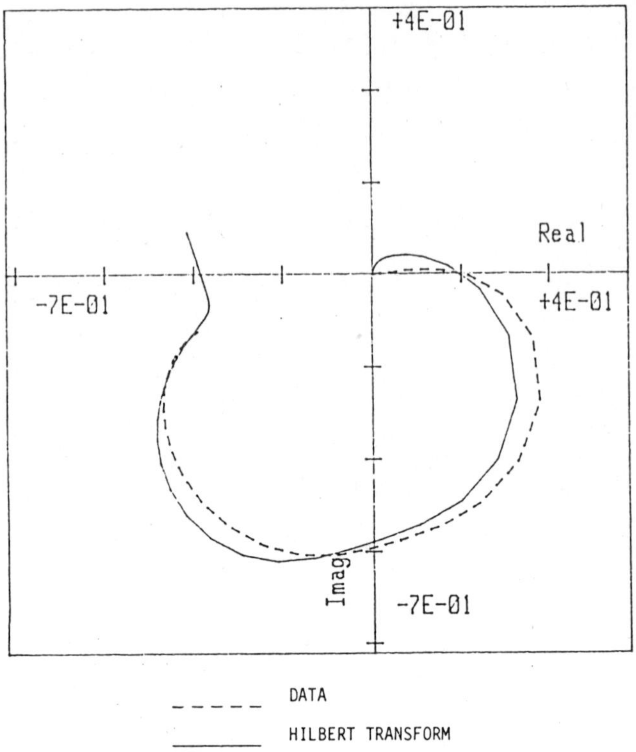

Fig 9 Effect of introducing square law damping into equation (15); non-linear effects are evident as the Hilbert transformed data shows a considerable deviation from the original frequency response data

C311/88

Improvement to prediction accuracy of stability limits and resonance amplitudes using instability-threshold-based journal bearing rotordynamic coefficients

M RASHIDI, BS, MS, PhD
Department of Mechanical Engineering, Cleveland State University, Cleveland, Ohio, United States of America
M L ADAMS, BS, MS, PhD, MASME, MASLE
Department of Mechanical and Aerospace Engineering, Case Western Reserve University, Cleveland, Ohio, United States of America

SYNOPSIS

Computer simulations, with postulated measurement errors, are presented on a controlled instability threshold based experimental approach to determine bearing rotordynamic coefficients. An imposed requirement that the retrieved coefficients exactly predict the experimentally observed test-rig instability threshold speed and self-excited vibration frequency is shown to translate into drastically improved threshold speed prediction accuracy when these coefficients are applied to multi-bearing flexible rotors. It is further shown that prediction accuracy of critical-speed vibration amplitudes, which are also quite damping sensitive, can be improved as well by using journal bearing rotordynamic coefficients obtained from such a controlled instability threshold experiment.

NOMENCLATURE

C_{ij} = bearing damping coefficients

c = bearing radial clearance

D = bearing bore diameter

E_{cyc} = cyclic vibration energy transferred from bearing film to system

F_i = ith component of bearing force

K_{ij} = bearing stiffness coefficients

L = bearing length

M = bearing mass

N = journal speed, rps

P = nonconservative portion of bearing force

R = D/2

S = Sommerfeld number = $(\mu NDL/W)(R/c)^2$

t = time

W = bearing static load

x = motion coordinate

$\bar{x}$ = single-peak amplitude of x-motion

X = dynamic radial position vector of journal relative to bearing referenced to static equilibrium

y = motion coordinate

$\bar{y}$ = single-peak amplitude of y-motion

α = real part of eigenvalue

ϕ = static attitude angle

λ = complex eigenvalue

μ = lubricant viscosity

θ = phase angle

Ω = imaginary part of eigenvalue

ω = journal speed

l = half of the shaft's length

K_s = dimensionless shaft stiffness = $\frac{EI}{l^3} \cdot \frac{c}{W}$

E = modulus of elasticity of shaft material

I = area moment of inertia

1. INTRODUCTION

The rotordynamic properties of fluid film journal bearings are commonly used in linearized lateral vibration analysis of rotor system dynamics. Determination of these fluid film properties, namely the stiffness and damping coefficients, has been pursued by researchers and machinery designers, both experimentally and analytically, for decades. Among the earliest works in this area are those of Hagg and Sankey [1] and Sternlicht [2]. Recently such work has been extended to characterize seals [3-5], and interactive fluid dynamical effects in complete turbomachinery stages [6-7]. Some experimental work has gone far beyond typical laboratory size scales, such as the impressive work reported by Morton [8] on full-size turbogenerator journal bearings.

Disagreement between the observed dynamical behavior of typical rotating machinery and the behavior predicted through theoretical analysis has motivated researchers and designers to construct elaborate dynamic-load test rigs that simulate application environments of journal bearings. In such experimental facilities, all dynamic characteristics of the system, except the bearing rotordynamic coefficients, are very accurately known a priori. The conventional approach to experimentally determine the rotordynamic coefficients of a journal bearing or seal is usually a linear transfer function (mechanical impedance) method. Typically, the system is excited by a controlled force input and the output motion is measured. Or, a controlled motion is prescribed as the input, and the resulting output force is measured. In either case, experimental data are used to invert the equations of motion of the system to evaluate the sought dynamic properties, namely, the K_{ij} and C_{ij} two-by-two stiffness and damping matrices respectively. The coefficients obtained from a transfer function or mechanical impedance approach (i.e., a forced vibration) are inherently quite prone to produce large errors when they are applied to predict an instability threshold (i.e., a self-excited vibration). Parameter locations of instability thresholds are inherently quite sensitive to damping errors in the model, since an instability threshold is a phenomenon in which the positive and negative damping factors of the system exactly cancel each other for a specific mode. Conversely, the predictions of critical speeds and off-resonance unbalance response are not so highly sensitive to reasonable damping errors in the mathematical model. The propensity for error in determining the individual stiffness and damping coefficients is obvious from the fact that these coefficients are actually derivatives of the measured forces, as shown by their definitions, as follows.

$$K_{ij} \equiv -\frac{\partial F_i}{\partial x_j} \quad , \quad C_{ij} \equiv -\frac{\partial F_i}{\partial \dot{x}_j} \qquad (1)$$

The significance of inaccuracies in journal bearing rotordynamic coefficients must be evaluated within the context of actual machinery and dimensional tolerances. Iwatsubo [9] concluded that the real part of system eigenvalues is strongly affected by errors, yielding potentially large errors in predicted instability threshold speeds and resonant peak amplitudes. This is consistent with many years of practical experience of rotating machinery builders. This is also consistent with vibration analysis in general, namely, those system characteristics which are sensitive to damping values are difficult to predict accurately because damping per se is difficult to predict accurately without appropriate experimental data.

2. INSTABILITY-THRESHOLD EXPERIMENTAL APPROACH

To circumvent this error sensitivity, an alternative approach has been proposed by Adams and Rashidi [10, 11], which in turn has its roots in the work presented by Adams and Padovan [12] and its supplement [13]. This approach makes use of the concept of cyclic energy imparted to the journal bearing system by the hydrodynamic fluid film, expressed in the following equation.

$$E_{cyc} = \oint P \cdot dX = -\pi \Big\{ \Omega [C^s_{xx} \bar{x}^2 + 2C^s_{xy} \bar{x}\bar{y} \cos(\theta_x - \theta_y) + C^s_{yy} \bar{y}^2] - 2K^{ss}_{xy} \bar{x}\bar{y} \sin(\theta_x - \theta_y) \Big\} \qquad (2)$$

Here, the superscripts "s" and "ss" stand for the "symmetric" and "skew symmetric" parts of the damping and stiffness matrices of a two-degree-of-freedom model containing a journal bearing as shown in Fig. 1.

As described in references [12] and [14], decomposition of bearing stiffness and damping matrices into their symmetric and skew-symmetric parts amounts to a separation of conservative and nonconservative bearing forces. Equation (2) expresses the cyclic energy imparted to the system due to the total nonconservative force components. At an instability threshold speed, the positive damping effect (C^s terms) and the negative damping effect (K^{ss} term for forward whirls) exactly cancel, yielding $E_{cyc}=0$.

A conceptual sketch for a test apparatus employing the instability-based approach is shown in Fig. 2 and described more fully in reference [10]. The equations of motion for this two-degree-of-freedom system can be expressed as follows.

$$\begin{aligned} M\ddot{x} + C_{xx}\dot{x} + K_{xx}x + C_{xy}\dot{y} + K_{xy}y = 0 \\ M\ddot{y} + C_{yy}\dot{y} + K_{yy}y + C_{yx}\dot{x} + K_{yx}x = 0 \end{aligned} \qquad (3)$$

Normalizing the coordinates, time and the coefficients in the following standard definitions,

$$x \equiv x/c \ , \ y \equiv y/c \ , \ t \equiv t\omega$$

$$M \equiv \frac{M\omega^2 c}{W} \quad , \quad C_{ij} \equiv \left(\frac{C_{ij}\omega c}{W}\right) \qquad (4)$$

$$\text{and } K_{ij} \equiv \left(\frac{K_{ij}c}{W}\right)$$

equations (3) become dimensionless and their eigenvalues become normalized by the rotational speed, ω. The dimensionless K_{ij} and C_{ij} coefficients are functions of Sommerfeld number (dimensionless speed). A controlled threshold of instability can be produced at any Sommerfeld number above the value below which the bearing is unconditionally stable. The harmonic portion of the measured x and y time base signals of the system at the threshold of instability can be captured by filtering out the growth portions of the total signals. The displacement signals utilized should be taken from the linear portion of the instability growth, i.e., before the orbital motion approaches the size of the bearing clearance and becomes highly non-linear.

At a threshold of instability there are four pieces of information provided by the two-time-based signals as follows.

$\alpha_f = 0$

Ω_f/ω_{TH} Subscript "f" = forward mode

$(\bar{x}/\bar{y})_f$

$\Delta\theta_f = \theta_x - \theta_y$

And of course these are the same parameter values provided from solution of the Eigen problem of equations (3) when the "correct" bearing rotordynamic coefficients are used.

To extract all eight stiffness and damping coefficients directly by inverting the Eigen problem would require knowing the complex eigenvalues and eigenvectors for both modes of the experimental setup (i.e., forward mode and backward mode). However, only the forward mode is potentially self-sustaining, and thus more readily detectable and measurable. The solution for the eight coefficients is not unique if based solely on the self-excited-mode eigenvalue and eigenvector information. For that reason, the number of unknowns is reduced from eight to four. To achieve this, the four bearing stiffness coefficients, K_{ij}, are measured through static loading, which can surely be made with relatively high accuracy. The following well known transformations yield the bearing stiffness in terms of the applied static load, W, the attitude angle, ϕ, and the eccentricity, e, shown in Fig. 1.

$$K_{xx} = \frac{dW}{de}\cos\phi \quad , \quad K_{xy} = \frac{dW}{de}\sin\phi$$

$$K_{yx} = -\frac{W}{e}\sin\phi - W\frac{d\phi}{de}\cos\phi \ ,$$

$$K_{yy} = \frac{W}{e}\cos\phi - W\frac{d\phi}{de}\sin\phi \qquad (5)$$

3. IMPLEMENTATION ON A TWO-BEARING FLEXIBLE ROTOR

Journal bearing rotordynamic coefficients were obtained using a computer simulated instability-threshold experiment with postulated displacement and phase-angle measurement errors. These bearing coefficients were then used to predict the instability threshold speed of a two-bearing flexible rotor (shown in Fig. 3), and the results compared to both the "exact" solution and to a solution employing bearing coefficients extracted from a simulated transfer function experiment having the same postulated measurement errors. The results are presented in a nondimensional form covering a wide range of shaft flexibility.

Changing the shaft flexibility, the critical Sommerfeld number (dimensionless instability threshold speed), S_{cr} may be determined from the solution of the eigenvalue problem. Fig. 4 shows the S_{cr} vs. the dimensionless shaft flexibility, $1/K_s$, where K_s is the dimensionless shaft stiffness (see nomenclature). S_{cr} is the Sommerfeld number at which the eigenvalue corresponding to the forward-whirl mode of the motion has a zero real part (i.e., threshold of instability). The graph of S_{sr} vs. $1/K_s$ may be developed based upon bearing dynamic coefficients extracted from measured motion parameters ($\Delta\theta = \theta_x - \theta_y$ and $\bar{x}/\bar{y}$), each containing plus 5% error. Here, the coefficients are determined first by using the proposed instability-based approach and then by the standard mechanical impedance approach, also with 5% error postulated in $\Delta\theta$ and $\bar{x}/\bar{y}$.

As shown in Fig. 4, the graph of S_{cr} vs. $1/K_s$ obtained from the new approach (with measured-parameter error) follows the exact-solution graph quite closely (within 1% error). Conversely, the S_{cr} vs. $1/K_s$ obtained from the results of the mechanical impedance approach (with the same measured-parameter error) does not follow the exact curve closely (i.e., up to 24% error). Therefore, the dimensionless instability threshold speed (S_{cr}) is predicted much more accurately using coefficients from the instability-based approach. Fig. 5 shows the percentage error in prediction of S_{cr} as a function of shaft flexibility, for both approaches.

4. CRITICAL-SPEED UNBALANCE-VIBRATION AMPLITUDE OF A FLEXIBLE ROTOR

The same simple flexible rotor system studied in the previous section is used here for steady-state unbalance response analysis. This is intended to study the resonance-peak amplitudes of the steady-state unbalance response at the first critical speed. The rationale is simply that the first lateral critical speed mode may be quite similar to the zero-damped mode at the instability threshold speed. Thus, the considerably enhanced accuracy of this mode's damping factors derived from the instability-based bearing coefficients, could be reasonably expected to yield a corresponding improvement for first-critical-speed vibration amplitude prediction. Dimensionless results are generated using both sets of bearing coefficients used in the previous section, and both sets of results are compared to the "exact" solution.

The equations of motion of this system are the same as those given by equations (3) except the null right hand side vector is replaced by an unbalance rotating force which is normalized by the static load (half of the weight of the system). The results were compared to both the "exact" solution and to a solution based upon a transfer function simulated experiment. The dimensionless displacement of the disc, X_2, for the three cases (exact, transfer function, and instability-based approach) are presented in Fig. 6 as functions of the shaft flexibility, $1/K_s$. Fig. 7 shows the percentage error in prediction of critical-speed unbalance vibration amplitude for each method. In this simple demonstration example, a considerable improvement is achieved in the prediction of first-critical-speed vibration amplitude. This is shown to translate into quite significant improvements in prediction accuracy of both instability threshold and critical speed analyses of multi-bearing flexible rotors.

5. SUMMARY AND CONCLUSIONS

Computer simulations are made to study the extraction of journal bearing rotordynamic coefficients from test-rig measurements. Two distinctly different experimental approaches are simulated and compared. One is the standard mechanical impedance approach and the other is a new approach based on motion measurements at a controlled instability threshold speed. In this new approach, the four bearing stiffness coefficients are presumed to be measured using static loading. Then, combining these with measured orbital motion parameters at an adjustable threshold speed, the bearing damping coefficients are extracted by inverting the associated eigenvalue problem with the constraint that the combined stiffness and damping coefficients exactly reproduce the experimentally observed instability threshold speed and self-excited orbital vibration frequency of the test rig.

A set of known errors were postulated in the test measurements of both approaches, to simulate the extraction of journal bearing rotordynamic coefficients with errors. The coefficients so obtained for both approaches (i.e., standard mechanical impedance and instability threshold based) were used to predict the instability threshold of a two-bearing flexible rotor. The results show that the new instability-based experimental approach has the potential to drastically improve instability threshold prediction accuracy. It is further shown that this new approach also has the potential to significantly improve prediction accuracy of critical-speed resonant-peak vibration amplitude.

REFERENCES:

(1) Hagg, A.C. and Sankey, G.O., "Some Dynamic Properties of Oil-Film Journal Bearings with Reference to the Unbalance Vibration of Rotors", ASME Journal of Applied Mechanics, Vol. 78, 1956, pp. 302-306.

(2) Sternlicht, B., "Elastic and Damping Properties of Cylindrical Journal Bearings", ASME Journal of Basic Engineering, Vol. 81, 1959, pp. 101-108.

(3) Childs, D.W., "Finite Length Solutions for the Rotordynamic Coefficients of Constant-Clearance and Convergent-Tapered Annular Seals", Proc. Third International Conference on Vibration in Rotating Machinery, I. Mech E. Confer. Publication 1984-10, Sept. 1984, pp. 223-231.

(4) Wright, D.V., "Labyrinth Seal Forces on a Whirling Rotor", ASME Applied Mechanics Division, Proc. of Symposium on Rotor Dynamical Instability, ASME Book AMD--Vol. 55, June 1983, pp. 19-31.

(5) Barrett, L.E., "Turbulent Flow Annular Pump Seals: A Literature Review", The Shock and Vibration Digest, Vol. 16, No. 2, Feb. 1984, pp.3-13.

(6) Flack, R.D., and Allaire, P.E., "Lateral Forces on Pump Impellers: A Literature Review", Shock and Vibration Digest, Vol. 16, No. 1, Jan. 1984, pp. 5-14.

(7) Chamieh, D.S., Acosta, A.J., Brennen, C.E., Caughey, T.K., and Franz R., "Experimental Measurement of Hydrodynamic Stiffness Matrices for a Centrifugal Pump Impeller", Workshop on Rotordynamic Instability Problems in High-Performance Turbomachinery, Texas A & M University NASA CP 2250, May 1982, pp. 382-398.

(8) Morton, P.G., "Dynamic Characteristics of Bearings-Measurement Under Operating Conditions", GEC Journal of Science and Technology, Vol. 42, 1975, pp. 37-47.

(9) Iwatsubo, T., "Influence of Errors on the Vibration of Rotor/Bearing Systems", Bulletin JSME, Vol. 24 (187), 1981, pp. 208-214.

(10) Adams, M.L., Rashidi, M., "On the Use of Rotor-Bearing Instability Thresholds to Accurately Measure Bearing Rotordynamic Properties", Journal of Vibration, Acoustics, Stress, and Reliability in Design, Trans. ASME, Vol. 107, Oct. 1985, pp. 404-409.

(11) Rashidi, M., "Accurate Measurement of Journal-Bearing Rotordynamic Proeperties Using Instability Threshold", Ph.D. Dissertation, Case Western Reserve Univ., Cleveland, Ohio, June 1987.

(12) Adams, M.L., and Padovan, J., "Insights into Linearized Rotor Dynamics", Journal of Sound and Vibration, Vol. 76, No. 1, 1981, pp. 129-142.

(13) Adams, M.L., "A Note on Rotor-Bearing Stability", Journal of Sound and Vibration, Vol. 86, 1983, pp. 435-438.

(14) Adams, M.L., "Insights Into Linearized Rotor Dynamics, Part 2", Journal of Sound and Vibration, Vol. 112, No. 1, 1987, pp. 97-110.

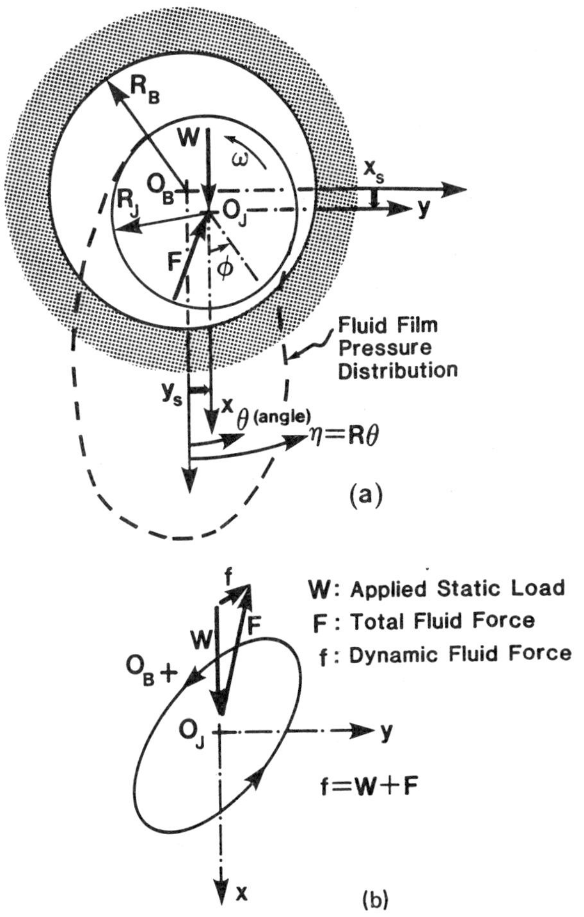

Fig 1 (a) Radial motion coordinates of journal relative to bearing, referenced to static equilibrium state (W = static load)
(b) Harmonic orbital motion of journal relative to bearing;
$x = \bar{x} \sin(\Omega t + \theta_x)$,
$y = \bar{y} \sin(\Omega t + \theta_y)$

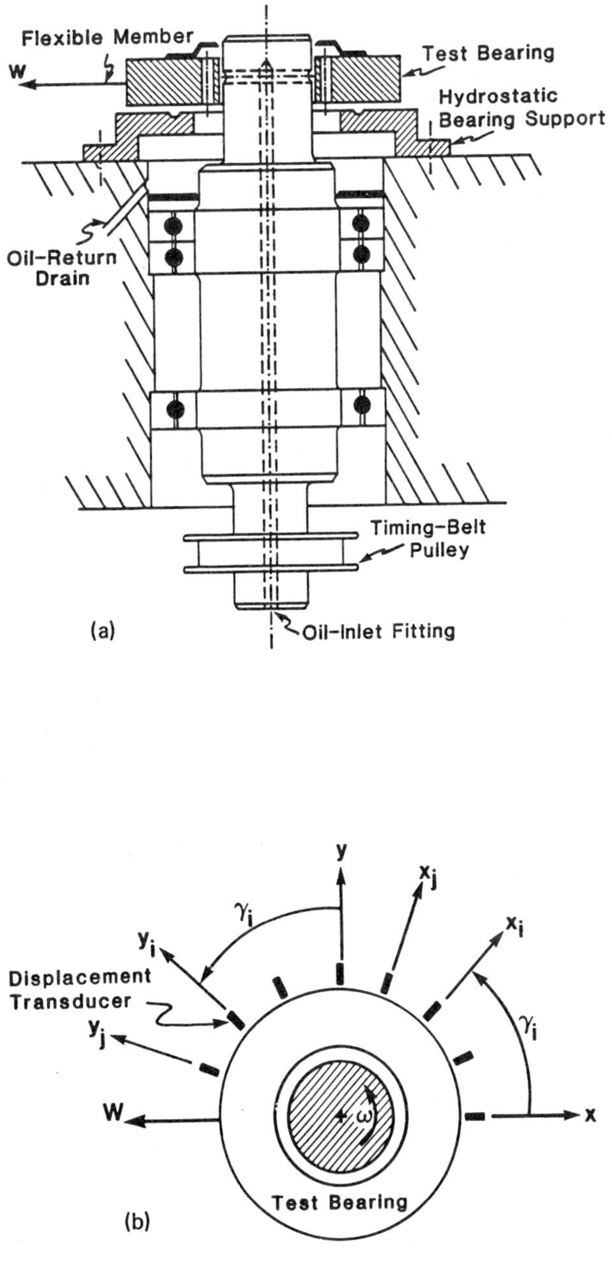

Fig 2 Conceptual sketch of a test apparatus embodying the instability-based approach for journal bearing rotordynamic coefficient extraction

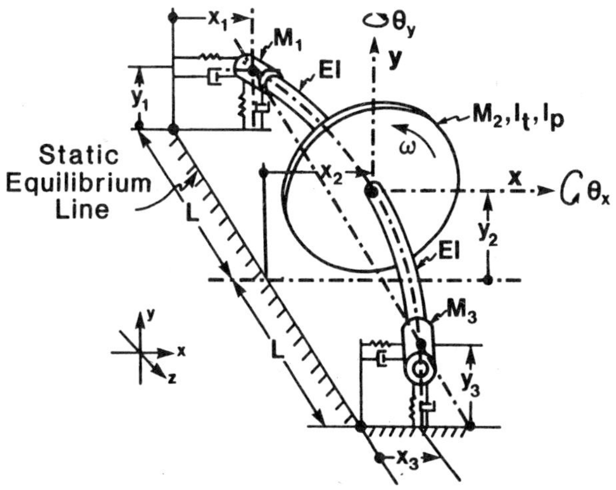

Fig 3 Flexible rotor supported by two identical journal bearings

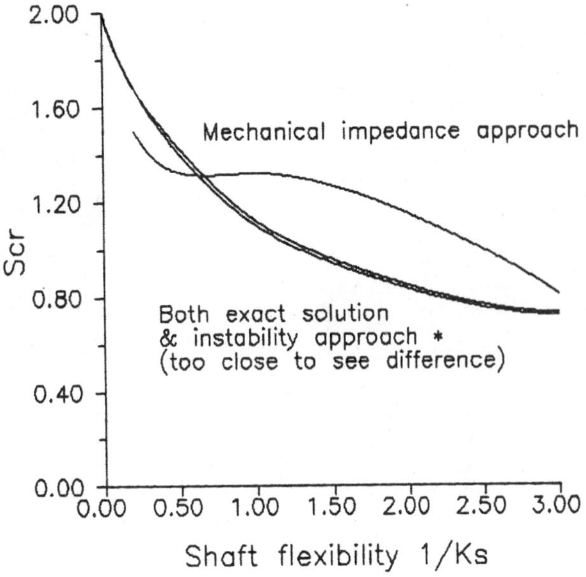

Fig 4 Critical Sommerfeld number, S_{cr} versus dimensionless shaft flexibility, $1/K_s$

* Based on postulated +5 per cent measurement error on $\overline{x}/\overline{y}$ and $\Delta\theta$

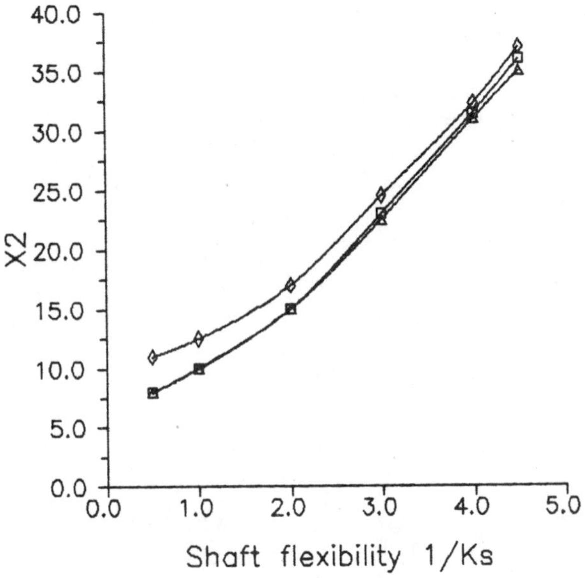

Fig 6 Dimensionless unbalance-excited critical-speed displacement, X2, at rotor disc versus dimensionless shaft flexibility, $1/K_s$

□ Exact bearing coefficients
△ Instability-based approach *
◇ Mechanical impedance approach *

* Based on postulated +5 per cent measurement error on $\overline{x}/\overline{y}$ and $\Delta\theta$

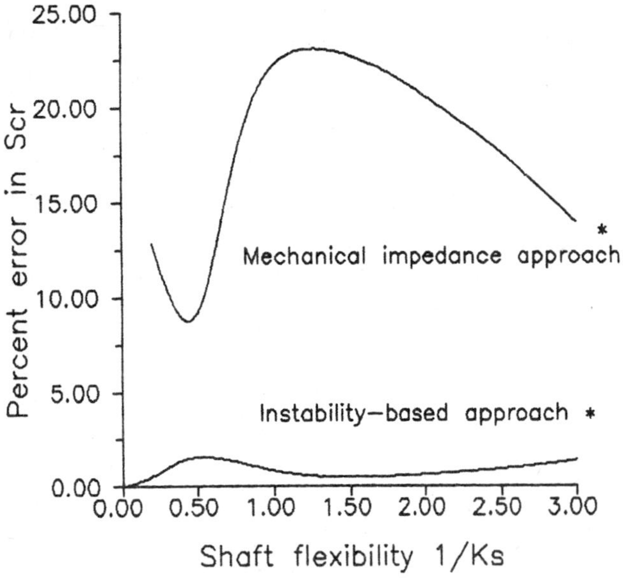

Fig 5 Percentage error in prediction of critical Sommerfeld number, S_{cr} versus dimensionless shaft flexibility, $1/K_s$

* Based on postulated +5 per cent measurement error on $\overline{x}/\overline{y}$ and $\Delta\theta$

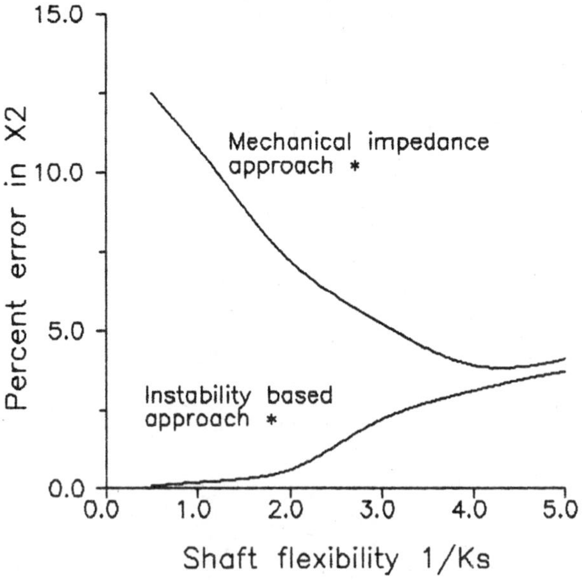

Fig 7 Percentage error in X2, unbalance-excited critical-speed displacement at rotor disc versus dimensionless shaft flexibility, $1/K_s$

* Based on postulated +5 per cent measurement error on $\overline{x}/\overline{y}$ and $\Delta\theta$

C240/88

The application of the model reference adaptive system in the identification of dynamic characteristics of oil-film bearings

W KANG MEng and **D JIN**, MCMES, MSVES
Department of Precision Instruments, Tsinghua University, Beijing, People's Republic of China

SYNOPSIS The model reference adaptive system(MRAS) is proposed for the identification of the eight linearized parameters which characterize the dynamics of a journal bearing oil film.Each of the bearing pedestals is considered as a mass-spring-damping system in vertical and horizontal directions respectively.The advantages of the method are given prominence in the problem of on-line identification and diagnosis of a rotor system with several bearings.The feasibility of the method is proved by computer simulation and laboratory experiments.

NOTATION

$(\dot{})$	derivation of ( )
Kxx,Kxy...	stiffness coefficients of oil film
Cxx,Cxy...	damping coefficients of oil film
Kx,Ky,Cx,Cy	stiffness and damping coefficients of bearing pedestal
Mx,My	modal masses of the pedestal
$\underline{K}$xx,$\underline{C}$xx...	ratios of stiffness and damping coefficients of oil film to the modal masses
X,Y	displacements of the pedestal
Ux,Uy	displacements of rotor relative to the pedestal
Z	output vector of bearing pedestal system
A	system matrix
B	input matrix
U	input vector
w	running frequency of rotor
c	clearance of the bearing
s	eigenvalue of the system
I	unit matrix
Sc	controllability matrix
So	observability matrix

1 INTRODUCTION

It has been well known that in the analysis of rotor bearing system,the journal bearing oil film can be represented by eight linearized parameters.However,such analyses as dynamic responses and stability are of limited value unless these parameters can be obtained experimentally.

The methods of measurement or identification of bearing oil film parameters can be classified as two categories--frequency domain and time domain.

Morton[1] has developed a technique which allows a step force to be applied to a rotating rotor. This is considerably more versatile than earlier attempts such as Glienicke's[2].It can be applied in-situ but suffers from the choice of step function and the high phase sensitivity of mechanical receptance.

Nordmann[3] has applied impact force as input to rotor bearing system and used curve-fitting technique to the frequency response functions obtained experimentally.The results of the method are better than those of other previous papers but suffer from some inconvenience in application under operating condition.

Burrows,Stanway[4] used PRBS (Pseudo Random Binary Sequence) as the input form of exciting force applied to the rotor bearing system and measured the responses.The dynamic state vector equations of the system have been erected on the consideration that the system is a one-span rotor with two identical bearings.The Least Squares Method (LSM) has been used as an identification algorithm.This is convenient for in-situ identification but is limited by the one-span rotor model and the biased properties of LSM.

Time domain methods have generally superior statistical properties to frequency domain methods and can be used to refine the results of frequency domain method,Lennart Ljung[5].Besides, time domain methods have another good property for they can be used to identify the nonlinear characteristics of oil film bearing if on-line identification algorithms are used.

Most rotor bearing systems working in industry are poly-span systems,which have several bearings.Many methods given in previous papers are based on the one-span model.It is necessary to analyse the dynamics of whole rotor when the methods are extended to poly-span rotor systems.Although the transfer matrix method by Lund[6] has been successfully used in many cases,there are still some difficulties such as the analyses of couplings.

Based on the considerations above,the MRAS is proposed as an identification algorithm in the paper.Each of bearing pedestals on which rotor is running in journal bearings is analysed individually.This model is applicable to the problem of on-line or in-situ identification and diagnosis of rotor system.The feasibility of the method is proved by computer simulation and laboratory experiments.

2 GENERAL CONCEPTS OF MRAS AND ITS APPLICATION

MRAS is a kind of identification method of a time domain. The system to be identified is taken as a reference model, the mathematical model of the system, as adjustable system. The parameters in the mathematical model are amended by an adaptive mechanism (identification algorithm) which is constructed by applying stability theory to make the difference between the output of the system to be identified and the adjustable system under the condition of same input vanish so that the real values of the parameters are obtained[7] Fig.1.

MRAS has some special properties compared with LSM and recursive LSM etc. It can give unbiased estimate results even if there is measurement noise providing that identification algorithm is constructed well and input signals have some properties given below.

2.1 Physical model and mathematical model

The dynamic characteristics of journal bearing and pedestal can be expressed by the system shown in Fig.2, among which oil film has eight stiffness and damping coefficients

Kxx Kxy Kyx Kyy

Cxx Cxy Cyx Cyy

and the pedestal has four stiffness and damping coefficients and two modal masses

Kx Ky Cx Cy Mx My

The coordinate oxy is erected with its initial point o attached at the center of journal when the pedestal is in static equilibrium position. The forces acted on the pedestal by rotor through oil film are considered as exciting forces. The dynamic equations of the system are

$$\begin{aligned} Mx\ddot{X}+Cx\dot{X}+KxX&=KxxUx+KxyUy+Cxx\dot{U}x+Cxy\dot{U}y \\ My\ddot{Y}+Cy\dot{Y}+KyY&=KyxUx+KyyUy+Cyx\dot{U}x+Cyy\dot{U}y \end{aligned} \quad (1)$$

define

$$\underline{K}ij=Kij/Mi \quad \underline{C}ij=Cij/Mi$$

$$\underline{K}i=Ki/Mi \quad \underline{C}i=Ci/Mi \quad (i,j=x,y) \quad (2)$$

Equ.(1) can be rewritten by substituting Equ.(2) into it

$$\ddot{X}=-\underline{K}xX-\underline{C}x\dot{X}+\underline{K}xxUx+\underline{K}xyUy+\underline{C}xx\dot{U}x+\underline{C}xy\dot{U}y \quad (3)$$

$$\ddot{Y}=-\underline{K}yY-\underline{C}y\dot{Y}+\underline{K}yxUx+\underline{K}yyUy+\underline{C}yx\dot{U}x+\underline{C}yy\dot{U}y \quad (4)$$

Equ.(3) can be non-dimensionalized to give

$$\dot{Z}=AZ+BU \quad (5)$$

where

$$Z=\begin{bmatrix}X/c & \dot{X}/wc\end{bmatrix}^T$$

$$A=\begin{bmatrix}0 & 1 \\ -\underline{K}x/w^2 & -\underline{C}x/w\end{bmatrix}$$

$$B=\begin{bmatrix}0 & 0 & 0 & 0 \\ \underline{K}xx/w^2 & \underline{K}xy/w^2 & \underline{C}xx/w & \underline{C}xy/w\end{bmatrix}$$

$$U=\begin{bmatrix}Ux/c & Uy/c & \dot{U}x/wc & \dot{U}y/wc\end{bmatrix}^T$$

The same treatment can be done to Equ.(4).

2.2 Stability, controllability and observability

In MRAS, the reference model must be stable, controllable and observable.

2.2.1 Stability

The characteristic equation of the system is

$$|sI-A|=\begin{vmatrix}s & -1 \\ \underline{K}x/w^2 & s+\underline{C}x/w\end{vmatrix}$$

$$=s^2+\underline{C}x/w\cdot s+\underline{K}x/w^2=0 \quad (6)$$

solve the equation, the eigenvalues of the system are obtained

$$s_{1,2}=-\underline{C}x/2w \pm \sqrt{(\underline{C}x/2w)^2-\underline{K}x/w^2}$$

In most cases, bearing pedestals are made of metal. It can be said that $\underline{K}x \gg \underline{C}x$, and $s_{1,2}$ have negative real parts.

Matrix A is a Hurwitz matrix and therefore verifies the Hurwitz stability criterion, so the system to be identified is stable.

2.2.2 Controllability

$$Sc=\begin{bmatrix}B & AB\end{bmatrix} \text{ and } RANK(Sc)=2$$

According to controllability criterion, the system to be identified is controllable.

2.2.3 Observability

$$So=\begin{bmatrix}I & IA\end{bmatrix}^T \text{ and } RANK(So)=2$$

According to observability criterion, the system to be identified is observable.

2.3 Establishment of MRAS identification algorithm

The state equation of the system is

$$\dot{Z}=AZ+BU \quad (Z(0)=Zo) \quad (8)$$

The identification model is

$$\dot{\underline{Z}}=As\underline{Z}+BsU \quad (\underline{Z}(0)=\underline{Z}o, As(0)=Aso, Bs(0)=Bso) \quad (9)$$

where $\underline{Z}$ is output vector, As, system matrix, and Bs, input matrix, of the mathematical model of the system. Zo, $\underline{Z}$o, Aso and Bso are the initial values of Z, $\underline{Z}$, As and Bs respectively.

The state generalized error vector is

$$e=Z-\underline{Z} \quad (10)$$

The identification algorithm is obtained by using Lyapunov stability theory. The details can be found in ref.(7).

$$As=\int_0^t F_A(Pe)\underline{Z}^T dt+As(0) \quad (11)$$

$$Bs=\int_0^t F_B(Pe)U^T dt+Bs(0) \quad (12)$$

$$PA+A^T P=-Q \quad (13)$$

where P is a positive definite matrix, and Q, F_A and F_B are arbitrary positive definite matrices. U should be piecewise continuous vector function.

2.4 Conditions of convergence of the algorithm

The bearing pedestal system is completely controllable. The components of U are linearly independent because U is the displacement and velocity of the rotor relative to pedestal. The Ux and Uy are signals of multiple components of frequency in the experiment described later in which the rotor is excited by impact force and in the in-situ conditions where the rotor is excited by various forces. The identification algorithm is convergent according to the conditions of convergence of MRAS[7].

2.5 Effects of measurement noise

Suppose that the signals Z and U be contaminated by zero mean noises α and β of finite amplitude respectively.

According to Equ.(11) and (12), it is obtained

$$As = \int_0^t F_A (P(e+\alpha))\underline{Z}^T dt + As(0) \qquad (14)$$

$$Bs = \int_0^t F_B (P(e+\alpha))(U+\beta)^T dt + Bs(0) \qquad (15)$$

Compared with noise α and β, e, $\underline{Z}$ and U vary slowly. Because $E(\alpha)=0$ and $E(\beta)=0$, where E( ) means the expectation of ( ), the additional terms of Equ.(14) and (15) will tend to become so small as to be neglected when F_A and F_B are small.

The frequencies of U and $\underline{Z}$ are not high even though step or impact forces are inputted to the bearing system. The efficient frequency band is less than 0.5KHz, which is much lower than the frequency of measurement noise. It may be said that this kind of MRAS can be used in bearing system identification with effect of noise.

3 PROGRAM OF THE IDENTIFICATION ALGORITHM

3.1 Flow chart

The program flow chart is given in Fig.3.

3.2 Selection of P matrix

P needs to be

a. positive definite and symmetric.

b. $PA + A^T P = -Q$

P can be derived from the conditions above

$$P = \begin{bmatrix} (\underline{C}x + 2\underline{K}x/\underline{C}x)/w & 1 \\ 1 & 2w/\underline{C}x \end{bmatrix}$$

Because of A being unknown, P is unknown. But there is a region in the parameter space for the conditions of P being satisfied in real situations. P can be calculated by use of the initial value of A. The demand is that the value of A must be in the region. The LSM is used to obtain the initial value of A in the paper.

F_A and F_B are called adaptation gains and can be selected as

$F_A = WI$, $F_B = VI$

where W and V, called adaptation gain factors, are positive values determined according to the noise signal ratio of real identification problems.

3.3 Considerations in program

3.3.1 Iteration method

The program can only treat finite number of sampling points of signals in a finite time interval. In order to obtain convergence, the adaptation gains must be small when there exists measurement noise. But small F_A and F_B can decrease the speed of convergence, so that the real values of parameters may not be achieved in the time interval. The solution to the problem is to use these sampling points repeatedly until the error vector reaches the error limit.

3.3.2 Error limit

When

$$E = \sum_{i=1}^{N} e_i^2 \,/\, (\max_{1\le i\le N} Z)^2 < \varepsilon$$

where E is error vector, ε, error limit, and N, number of sampling points, the parameters to be identified converge on the real values.

3.3.3 Solution of differential equation

The four-order Gill method is used to solve the Equ.(9) in order to decrease computing error and keep numerical stability.

4 COMPUTER SIMULATION OF MRAS

In order to show the feasibility of MRAS, a bearing pedestal system, parameters and inputs of which are taken from an actual generator in a power station, is investigated. The outputs are calculated by use of its parameters and inputs. The estimated results obtained are compared with its real values of the parameters. The calculation was carried out by DPS-8 computer system.

The real values of the parameters are

$Mx = 1000$ kg

$Kx = 5.0 \cdot 10^9$ N/m

$Cx = 1.0 \cdot 10^5$ N.s/m

$Kxx = 225\ 549\ 500.0$ N/m

$Kxy = -723\ 323\ 296.0$ N/m

$Cxx = 6\ 664\ 980.0$ N.s/m

$Cxy = -4\ 781\ 420.0$ N.s/m

$w = 285.6$ 1/s

$c = 0.1$ mm

$\dot{U}x$ and $\dot{U}y$ are derived by numerical derivations of Ux and Uy obtained from the measurement respectively. The output vector Z is obtained by solving the Equ.(8). The properties of the identification algorithm are investigated by giving different initial values of A and B, superposing constants (simulating measurement system errors) and random numbers (simulating measurement noises) on the input and output signals. The initial values of the parameters are given by multiplying the real values by a multiple.

4.1 Results of simulation

The results listed in table 1 show:

When there are no measurement system error and noise, the parameters to be identified converge on the real values and the error vector converges on zero with the iteration. The convergence speed increases with the increase of adaptation gains, No.1 and No.2.

When there are measurement system errors the parameters to be identified do not converge on the real values, and the E has a minimum value not equal to zero. When the measurement system error are large enough, the identification process overflows, No.3, No.4 and No.5.

When there is random noise on the signals, the E converges to a minimum value then vibrates around the value. The parameters to be identified converge approximately on real values then vibrate around the values, No.6, No.7 and No.8.

The precision of parameter convergence and the iteration number increase with the decrease of adaptation gains when noise is large. The error vector E has the properties shown in Fig.4.

5 EXPERIMENT AND RESULTS

5.1 Experiment system

An experiment was carried on a test rig with a rotor supported on two bearings. The diameter of the bearing investigated was 25 mm. The signals were measured by eddy current transducers and piezo-electric accelerometers which were mounted on the bearing pedestal in vertical and horizontal directions. Impact forces were given by use of a hammer with a force transducer on its top. All signals were recorded by a FM tape recorder, Fig.5.

5.2 Signal processing

In order to achieve the modal masses of test bearing pedestal Mx and My, impact forces were acted on the pedestal and its accelerations were measured in horizontal and vertical directions respectively when the rotor was not running. Modal analysis technique was used to obtain the modal parameters of the pedestal Kx, Ky, Cx, Cy, Mx and My using 7T17S signal processor.

5.3 Parameter identification by MRAS

The signals of rotor displacements relative to the bearing pedestal Ux, Uy, and the accelerations of the pedestal in horizontal and vertical directions at the rotor speed of 1200 r/min were converted from analog into digital form by the 7T17S, then inputted into DPS-8 computer.

Table 2 Results of identification

(stiffness: N/m damping: N.s/m mass: kg)

	Modal analysis technique	MRAS data 1	MRAS data 2
Mx	0.386		
My	0.9695		
Kx	$1.21\cdot10^6$	$1.03\cdot10^6$	$0.746\cdot10^6$
Ky	$1.23\cdot10^6$	$1.80\cdot10^6$	$3.22\cdot10^6$
Cx	34.6	56.7	30.6
Cy	48.8	472.	295.
Kxx		$1.90\cdot10^4$	$3.79\cdot10^4$
Kxy		$-4.83\cdot10^4$	$1.12\cdot10^6$
Kyx		$2.50\cdot10^4$	$-4.44\cdot10^4$
Kyy		$2.15\cdot10^5$	$1.96\cdot10^6$
Cxx		-61.4	66.2
Cxy		62.5	-39.6
Cyx		-53.1	-5.91
Cyy		-161.	17.1

The results are shown in table 2. The two groups of MRAS results are obtained under two different eccentricities of the bearing. The values of stiffness given by two kinds of methods are same in magnitude. This coincidence may be seen as good under the consideration that the measurement errors of rotor dynamics are often large and there is some bias in the identification by modal analysis technique. The damping coefficients exhibit considerably experimental scatter. This may be due to the phase error of measurement system and the error of numerical derivation and integration.

6 CONCLUSION

The MRAS is proposed for the identification of the eight linearized parameters of a journal bearing oil film by investigating every pedestal of a rotating system. The advantages of the method are given prominence in the problem of on-line identification, monitoring and diagnosis of a rotating system with several bearings because the model for analysing is based on a single bearing pedestal, and the excitation can be any kind of actions acted on the rotor by unbalance, asymmetry of the rotor and misalignments of the couplings etc.

By the MRAS identification algorithm, although the signals are contaminated by noise, the unbiased estimation could be achieved as long as the adaptation gains are small and the noise signal ratios of the signals are not too high. The signals should have no measurement system error or be pre-treated using such technique as filtering.

ACKNOWLEDGEMENT

The authors wish to acknowledge the assistance rendered by Zhengsong Zhang, Associate Professor, Tsinghua Vibration Laboratory, during the preparation of this paper.

REFERENCES

(1) MORTON, P.G. Dynamic characteristics of bearings--measurement under operating conditions.
GEC Journal of Science & Technology, 1975, 42, No.1, 37-47.

(2) GLIENICKE, J. Experimental investigation of the stiffness and damping coefficients of turbine bearings and their application to instability prediction.
Proceedings of the Institution of Mechanical Engineers, 1966-1967, 181, series 3B, 116-129.

(3) NORDMANN, R., SCHÖLLHORN, K. Identification of stiffness and damping coefficients of journal bearings by means of the impact method.
Vibrations in Rotating Machinery, IMechE 1980, 231-238.

(4) BURROWS, C.R., STANWAY, R. Identification of journal bearing characteristics.
Transactions of ASME, Journal of dynamic systems, Measurement, and Control, 1977, Sep., 167-173.

(5) LJUNG, L., GLOVER, K. Frequency domain versus time domain methods in system identification--a brief discussion.
Proc.5th IFAC Symposium on System Identification and Parameter Estimation, Pergamon(1979a), 1309-1322.

(6) LUND, J.W. Stability and damped critical

speeds of a flexible rotorin fluid-film bearings.
Transactions of ASME,Journal of Engineering for Industry,1974,May,509-517.

(7) LANDAU,Y.D. Adaptive Control--The model Reference Approach,1979,Marcel Dekker,Inc. ISBN 0-8247-6548-6.

Table 1 Results of computer simulation

No.		1	2	3	4	5	6	7	8
Initial value / Real value		2.0	2.0	2.0	1.0	1.0	1.0	2.0	2.0
Adaptation gain factor		10.0	100.0	100.0	100.0	100.0	100.0	10.0	10.0
Kinds of error		no error	no error	measurement system error	measurement system error	measurement system error	noise on Z	noise on Z	noise on Z,U
Noise / Signal (%)		0.0	0.0	5.0	5.0	10.0	5.0	5.0	5.0
Convergence error (%)		0.0009	0.0009	2.0	2.0	*	1.0	1.0	2.0
Iteration number		1256	258	754	455	*	425	3211	1112
Deviation of estimated value of parameters (%)	Kx	0.14	0.23	23.0	-19.0	*	2.4	-1.2	-0.86
	Cx	0.2	0.1	1.6	2.8	*	2.0	-1.4	2.8
	Kxx	0.52	1.2	101.0	-94.0	*	-15.0	-8.3	-7.6
	Kxy	0.47	1.1	89.0	-69.0	*	-5.1	-6.3	-1.5
	Cxx	0.15	0.09	-0.17	1.2	*	0.28	-1.3	0.37
	Cxy	0.35	0.09	-3.0	4.1	*	-5.7	-3.2	4.1

* algorithm overflow

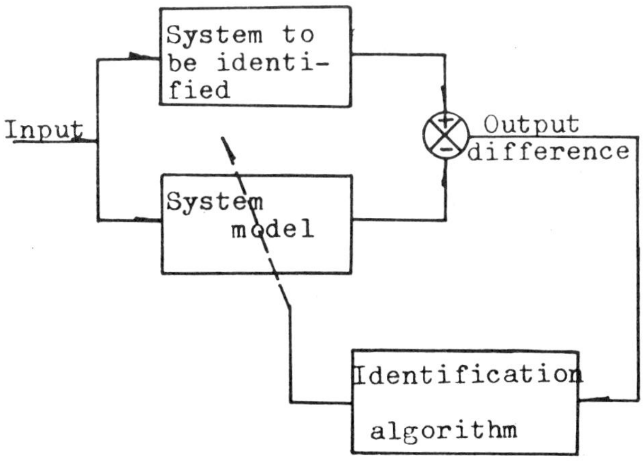

Fig 1 Basic configuration of MRAS

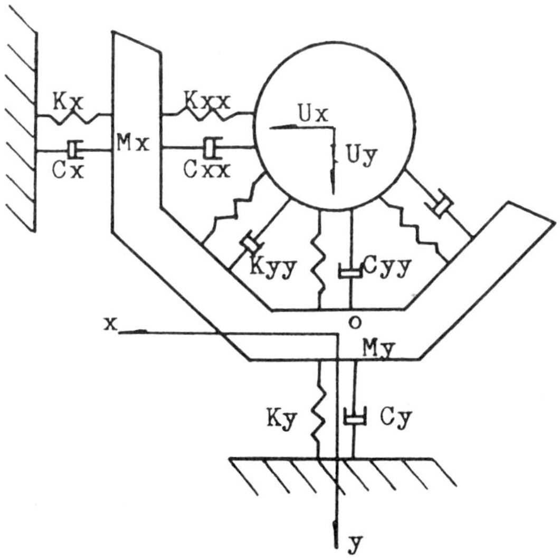

Fig 2 Bearing pedestal system model

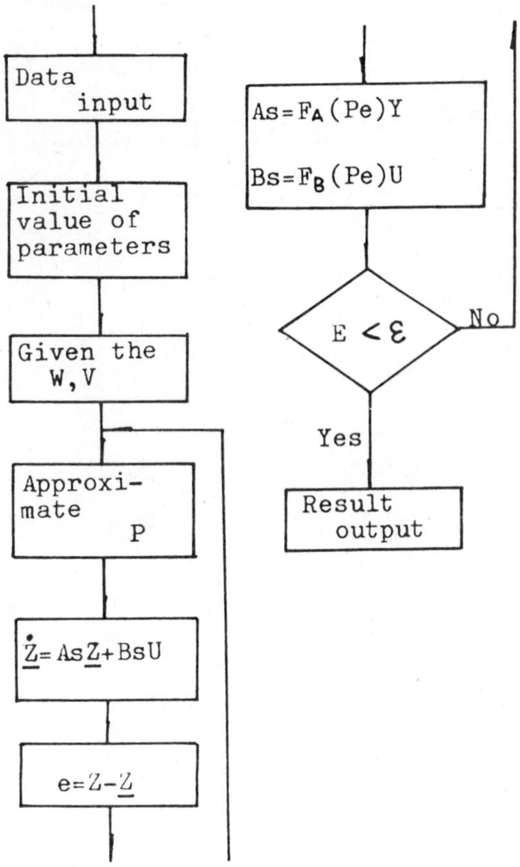

Fig 3 Simplified flow chart of the identification algorithm

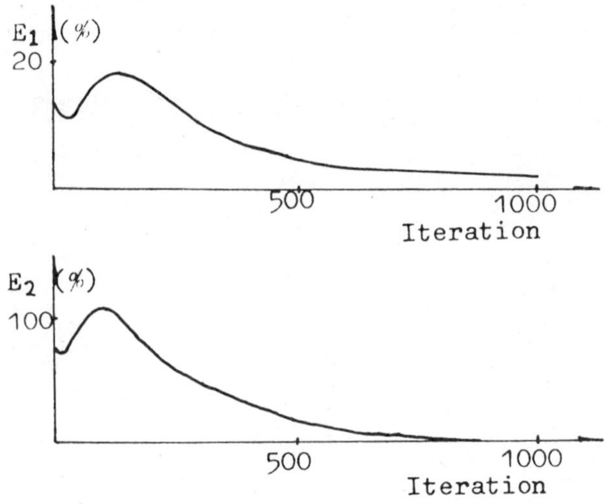

Fig 4 Changes of error vector with iteration

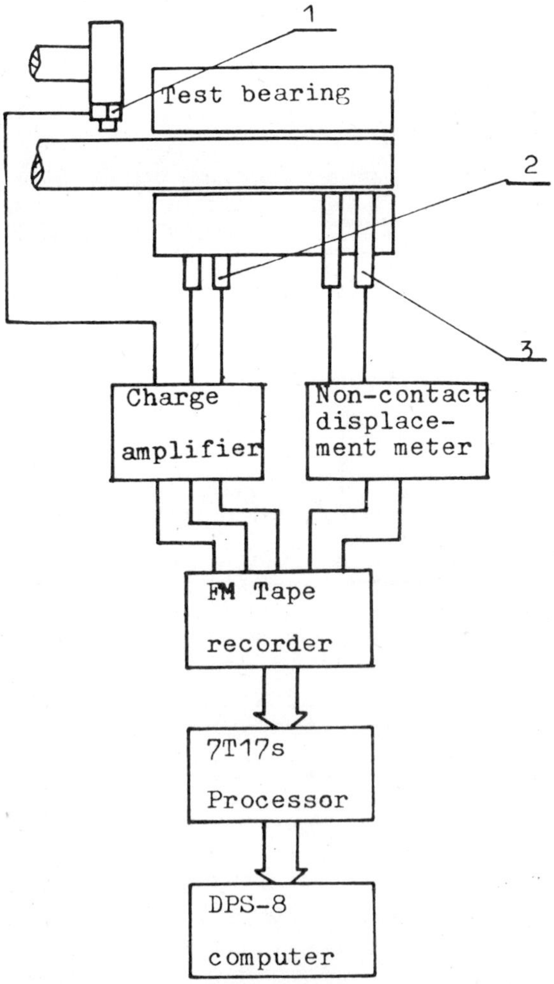

Fig 5 Simplified experiment system diagram
1 Force transducer
2 Piezo-electric accelerometers
3 Eddy current transducers

C274/88

Practical considerations on the rotordynamic behaviour and design of multi-stage pumps

J J VERHOEVEN
Byron Jackson BWIP, Etten-Leur, The Netherlands

SYNOPSIS. Operating speeds of multi-stage pumps tended to increase during the last decade from 4000-6000 RPM to 7000-10000 RPM, due to new super critical powerplant designs and new conversion processes in the hydro carbon industry. High speed pump selections are in general commercially attractive, due to smaller pumpsizes and especially when step down gears can be eliminated.
Reliability of high speed multi-stage pumps in boilerfeed, water injection and hydro carbon service is heavily related to the rotor dynamic behaviour of these pumps.

This paper reviews the influence of hydraulic impeller excitation forces and some design concepts on rotor dynamic behaviour. Real life high speed multi-stage pump examples are used to illustrate the tremendous influence of rotor dynamics.

NOTATION

F_t = Tangential force.
K_c = Cross coupled stiffness.
C^c = Direct stiffness term.
ω = Whirl frequency.
R = Rotor amplitude.

1 INTRODUCTION

Recently rotor dynamic models have been developed to simulate also the motion dependent forces generated by the impeller-casing hydrodynamic interaction. This in addition to the motion dependent forces generated in the annular clearance seals, which are included in pump rotor models for more than 15 years, as introduced by H.F. Black (ref. 19). Rotor response analysis with these models using maximum mechanical unbalance excitation forces always produces smaller amplitudes than the actual measured rotor amplitudes. The hydraulic impeller excitation forces responsible for the discrepancy, are much larger than the mechanical excitation forces. The paper shows a brief overview of these impeller excitation forces obtained from recent research results on impeller excitation forces. Rotor amplitude response predictions including both mechanical and hydraulic impeller excitation forces are shown to be very close to the actual rotor amplitudes measured.

Typical difference between an inline diffuser and opposed impeller volute style pump is the extra center bushing to seal the high pressure difference at the center. The influence of this typical center bushing on the rotor amplitude response is shown for excitation by mechanical and hydraulic forces.

Due to the load carrying capacity, developed in the internal annular clearance seals of these pumps, the static radial bearing forces will be influenced. This paper describes an iterative quasi-static procedure for evaluating the actual static bearing loads. An example on a 10 stage offshore pipe line reinjection pump running at 8900 RPM shows a tremendous decrease of the radial bearing force both in the vertical and horizontal direction. The paper reviews the negative effects of these low loaded pump journal bearings. By carefully machining a sag-bore in the pump casing, it is shown that the vertical static bearing load is almost again the rotor deadweight, creating improved rotor behaviour.

In high speed applications, centrifugal impeller growth has a tremendous influence on the impeller annular clearance seal geometry. The influence on rotor dynamic behaviour and leakage flow rates is reviewed.

Pump rotor stability is discussed in relation to smooth/grooved rotors, swirl brakes and annular clearance seal wear for both opposed impeller-volute style and inline diffuser style pumps.
When reverse running multistage pumps are used as hydraulic turbines the flow rates to speed ratio increases.
This has a negative effect on the stability of such a reverse running pump, especially at off design duties. This is shown for a 5 stage, double case, opposed impeller, turbine at 5700 RPM.
Subsynchronous unstable vibrations measured on the test floor disappeared after installation of "gap reducing rings" at the impeller tip. This supports the findings of Childs (18) and Bolleter (15), that large destabilizing forces are generated at the impeller shroud casing.

2 IMPELLER HYDRAULIC EXCITATION FORCES

The forces that act upon a pump rotor stem from mechanical and hydraulic causes. The mechanical forces due to unbalance, coupling misalignment, etc., are well-known and limited by international standards or vendor/user specifications. Hydraulic forces arise both at the impeller and the annular clearance seals in two distinctly different ways. Firstly when the

shaft center is fixed, radial forces occur as a result of the distribution of static pressure and fluid momenta, around the impellers. These forces are termed excitation forces and are independent of rotor motion. Secondly, when the shaft center also moves (as a result of vibration) both at the impeller tip, impeller shroud and the annular clearance seals additional forces are generated. These forces which are motion dependent can be represented by force coefficients in terms of stiffness, damping and mass. Except the impeller shroud forces which show a different behaviour for certain impeller whirl speeds and for large inlet tangential velocities of the impeller leakage flow (Childs 18). The latter forces received extensively research during the last decade (ref. -2-) and are better known as "lomakin effect" in the annular clearance seals and impeller/casing "hydrodynamic interaction" forces. This resulted in quite effective dynamic models of the pump rotors, which can be used to simulate vibration response with special purpose computer programs. In practice even simple unbalance response analysis, using the maximum mechanical unbalance allowed according factory standards, will never produce the results obtained during performance testing and field operation. It is at present quite clear that the vibration response of centrifugal pumps is mainly generated by impeller hydraulic excitation forces instead of mechanical excitation forces like unbalance, misalignment, disc skew, shaft bow, etc. Still little is known on the impeller hydraulic excitation forces, which makes accurate pump rotor response analysis cumbersome (ref. 1,2,3,4). A recently developed indirect method (5) was used to predict hydraulic impeller excitation forces in multistage pumps typically used for boilerfeed, water injection and hydro carbon feed pump service. Special test were also conducted on a large single stage horizontal process pump, to determine specific influences of hydraulic parameters and geometric deviations.
Correlation of the impeller excitation forces was made with following parameters:

* Flow rate.
* Hydraulic impeller/casing parameters.
* Geometric tolerances in impeller/casing.

Figure 1 shows ranges obtained for subsynchronous, synchronous and vane passing frequency impeller force components versus pump flow rates. The forces are made dimensionless per Stepanoff (7). More detailed results are discussed in reference 6. If one compares the synchronous, hydraulic force levels with synchronous mechanical force levels (unbalance), a tremendous difference is obtained.
Mechanical unbalance dimensionless force levels are much lower and typically range from .001 up to .009 for pumps balanced within Q = 2.5 per ISO 1944.
This demonstrates the importance of controlling these hydraulic excitation forces, and the need to include these forces in pump rotor response analysis.

3 MECHANICAL AND HYDRAULIC FORCED RESPONSE ANALYSIS OF PUMPS

From the review of the impeller hydraulic excitation forces it appears that predictions of pump vibration amplitudes cannot be done by considering only unbalance response. Large hydraulic force levels at subsynchronous, synchronous and vane passing frequency can seriously contribute to the total vibration amplitude level. This can be verified by additional non-synchronous harmonic response analysis using the above described hydraulic excitation forces together with the mechanical excitation forces.

The significance of the "hydraulic unbalance" on the synchronous response as well as the importance of evaluating non-synchronous response, is shown on a typical boilerfeed pump. Comprising a double case 4 stage feed pump design having an opposed impeller arrangement, with an axial split double volute inner casing (see figure 4). The main pump data is given below.

Impeller diameters	: 1st Stage : 350 mm double suction Serie stages : 420 mm
Operating speed range	: 4000 - 6000 RPM
Annular seals	: Impeller wearrings : Smooth Long center and balance bushing : Serrated
Capacity rated	: .44 m^3/S (6900 GPM)
Total head rated	: 2625 mtr. (8625 ft)

The rest unbalance of the rotor assembly was carefully measured on a dynamic balancing machine and was totally 2 OZin which represents a dimensionless force of K_R = .00116. Because the volute casing is axial split, there is no change in mechanical unbalance after complete assembly of the pump unit, which allows a direct comparison of field vibrations versus analysis vibration results. The pump vibrations were recorded over the total operating speed range, and figure 2 shows one proximity probe shaft measurement at the inboard bearing housing for 31% of rated pump capacity. A very detailed lumped mass elastic beam model comprising 48 stations was constructed for the rotor, pump casing and pedestal foundation mounting, using the RESP2V3 and PRS1V2 computer codes (8, 9). Both programs based on harmonic theory accept non-symmetric stiffness and damping matrices, gyroscopic forces, shaft hysteresis effects. The model comprises force coefficients for the journal bearings, annular clearance seals and impeller/volute interaction. The journal bearing (tilting pad) force coefficients were determined by finite element analysis. The annular seals were analysed either with a finite element approach per Schmauss (10) or based on Childs (11) "finite length" theory. The volute/impeller interaction forces were based on Jerry et al (12). The model was used in a synchronous unbalance response from 4000 to 6000 RPM and a non-synchronous analysis at 4300 and 5800 RPM. Unbalance response was done for pure mechanical unbalance of 2 OZin and for combined mechanical unbalance and synchronous hydraulic forces. The synchronous hydraulic force at the impellers was taken K_R = .01. Non-synchronous response was determined for a force spectrum consisting of a subsynchronous force K_R = .02, synchronous force K_R = .1 and vane passing force K_R = .15. The force spectrum contained a very small random background force level of K_R = .0005. The vane passing force is taken saw tooth shaped, while the other forces are pure Sinusoidal. Figure 2 shows a summary of the analysis. It is at once recognized that unbalance response analysis using only mechanical forces is producing much too low response results. Inclusion of both synchronous and non-synchronous hydraulic impeller forces produces quite accurate response amplitudes. The excitation of the rotor by the impeller hydraulic forces, is much larger than the mechanical excitation and occurs on predetermined locations. This allows for a direct comparison between the response sensitivity of an opposed impeller volute pump versus an inline diffuser style pump. For this purpose the boiler feed pump model described above was transformed into an equivalent inline diffuser style pump, by deleting the center stage piece, changing the volute/impeller force coefficients with diffuser/impeller force coefficients per Bolleter et al (13). The balance bushing was reanalysed for the full pump differential pressure. Figure 3 shows the synchronous and non-synchronous response results, which can be compared directly with the results of figure 2 since identical force levels were used. It can be concluded that the opposed impeller volute pump is less sensitivite for synchronous excitation forces and vane passing forces. This is explained by the extra (long) center bushing in an opposed impeller design which creates extra stiffening and damping just at those locations where the hydraulic excitation forces are acting. There is no substantial difference in the sensitivity for low frequency hydraulic excitation forces. The low frequency excitation sensitivity doesn't show large differences since small damping forces are generated due to very small rotor vibrational velocities at these low frequencies.

4 INFLUENCE OF ANNULAR SEAL LOAD CARRYING CAPACITY ON STATIC BEARING LOADS

Static bearing loads are used to determine a static equilibrium position of the rotor in a journal bearing oil film. Linearized force coefficients are then determined by perturbations around the static equilibrium position which are linearized. Since accurate rotor dynamic models are required, it is of prime importance to incorporate proper static loads in the specific journal bearing force coefficient evaluation. Without internal clearance seals, the bearings would react approximately half the rotor deadweight and half the static impeller radial force. The direction of the hydraulic radial impeller force depends on flow rates and volute geometric tolerances. Due to the internal clearance seals, which reflect load carrying capacity, the static bearing loads will decrease. Since the hydrodynamic oil film reflects a non linear force deflection behaviour, it is not possible to directly evaluate the force coefficients with the influence of the clearance seals. This requires either a transient non linear rotor dynamic analysis or an iterative quasi static procedure. The quasi static iterative approach can be done with a finite element beam model of the rotor supported by linear spring elements. The stiffness values of the spring elements are the stiffness force coefficients of the rotor dynamic model. For the bearings a starting value based on half the rotor deadweight and static impeller force is used. A static deflection analysis will result in a new bearing static load. By performing a new bearing force coefficients analysis, using the new static bearing load, new spring values for the finite element model can be calculated. This process is repeated until there is a convergence in static bearing load and deflections. The spring elements representing the annular clearance seals are not changed during the iterations, since they reflect a quite linear behaviour for eccentricities up to .8-.9. Because the seal force coefficients are speed dependent, this procedure must be repeated for each speed case. The analysis converges within 3-4 steps.

For smooth annular clearance seals, the static bearing load reduction is enormous and can be as low as 5% of the rotor static load. For grooved annular clearance seals, considerable static bearing load reductions are still obtained producing loads from 10 - 40% of the total static load. Small static bearing loads, generate low stiffness and damping values of the hydrodynamic oil films, which can have negative effects on rotor dynamic behaviour. The overall damping and rotor eigenvalues will be reduced, generating lower stability threshold speeds for the pump. As soon as the annular clearance seals yield a fully centered rotor journal, the load carrying capacity is reduced to zero. By machining the stationary wearring fits in the pump casing according to the static deadweight deflection of the rotor ("sag boring"), the rotor will not obtain internal deadweight reaction forces at the seals. This creates larger static bearing loads in vertical direction of appr. half the rotor deadweight and improves the rotor dynamic behaviour, both in vertical and horizontal direction. To illustrate the static bearing load reduction in a real life example, we used a modern 10 stage pump. The pump, used on an offshore platform in the North Sea for pipe line reinjection, is a double case, opposed impeller, double volute pump (see figure 4) driven by an electric motor and a combined fluid coupling/gearbox (variable speed turbo coupling).

Flow	: .0194 m^3/S (307 GPM)
Head rated	: 3300 mtr. (10827 ft)
Operating speed	: 4000 - 8900 RPM
Impeller diameters	: 190 mm
Shaft diameter	: 50 mm (between impeller)
Overall rotor length	: 2230 mm
Bearing span	: 1785 mm
Radial bearing type	: Tilting pad
Annular clearance seals	: All smooth with a few grooves on long center and balance bushing.

The pump has a very slender shaft, with a very

low 1st dry rotor eigenvalue.
This pump is also used as real life example subsequently in this paper.

A rotor dynamic model was developed with 50 stations, including clearance seal force coefficients and impeller/volute hydrodynamic interaction force coefficients, similar to the previous described model. The model was copied into a finite element beam model to perform an iterative quasi static bearing load analysis. Results are summarized in figure 5, in which we recognize a tremendous reduction of bearing loads and a changing static deflection of the rotor. The eccentricity of the bearing at operating speed changes from .43 to .003 at the pump outboard bearing, which emphasizes the very low loaded bearing condition. The force coefficients for the bearings are reduced to much lower values, which impacts the rotor dynamic behaviour.
Linear transient unbalance response analysis were conducted for 3 typical cases:

Case 1: Bearing force coefficients based on half static rotor loads.

Case 2: Bearing force coefficients based on corrected bearing loads due to internal load carrying capacity (pump without sag-bore).

Case 3: Bearing force coefficients based on corrected bearing loads due to internal load carrying capacity with a sag bore.

In a stable damped linear rotor system, the transient part of the rotor response dies out to yield the steady state response. By using an initial rotor position, it is possible to obtain a steady state solution after a certain time period when the transient part of rotor motion dies out. Stable operation is obtained when the rotor shows synchronous whirling. Subsynchronous whirling is obtained just before or on the stability threshold speed of the rotor. Unstable whirling can be readily observed from the transient amplitudes due to continuously growing amplitudes.
Figure 6 shows the rotor orbits and analog time signals resulting from transient analysis for station 47 of the model, being the inboard probe location, at 8900 RPM. It is at once recognized that for case 2 unstable whirling is fully developed, while for case 1 and 3 still stable synchronous response is obtained. Instability is caused by a decrease of net positive bearing damping which reduces the overall system damping. Destabilizing tangential forces at impellers, bearings and clearance seals are getting larger than the stabilizing damping forces for case 2 at 7500 RPM. This shows the positive effect of sag boring, which increases pump rotor stability. The results are for the quoted 10 stage reinjection pump with smooth clearance seals. Sag-boring is essential for good rotor dynamic behaviour of this 10 stage unit operating up to 8900 RPM. The above example also proves that the first pump rotor mode shapes at these high speeds are no longer having the character of a first bending mode, but are coupling overhung modes, which benefit from large damping forces generated due to relative large bearing motions at the inboard bearing. This behaviour is different with other rotating machines like multi-stage compressors, which develop relative large rotor motions at the rotor midspan, and have typical rotor bending mode shapes. It must be noted that in practice case 2 and 3 conditions can only be approached, due to tolerance generated during machining and manufacturing.

5 IMPELLER CENTRIFUGAL GROWTH

Operating speeds are constantly increased. New super critical powerplants are best equipped with auxiliary steam turbines with speeds of 8000-10000 RPM. Optimum selection, to provide high head are small directly driven boiler feed pumps running also 8000-10000 RPM. New processes in the hydrocarbon industry are designed to crack more and more heavier fractions. The new SHELL high conversion process, which becomes operational in 1988 is a typical example. For these processes, the feed pumps must supply larger differential heads and are operated at larger operating speeds than conventional hydrocarbon feed pumps. Accurate rotor dynamic models must incorporate all mechanical interactions generated in the machine. In this aspect the centrifugal growth of the pump impellers at high speeds appears to generate tremendous changes in annular seal geometry. Especially in the non-symmetric impellers, the straight annular clearance seals of the impeller at zero speed will become tapered and smaller at high operating speeds. Figure 7 shows a typical series impeller deformation due to centrifugal forces, impeller pressure loading and shrinkfit deformation calculated with a finite element analysis of a boiler feed pump impeller operating at 8000 RPM. The decrease of annular clearance will generate smaller leakage flows, improving efficiency of pump unit. For small impeller multi-stage pumps with relative small flow rates and many stages the leakage flow rate will heavily influence the pump efficiency. The 10 stage pump used on an offshore platform in the North Sea for pipe line reinjection is a typical example. Testing this unit at 2950 and 7000 RPM showed a difference in efficiency of almost 6%. Most of the efficiency improvement is generated by centrifugal impeller growth. Also important is the effect on rotor dynamic behaviour. Both the reduction in radial clearance as well as the tapered shape have effect on rotor dynamic characteristics of the pump. The eye wearring is divergent tapered and the backside wearring is convergent tapered. Still little is known on convergent and divergent tapered seals. Childs (14) performed a combined analytical computational finite length solution together with experimental testing. Large discrepancies were obtained between analysis and tests. It appears from testing that convergent tapered seals have substantially less damping on no greater direct stiffness than straight seals. From analysis and testing of divergent tapered seals the same tendency is observed. The total effect of tapered geometry and reduced clearance due to centrifugal growth on the impeller eye wearring dynamic force coefficients is shown in the table below. The table shows ratios obtained including centrifugal deformation and without centrifugal deformation effects of the impeller eye clearance seal of figure 7.

SPEED	K	kc	C
2000	1.1	1.1	.95
4000	1.2	1.2	.90
6000	1.3	1.4	.82
8000	1.42	1.7	.92
10000	1.9	2.2	1.07
12000	2.7	4.1	2.0

Ratio of direct stiffness K, cross coupled stiffness kc and direct damping term C.

The same is observed for the backside wearring, except that it is convergent tapered. This is all based on test data of Childs. Stability threshold speed is lowered by the centrifugal growth effect, which is best illustrated by comparing the tangential rotor force in the direction of rotation, which can be expressed as (15):

$$F_t = (K_c - C.\omega).R$$

Centrifugal deformations of the annular clearance seals of impellers are important for pump rotor stability evaluations. While many discrepancies between experiments and analysis are often sought in the analysis techniques, mechanical influences as described above are often forgotten. Modern high speed impellers should be designed to minimize the effects of the tapered deformation due to centrifugal forces. Tapered stationary wearrings, will eliminate the destabilizing effects of centrifugal deformation.

6 MULTI-STAGE PUMP ROTOR STABILITY

It is at present quite clear that multi-stage pumps, especially at high speeds, may suffer for rotor instability. Prior to complete rotor instability the pumps show subsynchronous whirling, with high amplitudes, in a typical frequency range of .65 up to .90 of operating speed (15, 16, 17). Most of the subsynchronous whirling is reported for boiler feed pumps, with light viscosity duty. In these cases most of the destabilizing tangential forces are generated by the impeller/casing hydrodynamic interaction, as reported by Pace et al (15).

To make a general review of the design factors affecting multi-stage pump instabilities in low viscosity duty two design concept are reviewed for following parameters:

* Smooth versus grooved annular seals.
* Swirl brakes.
* Influence of worn clearances.

Boiler feed pump designs can in general be divided into two major concepts, which are reviewed for above parameters:

* Diffuser, inline impeller style pump.
* Double volute, opposed impeller style pump.

The rotor model of the 10 stage pipe line reinjection pump, previous quoted in this paper for duty at 8900 RPM on an offshore platform in the North Sea is also transformed into an "equivalent" diffuser style pump, in the same fashion as the first boiler feed pump example of this paper. Both units incorporate the effects of sag boring and impeller seal deformation due to speed, pressure and shrink fit. The force coefficients for the impeller/diffuser hydrodynamic interaction were obtained from Bolleter et al (13). The center bushing, typical for the volute opposed impeller style pump was deleted from the model, while the balance piston was reanalysed for force coefficients with the total differential head instead of half pump differential pressure. Both linear damped stability analysis and transient linear unbalance response analysis were performed. Transient response analysis as done at constant operating speeds. The rotor was initially placed in its origin and the rotor motion was searched in time for either stable synchronous, subsynchronous or unstable whirling. The operating speed was gradually increased until subsynchronous and unstable whirling was obtained, for all cases studied. The synchronous forces used in the transient analysis were mechanical unbalance per ISO 1940 Q = 2.5 quality and hydraulic synchronous forces for K_R = .05. The results obtained from transient analysis agreed fairly well with linear damped stability analysis.

Table 1 shows all the results of the analysis including centrifugal growth effects and sag-boring. From the analysis it is seen that a double volute opposed impeller design is less sensitive for rotor instabilities. This is due to the extra center bushing typical for opposed impeller design. This concept increases the rotor eigenvalues and the stabilizing damping forces. Both for amplitude response and rotor stability the smooth annular seals are preferred.

Swirl brakes will reduce the inlet tangential velocities in the impeller eye annular selas, reducing cross coupled stiffness terms. This was modelled in the pump rotor models by reanalysing the force coeffcients of the grooved impeller eye wearrings, for zero tangential inlet velocities of the seals. For those cases when only grooved (serrated) annular seals can be used, the swirl brakes are an attractive solution for increasing instability threshold speed in high speed application. (Swirl brakes will affect however axial loads and leakage flow.)

By increased clearance of the annular seals generated by wear, the force coefficients of these seals are lowered. This results in lower rotor eigenvalues and less stabilizing damping forces. The two pump concepts were analysed for 100, 200 and 300% worn clearances for smooth clearance seal designs. Table 1 shows all subsynchronous whirl and stability threshold speeds analysed. Again the opposed impeller style pump reflects higher threshold speeds than the "equivalent" diffuser, inline impeller style pump. For this specific application, with operating speeds up to 8900 RPM, an opposed impeller, double volute pump with smooth clearance seals will give stable rotor response without any design modifications, while a grooved design concept will already suffer for subsynchronous and unstable shaft whirling with serious rotor/stator damage within a couple of hours of operation. Swirl brakes are necessary to obtain good rotor performance.

The diffuser inline concept is not suited for this application. Even introducing a smooth clearance seal design with sag-bore requires swirl brakes to prevent unstable and subsynchronous whirling.

The instability and subsynchronous whirl threshold speed results produced for the 10 stage reinjection pump in volute, opposed impeller style and diffuser inline impeller style are based on impeller-casing hydrodynamic interaction forces generated by the impeller tip-casing only. Recently it became evident that impeller shroud-casing interaction forces could be as large as the impeller tip-casing interaction forces (Childs, 18). This means that stability threshold speeds evaluated in the last part of this paper are in practice even lower. A review of impeller shroud-casing hydrodynamic force influence on the stability threshold speeds was made by assuming double values of the impeller tip-casing force coefficients. This is suggested by Bolleter (15) and Childs (18). Stability and subsynchronous whirl threshold speeds decreased with 8-10% for smooth clearance seals and 13-17% for grooved clearance seals using design clearances of the seals. It is of prime importance to focus research on the impeller shroud-casing force coefficients. Especially since there are indications that for large tangential fluid velocities at the entry the normally used force coefficients are no longer valid at certain whirl speeds (18). The reverse running pump example of next paragraph emphasizes the importance of impeller shroud-casing interaction forces.

7 IMPELLER SHROUD FORCES INFLUENCING HYDRAULIC TURBINE ROTOR STABILITY AT PART LOAD OPERATION.

The importance of impeller shroud forces on rotor stability can be seen from a real life example.

Pumps running in reverse as turbines are often used for energy recovery applications in the hydro carbon processing industry. A 5 stage double case, opposed impeller, volute type pump, reverse running as hydraulic turbine suffered for subsynchronous vibrations for zero load conditions. (Generator, driven by turbine not switched to utility net.)

The reverse running pump main water data is given below.

Impeller diameter		: 260 mm, 5 stages
Clearance seals		: Smooth, with a few grooves on center and balance bushing
Operating speed		: 5720 RPM
Rated flow	zero load	: 200 m^3/HR
	full load	: 500 m^3/HR
Rated head	zero load	: 1120 mtr.
	full load	: 1760 mtr.
Power recovered		: 1650 kW

During no load conditions the turbine rotor showed serious subsynchronous amplitudes at a frequency 77% of operating speed. At full load, the rotor showed pure synchronous whirling. After installation of gap narrowing rings at the impeller-tip subsynchronous whirling was completely eliminated.

Figure 8 shows rotor amplitude frequency spectrograms from proximity probe measurements near the inboard bearing before and after installation of the gap narrowing rings. The gap narrowing ring at the impeller tip shown in figure 9 have very large radial clearance (3x the impeller eye wearring clearance) and very small gap lengths (2 - 4 mm). This means that these clearances don't contribute as a flow restriction, producing smaller leakage flow rates over the wearrings. Tangential fluid velocities around the shroud are therefore also not influenced by this gap ring due to a reduced leakage flow rate.

It is presently assumed that the gap rings however will block secondary flow fields existing at the impeller tip and around the impeller shroud. The tangential velocity components of these secondary flow fields can act as destabilizing forces, which disappear due to the interference of the gap rings, see figure 9. The turbine at zero load conditions consumes 600 kW, of which only 5% is required to run the unloaded shafting at 5700 RPM. The rest are friction losses, which can only be so large by enormous tangential velocities at the impeller shrouds given disc friction losses. For normal pump operation, flow rates are smaller, while also the flow direction is reversed, resulting in smaller tangential velocities at the impeller shroud. In general it can be concluded that reverse running pumps develop much larger destabilizing forces compared to the normal pump operation. To simulate a reverse running pump in a rotor dynamic analysis is presently very difficult since the impeller tip-casing and impeller shroud-casing forces are unknown. No data on impeller/casing interaction forces for reverse running pumps are found in the public domain. Future research must be focussed on these impeller shroud forces.

CONCLUSIONS

In the design of high speed multi-stage pumps rotor dynamic aspects become very important, as well as control and reduction of the impeller hydraulic excitation forces. General conclusions are:

- Pump impeller hydraulic excitation forces are much larger than mechanical excitation forces present in a pump.
- Quantative ranges for subsynchronous, synchronous and vane passing force components are defined using research data obtained from 47 multi-stage pump tests, representing 327 different impellers.
- Synchronous and non-synchronous force predictions using both mechanical and hydraulic excitation forces yielded results quite close to actual measured vibrations.
- A comparison between an opposed impeller, volute style pump versus an "equivalent" inline impeller diffuser style pump showed much larger vibration amplitude sensitivity to synchronous and vane passing forces of the inline diffuser pump. For subsynchronous force excitation at frequencies 10% of operating speed no difference in sensitivity was obtained. This demonstrates the influence of the center bushing typical for opposed impeller arrangements.
- Internal clearance seals generate load carrying capacity which will decrease the static bearing loads to 5% of total rotor static loads for smooth seals and 10-40% for grooved seals, as long as the seals are concentric with the bearings. Introducing a casing "sag bore" following the static deflection of the rotor increases the vertical bearing load to almost half rotor deadweight. This increases rotor stability threshold speeds.
- Centrifugal deformation of impellers at high speeds, especially unsymmetrical impellers, create tapered seals, with smaller clear-

ances. Seal force coefficients change tremendously due to this geometric change, and based on test data of Childs (14) it appears that centrifugal deformations tend to decrease the stability threshold speed. Good high speed pump designs should consider use of tapered stationary seal parts, to obtain constant clearance seals at operating speeds.

- A stability comparison between a 10 stage opposed impeller, volute style pipe line reinjection pump and an "equivalent" inline impeller, diffuser style pump showed higher stability threshold speeds for the opposed impeller volute style pump. This was obtained both for smooth and grooved clearance seals as well as for worn clearances and for performance with special swirl brakes.
  This effect is generated by the long center bushing at the rotor midspan, typical for opposed impeller arrangements.
- A real life example of a reverse running pump used as hydraulic turbine, showed that gap-reducing rings solved subsynchronous whirling at no load operation. This further proves the importance of the impeller shroud forces as destabilizing force components, as recently published by Childs (18) and Bolleter (15).

ACKNOWLEDGEMENT

The author wishes to express his gratitude to Dr. S. Gopalakrishnan for his continued support. The author is grateful to the Byron Jackson management for permission to present this paper.

REFERENCES

1) Flack R.D., Allaire P.E., "Lateral forces on pump impellers". A literature review, Shock and Vibration Digest vol. 16, 5-14, 1984.
2) Verhoeven J.J., Gopalakrishnan S., "Rotor dynamic behaviour of centrifugal pumps" Shock and Vibration Digest of January 1988, Volume 20, no. 1.
3) Kanki, H., Kawata, J. and Kawatani, T., "Experimental Research on Hydraulic Excitation Force on the Pump Shaft", ASME Paper No. 81-DET-71.
4) Guelich, J., Jud, W. and Hughes, S.F., "Review of Parameters Influencing Hydraulic Forces on Centrifugal Impellers", Paper presented on Seminar "Radial Loads and Axial Thrusts on Centrifugal Pumps", Imech. E., London (5 Feb. 1986).
5) Verhoeven J.J., "Excitation force identification of rotating machines using operational rotor/stator amplitude data and analytical synthesized transfer function". 11th Biennial ASME design engineering division conference on Vibration and Noise, 27-30 September 1987, Boston, Mass.
6) Verhoeven J.J., "Unsteady hydraulic forces in centrifugal pumps". International Conference on "Part load pumping, operation, control & behaviour". 1-2 September 1988, Heriot-Watt University Edinburgh.
7) Stepanoff, A.J., "Centrifugal and Axial Flow Pumps - Theory Design and Application, 2nd Edition, Wiley, New York (1957).
8) "Unbalance response analysis of dual-rotor systems". RESP2V3.
   By: D.F. Li, Research Associate.
   E.J. Gunter, Professor.
   University of Virginia.
9) "Non synchronous response analysis of dual rotor systems" PRS1V2.
   By: J. Verhoeven, April 1986.
   Byron Jackson pumps, Parallelweg 6, Etten-Leur, Holland.
10) Schmaus, R.H. and Barrett, L.E., "Static and Dynamic Properties of Finite Length Turbulent Flow Annular Seals", School of Engineering and Applied Sciences, Univ. Virginia, Charlottesville, Rept. No. UVA/643092/MAE 81/178 (1981).
11) Childs, D.W., "Finite Length Solution for Rotor Dynamic Coefficients of Turbulent Annular Seals", ASME Paper No. 82-LUB-42 (1982).
12) Jerry, B., Brennen, C.E., Caughey, T.K. and Acosta, A., "Forces on Centrifugal Pump Impellers", Proc. of the Second Intl. Pump Symp., Texas A&M University, College Station, Texas (May 1985).
13) Bolleter, U., Wyn, A., Welte, I. and Stürchler, R., "Measurement of Hydrodynamic Matrices of Boilerfeed Pump Impellers", ASME Paper No. 85-DET-148 (1985).
14) Childs D.W., "Finite length solutions for the rotor dynamic coefficients of constant clearance and convergent tapered seals". Paper C276/84 3rd International Conference Vibrations in rotating machinery, University of York, England 11-13 September 1984.
15) Pace S.E., Florjancic S., Bolleter U., "Rotor dynamic developments for high speed multi-stage pumps". Paper presented at 3rd International Pump Symposium Texas A&M University, May 1986.
16) Brown R.D., "Vibration phenomena in boiler feed pumps originating from fluid forces". Proc. of Int. Conference on rotor dynamic problems in Power plants, Rome 1982.
17) Adams, M.L., Mackay E., "Development of advanced rotor/bearing systems for feed water pumps" EPRI research project 1266-7, phase 1, November 1979, final report.
18) Childs D.W., "Fluid-structure interaction forces at pump-impeller-shroud surfaces for rotor dynamic calculations" 11th Biennial ASME design engineering division conference on vibration and noise, 27-30 September 1987, Boston Mass.
19) Black H.F., "Effects of Hydraulic Forces in Annular Pressure Seals on the Vibrations of Centrifugal Pumps". Journal of Mechanical Engineering Science, Volume 11, no. 2, 1969.

Table 1 Subsynchronous whirl and stability threshold speed results of in-line diffuser style and opposed impeller volute style arrangement, for different conditions

CONDITION	OPPOSED IMPELLER, VOLUTE STYLE		"EQUIVALENT" INLINE IMPELLER, DIFFUSER STYLE	
	GROOVED	SMOOTH	GROOVED	SMOOTH
Normal design clearance	SSW 5100 RPM Unstable 5300 RPM	SSW 10200 RPM Unstable 10700 RPM	SSW 4600 RPM Unstable 4750 RPM	SSW 6200 RPM Unstable 6800 RPM
100% Worn clearance		SSW 9100 RPM Unstable 9400 RPM		SSW 5000 RPM Unstable 5300 RPM
200% Worn clearance		SSW 7600 RPM Unstable 7750 RPM		SSW 4150 RPM Unstable 4350 RPM
300% Worn clearance		SSW 5050 RPM Unstable 5200 RPM		SSW 2400 RPM Unstable 2500 RPM
Normal design clearance with swirl brakes	SSW 7650 RPM Unstable 7950 RPM		SSW 7500 RPM Unstable 7650 RPM	

SSW = Subsynchronous Whirling

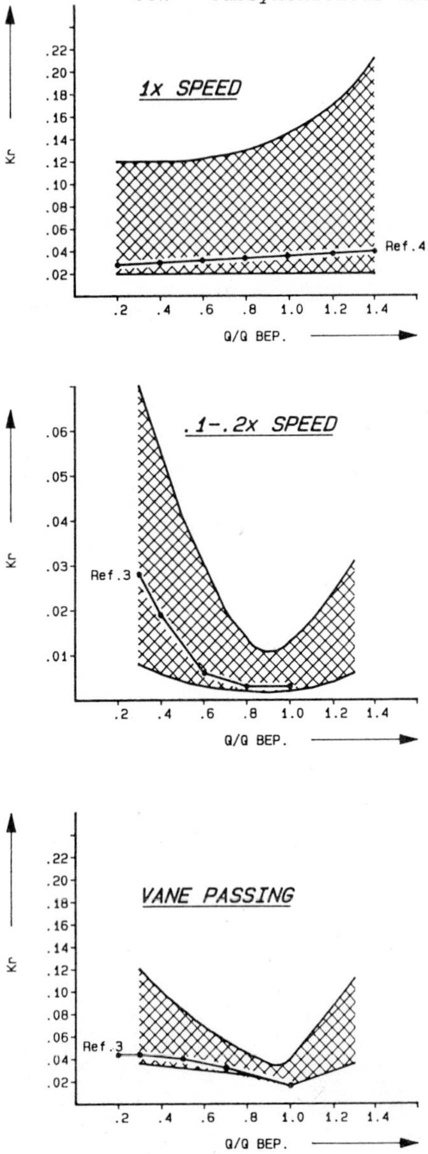

Fig 1 Ranges of hydraulic impeller excitation forces for synchronous, subsynchronous and vane passing frequency

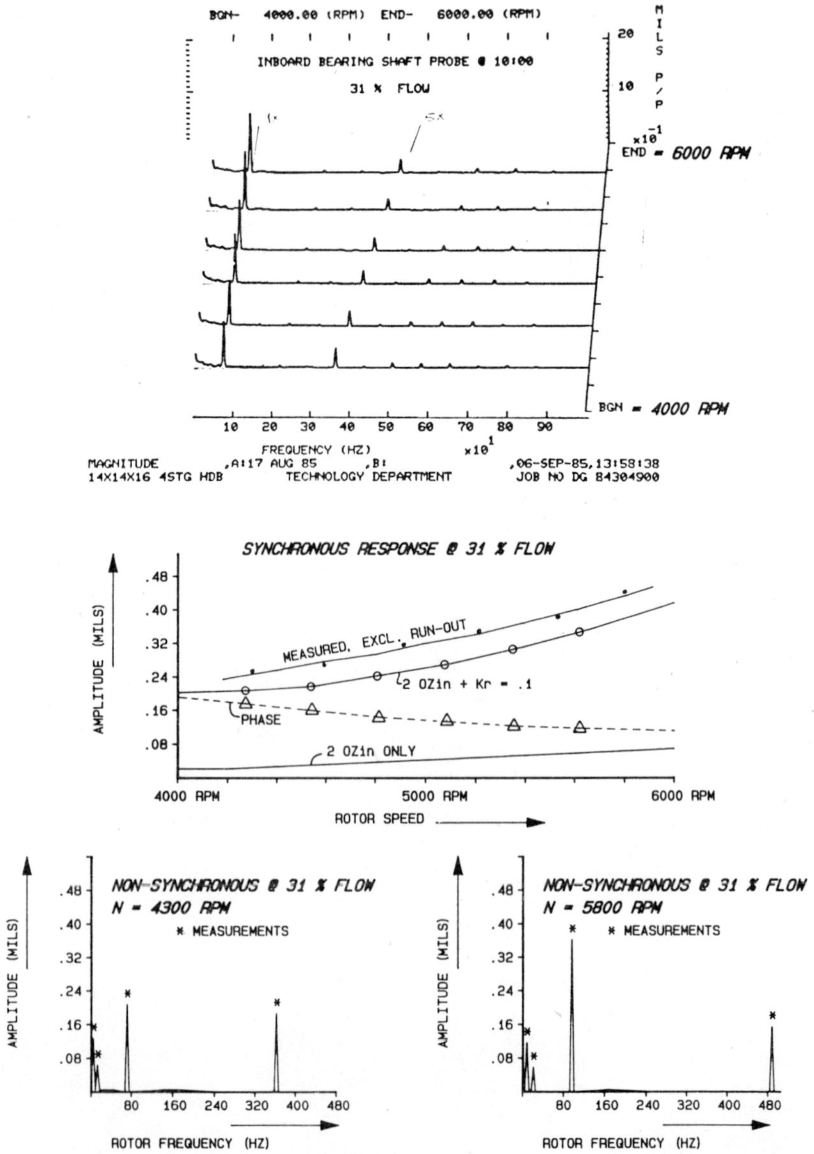

Fig 2 Four-stage boilerfeed pump; vibration measurements versus analysis

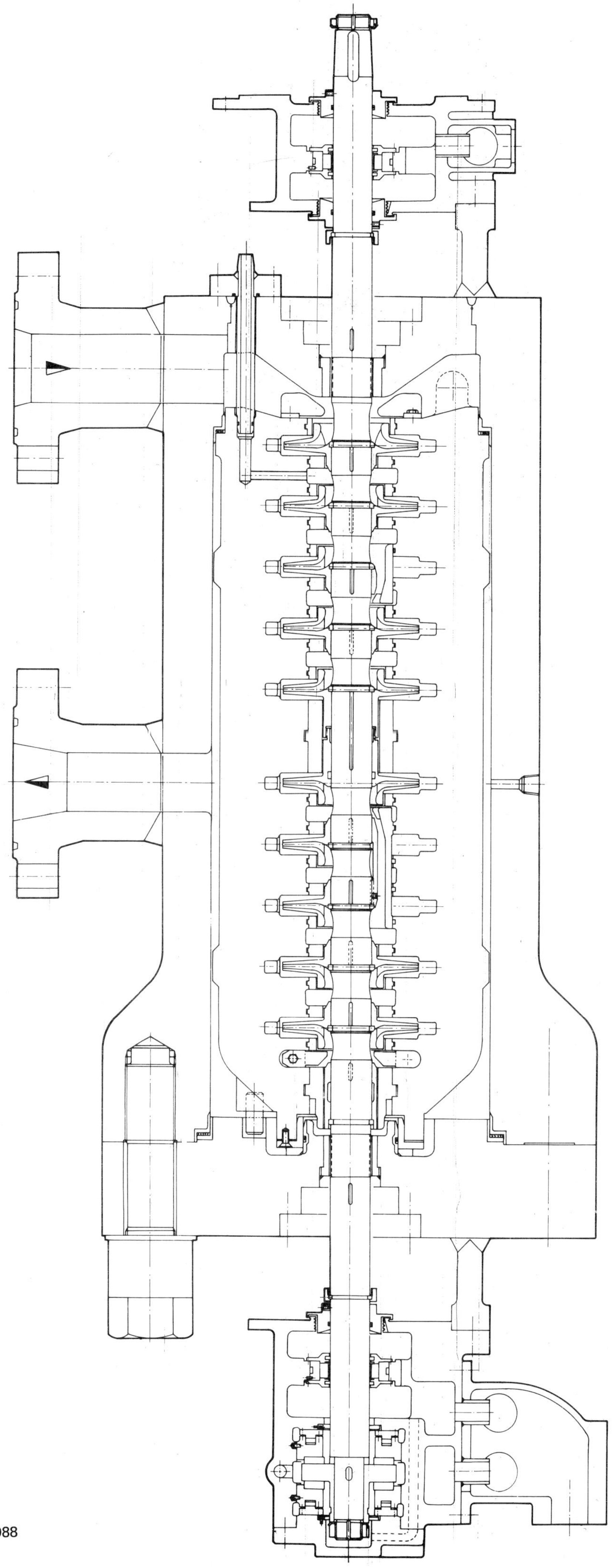

Fig 4 Horizontally opposed impeller, ten-stage axially split volute, double case pump for offshore reinjection service at 8900 r/min (bottom part shows the machine destaged to seven stages)

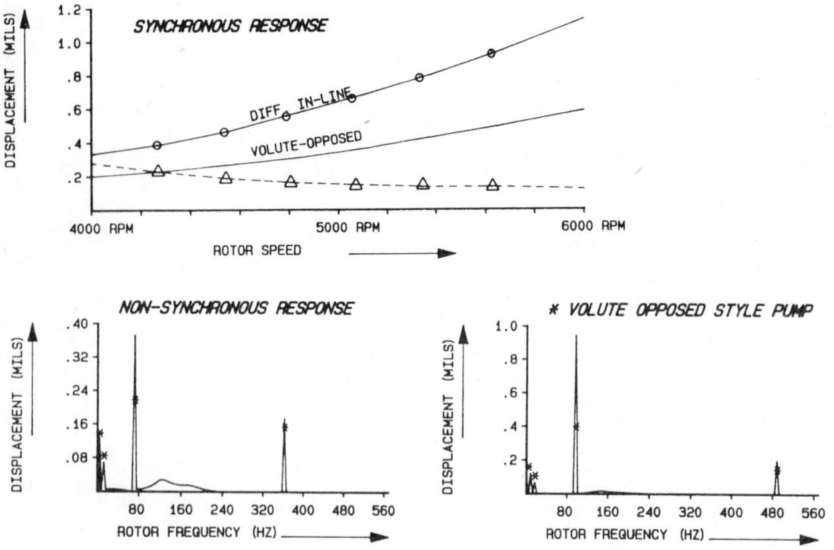

FREQ.	L.F.	1x	NZ
VOLUTE OPPOSED	1.17	2.18	.48
DIFFUSER IN-LINE	.93	6.16	.69

ROTOR MIDSPAN AMPL.
@ 5800 RPM IN MILS.

Fig 3 Synchronous and non-synchronous response results of 'equivalent' in-line diffuser-style pump

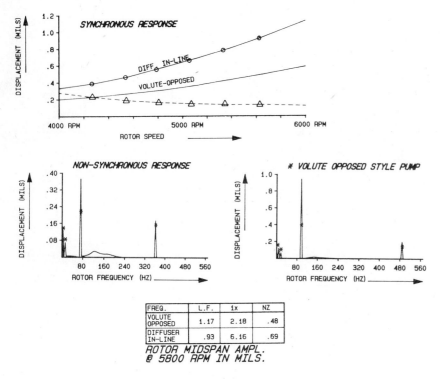

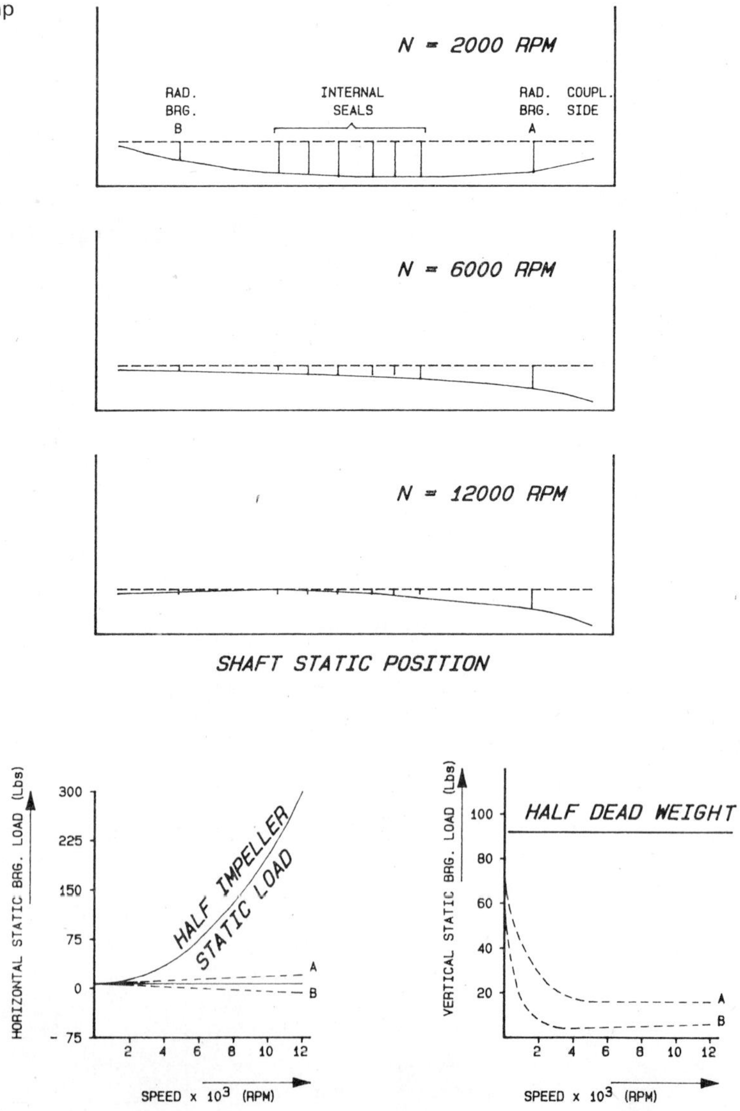

Fig 5 Static bearing load evaluation including clearance seal load carrying capacity of ten-stage reinjection pump

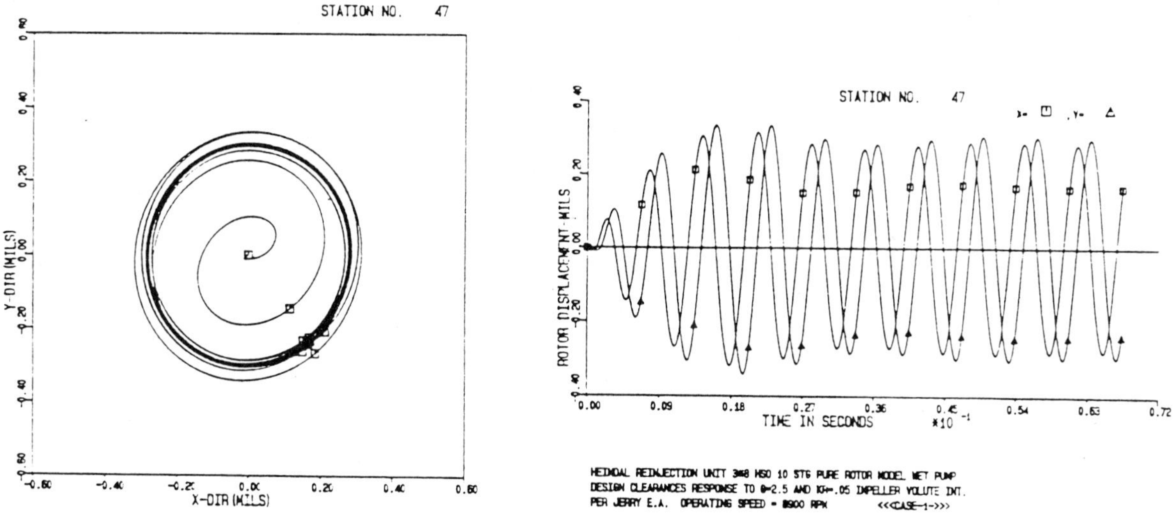

CASE -1- STABLE SYSTEM

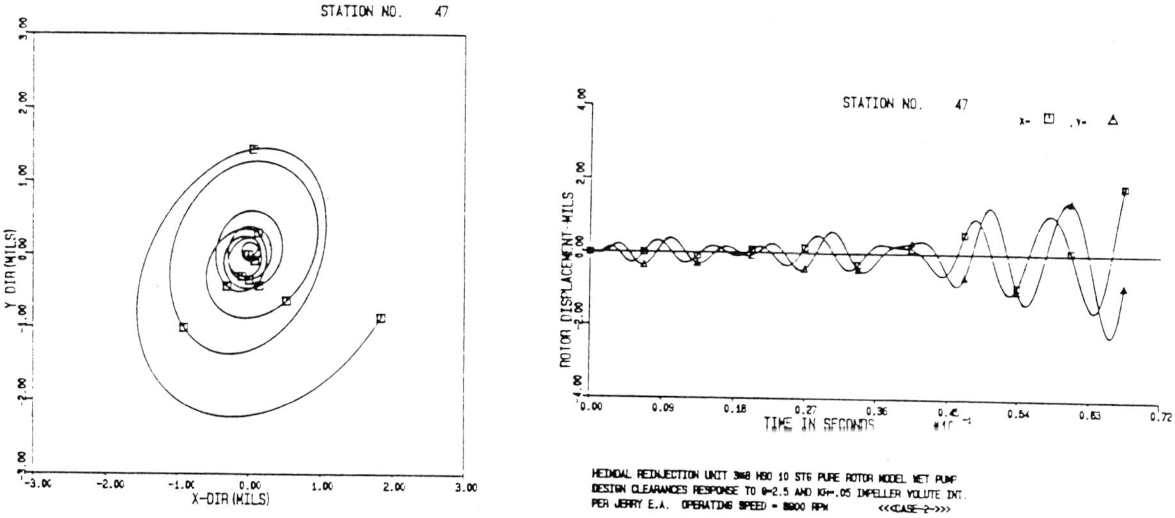

CASE -2- UNSTABLE SYSTEM

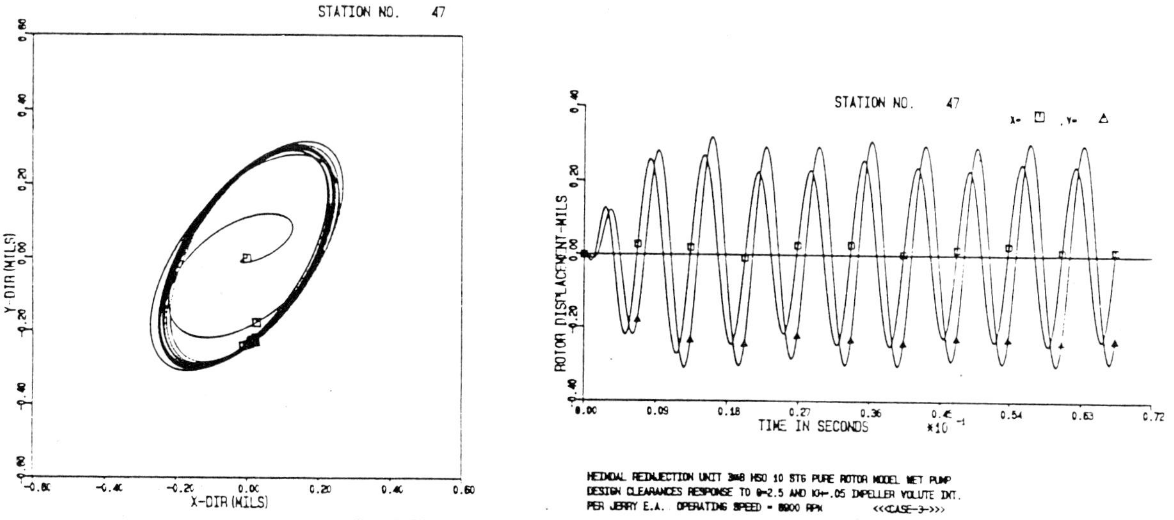

CASE -3- STABLE SYSTEM

CASE	1st Mode	MODE SHAPE
1	6955 RPM	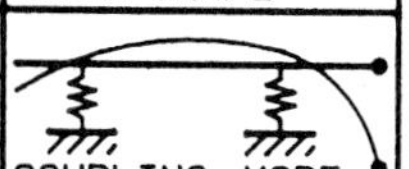
2	5780 RPM	
3	6513 RPM	COUPLING MODE

Fig 6 Transient response analysis results of synchronous excitation to show influence of corrected static bearing loads and pump case 'sag-boring'

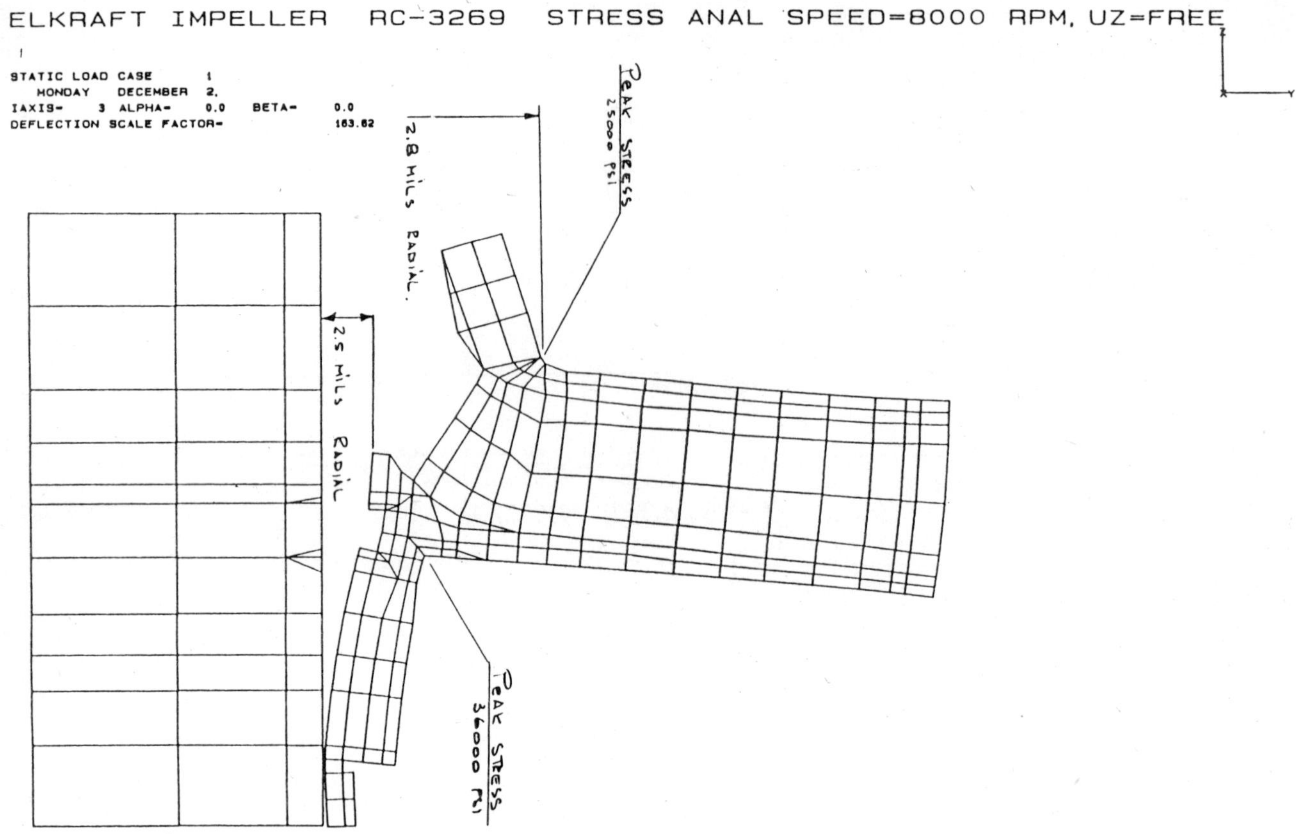

Fig 7 Asymmetrical impeller deformation due to centrifugal growth; pressure loading and shrinkfit at 8000 r/min

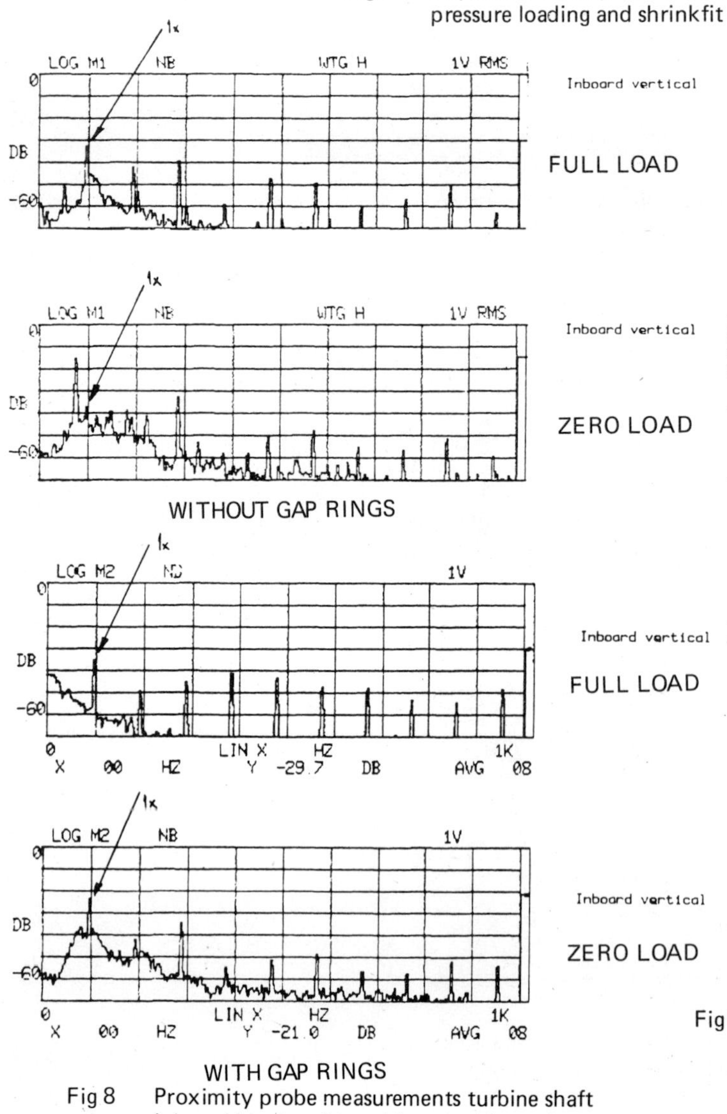

Fig 8 Proximity probe measurements turbine shaft inboard bearing side, with and without gap reducing rings

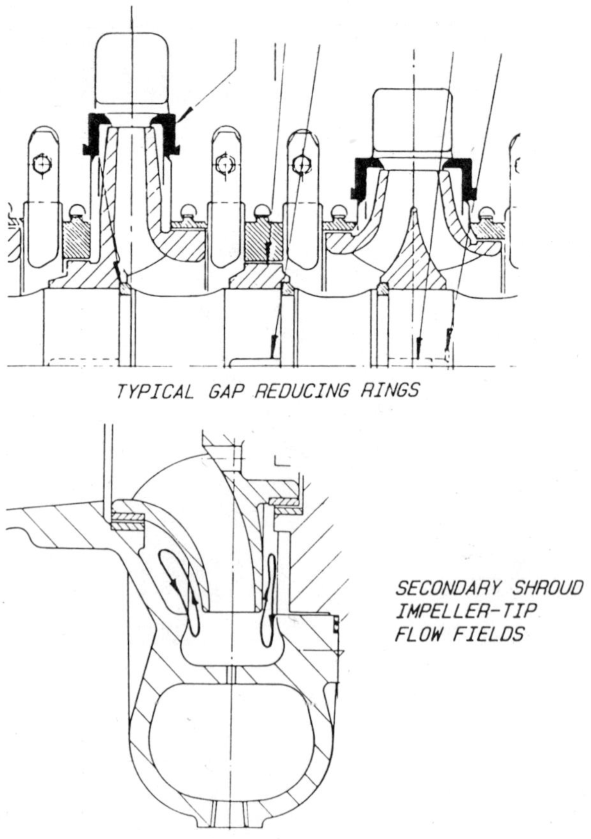

Fig 9 Gap reducing rings and secondary shroud flow fields

C244/88

The influence of seals on the dynamics of turbopumps

W DIEWALD, Dipl-Ing, R NORDMANN, Dr-Ing, C AST, Dipl-Ing, C HERZER, Dipl-Ing and C VOGEL, Dipl-Ing
Department of Mechanical Engineering, University of Kaiserslautern, Kaiserslautern, West Germany

ABSTRACT

One of the main problems in designing a centrifugal pump is to achieve a good efficiency while not neglecting the dynamic performance of the machine. The first aspect leads to the design of grooved seals in order to minimize the leakage flow. But the influence of these grooves on the dynamic behaviour is not known very well.
This paper presents experimental and theoretical results of the rotordynamic coefficients for different groove shapes and depths in seals. Finally the coefficients are applied to a simple pump model.

NOTATION

C	seal clearance
c_L	leakage coefficient = $\dot{V}/(2\pi R^2)\cdot(\rho/2\Delta p)^{.5}$
D, d	direct and crosscoupled damping
d_{ij}	dynamic coefficients for damping
F_x, F_y	external forces
H	groove depth
K, k	direct and crosscoupled stiffness
k_{ij}	dynamic coefficients for stiffness
L	seal length
M, m	direct and crosscoupled inertia
m_{ij}	dynamic coefficients for inertia
Δp	pressure drop
R	shaft radius
v	leakage velocity
$\dot{V}$	volume flux through the seal
x, y	coordinates for two orthogonal directions
α	damping coefficient (real part of the eigenvalue)
ξ	modal damping = $-\alpha/(\alpha^2+\omega^2)^{-.5}$
ρ	fluid density
ω	eigenfrequency (imaginary part of the eigenvalue)
ω_o	eigenfrequency of the "dry" shaft
Ω	rotational frequency

INTRODUCTION

Like all kinds of machines, centrifugal pumps shall operate with a good efficiency. Because of the high rotational speeds, contactless seals have to be used to seperate areas of different pressures in the pump. But the leakage flow through these seals reduces the efficiency of the machine (fig. 1).

As a result this aspect, pump manufacturers very often use grooved seal surfaces to give more resistance to the fluid flow and so to reduce the leakage.

Besides the efficiency, the dynamic behavior of the seals is also an important aspect for todays turbopumps. The operational speed usually is much higher than the first critical speed of the "dry" shaft. But a resonance problem is avoided because during operation the seals introduce a great amount of stiffness to the system. Therefore they are also called wear rings. But not only stiffness, also damping arises from the seals and the physical mechanisms can cause instability for the centrifugal pump.

So the influence of seals on the dynamic behaviour of turbopumps is very important, and has to be taken into account (ref. 1).
Especially the influence of grooved seal surfaces has to be investigated because these kind of seals are used for better effectiveness as mentioned before.

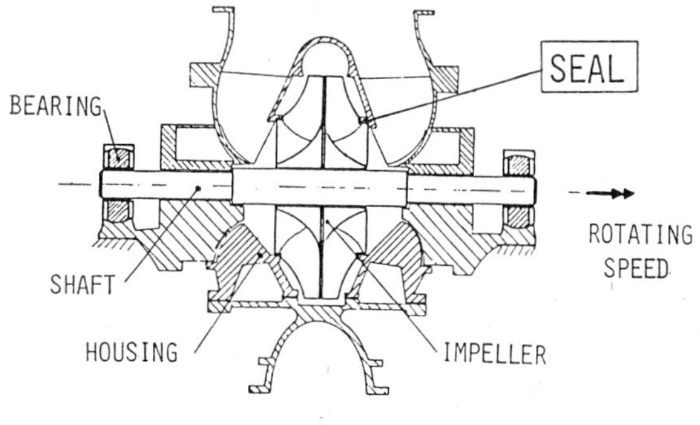

Fig 1 Cross-section of a turbopump

SEAL GEOMETRIES

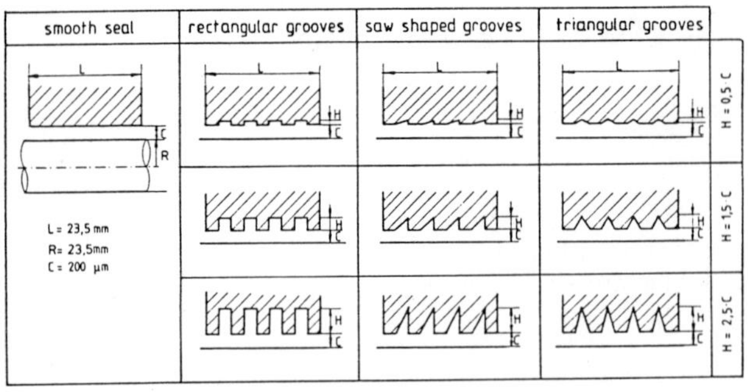

Fig 2 Seals with different groove shapes and depths

These influences are studied in this paper. For three different groove shapes with various depths (fig. 2) the dynamic coefficients of the seals are determined theoretically and experimentally. The results are discussed and the influence on the rotordynamic behaviour of a simple pump model is demonstrated.

THEORETICAL APPROACHES TO DETERMINE SEAL-COEFFICIENTS

It is the aim of the theoretical approaches to describe the relations between external forces acting on a seal and the motions of the shaft by a mathematical expression. The results can be easily included in a finite element procedure for the overall behaviour of a pump, if they are described in a matrix equation:

$$\begin{bmatrix} m_{11} & m_{12} \\ m_{21} & m_{22} \end{bmatrix} \begin{bmatrix} \ddot{x} \\ \ddot{y} \end{bmatrix} + \begin{bmatrix} d_{11} & d_{12} \\ d_{21} & d_{22} \end{bmatrix} \begin{bmatrix} \dot{x} \\ \dot{y} \end{bmatrix} + \begin{bmatrix} k_{11} & k_{12} \\ k_{21} & k_{22} \end{bmatrix} \begin{bmatrix} x \\ y \end{bmatrix} = \begin{bmatrix} F_x \\ F_y \end{bmatrix} \quad (1)$$

m_{ij}, d_{ij}, k_{ij} dynamic coefficients for inertia, damping and stiffness

x, y coordinates for two orthogonal directions

F_x, F_y external forces

In the following chapters two ways of calculating these dynamic seal coefficients are briefly described. Both theories come out with skewsymmetric matrices.

$m_{11} = m_{22} = M$ direct inertia

$m_{12} = -m_{21} = m$ crosscoupled inertia

$d_{11} = d_{22} = D$ direct damping

$d_{12} = -d_{21} = d$ crosscoupled damping

$k_{11} = k_{22} = K$ direct stiffness

$k_{12} = -k_{21} = k$ crosscoupled stiffness

Finite Length Theory

In 1982 CHILDS published a finite length theory (ref. 2,3) to calculate seal coefficients. It is based on a bulk-flow model (fig. 3a) for the

THEORETICAL APROACHES TO DETERMINE SEAL-COEFFICIENTS

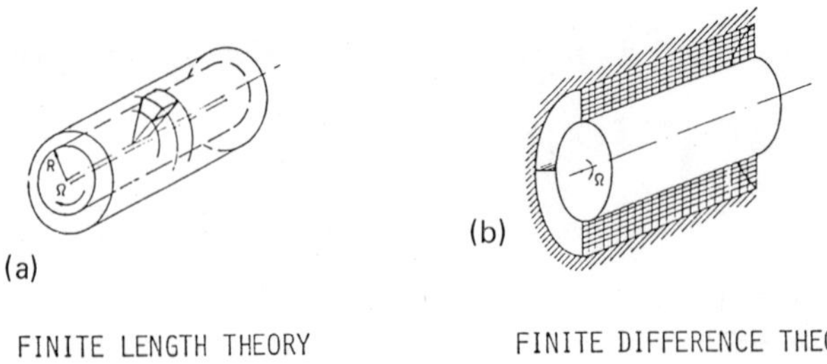

$$\begin{bmatrix} M & m \\ -m & M \end{bmatrix} \cdot \begin{bmatrix} \ddot{x}_1 \\ \ddot{x}_2 \end{bmatrix} + \begin{bmatrix} D & d \\ -d & D \end{bmatrix} \cdot \begin{bmatrix} \dot{x}_1 \\ \dot{x}_2 \end{bmatrix} + \begin{bmatrix} K & k \\ -k & K \end{bmatrix} \cdot \begin{bmatrix} x_1 \\ x_2 \end{bmatrix} = \begin{bmatrix} F_1 \\ F_2 \end{bmatrix}$$

Fig 3 Models for (a) finite length and (b) finite difference theory

fluid behaviour in the seal. The force equilibrium equations together with continuity equations can be solved by a perturbation method. Integrating the resulting pressure distribution over the seal surface leads to the seal-forces from which the dynamic coefficients can be extracted.

Enclosed in this calculation are empirical constants to include the friction at the rotor and stator surfaces, which have to be determined experimentally.

Regarding grooved seals it is possible to extend the model. The continuity equation in circumferential direction is enlarged by the additional cross-sections of the grooves, and the empirical friction factors are determined separately for circumferential and axial directions. Also the friction of the circumferential fluid flow at the groove walls is taken into account.

This extended bulk-flow theory was developed by NORDMANN (ref. 4) and is applicable to grooved seals, but some empirical constants have to be measured in addition.

Finite Difference Theory

In 1986 DIETZEN and NORDMANN (ref. 5) published a method to calculate seal-coefficients by means of a finite difference technique. The method is based on the Navier-Stokes-Equations, continuity equation and energy equation, so that every fluid flow can be modelled.
To calculate the fluid flow in a seal, the seal gap is described with a grid of calculation points (fig. 3b). Using also a disturbance method the fluid velocities and pressure values are calculated for all nodal points as a function of the values of the surrounding grid points.

The wall shear stresses are modelled by a logarithmical law, which only influences the first grid line at the wall. So it is possible to calculate the fluid flow for all seal geometries without using any empirical values. Again the resulting pressure distribution is integrated yielding the forces and finally the desired seal coefficients.

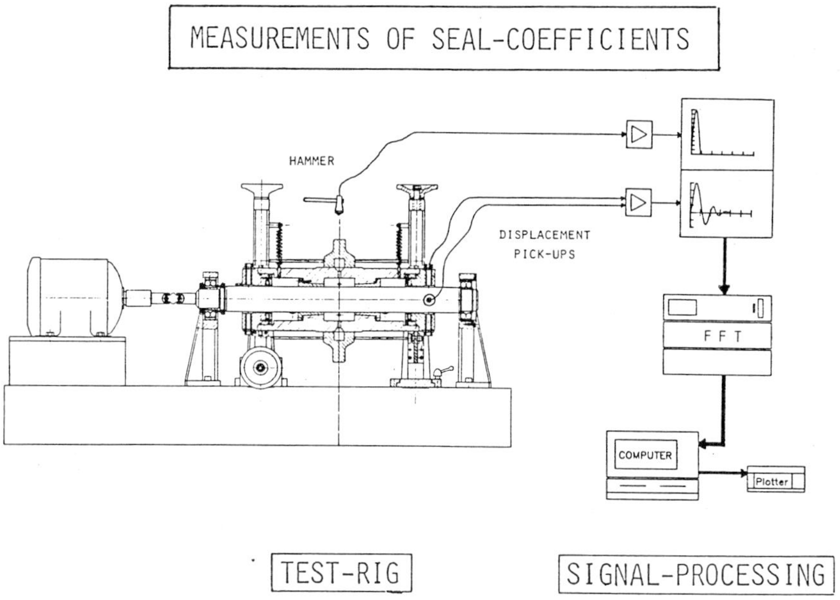

Fig 4 Test rig and signal processing

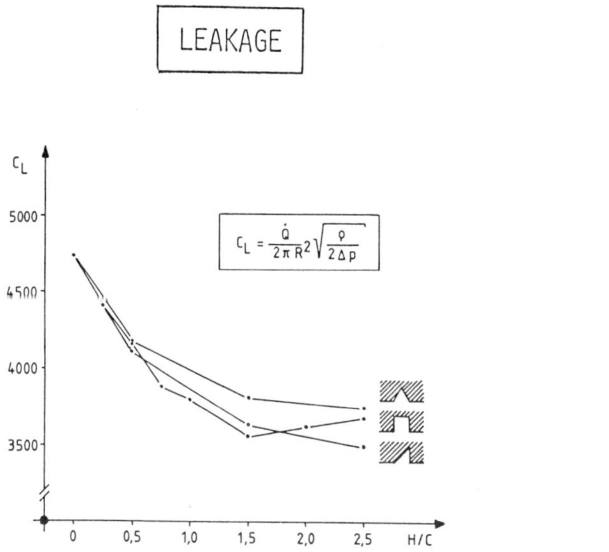

Fig 5 Leakage coefficients for different seal geometries versus groove depths

A disadvantage for this finite difference theory are the high computational times it takes, but the results are very good compared to measurements as the next chapter will show.

MEASUREMENTS OF SEAL-COEFFICIENTS

Test Rig

A test rig to measure rotordynamic seal coefficients was built at the University of Kaiserslautern by MASSMANN and NORDMANN (fig. 4), (ref. 6).

Between a stiff shaft which is rigidly supported and a stiff housing two symmetrical seal inlets are situated. During operation the shaft rotates and water is pumped through the seals. Their dynamic behaviour is measured by impacting the housing with a hammer and recording the input (force) and output signal (motion of the housing in relation to the shaft). A FFT-Analyser calculates the transfer function from wich the dynamic coefficients are extracted.

Because of the large mass of the housing, which is moved during measurements, the inertia coefficients cannot be evaluated by this test rig.

Leakage

A first result for the seal geometries shown in fig. 2 is the leakage performance. Fig. 5 presents the dimensionless leakage coefficient c_L for the different seal geometries versus groove depth.

It is obvious that the smooth seal has the highest leakage. For increasing groove depths, the coefficient goes down in general. Only the rectangular groove shape shows an increasing leakage for larger depths. Over all, rectangular and sawshaped grooves show the best performance.

Dynamic Coefficients

The resulting seal coefficients for one operational point are given in fig. 6. They are compared with theoretical results coming from the two theories mentioned before. The inertia coefficients are not shown in this figure, because they could not be measured as mentioned before.

For all geometries the values for the direct stiffness term are influenced very much by the various groove depths. They decrease with increasing depths. Cross coupled stiffnesses also decrease slightly, while cross coupled damping stays nearly constant. Especially for the rectangular groove shapes direct damping also goes down with increasing depths.

The comparison with the calculated data shows very good agreement for the finite difference theory in all parameters.
Finite length theory predicts higher direct stiffness terms, but shows good agreement for the other coefficients.

The results for other operational points are qualitatively similar.

STIFFNESS- AND DAMPING-COEFFICIENTS

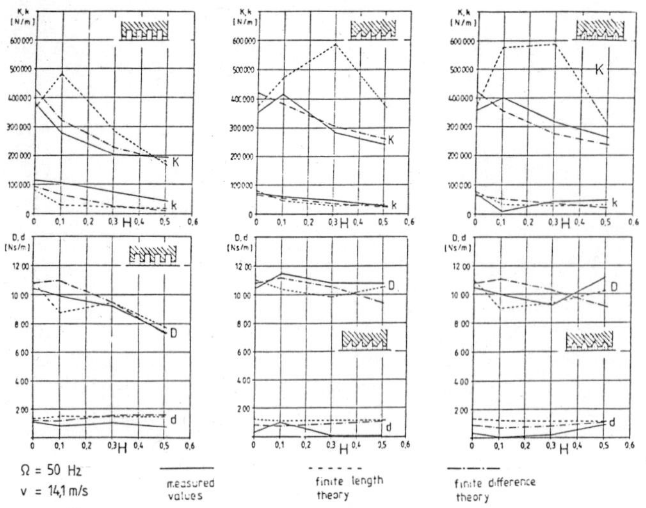

Fig 6 Dynamic stiffness (K, k) and damping (D, d) coefficients for different seal geometries and groove depths

Discussion

Using the finite difference technique it is possible to calculate accurate dynamic seal coefficients for all seal geometries, but it is very time consuming. Thus it is also of interest to find out when the finite length method works satisfactorily.

Direct damping is calculated quite accurately by finite length theory, but this may arise from the fact that measured friction factors have to be included in the calculation.

Also finite length theory shows reasonable results for rectangular groove shapes, but is very wrong for other geometries especially for the very important direct stiffness terms. This can be explained regarding the fluid flow in the seals calculated by the finite difference technique (fig. 7). One can see a recirculation inside rectangular grooves which does not affect the main flow very much. So the bulk flow model, with an extended circumferential continuity equation, can be applied successfully. For other groove geometries the main flow direction is very much influenced by the grooves, so that the bulk flow model is not applicable.

INFLUENCE ON EIGENVALUES

For real machines, it is not the single dynamic coefficients that are of direct importance but the resulting overall behaviour, for example critical speed, damping and stability behaviour. This information is given by the eigenvalues of a centrifugal pump, which consist of eigenfrequencies and damping values.

In order to demonstrate the effect of different seal geometries on the rotordynamics of turbopumps a simple rotor model is chosen. Fig. 8 shows a Jeffcott rotor as a model for a double suction feed pump.

The coefficients for the different seal geometries are applied to this model. The results at operational speed are shown in fig. 9.

FLOW FIELDS

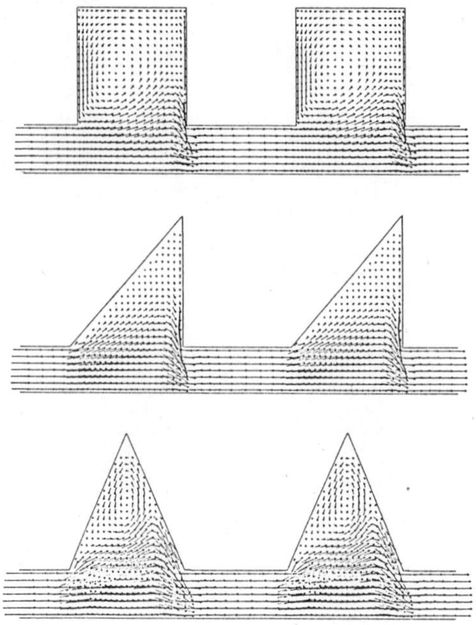

Fig 7 Flow fields for different groove shapes

TEST-MODEL

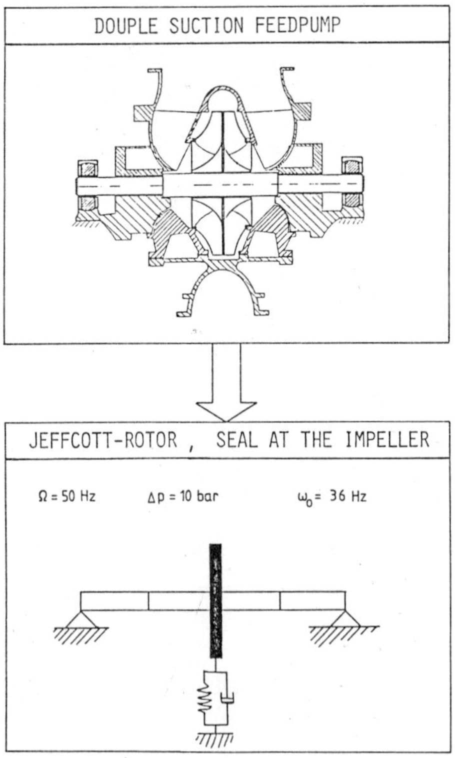

Fig 8 Simple model of a double suction feed pump

A smooth seal increases the first eigenfrequency of the dry shaft by 96 % and all grooved seals have a less stiffening effect, getting smaller with increasing groove depths. The best groove geometry to introduce stiffness to the system is saw shaped. Values for modal damping are improved using grooved seals, but a general trend cannot be observed. It has to be pointed out, that modal damping is determined by both,

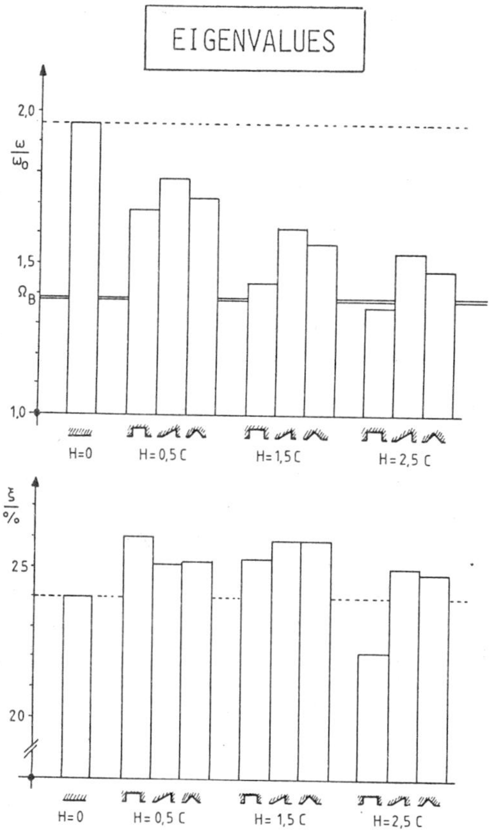

Fig 9 Eigenvalues of the pump model versus groove depths at operational speed for different seal geometries

the real and imaginary part of the eigenvalue. The real parts of the eigenvalues decrease for grooved seals (fig. 10).

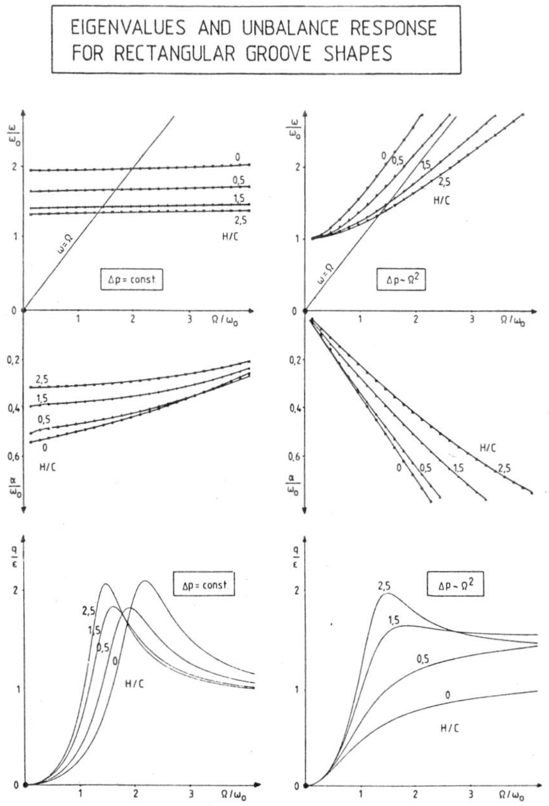

Fig 10 Eigenvalues and unbalance responses for rectangular groove shapes

In fig. 10 eigenvalues and unbalance responses are shown versus running speed for rectangular groove shapes. For a constant pressure drop versus speed, which is not realistic for a pump, eigenfrequencies do not change very much, damping slightly decreases and the unbalance responses show no dramatic differences for various groove depths. The pressure drop in a real machine would be nearly proportional to running speed squared. In this case the eigenfrequencies and damping values increase with running speed. The unbalance responses for grooved seals are now much worse compared to plain seals.

Finally the calculated eigenvalues of a multistage boiler feed pump are shown using the data for grooved seals and plain seals respectively (fig. 11). With one exception all eigenfrequencies are increased using plain seals. The results for modal damping show no uniform characteristics.

So from dynamic point of view it is hard to tell whether one should use plain or grooved seals. The special application has to be investigated.

CONCLUSIONS

The design of grooved seals for turbopumps reduces the leakage flow and so increases the efficiency.

The influence of these kinds of seals on the dynamic performance is investigated for different types of seals. Rotordynamic seal-coefficients are evaluated experimentally and the results are compared to predictions from finite length theory and a finite difference technique

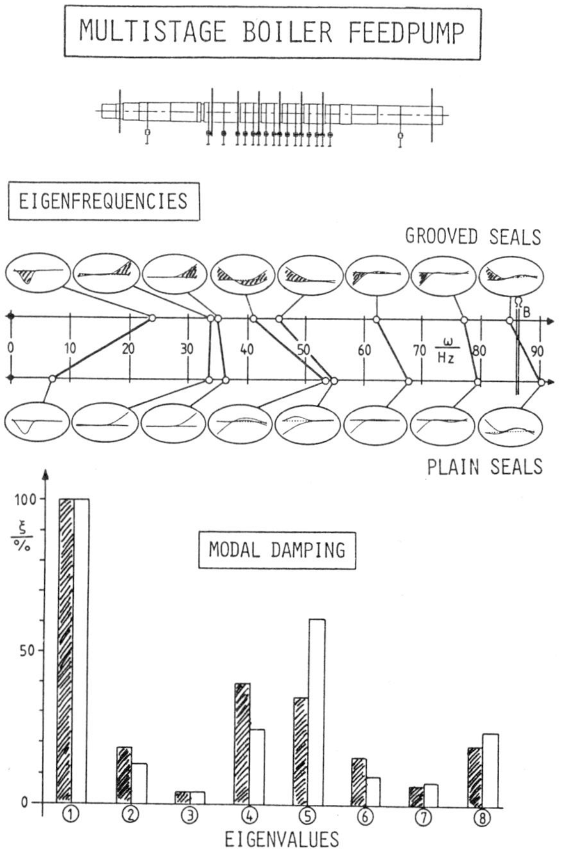

Fig 11 Eigenvalues of a multi-stage boiler feed pump

respectively. The latter gives the best results for all seal geometries, while finite length theory seems to be applicable for rectangular grooves only.

The influence on the eigenvalues of a simple pump model shows less stiffening effect of grooved seals compared to plain seals. Whether from a dynamic point of view grooved seals are better or worse compared to plain seals cannot be stated generally. An analysis of the special application seems to be necessary.

LITERATURE

1. W. DIEWALD, R. NORDMANN, "Dynamic Analysis of Centrifugal Pump Rotors with Fluid-Mechanical Interactions", 11th Biennal Conference on Mechanical Vibration and Noise, Boston, MA, September 27-30, 1987.

2. D.W. CHILDS, "Dynamic Analysis of Turbulent Annular Seals Based on Hirs' Lubrication Equation", Journal of Lubrication Technology, ASME-Paper No. 82 - Lub - 41, 1982.

3. D.W. CHILDS, "Finite-Length Solutions for Rotordynamic Coefficients of Turbulent Annular Seals", Journal of Lubrication Technology, ASME-Paper No. 82 - Lub - 42, 1982.

4. R. NORDMANN, "Rotordynamic Coefficients and Leakage Flow of Parallel Grooved Seals and Smooth Seals", the 4th Workshop on Rotordynamic Instability Problems in High Performance Turbomachinery, Texas A&M, June 2-4, 1986.

5. F.-J. DIETZEN, R. NORDMANN, "Calculating Rotordynamic Coefficients of Seals by Finite-Difference Techniques, the 4th Workshop on Rotordynamic Instability Problems in High Performance Turbomachinery, Texas A&M University, June 2-4, 1986.

6. R. NORDMANN, H. MASSMANN, "Identification of Stiffness, Damping and Mass Coefficients for Annular Seals", Vibrations in Rotating Machinery, IMechE, York, 1984.

C304/88

'Wet critical speeds' of boiler feed pump beyond the pump operational range

G MARENCO
Worthington Pompe Italia spa, Desio, Milan, Italy

SYNOPSIS

Modern boiler feed pump design philosophy considers the "stiff shaft" configuration a fundamental feature able to improve the dynamic behaviour of pumps by raising the rotor critical speeds over the operational range and therefore improving pump reliability.
Advantages and disadvantages of this fairly new philosophy are described, underlining technical difficulties to be overcome to achieve this objective.
The effect of possible design alternatives, i.e. the different geometry of shaft, tilting pad journal bearings, interstage seals, balance drum profile, taking into account casing and baseplate contribution, is emphasized in order to demonstrate that, only utilizing the maximum stiffening contribution of all aforesaid components, we can obtain a really "stiff shaft" pump configuration.
A case history is presented specifically describing different design alternatives examined for a 500 MW boiler feed pump.
The influence of cross coupled stiffness and damping coefficients, as well as considerations on volumetric efficiency variations, connected to different components geometry, are illustrated.
The excitation response diagrams confirm that a suitable combination of the main parameters can be found to avoid any interaction between shaft and/or casing natural frequencies and the well known exciting sources, multiple of the design pump rotational speed.
Experimental measurements will be carried out during a hot temperature, full speed pump test, to compare the real pump dynamic behaviour with the above mentioned analytical calculations.
Spectral vibration maps recorded on bearing housings during the pump run-up, not presently available, will be distributed at the Conference Meeting.

1. INTRODUCTION

"Boiler feed water pump failures are a major cause of power plant unavailability". This sentence, reported on EPRI Journal of July/August '87 (ref.1), confirms that the "Reliability" is becoming the most important feature to be pursued on feed-pump design and operation phases.

Besides, the current trend of using the fossil fuel units in cycling or peaking services involves an increase of the number of hours during which feed-pumps operate at reduced loads.
This undoubtly represents a potential source of failures, especially if we consider that the real pump behaviour at off-design conditions, remains nowadays, not entirely predictable.

The improvement of feed-pump reliability passes, therefore, through a better understanding of their hydraulic and dynamic behaviours.
Particularly, the pump-system rotordynamic analysis, taking into account the influence of all pump components, as well as their hydraulic interactions, is a proper mean to improve reliability.

As demonstrated by common experiences, the boiler feed pumps are capable to operate quietly, running through their critical speeds (sometimes in presence of subsynchronous vibrations) during the usual run-up or shut-down phases, as well as during the continuous operation at reduced loads.
A good system damping degree and a low level of interactions between resonant speeds and typical exciting forces (unbalance, recirculation flow, "vane" passing, etc.) permits infact low wear rates during the normal operating conditions.

This is no longer valid in presence of anomalous transient conditions, as it occurs during emergency start-up of the pump from hot stand-by, (ref.2) or when the pump occasionally operates at reduced loads, in correspondence of the onset recirculation flow (ref. 13).

These operating conditions, characterized by the presence of relevant exciting forces, drastically increase the risks of contact between the rotating and the stationary parts.

Reiterated contacts promote an increase of the rotor unbalance with a contemporaneous decrease of damping effects involving a consequent rapid increase of vibration levels.

For these reasons we believe that, the "Stiff Shaft" configuration, aiming at laying the wet critical speeds well over the pump operational range, maintains an unquestionable validity, mainly because this specific design tends to disengage the potential interactions between the a.m.exciting forces and the pump natural frequencies.

Efforts must be taken to achieve this objective still at off-design and "Worn" clearances conditions, restraining, as far as possible, the exciting forces amounts and defining adequately the seal characteristics in order to ensure a sufficient margin against subsynchronous vibration raising.

During the last 10 years, the major pump manufacturers, often in cooperation with widely known Universities, had made appreciable progress on this peculiar field by means of appropriate analytical and/or experimental research activities. Advanced analytical programs, able to predict the multistage pump dynamic behaviour, are presently worldwide available (ref. 3,4,5,6,7,8). However, as well underlined by W.D. Marscher (ref.13) and investigated in details by several authors (ref. 14,15,16,17,18,19,20,21), the centrifugal pump vibration levels are heavily influenced by the fluid forces actions.

The worthy part of these fluid forces is represented by the seals Lomakin effect which plays the well known positive dynamic role.

Other important hydraulic forces, instead, often constitute an undesired trouble source.
Within this latter group, the following shall be mentioned:

- "Vane pass" loads caused from the impeller-diffuser interaction (ref. 17,18)
- Internal recirculation flow (ref. 19,20)
- Bearings, seals and balancing drum Cross-Coupled forces when dominant on Direct values (ref.4, 21)

Therefore, the prediction of the pump dynamic behaviour, pursued using proven analytical tools, cannot absolutely neglect the presence of these potential vibration sources, to avoid, whenever possible, dangerous interactions between the hydraulic forces and the pump-system natural frequencies.

2. BOILER FEED PUMP ROTORDYNAMIC DESIGN

The dynamic behaviour of a 5 (*) stages, horizontal, Boiler Feed-Pump (fig.1), recently manufactured for a 500 MW Power Station, was determined through a series of integrated finite element programs, developed in cooperation with Milan Polytechnic.
(*) - including a kicker stage.

The Conditions of Service and the relevant pump Geometrical Data are reported on the following Table 1:

TABLE 1

B.E.P. Capacity	970	(m3/h)
Total dev. Head	2480	(m)
Design Speed	5800	(rpm)
Normal Operat.Range	400 ÷ 1150	(m3/h)
Run-Out Capacity	1350	(m3/h)
Bareshaft Pump Weight	4400	(Kgm/s2)
Overall Rotor Length	2403	(mm)
Bearing Span	1770	(mm)
Average Shaft Diam.	142	(mm)
Rotor Weight	442	(Kgm/s2)

Calculations were executed separately for casing and rotor, taking into account for this latter the casing mechanical impedances and the influence of oil-film bearings, interstage seals and balancing device (ref.9).

The following main assumptions were formulated for the analytical model description.

a) Barrel casing and Baseplate

A particular compact barrel casing, supported by a really stiff baseplate, constitute the basis to achieve higher system natural frequencies values.
Reasonable values of baseplate stiffnesses were considered referring to the experimental results carried out on similar baseplate configurations (direct stiffness values assumed = 2×10^9 N/m).

b) Rotor

The rotor analysis was performed simulating the "Worn" clearances configuration, i.e. assuming the diametral clearances in correspondence of impeller wear-rings and balancing drum equal to .7(mm) (twice the original value). Several design alternatives were investigated varying the following parameters:

- Shaft diameters in correspondence of impeller hub.

  Specifically, two calculations were executed with an average shaft diameter of 122 and 142 (mm).

- Tilting-pad journal bearings geometry

  Calculations were executed utilizing two different L/D bearing ratio (L/D = .4; L/D = .7)
  The relative dynamic characteristics, directly received from manufacturers, are listed on following Table 2:

TABLE 2

TILTING-PAD JOURNAL BEARINGS DYNAMIC CHARACTERISTICS

Design and Operating Data:

Small bearing clearances
Large preset configuration
Speed 5800 (rpm)
Load 2500 (N)

L/D = .7

Brg. Clear.	K_{xx} (N/m)	K_{xy} (N/m)	K_{yx} (N/m)	K_{yy} (N/m)
Min.	2.69×10^{8}	0.4×10^{6}	-0.4×10^{6}	2.6×10^{8}
Max.	1.16×10^{8}	0.5×10^{6}	-0.5×10^{6}	1.01×10^{8}

L/D = .4

Brg. Clear.	K_{xx} (N/m)	K_{xy} (N/m)	K_{yx} (N/m)	K_{yy} (N/m)
Min.	1.08×10^{8}	0.25×10^{6}	-0.25×10^{6}	0.64×10^{8}
Max.	0.8×10^{8}	0.26×10^{6}	-0.26×10^{6}	0.3×10^{8}

- Interstage Seals and Balancing Drum

  Different profiles, respectively straight and stepped for the front impeller wear-rings with 68 and 7 circumferential grooves for the balance drum, were taken into account on rotor analysis.

  Seals and balancing drum influence (Lomakin effect) were evaluated by means of analytical programs, validated through a series of experimental tests carried out on a full scale test-rig.

  Different straight seal geometries were infact in the past investigated varying L/D ratio, clearances, ΔP across seal and rotational speed (ref.10,11, 12).
  Affinities and discrepancies between the test results and analytical calculations, carried out using different evaluation methods (formulas, finite differences, finite elements), were in detail illustrated on ref.12.

  The final geometry of the impeller wearing-rings and the balance drum was selected taking into account the correlations indicated on ref.12 and the correspondent effect on leakage flow rate, which mainly depends by the surface profiles and seals clearances.

  The necessary compromise, derived from a series of iterative calculations, recommends infact the geometries described on following Tables 3 and 4, respectively for the frontal impeller wear-rings and balancing drum.

  On these tables are compared the selected impeller wear-ring and balancing drum characteristical data with the quite similar seal test-cases, shown on figs. 2a, 2b, 3a,3b.

TABLE 3

TEST SEAL AND IMPELLER WEAR-RING RELEVANT DATA

		Test Case	500 MW Imp.Wear-Ring
Length	L (mm)	40	33
Radius	r (mm)	80	113.5
L/D		.25	.145
Surface Profile		Straight	Straight
Radial Clear. C (mm) "Worn" condition		.36	.35
ΔP across Seal (MPa)		1.5	1.4
Speed (rpm)		4000	4000
Calc .Ax. Reynolds n°		24471	23653
Meas. Flow Rate (m3/h)		16.2	
Stiffness Exper.Data			
Kxx, Kyy	(N/m)	0.52×10^{7}	
Kxy, Kyx	(N/m)	0.15×10^{7}	
Calc.Flow Rate (m3/h)		22.1	30.4
Stiffness Anal.Data			
Kxx, Kyy	(N/m)	0.48×10^{7}	0.58×10^{7}
Kxy, Kyx	(N/m)	0.38×10^{7}	0.38×10^{7}

TABLE 4

TEST SEAL AND BALANCING DRUM RELEVANT DATA

		Test Case	500 MW Balanc.Drum
Length	L (mm)	120	212
Radius	r (mm)	80	106
L/D		.75	1.
Surface Profile		Straight	Grooved (7 circ.grooves)
Radial Clear. C (mm) "Worn" condition		.36	.35
ΔP across Seal (MPa)		2.2	9.5
Speed (rpm)		4000	4000
Calc.Axial Reynolds n°		19491	23459
Meas.Flow Rate (m3/h)		13.8	
Stiffness Exper.Data			
Kxx, Kyy	(N/m)	1.5×10^{7}	
Kxy, Kyx	(N/m)	$3.\times10^{7}$	
Calc.Flow Rate (m3/h)		17.6	28.1
Stiffness Anal.Data			
Kxx, Kyy	(N/m)	0.86×10^{7}	2.8×10^{7}
Kxy, Kyx	(N/m)	2.8×10^{7}	1.6×10^{7}

We can observe that the analytical program adopted to evaluate the impeller straight wear-ring characteristics leads to conservative analysis results, both from critical speed and stability points of view.
Infact the stiffness values used, which are respectively lower than the experimental direct and higher than the experimental cross-coupled coefficients, lead to obtain lower rotor critical speed values, increasing the calculation safety margin.

Unfortunately, for the balancing drum, insufficient experimental data are available, particularly for the grooved configuration profile, usually adopted on Boiler Feed-Pump.

The comparison with the tested straight profile configuration, moreover having lower P across the seal, is not particularly meaningful, thus indicating that the calculated stiffness values are of the same order of magnitude of the experimental ones.

Specifically, to evaluate the balance drum dynamic characteristics, we have utilized the guidelines indicated on the Atkins et al. paper (ref.6) which suggests to consider a grooved seal as a series of short plain seals subject to a pressure drop inversely proportional to the number of lands. The stiffness and damping values calculated for one land were multiplied by the number of lands to define the total drum dynamic characteristics.

3. ANALYTICAL RESULTS

The following results were obtained respectively for the barrel-casing and rotor analysis.

a) Barrel Casing and Baseplate

The calculated casing natural frequencies lie between the 9000 and 11500 (rpm), i.e. completely beyond the operational pump range.
Locally, the coupling side support shows an own frequency of about 31000 (cpm), which can be excited by the 7 impeller blades when the pump runs at 4400 (rpm).
We believe that the relatively low rotational speed and the adequate value of the gap existing between impeller and diffuser vanes (gap B) permit a consistent limitation of the hydraulic exciting forces able to contain the relative vibration levels.

b) Rotor Analysis

The various design alternatives examined lead to the results summarized on the following Table 5.
For data unavailability, the synchronous interaction existing between impeller and diffuser was neglected. Anyway this effect could decrease the calculated critical speed of about 5% (ref. 3).

All the data illustrated on Table 5 confirm the technical difficulties to reach even in "Worn" condition critical speeds above the pump design rotational speed.
An average residual margin of 7% is available in these conditions, between the lower critical and the design pump rotational speed.

On "New" clearances configuration we have, instead, as indicated on Table 6, an average calculated margin of 29%, still using the very conservative approach for the determination of seal and balancing drum effects.

The "Stiff Shaft" design affects moreover the volumetric efficiency as shown on the above mentioned Table 5.
We have estimated an efficiency difference of about 1.8% on "Worn" conditions depending by the seals and drum geometries.

On figs. 4a, 4b, we can see the undamped, uncoupled unbalance response diagrams obtained for the optimized configuration in correspondence of the coupling side tilting-pad bearing, respectively for vertical and horizontal directions.

Similar calculation results, obtained imposing the same unbalance load and introducing the cross-coupled coefficients still evaluated in "Worn" conditions, were illustrated in figs. 5a, 5b.
The introduction of these coefficients shows an evident tendency to modify, in terms of amplitudes, the unbalance

response diagrams, further on emphasized with "New" clearances configurations.
Finally, the obtained rotor natural frequencies denote in virtue of a disengaging policy, a low sensitivity to exciting forces having frequencies multiple of the rated rotational speed.

TABLE 5

COMPARISON AMONG DIFFERENT EXAMINED CONFIGURATIONS

Configuration	Direct.	Critical Speeds (cpm) 1st	2nd	3rd
Tilting Pad Journal Brg. L/D=.7				
Average Shaft Diameter 122 (mm)	Vert.	5850	7800	12000
Straight Frontal Wear-ring profile				
Balancing Drum (7 circum.grooves)	Horiz.	5800	7700	11600
Volumetric Eff.= .903				
Tilting Pad Journal Brg. L/D=.7				
Average Shaft Diameter 142 (mm)	Vert.	5100	8150	12750
Stepped frontal Wear-ring profile				
Balancing Drum (7 circum.grooves)	Horiz.	5000	7950	12350
Volumetric Eff. = .906				
Tilting Pad Journal Brg. L/D=.7				
Average Shaft Diam. 142 (mm)	Vert.	5500	7800	11500
Straight front.Wear-ring profile				
Balanc. Drum (64 circum.grooves)	Horiz.	5400	7400	11000
Volumetric Eff. = .917				
Tilting Pad Journal Brg. L/D=.4				
Average Shaft Diameter 142 (mm)	Vert.	5800	7200	8800
Straight frontal Wear-ring profile				
Balancing Drum (7 circum.grooves)	Horiz.	4400	6600	8200
Volumetric Eff. = .899				
Tilting Pad Journal Brg. L/D=.7				
Average Shaft Diameter 142 (mm)	Vert.	6200	8200	12400
Straight frontal Wear-ring profile				
Balancing Drum (7 circum.grooves)	Horiz.	6150	8100	12400
Volumetric Eff. = .899				

TABLE 6

FINAL CONFIGURATION "WET CRITICAL SPEEDS"

Design data
Tilting Pad Journal Brg. L/D = 0.7
Average Shaft Diam. = 142 (mm)
Straight frontal wear-ring profile
Balancing drum (7 circumferential grooves)

Seal Clear.	Direct.	Critical Speeds (cpm) 1st	2nd	3rd
Worn	Vert.	6250*	11200*	14900*
		6200	8200	12400
	Horiz.	6250*	11200*	14900*
		6150	8100	12400
New	Vert.	7600*	11300*	15400*
		7300	8200	13400
	Horiz.	7600*	11300*	15400*
		7250	8000	13400

* Min.Brg.Clear.

4. FLUID/MECHANICAL INTERACTIONS

- "Vane Pass" Loads
  The impeller-diffuser interaction can cause, as well known, the raise of consistent exciting forces, if the gap B in correspondence of interfacing blades results too low.
  The recommended value of 4% (ratio between impeller/diffuser diametral gap and the impeller diameter) was adopted to optimise both the dynamic and hydraulic performances.

- Recirculation Flow
  It was observed that the recirculation flow creates pressure and velocity fields rotating at a frequency ranging between .7 - .9 of the pump running speed.
  Subsynchronous natural frequencies can be easily excited by this phenomenon, which surely represents the major subsynchronous fluid exciting force. For the considered 500 MW Feed-pump we have chosen to place the onset recirculation flow below the normal pump capacity range adopting a particular hydraulic design in order to disengage any possibility of hydraulic/dynamic interaction.
  Visualization tests carried out on the more unfavourable single stage configuration (1st stage) confirm the design prediction; the onset impeller recirculation flow was measured at 40% of the pump B.E.P., which corresponds to the lower operational range limit.

- Journal Bearings and Rotor Stability

  Tilting-Pad Journal Bearings and swirl-breaks applied in correspondence of the balancing drum, were adopted to achieve a high rotor stability degree.

5. CONCLUSIONS

The analysis results suggest the following considerations:

- The undamped, uncoupled calculation model is appropriate to provide enough indications for a correct pump rotordynamic analysis.

- The cross-coupled stiffness and damping coefficients tend on one hand to closen the analytical results to the real pump behaviour, but on the other hand to confuse the resulting unbalance response causing sometimes unpleasant misunderstanding on obtained peaks.

- The possibility to adopt conservative seals contribution using different analytical programs, could be useful to increase the calculation safety margin.

- The above described calculation emphasizes the difficulties to reach a real "Stiff Shaft" configuration for the currently used multistage boiler feed-pumps.
  The maximum dynamic contribution is infact required for all pump components to increase the wet critical speeds beyond the normal operational range in "worn clearances" configuration.
  At this purpose, we remark the bearing relevant influence for the achievement of this target (see Table 5).

- The pump efficiency (i.e. volumetric and/or hydraulic efficiency) is normally negatively influenced by the "Stiff Shaft" design approach.
  Anyway, the Reliability must to be privileged versus the pump efficiency performances. On new clearances configuration the max. efficiency penalization was estimated at about .8%.

- The elimination of the fluid dynamic and mechanical interactions and subordinately the limitation of the impact of the exciting forces on casing and/or shaft natural frequencies represents in our opinion the main guideline to be followed during the pump rotordynamic design phase.

6. REFERENCES

1. Yeager K., "EPRI Journal", July/August 1987

2. Simon A., Eichhorn G., Vach R., Pace S., "Cycling Effects on Boiler Feed Pumps", Proceedings 1985 Fossil Plant Cycling Workshop, Miami Beach, Nov. 1985, pp. 4.21- 4.42

3. Pace S.E., Florjancic S., Bolleter U., "Rotordynamic Developments for High Speed Multistage Pumps", Proceedings of the Third International Pump Symposium, Texas A&M University, Houston, Texas, May 20-22 1986, pp. 45-54.

4. Massey I.C., "Subsynchronous Vibration Problems in High Speed Multistage Centrifugal Pumps", Proceedings of the 14th Turbomachinery Symposium, Turbomachinery Laboratories, Department of Mech.Engineering, Texas A&M University, College Station, Texas, Oct.22-24 1985, pp. 11-16.

5. France D., "Rotordynamics Considerations in the Design of High Speed Centrifugal Pumps", Proceedings of the 4th International Technical Seminar, Pumps for Developing Technologies, I.P.M.A., New Delhi, Feb. 10-11 1987.

6. Atkins, K.E., Tison J.D., Wachel J.C., "Critical Speed Analysis of an Eight-Stage Centrifugal Pump", Proceedings of the Second International Pump Symposium, Texas A&M University, College Station, Texas, Oct.22-24 1985, pp. 59-65.

7. Gopalakrishnan S., Fehlan R., Lorett J., "Critical Speeds in Centrifugal Pumps", ASME Paper 82-GT-277 (1982).

8. Kaneko S., Hori Y., Tanaka M., "Static and Dynamic Characteristics of Annular Plain Seals" Proceedings of the Third International Conference on Vibrations in Rotating Machinery, I.Mech.E., University of York, England, Sept. 11-13 1984, pp. 205-214.

9. Marenco G., Falco M., Gasparetto M., "On the Influence of the Dynamic Response of the Supports on the Critical Speeds of a Boiler Feed Pump", Proceedings of the Fluid Machinery Conference on Fluid Machinery Failures - Prediction, Analysis and Prevention, I.Mech.E., University of Sussex, England, April 15-17 1980, pp. 109-114.

10. Diana G., Falco M., Mimmi G., Marenco G., Saccenti P., "Experimental Research on the Behaviour of Hydrodynamic Plain Seals by Means of a Specific Testing Device", Proceedings of IFToMM Conference on Rotordynamic Problems in Power Plants, Rome, Sept. 1982, pp. 355-360.

11. Falco M., Mimmi G., Pizzigoni B., Marenco G., Negri G., "Plain Seal Dynamic Behaviour - Experimental and Analytical Results", Proceedings of the Third International Conference on Vibrations in Rotating Machinery, I.Mech.E., University of York, England, Sept. 11-13 1984, pp. 151-158.

12. Falco M., Mimmi G., Marenco G., "Effects of Seals on Rotor Dynamics", Proceedings of the International Conference on Rotordynamics, JSME-IFToMM, Tokyo, Sept. 14-17 1986, pp. 655-661.

13. Marscher W.D., "Subsynchronous Vibration in a Boiler Feed Pump" Proceedings of the Fifth Annual ROMAC Short Course on Vibrations in Compressors, Turbines and Pumps, Washington D.C., August 10-13 1987.

14. Lomakin, A.A. "Calculation of the Critical Speed and the Condition to Ensure Dynamic Stability of the Rotors in High Pressure Hydraulic Machines, Taking Account of the Forces in the Seals", Energo Mashinostroinie, 14 (4), pp. 1-5 (1958).

15. Black H.F., "Effects of Fluid-Filled Clearance Spaces on Centrifugal Pump ...Vibrations", Proceedings of the 8th Turbomachinery Symposium, Texas A&M University, 1979.

16. Childs D.W., "Finite Length Solutions for Rotordynamic Coefficients of Turbolent Annular Seals", ASME 82-LUB-42, 1982.

17. Agostinelli A., Nobles D., Mockridge C.R., "An Experimental Investigation of Radial Thrust in Centrifugal Pumps", ASME Trans., Journal of Engineering for Power, Vol.82, April 1960, pp.120-126.

18. Hergt P., Krieger P., "Radial Forces in Centrifugal Pumps with Guide Vanes" "Advance Class Boiler Feed Pumps" Proceedings, I.Mech.E., Vol. 184, Pt. 3N, 1969-1970, pp.101-107.

19. Fraser W.H., "Centrifugal Pump Hydraulic Performance and Diagnostics", Pump Handbook, McGraw-Hill, 1985.

20. Schiavello B., "On the Prediction of the Reverse Flow Onset at the Centrifugal Pump Inlet", Proceedings of the ASME Gas Turbine Conference and Fluids Engineering Conference, New Orleans, March 1980.

21. Bolleter U., Frei A., Florjancic D., "Predicting and Improving the Dynamic Behaviour of Multistage High Performance Pumps", Proceedings of the First International Pump Symposium, Texas A&M University, 1984.

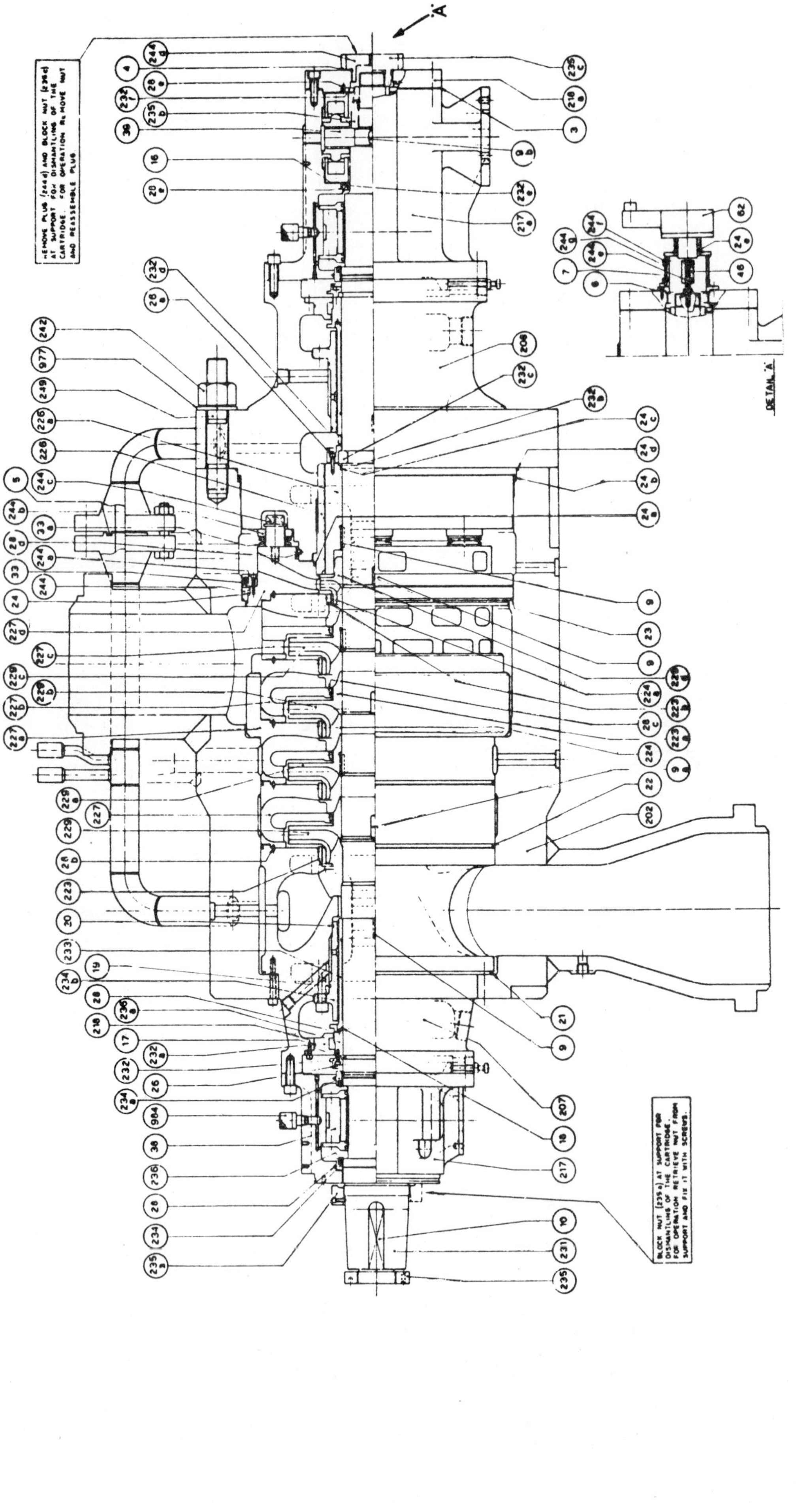

Fig 1

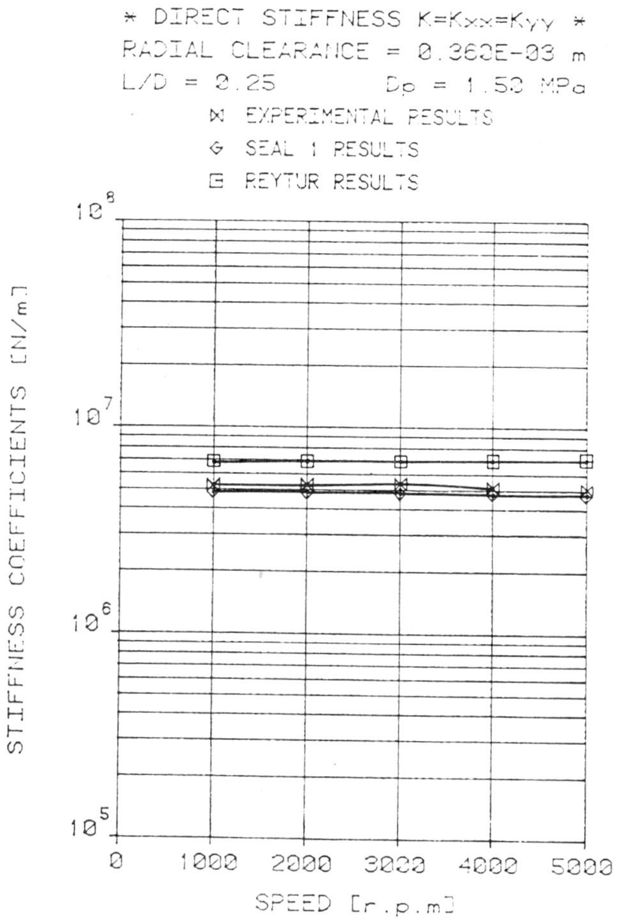

Fig 2a

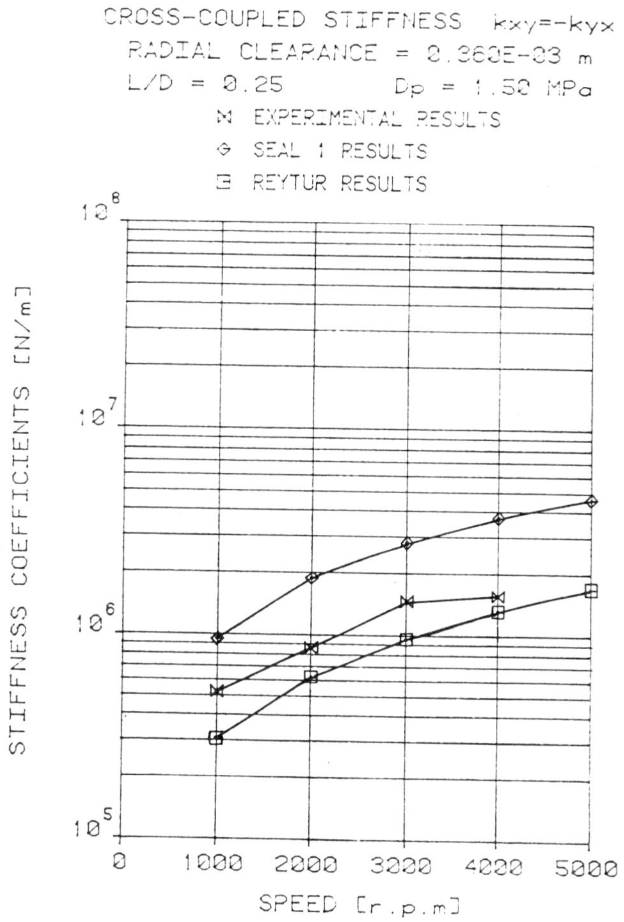

Fig 2b

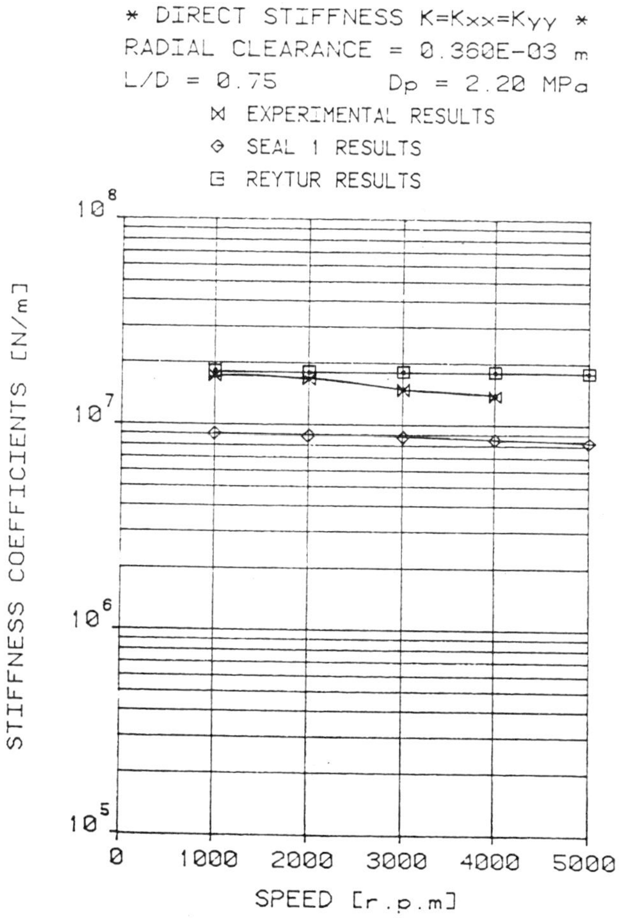

Fig 3a

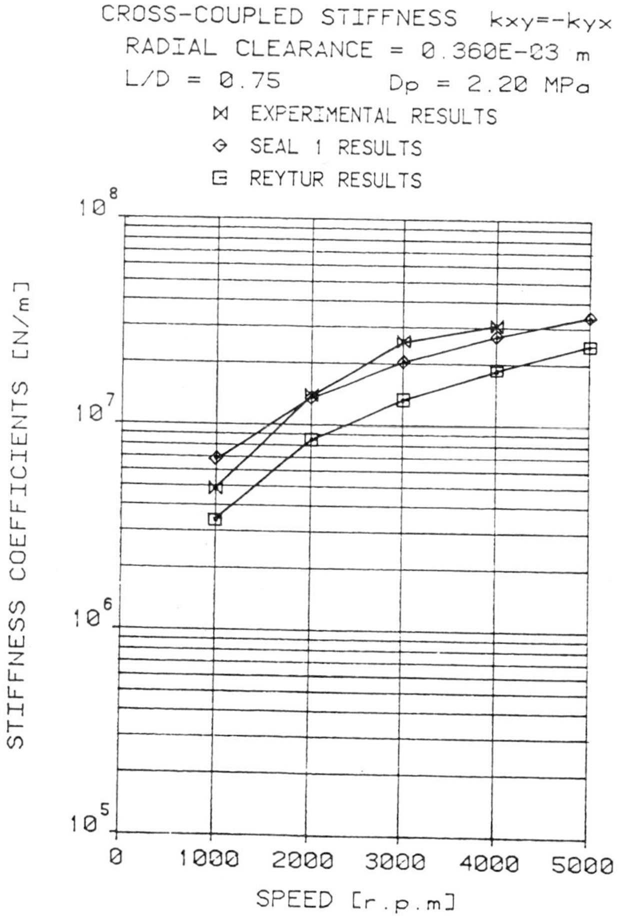

Fig 3b

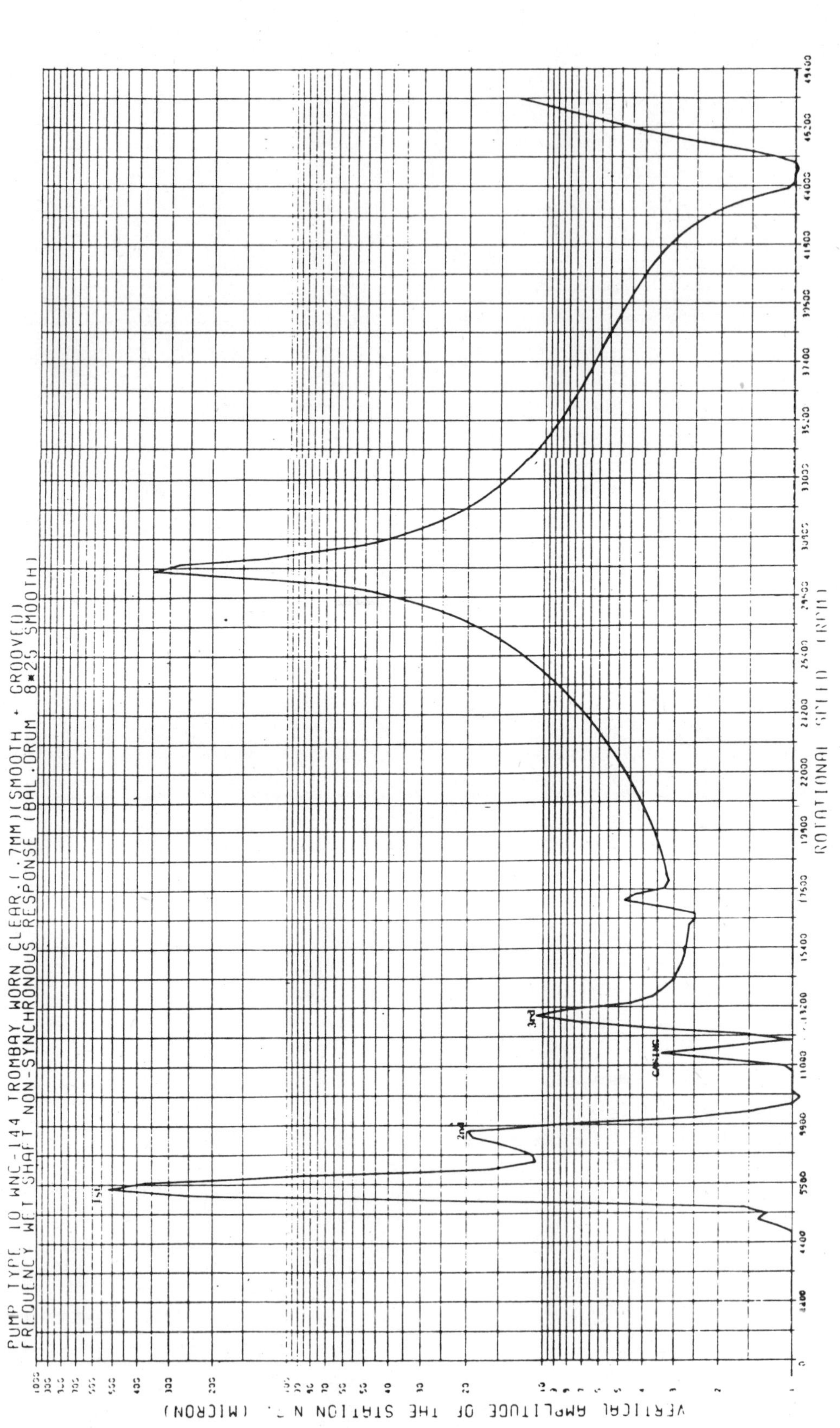
PUMP TYPE 10 WNC-144 TROMBAY WORN CLEAR.(.7MM)(SMOOTH + GROOVED)
FREQUENCY WET SHAFT NON-SYNCHRONOUS RESPONSE (BAL.DRUM 8*25 SMOOTH)
1st
2nd
3rd
ROTATIONAL SPEED (RPM)

Fig 4a

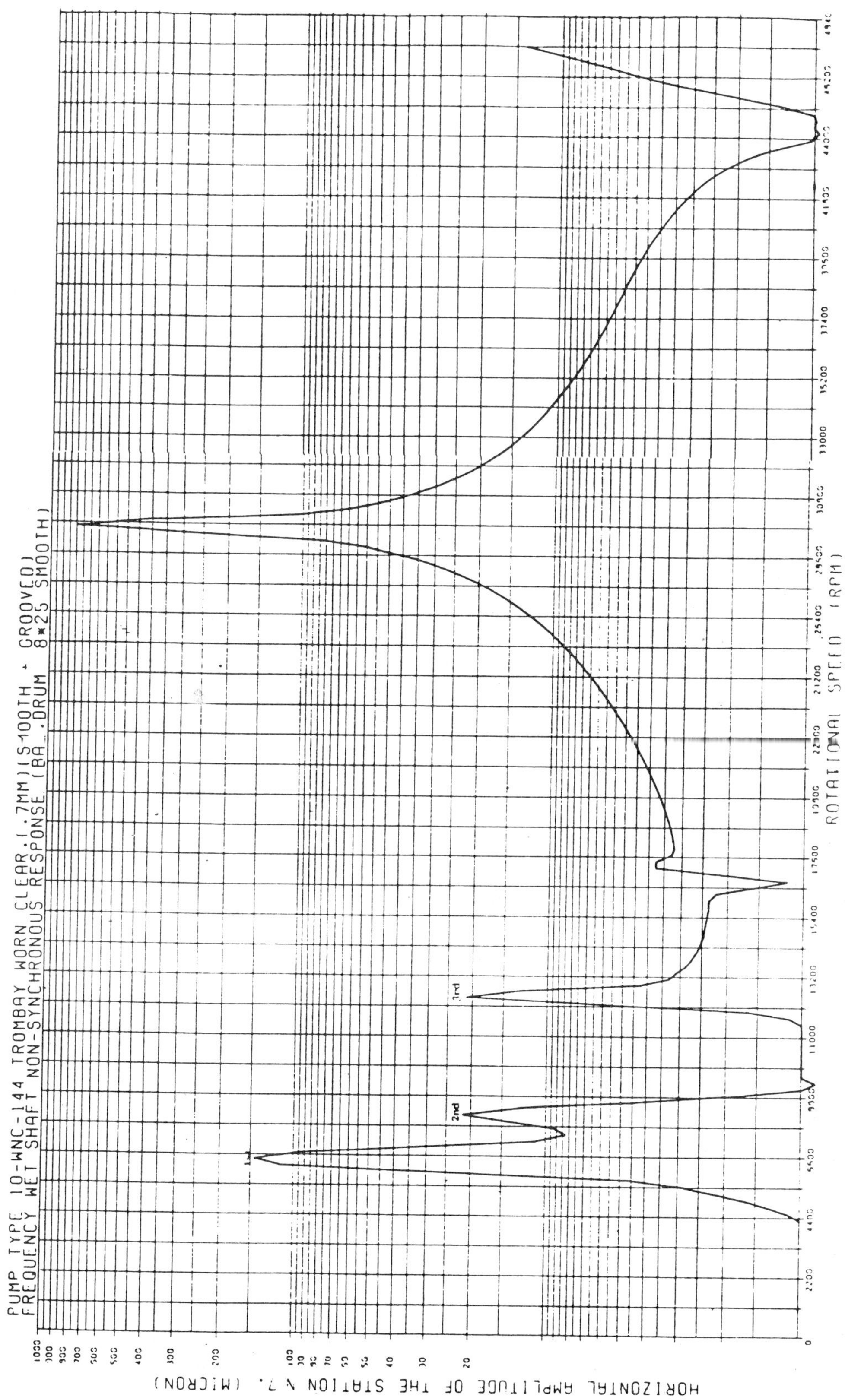
PUMP TYPE 10-WNC-144 TROMBAY WORN CLEAR.(.7MM)(S100TH + GROOVED)
FREQUENCY WET SHAFT NON-SYNCHRONOUS RESPONSE (BA .DRUM 8*25 SMOOTH)
1st
2nd
3rd
ROTATIONAL SPEED (RPM)
HORIZONTAL AMPLITUDE OF THE STATION N 7. (MICRON)

Fig 4b

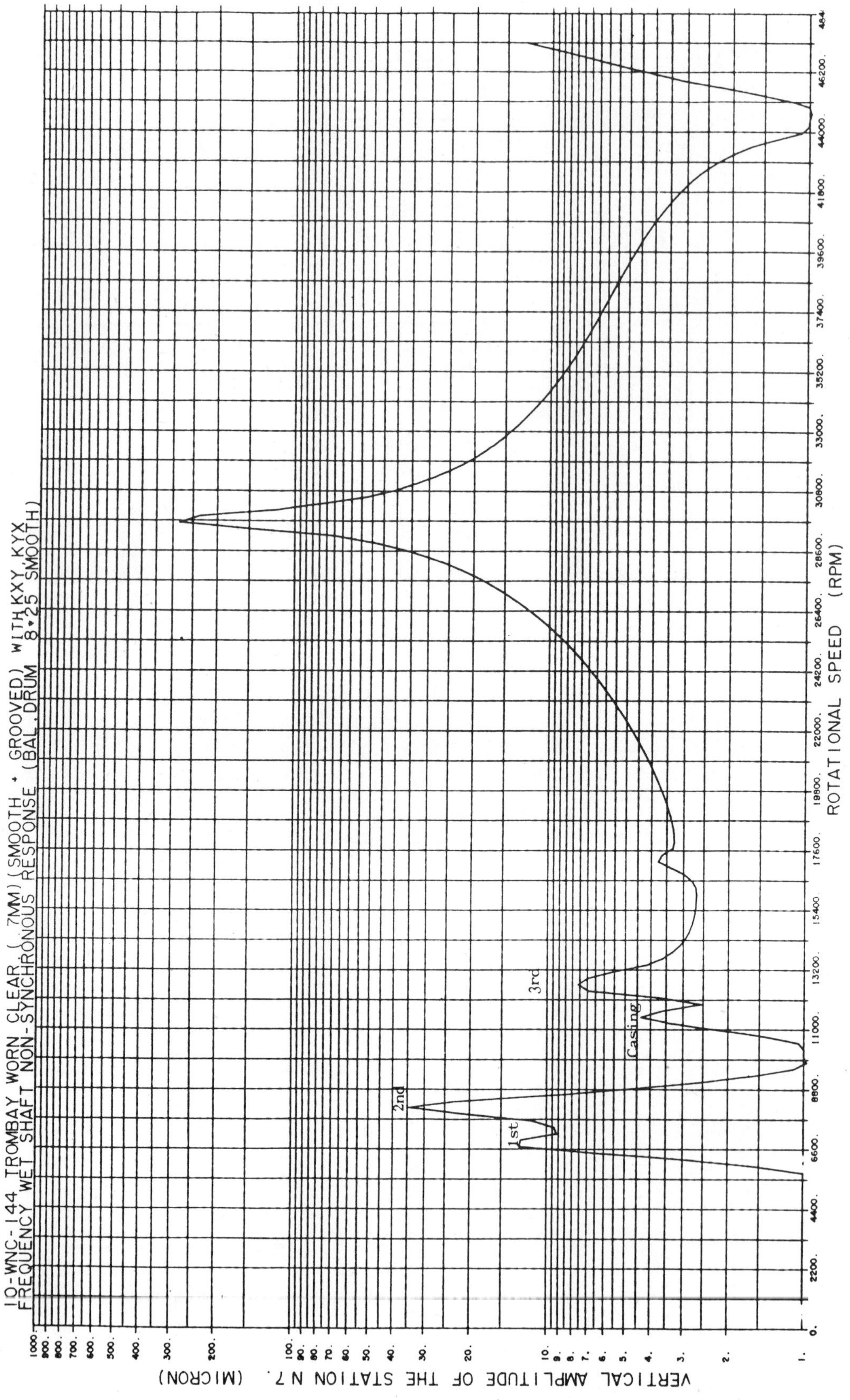
IO-WNC-144 TROMBAY WORN CLEAR (7MM) (SMOOTH + GROOVED) WITH KXY,KYX
FREQUENCY WET SHAFT NON-SYNCHRONOUS RESPONSE (BAL.DRUM 8.25 SMOOTH)
ROTATIONAL SPEED (RPM)
VERTICAL AMPLITUDE OF THE STATION N 7. (MICRON)
1st
2nd
3rd
Casing

Fig 5a

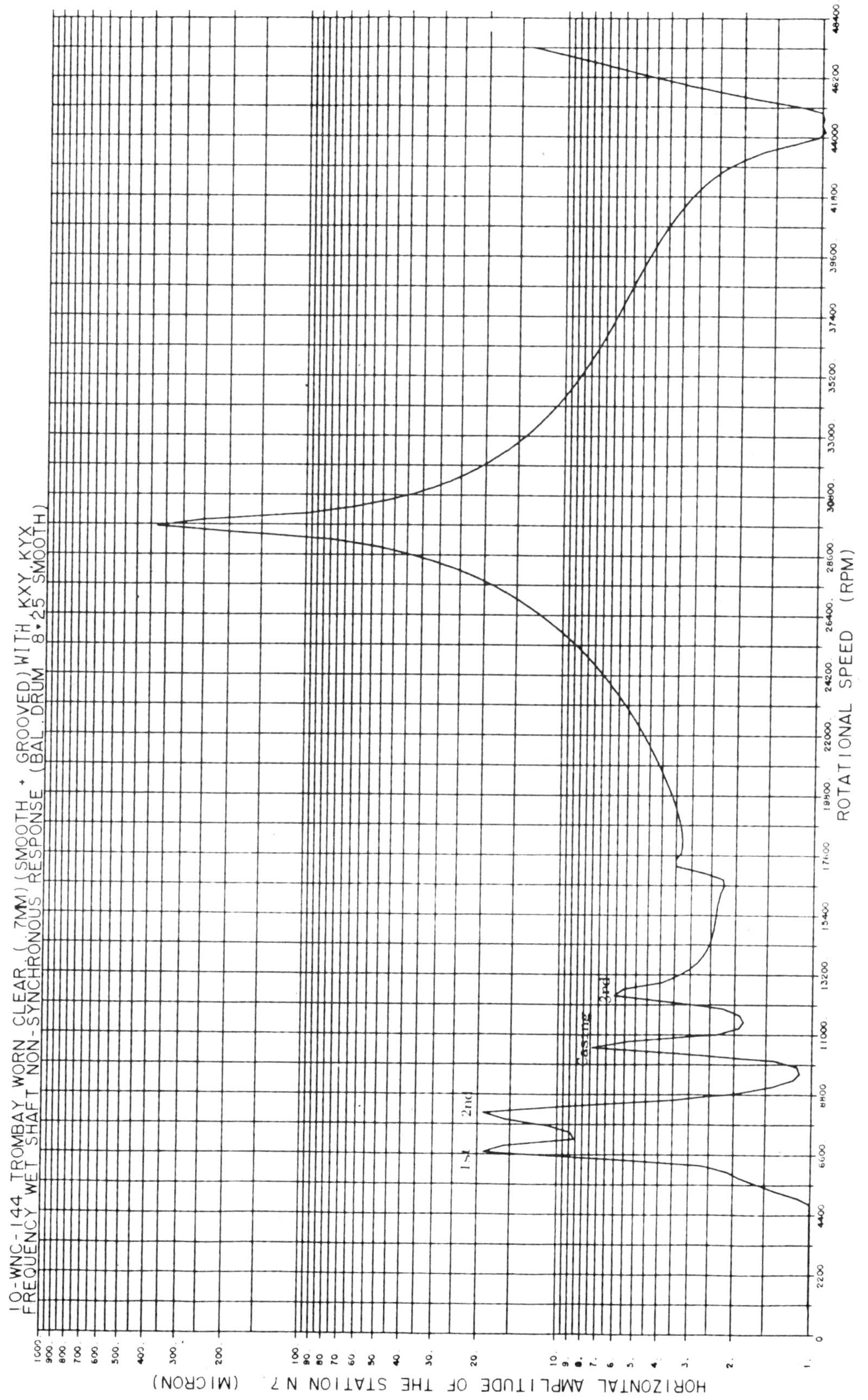
10-WNC-144 TROMBAY WORN CLEAR (7MM) (SMOOTH + GROOVED) WITH KXY KYX
FREQUENCY WET SHAFT NON-SYNCHRONOUS RESPONSE (BAL DRUM 8.25 SMOOTH)
ROTATIONAL SPEED (RPM)
HORIZONTAL AMPLITUDE OF THE STATION N 7 (MICRON)
1st
2nd
Casing
3rd

Fig 5b

C295/88

The influence of torsional – lateral – coupling in geared rotor systems on its eigenvalues, modes and unbalance vibrations

P SCHWIBINGER, PhD
Carl Freudenberg Company, Weinheim, West Germany
T NEUMER, A ZÜRBES and **R NORDMANN**, Dr-Ing
Department of Mechanical Engineering, University of Kaiserslautern, Kaiserslautern, West Germany

SUMMARY

In rotating machinery (turbines, compressors, pumps) for troublesome vibrations in particular the synchronous unbalance vibrations and the nonsynchronous selfexcited vibrations are known. This bending vibrations may cause expensive damage especially in high performance turbomachinery with high speeds and pressures. In recent times coupling mechanisms between torsional and lateral vibrations are considered as an additional influence, which is of special importance in systems with gears.

In gears a strong torsional-lateral coupling exists naturally due to the mechanism of power transmission. In this paper the influence of the torsional-lateral coupling on the eigenfrequencies and modes of a rotor test rig with gears is investigated analytically and experimentally. The studies are carried out at three different models: shafts without rotation, shafts rotating supported in ball bearings, shafts rotating in journal bearings. The identified modes have strong torsional and lateral components at the gear mesh. The measurements agree good with calculations.

The coupling does not only affect the eigenfrequencies and modes but also the forced vibrations, due to unbalance of the gear wheels. In the considered lower frequency range parametric excited vibrations caused by tooth meshing are neglected. Frequency spectra for torsion and bending in the uncoupled and the coupled case are discussed.

NOMENCLATURE

$\underline{D}$, $\underline{K}$, $\underline{M}$	Mass-, Damping-, Stiffness matrix
$\underline{f}$	Force vector
F	Tooth force
k_t	Tooth stiffness
$\underline{q}$	Displacement vector
q_{iG}	Displacement at gear wheels
r_i	Basic radii of gear wheels
f_m, f_G	Rotational frequency of motor, Generator shaft
α_j	Damping constant
ω_j	Circular natural frequency
Ω	Rotational circular frequency
x,y,z	Coordinate axes
$\tilde{( )}$	Complex variables

INTRODUCTION

The dynamic behaviour of many rotating machines, e.g. turbines, compressors, pumps is influenced by the mass and stiffness distribution of the rotor and by stiffness and damping characteristics of nonconservative effects like oil film forces, seals, clearing excitation. For bending vibrations the forced unbalance and the unstable nonsynchronous vibrations are important. Torsional vibrations of the shaft train are excited during startup and by transient electrical moments in motors and generators respectively.

Usually the dynamic analysis for the bending vibrations and for the torsional vibrations are carried out separately. However for geared rotor systems, Fig. 1, due to the offset centerlines of the shafts, the torsional and lateral vibrations are coupled. Till now we don't know much, how this coupling effects the dynamic behavior of the machine. In the literature we find several publications concerning torsional-lateral coupling in turbomachinery with gears.

WACHEL and SZENASI describe in /1/ a field problem, where in a geared system unstable lateral vibrations occured together with torsional oscillations. YAMADA, MITSUI deal in /2/ with a ship gear supported by oil film bearings, which runs unstable with partial load. A coupled torsional-lateral analysis shows, that the stability threshold is decisively influenced by the torsional stiffness of the rotor system.

IANNUZZELLI and ELWARD point out in /3/, that certain measured eigenfrequecies of a compressor train can be verified only by an analytical model considering the torsional-lateral coupling in a gear stage. SIMMONS and SMALLEY /4/ found by experimental and analytical investigations at a gas turbine/compressor train that torsional modes with a superposed bending component at the gear wheel can be damped significantly by the oil film bearings. IIDA et. al. discuss in /5/ the coupling effect at a simple rotor system.

In this paper the influence of the torsional-lateral coupling in a gear stage on the eigenfrequencies, modes and the vibration spectra of a simple test rig is investigated analytically and experimentally. The results for the uncoupled and the coupled system are presented.

MECHANICAL MODEL

Fig. 1 shows a typical turbomachine consisting of two elastic shafts connected by a reduction gear. The rotors are running in oil film bearings. Usually the lateral vibration analysis is carried out for both shafts seperately and independently from the torsional analysis. But in fact torsional and lateral vibrations of the rotors are coupled by the gear.

Gear Model

Fig. 2 shows the simplest model to describe the coupling of torsional and lateral displacements of spur gear wheels. The teeth are assumed as rigid compared to the elasticity of the shaft. Provided that both wheels maintain contact during operation the kinematic relationship in a gear stage is:

$$r_1 q_{1G} = r_2 q_{2G} \quad (1)$$

This is the model commonly used in rotor dynamics. If we consider lateral movement of the gear wheels the kinematic equation

$$r_1 q_{1G} + q_{3G} = r_2 q_{2G} + q_{4G} \quad (2)$$

implies a coupling of the torsional and lateral degrees of freedom.

If the gear teeth stiffness cannot be neglected compared to the shaft stiffness, the gear teeth are modelled as a spring with the constant elasticity k_t. Similiar to (1), (2) the tooth force F is a function of the torsional respectively the torsional and lateral displacements at the wheels.

uncoupled $\quad F = (r_1 \cdot q_{1G} - r_2 \cdot q_{2G}) \cdot k_t \quad (3)$

coupled $\quad F = (r_1 \cdot q_{1G} + q_{3G} - r_2 q_{2G} - q_{4G}) \cdot k_t \quad (4)$

It has to be noted that not all the effects of a meshing gear can be investigated with this simple models, e.g. the variation of gear mesh stiffness, gear mesh errors and backlash /7,8/. But we will concentrate on the coupling effect in the gear stage and its influence on the dynamics of turbomachinery in the lower frequency range. Therefore all important effects to model turbomachinery vibrations are included in the calculations: journal bearings, seals, stiffness and inertia of the shaft elements, gyroscopic effects and unbalance /9/.

A systematic investigation of the vibration behavior of a turbomachine with all these effects including torsional-lateral coupling is difficult to realize. In a first step we consider a simpler rotor system similiar to that of the turbocompressor. Fig. 3 shows this model with two elastic shafts connected by a gear. The rotors are running in ball bearings respectively journals. Shaft 1 is driven by a motor, shaft 2 drives a generator. The axes of the shafts are offset by the angle of mesh so that the tooth force acts in the vertical plane on the gear wheels. Both shafts are elastic for torsion and bending.

Equations of Motions

The equations of motion for the simple shaft system, Fig. 3, express the equilibrium of inertia, damping, stiffness, and external forces

$$\underline{M}\,\ddot{\underline{q}} + \underline{D}\,\dot{\underline{q}} + \underline{K}\,\underline{q} = \underline{f}(t) \quad (5)$$

$\underline{M}$ mass matrix
$\underline{D}$ damping matrix
$\underline{K}$ stiffness matrix
$\underline{q}$ vector of displacements (torsion and bending)
$\underline{f}$ vector of external forces

When we use journal bearings the matrices $\underline{D}$ and $\underline{K}$ contain the dynamic bearing coefficients. They are nonsymmetric and depend on the running speed of the rotors. Equation (5) contains torsional and lateral degrees of freedom. If we introduce equation (1) or (3) in (5) we get the equations of motion for the uncoupled system, the application of (2) respectively (4) yields the equations of motion for the torsional-lateral coupled system.

Natural Vibrations

From the homogeneous equations of motion ($\underline{f} = \underline{0}$) the natural vibrations can be calculated. Assuming a solution of the form

$$\underline{q}(t) = \underline{\varphi} \cdot e^{\lambda t} \quad (6)$$

we obtain a quadratic eigenvalue problem

$$(\lambda^2 \underline{M} + \lambda\,\underline{D} + \underline{K})\,\underline{\varphi} = 0 \quad (7)$$

with 2N eigenvalues λ_j and corresponding eigenvectors $\underline{\varphi}_j$. Eigenvalues as well as eigenvectors in most cases occur in conjugate complex pairs.

Eigenvalues: $\quad \lambda_j = \alpha_j + i\omega_j \quad \overline{\lambda}_j = \alpha_j - i\omega_j$

Eigenvectors: $\quad \underline{\varphi}_j = \underline{s}_j + i\underline{t}_j \quad \overline{\underline{\varphi}}_j = \underline{s}_j - i\underline{t}_j \quad (8)$

We consider only the part of the solution, which belongs to a conjugate complex pair

$$\underline{q}_j(t) = B_j e^{\alpha_j t} \{\underline{s}_j \sin(\omega_j t + \gamma_j) + \underline{t}_j \cos(\omega_j t + \gamma_j)\} \quad (9)$$

with ω_j as circular natural frequency of this part and α_j as damping constant. If the damping constant $\alpha_j > 0$ the natural vibrations increase and make the system unstable, for $\alpha_j < 0$ the natural vibration decrease and the system runs stable.

We define the expression in parantheses {...} of equation (9) as natural mode. For the torsional-lateral coupled system the natural modes are compared of torsional and lateral components.

Forced Vibrations

Mass eccentricity of the gear wheels excites the rotors to synchronous unbalance vibrations.

If we assume the unbalance of one shaft with the rotational circular speed Ω,

$$\underline{f}(t) = \underline{f}_c \cos\Omega t + \underline{f}_s \sin\Omega t \tag{10}$$

written complex,

$$\underline{f}(t) = \mathrm{Re}\ \{(\underline{f}_c - i\underline{f}_s \cdot (\cos\Omega + i\sin\Omega t)\}$$

$$\underline{f}(t) = \mathrm{Re}\ \{\tilde{\underline{f}} \cdot e^{i\Omega t}\}$$

a solution of the form

$$\underline{q}(t) = \underline{q}_c \cos\Omega t + \underline{q}_s \sin\Omega t$$

$$\underline{q}(t) = \mathrm{Re}\{(\underline{q}_c - i\underline{q}_s)\cdot(\cos\Omega t + i\sin\Omega t)\} \tag{11}$$

$$\underline{q}(t) = \mathrm{Re}\{\tilde{\underline{q}} \cdot e^{i\Omega t}\}$$

leads to a complex linear system of equations for the unknown displacements $\underline{q}$.

$$(\underline{K} - \Omega^2 \underline{M} + i\Omega \underline{D})\ \tilde{\underline{q}} = \tilde{\underline{f}} \tag{12}$$

In the coupled model the unbalance excites not only lateral but through the condition (2), (4) also torsional vibrations, which are synchronous to the rotational speed.

Besides mass eccentricity gears are not perfect. To mention a few errors: The mounting center is not the geometric center of the gear, the teeth are not spaced equally around the gear and the tooth profiles are not perfect involutes. Mounting errors cause synchronous vibrations, tooth space errors in addition vibrations of higher orders. The basic frequency of tooth profile induced vibrations is: number of teeth * rotational speed (= meshing frequency). In practice these transmission error induced vibrations of higher order can only be modelled satisfactory by a statistical approach /10/.

EXPERIMENTAL RESULTS

Models of the Test Rig

The experimental investigations are carried out at three different models of the test rig, Fig. 4:

a Both shafts are supported in ball bearings. The coupling effect is investigated without rotation of the shafts. A static preload by soft torsional springs prevents the teeth from losing contact.

b The shafts are rotating in ball bearings. The coupling effect can be discussed for different rotational speeds and transmitting torques.

c Shaft ① is supported in journal bearings, shaft ② is running in ball bearings.

How the uncoupled and the coupled model are realized at the test rig, shows Fig. 5:

- To measure the uncoupled bending vibrations both shafts are seperated. The gear wheels do not mesh.
- Uncoupled torsional vibrations are measured with meshing gears. To prevent lateral movements of the gear wheels, they are supported by rigid bearings.
- Without this rigid bearings the rotor system fulfills coupled torsional-lateral vibrations.

Results for Model ⓐ

In a first step the coupling effect is investigated at the test rig without rotation. Its eigenfrequencies and mode shapes are identified with experimental modal analysis using the impulse excitation technique. A torsional spring at the end of the motor and the generator shaft prevent backlash effects of the gear at standstill. Both shafts are supported by ball bearings.

Fig. 6 shows the measured eigenfrequencies and modes for the uncoupled and the coupled system. For the uncoupled system two bending modes (f_1 and f_4) and two torsional modes (f_2 and f_3) occur in the considered lower frequency range. In the torsional-lateral coupled system there are bending and torsional components in each mode shape. The coupling affects also the eigenfrequencies of the system.

In the first mode the lateral movement of the pinion acts in the same direction as the torsion of the wheel. Therefore the coupled system seems to be more flexible: The first eigenfrequency f_1 is 31 % lower than in the uncoupled system. The two torsional modes superimposes in the coupled case an opposite bending of shaft ①. This stiffening effect raises the eigenfrequencies f_2 and f_3 about 10 %. The forth eigenfre quency is a bit lower in the coupled model: The torsion of the pinion promotes the bending of shaft ②.

Results for Model ⓑ

Now shaft ① is run by a motor, shaft ② drives a generator. The rotor system is excited by unbalance forces and the gear mesh. The tooth force acts only in the vertical plane.

First we discuss again the eigenfrequencies and modes of the uncoupled and the coupled system, Fig. 7 and Fig. 9. Similiar to model ⓐ the coupling has a strong influence on this modes, which have components in the vertical tooth force plane. This are all torsional modes and the bending modes in the x-z plane. Fig. 9 shows two typical modes, where the coupling is very strong.

In Fig. 7 the calculated frequencies are also compared with the measurements. The agreement is good. Especially the shift from the uncoupled to the coupled model is predicted very well with the used gear model. It is difficult to identify the eigenfrequencies in the horizontal plane: the coulumb friction between the teeth depends strongly on the lubrication conditions and the transmitted load.

Fig. 8 shows spectra of the lateral and torsional vibrations at the gears. The motor runs with a speed of 960 rpm (f_M=16 Hz), the generator acts as a brake with M_G=2 Nm.

For the bending vibrations of the uncoupled model only the rotational frequencies f_M, f_G due to unbalance are significant. Higher orders in the spectra of wheel motion are caused by misalignment of the bearings. In the coupled case the gear mesh acts as an broadband excitation: all four eigenfrequencies can be identified as resonances in the considered frequency range. The peak amplitude of the synchronous components f_M, f_G do not differ from the uncoupled measurements. Geometric eccentricity of the gear wheels would change this peak when the gears are meshing. Therefore the mounting error does not play an important role at the test rig. As the static deflection widens the distance of the shafts the double frequency component ($2xf_G$) due to misalignment increases.

In the torsional response of the uncoupled system f_2+f_3 can be identified. The bending vibrations f_G, f_M are measured due to an unwelcome transverse sensitivity of the torsional pickup. In the coupled model all four eigenfrequencies can clearly be identified from the resonances, Fig. 7.

Results for Model ⓒ

In the next step the influence of the torsional-lateral coupling on the vibration behavior of the oil film supported rotor system is investigated, Fig. 4. Shaft ① is running in journals, shaft ② still in ball bearings. The oil film bearings are relatively stiff, compared to the elastic shaft ①. Therefore the eigenfrequencies and modes nearly do not change with running speed and are almost the same as in model ⓑ. Again the coupling strongly influences the response frequencies, Fig. 10. The first coupled eigenfrequency f_1 is 37 % lower than in the uncoupled case, the second 36 % and the third 13 % higher.

In Fig. 11a the rotational speed component of the pinion lateral vibration is plotted versus running speed: The calculated forced vibrations are compared with the measurements. The shaft is exited by an unbalance of 12 gmm at the pinion and additionally by a curvature of 13,5 μm having a phase angle of 55° to the unbalance.
For the uncoupled system resonance at 40 Hz occurs in the tooth force (x-z) and the horizontal (x-y) plane. In the coupled case this resonance occurs only in y-direction. In the z-direction the first resonance is shifted through the coupling to 25 Hz. The amplitude is about the same as in the uncoupled case. Additional resonances at 60 and 80 Hz occur belonging to the eigenfrequencies f_2 and f_3. The calculation predicts especially the occurrence of the resonances very well.

Fig. 11b shows the corresponding waterfall diagrams: The spectra for the y-direction show the uncoupled resonance at 40 Hz. In z-direction the three resonances at 25 Hz, 60 Hz and 80 Hz can be seen. There is an additional resonance at a rotating frequency between 55-58 Hz, when the speed of shaft ② (f_G) coincides with f_1. The excitation of bending vibrations of shaft ① by shaft ② confirms again the coupling effect.

In the waterfall plot of Fig. 11b for the uncoupled system the first natural vibration with $f_1 \simeq 40$ Hz gets unstable at a threshold speed f_M = 82 Hz. Unstable vibrations in the coupled system could not be investigated till now.

CONCLUSIONS

We have analyzed the torsional-lateral coupling in a simple shaft system with spur gears analytically and experimentally. The studies were carried out at three different models. We come to the following conclusions:

- The eigenfrequencies and modes of the geared system differ essentially, when a coupled instead of an uncoupled model for the gear mesh is used. The validity of the coupled model is confirmed by an experimental modal analysis.

- Experiments and calculations show: the forced vibrations due to unbalance and gear mesh excitation are strongly influenced by the used model. The coupled model is the more general one. If it is used resonance frequencies and amplitudes are different from the uncoupled model.

- The stability behavior of the coupled model is different from the uncoupled one. The experiments to this point are continuing.

ACKNOWLEDGEMENTS

We are grateful for the support of the "Deutsche Forschungsgemeinschaft".

REFERENCES

/1/ WACHEL, J.D.; SZENASI, F.R.: Field Verification of Lateral-Torsional Coupling Effects on Rotor Instabilities in Centrifugal Compressors.
NASA-Conference Publication No. 2147, 1980.

/2/ YAMADA, T.; MITSUI, J.: A Study on the Unstable Vibration Phenomena of a Reduction Gear System, Including the Lightly Loaded Journal Bearings for a Marine Steam Turbine
Bull. JSME, Vol. 22, No. 163, 1979.

/3/ IANNUZZELLI, R.J.; ELWARD, R.M.: Torsional-lateral Coupling in Geared Rotors.
ASME No. 84-GT-71, 1984.

/4/ SIMMONS, H.R.; SMALLEY, A.J.: Lateral Gear Shaft Dynamics Control Torsional Stresses in Turbine Driven Compressor Train. ASME No. 84-GT-28, 1984.

/5/ IIDA, H.; TAMURA, A.; KIKUCH, K.; AGATA, H.: Coupled Torsional-Flexural Vibration of a Shaft in a Geared System of Rotors. Bull JSME, Vol. 23, No. 1986, Dec. 1980.

/6/ SCHWIBINGER, P.; NORDMANN, R.: The Influence of Torsional-Lateral Coupling on the Stability Behavior of Geared Rotor Systems. Proc of 4th Workshop on Rotordynamics Instability Problems in High-Performance Turbomachinery. Texas A&M University, June 1986.

/7/ KÜCÜKAY, F.: Dynamic Behavior of High Speed Gears. Proc. IMEchE. C317/84, 1984.

/8/ DAVID, J.W.; PARK, N.C.: The Vibration Problem in Gear Coupled Rotor Systems. Proc. 11th ASME-Conf. Vibration and Noise, Boston, Mass., Sept. 1987.

/9/ SCHWIBINGER, P.: Torsionsschwingungen von Turbogruppen und ihre Kopplung mit den Biegeschwingungen bei Getriebemaschinen. VDI-Fortschrittsbericht, Reihe 11, Nr. 90, VDI-Verlag, Düsseldorf, 1986.

/10/ NERIYA, S.V.; BHAT, R.B.; SANKAR, T.S.: On the Dynamic Response of a Helical Geared System Subjected to a Static Transmission Error in the Form of Deterministic and Filtered White Noise Inputs. Proc. 11th ASME-Conf. Vibration and Noise, Boston, Mass., Sept. 1987.

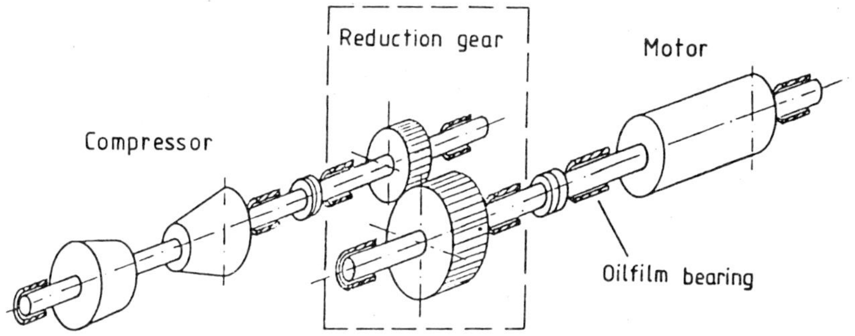

Fig 1   Reduction gear in a turbocompressor

Kinematic constraint without lateral displacement of gear wheels

$$r_1 q_{1G} = r_2 q_{2G}$$

Kinematic constraint with lateral displacement of gear wheels

$$r_1 q_{1G} + q_{3G} = r_2 q_{2G} + q_{4G}$$

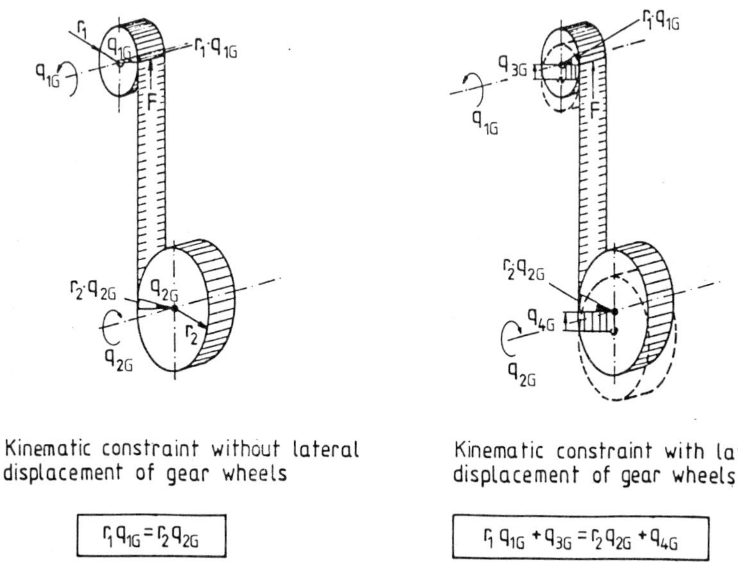

Fig 2   Kinematic constraints in a gear stage

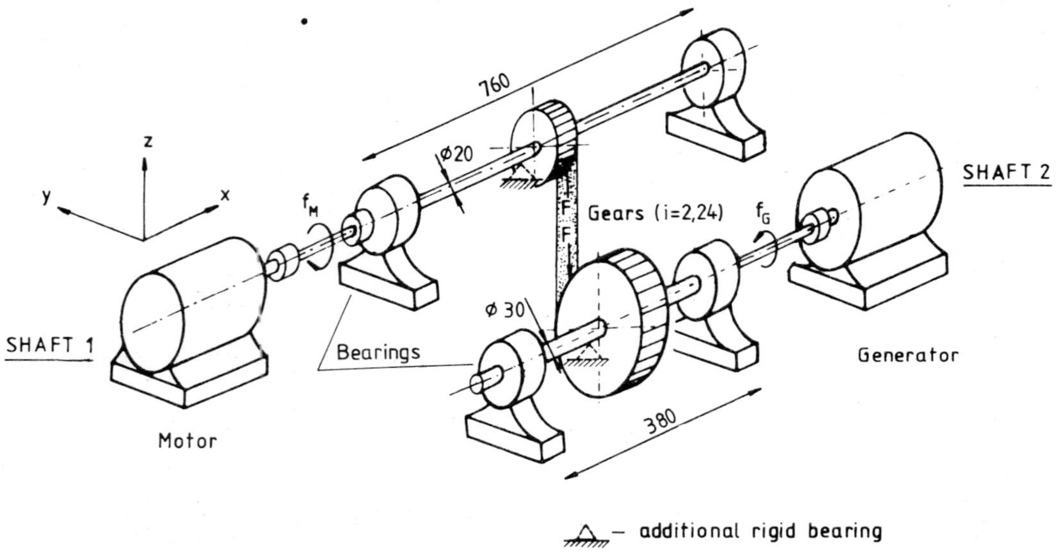

Fig 3 Test rig

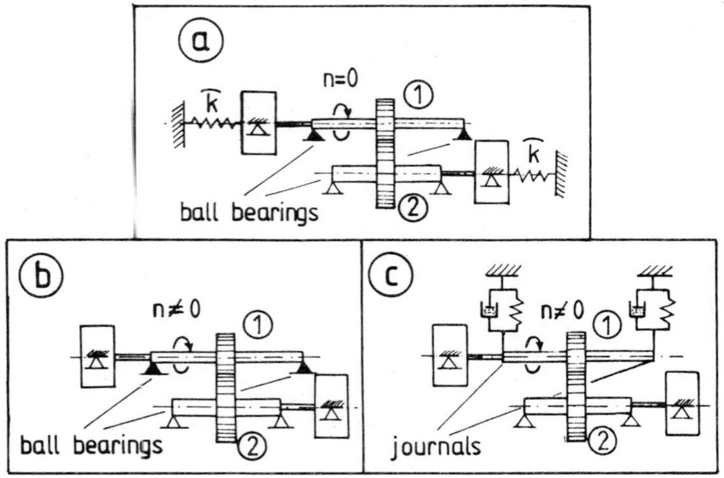

Fig 4 Test rig — models a, b and c

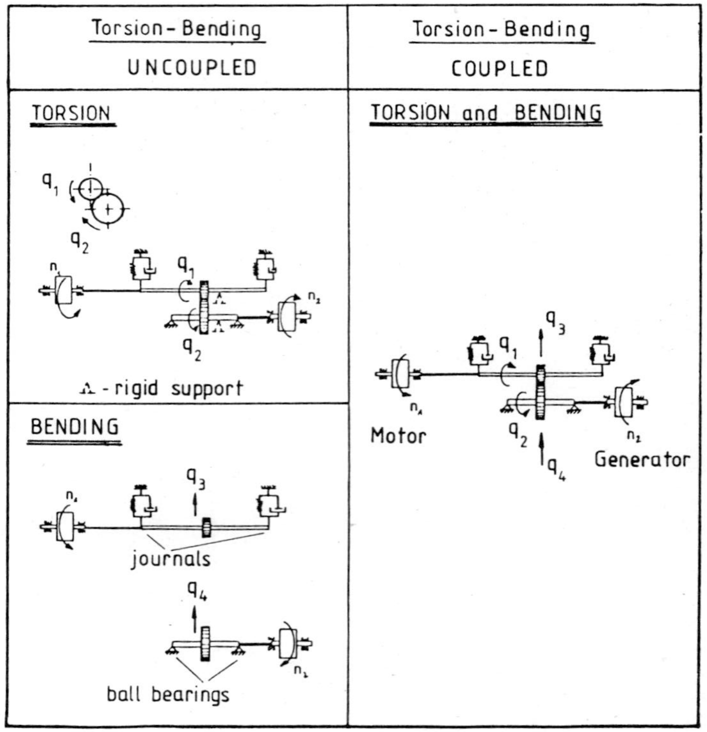

Fig 5 Measurement of the uncoupled and the coupled system (example c)

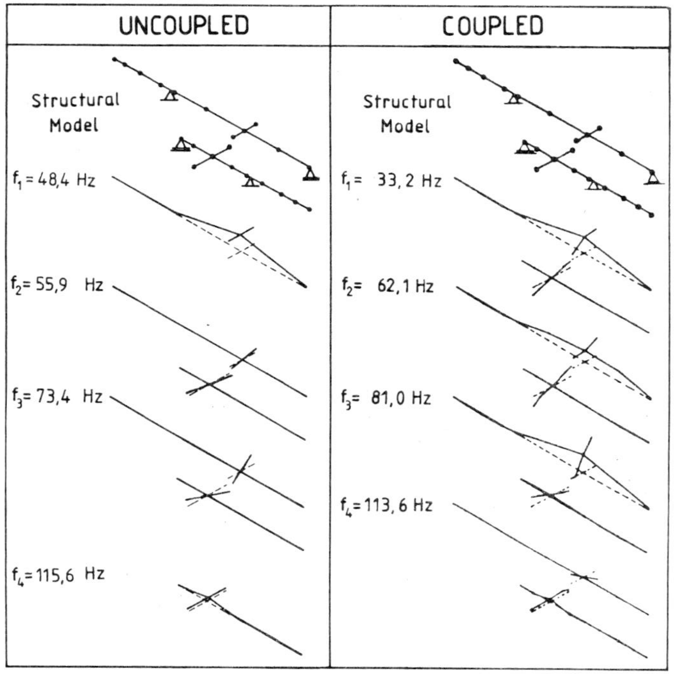

Fig 6 Identified modes and eigenfrequencies of test rig a

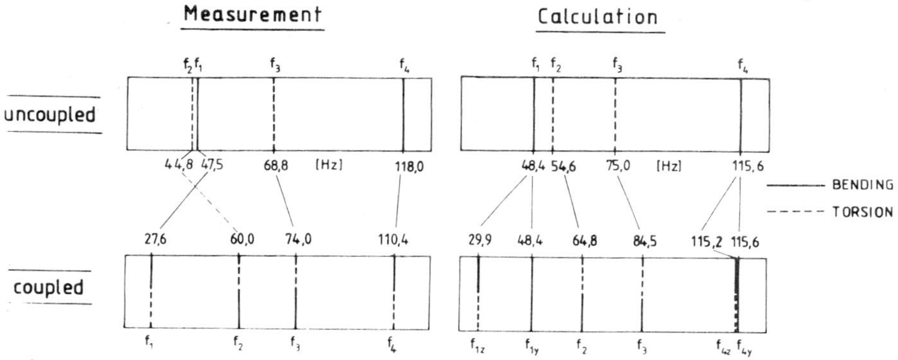

Fig 7 Eigenfrequencies of test rig b; measurement, calculation

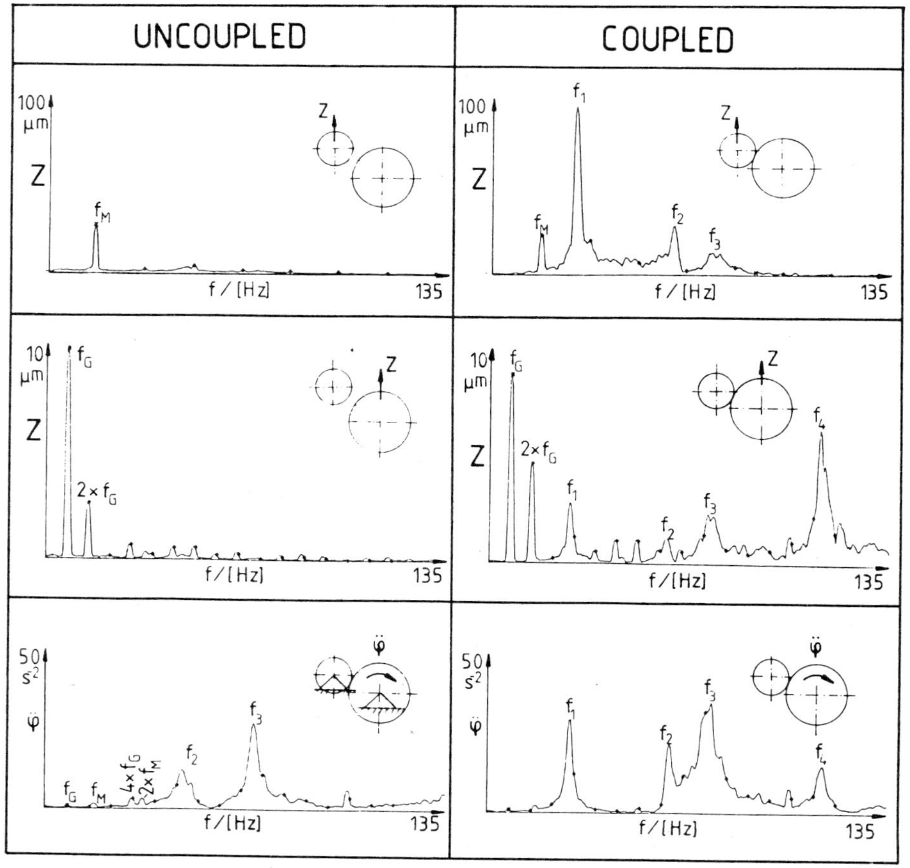

Fig 8 Spectra unbalance and gear excited vibration at test rig b (f_M = 16 Hz, f_G = 7.2 Hz)

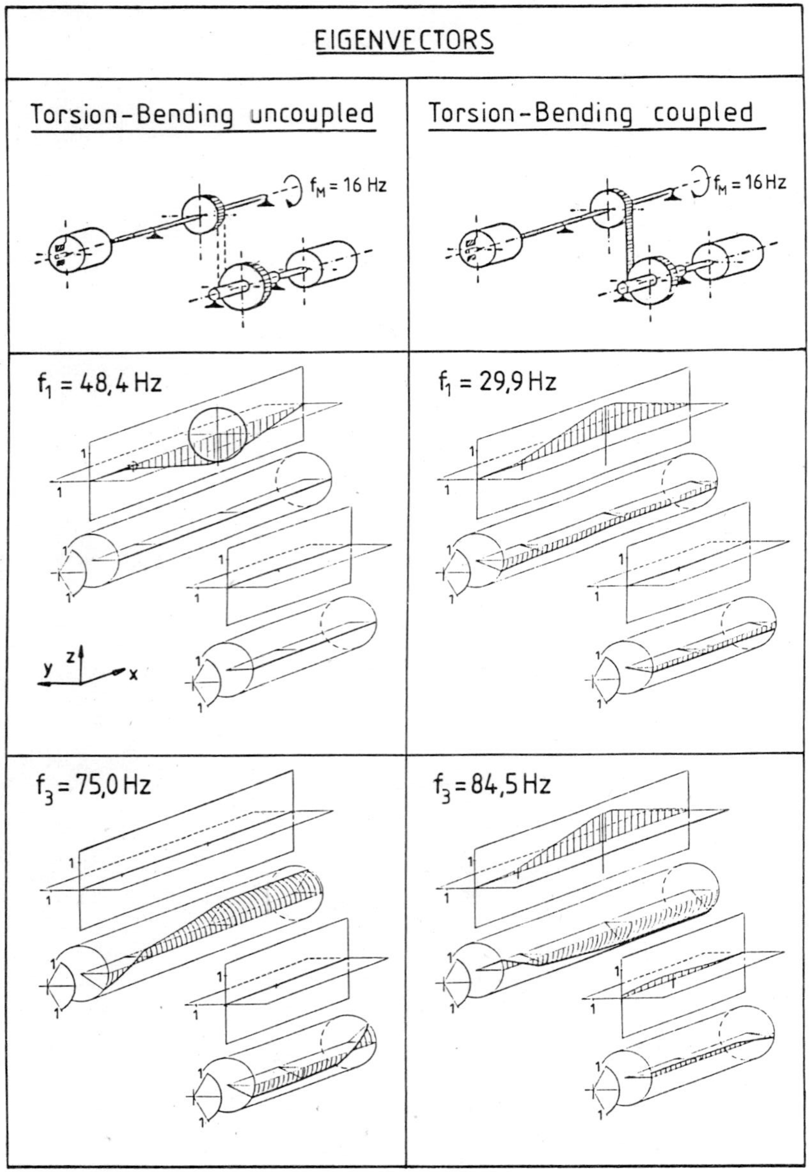

Fig 9 Calculated mode shapes and eigenfrequencies of test rig b

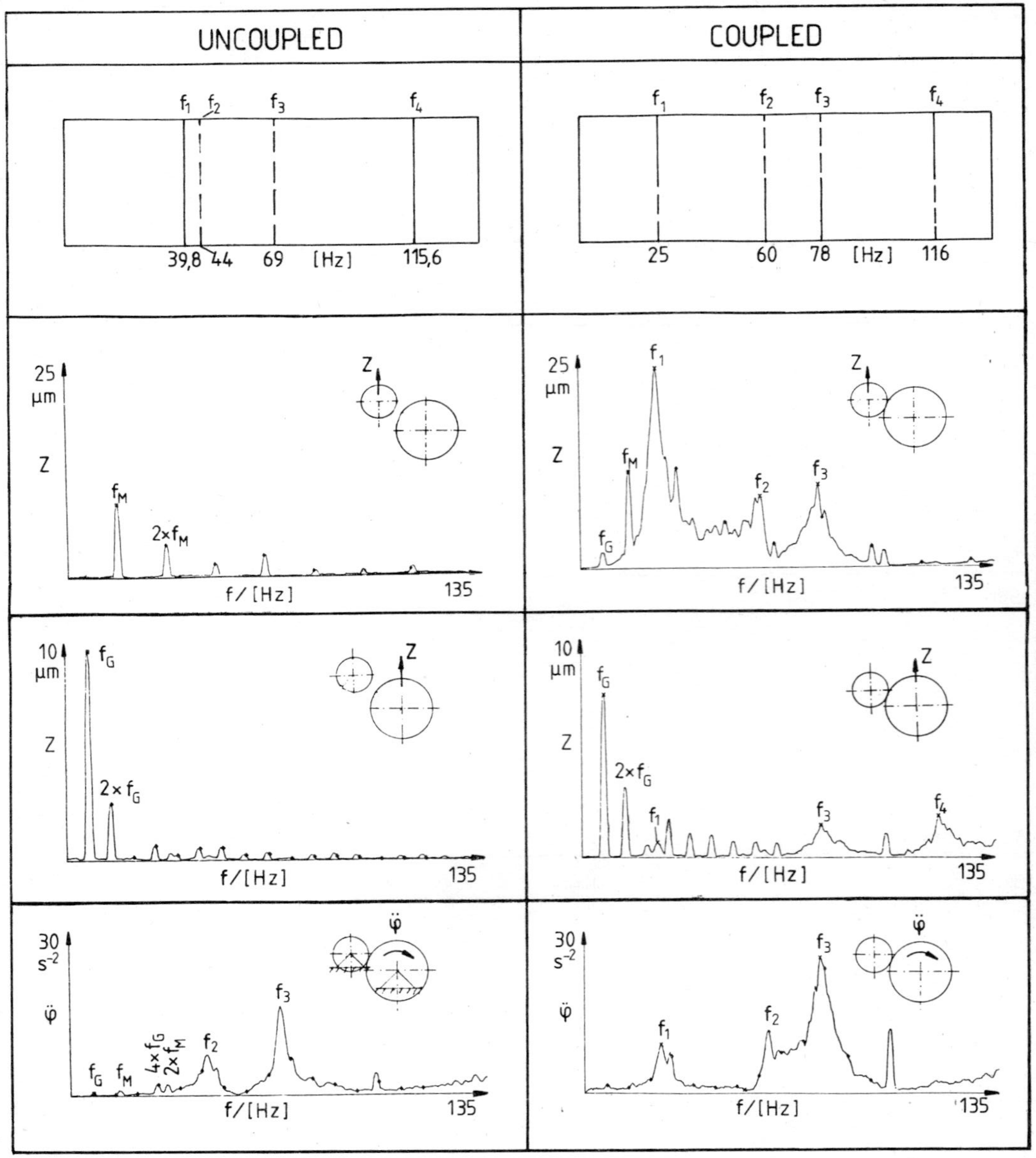

Fig 10 Spectra of unbalance and gear excited vibration at test rig c (f_M = 16 Hz, f_G = 7.2 Hz)

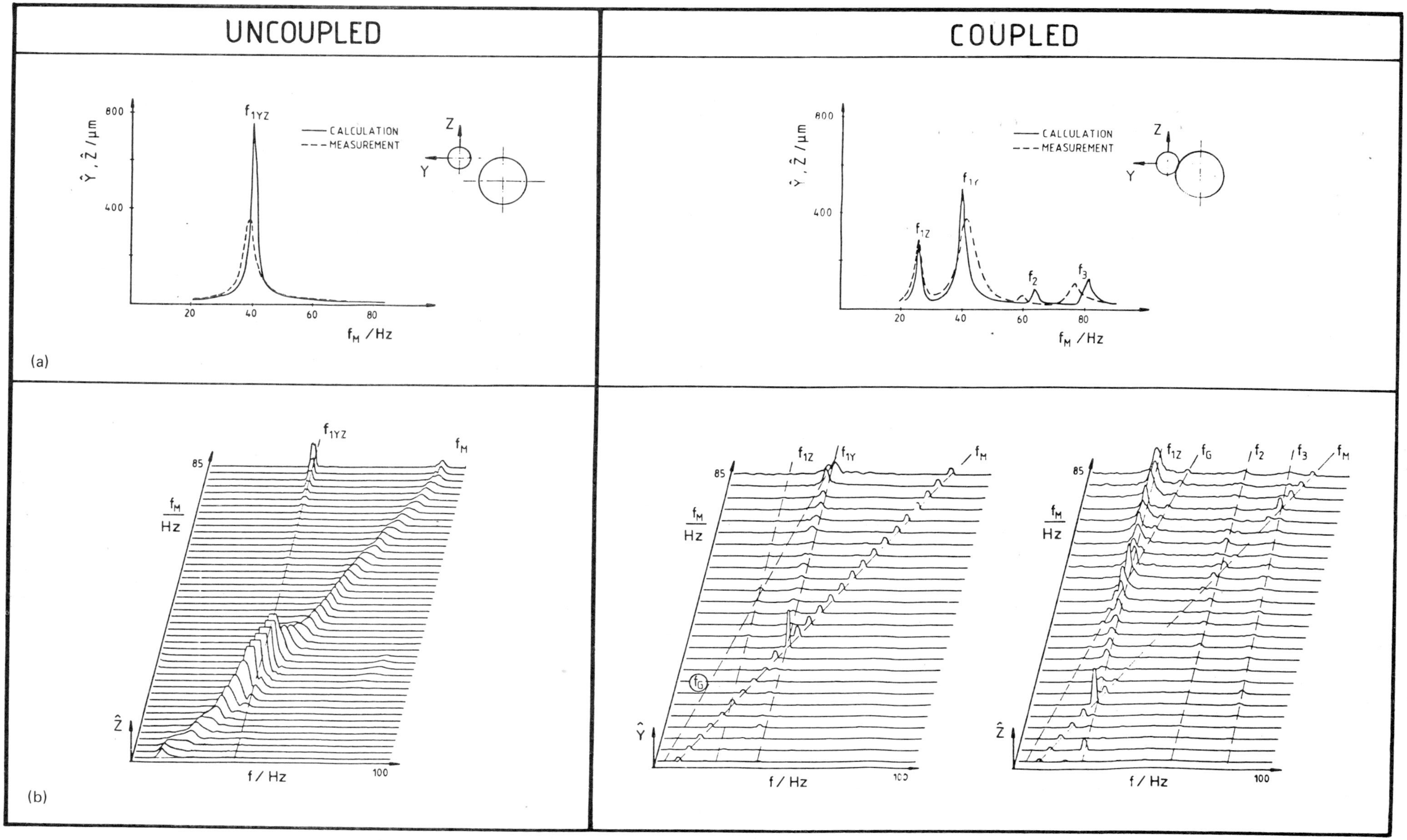

Fig 11 Forced vibrations at test rig c
(a) Measurement and calculations of the synchronous vibrations
(b) Spectra of the bending vibrations

C247/88

Vibration and noise of a geared rotor system —considered lateral and torsional motion

T IWATSUBO, DEng, MJSME
Faculty of Engineering, Kobe University, Kobe, Japan

SYNOPSIS This paper presents the vibration and noise obtained by the experimental results. The experimental test rig is constructed by a pair of spur gears, and both test rigs are prepared to compare the vibrational and noise characteristics of ball bearing and journal bearing supports. The amplitude and torsional acceleration of gear wheels, vibration of torque in the shafts and vibration of gear teeth are measured. These data are analyzed by FFT and the noise occurence mechanism is analysed. It is known that predominant gear noise occurs in the high frequency which coincides with intercross frequencies of the gear tooth resonant frequency and n multiplied gear mesh frequency. Tooth profile modification acts effectively on the vibrations of the geared-rotor system. Also low frequency noise of which frequency coincides with the natural frequency of the lateral vibration occurs.

1. INTRODUCTION

Gear system such as a power transmision is used in broad field of industry. Recently high speed and high performance gear is used in order to obtain high performance. Under these conditions requirement of vibration and noise reduction is increased.

There are many investigations on the vibration and noise analysis, and most of these works are concerned with tooth profile error, tooth deflection and dynamic force of gear tooth or noise reduction. In these analyses, torsional vibration of rotor system is considered and the effect of lateral vibration is neglected. Recently vibration and noise of geared rotor system considered torsional and lateral vibration is commenced. Iida et.al(1) investigated free vibration of the coupled lateral-torsional vibration of the geared rotor system and also investigated the forced vibration of unbalance due to geometric eccentricity. Hagihara et.al.(2) investigated the forced vibration by unbalance and the vibration due to torque fluctuation for geared rotor supported by journal bearings. Fukuma et.al.(5) investigated systematically the vibration and noise of the geared rotor system by theory and experiment. Kubo(6) and Masuda(7) investigated gear noise. However these investigations do not investigate the relation between lateral torsional vibration and noise in the gear coupled rotor system. Authers et.al.(3)(4) theoretically investigated stability and vibration due to tooth profile error by considering the coupled lateral torsional vibration of a geared rotor system.

This paper presents the effect of bearing type on the vibration and noise and noise occurence mechanism by experiment in order to compare the results with the former report. To investigate the effect of bearing type, rotor supported by ball bearings and supported by journal bearings are tested. To investigate the noise occurence mechanism three gear tooth profiles, which are optimum, over corrected and under corrected gear tooth profiles, are tested. And some important results are concluded.

2. EXPERIMENTAL SYSTEM

2.1 Experimental apparatus

Fig.1 shows the experimental apparatus and the measuring points of vibrations and Table 1 shows the details of the geared rotor system. Driving motor is a 3 phase induction motor and rotating speed is continuously changed by a variator and is measured by pulse senser. A generator is used as a driven part and the generated electricity is exhausted by resistor. The two types of bearing, that is, ball bearing and journal bearing are used for each bearing to investigate the effect of bearing on the vibration and noise. Diaphragm coupling and flexible coupling are used for driving and driven part, respectively. The diameter of both shafts is ϕ17mm and span lengths of driving and driven shafts are 270mm and 170mm for ball bearing and 405mm and 245mm for journal bearing, respectively. Each of the bending and torsional stiffnesses are shown in Table 2. The tested gears are spur gear with teeth surface finished to the JIS zero grade, that is, best finishing for standard gear, by grinding and is quenched by carburization. Its material is SNC M420. The three pairs of gears are tested which are different in tooth profile correction. Details of the tooth profile correction is illustrated in Fig.2. The correction lengths in the figure show the value which is projected on the line of action. The tooth profile (a) in Fig.2 is the optimum correction for load torque 29.4Nm and (b) is over corrected case and (c) is the true involute which means no profile correction and under corrected case. These three pairs of gears are named (a) Optimum correction gear, (b) over correction gear and (c) under correction gear.

2.2 Measuring system

Item and detail of measured vibrations of geared rotor system are shown in Table 3. Signals of item No.1, 3, 4 are transmitted to the Data recorder and Real time analyser by slip ring 6 in Fig.1 and analysed.

3. EXPERIMENTAL RESULTS AND DISCUSSIONS

3.1 Sound pressure distribution

Figs.3 and 4 show the sound pressure distributions of the test apparatus which is used journal bearing and optimum tooth profile correction gear, where rotating speed and load torque are 1200rpm and 29.4Nm, respectively. Fig.3 shows iso-sound pressure line on horizontal plane. It is known from this figure that the noise occurs from geared rotor system and noise from motor and generator is not large, so measured noise is mainly the noise from the geared rotor system. Fig.4 shows iso-sound pressure line on the cross section of gear wheels. It is known from this figure that the maximum noise occurs nearly on the line of action of gear force transmission. This sound pressure distribution is qualitatively similar for the case of ball bearing.

So microphone is located on the line of action to measure the gear noise.

3.2 Vibration and noise analysis of the case of ball bearing

Figs.5-7 are the experimental results for the optimum tooth profile correction and load torque 29.4Nm. Fig.5 shows the vibrational acceleration of radial direction of driven gear for various rotating speed. At the same time the vibrational acceleration of circumferential direction is measured, but this acceleration is about a half of the acceleration of radial direction. In Fig.5 meshing frequencies fz (rotating speed (rps))×(number of teeth) and 2fz are large, this is because the 3rd mode of which eigen frequency is 507Hz and this is mainly excited in lateral directed of driven shaft and is near the fz and 2fz. Fig.6 shows the lateral vibration of the driven shaft. From this figure the oscillation at 507Hz becomes large and fz lines become large. Fig.7 shows the noise which is occured in the geared rotor system. Comparing this figure with Fig.5, it is known that the acceleration characteristics and noise characteristics are very similar. The noise under 1KHz is mainly from the motor and the generator except for the fz and 2fz components. The peak value of the geared tooth noise is between 3KHz and 4KHz and this noise is very noisy for human. The vibrational mode of tooth deflection is 3502Hz, so this frequency and peak frequency coincide. From this discussion. It is known that the meshing frequency fz excite to the lateral vibration and high frequency vibration is coincident with tooth eigen-frequency. Figs.8 and 9 show the frequency analysis of the geared noise for rotating speeds and also show the cases of the over and under tooth profile correction for torque load 29.4Nm. The noise of these two cases is larger than the optimum tooth profile correction. This is the reason that if the gear teeth profile correction is optimum, the tooth meshes smooth, but it is not optimum some excitation force occurs, at the tooth meshing and it induces the tooth vibration. So it is known that the tooth profile correction is very important to reduce the gear noise.

3.3 Vibration and noise analysis of the case of journal bearing

Tests for the optimum, over and under tooth profile corrections are done and vibrations and noise are measured for the gear coupled rotor system supported by journal bearings. These characteristics were similar to these of ball bearing, so same conclusion is deduced. Comparing with the vibration of the rotor system supported the journal bearings with the ball bearings, the case of ball bearing is larger than the case of journal bearing in acceleration and sound pressure. To compare the sound pressure of both cases, the sound pressure of the rotor system supported by journal bearing is illustrated in Figs.10 and 11, where note that the scale is different. Large sound pressure is recorded in high rotating speed in Fig.10. In this case vibration due to gear backlash occurs and sound level increases, which is observed from the data of Fig.13.

3.4 Analysis of tooth vibration

In this section, tooth vibration, sound pressure and acceleration of gear are discussed. Fig.12 shows these vibration modes for the case of non tooth profile correction, 1440rpm, ball bearing and load torque 29.4Nm, where (a) and (b) show tooth vibrations of next tooth to each other. Overlapped range in two vibration modes means two pair meshing and non overlapped range means one pair meshing. (c) and (d) show sound pressure and acceleration in circumferential direction. The results of frequency analysis for one pair and two pair meshings are shown in Table 3. It is known from the table that frequencies of tooth vibration, gear noise and acceleration are almost same for each the two pair and one pair meshings. From this result noise in high frequency is induced by gear meshing, i.e. bending vibration of the tooth.

Fig.13 shows the vibration modes for the case of optimum tooth profile correction, 1980rpm, journal bearing and load torque 29.4Nm. As known from Fig.10 sound pressure is very large at this condition. This is due to the vibration induced by gear backlash, because tooth deflection becomes minus. In this case the results of frequency analysis are shown in Table 3, where frequencies of tooth vibration, gear noise and acceleration are almost same value as of as the case of ball bearing.

Fig.14 shows the relation between dedendum tooth vibration and the load torque. It is known from this figure that as the transmit torque increases, the tooth vibration increases and gear noise increases.

Fig.15 shows spectrum analysis of sound pressure for various torques. It is known from this figure that peak frequency is different for the high and low load torque. This characteristic is almost same for the rotating speed.

From the above discussions the following is summerized ; (1) high frequency noise induced by gear coupling occurs by the gear meshing, (2) vibration due to gear backlash occurs, tooth vibration amplitude becomes large and noise level increases, (3) As the load torque increases, the tooth vibration increases and noise increases.

3.5 Effect of rotating speed on the noise

Noise and vibration vary for the rotating speed. This effect is discussed in this section. Figs.16 and 17 show the sound pressure and circumferential acceleration for rotating speed where the load torques are a parameter. In these figures peaks shown by allow (A) and (B) do not change by load torque and other peaks change by load torque. It is known from this fact that the peaks (A) and (B) are the resonance frequencies of lateral vibration of both shafts. The other peaks which change by the load torques are related by the deflection of tooth excited vibration mode. Vibration and noise increases by rotating speed and load torque.

4. CONCLUSIONS

The gear coupled rotor supported by ball bearings and journal bearings are tested in order to investigate the lateral and torsional vibration and gear noise, where three types of gear tooth profile i.e. optimum, over and under tooth profile corrections, are tested and the following concluded;

(1) When natural frequency of the lateral vibration coincides with the gear meshing frequency, the lateral vibration is resonated and gear noise occurs.

(2) High frequency vibration correspond with the tooth eigenfrequency causes high frequency noise. This vibration and noise level is related to the tooth profile correction.

(3) The characteristics of the gear coupled vibration and noise is qualitatively same with no regard to the bearing type. But with regard to the vibration and noise level, gear coupled rotor supported by journal bearing is smaller than that by ball bearing.

(4) As the transmit torque increases or the rotating speed increases, the tooth vibration increases and gear noise increases.

(5) Vibration due to gear backlash occurs, tooth vibration amplitude becomes large and noise level increases.

REFERENCES

1. IIDA,H. & TAMURA,A., Coupled Torsional Flexural Vibration of a Shaft in a Geared System, Proc. of Vibration in Ratating Machinery, IME 1984-10, p.67.
2. HAGIHARA,N. et.al., Coupled Lateral-torsional Vibration of Gear System Supported by Jouranl Bearing, Trans. JSME Vol.49, No.445 (1983) p.1530.
3. FUKUMA,H., Fundamental Research of Vibration and Noise Gear, Doctor Thesis of Kyoto University, 1972.
4. IWATSUBO,T. et.al., Coupled Lateral-torsional Vibration of Rotor System Trained by Gears (Part1, Analysis by Transfer Matrix Method), Bulletin of JSME, Vol.27, No.227 (Feb. 1987), p.271.
5. IWATSUBO,T. et.al., The Coupled Lateral-torsional Vibration of a Geared Rotor System, Proc. of Vibration of Rotating Machinery, IME 1984-10, p.59.
6. KUBO,A. et.al., Load Transfer characteristics of Cylindrical Gear with Error, Trans. JSME, 43-371 (1977) p.2771.
7. MASUDA,T. et.al., Noise by Gear System (Effect of manufacturing method), Proc. of JSME 21th Symp., No.830-8 (1983) p.317.

Table 1 Details of gear system

Three-phase induction motor	
Capacity (kW)	5.5 (1740rpm)
Polar moment of inertia (kgm^2)	0.1666
D.C. generator	
Power (kW)	2.2 (1750rpm)
Polar moment of inertia (kgm^2)	7.413×10^{-2}
Spur gear	
Mass (kg)	1.7
Polar moment of inertia (kgm^2)	1.1598×10^{-3}
Module	4
Number of teeth	20
Pressure angle	20°
Face width (mm)	8
Gear ratio	1
Contact ratio	1.56

Table 2 Bending and torsional stiffness of the shafts

	Ball bearing	Journal bearing
K1 (N/mm)	8.31×10^3	3.15×10^3
K2 (N/mm)	1.99×10^4	1.53×10^4
K01 (Nmm/rad)	2.68×10^6	2.68×10^6
K02 (Nmm/rad)	2.03×10^5	2.03×10^5

Table 3 Measurement of vibrations of geared rotor system

Item	Location of Probe (No. in Fig.1	
(1) Acceleration of Circum-ferential Direction	8	Measured by acceleration pickups which are located at the symmetry for to the disc center.
(2) Disc Deflection	7	Measured by non-contacting displacement pickups which are located on the line of action of gear tooth and 90° phase shift.
(3) Torsional Torque	9	Measured by strain gauges attached at the shaft.
(4) Strain of Dedendum	10	Measured by strain gauge attached at the dedendum of two teeth.
(5) Rotating Speed	2	Measured at the coupling by pulse senser

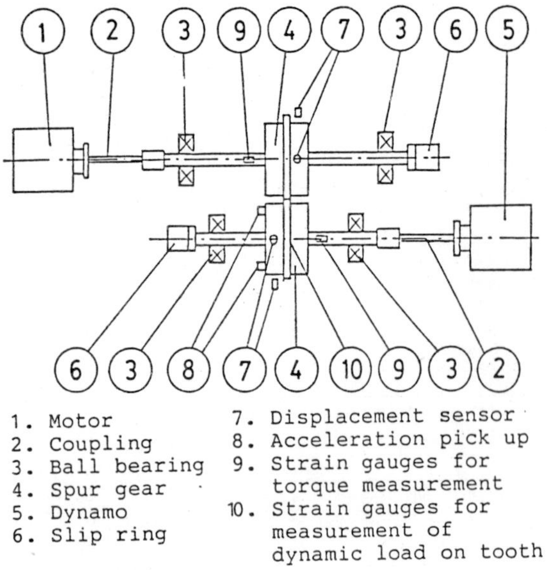

Fig 1 Experimental apparatus and vibration measuring location

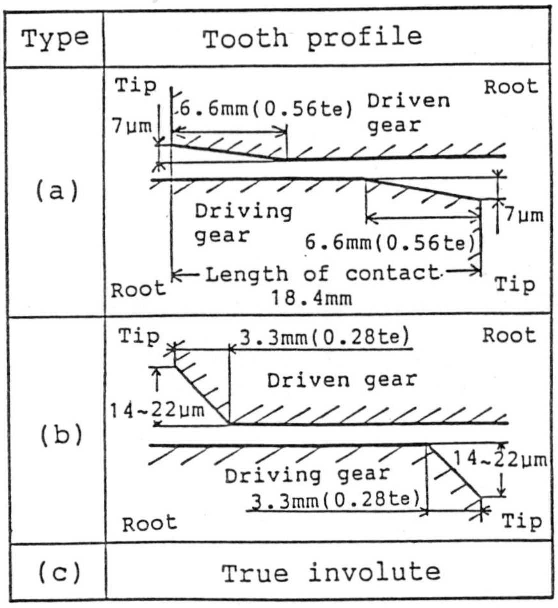

Fig 2 Tooth profile correction

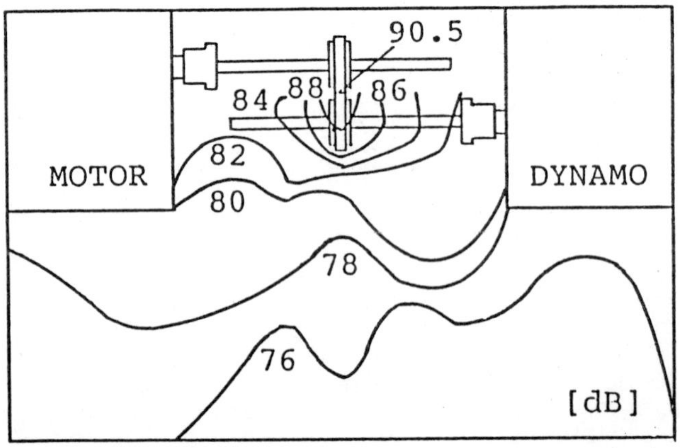

Fig 3 Iso-sound pressure curve of test rig (optimum tooth profile correction; journal-bearing; 1200 r/min; load torque 29.4 Nm)

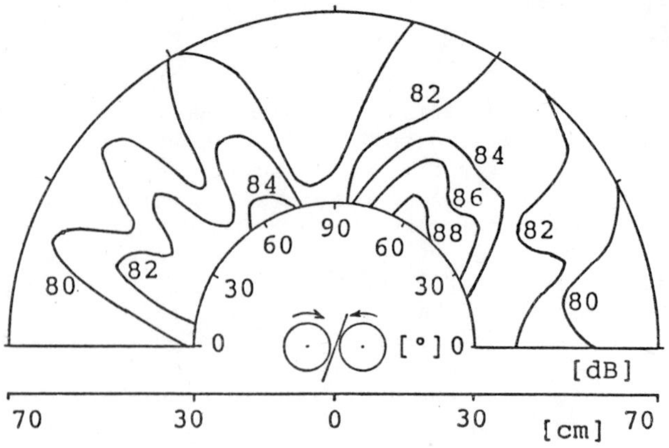

Fig 4 Iso-sound pressure curve of test rig (optimum tooth profile correction; journal-bearing; 1200 r/min; load torque 29.4 Nm)

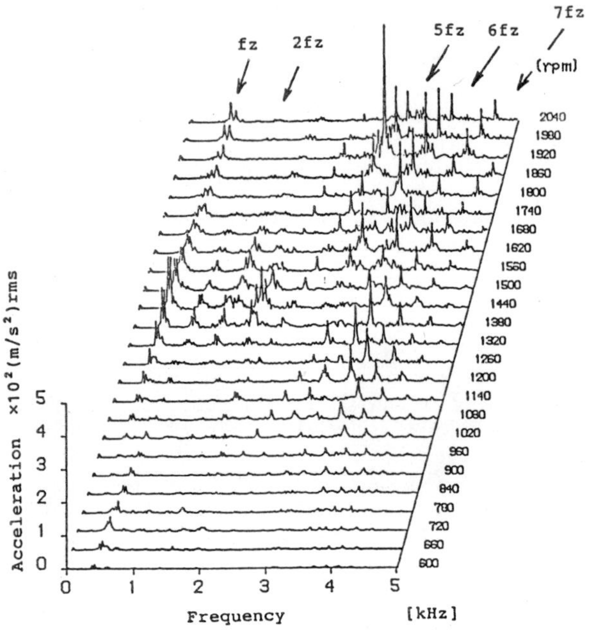

Fig 5 Spectrum analysis of acceleration of driven gear in radial direction (optimum tooth profile correction; ball-bearing; load torque 29.4 Nm)

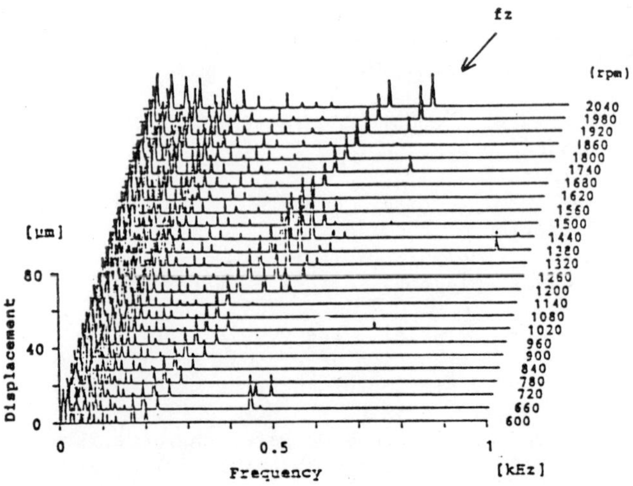

Fig 6 Spectrum analysis of lateral vibration of driven gear (same conditions as Fig 5)

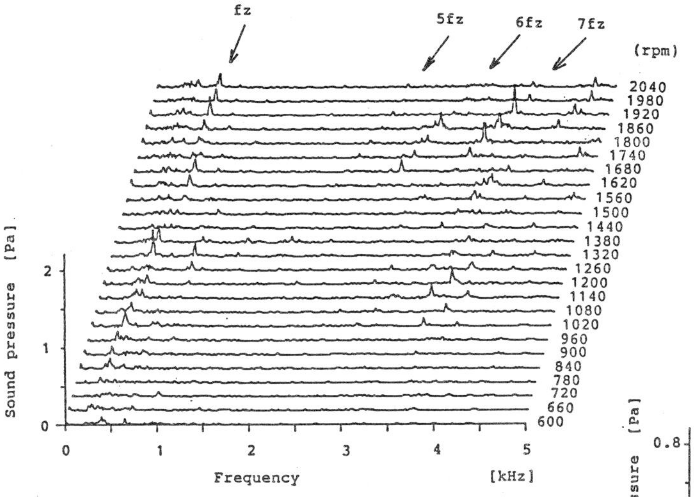

Fig 7 Spectrum analysis of gear noise (same condition as Fig 5)

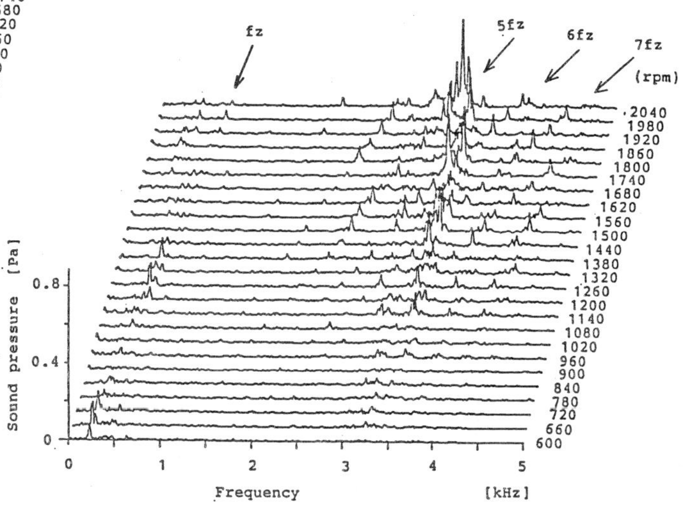

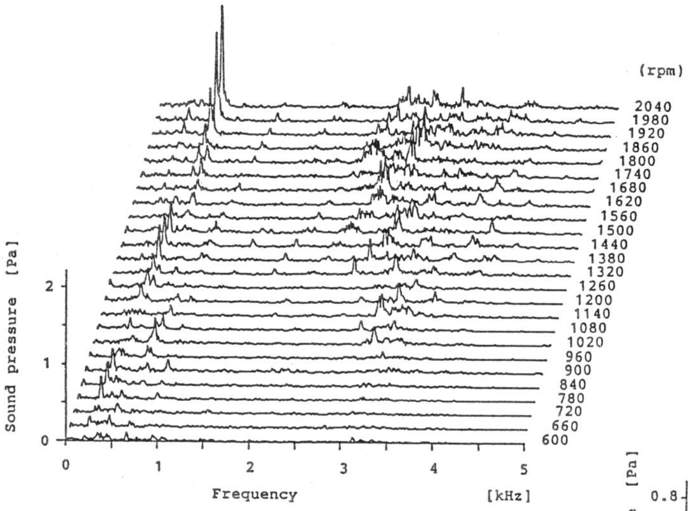

Fig 8 Spectrum analysis of gear noise (over tooth profile correction; ball-bearing; load torque 29.4 Nm)

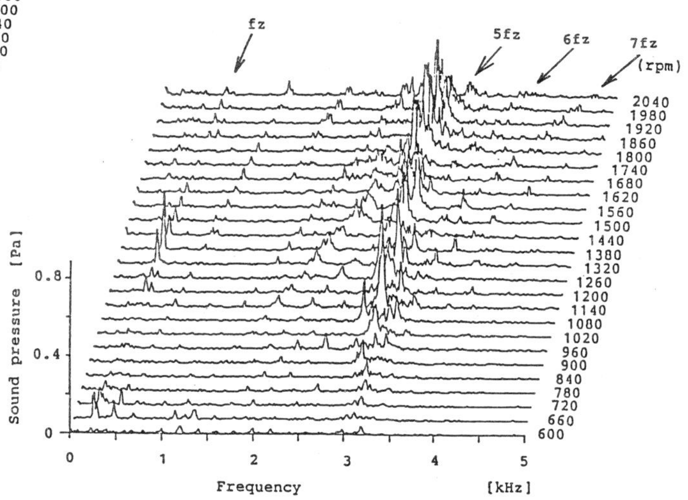

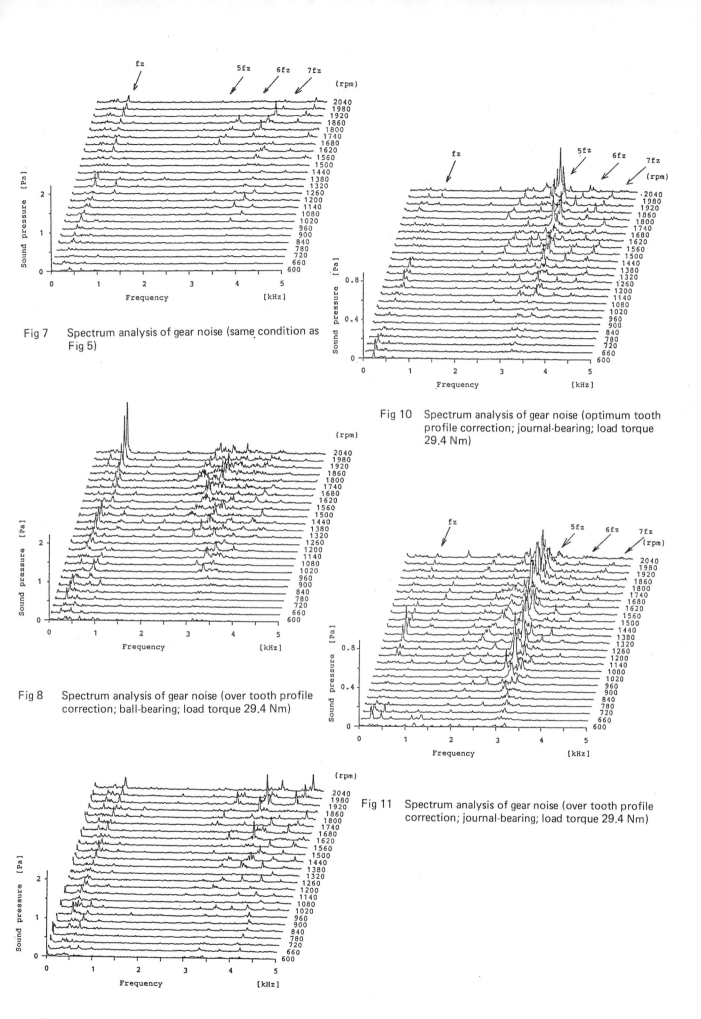

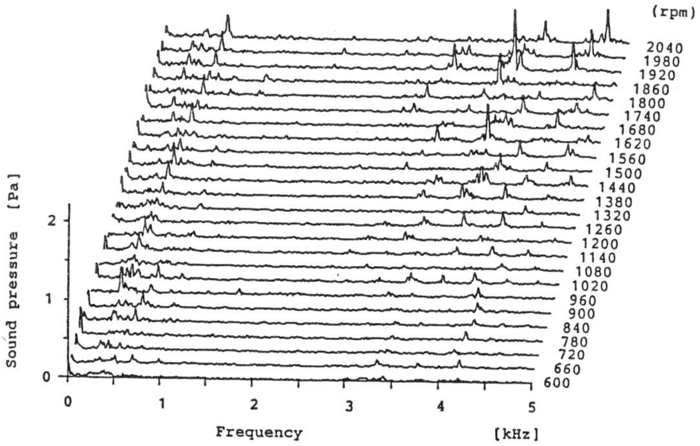

Fig 9 Spectrum analysis of gear noise (under tooth profile correction; ball-bearing; load torque 29.4 Nm)

Fig 10 Spectrum analysis of gear noise (optimum tooth profile correction; journal-bearing; load torque 29.4 Nm)

Fig 11 Spectrum analysis of gear noise (over tooth profile correction; journal-bearing; load torque 29.4 Nm)

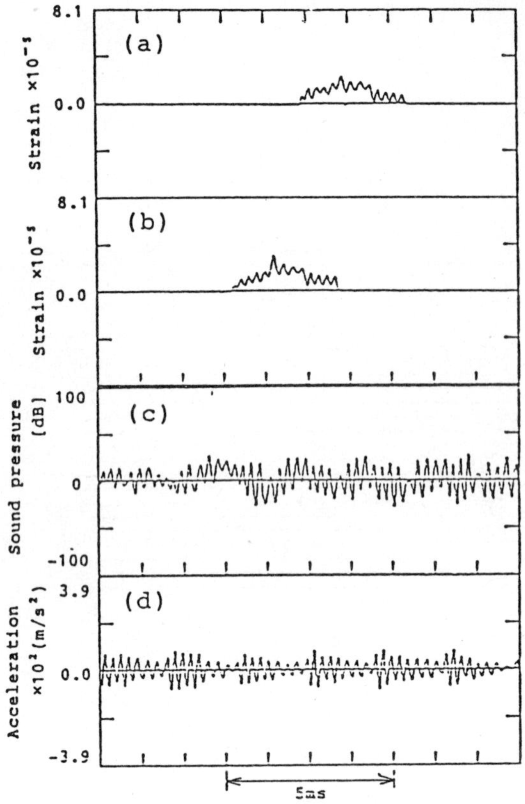

Fig 12 Vibration mode of dedendum strain; sound pressure and acceleration;
(a), (b) Dedendum strain
(c) Gear noise
(d) Circumferential acceleration of driven gear
(non-tooth profile correction; ball-bearing; 1440 r/min; load torque 29.4 Nm)

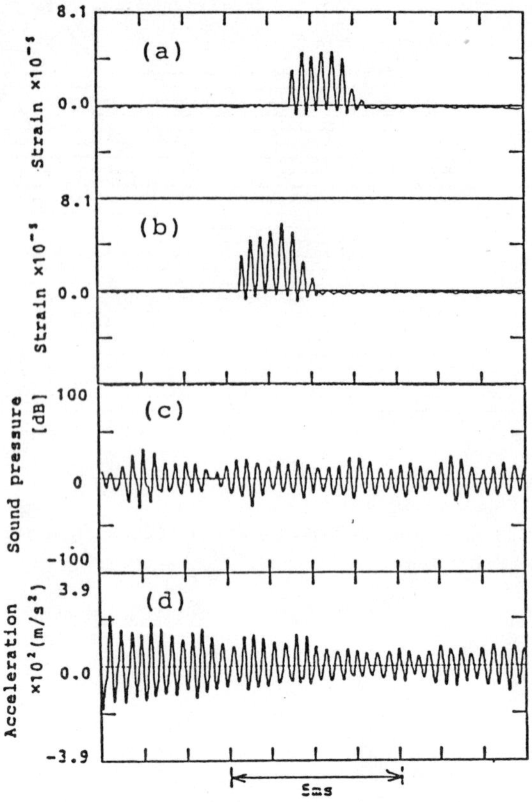

Fig 13 Vibration mode of dedendum strain; sound pressure and acceleration; (a)–(d) As for Fig 12 (optimum tooth profile correction; journal-bearing; 1980 r/min; load torque 29.4 Nm)

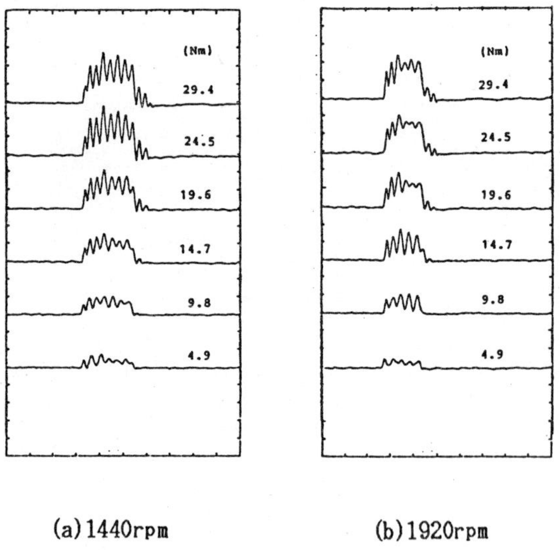

Fig 14 Dedendum tooth vibration for torques (optimum tooth profile correction; journal-bearing)

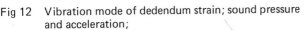

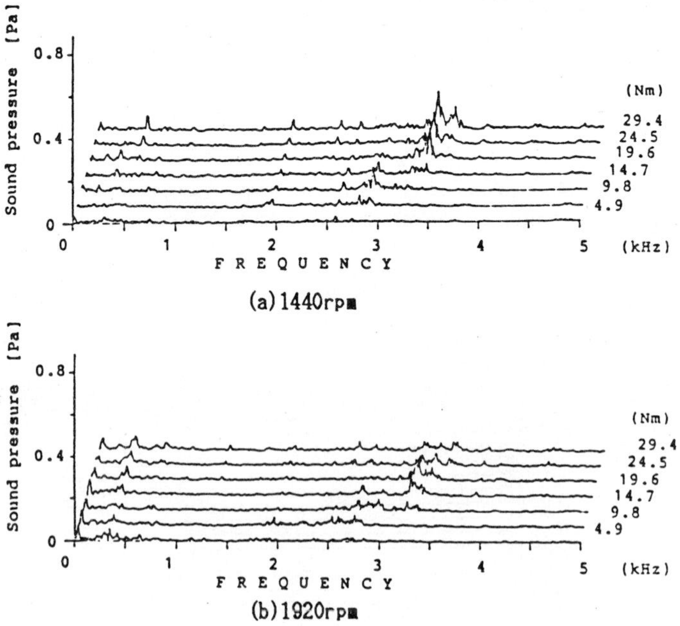

Fig 15 Spectrum analysis of sound pressure for torques (optimum tooth profile correction; journal-bearing)

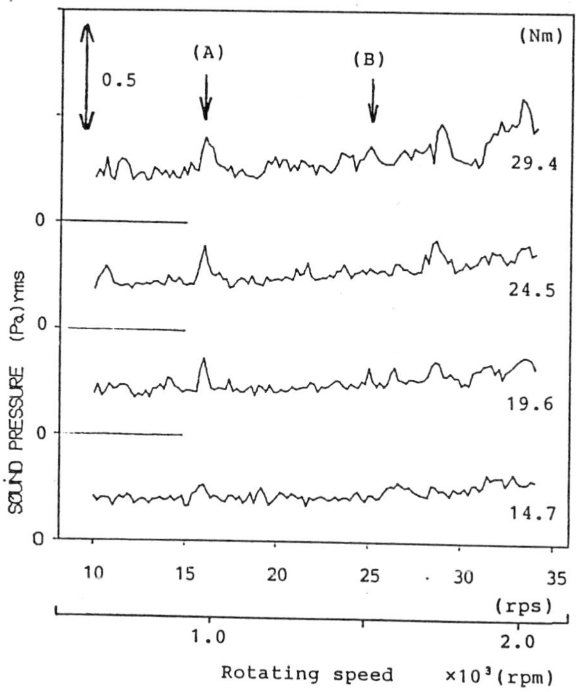

Fig 16 Sound pressure for rotating speed (non-tooth profile correction; ball-bearing)

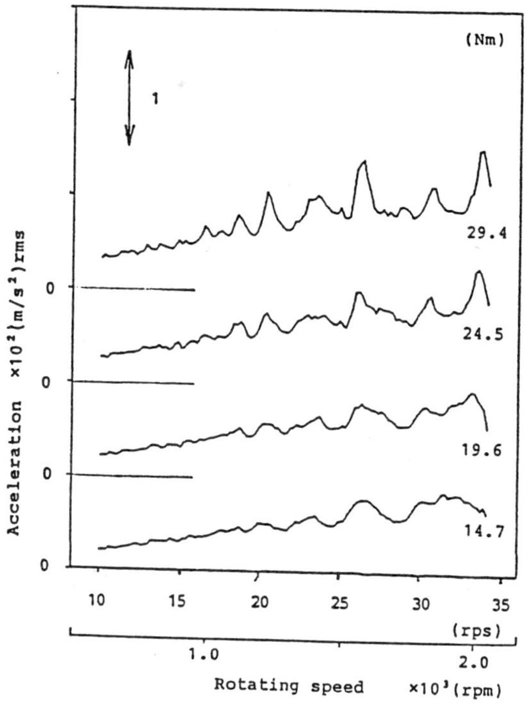

(a) Circumferential direction

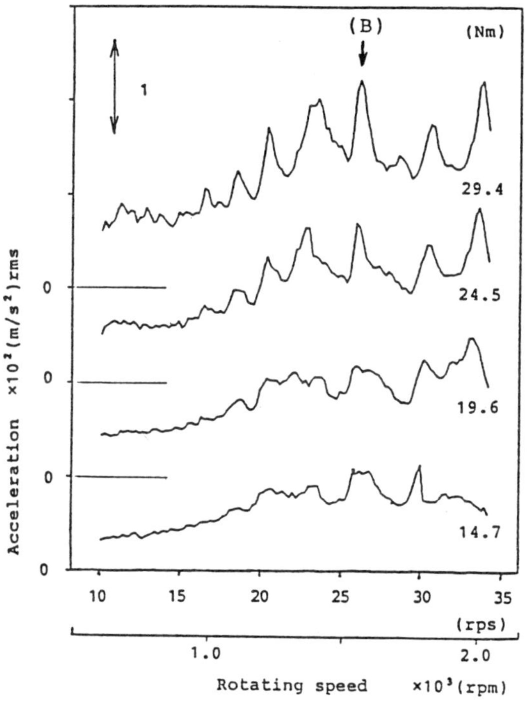

(b) Radial direction

Fig 17 Acceleration of driven gear for rotating speed (non-tooth profile correction; ball-bearing)

C284/88

The dynamics of a rotor with rubbing

D de KRAKER, BE, ME, PhD, M T M CROOIJMANS, ME, PhD and D H van CAMPEN, ME, PhD
Department of Mechanical Engineering, Eindhoven University of Technology, Eindhoven, The Netherlands

SYNOPSIS

In real rotor-bearing systems several tight clearances between the rotor and the fixed casing exist. Due to misalignment or excessive unbalances the rotor may rub against the casing. A simple model for this so-called rubbing is introduced. The model consists of a Laval rotor and a non-linear finite boundary stiffness with Coulomb friction.
It is known that rubbing is detected in the vibration spectrum by peaks at integer multiples of the rotorspeed. Therefore, a non-linear calculation technique is used that is based on a periodicity assumption. After discretization with respect to time and using finite differences, the non-linear equations of motion can be transformed into a set of non-linear algebraic equations, which can be solved by standard numerical techniques.
Using an arc continuation method a sensitivity analysis is performed for the system variables. For a particular static deflection of the rotor results are shown of rubbing near the system resonance. Phenomena such as jumping and backward whirl are discussed. Conclusions are drawn for the sensitivity of the phenomena with respect to the boundary stiffness, damping and friction coefficient.

LIST OF SYMBOLS

~	a tilde denotes a column matrix
.	a dot denotes the first derivative with respect to time
..	a double dot denotes the second derivative with respect to time
ODE	ordinary differential equation
A_0	maximum amplitude of rotor response predicted by linear models
C	clearance between rotating parts and the fixed housing
D	rotor diameter
$\vec{e}_x, \vec{e}_y, \vec{e}_z$	fixed Cartesian triad
F_c^*	dimensionless constant force
F_r^*	dimensionless coefficient in the expression of the restoring force
G	final set of descretized expressions
$\tilde{M}$	rotor mass
N	total number of discretization points
P_g	geometrical centre of disc
P_m	centre of disc mass
$\tilde{q}$	generalized co-ordinates
R	rotor radius
t	time
x,y	displacements of the rotor centre
ξ, η	dimensionless displacements of the rotor centre
$\tilde{z}$	column with unknowns after discretization
α	exponent in expression of the restoring force
δ	parameter that indicates whether the rotor touches the housing
ε	dimensionless mass eccentricity
ς	dimensionless damping of the rotor support
θ	angle defined by eq. (3)
τ	dimensionless time
ω	dimensionless rotor speed
ω_0	dimensionless frequency at A_0
Ω	constant rotational rotor speed

1 INTRODUCTION

In real rotors narrow clearances between the rotating parts and the fixed housing exist. In the ideal situation clearances between the fixed and rotating parts are circumferentially equal. Generally, this is not true due to misalignment; furthermore, the rotor is excited by mass unbalance and will vibrate. Due to this combination the rotating parts may rub against the housing and cause very different dynamic behaviour that is usually difficult to predict. However, it was recognized that due to rubbing the rotor vibrates periodically or quasi-periodically. In the vibration spectrum this can be identified by sharp peaks at a restricted number of frequencies and their multiples (lit. [2, 3, 4, 5]). In this paper we will assume that the rotor vibrates periodically.

We will analyse the dynamic behaviour of a simple model consisting of a Laval rotor and a massless boundary with a non-linear restoring force.

The mathematical model results in two coupled second order ODE's. The number of design variables can be reduced by using a non-dimensional form of the set of ODE's. We will assume that the rotor touches the static parts slightly and the sensitivity of the response with respect to the various design variables will be investigated.

2 ROTOR GEOMETRY, CONSTANT LOAD AND ISOTROPIC SUSPENSION

The Laval rotor has been modelled as a massless shaft mounted in two bearings at each end. Symmetrically between the bearings, it has an infinitely thin rigid disc perpendicular to the shaft. The disc has a mass M and radius R, whilst it rotates with a constant angular velocity Ω around the centre P_g of the disc. The centre of mass P_m of the disc does not necessarily coincide with P_g as indicated in figure 1a.

The bending vibrations of this rotor have been modelled with two generalized co-ordinates. In the reference configuration, P_g coincides with the origin P_o of the Cartesian triad ($\vec{e}_x$, $\vec{e}_y$, $\vec{e}_z$). The disc translates in the $\vec{e}_x$- and $\vec{e}_y$-directions and rotates around the direction of $\vec{e}_z$. The position of P_g with respect to P_o is given by the vector $\vec{r}_g$ and the position of P_m with respect to P_g by $\vec{r}_e$. $\vec{r}_g$ and $\vec{r}_e$ are expressed in terms of $\vec{e}_x$ and $\vec{e}_y$ as follows:

$$\begin{aligned} \vec{r}_g &= x\vec{e}_x + y\vec{e}_y \\ \vec{r}_e &= e\cos(\Omega t)\vec{e}_x + e\sin(\Omega t)\vec{e}_y \end{aligned} \tag{1}$$

Where x and y are generalized co-ordinates, e is the distance between P_m and P_g while Ωt is the rotation angle.

The disc is loaded with a constant force $\vec{F}^c$, a load $\vec{F}^l$ that is linearly dependent on the generalized co-ordinates, and a non-linear load $\vec{F}^r$. The constant load $\vec{F}^c$ with amplitude F_c points into the negative y-direction.

The linear force $\vec{F}^l$ results from isotropic stiffness, k_s, and isotropic damping, d_s, of the shaft and its bearings and it acts at P_g.

3 RUBBING FORCES

The non-linear load $\vec{F}^r$ is produced by the rubbing of the rotor against the stationary housing. We assume that during the motion of the rotor the housing only adds stiffness to the rotor. In the reference configuration, the clearance between the disc and the housing is constant and is called C. We can formulate a condition in which the disc contacts the housing:

$$\sqrt{(x^2+y^2)} \geqq C \text{ or } \delta = \sqrt{(x^2+y^2)}/C - 1 \geqq 0 \tag{2}$$

We assume that the disc and the housing make contact at one point P_r only if the contact criterium (2) is fulfilled. The position of point P_r relative to P_o is given by the vector $\vec{r}_p$:

$$\vec{r}_p = (R+C)\ (\cos\Theta\ \vec{e}_x + \sin\Theta\ \vec{e}_y) \tag{3}$$

with: $\tan\Theta = y/x$

The housing acts on the disc with a normal force $\vec{F}^t$ and a tangential or friction force $\vec{F}^w$ (figure 1b). We assume that the normal force depends on δ exponentially:

$$\vec{F}^t = \vec{F}^t_x + \vec{F}^t_y = \begin{cases} \vec{0} & \text{if } \delta<0 \\ -F_r\delta^\alpha(\cos\Theta\ \vec{e}_x + \sin\Theta\ \vec{e}_y) & \text{if } \delta \geqq 0 \end{cases} \tag{4}$$

In (4), we represent the normal force by means of a positive constant F_r and an exponent α. If $\alpha > 1$, the normal force as a function of δ has a continuous first derivative. Physically, the rotor presses against the housing.

Instantaneously, this will cause a local impression of the housing. The normal force needed for this impression increases non-linearly with the magnitude of the impression.

The friction force $\vec{F}^w$ is represented by Coulomb's law with the characteristic Coulomb friction coefficient f:

$$\vec{F}^w = \vec{F}^w_x + \vec{F}^w_y = \begin{cases} \vec{0} & \text{if } \delta<0 \\ f\, F_r \delta^\alpha (\sin\Theta\, \vec{e}_x - \cos\Theta\, \vec{e}_y) & \text{if } \delta \geq 0 \end{cases} \tag{5}$$

In this model, we have assumed that the relative velocity of the rotor contact point with respect to the housing contact point is always counterclockwise and unequals 0. This assumption was checked and appeared to be valid for the results that we will show later. With (4) and (5), we can find $\vec{F}^r$ from $\vec{F}^r = \vec{F}^t + \vec{F}^w$.

4 EQUATIONS OF MOTION

We are now able to formulate the equations of motion using Newton's law:

$$M\, \ddot{\vec{r}}_m = M\, (\ddot{\vec{r}}_g + \ddot{\vec{r}}_e) = \vec{F}^c + \vec{F}^1 + \vec{F}^r \tag{6}$$

This can be transformed in two coupled ODE's.

$$M\, (\ddot{x} - e\Omega^2 \cos(\Omega t)) = -d_s \dot{x} - k_s x + \underline{F_r \delta^\alpha (-\cos\Theta + f\sin\Theta)} \tag{7a}$$

$$M\, (\ddot{y} - e\Omega^2 \sin(\Omega t)) = -F_c - d_s \dot{y} - k_s y + \underline{F_r \delta^\alpha (-\sin\Theta - f\cos\Theta)} \tag{7b}$$

The underlined terms in the equation (7) have to be taken into account only when $\delta \geq 0$.

The number of design variables in equation (7) is eleven. We will reduce this number by writing these equations in a non-dimensional form. For this purpose, we must define a characteristic frequency ω_k as: $\sqrt{k_s/M}$. Furthermore, we will use the clearance C as a characteristic length of the problem. With C and ω_k we can define a number of non-dimensional variables:

$$\begin{aligned} (\xi,\eta) &= (x/C, y/C) \quad &; \varepsilon = e/C \\ \tau &= \omega_k\, t &; \omega = \Omega/\omega_k \\ \varsigma &= d_s/(2M\omega_k) & \\ F^*_r &= F_r/(MC\omega_k^2) &; F^*_c = F_c/(MC\omega_k^2) \end{aligned} \tag{8}$$

If we use (8) for rewriting equation (7), we obtain:

$$\xi'' + 2\varsigma\, \xi' + \xi + \underline{F^*_r \delta^\alpha (\cos\theta - f\sin\theta)} = \varepsilon\omega^2 \cos\omega\tau \tag{9a}$$

$$\eta'' + 2\varsigma\, \eta' + \eta + \underline{F^*_r \delta^\alpha (\sin\theta + f\cos\theta)} = -F^*_c + \varepsilon\omega^2 \sin\omega\tau \tag{9b}$$

with: $(\ )' = d(\ )/d\tau$, $(\ )'' = d^2(\ )/d\tau^2$

5 SOLUTION METHOD

A method is discussed which provides periodic solutions of the non-linear equations of motion (9) as a function of any of the design variables. This method is based on time discretization of the system equations combined with a numerical solution algorithm to solve the resulting non-linear algebraic equations.

Instead of a continuous periodic solution of the equations (9) an approximate solution at a discrete number of times will be determined. In order to discretise the equations (9) let τ_j (j=1,...,N) be an equidistant partition of one period of time T, yielding:

$$\tau_j = \frac{(j-1)}{N}\, T \tag{10}$$

with T the period of time that results from Ω ($T = 2\pi/\Omega$). The velocity and acceleration in the direction of $\vec{e}_x$ at τ_j are expressed with a 4th order central difference scheme. The velocity and the acceleration in the direction of $\vec{e}_y$ at τ_j can be expressed in a similar way.

Application of these discretization schemes in the equations of motion at the N discretization points yields 2N algebraic equations. The algebraic equations are denoted by:

$$\underset{\sim}{G}\, (\underset{\sim}{z}) = \underset{\sim}{0} \tag{11}$$

with the 2N unknowns:

$$\underset{\sim}{z}^t = [\xi(\tau_1),\ldots,\xi(\tau_N),\eta(\tau_1),\ldots,\eta(\tau_N)] \tag{12}$$

This set of algebraic equations can be solved by a standard multi-dimensional Newton-Raphson method. If we add one of the variables as an unknown we can solve the equations with an arc continuation method [1], and obtain solutions as a function of that variable.

The operation of the arc continuation method is based on a prediction step followed by correction steps until the solution at a next value of the design variable is reached.

In order to find an initial solution with which we can start the Newton Raphson procedure, a first calculation with the linearized set of equations can be made. The resulting solution will resemble the solution of the set of non-linear ODE's as long as the excitational forces in (9) are small.

6 DESCRIPTION OF THE BEHAVIOUR AS A FUNCTION OF ROTOR SPEED

If we neglect the stiffness of the housing a simple set of linear ODE's remain from (9). We will refer to this situation as the linear case whereas the non-linear case will mean that we do take into account the non-linear restoring force of the housing. Both the radial frequency, ω_o, and the corresponding vibration amplitude (A_o) at resonance can be solved analytically for the linear case.

We will assume that the static equilibrium position (ξ,η) of the rotor centre P_g due to F_c^* is at $P_2=(0,-0.9)$. Additional dynamic loads will cause the rotor to rub against its housing.

We have assumed a low damping value of $\varsigma=0.02$. The choice of $\varepsilon=0.008$ was made so that rub will occur. At this value the maximum amplitude of the rotor for the linear case is $A_o=0.2$ The housing characteristics were $F_r^*=10$, $\alpha=1.3$ and $f=0.20$.

Figure 2 shows the amplitude of the rotor vibration as a function of the rotor speed, in both the linear and non-linear cases. The amplitude of the non-linear periodic motion has been calculated as half of the maximum minus the minimum value of both the x- and the y-co-ordinates. The amplitude of the x- and y-co-ordinates of the linear curve are equal. At each of the symbols (Δ, o, etc.) the periodic solution has been determined using the arc continuation method. Some of the points at the amplitude curve of the x-co-ordinate have been numbered.

The differences between linear and non-linear forced response functions are obvious. The peaks of the x- and y-amplitudes are both at higher frequencies than the linear resonance frequency ω_0. The amplitudes are larger as well. An increase of the resonance frequency was to be expected from the increase of the stiffness produced by the housing.

In figure 3, the motion of P_g has been drawn in the x-y plane for the calculated points indicated in figure 2. The arrows indicate the whirl-direction of the rotor.

If we slowly increase the rotor speed, the rotor will hit the housing at (0,-1) for the first time. From that point the orbit in the x-y plane will deviate from the circular form that follows from linear analysis. The increasing y-amplitude is restricted by the housing. At points 8 to 17, the rotor centre P_g whirls in the same direction as the rotor speed.

At points 19 to 33, the rotor whirls backward, which is caused by the frictional force. In its final part, the motion will approache a circular form when the rotor no longer touches the housing.

In practice, the motions at the equilibrium points 23 to 43 will never occur because they appeared to be unstable as confirmed by numerical time integration. If we slowly increase the rotor speed, the motion will jump suddenly from near point 23 to a motion where the rotor does not touch the housing.

7 INFLUENCE OF THE DAMPING

The shape of a resonance peak of a linear system depends greatly on the dimensionless damping coefficient ς. When the damping increases, A_0 decreases, ω_0 increases and the peak becomes less sharp.

Starting from the non-linear results shown in figure 2 we can vary the damping ς at four levels: $\varsigma=0.02$, $\varsigma=0.05$, $\varsigma=0.10$ and $\varsigma=0.20$. The corresponding linear amplitude $A_0=0.2$ was kept constant by adjusting the eccentricity value ε; ε was 0.008, 0.01998, 0.0398 and 0.0784 respectively. Figure 4 shows the results of these successive cases.

With an increased damping, a larger unbalance ε was chosen in order to keep the same value of A_0. In this case, the magnitude of the unbalance forces increases as compared with the magnitude of the restoring forces of the housing. For large values of ς, and thus of ε, the solution resembles the linear solution more and more.

Also, after increasing the damping, its stabilizing effect increasingly dominates the destabilizing frictional forces. The unstable series of solutions will disappear entirely for large damping values. Also, the unstable backward whirl motion will not exist any longer for $\varsigma=0.10$.

The jump phenomenon that occurs when the rotor

speed is slowly increase, does not necessarily disappear when we increase the damping. In figure 4, we observe the largest jump at a damping level of $\varsigma=0.10$.

8 INFLUENCE OF HOUSING STIFFNESS AND FRICTION COEFFICIENT

The stiffness of the housing can be changed by varying both α and F_r^*. Figure 5a shows the effect of an increase of α for both the amplitudes of the x- and y-co-ordinates. A decreased stiffness leads to a reduced resonance frequency. It also diminishes the unstable series of solutions. Due to the increased stiffness, the contact time between the rotor and the housing decreases. Therefore, the friction forces will have less influence and the backward whirl will eventually disappear. The jump in the y-co-ordinate amplitudes decreases too, while varying F_r^* leads to similar results.

The influence of a variation in the Coulomb friction coefficient f on the stationary behaviour is shown in figure 5b. In this example, the frictional forces point mainly in the x-direction. Therefore, the largest influence can be expected in the x-direction too as can be observed in figure 5b. The frictional forces diminish with f and, thus, the amplitude of the x-co-ordinate increases.

9 CONCLUSIONS

A rotor model was introduced that took into account rubbing between the rotating part and the stationary part. The reaction of the housing was modelled by a non-linear restoring force and a friction force. In the dimensionless form, the design variables for the model were varied in order to investigate their influence. For all results the resonance frequencies increase compared to those in the corresponding linear case due to the added housing stiffness. In the spectrum of the periodic solutions higher harmonics are found in addition to the first harmonic that can be found from linear analysis. In slightly damped cases, a backward whirl is caused by the friction force. However, it is found that this backward whirl is never stable. With the aid of numerical time integration, we learn that increasing or decreasing the rotor frequency causes the rotor to jump from one rotor orbit to another. A series of unstable periodic solutions are then passed over.

Results obtained for a rotor that rubbed the housing slightly are sensitive to the damping. Therefore, we adjusted the mass unbalance in such a way that the system without the housing had a constant maximum amplitude for all damping levels. From these results, we conclude that when the damping increases, the influence of the housing decreases and the backward whirl finally disappears. However, a greater damping does not necessarily mean that the jump phenomenon is less pronounced.

An increased housing stiffness leads to somewhat higher resonance amplitudes and frequencies. An increased friction coefficient has little influence on the results.

ACKNOWLEDGEMENT

These investigations were supported (in part) by the Netherlands Foundation of Technical Research (STW). The authors are indebted to Professor W.L. Esmeijer for valuable discussions and contributions to this paper.

REFERENCES

[1] Fried I., Orthogonal trajectory accession to the nonlinear equilibrium curve, Computer Methods in Applied Mechanics and Engineering, Vol. 47, 1984, pp 283-297.

[2] Crooijmans, M.T.M., On the Computation of Stationary Deterministic Behaviour of Non-Linear Dynamic Systems with Application to Rotor-Bearing structures, Thesis, University of Technology Eindhoven, oktober 1987.

[3] Beatty R.F., Differentiating rotor response due to radial rubbing, Journal of Vibration, Acoustics, Stress and Reliability in Design, April 1985. Vol. 107, pp 151-159.

[4] Muszynska, A., Nonlinear excited and self-excited processional vibrations of symmetrical rotors, Dynamics of rotors (Symposium Denmark), 12-16 augustus 1974, Springer Verlag, Berlin 1975.

[5] Childs, D., Fractional-frequency rotormotion due to nonsymmetric clearance effects, Journal of Engineering for Power, july 1982, Vol. 104, pp 533-541.

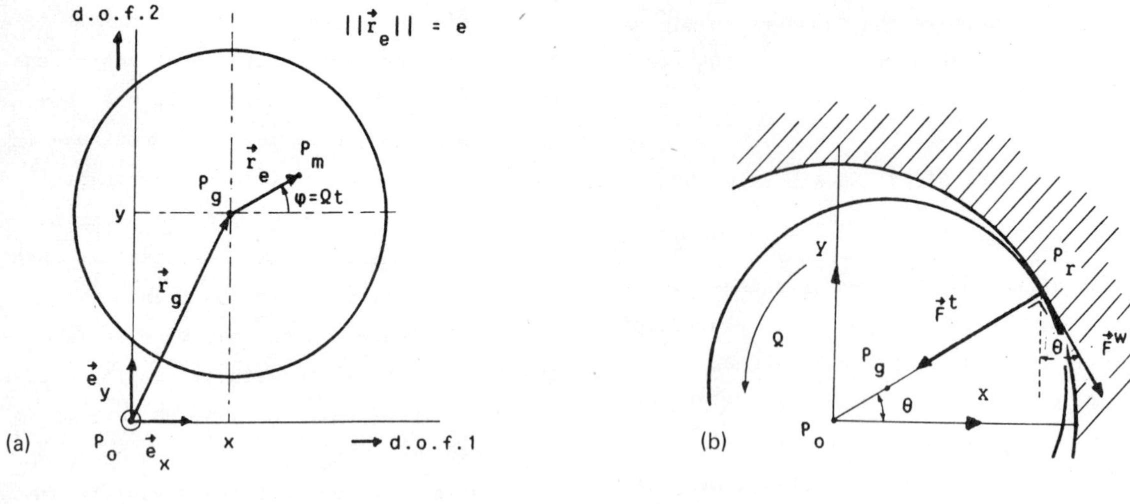

Fig 1 Degrees of freedom of the Laval rotor (a) and rubbing forces (b)

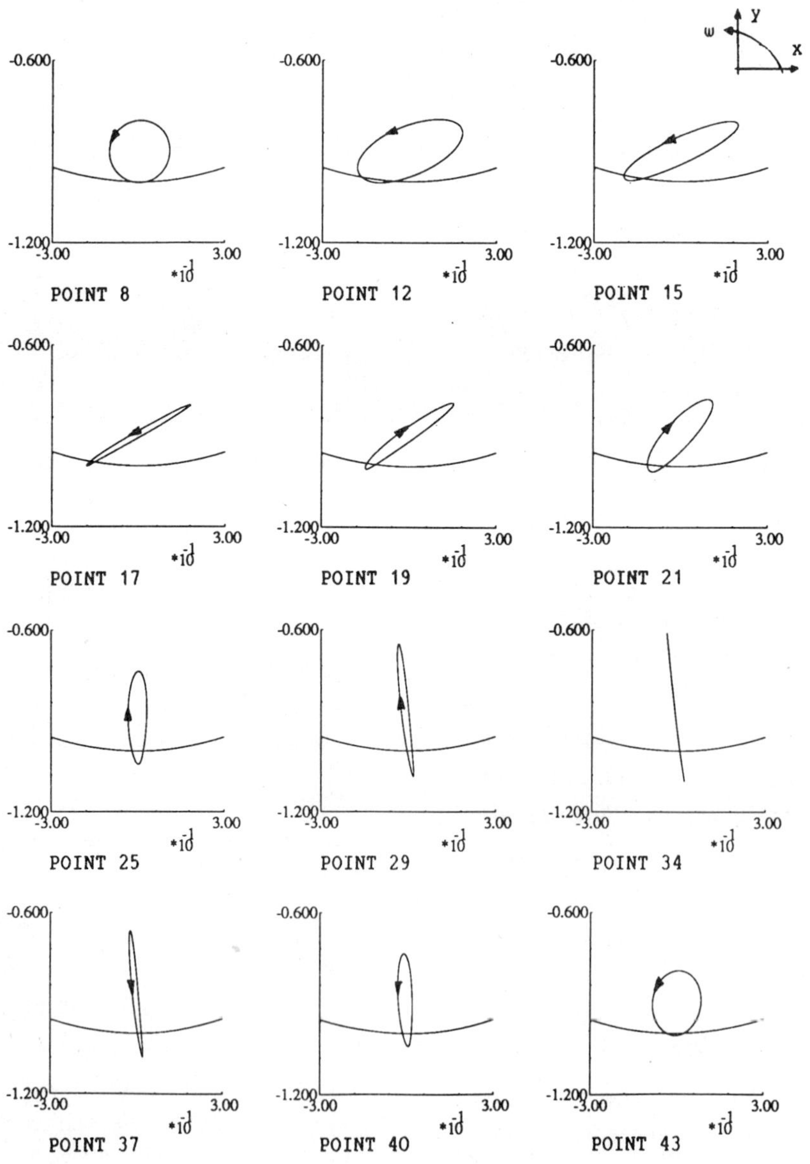

Fig 3 Motions described by the rotor centre P_g in the x–y plane for the calculations given in Fig 2

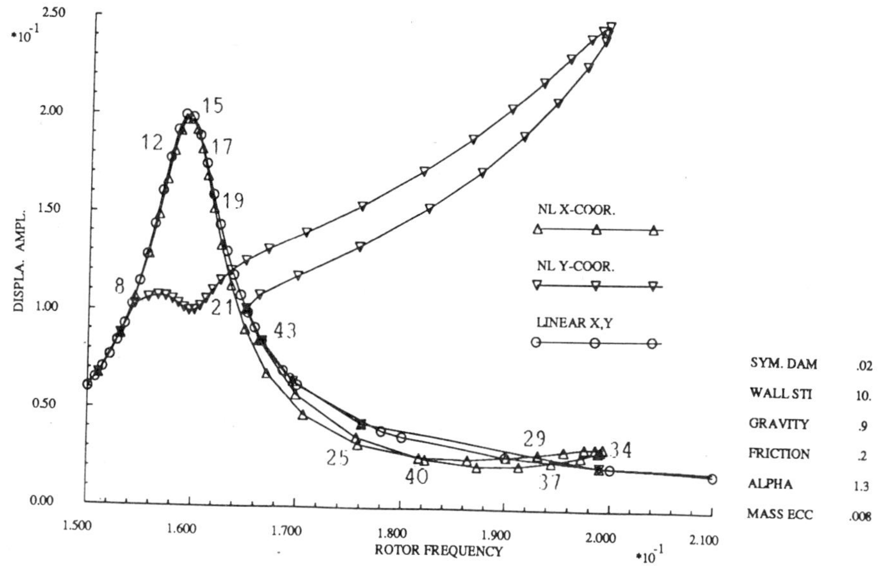

Fig 2 Amplitude of linear and non-linear orbits in the x and y directions

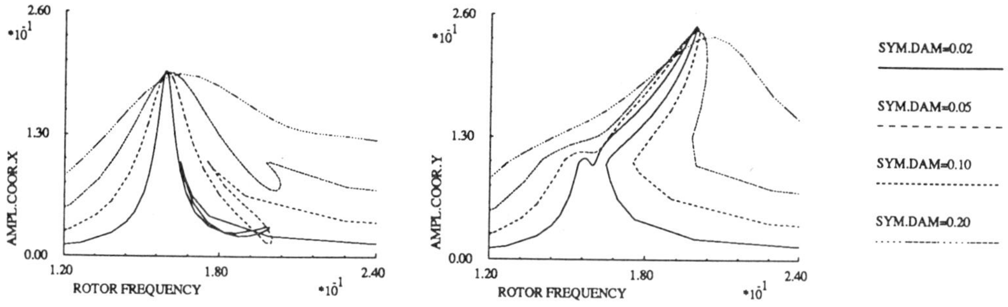

Fig 4 Influence of the damping ς on the amplitude of the coordinates in $\vec{e}_y$ directions for $F_r^* = 10$, $F_g^* = 0.9$, $f = 0.2$ and $\alpha = 1.3$

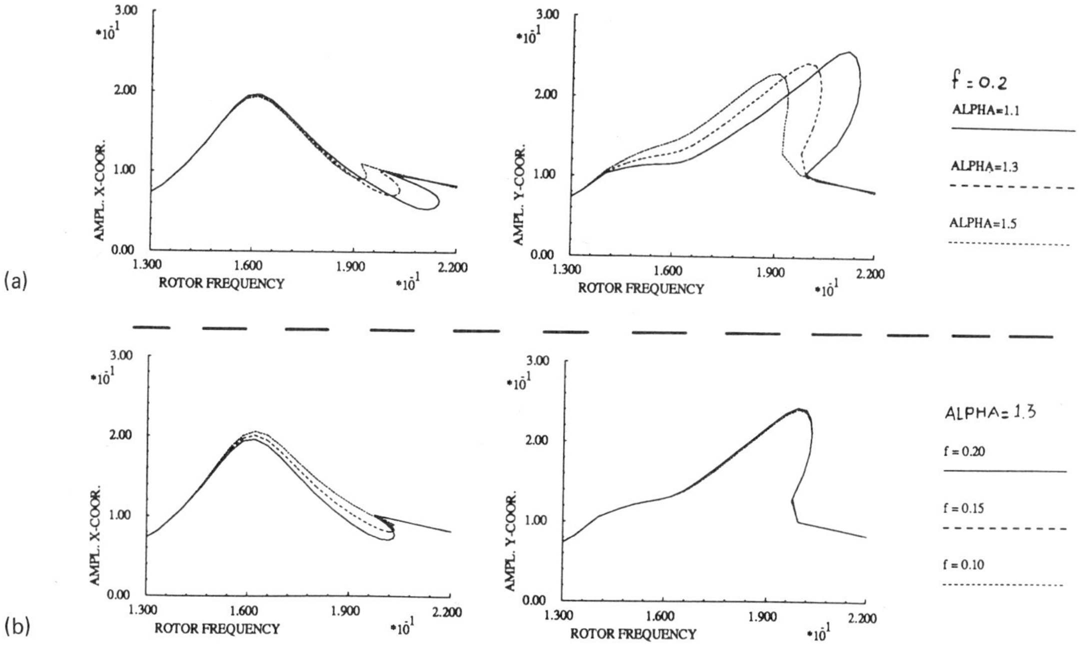

Fig 5 Influence of (a) α and (b) friction coefficient f on the amplitude of the coordinates in $\vec{e}_x$ and $\vec{e}_y$ directions for $\varsigma = 0.1$, $F_r^* = 1.0$, $F_g^* = 0.9$ and $\epsilon = 0.0398$

C252/88

Dynamic instability of multi-degree-of-freedom flexible rotor systems due to full annular rub

W ZHANG, CMES
Department of Applied Mechanics, Fudan University, Shanghai, People's Republic of China

SYNOPSIS The perturbed motion and its dynamic instability of multi-degree-of-freedom flexible rotor systems due to "full annular rub" is investigated. The general non-linear equation of motion is established. The approximate solution for complex rotor systems and the exact solution for a single disk rotor are both obtained. The dynamic instability is studied. The results show that the stability behaviour is dependent upon the initial condition. If the initial backward or forward perturbed precession velocity is larger than the threshold speed ω_1^* or ω_2^* respectively, the perturbation motion will be developed. The formulae for evaluating ω_1^* and ω_2^* are deduced in the paper.

1 INTRODUCTION *

There are many instances in which dynamic instability of rotor systems due to dry friction occurs. In fact, the optimum performance of many rotating machines lies in minimizing radial clearance between the rotor and stator to reduce leakage, which leads more frequently to rotor rubs. In contrast with other instability phenomena in rotor dynamics, perhaps the least studied and documented type of rotor whirl motion is that due to dry friction.

There are two cases of rotor radial rubs, the "full annular rub" and the "partial rub". In the former case, the rotor keeps in close contact with the stator during the whole cycle(360°) and the whirl motion is mainly due to the friction force. However, the latter case belongs in parameter vibration because of the partial contact and the consequent periodic "normal-loose" and "normal-tight" radial stiffness variation. It has been studied in some literature(Ref.1-3).

In this paper, the full annular rub is discussed. Den Hartog described the phenomenon in his well-known book(Ref.4) that, as soon as the rotor contacts with the stator, the tangential friction force will drive the rotor to whirl with increasing whirl speed and instability occurs. Although this qualitative explanation is accepted by many authors, it is a rare case in actual machinery. In practice, however, most rubbing rotors operate fairly well!

So, after Den Hartog, the problem is considered further by many authors. Though Ehrich(Ref.9) attempted to modify Den Hartog's model by introducing the flexibility in the stator, his definition of dynamic stability is quite unusual. Begg's suggestion(Ref.6) that the term "stable" should refer to a disturbed whirl motion which decays to zero is more correct. But the stability analysis in his work as well as in the most of the other literature known to the author are replaced by seeking the steady precession states so as to avoid the mathematical difficulty of the non-linear equation of motion. In fact, it is unavoidable to study the dynamic instability of a non-conservative system from its equation of motion.

Moreover, most of the previous works(Ref.4,7-9) only deal with the simplest rotor models such as the Jeffcott model with the limitation that the rubbing plane coincides with the disk plane. In the single disk system, the whirl amplitude of the disk must be bounded by the radial clearance. In a general multiple DOF rotor system, however, rubbing in one plane may cause a large amplitude whirl in another. But, up to now, few papers deal with multi-DOF rotor system. So the study of the problem is far from finished yet.

In this paper, the author presents an analytical approach to study the full annular rub problem for multi-DOF rotor systems. The non-linear equation of motion is established. The exact solution for single disk rotor is integrated out. The perturbation method is used to obtain the approximate solution for multi-DOF systems. The dynamic instability is discussed in detail. Begg's stability definition(Ref.6) is used. The results show that the behaviour of perturbed motion is largely dependent upon their initial state. If the initial backward and forward perturbed velocities are larger than the threshold speeds ω_1^* and ω_2^* respectively, the perturbation will be further developed and the system become unstable. The formulae of ω_1^* and ω_2^* are both deduced. In undamped case, ω_1^* and ω_2^* are equal respectively to the first critical speeds of the original rotor system and the constrained rotor system with the rubbing disk is constrained.

2 SINGLE DISK ROTOR

2.1 Equations of motion

Suppose that a single disk, supported by a light flexible shaft, rotates anticlockwise with high angular velocity Ω around its own axis. At a moment, the disk is temporarily perturbed from its steady state to come on the stator. The normal contact force R and the friction force μR then occur and act at the contact point A of the disk, where μ is the friction coefficient. Assuming that the disk keeps contact with the stator after that(the condition to realize it will be shown below), the sole generalized coordinate of the

* Projects supported by the National Natural Science Foundation of China

disk is then the polar angle θ (Fig.1). The equations of motion are as follows:

$$mh\ddot{\theta} + ch\dot{\theta} + \mu R = 0 \tag{1}$$

$$R = mh(\dot{\theta}^2 - \omega_o^2) \tag{2}$$

where $\omega_o^2 = k/m$ is the critical speed of the rotor, m is the disk mass, c is the external damping coefficient, k is the stiffness of the shaft at the disk position and h is the radial clearance.

The tangential linear velocity of the contact point A on the rotating disk is equal to $V_A = h\dot{\theta} + r\Omega$ (r is the radius of the disk). The direction of the friction force μR will be reversed only if $\dot{\theta} < -\frac{r}{h}\Omega$, which is a very large value. So the friction force μR usually is in the clockwise direction as shown in Fig.1.

Also it should be noticed that equations (1) and (2) hold only for R>0. Equation (2) shows that if $\dot{\theta}^2 < \omega_o^2$, the normal contact force R will be negative. The disk disengages from the stator and falls into the damped free vibration, which, of course, is stable.

2.2 Solution and dynamic stability

Substituting equation (2) into Eq.(1), we have

$$\ddot{\theta} = -\mu(\dot{\theta} - p_1)(\dot{\theta} - p_2) \tag{3}$$

where

$$p_1 = \omega_o(-\zeta/\mu + \sqrt{(\zeta/\mu)^2 + 1}) < \omega_o \tag{4}$$

$$p_2 = -\omega_o(\zeta/\mu + \sqrt{(\zeta/\mu)^2 + 1}) < -\omega_o \tag{5}$$

$$\zeta = c/2\sqrt{mk} \tag{6}$$

Assuming the initial value of $\dot{\theta}$ is $\dot{\theta}_o$, integrating eq.(3), we get

$$\dot{\theta}(t) = \frac{p_1 - p_2\xi e^{-\mu(p_1-p_2)t}}{1 - \xi e^{-\mu(p_1-p_2)t}} \tag{7}$$

where

$$\xi = (\dot{\theta}_o - p_1)/(\dot{\theta}_o - p_2) \tag{8}$$

ξ is an initial parameter.
Substituting eq.(7) into eq.(3), we obtain

$$\ddot{\theta}(t) = -\xi\mu\frac{(p_1 - p_2)^2 e^{-\mu(p_1-p_2)t}}{[1 - e^{-\mu(p_1-p_2)t}]^2} \tag{9}$$

which shows that $\ddot{\theta}$ and ξ are always of opposite signs.

From equations (7)-(9) we arrive at the following conclusions:

(I) If the initial perturbed whirl speed $\dot{\theta}_o$ falls into the following interval

$$-\omega_o < \dot{\theta}_o < \omega_o$$

then $R < 0$ is true. The disk disengages immediately from the stator after the initial contact. The system is stable.

(II) If $\dot{\theta}_o > \omega_o$, then $\ddot{\theta} < 0$, this means that $\dot{\theta}$ decreases from its initial value $\dot{\theta}_o$ monotonically and reaches ω_o in a finite time

$$T_1 = \frac{1}{\mu(p_1 - p_2)}\ln\frac{\xi(\omega_o - p_2)}{\omega_o - p_1}$$

After T_1, the disk comes off the stator and falls into the damped free vibration. The system is stable.

(III) If $p_2 < \dot{\theta}_o < -\omega_o$, then $\ddot{\theta} > 0$. $\dot{\theta}(t)$ increases monotonically from $\dot{\theta}_o$ and reaches $-\omega_o$ in a finite period of time

$$T_2 = \frac{1}{\mu(p_1 - p_2)}\ln\frac{\xi(\omega_o + p_2)}{\omega_o + p_1}$$

After T_2, the disk also disengages from the stator and falls into damped free vibration.

(IV) If $-\frac{r}{h}\Omega < \dot{\theta}_o < p_2$, then $\ddot{\theta} < 0$. The backward whirl speed $\dot{\theta}$ increases in magnitude up to $-\frac{r}{h}\Omega$ in a finite period of time

$$T_3 = \frac{1}{\mu(p_1 - p_2)}\ln\left(\xi\cdot\frac{\frac{r}{h} + p_2}{\frac{r}{h} + p_1}\right)$$

The system is unstable.

(V) If $\dot{\theta}_o < -\frac{r}{h}\Omega$, the sign of μR in eq.(1) will be reversed and the equation should be

$$\ddot{\theta} = \mu(\dot{\theta} - |p_2|)(\dot{\theta} + p_1)$$

By the same manner as above, it is not difficult to verify that in this region $\ddot{\theta} > 0$ is hold. So $\dot{\theta}$ increases to $-\frac{r}{h}\Omega$ in a finite time. But $-\frac{r}{h}\Omega$ is too large in magnitude to the initial tangential perturbation angular velocity, we do not pay any attention to this region.

In summary, the conclusion is that the stability behaviour of a rubbing disk is dependent upon its initial perturbation state. If $\dot{\theta}_o$ falls into the interval $(-\omega^*, \infty)$, the disk is always stable, where

$$\omega^* = \omega_o(\zeta/\mu + \sqrt{(\zeta/\mu)^2 + 1}) \tag{10}$$

Only in the case that the initial backward speed $|\dot{\theta}_o|$ is larger than ω^*, the rotor then may be unstable. This conclusion improves Den Hartog's description and reveals why most rubbing rotors operate fairly well. ω^* is a function of the damping to friction ratio ζ/μ. Formula (9) is depicted on Fig.2. ω^* is called as backward instability threshold speed for single disk rotor.

3 MULTI-DEGREE-OF-FREEDOM ROTOR SYSTEM

3.1 General equation of motion

Consider a multi-DOF lumped mass rotor system model (Fig.3). The mass of the jth disk is m_j. The displacements of its centre in the two transverse orthogonal directions are y_j and z_j respectively. The equations of motion are derived as follows:

$$\underline{M}\ddot{Y} + \underline{C}\dot{Y} + \underline{K}Y + \underline{R}Y - \mu\underline{RH}Z = 0$$

$$\underline{M}\ddot{Z} + \underline{C}\dot{Z} + \underline{K}Z + \underline{R}Z + \mu\underline{RH}Y = 0$$

or

$$\underline{M}\ddot{Q} + \underline{C}\dot{Q} + \underline{K}Q + (1 + \mu i)\underline{RH}Q = 0 \tag{11}$$

where

$$\underline{M} = \text{diag}[m_1, m_2, \ldots, m_n] \qquad \underline{K} = \begin{bmatrix} k_{11} & \cdots & k_{1n} \\ & \cdots\cdots & \\ k_{n1} & \cdots & k_{nn} \end{bmatrix}$$

$$\underline{C} = \text{diag}[c_1, c_2, \ldots, c_n]$$

$$\underline{R} = \text{diag}[R_1, R_2, \ldots, R_n]$$

$$\underline{H} = \text{diag}[h_1, h_2, \ldots, h_n]$$

$$Y^T = \lfloor y_1, y_2, \ldots, y_n \rfloor \qquad Z^T = \lfloor z_1, z_2, \ldots, z_n \rfloor$$

and

$$Q = Y + iZ$$

i is the imaginary unit. c_j, R_j and $h_j (j=1,\dots,n)$ are the external damping coefficient, normal contact force and radial clearance at the jth disk respectively. Most of them are zero if the associated disks are disengaged or undamped. The constrained conditions are

$$y_j^2 + z_j^2 = h_j^2 \quad (j = 1,2,\dots,r) \tag{12}$$

where r is the number of disks which are in an engaged state.

Assume that only one disk, say disk "1", comes on the stator, and let

$$Q = h \begin{bmatrix} 1 \\ \dots \\ Q_r(t) \end{bmatrix} e^{i\theta(t)} \tag{13}$$

where

$$h = h_1$$

$$Q_r(t) = A(t) + iB(t) = \frac{1}{h} \lfloor q_2, \dots, q_n \rfloor^T \tag{14}$$

The constrained condition (12) is then automatically satisfied. Substituting eq.(13) into eq.(11) and seperating the real and imaginary parts, we have

$$m_1\ddot{\theta} + c_1\dot{\theta} - \mu k_{11} + \mu m_1 \dot{\theta}^2 - \mu \underline{K}_{1r} A + \underline{K}_{1r} B = 0 \tag{15}$$

$$R + (k_{11} - m_1\dot{\theta}^2)h + h\underline{K}_{1r}A = 0 \tag{16}$$

$$\underline{M}_{rr}\ddot{A} - \dot{\theta}^2\underline{M}_{rr}A - 2\dot{\theta}\underline{M}_{rr}\dot{B} - \ddot{\theta}\underline{M}_{rr}B - \dot{\theta}\underline{C}_{rr}B + \underline{C}_{rr}\dot{A} + \underline{K}_{rr}h + \underline{K}_{rr}A = 0 \tag{17}$$

$$\underline{M}_{rr}\ddot{B} - \dot{\theta}^2\underline{M}_{rr}B + 2\dot{\theta}\underline{M}_{rr}\dot{A} + \ddot{\theta}\underline{M}_{rr}A + \dot{\theta}\underline{C}_{rr}A + \underline{C}_{rr}\dot{B} + \underline{K}_{rr}B = 0 \tag{18}$$

where

$$\underline{K} = \begin{bmatrix} k_{11} & \vdots & \underline{K}_{1r} \\ \cdots & & \cdots \\ \underline{K}_{r1} & \vdots & \underline{K}_{rr} \end{bmatrix} \qquad \underline{M}_{rr} = \mathrm{diag}\,[m_2, \dots, m_n] \qquad \underline{C}_{rr} = \mathrm{diag}[c_2, \dots, c_n]$$

$$R = R_1$$

Indeed, equations (15)-(18) are the equations of motion in the non-inertial coordinate system rotating with $\dot{\theta}(t)$.

3.2 Perturbation solution

Assume that μ and $c_j (j = 1,\dots,n)$ are all small quantities with the same order of magnitude. Set(Ref.11)

$$\begin{aligned} s &= \mu t \\ \dot{\theta} &= \omega_0(s) + \mu\omega_1(s) + \dots, \\ A(t) &= A_0(s) + \mu A_1(s) + \dots, \\ B(t) &= \mu B_1(s) + \dots . \end{aligned} \tag{19}$$

Substituting eq.(19) into (15)-(18) and remaining the terms of lowest order of magnitude in each equation, we get

$$m_1\omega_0' + \frac{c_1}{\mu}\omega_0 - k_{11} + m_1\omega_0^2 - \underline{K}_{1r}A_0 + \underline{K}_{1r}B_1 = 0$$

$$R = -h(k_{11} - m_1\omega_0^2 + \underline{K}_{1r}A_0)$$

$$A_0 = -\underline{L}_{rr}\underline{K}_{r1}$$

$$B_1 = -\underline{L}_{rr}(2\omega_0\underline{M}_{rr}A_0' + \omega_0'\underline{M}_{rr}A_0 + \frac{\omega_0}{\mu}\underline{C}_{rr}A_0)$$

where

$$\underline{L}_{rr}(\omega_0) = (\underline{K}_{rr} - \omega_0^2\underline{M}_{rr})^{-1}$$

"$'$" denotes the derivative with respect to s. After some manipulation, the first order approximate solutions are obtained as follows:

$$R = -h\,f(\dot{\theta}) \tag{20}$$

$$A = -\underline{L}_{rr}\underline{K}_{r1} \tag{21}$$

$$B = \underline{L}_{rr}[(4\dot{\theta}^2\underline{M}_{rr}\underline{L}_{rr}\underline{M}_{rr} + \underline{M}_{rr})\frac{d\dot{\theta}}{dt} + \frac{\dot{\theta}}{\mu}\underline{C}_{rr}]\underline{L}_{rr}\underline{K}_{rr} \tag{22}$$

$$\frac{d\dot{\theta}}{dt} = (\mu f(\dot{\theta}) - \dot{\theta}S^T\underline{C}S)/g(\dot{\theta}) \tag{23}$$

where

$$\underline{L}_{rr}(\dot{\theta}) = (\underline{K}_{rr} - \dot{\theta}^2\underline{M}_{rr})^{-1} \tag{24}$$

$$S(\dot{\theta}) = \begin{bmatrix} 1 \\ \dots \\ A \end{bmatrix} = \begin{bmatrix} 1 \\ \dots\dots \\ -\underline{L}_{rr}\underline{K}_{r1} \end{bmatrix} \tag{25}$$

$$f(\dot{\theta}) = k_{11} - m_1\dot{\theta}^2 - \underline{K}_{1r}\underline{L}_{rr}\underline{K}_{r1} \tag{26}$$

$$g(\dot{\theta}) = m_1 + A^T(\underline{M}_{rr} + 4\dot{\theta}^2\underline{M}_{rr}\underline{L}_{rr}\underline{M}_{rr})A \tag{27}$$

$\dot{\theta}(t)$ can be integrated from eq.(23). Then R(t), A(t) and B(t) can be obtained from eq.(20)-(22).

3.3 Dynamic instability

Dynamic stability can be analysed directly from eq.(23) without integrating it. For simplicity, neglect first all the external dampings. Eq.(23) reduces to

$$\frac{d\dot{\theta}}{dt} = \mu f(\dot{\theta})/g(\dot{\theta}) \tag{28}$$

The free vibration eigenvalue problem of the rotor system is

$$(\underline{K} - \omega^2\underline{M})X = 0 \text{ or } \begin{bmatrix} k_{11} - \omega^2 m_1 & \vdots & \underline{K}_{1r} \\ \dots\dots & & \dots\dots \\ \underline{K}_{r1} & \vdots & \underline{K}_{rr} - {}^2\underline{M}_{rr} \end{bmatrix} \begin{bmatrix} 1 \\ \dots \\ X_r \end{bmatrix} = 0 \tag{29}$$

from which, we have

$$X_r = -\underline{L}_{rr}(\omega)\underline{K}_{r1} \tag{30}$$

$$f(\omega) = k_{11} - m_1\omega^2 - \underline{K}_{1r}\underline{L}_{rr}\underline{K}_{r1} = 0 \tag{31}$$

So the characteristic equation of eq.(29) is equivalent to eq.(31). The eigenvalues are denoted by $\pm\omega_1$, $\pm\omega_2, \dots$, and the corresponding eigenvectors are $S(\omega_j)$. The curve $f(\dot{\theta})$ is qualitatively shown in the Fig.4(a). On the other hand, the following equation

$$\underline{L}_{rr}^{-1} X_r = (\underline{K}_{rr} - {}^2\underline{M}_{rr}) X_r = 0 \tag{32}$$

is the eigenvalue problem of the constrained system with the rubbing disk "1" constrained, whose eigenvalues are denoted as $\pm\omega_{r1}, \pm\omega_{r2}, \dots$. So the value of the determinant, $|\underline{L}_{rr}| = |\underline{K}_{rr} - \omega^2\underline{M}_{rr}|^{-1}$ tends to infinity as $\dot{\theta}$ tends to $\pm\omega_{rj}$, and so is $g(\dot{\theta})$. By the Courant-Fischer maximum and minimum theorem(Ref.10), it can be proved that $\pm\omega_j$ and $\pm\omega_{rj}$ are arranged alternately. So the function $f(\dot{\theta})/g(\dot{\theta})$ in eq.(28) may be shown qualitatively as in Fig.4(b). From these two figures we come to the following conclusions:

(I) If the initial perturbed value $\dot{\theta}_o$ is in the interval $(-\omega_1, \omega_1)$, in which we have $f(\dot{\theta})>0$, then $R<0$ due to eq.(20). The rotor disengages immediatly after the occasional contact moment. The system is stable.

(II) If $\dot{\theta}_o$ is in (ω_1, ω_{r1}), then $R>0$ and $\frac{d\dot{\theta}}{dt}<0$. $\dot{\theta}$ decreases from $\dot{\theta}_o$ monotonically and reaches ω_1, then falls into the disengaged state. The system is also stable.

(III) If $\dot{\theta}_o$ is larger than ω_{r1}, we have $\frac{d\dot{\theta}}{dt}>0$. $\dot{\theta}$ increases monotonically further. The system is unstable.

(IV) If $\dot{\theta}_o$ is in the region $(-\omega_{r1}, -\omega_1)$, $\frac{d\dot{\theta}}{dt}<0$, or if $\dot{\theta}_o<-\omega_{r1}$, $\frac{d\dot{\theta}}{dt}>0$. In both cases $\dot{\theta}$ monotonically approaches $-\omega_{r1}$. The amplitudes A and B also increase sharply according eq.(21) and (22). The system is unstable.

(V) Because the initial perturbed magnitude $\dot{\theta}_o$ could not much large, the other region of $\dot{\theta}_o$ may not be considered. The arrows in Fig.4(b) show the tendency of $\dot{\theta}(t)$ in various regions.

In summary, the rotor system will keep stable if its initial perturbation $\dot{\theta}_o$ satisfies

$$-\omega_1 < \dot{\theta}_o < \omega_{r1}$$

The above conclusion extents and improves Den Hartog's description to general complex rotor systems.

Let ω_1^* and ω_2^* denote the backward and forward instability threshold speeds respectively. Thus, in the undamped case, we have

$$\omega_1^* = -\omega_1 \qquad (33)$$

$$\omega_2^* = \omega_{r1} \qquad (34)$$

In the damped case, on the other hand, the ω_1^* should be found out from following equation

$$\mu f(\dot{\theta}) = \dot{\theta} S^T \underline{C} S \qquad (35)$$

Equation (35) comes from eq.(23). If $c/\mu \ll 1$, eq.(36) may be solved by perturbation method. Setting

$$\dot{\theta} = -\omega_1 + \varepsilon$$

where ε is a small quantity relative to ω_1, we obtain eventually from eq.(35) that

$$\omega_1^* = -\omega_1 - \frac{1}{2\mu} \frac{S_1^T \underline{C} S_1}{S_1^T \underline{M} S_1} \qquad (36)$$

where $S_1 = S(-\omega_1)$(see eq.(25)) is the first normal mode of the system. Eq.(34) is still valid in damped case.

4 ACKNOWLEDGEMENTS

The work is supported by the National Natural Science Foundation of China.

5 REFERENCES

1. BEATTY,R.F. Differentiating Rotor Response Due to Radial Rubbing, Journal of Vibration,Acoustics, Stress, and Reliability in Design, 1985,Vol.107, pp151-160.

2. BENTLY,D.E. Forced Subrotative Speed Dynamic Action of Rototing Mechinery, 1974, ASME paper 74-PET-16.

3. MUSZYNSKA,A. Partial Lateral Rotor to Stator Rubs, C281/84, Third International Conference on Vibrations in Rotating Mechinery, York, England, 1984.

4. DEN HARTOG,J.P. Mechanical Vibrations, 1965, Fourth Edition,(McGraw-Hill).

5. EHRICH,F.F. The Dynamic Stability of Rotor/ Stator Radial Rubs in Rotating Machinery, Journal of Engineering for Industry, 1965, Vol.91,pp1025-28.

6. BEGG,I.C. Friction Induced Rotor Whirl--A Study in Stability, Journal of Engineering in Industry,1974, Vol.96, pp450-454.

7. BILLETT,R.A. Shaft Whirl Induced by Dry Friction, The Engineer, 1965,Vol.29,pp713-714.

8. BLACK,H.F. Interaction of a Whirling Rotor With a Vibrating Stator Across a Clearance Annulus, Journal Mechanical Engineering Sc., 1968,Vol.10, pp1-12.

9. EHRICH,F.F. and O'CONNOR,J.J. Stator Whirl With Rotors in Bearing Clearance, Journal of Engineering for Industry, 1967, Vol.89,p381.

10. HUSEYIN,K. Vibrations and Stability of Multiple Parameter Systems, 1978, (Noordhoff International Publishing Alphen Aan Rijn).

11. NAYFEH,A.H. Perturbation Methods,1973,(Wiley, New York).

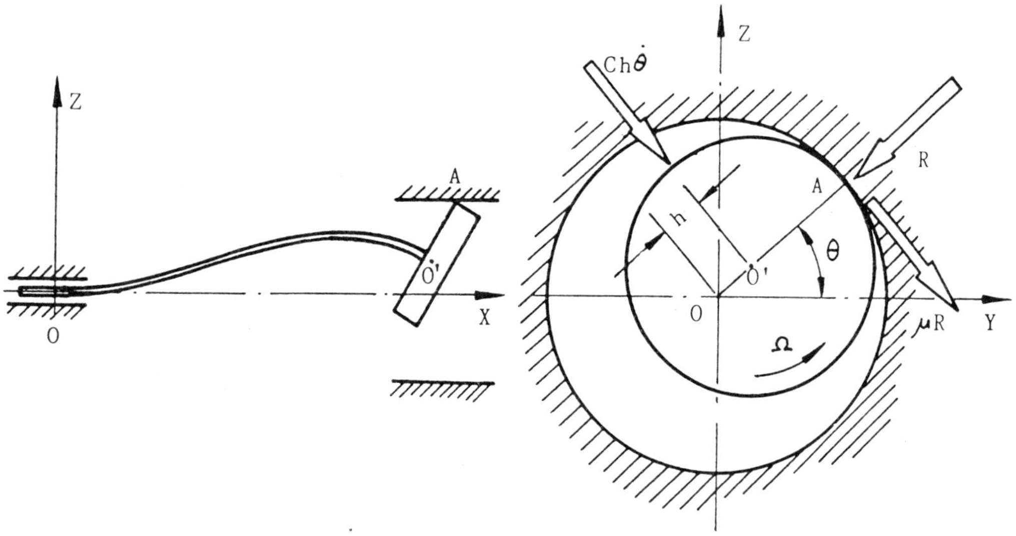

Fig 1 Single disk rotor

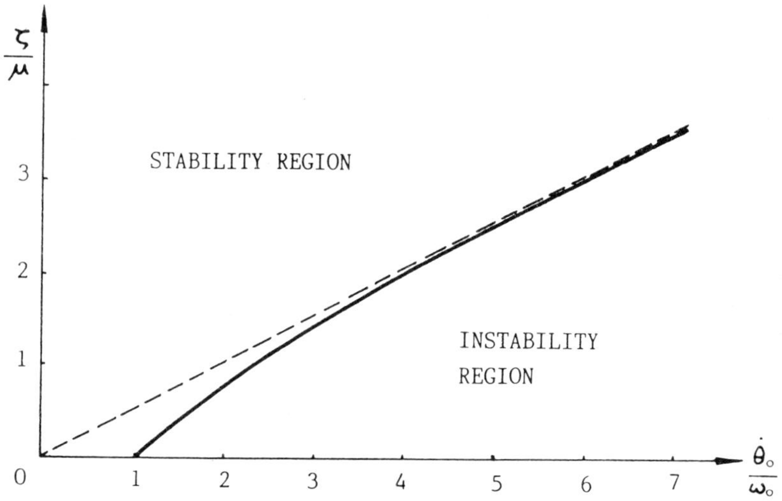

Fig 2 The threshold speed for single disk rotor

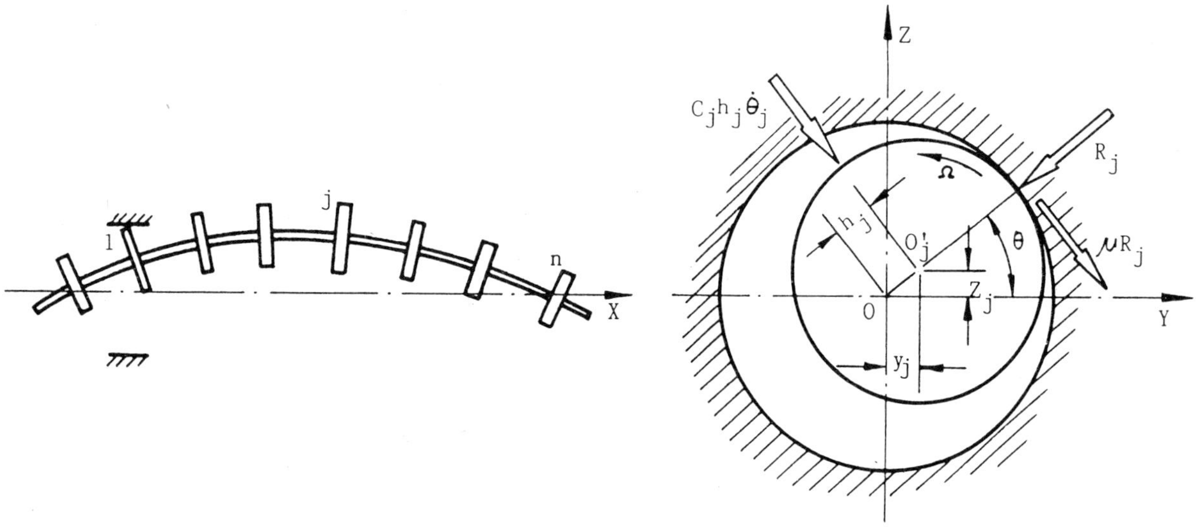

Fig 3 Lumped mass multi-DOF rotor system

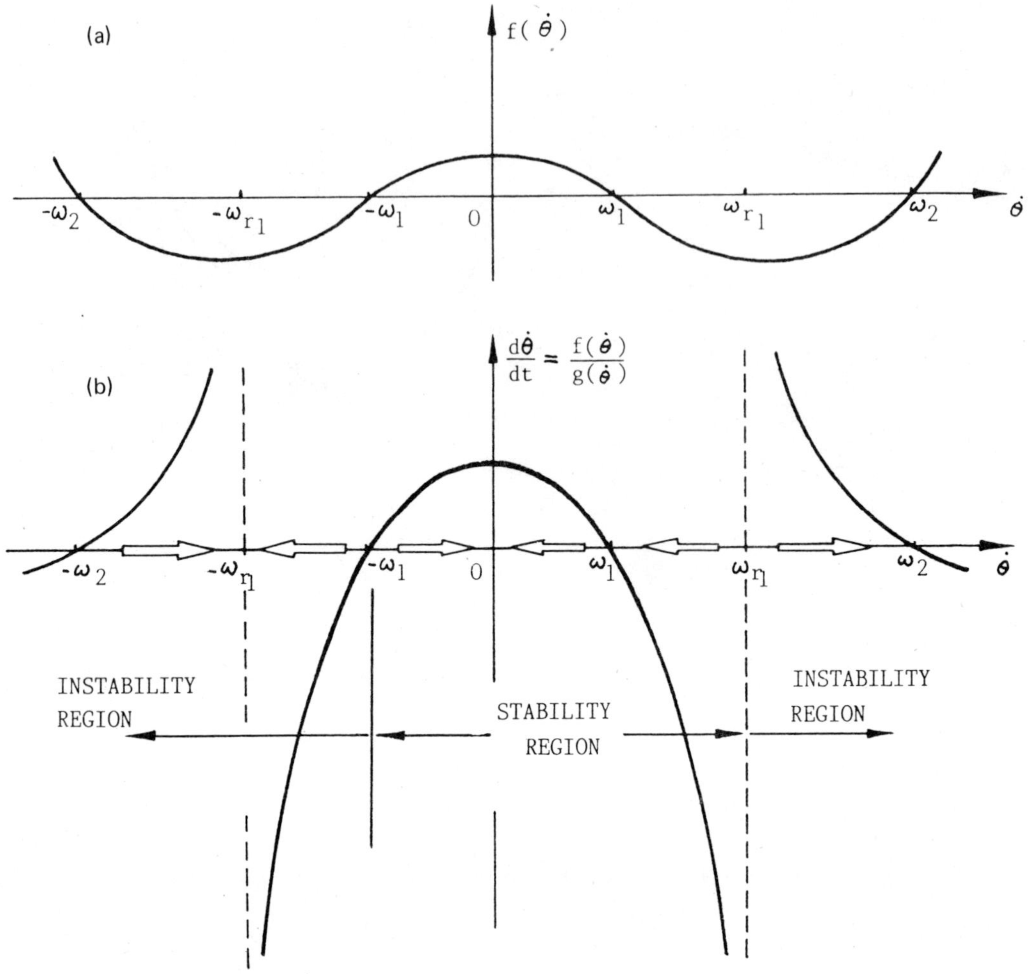

Fig 4 The illustrated curves of $f(\dot{\theta})$ and $f(\dot{\theta})/g(\dot{\theta})$

C263/88

Effect of bearing alignment on stability threshold and post-stability behaviour of rotor-bearing systems

Z A PARSZEWSKI, DrEngSc, CEng, FIMechE, K KRYNICKI, MEngSc and E D KIRBY, BEng
Department of Mechanical and Manufacturing Engineering, University of Melbourne, Parkville, Victoria, Australia

SUMMARY An approach to the system dynamics in terms of bearing alignment changes is presented. Such an approach allows for better understanding of machine performance at operating condition and has a number of advantages over classical methods. The effects of bearing alignment on the stability threshold and post stability operation of multi bearing-rotor systems is presented and analysed. In particular, the influence of bearing alignment on stability margin and post-stability behaviour of a test four bearing rotor system was studied.

1. INTRODUCTION

Most problems in machine and structural dynamics, involving the instability problem or system dynamic response, have been considered in the frequency or speed domain (Morton 1972, Lund 1974, Gasch 1976, Hori 1984). Information obtained in the speed domain is, however, of indirect type, giving for example the stability threshold speed or the critical speeds as compared with the operating speed. In particular it is not explicit as to how sensitive (and to what) the critical, unstable speeds or resonant frequencies are. Hence, additional information from analysis or experience is required.

For statically indeterminate systems, some additional conditions, which predetermine the system static and dynamic performance, are imposed to select the system configuration.

In the very important case of systems which use hydrodynamic oil bearings, any shifts or tilts introduce changes of the eccentricity ratio in the bearings. As the stiffness and damping of a journal bearing depend on the eccentricity, the dynamic characteristics of the system vary. Therefore, even a small variation of the bearing configuration may introduce important changes in the system critical speeds, instability threshold, and dynamic response.

In this paper some advantages of an approach carried out in terms of bearing configuration variations are shown with reference to design problems and vibrational diagnosis. The approach is used to investigate the effect of bearing alignment on the dynamics of a four-bearing test installation. The influence of (i) bearing reaction distribution, (ii) bearing transverse shifts on the stability margin, critical frequencies, and limit cycle development is assessed.

2. ADVANTAGES OF ANALYSIS IN TERMS OF BEARING ALIGNMENT CHANGES

An approach to dynamics of a machine operating with constant speed, which is considered in terms of bearing configuration changes, has a number of advantages over classical methods which deal in the frequency or speed domain. The advantages are evident when the approach is used:

(i) during diagnosis or identification of sources of vibration. Variations of the bearing configuration influence the dynamic characteristics of the system (non-homogeneous systems). The changes of dynamic properties may effect the stability threshold, reducing it or sometimes causing system instablility leading to the induction of self-excited vibrations in the system. Alteration of bearing configuration may also change the critical speeds and resonating (critical) frequencies. It may then happen, that the system running at some speed, due to a configuration change, may have a natural frequency (eigenvalue) which coincides with excitation from another machine or sub-system, which was previously harmless,

(ii) for definition of tolerances, with respect to manufacturing or assembly. These tolerances may be established, ensuring for example that the actual response (e.g. limit cycle or unbalance response) would not differ from the designed one more than an assumed percentage. The system configuration deviations from the optimal configuration can be treated as the necessary accuracy for manufacture and assembly,

(iii) for correction of the design to achieve the required dynamics (e.g. minimum of the amplitude of unbalanced or self-excited vibration). Limited optimization is possible, which involves for practical reasons limited number of bearing shifts.

Such an approach also allows for selection of the system configuration according to constraints imposed on the steady state and vibration response of the system. Such constraints can be introduced to:

(i) influence the steady state with respect

to bending moment distribution, rotor deflection, etc.,

(ii) select bearing operating conditions with respect to journal equilibrium position or distribution of bearing reactions,

(iii) to select (optimize) the dynamic response of the system to given forms of excitation and provide sufficient stability margin at working condition.

3. REPRESENTATION OF SYSTEM CONFIGURATION

The parameters used to represent a system configuration must be chosen to solve a problem in the most suitable way.

Two methods:

(i) bearing reaction method,
(ii) bearing displacement method,

which are of equal merit, are discussed and presented with their applications.

3.1 Bearing reaction method

Assuming the varying parameters are the reactions in the bearings, the rotor loading and deflection can be easily expressed as functions of these parameters. Thus, acheiving any additional constraints imposed on the steady state of the rotor (e.g. maximum bending moment or deflection) can be relatively easily achieved. Changing the bearing reactions results in a new equilibrium position of the rotor in the bearings. However, knowledge of the reactions allows the equilibrium position of the rotor to be found in each bearing separately. Therefore, the process of finding the equilibrium position in multi-bearing rotor systems is simplified by avoiding the statically indeterminate problem.

Moreover, knowledge of the bearing reactions allows for adjustment of each bearing orientation with reference to its recommended characteristics. Most modern bearings are designed such that they best carry a load in one direction. So, for instance, for a horizontal rotor-bearing system the best configuration can be arranged, when the bearing reactions are all vertical at steady state.

The static bearing load corresponds to an eccentricity ratio, from which the bearing stiffness and damping can be directly obtained. Thus, for an assumed set of reactions the bearing dynamic properties i.e. stiffness and damping are defined and can be assembled with the shaft and foundation characteristics, which allows for quick estimation of the system stability.

Once, the reactions are selected with respect to the static and dynamic requirements the deflection line of the shaft under (i) external loads and (ii) self-weight can be found relative to its unloaded line, which is used, from now on, as the reference line. The position of the bearing centers, relative to the reference line, can be found from the eccentricities which give the working position of the rotor in the bearings. On the other hand the casing and the foundation deformation along connecting co-ordinates under bearing reactions and thermal expansions can be also found.

The displacement compatibility between rotor-bearing sub-system and foundation must be satisfied for any set of reactions chosen. The rotor-bearing sub-system is fixed relative to the reference line for the assumed equilibrium imposed on it. Thus, the foundation levels need to be changed to join the two together. These changes will depend also on where the reference line is allocated which is optional. For example, the assembly of the rotor-bearing-foundation system can be achieved ensuring (i) a minimal number of adjustments or (ii) minimal correction.

3.2 Bearing displacement method

In some problems the additional constraints imposed on a system at equilibrium are given geometrically. Such cases are met e.g. (i) during the assembly, when admissible deviations of bearing alignment have to be specified, (ii) in the problems concerned with foundation settlement or (iii) thermal expansion of a supporting structure.

Knowledge of the influence of bearing alignment changes on the system static and dynamic behaviour may indicate the method for correction or improvement and the specified tolerances of an actual machine. The investigation of the vertical bearing shifts may provide significant information about the effects caused by foundation settlement or thermal expansion of a supporting structure.

In practice, due to cost, only a limited number of bearing shifts are recommended. Sensitivity analysis on bearing configuration variations allows for selection of the most influential bearings and their directions.

3.3 Conclusions

(i) Due to a direct correlation between the bearing load (Sommerfeld number) and journal position the bearing dynamic characteristics and therefore the characteristics of the whole system can be modified easily.

(ii) An advantage of the bearing displacement method is evident during diagnosis of the causes of vibration. Alteration of a bearing alignment allows for immediate observation of the influence of bearing shifts on the static and dynamic behaviour of the system. Thus, the assumptions made can be immediately verified. Also, the most influential bearing shifts can be selected in a straight forward way.

(iii) As bearing shifts are easily reflected in equations of motion, the bearing displacement method benefits during the simulation of limit cycles or the system non-linear unbalanced responses.

(iv) Sometimes, for practical reasons a part of the conditions superimposed on the system configuration is defined in terms of bearing reactions, while the rest is concerned with bearing alignment. Such a problem can arise, during the investigation of the effect of vertical bearing shifts on a turbogenerator's dynamics, where the restrictions regarding heavily-loaded bearing eccentricity ratios are imposed. Treating the bearing reactions as fixed parameters and applying them to the rotor as an external load the analysis is carried out on shifts of the remaining bearings.

4. DYNAMIC ANALYSIS OF TEST ROTOR SYSTEM

4.1 Four-bearing rotor system

A test rotor-bearing system, with a flexible shaft shown in Fig. 1, was used to study the effects of bearing configuration changes on the system dynamic behaviour. The bearing data is shown in Tab. 1.

Table 1 Journal bearing parameters

Diameter	Length	Rad.Clr.	Visc.	Max.Load
m	m	m	Ns/m2	N
0.050	0.0415	0.0003	0.02	150.0

Due to (i) very flexible coupling between the shaft and motor, thus transmitting only torque, and (ii) the use of short bearings, eight parameters (n=8) are used to describe the system configuration. However, between them, four linear relationships (m=4) exist. Thus, the system has four independant parameters (n-m=4). In the case where the reactions in the bearings R_i (i=1,2,...,8) are selected as configuration parameters, due to the four (m=4) force equilibrium conditions (two per plane), there are only four independant (n-m=4) reactions. When bearing's positions are selected, then eight co-ordinates define the bearing configuration. However, the introduction of a reference line i.e. alignment line (e.g. via two pairs of the co-ordinates) leaves four of the coordinates (two per plane) to be considered as independent.

4.2 System analysis in terms of bearing alignment

The results of a dynamic investigation of a test rotor-bearing system at working speed (3000 RPM) in terms of bearing reactions were presented by Parszewski, Krodkiewski (1986). The system response, caused by (i) machine excitation and (ii) environmental excitation transferred from other structures, and the predicted stability margin allowed for selection of the optimal system configuration.

The following is concerned with system analysis in terms of bearing displacements (Krynicki 1987).

The alignment line is understood as a line passing through the centres of the first and last bearings. The parameters which define the system configuration are the bearing displacements measured from this line and then normalized by the radial bearing clearance. Only vertical changes are considered. Such geometric conditions were deliberately imposed admitting only vertical movement of the supporting structure.

The effect of variation of the bearing position on the journal positions in the bearings is shown in Fig. 2. The journal positions in the bearings were obtained for fixed position of the third bearing.

It is apparent from Fig. 2 that due to the bearing shifts, the journal positions (curve A) are not located along the loci corresponding to the recommended load direction (curve B). Thus, the bearing capacities may be significantly reduced. In consequence, this may lead to more rapid bearing wear.

The dependence of (i) the system resonating frequencies and (ii) the system stability on bearing vertical shifts was recognized by solving the complex eigenvalue problem for the linearized system in the neighbourhood of the equilibrium position previously found. The first two eigenvalues were calculated for each set of bearing shifts. The contour maps (Fig. 3 and 4) illustrate the variation of the real parts and imaginary parts of eigenvalues on a plane of the two vertical bearing shifts. The solid lines join points at which real or imaginary parts of the eigenvalues have the same value. A symmetry of the contours about dashed line $x_2=x_3$ is justified by the symmetry of the rotor-bearing system. Vertical changes of the second bearing position cause a similar effect as that of the third vertical bearing position as expected.

By inspection of Fig. 3 it is apparent that the aligned system is stable. Its first two eigenvalues have negative real parts in the range of minus twenty and they change slowly with bearing shift, which provides a safety margin of stability. The system stability margin can be improved, if required, by shifting the bearings downward in the range of one radial bearing clearance.

Upward shifts in the range of one radial clearance of the bearings, however, reduce the stability margin, and if they exceed one diametral clearance, may in an extreme case, even destabilize the system. Thus, such a configuration may be responsible for vibration induced in the system.

A large value of upward bearing shifts restabilizes the system. But, referring to journal equilibrium locus (Fig. 2, curve B), such a type of stabilization is not recommended, because the journals operate away from the locus corresponding to recommended load direction.

From a point of view of system isolation, information provided by the eigenvalue analysis can be used (i) to recognize and eliminate frequencies transferred from the environment or alternatively (ii), if these frequencies cannot be eliminated, to tune the system away from them. Figure 4 shows the imaginary parts of eigenvalues due to the two vertical bearing shifts. Thus, it shows where the resonance is located. As it is apparent, the aligned system exhibits a tendency to amplify vibration of the frequency of 60 percent of rotational speed. A severe vibration with such a frequency component may then occur in the rotor-bearing system, resulting from any external excitation of the same frequency. The system can be tuned up toward higher frequencies by moving upward the bearing position, which, as mentioned above, also increases stability threshold. When the system is less stable, vibration with a half rotational frequency may occur as the effect of (i) self-excited phenomena or (ii) amplified vibration transferred from the environment. In some cases, they can be distinguished. The first one is asynchronous and the second is usually synchronous with rotation speed.

In order to estimate the level of self-excited vibrations induced, in operation, beyond the threshold of stability (i.e. amplitude of the limit cycles), non-linear simulation implementing the Newmark procedure for integration of the equations of motion was used. Figure 5 presents the trajectory of the first journal in its bearing. The trajectory is observed in the local co-ordinate system, attached to the centre of the bearing. The journal displacement was normalized by the bearing radial clearance. Limit cycles occur just after passing the threshold of stability (x_2 = +2.5, x_3= +1.5), which has been found on the basis of linear approach (solving eigenvalue problem).

Introduction of a higher value of misalignment in the range of 3.5 radial clearance reduced limit cycle amplitudes, causing even disappearance of self-excited vibration. This confirms the linear prediction concerning restabilization of the system. In the neighbourhood of instability threshold, the vibration amplitudes were small, however the bearings were heavily loaded, some of them outside recommended limits. The spectral analysis revealed the frequencies of self-excited vibration. They remained constant, mostly equal to 47 percent of rotational speed.

4.4 Conclusions

(i) Bearing displacement method was developed to reveal potential causes of the vibrations induced in the system. Method of bearing reactions being an alternative method was used to illustrate process of the optimal system configuration design (Parszewski 1986).

(ii) Vibration diagnosis in terms of bearing shifts showed that the straight line aligned system exhibited a safety margin of stability. It remained stable for bearing shifts up to two radil clearance.

(iii) Upward shifts of the middle bearings reduced stability margin. Upwards bearing shifts, exceeding one diametral clearance destabilized the system. Such a configuration may then cause the induction of self-excited vibrations in the system.

(iv) Vibration analysis revealed the resonances.

REFERENCES

LUND J.L., Stability and Damped Critical Speeds of a Flexible Rotor in Fluid-Film Bearings, J. of Eng. for Industry , Tran ASME, 1974, Ser.B, Vol.96, No.2, pp.509.

MORTON P.G., Analysis of Rotors Supported upon Many Bearings J. Mech. Eng. Science , 1972, Vol.14, No.1, pp.25.

GASCH R., Vibration of Large Turbo-Rotors in Fluid-Film Bearings on Elastic Foundation, J. of Sound and Vibration , 1976, Vol.47, No.1, pp.57.

HORI Y., UEMATSU R., Influence of Misalignment of Support Journal Bearings on Stability of a Multi-Rotor System, Tribology International , 1980, Vol.13, No.5, pp.249.

PARSZEWSKI Z.A., KRODKIEWSKI M.J., Machine Dynamics in Terms of the System Configuration Parameters, IFToMM JSME , Inter. Conf. of RotorDynamics, Tokyo, 1986.

KRYNICKI K., Limit Cycle in Rotor Bearing Instability, Ph.D. Thesis , University of Melbourne, 1987.

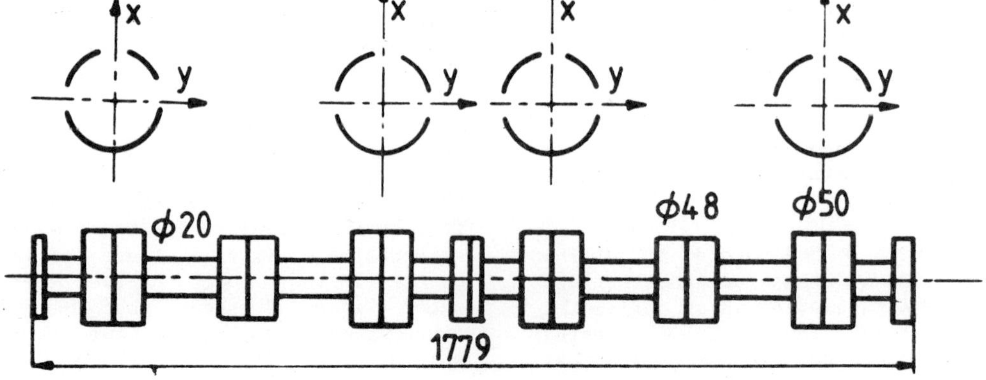

Fig 1 Test rotor

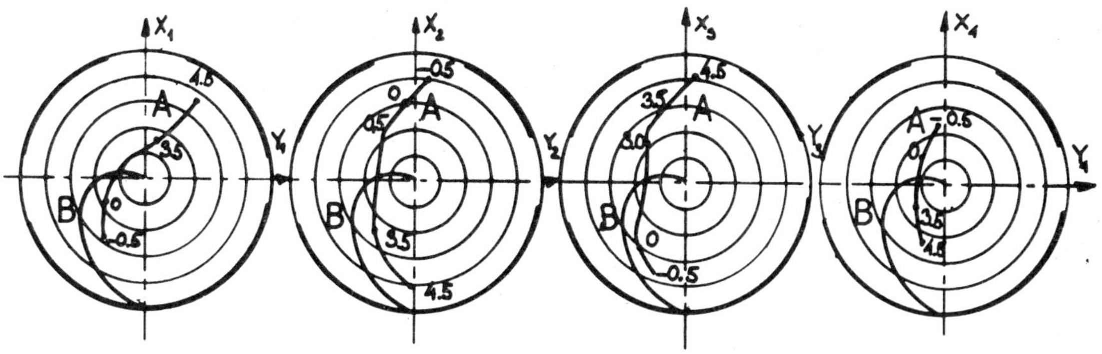

Fig 2 Locus of journal in bearings 1–4 for vertical misalignment of bearing 2

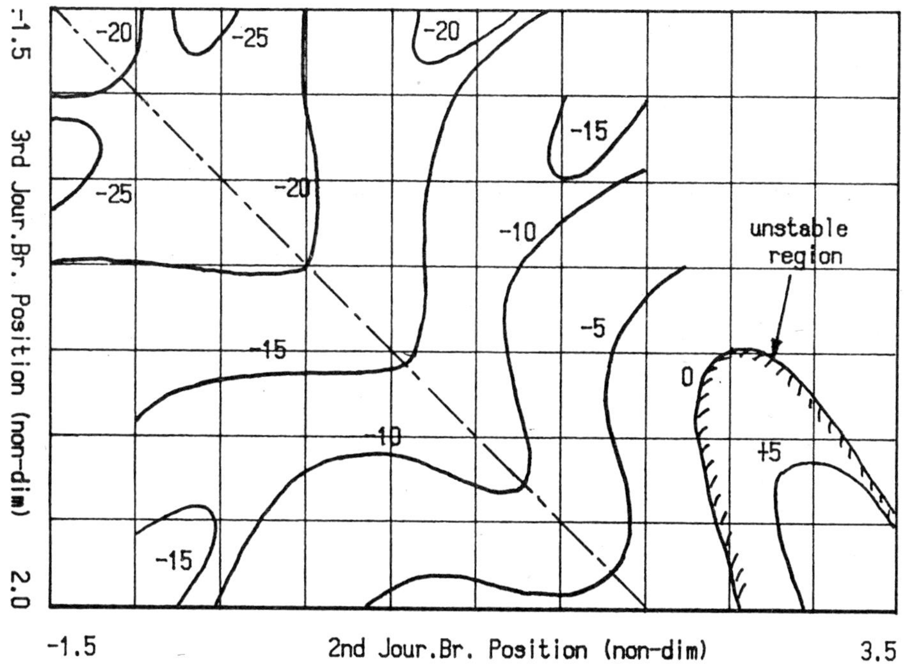

Fig 3 Stability map (real part of eigenvalue)

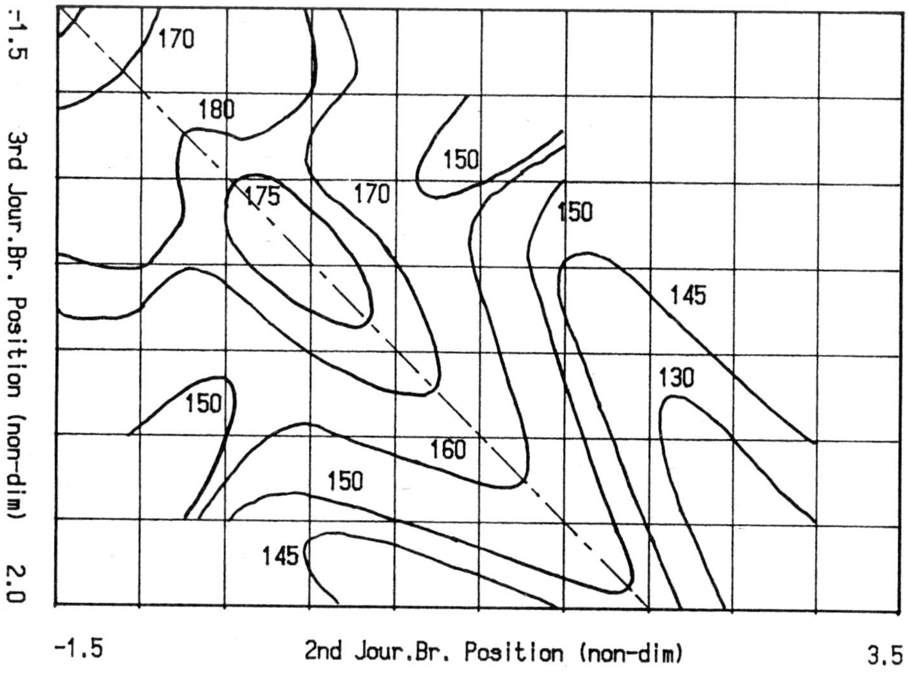

Fig 4 Resonance map (imaginary part of eigenvalue)

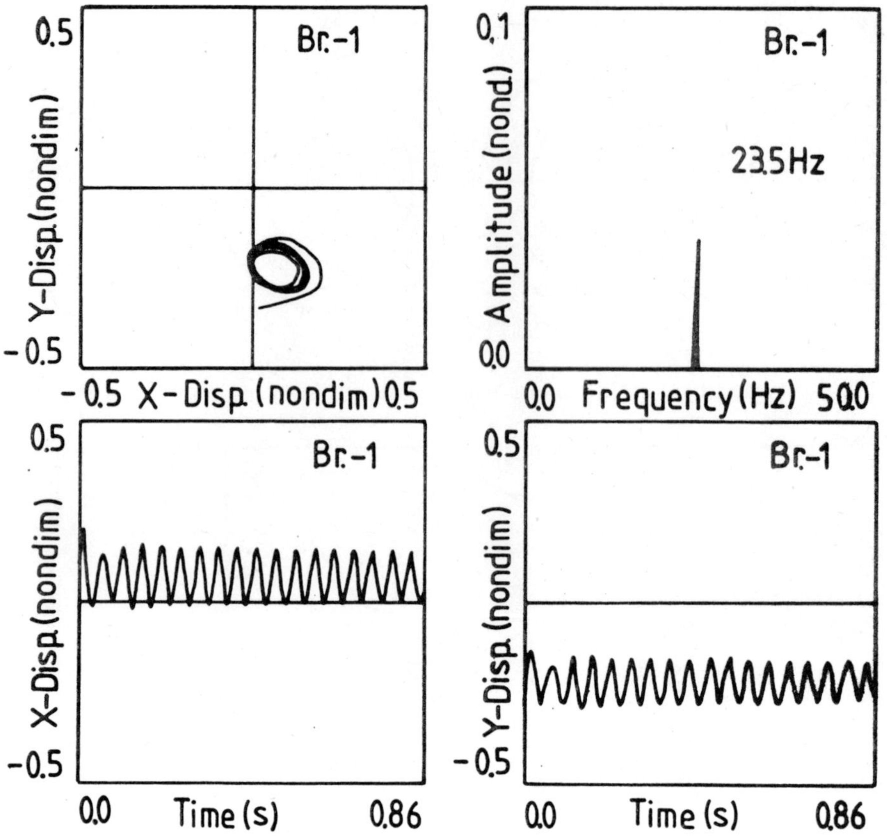

Fig 5 Development of limit cycle

C255/88

A study on non-steady stability of a rotor-journal bearing system—stability of the 360° short bearing

T INAGAKI, MEng, MJSME, T KAWAKAMI, BEng, MJSME and T OSADA, BEng, MJSME
Mitsubishi Heavy Industries Limited, Takasago City, Hyogo, Japan

Synopsis

The nonsteady stability evaluation means a judgement of rotor at an instant condition of $(x, y, \dot{x}, \dot{y})$ while the usual method evaluates the stability at the static equilibrium condition of $(x_0, y_0, \dot{x}=0, \dot{y}=0)$. In this paper, the 360 degree full bearing is studied. This bearing is always unstable by the usual method, but we find that it has the stable characteristics when it is evaluated by the new method, and show the stabilization effect of the external disturbance on the rotor time history motion.

1. Preface

The conventional method to evaluate the stability of the rotor-journal bearing system is to verify whether vibration will or will not be diverged after given disturbances at the static equilibrium bearing condition $(x_0, y_0, \dot{x}=0, \dot{y}=0)$. The 360 degree full bearing (non-cavitating full circumferential fluid film) is always unstable by the method wherever the rotor is located in the bearing. But we have experienced the large vertical pump rotor with the 360 degree bearing, which is evaluated as the unstable system by the usual complex eigenvalue analysis, can rotate without any trouble and has the unstable whirling component (N/2) whose amplitude, however, fluctuates unsteadily. From this view point, several studies about the nonlinear analysis have been carried out and reported the limit cycle behaviour and so on[1] [2], but they have no discussion about the stability characteristics, and also bearings in these studies are the partial types which have both stable and unstable regions for the rotor position and in these cases large vibrations could be expected to become in the limit cycles. The 360 degree full bearing, however, has no stable region as above mentioned and so the limit cycle phenomena could not be expected. In this paper, the nonsteady stability evaluation at the rotor's instant condition of $(x, y, \dot{x}, \dot{y})$ is studied for the 360 degree bearing under the short bearing approximation theory and it is found that the steady region arises, and also it is shown by the time history response calculations for the system that the rotor vibration can be stabilized by the external disturbances. This responds the phenomena mainly experienced with the actual vertical machines.

2. Stability Evaluation Formulas

In order to deal with the rotor and bearing in a simple manner, a single mass model is considered rotating inside the bearing, as shown in Fig.1. When the shaft center makes an infinitesimal movement from the state of $(x_0, y_0, \dot{x}_0, \dot{y}_0)$ to $(x, y, \dot{x}, \dot{y})$, the equation of motion is expressed as below.

$$\left.\begin{aligned} &m\ddot{x}+k_{xx}(x-x_0)+k_{xy}(y-y_0)+c_{xx}(\dot{x}-\dot{x}_0)\\ &\quad+c_{xy}(\dot{y}-\dot{y}_0)=F_{x0}+W_x\\ &m\ddot{y}+k_{yy}(y-y_0)+k_{yx}(x-x_0)+c_{yy}(\dot{y}-\dot{y}_0)\\ &\quad+c_{yx}(\dot{x}-\dot{x}_0)=F_{y0}+W_y \end{aligned}\right\} \quad \cdots\cdots(1)$$

where (F_{x0}, F_{y0}) is the bearing reaction force at $(x_0, y_0, \dot{x}_0, \dot{y}_0)$.
(W_x, W_y) ··· External force.

$$\left.\begin{aligned} k_{xx}&=-\frac{\partial F_x}{\partial x}\bigg|_{x_0, y_0, \dot{x}_0, \dot{y}_0}\\ k_{xy}&=-\frac{\partial F_x}{\partial y}\bigg|_{x_0, y_0, \dot{x}_0, \dot{y}_0}\\ c_{xx}&=-\frac{\partial F_x}{\partial \dot{x}}\bigg|_{x_0, y_0, \dot{x}_0, \dot{y}_0}\\ c_{xy}&=-\frac{\partial F_x}{\partial \dot{y}}\bigg|_{x_0, y_0, \dot{x}_0, \dot{y}_0}\\ &\vdots \end{aligned}\right\} \text{nonsteady spring and damping coefficient at } (x_0, y_0, \dot{x}_0, \dot{y}_0).$$

Presuming the free vibration at the right edge of (1) being zero and the root of its vibration equation being

$\lambda=a+i\omega$ (a, ω are real number)

the following condition to satisfy a=0 will give the stability boundary, i.e.,

$$\begin{aligned} M_c&=m\omega^2\frac{C}{W}=\frac{k^*_{xx}c^*_{yy}+k^*_{yy}c^*_{xx}-k^*_{xy}c^*_{yx}-k^*_{yx}c^*_{xy}}{c^*_{xx}+c^*_{yy}}\\ (\omega/\Omega)^2&=\frac{(k^*_{xx}-M_c)(k^*_{yy}-M_c)-k^*_{xy}k^*_{yx}}{c^*_{xx}c^*_{yy}-c^*_{xy}c^*_{yx}} \end{aligned} \quad \cdots\cdots(2)$$

where

ω/Ω······ natural frequency/rotating frequency

$k^*=kC/W$, $c^*=c\Omega(C/W)$······ nondimensional spring and damping coefficient

3. Oil film analysis on the 360 degree full bearing (short bearing approximation)

3.1 Fundamental equation

Reynolds equation

$$\frac{1}{R^2}\frac{\partial}{\partial\theta}\left(\frac{h^3}{\mu}\frac{\partial p}{\partial\theta}\right)+\frac{\partial}{\partial Z}\left(\frac{h^3}{\mu}\frac{\partial p}{\partial Z}\right)=\frac{6}{R}\frac{\partial}{\partial\theta}(R\Omega h)+12\frac{\partial h}{\partial t} \quad \cdots\cdots(3)$$

where

$$h=C(1+\varepsilon\cos\theta)$$

the boundary conditions are

$$\left.\begin{array}{ll} Z=\pm\frac{L}{2}, & p=0 \\ Z=0, & \frac{\partial p}{\partial Z}=0 \end{array}\right\} \quad \cdots\cdots(4)$$

By the short bearing theory of L≪D, the pressure fluctuation in the circumferential direction (θ) can be neglected unlike the monometric change in Z direction, therefore, the following equation can be obtained, i.e.,

$$\frac{\partial}{\partial Z}\left(\frac{h^3}{\mu}\frac{\partial p}{\partial Z}\right)=\frac{6}{R}\frac{\partial}{\partial\theta}(R\Omega h)+12\frac{\partial h}{\partial t} \quad \cdots\cdots(3)'$$

When the equation (3) is integrated by applying the boundary conditions, the pressure distribution p can be obtained which will be subsequently integrated on the whole circumferential bearing for obtaining the oil film reaction force, then the below equation is secured, i.e.,

$$p=-\frac{3\mu}{C^2(1+\varepsilon\cos\theta)^3}\{\varepsilon(\Omega+2\dot\theta)\sin\theta-2\dot\varepsilon\cos\theta\}\left(Z^2-\frac{L^2}{4}\right) \quad \cdots\cdots(5)$$

$$F_r=-WS\left(\frac{L}{D}\right)^2\frac{1}{\Omega}(-2\dot\varepsilon)J_{cc}$$

$$F_\theta=-WS\left(\frac{L}{D}\right)^2\frac{1}{\Omega}\varepsilon(\Omega+2\dot\theta)J_{ss}$$

where

$$J_{cc}=\pi(1-\varepsilon^2)^{-5/2}(1+2\varepsilon^2)$$
$$J_{ss}=\pi(1-\varepsilon^2)^{-3/2}$$
$$J_{sc}=0$$
$$\dot\theta=-\dot\varphi$$
$$S=\frac{\mu LD\Omega}{W}\left(\frac{R}{C}\right)^2,\ W\text{:sample load} \quad \cdots\cdots(6)$$

The oil film reaction forces in (x, y) axes are

$$\begin{vmatrix}F_x\\F_y\end{vmatrix}=\begin{vmatrix}\cos\varphi & -\sin\varphi\\ \sin\varphi & \cos\varphi\end{vmatrix}\begin{vmatrix}F_r\\F_\theta\end{vmatrix} \quad \cdots\cdots(7)$$

3.2 Spring and damping coefficients (non-steady)

As shown in Eq. (1), the spring and damping coefficients of the oil film are defined by the differential equation of the first order for the oil film reaction force (F_x, F_y), but the conventional method by which the characteristics in the vicintity of the static equilibrium points (x_0, y_0, $\dot x_0$, $\dot y_0$) were described, used to define the spring and damping coefficients from the pressure distribution and/or the oil film reaction force when the velocity at shaft center was zero, or, $\dot\varepsilon = \dot\varphi = 0$. In this case, however, the velocity at the shaft center is not made zero because the spring and damping coefficients are seeked for only when the rotor is in the arbitrary state of $(x, y, \dot x, \dot y)$ or $(\varepsilon, \varphi, \dot\varepsilon, \dot\varphi)$.

The spring and damping coefficients can be expressed in the differential equations as follows by the medium of ε, φ.

$$k_{xx}=\left[\frac{\partial F_r}{\partial\varepsilon}\cos^2\varphi-\frac{\partial F_r}{\partial\varphi}\frac{\sin\varphi\cos\varphi}{\varepsilon}+F_r\frac{\sin^2\varphi}{\varepsilon}-\frac{\partial F_\theta}{\partial\varepsilon}\sin\varphi\cos\varphi+\frac{\partial F_\theta}{\partial\varphi}\frac{\sin^2\varphi}{\varepsilon}+F_\theta\frac{\sin\varphi\cos\varphi}{\varepsilon}\right]\frac{1}{C}$$

$$k_{xy}=\left[\frac{\partial F_r}{\partial\varepsilon}\sin\varphi\cos\varphi+\frac{\partial F_r}{\partial\varphi}\frac{\cos^2\varphi}{\varepsilon}-F_r\frac{\sin\varphi\cos\varphi}{\varepsilon}-\frac{\partial F_\theta}{\partial\varepsilon}\sin^2\varphi-\frac{\partial F_\theta}{\partial\varphi}\frac{\sin\varphi\cos\varphi}{\varepsilon}-F_\theta\frac{\cos^2\varphi}{\varepsilon}\right]\frac{1}{C}$$

$$k_{yy}=\left[\frac{\partial F_r}{\partial\varepsilon}\sin^2\varphi+\frac{\partial F_r}{\partial\varphi}\frac{\sin\varphi\cos\varphi}{\varepsilon}+F_r\frac{\cos^2\varphi}{\varepsilon}+\frac{\partial F_\theta}{\partial\varepsilon}\sin\varphi\cos\varphi+\frac{\partial F_\theta}{\partial\varphi}\frac{\cos^2\varphi}{\varepsilon}-F_\theta\frac{\sin\varphi\cos\varphi}{\varepsilon}\right]\frac{1}{C}$$

$$k_{yx}=\left[\frac{\partial F_r}{\partial\varepsilon}\sin\varphi\cos\varphi-\frac{\partial F_r}{\partial\varphi}\frac{\sin^2\varphi}{\varepsilon}-F_r\frac{\sin\varphi\cos\varphi}{\varepsilon}+\frac{\partial F_\theta}{\partial\varepsilon}\cos^2\varphi-\frac{\partial F_\theta}{\partial\varphi}\frac{\sin\varphi\cos\varphi}{\varepsilon}+F_\theta\frac{\sin^2\varphi}{\varepsilon}\right]\frac{1}{C}$$

$$C_{xx}=\left[\frac{\partial F_r}{\partial\dot\varepsilon}\cos^2\varphi-\frac{\partial F_r}{\partial\dot\varphi}\frac{\sin\varphi\cos\varphi}{\varepsilon}-\frac{\partial F_\theta}{\partial\dot\varepsilon}\sin\varphi\cos\varphi+\frac{\partial F_\theta}{\partial\dot\varphi}\frac{\sin^2\varphi}{\varepsilon}\right]\frac{1}{C}$$

$$C_{xy}=\left[\frac{\partial F_r}{\partial\dot\varepsilon}\sin\varphi\cos\varphi+\frac{\partial F_r}{\partial\dot\varphi}\frac{\cos^2\varphi}{\varepsilon}-\frac{\partial F_\theta}{\partial\dot\varepsilon}\sin^2\varphi-\frac{\partial F_\theta}{\partial\dot\varphi}\frac{\sin\varphi\cos\varphi}{\varepsilon}\right]\frac{1}{C}$$

$$C_{yy}=\left[\frac{\partial F_r}{\partial\dot\varepsilon}\sin^2\varphi+\frac{\partial F_r}{\partial\dot\varphi}\frac{\sin\varphi\cos\varphi}{\varepsilon}+\frac{\partial F_\theta}{\partial\dot\varepsilon}\sin\varphi\cos\varphi+\frac{\partial F_r}{\partial\dot\varphi}\frac{\cos^2\varphi}{\varepsilon}\right]\frac{1}{C}$$

$$C_{yx}=\left[\frac{\partial F_r}{\partial\dot\varepsilon}\sin\varphi\cos\varphi-\frac{\partial F_r}{\partial\dot\varepsilon}\frac{\sin^2\varphi}{\varepsilon}+\frac{\partial F_\theta}{\partial\dot\varepsilon}\cos^2\varphi-\frac{\partial F_\theta}{\partial\dot\varphi}\frac{\sin\varphi\cos\varphi}{\varepsilon}\right]\frac{1}{C} \quad \cdots\cdots(8)$$

4. Characteristics in the equilibrium condition

In the equilibrium condition, it is $\dot\varepsilon = \dot\varphi = 0$, and the oil film reaction force and the eccentricity ratio are expressed as below, as widely known, i.e.,

Oil film reaction force:

$$\left.\begin{array}{l} F_r=0\\ F_\theta=-\pi\varepsilon(1-\varepsilon^2)^{-3/2}S\left(\frac{L}{D}\right)^2\cdot W\\ S=\left(\frac{R}{C}\right)^2\frac{\mu\Omega LD}{W}\end{array}\right\} \quad \cdots\cdots(9)$$

Eccentricity ratio:

$$S\left(\frac{L}{D}\right)^2=\frac{(1-\varepsilon^2)^{3/2}}{\pi\varepsilon} \quad \cdots\cdots(10)$$

Angular eccentricity:

$$\varphi=\gamma+\frac{\pi}{2} \quad (\gamma: \text{direction of load W}) \qquad \cdots\cdots(11)$$

Spring and damping coefficients:

$$k_{xx}=-\frac{W}{C}S\left(\frac{L}{D}\right)^2(1-\varepsilon^2)^{-5/2}\pi(-3\varepsilon^2)\sin\varphi\cos\varphi$$

$$k_{xy}=-\frac{W}{C}S\left(\frac{L}{D}\right)^2\{-\pi(1-\varepsilon^2)^{-5/2}\}\times\{1+\varepsilon^2(3\sin^2\varphi-1)\}$$

$$k_{yx}=-\frac{W}{C}S\left(\frac{L}{D}\right)^2\{\pi(1-\varepsilon^2)^{-5/2}\}\times\{1+\varepsilon^2(3\cos^2\varphi-1)\}$$

$$k_{yy}=-\frac{W}{C}S\left(\frac{L}{D}\right)^2(1-\varepsilon^2)^{-5/2}\pi\times 3\varepsilon^2\sin\varphi\cos\varphi(=-k_{xx})$$

$$c_{xx}=-2\frac{W}{C}S\left(\frac{L}{D}\right)^2\frac{1}{\Omega}\{-\pi(1-\varepsilon^2)^{-5/2}\times(1+\varepsilon^2\{3\cos^2\varphi-1)\}$$

$$c_{xy}=-2\frac{W}{C}S\left(\frac{L}{D}\right)^2\frac{1}{\Omega}\{-\pi(1-\varepsilon^2)^{-5/2}\times(3\varepsilon^2\sin\varphi\cos\varphi)$$

$$c_{yx}=c_{xy}$$

$$c_{yy}=-2\frac{W}{C}S\left(\frac{L}{D}\right)^2\frac{1}{\Omega}\{-\pi(1-\varepsilon^2)^{-5/2}\times(1+\varepsilon^2(3\sin^2\varphi-1)\} \qquad \cdots\cdots(12)$$

Therefore, when φ = 0, $\pi/2$, π, $3/2\pi$, 2π (the static load W coincides either with or direction), it becomes

$$k_{xx}=k_{yy}=c_{xy}=c_{yx}=0$$

and, on the other hand, when substituted by the stability boundary conditions of the equation (2) it becomes

$$m=0, \quad \omega/\Omega=0.5$$

and it is found that the system is always unsteady, with the centrifugal whirling vibrations being half of the rotational frequency.

For reference, both the floating track at the shaft center and the pressure distribution are illustrated in Fig. 3 and 4.

5. The nonsteady stability

The conventional evaluation of the stability boundary is conducted in the vicinity of the state of the static equilibrium, namely,

$$\varepsilon=\varepsilon_0, \quad \varphi=\varphi_0, \quad \dot{\varepsilon}=0, \quad \dot{\varphi}=0$$

therefore, as stated in chapter 4, as far as the case of the 360 degree full bearing is concerned, the system always becomes unsteady, irrespective of where the shaft center is located, accordingly, we have calculated the non-steady stability boundary for the two cases, i.e.,

(1) $\varepsilon=\varepsilon_0, \quad \varphi=\varphi_0, \quad \dot{\varepsilon}=0, \quad \dot{\varphi}\neq 0$

(2) $\varepsilon=\varepsilon_0, \quad \varphi=\varphi_0, \quad \dot{\varepsilon}>0, \quad \dot{\varphi}=0$

the serial calculation method is,

obtain, ε_0, φ_0 assume $\dot{\varepsilon}$, $\dot{\varphi}$.

calculate nonsteady spring and damping coefficients -------- Eq. (8)

calculate both the stability boundary and natural frequency --- Eq. (2)

in consequence of which, it was found that

(1) for the case of $\dot{\varphi} \neq 0$ ... always unsteady

$$k_{xx}=k_{yy}=C_{xy}=C_{yx}=0$$

(2) for the case of $\dot{\varepsilon}$ 0 ... stable region is formed

$$k_{xx}>0, \quad k_{yy}>0$$

In addition, the stability boundary $[M\Omega^2(C/W)]$ for the whole range of the shaft center ε_0 to the value of $\dot{\varepsilon}/\Omega$ = 0.00064 ∿ 0.064 is calculated and illustrated in Fig. 5. Fig. 5 reveals the horizontal axis of $S(L/D)^2$, the left vertical axis as $[m\Omega^2(C/W)]$ and the right vertical axis for eccentricity ratio ε_0, and shows the non-steady region existing in the upper portion of the stability boundary as shown by the curves of for the respective values, of $\dot{\varepsilon}/\Omega$. The eccentricity ratio is plotted by the broken line. The numerical values shown above the stability boundary line denote the frequency ratio ω/Ω at respective points.

The stability boundary belongs to the region revealed by the form of , prone to be stabilized at the eccentricity ratio of zero or near 1, whilst the stable region in the vicinity of ε_0 = 0.5, it becomes narrow. In general, the natural frequency is roughly 0.5Ω (half of the rotating frequency), but becomes below 0.4Ω for the range of ε_0 > 0.6 and $\dot{\varepsilon}/\Omega$ > 0.016.

6. Physical Interpretation

(1) Static equilibrium condition (ε_0, φ_0, 0, 0)

Along with the Fig. 4 in which the oil film pressure distribution is given, we have shown the same by the broken line in Fig. 6(a), where it is revealed that, in terms of pressure distribution, the static pressure is axisymmetrically on the positive pressure and in the reverse side, the negative pressure is distributed asymmetrically. This causes the shaft center eccentrically shift to the load in the orthogonal direction, besides tending to move eccentrically to the fluctuated load in the orthogonal direciton, namely,

direct spring : K_{xx}, K_{yy} = 0

coupling spring : K_{xy}, K_{xy} ≠ 0

causing the effect for stabilization not to occur.

(2) Nonsteady condition ($\dot{\varepsilon}$ > 0) of (ε_0, φ_0, $\dot{\varepsilon}$, 0)

The oil film pressure distribution in $\dot{\varepsilon}$ > 0 becomes the full line, as shown in Fig. 6(a), while the differential pressure between the oil film pressure distribution and the pressure distribution in the equilibrium status becomes as illustrated in Fig. 6(b). In other words, the restraint pressure arises in the same area of the eccentric velocity, in which case,

direct spring : K_{xx}, K_{yy} > 0 arises,

causing the stability effect to occur in consequence.

(3) Nonsteady condition ($\dot{\varphi}$ = 0) of (ε_0, φ_0, 0, φ)

The oil film pressure distribution in $\dot{\varphi}$ = 0 shown by the full line of Fig. 6 (c) is similar to the pressure distribution in the asymmetric type as shown by the broken line in the case of $\dot{\varphi}$ = 0 with a result that, from this distribution form,

direct spring : K_{xx}, K_{yy} will not occur, therefore, no stability effect will be generated.

7. Response calculation

7.1 Method of analysis

For a simple rotor shown in Fig. 7, the time history calculation is made. The equations of motion for the shaft center in the vicinity of (x_0, y_0) is expressed as below.

$$\left.\begin{aligned} &m\ddot{x}+K_{xx}(x-x_0)+K_{xy}(y-y_0)\\ &\quad+C_{xx}(\dot{x})-\dot{x}_0)\\ &\quad+C_{xy}(\dot{y}-\dot{y}_0)+K_s x+C_s\dot{x}\\ &\quad=F_{x0}+W_x\\ &m\ddot{y}+K_{yy}(y-y_0)+K_{yx}(x-x_0)\\ &\quad+C_{yy}(\dot{y}-\dot{y}_0)\\ &\quad+C_{xy}(\dot{x}-\dot{x}_0)+K_s y+C_s\dot{y}\\ &\quad=F_{y0}+W_y \end{aligned}\right\} \quad \cdots\cdots(13)$$

where

m: mass

Ks: shaft stiffness

Cs: shaft Damping Coefficient

(W_x, W_y)=External force

(F_{x0}, F_{y0})=Fluid film's reaction Force at (x_0, y_0)

K_{ij}, C_{ij}= Fluid film's dynamic characteristics at = (x_0, y_0) such as K_{xx}, C_{xx}

Presuming the initial condition $(x_0, y_0, \dot{x}_0, \dot{y}_0)$, the above formular is numerically integrated by the New Mark β method.

The dynamic characteristics of the oil film will be calculated at each fluctuation of the shaft center $(x, y, \dot{x}, \dot{y})$ and the nonsteady characteristics at each variance will be employed for calculation.

7.2 Calculation results of response

Giving an eye to the following points, response calculation is conducted as shown below i.e.,

(1) Relation between the frequency of the external excitation force and response frequency.
(2) Effect of the existence of unbalance.
(3) Effect of the rotor mass (m).

For these calculations, the rotor speed is made N, besides taking into consideration N/5, N/2, N (unbalanced excitation) and 2N in terms of the frequency for the external excitation force namely.

(1) Excitation by the resonance frequency of N/2 (Fig. 8(a)).
When excited by N/2, the system becomes strongly unstable, with vibration completely diverged, this is resulted from the whirl radius gradually developing to an extent of $\dot{\varepsilon} \simeq 0$.
(2) Excitation in direction and by N/5 external force (Fig. 8(b)(C))
As shown in Fig. 8(b), when excited in the direction of the initial eccentricity 0, namely in the direction of severe fluctuation of the eccentricity velocity of , the stabilization effect is large. For information, the principal frequency is N/2, not the frequency of the exciting force.
(3) 2N external force (Fig. 8(c)-(e))
As shown in Fig. 8(e), the system remains completely steady under the external force with 2N frequency. Even adding the 2N excitation to the response by the N/5 external force which is already cited in item (2), the stabilization effect can be recognized in Fig. 8(d). this fact precisely reveals the nonsteady stability effect.
(4) Effect of unbalance (Fig. 8(c)(f))
As revealed from the process of fluctuation from Fig. 8(c) to (f), unbalance is prone to physically spoil the stability;this is due to the fact that the unbalance promotes the round whirl, diminishing fluctuations of .
(5) Effect of the rotor mass (Fig. 8(g)(h))
When the rotor mass is reduced, the system becomes stable, this is due to that less in mass can yield more extended stable zone at less existence of $\dot{\varepsilon}$. (as shown in Fig. 5, i.e., when "m" becomes small the system tends turning toward the stable region.)

8. Conclusions

(1) We have analyzed the fluid film characteristics of the 360 full degree bearing (non-cavitating by the short-bearing approximation theory (L≪D). In this case, it is distinguished with having newly analyzed the dynamic characteristics of the shaft at an arbitary nonsteady condition $(x, y, \dot{x}, \dot{y})$, taking into consideration the velocity of the shaft center.
(2) According to the conventional stability evaluation method, the 360 full degree bearing is always unsteady irrespective of the location of the shaft center and/or the static load, with vibration diverging due to disturbance. Nevertheless, when considering the nonsteady condition of the 360 full degree bearing the following inherent and steady characteristics are brought to our fresh enlightenment, i.e.,
In case of the shaft center velocity exists: $\dot{\varepsilon} > 0$, $\dot{\varphi} = 0$ } Steady region arises. The larger $\dot{\varepsilon}$, the more the stable zone is extended.
In case the angular shaft center velocity exists: $\dot{\varphi} \neq 0$, $\dot{\varepsilon} = 0$ } stability is not change
(3) In consequence of the time history. Calculation of a simple rotor and of the model bearing by means of the nonsteady dynamic characteristics which we have conducted, the followings were brought to our knowledge, i.e.,
- According to either the amplitude or the frequency of the exciting force, self-excited vibration will be developed or restrained, yet, depending on the coupled exciting forces.
- Against distrurbances (with large $\dot{\varepsilon}$) having the high vibratory frequency, the system shows the positive stabilization.
- Unbalance is found prone to promote unstableness (because unbalance has the forward precession force in the equivalent

direction with self-excited vibration in whirling)
When the excitation force of (1/2)N works, vibration is apparently diverged.
When the rotor mass is small, the system is found inclined easily stabilized.

(4) The marvellous unsteady whirling phenomena experienced in the actual vertical pumps can be explained by this nonsteady stability theory. Based on this theory we could be accessible to bearings dynamically unstable by the conventional evaluation but meritorious in many other characteristics such as load capacity, toughness, simplicity, cost and so on.

9. Reference

1 . Malik, M.,Hori, Y., 'An approximate Nonlinear Transient Analysis of Journal Bearing Response in Unstable Region of Linearized System',the International Conference on Rotordynamics, IFTOMM, 1986.

2 . Adams, M.L., 'Non-Linear Dynamics of Flexible Multi-Bearing Rotors', J. Sound and Vibration 71 (1), 1980.

3 . Colsher,R., Anwar,I., Obeid, V., 'Nonlinear Bearing Effects on Rotor Dynamic Analysis', ASME paper 82-GT-291.

4 . Andres, L.S.,Vance, J.M., 'Effects of Fluid Inertia and Turbulence on the Force Coefficients for Squeeze Film Dampers',Trans. ASME Vol.108,1986.

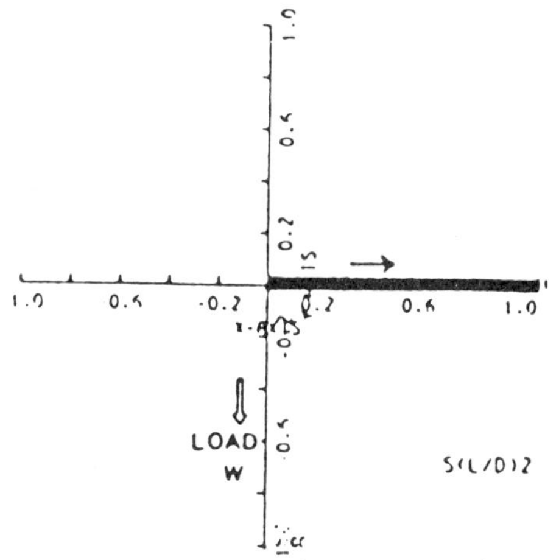

Fig 3 Floating track of the shaft in 360° full bearing

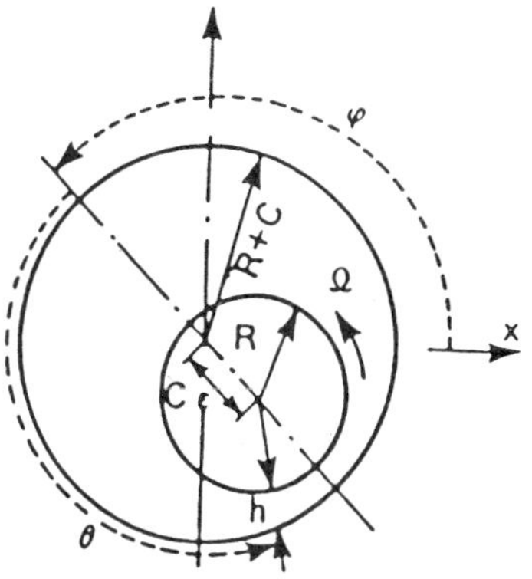

R = radius of journal
C = radius clearance
ε = eccentricity
P = pressure
μ = viscosity
θ = angular coordinate
φ = eccentric angle
Ω = angular speed of rotor
(x,y) = jounal center coordinate
L = bearing length
D = (R+C)×2, bearing diameter

Fig 1 Rotor and bearing

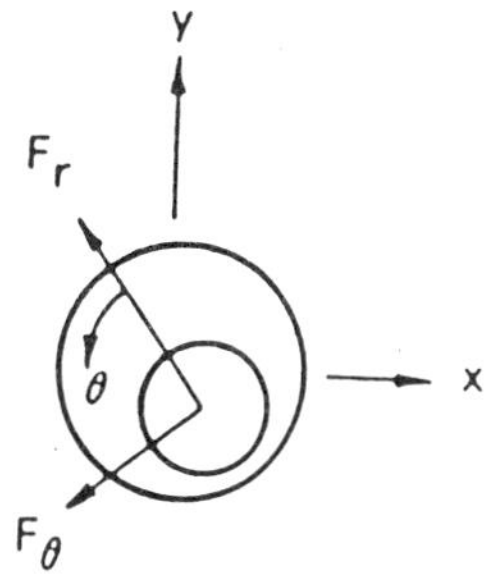

Fig 2 Coordinate of reaction force

pressure distribution

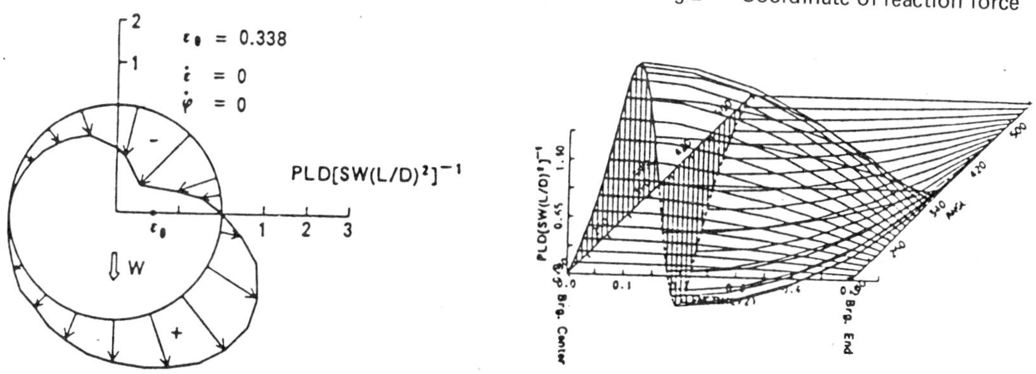

Fig 4 Oil film pressure distribution in 360° full bearing (at equilibrium of $\epsilon_0 = 0.338$)

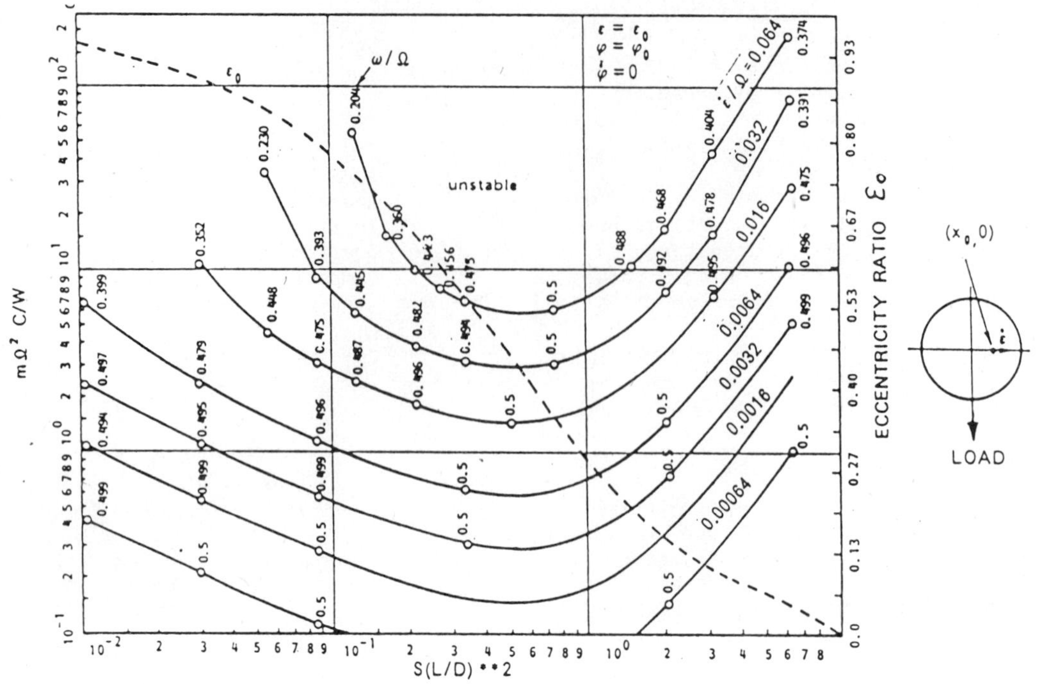

Fig 5 Unsteady stability threshold of 360° film bearing

1. $\dot{\epsilon}/\Omega = 0.08$ & $\dot{\epsilon} = 0$ ($\dot{\varphi} = 0$) 2. $\Delta p = p(\dot{\epsilon} > 0) - p(\dot{\epsilon} = 0)$ 3. $\dot{\varphi}/\Omega = 1$ & $\dot{\varphi} = 0$ ($\dot{\epsilon} = 0$)

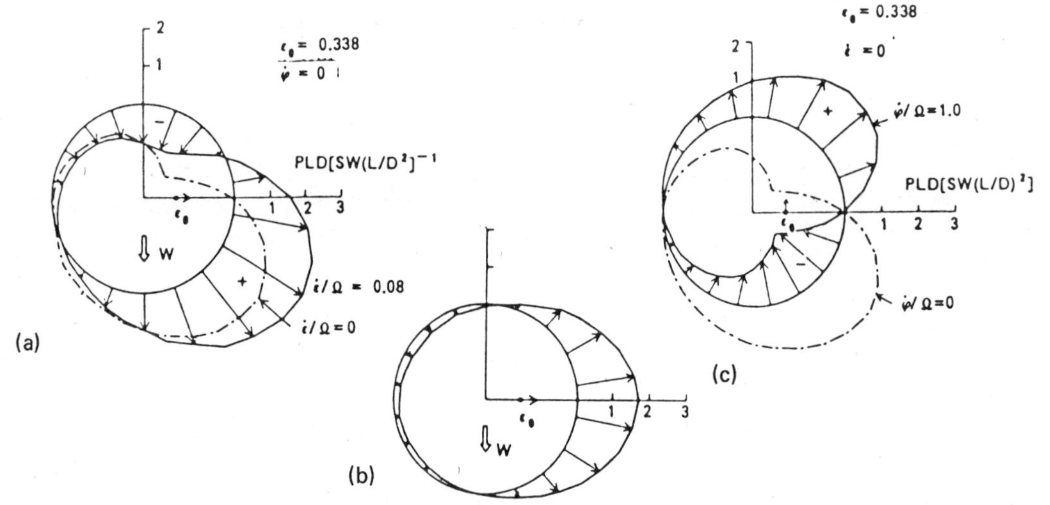

Fig 6 Oil-film pressure distribution of 360° full bearing at typical non-steady conditions ($\epsilon_0 = 0.338$)
(a) $\dot{\epsilon} > 0, \dot{\varphi} = 0$
(b) Differential pressure for $\dot{\epsilon} > 0$ and $\dot{\epsilon} = 0$
(c) $\dot{\epsilon} = 0, \dot{\varphi} = 0$

bearing length, dia.	L = D = 35.5cm
viscosity	$\mu = 45 \times 10^{-9}$ N·s/cm²
rotor weight	mg = 21717 or 9094N
static load	W_0 = 19600 N (-90° deg.)
radial clearance	C = 0.0241cm
rotor speed	$\Omega/2\pi$ = 20Hz
shaft stiffness	K_s = 29600 N/cm,
" damping	C_s = 0

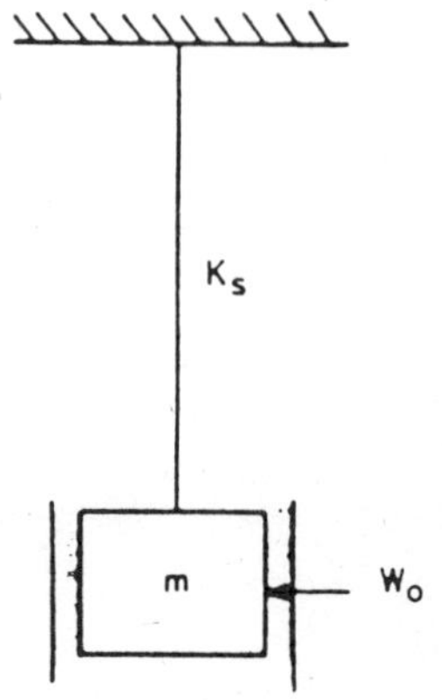

Fig 7 Response calculation model

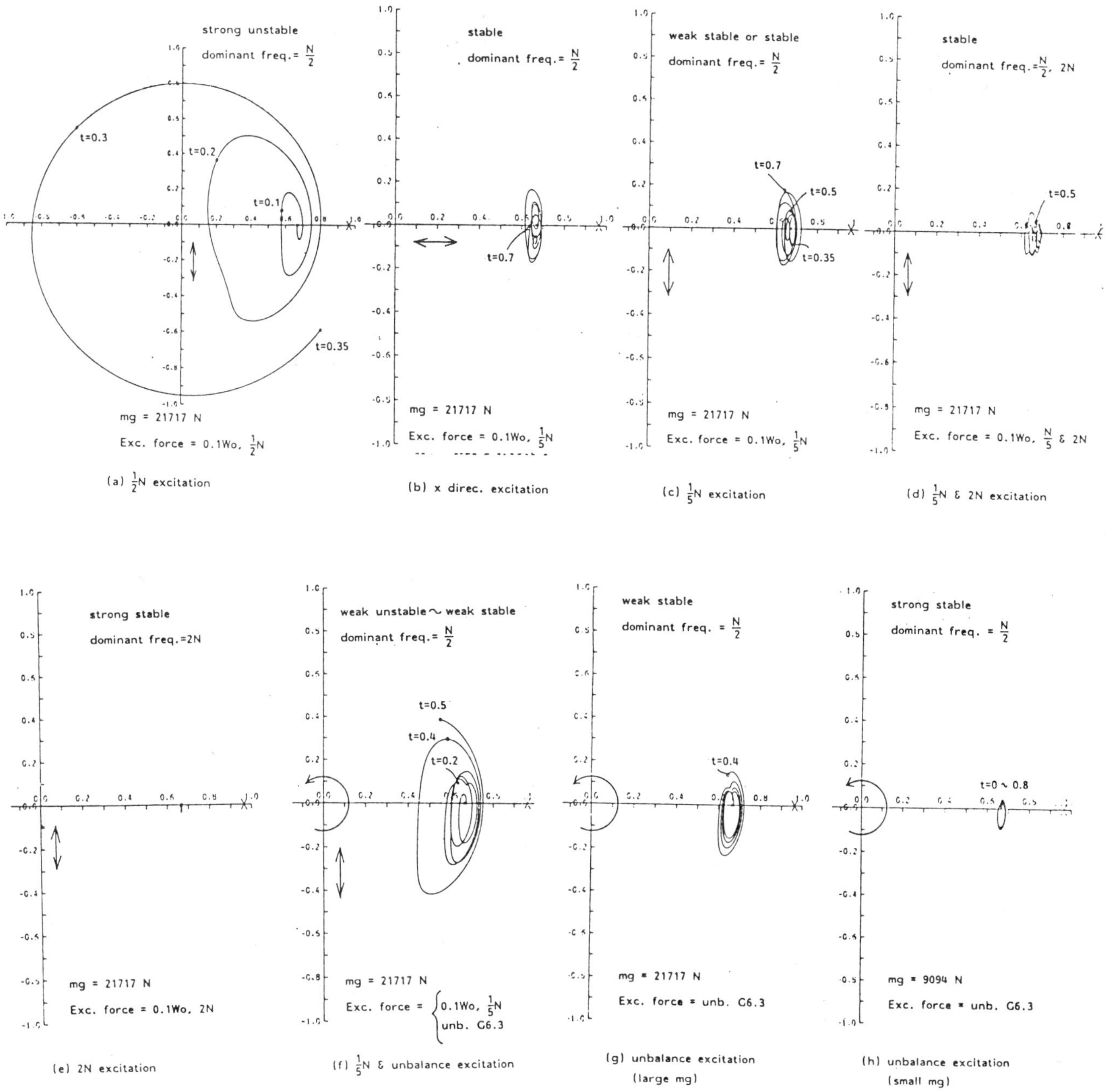

Fig 8 Results of response calculation

C285/88

Fluid-induced shaft vibrations in rotary atomizers

J COLDING-JØRGENSEN, MSc
Department of Machine Elements, Technical University of Denmark, Lyngby, Denmark

SYNOPSIS Subsynchronous shaft whirl in a rotary atomizer with vertical overhung shaft in anti-friction bearings has been studied experimentally and theoretically. A theoretical model is presented for the calculation of the fluid forces on the rotor. It shows that the fluid forces under certain conditions will produce self-exciting rotor whirl. A calculation of the rotordynamic coefficients is presented. The calculated coefficients of two different wheels for varying flowrate are consistent with whirl amplitudes and direction measured with these wheels mounted on the atomizer.

NOTATION

d	logarithmic decrement
$\underline{e}=e_x,e_y$	eccentricity of rotor center
$\dot{\underline{e}}=\dot{e}_x,\dot{e}_y$	velocity of rotor center
h	channel height
n	number of channels
r	radial distance from rotor center
r_1	channel inlet radius
r_2	wheel radius
r_d	liquid distributor exit radius
u	velocity in channel direction
u_{max}	maximum velocity in channel direction
$\bar{u}$	mean velocity in channel direction
$\underline{u}$	absolute vectorial fluid velocity
u_r	radial component of $\underline{u}$
u_{t0}	tangential component of $\underline{u}$ in centered wheel
u_{t20}	u_{t0} at wheel outlet
$\bar{u}_r$	radial component of $\bar{u}$
u_{r20}	$\bar{u}_r$ at outlet in centered wheel
w	channel width
(x,y)	stationary coordinate system perpendicular to rotor axis, with origin on wheel center.
B	damping coefficient
$\underline{F}$	fluid force on rotor
$\underline{\underline{B}} = \begin{Bmatrix} B_{xx} & B_{xy} \\ B_{yx} & B_{yy} \end{Bmatrix}$	damping matrix
$\underline{\underline{K}} = \begin{Bmatrix} K_{xx} & K_{xy} \\ K_{yx} & K_{yy} \end{Bmatrix}$	stiffness matrix
R_e	Reynolds number; $R_e = \frac{\bar{u}\cdot\varepsilon\cdot\rho}{\mu}$
U	free stream velocity in channel direction
U_0	inlet velocity
$\dot{V}$	total volumetric flowrate through wheel
$\frac{d\dot{V}}{d\theta}$	flowrate per angular segment of wheel
V_{dr}	radial velocity of flow disturbance in channel
δ	film thickness
ε	boundary layer thickness
θ	angular coordinate in (x,y) system
λ	angle between channel tangent and radial direction
μ	viscosity
ν	coordinate in channel direction
ξ	coordinate perpendicular to channel direction
ρ	fluid density
τ_0	shear at channel wall
ϕ	phase angle
ϕ_2	phase angle at outlet
ω	whirl angular frequency
Ω	rotational angular frequency

1 INTRODUCTION

Self excited lateral vibrations of high speed rotors is a phenomenon which has been known at least since the 1920's.

Most of this type of vibrations described in the early literature are caused by internal rotor damping or hydrodynamic bearing instability.

In recent years, much work has been devoted to investigate the rotordynamic stability effect of seals in turbomachinery.

Finally, it has been discovered that the working fluid in pumps and compressors can also create destabilizing forces on the rotor [1].

In this paper, a model of the fluid induced shaft vibrations of a rotary atomizer is presented and compared to measurements on such a machine under realistic process conditions.

A rotary atomizer works very much like a centrifugal pump at high speed and low flowrate.

The mixture of liquid and dispersed particles which is to be atomized is introduced into a rotating vaned wheel through a liquid distributor (Fig.1).

In the channels, the liquid is pressed against the wall, forming a thin film. At the exit, the liquid leaves the wheel as a spray jet.

There is no diffuser or volute around the wheel, since the function of the machine is not to build up pressure, but give the fluid a high velocity and spread it out in a thin sheet, which will disintegrate when it meets the air around the wheel exit.

Rotary atomizers are used for spray drying, in dry scrubbers, and other processes.

The present analysis and test results could have implications for a wider range of fluid handling machinery, since it shows that fluid induced unstable rotor vibrations can occur in a machine with an impeller, even in the absence of impeller/volute interaction, which has formed the basis for most of the work done on this type of rotordynamic destabilization so far [2], [3].

2 PHYSICAL MODEL

The fluid, which is assumed to be Newtonian and incompressible, enters the wheel through a circular ring gap in the distributor. The flow out of the distributor is uniform.

Then, when the wheel center is eccentric with respect to the distributor (Fig.2), the fluid will be distributed unevenly in the wheel inlet according to the expression:

$$\frac{d\dot{V}}{d\theta}(\theta)_{inlet} = \frac{\dot{V}}{2\pi}\left[1 - \frac{e_x}{r_d}\cos\theta - \frac{e_y}{r_d}\sin\theta\right] \qquad (1)$$

Second order contributions in e_x and e_y are neglected.

In the channels, the flow is treated as nearly parallel, with a free surface and a boundary layer growing from the wall. When the boundary layer meets the free surface, a fully developed film flow is formed (Fig.3).

The fluid is assumed to be guided perfectly by the channels.

3 FLUID FORCE AND ROTORDYNAMIC COEFFICIENTS

We disregard the pressure forces around the wheel, since an atomizer delivers the fluid to a large chamber filled with air at approximately atmospheric pressure.

Then, the momentum balance for a control volume gives the force excerted by the fluid on the wheel:

$$\underline{F} = -\left(\frac{\partial}{\partial t}\int_{C.V.} \underline{u}\rho dv + \int_{C.S.} \underline{u}\rho\underline{u}d\underline{A}\right) \qquad (2)$$

Where C.V. is a stationary control volume which contains the fluid at a given moment, from the inlet to the outlet of the wheel. C.S. is the surface of C.V. $\underline{dA}$ is normal to C.S., pointing outward.
In order to calculate these integrals, we need the velocity distribution in the wheel, giving the fluid velocity $\underline{u}$ in any point. In order to obtain the force as a function of rotor center position and velocity, which is required by the rotordynamic analysis, the fluid velocity distribution should be calculated for an impeller with a given eccentricity (e_x, e_y) and whirl velocity ($\dot{e}_x$, $\dot{e}_y$).

The mean velocity in the rotating channel direction is calculated in appendix A for a centered wheel. When the rotor center is whirling, two things are assumed to happen:

- The whirl velocity component normal to the channel is added to the fluid velocity, while the component tangential to the channel is assumed to have no effect.

- The uneven fluid distribution in an eccentric wheel, is assumed to alter the mean radial component of the fluid velocity, by the equation:

$$\bar{u}_r(\theta,r) = \frac{2\cdot\pi\cos\lambda}{h\cdot n\cdot\delta}\cdot\frac{d\dot{V}}{d\theta}(\theta,r) \qquad (3)$$

That is, the channels are assumed to be distributed continously over the circumference of the wheel with a "channel density" of $n/2\pi$.

In a wheel with radial channels, the tangential velocity is unaffected by this, since we assume the fluid to be guided perfectly by the channels.

In a wheel with curved channels, a variation of u_r will also affect u_t, since the term $u_r\cdot\sin\lambda$ is added to the tangential velocity. However, in the present investigation, the term $\Omega\cdot r$ from the wheel rotation is much larger than the term $u_r\cdot\sin\lambda$, so the tangential velocity is assumed to be unaffected by the variation in radial velocity.

The disturbances in the volume flow at the inlet of the whirling wheel, given by eq.(1), will propagate with a speed V_{dr} in radial direction, so the volume flow distribution is given by:

$$\frac{d\dot{V}}{d\theta}(\theta,r) = \frac{d\dot{V}}{d\theta}(\theta-\phi)_{inlet}$$

$$\text{where } \phi = \int_{r_1}^{r}\left(\frac{u_{t0}}{r^*} - \omega\right)\cdot\frac{1}{V_{dr}}\,dr^*$$

With these assumptions, the volume integral in (2) becomes:

$$\frac{\partial}{\partial t}\int_{CV}\rho\underline{u}dv = \frac{-\rho\cdot\dot{V}}{2r_d}\begin{Bmatrix}\int_{r_1}^{r_2}\cos\phi dr\cdot\dot{e}_x - \int_{r_1}^{r_2}\sin\phi dr\cdot\dot{e}_y \\ \int_{r_1}^{r_2}\sin\phi dr\cdot\dot{e}_x + \int_{r_1}^{r_2}\cos\phi dr\cdot\dot{e}_y\end{Bmatrix} \qquad (3)$$

In the calculations of the surface integral in eq.(2), the momentum transport into the wheel is neglected, since the fluid velocity at the inlet is much smaller than at the outlet of the wheel. Furthermore, at the outlet, the film is turbulent and fully developed for the flowrates considered. It can be shown from eq.(A4) that the mean radial velocity at the exit, $\bar{u}_{r2}$, is proportional to

$\dot{V}^{5/12}$ under this condition.

With these assumptions, the surface integral becomes:

$$\int_{C.S.} \underline{u}\rho\underline{u}d\underline{A} = -\frac{17}{24}\cdot\frac{\rho\dot{V}}{r_d}\cdot\bar{u}_{r20}\begin{pmatrix}e_x\cos\phi_2 - e_y\sin\phi_2\\ e_y\cos\phi_2 + e_x\sin\phi_2\end{pmatrix}$$

$$-\frac{1}{2}\cdot\left(\frac{\rho\cdot\dot{V}\cdot u_{t20}}{r_d}\right)\begin{pmatrix}-e_x\sin\phi_2 - e_y\cos\phi_2\\ -e_y\sin\phi_2 + e_x\cos\phi_2\end{pmatrix} \qquad (4)$$

In order to calculate ϕ, V_{dr} must be known.

According to Fulford [4], most experimental and theoretical work done on film flow indicates that disturbances will travel as surface waves on the film with a velocity between 1 and 3 times the mean velocity in the film. In this study, $V_{dr} = 2\cdot\bar{u}_r$ is used in the fully developed film.

In the inlet region, a mean disturbance velocity of $V_{dr} = \bar{u}_r(1 + \frac{\varepsilon}{\delta})$ is used, since it is reasonable to expect disturbances to propagate with the free stream velocity outside the boundary layer.

It should be noted that calculations of the integrals (3) and (4) for varying values of V_{dr} show that the influence of this parameter is small for high flowrates of water, but considerable for low flowrates or with high viscosity fluids.

The rotordynamic coefficients are defined by the equation:

$$\underline{F} = -\begin{pmatrix}B_{xx} & B_{xy}\\ B_{yx} & B_{yy}\end{pmatrix}\begin{pmatrix}\dot{e}_x\\ \dot{e}_y\end{pmatrix}-\begin{pmatrix}K_{xx} & K_{xy}\\ K_{yx} & K_{yy}\end{pmatrix}\begin{pmatrix}e_x\\ e_y\end{pmatrix} \qquad (5)$$

From eqs.(3), (4) and (5), we get:

$$K_{xx} = K_{yy} = \frac{-\rho\dot{V}}{r_d}\left(\frac{17}{24}\bar{u}_{r20}\cos\phi_2 - \frac{1}{2}u_{t20}\sin\phi_2\right)$$

$$K_{xy} = -K_{yx} = \frac{\rho\cdot\dot{V}}{r_d}\left(\frac{17}{24}\bar{u}_{r20}\sin\phi_2 + \frac{1}{2}u_{t20}\cos\phi_2\right) \qquad (6)$$

$$B_{xx} = B_{yy} = \frac{-\rho\cdot\dot{V}}{2r_d}\int_{r_1}^{r_2}\cos\phi dr$$

$$B_{yx} = -B_{xy} = \frac{\rho\cdot\dot{V}}{2r_d}\int_{r_1}^{r_2}\sin\phi dr$$

Using the results from Appendix A to calculate $\bar{u}_{r20}$, u_{t20}, $\phi(r)$ and $\phi_2 = \phi(r_2)$, the rotordynamic coefficients can now be calculated for any combination of flowrate $\dot{V}$, rotational frequency Ω and whirl frequency ω.

4 CALCULATION RESULTS

The rotordynamic stability influence of the fluid force is primarily given by the cross coupling stiffness $K_{xy} = -K_{yx}$ and the direct damping terms $B_{xx} = B_{yy}$.

The destabilizing effect of a negative direct damping term B_{xx} is of the same order of magnitude as a cross coupling stiffness of $\omega\cdot B_{xx}$.

Evaluating the expressions for K_{yx} and B_{xx} in eq.(6), it is seen that the destabilizing effect of B_{xx} is of the order of magnitude:

$$\omega B_{xx} \sim \frac{\omega}{\Omega}K_{xy}$$

In the present study, ω is one order of magnitude less than Ω, so the destabilizing influence is mainly given by K_{xy}.

When K_{xy} is positive, it will destabilize forward whirl. When negative, it will destabilize backward whirl.

In Fig.(4), calculated cross coupling coefficients K_{xy} are shown for two different wheels, with the stability effect on backward and forward whirl indicated. The coefficients are calculated separately for each type of whirl, since whirl frequency affects the results. The whirl frequencies are those observed in forward and backward whirl measured with each wheel mounted on the atomizer shaft.

Both wheels have a diameter of 210 mm. Wheel A has straight radial channels, wheel B forward curved channels.

The rotational frequency is fixed at Ω = 1900 rad/sec., and the flowrate is varied from 0 to 8 m^3/h. The fluid is water.

Fig.(5) shows an example with a fluid viscosity 100 times that of water.

5 SHAFT VIBRATION MEASUREMENTS

The radial shaft vibrations relative to the casing were measured, using two proximity probes placed at approximately right angles.

The signals were sent to a X-Y oscilloscope and a two channel FFT analyser, allowing frequency spectra of the individual channels, and the cross spectrum to be produced, displayed and stored digitally. The whirl direction was determined from the phase cross spectrum. The phase of the forward whirl component has the same sign as the syncronous component in the spectrum, and the phase of the backward component has the opposite sign.

Also, the forward whirl has a higher frequency, and the backward whirl a lower frequency, than the first natural frequency at no rotation, due to the gyroscopic moment [5].

Measurements were made on an atomizer with a vertical, overhung shaft in rolling element bearings, running at 1900 rad/sec., which is roughly 10 times the first critical speed, and well below the second critical.

The flowrate was varied from 0 to 8 m^3/h, requiring a power of up to 100 kW. Normally, this type of atomizer is used up to 4 m^3/h or 55 kW. The fluid used was water.

The whirl was measured between the lower bearing and the wheel, at a distance of approximately one third of the overhung shaft length, from the lower bearing.

Fig.(6) shows measured whirl amplitudes for the same two wheels as used in the calculations, and with a wheel with the same geometry as wheel B, except for an inlet shape to the channels, which has been observed to produce extensive backflow from the channel to the inner wheel (wheel C).

The observed whirl frequencies were:

Wheel A: {217 rad/sec (forward); 116 rad/sec (backward)}

Wheel B and C {167 rad/sec (forward); 78,5 rad/sec (backward)}

Also, by a simple impact test on the rotor standing still, a logarithmic decrement of the first critical of d = 0,07 was found for wheel B.

The impact test with wheel B also gave a first rotor natural frequency at no rotation of 123 rad/sec.

COMPARISON OF MEASUREMENTS AND RESULTS

In order to evaluate the order of magnitude of the calculated cross coupling stiffness coefficients an estimate of the damping in the rotor is necessary.

If we assume a simple one-degree of freedom model of the rotor, the measured logaritmic decrement is equivalent to a damping coefficient of

$$B = \frac{d \cdot m \cdot \omega_0}{\pi}$$

The mass of wheel B is 12,8 kg, so:

$$B = \frac{0.07 \cdot 12{,}8\,\mathrm{kg} \cdot 123\ \mathrm{rad/sec}}{\pi} = 35\ \mathrm{Nsec/m}$$

So, in order to create unstable whirl, the cross coupling coefficient must be:

$$|K_{xy}| > B \cdot |\omega|$$

Consequently, in order to create unstable whirl, K_{xy} must be:

Wheel A:

Forward whirl: $K_{xy} > 35\ \mathrm{Nsec/m} \cdot 217\ \mathrm{rad/sec} = \underline{7595\ \mathrm{N/m}}$

Backward whirl: $K_{xy} < -35\ \mathrm{Nsec/m} \cdot 116\ \mathrm{rad/sec} = \underline{-4060\ \mathrm{N/m}}$

Wheel B:

Forward whirl: $K_{xy} > 35\ \mathrm{Nsec/m} \cdot 167\ \mathrm{rad/sec} = \underline{5845\ \mathrm{N/m}}$

Backward whirl: $K_{xy} < -35\ \mathrm{Nsec/m} \cdot 78.5\ \mathrm{rad/sec} = \underline{-2747{,}5\ \mathrm{N/m}}$

Looking at fig.(4), it is seen that the calculated cross coupling terms should be large enough to create unstable forward whirl for wheel A for a flowrate higher than 5.5 m^3/h, and unstable backward whirl for wheel B for flowrates higher than 5 m^3/h.
Looking at fig.(6), it is seen, that wheel A does indeed show high forward whirl amplitudes, and wheel B high backward whirl amplitudes, as the flowrate exceeds 5 m^3/h. The whirl amplitudes increase further, as the flowrate is increased, as predicted by the calculation results displayed in fig.(4).

For intermediate flowrates, K_{xy} has a local minimum with negative value for wheel A and a local maximum with positive value for wheel B. Since the calculated stability limit is not exceeded, only moderate whirl amplitudes should be expected here, caused by random excitation. The forward component of this whirl will be amplified when K_{xy} is positive and the backward component will be amplified when K_{xy} is negative.

Looking at fig.(4), we should expect a local maximum of the backward component at 0.5 m^3/h for wheel A, and a local maximum for the forward component at 1.0 m^3/h for wheel B.

Looking at the experimental results in fig.6, we se indeed a local maximum of the backward component for wheel A, but at a higher flowrate than predicted (2.5 m^3/h). Also, wheel B exhibits a local maximum of the forward component, but again at a higher flowrate (5.2 m^3/h).

Generally, it is concluded that the theory predicts the whirl behaviour well for high flowrates, but only qualitatively correct for intermediate flowrates.

The measurements on wheel C with the high inlet backflow show a completely different whirl behaviour, with only small and moderate whirl amplitudes up to double the rated power level for this wheel.

This indicates that the destabilizing influence of the fluid force can be substantially reduced by designing the inlet to the channels so that a certain amount of fluid travels in and out of the inlet several times, before it is caught in a channel, when this is not forbidden by process requirements.

DISCUSSION

The comparison of calculated cross coupling stiffness coefficients and measured shaft whirl is very encouraging. However, there is room for some improvement of the theoretical model:

- K_{xy} is very sensitive to the disturbance speed V_{dr} at intermediate and low flowrates. A better estimate of V_{dr} could probably improve the prediction of whirl behaviour in the intermediate flowrate-range.

- The direct damping terms could be calculated from eq.(6) and included in the analysis. This would have a certain effect on the results, even when whirl frequency is small compared to rotational frequency, especially when K_{xy} is changing sign. This again should improve the predictions in the intermediate flowrate-range.

The measurements on wheel C show that modifications of the inlet to the channels is a very effective way to substantially reduce whirl problems in rotary atomizers. The optimal inlet shape should be determined as a compromise between process requirements and rotordynamic stability considerations.

Finally, it is possible that an eccentricity dependent inlet flow disturbance mechanism of a nature similar to the one presented in this paper, could influence the whirl behaviour of other fluid handling machinery, i.e. centrifugal pumps and compressors, especially at low flowrates combined with high power level, under which conditions such turbomachines are quite similar to a rotary atomizer.

AKNOWLEDGEMENT

The experimental part of this work was carried out, using the test facilities of Niro Atomizer A/S.

REFERENCES

(1) Ehrich, F. and Childs, D.: Self-excited Vibration in High-performance Turbomachinery. Mechanical Engineering, May 1984.

(2) Colding-Jørgensen, J.: Effect of fluid forces on Rotor Stability of Centrifugal Compressors and Pumps. NASA C.P.2133, Rotordynamic Instability Problems in High Performance Turbomachinery, Proceedings of a workshop held at Texas A&M University, 1980, pp.249-265.

(3) Chamieh, D.S., Acosta, A. J., Brennen, C. E., Caughey, T. K. and Franz, R.: Experimental Measurements of Hydrodynamic Stiffness Matrices for a Centrifugal Pump Impeller. NASA C.P. 2250, Proceedings of a workshop held at Texas A&M University 1982, pp.382-98.

(4) Fulford, G. D.: The Flow of Liquids in Thin Films, Advances in Chemical Engineering, Vol.5, Academic Press 1964.

(5) Den Hartog, J. P.: Mechanical Vibrations, McGraw-Hill 1956, pp.253-265.

(6) Schlicting, H.: Boundary Layer Theory, McGraw-Hill, 1968, pp.596-601.

(7) Lambert, J. D.: Computational Methods in Ordinary Differential Equations, John Wiley & Sons 1977, pp.85-114.

(8) Arpachi, V. D. and Scheel Larsen, P.: Convection and Heat Transfer, Chapter 8, Prentice Hall 1984.

APPENDIX A

Calculation of velocity distribution in atomizer wheel channel for rotating, centered wheel.

In the inlet, before the fully developed film is formed, we have a boundary layer and a free stream in the rotating channel (Fig.3). In the free stream, the Euler equation is used, giving the free stream velocity U in the channel direction as:

$$U\frac{dU}{d\nu} = \Omega^2 \cdot r \cdot \cos\lambda \qquad (A1)$$

ν is the coordiante in the channel direction and ξ is the coordinate perpendicular to the channel direction. Using an integral boundary layer formulation, the conservation of momentum and mass in the film gives [8]:

$$\frac{d}{d\nu}\int_0^\delta u^2 d\xi - u\frac{d}{d\nu}\int_0^\delta u d\xi = \delta\cdot\Omega^2\cdot r\cdot\cos\lambda - \frac{\tau_0}{\rho} \qquad (A2)$$

and

$$\int_0^\delta u d\xi - U_0\cdot\omega = 0 \qquad (A3)$$

For a laminar boundary layer, we assume [6]:

$$\tau_0 = \mu \cdot \left(\frac{\partial u}{\partial \xi}\right)_{\xi=0} \qquad (A4)$$

and a parabolic velocity distribution is assumed:

$$\frac{u}{U} = \begin{cases} 2\xi/\varepsilon - (\xi/\varepsilon)^2 & 0 \leqq \xi < \varepsilon \\ 1 & \varepsilon \leqq \xi \leqq \delta \end{cases} \qquad (A5)$$

For a turbulent boundary layer, we assume [6]:

$$\tau_0 = \rho \cdot U^2 \cdot 0{,}0225 \left(\frac{\mu}{\rho U\cdot\varepsilon}\right)^{1/4} \qquad (A6)$$

and

$$\frac{u}{U} = \begin{cases} (\xi/\varepsilon)^{1/7} & 0 \leqq \xi \leqq \varepsilon \\ 1 & \varepsilon \leqq \xi \leqq \delta \end{cases} \qquad (A7)$$

Inserting (A4) and (A5) into (A2) and (A3), we get for the laminar boundary layer:

$$\frac{d\varepsilon}{d\nu} = -\frac{9}{2}\frac{\Omega^2\cdot r\cos\lambda}{U^2}\varepsilon + \frac{\mu}{\rho}\cdot\frac{15}{U}\cdot\frac{1}{\varepsilon} \qquad (A8)$$

and

$$\bar{u} = \dot{V}/[(\tfrac{1}{3}\varepsilon + \frac{U_0}{U}\cdot w)\cdot n\cdot h] \qquad (A9)$$

Inserting (A6) and (A7) into (A2) and (A3), we get for the turbulent boundary layer:

$$\frac{d\varepsilon}{d\nu} = -3.29\cdot\frac{\Omega^2\cdot r\cos\lambda}{U}\cdot\varepsilon + 0.23\left(\frac{\mu}{\rho U\cdot\varepsilon}\right)^{1/4} \qquad (A10)$$

and

$$\bar{u} = V/[(\tfrac{1}{8}\varepsilon + \frac{U_0}{U}\cdot w)\cdot n\cdot h] \qquad (A11)$$

Following Fulford [4], the transition from laminar to turbulent flow is assumed to take place when $R_e = \frac{\bar{u}\cdot\varepsilon\cdot\rho}{\mu} = 400$.

At the inlet, the velocity in the direction of the channel is: $U_0 = \frac{\dot{V}\cos\lambda}{n\cdot h\cdot\omega}$.

The boundary layer thickness ε, and mean velocity $\bar{u}$ is calculated stepwise along the channel, solving eqs. (A1) and (A8) or (A10), depending on whether the flow is laminar or turbulent. The numerical method used in a simple trapez method with variable steplength controlled by an error estimate [7].

After a number of steps, the boundary layer meets the free surface, $\varepsilon = \delta$, and the fully developed film is formed. In this region, the shear forces must equal the centrifugal force giving:

$$\left.\begin{aligned} \rho\cdot\delta\cdot\Omega^2\cdot r\cdot\cos\lambda &= \mu\left(\frac{\partial u}{\partial \xi}\right)_{\xi=0} \quad \text{(laminar)} \\ \rho\cdot\delta\cdot\Omega^2\cdot r\cdot\cos\lambda &= \mu^2_{max}\cdot 0.0225\left(\frac{\mu}{\rho u_{max}\cdot\varepsilon}\right)^{1/4} \quad \text{(turbulent)} \end{aligned}\right\} \quad (A12)$$

Using the same velocity distribution as in the boundary layer, we get:

$$\bar{u} = \left(\frac{\rho\cdot\Omega^2\cdot r\cdot\cos\lambda\cdot\dot{V}^2}{3\cdot\mu\cdot n^2\cdot h^2}\right)^{1/3} \quad \text{(laminar)} \quad (A13)$$

and

$$\bar{u} = \left(\frac{\rho\cdot\dot{V}^5\cdot\Omega^8\cdot r^4}{0.028\cdot\mu\cdot n^5\cdot h^5}\right)^{1/12} \quad \text{(turbulent)} \quad (A14)$$

Both in the inlet region and in the fully developed film, the mean radial velocity is now given in any point as $\bar{u}_r = \bar{u}\cos\lambda$.

The tangential velocity is considered as constant through the film section:

$$u_{t0} = \Omega\cdot r + \bar{u}_r\cdot\sin\lambda$$

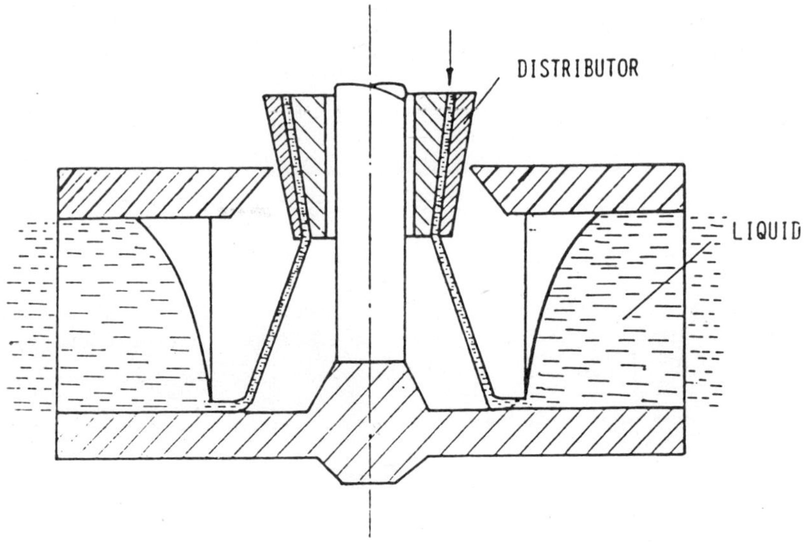

Fig 1

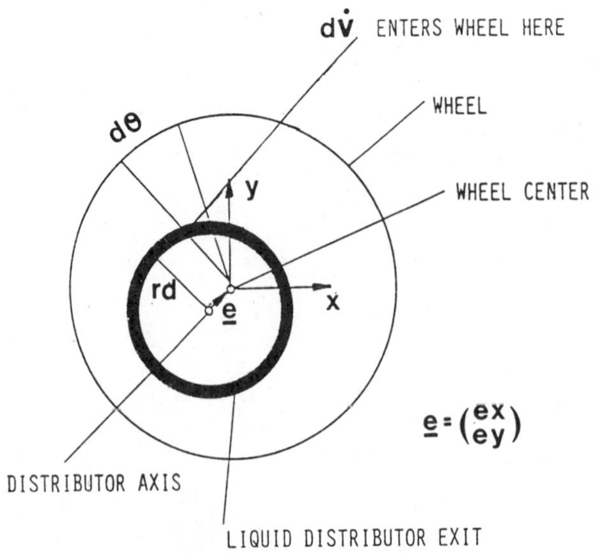

Fig 2

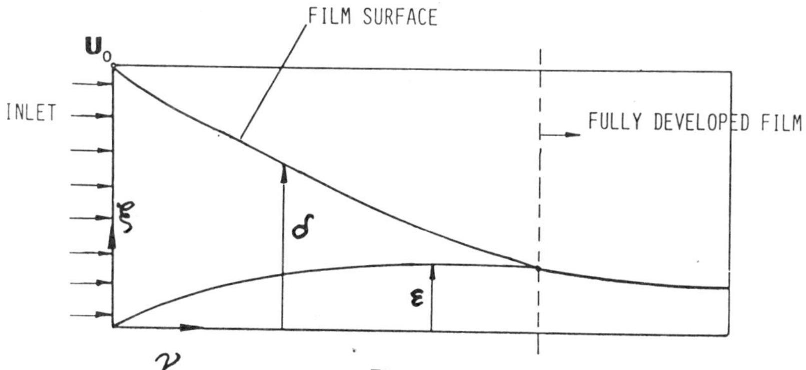
FILM SURFACE
U0
INLET
FULLY DEVELOPED FILM
δ
ε
Fig 3

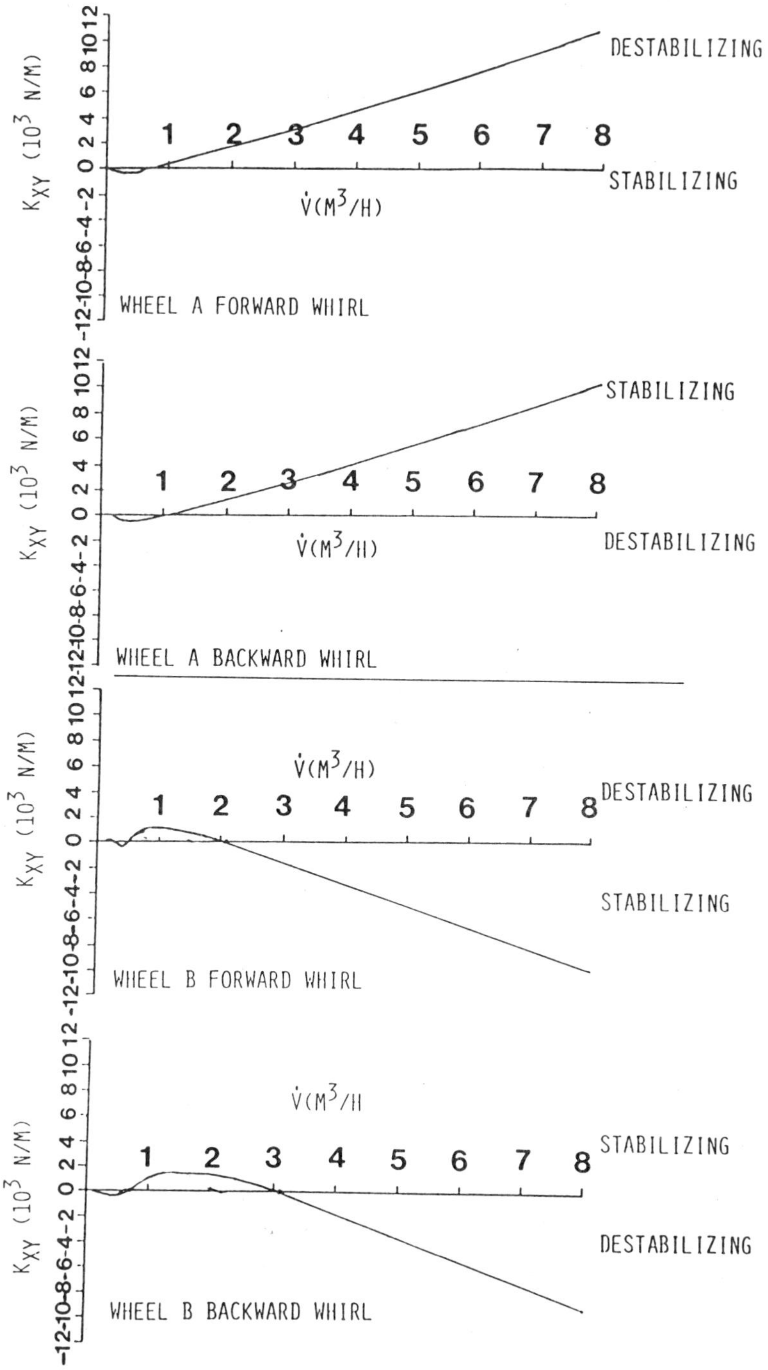
KXY (10^3 N/M)
DESTABILIZING
STABILIZING
V̇(M^3/H)
WHEEL A FORWARD WHIRL
STABILIZING
DESTABILIZING
WHEEL A BACKWARD WHIRL
DESTABILIZING
STABILIZING
WHEEL B FORWARD WHIRL
STABILIZING
DESTABILIZING
WHEEL B BACKWARD WHIRL
Fig 4

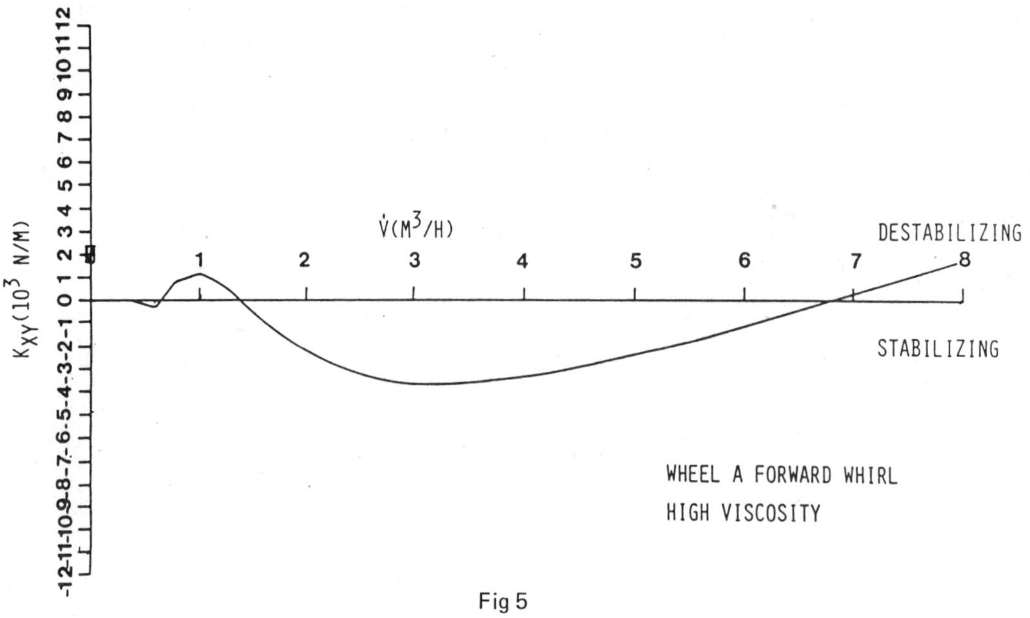

$K_{XY}(10^3$ N/M)
-12 -11 -10 -9 -8 -7 -6 -5 -4 -3 -2 -1 0 1 2 3 4 5 6 7 8 9 10 11 12
$\dot{V}(M^3/H)$
1 2 3 4 5 6 7 8
DESTABILIZING
STABILIZING
WHEEL A FORWARD WHIRL
HIGH VISCOSITY

Fig 5

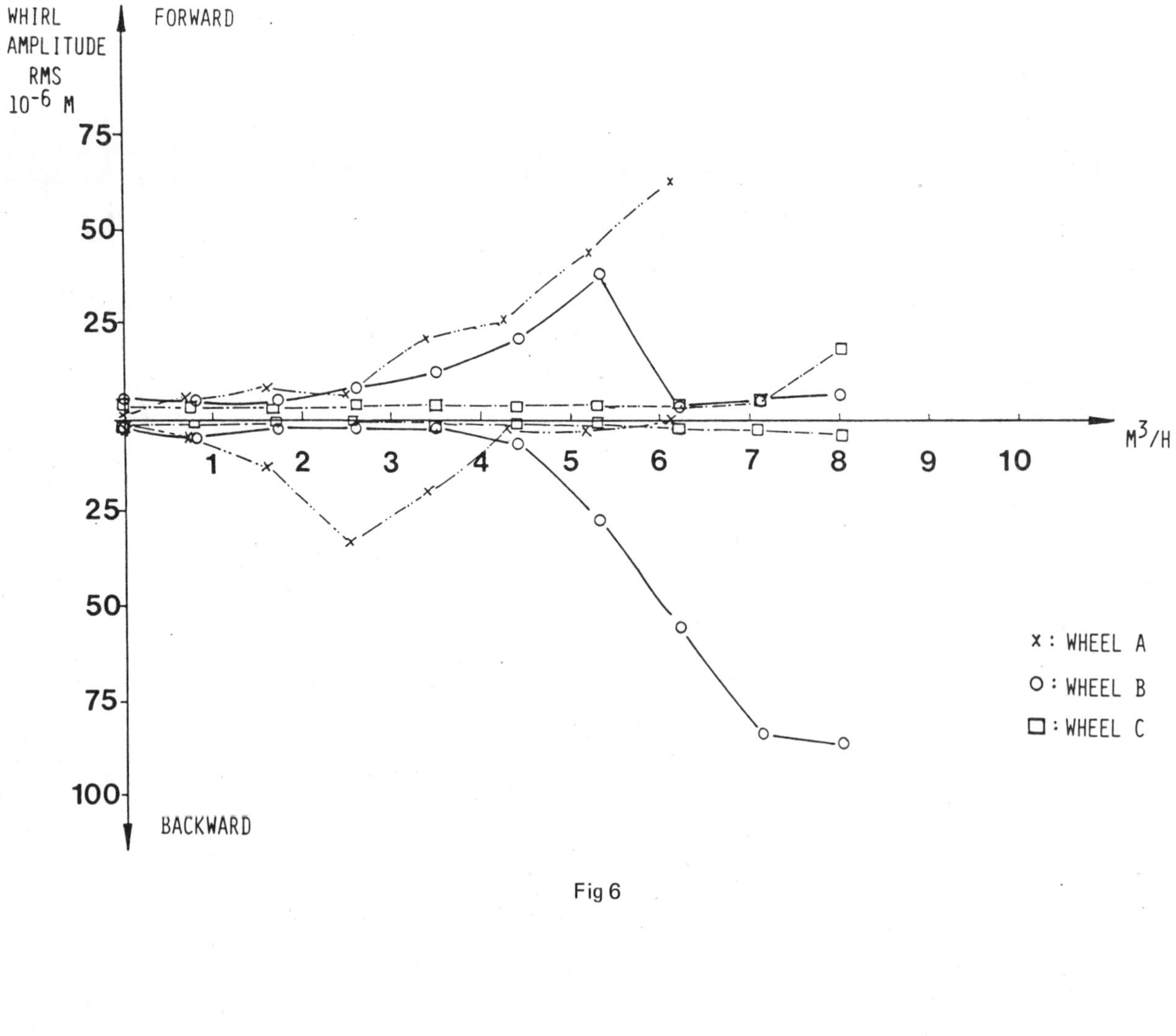

WHIRL
AMPLITUDE
RMS
10^{-6} M
FORWARD
75
50
25
25
50
75
100
BACKWARD
1 2 3 4 5 6 7 8 9 10
M^3/H
x : WHEEL A
O : WHEEL B
□ : WHEEL C

Fig 6

C309/88

On the natural frequencies and stability of a Laval rotor subjected to follower force

K CZOLCZYNSKI, MSc and **K P MARYNOWSKI**, PhD
Institute of Applied Mechanics, Technical University of Lodz, Lodz, Poland

SYNOPSIS The existance of regions of divergence and flutter instability for an elastically supported Laval-rotor under follower force is discussed. The compressive follower force acts only on the part of the rotor from a thrust bearing to the disk. The type of instability depends on the place where the disk is located and on the stiffness of the support. When gyroscopic effect is taken into consideration the dependence between angular velocity of the rotor and the critical follower force is observed.

NOTATION

$A_1 \ldots A_4, B_1 \ldots B_4$	constant terms
B	the mass moment of inertia of the disk with respect to its axis of bending
B_o	the mass moment of inertia of the disk with respect to its axis of rotating
C_A, C_B	stiffness coefficients of bearings
C	dimensionless stiffness coefficient of bearing
EI	bending stiffness of the shaft
P	follower force
$k = \sqrt{\frac{P}{EI}}\, l$	dimensionless follower force
l	length of the rotor
k_{cr}	critical dimensionless follower force
m	mass of the disk
$N = \begin{bmatrix} 0 & 1 \\ -1 & 0 \end{bmatrix}$	matrix for cross-coupled terms
r	radius of the disk
t	time
$y = \begin{Bmatrix} w \\ v \end{Bmatrix}$	vector of shaft's displacements for $0 \leqslant x \leqslant l_1$
$\tilde{y} = \begin{Bmatrix} \tilde{w} \\ \tilde{v} \end{Bmatrix}$	vector of shaft's displacements for $l_1 \leqslant x \leqslant l$
$\delta y = \begin{Bmatrix} \delta w \\ \delta v \end{Bmatrix}$	vector of shaft's virtual displacements for $0 \leqslant x \leqslant l_1$
$\delta \tilde{y} = \begin{Bmatrix} \delta \tilde{w} \\ \delta \tilde{v} \end{Bmatrix}$	vector of shaft's virtual displacements for $l_1 \leqslant x \leqslant l$
δU	the virtual work done by elastic forces of the rotor
δU_p	the virtual work done by follower force
δU_b	the virtual work done by inertia forces
ω	angular velocity of the rotor
$\omega^* = \omega \sqrt{\frac{ml^3}{EI}}$	dimensionless angular velocity
Ω	natural frequency of the rotor
$\Omega^* = \Omega \sqrt{\frac{ml^3}{EI}}$	dimensionless natural frequency
$\Theta = \begin{Bmatrix} \sin \Omega t \\ \cos \Omega t \end{Bmatrix}$	vector of harmonical functions
$\Theta_\omega = \begin{Bmatrix} \sin \omega t \\ \cos \omega t \end{Bmatrix}$	vector of harmonical functions
$\xi = \frac{l_1}{l}$	parameter of disk's location

1 INTRODUCTION

Rotors subjected to follower nonconservative forces can be found in many contemporary machines. For example longitudinal forces arising from the interaction between impellers and a working medium appear in axial compressors. The forces can be described as always tangential to deflected axis of the shaft and following its motion. The follower forces are able to change the natural frequencies of the rotor and in certain events cause an instability of rotor's motion.

2 THE MATHEMATICAL MODEL OF THE ROTOR

Fig.1 shows the Laval-rotor subjected to follower force P acting on the disk. The force compresses the part of the shaft between the disk and the thrust bearing A. The shaft of the rotor with constant cross-section is supported by two elastic isotropic bearings. The angular velocity of the rotor ω is constant. The axis x determines the static equilibrium position of the shaft. The considered system is undamped.

The mathematical model of the rotor

can be found by the principle of virtual work

$$\delta U = \delta U_p + \delta U_b \tag{1}$$

Using the vectorial notation, the equation (1) can be expressed as

$$EI\Big\{ y_{l1}''^{T}\delta y_{l1}' - y_o''^{T}\delta y_o'' - y_{l1}'''^{T}\delta y_{l1} + y_o'''^{T}\delta y_o + \int_0^{l_1} y^{IV}\delta y\,dx + \tilde{y}_1''^{T}\delta\tilde{y}_1' - \tilde{y}_{l1}''^{T}\delta\tilde{y}_{l1}' - \tilde{y}_{l1}''^{T}\delta\tilde{y}_{l1}'' - \tilde{y}_1'''^{T}\delta\tilde{y}_1 + \tilde{y}_{l1}'''^{T}\delta\tilde{y}_{l1} + \int_{l_1}^{l}\tilde{y}^{IV\,T}\delta\tilde{y}\,dx \Big\} + C_A y_o^T\delta y_o + C_B\tilde{y}_1^T\delta\tilde{y}_1 + P\Big\{ y_o'^{T}\delta y_o + \int_0^{l_1} y''^{T}\delta y\,dx \Big\} + m\,\ddot{y}_{l1}^T\delta y_{l1} + B\ddot{y}_{l1}'^{T}\delta y_{l1}' + B_o\omega N\dot{y}_{l1}'\delta y_{l1}' = 0 \tag{2}$$

The equation (2) is satisfied for all possible virtual displacements if the sum of work done by follower and inertia forces is equal to the change of elastic energy of the system for any virtual displacement. This condition leads to following eguations:

$$EI\,y_{l1}'' - EI\tilde{y}_{l1}'' + B\ddot{y}_{l1}' + B_o\omega N\dot{y}_{l1}' = 0 \tag{3}$$

$$EI\,y_o'' = 0 \tag{4}$$

$$-EI\,y_{l1}''' + EI\tilde{y}_{l1}''' + m\ddot{y}_{l1} = 0 \tag{5}$$

$$EI\,y_o''' + C_A y_o + P y_o' = 0 \tag{6}$$

$$EI\,\tilde{y}_1'' = 0 \tag{7}$$

$$-EI\,\tilde{y}_1''' + C_B\tilde{y}_1 = 0 \tag{8}$$

$$EI\,y^{IV} + P y'' = 0 \tag{9}$$

$$EI\,\tilde{y}^{IV} = 0 \tag{10}$$

The continuity conditions of bending shaft's line for the place where the disk is located are given by

$$y_{l1} = \tilde{y}_{l1} \tag{11}$$

$$y_{l1}' = \tilde{y}_{l1}' \tag{12}$$

(3) ÷ (8) and (11) , (12) give boundary conditions for the equations (9) and (10). (9) is the equation of deflection curve of the shaft in the partition $x=0 \div l_1$ and (10) in the partition $l_1 \div l$. The solution of the equation (9)

$$y(x,t) = \Theta\left(A_1\sin\frac{kx}{l} + A_2\cos\frac{kx}{l} + A_3x + A_4\right) \tag{13}$$

The solution of the equation (10)

$$\tilde{y}(x,t) = \Theta\left(B_1x^3 + B_2x^2 + B_3x + B_4\right) \tag{14}$$

Introduction (13) and (14) into (3) ÷ (8) and (11) , (12) gives homogeneous set of linear equations for the unknown constants A_1, A_2, A_3, A_4 and B_1, B_2, B_3, B_4.

The solution of the set of equations can exist only when its characteristic determinant is equal to zero. The equation (15) shows this condition.

$$\Delta D = 0 \tag{15}$$

This determinant is given by (16)

$MZk\,cs - k^2 sn$	$-MZk\,sn - k^2 cs$	MZ		$-6l_1$	-2		
$PZ\,sn - k^3 cs$	$PZ\,cs + k^3 sn$	PZl_1	PZ	-6			
	k^2						
	C_A	$\frac{k^2EI}{l^2}$	C_A				
				$6l$	2		
				$-6EI + C_Bl^3$	C_Bl^2	C_Bl	C_B
$k\,cs$	$-k\,sn$	1		$-3l_1^2$	$-2l_1$	-1	
sn	cs	l_1	1	$-l_1^3$	$-l_1^2$	$-l_1$	-1

(16)

where

$$MZ = \frac{\omega\Omega B_o - \Omega^2 B}{EI}$$

$$PZ = \frac{\Omega^2 m}{EI}$$

$$sn = \sin k\xi$$

$$cs = \cos k\xi$$

The frequency equation (15) admits to investigate natural frequencies and stability of the system.

3 NUMERICAL RESULTS

Fig.1 shows a flexibly supported Laval-rotor subjected to follower force. The bearings of the rotor were characterized by their dimensionless stiffness coefficient $C = 100$ ($C=C_Al^3/EI=C_Bl^3/EI$). In this investigated rotor $EI=5.7\,Nm^2$, $l=0.491m$. The parameter $\xi = l_1/l$ describes the place where the disk is located on the shaft.

3.1 The disk as a mass-point

At first, the rotor with the mass 0.72kg concentrated at a point on the shaft was investigated. Results of numerical calculations of natural frequencies of rotor's lateral vibrations under follower compressive force P at the mass-point are shown in Fig.2. The change of the values of natural frequencies is dependant on the value of the parameter ξ. For $0 \leqslant \xi \leqslant 0.03$ and $0.29 \leqslant \xi \leqslant 1$ increasing of follower force decreases the natural frequency of the rotor. It is shown in Fig.2 for $\xi = 0.75$. For a certain value of follower force $k=k_{cr}$ the natural frequency is equal to zero. In Fig.2 for $\xi=0.75$ $k_{cr}= 3.2$. Next for $k>k_{cr}$ the natural frequency has imaginary value. It means that for $k=k_{cr}$ the flexural buckling of the shaft apears there (instability in Euler's sense-divergence instability).

For $0.03 \leqslant \xi \leqslant 0.29$ increasing of

k causes also increasing of Ω^*. It is shown in Fig.2 for ξ = 0.15. In this region of ξ for $k=k_{cr}$ an abrupt change of the value of natural frequency apears there. It changes from real to imaginary value but doesn't pass through zero. This change means flutter instability of the rotor. In Fig.2 it apears for k_{cr}= 5.85.

Fig.3 shows the plot of the critical follower force k_{cr} versus the ξ value for the investigated rotor with mass-point.

3.2 Natural frequencies of Laval-rotor

The disk's parameters of considered rotor (Fig.1): B^*=0.01, B^*_o=0.02, m=0.81 kg, r=0.101m. The angular velocity of the rotor ω =160s^{-1} (ω^*= 20).

From computer calculations it was concluded that natural frequencies of rotor's lateral vibrations depend on the value of follower force. The change of Ω and stability of the rotor are dependant also on the place where the disk is located. Four different types of rotor's behaviour dependant on the disk's location were distinquished. In the first type for 0.34 $\leqslant \xi \leqslant$ 1.0 absolute values of all four natural frequencies of the rotor decrease with increasing of the follower force. It is shown in Fig.4 for ξ =0.8. For the k_m value of follower force the first natural frequency of retrograde precession is equal to zero. It is true for the range 0.5 $\leqslant \xi \leqslant$ 1.0. In the range 0.34 $\leqslant \xi \leqslant$ 0.5 for the k_m value of follower force, not the frequency of retrograde precession but the first natural frequency of direct precession is equal to zero. Both these cases mean rotation of the shaft about its immovable and deformed axis. For $k=k_{cr}$ both the least natural frequencies reach the same value and next they have conjugate complex values. It means instability of rotor's motion.

The second type of rotor's behaviour under follower force was distinquished for 0.075 $\leqslant \xi \leqslant$ 0.27. Fig.5 shows the plot of Ω^*versus k for ξ = 0.25.

The last types of rotor's behaviour (III and IV) can be observed in narrow ranges of ξ. They are shown in Fig.6. The changes of natural frequencies in these ranges can be described as a combination of changes in I and II. The continuous line in Fig.6 shows the plot of k_{cr} versus ξ for ω^*= 20.The broken lines in Fig.6 show the same function for other values of ω^*. It is worth to note in type II for ω^*= 10.3 the value of k_{cr} is equal to zero.

3.3 Critical speeds

Numerical calculations were conducted to observe critical speeds of Laval - rotor under follower force. To obtain the equation of critical speeds of the rotor the factor Θ in the solutions (13) and (14) was replaced by the factor Θ_ω. In this case ω is the unknown angular velocity of the rotor which is equal to its natural frequency.

Results of numerical calculations show the follower force changes the values of rotor's critical speeds. Results of investigations are shown in Fig.7 and Fig.8. Fig.7 shows the plot of ω^*versus k for ξ =0.4 as the example for the ranges 0 $\leqslant \xi \leqslant$ 0.03 and 0.3 $\leqslant \xi \leqslant$ 1.0. Fig.8 shows the same plot for ξ =0.2 as the example for the range 0.03 $\leqslant \xi \leqslant$ 0.3. It is worth to noting in Fig.8 for $k_m < k < k_{cr}$ increasing of rotor's angular velocity causes instability of rotor's motion without passing through its critical speed.

4. CONCLUSIONS

In this paper one particular case of Laval-rotor subjected to follower force was investigated. Any change of support's stiffness, the place of the disk's location or the disk's parameters gives different dependences between the natural frequencies of the rotor and the value of follower force.

The critical follower force is dependant on disk's location and on the stiffness of the support. Even small change of the place of disk's location may change the type of rotor's behaviour. In some ranges of the disk's location for particular rotor's velocities even small follower force may cause instability of rotor's motion. The follower force and gyroscopic effect of the disk may cause instability of rotor's motion without passing through any critical speed of the rotor.

REFERENCES

1 BOLOTIN, N.N. Nonconservative problems of the theory of elastic stability. Pergamon Press INC. N.Y. 1963.

2 KOUNDAIS, A.N. Divergence and flutter instability of elastically restrained structures under follower forces. Int.J.Eng.Sci,1981,19, 553-562

3 KOUNDAIS, A.N. The existence of regions of divergence instability for nonconservative systems under follower forces. Int.J.Solids Structures, 1983,19, 725-733.

4 HERMANN, G. BUNGAY, R.W. On the stability of elastic systems subjected to nonconservative forces. J. Appl. Mech. Vol. 31, Trans. ASME, 1964, 84, 435-440.

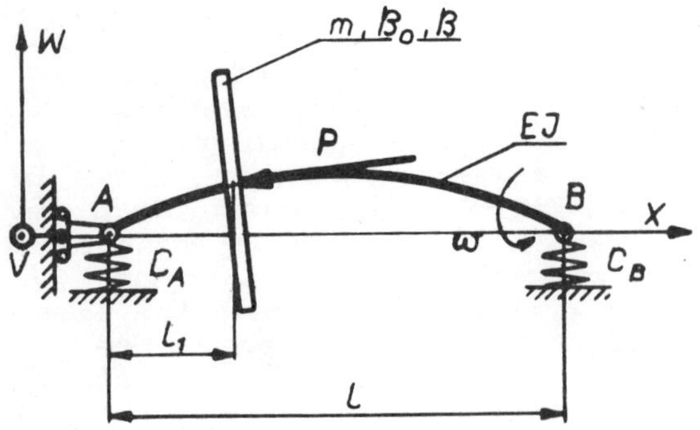

Fig 1 Model of the Laval rotor subjected to follower force

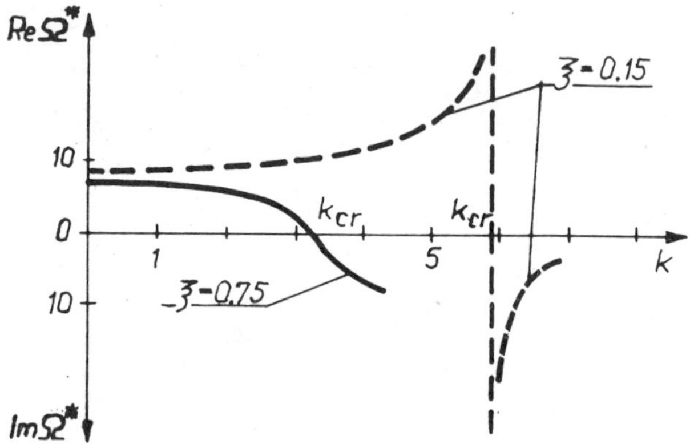

Fig 2 Natural frequencies of the rotor with mass-point

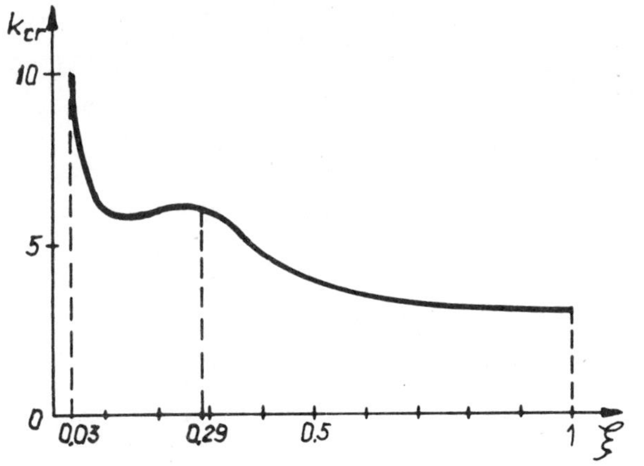

Fig 3 Critical follower forces of the rotor with mass-point

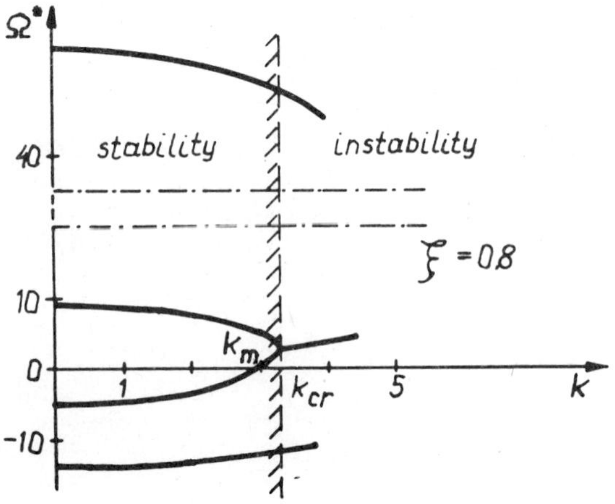

Fig 4 Natural frequencies of the Laval rotor (type I)

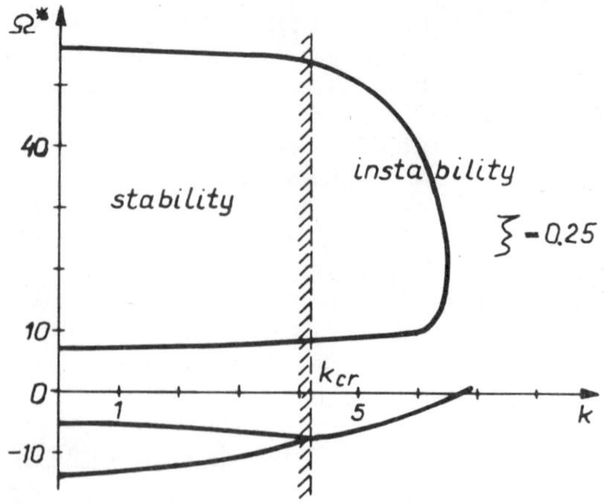

Fig 5 Natural frequencies of the Laval rotor (type II)

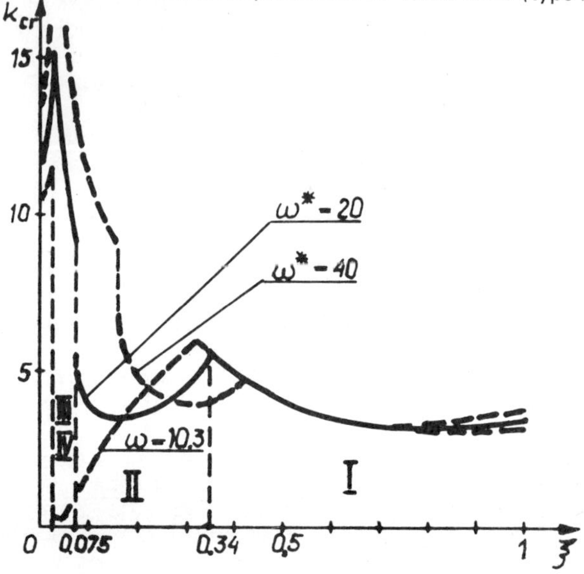

Fig 6 Critical follower forces of the Laval rotor

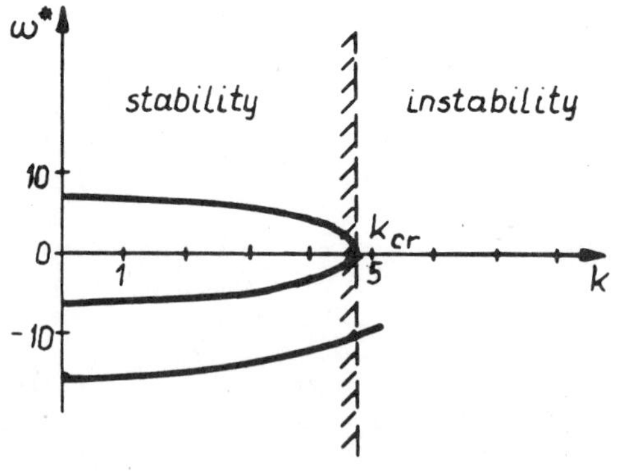

Fig 7 Critical speeds of the Laval rotor for $\xi = 0.4$

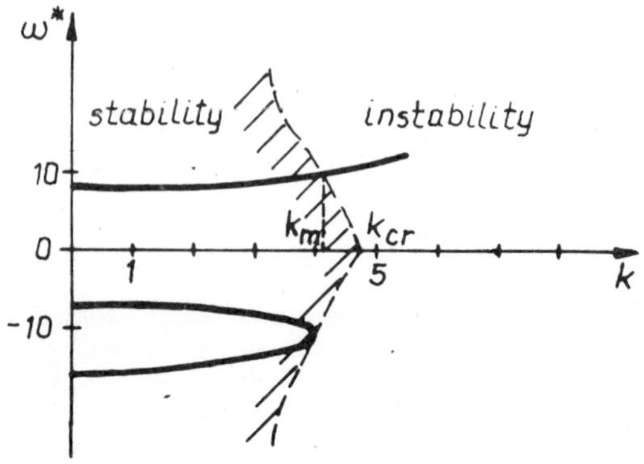

Fig 8 Critical speeds of the Laval rotor for $\xi = 0.2$

C243/88

Stability of the textile machine rotor

L CVETICANIN, PhD and **M ZLOKOLICA**, PhD
Faculty of Technical Sciences, University of Novi Sad, Novi Sad, Yugoslavia

The stability of rotation of the textile machine rotor is analysed. The model is a clamped-free rotor with variable mass. The non-linear rubbing force acts. The limits of instable rotation are denoted. They are the functions of the mass winding up on the rotor. During rubbing between the rotor and the housing the boundary moves in accordance to the geometrical and physical characteristics of the band and the shaft and clearance value.

1 INTRODUCTION

The fundamental working element of a lot of textile machines is the rotor with variable mass. Most of these rotors are clamped-free flexible rotors. These high speed rotors must be restricted to small deformations to insure safe operation. In spite of the smallness of the deformations there may be strong non-linear effects, which influence the dynamic behaviour of a rotor. One of the major sources of nonlinearity is rubbing of rotor elements against the stator. The interaction of a rotor with its housing can have a significant effect on the stability.

The paper |1| proposes a mathematical rationale of rubbing identification; defines the limits between benign contact and the initiation of a destructive instability based on harmonic spectral data.

The paper |2| gives a particular emphasis to rub phenomena and a detailed analysis of continuous rub during synchronous whirl.

In the paper |3| the stability of rotation of the rotor with variable mass is analysed. The influence of rubbing is taken into account. The model of the rotor is a shaft disc symmetrical system supported on the both ends.

The aim of the paper is to analyse the stability of rotation of the clamped-free rotor for the case of rubbing.

2 ROTOR DYNAMICS

The rotor is supposed as a shaft-disc system (see Fig. 1). The shaft is elastic. The elastic force in the shaft is non-linear. On the free end of the massless clamped-free elastic shaft a disc is mounted. The mass and the geometry of the dics are varying in time due to winding up of the band. The mass of the disc is varying (see p. |4|) according to

$$m(t) = m_0 + \rho L \Omega R_0 h t \tag{1}$$

where: m_0 is the mass of the empty disc, ρ is density of band, L is width of the band, Ω is angular velocity of rotor, R_0 is radius of the empty disc, h is thickness of the band and t is the time.

Also, the radius is varying as

$$R(t) = |R_0(R_0 + h\Omega t/\pi)|^{1/2} \tag{2}$$

The equations of motion for the rotor are developed using Lagranges equations. Three coordinates are defined: X,Y,Z - inertially fixed, x,y,z - spinning with angular velocity, x_1, y_1, z_1 - fixed in the disc. The second order Lagrange equation for the system with variable mass is after |5|

$$\frac{d}{dt}\frac{\partial T}{\partial \dot{q}_i} - \frac{\partial T}{\partial q_i} - Q_i - D_i = 0 \tag{3}$$

$$(i = 1,2, \ldots k)$$

where:

$$T = \sum_{\nu=1}^{n} \frac{m_\nu v_\nu^2}{2} \tag{4}$$

$$D_i = \sum_{\nu=1}^{n} \left(\frac{dm_\nu}{dt} \vec{v}_\nu \frac{\partial \vec{r}_\nu}{\partial q_i}\right) + R_i \tag{5}$$

T is kinetic energy, m_ν, $\vec{v}_\nu$, $\vec{r}_\nu$ are the mass, velocity and the position vector of the point ν, D_i is the added force, R_i is the reactive force, Q_i is the generalised force and q_i the generalised coordinate.

The eq. (3) can be written in another form using the principle of "rigidity" of the system. It means that at the moment the mass is supposed to be constant. The special operators for the "rigid" system

$$\frac{d}{dt}, \quad \frac{\partial^*}{\partial \dot{q}_i}, \quad \frac{\partial^*}{\partial q_i}$$

Then it is

$$\frac{d^*}{dt}\frac{\partial^* T}{\partial \dot{q}_i} - \frac{\partial^* T}{\partial q_i} = \frac{d}{dt}\frac{\partial T}{\partial \dot{q}_i} - \frac{\partial T}{\partial q_i} - \sum_{\nu=}^{n}\left(\frac{dm_\nu}{dt} \vec{v}_\nu \frac{\partial \vec{r}_\nu}{\partial q_i}\right) \tag{6}$$

Comparating the eqs. (3) and (6) it is

$$\frac{d^*}{dt}\frac{\partial^* T}{\partial \dot{q}_i} - \frac{\partial^* T}{\partial q_i} - Q_i - R_i = 0 \tag{7}$$

$$(0 = 1,2, \ldots k)$$

For the case when the conservative, dissipative forces and the rubbing forces act, it is

$$\frac{d^*}{dt}\frac{\partial^* T}{\partial \dot{q}_i} - \frac{\partial^* T}{\partial q_i} + \frac{\partial D}{\partial \dot{q}_i} + \frac{\partial V}{\partial q_i} - R_i - R_r = 0 \tag{8}$$

(i = 1,2, ... k)

where D is dissipative function and V is the potential energy. For the rotor with variable mass the equations of motion are then

$$\frac{d^*}{dt}\frac{\partial^* T}{\partial \dot{x}} - \frac{\partial^* T}{\partial x} + \frac{\partial D}{\partial \dot{x}} + \frac{\partial V}{\partial x} = R_x + F_{rx}$$
$$\frac{d^*}{dt}\frac{\partial^* T}{\partial \dot{y}} - \frac{\partial^* T}{\partial y} + \frac{\partial D}{\partial \dot{y}} + \frac{\partial V}{\partial y} = R_y + F_{ry} \qquad (9)$$

where x, y are the displacement coordinates and R_x, R_y are the projections of the reactive force, F_{rx}, F_{ry} are the projections of rubbing force.

The projections of reactive force are after |5|:

$$R_x = \frac{dm(t)}{dt}\cdot\frac{dR(t)}{dt} \ ; \ R_y = 0 \qquad (10)$$

and the projections of rubbing force |1| are:

$$F_{rx} = K\left(1 - \frac{\Delta}{\sqrt{x^2+y^2}}\right)(x-\mu^* y)$$
$$F_{ry} = K\left(1 - \frac{\Delta}{\sqrt{x^2+y^2}}\right)(y+\mu^* x) \qquad (11)$$

where:

Δ - radial clearance
K - coefficient of rigidity
μ^*- coefficient of rubbing

The kinetic energy, dissipation function and potential energy are according to the results of |6| respectively

$$T = \frac{1}{2}m(t)(1+\varepsilon d)^2|(\dot{x}-\Omega y)^2 + (\dot{y}+\Omega x)^2| + \frac{1}{2}I(t)\varepsilon^2(y+\Omega x)^2 + \frac{1}{2}I(t)\varepsilon^2(\dot{x}-\Omega y)^2 + \frac{1}{2}J(t)(\Omega+\varepsilon^2 xy) \qquad (12)$$

$$D = \frac{1}{2}c\ |(\dot{x}-\Omega y)^2 + (\dot{y}+\Omega x)^2| \qquad (13)$$

$$V = \frac{1}{2}k\ (x^2+y^2) + \frac{1}{4}k_1\ (x^2+y^2)^2 \qquad (14)$$

where ε is slope constant that measures how much the disc tilts as it runs out, d is the distance between the face surface of the disc and the centre of gravity, c is linear damping coefficient, k is coefficient of linear stiffness of the shaft, k_1 is the coefficient of non-linear stiffness of the shaft, I(t) is moment of inertia about the center of gravity, J(t) is the polar moment of inertia of the disc.

In order to highlight the important terms, the following non-dimensional quantities are introduced:

$$X = \frac{x}{R_o},\ Y = \frac{y}{R_o},\ T = \omega_o t,\ \omega_o = \sqrt{\frac{k}{m_o}},$$
$$\tau = \mu T,\ \mu = \mu_1\Omega_1\gamma,\ \Omega_1 = \frac{\Omega}{\omega_o},\ \gamma = \frac{\rho}{\rho_1},$$
$$\mu_1 = \frac{h}{R_o},\ d_1 = \frac{d}{R_o},\ \varepsilon_1 = \varepsilon R_o,\ \varepsilon' = \varepsilon_1 d_1,$$
$$C_1 = \frac{c}{m_o\omega_o},\ k_1' = \frac{k_1 R_o^2}{m_o\omega_o^2},\ K' = \frac{K}{m_o\omega_o^2},\ \Delta = \frac{\Delta}{R_o},$$
$$\omega^2(\tau) = \frac{A}{1+\tau},\ A = \frac{1}{(1+\varepsilon')^2+\varepsilon_1^2},\ m(\tau) = \frac{1}{\omega^2(\tau)} \qquad (15)$$

where ρ_1 is the density of disc.

The non-dimensional equations of motion are:

$$\ddot{X}+c_1\omega^2(\tau)\dot{X}-\Omega_1\left|2-\frac{1}{2}\varepsilon_1^2 A\right|\dot{Y}-\Omega_1 c_1\omega^2(\tau)Y+X|\omega^2(\tau)-\Omega_1^2|+k_1'\omega^2(\tau)X(x^2+y^2) = -K'\omega^2(\tau)\left(1-\frac{\Delta'}{\sqrt{x^2+y^2}}\right)(x-\mu^* y)$$
$$\ddot{Y}+c_1\omega^2(\tau)\dot{Y}+\Omega_1\left|2-\frac{1}{2}\varepsilon_1^2 A\right|\dot{X}+\Omega_1 c_1\omega^2(\tau)X+Y|\omega^2(\tau)-\Omega^2|+k_1'\omega^2(\tau)Y(x^2+y^2) = -K'\omega^2(\tau)\left(1-\frac{\Delta'}{\sqrt{x^2+y^2}}\right)(y+\mu^* x) \qquad (16)$$

The small parameter τ is varying in an interval

$$0 \leq \tau \leq \tau_{max} \qquad (17)$$

and the parameter $m(\tau)$

$$1 \leq m(\tau) \leq m_{max} \qquad (18)$$

3 STABILITY ANALYSES

The stability of rotation of the rotor of mass centre with zero deflection can be analysed by applying Lyapunov's second (direct) method. The Lyapunov function is supposed in the form

$$V = \frac{1}{2}m(\tau)(\dot{x}^2+\dot{y}^2) + \frac{1}{4}k_1'A(x^2+y^2)^2 + \frac{1}{2}m(\tau)(x^2+y^2) \qquad (19)$$

The function

$$W = \frac{1}{2}(\dot{x}^2+\dot{y}^2) + \frac{1}{4}k_1'A(x^2+y^2)^2 + \frac{1}{2}(x^2+y^2) \qquad (20)$$

is positive definite.

The function V satisfies the relations

$$V \geq W \text{ for } x,\dot{x},y,\dot{y} \neq 0 \quad T \neq 0$$
$$V = W = 0 \text{ for } x,\dot{x},y,\dot{y} = 0 \quad T \neq 0 \qquad (21)$$

So, the Lyapunov function V is positive definite, too. The first derivative of the function V along the integrating line of differential equations (16) is

$$\frac{dV}{dT} = \left(\frac{1}{2}\mu^*-c_1A\right)(\dot{x}^2+y^2) + (\Omega_1c_1-K'\mu^*)A(x\dot{y}-x\dot{y}) - \frac{K'A\Delta'\mu^*}{\sqrt{x^2+y^2}}(\dot{x}y-x\dot{y}) - (x\dot{x}+y\dot{y})|A(1+K') + m(\tau)(1-\Omega_1^2)| + \frac{K'\Delta'A}{\sqrt{x^2+y^2}}(x\dot{x}+y\dot{y}) + \frac{1}{2}\mu^*(x^2+y^2) \qquad (22)$$

The function $\frac{dV}{dT}$ is positive for

$$\frac{1}{2}\mu^* - c_1A > 0 \qquad (23)$$

$$x^2 + y^2 > R^2 \qquad (24)$$

where

$$R = \frac{K'^2A^2\Delta'^2(1+\mu^{*2})}{\frac{1}{2}\mu^*(\frac{1}{2}\mu^*-c_1A)-(\Omega_1c_1-K'\mu^*)^2-|A(1+K')+m(\tau)(1-\Omega_1^2)|^2}$$

The function V satisfies the Lyapunov's theorem of instability: If there is a positive definite function V whose first derivative along the integrating line of differential equations of motion is a positive function, the rotation is unstable. For the conditions (23) and (24) the rotation is unstable.

4 RESULTS

The region of instability is varying during the period of winding up of the textile band. In Fig. 2 the region of instability in function of mass $m(\tau)$ is plotted.

The region of unstable rotation is a function of the parameters $A, \gamma, c_1, \Omega_1, \mu_1, \Delta'$. In Fig. 3-8 the varying of the radius R in function of mass $m(\tau)$ for some values of these parameters is shown.

5 CONCLUSION

The textile machine rotor can work under rubbing conditions. For the case when the rubbing acts an unstable rotation is possible. At the moment of rubbing the width of the unstable region is a function of the quantity of the mass wound up on the rotor. A great influence on the boundary are the physical and geometrical characteristics of the textile band and the shaft, angular velocity of the rotor, damping characteristics and the clearance between the rotor and the stator.

When the band starts to wind up on the rotor the boundary of instability is high. After a period of time, more and more band is wound up and the limit decreases. At the moment when a large amount of band is on the rotor the boundary of instability region is high, in spite of rubbing.

The influence of the clearance is linear. For larger values of clearance the boundary of instability rises.

For the case when the rotor is rigid and the slope is small, the unstable region is smaller than for more elastic cases.

The position of unstable boundary rises for higher values of damping parameter.

The angular velocity has the tendency of increasing in the modern machinery. For the case of rubbing when the large amount of band is wound up the region of unstable rotation is quite small.

REFERENCES

|1| BEATTY, R.F. Differentiating rotor response due to radial rubbing. J.of Vibration, Acoustics, Stress and Reliability in Design, 1985, 107, 151-161.

|2| CRANDALL, S.H. Nonlinearities in rotor dynamics. XI International conference on nonlinear oscillations, 1987, Janos Bolyai Mathematical Society.

|3| CVETICANIN, L. The stability of rotating rotor due to radial rubbing, Publications of the School of Engineering Sciences, 1986, 17, 71-79.

|4| CVETICANIN, L. Vibrations of a textile machine rotor. J. of Sound and Vibration, 1986, 97, 181-187.

|5| BESSONOV, A.P. Osnovji dinamiki mehanizmov s peremennoj massoj zvenjev, 1967, Nauka

|6| HENDRICKS, S.L. Stability of a clamped-free rotor partially filled with liquid. J. of Applied Mechanics, 1986, 53, 166-172.

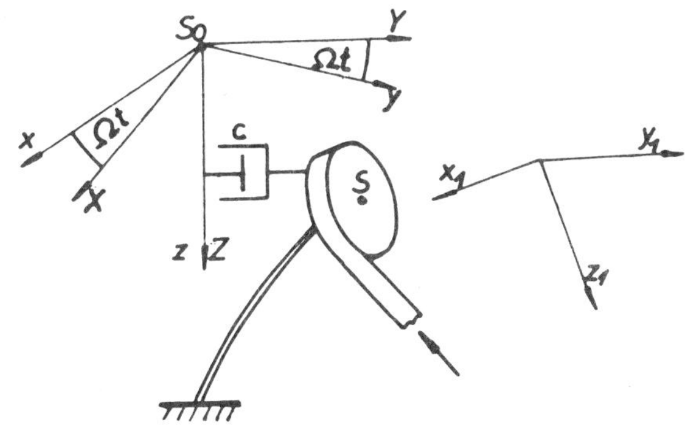

Fig 1 Model of the rotor

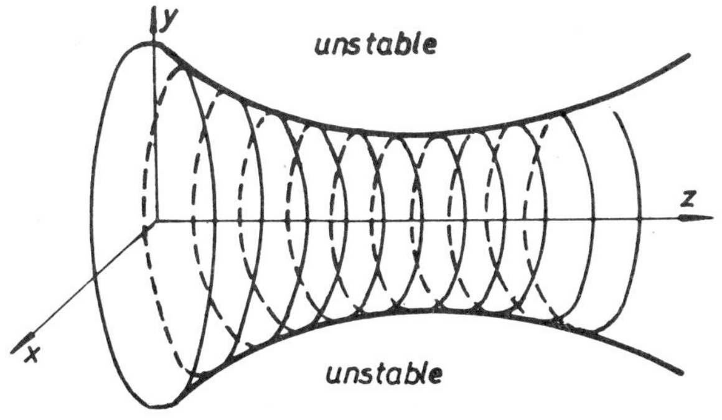

Fig 2 Instability region

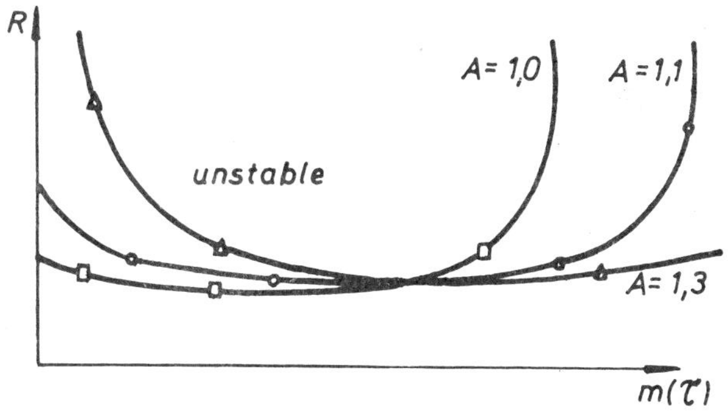

Fig 3 R versus $m(\tau)$ for some values of A

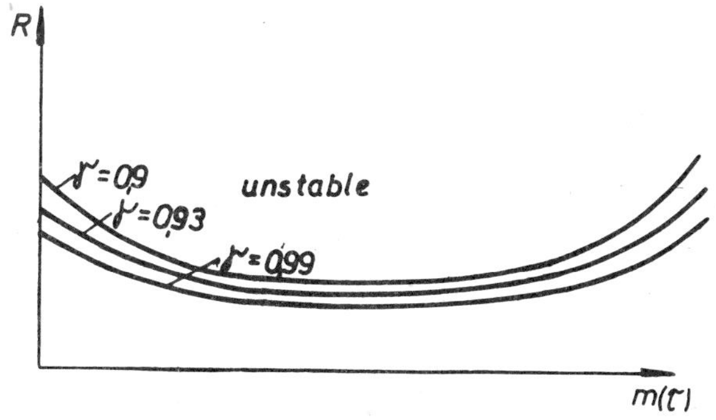

Fig 4 R versus $m(\tau)$ for some values of γ

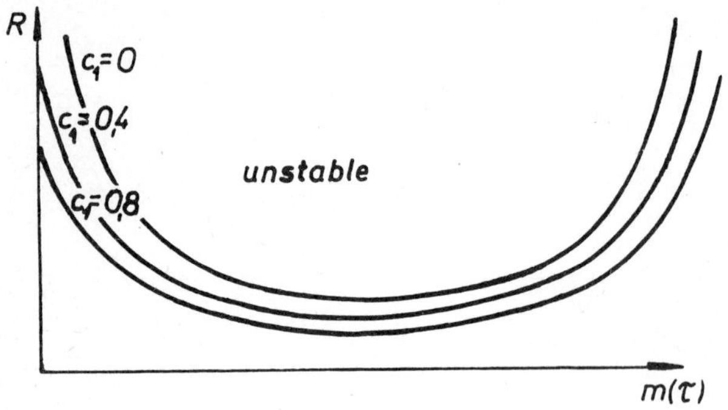

Fig 5 R versus $m(\tau)$ for some values of c_1

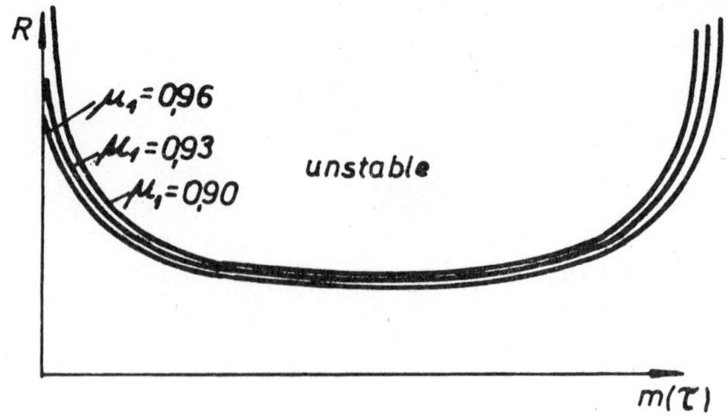

Fig 7 R versus $m(\tau)$ for some values of μ_1

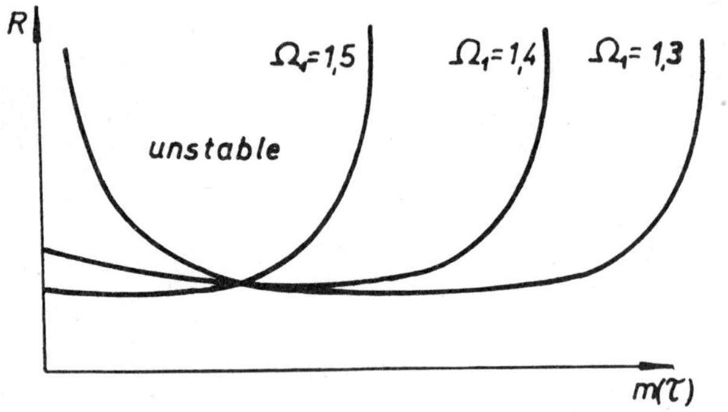

Fig 6 R versus $m(\tau)$ for some values of Ω_1

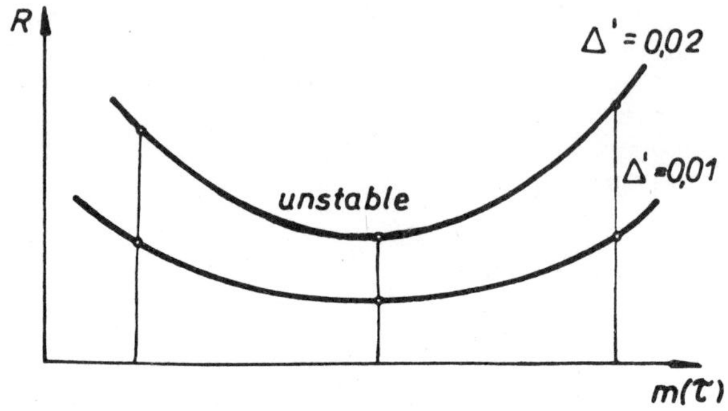

Fig 8 R versus $m(\tau)$ for some values of Δ

C259/88

Dynamic loading and response of reciprocating compressor bearings

A J SMALLEY, BSc, PhD, MASME
Southwest Research Institute, San Antonio, Texas, United States of America
M F WHITE, BSc, PhD, CEng, MIMechE, MIOA
Department of Marine Technology, University of Trondheim, Trondheim, Norway

ABSTRACT

This paper describes a theoretical and experimental investigation of dynamic loading and response in a small two stage reciprocating compressor with intercooler. Bond graph modelling techniques were used to develop a nonlinear model of the compressor. Models were developed for the different subsystems which were then assembled into a final overall bond graph for the thermo-fluid part of the system. Gas forces were predicted from the governing state equations which were set up directly from the overall system bond graph. Inertia forces were modelled to include second and fourth order oscillating components in addition to rotating components and the effects of the counterbalance mass.

A bearing orbital response analysis was used to calculate response of the shaft to various load conditions. The analysis solves the finite length Reynold's equation, in the time domain. It uses finite difference methods, and accounts for squeeze film effects and rupture of the fluid film. The rupture model employs the general equivalent of the π-film approach, in which all pressures below the rupture pressure are set equal to the rupture pressure in the caluculation of fluid film forces; the boundaries of the ruptured region are updated for each time step. Details of the lemon bore bearing geometry with circumferential groove in the upper half of the bearing are included in the model.

Predicted results are compared with the experimental data.

NOTATION

A_2, A_4	constants
B	damping matrix
cb, cv	control boundary, control volume
F	force
h	oil film thickness, high pressure cylinder
i, j	integer counters
k	counterbalance
L	length of con.rod, bearing width, low pressure cylinder
m	mass
o	oscillating component
P	pressure
q	flow vector
r	radius, rotating component
R	crank radius
s	length of control vol. boundary
v	vertical
x	vertical displacement
Y	horizontal displacement
Z	axial direction
β	crank angle for HP cylinder
η	horizontal coordinate
θ	crank angle
μ	dynamic viscosity
ξ	vertical coordinate
ϕ	crank angle for LP cylinder
ω	angular velocity

1. INTRODUCTION

Reciprocating compressor bearings are exposed to a complex dynamic load history resulting from pressure loads on the compressor pistons and from reciprocating inertia loads. These loads are transmitted via crosshead, connecting rod, and crank pin to the crankshaft where they must in turn be supported by the main bearings. On integral engine/compressors the main bearings must carry additional loads from the power piston pressures and inertia forces.

In general, the bearings of well-established compressor models have been successfully designed and built to carry expected loads reliably. However, failures do occur both on widely deployed compressor models and on advanced high performance compressors for such service as hypercompression for polyethylene manufacture and gas reinjection.

Contributors to these failures include dynamic over load, loss of lubricant, static misalignment of the compressor frame and foundation block, misalignment of the crankshaft within the bearing, and dynamic deflection of the compressor frame itself under the loads it must carry at each bearing, as shown by Smalley (1). Several of these factors can be contributors to bearing failure.

In evaluating the causes of bearing failures, and in designing future compressor bearings to carry design loads, it is desirable to have a predictive tool which can account for the lubrication physics, which can include the effects of complex load histories, and which can represent the detailed geometry of the bearing and its grooves in predicting the orbital motion of the journal within the clearance space.

The prediction of bearing response under reciprocating machinery dynamic loads has been addressed in a number of papers in the open literature. Worthy of note are: Horsnell and McCallion (2) who used finite difference, time domain modelling and a sophisticated film rupture model; Lloyd, Horsnell, and McCallion (3) who simplified the film rupture problem; the mobility method of Booker (4), and the work of Booker and Stickler (5) which considers the effects of crankshaft elastic deflection forces in determining bearing dynamic orbits.

One purpose of the present paper is to evaluate the effectiveness of the selected numerical technique in predicting dynamic journal motion as measured in a test rig described by White (6).

The predictive method selected accounts for a general grooving pattern in the bearing fluid film. It calculates journal velocities under instantaneous applied loads, and it advances the journal motion by small displacements. The small journal displacements are obtained by multiplying calculated velocities by a small time step. Fluid film rupture is handled by setting to zero all pressures which fall below zero in the total solution accounting for both journal rotation and squeeze film motion. For the fluid film analysis, the journal is treated as rigid and aligned, so that a single pair of orthogonal displacements describes the journal motion.

Dynamic loads on the journal are generated using a bond graph method for setting up the differential equations for the thermo-fluid model of the compressor as shown by White (6). Simulated pressures in both stages of the compressor are obtained as a function of crank angle by solving the model equations numerically. Gas and inertia forces in each stage are then decomposed into the vertical and horizontal forces acting on the main bearings.

Measured journal displacements were obtained in a 39 Horse Power two stage reciprocating compressor with a maximum delivery pressure of 35 bar. Two non-contacting displacement probes set in the center main bearing, each at 45° to the vertical, were used to measure journal displacements.

The paper describes the fluid film analysis, and the bond graph analysis for generating bearing loads, and presents the comparison of predicted and measured journal displacements.

2. BEARING LOADS

The thermo-fluid process in the compressor was modelled using the bond graph method. This enabled the state equations for the complete system to be set up by modelling the following four sub-systems independently: -

- slider-crank mechanism
- cylinder head
- valves
- heat exchanger

The bond graph shows all the mass and energy flow directions, causalities and the number of describing differential equations for the compressor model. A complete bond graph model of the total system is described by Engja [7].

Mechanical power input to the compressor is represented by a flow source which is transformed to the piston through the slider crank mechanism. This is represented by a modulated transformer, and the machine is assumed to be driven at constant speed, ω. The modulated transformer describes the relation between piston velocity, x_p, and shaft speed, ω: -

$$\dot{x}_p = \{R \sin \phi + \frac{\frac{L}{2}(\frac{R}{L})^2 \sin 2\phi}{(1-(\frac{R}{L})^2 \sin^2\phi)^{\frac{1}{2}}}\}\omega \quad (1)$$

ϕ, R and L are respectively crank angle, crank radius and connecting rod length. Flywheel and piston inertia are represented by I_{FW} and I_p.

Pressure and mass flow rate in the cylinder head are paired and indicated by single bonds which represent mass conservation. Energy conservation is represented by single thermal bonds which indicate effort and energy flow. Heat transfer between the gas and the cylinder walls is estimated at any instant from the instantaneous pressure, temperature and surface area. The total internal energy in the cylinder is then derived from instantaneous mass flow rate, temperature and specific heat of the gas.

Suction and discharge valves are treated as orifaces having isentropic flow. The mass flow rate through a valve is a function of valve area and can be computed from the well known relationship between upstream and downstream pressures and temperatures. The model also incorporates the case of choked flow.

The flow and thermal models for the interstage heat exchanger are used to calculate the energy flow rate assuming steady state turbulent heat transfer.

For the first stage compressor with intercooler the model is described by a set of 9 non-linear state equations. A similar analysis was carried out for the second stage of the compressor. Simulated pressures in the first and second stage cylinders as a function of crank angle were obtained by solving the model equations numerically.

Inertia forces were modelled using a two mass reduction for reciprocating components. This gave the following equations for force equilibrium in the two cylinders:

$$Y_L = \omega^2 m_{r1} \cdot r_v \cdot \sin\phi - \omega^2 r_k \; m_k \sin\phi \qquad (2)$$

$$X_L = \omega^2 r_v \; [(m_{r1}+m_{o1})\cos\phi + m_{o1} \; (A_2\cos 2\phi - A_4\cos 4\phi)]$$

$$-\omega^2 r_k m_k \cos\phi \qquad (3)$$

$$Y_h = m_{rh} r_v \sin\beta \qquad (4)$$

$$X_h = \omega^2 r_v \; [(m_{rh}+m_{oh})\cos\beta + m_{oh} \; (A_2\cos 2\beta - A_4\cos 4\beta)] \qquad (5)$$

m_r, m_o are rotating and oscillating masses and m_k is the counterbalance mass. r_v, r_k are crank and counterbalanse radii: X_L, Y_L, X_h, Y_h are vertical and horisontal components of inertia forces from 1st and 2nd stage cylinders.

Total bearing force history was found by summating gas and inertia forces vectorially for increments of shaft rotation.

3. FLUID FILM ANALYSIS

The Reynold's equation is formulated as follows:

$$\oint_{cb} \vec{q} \cdot \vec{ds} = \iint_{cv} \frac{dh}{dt} \, dx \, dz \qquad (6)$$

where the left-hand side represents the net flow into a control volume and the right-hand side represents the rate of increase in volume of that control volume.

$\vec{q}$ is the vector volume flow per unit width (flux)

$\vec{ds}$ has magnitude of an elemental length of control volume boundary and direction of the inwards normal

$\frac{dh}{dt}$ is the rate of change of film thickness and dx, dz are circumferential and axial elemental lengths.

The vector flow is defined as follows:

$$\vec{q} = \hat{i} q_x + \hat{j} q_z \qquad (7)$$

$$q_x = \frac{\omega R h}{2} - \frac{h^3}{12\mu} \frac{\Delta P}{\Delta X} \qquad (8)$$

$$q_z = \frac{-h^3}{12\mu} \frac{\Delta P}{\Delta z} \qquad (9)$$

$$\frac{dh}{dt} = \dot{\xi} \cos\theta + \dot{\eta} \sin\theta \qquad (10)$$

where $\dot{\xi}$ $\dot{\eta}$ are translational velocities with directions ξ, η as shown in Fig. 1a).

A finite difference formulation is obtained by making the control volume a rectangular shape located about a mesh point of interest. This is shown in Fig. 1b) for point i,j.

Values for q_x, q_z are written, in terms of the pressure at adjacent mesh points, to apply at the mid-sides of the control rectangle. For example, on Side 1:

$$(q_z)_1 = \frac{-\tilde{h}^3}{12\mu} \; (P_{i,j} - P_{i,j-1})/\Delta z_{j-1} \qquad (11)$$

$\vec{ds}$ takes the following values on Sides 1,2,3,4 of the rectangle in turn:

$$\text{Side 1, } \vec{ds} = \hat{j} \; (\Delta x_{i-1} + \Delta x_i)/2 \qquad (12)$$

$$\text{Side 2, } \vec{ds} = -\hat{i} \; (\Delta z_{j-1} + \Delta z_{j/2} \qquad (13)$$

$$\text{Side 3, } \vec{ds} = -\hat{j} \; (\Delta x_{i-1} + \Delta x_i)/2 \qquad (14)$$

$$\text{Side 4, } \vec{ds} = \hat{i} \; (\Delta z_{j-1} + \Delta z j)/2 \qquad (15)$$

The resultant series of algebraic equations in pressure is solved using Gauss-Seidel iteration with successive over-relaxtion.

The fluid film forces are found by numerically integrating the pressure distribution over the area of the finite difference mesh:

$$F_\xi = 2R\int_0^{L/2} \int_0^{2\pi} P \cos\theta \, d\theta \, dz \qquad (16)$$

$$F_\eta = 2R\int_0^{L/2} \int_0^{2\pi} P \sin\theta \, d\theta \, dz \qquad (17)$$

The equilibrium equation for the journal is:

$$-[B_s] \begin{Bmatrix} \dot{\xi} \\ \dot{\eta} \end{Bmatrix} = \begin{Bmatrix} R_\xi \\ R_\eta \end{Bmatrix} \qquad (18)$$

Equation 18 is solved numerically as follows:

$$\begin{Bmatrix} \xi \\ \eta \end{Bmatrix}_{t+\Delta t} = -\Delta_t [B_s]^{-1}_{t+\Delta t/2} \begin{Bmatrix} R_\xi \\ R_\eta \end{Bmatrix}_{t+\Delta t/2} + \begin{Bmatrix} \xi \\ \eta \end{Bmatrix}_t \qquad (19)$$

where $[B_s]_{t+\Delta t/2}$ is calculated using displacements extrapolated one half time step forward:

$$\begin{Bmatrix} \xi \\ \eta \end{Bmatrix}_{t+\Delta t/2} = \frac{3}{2} \begin{Bmatrix} \xi \\ \eta \end{Bmatrix}_t - \frac{1}{2} \begin{Bmatrix} \xi \\ \eta \end{Bmatrix}_{t-\Delta t} \qquad (20)$$

This procedure provides second order accuracy, with numerical errors of order Δt^2.

3.1 Boundary Conditions

The fluid film is subjected to fixed pressure boundary conditions at each end and at fluid supply regions within the film. Pressures accounting for journal rotation and translation which are predicted to fall below a specified film rupture pressure are reset to that rupture pressure. The velocities $\dot{\xi}$ and $\dot{\eta}$ are iteratively adjusted until the pressure distribution satisfies this film rupture condition for the equilibrium velocity vector. For the present analysis, pressures in grooves at the ends and in the film rupture zone have been treated as zero.

The damping matrix B_S is based on forces given by integrating pressures for unit $\dot{\xi}$ and $\dot{\eta}$ velocities over the non-ruptured film area.

3.2 Implied assumption

Clearly, the analysis is subject to several simplifying assumptions:

(1) Conventional lubrication theory applies.

(2) The half-Sommerfeld type film rupture boundary condition applies.

(3) The film is always adequately supplied with fluid; there is no tendency for starvation to occur.

(4) Journal inertia forces are negligible.

3.3 Verification

As partial verification, a comparison was made with the predictions of the Mobility analysis of J.F. Booker (4). Results are shown in Fig. 2 for a steadily loaded plain bearing. Characteristics of the two solutions are very similar and small quantitative differences are attributable to the fact that the particular mobility solution used was based on short bearing theory while the present analysis is finite length solution (of course finite length mobility analysis is also possible).

4. RESULTS AND DISCUSSION

The reciprocating compressor main bearing depicted in Fig. 3 has been modelled.

A finite difference mesh with variable mesh length is superimposed upon the bearing's surface, so that mesh lines match up with the grooving pattern. There are a total of 48 mesh intervals around the bearing and 8 mesh intervals across half the length of the bearing. A symmetry boundary condition is imposed at the bearing centerline. The bearing clearance varies circumferentially in the model to reproduce the elliptical or lemon bore configuration of the bearing. Table 1 presents the bearing model parameters.

The bearing is subjected to a load history as described above. This load history is repeated each revolution.

TABLE 1

RECIPROCATING COMPRESSOR BEARING MODEL

Length	50 mm
Diameter	63 mm
Nominal Diametral Clearance	0.10 mm
Diametral Clearance in Vertical Plane	0.05 mm
Drain/Supply Groove Orientation	Horizontal
Speed	996 rpm
Viscosity	20 cs
Specific Gravity	0.9
Number of Circumferential Intervals	48
Number of Axial Intervals (half bearing)	8

Table 2 presents load histories for two sets of compressor operating conditions, one at high delivery pressure (30 bar) and the other at low delivery pressure (7 bar).

TABLE 2

APPLIED LOAD HISTORY

Compress. cond.	Del. pressure = 7 bar		Del. pressure = 30 bar	
Cranksh. Rotation Degree	Vertic. Downw. Load ($F\xi$) Newtons	Horiz. Load ($F\eta$) Newtons	Vertic. Downw. Load ($F\xi$) Newtons	Horiz. Load ($F\eta$) Newtons
0,0	6969	0	7088	0
15,0	2276	298	2663	294
30,0	1564	681	1672	707
45,0	1920	957	2024	1026
60,0	2133	1234	2415	1314
75,0	2204	1298	2864	1558
90,0	1991	1468	3397	1773
105,0	1564	1404	4312	1981
120,0	1067	1255	6359	2252
135,0	498	1000	8055	2174
150,0	- 71	702	7706	1552
165,0	-427	383	7323	800
180,0	-569	0	7081	0
195,0	-782	- 340	3070	- 561
210,0	-284	- 660	337	- 736
225,0	356	- 894	427	- 920
240,0	1138	-1042	1288	-1075
255,0	2133	-1085	2275	-1100
270,0	3342	- 936	3407	- 970
285,0	4907	- 638	4948	- 649
300,0	6756	- 191	7075	- 176
315,0	7111	- 85	7437	- 71
330,0	6898	- 85	7316	- 66
345,0	6827	- 42	7175	- 44
360,0	6969	0	7088	0

For both compressor conditions the predominant load variation clearly occurs in the vertical plane. For the high pressure case (30 bar) the vertical bearing load, and corresponding polar load history, has two lobes corresponding to gas and inertia forces as shown in Fig. 4. These peak loads are approximately 180° out of phase with each other. In addition, the load contributions from each of the two cylinders are also 180° out of phase. These two effects are additive and therefore amplify the second order component. This effect predominates the vertical load characteristic more and more as compressor delivery pressure is increased from 7 bars and up towards the maximum pressure.

The predicted orbits under the loads are presented in Fig. 5.

As can be seen from the figure, the orbit response for the low load case (7 bar) is predominantly elliptical. Measured and predicted results show reasonable agreement. For the high load case (30 bar) the orbit responds to the two-loop load history with a lobed orbit which is flattened in the horizontal plane. The T-shape seen in the measured orbit was not predicted. This is most probably due to the limitation of the model which is for the one-bearing case. In the actual compressor the load from each cylinder will be shared between the three bearings in a statically indeterminate way.

Fig. 6 and 7 show displacement history vs crank angle for the 7 bar and 30 bar load cases respectively. Agreement between prediction and measurement is reasonable. Measured data shows a double peak per crank revolution. This is most marked in the high load case. Predicted results just show a flattering of the response curve in this region.

Fig. 8 presents the predicted variation of minimum film thickness. The results confirm the compressor manufacturers opinion that his bearings are overdimensioned.

5. CONCLUSIONS

The bearing load model predicted a two lobed polar load distribution with the largest amplitude in the vertical downwards direction as would be expected from the two cylinder reciprocating compressor.

Response analysis predicted an elliptical orbit for the case of low compressor delivery pressure. The shape related well to the elliptical bearing geometry. Both predicted response amplitudes and orbit shape showed good agreement with measured results for this low load condition.

At high delivery pressure the predicted orbit was flattened in the load direction into a banana shape. This corresponded well with the predicted load cycle. The measured orbit had a much more complex three lobed shape. Predicted and measured amplitudes of displacement history showed reasonable agreement, but measured results gave two peaks per crank revolution.

The selected numerical techniques predicted dynamic journal motion having amplitudes and general character which compared well with measured data. Discrepancies between prediction and measurement for the high delivery pressure case were probably due to load sharing between the three bearings, which gives a statically indeterminate result, for example, due to misallignment.

As a tool for predicting dynamic response of reciprocating compressor bearings the theoretical models presented in this paper give a good first approximation of dynamic journal motion. A better estimate may be possible if the model is extended to cover the three bearing case instead of one single bearing.

6. REFERENCES

(1) A.J. Smalley, Dynamic Forces Transmitted by a Compressor to its Foundation, to be presented at the Energy-Sources Technology Conference & Exhibition, January 10-13. 1988, New Orleans, L.A.

(2) R. Horsnell and H. McCallion, Prediction of Some Journal Bearing Characteristics Under Static and Dynamic Loading, Proc.Inst.Mech. Eng., 1963.

(3) T. Lloyd, R. Horsnell, and H. McCallion, An Investigation into the Performance of Dynamically Loaded Bearings: Design Study, Symp. on Journal Bearings for Reciprocating and Turbo Machinery, 1966, Proc.I.Mech.E., Part 3B, p. 181, 1966-67.

(4) J.F. Booker, Dynamically-Loaded Journal Bearings: Numerical Application of the Mobility Method, ASME Journal of Lubrication Technology, pp. 168-176, January 1971.

(5) J.F. Booker and A.C. Stickler, Bearing Load Displacement Determination for Multi-Cylinder Reciprocating Machinery, ASME 2nd International Computer Engineering Conference, San Diego, California, August 15-19, 1982.

(6) M.F. White, H. Engja, and M. Lærum, Rotor Dynamics of Reciprocating Compressors, Institution of Mechanical Engineers, Conference on Vibrations in Rotating Machinery, pp. 457-464, Sept. 1984, York, England.

(7) H. Engja, Bond Graph Model of a Reciprocating Compressor, The Journal of the Franklin Institute, 1984.

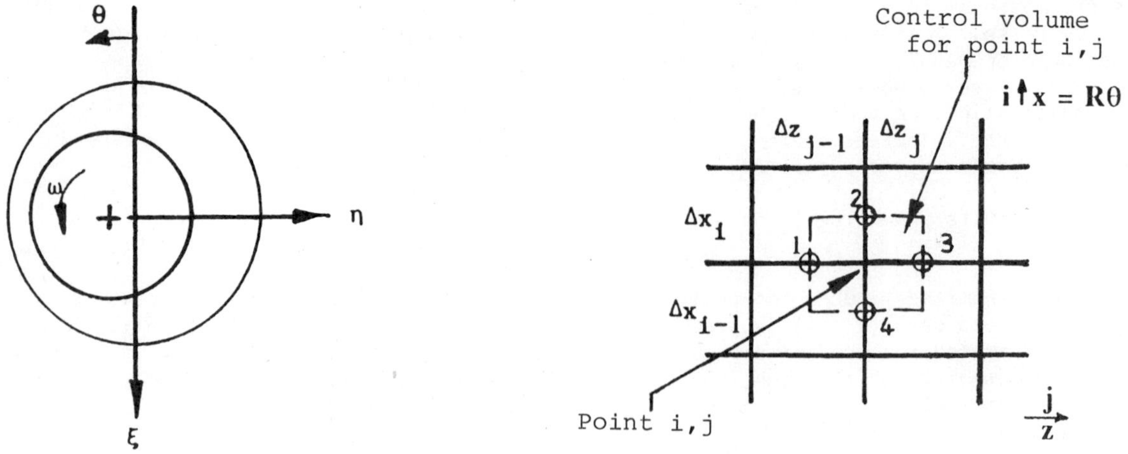

(a) Coordinate directions

(b) Typical finite difference mesh point

Fig 1 Journal coordinate directions and finite difference mesh

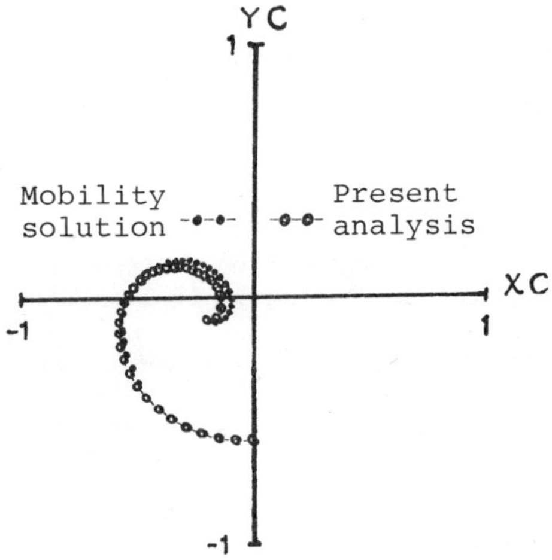

Fig 2 Comparison of predicted orbit for present analysis with predictions of short bearing mobility solution; steadily loaded plain bearing

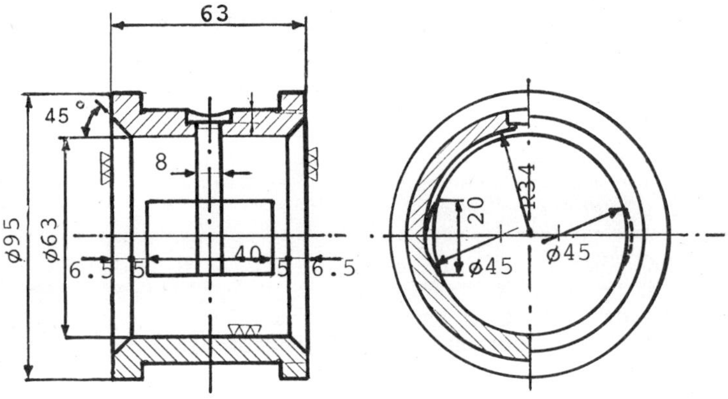

Fig 3 Elliptical main bearing with grooves

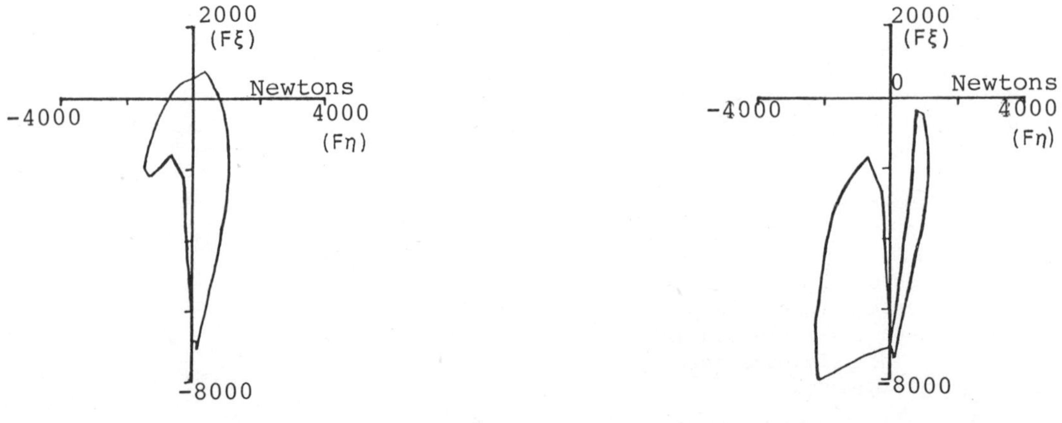

(a) 7 bar delivery pressure

(b) 30 bar delivery pressure

Fig 4 Polar load diagram (newtons)

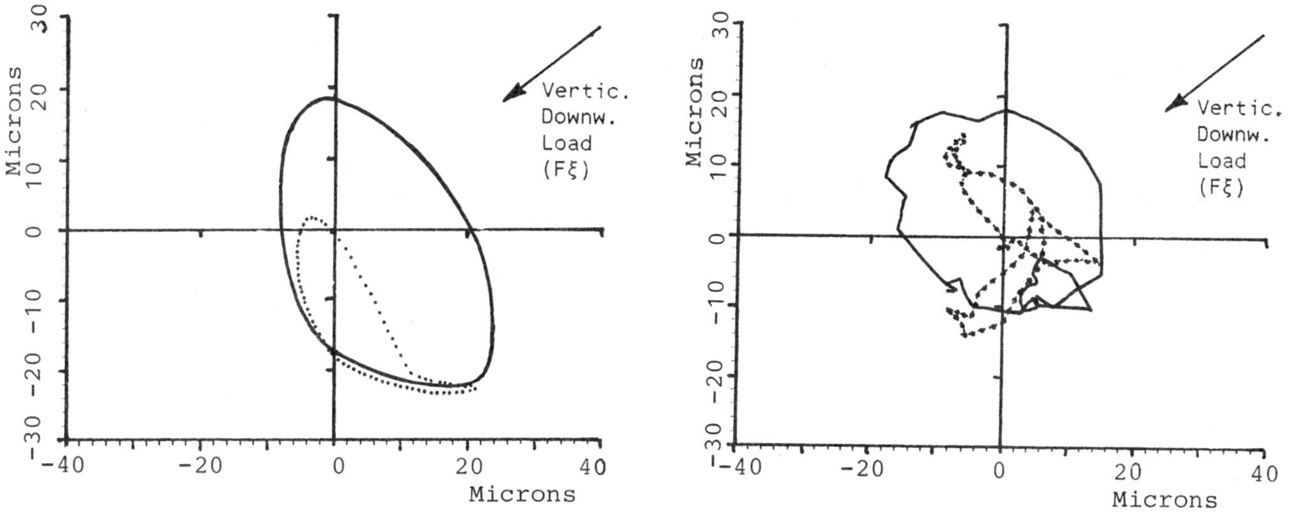

(a) Predicted response

(b) Measured response

Fig 5 Predicted and measured orbit response
(——— 7 bar - - - - - - - - 30 bar)

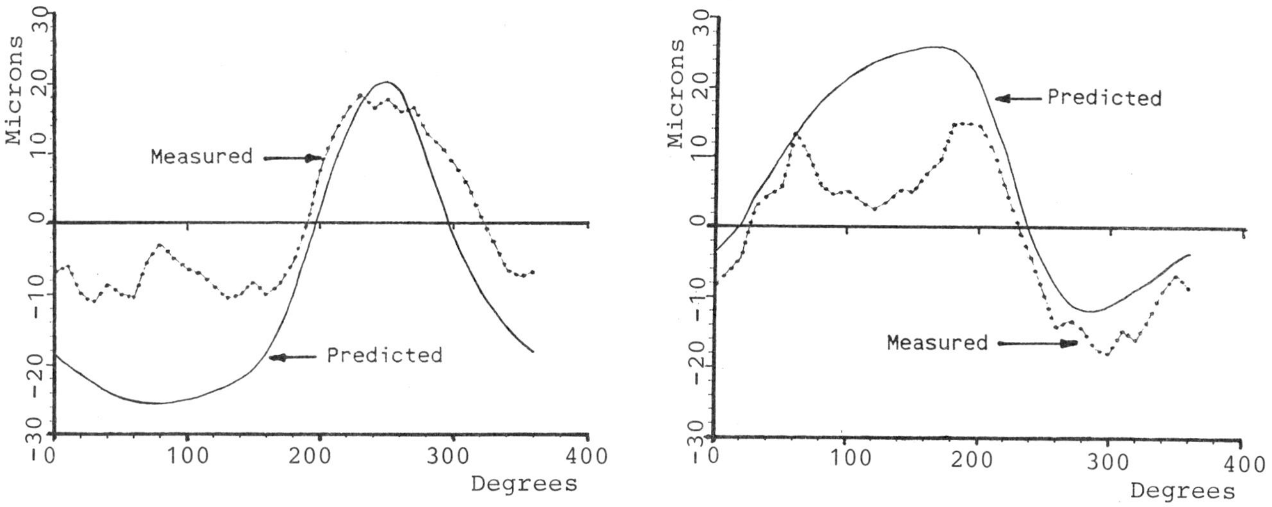

(a) Displacement towards probe 1 (microns)

(b) Displacement towards probe 2 (microns)

Fig 6 Displacement history for 7 bar pressure

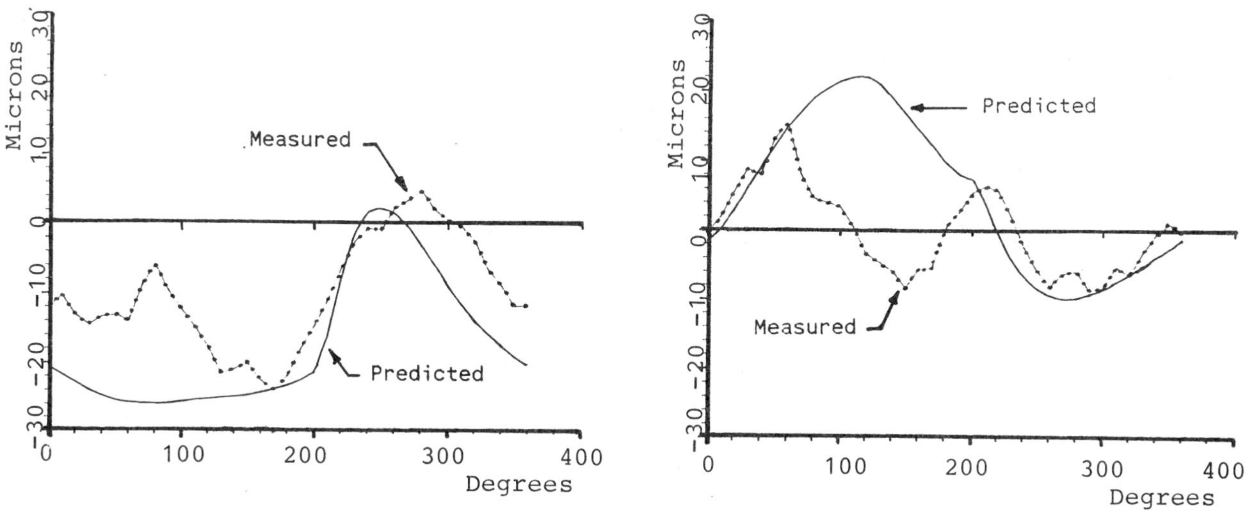

(a) Displacement towards probe 1 (microns)

(b) Displacement towards probe 2 (microns)

Fig 7 Displacement history for 30 bar pressure

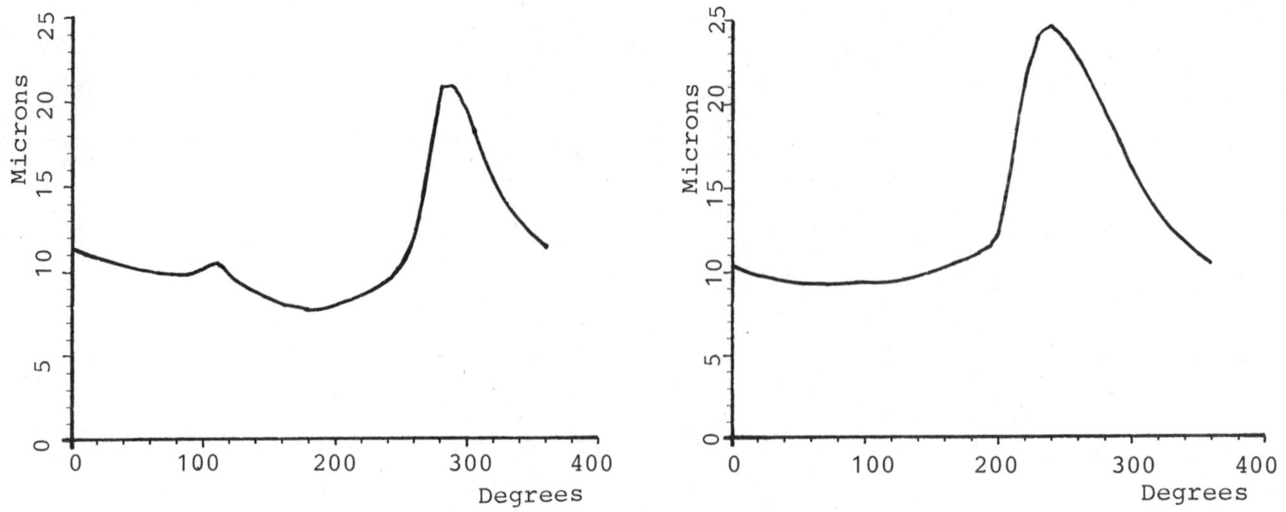

(a) For 7 bar pressure

(b) For 30 bar pressure

Fig 8 Predicted minimum film thickness (microns) versus crank angle

C242/88

An analysis of the steady state and dynamic characteristics of a cylindrical-spherical floating ring bearing

P S LEUNG, BSc, MSc and **I A CRAIGHEAD**, BSc, PhD, CEng, MIMechE
Department of Mechanical Engineering, Newcastle upon Tyne Polytechnic
T S WILKINSON, BSc
NEI Parsons Limited, Newcastle upon Tyne, Northumberland

SYNOPSIS An analysis of the steady state and dynamic characteristics of a novel type of floating ring bearing is performed. The journal and the inner surface of the ring are cylindrical whilst the outer surface of the ring and bearing are spherical providing axial location of the ring and allowing self alignment of the bearing. The effect of different bearing parameters is assessed and the bearing performance is compared with that of conventional bearings. A case study, based on a scaled turbine system, is also described.

NOTATION

C	film clearance
D	diameter
G_θ, G_z	turbulent coefficients
h	film thickness
$\bar{M}_1$	non-dimensional inner film viscous moment $= \dfrac{C_1}{\mu R_1^{\,3} L\ (\omega_1+\omega_2)} M_1$
$\bar{M}_2$	non-dimensional outer film viscous moment $= \dfrac{C_2}{\mu R_2^{\,4}\ \omega_2} M_2$
N	rotational speed in rev/s
P	pressure
R	radius
Re	Reynolds' number
S_1	inner film Sommerfeld number $= \dfrac{\mu\ (N_1+N_2)\ L\ D_1}{W_1} \left(\dfrac{R_1}{C_1}\right)^2$
S_2	outer film Sommerfeld number $= \dfrac{\mu\ N_2\ L\ D_2}{(W_1+W_2)} \left(\dfrac{R_2}{C_2}\right)^2$
S_b	bearing Sommerfeld number $\dfrac{\mu\ N_1\ L\ D_1}{W_1} \left(\dfrac{R_1}{C_1+C_2}\right)^2$
W	loading
Z	co-ordinate in longitudinal direction
α	(ring speed)/(rotor speed)
β	polar co-ordinate in longitudinal direction
γ	(weight of floating ring)/(bearing load)
ε	eccentricity ratio
θ	polar co-ordinate in circumferential direction
τ	viscous shear stress
μ	fluid viscosity
ω	rotational speed in rad/s
	Subscripts
1	inner fluid film or the journal
2	outer fluid film or the ring

1 INTRODUCTION

Modern designs of turbo-machinery usually aim at larger power output, improved efficiency and higher running speed. These design criteria demand an improved bearing performance. In fact, the industry is always seeking for a better bearing design. A new type of floating ring bearing is studied in this paper. The journal and inner surface of the ring are cylindrical whilst the outer surface of the ring and the bearing are spherical providing axial location of the ring and allowing self alignment of the bearing.

The floating ring bearing can be classified as a multi-film bearing. A floating ring is installed between the journal and the bearing surface dividing the bearing clearance into two separate fluid films. An annular ring with cylindrical surfaces on both sides of the ring was used in most early applications. The ring was allowed to move and rotate freely within the space available. This bearing design is known as the plain floating ring bearing. The earliest recorded use of these bearings can be dated back to 1912 where such a bearing was used in Leyland's vehicles [Campbell (1)]. This type of bearing was also used in a Parsons' steam turbine, and in the connecting rod of Bristol aircraft engines in the 1920's [Shaw and Nussdorfer(2)].

Shaw and Nussdorfer(2) presented possibly the first systematic analysis of the floating ring bearing in 1947. The long bearing assumption was used and the bearing was shown to be advantageous in reducing frictional power loss. Later (1955), Kettleborough(3) studied, experimentally, the frictional characteristics of the bearing. And, in the late 50's and early 60's, both Hill(4) and Dworski(5) reported favourable experience of the floating ring bearing applied to small gas turbines. The bearing was found capable of suppressing rotor vibration. In 1967, Orcutt and Ng(6) presented a theoretical and experimental study of the bearing. They were among the first to analyse the dynamic characteristics of the floating ring bearing and limited results were produced. The effect of turbulence was also studied and the long bearing assumption was retained.

A number of publications on the floating ring bearing were published in the 1970's. Short bearing theory was commonly used in these studies. Kahle(7) studied the floating ring bearing using the short bearing assumption. He also analysed the transient behaviour of the bearing. Tanaka and Hori(8) studied the linear stability of the bearing. They pointed out that the dynamic characteristics of the floating ring bearing were dependent on a number of bearing parameters and were very complicated. Nakagawa and Hiroshi(9) studied the unbalance vibration of the bearing whilst Nikolajsen(10) examined the effect of variable viscosity on its stability. In 1979, Ruddy(11) employed the finite difference method to analyse a floating ring bearing. Later (1981), Li and Rohde(12) pointed out that a rotor supported by floating ring bearings was able to whirl with large amplitude and in some circumstances, a limit cycle existed.

A number of studies were also made into various modifications of the plain floating ring bearing. The ring was either restricted from rotation, or in some cases, different holes and grooves were machined on the ring. Each of these individual designs is described in Tondl(13), Lund(14), Howarth(15), Chow(16) and Malik(17).

In the present study, the steady state analysis of the cylindrical-spherical floating ring bearing is analysed based on the finite difference method for bearings with finite length and incompressible lubricant. The dynamic characteristics of the bearing are represented by a set of eight linearised force coefficients obtained by perturbing the steady state solution. The effect of different parameters on the bearing performance is then examined and the bearing is compared with equivalent conventional bearings. The applicability of the new bearing to turbo-machinery is illustrated by a case study.

2. ANALYSIS

The configuration of the cylindrical-spherical floating ring bearing is illustrated in Figure 1. The inner fluid film supports the external load acting on the journal whilst the outer fluid film is required to support both the external load and the weight of the floating ring. Lubricant is supplied from the central circumferential grooves to the bearing surfaces. Hence, each fluid film may be analysed as having two shorter fluid films with the L/D ratio half the original value. The effective behaviour of each film is the sum of the shorter films.

The steady state solution of the floating ring bearing is mainly based on two equilibrium conditions,

(a) balance of forces: the sum of the hydrodynamic force of the inner film plus the weight of the floating ring should be equal to the hydrodynamic force of the outer film.

(b) balance of frictional moments on the ring: under steady state operation, the viscous moment acting on the inside surface of the ring should be equal to the viscous moment acting on the outside surface of the ring.

In order to establish the steady state solution of the bearing, it is necessary to calculate the bearing forces of each bearing film and the viscous moments acting at both sides of the ring. The procedure is described as follows.

2.1 The Reynolds' equation and bearing forces

The Reynolds' equation for the cylindrical inner film can be written as follows:

$$\frac{1}{R_1^2}\frac{\partial}{\partial\theta_1}\left(G_\theta\frac{h_1^3}{\mu}\frac{\partial P_1}{\partial\theta_1}\right)+\frac{\partial}{\partial Z}\left(G_Z\frac{h_1^3}{\mu}\frac{\partial P_1}{\partial Z}\right) = \frac{(\omega_1+\omega_2)}{2}\frac{\partial h_1}{\partial\theta_1}+\frac{\partial h_1}{\partial t} \qquad (1)$$

and for the spherical outer film, the Reynolds' equation is,

$$\frac{1}{R_2^2\cos^2\beta}\frac{\partial}{\partial\theta_2}\left(G_\theta\frac{h_2^3}{\mu}\frac{\partial P_2}{\partial\theta_2}\right)+\frac{1}{R_2^2}\frac{\partial}{\partial\beta}\left(G_Z\frac{h_2^3}{\mu}\frac{\partial P_2}{\partial\beta}\right) = \frac{\omega_2}{2}\frac{\partial h_2}{\partial\theta_2}+\frac{\partial h_2}{\partial t} \qquad (2)$$

where $G\theta$ and GZ are coefficients dependent on the local Reynolds' number [Constantinescu(18)], and the film thickness can be expressed as follows,

$$h_1 = C_1\,(1+\varepsilon_1\cos\theta_1) \qquad (3)$$

$$h_2 = C_2\,(1+\varepsilon_2\cos\theta_2\cos\beta) \qquad (4)$$

Equations (1) and (2) can be normalized by using selected non-dimensional groups [Craighead(19) and Leung et al(20)]. For a given inner and outer eccentricity ratio, the equations can then be solved by the finite difference method. Essentially, the pressure distribution of each fluid film is described by a rectangular mesh and the pressure at each nodal point is determined iteratively. Details of the computational scheme can be found in Craighead(19). The bearing forces may then be obtained by integrating the pressure field by a numerical integration method.

2.2 Frictional moments on the ring

The frictional forces acting on the inside and outside surfaces of the ring are dependent on the viscous shear force of the lubricant in each fluid film which is in turn dependent on the relative velocities of the corresponding solid surfaces enclosing the film. From the Navier-Stokes equation, the viscous shear stresses acting on the ring can be written as,

for the inside surface,

$$\tau_1 = Q\left[\frac{h_1}{2R_1}\frac{\partial P_1}{\partial\theta_1}-\frac{\mu}{h_1}(\omega_1-\omega_2)R_1\right] \qquad (5)$$

and for the outside surface,

$$\tau_2 = Q\left[\frac{h_2}{2R_2\cos\beta}\frac{\partial P_2}{\partial\theta_2}+\frac{\mu}{h_2}\omega_2R_2\cos\beta\right] \qquad (6)$$

Where Q is the ratio of turbulent friction to laminar friction, and is dependent on the mean Reynolds' number of each film. In the present study,

$Q = 0.039\ Re^{0.57}$ for turbulent flow

$Q = 1$ for laminar flow

The frictional moments can then be obtained by integrating the corresponding shear stress so that,

for the inside surface,

$$M_1 = 2\int_0^{L/2}\int_0^{2\pi} \tau_1 R_1^{\,2} d\theta_1 dZ \qquad (7)$$

and for the outside surface,

$$M_2 = 2\int_0^{\beta e}\int_0^{2\pi} \tau_2 R_2^{\,3} COS^2 \beta d\theta_2 d\beta \qquad (8)$$

The frictional moments may be obtained in non-dimensional form, from results of the normalized Reynolds' equation. The viscous friction in the cavitation region of the films is also included in the present study [Ruddy(11) and Leung et al (20)].

2.3 Calculation procedure for the steady state solution

The computational strategy to determine the steady stage solution of the bearing is described in this section. From the balance of forces, the ratio of Sommerfeld number for the inner and outer fluid films can be written as,

$$\frac{S_1}{S_2} - \frac{(1+\alpha)}{\alpha}\left(\frac{R_1}{R_2}\right)^3\left(\frac{C_2}{C_1}\right)^2(1+\gamma) = 0 = G_1 \qquad (9)$$

From the balance of frictional moments, the ratio of the non-dimensional moments acting at inside and outside surfaces of the ring can be written as,

$$\frac{\overline{M}_1}{\overline{M}_2} - \frac{\alpha}{(1+\alpha)}\left(\frac{R_2}{R_1}\right)^4\left(\frac{C_1}{C_2}\right)\left(\frac{R_1}{L}\right) = 0 = G_2 \qquad (10)$$

Equations (9) and (10) determine the relationship between the inner film eccentricity ratio, the outer film eccentricity ratio, and the (ring speed/rotor speed) ratio. For a given value of inner eccentricity ratio, the equations may then be used to search for the outer film eccentricity ratio and the speed ratio. A modified form of the Newton's iteration method [Gerald(21)] was used in this study, as follows,

$$\begin{bmatrix}\varepsilon_2\\ \\ \alpha\end{bmatrix}_{k+1} = \begin{bmatrix}\varepsilon_2\\ \\ \alpha\end{bmatrix}_k - R_f\begin{bmatrix}\dfrac{\partial G_1}{\partial \varepsilon_2} & \dfrac{\partial G_1}{\partial \alpha}\\ \dfrac{\partial G_2}{\partial \varepsilon_2} & \dfrac{\partial G_2}{\partial \alpha}\end{bmatrix}_k^{-1}\begin{bmatrix}G_1\\ \\ G_2\end{bmatrix}_k \qquad (11)$$

where k indicates the state of successive iterations and R_f is the relaxation factor.

The matrix containing the partial derivatives (the Jacobian matrix) is obtained by numerical perturbation of G1 and G2. The steady state solution is then obtained if a specified tolerance is satisfied. When R_f is equal to unity, the usual Newton's iteration method is used. It provides stable iteration for cases with laminar lubricant flow. However, the iteration may sometimes fail to converge if the lubricant flow is assumed turbulent. An under-relaxation factor of value from 0.5 to 0.75 was found satisfactory for all cases with either assumption. The application of the under-relaxation factor for laminar flow condition could reduce the computational time by up to 20% when compared with the case where R_f is equal to unity.

2.4 The dynamic characteristics of the bearing

The dynamic characteristics of the bearing are represented by a set of eight linearised force coefficients so that the bearing forces acting on the journal can be expressed by the following equations,

$$F_x = A_{xx}X + A_{xy}Y + B_{xx}\dot{X} + B_{xy}\dot{Y} \qquad (12)$$

$$F_y = A_{yx}X + A_{yy}Y + B_{yx}\dot{X} + B_{yy}\dot{Y}$$

The displacement coefficients, A's, are obtained by perturbing the displacement of the journal centre from its steady state position. The velocity coefficients, B's, are obtained by perturbing the velocity of the journal centre. These coefficients may then be used to study the stability and response characteristics of the bearing applied to different rotor systems.

3. PARAMETRIC INVESTIGATION OF BEARING PERFORMANCE

To ensure good design and reliable service of the new bearing, it is important to determine the effects of different parameters on the bearing performance. In this study, six bearing parameters were investigated. A bearing with typical parameter values was first selected as the reference case. Each parameter was then varied about the reference conditions and any changes in bearing characteristics have been noted. The results are described as follows,

3.1 The clearance ratio (C2/C1)

The ratio of the outer film clearance to the inner film clearance was found to be significant in determining many bearing characteristics. Design charts are given in this case as examples. In the present study, it is assumed that the total clearance of the inner and outer films is constant. Variation of C2/C1 is achieved by changing the inner film and outer film clearances at the same time.

For a given operating condition, it was found that increasing C2/C1 would cause a reduction in inner film eccentricity ratio, but an increase in outer film eccentricity ratio. As a result, the overall bearing eccentricity ratio, which is defined as the ratio of the journal displacement to the total clearance, is reduced (Fig. 2). This implies that the load carrying capacity of the bearing is increased by an increase of clearance ratio. The bearing attitude angle, however, is reduced (Fig. 3). Increasing the value of C2/C1 will also increase the (ring speed)/(rotor speed) ratio (Fig. 4). Consider that the inner film clearance is reduced as C2/C1 is increased, the shear force acting at the inside surface of the ring will then be increased. The speed ratio is therefore increased. The journal frictional loss (usually indicated by the friction factor) is also increased for similar reason (Fig. 5).

The dynamic coefficients of the bearing are shown in Figs. 6 and 7. These coefficients are used to calculate the limit of stability [Craighead(19)] of the bearing (Fig. 8). The effect of the clearance ratio on the dynamic stability can be divided into two regions : for low Sommerfeld numbers, the stability is reduced by increasing value of the clearance ratio; for high Sommerfeld numbers, increasing the value of the clearance ratio will increase the limit of stability. Large values of clearance ratio are therefore beneficial in high speed or light load operations. The ratio of the (whirl speed)/(rotor speed + ring speed) was found to be about 0.5 as expected [Orcutt and Ng (6)]. However, it is common to calculate the (whirl speed/ rotor speed) for bearings with a single film. For the floating ring bearing, the value of this ratio will be higher than 0.5 and within the range of 0.6 to 0.7.

3.2 The ratio of the ring radii (R2/R1)

The ratio of the ring outside radius to inside radius also affects the performance of the floating ring bearing. For an increasing value of R2/R1, the frictional moment acting at the outside surface of the ring will be increased. This reduces the (ring/rotor) speed ratio and hence reduces the load carrying capacity of the outer film. As a result, the overall load carrying capacity and attitude angle of the bearing is reduced. The frictional power loss, however, is increased because the reduction of ring speed will increase the frictional force acting at the journal surface. In terms of dynamic characteristics, increasing the value of R2/R1 increases the limit of stability of the bearing for the range of Sommerfeld Number examined.

3.3 The (L/D1) ratio

The effects of varying the L/D1 ratio on the floating ring bearing are similar to those found in conventional bearings. Increasing the value of L/D1 ratio will cause an increase in load carrying capacity and attitude angle of the bearing. However, for a given Sommerfeld Number, both the speed ratio and friction factor are not seriously affected by the change in L/D1 ratio. It should be noted that, if the variation of L/D1 ratio is achieved by simply varying the length of the bearing, a smaller L/D1 ratio will result in a smaller Sommerfeld number. In this case, reduction of L/D1 ratio will cause a slight reduction of the speed ratio and a significant reduction in frictional loss. In terms of dynamic behaviour, a reduction of L/D1 ratio will increase the limit of stability of the bearing. Hence, a bearing with a small value of L/D1 ratio is often preferred.

3.4 The mean bearing Reynolds' number

There are two fluid films in a floating ring bearing. The Reynolds' numbers are usually different for each film. In order to ensure comparable results, the mean bearing Reynolds' number of the floating ring bearing is defined herein to indicate the Reynolds' number of an equivalent single film bearing. In the present study, the bearing clearance of the single film bearing is equal to the total clearance of the floating ring bearing. The actual Reynolds' number for the inner and outer films is therefore less than the mean bearing Reynolds' number due to the existence of a rotating ring and a smaller clearance for each film.

The effects of increasing the bearing Reynolds' number on the floating ring bearing are quite similar to those found in single film bearings. However, the floating ring bearing is less sensitive to the change of bearing Reynolds' number, as would be expected. Increasing the value of bearing Reynolds' number will increase the load carrying capacity and the attitude angle of the bearing. The frictional loss is also increased as expected. It should be pointed out that the bearing Reynolds' number has a significant effect on the ring speed ratio. The ring speed is dependent on the viscous friction of the fluid films and the bearing Reynolds' number affects the viscous friction considerably. An increase in Reynolds' number will cause an increase in speed ratio. Also, increasing the value of the bearing Reynolds' number reduces the limit of stability of the bearing.

3.5 The supply pressure

The lubricant supply may sometimes be pressurized, for example, to increase the lubricant flow rate. It was found in this study that pressurizing the lubricant increased the load carrying capacity and attitude angle of the floating ring bearing. This was achieved at the cost of reducing bearing stability. It is generally known that the existence of a cavitation region in a bearing may have a stabilizing effect on the bearing performance. Pressurizing the lubricant, however, will reduce the cavitation region of the bearing.

3.6 The weight of the floating ring

The weight of the floating ring is usually small compared with the loading on the bearing (i.e. less than 5%), and is neglected in most analyses. The effect of the loading due to the ring was examined in this study. It was found that the weight of the floating ring had very little effect on both the steady state and the limit of stability of the floating ring bearing. The bearing performance was practically unchanged even when the weight of the ring was increased up to 10% of the bearing loading. This result confirms that the weight of the ring may be neglected without significant loss of accuracy. It also allows more flexibility in selecting the ring material without affecting the bearing performance.

4. COMPARISON WITH OTHER BEARINGS

The steady state and dynamic performance of the cylindrical-spherical floating ring bearing have been compared with an equivalent plain floating ring bearing, a conventional cylindrical bearing and an elliptical bearing. In order to ensure comparable results, all bearing parameters such as the L/D1 ratio, Reynolds' number, C2/C1 ratio and R2/R1 ratio, where applicable, are kept identical to those of the cylindrical-spherical floating ring bearing. For the single film bearings, the bearing clearance was equal to the total clearance of the inner and outer fluid films of the floating ring bearing. All bearings studied were analysed by the finite difference technique for finite bearings.

It was found that the behaviour of the cylindrical-spherical floating ring bearing is very similar to that found in a plain floating ring bearing. The load carrying capacity of the spherical outer film of the new bearing is slightly lower than that of the cylindrical outer film of the plain bearing [Leung et al(20)]. Despite some minor differences, the characteristics of the two floating ring bearings are so close that, for practical purposes, the differences may be ignored. The new floating ring bearing, hence, has the advantage over the plain floating ring bearing of being self aligning.

The load carrying capacity of the floating ring bearing was found to be higher than the cylindrical bearing but lower than the elliptical bearing. The separation of a single fluid film into two thinner films is thought to be responsible for the increase of the load carrying capacity of the floating ring bearing over the cylindrical bearing. However, the frictional power loss of the floating ring bearing, with laminar lubricant flow, is slightly higher than the loss of the cylindrical bearing but very close to that of the elliptical bearing. For similar reasons as above, the reduction of film thickness will increase the viscous friction in the bearing films and hence increase the frictional loss when compared with the cylindrical bearing. The geometry of the elliptical bearing is responsible for the high load carrying capacity and high frictional loss of the bearing.

The real advantage of the floating ring bearing in terms of energy reduction is its ability to reduce film turbulence when compared to a single film bearing. For a given operating condition (i.e. a given Sommerfeld No.), the degree of turbulence in the inner and outer films of the floating ring bearing is less than that of an equivalent cylindrical or elliptical bearing, especially with high Reynolds' numbers. As an example, when the bearing Reynolds' number is increased from laminar to about 3000, the power loss of the single film bearings is increased by 3 to 4 times the original value. The power loss of the floating ring bearing, however, is only increased by about 50% of the original value. As a result, the frictional loss of the floating ring bearing is only about 30% to 40% of that of the single film bearings. Potentially, a considerable amount of energy can be saved.

Stability analyses of the bearings with a rigid rotor show that the limit of stability of the floating ring bearing examined is higher than that of the cylindrical bearing but lower than that of the particular elliptical bearing. It should be pointed out that the ellipticity of the elliptical bearing studied is relatively high (ellipticity=0.4), which makes the bearing more stable. Although the behaviour of the reference floating ring bearing does reflect the general behaviour of this type of bearing, it is not optimised for high stability. More favourable result could be obtained with suitable selection of bearing parameters.

4.1 An alternative means of comparison

The results reported so far were concerned with a floating ring bearing having a total inner and outer film clearance equal to the clearance of the equivalent single film bearing : the total clearance was kept constant even when the C2/C1 ratio was varied. However, it is also possible to construct a floating ring bearing with the inner film clearance equal to that of the single film bearing. The inner film clearance can be kept unchanged and variation of C2/C1 achieved by varying the size of the outer film clearance. The total film clearance of the floating ring bearing is hence larger than that of the equivalent conventional bearing. Most early studies of the floating ring bearing are, in fact, concerned with this second arrangement. The performance of these two arrangements of the cylindrical-spherical floating ring bearing has also been examined in this study.

The first bearing, in which the total clearance of the bearing was kept constant, was found to have a higher load carrying capacity and higher limit of stability when compared with the second arrangement. However, the frictional loss of the second bearing was lower than that of the first arrangement. Because of the larger total clearance, the lubricant flowrate of the second bearing should be larger than the first bearing. Compared to conventional bearings, the load carrying capacity, frictional loss and limit of stability of the second bearing are all lower than those of the particular cylindrical bearing.

The first arrangement is more favourable when load carrying capacity and dynamic stability are important, which is the case for most turbomachines. The second arrangement is favourable only when frictional losses and lubricant overheating are serious problems.

5. A CASE STUDY WITH A SCALED LP TURBINE ROTOR

The application of the cylindrical-spherical floating ring bearing to turbo-machinery has been examined by a practical example. The floating ring bearing was used to support a scaled LP turbine rotor. The dynamic behaviour of the system was examined and the performance of the floating ring bearing has been compared with other bearings.

The scaled flexible rotor system constitutes to part of a long term research programme and is dynamically similar to a large steam turbine for a power plant [Leung(22)]. The bearing loading of a steam turbine is considered relatively high when compared to a fast speed gas turbine. The rotor was originally supported by a pair of elliptical bearings and the bearing Reynolds' number at operating speed was estimated to be about 3500.

The frictional loss of the cylindrical-spherical floating ring bearing at operating speed was found to be only 27% of the loss for elliptical bearings, and 40% of the loss for cylindrical bearings. Hence, considerable amounts of energy could be saved by using the floating ring bearing.

The stability of the rotor system was examined by eigen value analysis [Craighead(19)]. Rigid foundation was first assumed. The real and imaginary parts of the eigen values indicate the stability and the system (damped) natural frequencies. Results of the study showed that the highest threshold speed was achieved when the rotor was supported by the elliptical bearing. The threshold speed was about 1.75 times the operating speed. The floating ring bearing came second. The threshold speed was found to be 1.5 times the operating speed. The threshold speed of the rotor with cylindrical bearing was below the operating speed. The cylindrical bearing would be unsuitable for this application. The equivalent damping ratio of the lowest system frequency was 0.26 for the elliptical bearing, and 0.22 for the floating ring bearing. Despite a lower value of damping ratio, the damping in the floating ring bearing is sufficient to ensure safe and stable operation.

System response to mass unbalance was also investigated. It was found that least force was transmitted through the cylindrical bearing to ground. The elliptical bearing came second whilst the force transmitted through the floating ring bearing was the highest among the bearings examined. This result indicates that more attention is required for the supporting foundation if the floating ring bearing is to be employed. Damped flexible foundations are usually used in modern power plants. Results indicate that the forces transmitted to the ground were reduced by the introduction of a flexible foundation. The differences among the bearings in response analysis were also reduced to be very small.

The results of this case study suggest that each of the bearings examined offers some advantages over the others. The elliptical bearing is slightly favourable in terms of dynamic characteristics. However, the cylindrical-spherical floating ring bearing offers considerable energy saving and it is self aligning. The dynamic characteristics of the bearing are satisfactory, especially with damped flexible foundations. The overall performance of the cylindrical-spherical floating ring bearing makes it competitive with the elliptical bearing. Despite good response characteristics and simple geometry, the cylindrical bearing is least favourable in this application.

6. CONCLUSIONS

The steady state and dynamic characteristics of the cylindrical-spherical floating ring bearing have been studied. The factors affecting the bearing performance were also examined. Apart from the advantage of being self aligning, the bearing was shown to have very low frictional loss characteristics. A considerable amount of energy could be saved when compared with the other single film bearings, especially with high Reynolds' numbers. In terms of dynamic characteristics, the cylindrical-spherical floating bearing was found more stable than the cylindrical bearing. Compared to an elliptical bearing, the floating ring bearing may be less stable for heavy load applications. However, with the use of flexible foundations, favourable results could be obtained. Finally, the behaviour of the cylindrical-spherical floating ring bearing was found to be very close to the behaviour of a plain floating ring bearing.

The development of the cylindrical-spherical floating ring bearing is still in its early stage. The present study, however, demonstrates that the new bearing is an alternative and competitive design which is worthy of consideration.

ACKNOWLEDGEMENTS

The authors would like to acknowledge the support from the National Advisory Board and NEI Parsons Ltd. for the study described in this paper.

(1) CAMPBELL, C. Floating bush bearings. Chartered Mechanical Engineers, March 1987, pp 14.

(2) SHAW, M.C. & NUSSDORFER, T.J. An analysis of the full-floating journal bearing NACA report No. 866, 1947.

(3) KETTLEBOROUGH, C.F. Frictional experiments on lightly-loaded fully floating journal bearings. Australian Journal of Applied Science, 1955, pp 211-220.

(4) HILL, H.C. Slipper bearings and vibration control in small gas turbines. Transaction of A.S.M.E., 1958, Vol. 80, pp 1756-1764.

(5) DWORSKI, J. High speed rotor suspension formed by fully floating hydrodynamic radical and thrust bearings. Transaction of A.S.M.E. Journal of engineering for power, 1964, Vol. 86.

(6) ORCUTT, F.K. & NG, C.W. Steady-state and dynamic properties of journal bearing in laminar and superlaminar flow regimes. NASA report CR-733, June 1967.

(7) KAHLE, G.W. Analytical and experimental investigation of a full floating journal bearing. PhD thesis, University of Illinois, Jan. 1971.

(8) TANAKA, M. & HORI, Y. Stability characteristics of floating bush bearings. ASLE-ASME Joint Lubrication Conference, Oct. 1971.

(9) NAKAGAWA, E. & HIROSHI, A. Unbalance vibration of a rotor-bearing system supported by floating ring bearing. Bulletin of J.S.M.E., March 1973, Vol. 16, pp 503-512.

(10) NIKOLAJSEN, J.L. The effect of variable viscosity on the stability of plain journal bearings and floating-ring journal bearings. Journal of Lubrication Technology, Oct. 1973. pp 447-1973.

(11) RUDDY, A.V. The dynamics of rotor bearing systems with particular reference to the influence of fluid film journal bearings and the modelling of flexible rotors, PhD thesis, Dept. of Mech. Eng., University of Leeds, 1979.

(12) LI, C.H. & ROHDE, S.M. On the steady state and dynamic performance characteristics of floating ring bearings. Journal of Lubrication Technology, July 1981. pp 389-397.

(13) TONDL, A. Some problems of rotor dynamics. Chapman and Hall, 1965.

(14) LUND, J.W. Rotor-bearing dynamics design technology, part VII : the three lobe bearing and floating ring bearing. Mechanical Technology Inc., Feb. 1968.

(15) HOWARTH, R.B. A theoretical analysis of the floating pad journal bearing. Tribology Convention, 1970.

(16) CHOW, C.Y. Dynamic characteristics and stability of a helical-grooves floating ring bearing. Transaction of A.S.L.E., April 1983. Vol. 27, pp 154-163.

(17) MALIK, M. Externally-pressurized gas-lubricated floating-ring porous journal bearings, part 1 : steady state analysis. I. Mech. E. proceedings 1984. Vol. 1980, No. 16.

(18) CONSTANTINESCU, V.N. Basic relationships in turbulent lubrication and their extensions to include thermal effects. Transaction of A.S.M.E., paper no. 72-Lub-16.

(19) CRAIGHEAD, I.A. A study of the dynamics of rotor-bearing systems and related fluid-film bearing characteristics. PhD thesis, Dept. of Mech. Eng., University of Leeds, 1976.

(20) LEUNG, P.S., CRAIGHEAD, I.A. & WILKINSON, T.S. An analysis of the steady state and dynamic characteristics of a spherical hydrodynamic journal bearing. Paper submitted to the Journal of Tribology, A.S.M.E.

(21) GERALD, C.F. Applied numerical analysis. Addison-Wesley, 1980.

(22) LEUNG, P.S. The design of a flexible rotor for an investigation into the dynamic behaviour of floating ring bearings. Technical report, Newcastle Polytechnic, Aug. 1986.

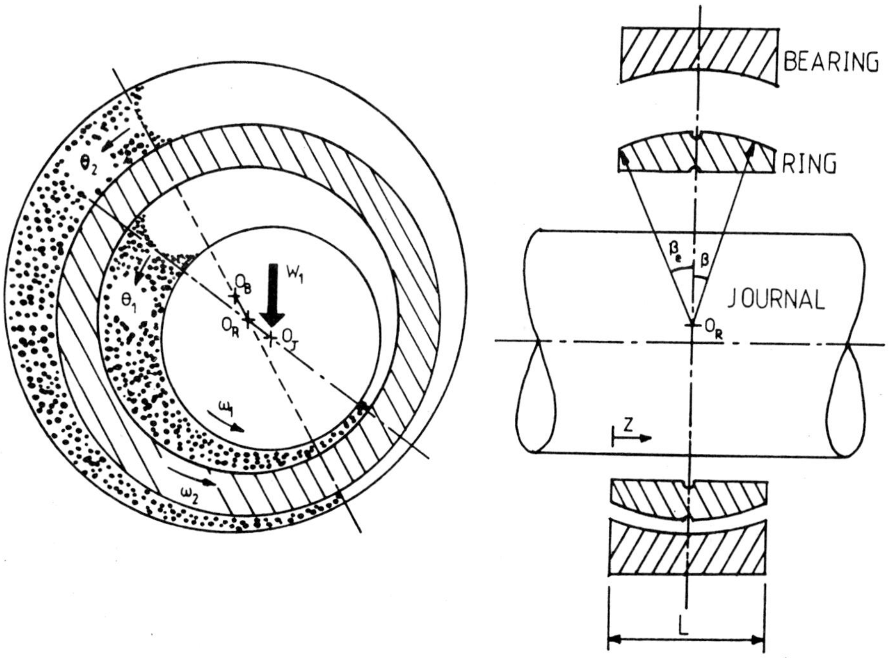

Fig 1 Configuration of the cylindrical-spherical floating ring bearing

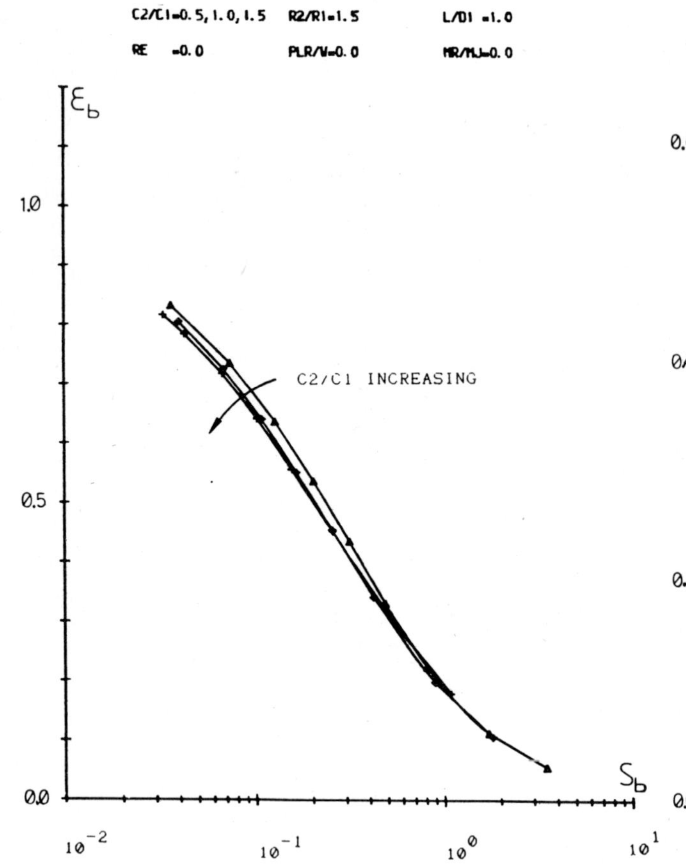

Fig 2 Bearing eccentricity ratio versus Sommerfeld number

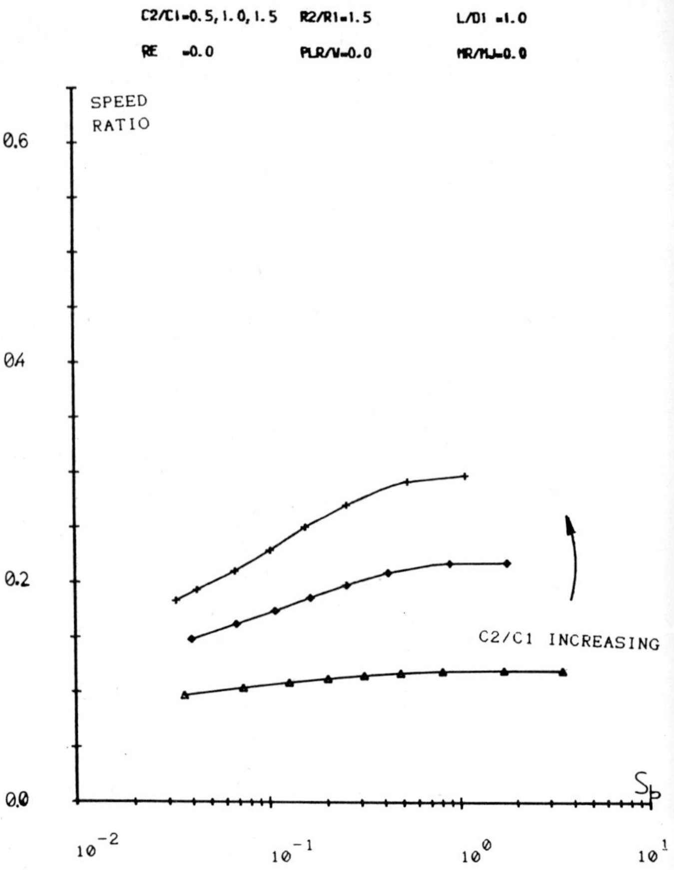

Fig 4 Speed ratio versus Sommerfeld number

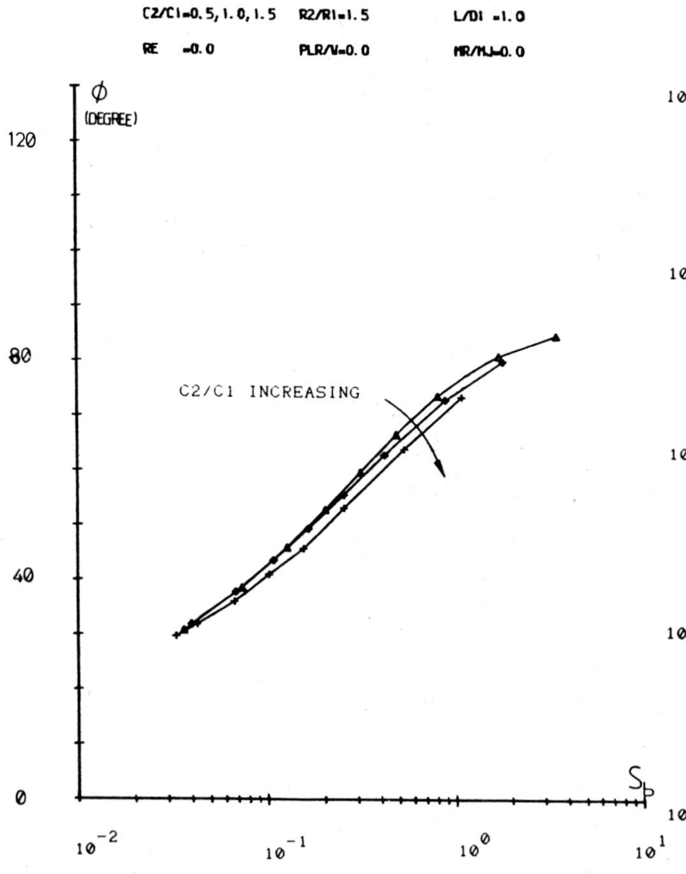

Fig 3 Attitude angle versus Sommerfeld number

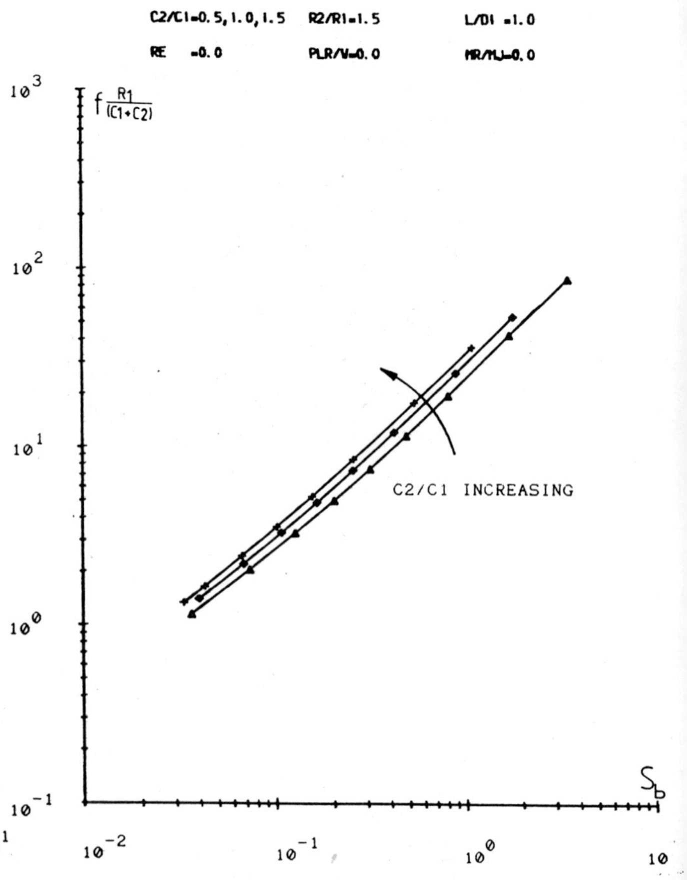

Fig 5 Friction factor versus Sommerfeld number

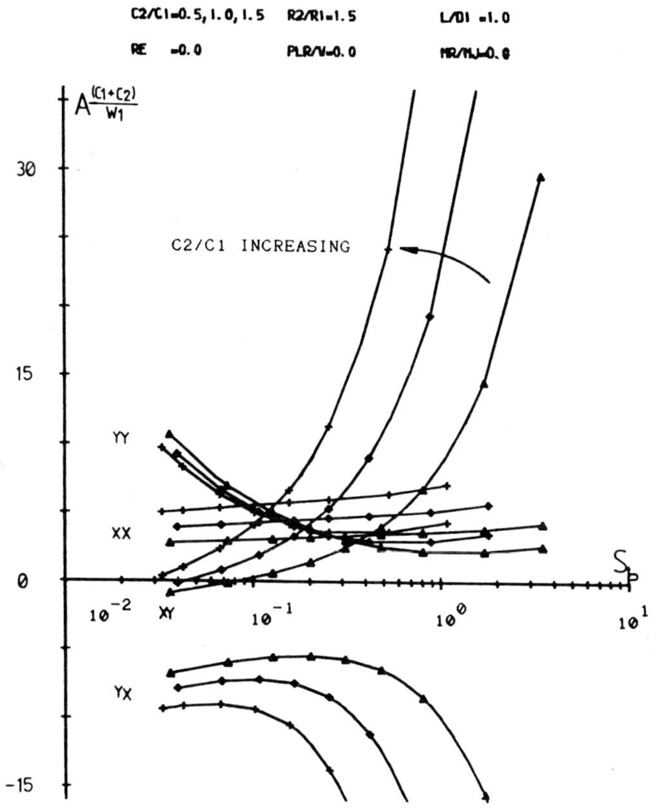

Fig 6 Displacement coefficients versus Sommerfeld number

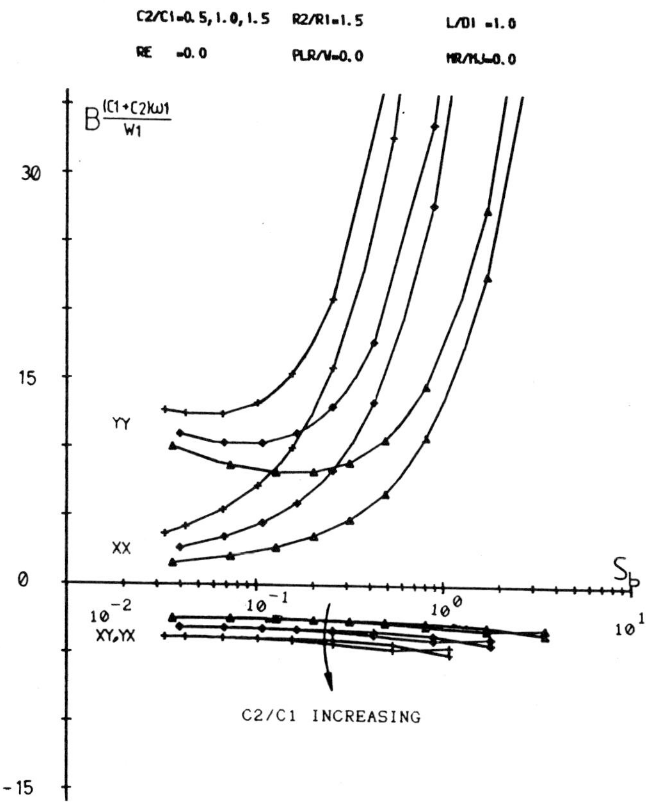

Fig 7 Velocity coefficients versus Sommerfeld number

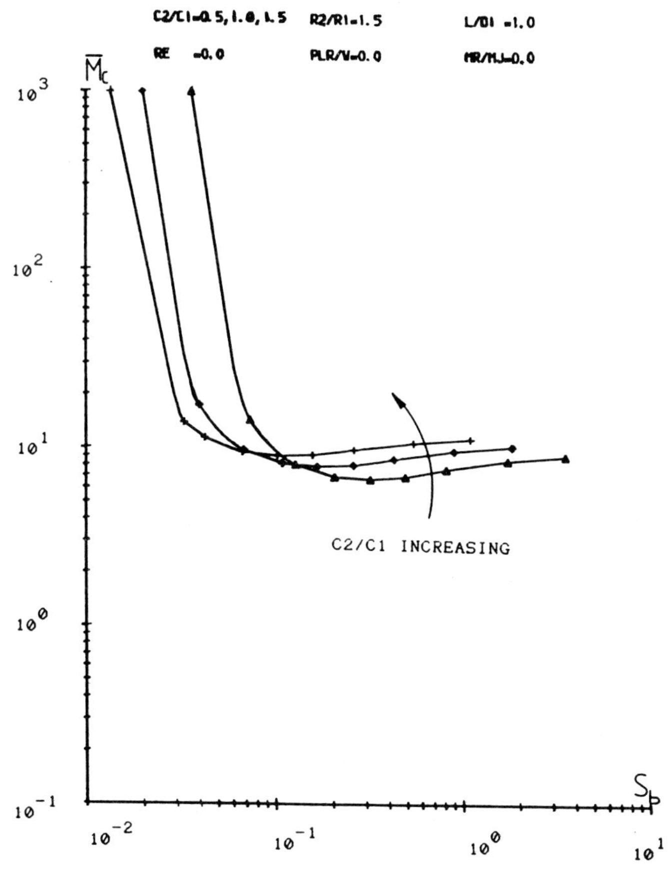

Fig 8 Critical mass versus Sommerfeld number

C293/88

One-dimensional finite element analysis of multiple-pad pressure dam bearings

T OKITA, MASSC
Mitsubishi Electric Corporation, Nagasaki, Japan
P E ALLAIRE, PhD, MASME, MASLE
Department of Mechanical and Aerospace Engineering, University of Virginia, Virginia, United States of America

SYNOPSIS Reynolds' equation is solved by the one dimensional finite element solution method under two assumptions. A variational method is employed to reduce the two dimensional pressure field problem to a one dimensional finite element formulation. The accuracy of this method is evaluated by comparing the results with ones of the two dimensional finite element solution method. Generally the results are quite close for load capacity, principal stiffness and damping but not so good for attitude angle and cross coupled coefficients.

NOTATION

R	radius of shaft (journal)
L	bearing length
L_d	axial dam length
c_p	bearing radial clearance
c_d	dam height
$\bar{c}_d$	dimensionless dam height, c_d/c_p
h_1	film thickness in dam region
h_2	film thickness in flat region
$\bar{h}_1$, $\bar{h}_2$	dimensionless film thickness, h_1/c_p, h_2/c_p
x_j, y_j	journal position in the x and y coordinates
θ	circumferential coordinate
z	axial coordinate
ω	rotational speed
G_θ, G_z	turbulence correction factors for governing equations $G_\theta = \dfrac{1}{12+0.0136Re^{0.9}}$ in circumferential; $G_z = \dfrac{1}{12+0.0043Re^{0.69}}$ in axial; where Re = Reynolds number $\rho R\omega h/\mu$, ρ = lubricant density
P	pressure
W	load
μ	viscocity
P_θ	centerline pressure
P_d	pressure before dam
P_f	pressure after dam
θ_l	location of leading edge measured with rotation from positive x-axis
θ_d	location of dam measured with rotation from positive x-axis
θ_t	location of trailing edge measured with rotation from positive x-axis
t	time
n	axial pressure exponent
ε	bearing eccentricity ratio
ϕ	attitude angle
$Q_{\theta d}$, $Q_{\theta f}$	circumferential flow rates before, after dam
$P_{\theta d}$	pressure at dam
F_{xi}, F_{yi}	bearing hydrodynamic forces
K_{ij}	dimensionless stiffness coefficients
C_{ij}	dimensionless damping coefficients
S	Sommerfeld number $\mu NRL(R/c_p)^2/(30W)$ where N = rotational speed (rpm)

1 INTRODUCTION

Pressure dam bearings have been used to improve the stability of horizontal turbines and compressors. A dam is cut in the upper half surface of the bearing to produce a hydrodynamic load on the journal. This load has the potential to suppress the vibration of vertical machinery due to instability when applying multiple pad pressure dam bearings with dams in each pad [1].

In recent years, the finite element method has been adapted to lubrication problems [2]-[4]. Reynolds' equation for pressure in a fluid film journal bearing is usually solved by a two dimensional finite element method. In the two dimensional method, any additional considerations are not necessary to handle the abrupt changes of film thickness across the dam.

Nicholas and Kirk calculated the characteristics of multiple pad bearings including pressure dam bearings by using the two dimensional method [5]. However, computer programs of two dimensional finite element method require large amounts of computer storage, and the execution time is quite long. They do not appear suited to micro computers. Knight and Barrett analysed multilobe bearings with a one dimensional finite element method which is based on the assumption of an approximate form of the pressure distribution in the axial direction [6]. They obtained good results which agreed with the experimental results.

The effects of turbulence and inertia forces at the dam and over the entire bearing have been considered by many authors [7]-[12]. Nicholas and Allaire determined that the step inertia effects may be neglected for a large class of oil bearings within the operating range [12]. They analyzed the pressure dam bearing neglecting step inertia effects but including the effects of turbulence over the entire bearing surface.

This paper describes a one dimensional finite element solution method for multiple pad pressure dam bearings. Reynolds' equation is solved under two assumptions. One is the assumption of a known pressure distribution in the axial direction and another is the pressure inclination at the step. The accuracy of this method is evaluated by comparing the results with ones of the two dimensional finite element solution method which is developed from Nicholas' two dimensional program [13].

2 REYNOLDS' EQUATION AND PRESSURE ASSUMPTIONS

The pad of a pressure dam bearing is illustrated in Fig. 1(a). Fig. 1(b) shows the side view of the shaft and pad. A fixed cartesian coordinate system is employed with respect to the load on the bearing.

The film thickness at the pocket and land are given by

$$h1 = cp + cd - xj\cos\theta - yj\sin\theta \qquad (1)$$

$$h2 = cp - xj\cos\theta - yj\sin\theta \qquad (2)$$

The governing equation for pressure in a fluid film journal bearing is Reynolds' equation [2]. When the viscosity of the fluid is assumed to be constant in the bearing, it can be written as

$$\frac{\partial}{\partial\theta}\left(\frac{G_\theta h^3}{R^2}\frac{\partial P}{\partial\theta}\right) + \frac{\partial}{\partial z}\left(\frac{\partial P}{\partial z}\right) = \mu\left(\frac{\omega}{2}\frac{\partial h}{\partial\theta} + \frac{\partial h}{\partial t}\right) \qquad (3)$$

To solve the Reynolds' equation and determine $P(\theta,z)$ the functional for it may be minimized by a finite element method in two dimensions.

$$J(P) = \int_{\theta l}^{\theta t}\int_{-L/2}^{L/2}\left\{\frac{G_\theta h^3}{R^2}\left(\frac{\partial P}{\partial\theta}\right)^2 + G_z h^3\left(\frac{\partial P}{\partial z}\right)^2 + 2\mu\left(\frac{\omega}{2}\frac{\partial h}{\partial\theta} + \frac{\partial h}{\partial t}\right)P\right\}dz\,d\theta \qquad (4)$$

When a bearing is axially symmetric with no shaft misalignment, the axial pressure distribution over the part including the dam and plain parts are assumed to be of the form

$$P_d(\theta,z) = \begin{cases} P_\theta, & 0 \leqq z \leqq L_d/2 \\ P_\theta\dfrac{L/2-z}{L/2-L_d/2}, & L_d/2 < z \leqq L/2 \end{cases}, \quad \theta_l \leqq \theta \leqq \theta_d \qquad (5)$$

$$P_f(\theta,z) = P_\theta\left[1 - \left(\frac{2z}{L}\right)^n\right], \quad \theta_d < \theta \leqq \theta_t \qquad (6)$$

where n is a chosen constant depending to some degree upon the length-to-diameter ratio of the bearing. Knight and Barret chose 2.0 as n, and got good results.

The functional (4) is conveniently broken up into integrals over each region by considering the axial pressure assumption. With the dimensionless film thickness, the integral reduces the functional to the one dimensional functional

$$\begin{aligned} J(P) = \int_{\theta l}^{\theta d}\Bigg\{&\left[\frac{G_{\theta 1}\bar{h}_1^3}{R^2}\left(\frac{L_d}{2}\right) + \frac{G_{\theta 2}\bar{h}_2^3}{R^2}\left(\frac{L/2-L_d/2}{3}\right)\right]\left(\frac{\partial P_\theta}{\partial\theta}\right)^2 \\ &+ G_{z2}\bar{h}_2^3\left(\frac{1}{L/2-L_d/2}\right)P_\theta^2 \\ &+ \frac{\mu}{c_p^2}\left(\frac{\omega}{2}\frac{\partial\bar{h}_2}{\partial\theta} + \frac{\partial\bar{h}_2}{\partial t}\right)\left(\frac{L}{2} + \frac{L_d}{2}\right)P_\theta\Bigg\}d\theta \\ + \int_{\theta d}^{\theta t}\Bigg\{&\frac{G_{\theta 2}\bar{h}_2^3}{R^2}\left(\frac{L}{2}\right)\left(1 - \frac{2}{n+1} + \frac{1}{2n+1}\right)\left(\frac{\partial P_\theta}{\partial\theta}\right)^2 \\ &+ G_{z2}\bar{h}_2^3\left(\frac{1}{L/2}\right)\frac{n}{2n-1}P_\theta^2 \\ &+ \frac{\mu}{c_p^2}\left(\frac{\omega}{2}\frac{\partial\bar{h}_2}{\partial\theta} + \frac{\partial\bar{h}_2}{\partial t}\right)\left(\frac{L}{2}\right)\left(2 - \frac{2}{n+1}\right)P_\theta\Bigg\}d\theta \end{aligned} \qquad (7)$$

This can be minimized by a simple finite element routine.

3 DAM CONDITION

Boundary conditions are normally zero pressure around each pad. In addition, the cavitation condition is usually met by setting the negative hydrodynamic pressures to zero. These conditions are also used in the two dimensional finite element program on multiple pad pressure dam bearings from Nicholas' [13].

Fig. 3 compares the pressure distributions along the centerline of a one pad bearing.

Table 1 Bearing geometry and lubricant properties used in pressure distribution comparison

R = 50 mm	L = 70 mm	L_d = 50 mm
c_p = 0.120 mm	c_d = 0.120 mm	
θ_l = 190°	θ_d = 290°	θ_t = 350°
μ = 0.12 x 10^{-8} kg·sec/mm^2		
N = 3600 rpm	ε = 0.50	ϕ = 45°

The pressure of one dimensional finite element solution with the general boundary conditions is much less than one of the two dimensional solution. There is a large difference in the slopes of the pressure just before and after the dam. The results of the two dimensional solutions show the same dam pressure distributions for various conditions as

$$\left.\frac{\partial P_\theta}{\partial\theta}\right|_{after\ dam} \approx -\left.\frac{\partial P_\theta}{\partial\theta}\right|_{before\ dam} \quad at\ \theta = \theta_d \qquad (8)$$

When assuming equation (8), the flow rates of dam and flat regions can be written as

$$Q_{\theta d} = -\frac{1}{12R\mu}\left(\left.\frac{\partial P_\theta}{\partial \theta}\right|_{before\ dam}\right)\left[h_1^3 L_d + h_2^3\left(\frac{L}{2} - \frac{L_d}{2}\right)\right] + R\omega\left[h_1\frac{L_d}{2} + h_2\left(\frac{L}{2} - \frac{L_d}{2}\right)\right] \quad (9)$$

$$Q_{\theta f} = \frac{1}{12R\mu}\left(\left.\frac{\partial P_\theta}{\partial \theta}\right|_{before\ dam}\right) h_2^3 L\left(1 - \frac{1}{n+1}\right) + R\omega h_2\frac{L}{2} \qquad at\ \theta = \theta_d \quad (10)$$

By considering the continuity of flow rate ($Q_{\theta d} = Q_{\theta f}$), the term ($\partial P_\theta / \partial\theta$ *before dam*) is obtained as

$$\left.\frac{\partial P_\theta}{\partial \theta}\right|_{before\ dam} = \frac{6R^2\omega\mu\bar{c}_d L_d}{c_p^2\left[\bar{h}_1^3 L_d + \bar{h}_2^3(L/2 - L_d/2) + \bar{h}_2^3 L\left(1 - \frac{1}{n+1}\right)\right]} \quad (11)$$

The boundary conditions applied to dam region ($\theta_l \leq \theta \leq \theta_d$) are set

$$\begin{cases} P_\theta = 0 & at\ \theta = \theta_l \\ \dfrac{\partial P_\theta}{\partial \theta} = \dfrac{6R^2\omega\mu\bar{c}_d L_d}{c_p^2\left[\bar{h}_1^3 L_d + \bar{h}_2^3(L/2 - L_d/2) + \bar{h}_2^3 L\left(1 - \frac{1}{n+1}\right)\right]} & \end{cases}$$

$$at\ \theta = \theta_d \quad (12)$$

After the pressure distribution has been decided in the dam region, the boundary conditions for the flat region ($\theta_d < \theta \leq \theta_t$) are

$$\begin{cases} P_\theta = P_{\theta d} & at\ \theta = \theta_d \\ P_\theta = 0 & at\ \theta = \theta_t \end{cases} \quad (13)$$

Fig. 4 shows the results of this one dimensional finite element solution by the above boundary conditions, which agrees with the two dimensional result quite well.

4 HYDRODYNAMIC BEARING FORCES AND LINEAR COEFFICIENTS

The multiple pad pressure dam bearing is usually made up of two or more pads which are arranged at equal intervals. Each pad has the same shape as shown in Fig. 1.

In the one dimensional analysis, the total hydrodynamic forces from each pad acting on the journal found by integrating equations (5) and (6) over the entire pad surface. When integrating with respect to z, the force components in x and y directions of a fixed cartesian coordinate system on the bearing are obtained as

$$F_{xi} = -\frac{1}{2}R(L + L_d)\int_{\theta_l}^{\theta_d} P_\theta \cos\theta\, d\theta - RL\left(1 - \frac{1}{n+1}\right)\int_{\theta_d}^{\theta_t} P_\theta \cos\theta\, d\theta \quad (14)$$

$$F_{yi} = -\frac{1}{2}R(L + L_d)\int_{\theta_l}^{\theta_d} P_\theta \sin\theta\, d\theta - RL\left(1 - \frac{1}{n+1}\right)\int_{\theta_d}^{\theta_t} P_\theta \sin\theta\, d\theta \quad (15)$$

These forces support the bearing load, when the journal operates at an equilibrium position. The solution of the journal equilibrium eccentricity requires iterating calculations. At first, the pressure distributions are calculated for an estimated eccentricity. Then, the calculations are repeated with prediction of a new journal eccentricity obtained by the Newton Raphson method. Finally, the total force on the bearing converges to the load within a reasonable error.

After the equilibrium position is obtained, stiffness and damping coefficients can be calculated by the same finite element method as that used to solve for the equilibrium position. These coefficients are given by linearizing the bearing forces for small displacements and velocities of the journal.

5 COMPARISON OF 1-D AND 2-D RESULTS

To evaluate the accuracy of the one dimensional solution to Reynolds' equation for the three pad pressure dam bearing shown in Fig. 5, results for equilibrium positions and characteristics are obtained by the one dimensional method were compared to those obtained by the two dimensional method [13]. These solutions include circumferential and axial turbulence corrections to solve for the hydrodynamic pressures. The two dimensional program was developed from Nicholas' [13], which calculates the characteristics of two lobe bearing with a pressure dam on the upper surface. The program was modified to calculate those of multiple pad bearing with pressure dam on every pad.

Figures 6 through 8 illustrate these comparisons. Fig. 6 shows the equilibrium positions of the journal. In the figure, the error of eccentricity ratio introduced from the assumption of axial pressure distributions in the form of equations (5) and (6) are very small over a wide range of eccentricity ratio. The error of attitude angle becomes larger when the eccentricity ratio becomes larger. Fig. 7 shows nondimensionalized stiffness coefficients, and Fig. 8 shows the nondimensionalized damping coefficients. In the two figures, the principal stiffness and damping coefficients of the one dimensional method agree with those of the two dimensional fairly well. The errors of cross-coupled coefficients are significant, but overall the results are good.

The errors of attitude angles and cross-coupled coefficients depend on the errors of the assumed axial pressure profile. The pressure distributions calculated by the two dimensional method show some differences to the assumed profile.

REFERENCES

(1) ALLAIRE, P. E. and OKITA, T. Multiple pad pressure dam journal bearings analysis and application to vertical machines. International Conference on Rotordynamics, Tokyo, 1986, Proceedings pp. 233-237.

(2) REDDI, M. M. Finite-element solution of the incompressible lubrication problem. Journal of Lubrication Technology, Trans. ASME, July 1969, pp. 524-433.

(3) BOOKER, J. F. and HUEBNER, K. H. Application of finite elements to lubrication: an engineering approach. Journal of Lubrication Technology, Trans. ASME, October 1972, pp. 313-323.

(4) ALLAIRE, P. E., NICHOLAS, J. C. and GUNTER, E. J. Systems of finite element for finite bearings. Journal of Lubrication Technology, Trans. ASME, April 1977, pp. 187-197.

(5) NICHOLAS, J. C. and KIRK, R. G. Theory and application of multipocket bearings for optimum turborotor stability. Trans. ASLE, April 1981, pp. 269-275.

(6) KNIGHT, J. D. and BARRETT, L. E. An approximate solution technique for multilobe journal bearings including thermal effects with comparison to experiment. Trans. ASLE, October 1983, pp.501-503.

(7) BURTON, R. A. and CARPER, H. J. An experimental study of annular flows with application in turbulent film lubrication Journal of Lubrication Technology, Trans. ASME, 1967, pp. 381-389.

(8) SMALLEY, A. J., VOHR, J. H., CASTELLI, V. and WACHMANN, C. An analytical and experimental investigation of turbulent flow in bearing films including convective fluid inertia forces. Journal of Lubrication Tehcnology, Trans. ASME, 1974, pp. 151-157.

(9) CONSTANTINESCU, V. N. and GALETUSE, S. Pressure drop due to inertia forces in step bearings. Journal of Lubrication Technology, Trans. ASME, January 1976, pp. 167-174.

(10) SALIMONU, D. O., KISTLER, A. I. and BURTON, R. A. Turbulent flows near steps in a thin channel with one moving wall. Journal of Lubrication Technology, Trans. ASME, April 1977, pp. 224-229.

(11) ALLAIRE, P. E., NICHOLAS, J. C. and BARRET, L. E. Analysis of step journal bearings--infinite length, inertia effects. ASLE Preprint No. 78-AM-3B-1.

(12) NICHOLAS, J. C. and ALLAIRE, P. E. Analysis of step journal bearings--finite length, stability. ASLE Preprint No. 78-LC-6B-2.

(13) NICHOLAS, J. C. A finite element dynamic analysis of pressure dam and tilting pad bearings. Ph. D. Thesis, 1977, Department of Mechanical and Aerospace Engineering, University of Virginia.

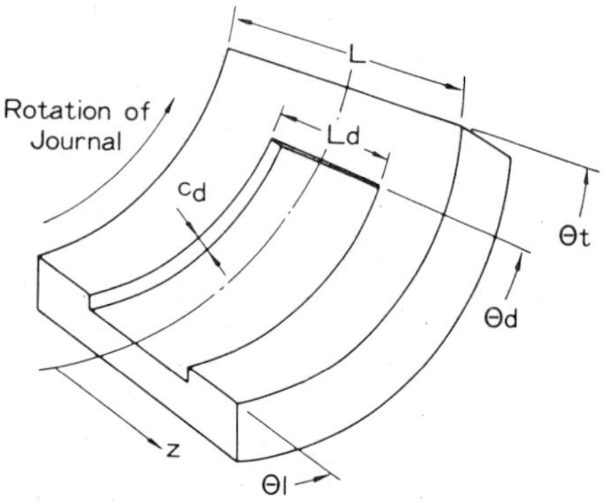

(a) Pressure dam bearing pad

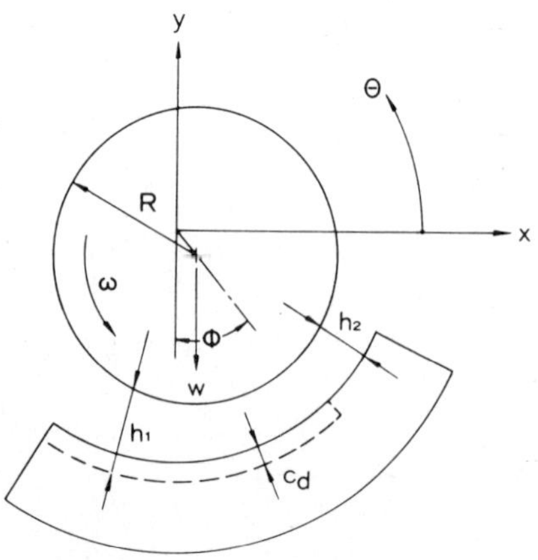

(b) Shaft and pad side view

Fig 1 Schematic of pressure dam bearing

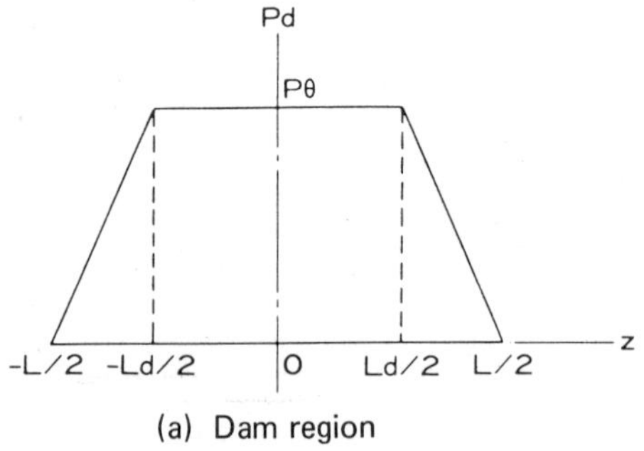

(a) Dam region

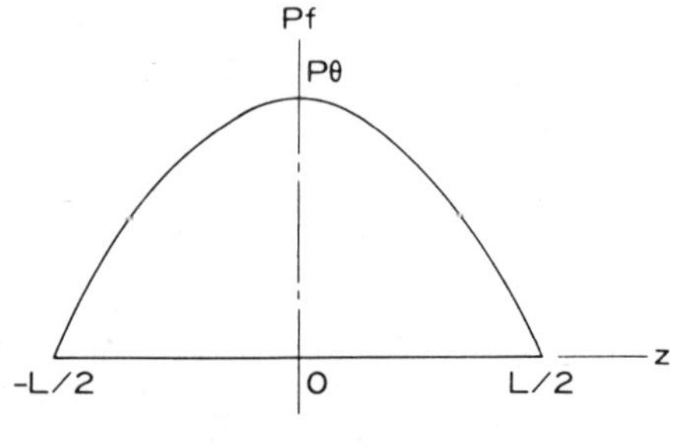

(b) Flat region

Fig 2 Pressure assumption in axial direction

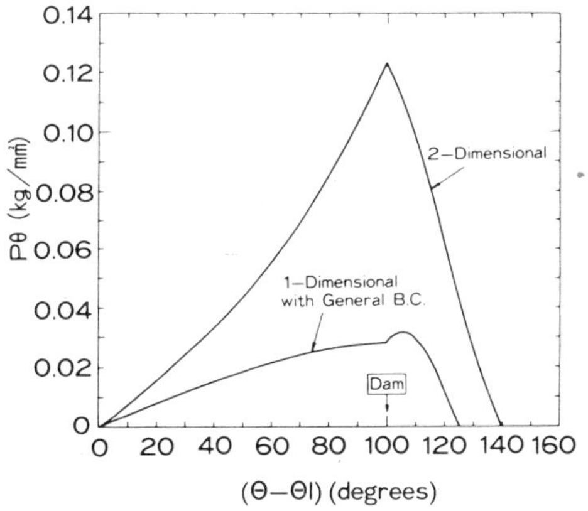

Fig 3 Comparison of centre-line pressure (one-dimensional with general BC)

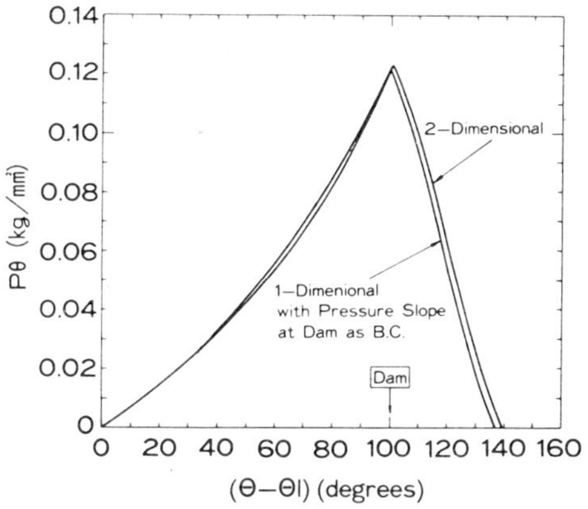

Fig 4 Comparison of centre-line pressure (one-dimensional with pressure slope at dam); L = 70 mm, Ld = 50 mm, ϵ = 0.50, ϕ = 45°

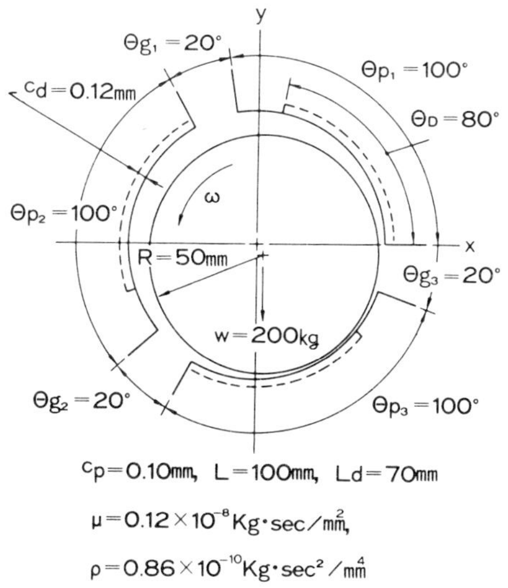

Fig 5 Three-pad pressure dam bearing

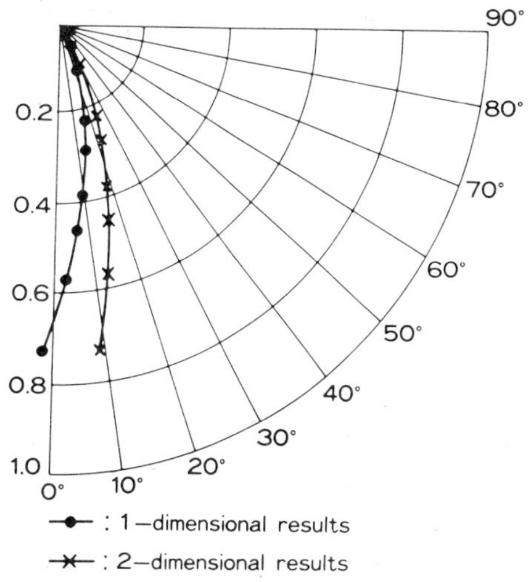

Fig 6 Journal equilibrium position comparison

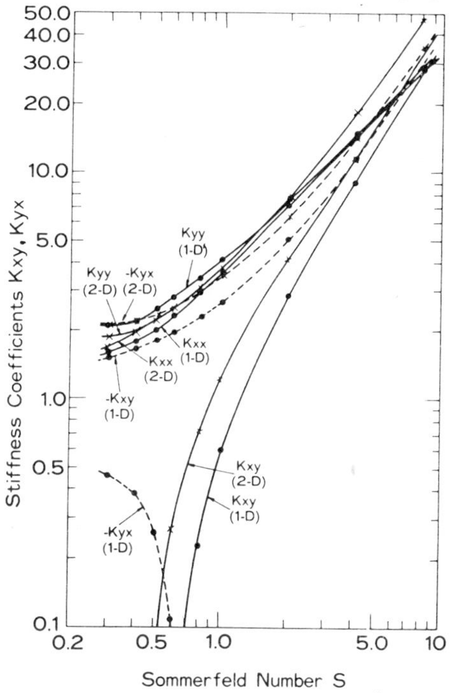

Fig 7 Non-dimensional linear stiffness coefficient comparison

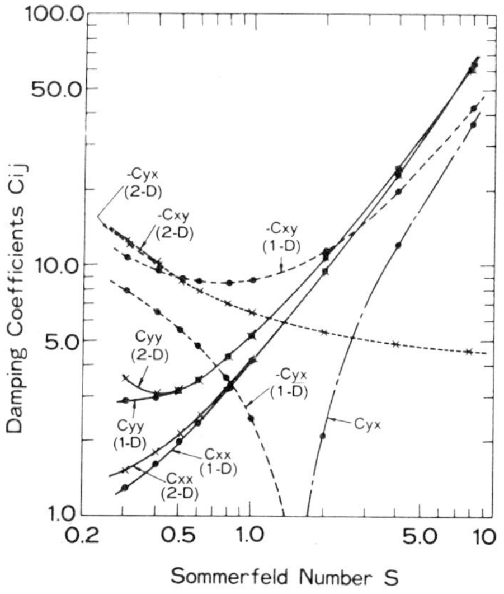

Fig 8 Non-dimensional linear damping coefficient comparison

Non-linear effects in lubricated bearings

G DIANA and **F CHELI**
Dipartimento di Meccanica, Politechno di Milano, Milan, Italy
A MANENTI
Dipartimento di Ingegneria Meccanica, Universita degli Studi di Brescia, Brescia, Italy
F PETRONE
Istituto di Macchine, Universita degli Studi di Catania, Catania, Italy

SYNOPSIS The aim of this paper is to illustrate a method to compute oil film stiffness and damping equivalent coefficients depending on the journal amplitudes of vibration. These coefficients can be used in a program for the simulation of the behaviour of a rotor-foundation system. The response to unbalance and the instability threshold obtained with these coefficients takes in some account the oil film non linearities.

NOTATION

$\underline{X}$	complex orbit amplitude along the vertical axis (x)
$\underline{Y}$	complex orbit amplitude along the horizontal axis (y)
Ω	journal rotational speed
ω_0	frequency of the orbital motion
$\underline{F}_v, \underline{F}_o$	complex amplitude of the oil film force components along the x and y axis
δ	radial clearance
R	journal radius, D = 2 R
L	journal bearing length
a	ratio between orbit amplitude and
χ	eccentricity ratio
$\underline{F}_v^I, \underline{F}_o^I$	film forces first harmonic components
$\underline{F}_v^{II}, \underline{F}_o^{II}$	film forces second harmonic components
$K_{xx}, \ldots$	oil film stiffness equivalent coefficients
$R_{xx}, \ldots$	oil film damping equivalent coefficients
$\underline{F}_{vL}, \underline{F}_{oL}$	'linearized' complex oil film force component along the x and y axis
I^I, I^{II}	non linearity coefficients related to I and II harmonic component
$A(\chi)$	journal bearing load capacity coefficient
Q	load on the journal
μ	lubricant viscosity
c_{xx}	non-dimensional direct stiffness
$\bar{\omega}$	frequency of system supported on journal bearings
g	acceleration of gravity
$\bar{\omega}_0$	$= \sqrt{g/\delta}$
$\alpha/\bar{\omega}$	instability coefficient

1 INTRODUCTION

The dynamic behaviour of a rotor on lubricated bearings is strongly affected by the fluid film characteristics.

The film forces are dynamically defined by means of the stiffness and damping coefficients, which are obtained by a linearization of the field of forces in the neighbourhood of the equilibrium position of the journal bearing. This linearization is no longer valid if significant vibration amplitudes of the journal occur: this may happen, for instance, if an important unbalance is present and/or in run-up or run-down conditions, when the critical speeds are passed through.

Big vibration amplitudes cause the non linear characteristics of the oil film forces to become important and cause some experimentally well known effects, such as the modification of the journal centre locus in respect to that defined for small amplitudes of vibration and the modification of the frequency response of a rotor, with the growing up of higher order harmonic components.

The instability phenomenona are in part connected to the non linear characteristics of the fluid film too: if the rotational speed of the journal is higher than twice the first flexural speed, the instability is put in evidence by the growing up of a sub-harmonic component with a frequency equal to the first critical speed (oil whip); if the rotational speed is lower, the sub-harmonic component frequency is one half of the rotor angular speed.

This last phenomenon is sometimes referred to as 'oil whirl' and some authors indicate its cause in the variation of the stiffness and damping coefficients along the journal orbits, and therefore they look at it as at a parametric instability [1] [2]. On the contrary, the oil whip phenomenon generally finds a linear justification at least in terms of incipient instability [3] [4].

An analytical simulation of all these problems could be obtained by means of a suitable non linear mathematical model: the fluid film forces would be computed in their complete non linear form (Reynolds equation integration) at any integration step of the motion equations relative to the system rotor + oil film + foundation. However this procedure would be greatly time consuming.

Another way to solve the problem could be the one described in this paper. A linear approach is adopted for the complete system, but fluid film stiffness and damping coefficients are computed taking into account the amplitude of the journal orbits.

Particular care will be devoted to the study of a rotor-foundation system response to unbalance, but some consideration about instability will be also considered.

The paper is organized as follows:

- Method to compute the oil film equivalent stiffness and damping coefficients
- Numerical applications and considerations
- Frequency response of a laboratory rotor model: comparison between experimental and analytical results obtained with the classical oil film stiffness and damping coefficients and with those computed with the method described
- Instability problems
- Conclusions

2 THE METHOD TO COMPUTE STIFFNESS AND DAMPING COEFFICIENTS OF THE OIL FILM DEPENDENT ON THE JOURNAL AMPLITUDE OF VIBRATION

Due to unbalance, the journal of a real rotor whirls with synchronous frequency in an orbit. Due to the journal movements, the film thickness is continuously changing and non linear modifications of the field of forces occur. Depending on the orbit amplitude, the non linear aspect of the problem is more or less significant. As an example, if the amplitude of vibration is of the same order of the journal eccentricity, the limits of the oil film convergent and divergent zones are continuously modified and so the boundary conditions of the Reynolds equation are changed. Moreover, an important role is also played, in these conditions, by the velocity terms associated to the journal motion: the minimum film thickness results to whirl with the same rotational speed associated to the orbital motion. It is clear that a linearization of the oil film field of forces is not suitable to reproduce such a situation.

The method hereafter described defines some oil film stiffness and damping parameters related to the amplitude of the journal orbit and it can be summarized in the following steps:

- all the journal bearing characteristics being defined, the journal is forced to whirl in a given elliptical orbit: the center of the orbit is the journal equilibrium position, depending on the values of the load and of the rotational speed. The orbit is defined by the two complex amplitudes along the vertical (x) and horizontal (y) axis:

$$\underline{x} = \overline{\underline{x}}\, e^{i\omega_o t}$$

$$\underline{y} = \overline{\underline{y}}\, e^{i\omega_o t}$$

  with associated velocities given by:

$$\dot{\underline{x}} = i\,\omega_o \overline{\underline{x}}\, e^{i\omega_o t}$$

$$\dot{\underline{y}} = i\,\omega_o \overline{\underline{y}}\, e^{i\omega_o t}$$

  where ω_o is the frequency associated to the orbital motion;
- the horizontal ($\underline{F}_o$) and vertical ($\underline{F}_v$) components of the fluid film forces are computed all along the orbit by integrating the Reynolds equation with the procedure described in [5] [6], which can take into account turbolence and temperature effects. In fig.1 the $\underline{F}_o$ component is reported as a function of the journal position along the orbit. The computation is relative to a cylindrical bearing characterized by:

  - radial clearance $\delta = 860\,\mu m$
  - ratio between δ and the journal radius R: $\delta/R = 4\ 10^{-3}$
  - ratio between the journal lenght L and diameter D: $L/D = 1$

  and the Reynolds equation is integrated, in this case, taking into account laminar flow and lubricant constant temperature. The orbit amplitude is given in terms of the non-dimensional parameter a, that is the ratio between the orbit amplitude and the radial clearance. Fig.1a refers to the case of an orbit amplitude defined by $a = .1$, with an initial position of the orbit center given by an eccentricity ratio $\chi = e/\delta$ equal to 0.373. Fig.1b refers to an orbit with $a = .55$ and the same initial position of the previous one;
- with an iterative procedure, a new position of the orbit center is found: the orbit with the new center represents a dynamic equilibrium condition for the journal bearing, in fact film forces components mean values equal the external loads: as an example, for the case of fig.1a this new position is characterized by $\chi = 0.360$, while for fig.1b we have $\chi = 0.157$;
- a Fourier analysis of the periodic functions representing the film forces along the orbit is performed and the first and second harmonic components are evaluated ($\underline{F}_o^{I}$, $\underline{F}_v^{I}$, and $\underline{F}_o^{II}$, $\underline{F}_v^{II}$). In fig.1a,b these values are also reported;
- if we indicate with K_{xx}, K_{xy}, K_{yx}, K_{yy}, and R_{xx}, R_{xy}, R_{yx}, R_{yy} the oil film coefficients in terms of equivalent stiffness and damping, the following expressions can be written:

$$\underline{F}_v^{I} = K_{xx}\overline{\underline{x}} + K_{xy}\overline{\underline{y}} + i\omega_o R_{xx}\overline{\underline{x}} + i\omega_o R_{xy}\overline{\underline{y}}$$

$$\underline{F}_o^{I} = K_{yx}\overline{\underline{x}} + K_{yy}\overline{\underline{y}} + i\omega_o R_{yx}\overline{\underline{x}} + i\omega_o R_{yy}\overline{\underline{y}} \qquad (1)$$

  where $\underline{F}_v^{I}$, $\underline{F}_o^{I}$, $\overline{\underline{x}}$ and $\overline{\underline{y}}$ are the known complex quantities previously defined.

  Expressions (1) are a system of 4 scalar equations with the 8 unknown K_{xx}, K_{xy}, K_{yx}, K_{yy}, R_{xx}, R_{xy}, R_{yx}, R_{yy}. The problem can be solved, by a deterministic approach, considering a second orbit for the journal. This orbit must be nearly equal but not similar to the previous one, in such a way that the number of equations is doubled and the equations themselves are not linearly dependent one on the other.

The stiffness and damping coefficients in this way obtained reproduce the non linear film forces first harmonic components.

The aforementioned first and second harmonic components of the oil film forces along the orbit can be compared with the values of the film forces obtained through expressions analogous to (1) in which the K and R coefficients are the stiffness and damping parameters derived from the linearization of the fluid film forces in the neighborough of static equilibrium position of the journal. We call $\underline{F}_{vL}$ and $\underline{F}_{oL}$ the forces in this way obtained and whose values are also reported in fig.1a,b.

The ratio between $\underline{F}_v^{I}$ and $\underline{F}_{vL}$ and between F_o^{I} and F_{oL} gives a non linearity index (I^{I}) related to first harmonic components, while the ratio between $\underline{F}_v^{II}$ and $\underline{F}_{vL}$ and between $\underline{F}_o^{II}$ and $\underline{F}_{oL}$ gives a non linearity index (I^{II}) related to second harmonic components. The bigger I^{I}, the larger the orbit. I^{II} is approximately zero for small orbit amplitudes, as the amplitudes increase this index approaches unity. These indices values are also reported in fig.1a,b.

A note must be made to this method: even if the only unbalance is taken into account and no significant second harmonic excitation components (such as those caused by a rotor with different stiffness along two axes or by a journal ovalization) are present, a journal, in real conditions, whirls in an orbit that would be elliptical only if the system rotor + oil film + foundation would be linear. Non linearities, in our case due to oil film, cause, as reported in fig.1a,b, second and higher order harmonic components in the oil film forces. These components, in turn, cause higher order harmonics in the journal vibrations and, therefore, non elliptical orbits. It is then clear that the method described better fits reality the less the higher harmonics contribute to vibrations in respect to the fundamental harmonic component: this happens, for instance, when a rotor is passing through one of its critical flexural speeds.

3 NUMERICAL APPLICATIONS AND CONSIDERATIONS

The modification of the various equilibrium positions of the already described cylindrical bearing in relation to different orbit amplitudes has been computed. Fig.2 shows the journal center loci, for the different cases considered, and the journal bearing load capacity coefficient $A(\chi)$. The continuous line represents the journal center locus computed for small amplitudes of vibration (a = 0). For the same rotational speed the various values of load on the journal are identified by arabian numbers (from 1 to 6). The dashed lines indicate the modification of the various equilibrium positions in relation to the orbit amplitudes considered.

As can be seen, the greater displacements take place for the higher χ values, and, in any case, all displacements are towards the bearing center. It can also be observed that these displacements become significant, for what concerns non linearities, for a greater than or equal to 0.2.

Fig.3 reports the direct non-dimensional stiffness in vertical direction $c_{xx} = K_{xx}\,\delta\ /\ Q$, where Q is the load on the journal, as a function of the parameter a, defining the vibration amplitude, and for different χ values characterizing the initial position of the orbit center.

Here it is considered the case of a cylindrical bearing with:
- δ = 225 μm; $\delta/R = 1.28\ 10^{-2}$; L/D = 1

and the Reynolds equation is integrated taking into account laminar flow and lubricant constant temperature.

As can be seen, with the increasing of a, the stiffness is significantly increasing for small χ values, while the same behaviour cannot be observed for higher χ values. However, to have a clearer evaluation of the oil film stiffening effect associated with the orbit amplitude, instead of the direct and cross coupled stiffness coefficients, it is possible to consider the variation of the natural frequency $\bar{\omega}$ of the system

$$[M]\,\ddot{\underline{X}} + [R]\,\dot{\underline{X}} + [K]\,\underline{X} = 0 \qquad (2)$$

where [R] and [K] are the fluid film damping and stiffness matrices and [M] is a diagonal matrix whose two terms value is given by the ratio Q/g, being g the acceleration of gravity. The solutions of this vector equation are of the type $\underline{X} = \bar{\underline{X}}\,e^{\lambda t}$ with $\lambda = \alpha + i\bar{\omega}$.

Four λ values are found: generally, two values are real and negative or complex conjugate with a very important and negative real part; the other two values are complex conjugate with a small real part that can be positive or negative: here we define as $\bar{\omega}$ the imaginary part of the eigenvalue in this last case.

In fig.4 $\bar{\omega}$ is reported in non-dimensional form ($\bar{\omega}/\bar{\omega}_o$, with $\bar{\omega}_o = \sqrt{g/\delta}$) as a function of a and for various values of the eccentricity ratio χ. As can be seen, the smaller is χ, the more significant is the increasing with a of $\bar{\omega}$, that is of the oil film stiffness. It can be observed that, for high χ values, the stiffness does not increase so much because large displacements of the journal center towards the bearing center occur for high a and χ values (see fig.2) and this fact counterbalances the oil film stiffening effect.

4 FREQUENCY RESPONSE OF A LABORATORY ROTOR MODEL: COMPARISON BETWEEN ANALYTICAL AND EXPERIMENTAL RESULTS

4.1 Experimental set up

The Sezione di Meccanica dei Sistemi del Politecnico di Milano has a small model consisting of a motor driven rotor supported by two bearings connected to a rigid base plate. This model is a modification of a rotor kit generously donated by Bently Nevada Co. A schematic drawing of this set up is shown in fig.5a.

Lubricated cylindrical bearings are mounted and the oil is supplied is an electric motor driven pump that is employed also to return the oil from the oil collectors to the reservoir.

The oil discharge is controllable and its maximum value is 5 l/min, which is more than required by the bearings until 6000 - 7000 r/min.

The bearing geometrical characteristics have been already reported describing fig.3.

A disk, called rotor mass, is fixed by two setscrews about at the middle of the rotor. This disk presents a series of axial threaded holes placed all at the same distance from its center. These holes can receive balancing/ unbalancing masses.

Relative vibrations between the rotor and the bearings are measured by a couple of probes (proximitor type) placed in vertical and horizontal directions in correspondence to the bearings and in the vicinity of the shaft midspan, near the rotor mass. Absolute bearing vibrations are measured by accelerometers placed on the bearing cases in vertical and horizontal directions.

The model rotational speed is evaluated by a proximitor, placed in horizontal direction, in front of a radial notch on the rotor. The signal from this probe is used also as phase reference for all the vibration signals.

A schematization of the model is reported in fig.5b, where there are shown also some peculiar geometric dimensions and the abovementioned probes position and denomination.

The signals from all the probes have been acquired by a PDP 11/34 Digital computer equipped with an A/D converter and suitable acquisition and elaboration software, that allows for a synchronous analysis of the signals with the determination of their first three harmonic components, in module and phase. The analysis can be performed both in transient and in steady state conditions.

4.2 Tests on the laboratory model

A number of tests have been performed on the laboratory set up. In particular, for what concerns the model response to unbalance, two configurations have been considered: the first one with no set screws used as balancing/ unbalancing masses in the rotor mass; the second with some set screws (total mass: $5.8\ 10^{-3}$ kg) placed in a sector of the rotor mass.

The lubricant temperature has been kept constant and equal to 48°C, in these conditions the oil viscosity being equal to 68 cSt.

A run-up (from 0 to about 6000 r/min) and a run-down have been recorded and then acquired.

As the first configuration has shown amplitudes of vibration negligible in respect to those of the second one, the results obtained in this last case can be considered the response of the system to unbalance.

Fig.6 a,b refers to the vertical and horizontal probes (no. 5 and 6 in fig.5b) relative to the journal bearing opposite to the motor.

As can be seen the first flexural speed of the rotor is put in evidence at about 3800 r/min. It is to be noted that the first natural frequency without oil film is about 6000 r/min.

4.3 Comparison between analytical and experimental results

The response of the rotor to unbalance has been first computed by a program generally used at the Sezione di Meccanica dei Sistemi del Dipartimento di Meccanica del Politecnico di Milano and described in [7] [8].

This program is based on a finite element schematization of the rotor and the oil film is taken into account by means of its stiffness and damping equivalent coefficients, which are obtained through the linearization of the film forces about the equilibrium position of the journal.

The supporting structure has been schematized by very stiff one- degree- of- freedom systems, in such a way to reproduce the real behaviour of the model rigid base.

The computation results are reported in fig.6 obviously for the same points to which the already reported experimental vibrations refer. It can be noted that the vibration amplitudes are considerably higher than the measured ones and that, while the mathematical model well enough reproduces a prevalently vertical critical speed not far from the experimental one, it introduces a prevalently horizontal critical speed at about 2000 r/min which does not find correspondence in the test model.

The presence in the mathematical simulation of a horizontal critical speed with a frequency lower than the experimental one has been quite frequently observed in computations performed with data relative to real turbo generator groups. Also in those cases 'linearized' film stiffness and damping coefficients had been used. This has been one of the reasons for which a research about a suitable schematization of the oil film in non linear terms has been examined closely [9].

The methodology described in this paper has been applied to the journal bearings of the laboratory model. The shape and dimension of the orbits necessary to run the method have been derived from the measurements made on the set up: as can be seen in fig.6, the orbits are nearly circular and a mean value of their amplitudes can be assumed around 80 μm, that corresponds, in non-dimensional terms, to a = 0.35.

Stiffness coefficients obtained with this data (that can be compared in terms of $\bar{\omega} / \bar{\omega}_o$ with the 'linear' ones in fig.4) have given the frequency responses reported in fig.6.

As can be noted there is a good fit between analytical and experimental results and this fact indicates the method validity, at least for this application. It can be pointed out, for instance, that the horizontal critical speed at about 2000 r/min is no more present and this is due to the oil film stiffening effect associated to non linearities.

5 INSTABILITY PROBLEMS

Using the laboratory set up, some investigations about the instability fields of the system have been carried out: various unbalance conditions and two different configurations for the supporting structure have been considered.

The first series of measures, with rigid foundation, has shown that, with small unbalance (orbits with a = 0.2), the instability threshold is about 3500 r/min and the subharmonic frequency is slightly lower than half the rotational speed, following the rotational speed itself during the run up (oil whirl).

Increasing unbalance, the instability threshold become higher and, with orbits having a = 0.5, it is over 6000 r/min.

A second series of measurements has been performed after having supported the base of the model on elastic elements. This fact introduces a critical frequency of the supporting structure at about 1500 - 2000 r/min (for further details see [10]). In this case a small unbalance (a = 0.2) makes the instability threshold to be around 3500 r/min as in the previous condition, but now the subharmonic frequency is equal to the foundation resonance, that is 1500 r/min, and does not change increasing the rotor speed (oil whip).

Increasing unbalance, the instability threshold become higher and, with orbits having a = 0.5, it is over 6000 r/min.

These tests demontrate two facts:

- the phenomenona here indicated as oil whip and oil whirl are not different, but it is only a problem connected to the natural frequencies: in one case it is an excited natural frequency prevalently dependent on the structure, in the other, the natural frequency involved in the phenomenon is related to the oil film and, therefore, it depends on the journal rotational speed;
- increasing the amplitude of the journal orbits (i.e. the unbalance), the system goes towards greater stability conditions.

Considering the already reported equation (2), that reproduces the behaviour of a rigid rotor on oil film, besides the ratio $\bar{\omega} / \bar{\omega}_o$ it is possible to compute also the instability coefficient $\alpha / \bar{\omega}$ (fig.8) and the ratio $\Omega / \bar{\omega}$ (fig.7), both reported as a function of χ and for different a values.

It can be observed that the journal rotational speed Ω is about twice the natural frequency $\bar{\omega}$ (fig.7), in all the field of the smaller χ , that is in all the field interested in instability conditions, where the $\alpha / \bar{\omega}$ values (fig.8) increase and can become positive (case of a = 0).

This indicates that the experimentally observed instability with frequency equal to about one half of the rotational speed is justified by a 'linear' theory [9] and it is caused by a flutter instability with frequency equal to the natural frequency of the rotor on oil film.

In other words, with rigid rotor and rigid supporting structure, the oil film stiffness, which depends on the rotational speed, gives the most important contribution to the natural frequency. This is approximately equal, in instability conditions (fig.7), to one half of the rotor speed (oil whirl). With flexible shaft and/or flexible supporting structure the prevailing terms become those associated to the structural parameters of the system and so the instability frequency is independent from the rotational speed (oil whip).

Observing fig. 7 and 8, it can be noted that, increasing a , $\alpha / \bar{\omega}$ becomes more negative, that is the instability threshold increases to higher rotational speeds as has been experimentally observed. This fact seems to be due, mainly, to an increase of the natural frequency $\bar{\omega}$ of the system, as clearly shown in fig.7.

6 CONCLUSIONS

As a conclusion of this paper, two facts can be put in evidence:

- the oil whip and the oil whirl phenomena are not two different instability problems, but they are both a flutter type instability;
- the linear approach here proposed, with oil film stiffness and damping equivalent coefficients dependent on the journal orbit amplitudes, seems to suitably reproduce some of the non linear phenomena that can be observed in the dynamic behaviour of a real rotor.

REFERENCES

[1] LUND, J.W. NIELSEN, H.B. Instability threshold of an unbalanced, rigid rotor in short journal bearings. II Int. Conf. Vibrations in Rotating Machinery, Cambridge 1980, Paper C263/80.

[2] LUND, J.W. SAIBEL, E. Oil whip whirl orbits of a rotor in sleeve bearings. Jrnl. Engineering for Industry, Trans. ASME, 1967, Vol. 89, pagg. 813-823.

[3] PORITSKY, H. Contribution to the theory of oil

whip. Trans. ASME, 1953, Vol. 75, pagg. 1153-1161

[4] HORI, Y. A theory of oil whip. Jrnl. Applied Mechanics, 1959, Vol. 26, pagg. 189-198

[5] RUGGIERI, G. Un metodo approssimato per la risoluzione dell'equazione di Reynolds. L'Energia Elettrica, 1976, Vol.LIII.

[6] DIANA, G. BORGESE, D. DUFOUR, A. Experimental and analytical research on a full scale turbine journal bearing. Proc. 2nd Conf. Vibration in Rotating Machinery, Cambridge, 1980.

[7] DIANA, G. P.A.L.L.A. A package to analyze the dynamic behaviour of a rotor supporting structure system. CISM Rotordynamics Conf., Udine, 1986.

[8] DIANA, G. MASSA, E., PIZZIGONI, B. A forced vibration method to calculate the oil film instability threshold of rotor-foundation systems. Int. Conf. Rotordynamic Problems in Power Plants, Rome, 1982.

[9] MANENTI, A. Effetti non lineari nei cuscinetti lubrificati. Tesi di Dottorato di Ricerca, 1987, Dipartimento di Meccanica, Politecnico di Milano.

[10] DIANA, G. CHELI, F. VANIA, A. A method to identify the foundation modal parameters through measurements of the rotor vibrations. 4th Int. Conf. on Vibration in Rotating Machinery, Edinburgh, 1988.

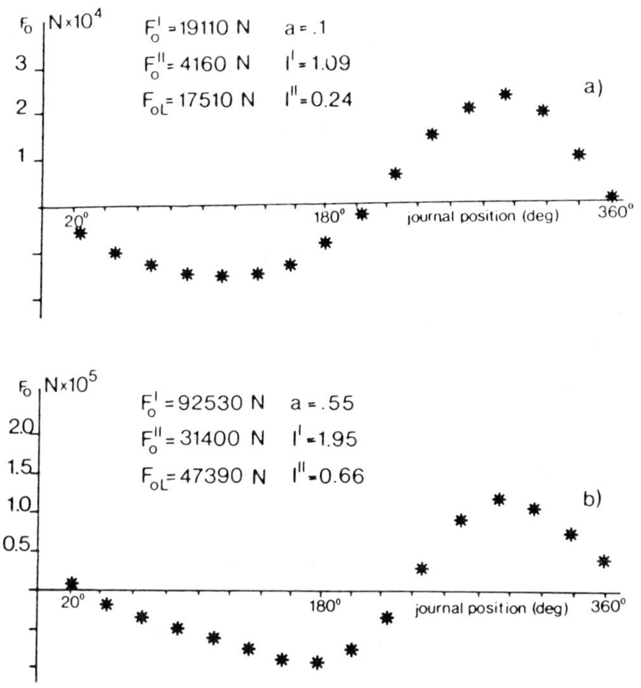

Fig 1 Horizontal oil—film force components calculated along two different orbits whose amplitudes are defined by the a parameter

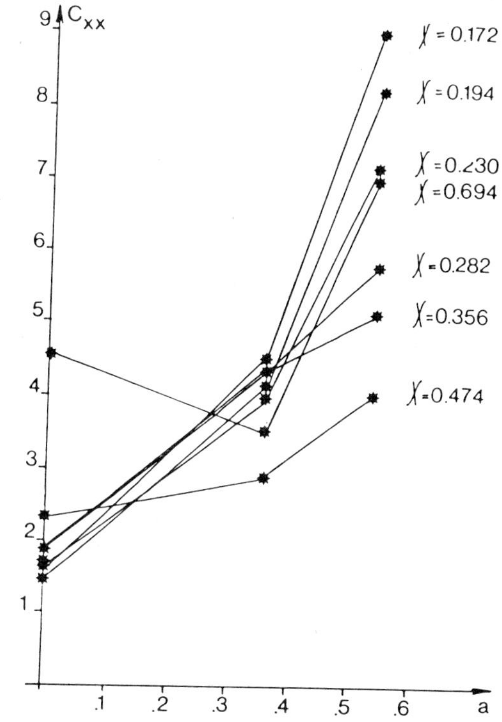

Fig 3 Direct non-dimensional stiffness (c_{xx}) in vertical direction versus orbit amplitude (a)

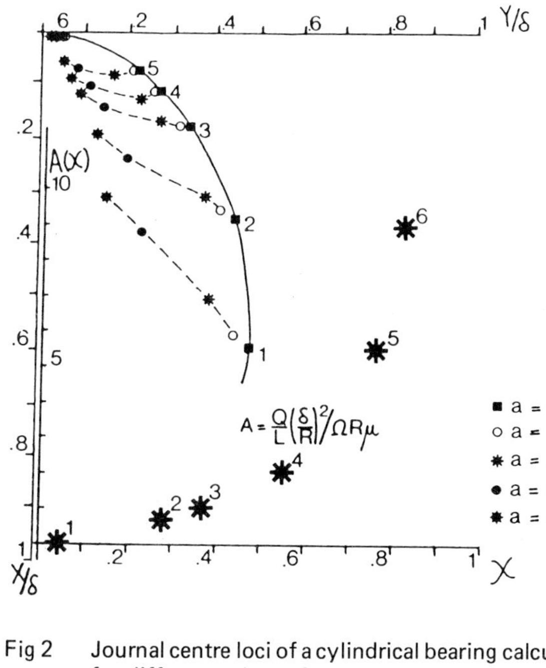

$A = \frac{Q}{L}\left(\frac{\delta}{R}\right)^2 / \Omega R \mu$

Fig 2 Journal centre loci of a cylindrical bearing calculated for different values of the orbit amplitude (a) and journal bearing load capacity coefficient (A)

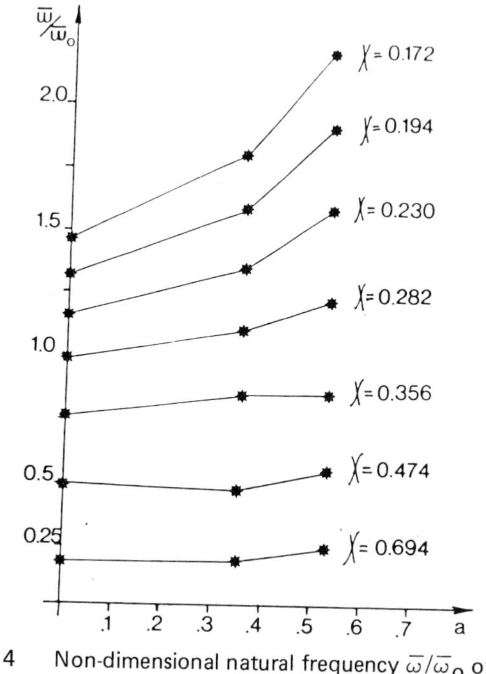

Fig 4 Non-dimensional natural frequency $\bar{\omega}/\bar{\omega}_o$ of the journal bearing system versus orbit amplitude (a) for different eccentricity ratios (χ)

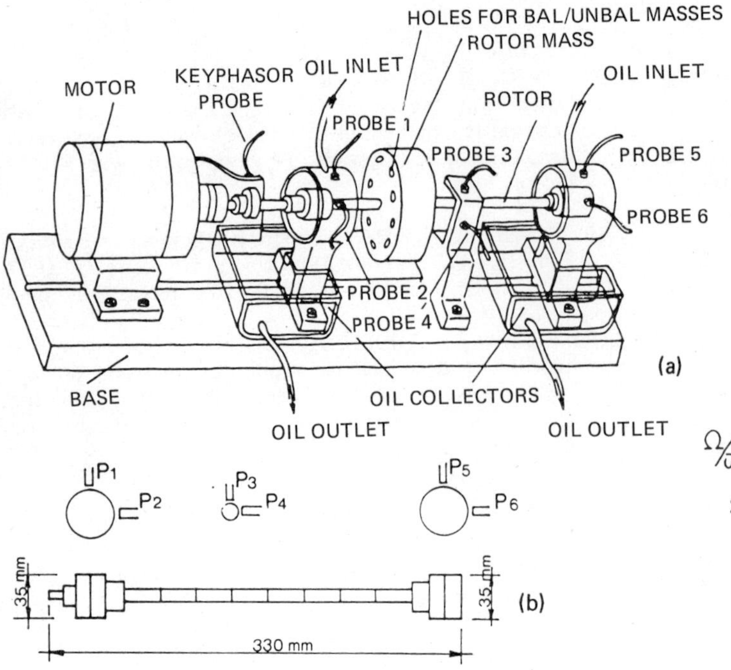

Fig 5 Laboratory set up (a) and its schematization (b)

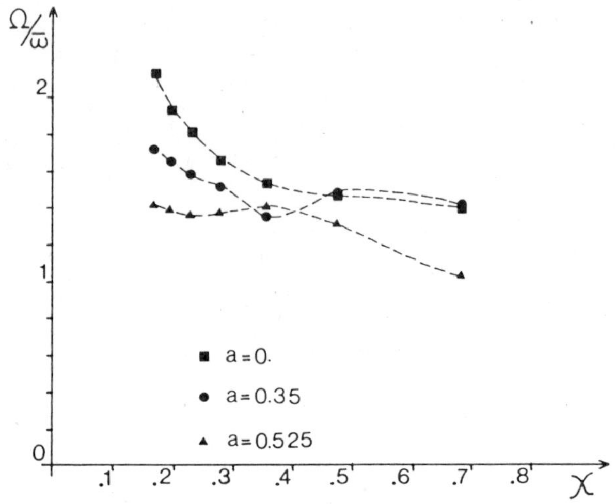

Fig 7 Calculated ratios between the journal rotational speed and the journal bearing natural frequency versus eccentricity ratio (χ) for different orbit amplitudes (a)

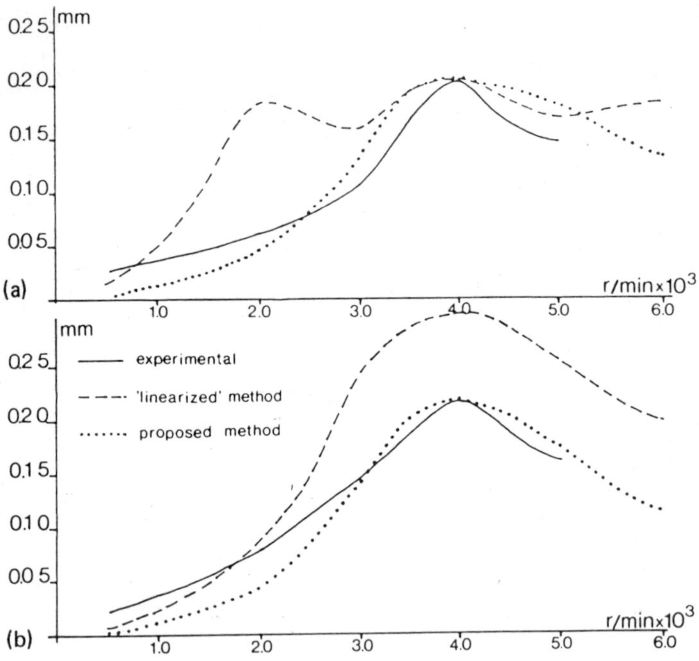

Fig 6 Frequency response of the laboratory rotor model in correspondence with the journal bearing opposite to the motor; (a) vertical vibrations, (b) horizontal vibrations

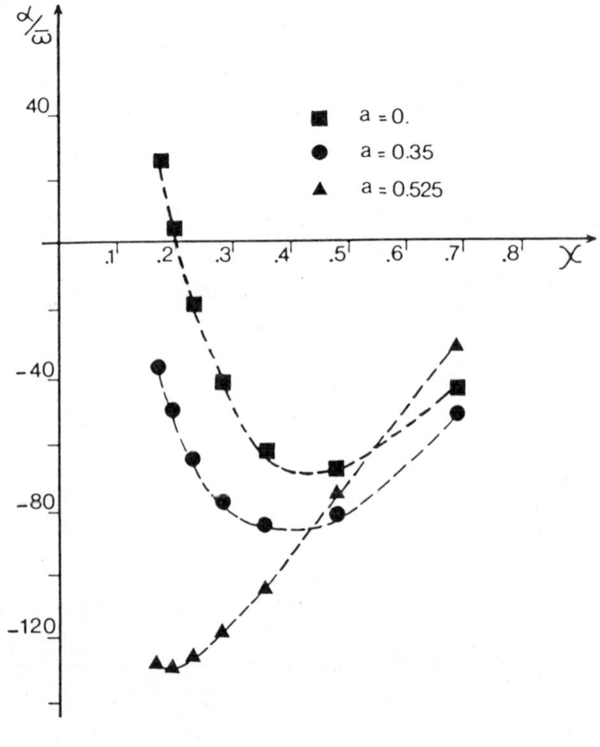

Fig 8 Calculated instability index of the journal bearing versus eccentricity ratio (χ) for different orbit amplitudes (a)

C315/88

Clearance effects on leakage and rotordynamic coefficients of smooth, liquid annular seals

D W CHILDS, BS, MS, PhD, PE, MASME
Department of Mechanical Engineering, Texas A and M University, Texas, United States of America
S NOLAN, MS, MASME and D NUNEZ, BS
Rockwell International, Canoga Park, California, United States of America

ABSTRACT

Test results are presented for smooth seals with $L/D = \frac{1}{2}$ and three clearance ratios: .005, .0076, .010. The results consist of leakage data, axial pressure gradients, and radial and tangential force coefficients. Measured axial pressure gradients yield friction factor data which are higher than Yamada's earlier measured results. Further, the decrease in friction factors with increasing Reynolds numbers is more rapid than would be expected from pipe data. Measured stiffness and damping coefficients compare favorably with theoretical predictions. Doubling the radial clearance yields an approximate 40% decrease in the stiffness and damping coefficients.

1. INTRODUCTION

From a rotordynamics viewpoint, seal analysis has the objective of predicting the coefficients for the following motion/reaction-force model

$$-\begin{Bmatrix} F_X \\ F_Y \end{Bmatrix} = \begin{bmatrix} K & k \\ -k & K \end{bmatrix} \begin{Bmatrix} X \\ Y \end{Bmatrix} + \begin{bmatrix} C & c \\ -c & C \end{bmatrix} \begin{Bmatrix} \dot{X} \\ \dot{Y} \end{Bmatrix} + M \begin{Bmatrix} \ddot{X} \\ \ddot{Y} \end{Bmatrix} \tag{1}$$

where X, Y are components of the seal-rotor displacement relative to its stator and F_X, F_Y are components of the reaction force. The diagonal and off-diagonal stiffness and damping coefficients are referred to, respectively, as "direct" and "cross-coupled." The cross-coupled coefficients arise due to fluid rotation within the seal. The coefficient M accounts for the seal's added mass.

Recent analyses which have been developed to define the rotordynamic coefficients of liquid annular seals use an expansion in the eccentricity ratio to yield zeroth and first-order equations. The zeroth-order equations define the flow and pressure fields for a centered position. The first-order equations depend on the zeroth-order solution and are used to define the rotordynamic coefficients. As a consequence of these developments, the rotordynamic coefficients depend on essentially static data for the centered position, viz., friction factor and inlet-loss correlations.

Childs [1] and Childs and Kim [2] use Hirs' turbulence model to define the wall-shear stress acting on a fluid element within the seal. This approach yields a Blasius friction-factor correlation. Nelson and Nguyen [4] use Moody's equation to define the friction factor and wall shear stress. The Blasius approach defines the friction factor as a power-law function of the Reynolds number; the two empirical coefficients used in the correlation change, depending on the relative roughness of the surfaces and the Reynolds number range. The Moody formula defines the friction factor as an explicit empirical function of Reynolds number and relative roughness. The data to support either of these correlations *in annular seals* are particularly sparse, consisting largely of Yamada's [5] data. The present work presents friction factor data for smooth seals for the following test configurations: (a) the inner (rotating) cylinder is concentric with the outer cylinder, and (b) the inner cylinder is nominally centered but orbits synchronously at its rotational speed. Data are presented for a range of clearances and Reynolds numbers.

Historically, the inlet-loss coefficient has been assumed to be constant in calculating rotordynamic coefficients. The present work presents inlet-loss versus Reynolds number data for three clearances over the axial Reynolds number range of 50,000 to 300,000.

Measured rotordynamic test data for smooth seals have generally been at C_r/R ratios on the order of .010

[1]This work was supported in part by NASA Contract NAS8-35824, George C. Marshall Space Flight Center, project monitor Hugh Campbell.

as compared to C_r/R ratios of .003 for pump applications, and the measured direct stiffness values from these test are larger than predicted: Childs and Dressman [7] and Childs [8]. There is an obvious question as to whether this discrepancy would increase or decrease as the C_r/R ratio is reduced; hence, dynamic tests were carried out in this study for the three C_r/R ratios: .005, .0076, .010.

In summary, results are reported and discussed from the following tests of smooth, annular seals:

(a) axial pressure gradients for four C_r/R ratios, with the seal rotor and sleeve concentric,
(b) axial pressure gradients for three C_r/R ratios with the inner cylinder orbiting at its rotational speed,
(c) inlet-loss coefficients for three C_r/R ratios, and
(d) dynamic force coefficients for three C_r/R ratios.

All seals are tested over a range of Reynolds numbers and running speeds.

2. TEST APPARATUS AND MEASUREMENTS

2.1 Dynamic Measurements

Figure 1 illustrates the dynamic seal-test apparatus. Fluid enters in the center and discharges axially across the two test seals. For dynamic tests, the rotor elements of the seals are mounted eccentrically to the bearing-enforced axial of rotation. Because of the eccentricity, shaft rotation generates a rotating pressure field. This transient pressure field is measured via five, axially-spaced, strain-gauge, pressure transducers. Transient measurements are also made of the seal-rotor motion in the horizontal and vertical planes. The pressure signals are integrated to yield force coefficients which are parallel and transverse to the rotating, seal-eccentricity vector, i.e., radial and circumferential force coefficients are calculated from test data.

The test section and flow loop achieve high-Reynolds numbers by pumping Halon (CB_rF_3) a Dupont-manufactured fire extinguisher fluid and refrigerant. Additional details of the test section and flow loop are provided by Childs et al. [9].

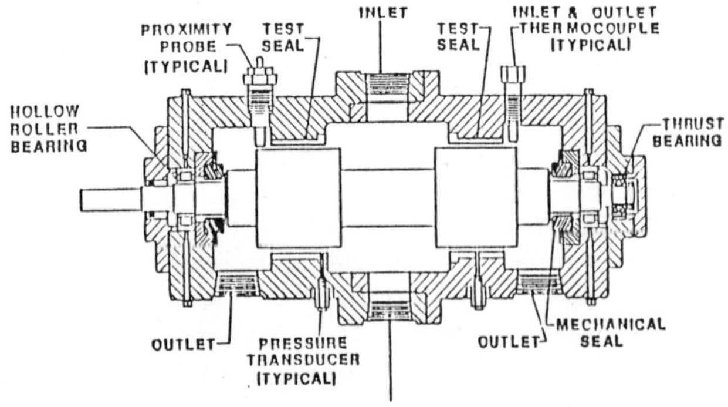

Fig 1 High Reynolds number seal test section

The dynamic test apparatus yields circular-orbital motion of the form

$$X = A\cos\omega t, \quad Y = A\sin\omega t \tag{2}$$

When substituted into Eq. (1), this motion yields the following definition of force coefficients which are, respectively, parallel and perpendicular to the rotating displacement vector

$$\begin{aligned} F_r/A &= -K - c\omega + M\omega^2 \\ F_\theta/A &= k - C\omega \end{aligned} \tag{3}$$

Observe that the cross-coupled-stiffness coefficient k yields a "driving" tangential contribution in the direction of rotation, while the direct damping coefficient develops a drag force opposing the tangential velocity.

The apparatus of Figure 1 can be used to measure F_r/A and F_θ/A versus Reynolds number and running speed. However it can not separately identify the coefficients of Eq. (3) because they depend on running speed. Eq. (3) can be used as the basis for a quantitative comparison between theory and experiment by assuming that k and c vary linearly with ω; the result is

$$\begin{aligned} f_\theta &= F_\theta/A = C_{ef}\omega \\ f_r &= F_r/A = K_{ef} + M_{ef}\omega^2 \end{aligned} \tag{4}$$

Test results are obtained for a fixed axial Reynolds-number over a range of running speeds and then curve-fitted to obtain K_{ef}, C_{ef}, and M_{ef}. K_{ef} is the zero-running-speed intercept of the $f_r(\omega)$ curve; C_{ef} is the slope of the $f_\theta(\omega)$ curve and is the "net-damping coefficient" resulting from the drag force $C\omega A$ and the forward-whirl excitation force kA. Quantitative comparisons can be made between measured and experimental values for K_{ef}, C_{ef}, and M_{ef}.

2.2 Static Measurements

Static measurements include upstream and downstream temperature and pressure measurements. Additionally, static results are provided by the five axially-spaced pressure transducers which are also used to measure the reaction force components. These transducers are located symmetrically about the midplane of the seal at .05, .23, and 0.5 of the seal length. The inlet-loss coefficient is defined by the pressure measurement upstream and immediately inside the seal. The axial pressure gradient is defined by the three centered pressure taps at $0.23L, 0.5L$, and $0.77L$. The seal length L is 5.08 cm.

3. STATIC MEASUREMENTS: THEORY VERSUS EXPERIMENTS

3.1 Hirs-Equation-Leakage Model

The zeroth-order equations which result from Hirs turbulent lubrication model define the pressure and circumferential velocity as a function of z for a centered, concentric position. For constant clearances, the equations are

$$\begin{aligned} \frac{dp_o}{dz} &= -(\sigma_s + \sigma_r)/2 \\ \frac{du_{\theta o}}{dz} &= [\sigma_s u_{\theta o} + a_{or}\sigma_r(u_{\theta o} - 1)] \end{aligned} \tag{5}$$

The σ parameters are related to the friction factor λ by

$$\sigma_r = \lambda_r \frac{L}{C_r} \ , \ \sigma_s = \lambda_s \frac{L}{C_r} \ , \tag{6}$$

and the friction factor is related to the empirical coefficients through the following friction-factor formulas

$$\begin{aligned} \lambda_r &= nrR_a^{mr}\{1 + [(u_{\theta o} - 1)/b]^2\}^{\frac{mr+1}{2}} \\ \lambda_s &= nsR_a^{ms}[1 + (u_{\theta o}/b)^2]^{\frac{ms+1}{2}} \end{aligned} \tag{7}$$

where

$$R_a = \frac{VC\rho}{\mu} \ , \ b = V/R\omega \tag{8}$$

The subscripts r and s denote rotor and stator respectively. the coefficients (ms, ns, mr, nr) characterize the friction factors for the rotor and stator and must be determined empirically.

Since equations (5) are coupled and nonlinear their solution is iterative. The initial conditions are

$$\begin{aligned} p(0) &= \Delta P/\rho V^2 \\ u_{\theta o}(0) &= U_{\theta o}(0)/R\omega \ , \end{aligned} \tag{9}$$

where V, which defines the leakage rate, is unknown.

For equal rotor and stator roughnesses $(mr = ms = m, nr = ns = n)$, the governing equations (5) have the following asymptotic $(z \to \infty)$ solution

$$\begin{aligned} \frac{dp_o}{dz} &= -\sigma \ , \ u_{\theta o} = \frac{1}{2} \\ a_{os} &= a_{or} = 1 \ , \ \sigma_r = \sigma_s = \sigma \\ \lambda_r &= \lambda_s = \lambda = nR_a^m(1 + \frac{1}{4b^2})^{\frac{m+1}{2}} \end{aligned}$$

The same solution results if $u_{\theta o}(0) = \frac{1}{2}$. However in the present test apparatus, there is no intentional prerotation of the fluid and a general solution results with coupled axial and circumferential-momentum equations (5).

The problem of immediate interest is: Given static leakage and axial-pressure measurements over a range of Reynolds numbers and running speeds, what values of m and n yield an optimum agreement between the differential equation model represented by Eq. (5) and the test results. This is, in fact, a dynamic parameter estimation problem; however, the following approximate algebraic solution approach is used to estimate m and n. The variable $u_{\theta o}(z)$ is replaced by its average

$$\overline{u}_{\theta o} = \int_0^1 u_{\theta o}(z)dz \tag{10}$$

Hence, Eq. (5.a) becomes

$$\frac{dp_o}{dz} = -\overline{\sigma}_c \ ,$$

where

$$\begin{aligned} \overline{\sigma}_c &= (\overline{\lambda}_s + \overline{\lambda}_r)L/2C_r \\ \overline{\lambda}_r &= nR_a^m\{1 + [(\overline{u}_{\theta o} - 1)/b]^2\}^{\frac{m+1}{2}} \\ \overline{\lambda}_s &= nR_a^m[1 + (\overline{u}_{\theta o}/b)^2]^{\frac{m+1}{2}} \end{aligned} \tag{11}$$

The first of Eq. (11) can be stated

$$2\overline{\sigma}_c\left(\frac{C_r}{L}\right) = \overline{\lambda}_c = n(C_1X^m + C_2Y^m) \tag{12}$$

where

$$\begin{aligned} C_1 &= [1 + (\overline{u}_{\theta o}/b)^2] \\ C_2 &= \{1 + [(\overline{u}_{\theta o} - 1)/b]^2\} \\ X &= R_aC_1 \ , \ Y = R_aC_2 \end{aligned} \tag{13}$$

The form of Eq. (12) isolates the unknowns m and n.

The parameters m and n are defined by minimizing the following objective function

$$E = \sum_{i=1}^{k}(2\lambda_{ci} - 2\hat{\lambda}_{ci})^2 = \sum_{i=1}^{k}[n(C_{1i}X_i^m + C_{2i}Y_i^m) - 2\hat{\lambda}_{2i}]^2 \tag{14}$$

which is the summation over the k data points of the squared difference between the predicted, λ_{ci}, and the measured, $\hat{\lambda}_{ci}$, friction factors. Minimization of E with respect to m and n yields the following equations

$$\begin{aligned} f &= \frac{\partial E}{\partial n} = \sum_{i=1}^{k}[n(C_{1i}X_i^m + C_{2i}Y_i^m) - 2\hat{\lambda}_{ci}](C_{1i}X_i^m + C_{2i}Y_i^m) = 0 \\ g &= \frac{\partial E}{\partial m} = \sum_{i=1}^{k}[n(C_{1i}X_i^m + C_{2i}Y_i^m) - 2\hat{\lambda}_{ci}]n(C_{1i}X_i^m \ln X_i \\ &\qquad + C_{2i}Y_i^m \ln Y_i)m = 0 \end{aligned} \tag{15}$$

These nonlinear equations are solved for n and m using a Newton-Raphson procedure. Figure 2 provides a flowchart for the solution algorithm used to obtain n and m for equal rotor and stator roughnesses. An iterative solution is required since the differential equations (5) include the unknown parameters n and m, and Eq. (15) contains C_1 and C_2 which depend on $\overline{u}_{\theta o}$.

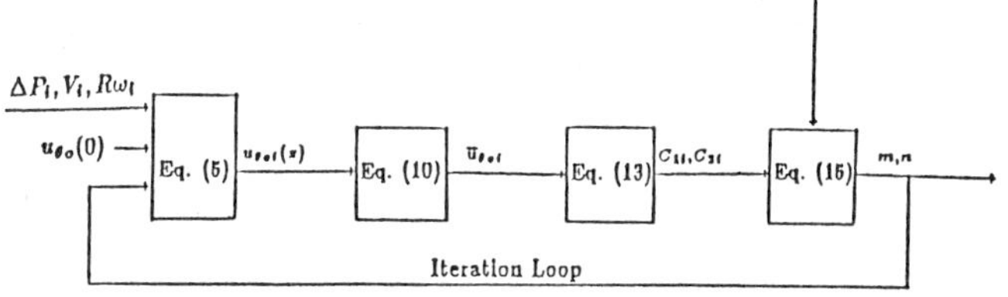

Fig 2 Flowchart for the solution of the empirical coefficients for equal rotor and stator surface roughnesses (the subscript i denotes the ith data point)

3.2 Friction Factor and Empirical-Coefficient Results for Smooth, Centered Seals

Four smooth stators were manufactured and tested in the apparatus of Figure 1 solely to obtain friction-factor versus leakage data. The test rotor was run at speeds from 1000 to 7200 *rpm*; however, the seal elements of the rotor had no eccentricity, unlike the dynamic tests with eccentric rotors which are reported subsequently. The stators all had measured surface roughnesses of approximately 0.81μm. The seals were tested at the following clearances: (a) .312 *mm*, (b) .375 *mm*, (c) .446 *mm*, and (d) .500 *mm*. Figure 3 illustrates representative test and theoretical results for $\overline{\sigma}_c$ as a function of running speed and Reynolds number. The theoretical results of Figure 3 are based on calculated m, n values which were obtained from the solution procedure of Figure 2. An inspection of these figures shows a generally good agreement between experiment and theory. A summary of the m, n results is provided in Table 1.

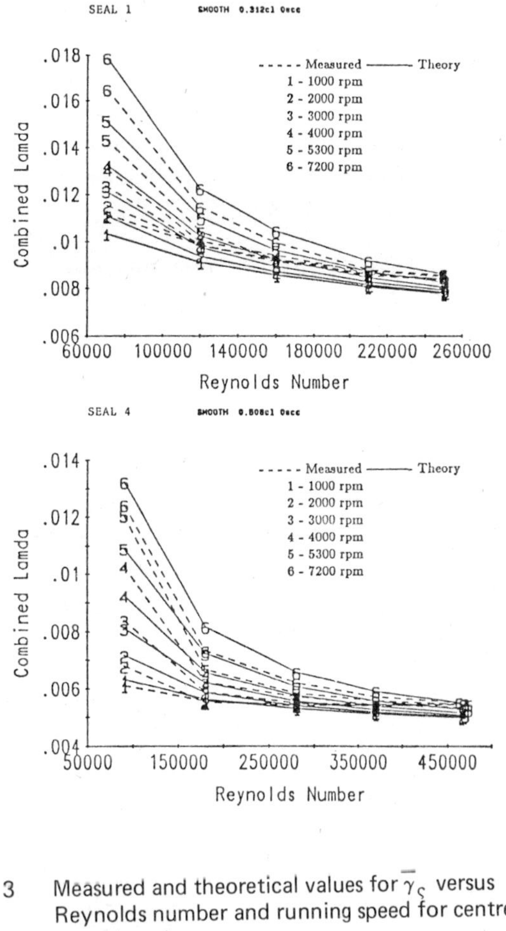

Fig 3 Measured and theoretical values for $\overline{\gamma}_c$ versus Reynolds number and running speed for centred smooth seals

Table 1. Hirs Coefficients for 0.81μm roughness.

No.	$C_r(mm)$	$D(mm)$	m	n	$R_a(max)$
1	0.312	101.00	-.205	.100	250,000
2	0.375	100.85	-.058	.011	360,000
3	0.446	100.71	-.139	.043	360,000
4	0.508	100.58	-.115	.022	475,000

These data indicate that the empirical parameters m, n can only characterize the friction-factor properties of the stator for the same clearance, roughness, and Reynolds-number range. The data can also be interpreted in terms of changes in *relative* roughness, $\epsilon = e/2C_r$, where e is the absolute roughness. As the clearance is reduced e remains fixed; hence, ϵ increases. These results support the use of a Moody friction-factor characterization as proposed by Nelson and Nguyen rather than a Hirs-based analysis.

3.3 Friction-Factor Data For Smooth Seals with Eccentric, Synchronously-Precessing Orbits

The experimental pressure gradients presented in this section were developed as peripheral data in the dynamic testing of smooth seals at the following clearances: 0.254, 0.381, and 0.508*mm*. For all tests the radial (dynamic) eccentricity was set at 89μm. Dynamic pressure measurements were averaged over several cycles of data to define nominal static pressure gradients. Measured surface roughness for the seals was 0.81μm.

The procedure of Figure 2 was used to estimate m and n values for the pressure gradient data. Calculated values for m and n yielded good agreement between measurements and predictions, comparable to the centered-seal results of Figure 3. Hirs coefficients for the three seals with orbiting rotors are presented in Table 2.

Table 2. Hirs coefficients for 0.81μm roughness seals with an orbiting rotor; seal diameter 101.6mm; $A = 89\mu m$.

No.	$C_r(mm)$	m	n	$R_a(max)$
5	0.254	-0.256	0.174	190,000
6	0.381	-0.098	0.019	350,000
7	0.508	-0.056	0.016	450,000

3.4 Seal Friction Factor Data in a Moody Diagram

The values of Hirs coefficients which are provided in Tables 1 and 2 are not of themselves particularly informative. To provide some perspective, the m and n values have been used to calculate friction factors which are illustrated in Figure 4. The λ friction factors of this report have been multiplied by a factor of 4 to yield the friction factor values for this figure. The numbers in this figure correspond to the seal numbers of tables 1 and 2.

Observe that the friction-factor data for concentric seals generally proceeds reasonably (f increases as C_r decreases) except for Seal 2. The friction factors corresponding to orbiting rotor cases do not proceed as expected since the friction factors are higher for Seal 7 than Seal 6. Note, that where the clearances are approximately the same in concentric and orbiting-rotor cases, that the dynamic friction factor is generally higher, but the difference decreases as the clearances decreases. A partial explanation for the difference between static and dynamic test results are provided by Tam et al. [10], whose computational results predict secondary flow developments in the orbiting-rotor test cases.

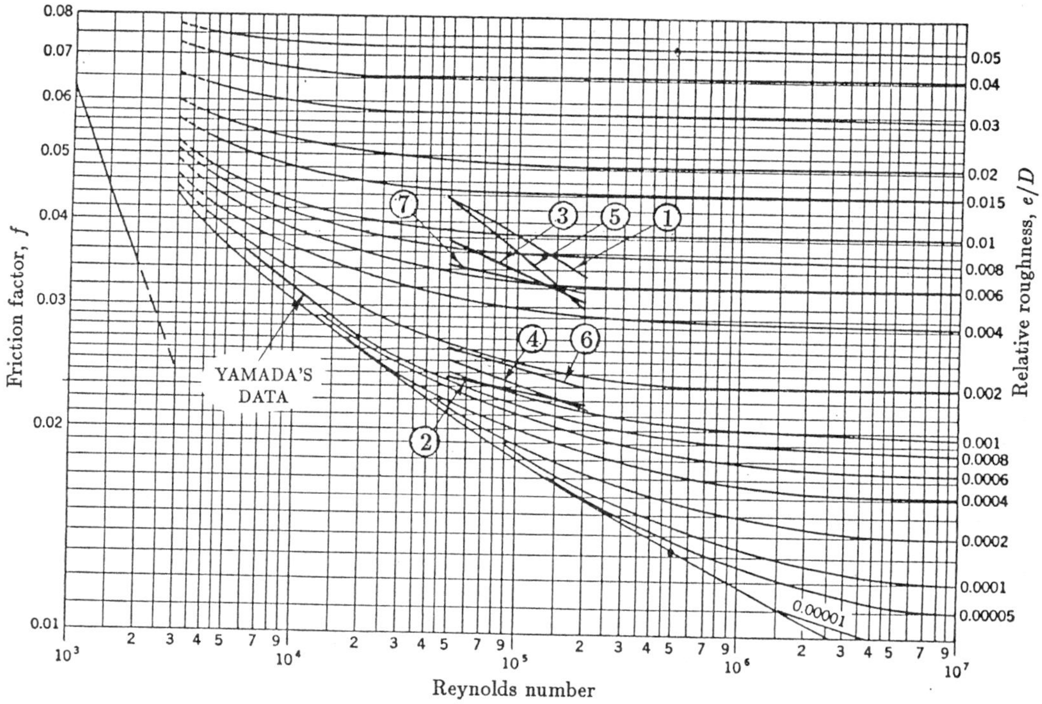

Fig 4 Measured friction-factor data plotted on a Moody diagram

The line at the lower left hand side of the turbulent friction factors illustrates Yamada's test-data correlation. Obviously, the "smooth" seals tested in this program have higher friction factors than those measured by Yamada. Differences in surface finishes between the current seals and Yamada's seals could possibly account for these differences.

3.5 Inlet-Loss Test Data

Figure 5 illustrates $(1+\xi)$ versus Reynolds number for the three orbiting-rotor, smooth, test seals. Observe that, in all cases, $(1+\xi)$ approaches an asymptote as R_a increases.

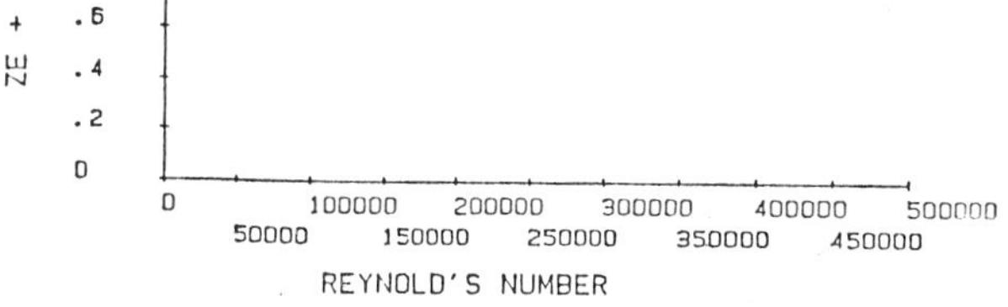

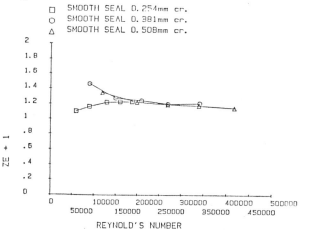

Fig 5 Inlet-loss-factors versus axial Reynolds number for three clearances

Table 3 provides the average value for $1+\xi$ over the Reynolds-number range.

Table 3. Entrance-loss data for smooth seals

$C_r (mm)$	$(1+\xi)$
0.254	1.21
0.381	1.25
0.508	1.18

4. DYNAMIC MEASUREMENTS: THEORY VERSUS EXPERIMENTS

The analysis procedure used to predict rotordynamic coefficients for annular seals is that of Childs and Kim [2]. Table 4 provides experimental values for K_{ef} and C_{ef} and a comparison between theory and experiment. The model slightly underpredicts K_{ef} at the larger clearances and over predicts at the minimum clearances. C_{ef} is well predicted for all clearances and Reynolds numbers.

Figure 6 illustrates K_{ef} for the three clearances versus Reynolds number. Note that doubling the clearances drops the stiffness by approximately 40%. Also illustrated is the nondimensionalized parameter

$$\overline{K}_{ef} = K_{ef}\left(\frac{C_r}{DL\Delta P}\right),$$

Observe that the nondimensionalization is generally effective in collapsing the data into a reasonably constant band.

Figure 7 illustrates C_{ef} for the three clearances and demonstrates that doubling the clearance also reduces damping by approximately 40%. The nondimensionalized parameter

$$\hat{C}_{ef} = C_{ef}\left(\frac{C_r T}{DL\Delta P}\right)$$

is also illustrated where $T = L/V$ is the transit time through the seal. The objective of this nondimensionalization would be to reduce the C_{ef} test results into a single curve which would be independent of both ΔP and C_r. While the ΔP dependency is eliminated, the C_r dependency remains.

Table 4. Measured values for K_{ef} and C_{ef} (SI units), a comparison between theory and experiment.

	K (Exp)	C_{ef} (Exp)	K_{ef} (Exp/Th)	C_{ef} (Exp/Th)
Smooth Seal 0.254 mm cr.				
R_a= 59,940	1,434,000	10,250	0.914	1.190
R_a= 89,980	4,703,000	13,440	1.460	0.954
R_a=130,000	5,426,000	20,830	0.823	0.982
R_a=160,000	8,033,000	25,980	0.784	0.982
R_a=188,500	10,696,000	28,900	0.758	0.928
Smooth Seal 0.381 mm cr.				
R_a= 90,140	737,200	4,971	1.370	0.954
R_a=149,900	1,754,000	8,507	1.677	0.947
R_a=210,200	3,324,000	12,670	1.113	1.017
R_a=270,100	4,953,000	17,620	1.023	1.085
R_a=341,400	7,743,000	21,740	1.000	1.060
Smooth Seal 0.508 mm cr.				
R_a=120,000	654,200	4,554	1.481	1.197
R_a=200,000	1,516,000	7,534	1.229	1.120
R_a=270,100	2,520,000	10,830	1.135	1.180
R_a=340,100	3,707,000	13,720	1.077	1.132
R_a=418,300	5,550,000	16,240	1.080	1.112

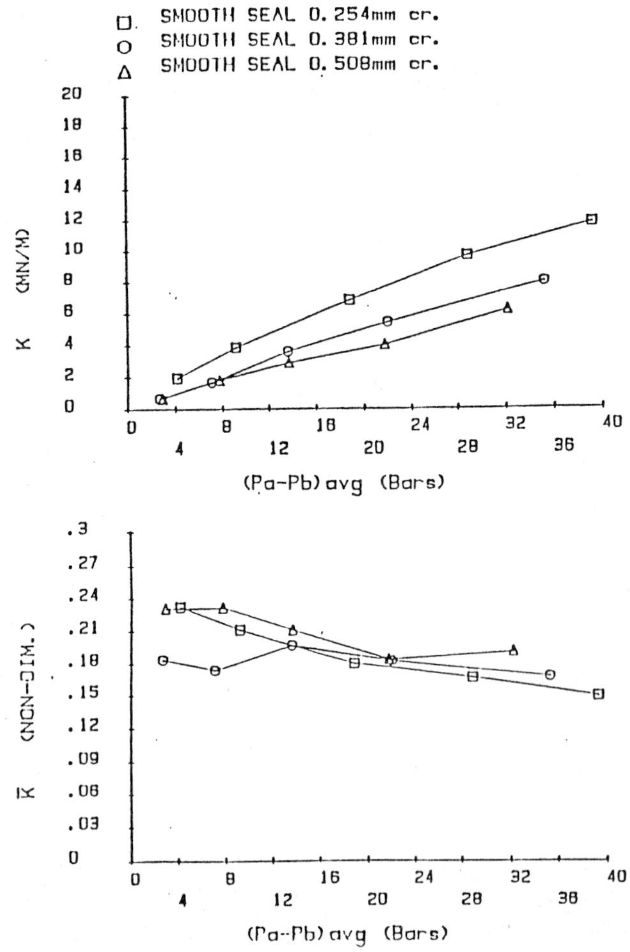

Fig 6 $\bar{K}_{ef}$ and $\hat{K}_{ef}$ versus ΔP for smooth seals at three different clearances

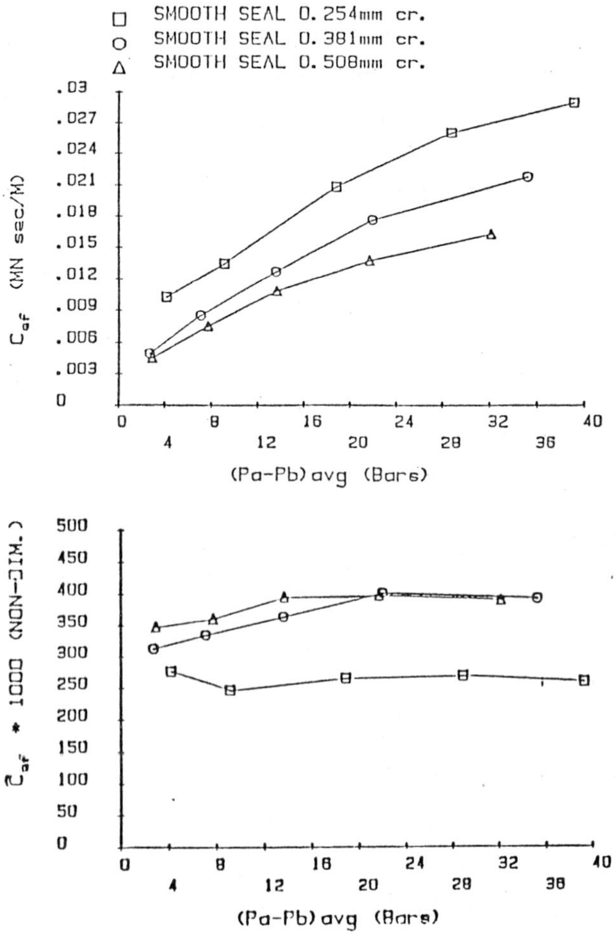

Fig 7 C_{ef} and $\hat{C}_{ef}$ versus ΔP for smooth seals at three different clearances

5. SUMMARY AND CONCLUSIONS

The test results presented here support the following conclusions:

5.1 Static Friction-Factor Data.

(a) The results reported here differ from Yamada's test results in that the friction factors are larger and tend to decrease more rapidly with increasing Reynolds numbers. These differences may be partially explained by differences in surface roughnesses and the fact that Yamada took pains (via a lengthy inlet section) to obtain fully developed flow, while the present tests apparatus clearly does not have fully-developed flow.

(b) The test results show consistently higher friction factors when the rotor is orbiting versus non-orbiting cases. These results are consistent with Tam et al.'s [10] numerical predictions of secondary flow due to the rotor's orbital motion.

(c) The friction-factor data demonstrate that the Hirs' wall-friction model is considerably more restricted than had previously been though. The empirical coefficients only characterize the seal surfaces at a fixed clearance and over a restricted Reynolds number range. The "Moody" approach of von Pragenau [12] and Nelson and Nguyen [4] needs to be examined to see whether it yields broader ranges of applicability.

(d) Given that the rotordynamic coefficients depend critically on the friction factor and inlet-loss data, considerably more data of this type needs to be developed.

5.2 Rotordynamic Coefficients

(a) The model does a good job of predicting C_{ef} for all clearances and Reynolds numbers.

(b) The model does a good job of predicting K_{ef} for the two larger clearances and the accuracy improved as R_a increases. However, for the smallest clearances measured, K_{ef} values are smaller than predicted and the accuracy is reduced as R_a increases.

(c) From a practical viewpoint, K_{ef} and C_{ef} both dropped by about 40% when the clearance was doubled.

REFERENCES

1. Childs, D., 1983, "Finite-Length Solutions for Rotordynamic Coefficients of Turbulent Annular Seals," *ASME Trans. J. of Lubrication Technology,* Vol. 105, pp. 437-444.
2. Childs, D. and Kim, C-H, 1985, "Analysis and Testing for Rotordynamic Coefficients of Turbulent Annular Seals with Different Directionally Homogeneous Surface Roughness Treatment for Rotor and Stator Elements," *ASME Trans. J. of Tribology,* Vol. 107, pp. 296-306.
3. Hirs, G. G., 1973, "A Bulk-Flow Theory for Turbulence in Lubricant Films," *ASME Trans. J. of Lubrication Technology,* Vol. 95, No.2, pp. 137-146.
4. Nelson, C. C. and Nguyen, B., 1987, "Comparison of Hirs' Equation with Moody's Equation for Determining Rotordynamic Coefficients of Annular Pressure Seals," *ASME Trans. J. of Tribology,* Vol. 109, pp. 144-148.
5. Yamada's, Y., 1962, "Resistance of Flow through Annulus with an Inner Rotating Cylinder," *Bulletin of Japanese Society of Mechanical Engineers,* Vol.5, No. 18, pp. 302-310.
6. Kim, C-H and Childs, D., 1987, "Analysis for Rotordynamic Coeffcients of Helically-Grooved Turbulent Annular Seals," *ASME Trans. J. of Tribology,* Vol. 109, pp. 136-143.
7. Childs, D. and Dressman, J., 1985, "Convergent-Tapered Annular Seals: Analysis and Testing for Rotordynamic Coefficients," *ASME Transaction Journal of Tribology Technology,* Vol. 107, pp. 307-317.
8. Childs, D., 1984, "Finite-Length Solutions for the Rotordyanmic Coefficients of Constant-Clearance and Convergent-Tapered Annular Seals," Third International Conference on Vibrations and Rotating Machinery, York, England.
9. Childs, D., Dressman, B., and Childs, B., 1980, "Testing of Turbulent Seals for Rotordynamic Coefficients," *Rotordynamic Instability Problems in High Performance Turbomachinery, NASA Conference Publication 2133,* pp. 121-138.
10. Tam, L. T., Przekwas, A. J., Muszynska, A., Hendricks, R. C., Braun, M. J., and Mullen, R. L., 1987, "Numerical and Analytical Study of Fluid Dynamic Forces in Seals and Bearings," *ASME Rotating Machinery Dynamics,* Vol. 1, pp. 359, 370.
11. von Pragenau, G., 1982, "Damping Seals for Turbomachinery," NASA Technical Paper #1987.

C317/88

Finite difference analysis of rotordynamic seal coefficients for an eccentric shaft position

R NORDMANN, Dr-Ing and F J DIETZEN, Dr-Ing
Department of Mechanical Engineering, University of Kaiserslautern, Kaiserslautern, West Germany

ABSTRACT

The dynamic coefficients of seals are calculated for shaft movements around an eccentric position. The turbulent flow is described by the Navier-Stokes equations in conjunction with a turbulence model. The equations are solved by a finite-difference procedure.

INTRODUCTION

To model the dynamic behaviour of turbopumps properly it is very important to consider the fluid forces which are developed in the seals. This has clearly been demonstrated by some authors, like for example Black (1969, 1972) and Diewald (1987). The fluid forces are normally described by the following equation:

$$-\begin{bmatrix} F_z \\ F_y \end{bmatrix} = \begin{bmatrix} K & k \\ -k & K \end{bmatrix}\begin{bmatrix} z \\ y \end{bmatrix} + \begin{bmatrix} D & d \\ -d & D \end{bmatrix}\begin{bmatrix} \dot{z} \\ \dot{y} \end{bmatrix} + \begin{bmatrix} M & 0 \\ 0 & M \end{bmatrix}\begin{bmatrix} \ddot{z} \\ \ddot{y} \end{bmatrix} \quad (1)$$

This equation is only valid for a shaft moving around the center of the seal, which is very seldom in reality, In most machines the shaft will orbit around an eccentric position, so that the fluid forces must be described by:

$$-\begin{bmatrix} F_z \\ F_y \end{bmatrix} = \begin{bmatrix} K_{zz} & k_{zy} \\ -k_{yz} & K_{yy} \end{bmatrix}\begin{bmatrix} Z \\ Y \end{bmatrix} + \begin{bmatrix} D_{zz} & d_{zy} \\ -d_{yz} & D_{yy} \end{bmatrix}\begin{bmatrix} \dot{Z} \\ \dot{Y} \end{bmatrix} + \begin{bmatrix} M_{zz} & 0 \\ 0 & M_{yy} \end{bmatrix}\begin{bmatrix} \ddot{Z} \\ \ddot{Y} \end{bmatrix} \quad (2)$$

The dynamic coefficients in such a case have been investigated by Jenssen (1970), Allaire et al. (1976) and recently in paper by Nelson and Nguyen (1987). To model the turbulent flow, all these have used so called "Bulk-Flow Theories" in which the shear stress at the wall is described as a function of the average fluid velocity relative to the wall.
Allaire used a relative simple approach, in which he neglected the inlet swirl and the influence of the circumferential pressure variation on the circumferential velocity. So he needed only the continuity and the axial momentum equation, in which the shear-stresses were modelled by a Blasius type turbulent friction factor, to describe the flow in the seal.
Nelson and Nguyen have taken into account all important effects, which effect the dynamic coefficients. So the influence of inlet swirl, different surface roughness on rotor and stator and seal taper are included. In this model the wall shear-stresses are described by the Moody equation. A special feature of this method is the use of the Fast Fourier Transform to describe the circumferential change of the flow, which allows a very efficient integration of the equations.

Some authors affirm that in the case of great eccentricities recirculation in circumferential direction occurs. This strongly effects the dynamic coefficients but can't be described by a bulk-flow model. Therefore we extended the theory of Dietzen and Nordmann (1986) and Nordmann (1987) to investigate the flow and the coefficients in the case of an eccentric shaft. In that theory the Navier Stokes equations in conjunction with the k-ϵ turbulence model were used to calculate the dynamic coefficients of incompressible and compressible seals for a shaft motion around the centric position.

GOVERNING EQUATIONS

To describe a turbulent flow by the Navier-Stokes equations, the velocities and the pressure are separated into mean and fluctuating quantities.

$$u = \bar{u} + u' \qquad v = \bar{v} + v'$$
$$w = \bar{w} + w' \qquad p = \bar{p} + p'$$

After a time-averaging of the Navier-Stokes equations, terms of the following form $\overline{u'v'}$, $\overline{v'w'}$, $\overline{u'w'}$ occur, which describe the turbulent stresses. These can be handled like laminar stresses, by using Boussinesq's eddy viscosity concept. For example

$$\overline{u'v'} = -\frac{\mu_t}{\rho}\left(\frac{\partial \bar{u}}{\partial y} + \frac{\partial \bar{v}}{\partial x}\right)$$

μ_t is the turbulent viscosity, which is not a fluid property, but depends strongly on the state of flow.
The turbulent viscosity must be described by a turbulence model. We use the k-ϵ model, but also much simpler mixing-length models will be appropriate for a straight seal. The turbulent viscosity is given by:

$$\mu_t = C_\mu \, \rho \, \frac{k^2}{\epsilon} \quad (3)$$

Summing up the laminar and turbulent viscosity we get an effective viscosity.

$$\mu_e = \mu_l + \mu_t \quad (4)$$

So we have the Navier-Stokes equations, the continuity equation and the equations of the k-ϵ model to describe the turbulent flow in a seal. These equations have the following form.

$$\frac{\partial(\rho\phi)}{\partial t} + \frac{\partial}{\partial x}(\rho u\phi) - \frac{\partial}{\partial x}(\Gamma\frac{\partial\phi}{\partial x}) + \frac{1\partial}{r\partial r}(r\rho v\phi) - \frac{1\partial}{r\partial r}(r\Gamma\frac{\partial\phi}{\partial r}) + \frac{1\partial}{r\partial\theta}(\rho w\phi) - \frac{1\partial}{r\partial\theta}(\Gamma\frac{1\partial\phi}{r\partial\theta}) = S_\phi \quad (5)$$

ϕ	Γ	S_ϕ
u	μ_e	$-\frac{\partial p}{\partial x} + \frac{\partial}{\partial x}(\mu_e\frac{\partial u}{\partial x}) + \frac{1\partial}{r\partial r}(r\mu_e\frac{\partial v}{\partial x}) + \frac{1\partial}{r\partial\theta}(\mu_e\frac{\partial w}{\partial x})$
v	μ_e	$-\frac{\partial p}{\partial r} + \frac{\partial}{\partial x}(\mu_e\frac{\partial u}{\partial r}) + \frac{1\partial}{r\partial r}(r\mu_c\frac{\partial v}{\partial r}) + - \frac{1\partial}{r\partial\theta}(r\mu_e\frac{\partial}{\partial r}(\frac{w}{r})) - \frac{2}{r^2}\mu_e\frac{\partial w}{\partial\theta} - \frac{2}{r^2}\mu_e v + \frac{\rho}{r}w^2$
w	μ_e	$-\frac{1\partial p}{r\partial\theta} + \frac{\partial}{\partial x}(\frac{1}{r}\mu_e\frac{\partial u}{\partial\theta}) + \frac{1\partial}{r\partial r}(\mu_e\frac{\partial v}{\partial\theta}) + \frac{1\partial}{r\partial\theta}(\frac{1}{r}\mu_e\frac{\partial w}{\partial\theta}) + \frac{1}{r^2}\mu_e\frac{\partial v}{\partial\theta} - \frac{w}{r^2}\frac{\partial}{\partial r}(r\mu_e) + \frac{1\partial}{r\partial\theta}(\frac{2}{r}\mu_e v) - \frac{\rho}{r}vw$
1	0	0
k	$\frac{\mu_e}{\sigma_k}$	$G - \rho\epsilon$
ϵ	$\frac{\mu_e}{\sigma_\epsilon}$	$C_1\frac{\epsilon}{k}G - C_2\,\rho\frac{\epsilon^2}{k}$

Table 1: The governing equations of the turbulent seal flow

(The constants of the k-ϵ model are given in appendix A)

PERTURBATION ANALYSIS

To describe the flow if the shaft is moving around an eccentric position we follow the procedure of Dietzen and Nordmann (1986) and use a similar transformation. (see Fig. 1)

$$\eta = r_a(\theta) - \frac{r_a(\theta) - r}{\delta(\theta, t)} C_0(\theta) \quad (6)$$

But now r_a and C_0 are functions of θ. This is not so if the shaft moves around the center. δ is the seal clearance, varying with angle θ and time t.

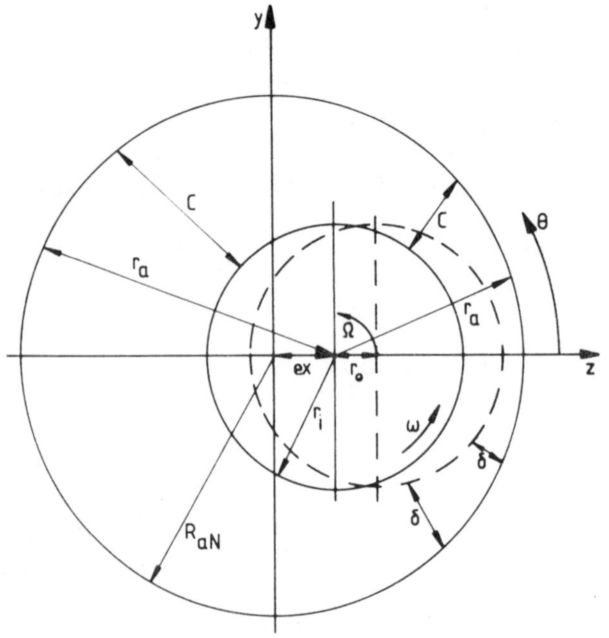

Fig 1 Geometry of the eccentric shaft

The radius r_a can be described by the following equation:

$$r_a(\theta) = \sqrt{R_{aN}^2 - ex^2 \sin^2(\theta)} - ex\cos(\theta) \quad (7)$$

As result of this transformation the shaft orbiting around an eccentric position is transformed to the stationary eccentric position. If we introduce this transformation into our equations, we must obey the following relations.

$$(\frac{\partial}{\partial\theta})_r = (\frac{\partial}{\partial\theta})_\eta + (\frac{\partial}{\partial\eta})_\theta(\frac{\partial\eta}{\partial\theta})_r$$
$$(\frac{\partial}{\partial t})_r = (\frac{\partial}{\partial t})_\eta + (\frac{\partial}{\partial\eta})_t(\frac{\partial\eta}{\partial t})_r \quad (8)$$

To calculate the rotordynamic coefficients we assume that the shaft moves around the eccentric position on small orbits, so that we are allowed to introduce a perturbation analysis.

$$\delta = C_0 - eh_1 \qquad u = u_0 + eu_1, \quad w = w_0 + ew_1, \quad v = v_0 + ev_1, \quad p = p_0 + ep_1 \quad (9)$$

If we introduce these expressions and the coordinate transformation in our governing equation, neglecting terms with power of e greater than 1 and seperating the parts with and without e we will get a set of zeroth order and first order equations. The zeroth order equations describe the stationary flow for the eccentric shaft, the first order equations the perturbation of this flow, if the shaft moves around the stationary position.

We assume that the shaft moves with the precession frequency Ω on a circular orbit with radius r_0 around the eccentric position. So the change of the clearance is given by:

$$eh_1 = r_0(\cos\Omega t\,\cos\theta + \sin\Omega t\,\sin\theta) \quad (10)$$

Because this change is periodic in time we introduce also periodic functions for the flow variables.

$$\begin{aligned} u_1 &= u_{1_c}\cos\Omega t + u_{1_s}\sin\Omega t \\ w_1 &= w_{1_c}\cos\Omega t + w_{1_s}\sin\Omega t \\ v_1 &= v_{1_c}\cos\Omega t + v_{1_s}\sin\Omega t \\ p_1 &= p_{1_c}\cos\Omega t + p_{1_s}\sin\Omega t \end{aligned} \tag{11}$$

By separating now in the first order equations the terms with $\cos\Omega t$ and $\sin\Omega t$ we obtain two real equations for every 1.order equation. These equations are then arranged in a new form by introducing complex variables.

$$\hat{u}_1 = u_{1_c} + iu_{1_s} \qquad \hat{v}_1 = v_{1_c} + iv_{1_s} \tag{12}$$
$$\hat{w}_1 = w_{1_c} + iw_{1_s} \qquad \hat{p}_1 = p_{1_c} + ip_{1_s}$$

Finally supplementary to our real zeroth order equations we have a set of complex first order equations. These equations have the following form.

$$\frac{\partial}{\partial x}(\rho u_o\phi) - \frac{\partial}{\partial x}(\Gamma\frac{\partial\phi}{\partial x}) + \frac{1\partial}{\eta\partial\eta}(\eta\rho v_o\phi) - \frac{1\partial}{\eta\partial\eta}(\eta\Gamma\frac{\partial\phi}{\partial\eta}) + \frac{1\partial}{\eta\partial\theta}(\rho w_o\phi) - \frac{1\partial}{\eta\partial\theta}(\Gamma\frac{1\partial\phi}{\eta\partial\theta}) = S_\phi \tag{13}$$

ϕ	Γ_ϕ	S_ϕ
u_o	μ_e	S_{u_o}
v_o	μ_e	S_{v_o}
w_o	μ_e	S_{w_o}
1	0	0
k_o	$\frac{\mu_e}{\sigma_k}$	S_{k_o}
ϵ_o	$\frac{\mu_e}{\sigma_\epsilon}$	S_{ϵ_o}
$\hat{u}_1$	μ_e	$-\frac{\partial\hat{p}_1}{\partial x} + \frac{\partial}{\partial x}(\mu_e\frac{\partial\hat{u}_1}{\partial x}) + \frac{1\partial}{\eta\partial\eta}(\eta\mu_e\frac{\partial\hat{v}_1}{\partial x}) + \frac{1}{\eta}\frac{\partial}{\partial\theta}(\mu_e\frac{\partial\hat{w}_1}{\partial x}) - C_{u_1} + i\rho\Omega\hat{u}_1 + D_{u_o}$
$\hat{v}_1$	μ_e	$-\frac{\partial\hat{p}_1}{\partial\eta} + \frac{\partial}{\partial x}(\mu_e\frac{\partial\hat{u}_1}{\partial\eta}) + \frac{1\partial}{\eta\partial\eta}(\eta\mu_e\frac{\partial\hat{v}_1}{\partial\eta}) + \frac{1\partial}{\eta\partial\theta}(\eta\mu_e\frac{\partial}{\partial\eta}(\frac{\hat{w}_1}{\eta})) - \frac{2}{\eta^2}\mu_e\frac{\partial\hat{w}_1}{\partial\theta} - \frac{2}{\eta^2}\mu_e\hat{v}_1 + \frac{\rho}{\eta}w_o\hat{w}_1 - C_{v_1} + i\rho\Omega\hat{v}_1 + D_{v_o}$
$\hat{w}_1$	μ_e	$-\frac{1\partial\hat{p}_1}{\eta\partial\theta} + \frac{\partial}{\partial x}(\frac{1}{\eta}\mu_e\frac{\partial\hat{u}_1}{\partial\theta}) + \frac{1\partial}{\eta\partial\eta}(\mu_e\frac{\partial\hat{v}_1}{\partial\theta}) + \frac{1\partial}{\eta\partial\theta}(\frac{1}{\eta}\mu_e\frac{\partial\hat{w}_1}{\partial\theta}) + \frac{1}{\eta^2}\mu_e\frac{\partial\hat{v}_1}{\partial\theta} - \frac{\hat{w}_1}{\eta^2}\frac{\partial}{\partial\eta}(\eta\mu_e) + \frac{1\partial}{\eta\partial\theta}(\frac{2}{\eta}\mu_e\hat{v}_1) - \frac{\rho}{\eta}v_o\hat{w}_1 - C_{w_1} + i\rho\Omega\hat{w}_1 + D_{w_o}$

Table 2: Source terms of zeroth and first order equations.

Only the first order continuity equation to calculate $\hat{p}_1$ has a slightly different form.

$$\frac{\partial}{\partial x}(\rho\hat{u}_1) + \frac{1\partial}{\eta\partial\eta}(\eta\rho\hat{v}_1) + \frac{1\partial}{\eta\partial\theta}(\rho\hat{w}_1) = D_{p_o} \tag{14}$$

You get S_{u_o}, S_{v_o}, S_{w_o}, S_{k_o}, S_{ϵ_o}, if you replace in table 1 in the corresponding terms r by η and u,v,w,p,k,ϵ by $u_o, v_o, w_o, p_o, k_o, \epsilon_o$. The terms C_{u_1}, C_{v_1}, C_{w_1} result of the perturbation of the convective terms in the Navier Stokes equations. D_{u_o}, D_{v_o}, D_{w_o}, D_{p_o} are constants resulting of the coordinate transformation, which are not functions of $\hat{u}_1$, $\hat{v}_1$, $\hat{w}_1$, $\hat{p}_1$. The terms with Ω represent the time dependent parts.

Because we assume that the viscosity μ_e remains constant for the small motions, we needtn't a $\hat{k}_1$ and $\hat{\epsilon}_1$ equation.

(D_{u_o}, D_{v_o}, D_{w_o}, D_{p_o}, C_{u_1}, C_{v_1}, C_{w_1} are given in appendix B)

BOUNDARY CONDITIONS

The zeroth order boundary conditions are:

stator	: $u_{o_s} = 0$	$v_{o_s} = 0$	$w_{o_s} = 0$
rotor	: $u_{o_r} = 0$	$v_{o_r} = 0$	$w_{o_r} = R_{iN}\omega$
entrance	: $p_{o_{En}} = p_{Res} - \frac{1}{2}\rho u^2_{o_{En}}(1 + \xi)$		
exit	: $p_{o_{Ex}} = 0$		

u_{En} is the average axial entrance velocity for every plane with θ constant. p_{Res} is the reservoir pressure and ω the rotational frequency of the shaft.

The first order boundary conditions are:

stator	: $\hat{u}_{1_s} = (0,0)$	$\hat{v}_{1_s} = (0,0)$
	$\hat{w}_{1_s} = (0,0)$	
rotor	: $\hat{u}_{1_r} = (0,0)$	
	$\hat{v}_{1_r} = \left[C_{o_N}(\Omega - \omega)\sin\theta,\ -C_{o_N}(\Omega - \omega)\cos\theta\right]$	
	$\hat{w}_{1_r} = \left[C_{o_N}\Omega\cos\theta,\ C_{o_N}\Omega\sin\theta\right]$	
entrance	: $\hat{p}_{1_{En}} = -\rho\, u_{o_{En}}(1 + \xi)\,\hat{u}_{1_{En}}$	
exit	: $\hat{p}_{1_{Ex}} = (0,0)$	

THE FINITE DIFFERENCE METHOD

For solving these equations a finite-difference procedure is used which is based on the method published by Gosman and Pun (1974). The seal is discretized by a grid (Fig.2) and the variables are calculated at the nodes. The velocities u_o, v_o, w_o ($\hat{u}_1, \hat{v}_1, \hat{w}_1$) are determined at points which lie between the nodes were the variables p_o, k_o, ϵ_o ($\hat{p}_1$) are calculated (Fig.3)

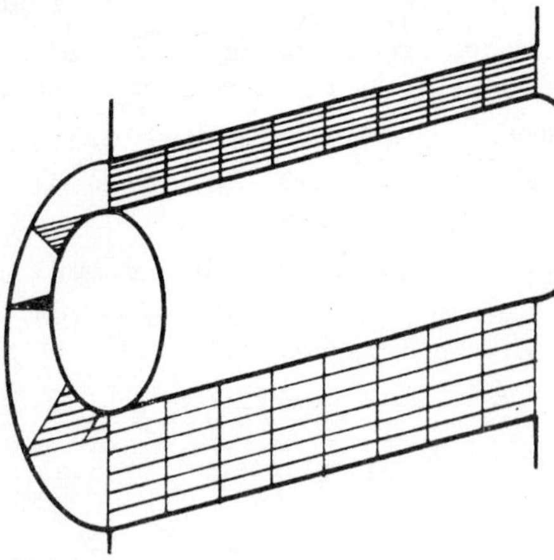

Fig 2 Three-dimensional grid in the seal

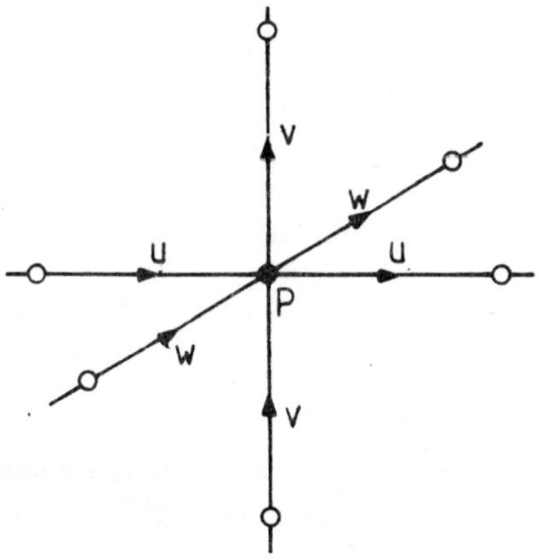

Fig 3 Velocities and pressure in a staggered grid

To calculate the flow we proceed as follow:

1. We start our procedure with guessed values for all variables.
2. First the velocities u_0, v_0, w_0 $(\hat{u}_1, \hat{v}_1, \hat{w}_1)$ are calculated.
3. Then the velocities and the pressure are corrected to satisfy the continuity equation. To do that we use a modified version of the 'PISO' procedure of Benodekar et al (1985)
4. After this k and ϵ are calculated (only for the zeroth order solution).

We repeat step 2 to 4 until we reach a convergent solution. First we solve the zeroth order equations and then the first order equations.

Of course we need a 3-dimensional finite-difference method to calculate the flow in the case of an eccentric shaft, while a two-dimensional method is sufficient for movements around the centric position.

DETERMINATION OF THE DYNAMIC COEFFICIENTS

By integration the pressure $\hat{p}_1$ around the shaft we get forces in z and y direction. Then we introduce $z, y, \dot{z}, \dot{y}, \ddot{z}, \ddot{y}$ resulting of our circular orbit into equation (2).
This gives us the following equations:

$$\int_0^L \int_0^{2\pi} p_{1c} \cos\theta \, R_{iN} \, d\theta dx = C_{oN} \, (K_{zz} + d_{zy}\Omega - M_{zz}\Omega^2)$$

$$\int_0^L \int_0^{2\pi} p_{1s} \cos\theta \, R_{iN} \, d\theta dx = C_{oN} \, (k_{zy} - D_{zz}\Omega)$$

$$\int_0^L \int_0^{2\pi} p_{1c} \sin\theta \, R_{iN} \, d\theta dx = C_{oN} \, (-k_{yz} + D_{yy}\Omega)$$

$$\int_0^L \int_0^{2\pi} p_{1s} \sin\theta \, R_{iN} \, d\theta dx = C_{oN} \, (K_{yy} + d_{yz}\Omega - M_{yy}\Omega^2)$$

If we calculate the forces for several precession frequencies Ω of the shaft, we can obtain the coefficients by a 'Least-Square-Fit'.

RESULTS

We compare our theory with the model of Nelson and Nguyen (1987) and some experimental results of Falco et al (1986) which also have been published in the paper of Nelson and Nguyen (1987).

In Fig. 4-10 dynamic coefficients are calculated as a function of the eccentricity. We compare our results with Nelson and Nguyen's theory and the stiffness coefficients also with experimental and theoretical results of Falco et al.

The seal data are:

Length	: $L = 40.0$ mm
Pressure drop	: 1.0 Mpa
Shaft radius	: $R_{iN} = 80.0$ mm
Shaft speed	: 4000 RPM
Nominal clearance	: $C_{oN} = 0.36$ mm
Density	: $\rho = 1000$ kg/m^3
Preswirl ratio	: $\bar{w}(0,\theta)/R_{iN}\omega = 0.3$
Viscosity	: $\mu_1 = 1.*10^{-3}$ Ns/m^2
Entrance lost-coefficient	: $\xi = 0.5$

The direct stiffnesses K_{zz}, K_{yy} remain nearly unchanged up to an eccentricity ratio of $e_x/C_o = 0.7$ while the cross-coupled stiffnesses increase with the eccentricity ratio. The damping and inertia coefficients are nearly constant up to $e_x/C_o = 0.5$.

From the three theories, the Finite-Element Method given by Falco yields the worst agreement and Falco only presents stiffness coefficients. In comparison with the measurements of Falco the results of Nelson, seem to be slightly better than the results of the Finite Difference method. We use the standard k-ϵ model to describe the turbulence, but we don't know if Nelson has chosen the roughness coefficients in the Moody formulas to get the best fit. Falco doesn't mention anything in his paper about the seal roughness.

We also don't know how exact the measurements are because Falco also doesn't give a uncertainty for his experimental values.

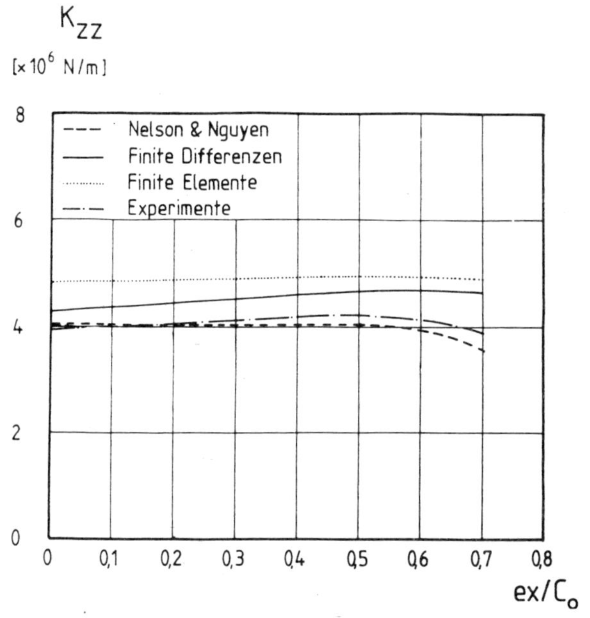

Fig 4 Direct stiffness K_{zz} versus eccentricity

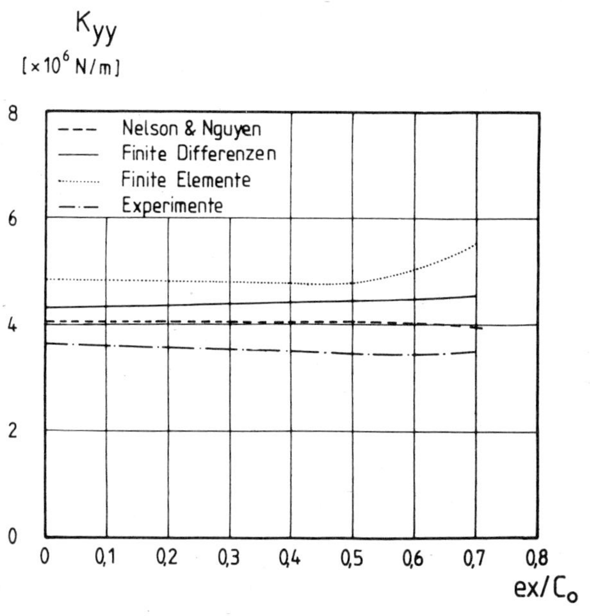

Fig 5 Direct stiffness K_{yy} versus eccentricity

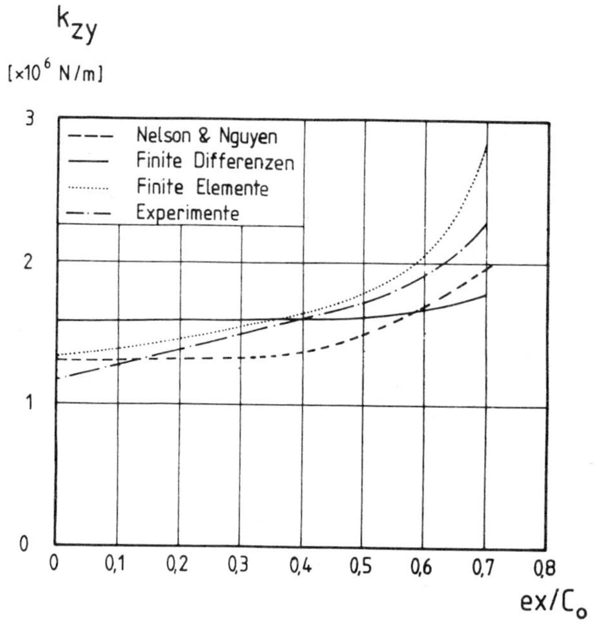

Fig 6 Cross-coupled stiffness k_{xy} versus eccentricity

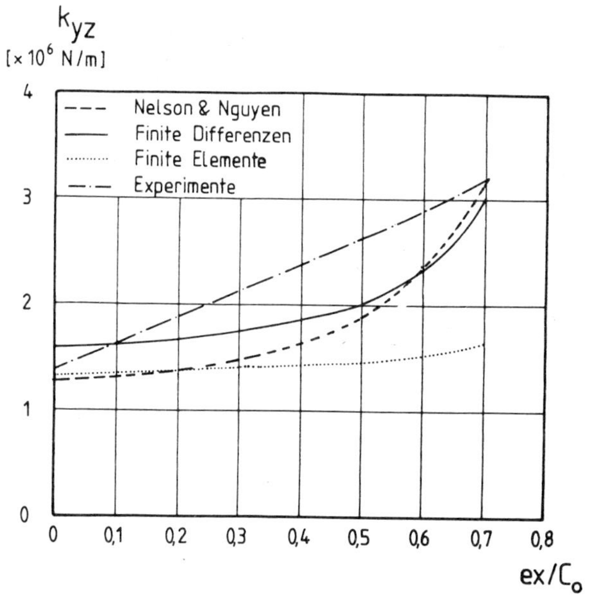

Fig 7 Cross-coupled stiffness k_{yz} versus eccentricity

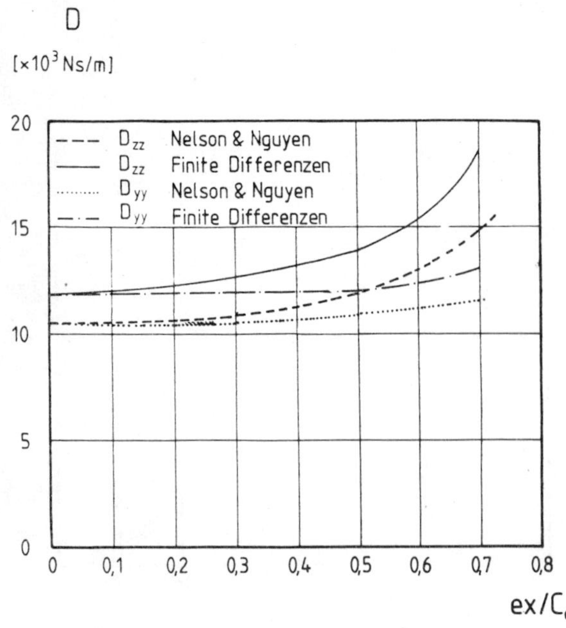

Fig 8 Direct damping versus eccentricity

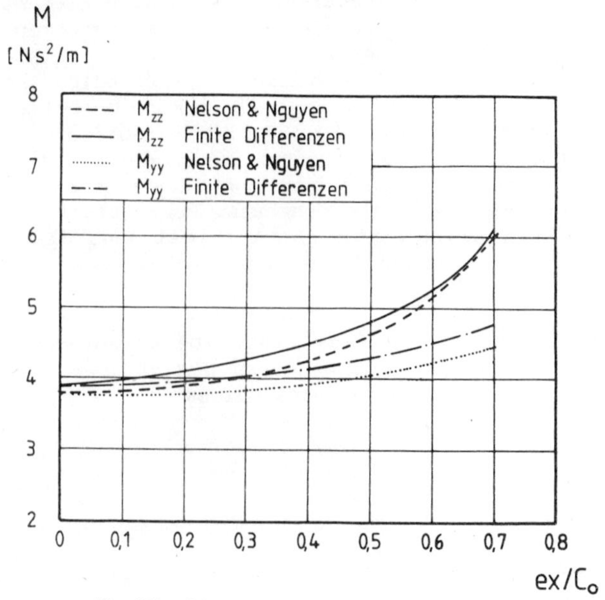

Fig 10 Direct inertia versus eccentricity

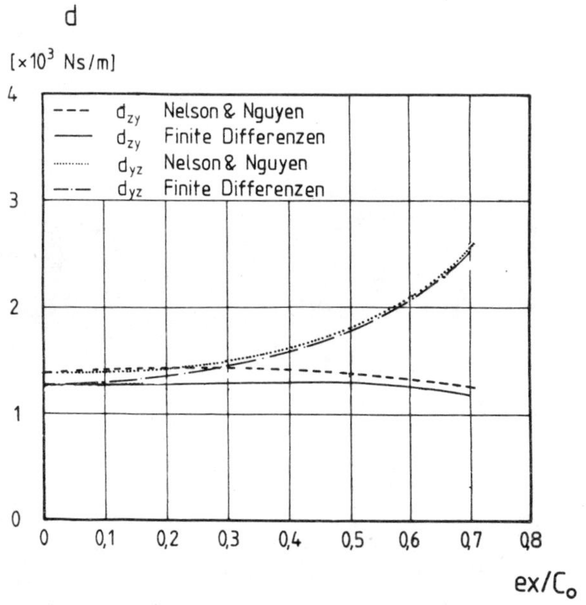

Fig 9 Cross-coupled damping versus eccentricity

CONCLUSION

We have shown that it is possible to calculate the dynamic coefficients of eccentric seals by a finite difference method based on the Navier Stokes equations. This method can also be extended to calculate the coefficients of eccentric gas seals by following the procedure of Nordmann (1987) and to calculate the dynamic coefficients of bearings by neglecting the turbulence model.

NOMENCLATURE

F_z, F_y	Forces on the shaft in z and y direction
K, k	direct and cross-coupling stiffness
D, d	direct and cross-coupled damping
M, m	direct and cross-coupling inertia
u,v,w	axial, radial and circumferential velocity
p	pressure
k	turbulence energy
ϵ	energy dissipation
μ_e, μ_l, μ_t	effective, laminar and turbulent viscosity
ρ	density
t	time
x, r, θ	axial, radial and circumferential coordinate
x,y,z	rectangular coordinate directions
Y,Z	rotor displacements from its steady-state position
η	radial coordinate after transformation
G	production term in k-ϵ-model
$\sigma_k, \sigma_\epsilon$	constants of the k-ϵ-model
C_μ, C_1, C_2	constants of the k-ϵ-model
Φ	general variable standing for u,v,w.p,k ϵ

S_ϕ	general source term
C_o	nominal seal clearance
C_{o_N}	steady-state clearance for an eccentric shaft
δ	varying seal clearance for orbiting shaft
r_o	radius of the precession motion of the shaft
$e = \frac{r_o}{C_{o_N}}$	perturbation parameter
ω	rotational frequency of the shaft $= RPM*\pi/30$
RPM	revolutions per minute
Ω	precession frequency of the shaft
ξ	entrance lost-coefficient
L	Length of the seal
r_a	stator radius (Fig. 1)
R_{a_N}	nominal stator radius (Fig. 1)
R_{i_N}	shaft radius (Fig. 1)
h_1	change in the clearance
ex	eccentricity
$\bar{w}$	average circumferential velocity
u_{En}	average axial velocity at the entrance
P_{Res}	sump pressure

REFERENCES

Allaire, P.E.; Gunter, E.J.; Lee, C.P.; and Barrett, L.E. (1976): The dynamic analysis of the space shuttle main engine-high pressure fuel turbopump. Part II - Load capacity and hybrid coefficients for the turbulent interstage seals. University of Virginia, Report NO UVA/528140/ME76/103

Benodekar, R.W.; Goddard, A.J.H.; Gosman, A.D.; Issa, R.I. (1985): Numerical prediction of turbulent flow over surfacemounted ribs. AIAA Journal Vol. 23, No 3, March 1985

Black, H.F.: Effects of Hydraulic Forces in Annular Pressure Seals on the Vibrations of Centrifugal Pump Rotors. Journal Mechanical Engineering Science, Vol. 11, No 2, 1969

Black, H.F.; Jenssen, D.N.: Effects of High Pressure Ring Seals on Pump Rotor Vibrations. ASME publication, 71-WA/FE-38, 1972

Dietzen, F.J.; Nordmann, R. (1987): Calculating rotordynamic coefficients of seals by 'Finite-Difference Techniques'. ASME Journal of Tribology 1987

Diewald, W.; Nordmann, R. (1987): Dynamic analysis of centrifugal pump rotors with fluid-mechanical interactions. 11th ASME Design Engineering Division Conference on Vibration and Noise, 27-30 Sept. 87 Boston.

Falco, M.; Mimmi, G.; Marenco, G. (1986): Effects of seals on rotor dynamics. Proceedings of the International Conference on Rotordynamics, September 14-17, 1986, Tokyo, Japan.

Gosman, A.D.; Pun, W. (1974): Lecture notes for course entitled: 'Calculation of recirculating flows'. Imperial College London, Mech. Eng. Dept., HTS/74/2.

Jenssen, D.N. (1970): Dynamics of rotor systems embodying high pressure ring seals. Ph.D. dissertation, Heriot-Watt University, Edingburg, Scotland, July 1970.

Nelson, C.C.; Nguyen, D.T. (1987): Analysis of eccentrically-positioned annular incompressible seals: Part 1 - A new solution procedure using the fast fourier transform to facilitate the solution for the hydrostatic force. Will be presented in ASME Journal of Tribology

Nelson, C.C.; Nguyen, D.T. (1987): Analysis of eccentrically-positioned annular incompressible seals: Part 2 - Investigating the effects of eccentricity on the rotordynamic coefficients. Will be presented in ASME Journal of Tribology

Nordmann, R.; Dietzen, F.J.; Weiser, H.P. (1987): Rotordynamic coefficients of turbulent gas seals. ASME Applied Mechanics, Bioengineering and Fluids Engineering Conference. June 14-17, 1987 Cincinnati, Ohio.

Patankar, S.V. (1980): Numerical heat transfer and fluid flow. Mc Graw Hill Book Company 1980.

APPENDIX A: Constants of the k-ϵ model.

$C_\mu = 0.09 \quad C_1 = 1.44 \quad C_2 = 1.92$

$\sigma_k = 1.0 \quad \sigma_\epsilon = 1.3$

and G is a given by:

$$G = \mu_e\left[2\left(\left(\frac{\partial v}{\partial r}\right)^2 + \left(\frac{\partial u}{\partial x}\right)^2 + \left(\frac{1\partial w}{r\partial\theta} + \frac{v}{r}\right)^2\right) + \left(\frac{\partial v}{\partial x} + \frac{\partial u}{\partial r}\right) + \left(\frac{1\partial v}{r\partial\theta} + \frac{\partial w}{\partial r} - \frac{w}{r}\right)^2 + \left(\frac{\partial w}{\partial x} + \frac{1\partial u}{r\partial\theta}\right)^2\right]$$

APPENDIX B: Terms of first order source term.

$$C_{u_1} = \frac{\partial}{\partial x}(\rho u_o \hat{u}_1) + \frac{1}{\eta}\frac{\partial}{\partial \eta}(\eta \rho u_o \hat{v}_1) + \frac{1}{\eta}\frac{\partial}{\partial \theta}(\rho u_o \hat{w}_1)$$

$$C_{v_1} = \frac{\partial}{\partial x}(\rho v_o \hat{u}_1) + \frac{1}{\eta}\frac{\partial}{\partial \eta}(\eta \rho v_o \hat{v}_1) + \frac{1}{\eta}\frac{\partial}{\partial \theta}(\rho v_o \hat{w}_1) - \frac{\rho}{\eta} w_o \hat{w}_1$$

$$C_{w_1} = \frac{\partial}{\partial x}(\rho w_o \hat{u}_1) + \frac{1}{\eta}\frac{\partial}{\partial \eta}(\eta \rho w_o \hat{v}_1) + \frac{1}{\eta}\frac{\partial}{\partial \theta}(\rho w_o \hat{w}_1) + \frac{\rho}{\eta} w_o \hat{v}_1$$

$$D_{u_o} = (((r_a-\eta)DFC+RCS)U1+CCC*U2+OCS*U3) + i(((r_a-\eta)DFS+RSS)U1+CCS*U2+OCC*U3)$$

$$D_{v_o} = (((r_a-\eta)DFC+RCS)V1+CCC*V2+OCS*V3) + i(((r_a-\eta)DFS+RSS)V1+CCS*V2+OCC*V3)$$

$$D_{w_o} = (((r_a-\eta)DFC+RCS)W1+CCC*W2+OCS*W3) + i(((r_a-\eta)DFS+RSS)W1+CCS*W2+OCC*W3)$$

$$D_{p_o} = (((r_a-\eta)DFC+RCS)P1-CCC*P2) + i(((r_a-\eta)DFS+RSS)P1-CCS*P2)$$

$$DFS = \frac{C_{o_N}}{C_o}\left[\cos(\theta) - \frac{ex}{C_o}*\sin(\theta)\sin(\theta) * \left(1 - \frac{ex\cos(\theta)}{\sqrt{R_{aN}^2 - ex^2\sin^2(\theta)}}\right)\right]$$

$$DFC = -\frac{C_{o_N}}{C_o}\left[\sin(\theta) + \frac{ex}{C_o}*\cos(\theta)\sin(\theta) * \left(1 - \frac{ex\cos(\theta)}{\sqrt{R_{aN}^2 - ex^2\sin^2(\theta)}}\right)\right]$$

$$RCS = ex\,\frac{C_{o_N}}{C_o}\left[\cos(\theta)\sin(\theta) * \left(1 - \frac{ex\cos(\theta)}{\sqrt{R_{aN}^2 - ex^2\sin^2(\theta)}}\right)\right]$$

$$RSS = ex\,\frac{C_{o_N}}{C_o}\left[\sin(\theta)\sin(\theta) * \left(1 - \frac{ex\cos(\theta)}{\sqrt{R_{aN}^2 - ex^2\sin^2(\theta)}}\right)\right]$$

$$CCC = \frac{C_{o_N}}{C_o}\cos(\theta) \qquad CCS = \frac{C_{o_N}}{C_o}\sin(\theta)$$

$$OCC = -\Omega\frac{C_{o_N}}{C_o}\cos(\theta) \qquad OCS = \Omega\frac{C_{o_N}}{C_o}\sin(\theta)$$

$$P1 = \frac{1}{\eta}\frac{\partial}{\partial\eta}(\rho w_o)$$

$$P2 = \frac{1}{\eta}\frac{\partial}{\partial\eta}(\eta\rho v_o) + \left(1 - \frac{r_a}{\eta}\right)\frac{1}{\eta}\frac{\partial}{\partial\theta}(\rho w_o) - \frac{r_a}{\eta^2}\rho v_o$$

$$U1 = \frac{1}{\eta}\left[\frac{\partial}{\partial\eta}(\rho u_o w_o) - \frac{\partial}{\partial\eta}\left(\frac{\mu_e}{\eta}\frac{\partial u_o}{\partial\theta}\right) - \frac{\partial}{\partial\eta}\left(\mu_e\frac{\partial w_o}{\partial x}\right)\right]$$

$$U2 = \frac{1}{\eta}\frac{\partial}{\partial\eta}\left(\eta\mu_e\frac{\partial v_o}{\partial x} + \eta\mu_e\frac{\partial u_o}{\partial\eta}\right) - \frac{r_a}{\eta^2}\mu_e\frac{\partial v_o}{\partial x} - \frac{r_a}{\eta^2}\mu_e\frac{\partial u_o}{\partial\eta} + \left(1 - \frac{r_a}{\eta}\right)\frac{1}{\eta}\frac{\partial}{\partial\theta}\left(\frac{\mu_e}{\eta}\frac{\partial u_o}{\partial\theta} + \mu_e\frac{\partial w_o}{\partial x}\right) - \frac{1}{\eta}\frac{\partial}{\partial\eta}(\eta\rho v_o u_o) + \frac{r_a}{\eta^2}\rho v_o u_o - \left(1 - \frac{r_a}{\eta}\right)\frac{1}{\eta}\frac{\partial}{\partial\theta}(\rho u_o w_o)$$

$$U3 = (r_a - \eta)\frac{\partial}{\partial\eta}(\rho u_o)$$

$$V1 = \frac{1}{\eta}\left[\frac{\partial}{\partial\eta}(\rho v_o w_o) - \frac{\partial}{\partial\eta}\left(\frac{\mu_e}{\eta}\frac{\partial v_o}{\partial\theta}\right) - \frac{\partial}{\partial\eta}\left(\mu_e\eta\frac{\partial}{\partial\eta}\left(\frac{w_o}{\eta}\right)\right)\right]$$

$$V2 = \left(1 - \frac{r_a}{\eta}\right)\left[\frac{1}{\eta}\frac{\partial}{\partial\theta}\left(\frac{\mu_e}{\eta}\frac{\partial v_o}{\partial\theta} + \mu_e\eta\frac{\partial}{\partial\eta}\left(\frac{w_o}{\eta}\right)\right) - \frac{2}{\eta}\left(\frac{\mu_e}{\eta}\frac{\partial w_o}{\partial\theta} + \frac{\mu_e}{\eta}v_o\right) - \frac{1}{\eta}\frac{\partial}{\partial\theta}(\rho v_o w_o)\right] - 2\frac{r_a}{\eta^2}\mu_e\frac{\partial v_o}{\partial\eta} + \frac{1}{\eta}\frac{\partial}{\partial\eta}\left(2\eta\mu_e\frac{\partial v_o}{\partial\eta}\right) - \frac{1}{\eta}\frac{\partial}{\partial\eta}(\eta\rho v_o v_o) + \frac{r_a}{\eta^2}\rho v_o v_o + \frac{\eta - r_a}{\eta}\frac{\rho}{\eta}w_o w_o - \frac{\partial p_o}{\partial\eta}$$

$$V3 = (r_a - \eta)\frac{\partial}{\partial\eta}(\rho v_o)$$

$$W1 = \frac{1}{\eta}\left[\frac{\partial}{\partial\eta}(\rho w_o w_o) - \frac{\partial}{\partial\eta}\left(2\frac{\mu_e}{\eta}v_o\right) - \frac{\partial}{\partial\eta}\left(2\frac{\mu_e}{\eta}\frac{\partial w_o}{\partial\theta}\right) + \frac{\partial p_o}{\partial\eta}\right]$$

$$W2 = \left(1 - \frac{r_a}{\eta}\right)\left[\frac{2}{\eta}\frac{\partial}{\partial\theta}\left(\frac{\mu_e}{\eta}\frac{\partial w_o}{\partial\theta} + \frac{\mu_e}{\eta}v_o\right) + \frac{1}{\eta}\left(\frac{\mu_e}{\eta}\frac{\partial v_o}{\partial\theta} + \mu_e\eta\frac{\partial}{\partial\eta}\left(\frac{w_o}{\eta}\right)\right) - \frac{\rho}{\eta}v_o w_o - \frac{1}{\eta}\frac{\partial p_o}{\partial\theta} - \frac{1}{\eta}\frac{\partial}{\partial\theta}(\rho w_o w_o)\right] + \frac{1}{\eta}\frac{\partial}{\partial\eta}\left(\eta^2\mu_e\frac{\partial}{\partial\eta}\left(\frac{w_o}{\eta}\right)\right) - \frac{r_a}{\eta}\mu_e\frac{\partial}{\partial\eta}\left(\frac{w_o}{\eta}\right) - \frac{r_a\mu_e}{\eta\,\eta^2}\frac{\partial v_o}{\partial\theta} + \frac{1}{\eta}\frac{\partial}{\partial\eta}\left(\mu_e\frac{\partial v_o}{\partial\theta}\right) - \frac{1}{\eta}\frac{\partial}{\partial\eta}(\eta\rho v_o w_o) + \frac{r_a}{\eta^2}\rho v_o w_o$$

$$W3 = (r_a - \eta)\frac{\partial}{\partial\eta}(\rho w_o)$$

C281/88

Evaluation of liquid and gas seals for improved design of turbomachinery

R G KIRK, BME, MME, PhDME, PE
Department of Mechanical Engineering, Virginia Polytechnic Institute and State University, Blacksburg, Virginia, United States of America
J F HUSTAK, BSME and K A SCHOENECK, BSME, MSME, PE
Dresser-Rand, Olean, New York, United States of America

SYNOPSIS

This paper reviews the types of seals used in modern turbomachinery and the methods of analysis that have been used to study their influence on rotor vibrations. Liquid seals used for high pressure compressors now use advanced design and analysis methods to prevent unwanted rotor instability. Results of a recent advanced analysis design study are given for a high pressure compressor.

Labyrinth gas seals are important for both efficiency and performance in addition to their interaction with the rotor system. Methods of analysis suitable for design studies are presented with summary results to guide advanced design of turbomachinery for optimum stability. The interaction of reduced clearance seals for improved efficiency with rotor shaft stability are evaluated for typical design conditions.

INTRODUCTION

The design of high speed, high performance turbomachinery demands the utilization of the best possible rotordynamic prediction capability. The many advances that have been made in recent years make it possible to conduct design evaluations with a high degree of confidence. The major emphasis in the 1950's was to rely upon undamped critical speed evaluations using the transfer matrix approach (1,2). This made it possible to evaluate the influence that support stiffness has on rotor critical speed placement. The 1960's introduced the capability to compute elliptic forced response (3) for linear rotor systems excited by rotating unbalance. This capability gave the designer the ability to evaluate the influence of damping on the response amplitude at resonance and at design speed. By the mid 1970's advances in calculation of damped systems eigenvalues (4,5,6,7) gave a new powerful capability for evaluation of possible destabilizing excitation from aerodynamic (8) or fluid-film bearing excitation.

The documentation of several classic compressor instability problems in the open literature that have occurred over the past twenty years (9,10,11,12,13) has given designers guidelines for advanced concept rotor systems (14). As the capability of the analyst improved (15), the demand to make more accurate predictions of rotor system response and stability has increased to the point that individual machine components must be treated in great detail. Some of the major potential sources of destabilizing forces are the seals in the machinery that have been responsible for the advances in machinery efficiency. The ability to predict the influence of liquid and gas seals on turbomachinery rotor dynamic stability is the major emphasis of the majority of the research and development currently in progress worldwide. This task is so important that the author's university has established an experimental laboratory to investigate the dynamic characteristics on full size rotor seal components. This will enable the current state-of-the-art prediction analysis tools of liquid and gas seals to be evaluated in greater detail than would be possible in actual industrial machinery installations. A 200-horsepower motor driven rotor dynamics seal tester will provide realistic evaluation of liquid and gas seals.

This paper will briefly discuss the current prediction capabilities for standard and advanced design liquid seals and gas labyrinth seals. Examples of the application of the new prediction programs to evaluate potential machinery component designs for improved stability will be given to illustrate the capabilities of the currently available component design codes.

Evaluation of Liquid Seal Characteristics

Liquid seals are essential for pumps and compressors that operate with liquids or gases that must be contained within the process. The analysis of turbulent liquid seals for pump wear rings and end seals was introduced by Black (16,17,18,19) and more recently extended by Allaire (20). Experimental and analytical work on turbulent liquid seals have been conducted by Childs (21), Kanki (22,23), and Takagi, (24).

The analysis of liquid end seals for compressors does not require turbulent seal theory and therefore the use of modified fluid-film bearing analysis has made it possible to predict liquid ring seal influence on turbocompressors (25,26,27,28).

Improved seal designs that are a combination of ring seals and tilting-pad bearing elements have been used in high pressure compressors with success. Some of the features of this type design are described in reference (29). The concept uses the stable tilting segment bearing to center the heavily grooved ring seal element that would otherwise have little if any ability to center the floating seal housing. Since seal leakage increases for greater eccentricity, this is highly undesirable from either leakage flow or dynamic stability considerations. A possible arrangement is shown in Figure 1. The analysis of this seal can be accomplished by combining the load calculations of a liquid ring seal analysis to the centering capacity of the tilting-pad portion of the seal. The recent automation of such a calculation has made it possible to predict the operating eccentricity, leakage flow, and dynamic damping and stiffness of the friction loaded floating cartridge. The advantage of the added stiffness inboard of the normal support bearings can easily increase the system stiffness and thereby extend the useful operating speed range of the compressor. The sealing elements are still producing destabilizing forces which mandate the prediction of the combined tilting-pad and ring-seal characteristics at the actual operating eccentricity. Consider a high pressure compressor operating at a design speed to first critical speed ratio of 4.0, considering only the bearing characteristics. When the stabilized seal cartridge is added to the rotor system the predicted stability log decrement increases from 0.118 to 0.324. This is in comparison to a log decrement of -0.899 for standard design ring seals. See Table 1 for further details of this example. This type seal is just one of many design features that can be used to extend the useful operating speed range of high pressure compressors.

Evaluation of Gas Labyrinth Seals

The importance of gas labyrinth seals for proper design evaluation of high pressure compressors has been documented by test stand and field experience on industrial high pressure compressors (30,31,32,33). The analysis of labyrinth seals was established by the work of Iwatsubo (34). Other researchers have also made substantial contributions to the method of analysis for labyrinth seals (35,36,37). The importance of labyrinth seal entry swirl velocity has been one common central factor in all these investigations. Methods of controlling the labyrinth seal entry swirl are therefore of great interest. It has been demonstrated (30,38,39) that gas injection in the leading edge teeth of a long balance piston by a series of flow nozzles directed against rotation can improve the stability of high pressure compressors (See Figure 2).

These results clearly demonstrate that the concept is real and that the labyrinth seals are a key factor in turbocompressor stability evaluations. Methods of controlling the entry swirl by mechanical vanes placed at the seal entry point or at the impeller tip represent possible design alternatives. It is therefore important to evaluate the influence of the placement of these mechanical vanes or flow control elements. The program developed to investigate disc leakage path influence on labyrinth entry swirl (39,40,41,42) can be used to evaluate the potential improvement in stability of high speed compressors.

Consider the influence of the flow control vanes placed at the impeller tip. Table 2 has example results for various swirl rates varying from the expected impeller tip swirl of 0.59 to a negative swirl of 30% rotor speed. The swirl rate of 2% is assumed to be close to the result of a high efficiency row of flow control vanes in the leakage path. The low pressure 1st stage cross-coupled stiffness, K_{xy}, is reduced by almost 50% while the damping is reduced by only 5%. For the high pressure stage results given in Table 3 the K_{xy} component reduces by only 32%, while the damping increases by 3.5%.

Similar results for the balance piston show a decrease of 72% in cross coupled stiffness and a 39% reduction in damping. These results are less favourable for a reduced pressure condition as shown in Tables 4 and 5. For the control of swirl at the labyrinth entrance, the results given in Table 6 indicate a total reversal of the cross-coupled stiffness when the swirl control reduces the swirl ratio to 0.02. By reducing the swirl to 30% at the seal entrance, the effective driving force, $Q_{eff} = K_{xy} - \omega C_{xx}$, is nearly zero.

These results illustrate that each seal location must be evaluated for influence of swirl control on the resultant cross-coupled stiffness and damping values.

Further consideration to seal running clearance should be given since one of the major goals of modern turbomachinery design is to increase the overall efficiency. One way to achieve this is to have minimum clearance labyrinth seal designs. Also to be considered is the variation from cold build clearance values to hot, at speed clearance values. Consider the reduction of clearance for the previous balance piston and last stage eye packing to a value of .0076 cm (0.003 in). Similar swirl condition results are given in Tables 7-9 where it is apparent that the placement of the swirl brake at the impeller tip is highly undesirable for this potential high efficiency seal design. The reduced leakage rate allows the swirl velocity to increase from the impeller tip down to the seal entrance much more than the higher leakage rate design given in Tables 2-6. The balance piston effective cross-coupling reduces by only 28% for the swirl brake at the impeller tip (i.e., swirl rate of 2%). The last stage eye seal effective cross-coupling reduces by only 6% for the same tip swirl brake condition of 2% swirl. These small reductions cannot be compared to the results of the flow control element located at or near the seal entrance (See Table 9.). The dimensions for the leakage path used for the results given in Tables 2-9 are shown in Table 10 for those interested readers that wish to compare their design conditions to the results given in this paper.

CONCLUSIONS

1. Influence of liquid and gas labyrinth seals must be considered in the design evaluations of all turbomachinery, with special concern for the analysis of high speed, high pressure turbomachinery.

2. Swirl control elements at the impeller tip can reduce destabilizing cross-coupled stiffness, but each design condition must be examined to determine the most desirable swirl velocity component.
3. Swirl control at the labyrinth entrance is most effective in giving known entry swirl velocities since the velocity variation from the impeller tip is eliminated.
4. Methods of adding stable load capacity to liquid seals, such as tilting-pad elements, reduce leakage and improve overall system stability of turbocompressors.

RECOMMENDATIONS

1. Analysis of seal elements must be considered in high speed, high performance turbomachinery.
2. Further experimental test results are needed to properly calibrate and verify the analytical prediction capability of computer programs that are to be used for design analysis.

REFERENCES

(1) MYKLESTAD, N. O. "A new method of calculating natural modes of uncoupled bending vibration of airplane wings and other types of beams." Journal of Aeronautical Sciences, April 1944, pp. 153-162.

(2) PROHL, M. A. "A general method for calculating critical speeds of flexible rotors." Journal of Applied Mechanics, Sept. 1945, pp. 142-148.

(3) LUND, J. W. "Rotor bearing dynamics design technology, part V." AFAPL-TR-65-45, Aero Propulsion Laboratory, Wright-Patterson AFB, OH, May 1965.

(4) RUHL, R. L. "Dynamics of distributed parameter rotor systems: transfer matrix and finite." Ph.D. dissertaion, Cornell University, Ithaca, NY, Jan. 1970.

(5) LUND, J. W. "Stability and damped critical speeds of a flexible rotor in fluid-film bearings." Journal of Engineering Industry, Trans. ASME, Vol. 96, No. 2, May 1974, pp. 509-517.

(6) BANSAL, P.N., and R. G. KIRK. "Stability and damped critical speeds of rotor-bearing systems." Journal of Engineering Industry, Trans. ASME, Series B, Vol. 98, No. 1, Feb. 1976.

(7) KIRK, R.G. "Stability and damped critical speed--how to calculate and interpret the results." CAGI, Technical Digest, Vol. 12, No. 2.

(8) ALFORD, J. S. "Protecting turbomachinery from self-excited rotor whirl." Journal of Eng. for Power, V. 87, Ser. A., No. 4, Oct. 1965.

(9) SMITH, K. J. "An operation history of fractional frequency whirl." Proceedings of the Fourth Turbomachinery Symposium, Texas A&M Univ., College Station TX, 1974, pp. 115-125.

(10) BOOTH, D. "Phillips' landmark injection project." Petroleum Engineer, Oct. 1975, pp. 105-109.

(11) KIRK, R. G., J. C. NICHOLAS, G. H. DONALD, and R. C. MURPHY. "Analysis and identification of subsynchronous vibration for a high pressure parallel flow centrifugal compressor." ASME Journal of Mechanical Design, Vol. 104, No. 2, April 1982, pp. 375-383.

(12) WACHEL, J. C. "Nonsynchronous instability of centrifugal compressors." ASME Paper 75-Pet-22, presented at Petroleum Mechanical Engineering Conference, Tulsa OK, Sept. 21-25, 1975.

(13) FERRARA, P. L. "Vibrations in very high pressure centrifugal compressors." ASME Preprint 77-DET-15, presented at Design Engineering Technical Conference, Chicago IL, Sept. 26-30, 1977.

(14) KIRK, R. G., and G. H. DONALD. "Design criteria for improved stability of centrifugal compressors." Rotor Dynamical Instability, AMD-Vol. 55, 1983, pp. 59-72.

(15) NICHOLAS, J. C., and R. G. KIRK. "Selection and design of tilting-pad and fixed lobe journal bearings for optimum turborotor dynamics." Proceedings of the 8th Turbomachinery Symposium, Texas A&M Univ., College Station TX, 1979, pp. 43-57.

(16) BLACK, H. F. "Effects of hydraulic forces in annular pressure seals on the vibrations of centrifugal pump rotors." Journal of Mechanical Engineering Science, Vol. 11, No. 2, 1969, pp. 206-213.

(17) BLACK, H. F., and D. N. JENSSEN. "Dynamic hybrid bearing characteristics of annular controlled leakage seals pressure seals." Institution of Mechanical Engineers, Vol. 184, Pt. 3N, Sept. 1970, pp. 92-100.

(18) BLACK, H. F., and D. N. JENSSEN. "Effects of high pressure ring seals on pump rotor vibrations." Fluids Engineering Division, ASME Paper 71-WA/FE-38, pp. 1-5.

(19) BLACK, H. F., and E. A. COCHRANE. "Leakage and hybrid bearing properties of serrated seals in centrifugal pumps." Paper G5, 6th Int. Conf. on Fluid Sealing, Munich, 1973, pp. 61-70.

(20) ALLAIRE, P. E., C. P. LEE, and R. C. FARRIS. "Turbulent flow in seals: load capacity and dynamic coefficients." University of Virginia, Report UVA/464761/ME76 /139, Dec. 1976.

(21) CHILDS, D. W. "Finite-length solution for rotordynamic coefficients of turbulent annular seals." ASME paper 82-Lub.-42, 1982.

(22) KANKI, H., et al. "Experimental study on the dynamic characteristics of pump annular seals." C297/84 I. Mech. E., 1984.

(23) KANKI, H., et al. "Study on the dynamic characteristics of a water-lubricated pump bearing." _Bulletin of JSME_, Vol. 29, No. 252, 1986.

(24) TAKAGI, M., et al. "Analysis and design of centrifugal pump considering rotor dynamics." _Proceedings of I. Mech. E._, 1980.

(25) KIRK, R. G., and R. G. MILLER. "The influence of high pressure oil seals on turbo-rotor stability." _ASLE Transactions_, Vol. 22, No. 1, Jan. 1979, pp. 14-24.

(26) KIRK, R. G., and J. C. NICHOLAS. "Analysis of high pressure oil seals for optimum turbo compressor dynamic performance." _Vibration in rotating machinery_, I. Mech. E. _Proceedings of Cambridge Conference_, 1980.

(27) KIRK, R. G. "Oil seal dynamics: considerations for analysis of centrifugal compressors." _Proceedings of 15th Turbomachinery Symposium_, Corpus Christi TX, Nov. 10-13, 1986.

(28) KIRK, R. G. "Transient response of floating ring liquid seals." to be published in _ASME Trans., J. of Tribology_, ASME preprint 87-Trib-23.

(29) KIRK, R. G. "_SEAL_" U.S. Patent No. 4,133,541, Jan. 9, 1979.

(30) KIRK, R. G., and M. SIMPSON. Full load testing of an 18000 Hp gas turbine driver centrifugal compressor for offshore platform service." NASA CP-2409 (1986).

(31) FULTON, J. W. "Full load testing in the platform module prior to tow-out: a case history of subsynchronous instability." NASA Conference Publication 2338, _Proceedings of a Workshop Held at Texas A&M_, College Station TX, May 1984.

(32) DESMOND, A. D. "A case study and redtification of subsynchronous instability in turbocompressors." I. Mech. E., Second European Congress on "Fluid Machinery for the Oil, Petrochemical and Related Industries," The Hague, The Netherlalnds, March 1984.

(33) BENTLY, D. E., and A. MUSZYNSKA. "The dynamic stiffness characteristics of high eccentricity ratio bearings and seals by perturbation testing." _Rotor dynamic Instability Problems in High-Performance Turbomachinery_, Texas A&M Univ., NASA C.P. 2338, 1984.

(34) IWATSUBO, T., N. MATOOKA, and R. KAWAI. "Spring and damping coefficients of the labyrinth seal." NASA CP 2250, 1980, pp. 205-222.

(35) JENNY, R. "Labyrinths as a cause of self-excited rotor oscillations in centrifugal compressors." _Sulzer Technical Review_, 4, 1980, pp. 149-156.

(36) BENCKERT, H., and J. WACHTER. "Flow induced spring coefficients of labyrinth seals for application in rotordynamics." NASA CP 2133 _Proceedings of a Workshop held at Texas A&M University_, Rotordynamic Instability Problems of High Performance Turbomachinery, May 12-14, 1980, pp. 189-212.

(37) CHILDS, D. W., and J. K. SCHARRER. "An Iwatsubo-based solution for labyrinth seals--comparison to experimental results." _Rotordynamic Instability Problems in High Performance Turbomachinery_, Texas A&M Univ., College Station TX, NASA CP 2338, May 28-30, 1984, pp. 257-279.

(38) MILLER, E. H. _Rotor Stabilizing Labyrinth Seal for Steam Turbines_. United States Patent #4,420,161, Dec. 1983.

(39) KIRK, R. G. "Labyrinth seal analysis for centrifugal compressor design--theory and practice." _Proceedings of the International Conference on Rotordynamics_, Tokyo, Sept. 14-17, 1986.

(40) KIRK, R. G., "A method for calculating labyrinth seal inlet swirl velocity." _Rotating Machinery Dynamics_, Vol. 2, ASME DE-Vol. 2, 1987.

(41) KIRK, R. G., "Evaluation of aerodynamic instability mechanisms for centrifugal compressors--Part I: current theory." to be published in _ASME Trans. J. of Vibration, Acoustics, Stress, and Reliability in Design_.

(42) KIRK, R. G., "Evaluation of aerodynamic instability mechanisms for centrifugal comrpessors--Part II: advanced analysis." to be published in _ASME Trans., J. of Vibrations, Acoustics, Stress, and Reliablity in Design_.

Table 1. Stability Summary for Liquid Seal Design Study

6-Stage Compressor
Design Speed = (10,000 RPM) 167 Hz
5-pad Tilting pad Bearing; L/D = 0.5; D = (4.0 in) 10.2 cm
Sealing Pressure = (945 lb/in^2) 651 N/cm^2

Type seal	Damped N_{cr} (RPM) Hz	Seal Ecc (Dim)	Sealing length (in) cm	Q (GPM)	δ*** (Dim)
None	(2592) 43.2	--	--	--	+0.118
2-Ring Outer Seal*	(2853) 47.6	0.79	(2.0) 5.08	10.9	-0.899
Tilt-pad Seal**	(3164) 52.7	0.55	(1.4) 3.56	10.3	+0.324

*2-rings; seal 1 and, 6 at (0.33 in) .84 cm each; C_D = (0.008 in) .02 cm

**L/D = 0.2; L = (1.0 in) 2.54 cm; D = (5.0 in) 12.7 cm; Load on Pad; W-Friction = (357 lb) 1587 N; Preload = 50%; C_D = (0.007 in) .018 cm; seal lands, 14 at (0.1 in) .254 cm each; C_D = (0.008 in) 0.020 cm

***log decrement = δ = stability parameter

Table 2. Dynamic Characteristics for Various Swirl Ratios for a Low Pressure Eye Packing Seal

1st Stage Eye Packing
Tooth on Stator

PS = (1188 psi) 819 N/cm^2
PE = (938 psi) 646 N/cm^2
MW = 43.86
6 teeth

H = (.094 in) .24 cm L = (.094 in) .24 cm
C = (.008 in) .020 cm
Leakage space = (0.1 in) .254 cm
N = (10 830 RPM) 180 Hz

Swirl at Impeller Tip	Swirl at Seal Entrance	K_{xy}(lb/in) N/cm	$C_{xx}\left(\frac{lb\text{-}s}{in}\right)\frac{N\text{-}s}{cm}$	Q_{eff}(lb/in) N/cm
.59	.73	(4137) 7240	(2.83) 4.95	(1943) 3400
.3	.59	(3134) 5485	(2.78) 4.87	(980) 1715
.02	.45	(2133) 3733	(2.7) 4.73	(41.54) 72.7
-.3	.28	(970) 1698	(2.6) 4.55	(-1042) -1824

Table 3. Dynamic Characteristics for Various Swirl Ratios for a High Pressure Eye Packing Seal

Last Stage Eye Packing
Tooth on Stator

PS = (2799 psi) 1929 N/cm^2
PE = (2550 psi) 1757 N/cm^2
MW = 43.86
6 teeth

H = (.094 in) .24 cm L = (.094 in) .24 cm
C = (.008 in) .020 cm
Leakage space = (0.1 in) .254 cm
N = (10 830 RPM) 180 Hz

Swirl at Impeller Tip	Swirl at Seal Entrance	K_{xy}(lb/in) N/cm	$C_{xx}\left(\frac{lb\text{-}s}{in}\right)\frac{N\text{-}s}{cm}$	Q_{eff}(lb/in) N/cm
.59	.69	(6775) 11 856	(4.89) 8.56	(2981) 5217
.3	.59	(5689) 9956	(4.99) 8.73	(1819) 3183
.02	.49	(4604) 8057	(5.06) 8.86	(684) 1197
-.3	.37	(3289) 5756	(5.09) 8.91	(-655) -1146

Table 4. Dynamic Characteristics for Various Swirl Ratios for a High Pressure Balance Piston Labyrinth Seal

Balance Piston Seal; Tooth on Stator
Full Pressure

PS = (2799 psi) 1929 N/cm^2 H = (.094 in) .24 cm L = (.094 in) .24 cm
PE = (938 psi) 646 N/cm^2 C = (.008 in) .020 cm
MW = 43.86 Leakage space = (0.1 in) .254 cm
13 teeth N = (10 830 RPM) 180 Hz

Swirl at Impeller Tip	Swirl at Seal Entrance	K_{xy}(lb/in) N/cm	$C_{xx}(\frac{lb\text{-}s}{in})\ \frac{N\text{-}s}{cm}$	Q_{eff}(lb/in) N/cm
.59	.79	(32 026) 56 046	(20.467) 3508	(16 166) 28 290
.30	.58	(19 539) 34 193	(15.7) 27.5	(7372) 12 901
.02	.36	(9002) 15 753	(12.5) 21.9	(-687) -1202
-.30	.11	(-1491) -2609	(10.47) 18.3	(-9608) -16 814

Table 5. Dynamic Characteristics for Various Swirl Ratios for a Medium Pressure Balance Piston Labyrinth Seal

Balance Piston: Tooth on Stator
Reduced Pressure

PS = (1800 psi) 1240 N/cm^2 H = (.094 in) .24 cm L = (.094 in) .24 cm
PE = (938 psi) 646 N/cm^2 C = (.008 in) .020 cm
MW = 43.86 Leakage space = (0.1 in) .254 cm
13 teeth N = (10 830 RPM) 180 Hz

Swirl at Impeller Tip	Swirl at Seal Entrance	K_{xy}(lb/in) N/cm	$C_{xx}(\frac{lb\text{-}s}{in})\ \frac{N\text{-}s}{cm}$	Q_{eff}(lb/in) N/cm
.59	.76	(19 886) 34 800	(12.5) 21.9	(10 197) 17 845
.3	.58	(13 206) 23 111	(10.1) 17.7	(5376) 9408
.02	.41	(7315) 12 801	(8.32) 14.6	(864) 1512
-.3	.19	(1129) 1976	(6.93) 12.1	(-4240) -7420

Table 6. Dynamic Characteristics for Various Swirl Ratios for a High Pressure Balance Piston Labyrinth Seal

Swirl Control at Seal Entrance
Balance Piston: Full Pressure

PS = (2799 psi) 1929 N/cm^2 H = (.094 in) .24 cm L = (.094 in) .24 cm
PE = (938 psi) 646 N/cm^2 C = (.008 in) .020 cm
MW = 43.86 Leakage space = (0.1 in) .254 cm
13 teeth N = (10 830 RPM) 180 Hz

Swirl at Impeller Tip	Swirl at Seal Entrance	K_{xy}(lb/in) N/cm	$C_{xx}(\frac{lb\text{-}s}{in})\ \frac{N\text{-}s}{cm}$	Q_{eff}(lb/in) N/cm
.59	.65	(23 910) 41 842	(17.45) 30.5	(10 385) 18 174
.3	.38	(9889) 17 306	(12.95) 22.7	(-142) -249
.02	.11	(-1483) -2595	(10.7) 18.7	(-9770) -17 098
-.3	-.2	(-12981) -22 717	(10.8) 18.9	(-21 339) -37 343

Table 7. Dynamic Characteristics for Various Swirl Ratios for a High Pressure Eye Packing Seal

Eye Packing - Last Stage
Reduced Clearance = (0.003 in) 0.076 cm

PS = (2799 psi) 1929 N/cm^2 H = (.094 in) .24 cm L = (.094 in) .24 cm
PE = (2550 psi) 1757 N/cm^2 C = (.008 in) .020 cm
MW = 43.86 Leakage space = (0.1 in) .254 cm
6 teeth N = (10 830 RPM) 180 Hz

Swirl at Impeller Tip	Swirl at Seal Entrance	K_{xy} (lb/in) N/cm	$C_{xx}(\frac{lb\text{-}s}{in})$ $\frac{N\text{-}s}{cm}$	Q_{eff} (lb/in) N/cm
.59	.55	(11 327) 19 822	(7.77) 13.6	(5302) 9278
.30	.54	(11 147) 19 507	(7.75) 13.5	(5136) 8988
.02	.54	(10 969) 19 196	(7.74) 13.5	(4974) 8705
-.3	.53	(10 760) 18 830	(7.71) 13.5	(4782) 8669

Table 8. Dynamic Characteristics for Various Swirl Ratios for a High Pressure Balance Piston Labyrinth Seal

Balance Piston: Full Pressure
Reduced Clearance = (0.003 in) 0.076 cm

PS = (2799 psi) 1929 N/cm^2 H = (.094 in) .24 cm L = (.094 in) .24 cm
PE = (938 psi) 646 N/cm^2 C = (.008 in) .020 cm
MW = 43.86 Leakage space = (0.1 in) .254 cm
13 teeth N = (10 830 RPM) 180 Hz

Swirl at Impeller Tip	Swirl at Seal Entrance	K_{xy}(lb/in) N/cm	$C_{xx}(\frac{lb\text{-}s}{in})$ $\frac{N\text{-}s}{cm}$	Q_{eff}(lb/in) N/cm
.59	.62	(18 595) 32 531	(7.26) 12.7	(12 970) 22 698
.30	.58	(16 427) 28 748	(6.86) 12.0	(11 109) 19 441
.02	.54	(14 316) 25 053	(6.49) 11.4	(9283) 16 245
-.3	.50	(11 808) 20 664	(6.08) 10.6	(7096) 12 418

Table 9. Dynamic Characteristics for Various Swirl Ratios for a High Pressure Balance Piston Labyrinth Seal

Balance Piston: Full Pressure
Reduced Clearance = (0.003) in) 0.076 cm
Brake at Entrance

PS = (2799 psi) 1929 N/cm^2 H = (.094 in) .24 cm L = (.094 in) .24 cm
PE = (938 psi) 646 N/cm^2 C = (.008 in) .020 cm
MW = 43.86 Leakage space = (0.1 in) .254 cm
13 teeth N = (10 830 RPM) 180 Hz

Swirl at Impeller Tip	Swirl at Seal Entrance	K_{xy}(lb/in) N/cm	$C_{xx}(\frac{lb\text{-}s}{in})$ $\frac{N\text{-}s}{cm}$	Q_{eff}(lb/in) N/cm
.59	.58	(16 551) 28 964	(6.95) 12.2	(11 166) 19 540
.30	.42	(7645) 1338	(5.51) 9.64	(3372) 5901
.02	.26	(-945) -1654	(4.45) 7.79	(-4391) -7684
-.3	.07	(-10 633) -18 608	(3.68) 6.44	(-13 487) -23 602

Table 10. Leakage Path Geometry for Labyrinth Seal Calculations

Impeller tip radius	=	(4.8 in) 12.2 cm
End wall uniform clearance	=	(0.1 in) 0.254 cm
Labyrinth radius	=	(2.9 in) 7.4 cm
Tooth spacing	=	(0.094 in) 0.24 cm
Tooth height	=	(0.094 in) 0.24 cm

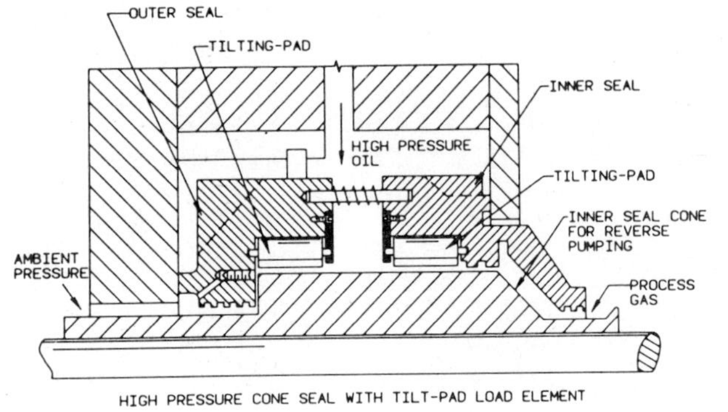

Fig 1 High-pressure liquid seal with tilting-pad load assist

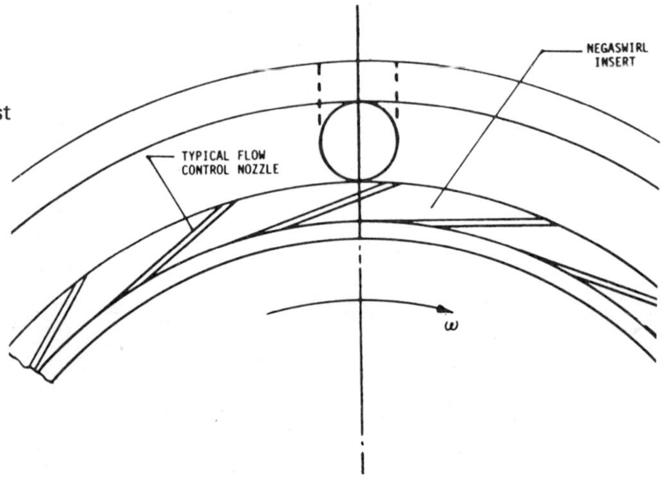

Fig 2a Cross-section of balance piston with flow control insert

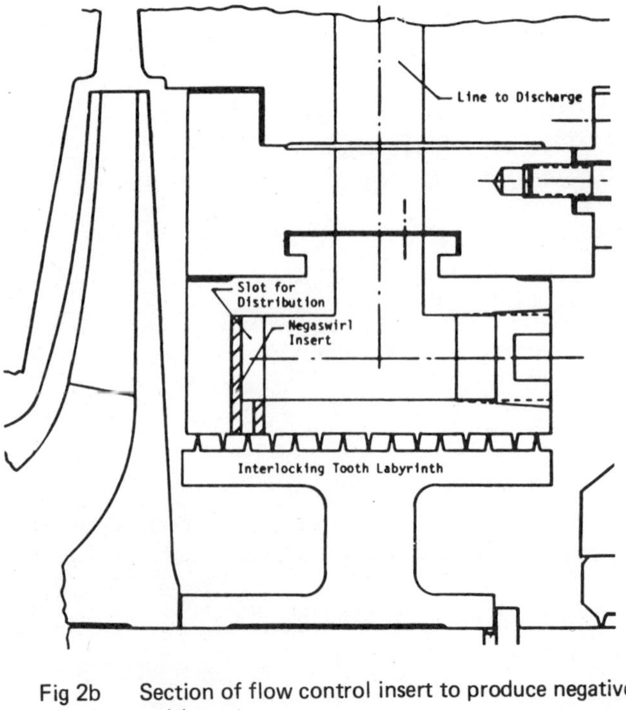

Fig 2b Section of flow control insert to produce negative swirl

C302/88

Some in-field experiences of non-repeatable behaviour in the dynamics of rotating machinery

C FRIGERI and G A ZANETTA
Ente Nazionale per l'Energia Elettrica, Cologno Monzese, Italy
A VALLINI
Ente Nazionale per l'Energia Elettrica, Pisa, Italy

SYNOPSIS The paper describes four different vibration problems recently occurred on large rotating machinery, all connected, to some extent, with non-repeatable dynamic behaviour of the machines due to thermal effects.

Each of these problems is presented in terms of rotor vibratory behaviour , by means of vibration recordings in the time or frequency domain. The specific cause of these faults is discussed and remedial action, if any, is reported.

The aim is to provide in-field experiences that can not be found in textbooks and might help a utility engineer in making a proper diagnosis.

1 INTRODUCTION

Digital data acquisition and processing have now been succesfully used for a decade and, nowadays, efforts are directed towards the search for the most suitable application for the large volume of information that has become available through these powerful means.

ENEL, too, is considering more efficient monitoring systems and data storage to assist early diagnosis, for this goal can be achieved by a suitable data processing, taking advantage of both signal analysis and data correlation techniques. In addition, mathematical models for the numerical simulation of dynamic behaviour of rotating machines have proved to be very useful in many cases.

In the case of a number of phenomena, enough experience has been built up to support pursuit of this goal. Nevertheless, especially in large rotating machinery in power plants, it is well-known that vibrations may undergo unexpected changes, even under seemingly the same operating conditions, at least to judge from the quantities monitored.

For some machinery, such as generator rotors, these phenomena may be considered usual and are, to some extent, easily recognized. In other cases, they are peculiar to a particular machine, or may appear at once in one machine only out of a number of machines of exactly the same design and are sometimes unsteady.

Moreover, there are many parameters acting together and it is not always possible to keep them apart from each other. Neither can the specialists always perform all the tests that would be needed.

This situation makes it rather difficult to obtain a correct diagnosis of these faults, which can cause very high vibration levels and eventually lead to sudden trips of the machinery, or even to unscheduled outages.

The paper deals with four such cases recently experienced at ENEL, which can be of some interest to the utility engineer. The first concerns a strong vibration difference measured on a generator rotor when passing through its first critical speed. The second problem was encountered in the HP rotor of a turboset, when the output was decreased from the maximum to a reduced load. The third example refers to noticeable and erratic increases in vibrations at the two bearings of an HP rotor in a 320 MW unit, with the machine operating in steady-state conditions. Lastly, the paper discusses severe vibrations in the L.P. rotors of two similar units after some hours with the machine on the turning gear.

2 SLACKENING OF AXIAL THRUST ON A GENERATOR ROTOR

The machine in the first case was the rotor of a generator used in short-circuit tests. The rotor, weighing about 500 kN, was supported by two cylindrical oil-film bearings 8.3 m apart. Its maximum speed was 3700 r/min (62 Hz).

At the time of the investigation, the machine was fitted with 2 eddy-current probes and 2 accelerometers at each bearing. The corresponding signals, together with a phase reference and a number of temperatures, were recorded on a magnetic tape for subsequent digital analysis.

Among other vibration problems, the machine showed high vibration differences when passing through its first critical speed, and these often prevented it from reaching the operating speed, especially after 1 or 2 days' stand-still.

In other such circumstances, a definite relationship was found between vibrations and the quantities (particularly rotor temperature) affecting them. Eventually, a satisfactory compromise was struck by means of balancing weights [1]. In this case, the vibration differences between different run-ups did not show any repeatable behaviour. On the contrary, the fundamental vibration component underwent substantial changes, with the phase ranging over an arc of 180 degrees. That made any compensation balancing impossible.

During two different shut-downs, a sudden variation in the vibration fundamental component was recorded at a speed slightly above the first critical speed. Comparing the vibration difference with a reference situation, an amplitude of 160 umpp was reached at the frequency of the first critical speed (Fig 1). Despite the marked effect on the fundamental component, the 2nd vibration harmonic did not react significantly (Fig 2). A frequency spectrum indicated a significant excitation of the rotor's first vibration mode, at the instant of the vibration step.

The experimental measurements were reproduced very effectively in a numerical simulation obtained by means of a finite element model [2], which assumed that there was an uneven axial thrust of about 80 kN between the coil ends, giving rise to shaft elastic bending. Such axial thrusts are due to thermal effects combined with friction. At lower speeds these effects decrease due to the reduced centrifugal force and the coils are able to slip in relation to the rotor body.

At first, the only remedial action possible was to keep the rotor spinning at low speed for about one hour before run-up, when the critical conditions occurred. Subsequently, in the course of the first scheduled overhaul, the problem was satisfactorily solved by adjusting the coil locking-bars, since when the trouble has no longer been experienced to the same extent.

3 LOAD-DEPENDENT VIBRATION CHANGES IN PARTIAL-ARC ADMISSION TURBINES

Load-dependent vibrations have long been the subject of reports from various authors ([3], [4], [5], [6], [7]).

In these works, attention was mainly focused on rotor instability due to steam forces exciting the H.P. rotor ([3], [4], [5]). In such cases, the problem grew worse when output was increased at given loads.

In this paper, we have to report experimental evidence of a marked increase in synchronous vibration in the H.P. rotor, when the load is decreased, in the case of partial-arc admission turbines. This kind of problem has been referred to in [6] and [4], in which of a theoretical approach and possible remedial actions are also suggested.

In our experience, the problem was first encountered on a 320 MW turboset. The shaft line consisted of the M.P.-H.P. rotor, the L.P. rotor and the generator, supported on 7 bearings. The M.P.-H.P. section, weighing about 250 kN, was equipped with 4 shoe tilting-pad bearings. In that unit, the load was reduced shutting off the upper nozzles, so that, at 200 MW, steam admission was in practice confined beneath the lower half of the shaft.

In these conditions, significant movement of the journal inside the bearing clearance was noticed, for both the M.P.-H.P. bearings (Fig 3), with the journal moving upwards and towards the bearing centre. A corresponding noticeable vibration change took place in the shaft-support relative vibration, especially at bearing No 1 (Fig 4). Only the fundamental vibration component was affected by the change. A slight effect on the 2nd harmonic was noticed, but was most likely due to the non-linear behaviour of the oil film. Furthermore, there were no sub-synchronous components, nor were any changes at all observed in the support vibrations.

At the same load of 200 MW, when changing from partial to full arc admission, the journal returned to the almost the same position as it had had at full load and the vibrations also returned to the values they had had at 320 MW.

Since throttle-controlled full-arc admission turbines have lower efficiency, solving the problem through a full-arc admission could not be considered. The most obvious solution of the problem lay in partial-arc admission through symmetrical nozzles. This action was indeed taken and gave the expected results, but it was not accepted by the manufacturer, who was afraid it might shorten the life of the first row of blades as the result of doubling the number of step excitations per turn. So closure of the lower nozzles, rather than the upper ones, was also tried, but though the results in terms of vibrations were fairly good, this solution had to be discarded because the white metal temperature in bearings 1 and 2 was too high.

Later on, the phenomenon was observed again in other turbosets, of either similar or different design and size, but all working at reduced loads with partial-arc admission.

In one case in particular, high vibrations took place at the first H.P. support within a narrow range around the two distinct load values of 120 MW and 285 MW, the maximum output being 320 MW. Oil pressure measurements taken on the four pads of the bearing showed a noticeable unbalance in the lower pad charges, precisely at those two loads. A check on bearing clearance indicated a somewhat higher value than usual. Reduction of the clearance, within tolerance limits, led to a significant improvement in terms of both pressure distribution and vibrations, without any substantial temperature rise.

In another 320 MW unit, one of the two lower pads of the second bearing of the H.P. rotor was strongly overcharged at the reduced load of 210 MW, with partial steam admission, with the consequence of a white metal temperature far beyond alarm limits. In that case, the operating conditions magnified the effects of a slight vertical and horizontal misalignment, which was finally found to be the source of the problem. The unit was kept in operation until a scheduled overhaul by changing the opening sequence of two admission valves, in order better to balance the total thrust on the bearing.

In both circumstances, a numerical model of the oil-film bearings turned out to be a very useful support for the diagnosis and for verifying the remedial action suggested.

4 SHORT-DURATION VIBRATION CHANGES ON TURBINES OF A 320 MW UNIT

As in the previous case, the machine, a 320 MW turboset, consisted of three shafts on 7 bearings: the H.P.-M.P. rotor, the L.P. stage and the generator.

From time to time, important changes were recorded in the absolute shaft vibrations at the turbine bearing locations. Several investigations were undertaken, but no relationship between the trouble and the various quantities of interest (temperatures, pressures, power, etc...) could be established. Measurements indicated that bearing No 2 was somewhat overcharged and that the L.P. rotor was in a poor state of balance.

But the phenomenon could not be ascribed to these problems: at most, they might magnify it. Its main characteristics were the following.

It was not load-dependent, because it appeared even at constant load and erraticly. It was unsteady, appeared even with sharp variations and developped vibration oscillations before disappearing. Frequency analysis showed that only the fundamental component was involved in the vibration changes. A typical record of both the amplitude and the phase of the fundamental component on the four turbine bearings in the presence of the phenomenon is shown in Fig. 5, in which absolute shaft motions are presented.

At first, attention was focused on L.P. bearing No 4, owing to the higher vibration levels reached on that bearing. But in fact the greatest changes took place at the two H.P.-M.P rotor bearings.

Finally, a steam loss from the glands on the side of bearing No 2 was found out, almost by chance. This loss was responsible for overheating on one side of the support of that bearing, which in turn caused distortion of the support itself. In such conditions, not only might the bearing reaction forces change, but also some slight friction had probably taken place. The resultant vibration counteracted the phenomenon in such a way that, after some time, it disappeared.

However, such events generally have a periodicity of their own. In this particular case no periodicity was observed and the erratic occurrence of the problem could not be explained.

However, at first a metal screen was put in-between the glands and the support of bearing No 2, as the result of which the problem disappeared. Afterwards, during a scheduled overhaul, the steam glands were checked. In addition, the support itself was lowered to compensate for the mentioned overload in hot machine conditions and, finally, the machine was balanced. So far, the problem has not reappeared.

5 NON-REPEATABLE BEHAVIOUR OF AN L.P. ROTOR WITH SHRUNK-ON DISKS

This problem was first encountered in an L.P. rotor of a 320 MW turboset, of the same design as that described at point 2 above.

The importance attached to the problem was not only due to its novelty and to the high vibration levels reached even at operating speed, but also to the large number of rotors of that type in operation at our power stations. Moreover, those particular machines were expected to operate with overnight stops, which were likely to cause the behaviour in question.

The rotor weighed about 520 kN and had 2 critical speeds at 22.5 and 47.5 Hz, the operating speed frequency being 50 Hz. The two disks, one on each side of the rotor, supporting the blades of the last stage, were shrunk on the shaft and had six centering plugs fitted on the external periphery of the disk, facing its inner rim. In more recent designs of the same machine, these disks are part of the rotor body.

The faulty behaviour first appeared during a run-up following a 7-hours wait of the machine on the turning gear (at about 3 r/min), when the vibrations at one of the two L.P. bearings reached about 300 umpp at a speed of 47.5 Hz. The previous shut-down had showed quite good vibration levels and could be considered absolutely normal. However, at 50 Hz, the load was increased to maximum output and, after 3 or 4 hours, the vibrations returned to the usual values.

The machine was then equipped with some more transducers and a digital acquisition and processing system, so as to be able to obtain a frequency analysis and phase information throughout the frequency range.

During that very period, a few stops of some hours were requested in the course of normal operation of the machine, thus reproducing the fault conditions. Briefly, examination of the data gave the following results: the phenomenon was present almost only in the two L.P. bearings; only the fundamental component was responsible for the vibration changes; the 2nd harmonic was repeatable and almost negligible; no subharmonic components were present; the fault excited strongly both the critical speeds, as was found in observing the phase of the vibration and taking the difference between a start-up in fault conditions and the previous run-down; though the amplitude was different as between one test and another, the phase of the vibration resulting from the fault was constant. Typical records taken in normal and fault conditions are given in Fig 6.

Additional tests were carried out at the first opportunity, without interfering with the regular operation of the machine. After a normal shut-down, the machine was started up again a few minutes later and the fault did not appear. After a run-up in the presence of the fault, the machine was immediately shut down: the malfunction was still present. Conversely, it did not recur after longer periods on the turning gear (more than 20 hours).

On the strength of this information, a crack was discounted as the cause of the problem with sufficient confidence to be able to go on operating the machine until the scheduled overhaul. In any case, the amplitude and phase of the fundamental component and 2nd vibration harmonic were continously monitored in order to detect possible significant changes that might be symptoms of cracks [8].

Thermal bending of the rotor was suspected. The dynamic behaviour of the shaft line in the presence of different forcing systems at various locations was then investigated, by means of a mathematical model [2]. The study confirmed the assumption and indicated the most likely section of the fault as being at the shrunk-on disk on the generator side. Yet it was not so clear what physical event might be responsible for this behaviour.

During the overhaul, the rotor was taken to the manufacturer's workshop. It was found permanently bent, the centering plugs of the turbine-side disk were broken over an arc of 180 degrees, non-repeatable behaviour in balancing tests was found and axial movements of the disks were noticed after overspeed tests. This rotor was then replaced.

After a few months, the same malfunction was recorded on a twin machine in service at another power station. Measurements confirmed the results of the previous case. After fine-tuning of the mathematical model, the absolute shaft vibrations at the two L.P. bearings around its two critical speeds were fitted satisfactorily (Fig 7), taking into account a forcing system arising from an angular deviation from the straight line of the shaft, corresponding to the inner rim of the shrunk-on disk on the turbine side.

It was therefore assumed that the shrinking on the surface between disk and shaft was uneven, thus giving rise to asymmetric differential expansions during thermal transients, also in conjunction with centering plugs broken over an arc of 180 degrees. Numerical evaluation of the fault (the angular deviation) indicated that the temperature differences, required to support the assumption, were within reasonable bounds.

Owing to the nature of the problem, it was not possible to devise any effective corrective action, by minor changes in the design of the rotor itself.

It was therefore decided to find out whether there was some procedure for operating the machine that might alleviate the trouble. But, unexpectedly, after an accidental overspeed, the fault disappeared completely. So further tests were no longer possible.

Finally, measurements were taken on two other similar sets, under the same transient conditions. In one of them, the malfunction was completely absent; in the other it was still present, though much less conspicuous.

6 CONCLUSIONS

Diagnosis of in-field vibratory problems in large rotating machinery is not always an easy task. Some times these problems are complex and cannot easily be reduced to known terms, bearing in mind, among other things, the fact that many factors have to be taken into account at the same time. Nor is it always so easy to arrive at the ultimate solution by means of corrective action on the machines.

The experience contributed here shows that a proper acquisition and analysis of vibration signals, sometimes in conjunction with analytical methods, gave enough confidence in diagnosis to operate the machinery even in the presence of high vibration levels, without stopping the machines for subsequent inspection, which would have proved to be of no use in cases like those described here. On the other hand, the kind of phenomena described here could not have been reproduced, and hence studied, by means of any test at the manufacturers' workshops.

ACKNOWLEDGMENTS

The authors wish to thank ENEL and CESI for their permission to publish the results of the investigations described in the paper.

REFERENCES

[1] Cadeddu,C. Frigeri,C. Gadda,E. Clapis,A. / Vibration Problems on Large Rotating Machinery: Some Cases Experienced in ENEL Power Stations./ IFToMM International Conference "Rotordynamic Problems in Power Plants", Rome-Italy, Sept.28-Oct.1, 1982.

[2] DYTS04-USER'S GUIDE-CEGB-RD/L/P 15/80 -JOB No. VE316

[3] Alford,J.S. /Protecting Turbomachinery from Self-excited Rotor Whirl./ ASME Journal of Engineering for Power, October 1965, pp 333-344.

[4] Pollman,E. Schwerdtfeger,H. Termuehlen,H. /Flow Excited Vibrations High Pressure Turbines (Steam Whirl)./ ASME Journal of Engineering for Power, April 1978, Vol 100, pp 219-228.

[5] Pollmann,E. Termuehlen,H. / Turbine Rotor Vibrations Excited by Steam Forces (Steam Whirl)./ ASME Paper 7 -WA/Pwr-11, December 1975.

[6] Greathead,S.H. Slocombe,M.D. /Further Investigations into Load Dependent Low Frequency Vibration of the High Pressure Rotor on Large Turbo-generator./ I.Mech.E. Conference "Vibrations in Rotating Machinery", Cambridge, September 2-4, 1980.

[7] Greathead,S.H. Slocombe,M.D. /Investigations into Output Dependent Rotordynamic Instability of the High Pressure Rotor on a Large Turbo-Generator./ IFToMM International Conference "Rotordynamic Problems in Power Plants", Rome-Italy, Sept.28-Oct.1, 1982.

[8] Nilsson,L.R.K. / On the Vibration Behaviour of a Cracked Rotor./ IFToMM International Conference "Rotordynamic Problems in Power Plants", Rome-Italy, Sept.28-Oct.1, 1982.

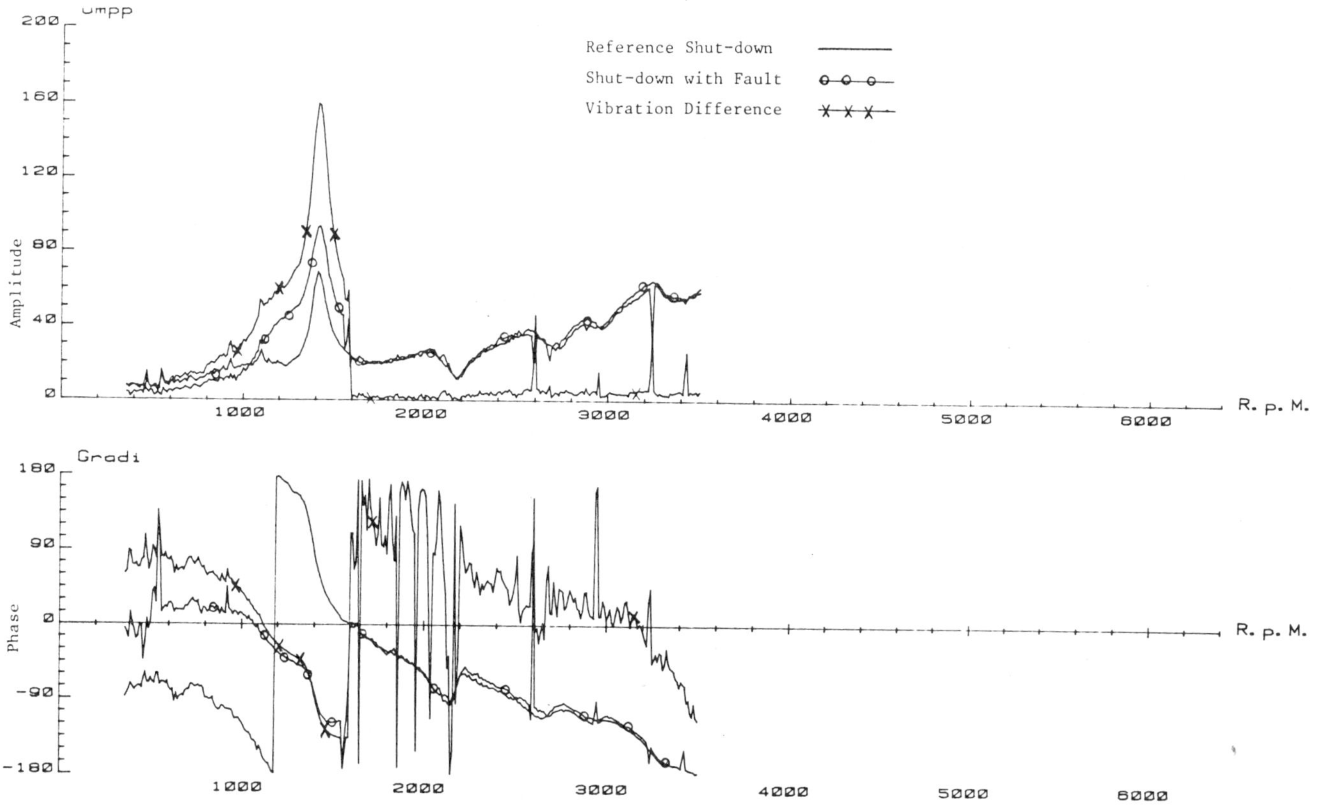

Fig 1 Fundamental component of the relative vibration at bearing number 1 of the alternator rotor for a reference shut-down, a shut-down with the fault and for the difference between the two tests

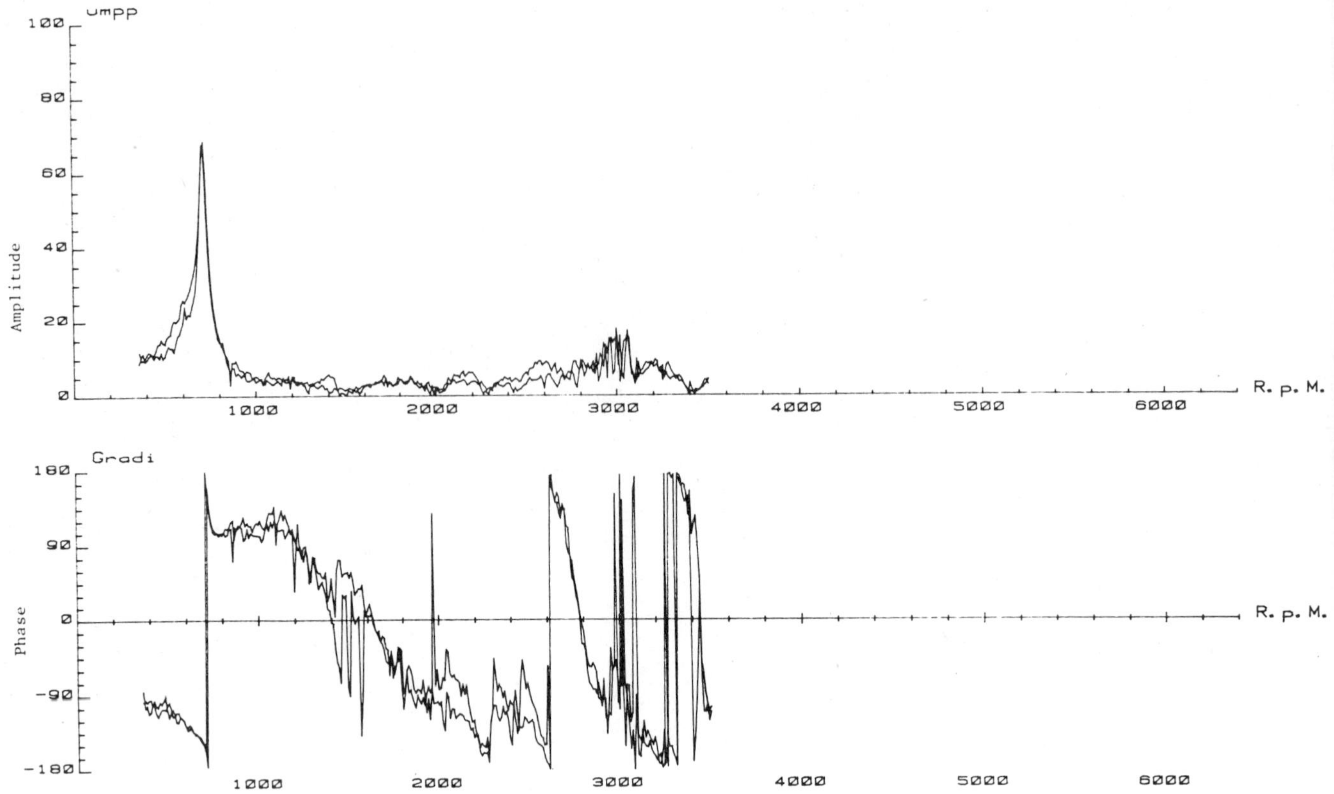

Fig 2 Second harmonic of the relative vibration at bearing number 1 of the alternator rotor during a normal shut-down and a shut-down with a sudden variation at 1600 r/min

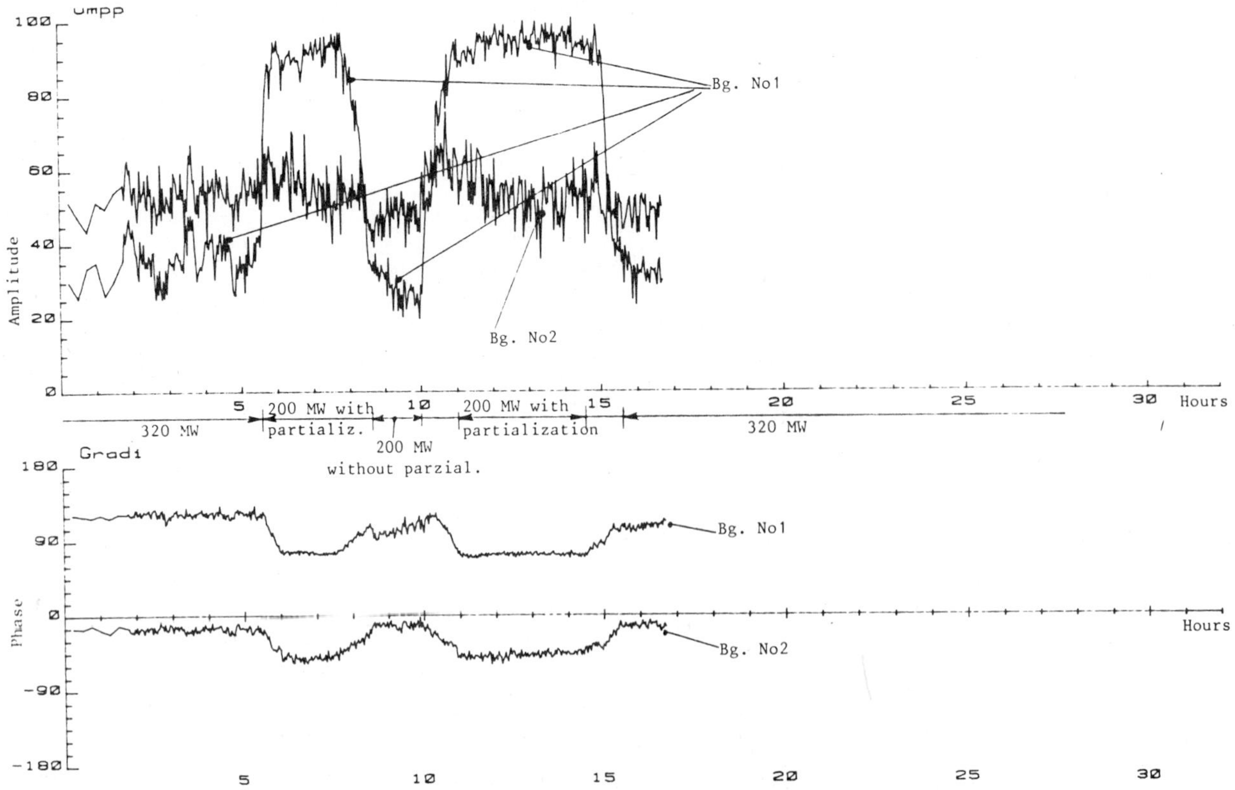

Fig 4 Fundamental component of the absolute shaft vibration at the two MP—HP turbine bearings during different operation conditions

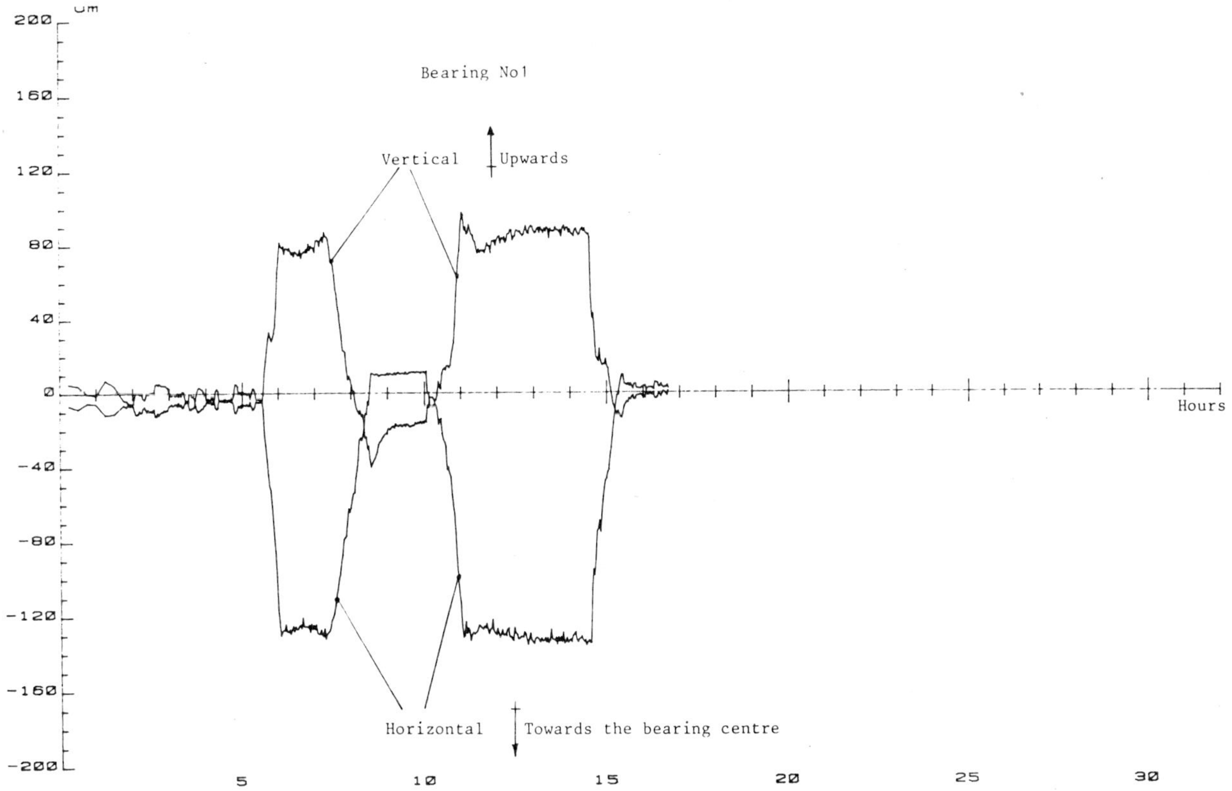

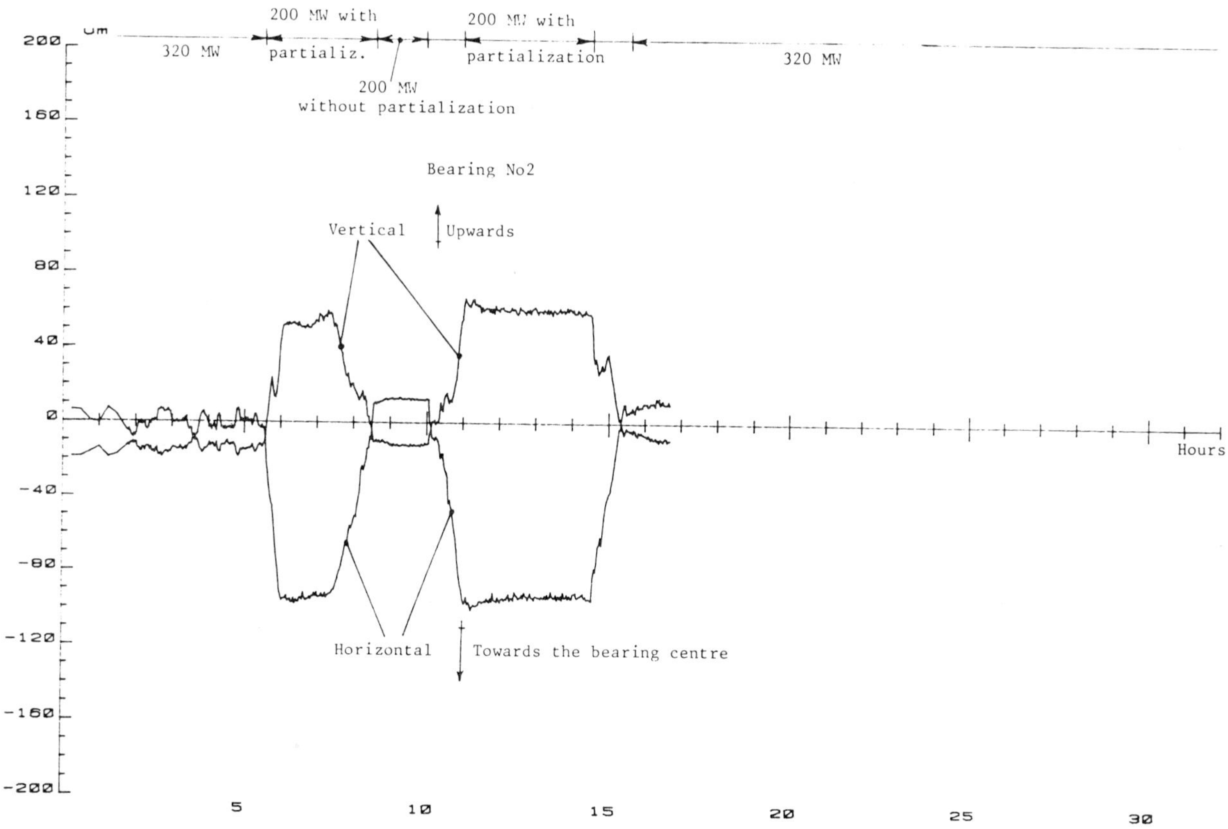

Fig 3 Journal movements in the two MP–HP turbine bearings for different operation conditions

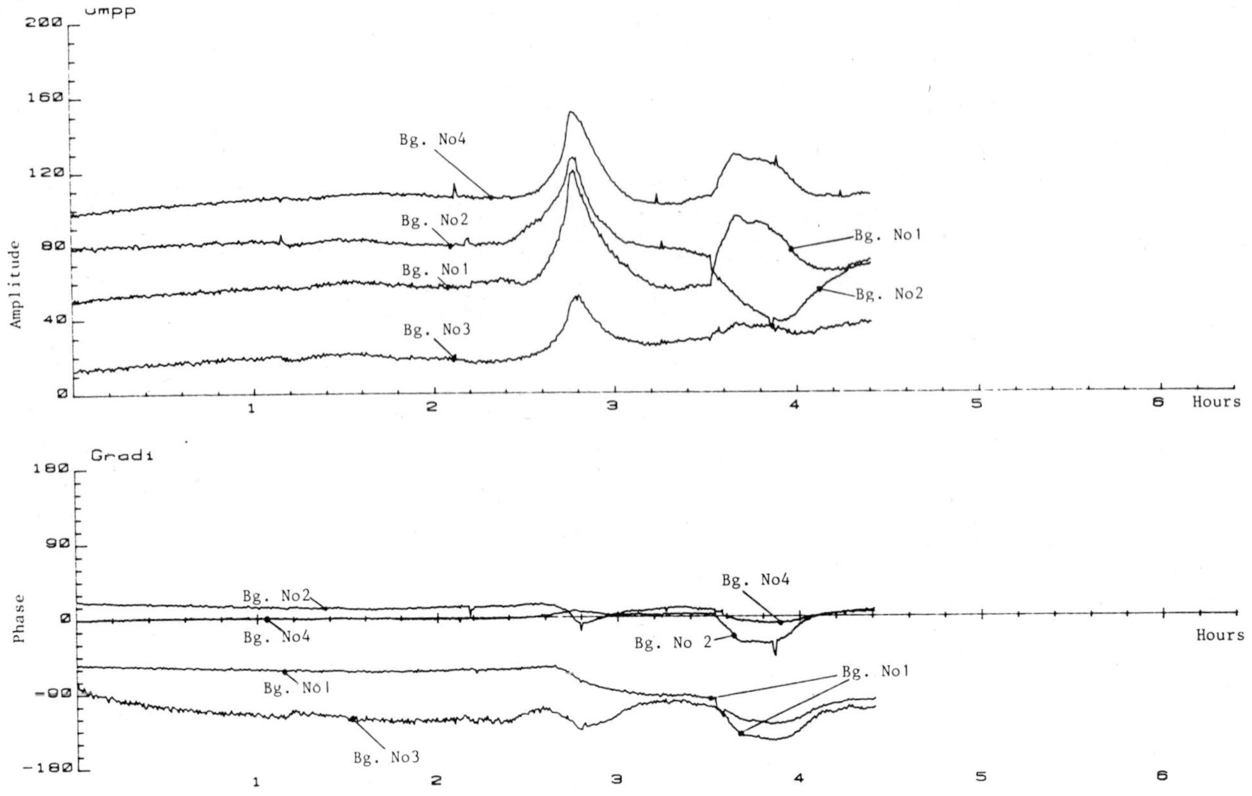

Fig 5 Fundamental component of the absolute shaft vibration at the four turbine bearings in the presence of the phenomenon

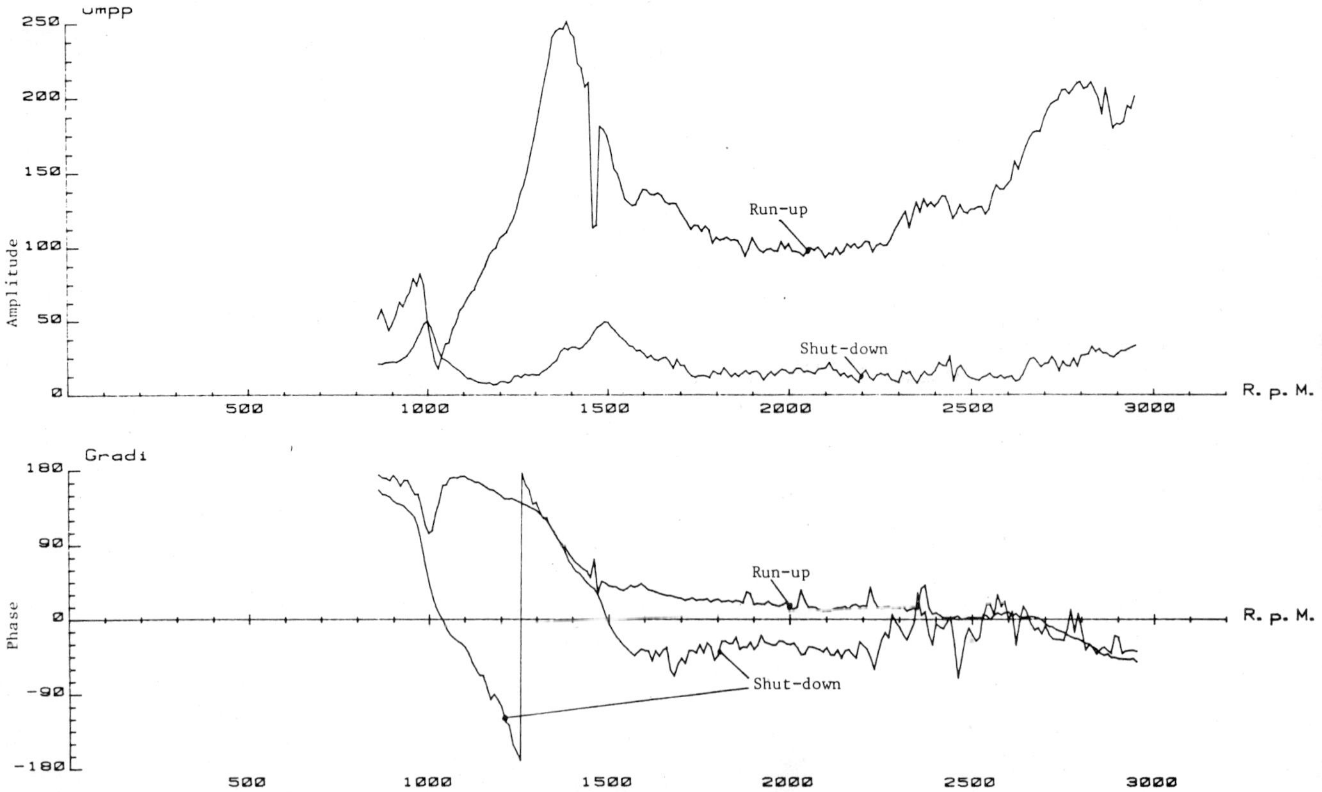

Fig 6 Fundamental component of the absolute shaft vibration at one LP turbine bearing during a normal shut-down and a run-up in the presence of the fault

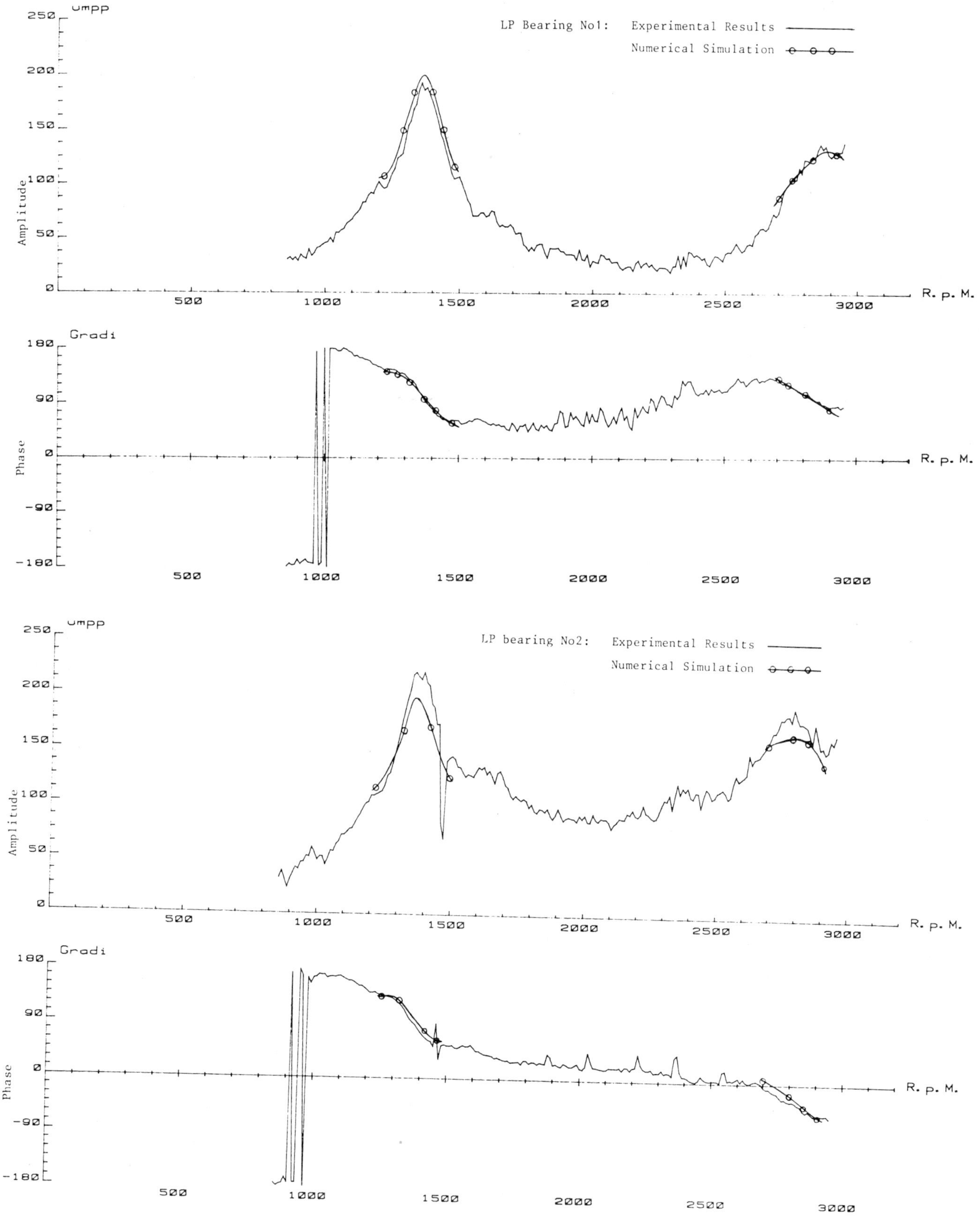

Fig 7 Experimental and numerical results for the fundamental component of the absolute shaft vibration at the two LP bearings obtained by difference between a run-up in the presence of the fault and a normal shut-down

C275/88

The diagnosis of faults of a turbine set

X W JIANG, F WANG and C F DONG
Harbin Institute of Technology, Harbin, People's Republic of China

ABSTRACT This paper discribes the diagnosis of the faults of turbine set by means of the vibration monitoring technique. The frequency spectrum analysis method is used for the analysing of the signals of vibration. The faults of turbine set are recognized by using Bayes method,pattern recognition method and fuzzy set. Finally some examples are given.

1 INTRODUCTION

The steam turbine generator set is the main equipment in a power plant. The running condition of turbine set directly effects the product of the power plant. So predicting the faults of a machine and avoiding serious accident is very important.The inside faults of structure is diagnosed by using outside vibration information. The practical application is simple and effective has been proved in these methods. As turbine set is a large and complex structure,its vibration information is very abundant. By common method it is difficult to diagnose the faults of structure. The method of the pattern recognition,Bayes method and fuzzy set are found new application in analysing the vibration spectrum of set.The suitable fault information can be gained.

These methods are realized in the examples.

2 PATTERN RECOGNITION

When the set is in operation the measured frequency spectrum lines are very abundant, its shapes are very complex as well.Therefore it is very difficult to find fault information in structure by means of conventional method. Consequently we adopt pattern recognition method to extract fault information. The key of pattern recognition is look for the discriminatory function. First we select the discriminatory function of diagram and then judge the sample of measured frequency spetrum by using discriminatory function. By this way we find out the fault information. The method of linear segmentation is basic. Its discriminatory function is straight and the linear discriminatory function can be written as:

$$g(x)=w_0+w^T \qquad (1)$$

where

$w=[w_1, w_2, \cdots]$ weight vector
$x=[x_1, x_2, \cdots]$ eigen vector
w_0 threshold value

When the number of eigen value is one, it is a simple straight line and discriminator is a single point. The function is:

$$g(x)=w_0+w_1 x \qquad (2)$$

The equation of discriminatory linear is:

$$g(x)=0 \qquad (3)$$

If $g(x)>0$ the x belong in the first sort. If $g(x)<0$ the x belong in the second sort. This is the case of two sorts discriminatiory function, it is shown in the Fig.1.

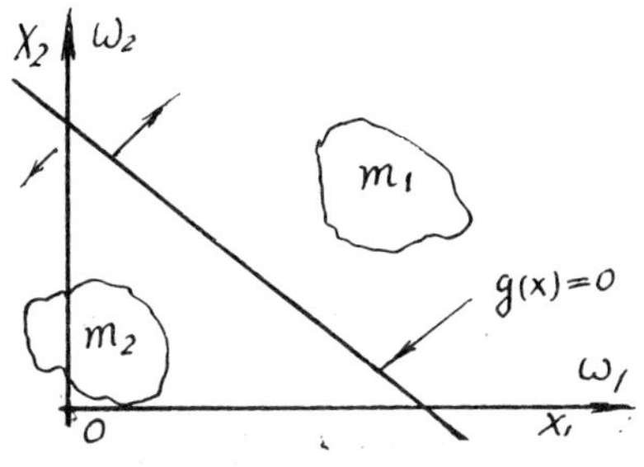

Fig 1 Linear discriminatory function

The process of finding discriminatory function is as same as the process of finding weight vector w that it can be find out using sequential iteration algorithem. At first we suppose one group value of w, and substitute the first sort eigen value into equ.(1).If $g(x)>0$ the weight vector does not change,otherwise increase by 1.And then, we check the second sort eigen value x again. If $g(x)<0$ the weight w does not change, otherwise reduce by 1. This process is continuously until that the weight is stabilizing.

Usually the picture is linear unsegmentation and the discriminatory function is not a straight line but a curve. such as:

$$g(x)=w_0+w_1 x+w_2 x^2 \qquad (4)$$

etc.

Now we still can find nonlinear segmentation method. Such as Fig.2.

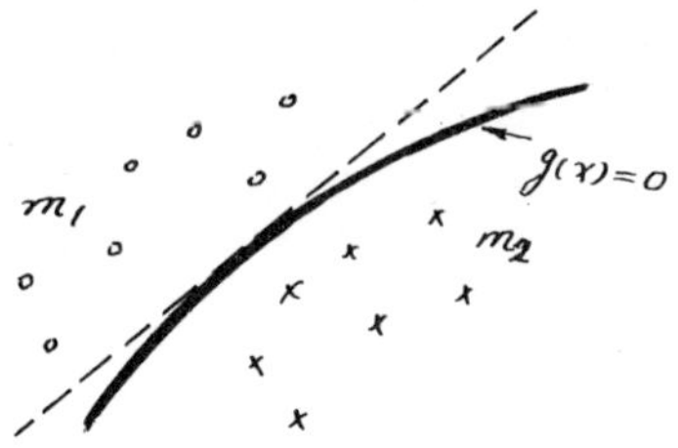

Fig 2 Non-linear discriminatory function

Then we make the sum of square of varition minimum.

$$\sum(G(x)-g(x))^2=\min \qquad (5)$$

and find out discriminatory function of linear unsegmentation. The weight function w_0, w_1, w_2 are the coefficient of discriminatory function of nonlinear segmentation. With this discriminatory function, we can perform a discriminator for the measured diagram. If $G(x)<0$ the eigen values belong to the first sort. If $G(x)>0$, the eigen values belong to the second sort. Sometimes we use the step by step linearation method to determine the discriminatory function in the nonlinear segmentation method. With this method the computing time can be saved. We can substitute several section of linear discriminatory function for the nonlinear discrlminatory function. Most of the judgement result is the final result. Other method in pattern recognition is the distance classiffication method, usually a least distance classiffication method is used in the case of multiclass picture, we assume:

$$\Omega=(\omega_1, \omega_2, \cdots \omega_n) \qquad (6)$$

for x class, we have

$$\mu_i=(\mu_{i1}, \mu_{i2}, \ \mu_{in})^T$$

If the recognized patten is

$$x=(x_1, x_2, \ x_n)^T \qquad (7)$$

the distance is $d(x, \mu_i)$. If

$$d(x, \mu_i)<d(x, \mu_j)$$

we classify it as its state, i.e. $x\sim\omega_i$.

$$\begin{aligned} d(x, \mu_i) &= |x-\mu_i|^2 \\ &= (x-\mu_i)^T(x-\mu_i) \\ &= x^T x - x^T \mu_i - \mu_i^T x + \mu_i^T \mu \qquad (8) \end{aligned}$$

Where $x^T x$ has no relationship with i. Thus the discriminatory function is:

$$g(x)=x^T \mu_i+\mu_i^T x-\mu_i^T \mu \qquad (9)$$

If $g_i(x)>g_j(x)$ $(i\neq j, j=1, 2, \ n)$

i.e. $d(x, \mu_i)$ is minimum.

For this reason x is judged to belong in ω_i. This method is simple and directly. It is widly used. Using above two methods, the faults of machinery have been recognized.

We have apply the pattern recognition to the diagnosis of faults of turbine set. The key is determining discriminatory function. Generally the discriminatory function is nonlinear. Our purpose is finding the discriminatory function. The work of diagnosis of fault is determining the border between the fault diagram and the normal diagram. Example, there are two group diagrams. One group is normal and other group is fault diagram. We should find the best curve between normal and fault diagram if the diagram were not very complex. Two frequency spectrum diagrams are shown in Fig.3.

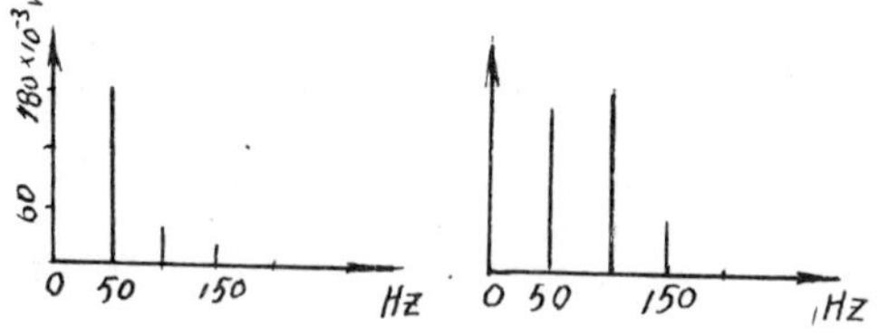

fault diagram normal diagram

Fig 3 Normal and fault diagram

This is a result of measurement on simulation testbed of rotor. One group is normal and other group is fault that it is placed disalignment. The average line between the two group diagrams that is finding discriminatory function.

3 THE APPLICATION OF BAYES EQUATION IN MACHINERY DIAGNOSIS

The random factor of distrib and noise is always existence in practice measurement. These factors influence result of tests so that it reduces the precision of result. The Bayes method is based on the statistical theory. For this reason the Bayes method is used widely for the fault diagnosis of machinery which operates in complex environment. It can reduce interference and noise, so the analysis precision is raised.

Bayes equation is:

$$p(\omega_i/x)=\frac{p(x/\omega_i)p(\omega_i)}{p(x)} \quad (10)$$

where:

$p(x)=\sum_{i=1}^{m} p(x/\omega_i)p(\omega_i)$ — total probability

$p(\omega_i)$ — prior probability of the event ω_i

$p(x/\omega_i)$ — with the event ω_i exists the condition probability of x

$p(\omega_i/x)$ — the condition probability of event ω_i, which is determined by feature x. It is called probability after test as well.

The equation (10) shows how it transform prior probability $p(\omega_i)$ into probability $p(\omega_i/x)$ after test. If any one observed value x, it makes $p(\omega_1/x)$ bigger than $p(\omega_2/x)$, then condition ω_1 is came to a decision. Otherwise, if $p(\omega_1/x)$ is lesser than $p(\omega_2/x)$, then condition ω_2 is came to a decision. It follows from equation (10) that: if $p(x/\omega_1)p(\omega_1)$ is more than $p(x/\omega_2)p(\omega_2)$ then condition ω_1 is came to a decision. Otherwise, if $p(x/\omega_2)p(\omega_2)$ is more than $p(x/\omega_1)p(\omega_1)$ then condition ω_2 is came to a decision. If logarithmically expressed, then it is given by

$$\log\frac{p(x/\omega_1)}{p(x/\omega_2)} > \frac{p(\omega_2)}{p(\omega_1)} \quad (11)$$

So ω_1 is came to decision. Otherwise, ω_2 is came to decision. A lot of problems of vibration follows normal distribution in engineering. The condition probability $p(x/\omega_i)$ can be written as:

$$p(x/\omega_i)=\frac{1}{(2\pi)^{n/2}|\Sigma|^{1/2}}\cdot \exp[-1/2(x-\mu^{(i)})\Sigma_i^{-1}(x-\mu^{(i)})] \quad (12)$$

$x=(x_1, x_2, \cdots, x_n)^T$

μ^i is n dimension vector;

Σ is $n\times n$ exponent positive symmetrical matrix;

$\mu^{(i)}=E_i[x]$;

$$\Sigma_i=E_i[(x-\mu^{(i)})(x-\mu^{(i)})^T] \quad (13)$$

General, a most pictures is normal distribution on machinery diagnosis. And the picture only has two sorts. Its covariance matrix are Σ_1 and Σ_2. It can be suppose equal. Then equation can be written as:

$$L(x)=\log\frac{p(x/\omega_1)}{p(x/\omega)} =(1/2)(x-\mu^{(2)})^T\Sigma^{-1}(x-\mu^{(2)})- (1/2)(x-\mu^{(1)})^T\Sigma^{-1}(x-\mu^{(1)}) \quad (14)$$

Thus $L(x)=x^T\Sigma^{-1}(\mu^{(1)}-\mu^{(2)})- (1/2)(\mu^{(1)}+\mu^{(2)})^T\Sigma^{-1}(\mu^{(1)}-\mu^{(2)})$ (15)

The prior probability $p(\omega_1)$ and $p(\omega_2)$ is known. Therefore they can be supposed equal 0.5.

Further we study the function L(x). Due to x is normal distribution as well. Its condition expectation value and variance can be derived as the expectation:

$$E[L|\omega_1]=(1/2)(\mu^{(1)}-\mu^{(2)})^T\Sigma^{-1}(\mu^{(1)}+\mu^{(2)}) =(1/2)J \quad (16)$$

$$E[L|\omega_2]=-(1/2)J \quad (17)$$

the variance:

$$\sigma_1^2=\sigma_2^2 =(\mu^{(1)}-\mu^{(2)})^T\Sigma^{-1}(\mu^{(1)}+\mu^{(2)}) =J \quad (18)$$

with covariance is equal, the misclassification probability R_0 as:

$$R_0=p_1\int_{\omega_2}p(x|\omega_1)dx+p_2\int_{\omega_1}p(x|\omega_2)dx \quad (19)$$

We know

$$p(L|\omega_1)\sim N(1/2J, J)$$
$$p(L|\omega_2)\sim N(-1/2J, J) \quad (20)$$

thus we can get

$$R_0=p_1G\left(\frac{\ln(p_2/p_1)-1/2J}{J^{1/2}}\right)+p_2G\left(\frac{-\ln(p_2/p_1)-1/2J}{J^{1/2}}\right) \quad (21)$$

where $G(x)=\frac{1}{(2\pi)^{1/2}}\int_{-\infty}^{x}e^{-x^2/2}dx$ (22)

when $p_2=p_1=1/2$, we obtain:

$$R_0=G(-1/2J^{1/2})=\int_{-\infty}^{-1/2J^{1/2}}\frac{1}{(2\pi)^{1/2}}e^{-x^2/2}dx \quad (23)$$

Thus it can be seen: if J is bigger, then misclassification R_0 is lesser; Otherwise if J is lesser, then R_0 is bigger. Practically the bigger J shows that the distance between mathematic expectation is bigger.

This shows that two pictures are easily separated. Otherwise it shows that distance is lesser, then they are not easily separated. When $\Sigma=I$, then J is general distance. J shows near degree between two sorts of phenomenon. Thus, it can be used as criterion of pictures classification. If we determined a threshold J_T, then the discriminant $J<J_T$ can be used sorting of phenomenon.

If $J \nless J_T$, two sorts of phenemenon are near, or can be called not fault. If $J<J_T$, then the fault has happened. when the fault is analysed, equation (14) and (15), or (16) and (17) can be selected.

x is amplitude or moment of zero point in frequency spectrum. It can be measured direcly. Equation μ is expectation of x, Σ is covariance matrix, its element $\sigma^{(1)}$ can be expressed as:

$$\sigma_{lk}^{1} = E_1[(x_l - \mu_l^{(1)})(x_k - \mu_k^{(1)})]$$
$$(l, k=1, 2, \quad n) \qquad (24)$$

4 THE APPLICATION OF FUZZY SET IN FAULT DIAGNOSIS OF MACHINERY

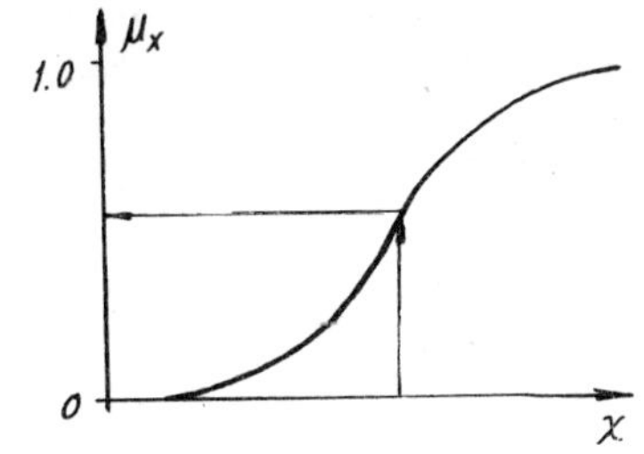

Fig 4 Curve of sub-degree of membership

It is difficult to say whether the intensity of a vibration is strong or weak quantitation, but one can really express it quantitatively by using the degree of membership in theory of fuzzy set. Figure 4 shows the curve of the relationship between the level of vibration and the sub-degree of membership. So with the help of membership, it looks much easy to express a complicated concept. The relationship between structure behaviours and fault is also fuzzy. Sometimes faults may cause several particular behaviour of a machine, meanwhile a certain particular behaviour may be caused by many faults. So it exists a quite complicated fuzzy relationship between vibration behaviour and fault. This relationship is predetermined by experiments. With the help of it the sub-degree of membership of vibration signal can transformed into the sub-degree of membership of the fault. Then the cause of fault of a machine can be found out by judging the sub-degree of membership of fault. If there are area $U=(u_1, u_2, - u_n)$ and area $V=(v_1, v_2, - v_n)$, the fuzzy relationship is $\underset{\sim}{R}$ and subarea $\underset{\sim}{A}=(a_1, a_2, - a_n)$ in U, subarea $\underset{\sim}{B}=(b_1, b_2, - b_n)$ in V, then they satisfy the equation

$$\underset{\sim}{B_1} = \underset{\sim}{A_1} \circ \underset{\sim}{R_1} \qquad (25)$$

where the fuzzy subset $\underset{\sim}{A}$ is composed of elements with a different weights:

$$A=[a_1, a_2, - a_n] \qquad (26)$$

and the fuzzy matrix $\underset{\sim}{R}$:

$$\underset{\sim}{R} = \begin{bmatrix} r_{11}, r_{12}, & r_{1n} \\ r_{21}, r_{22}, & r_{2n} \\ ---- & \\ & \\ r_{n1}, r_{n1}, & r_{nn} \end{bmatrix} \qquad (27)$$

In fault diagnosis of turbine generator set the vibration signal is analysed by frequency spectrum and area U is composed by component of frequency spectrum.

$$U_1 = (\text{subharmonic amplitude } x_{1/2}, \text{ primary harmonic amplitude } x_1, -) \qquad (28)$$

the fuzzy vector:

$$\underset{\sim}{A_1} - [\mu_{x1/2}, \mu_{x1}, - \mu_{xn}]^T \qquad (29)$$

is composed of sub-degree membership of every frequency spectrum element, where μ_{x1} is the sub-degree of membership.

Suppose fault area:

$$V_1 = (\text{initial unbalance } v_1, \text{disalignment } v_2, \text{whip of oil film } v_3, -) \qquad (30)$$

and fuzzy vecter:

$$B_1 = [\mu_{v1}, \mu_{v2}, \quad \mu_{vn}] \qquad (31)$$

consists of sub-degree of membership of synthesised judging equation of the first sort fuzzy relationship matrix R_1 is:

$$\underset{\sim}{B_1} = \underset{\sim}{A_1} \circ \underset{\sim}{R_1} \qquad (32)$$

The second synthesised judgement was chosen as:

A_2, R_2, B_2

The total sub-degree of membership of fault of a machine is composed as follows:

$$\underset{\sim}{B} = \underset{\sim}{B_1} + \underset{\sim}{B_2} = \underset{\sim}{R_1} \circ \underset{\sim}{A_1} + \underset{\sim}{R_2} \circ \underset{\sim}{A_2} \qquad (33)$$

If there are more factors, the synthesised judgement can be carried on further.

The synthesised judgement is on the basis of fuzzy relationship matrix. Thus it is very important. Its element can be determined by the test or experience. If there is not the matrix, then it can't realize synthesised judgement. Before we achieve the element of fuzzy relationship matrix, the data of similar machine can be quoted. After we gain data, it will be corrected.

5 THE APPLICATION OF DIAGNOSIS METHOD IN TURBINE SET

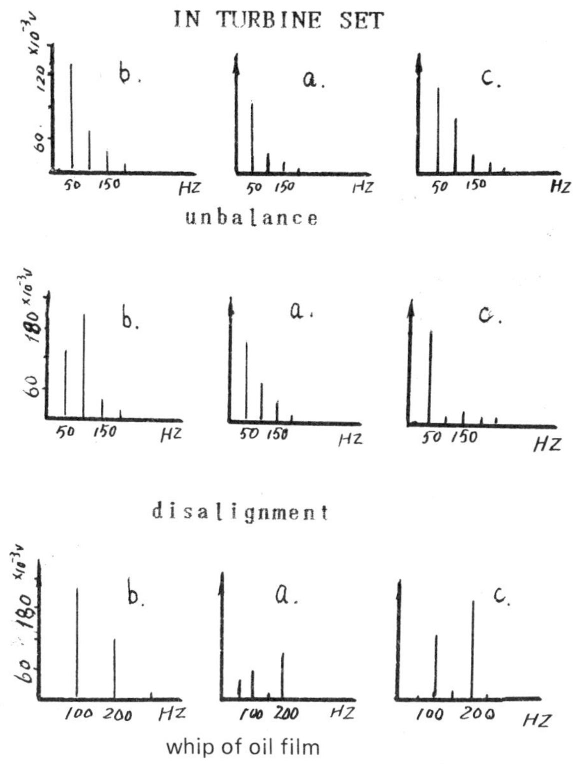

a. normal condition

b. unusual condition

c. the place without fault

Fig 5 Various frequency spectra

This paper carry on test study on similation testbed of rotor. We set up some faults on the modal, then these faults are measured and recorded. It is analysed by using various method of diagnosis.

There are unbalance, disalignment and whip of oil film on the modal. We measure vibration signal on some placesof the rotor at same time. One point of rotor has fault and other place of rotor has'nt fault. We call the point which is far from the place of fault. Meanwhile under normal condition and unusual condition signal are measured in modal.The result shows pattern recognition,Bayes method and fuzzy set that they have virtues and defects respectively. The frequency spectrum is shown in figure 5. From the spectrum line of Fig.5. it is seen that different condition has diffrent shapes of figure.

From these figures, we can see that there is condition of structure.
Pattern recognition can obviously judge the fault of unbalance of rotor. Not only it can judge fault of disalignment of rotor, but the place of fault on rotor can be found. It can effectively judge whip of oil film in modal. The Bayes method adopts the criterion J. It can judge the place of vibration source happened. It can clearly indicate the fault of whip of oil film as well. The fuzzy set employs sub-degree of membership. It can show the level of fault clearly. This method can show the order, grade and the place of fault. Thus it is very obrious and direct. The fuzzy set considers fuzzy property of the event. Therefore it reflects the essence of some events. It is most suitable for the events of diagnosis of disease to adopt the fuzzy set method. Someway the fuzzy set is adopted for the diagnosis of fault of machinery. It is very effective as well.

6 CONCLUSION

In this paper some method of fault diagnosis of machinery are studied. These methods are very useful . From above analysis we can seen that these methods possess common property of identification fault. But under various environment the methods are different from each other. The ability of identification is different. The pattern recognition possesses the ability of judging complex shapes of figure. The Bayes method possesses the ability that it reduces interferences and noise. The fuzzy set possesses ability of synthesizing fuzzy judgment. Thus various method can be selected depending on different working conditions.

REFERENCE

[1] S.Q.Chen, H.J.wei, "patten recognition thory and application", Chang du Electrical Engineering College Press, Sichuan, 1985.

[2] Y.J.Li, "pattern Recognition technique And Application", Mechinical In dustry Press, Bejing, 1986.

[3] G.X.Zhou, "Computation Pattern Recognition", Center China Institute of Technology Press, Hubin, 1986.

[4] Z.He, "fuzzy and Its Application", Taijin Science and Technology Press, taijin, 1985.

[5] C.X.Sao, "Application of Fuzzy Set Method in Turbine Set", CSMDT'-86, 1986,Shen Yang.

C262/88

Comparison between computed and measured vibrations of turbomachines

E KRÄMER, Dipl-Math, Dr and **L ECKERT**, Dipl-Ing
Department of Mechanical Engineering, Technische Hochschule Darmstadt, Darmstadt, West Germany

SYNOPSIS A project on the comparison between computed and measured vibrations of turbomachines is described. For computations rotor-models and rotor-foundation models were used. Measurements were made during start up, operation and run down. Four turbosets were investigated in this project. Problems of the comparison are discussed. For one machine computed results are presented and compared with measurements. By the results of the considered project and of known literature it may be concluded that the vibrational behaviour of turbosets can be computed accurately enough with available procedures. Resonance speeds may be predicted with less than five per cent deviation from measured values.

NOTATION

$\mathbf{D}$	damping matrix
D_k	k-th damping ratio acc. Eq.(4)
D_k'	k-th damping ratio acc. Eq.(5)
e	eccentricity
f	vector of exciting forces
f_k	k-th natural frequency
$\mathbf{G}$	gyroscopic matrix
h	horizontal
$\mathbf{K}$	stiffness matrix
$\mathbf{M}$	mass matrix
n	rotor speed
n_k;	k-th critical speed;
n_k'	corresponding resonance speed
t	time
U	unbalance
v	vertical
x	vector of displacements and rotations
$\hat{x}$	displacement amplitude
$\hat{x}_n$	displacement amplitude of n-vibration
$\underline{\lambda}_k$	k-th complex eigenvalue $\underline{\lambda}_k = \alpha_k + i\omega_k$
$\smile$	U-shaped mode (1st bending)
$\sim$	S-shaped mode (2nd bending)

1 INTRODUCTION

The technique of computation and measuring vibrations of turbomachines has reached a high level during the last decades. But only few efforts have been made to compare the results of measurements with those computed before. Such a comparison is however necessary to improve on the one side the accuracy of the computations and to reduce on the other side their extent to an acceptable size. The following questions arise:

- What kind of models are needed to represent the vibrational behaviour accurately enough?
- How should the foundation be considered in the model?
- Which computations should be carried out?
- Which unbalance cases should be assumed and which results should be plotted?

The problem of comparison is not easy, as will be described in the third chapter of this paper. Briefly the problem is the comparison of computed and measured

- resonance speeds,
- resonance amplitudes,
- sharpness of resonances and
- stability criteria.

In this paper a project for investigation of this problem is described. As example for one machine computed natural vibrations and responses to assumed unbalances are shown and compared with measured vibrations.

2 LITERATURE REVIEW

Only a few papers are published about the considered problem. In the following the papers /1-4/ are reviewed as examples.

In /1/ J.W. Lund and F.K. Orcutt considered test rotors with one, two and three discs and with two journal bearings. They compared computed and measured orbit radii due to unbalance excitation at several positions and due to different unbalances. In most cases good agreement was found. Partly this was explained by the bearings having only little influence in these cases. In other cases larger differences occurred because the assumed bearing coefficients were insufficiently accurate.

In /2, 3/ J.M. Vance, B.T. Murphy and H.A. Tripp describe a research project directed to improving computer programs for critical speeds of small turbomachines for petroleum industry.

The authors show in /2/ how the rotor model may be improved by investigations of the free-free natural frequencies. For improved models differences of 4 and 20 per cent for the first and third critical speed respectively were stated. In /3/ the influence of the journal bearings and the foundation on the critical speeds was studied. By improved technique the first and second critical speed of a single-stage steam turbine could be computed with 2 per cent accuracy.

A. Pons reports in /4/ on an important numerical and experimental study of a 900 MW turbogenerator unit with a rated speed of 1500 r/min. It was the aim to build up a model as good as possible for the rotor, the bearings and the supporting system. For this, the data derived from the drawings were complemented by measurements of the foundation with mounted pedestals and of the completely mounted unit. It may be mentioned that such an effort is not usual. The calculated mode shapes were characterized as mainly belonging to the supporting system, to the shaft line or to the whole system. Some of these mode shapes could also be observed during run down tests. A table shows predicted and measured critical speeds of main shaft line for horizontal and vertical direction. Differences between 0.6 and 2.7 per cent were stated for the first five critical speeds which are between 565 and 1035 r/min. The damping ratios of the corresponding eigenvalues were determined as 0.009 to 0.035. During the first period of operation vibration problems arose. They could be explained by the improved computation. With the computation model modifications of the supporting system could be studied before they were realized at the machine.

3 GENERAL REMARKS

For comparison between computations and measurements it is important

- that the same dimensions are compared and
- how accurate these dimensions are determined.

To explain this in more detail a short review is given about modern practice of computation and measuring of large turbomachine vibrations.

The computation model consists of the shaft line and the journal bearings. The supporting system mostly is approximated by equivalent mass-spring-damper systems with different parameters for horizontal and vertical direction, but uncoupled between the different bearings. Sometimes also the foundation is considered and modelled by one-, two- or threedimensional elements. For computation of these models several methods are available as described in many publications (see e.g. /4-11/). Normally the following quantities will be computed:

- Natural frequencies and modes,
- damping ratios,
- response to unbalance excitation and
- threshold parameters for instability.

Measurements normally are made at the pedestals and at shaft positions near the bearings, namely perpendicular to the shaft axis and in two orthogonal directions. From the possible measurement quantities the displacement amplitudes as a function of the rotor speed are of main interest. For judging the stability situation frequency analyses are necessary and cascade plots are useful.

Some problems arise with comparison of computations and measurements. Mostly the positions - and sometimes also the directions - of computations and measurements do not coincide. Additional computations are necessary in these cases. More serious is, that for the measurements normally the actual existing unbalance distribution is unknown. It should be known to assume the same distribution for the computation. Normally single unbalances are assumed for computations. To determine the corresponding response of the real machine, theoretically it is possible to add a trial mass, to measure the response and to determine the vector difference to the initial response. In practice however this procedure is often impractical because of uncontrollable variations of the unbalance due to thermal influences and other effects. Furthermore the number of such test runs often is limited due to economic reasons.

4 CONCEPT OF THE PRESENTED INVESTIGATION

As a contribution to the solution of the above general problem computations and measurements were carried out for four turbo-generator units, which have been in operation for some years. The considered machines have a nominal speed of 3000 r/min and a rated power from 125 to 660 MW. They have just one bearing between HP-, IP- ... part and are mounted on a reinforced concrete foundation, whose table top is spring supported in two cases (Fig 1). Further details see Table 1.

Table 1 Investigated Turbosets

Turbo-set	First operation	Rated power	Number of bearings		Shaft length	Shaft mass	Table top spring supported
		MW	1)	2)	m	tons	
M1	1967	125	6	7	26	81	no
M2	1967	125	6	7	26	81	no
E4	1982	315	5	6	30	129	yes
S4	1977	660	6	9	48	233	yes

1) Turbine and generator 2) Whole shaft line

4.1 Computations

For computation of vibrations the real unit was substituted by two kinds of models, called RO and RF, shown schematically in Fig. 2. For both models the shaft line was considered as beam with an appropriate number of beam elements. Gyroscopic moments were considered, damping was neglected. The journal bearings were considered as linear spring-damper systems with eight coefficients as tabulated functions of Sommerfeld number. The supporting system - pedestals and foundation - was approximated for model RO only as mass-spring systems at the bearings with normally different stiffness in horizontal and vertical direction but without coupling. For model RF the supporting system was considered as three dimensional framework of beam elements and rigid bodies. The housings were only modelled as concentrated masses. For the foundation proportional damping was assumed with damping ratio 0.02 for 1200 and 3000 c/min. In case of spring supported table top the foundation below the springs was assumed as rigid. The supporting systems were modelled originally by some hundreds of degrees of freedom and reduced afterwards to about 150 DOF. The models for the investigated turbosets had the following degrees of freedom:

Turboset	Model RO	Model RF
M1	106	310
M2	106	310
E4	172	274
S4	158	308

For the computations the program MADYN was used, which is described in /9,11/. It is based on the well known equation of motion

$$\mathbf{M}\ddot{x} + (\mathbf{D} + \mathbf{G})\,\dot{x} + \mathbf{K}\,x = f(t) \qquad (1)$$

for the displacements and rotations x_i.

The eigenvalues

$$\underline{\lambda}_k,\ \underline{\lambda}_k^* = \alpha_k \pm i\,\omega_k \qquad (2)$$

were computed by inverse vector iteration.

The natural frequency and damping ratio of order k are

$$f_k = \frac{1}{2\pi}\,\omega_k \quad (3) \qquad D_k = \frac{-\alpha_k}{|\underline{\lambda}_k|} \quad (4)$$

The eigenvalues and the corresponding conjugate complex pairs of eigenvectors lead after some transformations to the equation of the k-th natural vibration.

Due to the special kind of matrices the natural mode shapes are time-dependent. In the harmonic movement with natural frequency there are phase differences between various parts of the structure, thus the position of nodal points of the mode shape varies slightly during one period of natural vibration.

In case of unbalance excitation the solution of (1) is a system of algebraic equations for the response, which has to be solved for each speed. For each of the four considered turbo machines the following computations were carried out with model RO and RF.

Eigenvalues were computed for different speeds in the range from 600 to 3600 r/min. The natural frequencies and damping ratios were plotted as functions of speed. Mode shapes were plotted only for the rated speed 3000 r/min. For different unbalances the response at those points was computed, where the vibrations have been measured or at points in their neighbourhood. The unbalances were assumed with respect to the natural mode shapes. Those shapes with maximum displacement in a rotor span are of special interest. The deflection of this span normally is either ◡- or ∼- shaped. To excite mainly the corresponding natural vibration a single unbalance at midspan or a couple of single unbalances at 1/4 and 3/4 of span length was assumed.

4.2 Measurements

For comparison with the results of the computations the following measurements were carried out. The displacements for different positions at the pedestals and at the shaft were recorded as function of time during start up, run down and power operation. To be able to analyze these time histories also a once-per-revolution reference signal was recorded.

The pedestal vibrations were measured by two kinds of seismic pick-ups: Electrodynamic velocity transducers and piezoelectric accelerometers. The shaft vibrations were measured contactlessly either by the installed inductive pick-ups or by additionally mounted eddy current proximity displacement probes. For the once-per-revolution signal a dark mark was observed by a photoelectric sensor.

All signals were recorded after amplifying and filtering by tape recorders. A 14-track- and a 8-channel- tape recorder were used. The time histories of the 14-track-recorder were transferred afterwards to the 8-channel-PCM-recorder. By its digital Pulse Code Modulation technique all time histories were transferred as digital data to a large computer and interpreted by a program. The following results may be plotted:

- Time histories of displacements in one direction,
- orbits of points whose motion was measured in two directions,
- rotor speed as function of time,
- amplitudes and phase angles of the harmonics (n-, 2n-, ... vibration) as function of rotor speed or of time,
- amplitude- and phase-spectra,
- cascade diagrams as function of rotor speed or of time

and mean values, rms values and standard deviations.

Furthermore for n- and 2n-vibrations

- resonance speeds n_k',
- resonance amplitudes $\hat{x}_k$

and damping ratios D_k'

may be printed.

The damping ratio D_k' is determined in case of distinctly marked resonances as

$$D_k' = \frac{\Delta n_k}{2n_k'} \qquad (5)$$

with Δn_k as width of the resonance peak at $0.707\cdot\hat{x}_k$. It may be mentioned that the determination of this quantity sometimes is problematic for the observed kind of resonances.

5 RESULTS FOR TURBOSET S4

In this paper only results for turboset S4 are presented. The results of the whole project will be published later. The measurements of this turboset were concentrated mainly at positions at bearing 4 and 5 (Fig. 1). Therefore also the computations were focussed to these positions.

5.1 Computations

Computations were carried out for model RO and RF, as mentioned above. For model RO Fig. 3 shows natural frequencies and some of the corresponding damping ratios. The natural frequencies are numbered by k according to their order at rated speed. Mostly the natural frequencies depend only slightly on speed. In case of strong speed dependency the natural vibration is very highly damped. Fig. 4 shows natural mode shapes at rated speed as instant view in the moment of maximum deflection. Each mode shape is described by its deflection ($\smile$- or $\sim$-shaped) and by its dominating direction (horizontal or vertical). Those modes are shown which are relevant for vibrations at bearing 4 and 5. In Fig. 3 the corresponding curves are indicated by thick lines.

For model RF also eigenvalues and natural modes were computed for speeds between 600 and 3600 r/min. Up to 3600 c/min model RF has 53 natural vibrations, whereas model RO has 21. The mode shapes may be classified as

- rotor modes with mainly rotor deflections,
- foundation modes with mainly foundation deflections and
- combined modes.

Fig. 5 shows as example some natural modes, which are of special interest for this investigation.

Critical speeds are defined as speeds whose frequency is equal to a natural frequency. Table 2 shows some critical speeds and the corresponding damping ratios of the two models. They were selected with respect to the mode shapes and the damping. Cases with high damping were neglected.

Table 2 Turboset S4. Relevant critical speeds and corresponding damping ratios.

Model RO			Model RF		
k	n_k	D_k	k	n_k	D_k
	r/min	–		r/min	–
2	670	0.003	11	666	0.007
			18	1248	0.014
5	1320	0.011	19	1308	0.012
6	1410	0.010	20	1452	0.011
13	1924	0.006	32	1980	0.009

As mentioned above only the vibrations at the bearings 4 and 5 should be considered. In spite of this many cases of unbalance vibrations had to be studied. On the basis of the computed natural modes the following unbalance cases were assumed:

Case 1, single unbalance in LP1
Case 2, single unbalance in LP2
Case 3, single unbalance in Gen
Case 4, two unbalances in Gen.

The single unbalance was located at midspan. In case 4 one unbalance was located at 1/4, the other at 3/4 length of span and directed oppositely. In cases 1 to 3 the unbalance was U=me and in case 4 it was U/2 with m as the mass of the corresponding rotor and e=10 μm as eccentricity. Displacement amplitudes were computed in the speed range between 300 and 3600 r/min for both models. The results were plotted for pedestal and shaft vibrations at bearing 4 and 5 in horizontal and vertical direction.

As example Fig. 6 and 7 show the pedestal and shaft displacement amplitudes at bearing 4 in vertical direction. The resonances show up differently strong. This can be explained on the one hand by different damping ratios and on the other hand by the shape of the natural modes. As known by the modal analysis the magnitude of resonance amplitudes depends on the displacement of the corresponding mode at the position of the excitation and of the considered response (see e.g. /12/). The pattern of the results differs not essentionally for the two models. It is just remarkable that model RF has more important resonances than model RO. The computations showed that in most cases the horizontal resonance amplitudes are smaller than the vertical amplitudes. This is mainly due to different oil film damping in the two directions.

5.2 <u>Measurements</u>

For turboset S4 measurements were made during one start up, two runs down and during one minute of full power operation after stabilized conditions were obtained. Pedestal and shaft displacements at bearing 4 and 5 in horizontal and vertical direction were recorded. Analyses showed that the vibrations mainly are harmonic with rotational frequency – the n-vibration dominates. As example Fig. 8 and 9 show measured n-vibrations of pedestal and shaft at bearing 4, vertical for three runs. At run down 2 the resonance amplitudes were about 20 per cent larger than at run down 1. During start up the resonance amplitudes reached only about half the size of those during run down. Fewer distinct peaks occurred at start up compared to run down. This behaviour may be explained by different unbalance distributions due to thermal influences.

The measured resonance amplitudes in horizontal and vertical direction were nearly the same size, whereas the computed vertical amplitudes were larger than the horizontal. No satisfying explanation was found for this discrepancy.

5.3 <u>Comparison</u>

To show how accurate resonance speeds may be predicted by computation the results of Fig. 6 and 8 respectively Fig. 7 and 9 were compared. Such computed and measured resonance peaks were compared which have nearly equal resonance speeds and similar mode shapes. From the four assumed unbalance cases that one was considered, which causes mainly the measured resonance. In Fig. 8 and 9 the considered resonances are numbered by a to d.

Results of the comparison are given in Table 3. Values of associated resonances are written in the same row. The error for the resonance speeds is astonishing small for both models. The good results for the rather simple model RO is due to good assumptions for the horizontal and vertical bearing stiffnesses which are based on experiences with similar machines.

Table 3 also shows damping ratios. Those of measurements are between 0.019 and 0.023 and those of computations are between 0.003 and 0.013. The measured values were about double the corresponding computed ones. The damping ratios of model RO and model RF are nearly the same. The damping ratios of both models were computed by Eq. (4) and (5). Nearly identical values were achieved. Therefore resonance peaks may be determined approximately only by knowledge of the

Table 3 Turboset S4. Resonance speeds (r/min) and damping ratios

Measurements at run down 1			Model RO				Model RF				Error $\frac{n_k - n'_k}{n'_k}$ 100 (%)	
k	n'_k	D'_k	k	n_k	D_k Eq.(4)	D_k Eq.(5)	k	n_k	D_k Eq.(4)	D_k Eq.(5)	RO	RF
a	688	0.021	2	670	0.003	0.005	11	666	0.007	0.008	−2.6	−3.2
b	1330	0.019	5	1320	0.011	0.011	19	1308	0.012	0.013	−0.8	−1.7
c	1405	0.023	6	1410	0.010	0.010	20	1452	0.011	0.011	0.4	3.3
d	1940	0.021	13	1924	0.006	0.006	32	1980	0.009	0.011	−0.8	2.1

eigenvalues and eigenvectors /12/.

Finally it will be shown that patterns of resonance curves similar to those of measurements may be computed. As shown in Fig. 6 and 7 each resonance peak is mainly caused by only one or two of assumed unbalance cases 1 to 4 (chapter 5.1). With the knowledge of the mode shapes a set of unbalances was found by a linear combination of these unbalance cases and some variations of their phase angles. For the computation unbalances were assumed according to Table 4 which were found after seven trials. As example Fig. 10 shows computed and measured pedestal amplitudes at bearing 4 vertical. The figure shows that the main peaks, marked by A to F, coincide tolerably well. The computed resonance speeds deviate between −3 and 5 per cent from the measured ones.

Table 4 Set of unbalances for simulated existing unbalance distribution. Positions of unbalances at 0.25, 0.50, 0.75 length of span.

	LP1	LP2	Gen
U (kgmm)	150 750 150	150 350 150	280 350 280
Angle	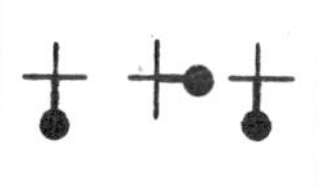	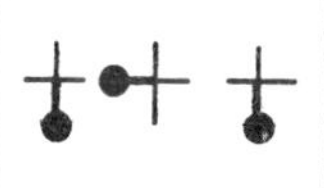	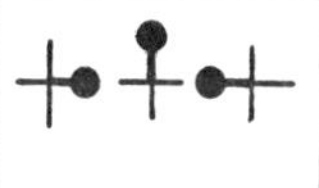

6 CONCLUSIONS

As part of a large project on comparison between computed and measured vibrations of turbomachines results of one machine were presented. From these results and others not presented here as well as from other published literature it may be concluded:

- For the computation of prototypes of important turbomachines the foundation should be considered as threedimensional framework with an appropriate number of degrees of freedom.

- With sufficient experience with similar machines the supporting system may be represented only by simple uncoupled mass-spring systems at the bearings.

- For all computations damping should be considered. For the journal bearings, in the usual manner, four speed dependent coefficients should be used. It is sufficient to assume proportional damping in the foundation. For the rotor damping may be neglected.

- The computation results in a lot of natural vibrations. Only a few of them are relevant for operation. These may be identified by their mode shapes and damping ratios.

- To investigate the unbalance behaviour the assumption of single unbalances is advantageous.

- For distinct resonance peaks the damping ratio estimated by the half power bandwidth is about the same as from the complex eigenvalue.
Therefore such resonance peaks may be determined approximately only on the basis of computed natural vibrations.

- For the presented example the measured damping ratios are approximately twice as large as the computed ones.

- The considered example shows a discrepancy between computed and measured resonances in horizontal and vertical direction. The computed peaks were higher for vertical than for horizontal direction. The measured peaks were approximately equal for both directions. For this discrepancy no satisfying explanation could be found.

- As a preliminary result one may assume that computed resonance speeds deviate less than five per cent from the real ones.

ACKNOWLEDGEMENTS

The authors wish to thank the companies Brown, Boveri & Cie AG, Mannheim and PreussenElektra Aktiengesellschaft, Hannover for their friendly support of this project.

REFERENCES

/1/ LUND, J.W., ORCUTT, F. K. Calculations and experiments on the unbalance response of a flexible rotor. *Journal of Engineering for Industry*, Trans. ASME, 1967, Vol. 89 No. 4, 785–796.

/2/ VANCE, J.M., MURPHY, B.T., TRIPP, H.A. Critical speeds of turbomachinery: Computer predictions vs. experimental measurements – Part I: The rotor mass-elastic model. *Journal of Vibration, Acoustics, Stress and Reliability in Design*, Trans. ASME, 1987, Vol. 109 No. 1, 1–7.

/3/ VANCE, J.M., MURPHY, B.T., TRIPP, H.A. Critical speeds of turbomachinery: Computer predictions vs. experimental measurements - Part II: Effect of tilt-pad bearings and foundation dynamics. *Journal of Vibration, Acoustics, Stress and Reliability in Design*, Trans. ASME, 1987, Vol. 109 No. 1, 9-14.

/4/ PONS, A. Experimental and numerical analysis on a large nuclear steam turbo-generator group. *Proceedings of International Conference on Rotordynamics* - IFToMM and JSME, Tokyo, Japan, 1986, 269-275.

/5/ GASCH, R. Vibration of large turbo-rotors in fluid-film bearings on an elastic foundation. *Journal of Sound and Vibration*, 1976, Vol. 47 No. 1, 53-73.

/6/ NORDMANN, R. Schwingungsberechnung von nichtkonservativen Rotoren mit Hilfe von Links- und Rechtseigenvektoren. *VDI-Berichte*, 1976, Nr. 269, 175-182.

/7/ JÄCKER, M. Vibration analysis of large rotor-bearing-foundation-systems using a modal condensation for the reduction of unknowns. *Second International Conference "Vibrations of Rotating Machinery"*, IMechE, C280/80, Cambridge, England, 1980, 195-202.

/8/ LI, D.F., GUNTER, E.J. Component mode synthesis of large rotor systems. *Journal of Engineering for Power*, Trans. ASME, Vol. 104, July 1982.

/9/ KRÄMER, E. *Maschinendynamik*. Berlin, Heidelberg, New York, Tokyo: Springer 1984.

/10/ LUND, J.W., WANG, Z. Application of the Riccati method to rotor dynamic analysis of long shafts on a flexible foundation. *Journal of Vibration, Acoustics, Stress and Reliability in Design*, Trans. ASME, 1986, Vol. 108, 177-181.

/11/ KLEMENT, H.D. MADYN - Ein Programmsystem für die Maschinenberechnung. *Unix/Mail 5* (1987) 1, 12-17.

/12/ KRÄMER, E. Approximative computation of unbalance vibrations of multi-bearing rotors. *Proceedings of International Conference on Rotordynamics* - IFToMM and JSME, Tokyo, Japan, 1986, 523-527.

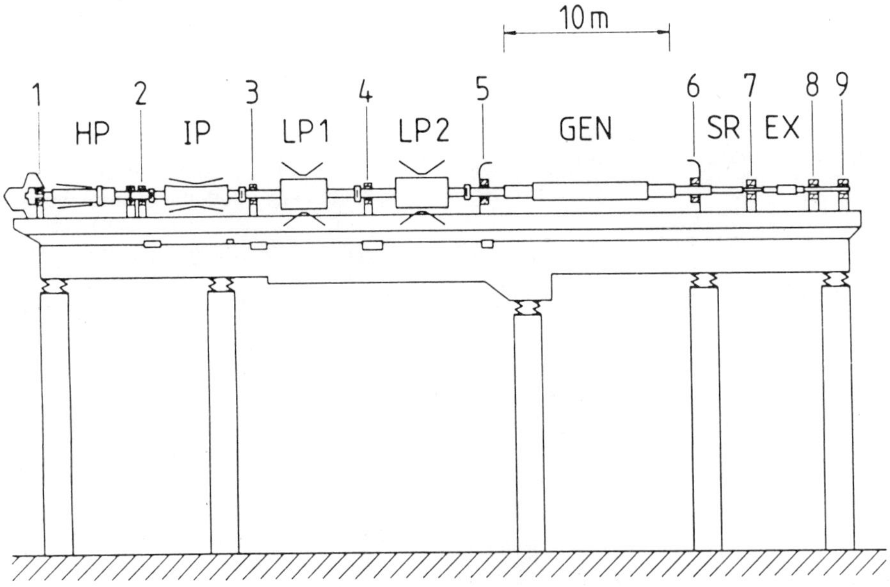

Fig 1 Schematic view of turboset S4

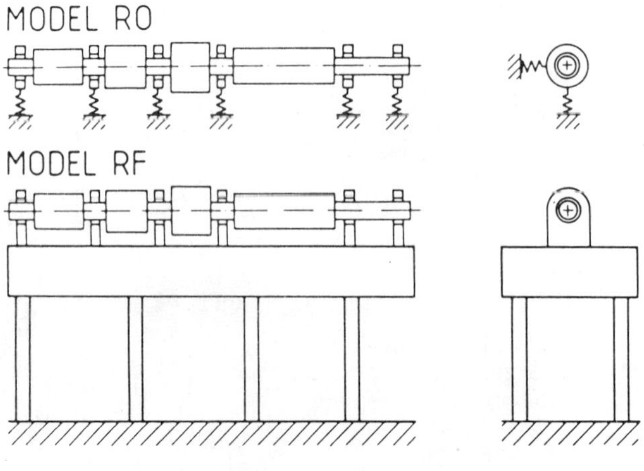

Fig 2 Computation models

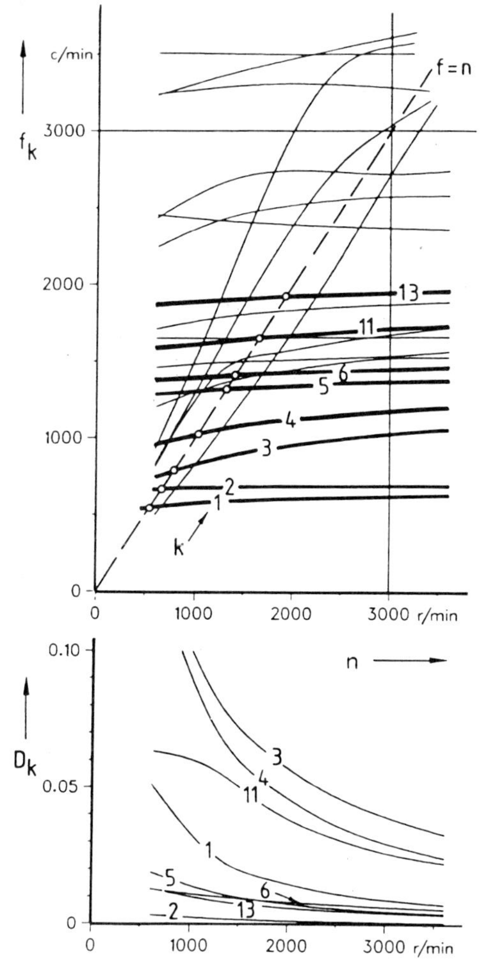

Fig 3 Model R0; natural frequencies and damping ratios

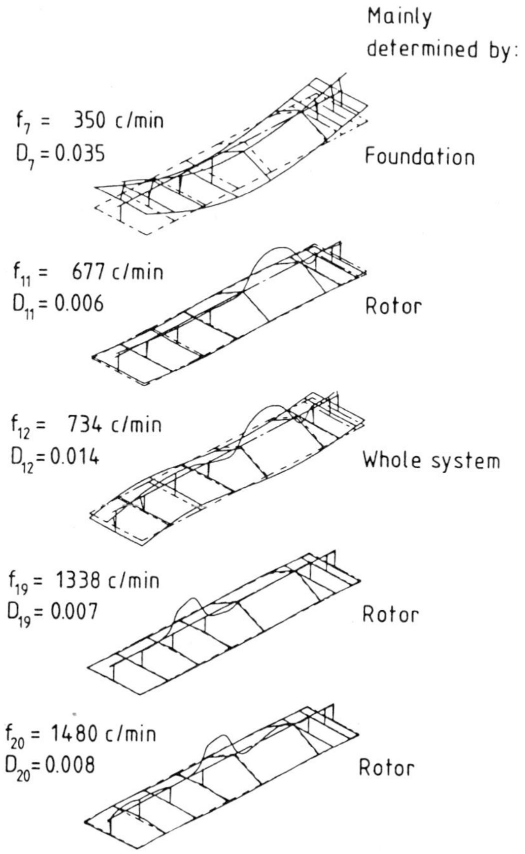

Fig 5 Model RF; some natural mode shapes at rated speed

k	Description	f_k c/min	D_k –
1	Gen◡h	623	0.009
2	Gen◡v	683	0.001
3	LP1◡h	1022	0.039
4	LP2◡LP1◡h	1169	0.030
5	LP1◡v	1359	0.005
6	LP2◡v	1452	0.006
11	Gen∽h	1708	0.026
13	Gen∽v	1944	0.004

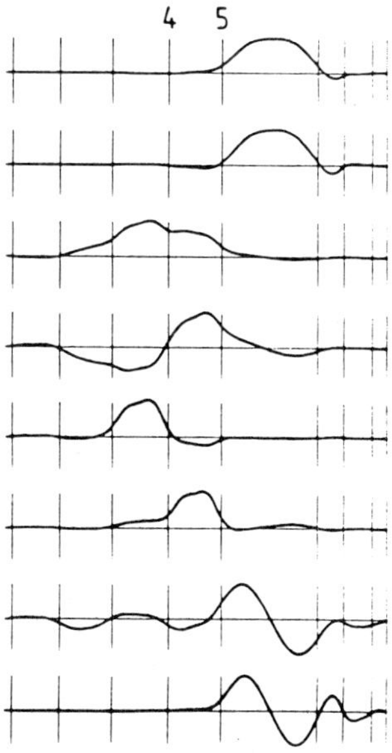

Fig 4 Model R0; natural modes at rated speed, relevant for vibrations at bearing 4 and 5

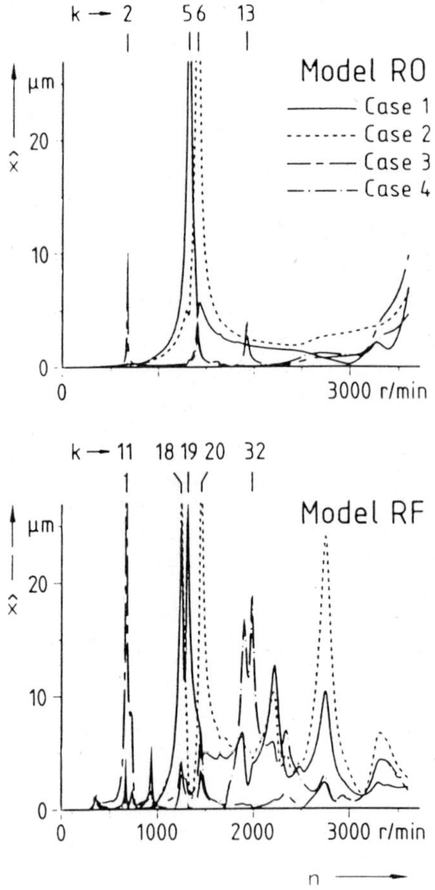

Fig 6 Computed pedestal amplitudes; bearing 4, vertical

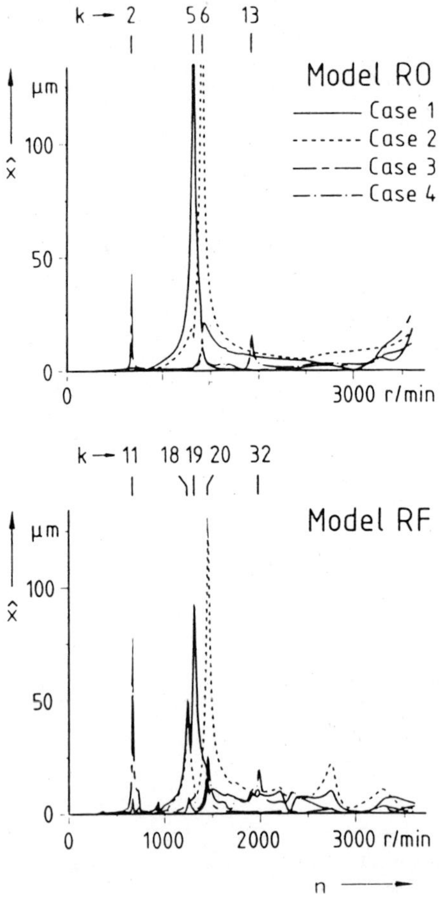

Fig 7 Computed shaft amplitudes; at bearing 4, vertical

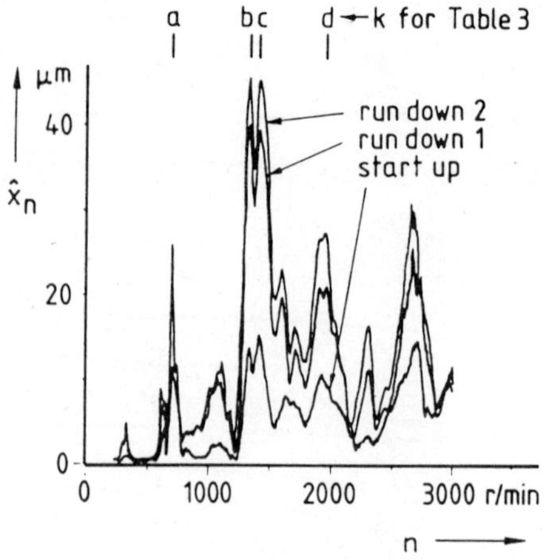

Fig 8 Measured pedestal amplitudes of n-vibration; bearing 4, vertical

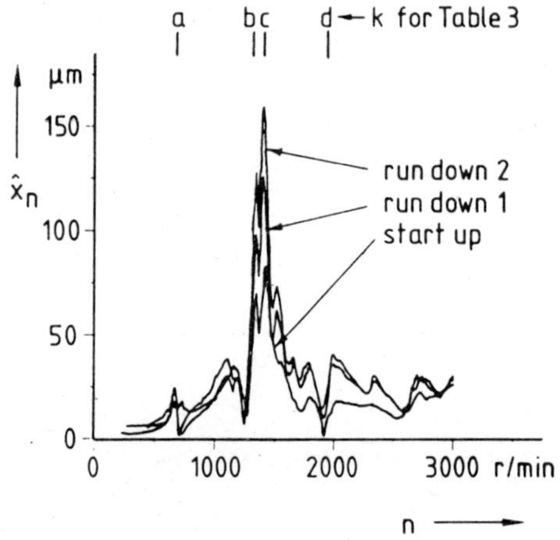

Fig 9 Measured shaft amplitudes of n-vibration; at bearing 4, vertical

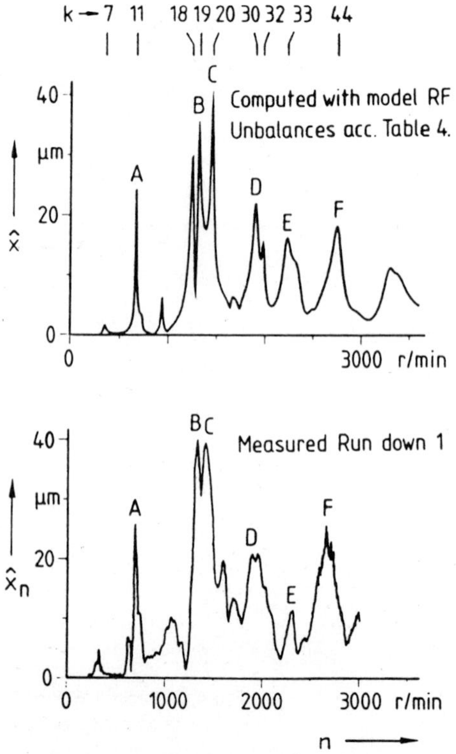

Fig 10 Computed and measured pedestal amplitudes; bearing 4, vertical

C286/88

In-service vibration levels on large turbine generators

H ELLAM, BSc and **D L THOMAS**, MA, PhD
Central Electricity Generating Board, Berkeley, Gloucestershire

SYNOPSIS The in-service vibration levels of large turbine generators on the CEGB system are compared with the quality bands of international standards. 14 incidents were identified in which a relatively high background level may have contributed to subsequent plant faults. In all these cases the pedestal vibrations prior to the fault exceeded 7.1 mms^{-1} rms.

1 INTRODUCTION

The Central Electricity Generating Board (CEGB) is responsible for electricity generation and transmission in England and Wales. It operates 78 power stations, and has a declared net capability of about 52GW. The majority of generation is from steam turbine-generators with outputs of 300 MW or greater.

The CEGB recently reviewed the in-service vibration levels on its large turbine-generators. The review considered the approach currently adopted on turbine generator vibration, particularly in relation to the standards of acceptable vibrations, and the implications of meeting these standards. This paper describes the distribution of vibration levels in CEGB machines, and discusses some of the implications.

The review was confined to bearing pedestal vibrations. It is recognised that shaft relative vibrations (eccentricities) may be more relevant parameters for certain parts of machines, particularly HP and IP rotors, but these are outside the scope of this paper. Moreover, the review was confined to absolute levels of vibration. The use of detailed monitoring to detect changes, is recognised to be of major importance, but is beyond the scope of this review. A strategy for detailed monitoring has been described previously (1,2).

In-service vibration levels are considered in the review. This excludes levels obtained during run-up and run-down. The levels of interest are the stable values obtained when running at normal operational loads. They refer to radial (horizontal or vertical) measurements, but not to axial vibrations.

The approach adopted for the review was to examine the existing in-service vibration levels of a large sample of machines for which data was available, and compare these levels to the ranges specified in existing standards. A survey of cases in which relatively high vibrations were considered to have caused subsequent problems was then carried out, and the costs of these problems estimated. This information enables improved operational policies to be formulated.

2 EXISTING AND PROPOSED STANDARDS

The existing relevant standards for measurement and evaluation of mechanical vibration in rotating machines are BS 4675: Part 1: 1976 (Mechanical vibration in rotating machines. Basis for specifying evaluation standards for rotating machines with operating speeds from 10 to 200 revolutions per second) and ISO 3945: 1977 (Mechanical vibration of large rotating machines with speed range from 10 to 200 rev/s. Measurement and evaluation of vibration severity in situ). Both of these standards specify 4 quality zones for vibration levels, shown below:

Vibration velocity v mms^{-1} r.m.s.	BS 4675 Quality judgement	ISO 3945 Quality judgement
$v<2.8$	A	Good
$2.8<v<7.1$	B	Satisfactory
$7.1<v<18.0$	C	Unsatisfactory
$v>18.0$	D	Unacceptable

The boundaries for the zones quoted above correspond to machines in Class IV of BS 4675, for machines mounted on foundations which are relatively soft in the direction of vibration measurements. All large CEGB turbine-generators come within this class.

A criticism of ISO 3945 has been the use of words such as 'satisfactory' and 'unsatisfactory' to define quality bands. The general view internationally is that whilst these are helpful they only give limited guidance. In future standards the intention is to provide more specific guidelines for interpreting the quality bands. This will be along the lines given in ISO 7919/1 which defines the 4 quality bands or zones as follows:-

<u>Zone A</u>

"Newly commissioned machines would normally fall within this zone."

Zone B

"Machines whose vibration levels fall within Zone B are considered acceptable for long term operation."

Zone C

"Once vibration levels are recorded within this zone, it is common practice to initiate alarms which provide a warning that remedial action is required. Generally, the machine may be operated for a limited period at these values until a suitable opportunity arises for remedial action to be taken."

Zone D

"Once vibration levels are recorded which are in excess of the upper value for Zone C, it is common practice to initiate a 'trip'. Vibration values which fall into this zone are considered to be of sufficient severity to cause damage."

It is also envisaged that future British and International standards will refer to criteria for assessing changes in vibration, in addition to the criteria on absolute levels described above.

It is normal practice for CEGB machines to have commercial operating limits for the initiation of alarms and trips, agreed between the manufacturers and the Board. These limits normally fall at some point within Zone C. A target vibration level of 2.8 mms^{-1} r.m.s. is specified for new plant, corresponding to Zone A.

3 REVIEW OF IN-SERVICE VIBRATION LEVELS

The in-service pedestal vibration levels for 43 500 and 660 MW units have been considered. Each unit has 14 bearings. In each case, measurements made at an arbitrary time at a typical high load were obtained for all the bearings on a machine. They therefore give a representative view of levels found in normal service on large machines in the CEGB. For individual bearings, the results, in terms of the zone boundaries from the standards are shown in Table 1. A more detailed analysis showed that the data approximately followed a Weibull distribution, as shown in Fig.1. This figure shows the data from Table 1 plotted on Weibull probability graph paper, together with a more detailed set of points for one type of machine, from all geographical areas. The data in Table 1 and Fig.1 can be used to estimate the likely number of bearings with radial vibrations above any proposed target values, and therefore to give some indication of the relative resources required to reduce levels below any particular target.

The survey showed that 37 per cent of bearing vibrations exceeded the 2.8 mms^{-1} r.m.s. level, 5 per cent exceeded the 7.1 mms^{-1} r.m.s. level, and there were no values in excess of 18.0. mms^{-1} r.m.s. Less than 1 per cent of the values exceeded 11.2 mms^{-1} r.m.s., which is the geometric mid-point of Zone C.

If complete machines are considered, rather than individual bearing pedestals, then 29 per cent of the machines had at least one bearing with a level greater than 7.1 mms^{-1}, and 6 per cent of the machines had at least one bearing with a level greater than 11.2 mms^{-1} r.m.s.

4 COSTS OF RUNNING WITH HIGH LEVELS

A question which can be posed is: 'Should an engineering policy objective of achieving in-service vibration levels of the order of the ISO 3945 'good' category be adopted?' The review of in-service levels described above shows that around 37 per cent of bearing vibrations lie outside this category, and would therefore require appropriate corrective action to implement the policy suggested in the question. The resources required to achieve this would clearly be very large. It was therefore decided that the costs incurred as a consequence of running machines with levels in excess of various figures should be considered, in order to give a yardstick against which the costs associated with programmes to reduce levels could be assessed.

To determine the costs mentioned above, operational records were examined, in order to provide case histories of situations where a relatively high background vibration level was implicated in causing subsequent problems, or hindered detection or diagnosis of some other problem of which vibration was a symptom. It should be emphasised that the intention was to consider those cases where an existing relatively high vibration led to subsequent problems, and to exclude the much larger number of cases where vibrations were a symptom of another problem. It was agreed that only units of 300 MW or larger nominal rating would be considered.

14 case histories were identified, covering a 5 year period from 1979 to 1984. Brief details are given in Table 2. In many cases there was uncertainty about the extent to which the existing vibrations had caused the subsequent problems. In these cases a probability has been assigned, and used to scale the resulting costs accordingly.

In order that the costs associated with each case could be assessed on a consistent basis, the costs for all the cases were calculated. The Appendix lists the assumptions made, and gives details of the calculations. Examples of the cases considered are described in the next Section.

5 EXAMPLES

5.1 Case 5 - LP turbine rubbing

Following an annual overhaul high vibration levels on the LP turbines were investigated. These investigations included the tape recording and analysis of machine run ups, run downs and on-load operation. The cause of the LP vibrations was subsequently diagnosed as being due to the combined effects of higher than desirable residual unbalance of the LP rotors and light rubbing within the LP cylinders. The background vibration level, in the absence of rubbing, was 7.7 mms^{-1}. This level was exceeded during excursions when rubbing took place. In-situ balancing was carried out and the LP vibrations were reduced to very low levels whereas previously they had prevented flexible operation of the machine. The balancing therefore removed the risks to safe operation of the machine arising from the high vibration levels and enabled the machine to be operated to system requirements, hence minimising generation costs.

5.2 Case 8 - bearing support ring pad failure

A generator was returned to service with a replacement generator rotor, and exhibited relatively high stable vertical vibration levels (about 11 mms^{-1} r.m.s.) on bearing 12 (the exciter end bearing). The rotor had not been high-speed balanced after remedial work. The vibrations were identified as the response of a vertical critical speed which occurs at 20 r/min above running speed, and is a known feature of the machine.

Because of operational requirements it was not possible to balance the rotor, and it continued to operate with vibration levels around 11 mms^{-1} r.m.s. After six months of satisfactory operation, an abnormal noise and vibration occurred around the exciter (adjacent to the generator). The (load dependent) vibration was investigated, and it was found that the exciter bearing 13, adjacent to the generating bearing 12, was experiencing large subsynchronous vibrations (see Figs. 2 and 3). Resonant oil whirl, at one of the exciter rotor critical speeds, was diagnosed. The whirl was suppressed by running the jacking oil pump.

A number of further incidents of exciter whirl occurred over the next four months, but they were controlled by operational expedients. However, the exciter bearing 13 then emitted rattling mechanical noises, followed by excessive and unusual vibration of both the exciter bearing 13 and the adjacent generator bearing 12 (Fig.4). The unit was shut down, and subsequent inspection showed that the pads locating the bearing 13 support ring had broken up.

It is not possible to be certain of the cause of this failure. However, the sequence of events of:

(i) high background vibrations
(ii) intermittent whirl
(iii) bearing support ring pad failure

may be related. A possible cause is a gradual deterioration (looseness/alignment changes) in bearing 13, induced by the high background vibration. The deterioration led to the onset of whirl, which encouraged further deterioration, until failure occurred. The probability that the initiating event was the high background vibration has been subjectively assessed as 0.5

6 POTENTIAL BENEFITS FROM A REDUCTION IN LEVELS

The calculated costs are shown in Table 2. The total cost incurred as a consequence of running with relatively high vibration levels is estimated to be £5M, and the average annual cost is £1M. This is equivalent to around £14K per machine year. The major proportion of these costs are "outage" costs (the costs of the unit's not being available for generation, in terms of the higher marginal cost of alternative sources of supply). The case histories account for 222 unit-days of outage over a 5-year period, equivalent to 44 unit-days outage per year, when account is taken of the probabilities.

The actual costs arising from vibration-induced problems give a guide to the financial benefits potentially achieveable from a policy of reducing levels, and hence allow a comparison with the extra costs which would be incurred by such a policy. One feature of the cases shown in Table 2 is that none of them involve pre-existing levels in Zone A or B. This implies that the benefit achieveable by reducing all levels to less than 2.8 mms^{-1} r.m.s., would be the same as that achieved by a lesser reduction to below 7.1 mms^{-1} r.m.s. However, problems have been identified from the bottom of Zone C upwards, demonstrating a consistency with the existing and proposed standards.

As the costs of reducing all levels to less than 2.8 mms^{-1} r.m.s. would be much greater than reducing them to less than 7.1 mms^{-1} r.m.s., and the identified benefits would be the same, there is no immediately quantifiable case for setting an overall target level for existing machines any lower than 7.1 mms^{-1} r.m.s., consistent with the draft ISO definition of Zone B quoted above. However, when detailed consideration is given to specific design features, there may well be benefits obtainable from lower targets in individual cases.

It is important to distinguish the concept of a target from that of a limit. Operating instructions specify limits, which should not be exceeded; operators must take immediate action to ensure limits are adhered to. A target is a desirable level, lower than the limit. Appropriate steps, consistent with the benefits obtainable, should be taken to ensure the target is met. If it is not, no short term operator intervention is called for. However, the cause should be investigated, and appropriate remedial measures identified. The machine may continue to operate at levels above the target until a suitable opportunity arises for the remedial measures to be implemented. These should then only be implemented if it is economically advantageous to do so.

The information on existing vibration levels, and the case histories described above, could be used to investigate whether targets are appropriate for turbine-generator pedestal vibrations. Two potential target levels are of particular interest, 7.1 mms^{-1} r.m.s. and 11.2 mms^{-1} r.m.s. The former corresponds to the Zone B/C boundary, and is just below the minimum level at which some problems were found to occur in the group's survey. The majority of those cases in the range 7.1 to 8.0 mms^{-1} r.m.s. involve rubbing caused by loss-of-clearance arising from imbalance. The second target level, 11.2 mms^{-1} r.m.s., is the geometric mid-point of Zone C.

The annual benefits obtainable on the CEGB system if <u>all</u> levels could be reduced below the possible targets of 7.1 and 11.2 mms^{-1} are £1M and £0.7M respectively. The costs required to achieve these benefits must also be considered when formulating a policy for acceptable vibration levels.

The figures suggest that although there are identifiable benefits to be obtained by reducing levels to less than 7.1 mms^{-1} r.m.s., the benefits are relatively small. Any policy must take the scale of the achievable benefits into account. Moreover, although no specific benefits have been identified for cases with stable on-load vibrations less than 7.1 mms^{-1} r.m.s., the sample size was relatively small. When specific design features or past history of a particular rotor are taken into account, similar benefits may be obtainable by setting a target less than 7.1 mms^{-1} r.m.s. in individual cases. The existence of significant day-to-day scatter in normal vibration levels could also be a justification for selecting a lower target.

7 CONCLUSIONS

(1) British and International Standards (BS.4675 and ISO 3945) exist to classify vibration levels. Whilst the quality bands are quantitatively the same, the verbal definitions of these bands differ between standards. The revised descriptions of the quality Zones A, B, C and D proposed in the draft ISO standard (ISO/DIS 7919/1) seek to rationalise these differences and are regarded as giving a significant improvement in guidance to operators.

(2) A review of in-service vibration levels showed that 63 per cent of all bearings fell into Zone A (<2.8 mms^{-1} r.m.s.), 32 per cent fell into Zone B (2.8 - 7.1 mms^{-1} r.m.s.), and 5 per cent fell into Zone C (7.1 - 18.0 mms^{-1} r.m.s.). No levels fell into the most severe category, Zone D (>18.0 mms^{-1} r.m.s.).

(3) For new plant, the CEGB specifies a target vibration level of 2.8 mms^{-1} r.m.s. consistent with Zone A of ISO 3945.

(4) 14 incidents were identified on machines of 300 MW and above over a five-year period (about 350 machine years), in which relatively high background vibration levels (falling within Zone C) may have contributed to subsequent plant faults.

(5) The probability of high background vibration being the root cause is a subjective assessment, but the annual cost was assessed as about £1M for the whole population of 69 machines, or around £14,000 per machine year. This mainly arises from an estimated 44 unit days of additional outages.

(6) It should be stressed that this paper has been concerned only with stable vibration levels. There is no departure from the requirement that any significant change in vibration level, even if below any target level, should be investigated, as it may be symptomatic of a fault on the machine.

(7) The review has been confined to bearing pedestal vibrations. Shaft relative vibrations (eccentricities) may be more relevant parameters for certain parts of machine, particularly HP and IP rotors.

REFERENCES

(1) THOMAS, D. L. Vibration monitoring strategy for large turbogenerators. 3rd Int.Conf. on Vibrations in Rotating Machinery, York, 1984. I.Mech.E. Conference Publications, 1984-10, pp.91-99.

(2) THOMAS, D.L. Turbine vibration monitoring in the CEGB. Int. Conf. on Rotordynamics, Tokyo, September 1986. pp.361-366.

ACKNOWLEDGEMENTS

The authors are grateful to many colleagues throughout the CEGB for providing the information used to compile this paper. The paper is published by permission of the Central Electricity Generating Board.

APPENDIX

Cost of Vibration Problems

The analysis was carried out using easily available information from CEGB information systems. Certain limitations of this are:-

1 Information on repair costs is usually difficult to identify; however in a few instances estimated costs are included.

2 The load limitations reported in cases 2 and 5 could not be easily evaluated, as these stations were operating at interim ratings.

3 Information on availability loss identifies the plant items involved but not the cause. For the purposes of this exercise all losses ascribed to the appropriate plant items during the relevant period are included.

4 It is not usually possible to identify availability losses due to late returns-to-service, following repair or maintenance, although a nominal (identified) amount has been included.

The cases were divided into three categories:-

(a) Those where one or more actual failures are involved (cases 3, 4, 8, 10, 12, 13, 14).

(b) Those where no actual failure occurred but where there were load limitations for a substantial period (2, 5).

(c) Those where vibration problems occurred but where these were surmounted without incurring identifiable costs (cases 1, 6, 7, 9, 11).

Cases Where Actual Failures Occurred

The information is summarised in Table 3. The information is summarised and where it was not available the assumed availability loss is shown.

The monetary value of this loss was calculated by assuming average forecast replacement costs for 500/600 and 300/350 MW units. All units were assumed to have an output of 93 per cent of gross capability, and an availability of 89 per cent.

Cases Resulting in Load Restrictions/Loss of Flexibility

These are summarised in Table 4. Each of these incidents was repeated at other units on the same site, in total an 'expected' value of 2.25 incidents over the five year period.

Each incident was evaluated as a 10 MW output restriction for a period of 12 months.

Evaluating as in Section 1 above thus gives a value of about £0.6M.

No value has been ascribed to loss of flexibility (ie a reduction in the ability to respond to changing local requirements), as no records are available but it is believed that this would be small as long as only one or two units at each station were involved simultaneously, leaving the others to respond to changing requirements.

(This is not the same as saying that there was no benefit in tackling the problems; if such problems were allowed to persist then eventually all units at the stations concerned would have become inflexible, at considerable cost).

Other Cases

Table 5 lists the cases where no easily identifiable costs were ascribed. Reasons for this include:

Cure effected by balancing, an assumption is that balancing would have been required anyway.

Repair effected during plant outage, required for other purposes, eg statutory boiler inspection, breakdown of other components at no loss of available output.

Costs only in terms of flexibility, see above.

Total Costs

Costs identified over a five year period were:-

£3.8M in outage costs (see Table 3).

£0.9M in repair costs (see Table 3).

£0.6M from load restrictions at two stations (see Table 4.
Total £5.3M or just over £1M per annum.

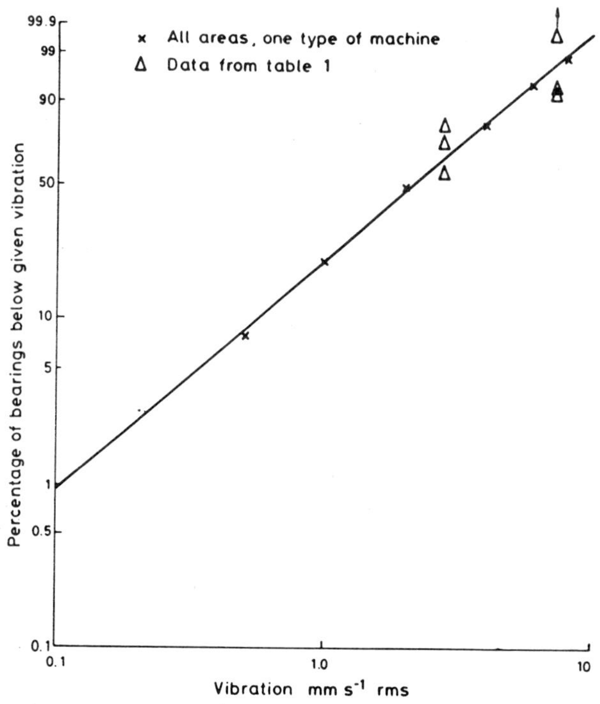

Fig 1 Weibull plot of distribution of stable on-load bearing vibration levels

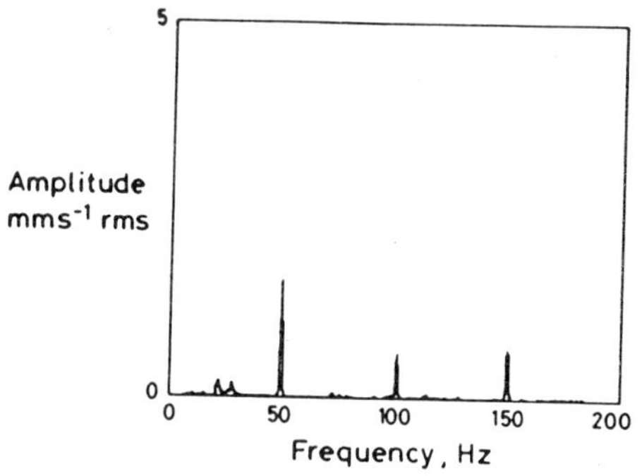

Fig 2 Bearing 13 vibration spectrum prior to excursion

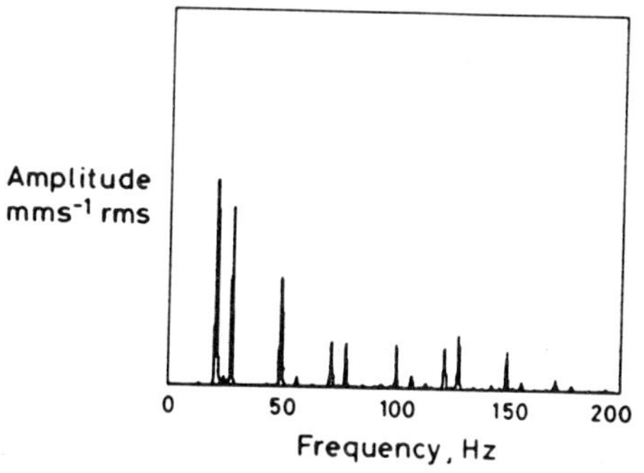

Fig 3 Bearing 13 vibration spectrum during excursion; peak is at 21.5 Hz

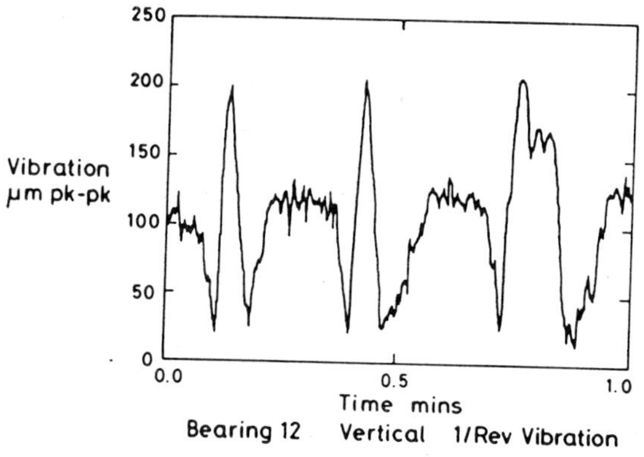

Fig 4 Generator bearing vibration during de-load prior to shut-down

Table 1
Percentage of in-service pedestal vibrations falling within quality zone

Zone	Limits mms^{-1} r.m.s.	Geographical Area W (17 units) (238 bearings)	X (10 units) (140 bearings)	Y (8 units) (112 bearings)	Z (8 units) (112 bearings)	All 500/660 (43 units) (602 bearings)
A	<2.8	55	54	72	81	63
B	2.8 - 7.1	38	39	22	19	32
C	7.1 - 18	6	7	6	-	5
D	>18.0	-	-	-	-	-

Table 2
Cases in which relatively high background vibration levels were implicated in causing subsequent problems and/or hindered detection or diagnoses

Case No	Year	Background Level mms^{-1} r.m.s.	Problem	Cost £M	Probability Caused by Vibration	Potential Benefit £M
1	1981	7.7	LP excursions - rubbing and Tilting	-	0.2	-
2	1979	7.7	High eccentricity excursions - rubbing	**	0.25	0.2
3	1981	7.8	Wedge failure - major damage	10.0	0.1	1.0
4	1983	up to 7.8	Rubbing	0.7	0.3	0.2
5	1979	7.8	Rubbing	**	0.5	0.4
6	1984	8.0	Excursions - rubbing and Tilting	-	0.75	-
7	1981	9.3	Pedestal support looseness	-	0.75	-
8	1984	11.0	Bearing support ring pad failure	0.4	0.5	0.2
9	1984	12.0	Weld failures	-	1.0	-
10	1983	12.0	Rubbing - bearing damage	1.6	0.5	0.8
11	1984	17.0	Pedestal support looseness - under review	-	0.75	-
12	1981	17.9	Hydrogen cooler leak Insulation degradation	0.3 1.2	1.0 0.75	0.3 0.9
13	1982	up to 35.0	Pedestal looseness. Bearing damage. Bushgear failure	0.2	0.95	0.2
14	1979	*	Bent HP rotor	2.2	0.5	1.1

* Eccentricity 500µm pk-pk ** See Table 4

Table 3
Cases involving outages

Case No	Description of Loss	Data Available On Outage Duration/Cause	Outage Cost (£M)	Estimated Repair Cost (£M)	Probability Vibration Responsible	Total Cost x Probability (£M)
3	Major Failure of Generator Rotor	260 Unit days for Generator Replacement	7.0	3.0	0.1	1.0
4	(a) Delayed RTS+ arising from LP2 bearing vibrations	No records 12 Unit days assessed	0.7	-	0.3	0.2
	(b) Unplanned Outages to investigate LP bearing vibrations	12 Unit days on 'LP cylinder bearings' 'Turbine LP rotor shaft'				
8	Failure of Bearing 13 Support Ring Pads	12 Unit days on 'Bearing 13 vibration'	0.4	-	0.5	0.2
10	Loss of blades, + restrictions and minor outages due to blade cropping	130 Unit days in addition to the 10 that would have been required for immediate repair	1.1	0.5	0.5	0.8
12	1. Failure of Hydrogen Pipe leading from cooler	10 Unit days as 'Generator'	0.3	-	1.0	0.3
	2. Extended Repair of Generator Rotor	21 Unit days outage	0.6	0.6	0.75	0.9
					Sub Total	1.2
13	Miscellaneous problems related vibration at main and pilot exciter bearings	Outages identified as 'Slip Rings and Brush Gear' 'Pilot Exciter' 'Main Exciter' 'H_2 Seals'	0.2	-	1.0	0.2
14	1. Delays to RTS+ over a number of years	15 Unit days (estimated)	0.4	-	0.5	0.2
	2. 68 Unit day for 'Turbine HP Rotor Shaft'	68 Unit days on 'Turbine HP Rotor Shaft'	1.8	-	0.5	0.9

\+ RTS - Return to Service

Table 4
Cases involving load limitations and lack of flexibility

Case No	Total No of Incidents	Probability Caused by Vibration	Expectation of Incidents	Cost £M
2	3	0.25	0.75	0.2
5	3	0.5	1.5	0.4

Table 5
Cases without identifiable major costs

Case No	Nature of Incident	Comment
1	Reduced Loading Rate (also encountered on other units at same station.	Cost of lack of flexibility would be low as long as only one unit involved simultaneously.
6	High IP Vibrations during return to service, following plant outage	Cured by balancing. No cost as balancing would have been required anyway.
7	Repairs carried out during existing outage.	No outage cost
9	Weld failures	Weld repairs in existing outage
11	Damage to turbine pedestal arising from high vibrations.	Repairs in parallel with other work.

C250/88

Turbogenerator layout for optimal dynamic response—a study and a case history

Z A PARSZEWSKI, DrEngSc, MEngSc, CEng, FIMechE, **T J CHALKO**, DrEngSc, MEngSc and **D-X LI**, MEngSc
Department of Mechanical and Manufacturing Engineering, University of Melbourne, Parkville, Victoria, Australia

SYNOPSIS Application of the method of optimisation of the system dynamic behaviour in The System Configuration Domain to a 200 MW turbogenerator set is presented. Bearing positions (system layout) are selected as system configuration parameters. This enabled optimisation of the system dynamics without changing any component of the system. Optimum system layout is calculated for the turbogenerator set , providing its optimal dynamic performance , that is a minimum vibration level at working speed. Two optimisation goals : minimization of maximum amplitude of vibration and minimization of root mean square amplitude of vibration ,at selected positions along the shaftline, are considered and results compared. Dynamic performance of the system is compared for conventional catenary based (turbogenerator manufacturer recommended) alignment and obtained optimum layout. It is shown that by selecting the optimal layout in the system considered, the maximum vibration level could be reduced by 40 percent , for the same amount of residual unbalance of the shaftline . It is confirmed , that catenary based alignment is not necessarily the best, from the point of view of the turbogenerator dynamics. The approach presented may be adopted at (layout) design stage , but in the case described in this paper it is used to improve vibration response of an existing 200 MW turbogenerator on site.

1. INTRODUCTION

This paper presents an application of Parszewski and Krodkiewski (1986) approach , machine dynamics in terms of the system configuration parameters, to system dynamic performance optimisation. Such configuration parameters should be selected , which can be easily modified at design or assembly stage , without changing the subsystems . Furthermore, small variations of selected parameters should also have significant influence on the system dynamics ,being the subject of optimisation. Considered turbogenerator was a statically indeterminate system (6 journal bearings) , with dynamic performance depending on its steady state equilibrium position. Bearing pedestals positions were taken as configuration parameters , which enabled to change the system equilibrium position ,therefore modifying its dynamic performance and the level of vibration at the working speed in particular. Optimisation of system dynamics was therefore possible by adjusting the bearing pedestals positions (alignment) without changing the existing design of the system components or adding any new components.

The formulation of the dynamic response of the considered turbogenerator unit in terms of selected configuration parameters is presented first. Then ,results of the optimisation process are presented and discussed.

2. NOTATION

- A – vibration amplitude vector
- C_j – j-th bearing eccentricity ratio
- c,d – axes of elliptical vibration orbit,
- $\{F\}$ – excitation forces vector,
- F,R,B – subscripts for Foundation,Rotor,Bearing
- f_x,f_y – hydrodynamic force components,
- $[M],[K],[C]$ – mass,stiffness ,damping matrices
- $[R]$ – receptance matrix
- n_B – number of bearings
- n_R, nR – number of nodes on the rotor
- R_x,R_y – bearing reaction components,
- X,Y,Z – global inertial coordinate system ,
- X_B,Y_B – bearing centre coordinates,
- X_J,Y_J – journal centre coordinates(equilibrium)
- X_{JB},Y_{JB} – relative journal position with respect to bearing centre (equilibrium),
- x,y – journal centre displacement from equilibrium position,
- θ,Ψ – shaft nodal cross-section rotations.
- ϕ,β – phase angles
- Ω – working speed
- ω – excitation frequency,

3. TURBOGENERATOR MODELLING

The turbogenerator considered was of C.A.Parsons Co. Ltd. (U.K) production and had the rating of 200 MW . The shaftline consisted of high pressure turbine rotor, low pressure turbine rotor and generator rotor ,all coupled rigidly, and is shown in Fig.1 , with indicated main dimensions. Auxillary equipment (turning gear and motor) axially connected to the main shaftline by means of an elastic coupling are considered as dynamically isolated from the main shaftline and neglected in the model. Each of the three main rotors was supported by two journal bearings , so there were six journal bearings in the system. Thrust bearings were not incorporated in the model, since the transverse vibration only as considered . The bearings were of split , two cylindrical concentric sleeves type and had the following dimensions:

No.	location	nominal diameter mm(inches)	length mm(inches)	
1.	H.P. front	254 (10)	152.4 (6)	*2
2.	H.P. rear	355.6 (14)	254 (10)	
3.	L.P. front	355.6 (14)	406.4 (16)	
4.	L.P. rear	381 (15)	381 (15)	
5.	Gen.inboard	381 (15)	381 (15)	
6.	Gen.outboard	381 (15)	381 (15)	

The radial clearance in all bearings was on average 0.0015.

The first three critical speeds , specified by the manufacturer were:

	r/min	rad/s	Hz
1	1385	145.0	23.08
2	2400	251.3	40.00
3	3700	387.4	61.66

Operating speed was 50 Hz ,which means that the unit was designed to operate between second and third critical speeds.

3.1 Rotor modelling

The rotor was assumed to be rotary symmetric. It was modelled using the finite element method with the beam element formulation including shear deformations. (Parszewski and Li 1987). Thermal effects on Young modulus along the shaftline are also taken into account. After the shaftline model condensation ,using FEM superelements , mass and stiffness matrices were obtained at 13 selected nodal cross-sections , which included all bearing positions. (Fig.1) Four displacement components were considered at each node : along 2 cartesian translational (x,y) and 2 angular (θ,Ψ) coordinates.

3.2 Bearing and foundation modelling

Journal bearing characteristics were obtained using the finite element fluid film model including thermal effects. Bearing hydrodynamic forces were evaluated by integrating the pressure distribution on individual bearing sleeves , found as the numerical solution of the Reynold's equation using FEM (Parszewki Nan and Li 1987). The hydrodynamic force of the bearing was obtained by summing the corresponding force components of all its sleeves. Geometry of the bearing used in the turbogenerator considered is shown in Fig.2. Hydrodynamic force was calculated as a function of position and velocity components of the journal centre with respect to the bearing centre:

$$f_{jx} = f_{jx}(X_{jJB}+ x, Y_{jJB}+ y, \dot{x}, \dot{y}) \quad (1)$$

$$f_{jy} = f_{jy}(X_{jJB}+ x, Y_{jJB}+ y, \dot{x}, \dot{y}) \quad (2)$$

Bearing static characteristics (equilibrium positions) were calculated as :

$$f_{jx0}= f_{jx}(X_{jJB}, Y_{jJB}, 0, 0) \quad (3)$$

$$f_{jy0}= f_{jy}(X_{jJB}, Y_{jJB}, 0, 0) \quad (4)$$

For small vibration x ,y ,$\dot{x}$ and $\dot{y}$ in the vicinity of the above equilibrium position the hydrodynamic forces were linearised as follows:

$$f_{jx}= f_{jx0} - K_{jxx}x_j - K_{jxy}y_j - C_{jxx}\dot{x}_j - C_{jxy}\dot{y}_j \quad (5)$$

$$f_{jy}= f_{jy0} - K_{jyx}x_j - K_{jyy}y_j - C_{jyx}\dot{x}_j - C_{jyy}\dot{y}_j \quad (6)$$

where , for example , coefficients Kxx and Cxy are of the form :

$$K_{jxy} = - \left[\frac{\partial f_{jx}}{\partial Y_{jJB}}\right]_{(X_{jJB}, Y_{jJB}, 0, 0)} = - \frac{f_{jx}(X_{jJB}+\Delta X_{jJB}, Y_{jJB}, 0, 0) - f_{jx0}}{\Delta Y_{jJB}} \quad (7)$$

$$C_{jxy} = - \left[\frac{\partial f_{jx}}{\partial \dot{Y}_{jJB}}\right]_{(X_{jJB}, Y_{jJB}, 0, 0)} = - \frac{f_{jx}(X_{jJB}, Y_{jJB}, 0, \Delta\dot{Y}_{jJB}) - f_{jx0}}{\Delta\dot{Y}_{jJB}} \quad (8)$$

Bearing linearised stiffness and damping coefficients were calculated numerically for each equilibrium position of the system , using finite difference method shown by equations (7) and (8).

The j-th bearing receptance was calculated for the working speed Ω ,using coefficients (9).

$$R_{Bj} = \left[\begin{bmatrix} K_{jxx} & K_{jxy} \\ K_{jyx} & K_{jyy} \end{bmatrix} + i\Omega \begin{bmatrix} C_{jxx} & C_{jxy} \\ C_{jyx} & C_{jyy} \end{bmatrix}\right]^{-1} \quad (9)$$

Short bearings were assumed. The receptance matrix for all bearings was assembled from receptances of the individual bearings :

$$[R]_B= \mathrm{diag}\left[R_{Bj} \right] \quad (10)$$

Flexible casing, pedestals and foundation can be introduced in the presented approach by the receptance matrix of the foundation

$$[R]_F$$

along coordinates connecting it with the bearings fluid film. It was approximated as rigid in the considered case , as no corresponding data was available. Foundation influence will be considered in detail in the future investigations.

3.3 Subsystem composition

The size of the matrices in the dynamic response calculation algorithm were minimised by the use of receptance composition method (Parszewski 1982) for the system containing the shaftline, linearised bearings and foundation. Using this technique, the system composition was done for the working speed only. The receptance of the combined foundation and massless bearings was calculated from the relationship:

$$[R]'_{BF} = [R]_B + [R]_F \tag{11}$$

along connecting coordinates.

The obtained bearing - foundation receptance matrix was next extended so that it included additional shaft nodes and angular coordinates. Then, the receptance of the entire system [R] was calculated :

$$[R] = [\,[R]^{-1}_{BF} + [K]_R - \Omega^2 [M]_R]^{-1} \tag{12}$$

The coordinates $[x, y, \theta, \Psi]$ along which the final receptance matrix was obtained included cartesian translations (x,y) and angular (θ, Ψ) components of the shaft nodal cross-sections displacements. These were taken at the bearings and at a few additional locations of interest along the shaftline. For the system considered, the final number of nodes of the composed system was 13, with 4 coordinates per node, so the size of the final receptance matrix was 52x52.

3.4 Dynamic response of the system

Receptance [R] gives the relationship between displacement and force vectors along the selected coordinates. In particular, the motion along the cartesian coordinates may be calculated using equation (13).

$$[x, y]^T = [Rc]\ \{F\} \tag{13}$$

where [Rc] contains elements of [R] corresponding to cartesian translations only and {F} is the force excitation vector along these coordinates.

Single frequency harmonic excitation synchronous with the working speed $(\omega = \Omega)$, corresponding to some rotor unbalance, was assumed of the form :

$$F_{xi} = F_{0xi} \cos(\omega t + \beta_{xi}) \tag{14}$$

where $x = x$ or y, $i = 1, 2, \cdots, n_R$,

By using equation (13), the dynamic responses of the system can be obtained along the considered coordinates of the form :

$$x_i = a_{xi} \cos(\omega t - \phi_{xi}) \tag{15}$$

The trajectory of the i-th shaft node given by (x_i, y_i) is an ellipse, and the lengths of its principal axes can be written in the form :

$$c_i = \sqrt{a^2_{xi} \cos^2(\phi_{ci} - \phi_{xi}) + a^2_{yi} \cos^2(\phi_{ci} - \phi_{yi})}\ , \tag{16}$$

$$d_i = \sqrt{a^2_{xi} \cos^2(\phi_{di} - \phi_{xi}) + a^2_{yi} \cos^2(\phi_{di} - \phi_{yi})}\ , \tag{17}$$

where,

$$\phi_{ci} = \frac{1}{2} \tan^{-1} \frac{a^2_{xi} \sin 2\phi_{xi} + a^2_{yi} \sin 2\phi_{yi}}{a^2_{xi} \cos 2\phi_{xi} + a^2_{yi} \cos 2\phi_{yi}}\ , \tag{18}$$

$$\phi_{di} = \phi_{ci} + \pi/2\ . \tag{19}$$

(see Parszewski and Li 1987). Values c and d were used to form a vibration amplitude vector A, being the argument of the goal function (described in Section 4.1) :

$$A = \{A_i\} = \{c_1 .. c_{n_R}, d_1 .. d_{n_R}\} \tag{20}$$

3.5 Catenary based alignment

It was decided to start the optimisation algorithm from the bearing alignment recommended by the turbogenerator manufacturer, in an attempt to find the local optimum closest to it. The manufacturer recommended alignment procedure was based on the minimisation of the step and gap between the shaftline couplings at the assembly stage. The following procedure was applied to calculate the corresponding equilibrium of the system and bearings eccentricities for the steady state operation :

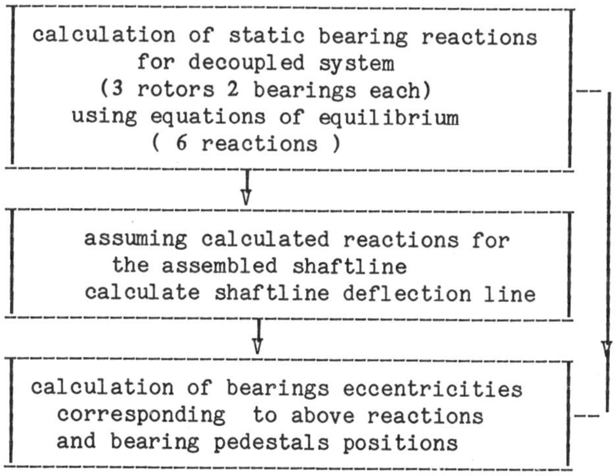

Fig 3 Calculation of steady state bearing eccentricities corresponding to initial alignment

Initial layout calculated and presented in Table 1. agreed closely with the manufacturer alignment data.

4. THE METHOD

Bearing eccentricities were taken as configuration parameters. For bearings, 4 bearing eccentricities could be assumed and the remaining two calculated from equilibrium conditions.

The calculation algorithm is presented in Fig.4.

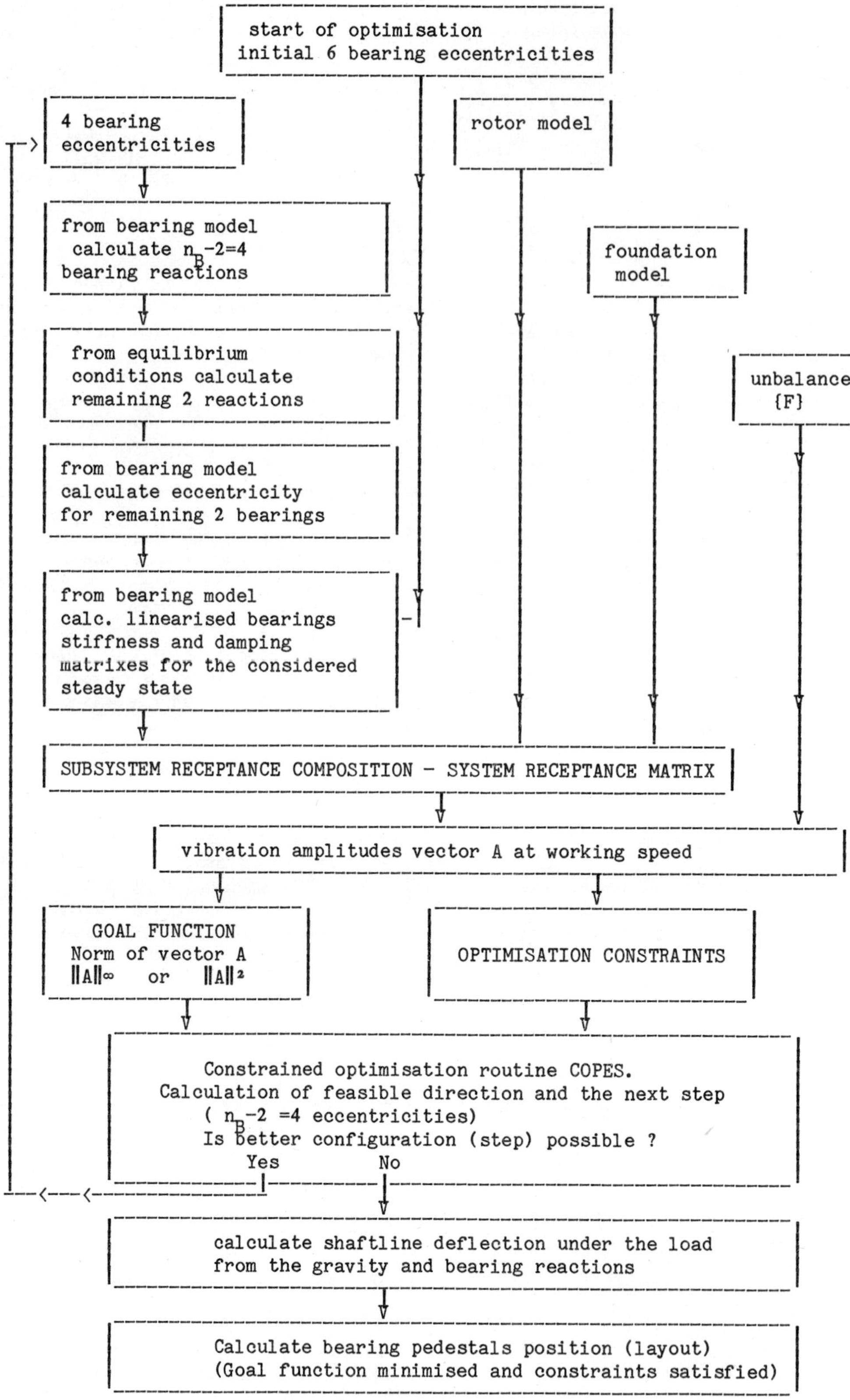

Fig 4 Calculation flowchart

The IBM PC-AT ,FORTRAN77 version of the mainframe based optimization program COPES (Control Program for Engineering Synthesis - in FORTRAN-IV, Vanderplaats, 1979) was compiled and applied by the authors. COPES uses constrained function minimisation routine CONMIN which employs both the Fletcher-Reeves Conjugate Direction Method and Zoutendijks Method of Feasible Directions.

The optimization procedure requires the user to supply a subprogram ANALIZ which initializes design variables (in our case, the described configuration parameters), evaluates the values of the goal function and constrained variables , and outputs the results. The content of ANALIZ is presented in sections 3 and 4. Fig.4 shows the interface of the goal function calculations (the ANALIZ part) with COPES.

4.1 Goal function

As the optimisation criterion, the norm of the vector A of the form

$$\|A\|_p = \left(\sum_{i=1}^{2n_B} a_i^p \right)^{1/p} \qquad (21)$$

$$p=1,2\ldots\infty$$

was introduced. The following two cases : p=2 and p=∞ were found as useful criteria for optimisation, with the added advantage of having meaningful practical interpretation. In the case of p=2 - minimised was the root mean square of vibration amplitudes along the shaftline. With p=∞, the maximum vibration amplitude was minimised.

4.2 Optimisation constraints

Non-dimensional eccentricity ratios were restricted at each bearing:

$$0.2 < C_j < 0.8 \qquad (22)$$

$$j = 1,2,\cdots,n_B,$$

Eccentricity ratios smaller than 0.2 were considered impractical from the point of view of possible instability, resulting from insufficient bearing load. On the other hand, an eccentricity ratio bigger than 0.8 would practically mean a bearing overload. Another optimisation constraint was a limit imposed on directions of the resultant bearing reaction forces:

$$\alpha_{R1j} \leq \tan^{-1} \frac{R_{yj}}{R_{xj}} \leq \alpha_{R2j} \qquad (23)$$

$$j = 1,2,\cdots,n_B.$$

Bearings, their pedestals and foundation are actually designed for mainly vertical load. Constraint (23) restricts the direction and sense of each bearing reaction to vertical and opposing the gravity if :

$$\alpha_{R1j} = \alpha_{R2j} = 0 \qquad (24)$$

$$j = 1,2,\cdots,n_B.$$

5. RESULTS

5.1 Calculation

Optimisation was performed for assumed unbalance force distribution along the shaftline. It was found that the amount and distribution of unbalance was not affecting the optimisation result i.e the optimum layout obtained. The results are presented in Table 1. and the goal function history in Table 2.

Initial (manufacturer recommended) and optimal catenary lines of the shaftline are compared graphically in Fig.5. Indicated are bearing centres positions.

It was found that the goal function taken as the norm $\|A\|_\infty$, as compared with $\|A\|_2$, was in the case considered improving significantly convergence of the optimisation algorithm during initial few iterations, and for further iterations, close to the final optimum, the norm $\|A\|_2$ became more sensitive criterion for optimisation.

Table 1. Initial and optimal configuration

	Bearing Number	Initial Values	Optimal Values
Vertical Pedestal Positions (Bearing Centres) X_B (mm)	1	1.762	7.361
	2	0.230	0.883
	3	0.000	0.000
	4	0.000	0.000
	5	0.191	2.004
	6	5.377	29.406
Horizontal Pedestal Positions (Bearing Centres) Y_B (mm)	1	0.019	-0.046
	2	0.036	0.005
	3	0.000	0.000
	4	0.000	0.000
	5	0.004	0.049
	6	-0.032	0.105
Journal Eccentricity Ratios (Non-Dimensional) C	1	0.448	0.498
	2	0.506	0.453
	3	0.310	0.386
	4	0.367	0.208
	5	0.419	0.417
	6	0.390	0.431
Vertical Bearing Reactions R_X (kN) ($R_y = 0$)	1	107.84	133.34
	2	159.84	128.03
	3	170.51	231.44
	4	173.02	83.75
	5	212.16	210.87
	6	203.02	238.98
$\|A\|_\infty / \|A\|_0^\infty$		1.0	0.6099
$\|A\|_2 / \|A\|_0^2$		1.0	0.8477

Table 2. Goal functions history

Iteration	$\|A\|_\infty / \|A\|_0^\infty$	$\|A\|_2 / \|A\|_0^2$
0	0.1966	0.2968
1	0.6607	0.8541
2	0.6607	0.8538
3	0.6607	0.8598
4	0.6119	0.8568
5	0.6099	0.8518
6	0.6099	0.8477 *
7	0.6099	0.8578
8	0.6099	0.8575

In the case considered, the optimum configuration was obtained with sufficient accuracy after only 5 iterations (see Table 2). Table 1 compares initial system layout with the one corresponding to iteration 6 in Table 2. This final bearing layout presented corresponds to local minima of both goal functions $\|A\|_2$ and $\|A\|_\infty$. Fig.7 presents frequency responses of system with catenary and optimal alignment for nodes 1 and 5. Experimentally obtained 2-nd critical (Fig.6) is indicated in Fig.7 for comparison. System stability was checked for the optimum configuration.

5.2 Experiment

The rundown test was performed for the tubogenerator with conventional layout and vibrations of all bearing pedestals were measured as a function of frequency (rotating speed) . Example result of the measurement at bearing No.6 in the vertical direction is presented in Fig.6.

The first two critical measured speeds of the unit were 19.1 Hz and 44.8 Hz.

The unit was therefore operating close to the second critical , which can be seen in Fig.6., where the plot ends at 50 Hz (working speed).

The obtained optimal layout has not yet been implemented (negotiations are underway), hence no direct comparison of measurements given in Fig.6. ,corresponding to the optimal alignment , can be presented.

However , important conclusions can be readily derived.

6. CONCLUSIONS

Results of the optimisation of turbogenerator dynamics using its configuration parameters in the case considered indicate , that vibration level , understood as the norm $\|A\|_\infty$ of the vibration amplitude vector A , may be reduced by up to 40 percent for a given unbalance - by changing the configuration of the system , i.e modifying the bearing alignment (layout of the system)

Proposed bearing shifts from the original layout are a few millimeters for 3 bearings and about 24 mm for bearing No.6. (Table 1).

Only vertical bearing shifts were found significant in the case considered.

Proposed layout changes the bearings static (steady state) loads , with the most significant effect being the reduction of bearing No.4 load by a half.

The effectiveness of the layout optimisation procedure in the case considered is enhanced by the fact , that the system operated on a steep slope of the frequency response curve just above second critical (see Fig.6). Changes in the layout effectively modified the stiffness distribution and the natural frequencies of the system. In particular , softening bearing No.4 resulted in the second critical being shifted away from working speed. Because the amplitude of the forced vibration (vector A) was very sensitive to the location of the second critical (steep slope) - a small shift of the second critical could result in substantial reduction in the vibration level in the system. Also the damping in the vicinity of a working speed (50 Hz) was increased . Both effects can be observed in Fig.7.

The method is shown to have a significant potential in optimisation of dynamic performance of turbogenerators and other machinery.

7. ACKNOWLEDGEMENTS

This work was supported by the Australian Electrical Research Board and the State Electricity Commission of Victoria , Australia. Authors express their thanks to M.H.Melksham ,G.E.Pleasance , S.Dennis and other staff of the SECV for enabling the site data acquisition , providing the measurement data ,valuable discussions and help.

8. REFERENCES

[1] Parszewski, Z. A., 1982, "Vibration and Dynamics of Machines," (Polish), WNT, Warsaw.

[2] Z.A.Parszewski ,E.Kirby , K.Krynicki , F.E.M in turbo-generator dynamics - rotor modelling. Proc. of 5-th International Conference in Australia on Finite Element Methods , Melbourne , 19-21 August 1987 p.192-196

[3] Z.A.Parszewski , X.Nan , D.X.Li , FEM for thermo-fluid bearing dynamic characteristics. Proc. of 5-th International Conference in Australia on Finite Element Methods , Melbourne , 19-21 August 1987 p.205-210

[4] Z.A.Parszewski , Dynamics in terms of the system configuration parameters. Proc. of IFToMM VII World Congress on the Theory of Machines and Mechanisms, Sevilla, Sept, 17-22 , 1987 , Spain.

[5] Z.A.Parszewski , System Configuration and its Vibration Response , Proc. ASME Bien. Design Conference on Vibration and Noise , Boston , Sept. 1987

[6] Z.A.Parszewski, X.Nan, Donx Xu Li, Steady and unsteady bearing forces with thermal effects. Proc. of International Conference on Mechanical Dynamics , Shenyang , China, August 1987.

[7] Z.A.Parszewski , Dong Xu LI , Multi-bearing Rotor Response Presentation for system optimization . Proc. of International Conference on Mechanical Dynamics , Shenyang , China, August 1987.

[8] Z.A.Parszewski , K.Krynicki , Rotor-bearing system synthesis for Stability Analysis. Proc. of International Conference on Mechanical Dynamics , Shenyang , China, August 1987.

[10] Vanderplaats, G. N., 1979, "COPES - A FORTRAN Control Program for Engineering Synthesis,"

[11] Vanderplaats, G. N., May 1979, "CONMIN User's Manual Addendum,"

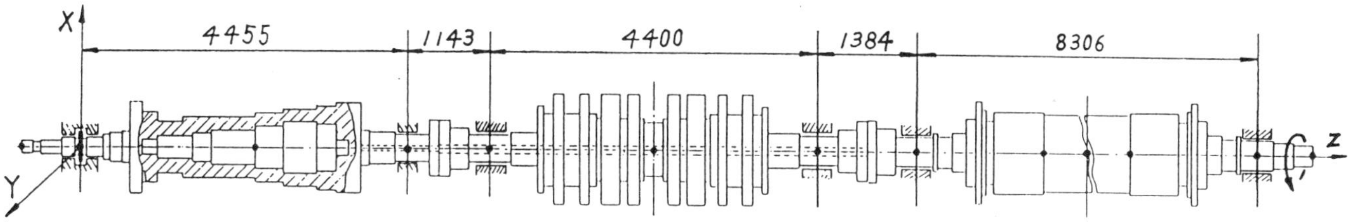

Fig 1 Turbogenerator shaftline; nodal points are indicated

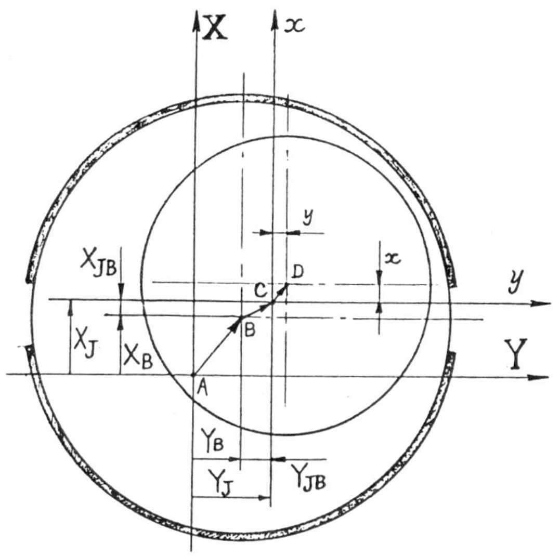

Fig 2 Bearing diagram and coordinates

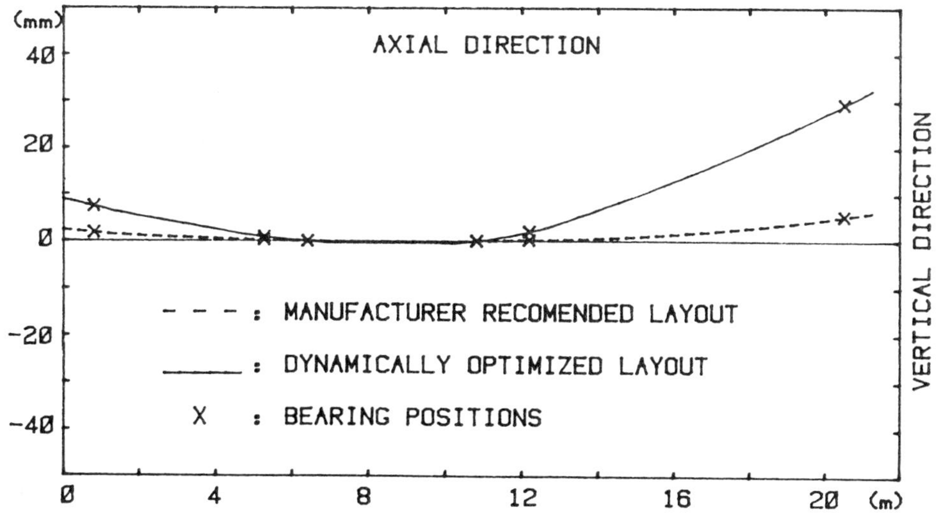

Fig 5 Initial and optimal bearing positions and shaft deflection in XZ plane

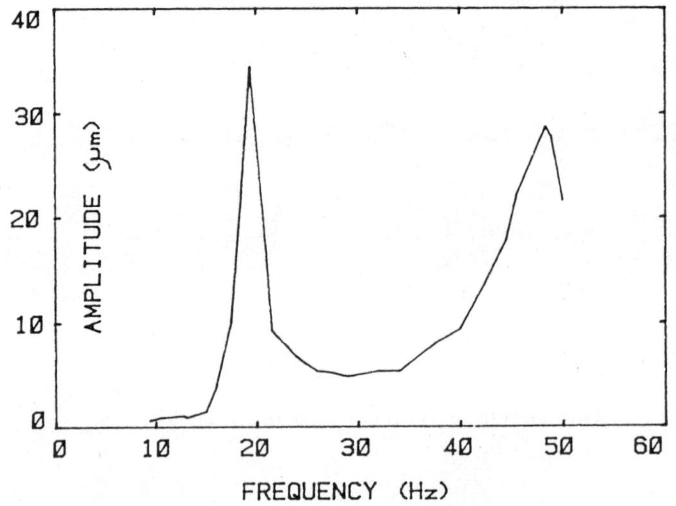

Fig 6 Peak to peak vibration at bearing number 6 pedestal (vertical)
[Courtesy of the State Electricity Commission of Victoria, Australia]

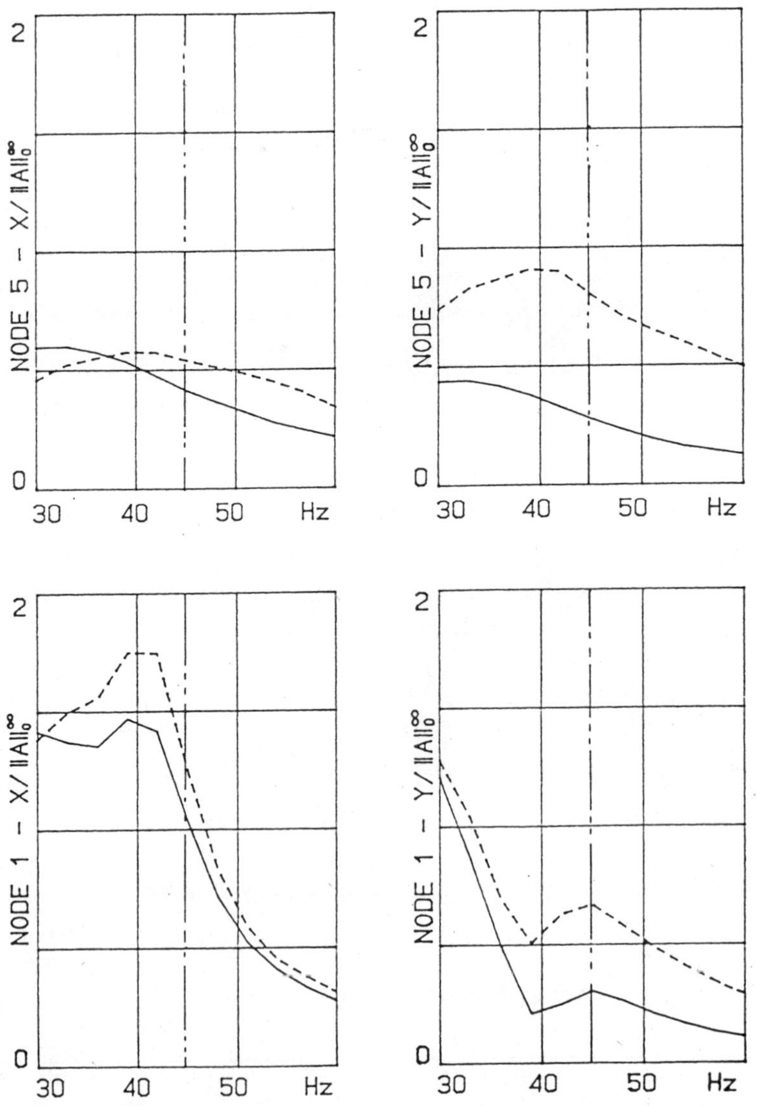

Fig 7 Frequency response for initial (– – –) and optimal (——) configuration

C289/88

Prediction and measurement of vibrations of pump rotors

E T BÜEHLMANN, MSME, PhD, **R STÜERCHLER**, **A FREI**, Dipl-Ing and **E LEIBUNDGUT**
Sulzer Brothers Limited, Winterhur, Switzerland
S E PACE, BS, MSME
Electric Power Research Institute, Palo Alto, California, United States of America

SYNOPSIS The rotor model of a high performance boiler feed pump comprises asymmetric stiffness and damping matrices for the hydrodynamic bearings, and rotationally symmetric stiffness, damping and mass matrices for impeller interaction and the gaps in the seals and balance piston. The coefficients for the impeller interaction are determined experimentally. The nondimensionalized radial and tangential forces are characteristic, for the centering of the rotor and the stability resp. Here centering and stability factors are defined for forces due to impeller interaction. These numbers are relatively independent on the Reynolds number and flow rate ratio and typical for a chosen design, e.g. with and without a "swirl break". The effect of the swirl break is also shown on a computation of a full scale boiler feed pump where the damping ratio increases substantially.

NOTATION

B_2	impeller discharge width
$CF, \overline{CF}$	centering factor actual and averaged resp.
C_{xx}, C_{yy}, C_{xy}, C_{xy}	damping coefficients of bearing
c, c_c	damping coefficients of seal gap
C, C_c	damping coefficients of impeller interaction
D	critical damping ratio
f	frequency
f_e	eigenfrequency
f_0	frequency of rotation at BEP
F_y, F_z	forces in the Y-, Z-direction
$F_{y,s}, F_{z,s}$	static forces in the Y- and Z-direction
F_r, F_t	radial and tangential forces for circular orbit
F_r^*, F_t^*	normalized radial and tangential forces $F^* = F/r \cdot \pi \cdot r_2^2 \cdot B_2 \cdot \rho \cdot \Omega^2$
K_{xx}, K_{xy}, K_{yx}, K_{yy}	stiffness coefficients of bearing
k, k_c	stiffness coefficients of seal gap
K, K_c	stiffness coefficients of impeller interaction
m, m_c	mass coefficients of seal gap
M, M_c	mass coefficients of impeller interaction
r_2	outer radius of impeller
r	radius of circular orbit
Q, Q_{REF}	actual flow rate and at BEP resp.
Re	Reynolds number
$SF, \overline{SF}$	stability factor, actual and averaged resp.
U	unbalance (hydraulic and mechanical)
$X,Y,Z,y,z,\dot{y},\dot{z}$	coordinates, displacements, velocities resp.
ω	circular frequency of the excitation
Ω	circular frequency of the shaft rotation
ρ	density of water
*	(superscript): normalized

INTRODUCTION

The trend in the design of high performance boiler feed pumps is towards smaller and fewer impellers, shorter and lighter rotors and increased speed. Heads of 800 m per impeller is not unusual anymore. At the same time the dynamics of the rotor and especially the stable behaviour becomes more and more important as high vibrations reduce the life span.

Several dynamic forces are acting on the rotor as shown in Fig. 1. They may be divided into forces a) due to excitation such as hydraulic excitation, unbalance and the coupling, b) due to deflection of the rotor such as reactances at the journal bearings, in the thrust bearings, in the seal gaps and balance piston, and interaction forces at the impeller-diffusor, and in the coupling. In order to predict the dynamic behaviour of the rotor in the design stage, the forces due to the deflection should be known accurately. It is common practice to analyze the torsional and lateral vibrations separately since they are not strongly coupled. The stability of a rotor is mainly dependent on the lateral eigenvalues and mode shapes. For that reason this paper focuses on lateral vibrations only. The forces in bearings, in the gaps, and at the impeller diffuser are nonlinear in nature, however they may be linearized for most practical purposes. In that form the forces may be given in the following mathematical form.

1) for the bearings

$$\begin{Bmatrix} F_y \\ F_z \end{Bmatrix} = - \begin{bmatrix} K_{yy} & K_{yz} \\ K_{zy} & K_{zz} \end{bmatrix} \begin{Bmatrix} y \\ z \end{Bmatrix} - \begin{bmatrix} C_{yy} & C_{yz} \\ C_{zy} & C_{zz} \end{bmatrix} \begin{Bmatrix} \dot{y} \\ \dot{z} \end{Bmatrix}$$

2) for the gaps

$$\begin{Bmatrix} F_y \\ F_z \end{Bmatrix} = - \begin{bmatrix} k & k_c \\ -k_c & k \end{bmatrix} \begin{Bmatrix} \dot{y} \\ \dot{z} \end{Bmatrix} - \begin{bmatrix} c & c_c \\ -c_c & c \end{bmatrix} \begin{Bmatrix} \ddot{y} \\ \ddot{z} \end{Bmatrix}$$

3) for the impeller

$$\begin{Bmatrix} F_y \\ F_z \end{Bmatrix} = \underbrace{+ \begin{Bmatrix} F_{y,s} \\ F_{z,s} \end{Bmatrix} + U\omega^2 \begin{Bmatrix} \cos \omega t \\ \sin \omega t \end{Bmatrix}}_{\text{Excitation}}$$

$$\underbrace{- \begin{bmatrix} K & K_c \\ -K_c & K \end{bmatrix} \begin{Bmatrix} \dot{y} \\ \dot{z} \end{Bmatrix} - \begin{bmatrix} C & C_c \\ -C_c & C \end{bmatrix} \begin{Bmatrix} y \\ z \end{Bmatrix} - \begin{bmatrix} M & M_c \\ -M_c & M \end{bmatrix} \begin{Bmatrix} \ddot{y} \\ \ddot{z} \end{Bmatrix}}_{\text{Interaction}}$$

The coefficients of the bearings K_{ij} and C_{ij} are well established and documented in the literature. Although the later ones are not yet completely investigated. The coefficients of the gaps in the seals and balance piston k_i, c_i, m_i can be predicted by the theories of Childs [1] and Nordmann [2].

The investigation of the coefficients K, K_c, C, C_c, M, M_c of the impeller interaction has been a major part of a research programm sponsored by EPRI (Energy Power Rerearch Institute) at the Pump Division of Sulzer Bros. The measurement of these coefficients is based on a rotationally symmetric model [3].

THE MEASUREMENT OF THE COEFFICIENT OF THE IMPELLER INTERACTION

a) Instrumentation

Bolleter et al [3] have made a great effort to determine the coefficients of the impeller interaction. A special rig was built as shown in Fig. 2 and 3. It mainly consists of a rigid bedplate, a single stage pump casing, a swinging arm, and a hydraulic exciter. The impeller is fixed on a shaft which rotates in anti-friction bearings in the arm. The arm is moved by the exciter in vertical direction only. The displacements in Z- and Y-directions are measured between the casing and the impeller. The forces and moments at the impeller are picked up by a transducer specially built for this purpose. The input displacement signal at the swinging arm is a chirp (rapid sweep) with duration of 32 revolutions of the impeller, a frequency band of $0,1\ f_0 < f < 1.5\ f_0$, and max. amplitude of 0.1 mm only. The output signals, displacements in the Y- and Z-direction, and the radial and tangential forces contain the effects of the hydraulic forces which are turbulent in nature. As the input signal is continuously repeated and the repetition period is equel to the time window of the data aquisition system, the averaging is done in the time domain. Averaging of over 100 triggered samples recoveres the coherent signals resulting from the excitation and makes it possible to calculate the coefficients. For simplicity they are plotted in nondimensional form, following [4].

$$K^*,\ Kc^* = \frac{1}{\pi r_2^2 B_2 \rho \Omega^2} \qquad (K,\ K_c)$$

$$C^*,\ C_c^* = \frac{1}{\pi r_2^2 B_2 \rho \Omega} \qquad (C,\ C_c)$$

$$M^*,\ M_c^* = \frac{1}{\pi r_2^2 B_2 \rho} \qquad (M,\ M_c)$$

These normalized coefficients depend on different parameters: flow rate ratio Q/Q_{REF}, Reynolds number Re, speed of rotation Ω, design of the intermediate spaces between impeller-casing, impeller-diffusor and swirl break etc.

b) Results

The normalized coefficients in Fig. 4 are given as functions of the Reynolds number Re and the flow rate ratio Q/Q_{REF} for a design without any stabilizing design features. Fig. 5 shows similar diagrams however, the design of the pump casing was improved by a "swirl break". This feature changes the flow pattern in the space between the impeller and casing. Therefore it tends to destabilize the rotor. This is demonstrated in these figures by the increased direct damping coefficients and reduced crosscoupled stiffnesses.

EFFECT OF THE COEFFICIENTS ON THE RADIAL AND TANGENTIAL FORCE

Radial and tangential force for circular orbits

The force acting at the impeller may be split into a radial and tangential force F_r^* and F_t^* (Fig. 6). Both of these forces influence the dynamic behaviour of the rotor in different ways. The radial force F_r^*, depending on the sign, forces the rotor back into or away from the original position (center); the tangential force F_t^*, stabilizes or destabilizes the rotor. The normalized forces are given for circular orbits by the coefficients as follows:

$$F_r^* = -K^* - C_c^*\left(\frac{\omega}{\Omega}\right) + M^*\left(\frac{\omega}{\Omega}\right)^2$$

and

$$F_t^* = K_c^* - C^*\left(\frac{\omega}{\Omega}\right) - M_c^*\left(\frac{\omega}{\Omega}\right)^2$$

From the sign convention one may observe:

(a) The radial force may push back the rotor to its origin (center) when it is negative e.g. K^* and C_c positive. Note M^* is always positive and decenters the rotor.

(b) The tangential force will stabilize the rotor when it is negative again, e.g. C^* and M_c^* positive and K_c negative

Fig. 7 and 8 give the radial and tangential forces F_t^* and F_r^* as functions of the Reynolds number Re, the flow ratio Q/Q_{REF} and frequency ratio $\omega/\Omega = 1.0$. Fig. 7 shows the forces of the conventional design and Fig. 8 the one with the swirl break.

Stability and centering factor

Prediction of the stability and centering of the rotor form a fundamental part in the analysis of rotor dynamics. The radial and tangential forces are influenced differently by the measured coefficients. It is convenient to have a number which expresses the tendencies toward the stability and centering. The radial force is composed by centering and decentering components. The ratio of this forces may be defined as centering factor CF and given mathematically (for $\omega > 0$).

$$CF = \frac{\text{centering parts of } F_r^*}{\text{decentering parts of } F_r^*} =$$

$$\frac{C_c^*\left(\frac{\omega}{\Omega}\right) + K^*}{M^*\left(\frac{\omega}{\Omega}\right)^2} \quad \text{when } K^* > 0$$

or

$$CF = \frac{C_c^*\left(\frac{\omega}{\Omega}\right)}{M^*\left(\frac{\omega}{\Omega}\right)^2 + |K^*|} \quad \text{when } K^* < 0$$

A rotor may be selfcentering when CF is larger than 1. Similarly the tangential force is composed by stabilizing and destabilizing forces. Also here one may define a stability factor SF (for $\omega > 0$).

$$SF = \frac{\text{stabilizing parts of } F_t^*}{\text{destabilizing parts of } F_t^*} =$$

$$\frac{C^*\left(\frac{\omega}{\Omega}\right) + M_c\left(\frac{\omega}{\Omega}\right)^2}{K_c} \quad \text{when } M_c > 0$$

or

$$SF = \frac{C^*\left(\frac{\omega}{\Omega}\right)}{K_c + |M_c|\left(\frac{\omega}{\Omega}\right)^2} \quad \text{when } M_c < 0$$

A rotor may behave stable when SF is larger than 1 and unstable when SF is smaller than 1.
In Fig. 9 the centering and stability factors are given as functions of the Reynolds number Re and flow rate ratio Q/Q_{REF} for the conventional pump. One may observe CF and SF are relatively independent of the Reynolds number and flow rate ratio. This facts leads to the definition of averaged values $\overline{CF}$, $\overline{SF}$. They are calculated and given in Fig. 9 for all the measured values of Re and for $0.875 \le Q/Q_{REF} \le 1.25$.
The averaged values do not vary substantially. As the stability factor is an indicator of the stability of the rotor it may also show the limit of the stability or indicate the onset of instability. This is so when $F_t = 0$. Fig. 10 shows this limit for $Q/Q_{REF} = 1.0$.
Excitation at very low frequencies are instable in any case. e.g. they may be amplified by the internal forces. However these excitations are turbulent and small and do not coincide with natural frequencies of the rotors, so that there is no danger for the system in general; except at low flow rate, when the hydraulic forces at the impeller are strong and when the vibrations occur at low frequencies.
As mentioned before the averaged stability and centering factors ($\overline{SF}$, $\overline{CF}$) depend significantly on the design of impeller and casing. Therefore these two numbers are characteristic for a design configuration. They can be used for the evaluation and comparisons of designs. Such a comparison is given in Fig. 11 where the ratio of the averaged stability and centering factors as functions of the frequency are given. The change of the design, implementation of the swirl break increases the centering as well as the stability factors, which means the swirl break improves the overall dynamic behaviour of the impeller.

PREDICTION OF UNSTABLE NATURAL FREQUENCIES

The natural frequencies and the mode shapes of rotors of pumps may be estimated by computer codes. Often calculations considered neither damping nor mass matrices as postulated by [3]. Such calculations provided undamped eigenvalues independent of the speed of rotation as long as the speed dependent direct stiffness (K_{xx}, K_{yy}) was not considered. Consequently these values deviated very much from the measured ones. In advanced rotor dynamics the model of a rotor includes all the necessary stiffness, damping, mass, and gyrosgopic moment matrices. The predicted natural frequencies and mode shapes are damped and increase with the speed of rotation. Often the speed coincides with eigenvalues. Eigenvalues with high damping rarely harm the rotors as practice shows. The mode shapes of such eigenvalues usually have large amplitudes in the bearings, where the vibrations are easily damped and the energy removed. These facts are demonstrated in an example. The model of the

rotor comprises all the matrices in question. The gap in the seals is twice the design clearance. The calculation reveals 5 eigenvalues in the frequency band $0 < f < 1.25\ f_0$ (f_0 = frequency of rotation at design point). In the Campbell diagramm Fig. 12 [5] the dependence of the eigenvalues on the frequency of rotation is clearly demonstrated. At resonance the critical damping coefficient is given, D = 12, 48, and 29 % when resonance exists with the first, second and third natural frequency resp. These values are rather large, so that no problems are expected. The limit of stability is given, when D = 0 for the first natural frequency. Note the critical damping decreases successively with increasing speed. This diagram indicates no problems are to be expected in the entire speed range of pump, except when the damping of the first mode decreases further. In reality this may be due to an additional increase of the seal clearance. In such a case the rotor may become unstable and vibrate at a subsynchronous frequency (at the first natural frequency of the rotor).
The influence of extending the rotor model by using the stiffness, damping, and mass matrices is shown in Fig. 13. The model is made for design speed f_0. The conventional calculation of a conservative system estimates a normalized natural frequency $f/f_0 = 1$ and D = 0. Inclusion of the matrices for designed seal clearance (100 %), reduces the critical natural frequency to $f/f_0 = 0.9$ and increases the damping to a safe value D = 13 %. Such a rotor will certainly have a stable behaviour. Increasing the seal clearance (e.g. wear) reduces the natural frequency as well as the damping. This predicted value $f/f_0 = 0.88$ and D = 1.5 % compares very well with the measured ones $f/f_0 = 0.75$ and D = 0.5 % resp. Generally pumps are designed for increased seal clearance to account for wear in service. To maintain high stability for increased seal clearance the design may be changed and a swirl break may be included. Such a change may reduce the natural frequency further to $f/f_0 = 0.74$ but increases the damping substantially to D = 25 %. Thus the rotor always will behave stable.

CONCLUSIONS

The limit of the stability of the rotor of a high performance boiler feed pump can be well estimated when the rotor models comprised of linearized stiffness, damping and mass balance piston and the impeller interaction are used.
The matrices for the interaction cannot be calculated yet and must be measured. The coefficients of the matrices depend on the design of the spaces, the flow rate ratio, and the Reynolds number. They are the coefficients for the normalized radial and tangential forces. These forces form the basis of the centering and stability factor CF and SC respectively, which are mainly independent on the flow rate and the Reynolds number.
The averaged values of CF and SF are characteristic for a chosen design configuration and may be useful when comparing them.

REFERENCES

(1) CHILDS, D.W. Finite-Length solutions for the rotordynamic coefficients of constant-clearance and convergent-tapered annular seals. *Third International Conference on Vibrations and Rotating Machinery, York, England*, 1984, Sept. 10-12.

(2) NORDMANN, R.; DIETZEN, F.J.; JANSON, W.; FREI, A.; FLORIJANCIC, S. Rotordynamic coefficients and leakage flow for smooth and grooved seals in turbopumps. *International Conference on Rotordynamics, 1986, Sept. 14-17.*

(3) BOLLETER, U.; WYSS, A.; WELTE, Y.; STUERCHLER, R. Measurement of hydrodynamic interaction matrices of boiler feed pump impellers. *Trans. ASME, Journ. Vibrations and Acoustics, Stress, Reliab. in Design. Vol. 109*, 1987, April, pp. 144-151.

(4) OHASHI, H.; SHOZI, H. Lateral fluid forces acting on a whirling centrifugal impeller in vaneless and vaned diffusor. *Proceedings of Workshop on Rotordynamic Instability Problems in High Performance Turbomachinery, Texas A & M University College Station, TX, NASA Conference Publication 2338*, pp. 109-122.

(5) PACE S.E.; FLORIJANCIC S.; BOLLETER, U. Rotordynamic developments for high speed multistage pumps. *Third International Pump Symposium*, 1986, May, Houston Texas.

ACKNOWLEDGEMENT

This work was made possible by the contract between the Electric Power Research Institute Paolo Alto California and the Pump Division of Sulzer Bros. Ltd. The authors like to thank both organizations for their support and the permission to publish these results.

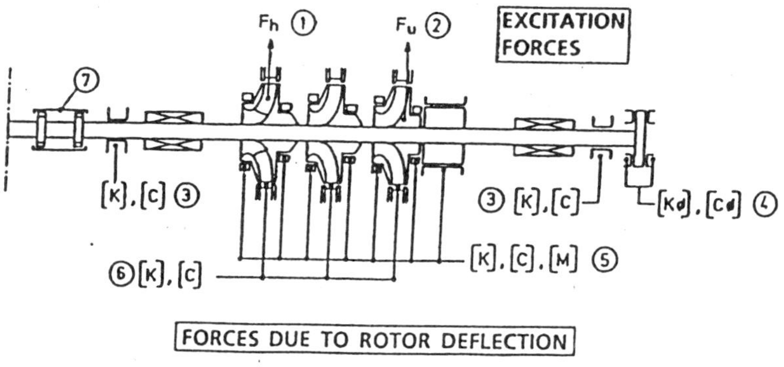

Fig 1 Forces on a pump rotor due to excitation forces, (1) hydraulic, (2) unbalance, (7) coupling and due to rotor deflection, (3) journal bearing, (4) thrust bearing, (5) gaps in the seal and balance piston and (6) impeller interaction

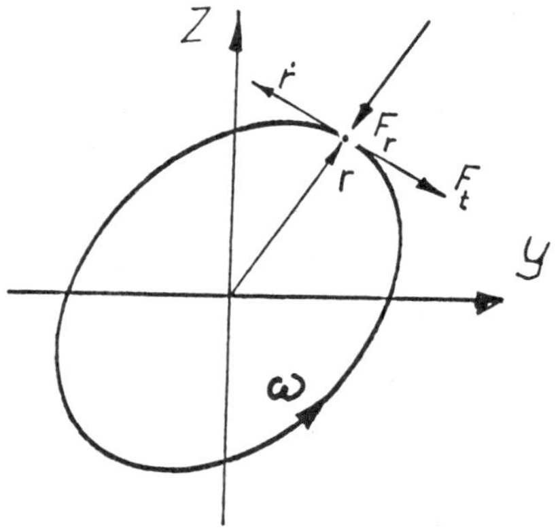

Fig 6 Radial and tangential forces on a forward orbit when they centre and stabilize a rotor

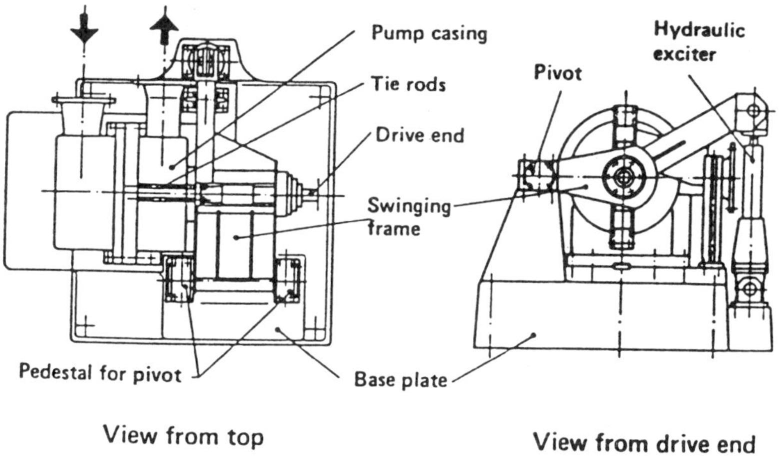

Fig 2 Test rig

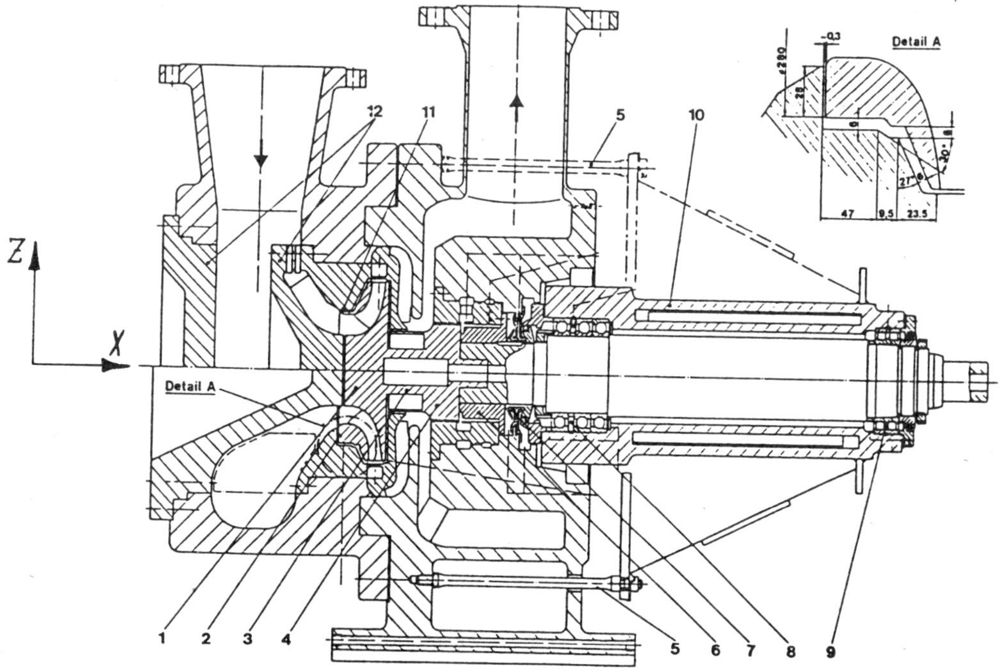

Fig 3 Horizontal (top half) and vertical (bottom half) section through the test arrangement; (1) impeller, (2) face seal, (3) measuring section, (4) balancing piston, (5) tie rods, (6) mechanical seal, (7) oil-water separation, (8) radial and axial bearings (angular ball-bearings), (9) cylindrical roller bearings, (10) swinging frame, (11) axial seal (alternative to radial seal), (12) insert for rotationally symmetric approach flow (shown in top half only, bottom half shows normal intake)

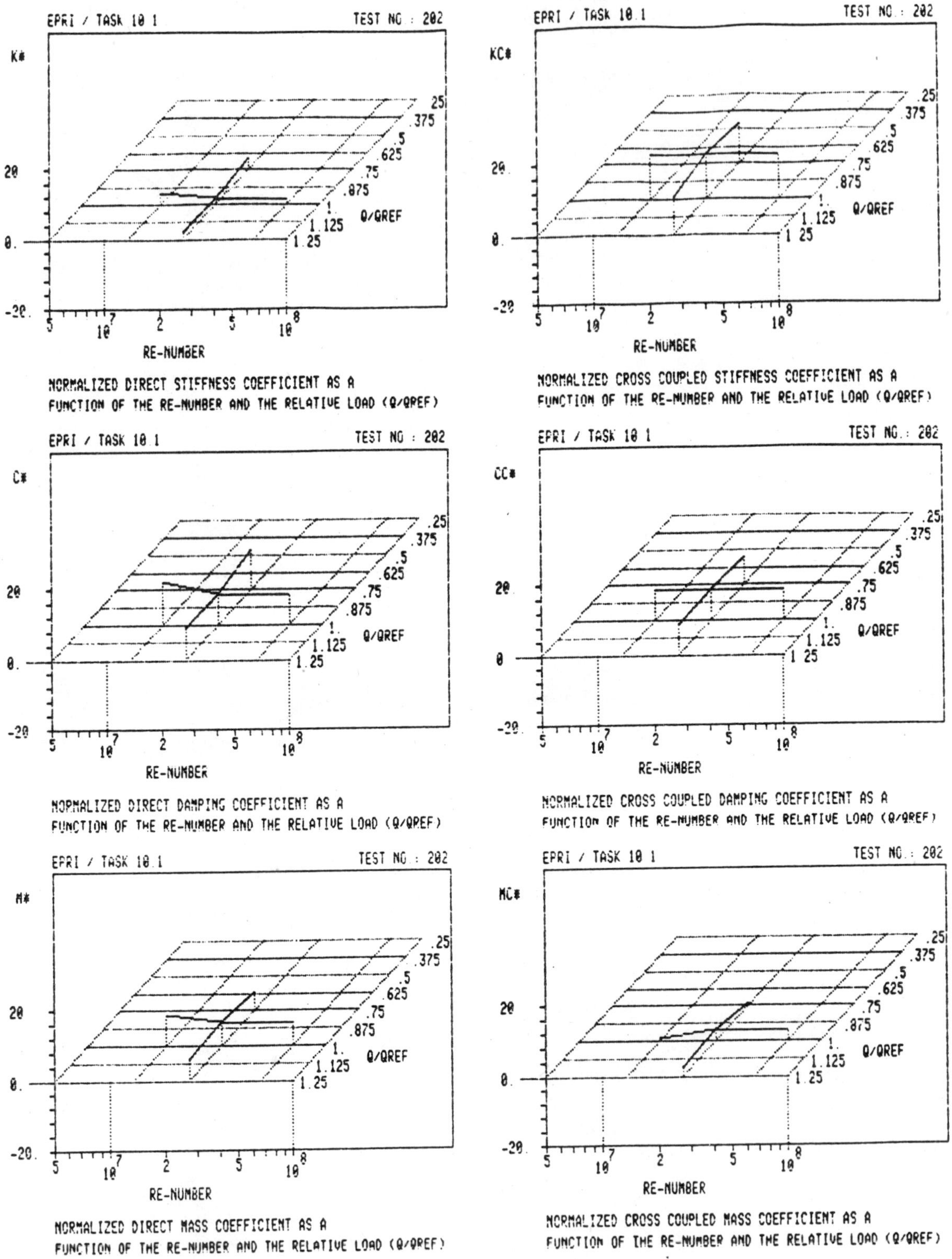

Fig 4 Normalized coefficients of the stiffness, damping and mass matrices of the conventional design

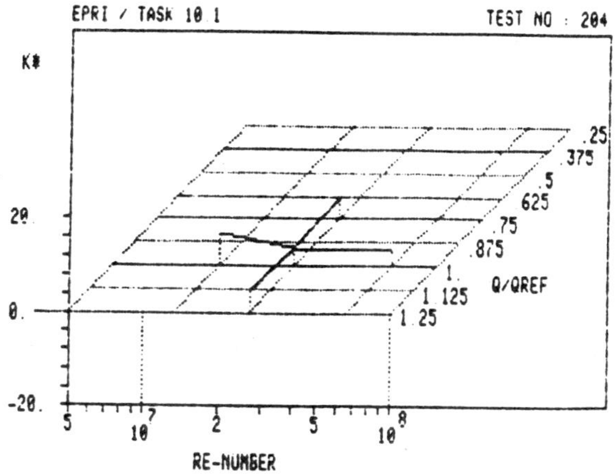

NORMALIZED DIRECT STIFFNESS COEFFICIENT AS A
FUNCTION OF THE RE-NUMBER AND THE RELATIVE LOAD (Q/QREF)

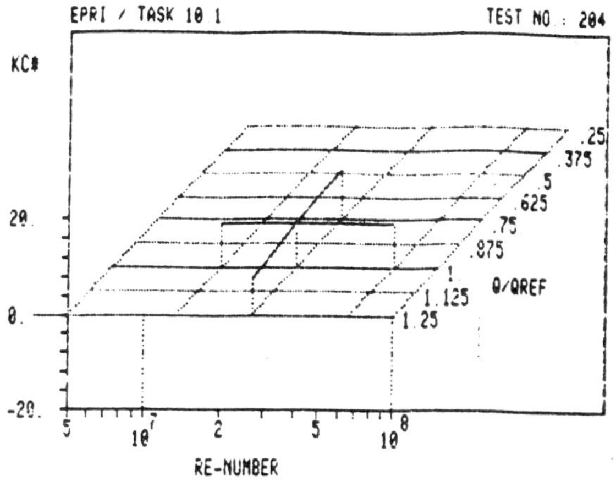

NORMALIZED CROSS COUPLED STIFFNESS COEFFICIENT AS A
FUNCTION OF THE RE-NUMBER AND THE RELATIVE LOAD (Q/QREF)

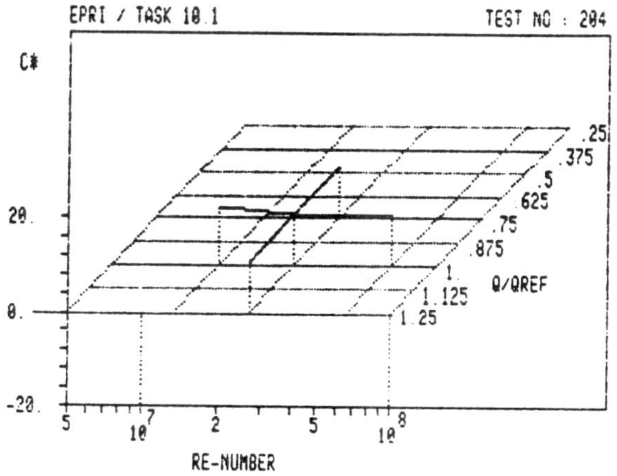

NORMALIZED DIRECT DAMPING COEFFICIENT AS A
FUNCTION OF THE RE-NUMBER AND THE RELATIVE LOAD (Q/QREF)

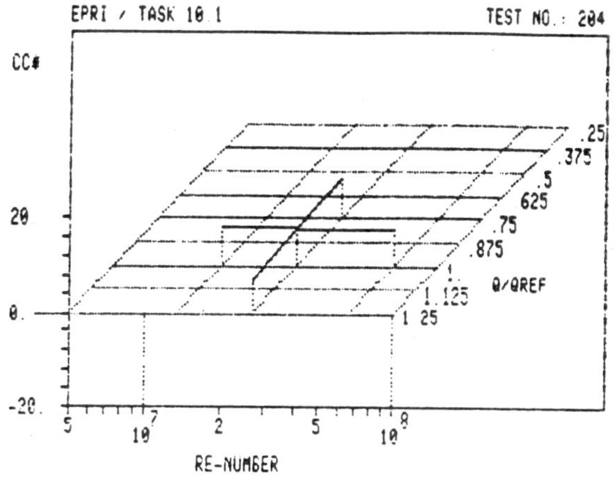

NORMALIZED CROSS COUPLED DAMPING COEFFICIENT AS A
FUNCTION OF THE RE-NUMBER AND THE RELATIVE LOAD (Q/QREF)

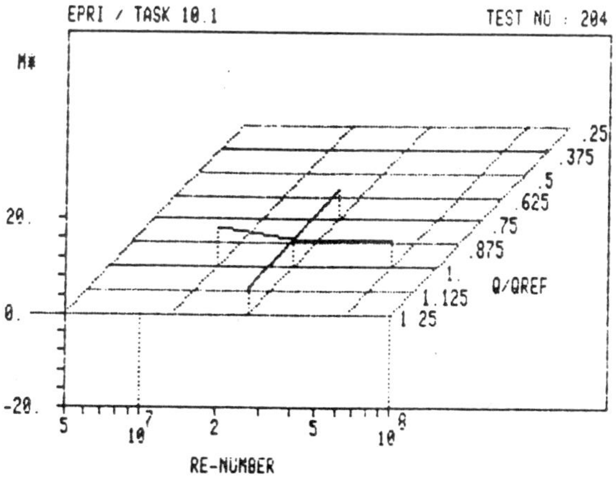

NORMALIZED DIRECT MASS COEFFICIENT AS A
FUNCTION OF THE RE-NUMBER AND THE RELATIVE LOAD (Q/QREF)

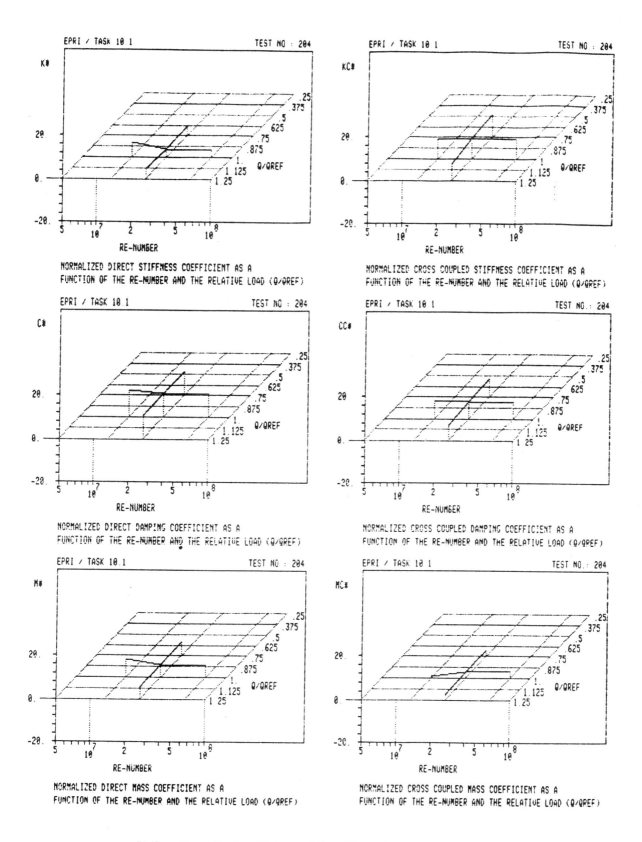

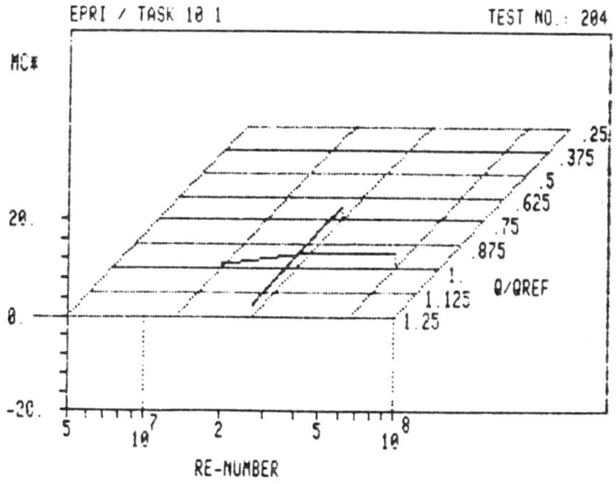

NORMALIZED CROSS COUPLED MASS COEFFICIENT AS A
FUNCTION OF THE RE-NUMBER AND THE RELATIVE LOAD (Q/QREF)

Fig 5 Normalized coefficients of the stiffness, damping and mass matrices of the design with 'swirl break'

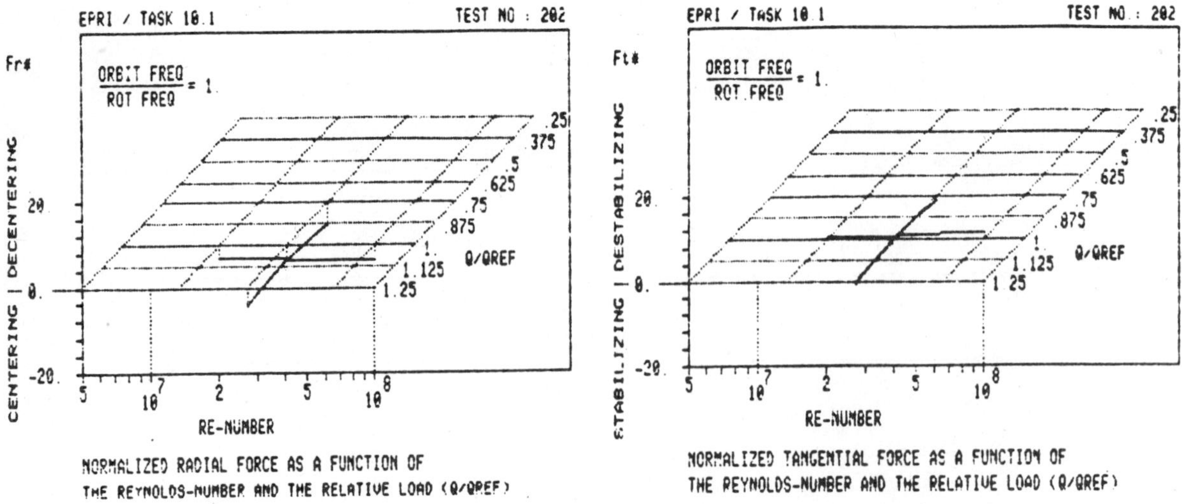

Fig 7 Normalized radial and tangential forces as functions of Reynolds number Re and flow rate ratio Q/Q_{REF} for frequency ratios ω/Ω = 1.0 of the conventional pump design

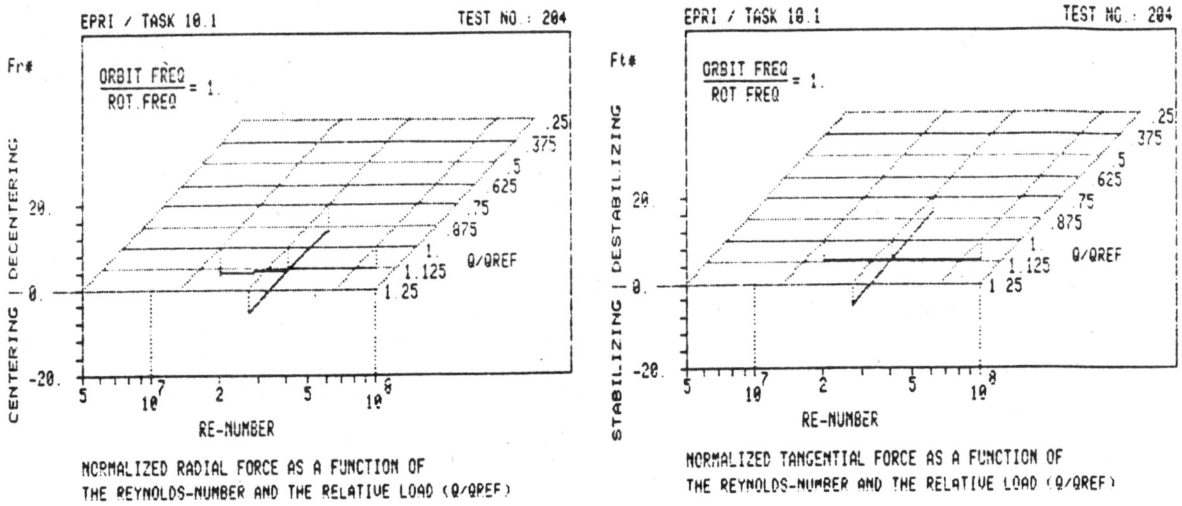

Fig 8 Normalized radial and tangential forces as functions of Reynolds number Re and flow rate ratio Q/Q_{REF} for frequency ratio ω/Ω = 1.0 for a pump with swirl break

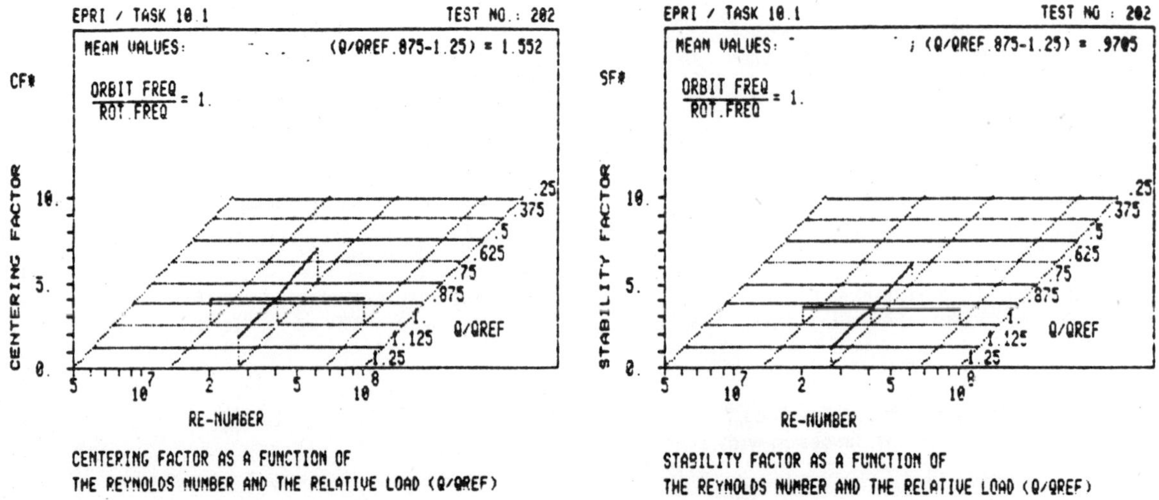

Fig 9 Centering and stability factor as functions of Reynolds number Re and the flow rate ratio Q/Q_{REF} for frequency ratio ω/Ω = 1.0 of a conventional pump design

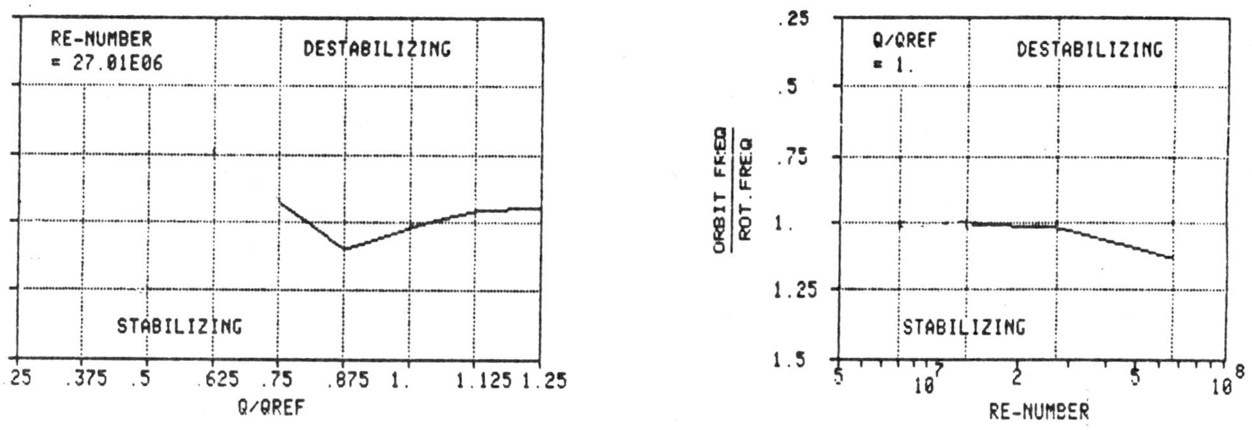

Fig 10 Frequency ratio ω/Ω as a function of the stability limit when $F_t^* = 0$ for constant Reynolds number Re and flow rate ratio Q/Q_{REF} of a conventional pump design

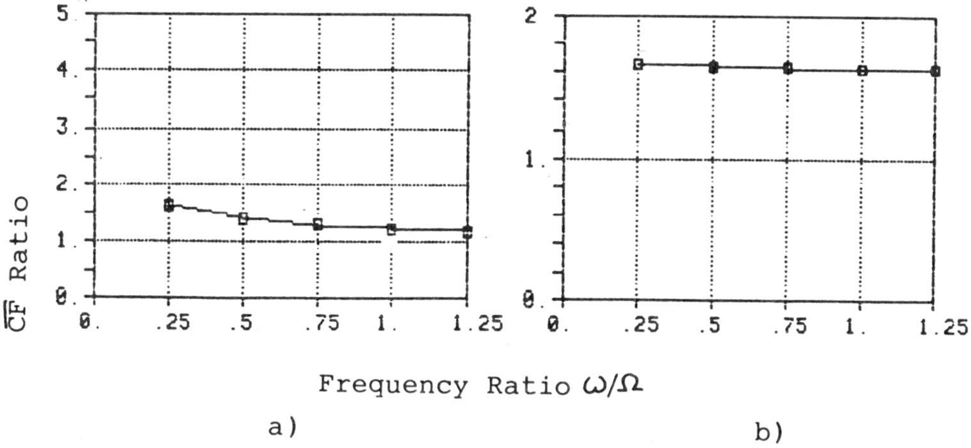

Fig 11 Effect of the insertion of a swirl break on the (a) centering and (b) stability factor as a function of the frequency ratio ω/Ω (ratio built by the $\overline{CF}$, $\overline{SF}$ averaged over all Re numbers and a flow rate ratio $0.875 = Q/Q_{REF} = 1.25$)

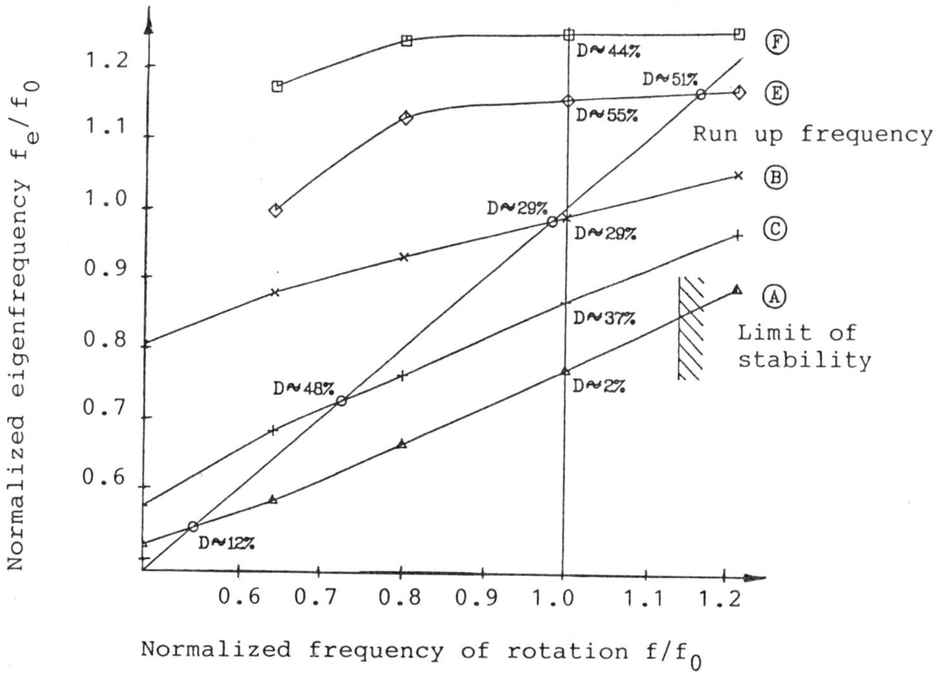

Fig 12 Campbell diagram of the five lowest normalized eigenfrequencies as a function of the normalized rotational speed

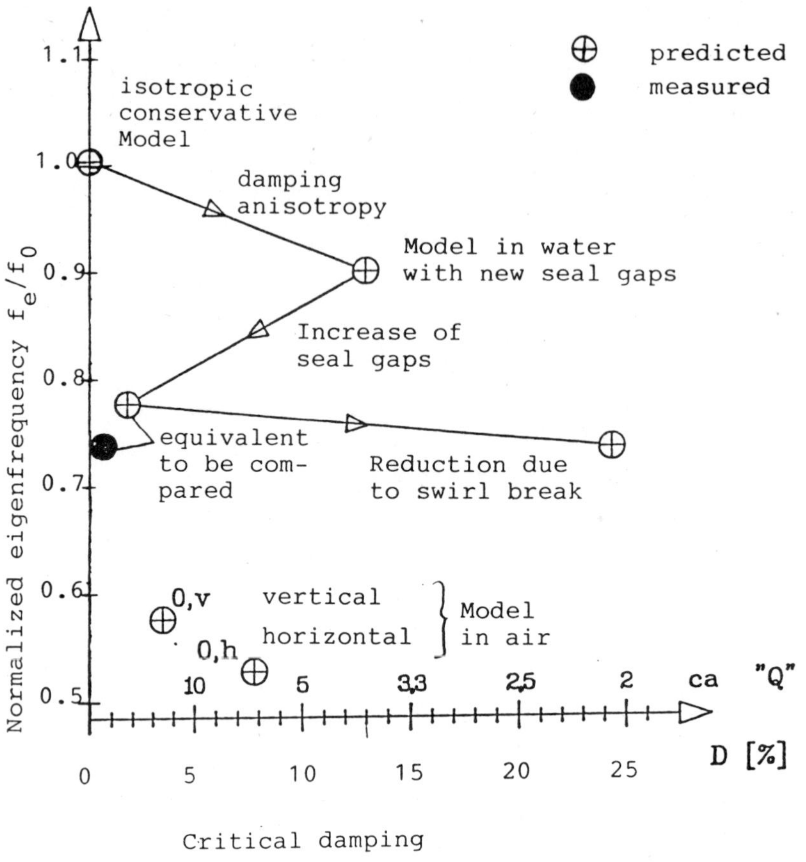

Fig 13 Influence of the insertion of the dynamic stiffness, damping and mass matrices and the swirl break on the normalized eigenvalue of the least damped eigenmode of a pump with worn gaps (200 per cent of the original design) and at full speed

C256/88

A method for overall condition monitoring by controlling the efficiency and vibration level of rotating machinery

E H GÖTTLICH, Dr-Ing
Vibro-Meter, Fribourg, Switzerland

SYNOPSIS The method described in this paper, shows a way of monitoring a machine by looking at the vibration level and the off-design performance map. The off-design performance map is calculated by a theoretical model using, as input, the point with the maximum efficiency. This point will be found by an iteration of measured values like mass flow, pressure-ratio, air-intake temperature and rotating speed.

NOTATION

k isotropic coefficient
$\dot{m}$ mass flow
n speed
P total pressure
T total temperature
ϵ pressure factor
η isotropic efficiency
μ corrected mass flow
ν corrected speed
π pressure ratio
φ flow factor
ψ work factor

subscripts:

$_o$ point of max. efficiency
$*$ related to max. efficiency
DP design point
E related to compressor entrance

1 INTRODUCTION

Today there is a lot of equipment for condition monitoring of rotating machinery on the market. This equipment can be separated in three groups. The first group are systems with static signal input, like pressures, temperatures etc. With this group a trend on the behaviour of the machine can be done. The second group is a little more intelligent, while treating dynamic and static signals. The dynamic signals, like measured vibration, in combination with the static signals, give a far better impression of the actual state of the machines. Big and expensive systems called "Expert Systems", representing the third group, try to give information about the actual aerothermal situation of the machine, as well as the static and dynamic values of the operating condition.
A handicap for the last mentioned group is calculation of the off-design performance map. It would be very useful to give information about the present efficiency of the machine. It is imaginable, that a aerothermal machine is not vibrating at a critical level, but working at lower efficiency, e.g. because of pollution of the compressor blades.
To give the possibility of showing the actual efficiency of a machine, the method described herein, shows a way of a synthetically calculating the off-design performance map of rotating machinery. The combination with measurement equipment for static and dynamic signals and a synthetical calculation of the machines present efficiency will decrease the operating costs and prolong the life of the machine. In former times measuring of the actual efficiency caused a lot of expenditure, but today, with the high performance of computers and computer controlled measurement equipment, an online calculation of the efficiency can be well done.

2 OFF-DESIGN PERFORMANCE MAP CALCULATION

The theory of the off-design performance map calculation is based on the behaviour of a single stage. Single stage characteristics are super-imposed with a stage-stacking method for multi-stage machines. The basic ideas are to be found at (1,2) for a multi-stage machine and at (3) for single stage compressors or fans.
All values are related to the point with the maximum efficiency of the machine. With this relationship nondimensional factors are the result. These factors are shown with a star as superscript. The values of the maximum efficiency point are written with a 0 as subscript. The following figure shows the characteristic of a single stage, expressed by these to the maximum efficiency related values, for a variation of the flow factor.

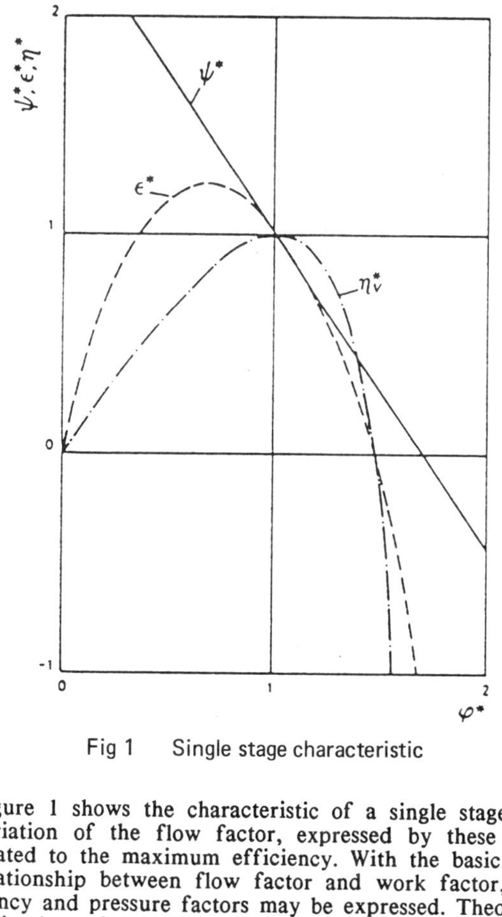

Fig 1 Single stage characteristic

Figure 1 shows the characteristic of a single stage for a variation of the flow factor, expressed by these values related to the maximum efficiency. With the basic linear relationship between flow factor and work factor, efficiency and pressure factors may be expressed. Theory and derivation of this basic linear relationship are given at (1,2). The following equations give the different relationships of the shown factors.

$$\mu^* = \frac{\mu}{\mu_o} = \dot{m}^* \sqrt{T_E} / P_E^* \qquad \nu^* = \frac{\nu}{\nu_o} = \frac{n/\sqrt{T_E}}{n_o/\sqrt{T_E}}$$

$$\varphi^* = \frac{\mu^*}{\nu^*}$$

$$\psi^* = f(\varphi^*) \qquad \eta^* = f(\psi^*) \qquad \epsilon^* = f(\psi^*)$$

3 ITERATION OF THE POINT WITH MAXIMUM EFFICIENCY

Paragraph 2 described a method for a synthetical off-design preformance map calculation. In principal there is no problem in using this method, but it is essential to know the point with the maximum efficiency of the machine. This point is a theoretical one, therefore an iteration must be used to find this point as input for the calculation.
The procedure to find this critical point is divided in two parts. First, using measurements, the speed and mass flow of this point must be specified. In addition to this the pressure ratio can be measured but is not essential because pressure ratio may be calculated with the related speed and mass flow. The second part is the iteration, using the measured values as input, to find the work factor for this point with maximum efficiency.

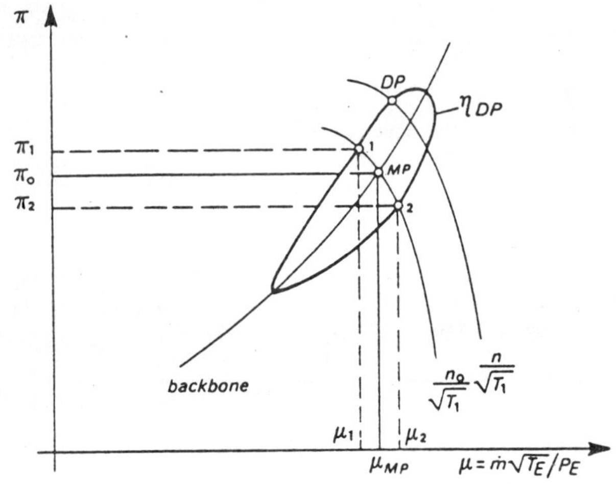

Fig 2 Measurement of the maximum efficiency point

Figure 2 shows the different points of measurement. At the design point the efficiency is given by the manufacturer of the machine. For nearly all machines the design point **DP** is situated in the off-design performance map at higher pressure ratio and speed, but with a lower efficiency, as the maximum. Reducing the speed will move the working point to the direction of the point with the highest efficiency **MP**. With a variation of speed and a throttle or dethrottle at a certain constant speed it is possible to find the point MP. In figure 2 the shell characteristic of the line with constant efficiency for the point DP is shown. This characteristic of the curve explains the necessity of a throttle or dethrottling at a certain constant speed. Every speed line has only one point with the highest efficiency. All other efficiencies exist twice for a certain speed. The connection of all points with the maximum efficiency of each speed is shown in figure 2 as the backbone line. The measurement of these points requires a fast data acquisition system, which is fast enough to handle the data acquisition and to calculate the actual efficiencies online.
After the values of the point with maximum efficiency is known, an iteration for the work factor of this point is necessary. It is impossible to calculate this work factor without iteration, because the theory uses this work factor for all losses of a single stage or the complete machine respectively. The equation below shows the basic relation of the iteration.

$$\psi^* = \frac{\psi}{\psi_o} = \frac{1}{\nu^{*2}} \left(\frac{\pi^{\frac{k-1}{k}} - 1}{\pi_o^{\frac{k-1}{k}} - 1} \right) \frac{1}{\eta^*}$$

All points of the calculation are related to the point MP. As shown in figure 1 all factors, like flow, work, pressure and efficiency, are "1". For this reason it is impossible to use for the iteration the relationship of the speed with the highest efficiency. The equation above gives the right work factor, if a variation of this factor and the related speed of point DP result in the same pressure ratio and efficiency as the well known values of the design point DP. Except for the work factor, all other values are known by measurement or given by the manufacturer.
Figure 3 shows an example of a work factor variation and a comparison with measured values. The measured values are taken from Eckert, Schnell (4).

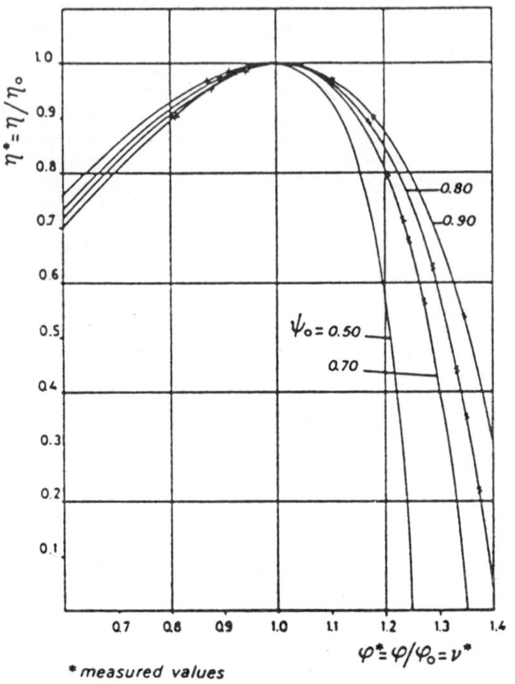

Fig 3 Variation of the work factor

The comparison with measured values shows very well that the estimated losses expressed by the work factor correspond to reality.
With the nondimensional factors and the functionalities shown in Figure 1, all points of the off-design performance map may be calculated, using the equation below and the one for the work factor iteration. The determination of the surge- and choke-line is not mentioned within this paper, but the theory and examples are given at (1,2,3).

$$\epsilon^* = \frac{\epsilon}{\epsilon_o} = \frac{1}{\nu^{*2}} \left(\frac{\pi^{\frac{k-1}{k}} - 1}{\pi_o^{\frac{k-1}{k}} - 1} \right)$$

4 CONDITION MONITORING

The above paragraphs showed a way of calculating the efficiency of rotating machinery. For the overall condition monitoring this calculation is only one part of the system. Efficiency calculation may have different functions. First the system gives online information about the actual efficiency of the machine. On an initial run of a new machine, all measurements for the off-design performance map calculation are taken. With these values the map can be calculated and stored away as a screen mask of a graphic terminal. During operation the actual working point is visualized in the off-design performance map.
A screen mask generation gives some advantages compared with a online calculation of the actual efficiency. Online efficiency calculation uses three parameters to determine the efficiency, mass flow, speed and pressure ratio. Using a screen mask the working point calculation is reduced to two parameters, pressure ratio and mass flow. Entering these actual measured values to the map, the working point is described without a calculation of the efficiency. The efficiency is visualized with the shell curves of the map. On demand it is possible to give the actual efficiency.

Considering the working region of the design point, it is not obvious, why a online calculation of the efficiency should not be done. For example a gas turbine at a power generation plant is working at the design point most of the time, therefore there will be enough time to calculate the efficiency online. However if a run-up or run-down of such a machine is controlled, an online calculation will take too much time. Measurements of pressure and mass flow are taken very quickly during this period and the operator can follow the run-up or run-down in the off-design performance map. Besides the information of the efficiency the operator will see the information about the surge- and choke-line. If someone wants to run-up a machine manually, such a system will avoid the passage of the surge-line.
Besides the control of the efficiency and visualisation of the off-design performance map, the efficiency calculation is imbedded in the condition monitoring. As mentioned in the introduction a lot of equipment corresponding to the group two is on the market. A calculation of the efficiency in combination with a surveillance of dynamic and static signals represent a monitoring system of group three. Such a system should fullfill the following specifications:

- The data acquisition system must have a fast analogue/digital conversion. To have the highest throughput rate on such a system, it has to run with a multi-tasking operating system. Multi-tasking gives the possibility to calculate a dynamic signal (Fast Fourier Transformation), while the CPU is controlling a new set of 1024 measurements, for another FFT.
- As many simultaneous channels as possible should be used for the measurement. Simultaneous measurements give the advantage of looking at the vibration level of the shaft in a reliable manner. A measurement of a single shaft should have at least four simultaneous measurement channels, e.g. horizontal and vertical vibration of every bearing.
- Beside the dynamic values, it is essential to have a certain capacity of DC-input channels. DC-inputs, like pressures, temperatures or mass flows, are used to have the information about the aerothermal process of the machine.
- The data acquisition system must have a high speed communication interface. Without such an interface the amount of data can not be handled by the host computer in a sufficient manner, without blockading the bus for the other control functions.
- On the host computer side it is absolutely essential to have a multitasking operating system. A condition monitoring system on a PC with a single-tasking system will give no reliable results. The monitoring software on the host should fullfill the following specifications:
  * Trending of DC-values and vibration signals. Vibration signals are order tracks with amplitude and phase information. Trending of phase is very important, because certain failures, like unbalance, produce a phase shift at the beginning. The trend should have different time-bases, so that a trouble-shooting is possible in different time intervals. Imaginable are ranges of last year(s), last month(s) and the last one or two days in very short time intervals.
  * Before a cyclic measurement is stored to the trend data-base, all incoming signals are compared with base-lines or initial measurements. This will give information about critical signal levels of the system, like "Alarm" or "Danger" mesages.
  * Beside the trending, cyclic measurements, an interactive communication with the data acquisition system should be possible. With this function online information about the machine is given to the operator.
  * The aerothermal process of the machine is monitored using the DC-channlels of the system. Process-data and efficiency is shown and may be displayed with bargraphs, including the critical levels of the machine.

5 CONCLUSIONS

The paper gives a new approach to a condition monitoring system. The described monitoring is based on a new theory of calculating the actual efficiency of the machine. Existing efficiency calculating software use isotropic equations. With this new method of estimating the losses with the work factor results are closer to reality.
Using an existing data acquistion hardware this new approach can be realised. Because of the well known characteristics of dynamic and static signal processing units, that part of the monitoring is not described in detail within this paper.
At the end, after all these theoretical and technical specifications it may sound ridiculous to mention the reliability of the used hardware. But this fact is not negligible. More and more the industry is using critical machines to have better efficiencies and a better cost profit relationship. Maintenance costs are increasing and unexpected catastrophic failures cause both loss of production and large repair bills. A permanently installed monitoring system can avoid such failures and reduce maintenance costs, but it must have guarenteed reliable monitoring hardware. Saving some money on this data acquisition and host computer hardware can produce higher cost, if a failure is not detected or the missing data-base prevents a retracing of the failure. A monitoring system is in operation 24 hours a day and 365 days a year.

6 REFERENCES

1. GASPAROVIC, N., STAPERSMA, D.:' Berechnung der Kennfelder mehrstufiger axialer Turbomaschinen.'. *Forsch. Ing. Wes. 39 (1973) Nr.5, pp.133-143.*
2. GÖTTLICH, H.: 'Verfahren zur Berechnung der Kennfelder von axialen bzw. radialen, ein- oder mehrstufigen thermischen Strömungsmachinen, auch unter Berücksichtigung des Einflusses von verstellbaren Leitapparaten.' *Phd. Thesis, Technische Universität Berlin, 1984, D83.*
3. GASPAROVIC, N., GÖTTLICH H.: 'Einfaches Verfahren zur Simulation des Betriebsverhaltens einstufiger Axialverdichter und Gebläse mit fester Geometrie.', *Brennstoff Wärme Kraft 37 (1985) 3 pp.102-105, reprinted MTU-Informationsdienst 15(1985) Heft 2.*
4. ECKERT B., SCHNELL, E.: 'Axial- Radialkompressoren.', *2.Edition, Springer-Verlag, Berlin-Heidelberg-New York, 1980.*

C307/88

Application of an expert system to rotating machinery health monitoring

J W HILL, BSc(Eng), ACGI, CEng, MIMechE, MIOA
Acoustic Technology Limited, Southampton, Hampshire
N C BAINES, BSc, PhD, MIOA
MJA Dynamics Limited, Southampton, Hampshire

SYNOPSIS Condition based maintenance is now common practice with operators of major rotating equipment. However, as condition monitoring systems become more complex so does interpretation of the measured data. Expert systems embodying the knowledge and experience of key personnel provide a cost effective means of automating that interpretation and recommending the actions required. This paper discusses the design of such a system to accommodate the requirements of operators, maintenance managers and specialist machinery engineers.

1 INTRODUCTION

The practice of condition monitoring is now well established with most operators of major rotating equipment. Vibration measurement can provide valuable information regarding the mechanical health of a machine. Supplementary data is obtained from lube oil analysis and by monitoring bearing temperatures. In the particular case of turbo machinery, performance can be determined by monitoring fluid pressures, temperatures and flow rates. The major objective of applying condition monitoring is the reduction of operating costs. This is achieved in a number of ways:

(a) By monitoring the progressive development of machine faults remedial action can, if and when necessary, be taken and serious failure avoided.
(b) Shutdown time is reduced when the fault can be diagnosed prior to strip down.
(c) Time between overhauls can be increased by adopting a condition based maintenance strategy, rather than routinely overhauling after an elapsed running period.
(c) Direct fuel cost savings are achieved by ensuring that machinery operates at optimum efficiency.

Development of condition monitoring systems has undoubtedly been influenced by modern developments in digital electronics, both micro-processor based data acquisition equipment and mini/micro computers for subsequent data management. Each is continually becoming both cheaper and more powerful. Once suitable equipment has been installed, acquisition of monitored data, often in excessive volume, is almost too easy. Problems arise however with interpretation of that data, since this usually requires the services of specialists in vibration and aerothermal analysis. This is of little consolation to the machine operator, who wishes to know the current state of the machine, or the maintenance manager who wishes to plan maintenance schedules. By integrating the data acquisition system with an expert system, data analysis and interpretation can be undertaken using the host computer. Data is processed by reference to a 'knowledge base' which embodies all the 'rules' for interpreting the data from a given machine. The result is a condition monitoring system that can communicate the information required by each of the different levels of user viz plant operator, maintenance manager and specialist machinery engineer. This paper describes a project currently being undertaken by Acoustic Technology Limited and MJA Dynamics Limited to develop an expert condition monitoring system for a gas turbine driven compression station operated by a major international oil company.

2 WHAT IS AN EXPERT SYSTEM?

An expert system is essentially a computer program that is able to produce decisions in much the same way that an expert would, given the same information. They have found use in many different application areas. Two of the best known expert systems are MYCIN and PROSPECTOR. The first of these two was developed to provide Doctors with consultative advice about patients with meningitis and bacteremia (infections that involve bacteria in the blood). MYCIN diagnoses the cause of the infection, and offers recommended drug treatments. PROSPECTOR is designed to help geologists evaluate the mineral potential of a region during exploration for new mineral deposits. Based on information supplied by the Geologist, the program identifies the most likely mineral deposits in the area.

An expert system for condition monitoring purposes consists of 3 key components as follows:

(a) Knowledge base
(b) Inference engine
(c) Database

These components are linked together to produce a system that advises, informs and solves problems in a manner similar to that of one or more human experts.

The Knowledge base should embody all the information that the human expert(s) would use to solve a problem. In general this knowledge will either be based upon heuristics (ie. observational evidence which cannot be rigorously proved), or facts (ie. knowledge which is in some way "provable"). It is interesting to note that heuristic knowledge is not necessarily possessed only by highly trained, senior personnel, since it will in general be based upon experience in carrying out a particular task. In this respect it is often most important to encapsulate such information in an expert system before an employee leaves a particular post. Otherwise it may be a considerable period of time before his replacement is able to diagnose faults with comparable skill.

By far the most common way of representing knowledge in condition monitoring expert systems is as rules. The general form of a rule is:

IF (condition) THEN (consequence)

If the conditional part of the rule is satisfied, then the consequence follows.

The initial phase in the construction of an expert system involves gathering together all the available knowledge. This process is often referred to as 'Knowledge engineering'. Since an expert system can only ever be as good as the knowledge that it embodies, the importance of this process cannot be overstressed.

For a condition monitoring expert system, the database contains all the acquired data that is relevant to making a diagnosis about a particular machine. Typically, the majority of the data would derive from sensor measurements, although visual observations and manual measurements might also be included. Since some rules will be based upon trends in the machine's behaviour, it is important that historical data is maintained in the database as required.

The Inference engine provides the expert system with its ability to arrive at a diagnosis. It is basically a program that uses the knowledge base to operate upon the information in the database. The two most common strategies employed to perform the inference process are termed forwards and backwards chaining. In the first of these, the inference process starts from a new piece of data and searches through the rules to find which ones are satisfied, and hence identify if there are any faults present. The alternative strategy of backwards chaining starts from a particular fault and then works 'backwards' through the rules to the point where the implied conditions satisfy the information available in the database. If all the conditions are satisfied then the fault is judged to exist.

Why Use an Expert System? -

There are a large number of reasons why expert systems are a very effective part of modern condition monitoring systems. These include:

(a) Repeatable diagnosis given the same data.
(b) The knowledge from more than one expert can be incorporated.
(c) The expert system is available at any hour of the day throughout the year.
(d) The expert system can be available at a number of sites simultaneously.
(e) Because the knowledge base is not part of the inference program, changes can be made to the knowledge base far more easily than with conventional programming techniques. Hence as more knowledge becomes available, the system can easily be updated.

What are the Drawbacks? -

The current generation of systems is unable to have the 'flashes of inspiration' that a human expert might have. Practical systems that learn new rules as more data becomes available have yet to be developed. However, because of their structure, the current generation of expert systems lend themselves to modification as the human expert(s) discover new rules.

One of the perennial problems with condition monitoring systems is that of high false alarm rates. This problem is just as prevalent with expert systems, and great care must be taken in the system design to ensure a high degree of integrity. The next sections of this paper will detail the strategies being used for one particular expert system for on-line condition monitoring.

3 SYSTEM FUNCTIONAL DESIGN

3.1 System overview

The on-line expert condition monitoring system which is currently being developed can be broadly described by the function block diagram given in Figure 1. The blocks in this figure are considered in more detail in the following sections of this paper.

3.2 Signal conditioning and data acquisition subsystems

The system is designed to enable both analogue and digital signals to be acquired, sampled and conditioned.

The inputs are a combination of static and dynamic signals which either singularly or in combination can be used to provide relevant information on the mechanical, aerothermal and performance condition of the gas compression train. The input data is typically:

Vibration
- acceleration
- velocity
- displacement

Pressures
Temperatures
Flow rates
Speed
Position indication (valves, guide vanes)
Alarm states (digital)

The data acquisition subsystem operates in two modes controlled by the system management software. Rapid scan mode allows the monitoring of gross parameters to allow alarm sequencing to be accomplished. Periodic scan mode enables the collection of data for long term analysis trends.

The major functions of the signal conditioning and data acquisition subsystems are as follows:

(a) Signal validation/transducer calibration checks.
(b) Signal processing (FFT, Enveloping, Fault specific parameter extraction).
(c) Alarm/threshold comparison.
(d) Buffer memory.
(e) Data transfer to database.
(f) First level alarm validation.
(g) Service interrupt to host computer upon validated alarm/threshold exceedence.

The software required to execute these functions is self contained within each data acquisition/signal conditioning unit. Direct access and control are available to the system management software installed in the host computer. Standard industrial grade hardware with a proven in-field track record is utilised throughout.

3.3 Database

The database function is to accept, manage and manipulate conditioned and processed data from the data acquisition subsystem and from the keyboard. The database is accessed directly by the system management software and by the expert system. This is shown schematically in Figure 2.

3.4 System management software

The primary functions of the system management software are:

(a) Interface with operator (monitor/keyboard)
(b) Direct control of system hardware configuration (eg scan rates).
(c) Direct interrogation of all real time parameters.
(d) Direct access to database and expert system.
(e) Overall system control (passwords, file handling).
(f) Modification of expert system decision tools.

3.5 Expert system software

The primary function of the expert system software is to convert monitored data into meaningful information.

As mentioned previously there are three main types of system user, each with its own requirements:

(a) The plant operator whose prime requirement will be immediate on-line information to warn of current alarms.
(b) The maintenance manager who will require longer term, more detailed information in order to plan maintenance effectively.
(c) The specialist machinery engineer who will need much more detailed information on predicted faults in order to establish the causes of alarms and hence identify the necessary action.

This three level system is typical of many maintenance management structures.

4. DESIGN CONSIDERATIONS

4.1 Rule structure

Up to this point the description has been of a monitoring system that could be applied to many pieces of industrial plant. This section will detail the application of the general principles to a gas turbine driven compressor string as shown in Figure 3.

Compressor strings typically consist of a gas turbine driver, and one or more process compressors and gearboxes. However the physical construction of the component machines, and the instrumentation fitted varies tremendously between compressor strings. A troubleshooting engineer makes use of a lot of this information when diagnosing a machinery fault without perhaps giving the matter much conscious thought. Encapsulating this process in a computer system in a manner which allows any gas turbine driven compressor string to be accommodated is not a trivial problem, since the type and number of machines within the string may vary, as will the sensor fit.

The first step in tackling this problem is to subdivide the sensors fitted to each of the machines into a number of subgroups. Typical subdivisions for a gas turbine are shown in Figure 4. Each subdivision contains rules that are associated with that particular class of sensors. However, this process is not as straightforward as it may at first appear, and the final success of the system depends upon it. A simple example will be discussed in order to illustrate some of the principles involved.

For a gas turbine the major faults that are normally diagnosed from vibration data are:

(a) Unbalance
(b) Misalignment
(c) Bearing problems

At first sight the differentiation between these three may appear to be an easy problem. However, there are a number of key features that need to be incorporated in an expert system in order to achieve robust diagnoses. These are as follows:

(a) A knowledge of the construction of the particular type of machine.
(b) A knowledge of the detailed sensor fit.
(c) Corroboration.

Considering first the diagnosis of unbalance one might produce the prototype rule (triggered after a high level has been detected in a spectrum analysis).

> If frequency is 1R then fault is unbalance.
> (where R is the rotational speed of the shaft)

However, a large number of misalignment faults also produce once per rev. vibration. The principle difference is that for misalignment the vibration is more often axial (1). Hence in order to produce an on-line diagnosis of misalignment, the sensor orientation must be known. Modified rules for differentiation of misalignment and unbalance might be:

> If frequency is 1R and sensor direction is radial then fault is unbalance.

> If frequency is 1R and sensor direction is axial then fault is misalignment.

However, there are other ways of diagnosing misalignment from vibration spectra (2). These require a knowledge of the couplings, an example of one such simple rule would be:

> If frequency is 2R and coupling type is gear then fault is misalignment.

From the above it is apparent that in order to achieve successful differentiation between unbalance and misalignment, the expert system requires a knowledge of both the sensor orientations, and the machinery involved. This latter point extends to the diagnosis of bearing faults, since rolling element and journal bearings require completely different types of analyses. In principle rolling element bearing failures are easy to diagnose from vibration sensors, since the normal failure modes give rise to surface imperfections which cause the impulsiveness of the signal (as measured by, say, Kurtosis, or enveloping techniques) to increase. Sophisticated techniques such as enveloping require a detailed knowledge of the bearing geometry for an accurate diagnosis to be made. The simpler impulse detection techniques (such as Kurtosis) are non-discriminatory in that they indicate that impulses are being generated without identifying their precise cause. In such cases corroboration, either with another measure from the same sensor (e.g. the envelope spectrum) or with another sensor (e.g. magnetic chip detector) is highly desirable for error free diagnosis.

Journal bearings generally produce little vibration as a precursor to failure. For such bearings, a well placed temperature sensor may prove far more valuable. Journal bearing systems may however suffer from oil whirl, for which vibration sensors are ideal. Typically oil whirl produces a vibration component at a frequency just less than half the shaft frequency. Such vibrations are also observed in non-journal bearing systems (1), when the fault is normally attributed to 'looseness induced vibration'. Again a knowledge of both the detailed machinery configuration and the sensor locations are essential for accurate diagnosis of bearing related faults.

The above example demonstrates some of the principles involved in forming a robust rule-set for diagnosing gas turbine faults from vibration measurements. In practice, further rule-sets must be identified, both for the other sensor sets for the gas turbine module, and also for the other modules that comprise the machinery string.

4.2 Knowledge engineering

From the foregoing discussions it is clear that the performance of an expert system is critically dependent upon the knowledge that is embodied within it. Rules and procedures must be established which cover the known or anticipated behaviour of a machine in a given set of circumstances. The major sources of input to the knowledge base are therefore:

- Operational experiences of key personnel
- Maintenance manuals/operating handbooks and procedures
- Fault history/measurement records
- Equipment manufacturer/designer
- Cause and effect relationships.

This information is encapsulated in the knowledge base after the knowledge engineering phase. Generally an initial knowledge base is developed and then refined during operation. The ease with which knowledge bases can be altered plays a vital part in this refinement process.

4.3 Monitored parameters

Once the machine faults to be included in the knowledge base have been established, suitable measurable parameters must be determined which effectively detect development of those faults. Parameters must be related to the associated symptoms. In addition, they need to be stable and repeatable as well as being accessible from available sensors.

Some symptoms are more reliably determined by corroboration of more than one measurement (eg misalignment) while others rely on subsequent digital signal processing and parameter extraction (eg gear faults). The parameters to be monitored define the system's transducer and signal conditioning requirements.

5. CONCLUSIONS

This paper has described the design principles and the considerable advantages that can arise from the incorporation of an expert system within the framework of a condition monitoring system. Great emphasis however must be placed on the detailed design of the system, if any potential cost savings are not to be destroyed by high false or missed alarm rates.

The key features in the design process can be summarised:

(a) An efficient knowledge engineering exercise which defines both the knowledge base structure, and the parameters to be measured.
(b) A system hardware design that enables the required parameters to be accurately and repeatably measured, and the results passed to the database.
(c) The development of an expert system structure which uses the interrelationships between the parameters, in order to reduce the number of false alarms.
(d) The incorporation of procedural func-tions that can be controlled by the expert system. These functions include signal processing routines to extract more information from a raw signal, and the use of analytic models to perform complex computations involving a number of the measured parameters. Both of these techniques yield diagnoses that are more reliable than simple overall level monitoring.
(e) The rapid refinement and enhancement of the systems performance during its initial operating phase.

All of the above features have been incorporated into the current joint Acoustic Technology - MJA Dynamics projects.

References

1. J.S. Sohre. Turbomachinery problems and their correction. Sawyers' Turbomachinery Handbook. Turbomachinery Publication, Norwolk CT.

2. D.L. Dewell and L.D. Mitchell. Detection of a misaligned disk coupling using spectrum analysis. Journal of vibration, acoustics, stress and reliability in design. 1984, 106, (9) 9-16.

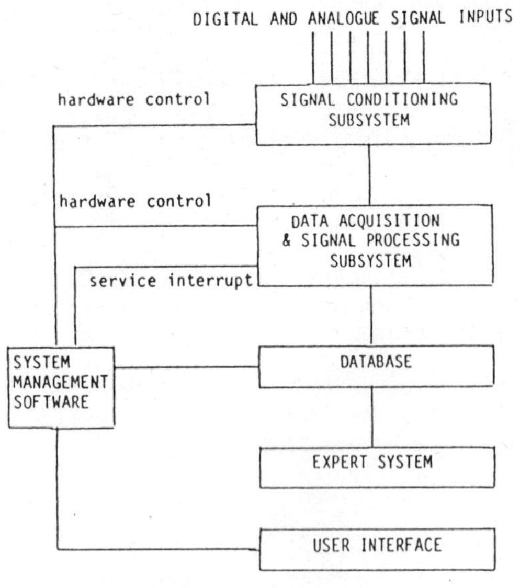

Fig 1 System functional blocks

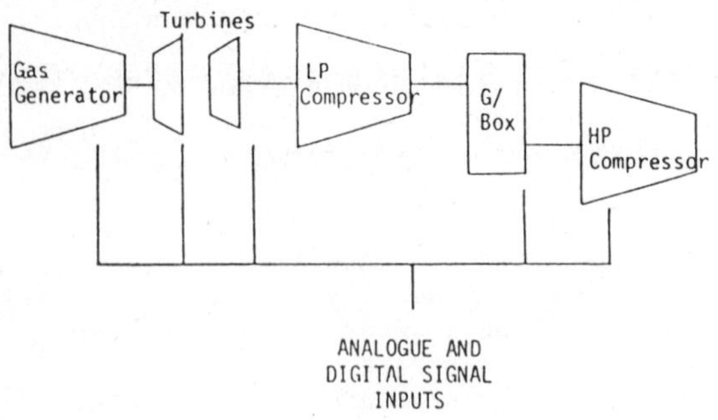

Fig 3 Typical gas turbine driven compressor string

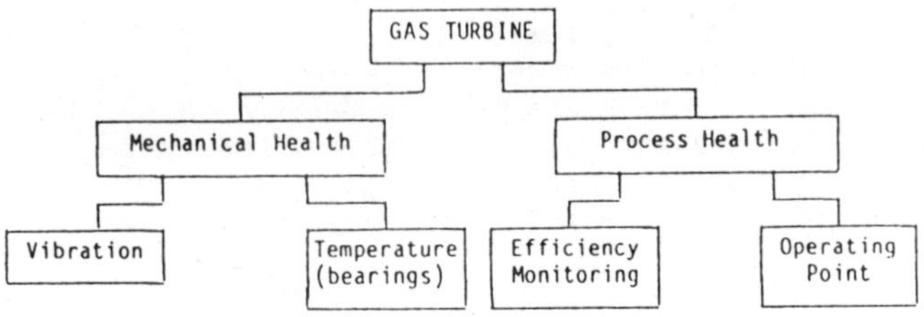

Fig 4 Generic rule groupings (based on sensor types)

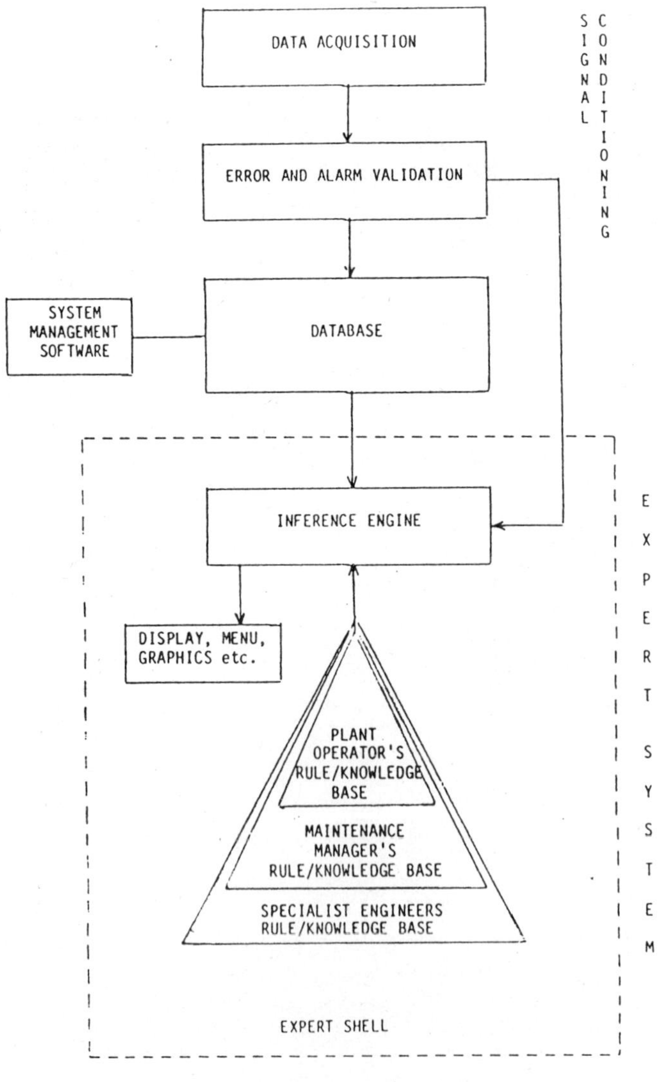

Fig 2 Schematic diagram of software structure

C254/88

A microcomputer-based vibration diagnostic system for steam turbines and generators

Y W KUO, MEng, MJSME
China Steel Corporation, Hsiao Kang, Kaohsiung, Taiwan, Republic of China

ABSTRACT: This paper describes a microcomputer-based vibration diagnostic system for turbine and generator in CSC (CHINA STEEL CORPORATION).

This on-line system can be served in many functions for turbine generator, such as the monitoring for a health condition, running-up and coasting-down and load operation status. Therefore, a precise prediction and diagnosis are easily achieved. Two successful examples are illustrated in this paper; one is generator unbalance and the other one is rubbing problem.

1. INTRODUCTION

In CSC, the power sources of all facilities and equipments are totally dependent upon the electric power supplied from power plant. Any trouble, such as the emergent shutdown of the steam turbine generator, might cause a great loss to the whole company, therefore, a good operation status for the generator is absolutely important.

Due to the improvement of technology in recent years, a new turbine generator is required to be operated in a higher temperature and pressure, larger capacity, better performance, multi-shaft, auto-control, and computerized. In other words, the goal of "High efficiency, low cost" will be the dominant tendency even in the future.

Consequently, the maintenance work becomes more challenging than before. Above all, it is even more harder to analyse and to solve the vibration problem, CSC spent a lot of money and tried to make every effort to study all the possible means to get rid of the unnecessary loss. Again, CSC provides that the previous breakdown maintenance can no longer be available, especially under the situations for lack of some supporting data. Furthermore, if the vibration conditions can be monitored and kept within the threshold value all the time, the power failure loss can easily be minimized. Therefore it is necessary to collect continuously the vibration data of the turbine generator in order to analyse or to diagnose at any time. In other words, it is essential to set up the computer-based vibration monitoring & diagnosis system.

The system in CSC power plant can provide a prediction for almost all accidental turbine generator shutdowns caused by abnormal shaft vibration during the load operation. Abnormal vibration behaviour can be detected at an early stage by continuously monitoring. It is also possible to judge automatically between types of abnormal vibration by comparing actual vibration characteristics with predetermined malfunction patterns which are memorized by the microcomputer. Furthermore, it can provide the one run balancing method to determine the correct weight and location of the unbalance.

2. SYSTEM STRUCTURE

This system consist of vibration sensor & phase meter, monitor racks, data acquistion instruments, central computer & periphery and software programs. See Fig. (1) for an overall block diagram.

2.1 Vibration block sensor & phase sensor

The vibration sensor is used to pick up vibration amplitude while the phase sensor is to pick up the phase. All the signal of amplitude and phase will be transfer to monitor rack.

2.2 Monitor rack

The monitor rack consists of indicator & backplane electronics. The indicator will indicate turbine generator vibration, such as ok, danger, alarm and vibration amplitube & phase etc. The backplane electronics which include multiplexing circuitry and buffered transducer output can provide the interface with the data acquistion instrument.

2.3 Data acquistion instrument

Data acquistion instrument is used to collect data from monitor rack, store the information and transfers this data to host computer. The collected data include steady-state, dynamic vibration data which are acquired while the turbine generator is operating at constant speed, and transient dynamic vibration informations, which are acquired during start-up and shutdown, as well as static vibration data which are presented in formats that show what is happening on the turbine generator.

The data acquistion instrument captures vector data and waveform data. All data can be loaded into the host computer system via a RS-232 interface for data reduction, display, calculation and comparing.

2.4 Computer

The computer system being offered for this system is based a MICRO VAX II computer, a basic host computer should include:

(a) MICRO VAX II computer.
(b) 5M Bytes RAM.
(c) 55M Bytes winchester disk.
(d) Floppy disk.
(e) Printer & Plotter.
(f) CRT & terminal.

The data acquistion instruments are used by the host computer to gather information from the monitor racks in the system, this takes the communications burden from the MICRO VAX. Allowing the MICRO VAX to dedicate its resources for storing calculating, comparing and retrieving data and servicing user requests.

2.5 Software

(a) Data acquistion program.
(b) Data trend program.
(c) Diagnosis & comparing program.
(d) Graphics program.
(e) Balancing support program.
(f) Reporting program.

3. SYSTEM FUNCTION

A microcompter is the center of the system, an overall block diagram of the function is shown in Fig. (2), which is used for the abnormal omen detection and cause diagnosis, the main functions of system are as following:

3.1 Pre-processing for auto-acquistion of data

There are four kinds of data to be used for monitoring and diagnosis, namely overall amplitude, waveform, spectrum, and vibration vector, which are processed by computer, those four kinds data are acquired during machine run up/ run down or load operation.

3.2 Detection for abnormal vibration

This subsystem is used for monitoring if there were abnormal omen

(a) Overall amplitude survey

Compare the vibration amplitude between the actual measured value and the limited value for each bearing point, it alarms when the former is higher than the latter.

(b) Frequency component survey

Refer to Fig. (3), it alarms when the component level of vibration spectrum exceeds that of limited spectrum window.

(c) Trend survey

Compare the trend growth rate between measured value and the limited value, it alarms as the measured trend growth rate exceeds the threshold value.

(d) Vector survey

Compare the vibration vector and normal vibration vector, it alarms as the difference of vector exceeds the limited region.

3.3 Reporting the status of turbine generator

It is divided into two categories regarding to the report of turbine generator

(a) On-line report

It is essential to have on-line report under continuous monitor and inspection condition which include (1) overall amplitude and trend (2) spectrum (3) waveform (4) vector in a form of map or value type are available as required.

(b) Periodic report

The printer prints out data automatically at certain selected period such as daily, weekly, monthly etc.

3.4 Diagnosis

This is the heart of whole system, including auto-diagnosis and interactive diagnosis.

(a) Auto-diagnosis

It executes auto-diagnosis as abnormality has been occurred and list the possible causes, because it judges each element by Table 1-(a) & Table 1-(b).

Table 1-(a): Seven (7) categories of vibration phenomena can be detected

(1) Operating status (Status Analysis)
Standing letter is "A", normally the operating status for generator can be divided into four (4) different situations:
1. Speed up or speed down
2. Constant speed
3. Oven speed
4. Turning
Example: "A1" Stands for the vibration signal of the machine when it is speeding up or coasting down.

(2) Vibration phase (Phase Analysis)
Standing letter is "B", there are two kinds of vibrating phase
1. Stable phase
2. Unstable phase
Example: "B1" Stands for a stable phase reading.

(3) Frequency (Frequency-Amplitude Spectrum)
Standing letter is "C", there are seven (7) types of frequency for generator
1. 1 x r/m synchronous frequency
2 Shaft natural frequency
3. 2 x r/m synchronous frequency
4. Vibration frequency of 120 Hz
5. N x r/m synchronous frequency
N = integer or N= fractional
6. Specified N x r/m synchronous frequency
7. Tortional natural frequency.
Example: "C1" Stands for the synchronous frequency of generator.

(4) Vibration location
Standing letter is "D". There are three (3) location in vibration measurement
1. Journal
2. Structure
3. Foundations
Example: "D1" Stands for a signal source from Journal.

(5) Time state (Time Wave Analysis)
Standing letter is "E" usually, the relationships between vibration signal and time can be classified into three types
1. Steady-state
2. Non-steady state
3. Special period status
Examples: "E1" Stands for a steady-state base.

(6) Correlated parameters
Standing letter is "F". There are seventeen (17) parameters in total
1. Steam temperature change
2. Steam pressure change

3. Alignment change
4. Load change
5. Generator voltage change
6. Generator current change
7. Exciter current change
8. Oil temperature change
9. Oil pressure change
10. Orifice change
11. Vacuum change
12. Seal oil pressure change
13. Turbine vibration change
14. Turbine-generator coupling change
15. Generator-exciter coupling change
16. Gas cooler single side stop change
17. Start-stop change.

Example: "F1" is a condition which is related with steam temperature deviation.

(7) Vibration category
(Vibration Mode Analysis)
Standing letter is "G" basically there are four (4) types of vibration
1. Force vibration
2. Resonance
3. Self-excited vibration
4. Tortional vibration.

Example: "G1" is a force vibration caused by pre-load, such as unbalance.

Table 1-(b): This is a basic supporting data for auto-diagnosis, it is based on the results of that seven (7) vibration analysis listed in Table 1-(a).

A scanning comparison is processed in Table 1-(b), the computer will automatically indicate the compared results and print out the "cause of vibration", a manual analysis is required for further identification. For instance, if the scanning results from computer are A1, B1, C1, D1, E1 and G1, respectively, a printing cause of vibration is definitely "unbalance".

(b) Interactive diagnosis

This is for detailed judgement from the result of auto-diagnosis. It executes amplitude measurement, waveform analysis, spectrum analysis, vector analysis, and then identifies by manual analysis.

3.5 Computer-aided balancing system

Almost 80% of the vibration causes are due to unbalance; therefore the computer-aidedbalancing system will be helpful for auto-diagnosis.

The lag angle between high spot and heavy spot is defined as characteristic angle, and the relation between balancing weight and vibration amplitude are defined as influence coefficient for each bearing of turbine generator can be determined by previous experience and pre-stored in computer for application, this is so called "one run balancing technique" [1] and provides to be a successful example for the computer-aided balancing system.

4. EXAMPLES

The following two examples are illustration of the vibration diagnostic system, describing about No. 3 turbine generator at power plant, with the specifications given in Table (2).

Table (2)

Item	Specification
Rated load	50 MW
Rated speed	3600 rpm
Rotor system	Tandem compound type consisted of one turbine and one generator

4.1 Unbalance

Fig (4) Shows the shaft vibration monitored by the trend function system, the vibration amplitude at No. 4 Brg. exceeds the limited value and alarms, meanwhile, the diagnostic function system shows a diagnostic result of "unbalance" and gives a waveform & spectrum diagram as Fig. (5), a vector diagram as Fig. (6) and the computer-adid balancing system gives a correction weight 105 grams and the angle 130 degrees on the correction plane to indicate where it should be attached. Fig. (7) show the polar plot for "before" and "after" balancing.

4.2 Rubbing

Another example is an illustration of the same turbine generator that was tested during an annual inspection under turbine-startup conditions after replacement of oil retainer.

Figure (8), (9), (10) show the waveform, spectrum, vector diagram of the amplitude and phase at No. 2 Brg. The diagnostic function system gives that this abnormal vibration is due to a rubbing malfunction, because the phase is untable, and move in a direction opposite to the rotor rotation, and the spectrum have ½ x r/m component. After, this auto-diagnosis result is given, and manual inspection confirms that the oil retainer has rubbed the rotor shaft.

5. CONCLUSION

This paper describes a microcomputer-based vibration diagnostic system which was effectively used in CSC. When a significant change in vibration is noticed by the continuous monitoring system, the diagnostic program functions automatically, the program performs a sweep on all elements of the pre-memorized causes, and identify which one is the most possible.

The one run balance method which has been applied in this system is very accurate because the correction weight and location of unbalance are obtained from previous experience on the features and characteristics of each turbine generator.

6. REFERENCE

1. K. M. Wu. "Field Balancing in One Run" Technique in CSC, China Steel Technical Report Vol. 1, 1987.
2. Y. W. Kuo. "Automatic Monitoring and Diagnostic on Vibration in Rotating Machinery", Journal of China Mechamical Engineering Vol. 130, 1985.
3. Toshio Toyota, Kanji Maekawa, Satoshi Nakashima, "Automatic Diagnosis for Malfunctions of Rotating Machinery By Falt-Matrix, Japan Soc. of Prec. Engg. Vol. 16 No. 2, 1982.

Table 1-(a). Diagnostic chart

	A Operating Status	B Vibration Phase	C Frequency	D Vibration Location	E Time State	F Correlated Parameter	G Vibration Category
1	Speed Up or Speed Down	Stable Phase	1 × r/m	Journal	Steady-State	Steam Temp. Change	Force Vibration
2	Constant Speed	Unstable Phase	Shaft Natural Frequency	Structure	Non-Steady	Steam Pressure Change	Resonance
3	Over Speed		2 × r/m	Foundation	Special Period	Alignment Change	Self-Excited Vibration
4	Turnning		120 Hz			Load Change	Tortional Vibration
5			n × r/m			Generator Voltage Change	
6			Special Constant r/m			Generator Current Change	
7			Tortional Natural Freq.			Exciter Current Change	
8						Oil Temp. Change	
9						Oil Pressure Change	
10						Orifice Change	
11						Vacuum Change	
12						Seal Oil Pressure Change	
13						Turbine Vibration Change	
14						T. G. Coupling Change	
15						G. Ex. Coupling Change	
16						Gas Cooler Single Side Stop Change	
17						Start-stop Change	

Table 1-(b) Diagnostic chart

Vibration Phenomenon							
Operation Status	Phase	Frequency	Vibration Location	Time State	Correlated Parameter	Vibration Category	Cause of Vibration
A1 A2 A3	B1	C1	D1	E1		G1	Unbalance
A1 A2 A3	B1	C1	D1 D2	E1	F1 F2	G1	Structure is too weak
A1 A2 A3	B1	C1 C6	D2	E1	F2 F3	G2	Structure is resonance
A1 A2 A3	B1	C6	D1 D2	E1		G1	Foundation is too weak
A1 A2 A3	B1	C1 C6	D3	E1		G2	Foundation is resonance
A1 A3	B1	C2	D1	E1	F10 F11	G1	Misalignment
A1 A2 A3	B1	C1	D1	E1	F4 F14 F15	G1	Coupling is bad
A2		C4	D2	E1	F2 F3 F5	G1	Spring support is bad
A1 A2			D2	E2	F1 F3	G1	Frame expansion is constrained
A1 A2 A3	B1	C1	D1	E1	F1 F2 F3 F16	G1	Frame is deformation
A1 A2 A3	B1	C3	D1	E1		G1	Cross-Slot arrangement is bad
A2 A3			D2	E2	F3 F8 F9 F10 F11	G3	Oil supply is lacking
A2 A3	B1	C2	D1	E1	F3 F8 F9 F10 F11	G3	Oil supply is over
A2		C4	D2	E1	F1 F5	G1 G2	Core struture is bad
A2		C4	D2	E1	F2 F3 F5	G1 G2	Frame stiffness is too weak
A1 A2 A3	B2	C1	D1	E2 E3	F2 F12	G1	Seal lubrication is bad
A1 A2 A3	B2	C1	D1	E2 E3	F2 F3	G1	Oil retainer is rubbing
A1 A2 A3	B1	C1	D1	E1	F14 F15	G1	Run out is too large
A1 A2	B1	C1	D1	E1	F1 F8 F9	G1	Shaft material is bad
A1 A2 A3	B1	C5	D1	E1		G1	Journal is deformation
A4		C7		E1	F3 F8 F9	G3 G4	Turning lubrication is bad
A1 A2 A3	B2	C1	D1	E1	F17	G1	Floating part is used
A1 A2 A3	B2	C1	D1	E1	F1 F5 F17	G1	Thermal interference is bad
A1 A2 A3	B2	C1	D1	E1	F17	G1	Contaminant is adhesive
A2	B1	C4	D1	E1	F1 F3 F5	G1	Comperession ring arrangement is bad
A2	B1	C1		E1	F7	G1	Gap unbalance
A2	B1	C1	D1	E1	F7	G1	Rotor coil is layer short
A2	B1	C1	D1	E1	F7	G1	Rotor coil is earthing
A2	B1	C1	D1	E1	F4 F6	G1	System load is unbalance
A2	B1	C1	D1	E1	F4 F6	G1	Connection is mistake
A2	B1	C1	D1	E1	F6	G1	Surface loss is unbalance
A2	B1	C1	D1	E1	F7	G1	Rotor wedge cooling is unbalance
A2	B1	C1	D1	E1	F1 F3	G1	Stator cooling is unbalance
A2	B1	C1	D1	E1	F7 F1	G1	Rotor coil sliding is unbalance
A2	B1	C1	D1	E1	F7 F1	G1	Clippage block sliding is unbalance
A2	B1	C1	D1	E1	F6 F7	G1	Rotor surface varnish is non-uniform
A2	B1	C1	D1	E1	F7	G1	Rotor coil hole is plugged

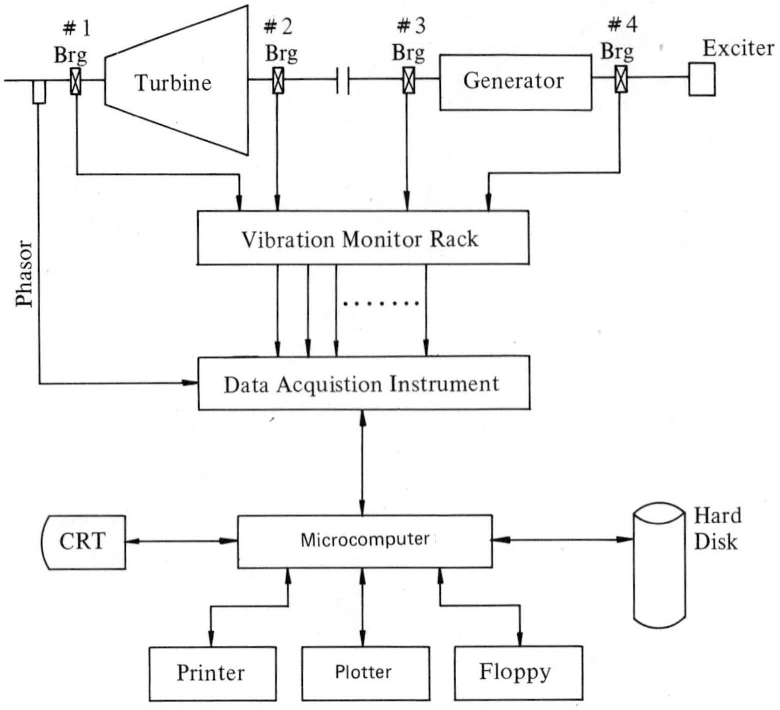

Fig 1 System structure

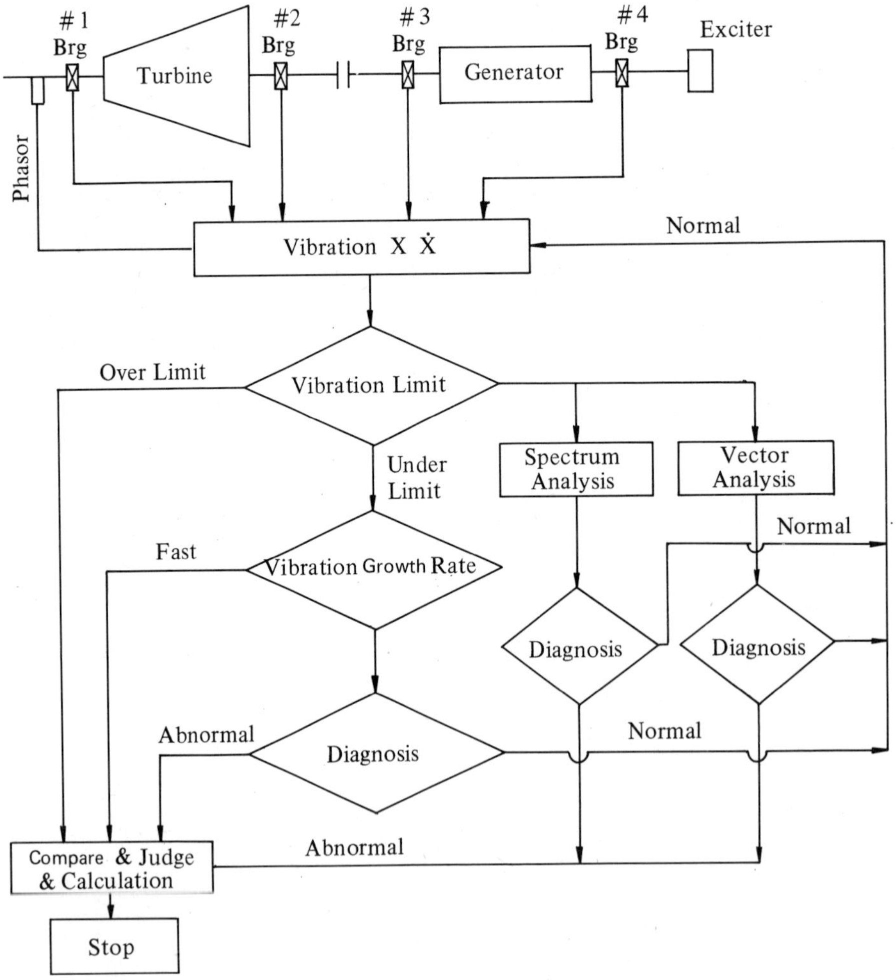

Fig 2 System function

(Exciter is a part of the generator, which can excite the field windings of the generator, and provide a magnetic field for electric generation)

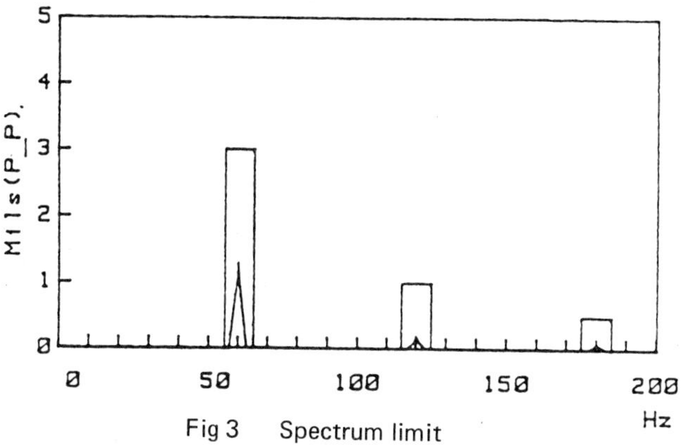

Fig 3 Spectrum limit

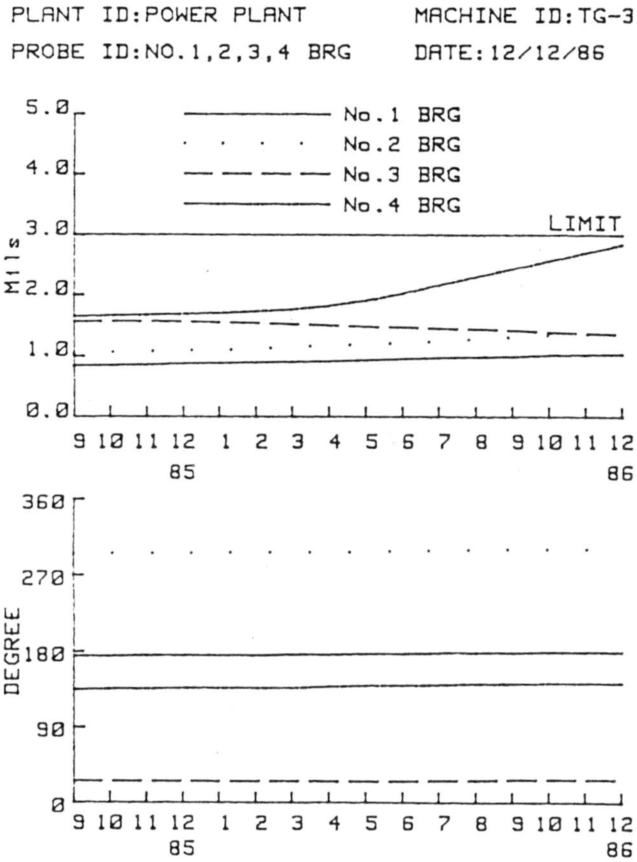

Fig 4 Trend for TG-3

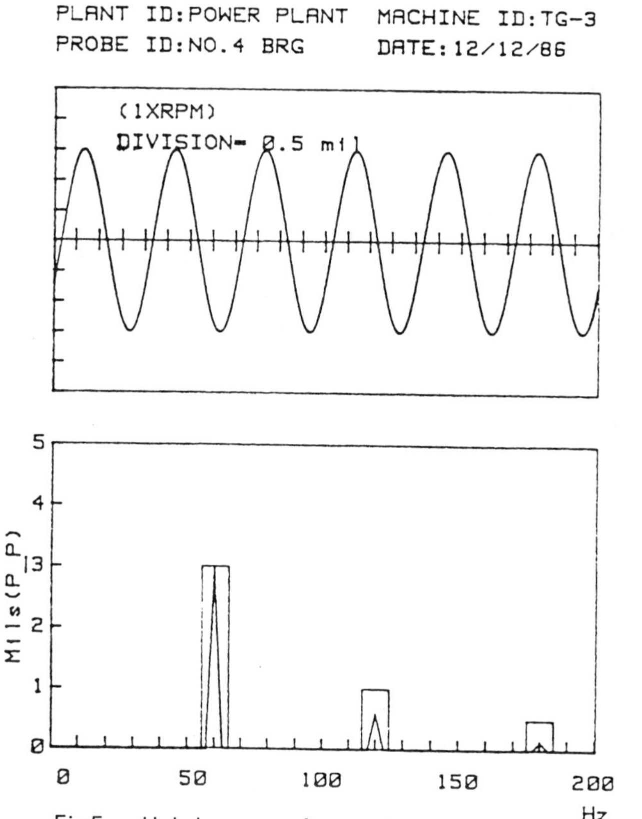

Fig 5 Unbalance waveform and spectrum

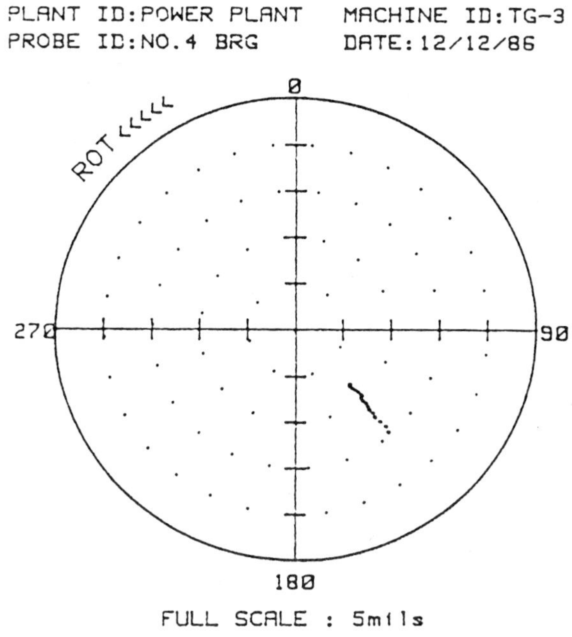

Fig 6 Vector diagram for TG-3

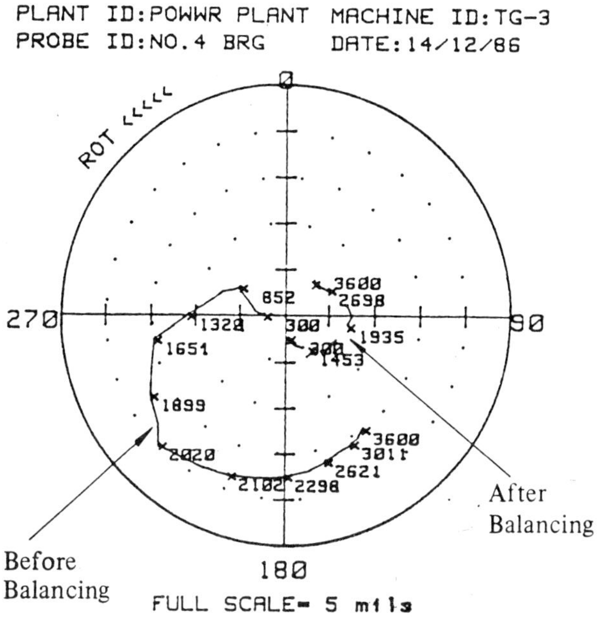

Fig 7 Polar plot for TG-3

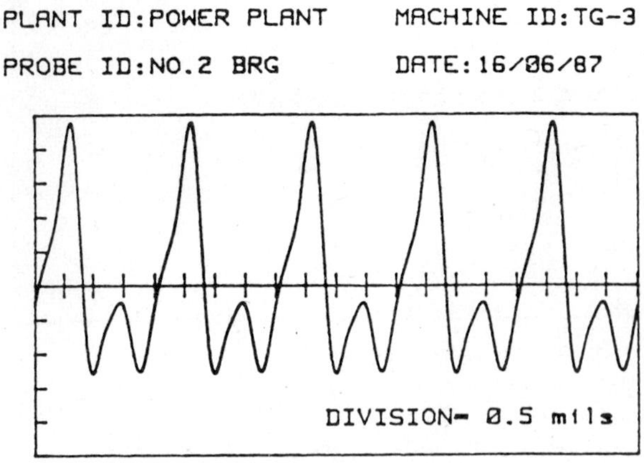

Fig 8 Rubbing waveform for TG-3

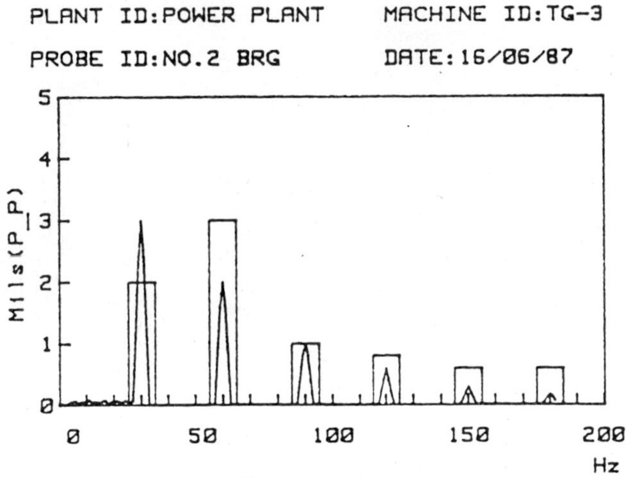

Fig 9 Rubbing spectrum for TG-3

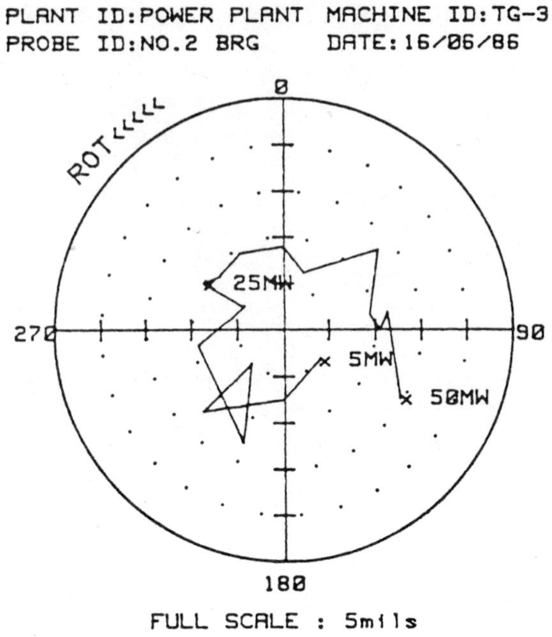

Fig 10 Rubbing vector diagram for TG-3

C314/88

Dynamic behaviour of the Laval rotor with a cracked hollow shaft—a comparison of crack models

R GASCH, Dr-Ing, **M PERSON**, Dr-Ing and **B WEITZ**, Dipl-Ing
Technical University of Berlin, West Berlin

1 INTRODUCTION

In 1976 three papers dealing with the dynamic behaviour of a simple Laval rotor with a transverse crack were presented at the Cambridge Conference on Vibrations in Rotating Machinery (Mayes and Davies |1|, Henry and Okah-Avae |2|, Gasch |3|. Nearly all phenomena of this system were revealed at that time

- the wide range of instability near the Laval-critical speed, $\Omega \simeq \omega_o$, and the unstable speed range near $\Omega \simeq 2\,\omega_o$, see |5|,
- the parameter resonances (further unstable zones) at Ω/ω_o = 2/3; 2/4; 2/5 etc.
- the effect that - according to its angular direction - the mass unbalance may keep the crack close or open,
- the fact that there are speed ranges with hardly any difference in the dynamic behaviour of a rotor with and a rotor without crack.

Later papers laid emphasis on the finite element analysis of large turborotors with a crack |6-13|. But nearly all authors used simplified crackmodels ignoring the details of opening and closing during one revolution of the shaft. The finite element analysis was made easier by the discovery of the weight dominance |7, 8, 10| with horizontal rotors. Assuming small vibrations compared to the deflection due to weight the originally non-linear equations of motion change to linear, periodically time-variant equations.

Nearly all authors use this simplification for the calculation of the f o r c e d v i b r a t i o n s in the linearized system. Only in |14| a systematic stability-analysis of the Laval-rotor was intended using a method similar to Floquets. Unfortunately the crack model used there is somewhat strange.

Taking the classical Laval-rotor with a c r a c k e d t h i n - w a l l e d h o l - l o w s h a f t this paper tries:

- to illuminate in detail the opening and closing procedure in the crack surface (weight dominance, classical beam theory),
- to obtain the local crack flexibilities from a stress analysis of the residual problem via Vlassouv's theory of thin walled shells,
- to carry out a systematic stability analysis using Floquet's method in order to get a survey on the influence of c r a c k - d e p t h , d a m p i n g and r o t a - t i o n a l s p e e d .
- to present a systematic stability analysis even for the simplified crack models,
- to compare the results of the three different crack models analysed.

2 THE EQUATIONS OF MOTION OF THE LAVAL-ROTOR WITH A CRACK

As we know from |1-5| the Laval-rotor with a cross-sectional crack in its shaft can be described by equations of motion of the following type

$$\begin{bmatrix} m & \\ & m \end{bmatrix}\begin{bmatrix} \ddot{w} \\ \ddot{v} \end{bmatrix} + \begin{bmatrix} d & \\ & d \end{bmatrix}\begin{bmatrix} \dot{w} \\ \dot{v} \end{bmatrix} + \begin{bmatrix} s_{11} & s_{12} \\ s_{21} & s_{22} \end{bmatrix}\begin{bmatrix} w \\ v \end{bmatrix} =$$

$$\underline{M}\;\ddot{\underline{u}} + \underline{D}\;\dot{\underline{u}} + \underline{S}(\underline{u},t)\;\underline{u} = \tag{1}$$

$$\begin{bmatrix} mg \\ 0 \end{bmatrix} + \varepsilon\, m\Omega^2 \begin{bmatrix} \cos(\beta+\Omega t) \\ \sin(\beta+\Omega t) \end{bmatrix}$$

$$\underline{p}_o \quad + \quad \underline{p}_u$$

w and v are the displacement coordinates of the center of the shaft in the plane of the disc, fig. 1. As parameters we have mass m or weight mg, damping coefficient d, the eccentricity ε of the center of gravity and its direction β (related to the crack centerline). The stiffness s_{ik} of the shaft is time depentent and non-linear.

The stiffness-matrix $\underline{S}(\underline{u},t)$ may be split up into

$$\underline{S}(\underline{u},t) = \underline{S}_o + \underline{\Delta S}(\underline{u},t) \tag{2}$$

where $\underline{S}_o$ is the linear, time independent stiffness-matrix of the uncracked shaft. $\underline{\Delta S}$ - the aditional (negative) crack stiffness - depends on the deflections $\underline{u}(t)$ and the position angle $\psi = \Omega t$. The rotor response $\underline{u}(t)$ also can be split up into

$$\underline{u} = \underline{u}_o + \underline{\Delta u}(t) \tag{3}$$

where

$$\underline{u}_o = \underline{S}_o^{-1}\,\underline{p}_o = \begin{bmatrix} mg/s_o \\ 0 \end{bmatrix} \tag{4}$$

is the deflection of the uncracked shaft due to weight and $\underline{\Delta u}(t)$ is the vector of the vibrational behaviour. Assuming the dominance of weight, i.e. $\Delta\underline{u}(t) << \underline{u}_o$, as suggested by Grabowski, Mayes and Bachschmid, we find the modified equations of motion by introducing (4), (3) and (2) into (1)

$$\underline{M}\,\Delta\underline{\ddot{u}} + \underline{D}\,\Delta\underline{\dot{u}} + [\underline{S}_o + \Delta\underline{S}(t)]\,\Delta\underline{u} = -\Delta\underline{S}(t)\,\underline{u}_o + \underline{p}_u \qquad (5)$$

These equations are linear, but periodically time-variant. On the right hand side in addition to the unbalance excitation we find the weight excitation $-\Delta\underline{S}(\underline{u}_o,\Omega t)\underline{u}_o$. As we introduced the deflection $\underline{u}_o$ instead of the general deflection $\underline{u}(t) \simeq \underline{u}_o$ this is a linear expression. The time-dependance is governed by the angle of rotation $\phi = \Omega t$.

The stability of this system can be analyzed using Floquet's theory. In case of stable solutions the calculations of forced vibration may be found from the time-invariant system

$$\underline{M}\,\Delta\underline{\ddot{u}} + \underline{D}\,\Delta\underline{\dot{u}} + \underline{S}_o\,\Delta\underline{u} = -\Delta\underline{S}(\underline{u}_o,t)\underline{u}_o + \underline{p}_u \qquad (6)$$

-provided that the resonance peaks are significantly smaller than the weight deflection $\underline{u}_o$.

The stiffnes matrix $\underline{S}(t) - \underline{S}_o + \Delta\underline{S}(t)$ of three different crack-models is presented in fig. 8.

3 STABILITY ANALYSIS FOLLOWING FLOQUET

The linear homogeneous differential-equation (5)

$$\underline{M}\,\Delta\underline{\ddot{u}} + \underline{D}\,\Delta\underline{\dot{u}} + \underline{S}(t)\,\Delta\underline{u} = \underline{0} \qquad (7)$$

reveals its behaviour concerning stability within one period T.

For the Floquet-analysis we first have to calculate the transition matrix $\underline{\Phi}(T)$ of the system. This matrix $\underline{\Phi}(T)$ tells us how the state vector $\underline{x}^T = \{\Delta\underline{u}^T, \Delta\underline{\dot{u}}^T\}$ of the system has changed after one period ($t = T$):

$$\begin{bmatrix} \Delta w \\ \Delta v \\ \Delta\dot{w} \\ \Delta\dot{v} \end{bmatrix}_{t=T} = \underline{\Phi}(T) \begin{bmatrix} \Delta w \\ \Delta v \\ \Delta\dot{w} \\ \Delta\dot{v} \end{bmatrix}_{t=0} \qquad (8)$$

$$\underline{x}_T = \underline{\Phi}(T)\quad \underline{x}_o$$

$\underline{x}_o$ are the initial conditions. $\underline{x}_T$ is found by numerical integration of the equations of motion.

Following Floquet we then ask of which kind is the proportionality between $\underline{x}_T$ and $\underline{x}_o$,

$$\underline{x}_T = \mu\,\underline{x}_o\ . \qquad (9)$$

Introducing (9) in (8) we get an eigenvalue problem

$$(\Phi(T) - \mu\underline{I})\underline{x}_o = o \qquad (10)$$

which gives us an answer.

The vibrations fade away if all eigenvalues of (10) have a magnitude less than one, $|\mu_k| < 1$. Then the system is stable. In case there are one or more eigenvalues with a magnitude larger than one, $|\mu_k| > 1$, the system is unstable. $|\mu_k| = 1$ is the borderline of stability. $|\mu_k|$ is the factor of increase or decrease of the amplitudes during one period T, that is one revolution of the shaft.

4 CRACK MODELS

4.1 The hinge model

Instead of the real crack in the shaft a hinge mechanism is introduced. When the hinge is closed the shaft has its original flexibility h_o. When the hinge is open the additional spring adds the (maximum) crack flexibility Δh_ζ, which depends on the crack depth and can be determined by experiments or by analytical calculations, fig. 2.

In rotating coordinates the hinge leads to the flexibility matrix

$$\begin{bmatrix} w_\zeta \\ v_\eta \end{bmatrix} = \begin{bmatrix} h_o+\Delta h_\zeta & 0 \\ 0 & h_o \end{bmatrix} \begin{bmatrix} F_\zeta \\ F_\eta \end{bmatrix} \qquad (11)$$

$$\underline{\underline{u}} \;=\; \underline{\underline{H}} \qquad \underline{\underline{f}}$$

with $\Delta h_\zeta = 0$ as long as $F_\zeta \leq 0$ ($w_\zeta \leq 0$) and $\Delta h_\zeta \neq 0$ for $F_\zeta > 0$($w_\zeta > 0$). The flexibility matrix is b i - l i n e a r . Whether Δh_ζ is zero or not depends on the direction of F_ζ.

Transferring the deflections $\underline{u}^T = \{w_\zeta, v_\eta\}$ from the rotating coordinates to fixed coordinates with the deflections $\underline{u}^T = \{w_z, v_y\}$ using the transformation matrix

$$\underline{T} = \begin{bmatrix} \cos\Omega t & -\sin\Omega t \\ \sin\Omega t & \cos\Omega t \end{bmatrix} \qquad (12)$$

we find the flexibilities in an inertial frame $\underline{H} = \underline{T}^{-1}\,\underline{H}\underline{T}$. Inverting $\underline{H}$ we get the stiffness matrix $\underline{S}(\underline{u},t) = \underline{H}^{-1}$ needed for the equation of motion

$$\underline{S}(\underline{u},t) = \begin{bmatrix} 1 + \frac{h_\zeta}{h_o}\sin^2\Omega t & -\frac{\Delta h_\zeta}{h_o}\sin\Omega t\,\cos\Omega t \\ -\frac{\Delta h_\zeta}{h_o}\sin\Omega t\,\cos\Omega t & 1 + \frac{\Delta h_\zeta}{h_o}\cos^2\Omega t \end{bmatrix} \frac{1}{h_o(1+\frac{\Delta h_\zeta}{h_o})} \qquad (13)$$

With the assumption of weight dominance in the shaft deflections the non-linear properties are lost, because $\underline{u}(t) \simeq \underline{u}_o$, thus $\underline{S}(\underline{u},t) \rightarrow \underline{S}(\underline{u}_o,t)$. The hinge opens at 90° and closes at 270°. In fig. 8 the stiffness-coefficients s_{ik} for the hinge model are presented.

Although we do not intend to discuss here the forced vibrations of the system caused by the weight involuence, eq. (5), we will split up the stiffness matrix according to equations (2).

$$\underline{S} = \begin{bmatrix} 1/h_o & 0 \\ 0 & 1/h_o \end{bmatrix} + \frac{\Delta h_\zeta}{h_o^2}\left(\begin{bmatrix} -1 & 0 \\ 0 & -1 \end{bmatrix} + \begin{bmatrix} \sin^2\Omega t & -\sin\Omega t\cos\Omega t \\ -\sin\Omega t\cos\Omega t & \cos^2\Omega t \end{bmatrix} \right)$$

uncracked shaft   crack component

$$\underline{S} = \underline{S} \quad + \quad \Delta\underline{S}(\underline{u},t)$$

4.2 Improved crack models

Concerning the opening and closing of the crack surface during one revolution authors like |6, 7, 8, 13| made more or less plausible assumptions. In general it was felt that the closure of the hinge model is too abrupt for deeper cracks. Mayes suggested in |9| to assume an opening procedure following the (steady) law $K(t) = (1-\sin\Omega t)/2$ instead of

$$K(t) = \begin{cases} 1 \text{ for } 0 < t < 90^{\circ} \text{ and } 270^{\circ} < \Omega t \quad 360^{\circ} \\ 0 \text{ for } 90^{\circ} < \Omega t < 270^{\circ} \end{cases}$$

following the hinge model. Bachschmid |10| recommended a more consistent way: to balance for each time step the bending moment in the shaft with the stresses in that area of the cracked cross section that is able to bear. However, for a circular cross-section this leads to an iterative (time-consuming) numerical procedure.

Concerning the additional (local) crack-flexibilities many authors made more or less plausible assumptions of an "equivalent slot-length and slot-depth". Mayes |9| and Dimarogonas |14| proposed a more consistant way: to take the residual flexibilities of the cracked shaft from a 3-dimensional stress-strain analysis or from experimental results.

5 THE HOLLOW SHAFT WITH A CRACK

5.1 Opening and closing of the cracked cross-section

Starting with the usual assumptions of the beam theory we consider a hollow shaft (wall-thickness t, mean radius a, t << a) that has to transfer the bending M_g caused by weight over the cracked cross-section, fig. 3. The ζ^*-η^*-coordinates are based at the center of gravity C of the symmetric cross-section; the ζ-η-coordinates are based at M in the middle of the circle.

Turning now the cross-section into the direction of a larger angle φ the crack begins to close as soon as its leading edge passes from the zone of tension to the zone of compression. The angle φ where the stress at the leading edge becomes zero, $\sigma_x = 0$, can be found with the aid of beam theory. For a symmetric profile under bending moments the stress is

$$\sigma_x = -\frac{M_\eta^*}{I_\eta^*}\zeta^* + \frac{M_\zeta^*}{I_\zeta^*}\eta^* \tag{15}$$

The bending moments caused by the weight are

$$M_\zeta^* = M_\zeta = -M_g \sin\varphi, \quad M_\eta^* = M_\eta = -M_g \cos\varphi \tag{16}$$

The (area) moments of inertia (in ζ-η-coordinates), the area F and the distance s between M and C are:

$$I_\eta = \int \zeta^2 dF = ta^3 \left[\pi-\alpha_o-0.5\sin 2\alpha_o\right]$$

$$I_\zeta = \int \eta^2 dF = ta^3 \left[\pi-\alpha_o+0.5\sin 2\alpha_o\right] \tag{17}$$

$$F = 2(\pi-\alpha_o)t\,a$$

$$s = a\sin\alpha_o/(\pi-\alpha_o)$$

Thus we find $I_\zeta^* = I_\zeta$ and $I_\eta^* = I_\eta - s^2F$. The leading edge's coordinates are

$$\zeta^* = a\cos\alpha_o + s; \quad \eta^* = a\sin\alpha_o. \tag{18}$$

Introducing the equations (16), (17) and (18) into (15) and demanding that $\sigma_x = 0$, we get an equation for the angle φ at which the crack begins to close

$$\tan\varphi = \left(\frac{\cos\alpha_o}{\sin\alpha_o} + \frac{1}{\pi-\alpha_o}\right) \times \left(\frac{\pi-\alpha_o+0.5\sin 2\alpha_o}{\pi-\alpha_o-0.5\sin 2\alpha_o-(2\sin^2\alpha_o/(\pi-\alpha_o))}\right) \tag{19}$$

After the beginning of the closure ($\alpha < \alpha_o$) the profile is still a symmetrical one. So we may again find the position of the new centerline $\varphi = \varphi(\alpha)$ of the diminished crack from our formula, eq. (19), see fig. 4.

For our rotor dynamic analysis the reference angle of rotation $\psi = \Omega t$ refers to the old centerline. So we have to take into account the shift δ between old and new centerline

$$\psi = \varphi + \delta \quad \text{or} \quad \psi = \varphi(\alpha) + (\alpha_o-\alpha) \tag{20}$$

Fig. 5 shows the graph of the "breathing" of the crack during rotation as found from eq. (19) and (20).

From this diagram we learn already:

- o Whilst the hinge model closes abruptly at $\psi=90^{\circ}$ and opens abruptly at 270° the "real" crack closes and opens steadily over a wide range. But for small cracks the assumption of the hinge model is not so bad.
- o The principal axes of inertia are not fixed in the rotating coordinates ζ-η. They are swaying within an angle of $\alpha_o/2$.

5.2 The flexibilities of the open crack

The flexibilities of the cracked hollow shaft due to the bending moments X are found by superimposing two problems, fig. 6. First we have to find the deflections of the shaft due to the influence of external moment X and an additional internal moment X* that keeps the crack closed; this problem I is solved by the classical beam theory. Next we have to solve the residual problem II: We have to find the inclinations of the stress-free boundaries of the cracked shaft caused by the internal moments X*. We did this with the aid of Vlassow's theory of thin-walled shells |15|. As these flexibilities $\hat{h}_{main}$, $\hat{h}_{cross}$ are local (St. Venant) they do not depend on the beam length but only on the ratio t/a and Poisson's ratio ν. The results presented in the diagram, fig. 6b, were confirmed by some little experiments. These dimensionless flexibilities of the cracked hollow shaft are similar to those given by I. Mayes in |9| for the full circular cross-section.

5.3 Transformation of the crack flexibilities h_{main} and h_{cross} into the inertial frame

For the open crack the principal axes of inertia are the ζ-η-axes, see fig. 6b. To transform the (dimensionless) crack flexibilities $\hat{h}_{main}(\alpha_o)$, $\hat{h}_{cross}(\alpha_o)$

$$\underline{\underline{H}}_{crack} = \begin{bmatrix} \hat{h}_{main} & \\ & \hat{h}_{cross} \end{bmatrix} \qquad (21)$$

into the inertial coordinates we need the transformation matrix $\underline{T}$, eq. (12), and get

$$\underline{H}_{crack} = \underline{T}^{-1}\,\underline{\underline{H}}_{crack} \cdot \underline{T}\,. \qquad (22)$$

The argument of the sine and cosine terms in the transformation matrix $\underline{T}$ is the angle $\phi = \Omega t$.

This changes as soon as the crack commences to close, $\alpha < \alpha_o$.

- We now have to take the flexibilities $\hat{h}_{main}(\alpha)$, $\hat{h}_{cross}(\alpha)$ for the actual opening angle α.
- And we have to take the angle φ instead of ϕ as transformation angle because the principal axis swayed from ζ to ζ' which is the new centerline.

In fig. 7a, b, c the dimensionless flexibilities $\hat{h}_{ik}$ of the cracked shaft are presented for the "crack-depth" $\alpha_o = 30^o$, 60° and 90°. For comparison the crack flexibilities $\hat{h}^s_{ik}$ of the hinge model - making use of only the main flexibility of the fully open crack $\hat{h}_{main}(\alpha_o)$ - are shown by the dashed lines.

From these figures we learn: The brutal simplification of the hinge model mainly affects the coefficient with subscript 22 (main horizontal flexibility). With the hinge model there is a jump of this coefficient h_{22} at 90° and 270° degrees. With the more sophisticated crack model $\hat{h}_{22}$ is steady at 90° and 270° degrees but there is still a very sharp drop.

5.4 The complete flexibility matrix $\underline{H}(t)$ of the rotating hollow shaft with a breathing crack and its inverse $\underline{H}^{-1} = \underline{S}(t)$

We assume that both the disc mass m and the crack are situated in the middle of a (uniform) elastic shaft with the length 2 ℓ. The deflections in the middle of the shaft are superimposed from the deflections of the uncracked shaft plus the additional contribution of the crack flexibilities. But before superposition we have to bear in mind that the crack flexibilities were defined as "inclination angle ρ caused by bending moment X", see fig. 6, so we first have to transform these flexibilities into "deflections caused by forces" in the middle of the shaft. Doing so, we finally get

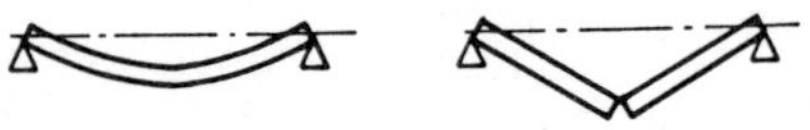

$$\begin{bmatrix} \Delta w \\ \Delta v \end{bmatrix} = \underbrace{\left(\frac{\ell^3}{6EI}\begin{bmatrix} 1 & 0 \\ 0 & 1 \end{bmatrix} + \frac{\ell^2}{4Eta^2}\begin{bmatrix} \hat{h}^{II}_{11} & \hat{h}^{II}_{12} \\ \hat{h}^{II}_{12} & \hat{h}^{II}_{22} \end{bmatrix}\right)}_{\underline{H}(t)} \begin{bmatrix} \Delta F_z \\ \Delta F_y \end{bmatrix} \qquad (23)$$

where $I = \pi t a^3$. The dimensionless flexibilities $\hat{\underline{H}}$ of the crack are shown in fig. 7. The stiffnes matrix $\underline{S}(t)$ is now found by inversion $\underline{S}(t) = \underline{H}^{-1}(t)$. In fig. 8 we find the time varying coefficients of the stiffness-matrix for

- the hinge model c_{ik}
- the hollow shaft with a crack s_{ik} and
- the model suggested by I. Mayes k_{ik}.

In order to keep the presentation dimensionless the flexibilities $\underline{H}(t)$ were multiplied by $6\,EI/\ell^3$ before the inversion. The ratio half-length of the shaft to radius was fixed to $\ell/a = 15$.

6 RESULTS OF THE STABILITY ANALYSIS

A survey of the stability behaviour of the Laval rotor with a cracked hollow shaft is presented in fig. 9. In fig. 9a, D = 0, we clearly recognize three ridges of instability starting from

$$\Omega/\omega_o = 2;\ 1 \text{ and } 2/3\ .$$

The small ridges at $\Omega/\omega_o = 2/4$; 2/5 etc. are drowned by the graphic-program but do exist, see fig. 10.

There is a strikingly broad unstable area near to $\Omega/\omega_o \simeq 2$ that was already discussed in detail in |5, 6|. At $\Omega/\omega_o \simeq 1$ we find the unstable zone well known from |1, 2, 3|. The unstable ridge starting from $\Omega/\omega_o = 2/3$ is small but very violent, $|\mu| > 2$. In general the amplitudes grow very rapidly in the unstable zones. $|\mu|$ is the factor of increase during one perod T. Thus $|\mu| = 2$ means an increase of over 1000 during 10 revolutions of the shaft! (Of course our linear theory can not decide whether the amplitudes grow to "infinity" or yield into a non-linear limit cycle).

The strong influence of damping is shown in fig. 9b, D = 0.05. The mountain chains of instability lose their height, the unstable zones get smaller, see fig. 10b.

7 A COMPARISON OF CRACK MODELS

The more sophisticated crack model ("real crack") used for the hollow shaft permits to test the quality and tolerances of the simplified models.

7.1 A comparison of the crack flexibilities and stiffness-matrices $\underline{S}(t)$

Comparing the pure crack flexibilities, fig. 7, we find an excellent agreement between the hinge model and the "real" crack as long as the crack-

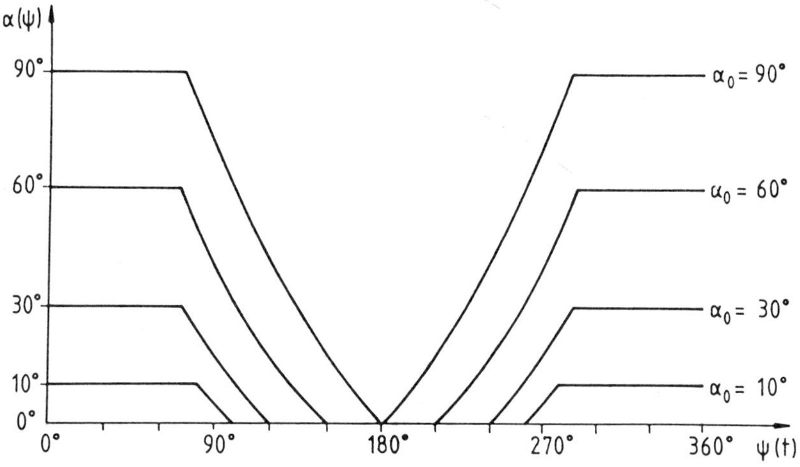

Fig 5 Crack opening α versus rotational angle $\Psi = \Omega t$

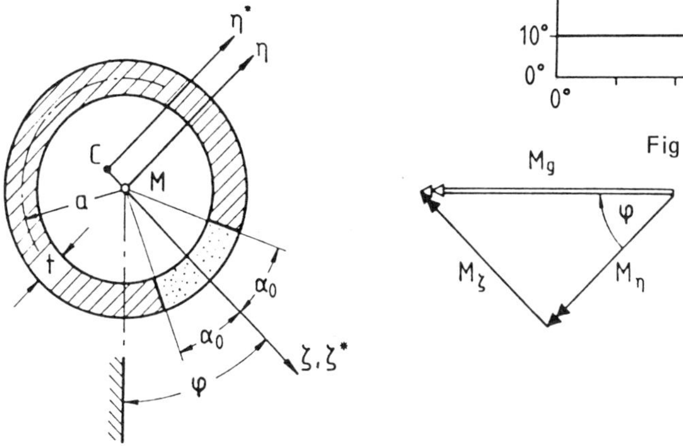

Fig 3 Angle of crack opening $2\alpha_0$; coordinates; bending moment M_g due to weight

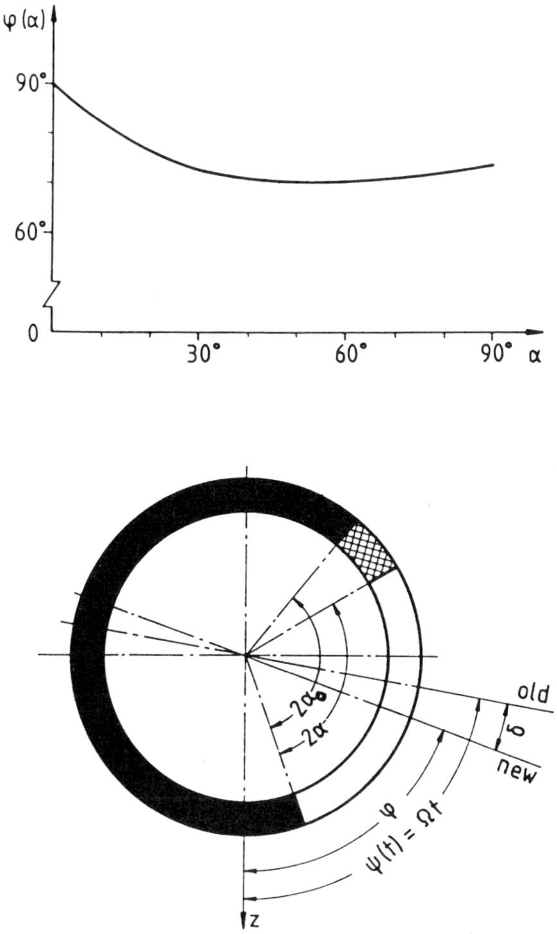

Fig 4 Crack centre position φ versus actual angle α of opening ($\alpha < \alpha_0$)

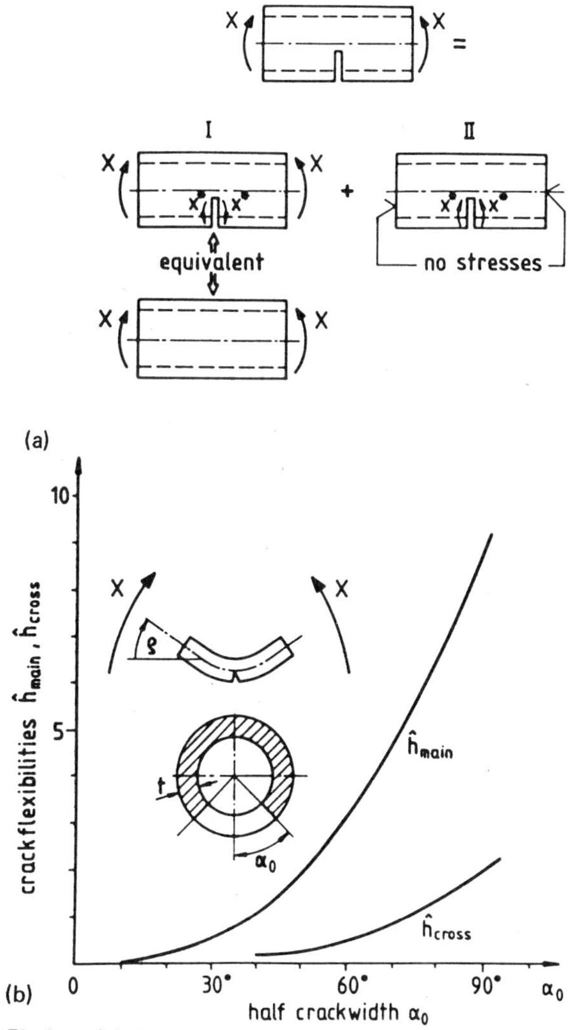

Fig 6 (a) Crack flexibility from the residual problem
(b) Crack flexibilities (dimensionless) $\hat{h}_{main}$, $\hat{h}_{cross}$ versus crack opening angle α_0
$t/a = 0.1$; $\nu = 0.3$; angle $\rho = \hat{h}_{main} X/2Eta^2$

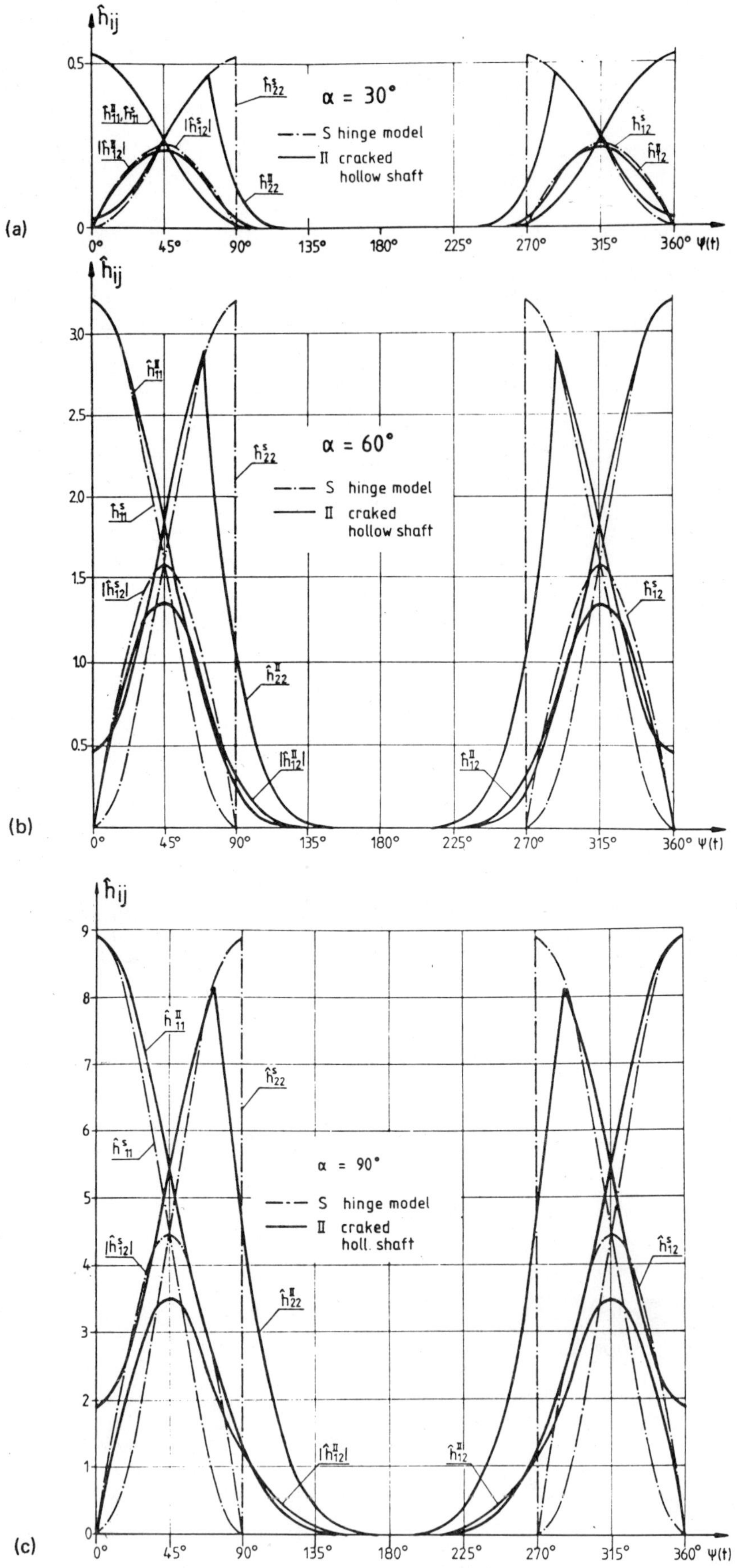

Fig 7 Crack flexibilities of the rotating shaft in an inertial frame. Crack depth
(a) α_o = 30°
(b) α_o = 60°
(c) α_o = 90°
hollow cracked shaft ————————
hinge model ——— · ——— · ———

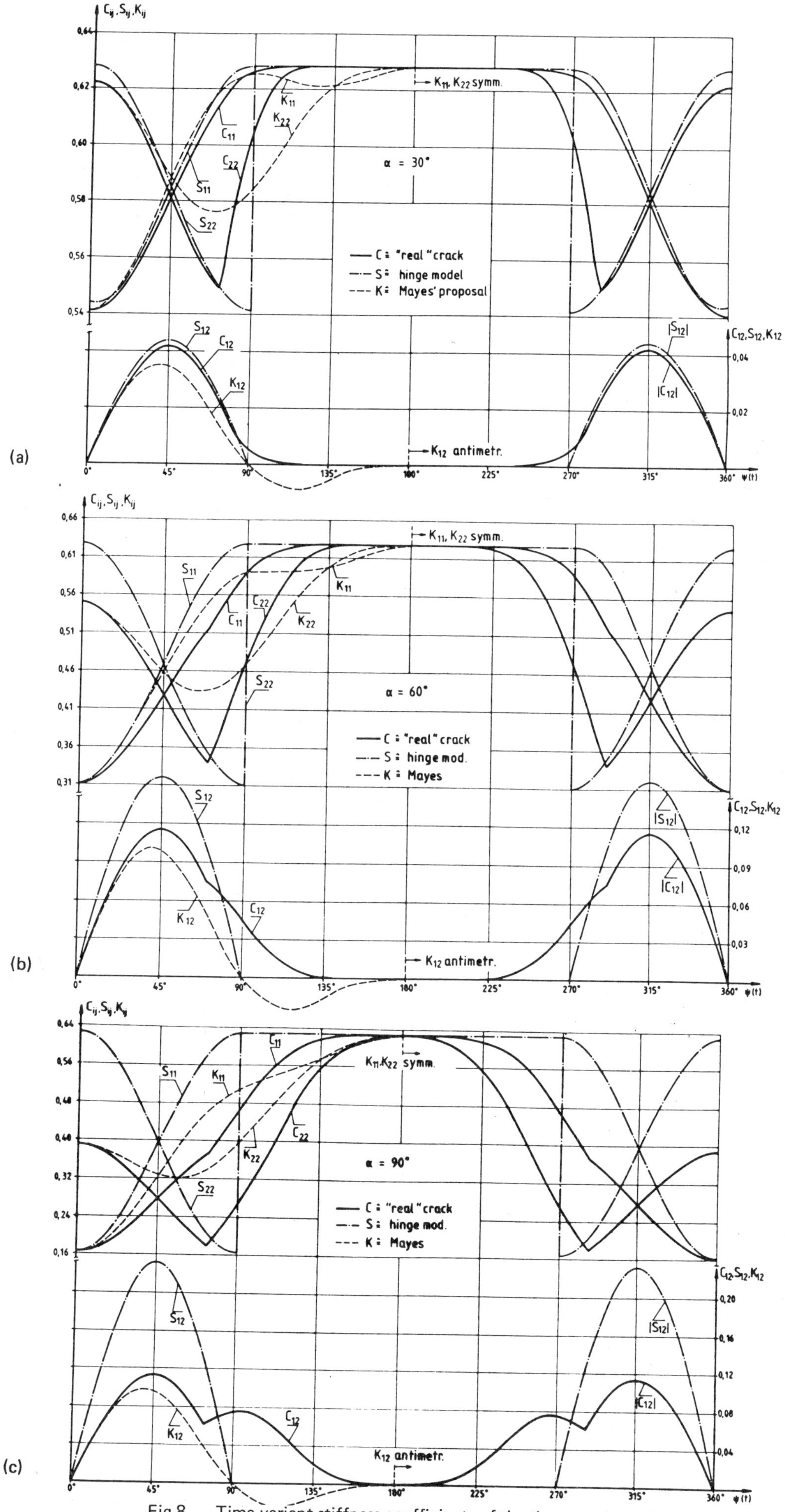

Fig 8 Time-variant stiffness coefficients of the three crack models. Crack depth (a) $\alpha_0 = 30°$; (b) $\alpha_0 = 60°$; (c) $\alpha_0 = 90°$
$\ell/a = 15$, $t/a = 0.1$, $\nu = 0.3$

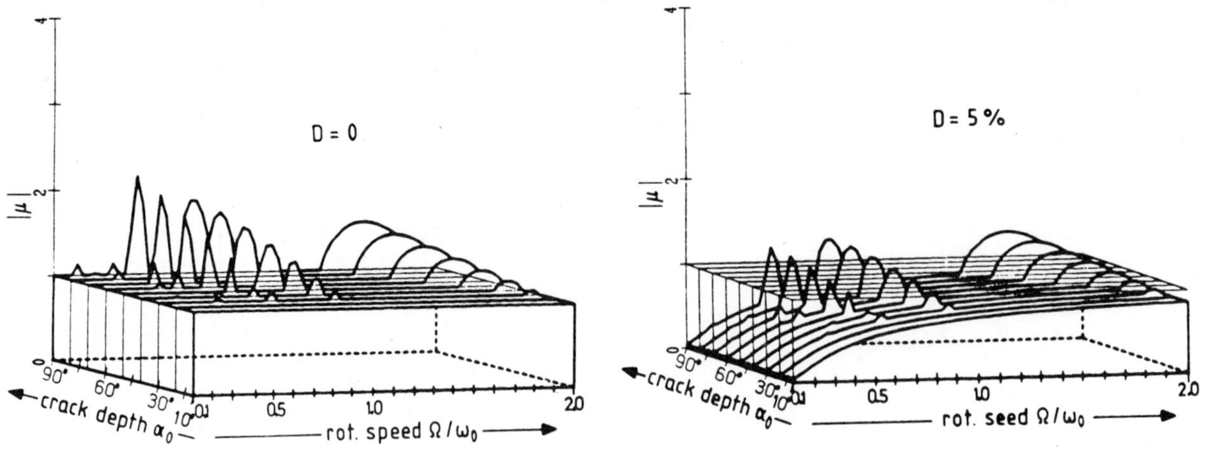

Fig 9 Stability of the cracked hollow shaft versus crack depth α_0 and rotational speed Ω/ω_0; damping (a) D = 0 and (b) D = 0.05; unstable regions $|\mu| > 1$

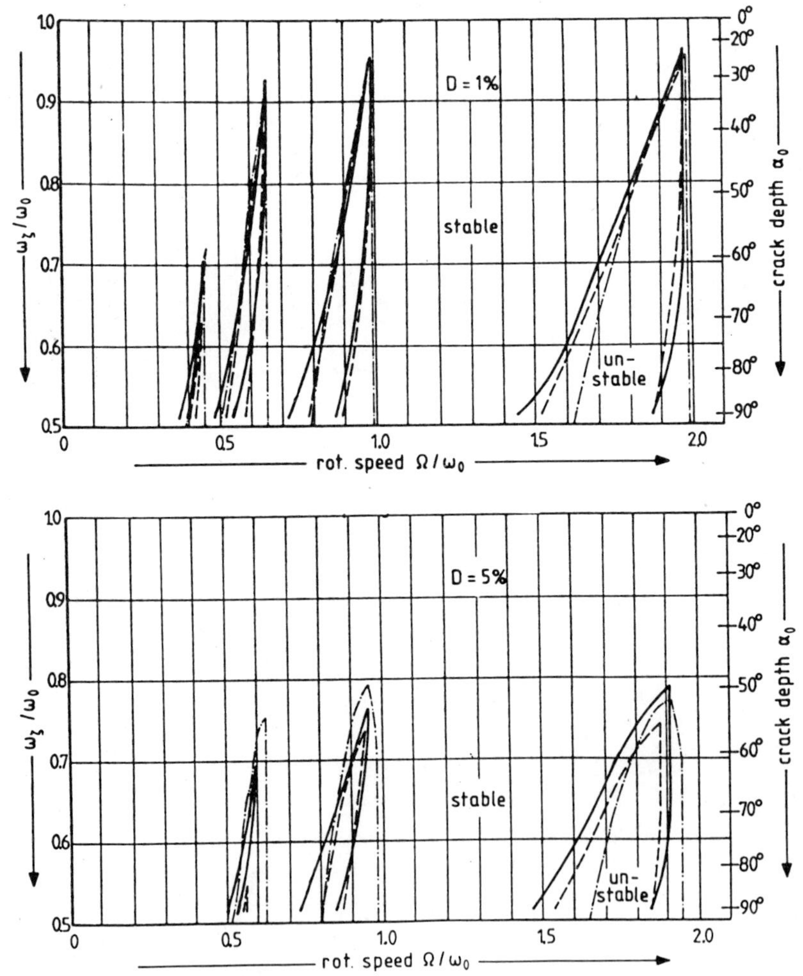

Fig 10 Borderlines of stability versus crack depth and rotational speed Ω/ω_0
cracked hollow shaft ————
hinge model —— · —— · ——
Mayes' proposal —— —— ——
α_0 crack-half-width
$\omega\zeta/\omega_0$ frequency ratio for open to closed crack

C301/88

Theoretical research, calculation and experiments of cracked shaft dynamic response

B C WEN, MCMES and **Y B WANG**
Department of Mechanical Engineering, Northeast University of Technology, Shenyang, Liaoning, China

SYNOPSIS The responses of the cracked shafts are analysed by using perturbation and numerical methods. A vast amount of calculation has been carried out for an experimental model rotor system under different crack depths. Some experimental results are presented in this paper.

NOTATION

m = mass of the disk
k_o = stiffness of the uncracked shaft
e = eccentricity of the rotor
c = damping coefficient
k_ξ, k_η = stiffness at ξ and η directions of cracked shaft
β = orientation of crack
$\Delta k_\xi = k_o - k_\xi$
$\Delta k_\eta = k_o - k_\eta$
$\omega_o^2 = k_o/m$
$c/m = 2\zeta\omega_o$
$\lambda^2 = 1 - \varepsilon_1\delta$
$y_o = -g/(1-\varepsilon_1\delta)\omega_o^2$
$\varepsilon_1 = (1/4)(\Delta k_\eta - \Delta k_\xi)$
$\delta = (\Delta k_\eta + \Delta k_\xi)/(\Delta k_\eta - \Delta k_\xi)$
$F(\Omega T) = 1 - f(\lambda_o\tau)$
$\tau = \omega_o t$
$\varepsilon_1 E = e\lambda_o^2$
$\lambda_o = \Omega/\omega_o$

1 INTRODUCTION

For large rotating machinery, such as turbine generators and compressors, the expansion of cracks in their rotating shafts cause severe accidents. How to make early diagnostics of generation and development of those cracks has received great attention by the scientists.
Recently, many scholars have published papers on the effects of cracks in rotors upon responses of the system, and received research results. In this paper, perturbation and numerical methods have been used to analyse responses of shafts with cracks and large amount of experiments have been conducted to obtain a variety of situations of responses of shaft with different-depth cracks. It is shown that analytical results fit experimental ones very well.

2 ANALYSIS OF RESPONSES OF CRACKED SHAFTS BY PERTURBATION METHOD

Fig. 1 and 2 show a simple rotor with surface transverse crack in the middle of span and the section of the rotor shaft respectively. The equations of motion of the system are as follows:

$$X'' + 2\zeta X' + \lambda^2 X = -\varepsilon_1 E\sin(\lambda_o\tau-\beta) + \varepsilon_1\{-\delta\cdot f(\lambda_o\tau)X - [1-f(\lambda_o\tau)][\cos 2\lambda_o\tau\cdot X + \sin 2\lambda_o\tau(Y+y_o)]\}$$
$$Y'' + 2\zeta Y' + \lambda^2 Y = \varepsilon_1 E\cos(\lambda_o\tau-\beta) + \varepsilon_1\{-\delta f(\lambda_o\tau)(Y+y_o) - [1-f(\lambda_o\tau)][\sin 2\lambda_o\tau\cdot X - \cos 2\lambda_o\tau(Y+y_o)]\} \quad (1)$$

where

$$f(\lambda_o\tau) = \frac{4}{\pi}\cos\lambda_o\tau - \frac{4}{3\pi}\cos 3\lambda_o\tau + \frac{4}{5\pi}\cos 5\lambda_o\tau - \frac{4}{7\pi}\cos 7\lambda_o\tau \quad (2)$$

The equation is a forced equation with parametric excitation.

2.1 The Solution in Non-resonant Case

According to perturbation method, we can get the improving approximation solution:

$$X_1 = X + \varepsilon_1 u_1(\lambda_o\tau), \qquad Y_1 = Y + \varepsilon_1 v_1(\lambda_o\tau) \quad (3)$$

$$X = a_r(\tau)\cos\lambda\tau + b_r(\tau)\sin\lambda\tau$$
$$Y = a_i(\tau)\cos\lambda\tau + b_i(\tau)\sin\lambda\tau \quad (4)$$

or

$$X = e^{-\zeta\tau}(c_1^r\cos\lambda\tau + c_2^r\sin\lambda\tau)$$
$$Y = e^{-\zeta\tau}(c_1^i\cos\lambda\tau + c_2^i\sin\lambda\tau) \quad (5)$$

$$u_1 = \frac{-E}{\lambda^2-\lambda_o^2}\sin(\lambda_o\tau-\beta) - \frac{y_o}{\lambda^2-4\lambda_o^2}\cdot\sin 2\lambda_o\tau + \sum_{n=1}^{5}\frac{y_o}{\lambda^2-(2n-1)^2\lambda_o^2}A_{2n-1}\cdot\sin(2n-1)\lambda_o\tau$$

$$v_1 = \frac{E}{\lambda^2-\lambda_o^2}\cos(\lambda_o\tau-\beta) + \frac{y_o}{\lambda^2-4\lambda_o^2}\cdot$$

$$\cdot \cos 2\lambda_o \tau + \sum_{n=1}^{5} \frac{y_o}{\lambda^2 - (2n-1)^2 \lambda_o^2} B_{2n-1} \cdot$$

$$\cdot \cos(2n-1)\lambda_o \tau \qquad (6)$$

A_1	$8/3\pi$	B_1	$-4(\delta+1/3)/\pi$
A_3	$8/5\pi$	B_3	$4(\delta-9/5)/3\pi$
A_5	$-8/21\pi$	B_5	$-4(\delta-25/21)/5\pi$
A_7	$2/5\pi$	B_7	$4(\delta-7/10)/7\pi$
A_9	$-2/7\pi$	B_9	$2/7\pi$

where, c_1^r, c_2^r .... are constants which are determined according to initial conditions.

From formula (6), we can find that the main parts of the responses in the non-resonant case are the first and second order, and the amplitudes of the higher harmonics are rather small.

Table 1 shows the amplitudes of each harmonic under three non-resonant frequencies which are obtained when $\varepsilon_1=0.018$, $\delta=1.1927$, e=1mm and $\beta=0$.

Table 1 The amplitudes of each harmonic under three non-resonant frequencies

in X direction

	$\lambda_o=0.29$	$\lambda_o=0.4$	$\lambda_o=0.6$
$X_m^{(1)}$	0.0784	0.1784	0.5595
$\sin 2\lambda_o\tau$	0.0256	0.0486	0.0356
$\sin 3\lambda_o\tau$	0.0378	0.0181	0.0037
$\sin 5\lambda_o\tau$	0.0018	0.0007	0.0002
$\sin 7\lambda_o\tau$	0.0007	0.0003	0.0001
$\sin 9\lambda_o\tau$	0.0003	0.0001	0.0000

in Y direction

	$\lambda_o=0.29$	$\lambda_o=0.4$	$\lambda_o=0.6$
$X_m^{(1)}$	0.0660	0.1648	0.5415
$\sin 2\lambda_o\tau$	0.0256	0.0486	0.0356
$\sin 3\lambda_o\tau$	0.0191	0.0092	0.0019
$\sin 5\lambda_o\tau$	0.0000	0.0000	0.0000
$\sin 7\lambda_o\tau$	0.0005	0.0002	0.0001
$\sin 9\lambda_o\tau$	0.0003	0.0001	0.0000

From above table, we can see that under non-resonant case the main part is harmonic, next part is second order harmonic and other harmonics are rather small in the responses of forced vibration.

2.2 The Solution in Resonant Case

According to the perturbation method, the improving first approximation solution can be expressed as follows:

$$X = a_r(\tau)\cos\frac{p}{q}\lambda_o\tau + b_r(\tau)\sin\frac{p}{q}\lambda_o\tau + \varepsilon_1 u_1$$

$$Y = a_i(\tau)\cos\frac{p}{q}\lambda_o\tau + b_i(\tau)\sin\frac{p}{q}\lambda_o\tau + \varepsilon_1 v_1 \qquad (7)$$

where a_r, b_r, a_i and b_i can be determined according to the following formulae:

$$\frac{da_r}{d\tau} = \frac{\varepsilon_1\Delta q}{2p\lambda_o} b_r - \zeta a_r - \frac{\varepsilon_1 q}{2p\lambda_o}(\Sigma_{1s})$$

$$\frac{db_r}{d\tau} = \frac{\varepsilon_1\Delta q}{2p\lambda_o} a_r - \zeta b_r + \frac{\varepsilon_1 q}{2p\lambda_o}(\Sigma_{1c})$$

$$\frac{da_i}{d\tau} = \frac{\varepsilon_1\Delta q}{2p\lambda_o} b_i - \zeta a_i - \frac{\varepsilon_1 q}{2p\lambda_o}(\Sigma_{2s})$$

$$\frac{db_i}{d\tau} = \frac{\varepsilon_1\Delta q}{2p\lambda_o} a_i - \zeta b_i + \frac{\varepsilon_1 q}{2p\lambda_o}(\Sigma_{2c}) \qquad (8)$$

where, Σ_{1s} and Σ_{1c} are the coefficients of $\sin (p/q)\lambda_o\tau$ and $\cos(p/q)\lambda_o\tau$, which are the averaging values of the right parts of first equation of (8), and Σ_{2s} and Σ_{2c} are of second equation. u_1 and v_1 are very complex, so they are not written out here.

When p=q=1 and let $\frac{da_r}{d\tau}=0$, ..., we can obtain following equations

$$a_r = [(1-\varepsilon_1\delta-\varepsilon_1-\lambda_o^2)(e\lambda_o^2\sin\beta+\frac{\varepsilon_1}{2}\alpha_1 K)+2\zeta\lambda_o\cdot (e\lambda_o^2\cos\beta+\frac{\varepsilon_1 y_o}{3\pi}+\frac{\varepsilon_1}{2}\alpha_2 K)]/[(1-\varepsilon_1\delta+\varepsilon_1-\lambda_o^2)\cdot (1-\varepsilon_1\Delta-\varepsilon_1-\lambda_o^2)+4\zeta^2\lambda_o^2]$$

$$b_r = [(1-\varepsilon_1\Delta+\varepsilon_1-\lambda_o^2)(e\lambda_o^2\cos\beta+\frac{\varepsilon_1 y_o}{3\pi}+\frac{\varepsilon_1}{2}\alpha_2 K) -2\zeta\lambda_o(e\lambda_o^2\sin\beta+\frac{\varepsilon_1}{2}\alpha_1 K)]/[(1-\varepsilon_1\delta+\varepsilon_1-\lambda_o^2)\cdot (1-\varepsilon_1\delta-\varepsilon_1-\lambda_o^2)+4\zeta^2\lambda_o^2]$$

$$a_i = [(1-\varepsilon_1\delta+\varepsilon_1-\lambda_o^2)(e\lambda_o^2\cos\beta-\frac{4}{\pi}(\delta-\frac{1}{3})\cdot\varepsilon_1 y_o - \frac{\varepsilon_1}{2}\alpha_2 K)-2\zeta\lambda_o(e\lambda_o^2\sin\beta-\frac{\varepsilon_1}{2}\alpha_1 K)]/[(1-\varepsilon_1\delta+\varepsilon_1-\lambda_o^2)(1-\varepsilon_1\delta-\varepsilon_1-\lambda_o^2)+4\zeta^2\lambda_o^2]$$

$$b_i = [(1-\varepsilon_1\delta-\varepsilon-\lambda_o^2)(e\lambda_o^2\sin\beta-\frac{\varepsilon_1}{2}\alpha_1 K)+2\zeta\lambda_o^2\cdot (e\lambda_o^2\cos\beta-\frac{4}{\pi}(\delta-\frac{1}{3})\varepsilon_1 y_o-\frac{\varepsilon_1}{2}\alpha_2 K)]/[(1-\varepsilon_1\delta+\varepsilon_1-\lambda_o^2)(1-\varepsilon_1\delta-\varepsilon_1-\lambda_o^2)+4\zeta^2\lambda_o^2] \qquad (9)$$

In the above formulae, α_1 and α_2 are as

$$\alpha_1 = \frac{2\zeta\lambda_o}{(1-\varepsilon_1\delta-\lambda_o^2)^2+4\zeta^2\lambda_o^2}$$

$$\alpha_2 = \frac{-(1-\varepsilon_1\delta-\lambda_o^2)}{(1-\varepsilon_1\delta-\lambda_o^2)^2+4\zeta^2\lambda_o^2}$$

$$K = \frac{1}{\pi}(4\delta-1)\varepsilon_1 y_o \qquad (10)$$

When p=2 and q=1, the first approximate solution in directions x and y are as

$$X = a_r(\tau)\cos 2\lambda_o\tau + b_r(\tau)\sin 2\lambda_o\tau$$

$$Y = a_i(\tau)\cos 2\lambda_o\tau + b_i(\tau)\sin 2\lambda_o\tau \quad (11)$$

From equation (7), we obtain

$$a_r(\tau) = \frac{-4\varepsilon_1\lambda_o g\zeta}{(1-\varepsilon_1\delta)\omega_o^2[(1-\varepsilon_1\delta-4\lambda_o^2)^2+16\zeta^2\lambda_o^2]}$$

$$b_r(\tau) = \frac{\varepsilon_1(1-\varepsilon_1\delta-4\lambda_o^2)g}{(1-\varepsilon_1\delta)\omega_o^2[(1-\varepsilon_1\delta-4\lambda_o^2)^2+16\zeta^2\lambda_o^2]}$$

$$a_i(\tau) = \frac{-\varepsilon_1(1-\varepsilon_1\delta-4\lambda_o^2)g}{(1-\varepsilon_1\delta)\omega_o^2[(1-\varepsilon_1\delta-4\lambda_o^2)^2+16\zeta^2\lambda_o^2]}$$

$$b_i(\tau) = \frac{-4\varepsilon_1\lambda_o g\zeta}{(1-\varepsilon_1\delta)\omega_o^2[(1-\varepsilon_1\delta-4\lambda_o^2)^2+16\zeta^2\lambda_o^2]} \quad (12)$$

Equations (11)can also be written as

$$X = X_m^{(2)}\cos(2\lambda_o\tau-\psi_2)$$

$$Y = Y_m^{(2)}\sin(2\lambda_o\tau-\psi_2) \quad (13)$$

in which

$$X_m^{(2)} = Y_m^{(2)} = \frac{\varepsilon_1 g}{(1-\varepsilon_1\delta)\omega_o^2}\cdot\frac{1}{\sqrt{(1-\varepsilon_1\delta-4\lambda_o^2)^2+4\zeta^2\lambda_o^2}}$$

$$\psi_2 = \mathrm{arctg}\left(-\frac{1-\varepsilon_1\delta-4\lambda_o^2}{4\zeta\lambda_o}\right) \quad (14)$$

The $\frac{1}{2}$ subcritical speed is

$$\Omega_s = \frac{1}{2}\omega_o\sqrt{1-\varepsilon_1\delta} \quad (15)$$

When p=3 and q=1, we have

$$X = X_m^{(3)}\cos(3\lambda_o\tau-\psi_3)$$

$$Y = Y_m^{(3)}\sin(3\lambda_o\tau-\psi_3) \quad (16)$$

where

$$X_m^{(3)} = \frac{8\varepsilon_1 g}{5\pi(1-\varepsilon_1\delta)\omega_o^2}\;\frac{1}{\sqrt{(1-\varepsilon_1\delta-9\lambda_o^2)^2+36\zeta^2\lambda_o^2}}$$

$$Y_m^{(3)} = \frac{4(\delta-9/5)\varepsilon_1 g}{3\pi(1-\varepsilon_1\delta)\omega_o^2}\;\frac{1}{\sqrt{(1-\varepsilon_1\delta-9\lambda_o^2)^2+36\zeta^2\lambda_o^2}}$$

$$\psi_3 = \mathrm{arctg}\left(-\frac{1-\varepsilon_1\delta-9\lambda^2}{6\zeta_o\lambda}\right) \quad (17)$$

When p=5 and q=1, we have

$$X = X_m^{(5)}\cos(5\lambda_o\tau-\psi_5)$$

$$Y = Y_m^{(5)}\sin(5\lambda_o\tau-\psi_5) \quad (18)$$

where

$$X_m^{(5)} = \frac{8\varepsilon_1 g}{21\pi(1-\varepsilon_1\delta)\omega_o^2}\;\frac{1}{\sqrt{(1-\varepsilon_1\delta-25\lambda_o^2)^2+100\zeta^2\lambda_o^2}}$$

$$Y_m^{(5)} = \frac{4(\delta-25/21)\varepsilon_1}{5\pi(1-\varepsilon_1\delta)\omega_o^2}\;\frac{1}{\sqrt{(1-\varepsilon_1\delta-25\lambda_o^2)^2+100\zeta^2\lambda_o^2}}$$

$$\psi_5 = \mathrm{arctg}\left(-\frac{1-\varepsilon_1\delta-25\lambda_o^2}{10\zeta\lambda_o}\right) \quad (19)$$

When p=7 and q=1, we have

$$X = X_m^{(7)}\cos(7\lambda_o\tau-\psi_7)$$

$$Y = Y_m^{(7)}\sin(7\lambda_o\tau-\psi_7) \quad (20)$$

where

$$X_m^{(7)} = \frac{2\varepsilon_1 g}{5\pi(1-\varepsilon_1\delta)\omega_o^2}\;\frac{1}{\sqrt{(1-\varepsilon_1\delta-49\lambda_o^2)^2+156\zeta^2\lambda_o^2}}$$

$$Y_m^{(7)} = \frac{4(\delta-7/10)\varepsilon_1 g}{7\pi(1-\varepsilon_1\delta)\omega_o^2}\;\frac{1}{\sqrt{(1-\varepsilon_1\delta-49\lambda_o^2)^2+156\zeta^2\lambda_o^2}}$$

$$\psi_7 = \mathrm{arctg}\left(-\frac{1-\varepsilon_1\delta-49\lambda_o^2}{14\zeta\lambda_o}\right) \quad (21)$$

According to above results, we can calculate the responses of the cracked shafts in non-resonant or resonant case. Some results by means of analytical method are shown in tables 2 and 3.

3 RESPONSES OF CRACKED SHAFTS CALCULATED BY MEANS OF FINITE-ELEMENT METHOD

Fig. 3 shows a mechanical model of the single-disc rotor system. The rotor consists of a mass-lumped disc and a circular cross, uniform mass shaft. The horizontal shaft is mounted on rigid supports. Cracks on shaft can be anywhere in axial direction.

The XYZ has been selected as a fixed coordinate system and $\xi\zeta\gamma$ - rotating coordinate system (see Fig. 3).

The following method is used to generate the elements. Suppose that the lumped-mass disc is positioned at a node, letting its node number be d, the crack is positioned in an element, letting its element number be c and remaining elements are generated by the finite-element method. Assembling every elements according to their node codes, we can get the differential equation of motion of the rotor system.

$$[M]\{\ddot{S}\} + ([C]-\Omega[G])\{\dot{S}\} + ([K]-\varepsilon_1[B_1])\{S\} =$$

$$= \{Q_g\} + \varepsilon_1\{Q_R^{\varepsilon}1\}e^{i\Omega t} + \varepsilon_1[-[B]\{S\}f(\Omega t)-(1-f(\Omega t))\cdot$$

$$\cdot[B_2]\{\bar{S}\}e^{i2\Omega t} \quad (22)$$

where

$$f(\Omega t) = (2/\pi)e^{i\Omega t}-(2/3\pi)e^{i3\Omega t}+(2/5\pi)e^{i5\Omega t}-(2/7\pi)\cdot$$

$$\cdot e^{i7\Omega t}+(2/\pi)e^{-i\Omega t}-(2/3\pi)\,e^{-i3\Omega t}+(2/5\pi)\cdot$$

$$\cdot e^{-i5\Omega t}-(2/7\pi)e^{-i7\Omega t} \quad (23)$$

We assume that the solution of above equation has following form:

$$\{S\} = \{S_o\}+\varepsilon_1\{S_1\}+\varepsilon_1^2\{S_2\}+\ldots \quad (24)$$

From equation (22), we can get

$$[M]\{\ddot{S}_o\}+([C]-\Omega[G]\{\dot{S}_o\}+([K]-\varepsilon_1[B_1])\{S_o\}=\{Q_g\} \quad (25)$$

$$[M]\{\ddot{S}_1\}+([C]-\Omega[G])\{\dot{S}_1\}+([K]-\varepsilon_1[B_1])\{S_1\}= \quad (26)$$

$$=\{Q_R^{\varepsilon_1}\}e^{i\Omega t}-[B_1]\{S_o\}f(\Omega t)-(1-f(\Omega t))[B_2]\{\bar{S}_o\}e^{i2\Omega t}$$

$$[M]\{\ddot{S}_j\}+([C]-\Omega[G])\{\dot{S}_j\}+([K]-\varepsilon_1[B_1])\{S_j\}=$$

$$=-[B_1]\{S_{j-1}\}f(\Omega t)-(1-f(\Omega t))[B_2]\{\bar{S}_{j-1}\}e^{i2\Omega t} \quad (27)$$

$$j=2,3,4...$$

From equation (25), we can obtain zero-order approximate solution S_o and from equation (26) and equation (27), the first and second solutions S_1 and S_2 can be obtained.

According to above formulae, we can make use of finite element method of rotor dynamics to analyse responses of cracked shaft system. A computer program, called CRRE, has been made by authors. A great amount of calculation on different depth of crack have been conducted. Table 2 and 3 show some calculation results. From these tables, we can see that the results by means of Finite Element Method approach to those from approximate analytical solutions. Therefore, the CRRE is suitable to analyse the responses of cracked shaft system. Because the program has generality, it can be used to analyse a cracked shaft system with multi-lumped-mass or distributed-mass and multi-supports.

Table 2. The calculation results of the system responses when the system in the vicinity of 1/2 subcritical speed(when Ω=48.17 rad/s, $\zeta=\beta=0$)

T	Frequencies	Analytic solution Verti.	Analytic solution Hori.	F.E.M solution Verti.	F.E.M solution Hori.
T=1/7d	Fundamental	0.3260	0.3298	0.2841	0.2833
	2-order	0.0307	0.0307	0.0415	0.0415
	3-order	7.6367 E-03	1.2540 E-03	7.7041 E-04	1.4816 E-03
	4-order	3.4924 E-05	3.4924 E-05		
	5-order	1.8922 E-05	7.1089 E-05	2.0408 E-05	7.2187 E-05
T=2/7d	Fundamental	0.2950	0.3143	0.2894	0.2852
	2-order	0.1663	0.1663	0.1706	0.1706
	3-order	1.7157 E-03	7.0045 E-03	1.7596 E-03	7.1985 E-03
	4-order	1.0196 E-03	1.0196 E-03		
	5-order	2.5875 E-04	3.9709 E-04	2.3628 E-04	3.7404 E-04
T=3/7d	Fundamental	0.2473	0.2873	0.2985	0.2876
	2-order	0.3969	0.3969	0.4225	0.4225
	3-order	0.0185	0.0198	0.0190	0.0211
	4-order	5.6666 E-03	5.6666 E-03		
	5-order	1.5078 E-03	1.1206 E-03	1.8880 E-03	1.0648 E-03

Table 3. The calculation results of higher-order harmonic amplitudes at several fraction critical speeds in the different crack depths

when Ω=32.117rad/s, $\zeta=\beta=0$

T	Analytic solution: 3-order harmonic amplitude Verti.	Analytic solution: 3-order harmonic amplitude Hori.	F.E.M. solution: Harmonic amplitude Verti	F.E.M. solution: Harmonic amplitude Hori	F.E.M. solution: 3-order harmonic amplitude Verti.	F.E.M. solution: 3-order harmonic amplitude Hori.
1/7d	0.0107	6.1934 E-03	0.107	0.107	0.0119	6.4532 E-03
2/7d	6.5395 E-02	3.3095 E-02	0.108	0.112	6.6981 E-02	3.6916 E-02
3/7d	0.2224	0.1512	0.109	0.119	0.2114	0.1907

when Ω=19.270rad/s, $\zeta=\beta=0$

T	Analytic solution: 5-order harmonic amplitude Verti.	Analytic solution: 5-order harmonic amplitude Hori.	F.E.M. solution: Harmonic amplitude Verti	F.E.M. solution: Harmonic amplitude Hori.	F.E.M. solution: 5-order harmonic amplitude Verti.	F.E.M. solution: 5-order harmonic amplitude Hori.
1/7d	2.6760 E-03	1.5505 E-03	3.589 E-02	3.649 E-02	2.5380 E-03	1.2299 E-03
2/7d	1.6346 E-02	1.1502 E-02	3.697 E-02	4.025 E-02	1.7381 E-02	1.2692 E-02
3/7d	5.5583 E-02	6.5728 E-02	3.790 E-02	4.648 E-02	5.0982 E-02	4.1100 E-02

when Ω=13.764rad/s, $\zeta=\beta=0$

T	Analytic solution: 7-order harmonic amplitude Verti.	Analytic solution: 7-order harmonic amplitude Hori.	F.E.M. solution: Harmonic amplitude Verti	F.E.M. solution: Harmonic amplitude Hori.	F.E.M. solution: 7-order harmonic amplitude Verti.	F.E.M. solution: 7-order harmonic amplitude Hori.
1/7d	2.5486 E-03	4.5265 E-03	0.018	0.019	2.6966 E-03	4.6922 E-03
2/7d	1.5663 E-02	2.4226 E-02	1.918 E-02	2.239 E-02	1.6192 E-02	2.1837 E-02
3/7d	5.2937 E-02	3.7532 E-02	2.003 E-02	0.028	4.8560 E-02	4.0093 E-02

4 EXPERIMENTAL RESEARCH

In order to verify the theoretical results, an experimental table has been designed.

The experimental results of the principal critical rotating speed of the system in the case of no crack in shaft are shown in table 4. The theoretical value of critical speed is very close to the measured results.

Table 5 lists the theoretical calculation values and measured results of the principal critical speeds of the cracked rotor system. From the theoretical results, we can see that the varying region of the principal critical speeds is 0-5.218 %. The change is very small. From the measured results, we can see that every measured results have a little bit of difference and the critical speeds vary from 995 to 1000 rpm, there is no obvious change. In practice, there are many factors to affect the change of the critical speeds, and in general, it is very difficult to determine exactly the critical speed values, either from theoretical calculation or from practical measure. Thus 5 % change is not enough to mean there exists a crack in shaft. The conclusion can be drawn that change of principal critical speeds is not sensitive to existence of cracks.

From experimental results, we can see that when the speed of the rotor system is steady at near 1/2 subcritical speed, the second harmonic component in responses is very obvious and there exist other higher harmonics, which are more and more obvious as the crack deepens. Fig.

4 shows the responses and power spectrum of the system at the different depths of the crack. When the speed of the system passes through 1/3, 1/5, 1/7 ... critical speed, the higher harmonic responses will appear, but their amplitudes are less. Fig. 5 shows the responses of the system, when it passes through 1/3, 1/5, 1/7 ... fractional critical speed. The second harmonic is very obvious at near 1/2 subcritical speed. When the speed is getting away from this speed, the second harmonic is getting less and less. Fig. 6 shows the curves about the amplitude of second harmonic and time t. Fig. 7 shows some speed power spectrum during starting (see Fig. 7 a,b,c and d) at different depths of crack.

Table 4. Theoretical principal critical speed and measured critical speed

Theoretical critical speed	Measured critical speed
997.6793 rpm	995-1000 rpm

Table 5. Theoretical calculation results and measured results (rpm)

Theoretical calculation critical speed	Measured critical speed
0 997.6793 1/7d 994.1012,997.4997 2/7d 977.7926,995.9475 3/7d 945.6244,987.0323	995-1000

5 CONCLUSIONS

According to the calculation results by perturbation method and finite element method and experimental results, some conclusions may be summarized as follows:

1). The principal critical speed will decrease with the increase of crack depth. But the changed values are very small and it can not be found easily. Thus it is difficult to determine the existence of crack on shaft according to the change of critical speed.

2). In the responses of cracked shaft system, the amplitude of fundamental harmonic is associated with the depth of crack and with relative unbalanced direction of crack. In general, the amplitude is maximum when $\beta=0$ or 180 and minimum when $\beta=\pm 90$. The phase angle difference of fundamental harmonic is affected by β very obviously.

3). In the responses of cracked shaft system, compared to those of rotor system without crack, the second order harmonic component is very obvious and the 1/2 subcritical speed appears in the system. When the rotating speed of the system approaches 1/2 subcritical speed, second order harmonic resonance will occur.

4). In the transient process of starting or stopping when the system passes through 1/2 subcritical speed, the maximum amplitude, in general, is less than the maximum steady-state amplitude. The faster the starting speed, the smaller the maximum amplitude. After passing through the resonance the beating phenomenon appears.

5). For open-close crack, there exists not only second order harmonic but also many higher order and corresponding fractional harmonic in the system responses. When the rotating speed of the system is in the vicinity of various fraction cratical speeds, the system will produce higher order harmonic resonance, such as 3, 5 and 7-order harmonics. Besides, there are some generating higher harmonics, such as 4 and 6-order harmonics, etc, which, in general, do not cause resonance, and are very small.

6). The system without the crack in the shaft is stable. The existence of a crack may make the system produce the unstable region in the vicinity of the principal critical speed. The unstable region will expand with the deepening of the depth of the crack. However, when the damping of the system is large enough the unstable region will disappear.

7). The asymptotic formulae of finding out responses of rotor system with crack, set up in this paper, can be used to analyse the responses of the similar rotor system with crack. The computer program CRRE set up by using finite element method of rotor dynamics, can be used to analyse the responses of a general single disc rotor system with transverse cracks.

Crack diagnosis is very difficult and complicated problem, work done in this paper is elementary. In order to diagnose cracks on rotor much more information should be synthesized.

REFERENCES

(1) BACHCHMID, N and DIANA, G. The influence of unbalance on cracked rotors, The Proceedings of International Conference on Vibrations in Rotating Machinery, C304/84, 1984.

(2) GASCH, R. Dynamic behaviour of a simple rotor with a cross-sectional crack, The Proceedings of International Conference on Vibrations in Rotating Machinery, C178/78, 1976.

(3) GRABOWSKI, B. The vibrational behaviour of a turbine rotor containing a transverse crack, Paper of ASME Conference, 79-DET-67, 1979.

(4) HENRY, T.A. and OKAHAVAE, B.E. Vibration in cracked shafts, The Proceedings of International Conference on Vibrations in Rotating Machinery, C162/76, 1976.

(5) INAGAKI, T., KANKI, H. and SHIRAK, K. Transverse vibrations of a general cracked rotor bearing system, Paper of ASME Conference, 81 -DET-45, 1981.

(6) MAYES, I.W. and DAVIES, W.G.R. The vibrational behaviour of a rotating shaft system containing a transverse crack, The Proceedings of International Conference on Vibrations in Rotating Machinery, C168/76, 1976.

(7) MAYES, I.W. and DAVIS, W.G.R. Amethod of calculating the vibrational behaviour of coupled rotating shafts containing a transverse crack, The Proceedings of International Conference on Vibrations in Rotating Machinery, C254/80, 1980.

(8) NELSON, H.D. and HATARAJ, C. The dynamics of a rotor system with a cracked shaft, Paper of ASME Conference, 85-DET-31, 1985.

(9) NILSSON, L.R.K. On the vibration behaviour of a cracked rotor, The Proceedings of the World Congress on ToMM, Sept., 1982 Rome.

(10) SCHMIED, J. and KRAMER, E. Vibrational behaviour of a rotor with a cross-sectional crack, The Proceedings of International Conference on Vibration in Rotating Machinery, C279/84. 1984.

(11) IMAM, I., SCHEIBEL, J. AZZARO, S.H. and BANKERT, R.J. Development of an on-line rotor crack detection and monitoring system, Paper of ASME Conference, DE-Vol 2, 1987.

(12) HERBERT, R.G. Turbine-alternator run-down vibration analysis: automated crack detection, Paper of ASME Conference, DE-Vol 2, 1987.

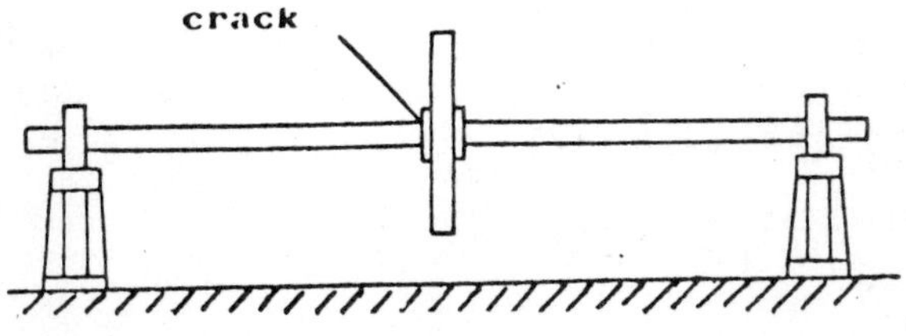

Fig 1 Simple rotor with surface transverse crack — middle of span

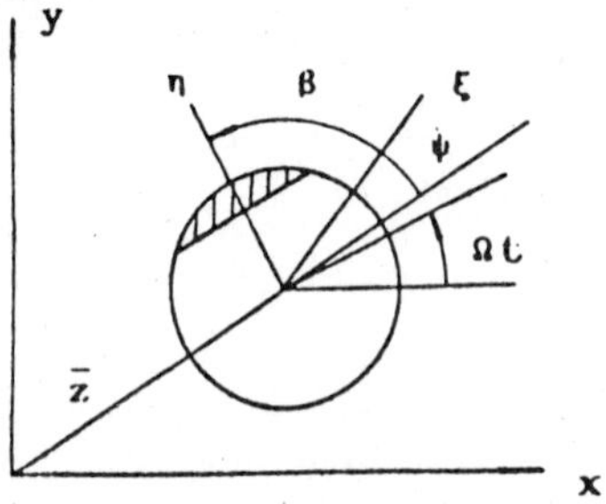

Fig 2 Simple rotor with surface transverse crack — section of the rotor shaft

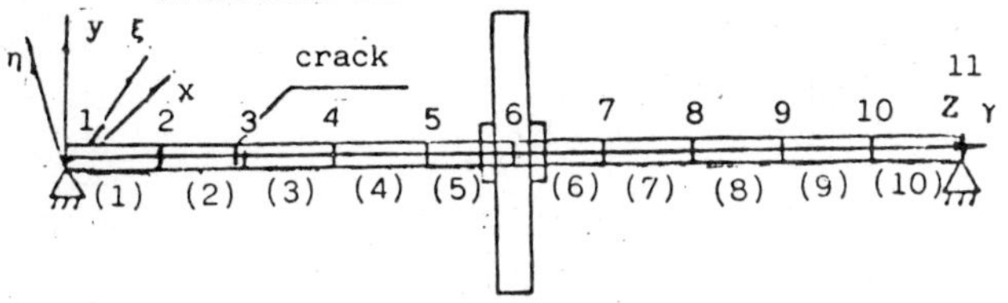

Fig 3 Mechanical model of the single-disc rotor system

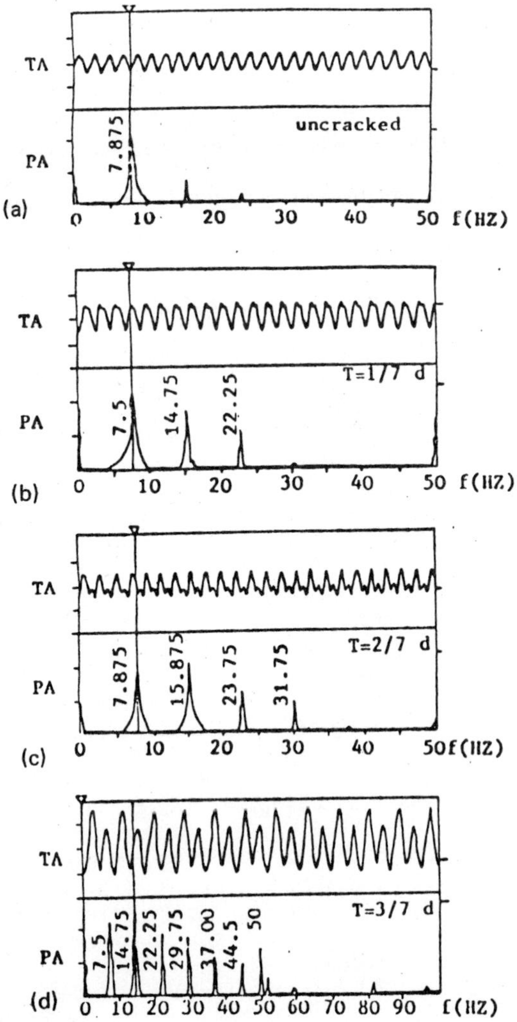

Fig 4 Responses and power spectra of the rotor system at different depths of crack

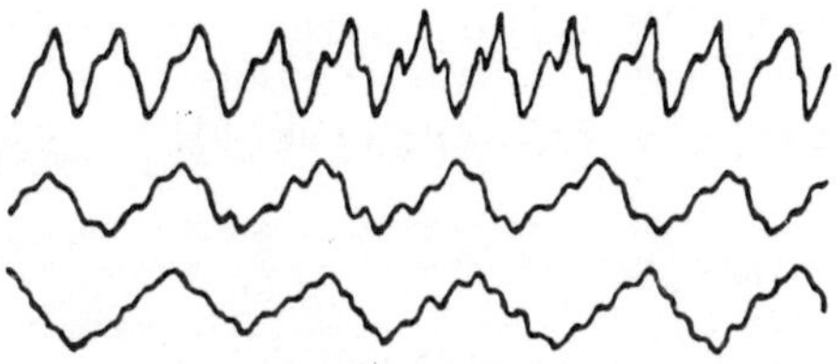

Fig 5 Responses of the rotor system when the speed of the system passes through 1/3, 1/5, 1/7 . . . fractional critical speed

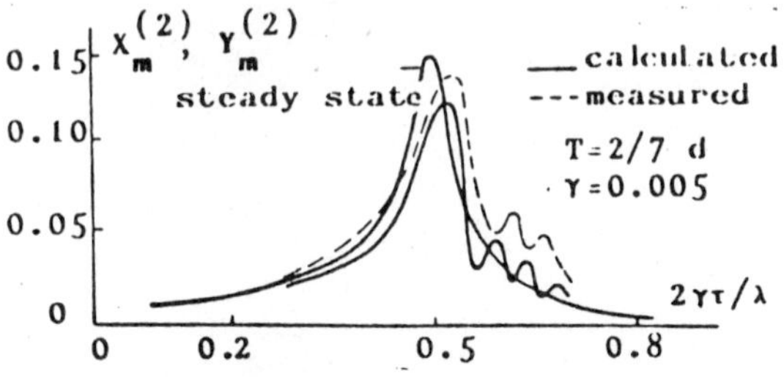

Fig 6 Curves about the amplitude of the second harmonic and time t

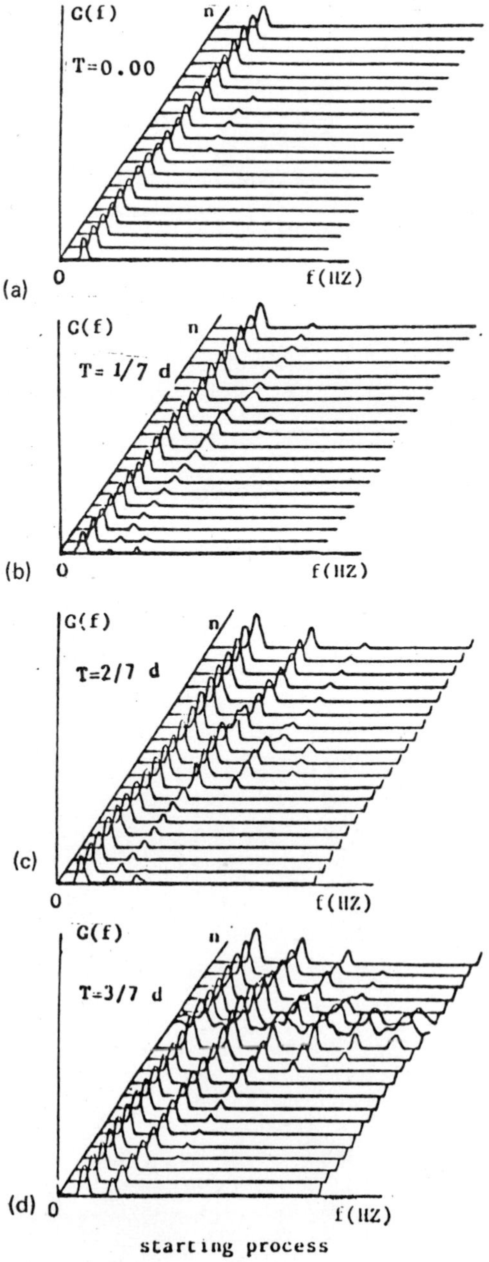

Fig 7 Speed—power spectra during starting at different depths of crack

C292/88

Bimodal (complex) balancing of large turbogenerator rotors having large or small unbalance

W KELLENBERGER, PhD
BBC Brown Boveri Limited, Birr, Switzerland
P RIHAK, PhD
BBC Brown Boveri Limited, Baden, Switzerland

Abstract

Bimodal balancing, in contrast to modal balancing, also applies theoretically to rotors with plain journal bearings. Using as an example a larger turbogenerator rotor having an initial unbalance, balancing sets are calculated. Four cases are considered, two with and two without large unbalance. "Large unbalance" means that because of excessive vibration amplitudes, the rotor cannot be run at or near the critical speeds. In this case, it is helpful to apply a preliminary balancing of the "rigid" rotor at low speeds, and as a result the subsequent "elastic" balancing sets will be considerably smaller. The first two cases illustrate the conventional bimodal balancing, and the other two illustrate bimodal "forward" balancing.

Introduction

Rotors having unbalance reduce the operational reliability and shorten the useful life of the rotating machine. They also disturb the surroundings because of the undesirable vibration. In spite of the current improved methods of fabrication, however, it is still not possible to produce rotors without unbalance, and in the foreseeable future rotors will still have to be balanced. Whereas two balancing planes are sufficient for rigid rotors, flexible rotors theoreticaly require compensation unbalance to counteract the initial unbalance wherever it occurs, i.e. along the full length of the rotor axis, and this would require infinitely many balancing planes. In practice, of course, there are only a finite number of balancing planes available, so that - with few exceptions - a flexible rotor can never be completely balanced. Complete balancing is also not necessary. The balancing must only be good enough to satisfy the tolerance specifications as given in the standards or as agreed with the customer. The classical approach to the balancing of flexible rotors is based on two methods: modal balancing, and balancing with the help of influence coefficients.
Balancing with the help of influence coefficients in particular requires numerous test runs, first in order to determine the coefficients and then to perform the balancing with them. The procedure is particularly suitable where many identical rotors (series) have to be balanced. For single rotors (prototypes), either modal or bimodal balancing is to be recommended.

Modal balancing assumes symmetrical mass, damping, and stiffness matrixes ($\underline{\underline{M}}$, $\underline{\underline{D}}$ and $\underline{\underline{S}}$) in the differential equation for the rotor reduced in finite elements

$$\underline{\underline{M}}\,\ddot{\underline{p}} + \underline{\underline{D}}\,\dot{\underline{p}} + \underline{\underline{S}}\,\underline{p} = \underline{F}(t) \qquad (1)$$

In addition, $\underline{\underline{D}}$ must satisfy the compatibility conditions, i.e. it must be proportional to $\underline{\underline{M}}$ and/or $\underline{\underline{S}}$, so that theoretically the journal bearings, gyroscopic effects, and internal damping are eliminated. In spite of this, however, modal balancing has been in practical use for some decades.

For bimodal balancing, matrixes $\underline{\underline{D}}$ and $\underline{\underline{S}}$ in (1) need not be symmetrical, and the compatibility condition does not apply, so that the journal bearings, gyroscopic effect (polar inertia), rotary inertia, shear, inner damping, and damping-free foundation can be expressed.

Turbogenerator Rotor and Its Model

We have investigated the bimodal balancing of the rotor of an actual turbogenerator rated at 723 MVA, 615 MW, 50 Hz, 3000 rpm, of mass 77 733 kg and length 15.275 m, as shown in Fig. 1. There are three different journal bearings to be considered, each with eight coefficients (DE = drive end bearing, NDE = non - drive end bearing, AUX = auxiliary bearing, V = vertical, H = horizontal):

$$\begin{Bmatrix} F_V \\ F_H \end{Bmatrix} = \overset{\mathbf{C}}{\begin{bmatrix} c_{11} & c_{12} \\ c_{21} & c_{22} \end{bmatrix}} \begin{Bmatrix} x_V \\ x_H \end{Bmatrix} + \overset{\mathbf{D}}{\begin{bmatrix} b_{11} & b_{12} \\ b_{21} & b_{22} \end{bmatrix}} \begin{Bmatrix} \dot{x}_V \\ \dot{x}_H \end{Bmatrix}$$

At 3000 rpm, the numerical values are:

$$\text{DE-B.:}\ \overset{\mathbf{C}}{\begin{vmatrix} 1.06 & 0.30 \\ 0.25 & 1.63 \end{vmatrix}} 10^9\ \frac{\text{N}}{\text{m}}; \quad \overset{\mathbf{D}}{\begin{vmatrix} 9.00 & -1.90 \\ -4.44 & 23.8 \end{vmatrix}} 10^5\ \frac{\text{Ns}}{\text{m}}$$

$$\text{NDE-B.:}\ \overset{\mathbf{C}}{\begin{vmatrix} 1.32 & 0.36 \\ 0.55 & 2.26 \end{vmatrix}} 10^9\ \frac{\text{N}}{\text{m}}; \quad \overset{\mathbf{D}}{\begin{vmatrix} 19.8 & -0.30 \\ -1.10 & 38.4 \end{vmatrix}} 10^5\ \frac{\text{Ns}}{\text{m}}$$

$$\text{AUX-B.:}\ \overset{\mathbf{C}}{\begin{vmatrix} 4.66 & 2.17 \\ 0.31 & 4.74 \end{vmatrix}} 10^8\ \frac{\text{N}}{\text{m}}; \quad \overset{\mathbf{D}}{\begin{vmatrix} 0.65 & -2.51 \\ -7.77 & 4.29 \end{vmatrix}} 10^5\ \frac{\text{Ns}}{\text{m}}$$

The bearings used in our example are tilting-pad bearings. Their stiffness and damping coefficients have been determined experimentally by Glienicke [Gℓ], in dimensionless form, for a test bearing of the same design.
For sake of simplicity, the dependence of the dynamic bearing coefficients on speed has been neglected. This neglect has no effect on the main results of our investigation.

Four alternative cases were calculated. The results may be taken as representative for all generator rotors of medium to high rating. The model used for the finite-element calculations for the rotor is shown in Fig. 2. All natural modes are excited by the theoretically assumed initial unbalance. Four balancing planes are provided in the element boundaries (nodes) 7, 11, 15 and 25; for simplicity, these are denoted A, B, C and D.

Bimodal Balancing

In the differential equation (1)

$$\underline{F}(t) = \Omega^2 \begin{bmatrix} \mathrm{Re}[(\underline{U}+\underline{U}_\varepsilon)e^{i\Omega t}] \\ \mathrm{Im}[(\underline{U}+\underline{U}_\varepsilon)e^{i\Omega t}] \end{bmatrix} = \Omega^2 \begin{bmatrix} \mathrm{Re}[(\underline{U}+\underline{U}_\varepsilon)e^{i\Omega t}] \\ \mathrm{Re}[-i(\underline{U}+\underline{U}_\varepsilon)e^{i\Omega t}] \end{bmatrix}$$

$\underline{\rho} = [\underline{x}_2, \underline{x}_3, \underline{\Phi}_2, \underline{\Phi}_3]^T$; $\underline{x}_2$=horizontal deflections, $\underline{x}_3$=vertical deflections, $\underline{\Phi}_2$, $\underline{\Phi}_3$=slopes

$\underline{x}_n = [x_{1n}, x_{2n}, \ldots x_{Kn}]^T$; K=number of nodes

$\underline{U} = [U_1, U_2, \ldots U_N]^T$; $U_j = U'_j + iU''_j$

$\underline{U}_\varepsilon = [U_{\varepsilon 1}, U_{\varepsilon 2}, \ldots U_{\varepsilon M}]^T$; $U_{\varepsilon j} = U'_{\varepsilon j} + iU''_{\varepsilon j}$

$\underline{U}$ = vector of compensation unbalance (CU)

$\underline{U}_\varepsilon$ = vector of initial unbalance (IU)

N is the number of balancing planes; in our example, N = 4.

M is the number of planes with initial unbalances. The calculation of the eigenvalues and of the forced vibration are given in the Appendix.

The calculation of the forced vibrations from the unbalance is, of course, more laborious. Decoupling of differential equation (1) requires a coordinate transformation, which leads to higher-order matrixes and vectors, see [No], [Ke].

It is necessary to calculate only the eigenvalues which lie within and just above the operating range. When the number is 4N, then the 2N conjugate complex eigenvalues $\lambda_n^* = \alpha_n - i\omega_n$ are eliminated with the assumed positive sense of rotation Ω (angular velocity) for the rotor. Only the 2N eigenvalues λ_n remain, which occur in pairs having practically equal ω_n and similar eigenvectors, see Table 1.

Table 1 COMPLEX EIGENVALUE $\lambda = \alpha + i\omega$

	EIGENVALUE N	α	ω[rad/s]	f[rps]	n[rpm]
2N =8	1	-4.8152D-01	6.6763D+01	10.63	638
	3(•)	-2.3377D-01	6.9106D+01	11.00	660
	5	-6.7020D+00	1.9268D+02	30.67	1840
	7(•)	-4.2785D+00	2.0823D+02	33.14	1988
	9	-8.0719D+00	2.4782D+02	39.44	2367
	11(•)	-3.5589D+00	2.5640D+02	40.81	2448
	13	-1.9748D+01	3.5412D+02	56.36	3382
	15(•)	-1.5215D+01	3.7500D+02	59.68	3581

For balancing, only one eigenvalue per pair is required, preferably the one having the smaller damping α_n. Hence in the operating range, only N eigenvalues remain, with their corresponding vectors. In the present example, the N = 4 eigenvalues are n = 3, 7, 11, 15. The corresponding right eigenvectors (without slope) $\underline{r}_n = [\underline{x}_2, \underline{x}_3]_R^T$ are shown in Fig. 3. The components do not lie in a single plane through the rotor axis, but are twisted.

For balancing, only left eigenvectors $\underline{\ell}_n = [\underline{x}_2, \underline{x}_3]_L^T$ (n = 3, 7, 11, 15) are required. They are shown in Table 2, for nodes 7, 9, 11, 13, 15, 25.

Table 2 LEFT EIGENVECTORS $\underline{\ell} = [\underline{x}_2, \underline{x}_3]_L^T ; \begin{vmatrix} \mathrm{Re} \\ \mathrm{Im} \end{vmatrix}$

NODE	ℓ_3	ℓ_7	ℓ_{11}	ℓ_{15}
7	0.37283 -0.18935	0.39330 -0.61093	-0.00044 -0.01843	0.00207 0.01494
9	0.41707 -0.21068	0.20927 -0.32609	0.00140 -0.00796	-0.00566 0.03531
11	0.43024 -0.21756	-0.01186 0.01006	0.00326 0.00254	-0.00601 0.02715
13	0.41010 -0.20904	-0.22700 0.34287	0.00550 0.00980	-0.00010 -0.00310
15	0.35942 -0.18623	-0.39431 0.61858	0.00900 0.01086	0.00826 -0.04147
25	-0.10584 0.03832	0.29485 0.22654	0.44574 -0.41349	0.00892 0.03711
7	0.86130 0.00015	0.98268 0.00862	0.01707 -0.01097	-0.01670 0.00858
9	0.96652 0.00008	0.53079 0.00526	0.00968 -0.00433	-0.06210 0.00638
11	1.00000 0.00000	-0.01269 -0.00044	0.00229 0.00260	-0.05235 0.00229
13	0.95650 -0.00005	-0.55115 -0.00572	-0.00167 0.00804	0.00248 -0.00232
15	0.84233 -0.00013	-0.98922 -0.00782	0.00167 0.01068	0.07532 -0.00651
25	-0.23107 -0.00276	0.06279 0.09266	1.00000 0.00000	-0.06001 0.01576

When the 4 left eigenvectors are combined in a matrix,

$$\underline{\underline{L}} = \overset{\rightarrow N}{\downarrow 2M}[\underline{\ell}_3, \underline{\ell}_7, \underline{\ell}_{11}, \underline{\ell}_{15}] \tag{12}$$

we obtain the following balancing equation:

$$\overset{\rightarrow 2N}{\downarrow N\boxed{\underline{\underline{L}}^T_N}}\,[U_7, U_{11}, U_{15}, U_{25}, -iU_7, -iU_{11}, -iU_{15}, -iU_{25}]^T_{\rightarrow 2N}$$
$$= -\overset{\rightarrow 2M}{\downarrow N\boxed{\underline{\underline{L}}^T_M}}\,[U_{\varepsilon 7}, U_{\varepsilon 9}, U_{\varepsilon 11}, U_{\varepsilon 13}, U_{\varepsilon 25}, -iU_{\varepsilon 7}, \ldots -iU_{\varepsilon 25}]^T_{\rightarrow 2M} \tag{13}$$

Index N signifies that, for the left eigenvectors in L, only the 4 nodes 7, 11, 15 and 25 (with balancing planes) are to be considered. Index M signifies that, for the left eigenvectors in L, all nodes 7, 9, ... 25 with initial unbalance must be considered.

The four individual complex equations in (13) permit the unique determination of the four complex compensation unbalances U_7, U_{11}, U_{15} and U_{25}. Equation system (13) can be further simplified as follows.

From matrices

$$\begin{matrix}\underline{\underline{\Theta}}_N \\ \underline{\underline{\Theta}}_M\end{matrix} = [\underline{\theta}^T_3, \underline{\theta}^T_7, \underline{\theta}^T_{11}, \underline{\theta}^T_{15}]_{\substack{N\\M}} \; ; \; \underline{\theta}_n = \underline{x}_{2n} - i\,\underline{x}_{3n} \tag{14}$$

and equation (13), we obtain

$$\overset{\rightarrow N}{\downarrow N\boxed{\underline{\underline{\Theta}}^T_N}}\;\underline{U} = -\overset{\rightarrow M}{\downarrow N\boxed{\underline{\underline{\Theta}}^T_M}}\;\underline{U}_\varepsilon \tag{15}$$

$$\underline{U} = [U_7, U_{11}, U_{15}, U_{25}]^T_{\rightarrow N} \; ; \; \underline{U}_\varepsilon = [U_{\varepsilon 7}, U_{\varepsilon 9}, U_{\varepsilon 11}, U_{\varepsilon 13}, U_{\varepsilon 25}]^T_{\rightarrow M}$$

This is broken down into the individual balancing sets or natural modes by subdividing the right-hand side of (15) and solving the N = 4 individual systems

$$\underline{\underline{\theta}}^T_N \underline{U} = -\begin{bmatrix}\underline{\theta}^T_{3M}\\0\\0\\0\end{bmatrix}\underline{U}_\varepsilon,\; -\begin{bmatrix}0\\\underline{\theta}^T_{7M}\\0\\0\end{bmatrix}\underline{U}_\varepsilon,\; -\begin{bmatrix}0\\0\\\underline{\theta}^T_{11M}\\0\end{bmatrix}\underline{U}_\varepsilon,\; -\begin{bmatrix}0\\0\\0\\\underline{\theta}^T_{15M}\end{bmatrix}\underline{U}_\varepsilon$$

bal. set 1   bal. set 2   bal. set 3   bal set 4

Case 1: Bimodal balancing in the four balancing planes (BP): A (node 7), B (node 11), C (node 15), and D (node 25) with four balancing sets for the four mode shapes to be balanced in the operating range and somewhat above. None of the balancing sets excite any of the other modes. The balancing sets are shown in Table 3. For clarity, they are given in polar form $U = |U|e^{i\phi}$ (the first number is $|U|$ in kg dm; the second number, below it, is the angle ϕ in degrees). The balancing sets do not lie in planes through the rotor axis, but are twisted.

In practice, the size and angular position of the balancing masses must be determined by test runs at or near the critical speeds, as in modal balancing. If this is not possible, due to excessive initial unbalance, then Case 2 will apply.

Table 3, Compensation unbalances, their sums, and gross material requirement

$$U = |U|e^{i\phi} \left\{ \begin{matrix}|U|, & \text{kg dm} \\ \phi, & \text{degrees}\end{matrix} \right\}$$

		Bimodal balancing		Bimodal "forward" balancing	
		Case 1	Case 2	Case 3	Case 4
"rigid" bal.set	U_A	-	4.3214 +180	-	4.3214 +180
	U_D	-	1.6786 +180	-	1.6786 +180
Bal .set 1 (Mode 1)	U_A	1.6046 - 1.84+180	0.3901 - 1.51+180	-	-
	U_B	1.7196 + 3.09+180	0.4181 + 3.33+180	4.4311 - 0.09+180	1.0774 + 0.23+180
	U_C	1.5688 - 2.32+180	0.3814 - 2.00+180	-	-
	U_D	0.0319 + 9.37	0.0078 + 9.58	-	-
Bal .set 2 (Mode 2)	U_A	2.0312 + 3.27+180	1.4676 + 3.32	1.2561 + 5.82+180	0.8747 + 5.56
	U_B	1.4016 - 0.81	1.0127 - 0.76+180	-	-
	U_C	0.4472 +17.84	0.3231 +17.88+180	1.2853 + 6.11	0.8951 + 5.85+180
	U_D	0.0589 -40.73	0.0426 -40.62+180	-	-
Bal .set 3 (Mode 3)	U_A	0.6652 -35.32+180	0.4485 -36.61	0.1961 -38.58+180	0.1329 -38.55
	U_B	0.7982 -33.49	0.5382 -34.78+180	-	-
	U_C	0.5278 -13.13+180	0.3559 -14.42	0.1770 +49.79+180	0.1199 +49.84
	U_D	1.0522 - 0.97+180	0.7095 - 2.26	1.0202 - 0.43+180	0.6909 - 0.40
Bal .set 4 (Mode 4)	U_A	1.7024 -10.92	0.4194 +23.85	1.0999 - 0.13+180	0.8442 - 0.29
	U_B	2.9368 -10.90+180	0.7235 +23.87+180	1.8974 - 0.12	1.4563 - 0.28+180
	U_C	1.7311 -10.68	0.4265 +24.09	1.1184 + 0.10+180	0.8584 - 0.06
	U_D	0.0305 -31.87+180	0.0075 + 3.05+180	0.0197 -21.10	0.0151 -21.31+180
$\Sigma\ U_A$		2.5028 +0.05+180≈180	2.5027 +0.06+180≈180	2.5028 +0.06+180≈180	2.5027 +0.05+180≈180
$\Sigma\ U_B$		2.5337 -0.06+180≈180	2.5337 -0.07+180≈180	2.5337 -0.06+180≈180	2.5337 -0.06+180≈180
$\Sigma\ U_C$		0.0453 + 0.51≈0	0.0453 - 0.51≈0	0.0453 - 0.63≈0	0.0453 - 0.63≈0
$\Sigma\ U_D$		1.0018 -0.04+180≈180	1.0018 -0.03+180≈180	1.0018 -0.03+180≈180	1.0018 -0.04+180≈180
$\sum_A^D (\Sigma U_j)$		5.9930 -0.01+180≈180	5.9926 -0.01+180≈180	5.9930 -0.00+180 =180	5.9929 -0.00+180=180
Gross material $\Sigma\lvert U\rvert/\Sigma(IU) = \frac{\Sigma\lvert U\rvert}{6}$		3.05	2.28	2.08	2.16

The final compensating unbalances and their sum are also shown in Table 3. The sum is, with remarkable exactness, opposite the sum of the initial unbalances. The "gross material" is to be understood as the sum of the individual balancing masses (assumption: all at same radius 1) which in practice must be handled and installed for the total number of balancing sets. This gross material figure is also shown in Table 3. The resultant balance - horizontal amplitude versus speed - is shown in Fig. 4 for the DE bearing (shaft) and for the individual steps in balancing.

Case 2: as for Case 1, except with preliminary balancing of the "rigid" rotor at low speeds. In this way, the initial unbalance is reduced, and the rotor can be run at the critical speeds. The individual balancing sets for the subsequent "flexible" balancing become smaller. For the "rigid" balancing, two balancing planes are sufficient. We selected planes A and D, see Fig. 2. The "rigid" balancing set yields:

$$U_A = 4.3214 \quad \angle +180$$
$$U_D = 1.6786 \quad \angle +180$$
$$\Sigma = 6.0000 \quad \angle +180$$

The subsequent bimodal balancing yields the four balancing sets for the four mode shapes, as shown in Table 3. The final balancing sets and the gross material are also shown in the same table. Fig. 5 shows, for each balancing step, the horizontal amplitude versus the speed, for the DE bearing (shaft).

Case 3: Bimodal forward balancing of the four mode shapes. In the first step, only the first mode shape is balanced, at the first critical speed, and without regard to the higher mode shapes. For this procedure, a single balancing plane is sufficient. We selected balancing plane B. The compensation unbalance is

$$U_B = 4.4311 \quad \angle -0.09+180$$

This may excite the higher mode shapes either favourably or unfavourably.
In the second step, only the second mode shape is balanced, but in such a way, however, that the balancing set no longer excites the first mode shape (rundown through the first critical speed). For this procedure, two compensation unbalances, in two balancing planes, are required. Here we selected planes A and C. The compensation balances obtained are:

$$U_A = 1.2561 \quad \angle +5.82+180$$
$$U_C = 1.2853 \quad \angle +6.11$$

In the third step, only the third mode shape is balanced, but in such a way as not to excite the second and first mode shapes (rundown).For this procedure, three compensation unbalances in three balancing planes are required. We selected balancing planes A, C, and D. The compensation unbalances are shown in Table 3.

In the fourth step, only the fourth mode shape is balanced, but without exciting the third, second, and first. For this, four compensation unbalances, in all four balancing planes, are required. The compensation balances are shown in Table 3.

With this, the rotor is now balanced in the operating range. The final balancing sets and the gross material are also shown in Table 3. Fig. 6 shows, for each balancing step, the horizontal amplitude versus the speed, for the DE bearing (shaft).

Case 4: As for Case 3, except with initial balancing of the "rigid" rotor at low speeds. In this way, the initial unbalance is reduced, and the rotor can be run at the critical speeds. The individual balancing sets for the subsequent "flexible" balancing become smaller.

The "rigid" balancing sets for balancing planes A and D are the same as for Case 2.

The subsequent forward balancing yields the following for the first mode shape in balancing plane B:

$$U_B = 1.0774 \quad \angle 0.23+180$$

and for the second mode shape in balancing planes A and C:

$$U_A = 0.8747 \quad \angle 5.56$$
$$U_C = 0.8951 \quad \angle 5.85+180$$

The balancing sets for the third mode shape in balancing planes A, C, and D and the fourth mode shape in all four planes are shown in Table 3; this table also contains the final balancing sets and the gross material. Fig. 7 shows, for each balancing step, the horizontal amplitude versus speed, for the DE bearing (shaft).

Comment

The final balancing sets are the same for all four cases. The individual compensation unbalances U_C practically cancel each other out; their sum ΣU_C is almost zero. The balancing could be done sufficiently accurately in only three balancing planes A, B, and D. The individual compensation unbalances U_C are necessary, however, in order to make up the balancing sets, and thus to be able to balance each mode shape alone. Case 2, compared with Case 1, and Case 4, compared with Case 3, yield substantially smaller "flexible" balancing sets. With proper selection of the two balancing planes, the initial "rigid" balancing is advantageous in both cases (and in many places it is standard practice). With higher values of initial unbalance it is an effective means of avoiding excessive vibration and thus to reach the critical speeds without damage.

The bimodal forward balancing (Cases 3 and 4) is simpler, because at the beginning only one compensation unbalance and one balancing plane have to be treated. It is necessary here, however, to establish at the start the N balancing planes for the N mode shapes. The gross material, as compared with bimodal balancing (Cases 1 and 2), is not significantly less.

References

[Gℓ] Glienicke, J.: FVV Forschungsbericht Nr. 211/17, 1969.

[No] Nordmann, R.: Schwingungsberechnung von nicht konservativen Rotoren mit Hilfe von Links- und Rechts-Eigenvektoren. VDI Ber. 269(1976), 175-182.

[Ke] Kellenberger, W.: Elastisches Wuchten. Springer, Berlin 1987.

Appendix

1. Natural vibration

Differential equation (1) with $\underline{F}(t) = 0$ can be solved with the expression

$$\underline{p}(t) = \underline{p}\, e^{\lambda t} \tag{2}$$

When substituted in (1), the result is the eigenvalue problem

$$(\lambda_n{}^2 \underline{\underline{M}} + \lambda_n \underline{\underline{D}} + \underline{\underline{S}})\underline{p}_n = 0 \tag{3}$$

The associate eigenvalue problem

$$(\lambda_n{}^2 \underline{\underline{M}}^T + \lambda_n \underline{\underline{D}}^T + \underline{\underline{S}}^T)\underline{q}_n = 0 \tag{4}$$

contains the same eigenvalues λ_n, but different eigenvectors $\underline{p}_n$ and $\underline{q}_n$.
With the matrices $\underline{\underline{A}}$, $\underline{\underline{B}}$, and eigenvectors $\underline{r}_n$, $\underline{\ell}_n$:

$$\underline{\underline{A}} = \begin{bmatrix} \underline{\underline{M}} & 0 \\ 0 & -\underline{\underline{S}} \end{bmatrix}; \quad \underline{\underline{B}} = \begin{bmatrix} 0 & \underline{\underline{M}} \\ \underline{\underline{M}} & \underline{\underline{D}} \end{bmatrix}; \quad \underline{r}_n = \begin{bmatrix} \lambda_n \underline{p}_n \\ \underline{p}_n \end{bmatrix}; \quad \underline{\ell}_n = \begin{bmatrix} \lambda_n \underline{q}_n \\ \underline{q}_n \end{bmatrix}$$

(3) and (4) yield the new eigenvalue problems

$$[\underline{\underline{A}} - \lambda_n \underline{\underline{B}}]\underline{r}_n = 0; \qquad [\underline{\underline{A}}^T - \lambda_n \underline{\underline{B}}^T]\underline{\ell}_n = 0 \tag{5, 6}$$

with right eigenvectors $\underline{r}_n$ and left eigenvectors $\underline{\ell}_n$.
If one combines, column-wise, all right eigenvectors in matrix $\underline{\underline{R}}$ and all left eigenvectors in matrix $\underline{\underline{L}}$, then with identity matrix $\underline{\underline{E}}$ and $\underline{\underline{\Lambda}} = \text{diag}\,(\lambda_n)$ the following orthogonality relations apply:

$$\underline{\underline{L}}^T \underline{\underline{A}}\, \underline{\underline{R}} = \underline{\underline{\Lambda}}; \qquad \underline{\underline{L}}^T \underline{\underline{B}}\, \underline{\underline{R}} = \underline{\underline{E}} \tag{7}$$

These permit decoupling of the equations of motion for the enforced vibration.

2. Forced vibration due to Unbalance

We transform the 2nd-order system (1) into a larger 1st-order system with $\dot{\underline{p}}(t)$ as auxiliary co-ordinate:

$$\underline{\underline{A}}\,\underline{\nu} - \underline{\underline{B}}\,\dot{\underline{\nu}} = \underline{F}(t); \qquad \underline{\nu}(t) = \begin{bmatrix} \dot{\underline{p}} \\ \underline{p} \end{bmatrix}; \quad \underline{F}(t) = \begin{bmatrix} 0 \\ -\underline{f}(t) \end{bmatrix} \tag{8}$$

and develop the new vector $\underline{\nu}(t)$ from the right eigenvectors

$$\underline{\nu}(t) = \underline{\underline{R}}\,\underline{\eta}(t) \tag{9}$$

By substituting (9) in (8), we obtain

$$\underline{\underline{A}}\,\underline{\underline{R}}\,\underline{\eta} - \underline{\underline{B}}\,\underline{\underline{R}}\,\dot{\underline{\eta}} = \underline{F}(t)$$

which when multiplied by $\underline{\underline{L}}^T$ gives the system

$$\underbrace{\underline{\underline{L}}^T \underline{\underline{A}}\,\underline{\underline{R}}}_{\underline{\underline{\Lambda}}}\,\underline{\eta} - \underbrace{\underline{\underline{L}}^T \underline{\underline{B}}\,\underline{\underline{R}}}_{\underline{\underline{E}}}\,\dot{\underline{\eta}} = \underline{\underline{L}}^T \underline{F}(t) \tag{10}$$

from (7)

thus obtaining a decoupling of the individual equations (bimodal solution).
For a complete mass balance, $\underline{p}$ must disappear, and hence also $\underline{\nu}$ and $\underline{\eta}$. Thus from (10) we obtain the balance equation

$$\underline{\underline{L}}^T \underline{F}(t) = 0 \tag{11}$$

from which one can see that for complete mass balance, only left eigenvectors of system (6) are significant.

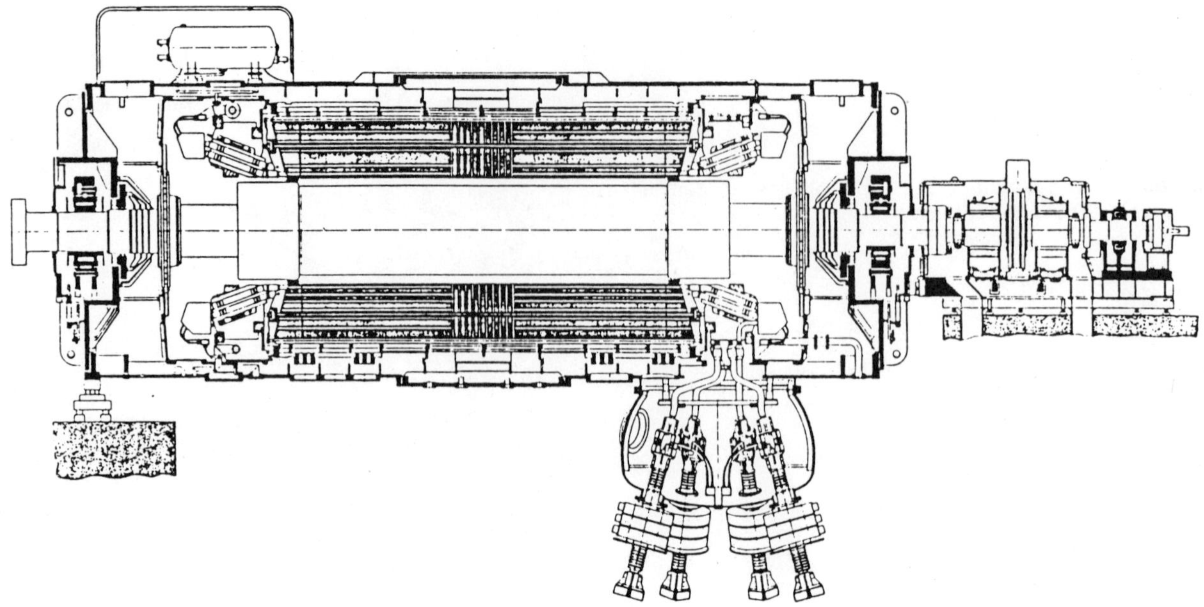

Fig 1 Turbogenerator (723 MVA, 3000 r/min)

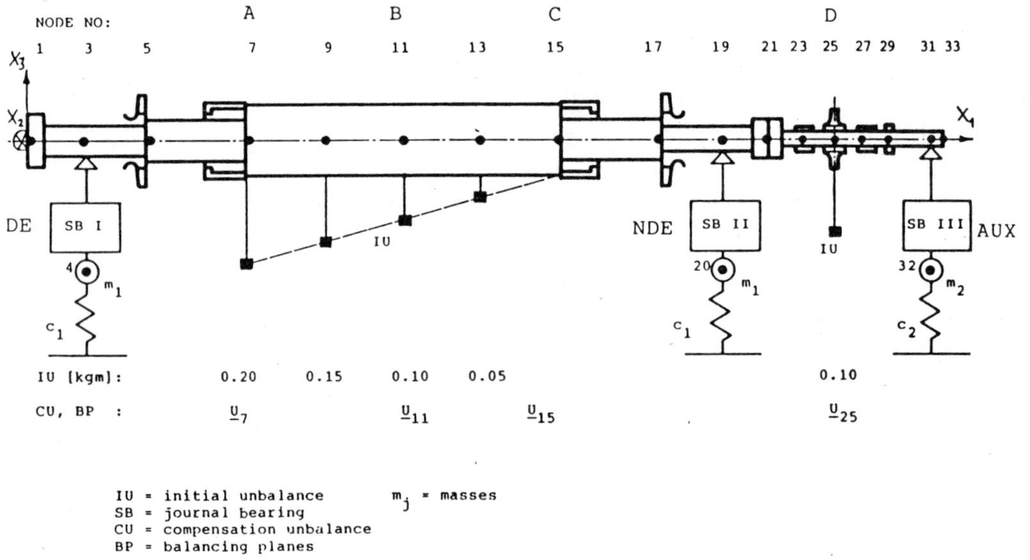

Fig 2 Model

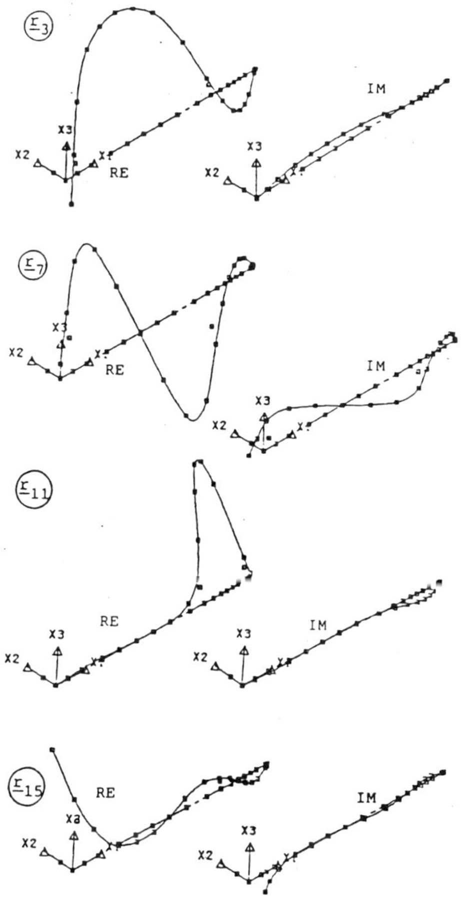

Fig 3 Right eigenvectors (bimodal);
$\underline{RE} = Re(\underline{x}_2) + iRe(\underline{x}_3)$;
$\underline{IM} = Im(\underline{x}_2) + iIm(\underline{x}_3)$;
$\underline{r} = \underline{RE} + i\underline{IM} = \underline{x}_2 + i\underline{x}_3$

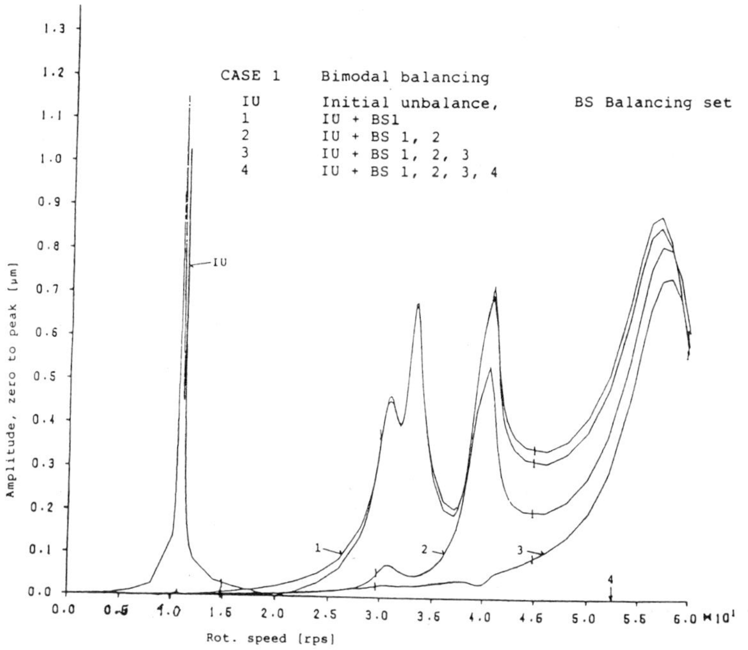

Fig 4 DE-bearing (shaft)

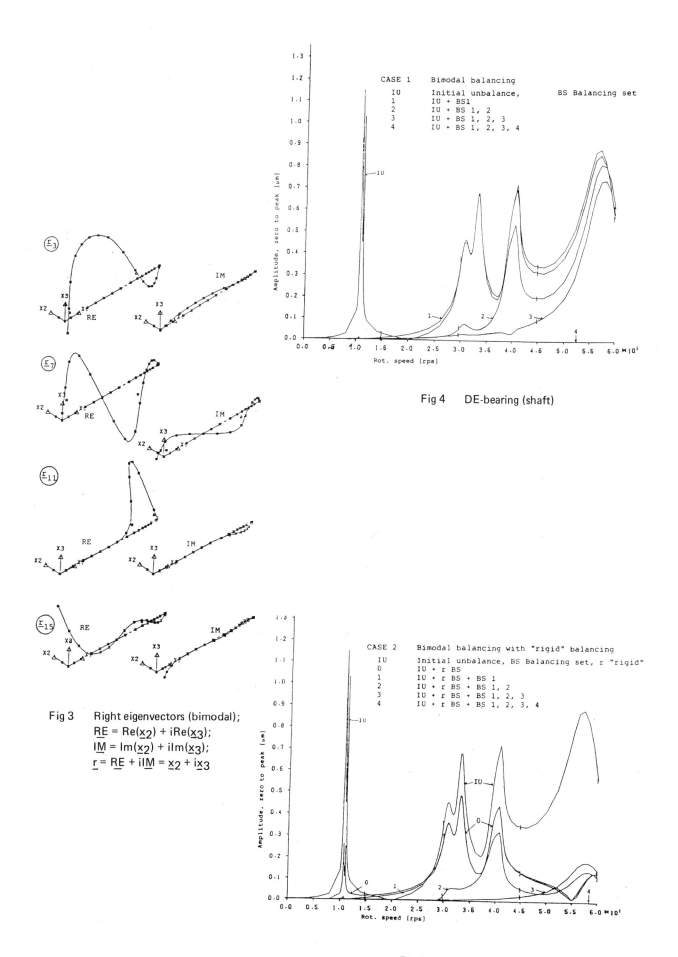

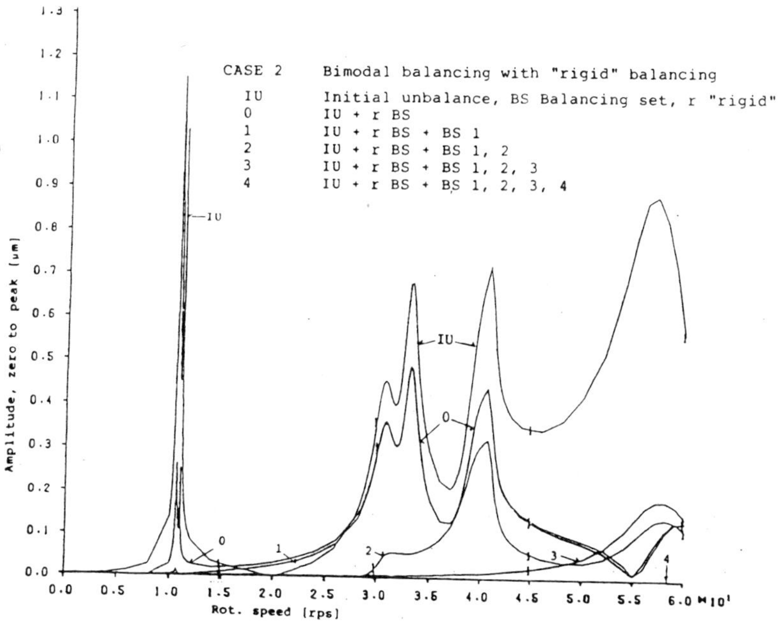

Fig 5 DE-bearing (shaft)

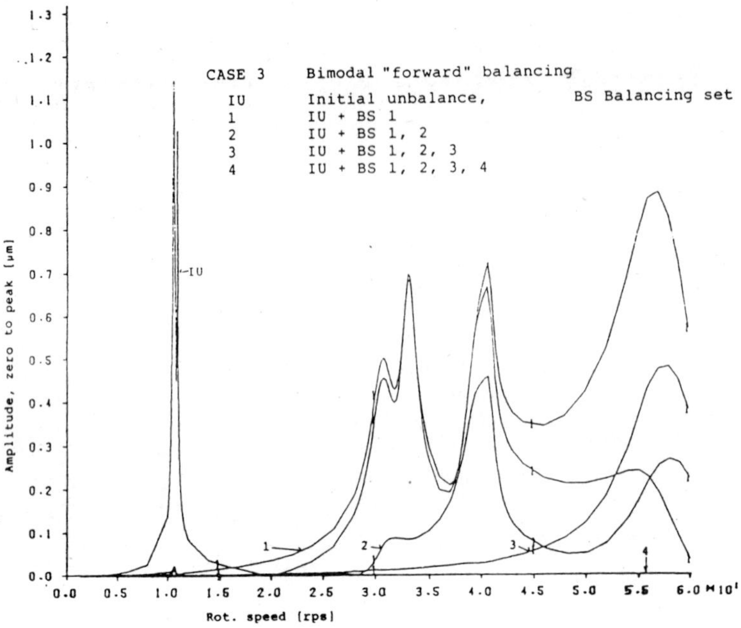

Fig 6 DE-bearing (shaft)

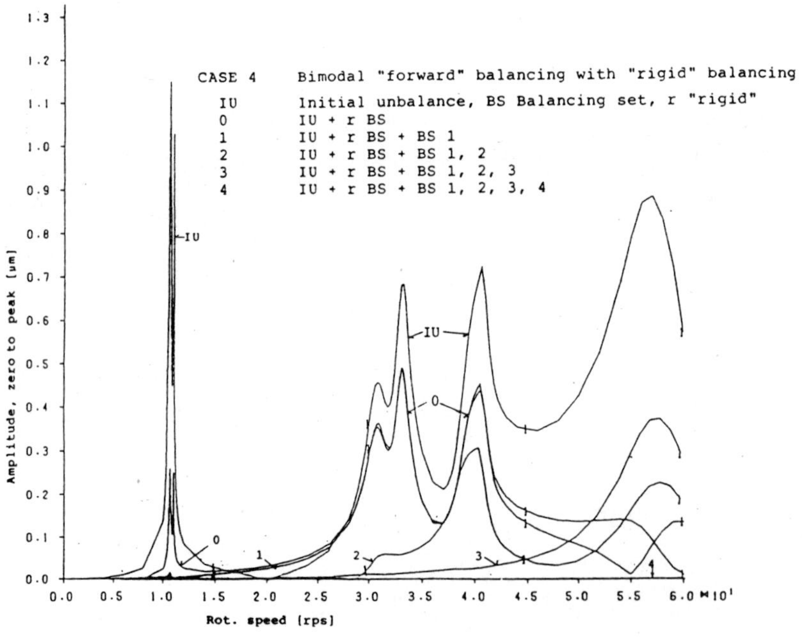

Fig 7 DE-bearing (shaft)

C265/88

Balancing turboset rotors as evolutionary systems

M BALDA, DrSc, MCSSM
Central Research Institute of SKODA, Plzen, Czechoslovakia

SYNOPSIS The paper deals with the problem of balancing flexible rotors supported by many slide bearings, which cause the dynamic properties of a system to vary with speed. A new iterative method developed by the author makes it possible to find optimum balancing weights from a set of measured data collected during different balancing runs at independently chosen speeds.

NOTATION

$\mathbf{A}$ matrix of modal coefficients
$\mathbf{A}_n$ submatrix of modal coefficients corresponding to n-th eigenvalue
$\mathbf{B}$ damping matrix
b, c coefficients
d_{kn} dynamic coefficient (Eq.(15))
$f(x,\omega)$ force distribution acting on a rotor
$\mathbf{G}(x,\xi,\omega)$ matrix of Green's resolvents of a rotor at rotational frequency ω
$\mathbf{G}(\omega) = [\mathbf{g}_j(\omega)]$, matrix of frequency responses of a rotor
i imaginary unity, $(-1)^{1/2}$
$\mathbf{K}$ stiffness matrix
ℓ length of a rotor
n an order of eigenvalue, $n=1, \ldots, N$
N number of modes to be balanced
$\mathbf{M}$ mass matrix
$\mathbf{p}_j$ frequency independent vector of rotor sensitivities to an applied unbalance (Eq.(18))
$\mathbf{P}$ $=[\mathbf{p}_j]$, matrix of sensitivities of a rotor to unbalances applied at different runs
$\mathbf{q}(x,\omega)$ vector of complex amplitudes
$\mathbf{s}$ vector of spectral parameters (Eq.(20))
$\mathbf{S}$ $=\mathrm{diag}\,[s_n]$, matrix of eigenvalues
u(x) complex unbalance density with respect to the reference axis v
$\mathbf{u}(x)$ $= \begin{bmatrix} u(x) \\ -iu(x) \end{bmatrix}$, vector of unbalance density
$\mathbf{u}_j$ vector of complex point unbalances u_{ij} at selected points (balancing planes) x_i
$\mathbf{U}$ $=[\mathbf{u}_j]$, matrix of effective unbalances in runs "j"
$v(x,\omega)$, $w(x,\omega)$ complex components of rotor deviations from equilibrium
$\mathbf{v}_n$, $\mathbf{w}_n$ n-th right- and left-hand eigenvectors of a rotor
$\mathbf{V}$ $=[\mathbf{v}_n]$, modal matrix of right-hand eigenvectors
$\mathbf{W}$ $=[\mathbf{w}_n]$, modal matrix of left-hand eigenvectors

Superscripts

C complex conjugate
H Hermitian transpose (e.g. transpose and complex conjugate)
T transpose
(ℓ) ℓ-th iteration step
+ matrix pseudoinverse

1 INTRODUCTION

Much has been written on the problem of balancing flexible rotors. Nevertheless, there are a few aspects that have not been mentioned so far. One of them is the dependence of dynamic properties of bearings and foundation on rotor speed. These phenomena cause the observed system to vary in speed interval of a machine. Consequently, both natural frequencies and modes are speed dependent. While conditions of generalized orthogonality are valid among the natural modes corresponding to the dynamic properties of the system at a particular speed ω, they do not hold among modes corresponding to different speeds (say critical) of a rotor. This fact affects negatively the balancing process. A partial solution to the problem will be described in the following paragraphs.

2 FORCED VIBRATIONS OF A ROTOR

Let horizontal and vertical components of a spatial density of unbalance at time t=0 be $u_w(x)$ and $u_v(x)$ respectively. The complex unbalance density, as a function of the longitudinal coordinate, is then expressed with respect to the vertical axis v by the formula

$$u(x) = u_v(x) + i.u_w(x) \tag{1}$$

When rotating, the unbalance generates centrifugal loading distribution

$$f(x,\omega) = \omega^2.u(x) , \tag{2}$$

whose influence decomposed into coordinate axes v, w, may be expressed by a complex vector

$$\mathbf{f}(x,\omega) = \begin{bmatrix} f(x,\omega) \\ -if(x,\omega) \end{bmatrix} = \omega^2 \begin{bmatrix} u(x) \\ -iu(x) \end{bmatrix} = \omega^2.\mathbf{u}(x) \tag{3}$$

The forces acting on a continuous rotor of length ℓ generate lateral deflections. Their steady state is characterized by vector $\mathbf{q}(x,\omega)$ composed of two complex elements, i.e. the deviations of the shaft centre line from the straight line connecting bearing centres. It has the form

$$\mathbf{q}(x,\omega) = \omega^2.\int_0^\ell \mathbf{G}(x,\xi,\omega).\mathbf{u}(\xi).d\xi \tag{4}$$

Its discrete equivalent is

$$\mathbf{q}(\omega) = \omega^2 \mathbf{G}(\omega).\mathbf{u} \tag{5}$$

where Green´s resolvent $\mathbf{G}(x,\xi,\omega)$ has taken the form of a complex matrix of dynamic flexibilities (or receptances), e.g. frequency responses of the rotor to the steady harmonic excitation $\mathbf{f}(\omega)$ applied at fixed points of the rotor.

The methods of balancing flexible rotors used so far are based on evaluating the effective unbalances $\mathbf{u}$ from Eq.(5) and on inserting the balancing group $(-\mathbf{u})$ into the rotor. In calculating $\mathbf{u}$, an important role plays $\mathbf{G}(\omega)$ which includes all information on the dynamics of the particular system. It is well known that for discrete systems it may be calculated as

$$\mathbf{G}(\omega) = [-\omega^2\mathbf{M} + i\omega\mathbf{B} + \mathbf{K}]^{-1} \tag{6}$$

For our purposes, however, another formula is more convenient. It may be derived with the use of Laplace transform

$$\mathbf{G}(\omega) = \mathbf{V}.[i\omega\mathbf{I} - \mathbf{S}]^{-1}.\mathbf{W}^H + \mathbf{V}^C.[i\omega\mathbf{I} - \mathbf{S}^C]^{-1}.\mathbf{W}^T, \tag{7}$$

where superscripts C and H represent the complex conjugate and Hermitian transpose respectively. Inverses of the matrices on the right-hand side of Eq(7) may be done rather easily because of the diagonal form of the spectral matrix $\mathbf{S}$, provided all eigenvalues are simple. Hence, the matrix of frequency responses $\mathbf{G}(\omega)$ may be rewritten into the form

$$\mathbf{G}(\omega) \doteq \sum_{n=1}^{N} (i\omega - s_n)^{-1}.\mathbf{V}_n.\mathbf{W}^H, \tag{8}$$

where $\mathbf{V}_n$ is the zero matrix except for the n-th column containing the n-th natural mode $\mathbf{v}_n$. Note that the second part of $\mathbf{G}(\omega)$ containing complex conjugate matrices has been omitted because of its low contribution to the resulting $\mathbf{G}(\omega)$. According to the special form of $\mathbf{V}_n$, Eq.(8) may also be written as

$$\mathbf{G}(\omega) \doteq \sum_{n=1}^{N} (i\omega - s_n)^{-1} \mathbf{v}_n.(\mathbf{w}_n)^H = \sum_{n=1}^{N} (i\omega - s_n)^{-1} \mathbf{A}_n, \tag{9}$$

where $\mathbf{v}_n$ and $\mathbf{w}_n$ are n-th columns of modal matrices $\mathbf{V}$ and $\mathbf{W}$, respectively.

Approximation (9) to the matrix of the rotor frequency responses is advantageous especially in systems in which the number of degrees of freedom is much higher than N. In this case the rank of approximation to matrix $\mathbf{G}(\omega)$ is not greater than N.

Let us assume, in agreement with experience, that the variability of ω influences eigenvalues s_n, while natural modes $\mathbf{v}_n$ and $\mathbf{w}_n$ are influenced only slightly. Since the real dependence of s_n is very complicated, let us approximate it by the complex polynomial

$$s_n(\omega) = \sum_i s_{in}\omega^i = s_{on}.(1 + b_{1n}\omega + b_{2n}\omega^2 + \dots) \tag{10}$$

When neglecting the influence of ω on natural modes, the vibration of the rotor caused by unbalance $\mathbf{u}$ may be calculated from the formula

$$\mathbf{q}(\omega) = \omega^2 \sum_{n=1}^{N} [i\omega - s_n(\omega)]^{-1}.\mathbf{A}_n.\mathbf{u} \tag{11}$$

Unfortunately, the dynamic properties of the rotor to be balanced, eigenvalues $s_n(\omega)$ and matrices $\mathbf{A}_n$ are not known at the beginning of balancing. They have to be determined from the measured data acquired during the balancing runs of the rotor by means of identification.

3 IDENTIFICATION

It is clear from Eq.(11) that the measured data depend on the unknowns nolinearly. This fact complicates the identification procedure for finding s_n and $\mathbf{A}_n$.

The formula given by Eq.(11) may be rewritten for the j-th run and k-th speed into

$$\mathbf{q}_j(\omega_k) = \mathbf{D}_k.\mathbf{A}.\mathbf{u}_j, \tag{12}$$

where

$$\mathbf{D}_k = \left[\frac{\omega_k^2}{i\omega_k - s_1(\omega_k)}\mathbf{I}, \dots, \frac{\omega_k^2}{i\omega_k - s_N(\omega_k)}\mathbf{I}\right] = \mathbf{d}_k^T \otimes \mathbf{I} \tag{13}$$

and

$$\mathbf{A} = \begin{bmatrix} \mathbf{A}_1 \\ \vdots \\ \mathbf{A}_N \end{bmatrix} \tag{14}$$

Symbol $\otimes$ in Eq.(13) is the operator of the Kronecker product of matrices (see Ref.(1)). Vector $\mathbf{d}_k^T$ contains elements

$$d_{kn} = \frac{\omega_k^2}{i\omega_k - s_n(\omega_k)}, \tag{15}$$

which are dynamic amplifications of the n-th modal component. After ordering vectors $\mathbf{q}_j(\omega_k)$ for different balancing speeds ω_k, k = 1,..,K into the vector

$$\mathbf{q}_j = [\mathbf{q}_j^T(\omega_1), \dots, \mathbf{q}_j^T(\omega_K)]^T \tag{16}$$

Eq.(12) takes the form

$$\mathbf{q}_j = [\mathbf{D}_k]_j.\mathbf{A}.\mathbf{u}_j, \tag{17}$$

from which it follows

$$\mathbf{A}.\mathbf{u}_j = [\mathbf{D}_k]_j^+.\mathbf{q}_j = \mathbf{p}_j \tag{18}$$

The matrix

$$[\mathbf{D}_k]_j^+ = ([\mathbf{d}_k]_j \otimes \mathbf{I})^+ = [\mathbf{d}_k]_j^+ \otimes \mathbf{I} \tag{19}$$

is the pseudoinverse matrix to $[\mathbf{D}_k]_j$ in the sense of Moore and Penrose (cf (2), (3)).

The formula (18) is of major importance for the new method of balancing. While types of matrix $[\mathbf{D}_k]_j$ and vector $\mathbf{q}_j$ are dependent on the number of measurements in the j-th run, the dimension of vector $\mathbf{p}_j$ is given only by a fixed type of matrix $\mathbf{A}$ and is independent of the values and number of balancing speeds. It is a great advantage of this approach that it is not necessary to measure the responses of the rotor at the same speeds in different runs. The advantage is apparent at the moment when it is impossible to exceed a certain critical speed, and consequently, data corresponding to higher speeds are not available in that particular balancing run. Nevertheless, even measurements with reduced information may be exploited in the balancing process.

The identification process consists of the following steps:

(a) Set the counter of balancing runs, j = 0.
(b) Make the initial run, measure responses $\mathbf{q}_o(\omega_k)$ to be eliminated, assemble vector $\mathbf{q}_o$ and calculate vector $\mathbf{p}_o$ using Eq.(18).
(c) Estimate eigenvalues s_{on} either by numerical analysis of the rotor-bearing system, or from the data of the initial run, set $b_{1n} = b_{2n} = \ldots = 0$. Use them to estimate the vector of spectral parameters

$$\mathbf{s} = [s_{o1}, b_{11}, \ldots, s_{o2}, b_{12}, \ldots, s_{oN}, b_{1N}, \ldots]^T \qquad (20)$$

(d) Increment counter of runs, j := j+1.
(e) Insert trial balancing group $\Delta\mathbf{u}_j$, run the rotor, measure and assemble vector of (total) responses $\mathbf{q}_j$.
(f) Start the cycle of iterations, $\ell := 0$.
(g) Increment the iteration counter $\ell := \ell + 1$, and set $\mathbf{s}^{(\ell)} = \mathbf{s}$.
(h) Assemble the matrix $[\mathbf{d}_k]_j$ as a function of $\mathbf{s}^{(\ell)}$, calculate its pseudoinverse and the vector $\mathbf{p}_j$ using Eq.(18).
(i) Evaluate estimate $\tilde{\mathbf{A}}$ of matrix $\mathbf{A}$ from modified Eq.(18) as

$$\tilde{\mathbf{A}} = \Delta\mathbf{P}_j \cdot \Delta\mathbf{U}_j^+ \qquad (21)$$

where

$$\mathbf{P} = [\mathbf{p}_o, \ldots, \mathbf{p}_j] \; , \quad \mathbf{U} = [\mathbf{u}_o, \ldots, \mathbf{u}_j] \qquad (22)$$

$$\Delta\mathbf{P} = \mathbf{P}.\mathbf{R} \qquad \Delta\mathbf{U} = \mathbf{U}.\mathbf{R} \qquad (23)$$

User chosen matrix $\mathbf{R}$ ensuring the generation of difference matrices $\Delta\mathbf{P}$ and $\Delta\mathbf{U}$ has the form

$$\mathbf{R} = \begin{bmatrix} \mathbf{o}^T \\ \mathbf{I} \end{bmatrix} - \bar{\mathbf{I}} \qquad (24)$$

where $\bar{\mathbf{I}}$ is a (rectangular) matrix having in every column only one nonzero element equal to one. Thus matrix

$$\mathbf{R} = \begin{bmatrix} -1 & -1 & \ldots & -1 \\ 1 & & & \\ & 1 & & \\ & & \ddots & \\ & & & 1 \end{bmatrix} \quad \text{generates} \qquad (25)$$

$$\Delta\mathbf{U} = [\mathbf{u}_1 - \mathbf{u}_o, \ldots, \mathbf{u}_j - \mathbf{u}_o]$$

while matrix

$$\mathbf{R} = \begin{bmatrix} -1 & & & \\ 1 & -1 & & \\ & \ddots & \ddots & \\ & & 1 & -1 \\ & & & 1 \end{bmatrix} \quad \text{generates} \qquad (26)$$

$$\Delta\mathbf{U} = [\mathbf{u}_1 - \mathbf{u}_o, \ldots, \mathbf{u}_j - \mathbf{u}_{j-1}]$$

The other forms of matrix $\mathbf{R}$ may generate arbitrary difference matrices $\Delta\mathbf{U}$ and $\Delta\mathbf{P}$.
(j) Find vectors $\mathbf{v}_n$ and $\mathbf{w}_n$ from $\tilde{\mathbf{A}}_n$ using formulae derived from the definition of $\tilde{\mathbf{A}}_n$ in Eq.(9)

$$\tilde{\mathbf{v}}_n \cdot \tilde{\mathbf{w}}_n^H = \tilde{\mathbf{A}}_n \qquad (27)$$

If $\tilde{\mathbf{a}}_j$ is a vector of $\tilde{\mathbf{A}}_n$ with maximum length, then it holds

$$\tilde{\mathbf{v}}_n \cdot \tilde{\mathbf{w}}_n^H \cdot \tilde{\mathbf{a}}_j = \tilde{\mathbf{A}}_n \cdot \tilde{\mathbf{a}}_j = \bar{\mathbf{v}}_n \; , \qquad (28)$$

from where the normalized eigenvector is

$$\mathbf{v}_n = \frac{1}{\| \bar{\mathbf{v}}_n \|_2} \cdot \bar{\mathbf{v}}_n \qquad (29)$$

Hence, it follows from Eq.(27)

$$\mathbf{w}_n = \tilde{\mathbf{A}}_n^H \cdot \mathbf{v}_n \qquad (30)$$

Having vectors $\mathbf{v}_n$ and $\mathbf{w}_n$, the averaged matrix $\mathbf{A}_n$ is given by Eq.(9).
(k) Assemble new estimate of matrix $\mathbf{A}$ from Eq.(14).
(l) Get a new vector $\mathbf{s}$ of the parameters of the eigenvalues by means of a nonlinear regression starting from the point $\mathbf{s}^{(\ell)}$ using modified Eq.(16):

$$[\mathbf{D}_k]_j \cdot \Delta\mathbf{P}_j - \Delta\mathbf{Q}_j = \mathbf{O} \; , \qquad (31)$$

where

$$\Delta\mathbf{Q}_j = [\mathbf{q}_o, \mathbf{q}_1, \ldots, \mathbf{q}_j] . \mathbf{R} \qquad (32)$$

(m) If $| \|\mathbf{s}\| - \|\mathbf{s}\|^{(\ell)} | > \varepsilon$, then repeat from (g)
(n) The end of the iteration cycle of the j-th run. Repeat from (d) for the next balancing group, until j=N, i.e. the required number of modes to be balanced.

4 BALANCING

As soon as the identification procedure is successfully completed, the optimum balancing group of weights may be found from a slight modification of Eq.(18). Let the matrix of unbalances $\mathbf{U}$ be decomposed in such a way that it holds

$$\mathbf{A} . [\mathbf{U}_o + \Delta\mathbf{U}] = \mathbf{P} \qquad (33)$$

Hence, matrix $\mathbf{U}_o$ may be obtained simply from Eq.(33) as

$$\mathbf{U}_o = \mathbf{A}^+ . \mathbf{P} - \Delta\mathbf{U} \qquad (34)$$

If the measurement were without any errors and the observed system were linear, matrix $\mathbf{U}_o$ would contain j+1 identical columns $\mathbf{u}_o$. In the real case, however, the columns will be different due to measurement noise and nonlinearities. Then, the effective unbalance vector may be found by an averaging process. Its general form is

$$\mathbf{u}_o = \mathbf{U}_o . \mathbf{c} \qquad (35)$$

where $\mathbf{c}$ is such that $\sum_{i=0}^{j} c_i = 1$.

A special case occurs, if a certain "coefficient of forgetting", $\alpha \leq 1$, is chosen and used for the generation of weight c_i in the form

$$c_i = \frac{\alpha^{j-i}}{\sum_{k=0}^{j} \alpha^k} \qquad (36)$$

The recent information has then a higher weighing coefficient expressing thus that the earlier data are less important in the final stage of balancing.

5 CONCLUSIONS

The above described procedure of balancing flexible rotors makes it possible to take into account changes of eigenvalues of rotor-bearing systems due to the speed-dependent properties of the supporting elements. The procedure becomes simpler and remains equally useful when applied to systems with constant parameters.

The main advantage of the method is that balancing may proceed without any requirements on the number and values of balancing speeds in particular trial runs. Even measurements which are not complete from the point of view of the other methods are fully exploitable. The described method may thus save a number of balancing runs.

REFERENCES

(1) GRAHAM, A. Kronecker Products and Matrix Calculus With Application. Ellis Horwood Ltd., Chichester, 1981

(2) NOBLE, B., DANIEL, J.W. Applied Linear Algebra. Prentice-Hall, Englewood Cliffs, 1977

(3) BUSINGER, P.A., GOLUB, G.H. Singular Value Decomposition of a Complex Matrix. Communications of the ACM, vol. 12, pp. 564-565, Oct. 1969

C308/88

Turbine generator trim balancing using optimized least squares methods

A F P SANDERSON, CEng, MIMechE
Ontario Hydro, Toronto, Ontario, Canada

SYNOPSIS

In-place trim balancing of large steam turbine generator sets is periodically required in order to control vibration during run up through the critical speed regions and at various load conditions at synchronous speed. The effects of slight thermal distortions preclude the elimination of unbalance under all operating conditions. Even when extensive, reliable balance plane calibration data are available, the chosen distribution of balance weights will anticipate residual vibrations. This paper is concerned with that residual vibration distribution. The basic concept involves normalization of influence coefficients and vibration data in terms of desired limits at each point of measurement. Examples are given illustrating the application of the method where vibration limits vary and where the effects of statistical influence coefficient scatter are considered.

INTRODUCTION

Steam turbine generators used in the power generation industry require periodic trim balancing in order to control bearing vibrations and to limit rotor motions within seal clearances. This is achieved by measuring the effect of weight adjustments on the running vibration vectors. The calibration information is then used to determine an appropriate set of corrective weights that will reduce the observed vibrations. Once it has been measured, the calibration data remain available for future use on all similar machines. Minor variations in dynamic behaviour, due to changes in alignment and bearing clearance, do influence the results. Evaluation of the statistical variations in the calibration data for a group of similar machines shows that repeated use and transportation of the data can lead to significant errors in the balancing solution. In practice, this limits the number of calibrations that should be combined in any one balancing calculation. The trim balancing procedure must address these limitations. Computer programmes for balancing are capable of solving the most complex of problems. As a result, the limitation in number of balance planes used for a single weight change are those arising from influence coefficient validity and the economics of manpower and time to do the work.

On those machines where trim balancing has been done many times before, it has become routine at Ontario Hydro to install weights during a single shutdown with the expectation that completely satisfactory performance will result. Often this can be combined with an outage for maintenance to another part of the plant so that the actual cost of performing the balancing is virtually nil. In order to achieve this, a new computer programme has been written which is developed from the work of Goodman (1). The procedure used is called "the normalized least squares method." Its most important features include the capability of dealing with a variety of severity limits and the assessment of errors due to influence coefficient variations. The programme may be used to resolve the often conflicting requirements for weight adjustment considering vibration on run up and at various conditions of load.

TRIM BALANCING OBJECTIVES

Balancing during a concurrent outage requires that the scope of the corrective work be limited to only those balance planes that can reasonably be accessed without extending the critical path. Only those most effective to the solution of

each problem are used. To date, this has not exceeded seven planes during a single trim balance. To permit evaluation of the many permutations of possible planes included in the solution, the balancing programme must allow convenient reformulation of the influence coefficient matrix from the available data base.
The data used to define the unbalance state of the machine will represent all bearings that are significantly effected by the balance planes used. Both extremes of generator output and also selected points at an intermediate speed may be included.

In practice, the operating limits of vibration level may not be the same at all bearings. Generally, the high pressure and intermediate pressure turbines have tighter internal clearances and correspondingly lower limits. Transient conditions such as run up and passage through the initial loading range require consideration in the solution, but are less important than the normal operating point. If a limit is set for each of these data points, then the objective in trim balancing is to provide the greatest practical margin until any limit is exceeded. With no prior knowledge of how the subsequent in service deterioration will be distributed, this strategy gives the solution consistent with the longest cycle time between repeat balances.

Where the balance weights are installed during an outage for other maintenance work, the balance run to assess the effect of weight moves is actually the return to service. The weight adjustments made must, therefore, be highly reliable otherwise there is a risk of extending the outage. Generally, the more balance planes that are included in any balancing calculation, the lower the predicted residual vibration becomes. However, as the solution becomes more complex, the confidence level becomes lower, since the propagation of influence coefficient error and statistical scatter becomes a greater factor. Means of assessing these factors in order to avoid high risk solutions are required.

NORMALIZED LEAST SQUARES

The basic least squares procedure provides a means of solving the general problem where there are more points of measurement than planes available. The solution obtained is the balance weight distribution for minimum root mean square residual vibration. This does not satisfy our objective of providing the greatest margin for deterioration since invariably the residual at some data points will be higher than at others in an unpredictable pattern. Goodman proposed an iterative scheme to reduce the highest residuals at the expense of the lower ones, however, in practical problems, the procedure sometimes fails to converge. A requirement to deal with various limits for residual vibration at the different locations and conditions of measurement exists. The work of Goodman is developed here in order to overcome practical difficulties.

Notation:

$[\alpha]$ rectangular matrix of complex influence coefficients formatted to reflect the planes selected for use.

$[x]$ vector of complex vibrations at the data point used to define the unbalance state.

$[\rho]$ vector of complex residual vibrations associated with a calculated balance weight distribution.

$[\omega]$ vector of complex corrective weights.

$[\lambda 1]$ vector of real convergence coefficients used to stabilize the iteration process.

$[\lambda 2]$ vector of real mini-max coefficients used to equalize the residual vibrations.

$[\lambda 3]$ vector of real normalizing coefficients used to transform the equations to units of equal severity at all measurement points. After reaching a converged solution, the vector of residual vibration magnitudes will be approximately proportional to $[\lambda 3]$.

$[\lambda]$ vector of real composite weighting coefficients used to drive the iteration process.

$[\beta]$ rectangular matrix of weighted influence coefficients obtained by multiplying the elements of $[\alpha]$ by the corresponding terms of $[\lambda]$.

$[\sigma]$ vector of unbalance severities obtained by multiplying the terms of $[x]$ by the corresponding terms of $[\lambda]$.

$[\upsilon]$ matrix of real standard deviations with each element specifying the data base scatter for the corresponding element in the influence coefficient matrix $[\alpha]$.

$[\varepsilon]$ vector of real values each element of which represents the calculated uncertainty of the predicted effect at each data point.

The normalized least squares process is started by setting $[\lambda 1]$ and $[\lambda 2]$ to unity and then assigning values to $[\lambda 3]$. Judgement is required in assessing a once per revolution vibration level for each data point that constitutes a common measure of severity. If the data points include both shaft readings and casing readings, or if a variety of limits apply, then those factors need consideration in filling out $[\lambda 3]$. In some situations, a significant contribution to the overall vibration level is due to second harmonic or nonsynchronous motion. Here the measures of severity should anticipate that this will add to the residual unbalance vibrations. The corresponding value of $\lambda 3$ will be given by the local vibration limit minus the magnitude of vibration not caused by unbalance.

The composite weighting vector is given by:

$$[\lambda] = [\lambda 1 \times \lambda 2 \div \lambda 3] \qquad (1)$$

The weighted influence coefficient matrix $[\beta]$ and weighted unbalances $[\sigma]$ are determined by multiplying the terms of $[\alpha]$ and $[\chi]$ by the corresponding elements of $[\lambda]$. Corrective weights are calculated using the least squares solution:

$$[\omega] = -([\beta]^T[\beta])^{-1}[\beta]^T[\sigma] \qquad (2)$$

The corresponding residual vibrations are calculated in units of equal severity as follows:

$$[\rho] = [\beta][\omega] + [\sigma] \qquad (3)$$

The convergence coefficients are calculated as the average of the residual severity and the value of $\lambda 1$ at the previous iteration:

$$[\lambda 1'] = ([\lambda 1 + |\sigma|] \div 2) \qquad (4)$$

The convergence of $[\lambda 1]$ is tested and unless satisfied, the programme loops back to equation (1). If the convergence criteria are satisfied then an intermediate solution has been obtained. At this point, it is necessary to update the mini-max coefficients in order to continue the iteration process towards its final solution. This is described by the following expressions.

$$[\lambda 2'] = [\lambda 2 \times |\sigma|] \qquad (5)$$

$$[\lambda''] = [\lambda 2' \div \lambda 2'(\max)] \qquad (6)$$

The convergence of $[\lambda 2]$ is tested and unless satisfied, the program loops back to equation (1). If the convergence criteria are satisfied, then the solution is complete. The final residual vibrations may now be determined using the expression:

$$[\rho] = [\alpha][\omega] + [\chi] \qquad (7)$$

CUMULATIVE ERRORS

A common problem with computer balancing solutions occurs when two sets of quite similar influence coefficients are used together in the same solution. Small differences between them can be exploited for reduction of the residual vibration by the application of large weights. The problem with this is that errors in the influence coefficients can eventually swamp the theoretical improvement and produce unpredictable results. The effect will be minimized by an intelligent selection of balance planes or combination of planes for each calibration . In practice, this is achieved by using "couples" and "in phase pairs" whenever possible, so that modal insight is helpful in recognizing which planes or groups of planes should be included in the solution. Assuming that the balance planes chosen produce quite distinct effects, there is still a need to evaluate the potential errors in the various solutions considered. Clearly a convoluted solution in which the resultant improvement is brought about by the sum of several large weight effects at each bearing is sensitive to individual influence coefficient errors. This can be assessed by considering the variations in the measured influence coefficients with repeated use. Processing of the accumulated data base includes calculation of the standard deviation for use in error assessment. Figure 1 shows all elements in the data base population for one influence coefficient set for a group of four similar machines. The vector average is calculated and the standard deviation (shown as a dotted circle) is used to measure the scatter. For each measurement point, the potential error in any solution considered can be estimated by the expression.

$$\varepsilon_j = \sum_{i=1,n} \upsilon_{ij} \times |\omega_i| \qquad (8)$$

where i is the balance plane index and j is the measurement point index.

This is a pessimistic assessment since it assumes that all errors are cumulative at the same angle. However, it provides a means by which the balancing programme can identify high risk solutions. Usually, the difference between predicted and measured residuals is much lower than the value given by (8).

EXAMPLES

The examples chosen here have been selected to illustrate the application of the techniques described in this paper. In both cases, they represent actual case histories and real machines.

The first example is a 510 MW unit with five rotors supported in ten bearings which runs at 3 600 r/min. Vibration at all ten bearings is shaft absolute. The data base for this group of machines is quite extensive and it is required to make use of a short planned outage to make a reliable balance adjustment. The initial calculation includes all balance planes as shown at the lower right of Figure 2, and uses the basic least squares procedure. Although generally low residual vibrations are predicted, there is significant variation from one bearing to the next. It is noted that some planes require small weights and it is doubtful that use of the high pressure to reheat coupling (A CPLG) or reheat mid-span (RHT MID) can be justified. The error indicator (ERR) is quite large particularly at bearings five and six suggesting that this solution is high risk. Figure 3 shows the normalized least squares result for the same problem with all planes included and an equal severity at all bearings. The predicted residual at all bearings is now equal at 23 μm, however, in this case, the error indicators are significantly greater than for the basic least squares. (It should be noted that equal residuals are not generally predicted and that the error indicators are not necessarily larger for the normalized least squares method.) Figure 4 shows how [λ3] can be used to shape the residual vibration distribution. Note that the predicted residuals at bearings one to four are now about half the residuals at the remaining bearings. This distribution of residual vibration is judged to more closely represent equal severity of residual vibration at all bearings. It was obtained by setting λ3 for bearings one to four to 25 μm and using 50 μm at bearings five to ten. Figure 5 shows the solution actually selected which includes only three influence coefficient sets, that is couples on both low pressure rotors and a weight on the generator coupling. The calculation procedure is the normalized least squares with equal severity at all bearings. It was judged here that the planes chosen were not consistent with assigning lower severity numbers to the high pressure and reheat rotor bearings (one to four) as is sometimes done. The important consideration is that the error indicators are considered reasonable and as such, the probability of success is good. The results are also shown in Figure 5.

The second example is included in order to illustrate the effectiveness of the normalized least squares method when both shaft and casing readings are required. This machine is the high pressure line of a 300 MW cross compound unit and comprises three rotors running on six bearings at 3 600 r/min. The design of the generator is somewhat unusual in that there is no steady bearing outboard of the slip rings. In order to control vibration of the brush gear, it is necessary that the shaft vibration near the slip rings be considered in the balancing solution. In the example problem, three balance planes on the generator are available to reduce the vertical and horizontal shaft vibration in the slip rings. It is required that this be done without producing excessive vibration at the main generator bearings numbered five and six. Calculation results are presented in Figure 6, 7 and 8. Note that the basic least squares solution (Figure 6) is very effective in reducing the slip ring vibration, but increases the horizontal vibration at Bearing 6 to 77 micrometres which is an unacceptable level. A better solution is obtained if iteration is used to minimize the maximum residuals (Figure 7), but clearly the 57 micrometres remaining at the slip rings could be allowed to increase if lower residuals at the main bearings could be attained. This is satisfied by the normalized least squares method described in this paper by assigning realistically weighted severity levels to the shaft and main bearing casing measurement points. In this example, it was considered that a ratio of five to one for shaft and casing residual vibration would represent a fair estimate of equal severity. The normalized least squares calculation for this assumption gives reasonable residual vibrations at all points. Further improvement could only be achieved by the inclusion of additional balance planes and is not an economically viable option.

CONCLUSIONS

1. The combined influence coefficient data base from a family of similar turbine generators has been used to compute a reliable balance correction for one of the machines. Potential errors introduced by influence coefficient scatter suggested that in this case, no more than three influence coefficient sets should be combined in the solution.

2. The normalized least squares procedure provides an effective method of control to shape the residual vibration distribution. This is useful where shaft and pedestal data must be used together or whenever varying limits apply.

REFERENCE

(1) Goodman, TP. A least squares method for computing balancing corrections.

ACKNOWLEDGEMENTS

The author wishes to thank Ontario Hydro for permission to present this paper. The contents of the paper represent the personal views of the author and are not implied to be the official views of Ontario Hydro.

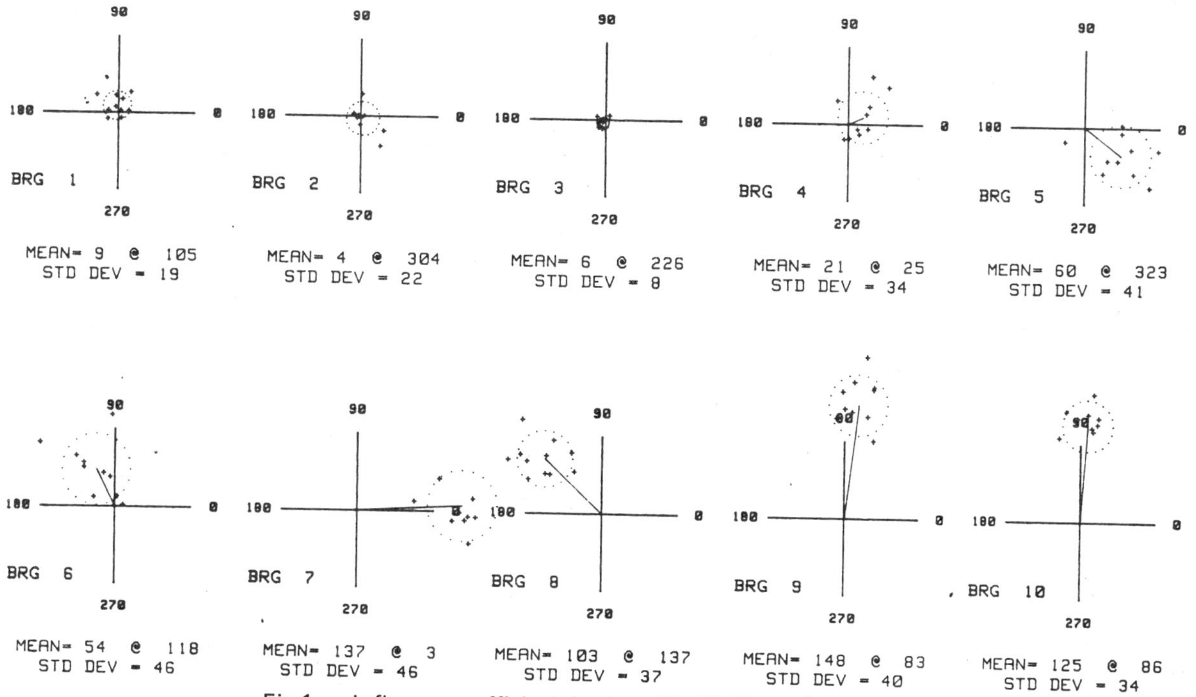

Fig 1 Influence coefficient database 'D CPLG' um/kg population, mean and standard deviation

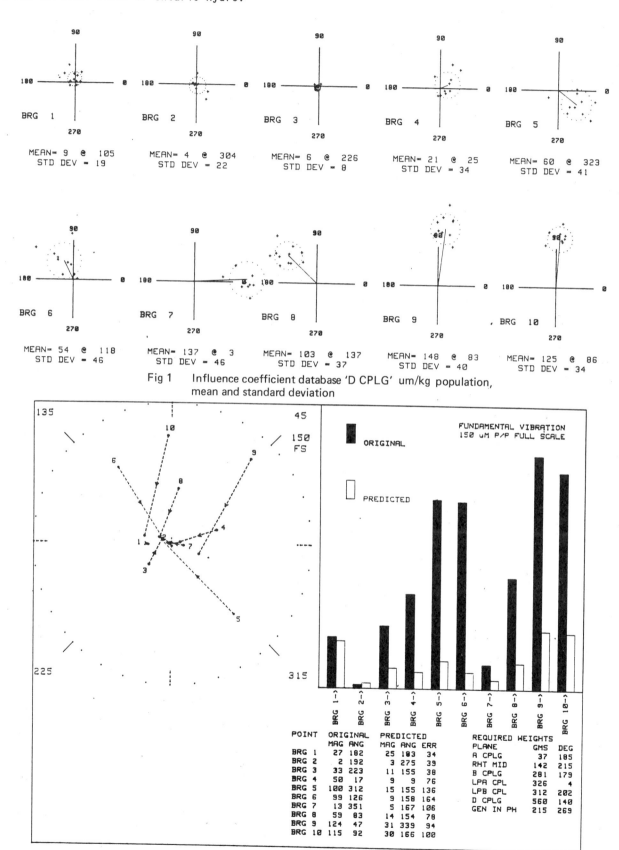

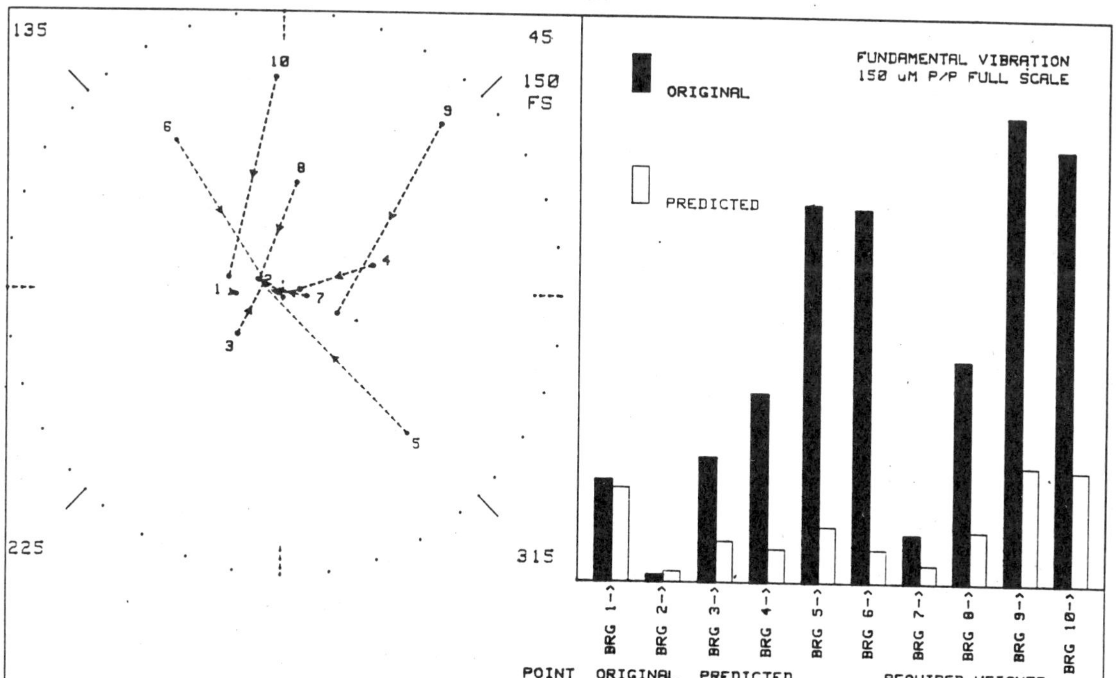

POINT	ORIGINAL MAG	ORIGINAL ANG	PREDICTED MAG	PREDICTED ANG	PREDICTED ERR
BRG 1	27	182	25	183	34
BRG 2	2	192	3	275	39
BRG 3	33	223	11	155	38
BRG 4	50	17	9	9	76
BRG 5	100	312	15	155	136
BRG 6	99	126	9	158	164
BRG 7	13	351	5	167	106
BRG 8	59	83	14	154	78
BRG 9	124	47	31	339	94
BRG 10	115	92	30	166	100

REQUIRED WEIGHTS

PLANE	GMS	DEG
A CPLG	37	185
RHT MID	142	215
B CPLG	281	179
LPA CPL	326	4
LPB CPL	312	202
D CPLG	560	140
GEN IN PH	215	269

Fig 2 Example number 1 least squares solution for all planes

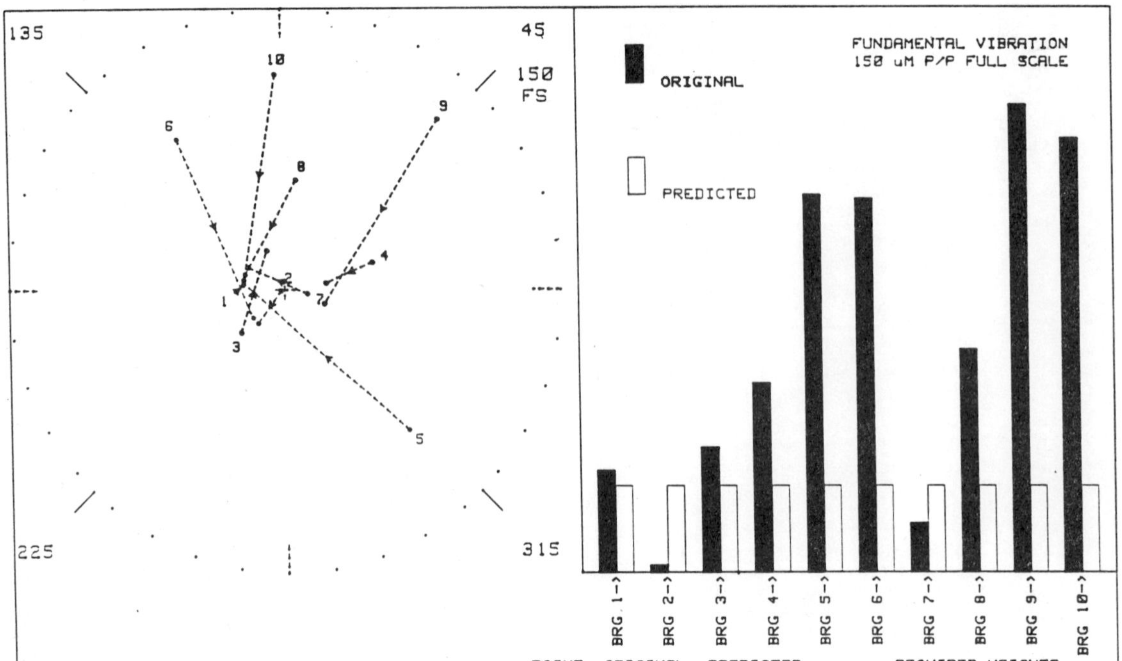

POINT	ORIGINAL MAG	ORIGINAL ANG	PREDICTED MAG	PREDICTED ANG	PREDICTED ERR
BRG 1	27	182	23	174	54
BRG 2	2	192	23	230	68
BRG 3	33	223	23	115	71
BRG 4	50	17	23	9	141
BRG 5	100	312	23	171	206
BRG 6	99	126	23	220	243
BRG 7	13	351	23	149	179
BRG 8	59	83	23	160	132
BRG 9	124	47	23	341	163
BRG 10	115	92	23	168	172

REQUIRED WEIGHTS PLANE	GMS	DEG
A CPLG	208	177
RHT MID	494	227
B CPLG	531	198
LPA CPL	419	13
LPB CPL	354	188
D CPLG	953	127
GEN IN PH	435	289

Fig 3 Example number 1 least squares solution for all planes; equal severity at all bearings

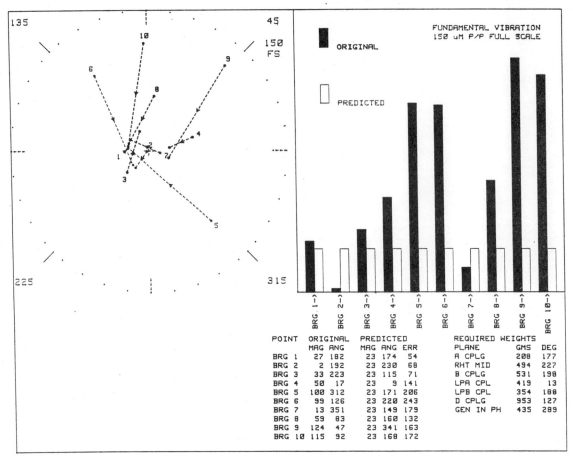

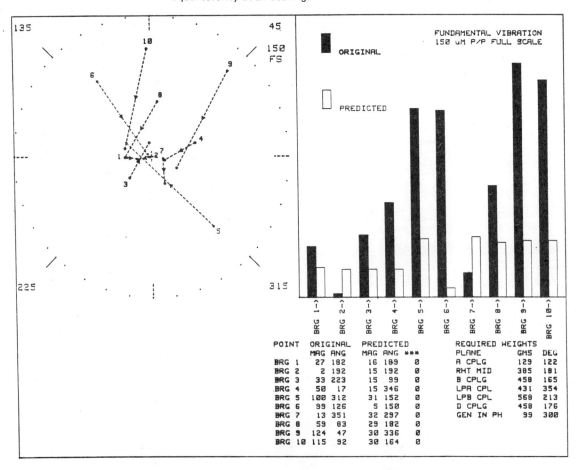

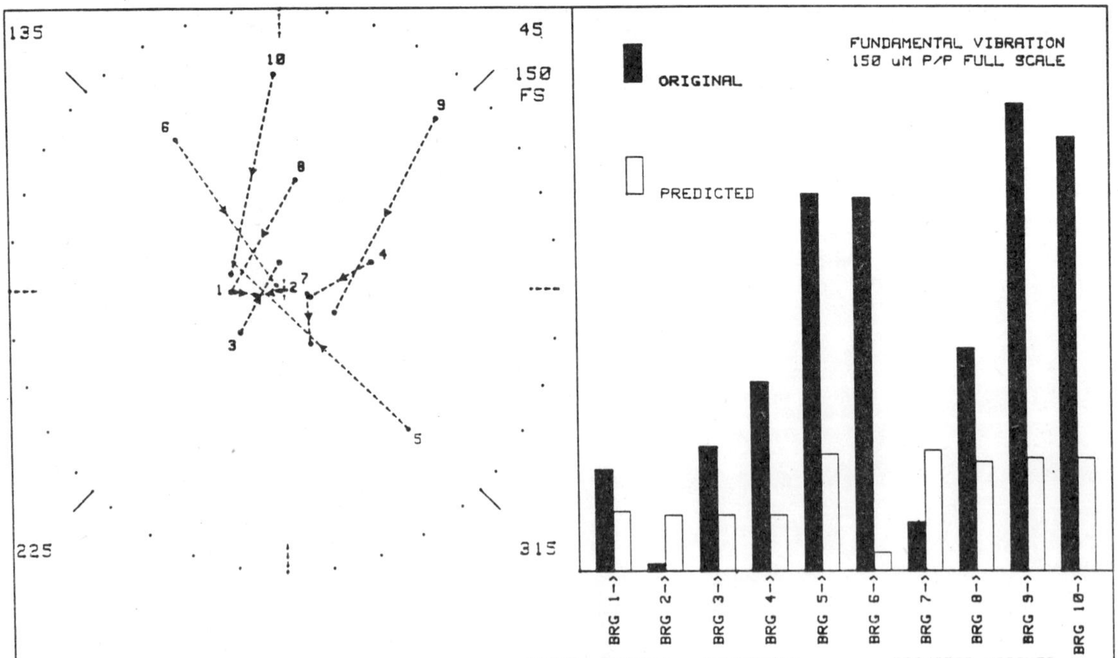

POINT	ORIGINAL MAG	ORIGINAL ANG	PREDICTED MAG	PREDICTED ANG	***
BRG 1	27	182	16	189	0
BRG 2	2	192	15	192	0
BRG 3	33	223	15	99	0
BRG 4	50	17	15	346	0
BRG 5	100	312	31	152	0
BRG 6	99	126	5	150	0
BRG 7	13	351	32	297	0
BRG 8	59	83	29	182	0
BRG 9	124	47	30	336	0
BRG 10	115	92	30	164	0

REQUIRED WEIGHTS PLANE	GMS	DEG
A CPLG	129	122
RHT MID	385	181
B CPLG	458	165
LPA CPL	431	354
LPB CPL	568	213
D CPLG	458	176
GEN IN PH	99	300

Fig 4 Example number 1 least squares solution for all planes; bearings 1–4 severity = 25, bearings 5–10 severity = 50

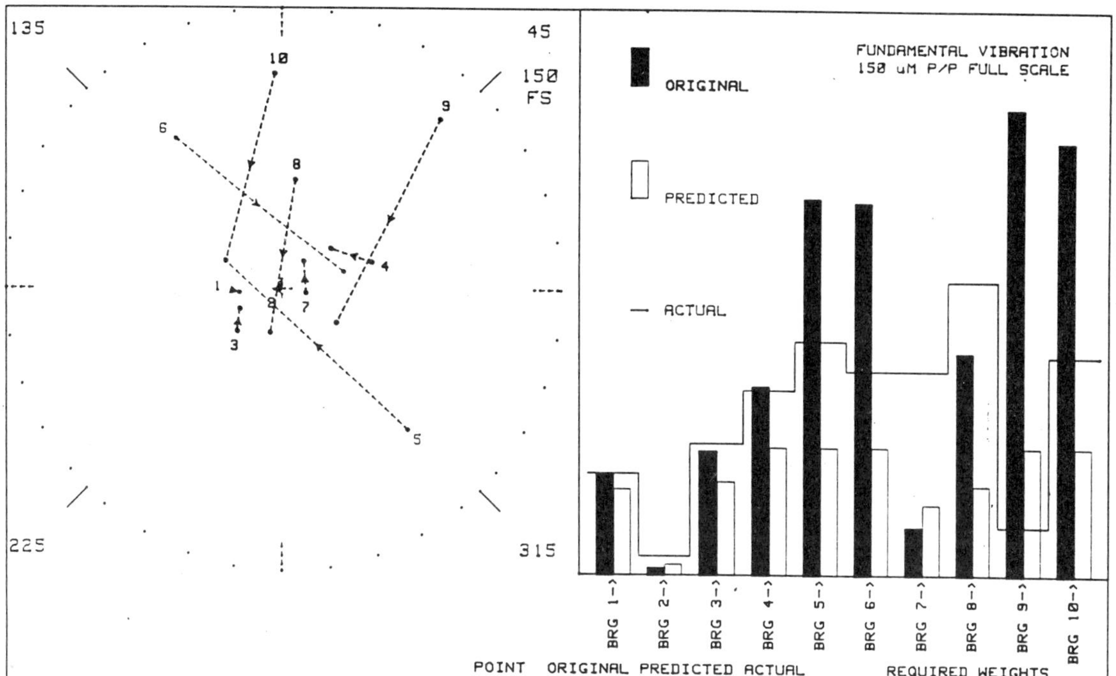

POINT	ORIGINAL MAG	ORIGINAL ANG	PREDICTED MAG	PREDICTED ANG	ERR	ACTUAL MAG	ACTUAL ANG
BRG 1	27	182	23	185	26	27	180
BRG 2	2	192	3	102	24	5	152
BRG 3	33	223	25	205	16	35	208
BRG 4	50	17	34	40	34	49	38
BRG 5	100	312	34	154	104	62	50
BRG 6	99	126	34	16	135	54	224
BRG 7	13	351	19	52	71	54	320
BRG 8	59	83	24	255	52	78	166
BRG 9	124	47	34	328	61	13	210
BRG 10	115	92	34	154	59	58	158

REQUIRED WEIGHTS

PLANE	GMS	DEG
A CPLG	0	0
RHT MID	0	0
B CPLG	0	0
LPA CPL	305	16
LPB CPL	412	232
D CPLG	491	171
GEN IN PH	0	0

Fig 5 Example number 1 normalized least squares solution for limited planes; equal severity at all bearings (results are included)

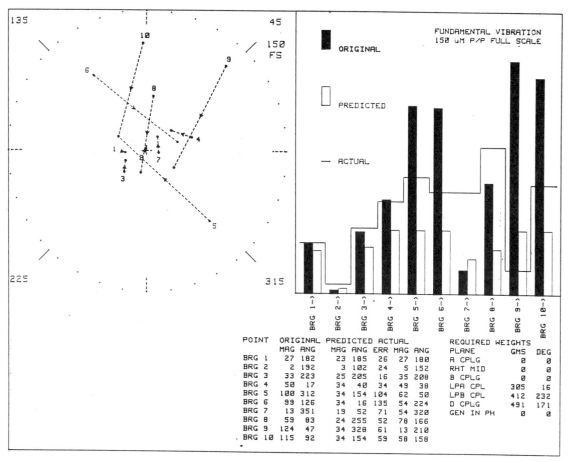

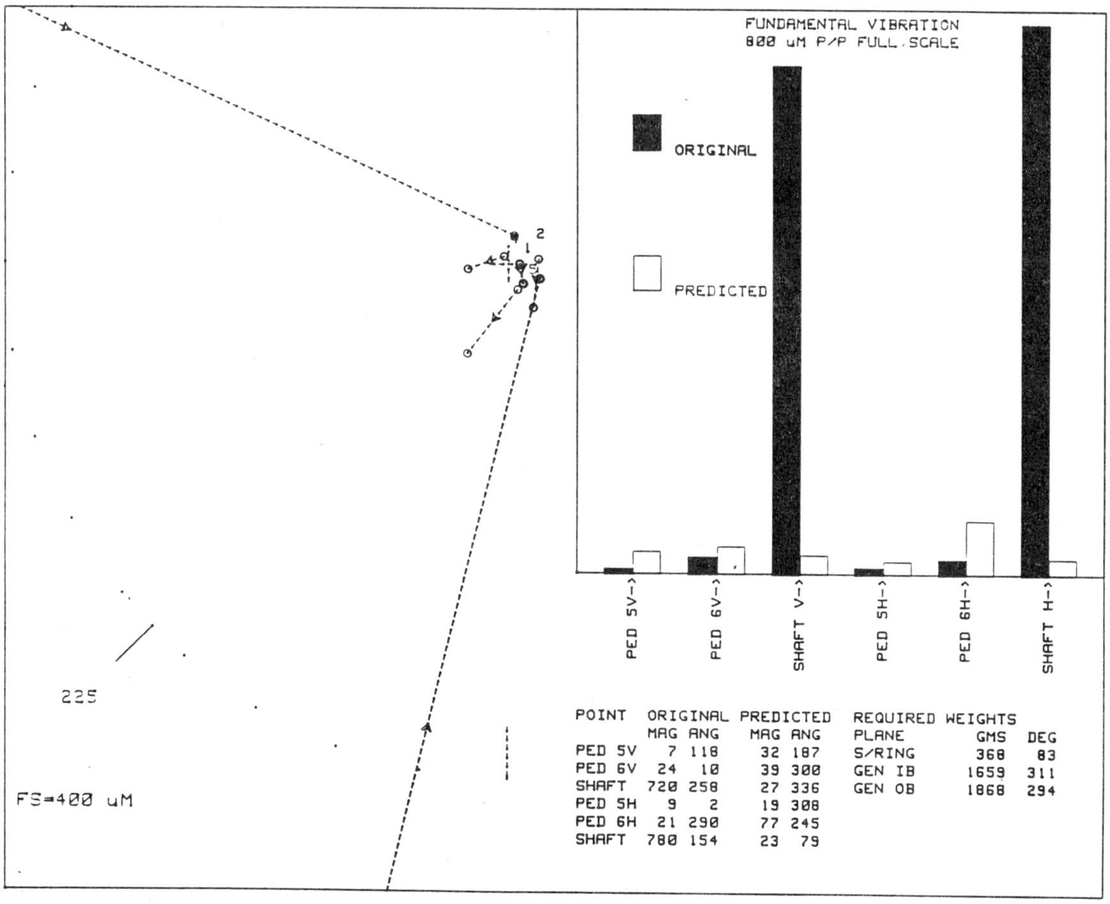

POINT	ORIGINAL MAG	ORIGINAL ANG	PREDICTED MAG	PREDICTED ANG
PED 5V	7	118	32	187
PED 6V	24	10	39	300
SHAFT	720	258	27	336
PED 5H	9	2	19	308
PED 6H	21	290	77	245
SHAFT	780	154	23	79

REQUIRED WEIGHTS

PLANE	GMS	DEG
S/RING	368	83
GEN IB	1659	311
GEN OB	1868	294

Fig 6 Example number 2 least squares method

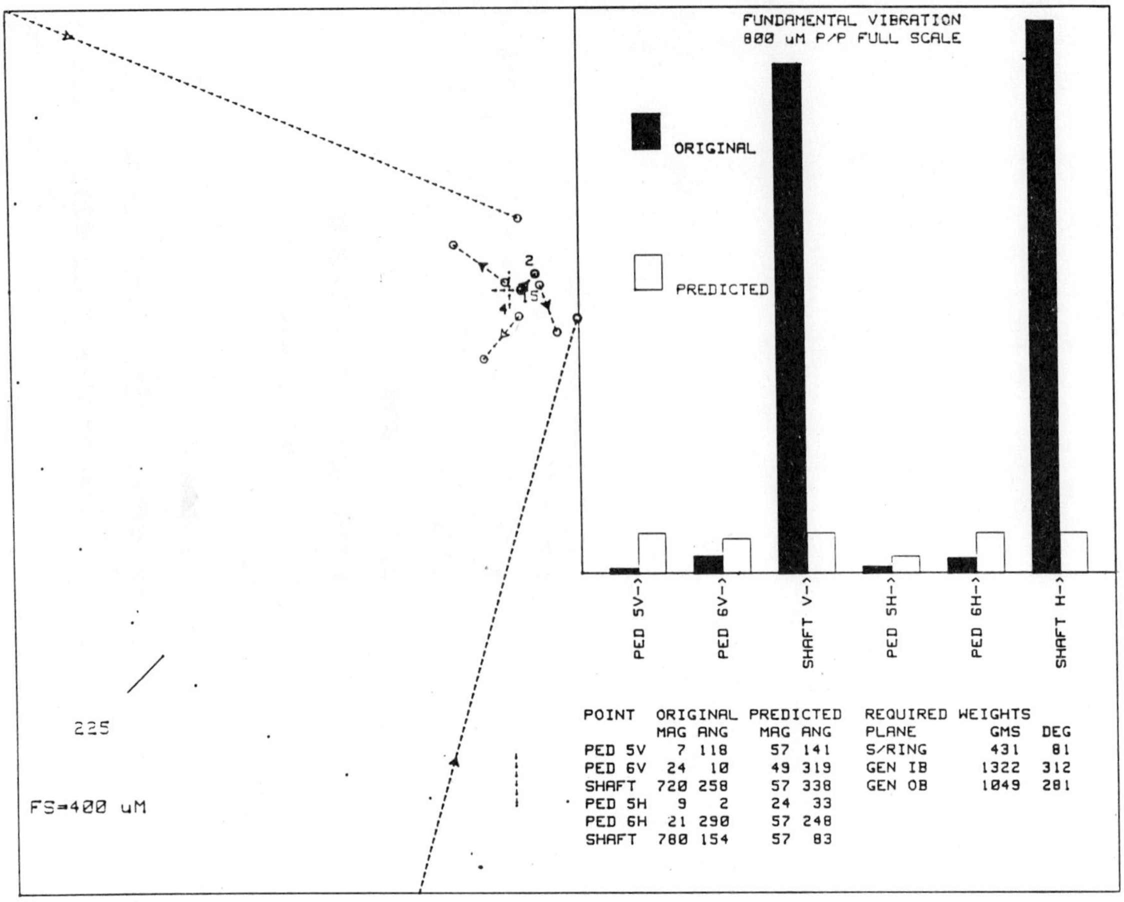

Fig 7 Example number 2 normalized least squares

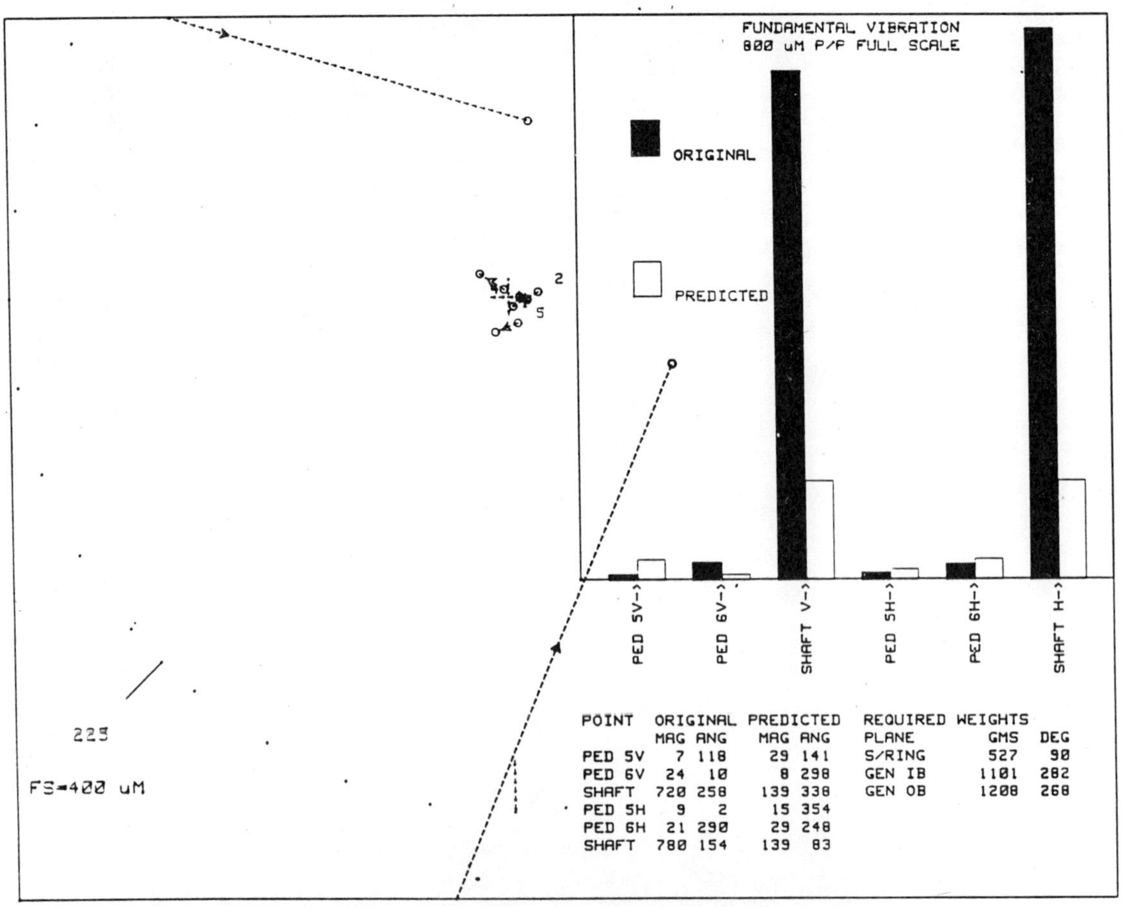

Fig 8 Example number 2 normalized least squares; unequal severity

C249/88

An improved least squares method for calculating balancing correction masses

H GU, H R REN and **J M YANG** MSc
Department of Thermoscience and Engineering, Zhejiang University, Hangzhou, People's Republic of China

SYNOPSIS In this paper an improved least squares method for calculating balancing correction masses is developed. This improved method can provide the possibility for balancing engineers to make the best choice of magnitudes and phase angles of correction masses according to the needs of an acceptable level of residual vibrations. The correction masses calculated by using this improved least squares method are more reasonable than those obtained by using Goodman's method and Darlow's method for identifying and eliminating non-independent balance planes.

1 INTRODUCTION

During the past two decades the least squares method developed by Goodman (1) for calculating balancing correction masses has received wide applications in the manufacture and balancing of field rotors. However, owing to the physical meaning of balancing being ignored and overemphasis on the minimum of residual vibrations, in the case of inadvertent selection of balance planes, especially in multi-bearing multi-plane rotor balancing, there may exist non-independent or nearly non-independent columns in the influence coefficient matrix. In this case the calculation may result in impractically large, mutually counteracting correction masses (2) and the errors in calculating correction masses due to measurement errors are generally also very large in practice.

In ref. (3) Darlow presented a method to identify and eliminate the non-independent balance planes using a Gram-Schmidt orthogonalization procedure. Darlow defined a "significance factor" as a criterion to identify the non-independence of the balance plane. When the value of the "significance factor" is less than 0.2 this balance plane is considered as a non-independent plane and has to be eliminated. After the non-independent plane has been eliminated the further calculation is still the Goodman's method.

In this paper based on the theory of rotor dynamics the condition of non-independent planes is analysed and an improved least squares method for calculating correction masses by using successive iteration is developed. Three numerical examples for rotor balancing are presented to explain the process of iterative calculations and to compare the improved method with Goodman's method and Darlow's identification method. The comparison shows the significance and advantage of the improved least squares method.

2 CHARACTERISTICS OF THE NON-INDEPENDENT BALANCE PLANES

2.1 Influence coefficient

The influence coefficient is defined as the vibration resulting from placing a unit trial mass on a balance plane. For the nth balance plane and the ith vibration reading the influence coefficient α_{in} can be expressed as follows:

$$\alpha_{in} = \frac{A_{in} - A_{io}}{W_n} \qquad (1)$$

Here, A_{io}, A_{in} are vibration readings before and after the trial mass W_n is applied on the nth balance plane respectively.

According to the theory of rotor dynamics the unbalanced response of a rotor can be expressed in the form:

$$Z(s,t) = \sum_{r=1}^{\infty} \frac{E_r \phi_r(s)\, f_r^2}{\sqrt{(1-f_r^2)^2 + (2\zeta_r f_r)^2}} e^{I(\Omega t - \varphi_r)} \qquad (2)$$

Here

$Z(s,t)$ is the deformation of a rotor at an axial location s and time t

$\phi_r(s)$ — r^{th} modal function

$f_r = \frac{\Omega}{\omega_r}$ — r^{th} modal frequency ratio

Ω — angular speed

ω_r — r^{th} critical angular speed

ζ_r — r^{th} modal damping ratio

$$\varphi_r = tg^{-1}\left(\frac{2\zeta_r f_r}{1-f_r^2}\right)$$

$$E_r = e_r + \sum_{n=1}^{N} U_n \delta(s-s_n)\phi_r(s_n)\frac{1}{M_r}$$

e_r — amplitude of rth modal continuous unbalance

U_n — isolated unbalance

$\delta(s-s)$ — delta function

$M_r = \int_0^{\ell} m(s)\varphi_r^2(s)ds$ — r^{th} modal mass

ℓ — bearing distance

Therefore the vibration variation at the measuring point i after and before the trial mass is applied on the nth balance plane is:

$$A_{in} - A_{io} = \sum_{r=1}^{\infty} \frac{W_n e^{I\Omega t}\phi_r(S_n)\phi_r(S_i)}{\sqrt{(1-f_r^2)^2 + (2\zeta_r f_r)^2}} \frac{f_r^2}{M_r} e^{-I\varphi_r}$$

and the corresponding influence coefficient will be:

$$\alpha_{in} = \frac{A_{in} - A_{io}}{W_n e^{I\Omega t}} = \sum_{r=1}^{\infty} \frac{\Phi_r(S_n)\Phi_r(S_i)}{\sqrt{(1-f_r^2)^2 + (2\varsigma_r f_r)^2}} \cdot \frac{f_r^2}{M_r} e^{-I\varphi_r} \qquad (3)$$

In expression (3) each term expresses the contribution of the corresponding modal component. Thus, we define the influence coefficient as follows:

$$\alpha_{in}^{(r)} = \frac{\Phi_r(S_n)\Phi_r(S_i)}{\sqrt{(1-f_r^2)^2 + (2\varsigma_r f_r)^2}} \frac{f_r^2}{M_r} e^{-I\varphi_r}$$

Thus, $\alpha_{in} = \sum_{r=1}^{\infty} \alpha_{in}^{(r)}$

From the above we see that the influence coefficient is the vector sum of various modal influence coefficients. For the given rotor system all the modal parameters are definite. The values of various components included in the general influence coefficient depend not only on the locations of the balance planes and the measuring points, but also on the modal frequency ratio. In general, when the selected balancing speed is near the critical speed, the larger the contribution of this modal component will be. Therefore for the influence coefficient usually there are only a few modal components which are dominating.

2.2 Non-independence of balance planes

Correction masses are added to a rotor to reduce vibrations. For a rotor having the initial vibrations (Ao) after the correction masses W are added, the residual vibrations $\{\varepsilon\}$ of a rotor will be:

$$\{\varepsilon\} = \{A_o\} + [\alpha]\{W\} \qquad (4)$$

According to Goodman's method in order to minimize the sum of squared residual vibrations the correction masses can be found as follows:

$$\{W\} = -([\bar{\alpha}]^T[\alpha])^{-1}[\bar{\alpha}]^T\{A_o\} \qquad (5)$$

From the mathematical theory when there exist non-independent columns in an influence coefficient matrix $[\alpha]$, $[\bar{\alpha}]^T[\alpha]$ is singular and the solution vector becomes infinite. When there exist nearly non-independent columns in matrix $[\alpha]$, Equation (5) is ill conditioned. Moreover, the more the non-independence, the more serious the ill condition will be. Mathematically the existence of non-independent columns in a matrix means that a certain column can be obtained by the linear combination of other columns. In the influence coefficient balancing method each column in matrix $[\alpha]$ expresses the influence coefficients of a balance plane with respect to all measuring points. Therefore mathematical non-independence of columns is equal to the non-independence of balance planes. The physical meaning of non-independence of balance planes means that among the selected balance planes a certain plane can be replaced by other balance planes.

Unbalanced vibrations of a rotor consist of various modal unbalance vibrations. Theoretically for a certain modal vibration to vanish it is necessary to use only one balance plane. If two balance planes are used we can consider two balance planes as one equivalent balance plane in calculation and place correction masses. If we use two balance planes separately calculations will be prone to certain difficulties due to the non-independence.

In the balancing practice of using the influence coefficient method it is almost impossible for complete non-independence of balance planes to exist due to the simultaneous influences of various modal components. However under the selected balancing speed it is always true that contributions of some modal components are larger than those of others. The less the number of dominant modal components are excited at balancing speed the more the non-independence will be. There must exist non-independent balance planes especially when the number of balance planes is greater than the number of dominant modal components (4).

The non-independence of balance planes is a relative concept which relates to the balancing speed and selection of balance planes. The more the content of modal components at balancing speed with no one modal component predominant, the less the possibility of the existence of non-independence.

3 IMPROVED LEAST SQUARES METHOD FOR CALCULATING THE BALANCING CORRECTION MASSES

Suppose that N represents the number of balance planes and M the number of vibration readings. According to Equation (4) we have the sum of the squared residual vibrations S in the following form:

$$S = \sum_{i=1}^{M} \varepsilon_i \bar{\varepsilon}_i = \sum_{i=1}^{M} (A_{io} + \sum_{j=1}^{N} \alpha_{ij} W_j)(\bar{A}_{io} + \sum_{j=1}^{N} \bar{\alpha}_{ij} \bar{W}_j) \qquad (6)$$

Next, we express the correction mass W_j by the product of magnitude $|W_j|$ and phase angle q_j i.e. $W_j = |W_j| e^{Iq_j}$, then Equation (6) becomes:

$$S = \sum_{i=1}^{M} (\bar{A}_{io} + \sum_{j=1}^{N} \bar{\alpha}_{ij} |W_j| e^{-Iq_j}) \cdot (A_{io} + \sum_{j=1}^{N} \alpha_{ij} |W_j| e^{Iq_j}) \qquad (7)$$

Here, I is an imaginary number unit

In matrix form Equation (7) becomes:

$$S = (\{\bar{A}_o\} + [\bar{\alpha}][|W|]\{\bar{Q}\})^T(\{A_o\} + [\alpha][|W|]\{Q\}) \qquad (8)$$

or

$$S = (\{\bar{A}_o\} + [\bar{\alpha}][\bar{Q}]\{|W|\})^T(\{A_o\} + [\alpha][Q]\{|W|\}) \qquad (8')$$

Here

$$[Q] = \begin{bmatrix} e^{Iq_1} & & & \\ & e^{Iq_2} & & \\ & & \cdot & \\ & & & \cdot \\ & & & & e^{Iq_N} \end{bmatrix} \qquad \{Q\} = \begin{Bmatrix} e^{Iq_1} \\ e^{Iq_2} \\ \cdot \\ \cdot \\ e^{Iq_N} \end{Bmatrix}$$

$$[|W|] = \begin{bmatrix} |W_1| & & & & & \\ & |W_2| & & & & \\ & & \cdot & & & \\ & & & \cdot & & \\ & & & & \cdot & \\ & & & & & |W_N| \end{bmatrix} \qquad \{|W|\} = \begin{Bmatrix} |W_1| \\ |W_2| \\ \cdot \\ \cdot \\ \cdot \\ |W_N| \end{Bmatrix}$$

In the following instead of finding the minimum of Equation (6) with respect to $\overline{W}_j$ or W_j by Goodman's method we use Equation (7) and find the minimum of S with respect to $|W_j|$, q_j separately and successively.

Under the assumed constant angles of correction masses the optimum magnitudes of correction masses can be determined by using $\frac{\partial S}{\partial |W_j|} = 0$ with the following expression:

$$\sum_{i=1}^{M} [Re(A_{io}\bar{\alpha}_{ij}e^{-Iq_j}) + Re(\bar{\alpha}_{ij}e^{-Iq_j}\sum_{k=1}^{N}\alpha_{ik}|W_k|e^{Iq_k})]$$
$$= 0 \qquad j=1,2......N \qquad (9)$$

or in matrix form:

$$Re([\bar{Q}]^T[\bar{\alpha}]^T\{A_o\} + [\bar{Q}]^T[\bar{\alpha}]^T[\alpha][Q]\{|W|\}) = 0 \quad (10)$$

After the magnitudes of correction masses $|W_j|$ have been obtained we substitute them into Equation (7) and using the optimum condition $\frac{\partial S}{\partial q_j} = 0$ we get:

$$I_m[\sum_{i=1}^{M}(\bar{A}_{io}\alpha_{ij}|W_j|e^{Iq_j})] + $$
$$+ I_m[\sum_{i=1}^{M}(\alpha_{ij}|W_j|e^{Iq_j}\sum_{\substack{k=1\\k\neq j}}^{N}\bar{\alpha}_{ik}|W_k|e^{-Iq_k})] = 0$$
$$j=1,2......N \qquad (11)$$
$$k \neq j$$

In order to ensure that S is a minimum, rather than a maximum, $\frac{\partial^2 S}{\partial q_j^2}$ must be >0. Using this condition with Equation (7) we get the following expression:

$$-\{Re(\sum_{i=1}^{M}(\bar{A}_{io}\alpha_{ij}|W_j|e^{Iq_j}))$$
$$+ Re(\sum_{i=1}^{M}\alpha_{ij}|W_j|e^{Iq_j}\sum_{\substack{k=1\\k\neq j}}^{N}\bar{\alpha}_{ik}|W_k|e^{-Iq_k})\} > 0$$
$$j=1,2......N$$
$$k \neq j \qquad (12)$$

Therefore in order to ensure that S is a minimum, the phase angles must satisfy Equations (11) and (12). When the imaginary part of a complex is zero and the real part is less than zero, the argument of this complex must equal 180°. Thus, we have:

$$Arg[\sum_{i=1}^{M}(\bar{A}_{io}\alpha_{ij}|W_j|e^{Iq_j})$$
$$+ \sum_{i=1}^{M}(\alpha_{ij}|W_j|e^{Iq_j}\sum_{\substack{k=1\\k\neq j}}^{N}\bar{\alpha}_{ik}|W_k|e^{-Iq_k})] = 180°$$
$$j=1,2......N$$
$$k \neq j \qquad (13)$$

Here, Arg[ ] is the argument of complex.

On substituting the values of the phase angles of correction masses obtained from Equation (13) into Equation (9) we can calculate the next value of magnitude of the correction masses.

From the above we see that the main feature of the improved least squares method is in the use of Equation (9) and (13) successively to find the optimum magnitudes and phase angles of correction masses.

Theoretically the above iterative calculations can begin either with assumed arbitrary phase angles or assumed arbitrary magnitudes of correction masses. As a matter of experience, in order to get a faster convergence we usually choose the optimum phase angles for each balance plane, neglecting the mutual influence of balance planes, as the starting phase angles of correction masses.

Let the partial differential (Equation (7)) $\frac{\partial S}{\partial q_j} = 0$ then the optimum phase angles for each balance plane, neglecting the mutual influences of balance planes can be derived in the form:

$$q_j = -\pi + Arg(\sum_{i=1}^{M} A_{io}\bar{\alpha}_{ij}) \quad j=1,2...N \qquad (14)$$

4 NUMERICAL EXAMPLE AND COMPARISON

In order to explain the practical significance of the iterative calculation method presented in this paper, the authors have composed a computer program for iterative calculations.

For evaluating the balancing quality the total correction masses and root-mean-square (R.M.S.) of residual vibrations are used.

R.M.S. is defined as:

$$R.M.S. = \sqrt{\frac{1}{M}\sum_{i=1}^{M}\bar{\varepsilon}_i\varepsilon_i}$$

Case 1. This is a practical example for balancing a 100 MW steam turbine-generator rotor, having three balance planes and three vibration readings N=M=3. The initial data are presented in Table 1. As shown in ref. (4) for this example there is a nearly non-independent balance plane. Thus, by using Goodman's method the theoretical vibrations can be reduced to zero, but the correction masses are impractically large (in total 1.328 kg).

In Table 1 also are shown the values of "significance factors" of the three planes calculated by using Darlow's method. The "significance factor" of the third plane is less than 0.2, thus according to Darlow's method this third plane has to be eliminated.

According to the improved least squares method presented in this paper, without eliminating the third plane, with the initial data from Table 1 under the variously assumed starting values of magnitudes and phase angles including the use of Equation (14), the results of successive iterative calculations are summarized and plotted in Table 2, Figure 1.

From the curves in Fig.1 it can be clearly shown that with the decrease of R.M.S. of residual vibrations the magnitudes of total correction masses increase rapidly. The convergence of iterative calculation is fast and no matter whether the calculations begin with the assumed starting values of phase angles or with magnitudes of correction masses, the solution finally approaches Goodman's solution.

From Table 2 we see that the improved least squares method, by using successive iterative calculations, can provide the possibility from many intermediate iterative results for balancing engineers to make the best choice of magnitudes and phase angles of correction masses. With respect to this example the second or fourth set of iterative results may be the best choice, since the magnitude of the total correction masses is small and the R.M.S. is still an acceptable low level. Although in this case, according to Goodman's method, the R.M.S. can be reduced to zero, this solution is of no practical significance because the total correction masses (1.328 kg) are too large due to the existence of one nearly non-independent balance plane. If too large correction masses are applied on a rotor, unexpected stress may occur.

Case 2. In Table 3 is presented another example cited from Darlow's paper (3). Since the "significance factors" of all three planes are greater than 0.2, Darlow considered that all three planes are independent and the solution by using Goodman's method is reasonable.

The calculated results using the improved least squares method presented in this paper and Goodman's solution are plotted in Figure 2.

The curve in Figure 2 shows that the R.M.S. of residual vibrations by using the improved method after a second iteration is already quite close to that of Goodman's solution. The numerical comparison between the results of a second iteration and Goodman's solution is also presented in Table 4. From Table 4 we see that after a second iteration the R.M.S. of both methods are quite close but the correction masses by using the improved least squares method are only 73% of that by using Goodman's method, moreover, the maximum residual vibration by using the improved method is lower than that obtained by Goodman's method. Therefore this example indicates that even if the value of the "significance factor" is much greater than 0.2, Goodman's solution is still not the best choice. Perhaps this is really the disadvantage of Darlow's identification method.

Case 3. This example is also cited from Darlow's paper (3). The initial data are presented in Table 5. The value of the "significance factor" of the second balance plane, 0.110, is less than 0.2. Thus, according to Darlow's method, the second plane is a non-independent plane and has to be eliminated. However, by using the improved least squares method presented in this paper, this second plane can still be retained. The successive iterative results are plotted in Figure 3. The iterative curve shows that after a second iteration a large portion of the vibrations has been dropped and the correction masses of the second iteration are much smaller than that (2.9005) obtained by using Darlow's method after eliminating the second plane (3).

5 CONCLUSION

(1) Goodman's method overemphasizes the minimum of sum of squared residual vibrations. If there exists a non-independent or nearly non-independent balance plane, the calculation may result in impractical large correction masses.

(2) Gompared with the Goodman's method, the improved least squares method developed in this paper can reduce the non-independence of balance planes and the calculated correction masses are generally reasonable. The more the iterations are carried on, the less the R.M.S. and the more the total correction masses will be. Finally the solution approaches to that of Goodman's method.

(3) Darlow suggested a value of the "significance factor" 0.2 as a criterion to identify and eliminate the non-independent balance plane. However, the calculations of some examples show that whether the value of the "significance factor" is greater or less than 0.2, the improved least squares method presented in this paper can provide better results than Darlow's method.

(4) The improved successive iterative method can provide the possibility for balancing engineers to choose the optimum total correction masses and residual vibrations according to the needs of an acceptable level of residual vibrations.

REFERENCES

(1) GOODMAN, T.P. A least squares method for computing balancing corrections. ASME Transaction, Journal of Engineering for Industry, 1964, vol.86, 273-279.

(2) ZHANG, H.S. Calculation method for the optimal on-site dynamic balancing for a steam turbine-generator, CASE, Power Engineering, 1983.5,1-10.

(3) DARLOW, M.S. The identification and elimination of non-independent balance planes in influence coefficient balancing. ASME paper, 82-GT-269.

(4) YANG, J.M. Study on improving balancing method of flexible rotors. Msc Thesis, Jan. 1985, Zhejiang University, P.R.China.

Table 2 Iterative results for case 1

Initial vibration	Measuring point	1	2	3
	values $\mu m \angle °$	54 $\angle 135$	35 $\angle 250$	49 $\angle 342$

Goodman's solution:	Correction masses (kg)	Phase (°)	R.M.S. (μm)
	.2844 .261 .7832	28.0 162.0 346.0	0.

A. Starting values of phase angles of correction masses are given by using Eq. (14) and the corresponding curve in Fig. (1) is A

No: Itera-tion	Balance planes	Correction masses Mass(kg)	Correction masses Phase(°)	Residual vibration (μm)	R.M.S. (μm)
0	1 2 3		347.3 7.9 12.2		
1	1 2 3	.113 .06 .247	347.3 7.9 12.2	7.884 17.491 12.976	13.37
2	1 2 3	.113 .06 .247	19.5 7.8 .9	11.294 12.338 9.28	11.043
: :		: :	: :		
9	1 2 3	.257 .218 .71	26.7 158.5 346.7	.432 2.14 2.541	1.934
10	1 2 3	.257 .218 .71	27.2 160.0 346.7	1.263 2.157 1.22	1.606
: :		: :	: :		
19	1 2 3	.282 .258 .778	28.0 162.2 346.6	.112 .171 .263	.192
20	1 2 3	.282 .258 .778	28.1 162.4 346.5	.178 .172 .131	.162

B. Starting values of phase angles of correction masses are given arbitrarily and the corresponding curve in Fig. (1) is B

No:	correction masses Mass(kg)	correction masses Phase(°)	Residual vibration (μm)	R.M.S. (μm)
0		372.7 187.1 12.2		
1	.135 .009 .336	372.7 187.1 12.2	4.797 12.403 15.191	11.656
2	.135 .009 .336	30.2 307.8 2.6	8.752 10.497 7.811	9.088
: :	: :	: :		
13	.203 .112 .543	387.9 517.7 712.8	1.976 5.736 7.477	5.559
14	.203 .112 .543	28.6 163.3 352.2	4.924 5.661 4.428	5.03
: :	: :	: :		
25	.282 .256 .784	390.8 529.2 710.3	1.226 .694 1.185	1.063
26	.282 .256 .784	30.7 168.9 350.0	1.175 .68 1.115	1.014

C. Starting values of the magnitudes of correction masses are given arbitrarily and the corresponding curve in Fig. (1) is C

No.	Correction masses Mass(kg)	Correction masses Phase(°)	Residual vibration (μm)	R.M.S. (μm)
0	.109 .101 .186			
1	.109 .101 .186	18.1 7.4 358.6	12.458 11.46 11.099	11.686
2	.223 .111 .515	18.1 7.4 358.6	8.845 7.239 6.698	7.648
: :	: :	: :		
7	.237 .178 .64	26.1 156.4 346.5	2.601 3.555 2.57	2.944
8	.249 .202 .682	26.1 156.4 346.5	.783 2.721 3.294	2.508
: :	: :	: :		
13	.268 .237 .742	27.5 160.8 346.4	.686 1.324 .665	.943
14	.274 .246 .758	27.5 160.8 346.4	.192 .835 .951	.739

Table 1 Initial data for case 1

sensor	influence coefficient μm/° / g/° plane1	plane2	plane3	initial vibration μm/°
1	0.0465 /325.71°	0.2174 /290.18°	0.1242 /299.97°	54 /135°
2	0.1409 /43.36°	0.0629 /88.45°	0.0145 /79.55°	35 /250°
3	0.1158 /209.07°	0.2106 /167.84°	0.1325 /151.02°	49 /342°
"significance factor"	0.78049	1.0	0.17795	RMS=46.7

Table 3 Initial for case 2

sensor	influence coefficient plane1	plane2	plane3	initial vibration
1	1.414 /45°	2.236 /26.6°	2.606 /33.7°	3.162 /71.6°
2	3.162 /71.6°	4.472 /26.6°	2.236 /26.6°	3.162 /18.4°
3	2.828 /45°	2.236 /26.6°	5.0 /36.9°	4.123 /14°
4	3.162 /18.4°	3.606 /37.7°	4.472 /26.6°	5.385 /68.2°
"signif. factor"	0.331	0.502	1.0	RMS= 4.062

Table 4 Result comparison for case 2

correction masses			Residual vibration		
plane	Goodman's method	Improved method	Sensor	Goodman's method	Improved method
1	1.394 /356.4°	0.945 /354°	1	2.159 /165.6°	1.765 /156.1°
2	1.249 /216.3°	0.95 /206.1°	2	0.423 /267°	0.375 /285.1°
3	0.98 /167.6°	0.767 /177.6°	3	1.525 /323.3°	2.013 /324.1°
Total	3.623	2.662	4	0.945 /60.4°	1.278 /87°
			RMS	1.42	1.49

Table 5 Initial data for case 3

sensor	influence coefficient plane1	plane2	plane3	initial vibration
1	1.414 /45°	3.606 /33.7°	3.606 /33.7°	3.162 /71.6°
2	3.162 /71.6°	2.236 /26.6°	2.236 /26.6°	3.162 /18.4°
3	2.828 /45°	5.0 /36.9°	5.0 /36.9°	4.123 /14°
4	3.162 /18.4°	3.606 /33.7°	4.472 /26.6°	5.385 /68.2°
"signif. factor"	0.466	0.110	1.0	R.M.S.= 4.062

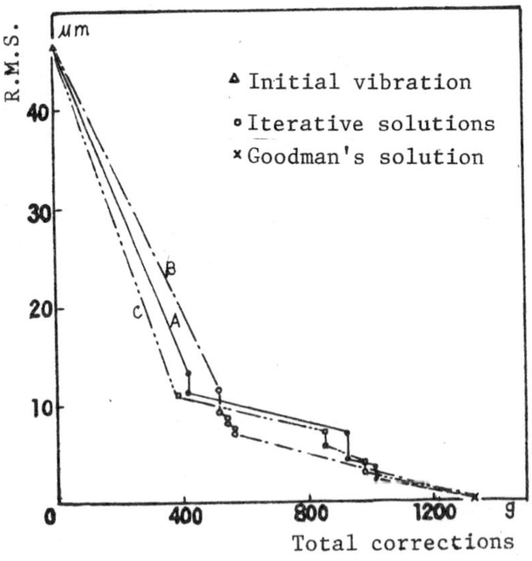

Fig 1 Iterative results for case 1

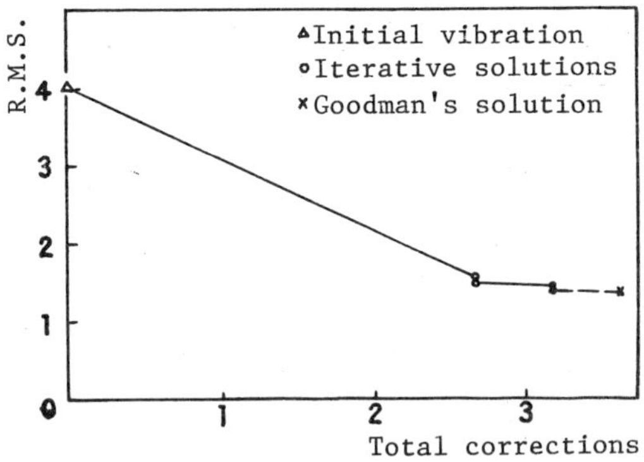

Fig 2 Iterative results for case 2

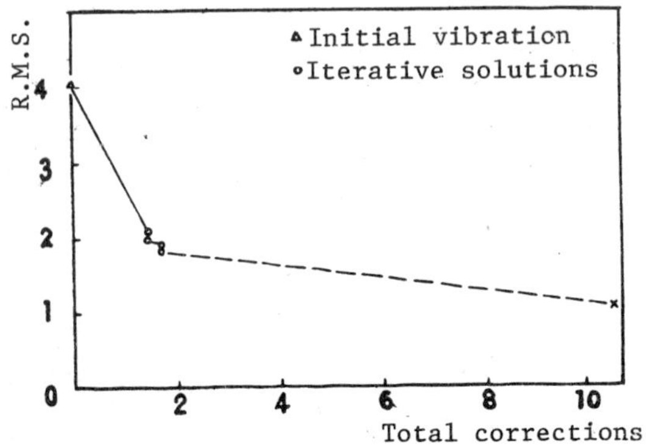

Fig 3 Iterative results for case 3

C279/88

Experimental study on balancing of flexible rotor by torsional excitation

K ONO, PhD, MJSME, MJSLE, MJSPE and **Y NAKAYAMA**
Department of Mechanical Engineering, Tokyo Institute of Technology, Tokyo, Japan

SYNOPSIS The fundamental experimental study on the torsionally excited unbalance vibration and modal balancing were carried out for the simple flexible rotor at the frequency of 96.4 percent of the first resonance frequency. As a result, it was found that the torsionally excited unbalance bending vibration was perpendicular to unbalance vector and proportional to the unbalance amount as predicted from the theory. It was also found that the initial unbalance of the test rotor was mainly due to the initial bending of the shaft and that the bending and unbalance changed slightly according to the angular position of the rotor. The sensitivity of mass eccentricity to lateral vibration was 0.3 under the torsional excitation with the frequency of 96.4 percent of the resonance frequency and the amplitude of 0.0481 rad. This experimental value was in good agreement with the theoretical one.

1. Introduction

Measurement of a rotor unbalance has been carried out by measuring the unbalance force or vibration due to the unbalance force when the rotor is rotating. Since the modal unbalance of a flexible rotor can not be measured unless the rotor is rotating at resonance frequency, it is very hard to balance the modal unbalance of a flexible rotor without supporting it by high damping bearings.

Taking this difficulty of the present technology into consideration, one of the authors has proposed a new measuring and balancing method of modal unbalance by means of torsional excitation. This method is regarded to be applicable to the balancing of a rigid body which is hard to rotate, if the rigid body is torsionally excited through a flexible shaft. Following to the theoretical study, the possibility of measurement and balancing of modal unbalance by means of the torsional excitation are investigated experimentally in this paper. A simple flexible rotor which has two equivalent disks is torsionally excited at its one end by DC motor at zero rotational speed. The general characteristics of the torsionally excited unbalance bending vibration and complicated behaviors of unbalance due to initial shaft deflection are also revealed.

2. Theory on Measurement and Balancing of Modal Unbalance by Torsional Excitation

Although the details of the theory has been presented in the previous paper[1], the relation of torsionally excited bending vibration to the first modal unbalance, torsionally exciting frequency and amplitude is described briefly here for the better understanding of the experiment.

The physical model of a flexible rotor and coordinates system are shown in Fig. 1.

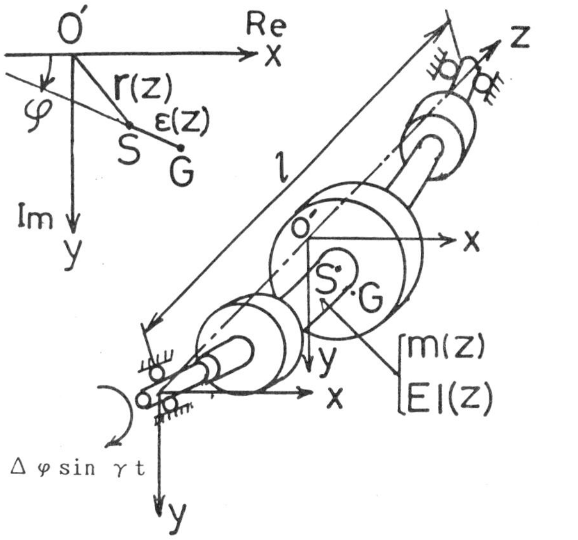

Fig 1 Analytical model of flexible rotor under torsional excitation

Although the rotor in this figure is simply supported at both ends, one end can be fixed as shown later in the experiment. The equation of motion for the local mass center of unit length at axial position z is written as

$$m \frac{\partial^2}{\partial t^2}\left(\mathbf{r}+\varepsilon e^{j\varphi}\right) + \frac{\partial^2}{\partial z^2}\left\{EI \frac{\partial^2 \mathbf{r}}{\partial z^2}\right\} = 0 \qquad (1)$$

where $\mathbf{r}$ is the displacement vector, m is the linear mass density, EI is the bending stiffness, ε is local mass eccentricity. All these quantities are the functions of z. j is the unit of the imaginary number. Damping of the system is neglected here. If the rotor is excited torsionally as $\varphi = \varphi_0 + \Delta\varphi \sin \gamma t$ at zero rotational speed, putting this equation into Eq. (1) and neglecting the higher order of $\Delta\varphi$ yield

$$m\frac{\partial^2 \mathbf{r}}{\partial t^2}+\frac{\partial^2}{\partial z^2}\left\{EI\frac{\partial^2 \mathbf{r}}{\partial z^2}\right\}$$

$$= j\gamma^2 \mathbf{u}\Delta\varphi\sin\gamma \qquad (2)$$

where $\mathbf{u}$ is the local unbalance vector of the rotor of unit length and given by

$$\mathbf{u} = m\varepsilon\exp(j\varphi_o) \qquad (3)$$

We assume here that the lowest torsional resonance frequency is high enough compared with the first bending resonance frequency and that $\Delta\varphi$ thus can be regarded as constant.

This displacement vector of the rotor can be expressed as,. by using the mode function Φ_1,

$$\mathbf{r} = \sum_{i=1}^{\infty} \mathbf{a}_i\phi_i\sin\gamma t \qquad (4)$$

Substituting Eq. (4) into Eq.(2), multiplying Eq. (2) by $\Phi_1 dz$ and integrating it from z=0 to l, we can get the amplitude vector of the first mode vibration of the form

$$\mathbf{a}_1 = \frac{\gamma^2\Delta\varphi\int_0^l \mathbf{u}\,\phi_1 dz}{\overline{m}_1(\Omega_1^2-\gamma^2)}j \qquad (5)$$

where $\overline{m}_1$ is the modal mass of the first mode and given by

$$\overline{m}_1 = \int_0^l m\phi_1^2 dz \qquad (6)$$

Since the higher modes effect to unbalance response at the vicinity of the first mode resonance can be neglected if the system has small damping, the torsionally excited unbalance vibration response is expressed with good approximation, except resonance, as,

$$\mathbf{r} = \frac{\gamma^2\Delta\varphi\,\phi_1\int_0^l \mathbf{u}\,\phi_1 dz}{\overline{m}_1(\Omega_1^2-\gamma^2)}j\sin\gamma t \qquad (7)$$

The torsionally excited response just at the resonance is subject to the damping of the system and given by

$$\mathbf{r} = \frac{\gamma^2\Delta\varphi\,\phi_1\int_0^l \mathbf{u}\,\phi_1 dz}{\overline{m}_1(\Omega_1^2+2\zeta_1\Omega_1\gamma j-\gamma^2)}j\sin\gamma t \qquad (8)$$

where ζ_1 is the modal damping ratio of the first mode. The term

$$\int_0^l \mathbf{u}\,\phi_1 dz$$

in Eqs. (7) and (8) is the modal unbalance which excites the first mode vibration. It is noted from these equations that the vibration amplitude is proportional to the modal unbalance and the torsionally exciting amplitude $\Delta\varphi$.

Next suppose the concentrated unbalance at $z=l_1$ equivalent to the first modal unbalance as **U**, then we have the relation of the form

$$\int_0^l \mathbf{u}\,\phi_1 dz = \int_0^l \mathbf{U}\,\delta(z-l_1)\phi_1 dz$$

$$= \mathbf{U}\,\phi_1(l_1) \qquad (9)$$

From this equation, we have

$$\mathbf{U} = \int_0^l \mathbf{u}\,\phi_1 dz/\phi_1(l_1) \qquad (10)$$

If we denote the torsionally excited vibration vector by **A**, we get from Eqs. (7) and (10)

$$\mathbf{A} = \frac{\gamma^2\Delta\varphi\,\phi_1(l_1)\,\phi_1(l_2)\,\mathbf{U}}{\overline{m}_1(\Omega_1^2-\gamma^2)}j \qquad (11)$$

where $\Phi_1(l_2)$ is the first mode amplitude at measuring position. It is seen from Eq. (11) that the direction of the vibration is perpendicular to the unbalance vector and the vibration amplitude is proportional to the amount of unbalance.

3. Experimental Apparatus and Method

Experimental set up and measuring system are shown in Fig. 2. One end of the flexible shaft with 10 mm diameter is connected to the driving DC motor shaft by means of the frictional coupling element. The other end of it is simply supported by outer races of two ball bearings. The two disks with 100 mm diameter and 20 mm width are mounted concentrically on the shaft at z=539.5 and 560.5 mm by means of frictional fixing elements with 5 mm contact region. Mass of the disk is 1.29 kg including the fixing element. The length of the shaft between two supporting points is 900 mm. Sinusoidal signal for the torsional excitation is obtained from the sweep oscillator in the FFT analyser or another oscillator with DC offset. The relation between torsional amplitude and input voltage amplitude is previously measured at a certain frequency by frictionally contacting 1024 pulses optical encorder with the outer surface of the frictional coupling. The torsional amplitude is calcutated from this calibration data. The lateral vibration of the shaft is measured by the eddy-current-type noncontact displacement detectors in the vertical and horizontal directions. Since the interference between two probes appears when they approach too close each other, the two probes are separated each other at a distance of 20 mm in the axial direction. Since the initial shaft deflection was found to be the main cause of the initial unbalance, the most straight one

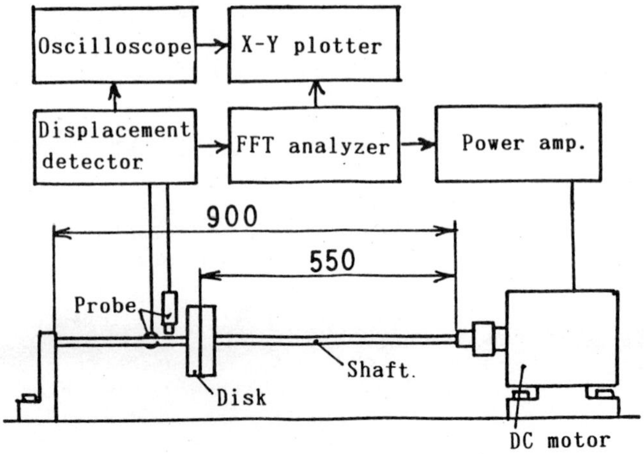

Fig 2 Experimental set-up

was chosen from more than ten shafts. The misalignment between the shaft and driving motor is another important factor to the unbalance. Then the frictionally fixing coupling was chosen and the runout of the shaft at the coupling was reduced to less than 10 μm.

4. Experimental Results and Discussions

The steady frequency response of torsionally excited bending vibration was first measured by sweeping slowly the exciting frequency near the first bending natural frequency. Figure 3 shows the typical data of the torsionally excited resonance curve. The first natural bending frequency of the test rotor is 11.2 Hz, whereas Q factor at the resonance is about 160. Then it is considered that the unbalance can be detected with high sensitivity at the resonance. The amplitude distribution of the test rotor at resonance, that is, the first natural mode, is calculated by using the finite element method. Figure 4 shows the typical calculated results of the assumed unbalance vectors at the disks and the resulting amplitude distributions at 11 and 12 Hz, just before and after the resonance. The calculated natural frequency is 11.18 Hz.

Before the measurement and balancing at resonance, the possibility of measurement and balancing below the resonance is investigated here by choosing 10.5 Hz, 96.4 percent of the resonance frequency, for the exciting frequency. When the small DC voltage is superposed to the 10.5 Hz exciting voltage signal, the rotor rotates very slowly and exhibits the whirling motion due to the initial shaft deflection and the torsionally excited lateral vibration. The whirling motion and the lateral vibration at twelve representative angular positions measured at z=600 mm are shown in Fig. 5. The number in this figure means the rotor position where the rotor position denoted by this number takes the uppermost vertical position.

The reference scale of 50 μm is valid for the whirling motion trajectry. The size of each vibration orbit is reduced to 33.8 percent of the original one, for the better illustration of the orbit without intersection. It is seen from this figure that the shaft center exhibits not a complete circle but a somewhat deformed circle. If we assume that the disks have no unbalance with respect to the shaft center, the unbalance due to the initial shaft deflection is the product of the shaft whirling radius at the disk and the disk mass. If we further assume that the whirling trajectory measured at z=600 mm is approximately equal to those of the two disk centers, the deflection vector drawn from the quasi center of the trajectory to the each number point shown in Fig. 5 is considered to be the mass eccentricity vector of the disks due to the initial shaft deflection. Since the vibration vector is approximately perpendicular to the deflection vector, it is found that the initial shaft deflection is the main cause of the unbalance.

Accordingly, the whirling motion of the shaft while rotating is next measured at equally separated nine positions from z=50 to 850 mm.

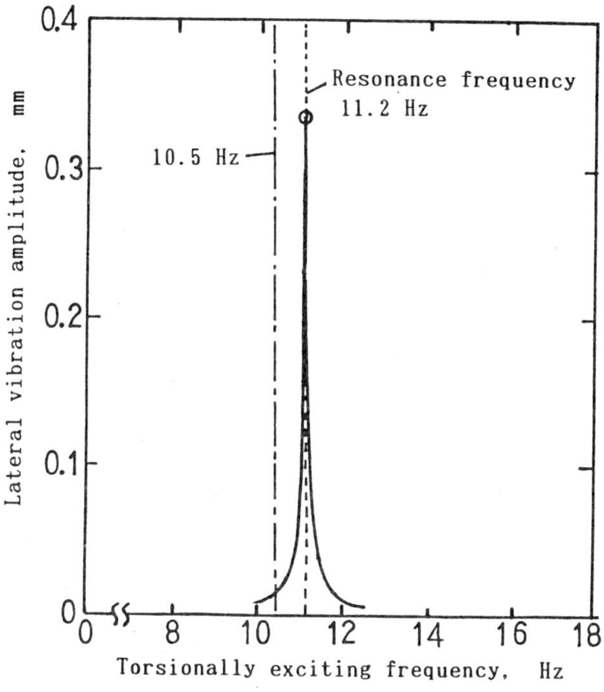

Fig 3 Resonance response of torsionally excited bending vibration

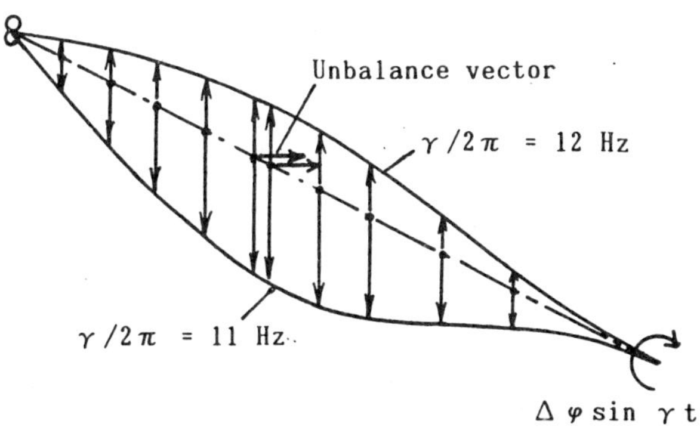

Fig 4 Calculated result of torsionally excited bending vibration

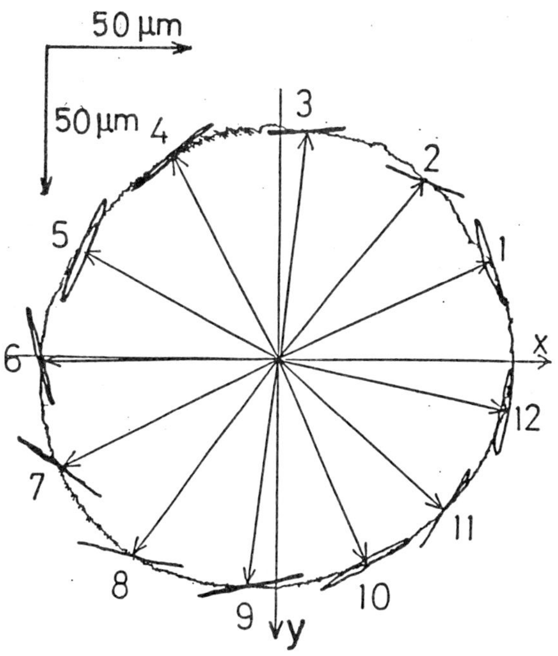

Fig 5 Whirling motion and torsionally excited vibration of the rotor ($\Delta\varphi$ = 0.042 rad, $\gamma/2$ = 10.5 Hz, z = 600 mm)

The examples of the measured trajectories at z=450 and 650 mm are shown in Fig. 6(a) and (b), respectively. From these data of the trajectories, the initial shaft deflection distribution is calculated. Fig.6(c) shows the two typical deflection distributions when the rotor positions of number 4 and 10 take the uppermost vertical position. Although the effective local mass eccentricity vector which results in unbalance at each position is actually the radius of curvature of the whirling trajectory, deflection vector from the common center of the trajectory is regarded as the local mass eccentricity. As seen from Fig. 6, it is found that the initial shaft deflection changes slightly according to the rotor position. The reason of the deflection change is not clear but seems to result from the interaction between the initial deflection and supporting conditions.

By multiplying the measured shaft deflection by the effective rotor mass m_i and the first natural mode function $\Phi_1(z)$ shown in Fig. 4 and integrating the product along the rotor, we can get the concentrated first mode unbalance vector $\mathbf{U}_1$ at disk 2 position from Eq. (10). The first mode unbalance components values U_{1x} and U_{1y} in the x and y directions and its absolute value $|U_1|$ ($=\sqrt{U_{1x}^2 + U_{1y}^2}$) at z=560.5 mm thus calculated from the measured shaft deflection shown in Fig. 6(c) is listed in the Table 1. In this table, the corresponding values U_{1xm}, U_{1ym} and $|\mathbf{U}_{1m}|$ calculated from the torsionally excited vibration at 4 and 10 rotor positions shown in Fig. 5 are also shown for comparison. It is seen from this table that the component and absolute values of the first mode unbalance calculated from the different data are nearly equal each other. Then, it can be said that the initial shaft deflection is the main cause of the initial unbalance of the test rotor.

In order to reduce the initial unbalance due to the initial shaft deflection, the balancing masses described in Table 2 are put on the disk 2 for the two cases where the rotor positions are in 4 and 10. After the balancing, the torsionally excited vibration amplitude A_x and A_y in the x and y direction are measured by putting the known unbalance U_y and U_x in the y and x directions, respectively. Figures 7(a) and (b) show A_y versus mass eccentricity ε_x and A_x versus mass eccentricity ε_y, respectively, for the rotor position of 4. Figure 8(a) and (b) are the similar data for the rotor position of 10. In these figures the mass of the two disks, 2.58 kg, are used for M. It is seen from this figures that the torsionally excited vibration amplitude is in good proportion to the additional mass eccentricity in the perpendicular direction. The crossing point of the abscissa means the opposite sign value of the residual mass eccentricity.

Then, the component and absolute values of the residual unbalance, U_{2xm}, U_{2ym}, $|\mathbf{U}_{2m}|$ obtained from Figs. 7 and 8 are listed in Table 3. The residual unbalance calculated from the initial shaft deflection, $U_{1x}-U_{xb}$, $U_{1y}-U_{yb}$, $|\mathbf{U}_1 - \mathbf{U}_b|$, and those calculated from the first unbalance vibration data, $U_{1xm} - U_{xb}$, $U_{1ym} - U_{yb}$, $|\mathbf{U}_{1m} - \mathbf{U}_b|$, are also presented in Table 3 for comparison. From this table it is seen that the residual unbalance U_{2xm}, U_{2ym} and $|\mathbf{U}_{2m}|$, obtained from Figs. 7 and 8, are in good agreement with $U_{1x} - U_{xb}$, $U_{1y}-U_{yb}$

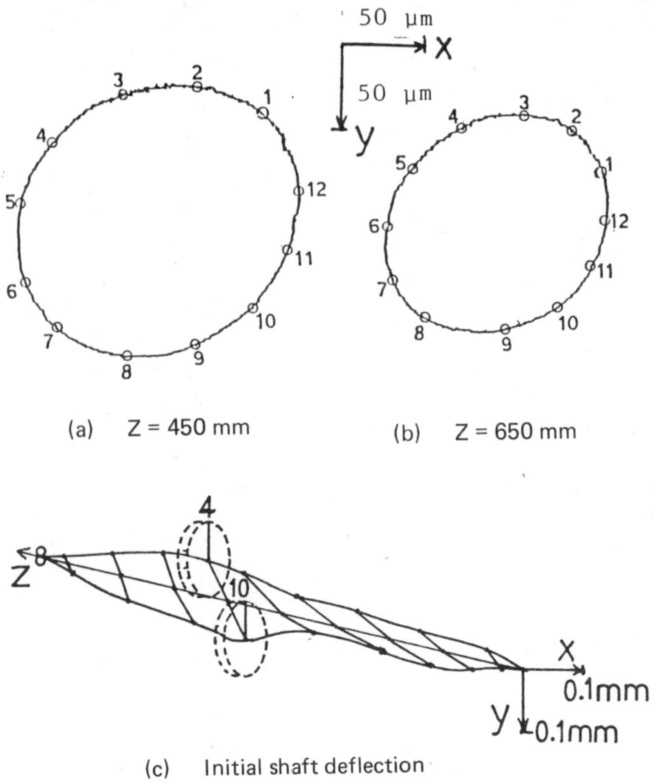

(a) Z = 450 mm (b) Z = 650 mm

(c) Initial shaft deflection

Fig 6 Shaft whirling motion and initial shaft deflection

Table 1 Initial unbalances calculated from initial deflection and torsionally excited vibration (g·mm)

Rotor position		4	10
Calculated from initial shaft deflection	U_{1x}	-80.7	92.6
	U_{1y}	-176.1	168.3
	$\|\mathbf{U}_1\|$	193.7	192.1
Calculated from torsionally excited vibration	U_{1xm}	-109.0	92.2
	U_{1ym}	-150.9	159.2
	$\|\mathbf{U}_{1m}\|$	186.1	184.1

Table 2 Balance amount (g·mm)

Rotor position	U_{xb}	U_{yb}	$\|\mathbf{U}_b\|$
4	68.1	152.0	166.6
10	-103.8	-132.2	168.1

and $|\mathbf{U}_1 - \mathbf{U}_b|$, respectively, whereas some large difference of 32.1 g mm appears between U_{2ym} and $U_{1ym}-U_{yb}$. This unbalance amount corresponds to mass eccentricity of 12μm. The reason of this discrepancy is not clear, but may result from the measurement error and the change of shaft deflection during the different measuring occasions.

Lastly, let us consider the ability of this method for unbalance measurement. Since the measurement accuracy of the present method can

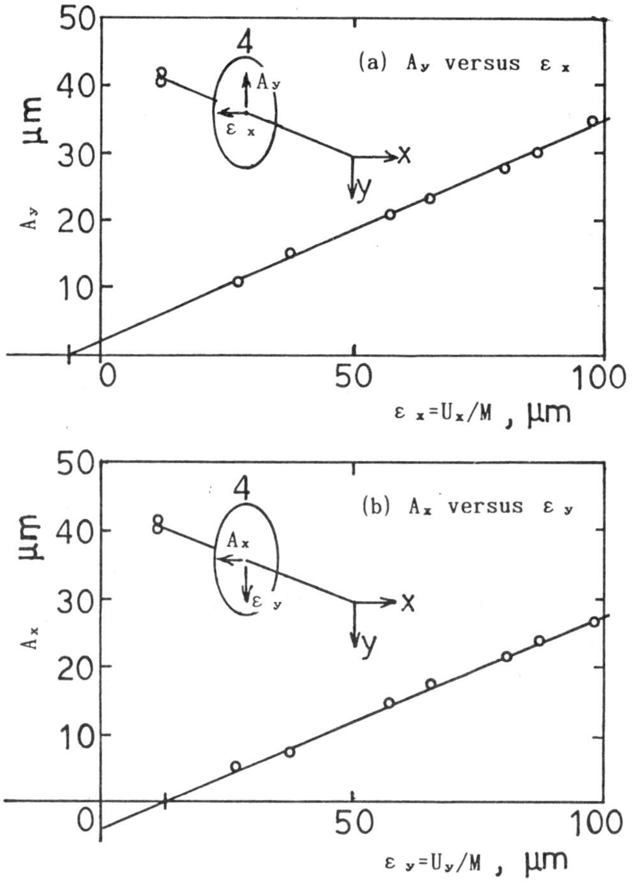

Fig 7 Vibration amplitude versus additional mass eccentricity (rotor position ten, $\Delta\varphi$ = 0.0481 rad, $\gamma/2$ = 10.5 Hz)

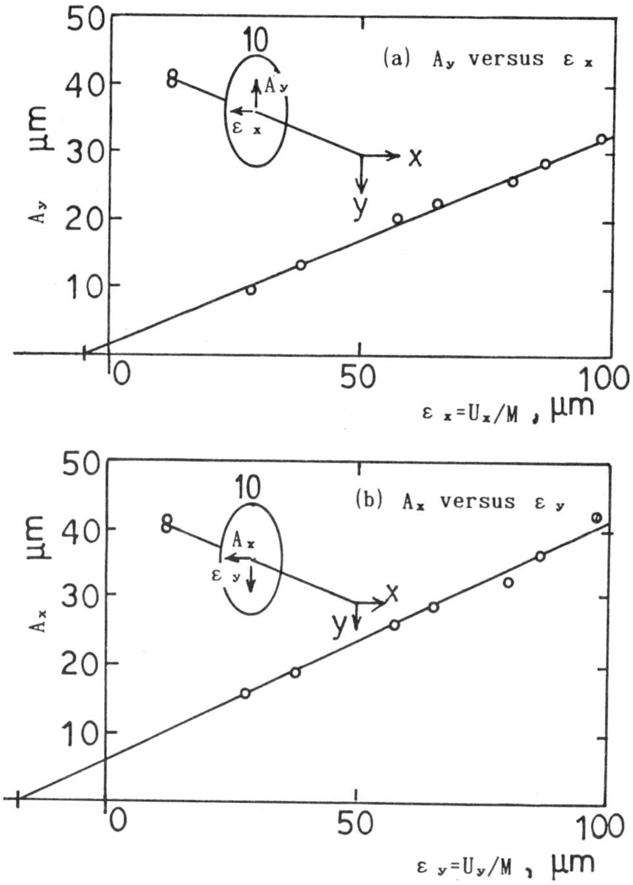

Fig 8 Vibration amplitude versus additional mass eccentricity (rotor position ten, $\Delta\varphi$ = 0.0481 rad, $\gamma/2$ =10.5 Hz)

Table 3 Residual unbalance after balancing (g·mm)

Residual unbalance \ Rotor position		4	10
U_{2x}	U_{2xm}	-14.2	-12.2
	$U_{1x}-U_{xb}$	-12.6	-11.2
	$U_{1xm}-U_{xb}$	-40.9	-11.6
U_{2y}	U_{2ym}	-31.0	44.2
	$U_{1y}-U_{yb}$	-24.1	36.1
	$U_{1ym}-U_{yb}$	1.1	27.0
$\lvert \mathbf{U}_2 \rvert$	$\lvert \mathbf{U}_{2m} \rvert$	34.1	45.9
	$\lvert \mathbf{U}_1-\mathbf{U}_b \rvert$	27.2	37.8
	$\lvert \mathbf{U}_{1m}-\mathbf{U}_b \rvert$	40.9	29.4

be evaluated by the sensitivity of mass eccentricity to vibration amplitude. From Eq. (11), the ratio of vibration amplitude to mass eccentricity S is given by

$$S = \frac{|\mathbf{A}|}{(\mathbf{U}/M)} = \frac{\Delta\varphi\,\gamma^2}{(\Omega_1^2-\gamma^2)} \times \frac{M\phi_1(l_1)\,\phi_1(l_2)}{\sum_{i=1}^{9} m_i\,\phi_1(z_i)} \quad \cdots\cdots \quad (12)$$

where $\Phi_1(l_1)$ and $\Phi_1(l_2)$ are the first mode amplitude at the unbalance point and the measuring point, respectively. The theoretical value of the sensitivity S for this experiment is calculated from Eq. (12) by using the parameter values that l_1=560.5mm, l_2=600mm , $\Delta\varphi$ =0.0481 rad, $\gamma/2\pi$=10.5 Hz, and listed in Table 4 in comparison with the experimentalvalues S_x and S_y calculated from Figs. 7 and 8. It is found from this table that the theoretical value is in good agreement with the experimental ones, and that the mass eccentricity of 1 μm can be detected as the lateral vibration amplitude of 0.3μm, when the torsional frequency is 96.4 percent of the resonance frequency and the torsional amplitude is 0.0481 rad.

Table 4 Sensitivity of mass eccentricity to vibration amplitude ($\Delta\varphi$ =0.0481rad, $\gamma/2\pi$ = 10.5 Hz)

		Rotor position	
		4	10
Experiment	$S_x=A_x/\varepsilon_y$	0.31	0.34
	$S_y=A_y/\varepsilon_x$	0.33	0.31
Theory S		0.35	

5. Conclusion

The general characteristics of the torsionally excited unbalance vibration and the measuring method of modal unbalance of a simple flexible rotor are experimentally investigated at the exciting frequency of 96.4 percent of its first resonance frequency. The results of this study are summarized as follows:

(1) As predicted from the theory, the torsionally excited bending vibration is perpendicular and proportional to the unbalance vector.

(2) The initial unbalance of the test rotor results mainly from the initial shaft deflection. The shaft deflection and, thus, the initial unbalance change slightly according to the angular position of the rotor, probably because of the misalignment between the test shaft and the driving motor.

(3) One third of the mass eccentricity can be detected as the lateral vibration amplitude if the exciting frequency is 96.4 percent of the resonant frequency and torsional amplitude is 0.0481 rad. This experimental value of sensitivity agrees well with the theoretical one.

Reference

[1] K. Ono, "Theory on Balancing of a Flexible Rotor by Driving Torque Excitation", Vibration in Rotating Machinery, I. Mech E. , 1984, p. 475

C305/88

Balancing of flexible rotors with one transient run—an experimental and theoretical investigation

R de SILVA, Dipl-Ing
Technical University of Berlin, West Berlin

1 INTRODUCTION

All classical balancing processes, modal balancing and influence coefficient method, are based on a data acquisition at constant rotational speeds (stationary states). By modal balancing methods, the unbalance state of a flexible rotor is identified from stationary state measurement of unbalance vibration in the neighbourhood of the critical speeds. Apart from the fact, that these standard methods usually require a large number of time consuming test runs, problems can also arise by measuring the vibrational response close to critical speeds, due to heavy vibrations which might develop depending on the unbalance and damping state of the rotor.

Therefore, in this paper the concept of stationary state measurement is left behind in favour of a t r a n s i e n t balancing procedure, that needs (for data acquisition) only o n e highly accelerated and decelerated test run through all critical speeds of interest (see Fig 1). The basic elements of this transient balancing method are: modal theory, FFT-Analysis and an identification procedure. The identification procedure requires an estimation of the mode shapes and generalized masses of the critical speeds to be passed, as in the modal balancing methods suggested by GNIELKA |1|, GASCH and DRECHSLER |2|. The signal analysis needs the measurement and sampling of the angle of rotation and the shaft vibrations in orthogonal directions (horizontal and vertical) throughout a transient test run. The time functions of the vibrations (output of the mechanical system) and the sine and cosine of the time dependent angle of rotation (input of the mechanical system) are transformed by FFT into the frequency domain where the frequency response function of the mechanical system is evaluated. A relatively simple linear mathematical model, based on modal theory, is used to describe the bending vibration (by low damping) due to transient excitation of rotor unbalance, whereby an initial bow for the rotor axis is admitted in the model. The analytically derived frequency response function of the mechanical system contains the generalized unbalances, eigenfrequencies, damping coefficients and the constrained bow (due to initial bow) as the only unknown parameters (the generalized masses and mode shapes are assumed to be known by precalculation). The contributions of the constrained bow and the different mode shapes to the frequency response function can be elegantly seperated from each other by considering different regions in the frequency domain. The contribution of the constrained bow dominates in the range of very low frequencies, whereas the i-th mode influence is the greatest in the vicinity of the i-th eigenfrequency. Comparision of the analytical frequency response function with the evaluated function, in these specific regions, leads usually to overdetermined equation systems for the respective parameters. Thus, using least squares procedures, linear equation systems are obtained, which can be solved easily with respect to the unknown parameters.

In this way the identification of the generalized unbalances is possible and the rotor can be balanced by using orthogonal weight sets.

2 BENDING VIBRATION OF FLEXIBLE ROTORS DUE TO TRANSIENT EXCITATION OF UNBALANCE

The objective of this section is to obtain a mathematical model describing the vibrational response of an unbalanced flexible rotor due to non stationary rotational speeds. The model should, of course, contain all the essential features of the behaviour of the real system, but it should not be too complicated for practical purposes such as delivering a solid basis for the unbalance identification explained in the next section.

In view of this the rotor will be idealized to a round elastic shaft with arbitrary mass and bending stiffness distribution ($\mu(x)$ and $B(x)$). The effects of distributed rotary inertia with respect to the cross sectional plane as well as the deformations due to shear and torsion will be neglected. The rotor is considered to be mounted in elastic bearings assumed to have isotropic stiffness (s_i), no mass and no damping. A space-fixed co-ordinate system (x y z) is chosen so that the x-axis passes through the centers of all bearings in their initial state before the rotor is mounted. Further, the gravitational field effects will be neglected.

Figure 2 shows this idealized mechanical system, whereby a constrained bow of the mounted shaft resulting from an initial bow before mounting has been taken into consideration.

For kinematic analysis (see Fig 3) all deflections along the z- and y-axes are superimposed into a complex deflection. Thus the constrained bow at the beginning (t=o) is represented by the complex function $c(x)=w_c(x)+jv_c(x)$ and $\varepsilon(x)$ is the complex function of distributed mass eccentricity.

Let the function $\varphi(t)$ represent the angle of rotation of the shaft assumed to be of infinite torsional stiffness. If the inertial forces were absent ($\mu(x)=o$), the center **S** of any cross section would describe a circular path, given by the equation (1).

$$r(x,t) = r_c(x,t) = c(x)\, e^{j\varphi(t)} \tag{1}$$

In the presence of inertial forces, however, the path of the geometrical center **S** will differ to that of equation (1). With $r_d(x,t)$ being the dynamic deflection representing this difference, one obtains equation (2) for the time dependent co-ordinates of the center.

$$r(x,t) = r_c(x,t) + r_d(x,t) = c(x)\, e^{j\varphi(t)} + r_d(x,t) \tag{2}$$

Further, the path of the center of mass M can be represented by the following kinematic equation:

$$r_M(x,t) = (c(x)+\varepsilon(x))\, e^{j\varphi(t)} + r_d(x,t) \tag{3}$$

The resulting distributed inertial forces of the motion will be proportional to the acceleration

$$\ddot{r}_M(x,t) = (c(x)+\varepsilon(x))\left(e^{j\varphi(t)}\right)^{\bullet\bullet} + \ddot{r}_d(x,t) \tag{4}$$

For the special case of constant rotational speed $\dot{\varphi}(t)=\Omega$ GNIELKA |1| derives the equation of dynamic equilibrium by applying the principle of virtual work. He shows that the mode shapes $\Phi_i(x)$ are the same for a shaft with or without initial bow. In the case of the bowed rotor the "reference line" of the mode shapes will be the center line of the constrained bow and not the x-axis. GNIELKA expands the dynamic deflection $r_d(x,t)$ into a modal series.

$$r_d(x,t) = \sum_{i=1}^{\infty} \phi_i(x)\, f_i(t) \tag{5}$$

The generalized co-ordinate $f_i(t)$ has to satisfy the generalized equation of motion

$$m_i \ddot{f}_i(t) + c_i f_i(t) = -u_i \left(e^{j\Omega t}\right)^{\bullet\bullet} \tag{6}$$

The generalized mass m_i, stiffness c_i and unbalance u_i are given by the following relations:

$$m_i = \int_0^{\ell} \mu(x)\, \phi_i^2(x)\, dx \tag{7a}$$

$$c_i = \int_0^{\ell} B(x)\, \phi_i''^{\,2}(x)\, dx + \sum_{\ell=1}^{L} S_\ell\, \phi_i^2(x_\ell) \tag{7b}$$

$$u_i = \int_0^{\ell} (c(x)+\varepsilon(x))\, \mu(x)\, \phi_i(x)\, dx \tag{7c}$$

GNIELKA |1| improves the equation (6) by admitting modal damping d_i

$$m_i \ddot{f}_i(t) + d_i \dot{f}_i(t) + c_i f_i(t) = -u_i \left(e^{j\Omega t}\right)^{\bullet\bullet} \tag{8}$$

In this paper, the equation (8) will be extended for the general case of transient excitation as follows:

$$m_i \ddot{f}_i(t) + d_i \dot{f}_i(t) + c_i f_i(t) = -u_i \left(e^{j\varphi(t)}\right)^{\bullet\bullet} \tag{9}$$

In this general case, the time dependency of the unbalance excitation is represented by the complex function $(e^{j\varphi(t)})^{\bullet\bullet} = (-\dot{\varphi}^2(t) + j\cdot\ddot{\varphi}(t))\cdot e^{j\varphi(t)}$ (which can be considered as a harmonic function only for the special case of constant rotational speed $\dot{\varphi}(t)=\Omega$).

Now, using the well known unit-impulse response function $h_i(t)$ or the frequency response function $H_i(j\omega)$ of a damped system, the general solution of equation (9) can be expressed in the form of a convolution integral (in time domain) or as an algebraic product (in frequency domain).

$$f_i(t) = -u_i \int_{-\infty}^{+\infty} \left[e^{j\varphi(\tau)}\right]^{\bullet\bullet} h_i(t-\tau)\, d\tau \tag{10a}$$

$$f_i(t) = -u_i \int_{-\infty}^{+\infty} e^{j\varphi(\tau)} \left[\dot{h}_i(0)\,\delta(t-\tau) + \ddot{h}_i(t-\tau)\right] d\tau \tag{10b}$$

$$h_i(t) = \begin{cases} 0 & \text{for } t<0 \\ \dfrac{e^{-\frac{d_i}{2m_i}t}}{m_i\,\omega_{d_i}} \sin(\omega_{d_i} t) & \text{for } t \geq 0 \end{cases} \tag{10c}$$

with $\omega_{d_i}^2 = \omega_i^2 - \left(\frac{d_i}{2m_i}\right)^2$ and $\omega_i^2 = \frac{c_i}{m_i}$

$$F_i(j\omega) = \int_{-\infty}^{+\infty} f_i(t)\, e^{-j\omega t}\, dt = -u_i \left\{(j\omega)^2 E(j\omega)\right\} H_i(j\omega) \tag{11a}$$

$$E(j\omega) = \int_{-\infty}^{+\infty} e^{j\varphi(t)}\, e^{-j\omega t}\, dt \tag{11b}$$

$$H_i(j\omega) = \int_{-\infty}^{+\infty} h_i(t)\, e^{-j\omega t}\, dt = \frac{1}{m_i\left(\omega_i^2 - \omega^2 + j\,\frac{d_i}{m_i}\,\omega\right)} \tag{11c}$$

with $\omega_i^2 = \frac{c_i}{m_i}$

Note, that the rules for convolution and differentiation of the Fourier transformation have been applied in obtaining (11a) from (10a). Further, to derive (10b) from (10a), integration by parts was used, whereas $\delta(t)$ represents the Dirač-impulse function. In view of the equations (2), (5), (10) and (11), the dynamical response of a bowed flexible rotor due to transient excitation of its balance can be described by the following mathematical model, which is linear, time invariant and deterministic:

$$r(x,t) = w(x,t) + j\,v(x,t) = \int_{-\infty}^{+\infty} e^{j\varphi(\tau)}\, h(x,t-\tau)\, d\tau \tag{12a}$$

$$R(x,j\omega) = \int_{-\infty}^{+\infty} r(x,t)\, e^{-j\omega t}\, dt = E(j\omega)\; H(x,j\omega) \tag{12b}$$

$$h(x,t) = c(x)\,\delta(t) - \sum_{i=1}^{\infty} \phi_i(x)\, u_i \left\{\dot{h}_i(0)\,\delta(t) + \ddot{h}_i(t)\right\} \tag{12c}$$

$$H(x,j\omega) = \int_{-\infty}^{+\infty} h(x,t)\, e^{-j\omega t}\, dt = c(x) + \sum_{i=1}^{\infty} \phi_i(x)\, \frac{u_i}{m_i}\left(\frac{\omega^2}{\omega_i^2 - \omega^2 + j\,\frac{d_i}{m_i}\,\omega}\right) \tag{12d}$$

This general model includes, of course, the stationary state response as a special case for $\varphi(t)=\Omega\cdot t$. In the stationary case one obtains from (12)

$$r(x,t) = \int_{-\infty}^{+\infty} e^{j\Omega\tau} h(x,t-\tau)\, d\tau$$
$$= \int_{-\infty}^{+\infty} e^{j\Omega(t-\tau_1)} h(x,\tau_1)\, d\tau_1$$

(variable substitution: $\tau_1 = t-\tau$)

$$= e^{j\Omega t} \int_{-\infty}^{+\infty} h(x,\tau_1)\, e^{-j\Omega\tau_1} d\tau_1 \quad (13)$$
$$= e^{j\Omega t}\, H(x,j\Omega)$$
$$= e^{j\Omega t}\left\{c(x) + \sum_{i=1}^{\infty} \phi_i(x) \frac{u_i}{m_i} \frac{\Omega^2}{\omega_i^2 - \Omega^2 + j\frac{d_i}{m_i}\Omega}\right\}$$

which is the same as equation (21) from GNIELKA.

3 PARAMETER IDENTIFICATION BY SPECTRUM ANALYSIS

Figure 4 gives a schematic presentation of the model derived in the last section. The time functions $e^{j\varphi(t)}$, $r(x,t)$ or their Fourier transforms $E(j\omega)$, $R(x,j\omega)$ are considered to be the input and output of the mechanical system. The impulse response function $h(x,t)$ or its Fourier transform $H(x,j\omega)$, the frequency response function, represent the mechanical features of the rotor mounted in bearings. They contain the constrained bow $c(x)$, eigenmodes $\Phi_i(x)$, generalized masses and unbalances m_i, u_i, eigenfrequencies ω_i and the modal damping coefficients $d_i (i=1,2..., \infty)$ as parameters. Equations (12a) and (12b) describe the input-output relations in time and frequency domains. Analytical expressions for $h(x,t)$ and $H(x,j\omega)$ are given by the equations (12c), (12d).

If the Fourier transforms $E(j\omega)$ and $R(x,j\omega)$ were known, the frequency response function $H(x,j\omega)$ could be evaluated easily.

$$H(x,j\omega) = R(x,j\omega) / E(j\omega) \quad (14)$$

Note, that the equation (14) (which has been derived from the relation (12b)) is only valid for such frequencies, where the input spectrum $E(j\omega)$ has non-zero values. This means, that, to evaluate $H(x,j\omega)$ from (14) for a certain frequency range, the input $e^{j\varphi(t)}$ must also contain non vanishing frequency components in the desired range.

Once the frequency response function is determined, however, the problem of parameter identification can be tackled as shown later in this section. First, the approach used by the numerical evaluation of the FOURIER transformation of the input-output signals will be explained.

3.1 Spectrum formation with fast Fourier Transformation (FFT)

The balancing procedure presented in this paper requires the measurement and sampling of the angle of rotation $\varphi(t)$ and the vertical and horizontal vibrations $w(x,t)$, $v(x,t)$ of the shaft center throughout a transient run of the rotor.

Using a sampling frequency $f_s = 1/\Delta T$, the input-output functions will be computed at discrete times $t_k = k\cdot\Delta T (k=0,1,...,N-1)$ during the finite observation time $T = N\cdot\Delta T$.

$$e^{j\varphi(t_k)} = \cos\varphi(t_k) + j\sin\varphi(t_k) \quad (15a)$$
$$(k=0,1,...,N-1)$$
$$r(x,t_k) = w(x,t_k) + j\,v(x,t_k) \quad (15b)$$

The continuous Fourier transforms,

$$E(jf) = \int_{-\infty}^{+\infty} e^{j\varphi(t)}\, e^{-j2\pi ft} dt$$

and

$$R(x,jf) = \int_{-\infty}^{+\infty} r(x,t)\, e^{-j2\pi ft} dt$$

will be approximated by the discrete Fourier transforms (DFT), $\tilde{E}(jf_k)$ and $\tilde{R}(x,jf_k)$, using the fast Fourier transform algorithm (FFT).

$$\tilde{E}(jf_k) = \tilde{E}(j\tfrac{k}{T}) = \Delta T \sum_{n=0}^{N-1} e^{j\varphi(t_n)}\, e^{-j\frac{2\pi kn}{N}} \quad (16a)$$
$$\tilde{R}(x,jf_k) = \tilde{R}(x,j\tfrac{k}{T}) = \Delta T \sum_{n=0}^{N-1} r(x,t_n)\, e^{-j\frac{2\pi kn}{N}} \quad (16b)$$
$$(k=0,1,...,N-1)$$

Both, aliasing and leakage, the well known major errors of this approximation, can be completely eliminated in the case of periodic signals with band limited spectra, if the used sampling frequency is at least twice the highest frequency present in the signals (Shannon's theorem: $f_s \geq 2\cdot f_{max}$) and if the signals were cut off exactly after one period (or multiples of periods).

Therefore, the approach emphasized in this paper is to use such transient runs of the rotor, which makes it possible to consider the input-output signals as band limited periodics (at least approximately for practical purposes). In view of this, the rotor will be accelerated from an initial stationary state, given by the rotational speed Ω_o (which can also be zero, $\Omega_o = o$), through all the critical speeds of interest to reach a maximum rotational speed Ω_{max} and then retarded back to the same (initial) state during the observation time T (Fig 1). The sampling frequency will be chosen to fulfill the condition $f_s > \Omega_{max}/\pi$. Within the time interval T the non-stationary vibrations should completely decay and, strictly speaking, the rotor should also complete a natural number of revolutions.

Now, the signals $e^{j\varphi(t)}$ and $r(x,t)$ can be considered as periodics (period = T), because the transient run can be repeated without causing any change in the system response, according to the principle of causality. Further, the frequency range $\Omega_o \leq |\omega| \leq \Omega_{max}$ will contain the major frequency components of the input signal $e^{j\varphi(t)}$, whereas the rest of the input stectrum may be practically neglected. The band limited input spectrum $E(j\omega)$ will also lead to a band limited output spectrum $R(x,j\omega)$ in view of the relation (12b).

The use of antialiasing filters (low pass filters) may still prove to be necessary to cut off higher frequencies caused by measuring disturbances, but the undesirable filter effects should also be taken into consideration. For example, the distortion of phase information by using Butterworth filters can result in significant spectrum errors, if not corrected. A numerical correction will be possible, as the filter coefficients are usually known.

However, the use of other windows (Hanning, Hamming, Gaussian, etc.) to minimize leakage can be avoided, as the rectangular window is best suited for periodic signals.

3.2 Identification of system parameters

To demonstrate the identification procedure, the evaluated frequency response function $H(x,j\omega)$ will be analytically expressed as follows:

$$H(x,j\omega) = c(x) + \sum_{i=1}^{\infty} \varepsilon_i(x) \frac{\omega^2}{\omega_i^2 - \omega^2 + j\delta_i\omega} \tag{17a}$$

with

$$\varepsilon_i(x) = \phi_i(x) \frac{u_i}{m_i} \tag{17b}$$

$$\delta_i = \frac{d_i}{m_i} \tag{17c}$$

The new parameters, the modal eccentricity function $\varepsilon_i(x)$ and the mass proportional (modal) damping coefficient δ_i, have been introduced, because they can be directly identified from the spectrum $H(x,j\omega)$. The eigenmodes $\Phi_i(x)$ and generalized masses will be assumed to be known (for example by pre-calculation), so that the identification of the generalized unbalance u_i will be possible from equation (17b).

The identification of the constrained bow c(x):

If the initial stationary state of the rotor is chosen to be the zero state $\Omega_o = o$ (see section 3.1), the lower frequency range ($\omega \approx o$) will be present in the spectrum $H(x,j\omega)$. This part of the spectrum is best suited for the identification of the constrained bow $c(x)$. In this range of very small frequencies ($\omega \approx o$) the frequency response is dominated by the constrained bow $c(x)$ and the contributions of all generalized eccentricities $\varepsilon_i(x)$ may be neglected, as they are proportional to the second power of the frequency ω^2 (see eq. (17a)).

Therefore, one obtains from (17a):

$$H(x,j\omega) = c(x) + q_o(x,j\omega) \qquad (\text{for } o \leqq |\omega| \leqq \Delta\omega) \tag{18}$$

The complex error function $q_o(x,j\omega)$ may contain all the errors such as approximation error (caused by neglecting the $\varepsilon_i(x)$-contributions), model error (assumption of a round shaft), measurement error, etc. For (low) discrete frequencies $\Omega_p (p = 1,2,\ldots,P)$ one obtains from (18) the overdetermined equation system

$$\begin{bmatrix} 1 \\ \vdots \\ 1 \\ \vdots \\ 1 \end{bmatrix} c(x) - \begin{bmatrix} H(x,j\Omega_1) \\ \vdots \\ H(x,j\Omega_p) \\ \vdots \\ H(x,j\Omega_P) \end{bmatrix} = \begin{bmatrix} q_o(x,j\Omega_1) \\ \vdots \\ q_o(x,j\Omega_p) \\ \vdots \\ q_o(x,j\Omega_P) \end{bmatrix} \tag{19}$$

Now, using a least squares procedure, the constrained bow $c(x)$ will be found.

$$c(x) = \frac{1}{P}\sum_{p=1}^{P} H(x,j\Omega_p) \qquad (o \leqq \Omega_p \leqq \Delta\omega) \tag{20}$$

Identification of the parameters $\varepsilon_i(x)$, ω_i and δ_i:

Once $c(x)$ is identified, it can be eliminated from the original spectrum

$$\tilde{H}(x,j\omega) = H(x,j\omega) - c(x) = \sum_{i=1}^{\infty} \varepsilon_i(x) \frac{\omega^2}{\omega_i^2 - \omega^2 + j\delta_i\omega} \tag{21}$$

In the following it will be assumed that

(a) the eigenfrequencies are located on the frequency axis well apart from each other

(b) the measuring station x does not coincide with a node of the i-th eigenmode ($\Phi_i(x) \neq o$), and

(c) all generalized eccentricities $\varepsilon_i(x)$ are of the same order of magnitude.

Now, the frequency range in the neighbourhood of the i-th eigenfrequency will be considered. In this region the contribution of the i-th eccentricity function $\varepsilon_i(x)$ will dominate the spectrum $\tilde{H}(x,j\omega)$, because the denominater $(\omega_i^2-\omega^2+j\delta_i\omega)$ of the i-th term of the sum in (21) will be a minimum for $\omega \approx \omega_i$. Therefore, neglecting the other contributions, the following approximation will be made:

$$\tilde{H}(x,j\omega) \approx \varepsilon_i(x) \frac{\omega^2}{\omega_i^2 - \omega^2 + j\delta_i\omega} \qquad (\omega_i - \Delta\omega \leqq \omega \leqq \omega_i + \Delta\omega) \tag{22}$$

Rearrangement of (22) leads to the equation

$$-\omega^2\varepsilon_i(x) + \tilde{H}(x,j\omega)\omega_i^2 + j\omega\tilde{H}(x,j\omega)\delta_i - \omega^2\tilde{H}(x,j\omega) = q_i(x,j\omega) \tag{23}$$

Also here, the error function $q_i(x,j\omega)$ contains all the errors made.

Evaluation of the equation (23) for discrete frequencies $\Omega_k (k=1,2,\ldots,K)$ in the vicinity of ω_i results in the equation system (24) which is usually overdetermined (for $K > 3$).

$$
\begin{bmatrix}
-\Omega_1^2 & -j\Omega_1^2 & \tilde{H}(x,j\Omega_1) & j\Omega_1\tilde{H}(x,j\Omega_1) \\
\vdots & \vdots & \vdots & \vdots \\
-\Omega_k^2 & -j\Omega_k^2 & \tilde{H}(x,j\Omega_k) & j\Omega_k\tilde{H}(x,j\Omega_k) \\
\vdots & \vdots & \vdots & \vdots \\
-\Omega_K^2 & -j\Omega_K^2 & \tilde{H}(x,j\Omega_K) & j\Omega_K\tilde{H}(x,j\Omega_K)
\end{bmatrix}
\cdot
\begin{bmatrix}
\overset{re}{\varepsilon}_i(x) \\ \overset{im}{\varepsilon}_i(x) \\ \omega_i^2 \\ \delta_i
\end{bmatrix}
-
\begin{bmatrix}
\Omega_1^2\tilde{H}(x,j\Omega_1) \\ \vdots \\ \Omega_k^2\tilde{H}(x,j\Omega_k) \\ \vdots \\ \Omega_K^2\tilde{H}(x,j\Omega_K)
\end{bmatrix}
=
\begin{bmatrix}
q_i(x,j\Omega_1) \\ \vdots \\ q_i(x,j\Omega_k) \\ \vdots \\ q_i(x,j\Omega_K)
\end{bmatrix}
\qquad (24)
$$

$$
\underline{A}\cdot\underline{a} - \underline{b} = \underline{q}(\underline{a})
$$

Now, using the least squares procedure

$$
\underline{q}^{*T}\cdot\underline{q} = \text{minimum} \rightarrow \frac{\partial}{\partial\underline{a}}\left(\underline{q}^{*T}\cdot\underline{q}\right) = \underline{0} \qquad (25)
$$

($\underline{q}^{*T}$ is the transposed and conjugate-complex vector $\underline{q}$)

one finally obtains the symmetrical linear equation system

$$
\begin{bmatrix}
A_9 & 0 & -A_1 & A_5 \\
0 & A_9 & -A_2 & -A_4 \\
-A_1 & -A_2 & A_3 & 0 \\
A_5 & -A_4 & 0 & A_8
\end{bmatrix}
\cdot
\begin{bmatrix}
\overset{re}{\varepsilon}_i(x) \\ \overset{im}{\varepsilon}_i(x) \\ \omega_i^2 \\ \delta_i
\end{bmatrix}
=
\begin{bmatrix}
-A_6 \\ -A_7 \\ A_8 \\ 0
\end{bmatrix}
\qquad (26)
$$

with the coefficients

$$
A_1 = \frac{1}{K}\sum_{k=1}^{K}\Omega_k^2\,\overset{re}{\tilde{H}}(x,j\Omega_k)
$$

$$
A_2 = \frac{1}{K}\sum_{k=1}^{K}\Omega_k^2\,\overset{im}{\tilde{H}}(x,j\Omega_k)
$$

$$
A_3 = \frac{1}{K}\sum_{k=1}^{K}\left|\tilde{H}(x,j\Omega_k)\right|^2
$$

$$
A_4 = \frac{1}{K}\sum_{k=1}^{K}\Omega_k^3\,\overset{re}{\tilde{H}}(x,j\Omega_k)
$$

$$
A_5 = \frac{1}{K}\sum_{k=1}^{K}\Omega_k^3\,\overset{im}{\tilde{H}}(x,j\Omega_k) \qquad (27)
$$

$$
A_6 = \frac{1}{K}\sum_{k=1}^{K}\Omega_k^4\,\overset{re}{\tilde{H}}(x,j\Omega_k)
$$

$$
A_7 = \frac{1}{K}\sum_{k=1}^{K}\Omega_k^4\,\overset{im}{\tilde{H}}(x,j\Omega_k)
$$

$$
A_8 = \frac{1}{K}\sum_{k=1}^{K}\Omega_k^2\left|\tilde{H}(x,j\Omega_k)\right|^2
$$

$$
A_9 = \frac{1}{K}\sum_{k=1}^{K}\Omega_k^4
$$

The system (26) can be easily solved, the solution will be given as follows:

$$
\omega_i^2 = \frac{1}{D}\left[\left\{A_8 - \frac{(A_4^2+A_5^2)}{A_9}\right\}\left\{A_8 - \frac{(A_1A_6+A_2A_7)}{A_9}\right\} + \frac{1}{A_9^2}\left\{-A_1A_5+A_2A_4\right\}\left\{A_5A_6-A_4A_7\right\}\right] \qquad (28a)
$$

$$
\delta_i = \frac{1}{D}\left[\frac{\{-A_1A_5+A_2A_4\}}{A_9}\left\{A_8 - \frac{(A_1A_6+A_2A_7)}{A_9}\right\} + \left\{A_3 - \frac{A_1^2+A_2^2}{A_9}\right\}\frac{\{A_5A_6-A_4A_7\}}{A_9}\right] \qquad (28b)
$$

$$
\overset{re}{\varepsilon}_i(x) = \frac{1}{A_9}\left[-A_6 + \omega_i^2A_1 - \delta_iA_5\right] \qquad (28c)
$$

$$
\overset{im}{\varepsilon}_i(x) = \frac{1}{A_9}\left[-A_7 + \omega_i^2A_2 + \delta_iA_4\right] \qquad (28d)
$$

with

$$
D = A_3A_8 + \frac{1}{A_9^2}\left(A_1A_4+A_2A_5\right)^2 - \frac{1}{A_9}\left\{A_3(A_4^2+A_5^2) + A_8(A_1^2+A_2^2)\right\} \qquad (28e)
$$

Unbalance identification:

In view of the equation (17b) one may make the following approximation:

$$
\frac{1}{m_i}\phi_i(x)\,u_i \approx \varepsilon_i(x) \qquad (29)
$$

If $\varepsilon_i(x)$ has been identified for M measuring locations $x = x_m (m = 1,2,\ldots M)$, the approximation (29) will be valid for each location x_m. Thus, using a least squares procedure, one obtains

$$u_i = m_i \frac{\sum_{m=1}^{M} \phi_i(x_m)\, \varepsilon_i(x_m)}{\sum_{m=1}^{M} \phi_i^2(x_m)} \tag{30}$$

Once the modal unbalances u_i are identified, the rotor can be balanced by making use of orthogonal test weight sets $\underline{t}_i$ (see |1|), which are precalculated from

$$\underline{t}_i = -\underline{\phi}^{-1} \cdot \underline{e}_i$$

with (31)

$$\underline{\phi} = [\phi_{ij}] = [\phi_i(x_j)]$$

$$\underline{e}_i^T = [0; 0; \ldots 0; \underset{\uparrow}{1}; 0; \ldots .0]$$

i-th element

The i-th mode contribution to the unbalance vibration will be compensated by mounting the unbalance set $\underline{w}_i = u_i \underline{t}_i$ in the balancing planes $x = x_j (j = 1,2,\ldots,N)$.

4 THE FIRST EXPERIMENTAL RESULTS

4.1 Test rotor in roller bearings

To test the efficiency of the newly introduced identification method, experiments are carried out on a rotor supported in three roller bearings, which are mounted on a heavy steel foundation with the aid of bearing blocks. The test rotor, an initially bowed shaft, is 1.63 m long and has a diameter of 1.5 cm. In order to increase the mass distribution of the mechanical system nine discs are mounted on the shaft. Each disc (of weight 387 g) can be used as a balancing plane. Balancing planes, measuring stations and precalculated modal information (mode shapes, eigenfrequencies, generalized masses) are shown in Fig 5. Although the bearing blocks are a l m o s t rigid, they are considered to be springs of very high stiffness for the sake of generality (as shown in the sketch of the system).

The shaft is driven by a 2 kW d.c. motor (see Fig 6). The rotational speed is controlled by a feedback system and can be regulated continuously up to 6000 rev/min. The manual or automatic control of the ideal speed value is possible. The motor speed is transmitted to the shaft by a flat belt. A transmission ratio of 1 : 1 makes high accelerations possible, which are needed to pass the critical speeds without developing heavy vibrations.

4.2 Measurement System

The measuring system is shown in Fig 6.

The shaft vibrations are measured in two orthogonal directions (horizontal and vertical) at three locations close to the bearings with six non-contact pickups (SV 101/SCHENCK). The output voltage of a sensor (pickup) is an analog signal consisting of constant and alternating parts. The signal preparation device reduces the d.c. voltage approximately to zero and amplifies the a.c. voltage (which is proportional to the time dependent deflection of the shaft center) to a suitable level for further processing. These signals, which can be filtered if necessary, are sampled by a spectral analyser (IN 90 S/INTERTECHNIQUE) for further analysing.

The angle of rotation is measured at the end of the shaft with an incremental, photoelectric angle encoder (ROD 426/HEIDENHAIN). The output signal trains (square-wave pulses) are subjected to counting and digital-to-analog conversion by the signal preparation device. The resulting, angle proportional analog signal will also be sampled by the analyser. However, this signal will not be filtered before sampling, because it is not proportional to the input $e^{j\varphi(t)}$ of the mechanical system.

Computer programs are implemented in the spectral analyser, which make the computation of the input-output signals, the fast Fourier transformation and finally the parameter identification possible. Further, these interactive programs enable the user to neglect such sampled signals which seem to be incorrect or to vary parameters such as the sampling frequency, the observation time, etc. .

4.3 Experimental concept

The following approach will be used to examine the accuracy of the transient balancing procedure:

(a) First, the rest rotor is balanced to a 'zero-shaft' (with almost no residual unbalance in the frequency range of interest) by means of a standard balancing method.

(b) Next, a selected unbalance state $u_i (i = 1,\ldots,N)$ affecting the first N modes is created in the well balanced rotor by mounting the modal unbalance set $\underline{w} = -\Sigma u_i \underline{t}_i$ whereas $\underline{t}_i (i = 1,\ldots,N)$ represent the orthogonal weight sets given by the equation (31).

(c) A transient run through N critical speeds of interest is used for data acquisition.

(d) The computer program idenfifies all parameters including the N modal unbalances, and calculates the balancing weights as described in section 3.

(e) The identified unbalance $\tilde{u}_i$ and the initially set unbalance u_i are compared to find the unbalance reduction ratio $R_i = 1 - |u_i - \tilde{u}_i| / |u_i|$ according to ISO 1925. This ratio eliminates the influence of inaccurate weight sets and it quantifies the accuracy of the transient balancing method.

(f) Finally, the balancing set $\tilde{\underline{w}} = \Sigma \tilde{u}_i \underline{t}_i$ is mounted to compensate the contributions of the first N modes and a control run is made.

Table 1

Results for transient runs of type (a)

Initially set generalized unbalance-for first mode (g cm)	Measuring station 1	2	3	Identified unbalance $\tilde{u}_1^{re}$ (g cm)	$\tilde{u}_1^{im}$ (g cm)	Unbalance reduction ratio R_1 (%)
u_1^{re} = -5	x			-2.700	-0.574	52.6
		x		-4.346	0.735	80.3
u_1^{im} = 0			x	-5.792	1.173	71.7
	x	x	x	-4.833	0.755	84.53
u_1^{re} = -3.53	x			-1.825	-2.581	60.9
		x		-4.418	-3.000	79.3
u_1^{im} = -3.53			x	-3.642	-2.100	71.3
	x	x	x	-3.610	-2.464	78.6
u_1^{re} = 0	x			0.273	-3.911	77.5
		x		-0.902	-4.408	78.4
u_1^{im} = -5			x	-0.705	-4.603	83.8
	x	x	x	-0.614	-4.430	83.2

Table 2

Results for transient runs of type (b)

Initially set generalized unbalance-for first mode (g cm)	Measuring station 1	2	3	Identified unbalance $\tilde{u}_1^{re}$ (g cm)	$\tilde{u}_1^{im}$ (g cm)	Unbalance reduction ratio R_1 (%)
u_1^{re} = -5	x			-2.184	-0.315	43.4
		x		-4.955	0.852	83.0
u_1^{im} = 0			x	-5.440	1.083	76.6
	x	x	x	-4.770	0.787	83.6
u_1^{re} = -3.53	x			-3.231	-3.640	93.6
		x		-4.327	-2.885	79.5
u_1^{im} = -3.53			x	-4.718	-3.333	76.0
	x	x	x	-4.356	-3.236	82.5
u_1^{re} = 0	x			1.014	-4.962	79.7
		x		-0.800	-5.538	80.7
u_1^{im} = -5			x	-0.874	-6.080	72.2
	x	x	x	-0.552	-5.727	81.7

Table 3

Results for transient runs of type (c)

Initially set generalized unbalance-for first mode (g cm)	Measuring station 1	2	3	Identified unbalance $\tilde{u}_1^{re}$ (g cm)	$\tilde{u}_1^{im}$ (g cm)	Unbalance reduction ratio R_1 (%)
u_1^{re} = -5	x			-3.771	-0.103	53.7
		x		-4.700	1.024	78.6
u_1^{im} = 0			x	-4.793	0.568	88.0
	x	x	x	-4.600	0.610	85.4
u_1^{re} = -3.53	x			-2.870	-4.842	70.6
		x		-3.718	-3.242	93.1
u_1^{im} = -3.53			x	-5.420	-3.304	62.0
	x	x	x	-4.463	-3.526	81.3
u_1^{re} = 0	x			0.636	-5.427	84.7
		x		-1.063	-4.980	78.7
u_1^{im} = -5			x	-0.254	-6.510	69.3
	x	x	x	-0.376	-5.842	81.5

4.4 Preliminary experiments and results

The experimental part of the investigation is still at an early stage. During the preliminary experiments only the first mode was considered. Three different initial unbalance states were created in the well balanced rotor:

$$u_1^{(1)} = -5.00 \text{ gcm} + j\,0$$

$$u_1^{(2)} = -3.53 \text{ gcm} - j\,3.53 \text{ gcm}$$

$$u_1^{(3)} = 0 \quad - j\,5.00 \text{ gcm}$$

Each unbalance state was identified for three different types of transient test runs:

Typ (a)
The rotor was a c c e l e r a t e d from the standstill position (f_o = 0 Hz) through the first critical speed ($f_1 \approx 27$ Hz) to reach a maximum rotational speed ($f_{max} \approx 35$ Hz) and then r e t a r d e d back to the initial zero-state. The evaluated, b a n d l i m i t e d input-output spectra and the frequency response of the mechanical system are shown in Fig 7 and Fig 8.

Type (b)
The rotor was o n l y a c c e l e r a t e d from the zero-state to reach the stationary state of the maximum rotational speed ($f_{max} \approx 35$ Hz).

Type (c)
The rotor was o n l y d e c e l e r a t e d from the stationary state of the maximum rotational speed ($f_{max} \approx 35$ Hz) to the standstill position.

Note, that, by using test runs of type (b) and type (c), the output of the mechanical system cannot be considered as a periodic signal which makes physical sense. However, curiosity and intuition influenced the author to use these types of transient runs, and the experimental results imply that they are quite suitable to minimize the errors made by the FFT.

The sampling frequency f_s = 128 Hz and the observation period T = 16 s were used throughout all experiments. Antialiasing filters and windows (other than the rectangular window) were completely avoided. The unbalance vibration of the rotor was measured at three locations (see Fig 5) in orthogonal directions.

The parameter identification was carried out for each measuring location seperately, and the unbalance parameter was identified also by taking measurements from all three locations into account, according to equation (30).

Although the other parameters were successfully identified, only the results of the unbalance identification are given in Tables 1, 2 and 3, which correspond to the used transient test runs of type (a), (b) and (c). Each table is split into three sub-tables, one for each unbalance state. From the first three lines of each sub-table , the result for each particular measuring location can be seen. In the last line the result of identification is shown, when the maximum available information was taken into account.

The three unbalance states were well identified for all the used types of transient test runs, when the measurements of all locations were integrated in the calculation. The unbalance reduction ratio R_1 lies between 78,6 % and 85.4 %, which can be considered as a very good result. Each time the rotor was balanced very successfully.

The results of these early stage experiments are very encouraging. The author has no doubts, that the transient balancing procedure will prove to be a very efficient method for practical applications.

ACKNOWLEDGEMENT

The author wishes to thank the DEUTSCHE FORSCHUNGSGEMEINSCHAFT (DFG) for supporting this investigation.

REFERENCES

|1| GNIELKA, P., 'Modal balancing of flexible rotors without rest runs: an experimental investigation', Journal of Sound and Vibration, 1983, 90(2), 157-172.

|2| GASCH, R., DRECHSLER, J., 'Modales Auswuchten elastischer Läufer ohne Testgewichtssetzungen - Darstellung der gemischt analytisch-experimentellen Vorgehensweisen, erste Versuchsergebnisse', VDI-Berichte Nr. 320, 1978, 445-454.

|3| CICHON,D., 'Transientes Auswuchten elastischer Rotoren ohne Testgewichtssetzungen', ILR, TU-Berlin: Diplomarbeit, 1983.

|4| MARKERT, R., 'System- und Unwuchtidentifikation von elastischen Rotoren aus Anfahrmessungen', VDI-Berichte Nr. 536, 1984, 121-139.

|5| BRIGHAM, E.O., 'Schnelle Fourier-Transformation', München, Wien: R. Oldenbourg, 1985.

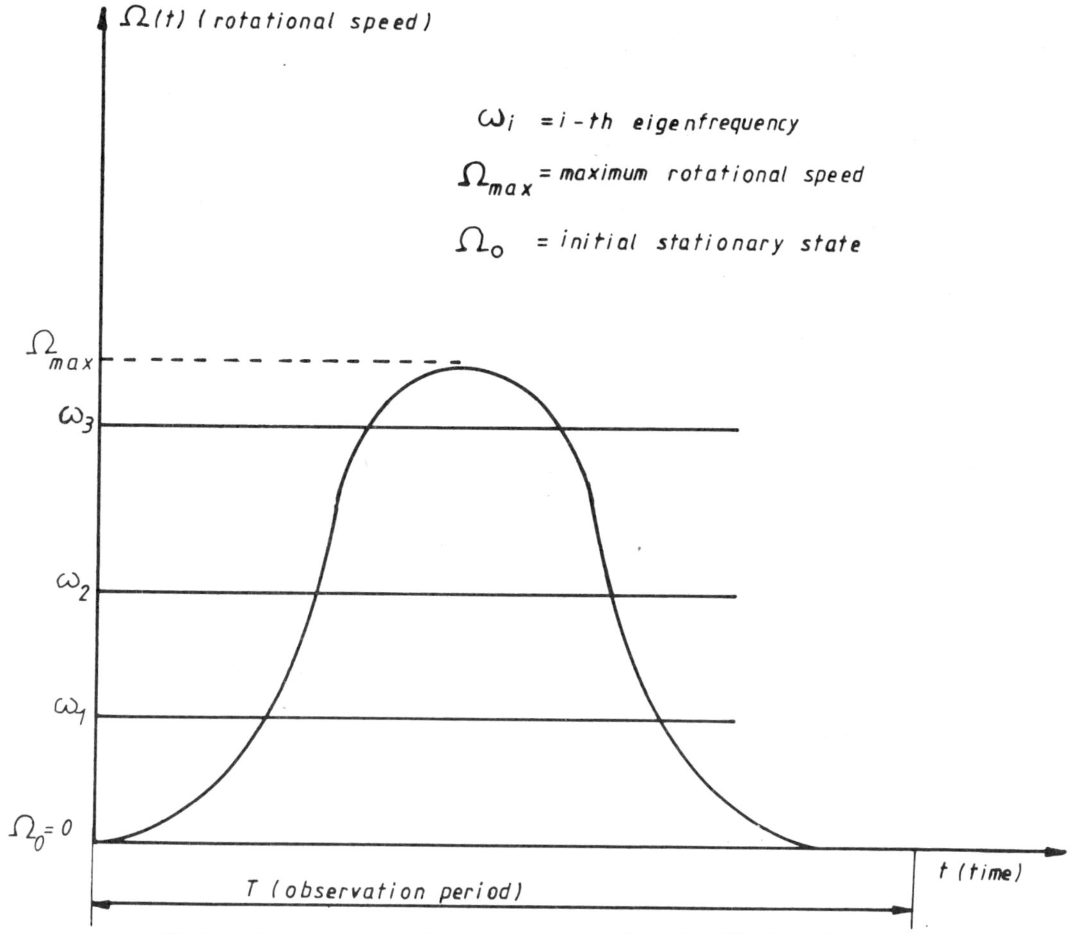

Fig 1 Accelerated and decelerated test run through critical speeds for data acquisition

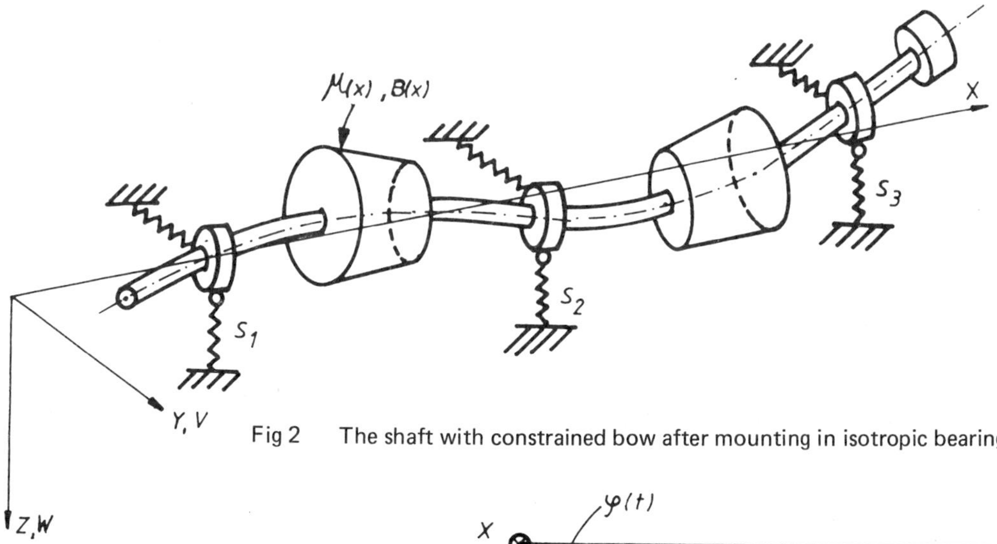

Fig 2 The shaft with constrained bow after mounting in isotropic bearings

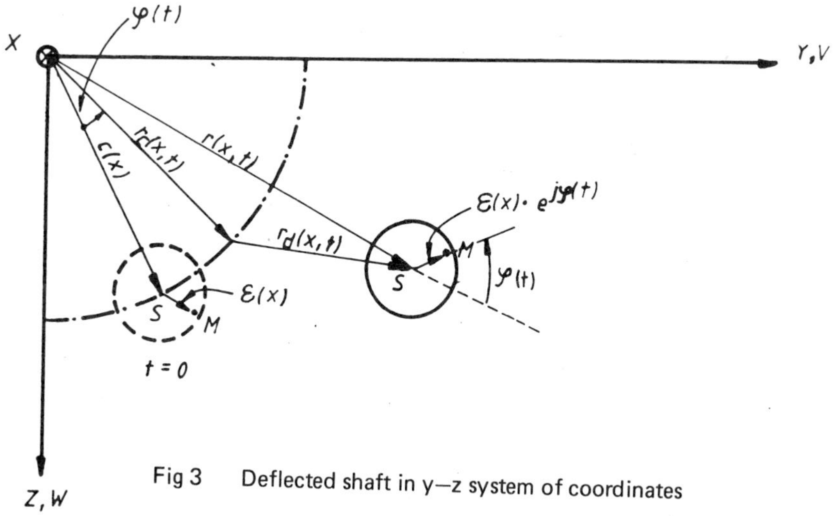

Fig 3 Deflected shaft in y–z system of coordinates

time domain INPUT $e^{j\varphi(t)}$ $E(j\omega)$

$$r(x,t) = \int_{-\infty}^{+\infty} e^{j\varphi(\tau)} \cdot h(x, t-\tau) d\tau$$

$$R(x, j\omega) = E(j\omega) \cdot H(x, j\omega)$$

parameters: $c(x), \phi_i(x),$ u_i, m_i, ω_i, d_i $(i=1,2,\ldots,\infty)$

time domain $r(x,t)$ OUTPUT $R(x, j\omega)$

frequency domain

frequency domain

SYSTEM (linear; time invariant; deterministic)

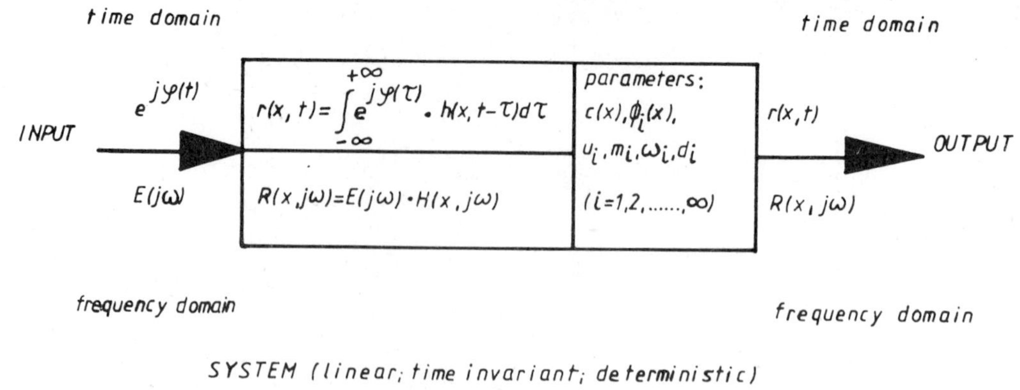

Fig 4   Mathematical model of mechanical system

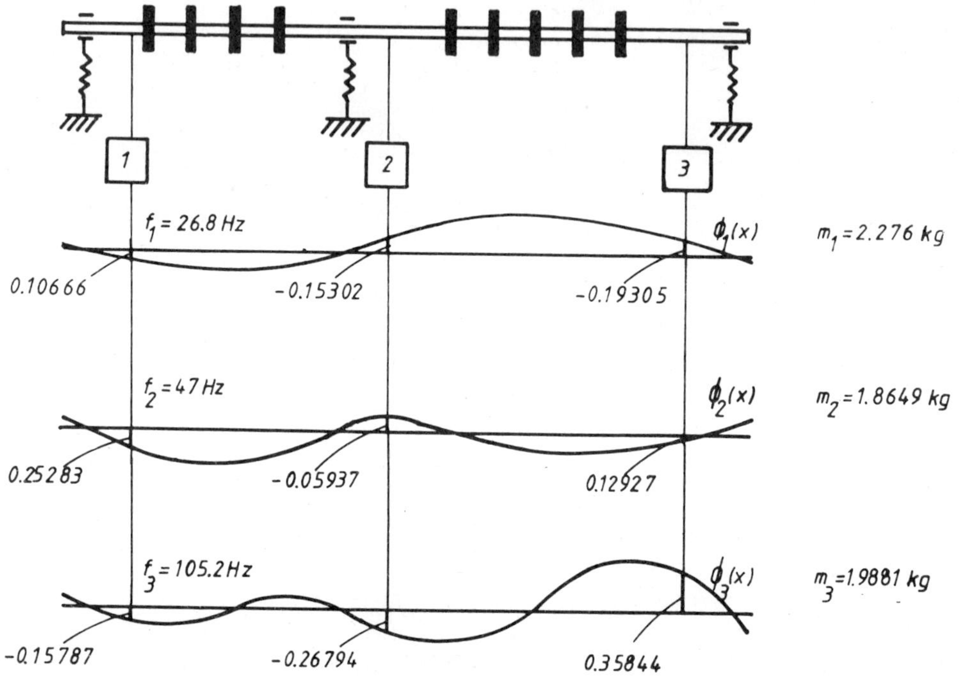

Fig 5   Test rotor in roller bearings with pre-calculated mode information

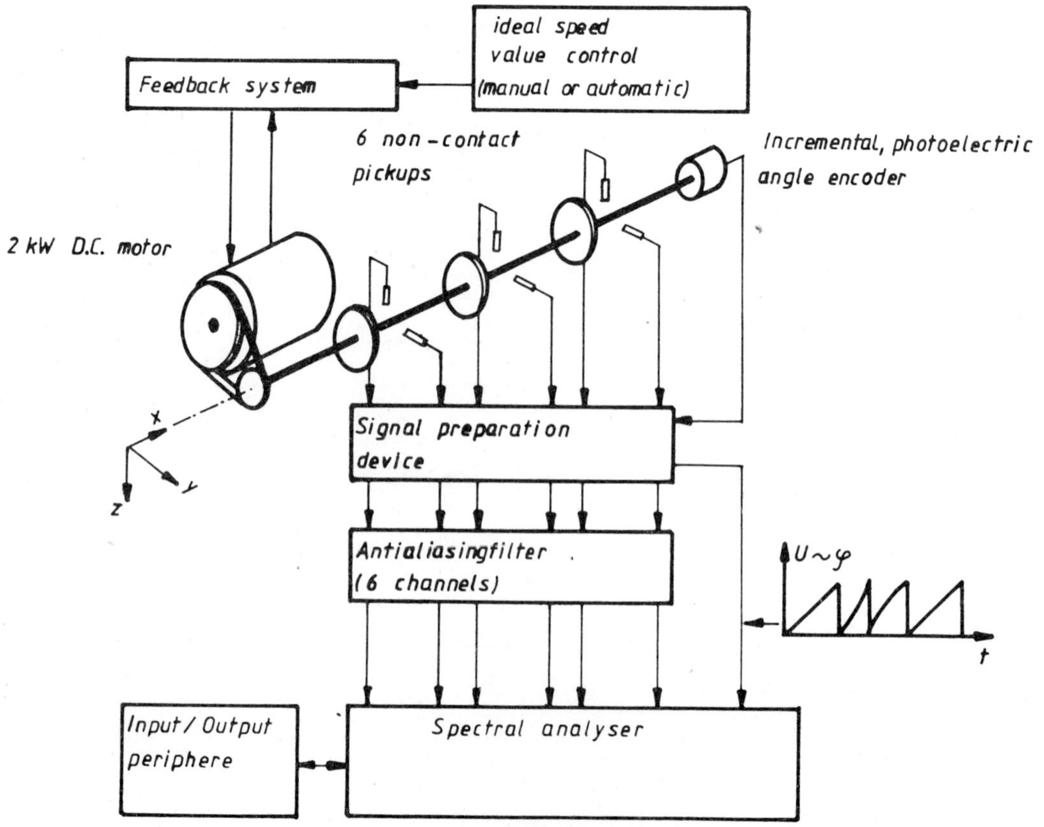

Fig 6   The measurement system

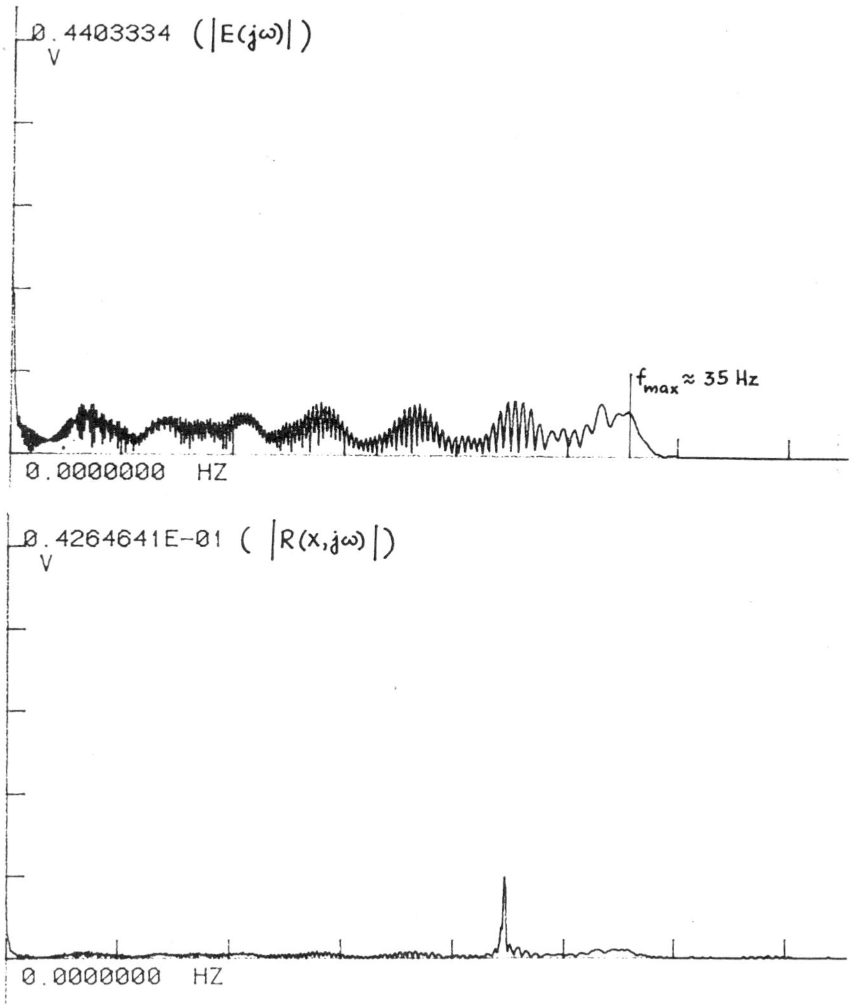

Fig 7 Input–output spectra for transient runs of type (a)

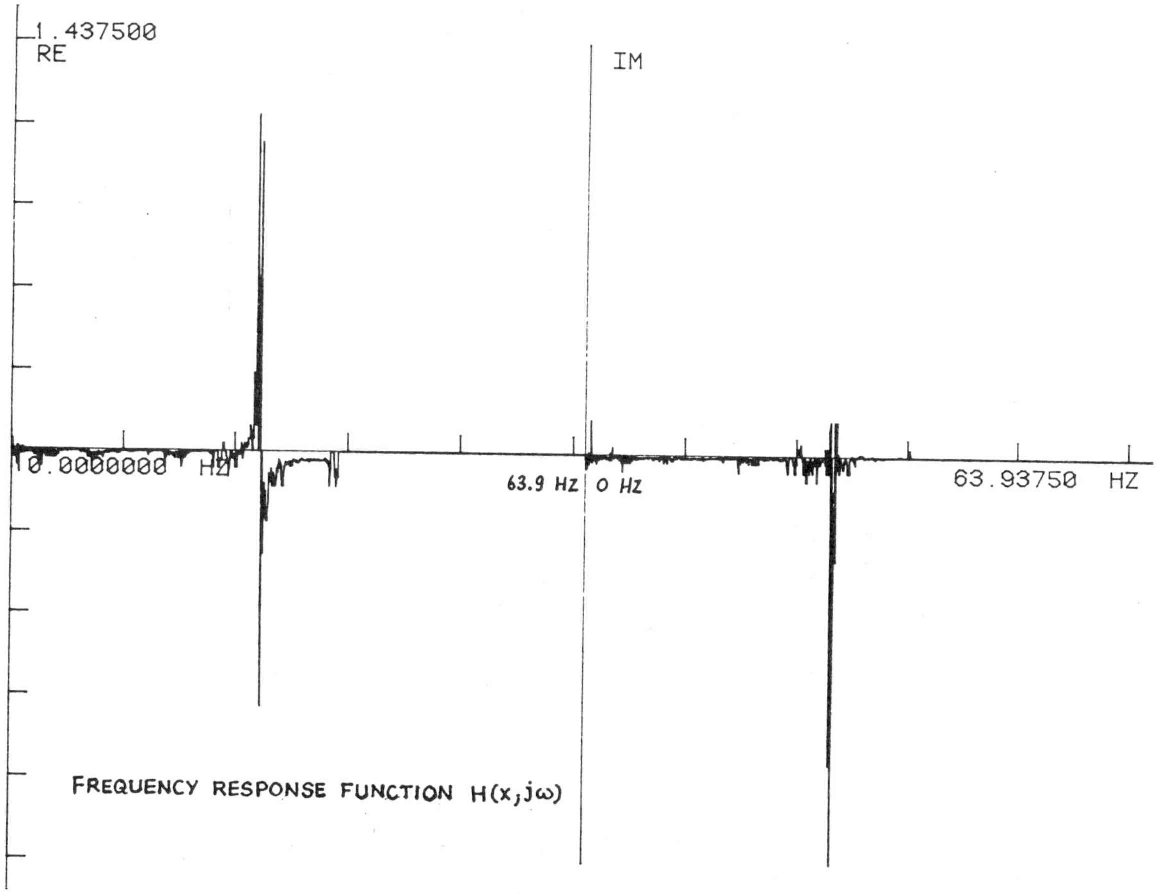

Fig 8 Frequency response of mechanical system

C316/88

Balancing of jet engine modules

H SCHNEIDER, Dipl-Ing
Carl Schenck AG, Darmstadt, West Germany

SYNOPSIS A theoretical approach is used to consider and compare different balancing procedures and arrangements of correction planes in order to optimize the low speed, multi-step balancing of jet engine modules.

NOTATION

A B	module A, B
L, L_A, L_B	length of assembly, modules
L_{AC}, L_{BC}	distance of center of gravity of modules A, B from bearings
L_P	distance between correction planes
m_A, m_B	masses of modules A, B
J_{Az}, J_{Bz}	moments of inertia of modules A, B around shaft axis
J_{Ax}, J_{Bx}	moments of inertia of modules A, B around transverse principal axis
α, β	angle errors at mating plane
$\gamma_{A\alpha}$, $\gamma_{B\alpha}$	inclination of main axis of modules due to error α
$\gamma_{A\beta}$, $\gamma_{B\beta}$	inclination of main axis of modules due to error β
U_S, U_C	static, couple unbalance
R	radius of zylinder or disc
①,②,③,④	correction planes 1 to 4
(1a), (3a), (4a)	alternative correction planes

1 INTRODUCTION

In order to facilitate service and reduce downtime some jet-engines are designed and built to allow for the exchange of complete modules.

As far as rotating elements (parts) are concerned, this naturally influences the balancing procedure to quite a large extent.

The dynamic behaviour of some rotors is a flexible one, therefore class 2 (low speed) balancing procedures according to ISO 5406 (1) are applied. The following addresses not a particular engine but these rotors in general.

2 CONFIGURATIONS

2.1 Single module with two bearings

A single module with two bearings - either at each end or both on one end - doesn't seem to impose too many difficulties, apart from the problems with transfer-unbalances (2) which sometimes seem to be missed.

2.2 Two modules with two bearings

This for instance could be a fan/booster system (figure 1), where both the bearings are on the fan shaft, or on HPC / HPT system, where each module has one bearing (figure 2).

Since the latter would be the more critial case, it will be the basis for our theoretical approach, but some main principles and findings could also be applied to the other case.

2.3 Two modules with one bearing each

There would be no difference in balancing procedure between two individual modules with two bearings each, and this afore mentioned system, if the mating surfaces are accurate enough to meet the balancing tolerances. In this case both the modules would be balanced individually- using an auxiliary stub shaft on each - and could then be mated and used without the need for an additional balancing step.

But even with today's sophisticated manufacturing and assembly methods the geometric errors - axial and radial runout - at the mating surfaces are too large (compared to balancing tolerances) to be neglectable.

Whilst axial and radial runout have different influences on the geometric performance of the assembly - runout, gaps at blade tips etc. - for the purpose of balancing both the errors can be expressed just as one error: an axial runout. One can assume the shaft-axis of a module itself to be the axis connecting the bearing center on one side with the center of the mating plane on the other side. Then only the axial runout determines the position of the module-axis after assembly. For calculation purposes it is easier to use the tilt angle of the mating plane in preference to the runout.

3 BALANCING PROBLEM

Two well balanced modules (relative to their shaft axis) (figure 3) will show new unbalances after assembly, due to the angle errors at the mating plane α of module A and β of the module B.

α and β will have different magnitudes and different angular location (circumferential), and will sum up vectorially. Obviously due to angle α, the principal axis of either module is inclined to the shaft axis of the assembly and the centers of gravitiy of both modules are displaced from this new shaft axis.

For small angles one can state:

$$\gamma_{A\alpha} = \alpha \frac{L_B}{L} \quad (1)$$

$$\gamma_{B\alpha} = \alpha \frac{L_A}{L} \quad (2)$$

$$e_{A\alpha} = \gamma_{A\alpha} \cdot L_{AC} \quad (3)$$

$$e_{B\alpha} = \gamma_{B\alpha} \cdot L_{BC} \quad (4)$$

For an angular error β of module B similar equations apply.

Static and couple unbalances will be:

$$U_{AS} = e_{A\alpha} \cdot m_A \quad (5)$$

$$U_{AC} = \gamma_{A\alpha} (J_{Ax} - J_{Az}) \quad (6)$$

with corresponding equations for module B (due to error α) and modules A and B due to error β .

If these unbalances exceed (or are too large a portion of) the tolerances an additional balancing procedure is necessary.

3.1 Indexing Procedure

In order to arbitrarily match modules in arbitrarily chosen angles, an indexing procedure is applied:

- The unbalance of the assembly is measured
- One module is rotated (if possible through 180°) against the other
- The unbalance is measured again

By numerical or graphical evaluation all unbalances caused by module A can be related to A, all unbalances caused by B, are related to B. If the prebalancing of the single modules is included, this means:

	Unbalance of A around its shaft axis
+	Unbalance of A due to error α
+	Unbalance of B due to error α
=	Sum of all unbalances corrected at Module A.

	Unbalance of B around its shaft axis
+	Unbalance of A due to error β
+	Unbalance of B due to error β
=	Sum of all unbalances corrected at module B.

This procedure works well - even though it sometimes is a difficult and time consuming task on jet-engine rotors - but there is a further problem: By correcting unbalances at the module A, which really occur at module B, the basic rule of a ISO class 2 - rotor*) is broken: All unbalances have to be corrected close to those planes from which they result.

*) ISO 5406 (1) describes flexible rotors, which can be balanced at low speeds using special balancing procedures as class 2 (quasi-rigid) rotors.

To examine more carefully what this could mean, we have to simplify the rotors and then look for typical unbalances:

Figure 3 shows that both modules can be simplified to cylinders or discs of different diameters and lengths.

Unbalances of module A are, for example:

$$U_{S1,2} = \gamma_A \cdot m \frac{L_{AC}}{2} \quad (7)$$

$$U_{C1,2} = \mp \gamma_A \cdot m \frac{L_A^2/3 - R^2}{4 L_P} \quad (8)$$

3.2 HP compressor

This typically is a cylindrical rotor with a length larger than the radius. Keeping L_A constant and assuming the c of g to be in the middle ($L_{AC} = 1/2 L_A$) and correction planes to be the end planes, figure 4 shows the relative unbalances in planes 1 and 2 caused by an angle error α (or β).

- The static unbalance outweights the couple unbalances
- Within the range of interest, $0{,}15 > \frac{R}{L_A} > 0{,}5$ U_{C1} is always negative and substracts from U_{S1} , U_{C2} is always positive, and adds to U_{S2} .

This typically means: the unbalances in both planes have the same sign, the one in plane 1 being smaller than the one in plane 2.

3.3 HP turbine

This module typically is disc-shaped, with the c of g remote from the compressor and near to the bearing. Assuming this distance from the bearing to be 1/4 of the module length and the correction planes to be located symmetrically to the c of g and 1/4 of the module length apart, typical relative unbalances for planes 3 and 4 are shown in figure 5.

- the relative static unbalances are smaller than on the HPC, couple unbalances can be much larger
- the relative couple unbalance substracts from the static portion in plane 3, in plane 4 they are additive

3.4 Typical unbalances

Typical unbalances (based on the above) in planes 1 to 4 caused by an angular error α or β are shown in figure 6: smaller or larger angular errors change all unbalances in proportion. Unbalances due to errors α and β are added vectorially, but with proper angular matching of the modules, only adding or substracting of the errors and the unbalances is necessary.

4 FLEXIBLE ROTORS

4.1 Bending Mode

We restrict our considerations to the first critical speed and therefore to the first flexible mode.

If there were no appreciable bending between the left hand and the right hand bearing, the assembly would be a rigid rotor - class 1 according to ISO (1) definition - and could be balanced with typical procedures used for rigid rotors, if necessary, including index balancing.

In the case, where the mode shape between correction planes 1 and 4 is a straight line with some appreciable bending between the bearings and the outer correction planes, the assembly would be considered as a quasi-rigid rotor of class 2c, but one would nevertheless allow for index balancing the modules.

But sometimes bending between correction planes 1 and 4 appears to be unavoidable and an assumed bending line is shown on figure 6 (if this shape doesn't coincide with that of an actual rotor, it doesn't matter: all further explanations will none the less apply).

4.2 Rotor deflection

All unbalances will cause deflections proportional to the ordinate of the mode shape at their planes. This means e.g. an unbalance in plane 2 is much more effective than the same magnitude of unbalance in planes 1 or 4. Deflections add up or substract from each other. In order to compare different unbalance distributions one can simply multiply the unbalance units in each plane with the corresonding ordinate and build the sum over all corection planes.

The unbalances of figure 6 (according to the above) will cause a total deflection of 130. The compressor alone contributes 115 (88 %), the turbine only 15 (12 %).

If error α (of module A) has caused these unbalances, using the index balancing procedure <u>all</u> unbalances are corrected using only planes 1 and 2 (figure 7): Now for low speeds (rigid behaviour) the assembly is properly balanced, but the figure for the deflection is calculated to be 109, which is only a slight improvement.

If error β (of module B) has caused these unbalances, all unbalances are to be corrected in planes 3 and 4 (figure 8):

The low speed condition again is good, but the deflection is calculated to be 145, i.e. it is even worse. In addition, the amount of correction (74 units) is quite large, which might be difficult or impossible to perform under typical correction conditions on jet engines.

An improvement would be a correction in planes 3a and 4 (together with the small unbalance in plane 3), figure 9: Again low speed is o.k., deflection amounts to 91, correction is performed with only 12 units.

A further improvement would be a correction of unbalances in module B in planes 3 and 4 and of unbalances in module A as near as possible to A, e.g. in planes 3a and 4a (figure 10): Low speed is o.k., deflection amounts to 57 (a clear improvement) correction is 38 units, worse than in figure 9 due to small distance between correction planes.

In a similar manner deflection can be reduced, but the number of correction units will increase if for error α at module A correction planes 1, 1a and 2 are used.

4.3 Assembly

One would assume that two equal errors α and β would give a well balanced assembly if they are oriented in opposite directions. This would of course hold true, if both modules are only balanced on their own, but with index balancing (based on the corrections of figure 7 and 9), figure 11 shows: low speed again is o.k., but deflection amounts to 38. The total number of correction units is 27.

Surprisingly deflection becomes nearly zero under same general conditions) if error β is 1,4 times error α.

5 POSSIBLE VARIATIONS

From the above we can conclude:

- If the assembly exhibits flexible behaviour, there is no possibility of balancing and matching the modules arbitrarily: The deflection at high speed will not be tolerable.
- A straight rotor can reduce the problems, but sometimes the errors α and β should be different if an optimum assembly state is to be achieved.
- Index balancing keeps the low speed unbalance at zero, but sometimes worsens the high speed condition.
- If a dummy module B is used whilst balancing module A (and vice versa), the mass properties could be varied to achieve better results.

In summary this means that for a particular rotor we have to compare different procedures in order to find an optimal solution.

I - balancing the modules individually (without displacement - due to errors α and β)

II - correcting for the displacement too, but without dummy mass and moment of inertia

III - correcting for the displacement too, with dummies of reduced mass and moment of inertia

IV - correcting for the displacement too, with full dummy properties

Additionally three different positions of correction planes (referred to as cases) can be considered:

Table 1 Results for different procedures and cases (correction planes)

	procedure	without displacements	with displacements due to errors α and β							
		I individual modules	II wo. dummies		III half dummies			IV full dummies		
	case	1	1	2	1	2	3	1	2	3
flex. behaviour	no excitat. $\alpha : \beta$ 1	1:1	8:1	7:1	1:4	1:5	1:1,1	1,1:1	1:1,4	1:1,4
flex. behaviour	assembly 2	↑↓	↑↓	↑↓	↑↓	↑↑	↑↑	↑↓	↑↓	↑↓
flex. behaviour	sensitivity 20% 3	26	23	22	15	10	6	26	18	11
flex. behaviour	sum of correction units 4	0	70	53	198	72	46	90	31	100
rigid behaviour	static unbalances 5	0	24	22	14	22	6	0	0	0
rigid behaviour	couple unbalances 6	0	31	34	8	5	0.5	0	0	0
rigid behaviour	sensitivity static 20 % 7	2.4	6	6	4	4	0.3	0	0	0
rigid behaviour	sensitivity couple 20% 8	0.3	4.5	6	2	1	0.3	0	0	0

Case 1: planes 1, 2, 3 and 4

Case 2: planes 1, 2, 3a and 4

Case 3: planes 1 and 2 for unbalance of module A due to α, 3, 4 for unbalances of module B due to β, 1a and 2 for unbalance of B due to α, 3a, 4a for unbalances of A due to β.

6 RESULTS

Some theoretical results are listed in table 1:

Line 1: ratio $\alpha:\beta$ with no excitation of mode

Line 2: assembly: ↑↑ errors α and β add up
↑↓ errors α and β substract from each other

Line 3: deflection of assembly at high speed if α or β deviates by 20 % from proper ratio

Line 4: sum of correction units

Line 5: static unbalance at low-speed

Line 6: couple unbalance at low speed

Line 7: change of static unbalance if α or β deviates by 20 % from proper ratio

Line 8: change of couple unbalance if α or β deviates by 20 % from proper ratio

7 EVALUATION

From the beginning we can delete all ratios $\alpha : \beta$ which do not meet practical considerations:

Since the smaller value can't be reduced any more, one would have to build into the other module a much larger angle error than necessary to satisfy the ratio. If in practice errors α and β have the same order of magnitude, we have to delete the posssibilities II,1 ; II,2 ; III,1 and III,2.

The remaining 5 possibilties show different characteristics:

With reference to column I:
Clearly this procedure needs identical errors α and β, a straight build, and is very sensitive to deviations from the ratio 1:1 (26 for 20 %). There is no correction for a (non existing) displacement (0).

Low speed condition is good (0), sensitivity against deviations from ratio 1:1 seems to be acceptable (2,4 : 0,3).

With reference to column III,3:
Balanced with dummies of half the value of the real modules, the modules show no excitation at high speed for $\alpha : \beta$ = 1:1,1 and both errors adding up! Sensitivity agains deviations of ratio is very small (6), but a lot of correction is necessary (46). Critical could be the static unbalance at low speed (6) whereas couple unbalance and sensitivities against deviations are very small.

With reference to column IV,1 to 3:
All error ratios are around 1:1, α and β to substract, but sensitivities to deviations in ratio are large, and a lot of correction has to be done (except IV,2).

Low speed balancing conditions and sensitivity are naturally ideal.

8 CONCLUSION

- Beyond a certain limit of flexibility, there is no possibility of exchanging modules without taking account of the angle errors of the modules and matching them properly.

- Depending on the degree of flexibility, the dimensions, masses and moments of inertia of the modules and the typical errors, there is a good chance of finding a procedure and correction planes, which are suitable and lead to a satisfactory compromise between low speed and high speed unbalance conditions.

- Balancing machines can be used to measure the unbalance caused by the errors of the mating plane - and hence to quantify these errors - in order to select modules and match them angularly in an optimal relationship.

REFERENCES:

(1) ISO 5406:
The mechanical balancing of flexible rotors
Nov. 1980

(2) SCHNEIDER, H.
Balancing of integral gear-driven centrifugal compressors.
15th Turbomachinery Symposium,
November 1986, Texas University

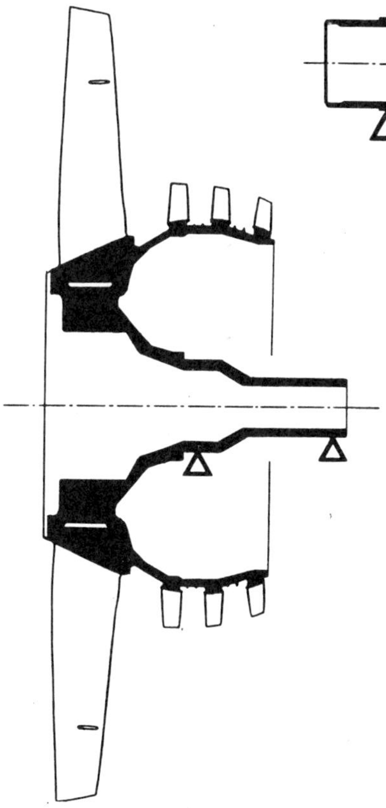

Fig 1 Fan with two bearings and booster

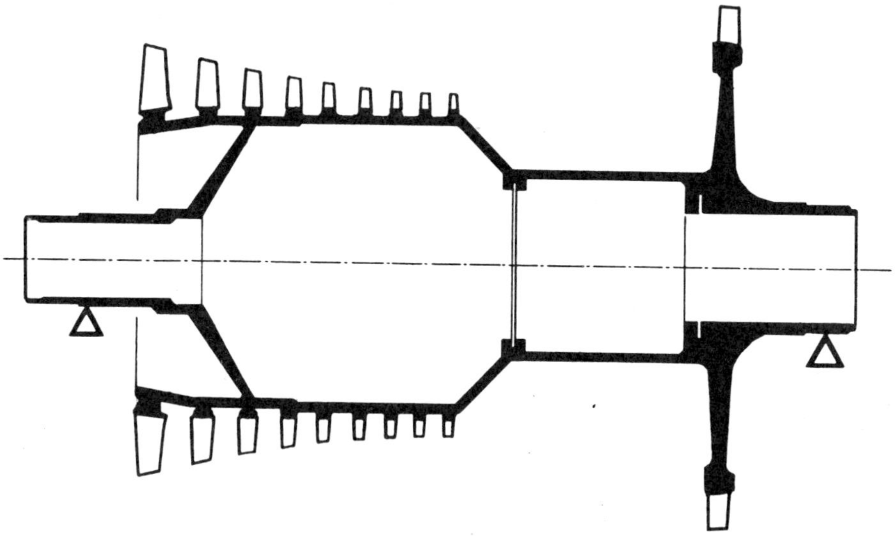

Fig 2 HP compressor/HP turbine with one bearing each

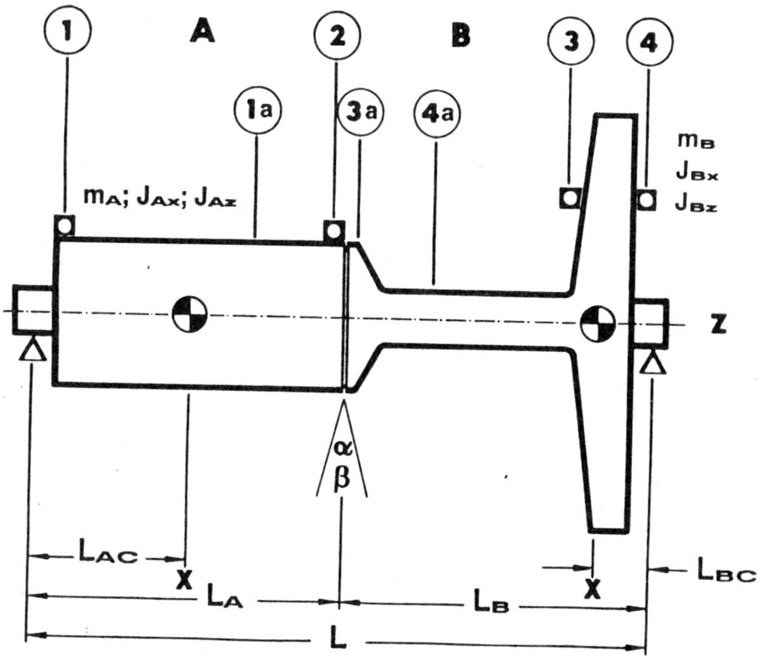

Fig 3 Assembly with main physical data

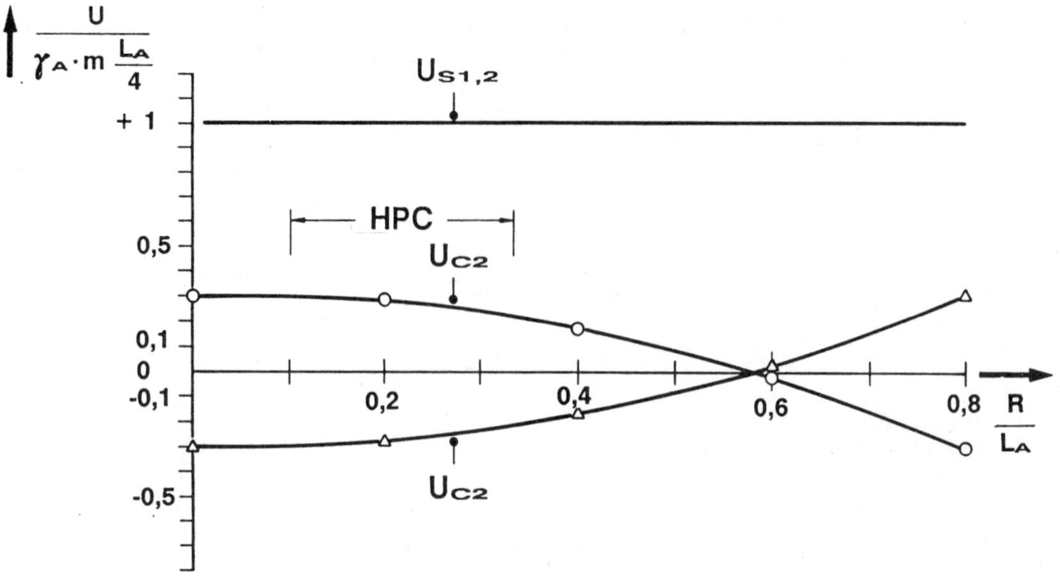

Fig 4 Typical relative unbalance (static and couple) on an HP compressor plotted against R/L_A

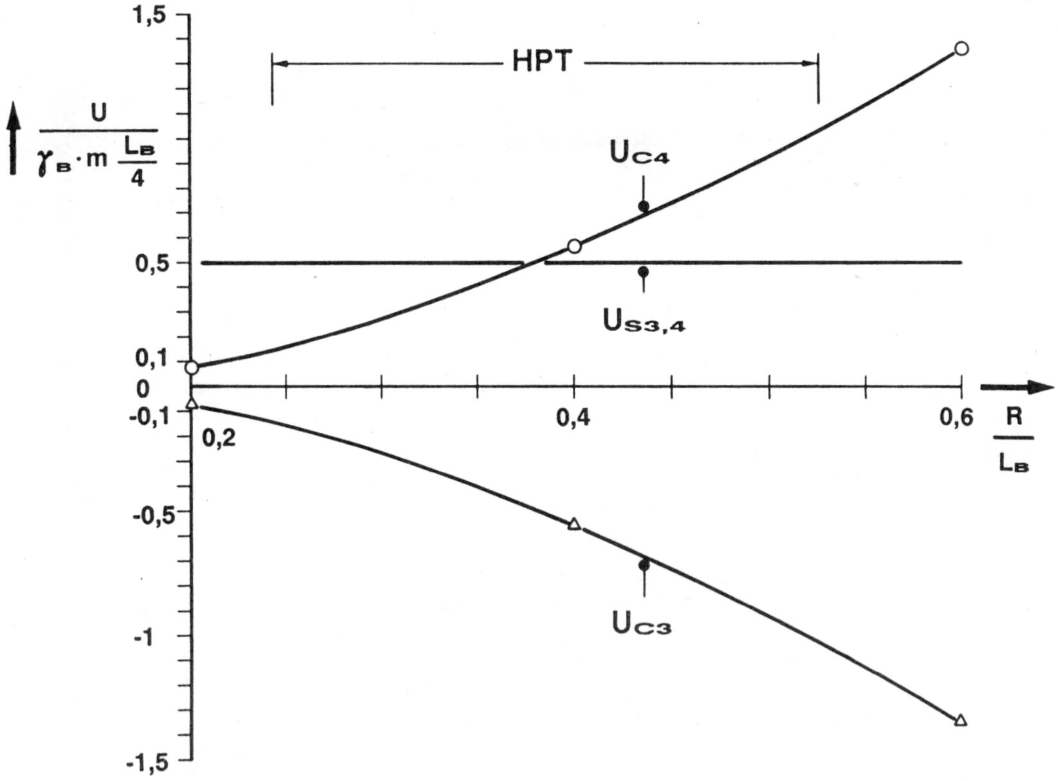

Fig 5 Typical relative unbalance (static and couple) on an HP turbine plotted against R/L_B

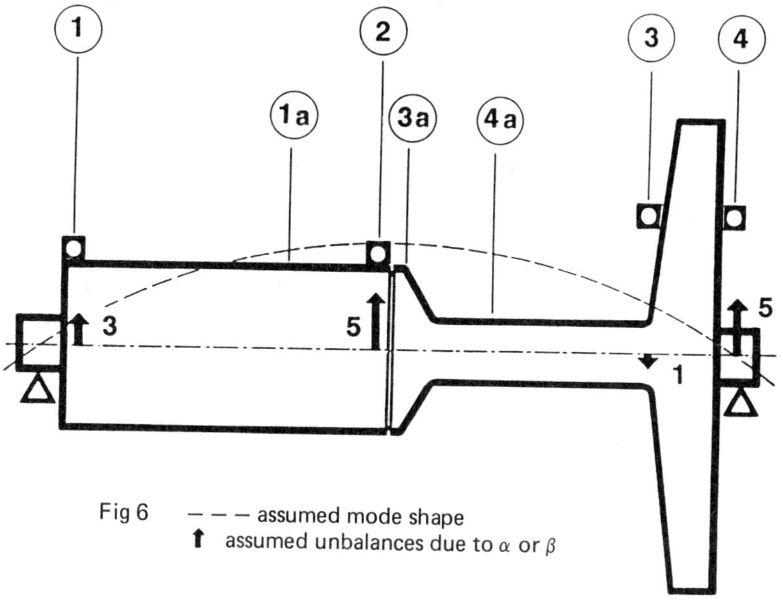

Fig 6 — — — assumed mode shape
⬆ assumed unbalances due to α or β

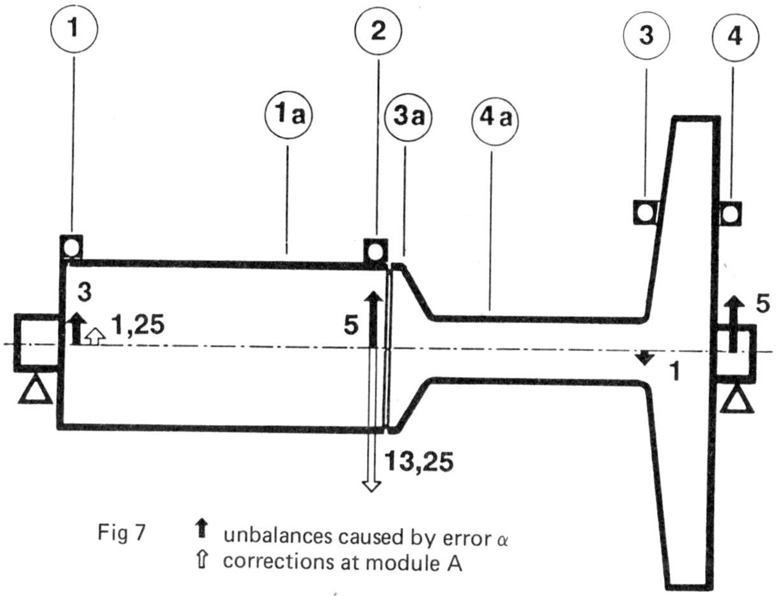

Fig 7 ⬆ unbalances caused by error α
⇧ corrections at module A

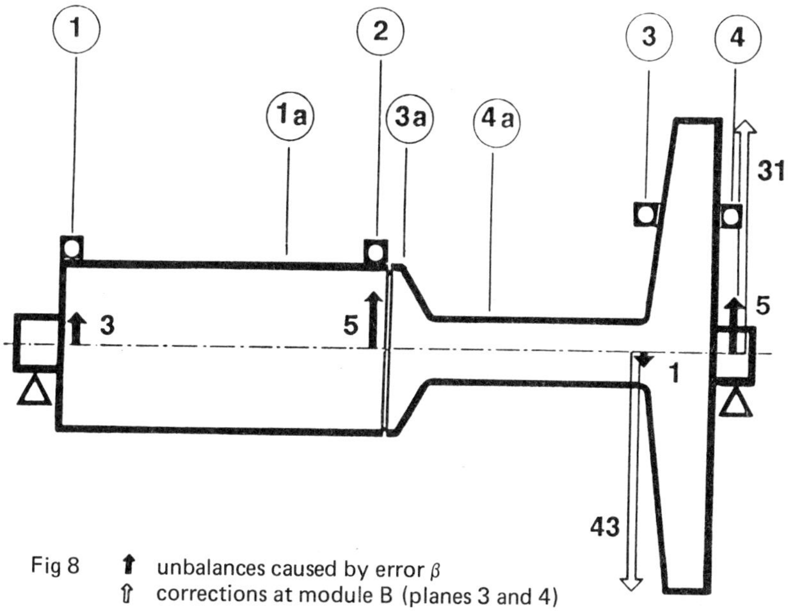

Fig 8 ⬆ unbalances caused by error β
⇧ corrections at module B (planes 3 and 4)

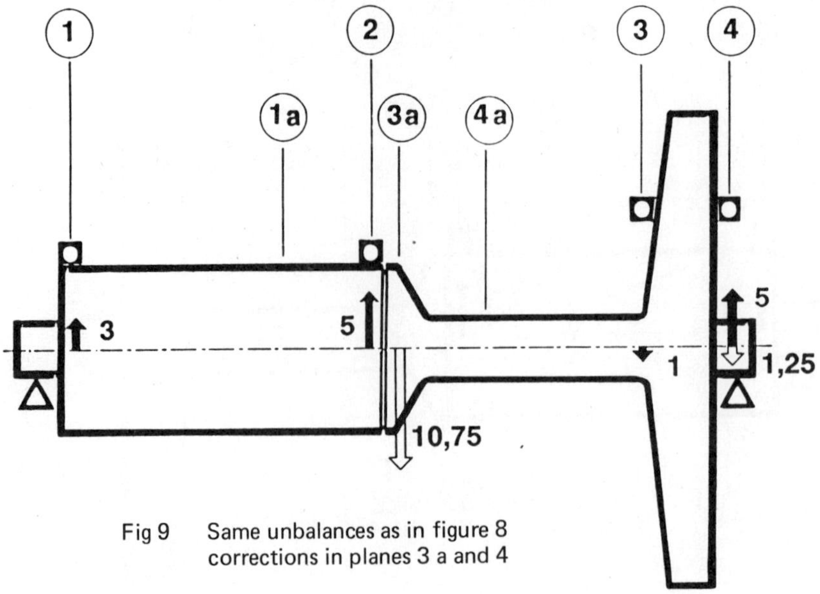

Fig 9 Same unbalances as in figure 8
corrections in planes 3 a and 4

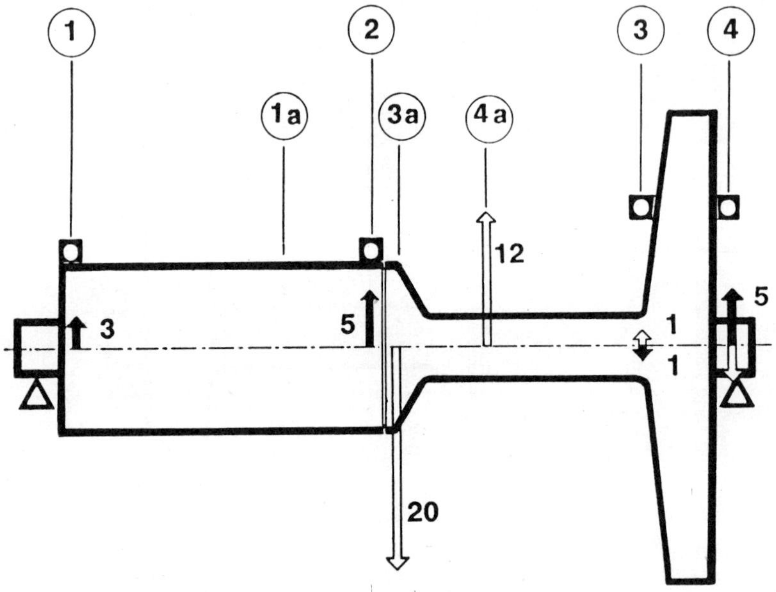

Fig 10 Same unbalances as in figure 8
⇧ corrections in planes 3, 4 and 3 a, 4 a

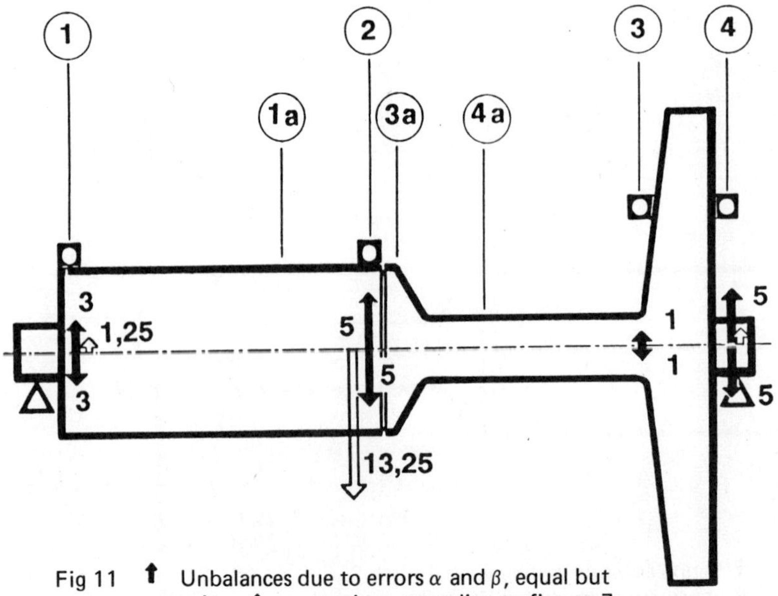

Fig 11 ⬆ Unbalances due to errors α and β, equal but opposite ⇧ corrections according to figures 7 and 9

C260/88

Calculation of unbalance sensitivity in complex rotor systems

H-L OLAUSSON, LicEng and A KLANG, MSc
Department of Mechanical Engineering, ASEA STAL, Finspong, Sweden

SYNOPSIS A modal approach to an approximative calculation of sensitivity to unbalance in rotating machinery is presented. The approach is purely analytical, is cost effective, simple to use and gives a unique number for each mode. To demonstrate the usability of being able to easily estimate unbalance sensitivity in a rotating machine, a calculation from an optimization of a squeeze film damper in a twin-shaft gas turbine is shown.

NOTATION

e	Mass eccentricity
f_x	Force along the x axis
f_y	Force along the y axis
i	Imaginary number, $(-1)^{1/2}$
m	Mass
r_{ij}	Transfer function
x,y	Lateral displacement
δ_i	Displacement at point i
ϕ	Phase angle
λ_k	Eigenvalue number k
μ_k	Real part of eigenvalue λ_k
σ_k	Imaginary part of eigenvalue λ_k
ω	Circular frequency of vibration
Ω	Angular speed of shaft rotation
a_k	Normalization constant for mode number k
diag($\mathbf{a}$)	Diagonal $\mathbf{A}$ matrix
$\mathbf{A,B}$	System dynamic matrices in state space form (complex)
$\mathbf{C}$	System damping matrix
$\mathbf{f}(t)$	Force vector
$\mathbf{H}$	System right-hand eigenvector matrix
$\mathbf{h}_j$	Right-hand eigenvector associated with eigenvalue λ_j
$\mathbf{I}$	Identity matrix
$\mathbf{K}$	System stiffness matrix
$\mathbf{M}$	System mass matrix
$\mathbf{q}$	System state space vector
$\mathbf{U}$	System left-hand eigenvector matrix
$\mathbf{u}_j$	Left-hand eigenvector associated with eigenvalue λ_j
$\mathbf{v}$	Real velocity vector
$\mathbf{x}$	Real variable vector
ζ	Complex normal coordinates
γ	State space load vector
Λ	Diagonal eigenvalue matrix
ϕ_j	Left-hand mode form associated with eigenvalue λ_j
ψ_j	Right-hand mode form associated with eigenvalue λ_j

1 INTRODUCTION

The term unbalance sensitivity has become more topical now that rotors and foundations have become so flexible that it is unrealistic to design a machine which is free from resonance in the operating area. It is not even technically justifiable to assert freedom of resonance.The unbalance sensitivity is often connected with the damping of the resonance, which is only one of several parameters which are important when describing sensitivity to unbalance. Two other parameters have proved significant. The first is the eigenvector's amplitude in relation to its normalization. The second is the left-hand eigenvector's whirl orbit. Even modes with low damping can be ignored if unbalance sensitivity is correctly calculated.

The method to calculate unbalance sensitivity described in this Paper is to be used purely as an analysis tool. The method does not strictly maximize the unbalance distribution along the rotor , instead the approach is to find a unique distribution for each modeshape with the property that all modes are excited in a similar way.

To demonstrate the method a sample calculation has been done where the minimization of the calculated unbalance sensitivity has been used as the objective function to optimize a squeeze film damper in a twin-shaft gas turbine.

2 UNBALANCE FORCE

In Appendix, the transfer function r_{ij} for displacement at points i caused by an applied force at j can be shown thus:

$$r_{ij} = \sum_k \frac{\psi_{ik}\phi_{jk}}{a_k(\lambda - \lambda_k)} + \frac{\bar{\psi}_{ik}\bar{\phi}_{jk}}{\bar{a}_k(\lambda - \bar{\lambda}_k)} \quad (1)$$

where k is the number of modes, ϕ and ψ are the left- and right-hand eigenvectors, a_k is the normalization constant, λ_k is the eigenvalue and λ is the excitation frequency. This means that the displacement δ_i, in an arbitrary point i depending on forces f_j along the rotor can be shown:

$$\delta_i = \sum_j \left[\sum_k \left[\frac{\psi_{ik}\phi_{jk}}{a_k(\lambda - \lambda_k)} + \frac{\bar{\psi}_{ik}\bar{\phi}_{jk}}{\bar{a}_k(\lambda - \bar{\lambda}_k)} \right]\right] f_j \quad (2)$$

where j is the number of the lumped mass rotor station. The contribution from mode k can be shown for a harmonic excitation $\lambda = i\omega$

$$\delta_i = \Psi_{ik} \sum_j \frac{\phi_{jk} f_j}{a_k(i\omega - \lambda_k)} + \bar{\Psi}_{ik} \sum_j \frac{\bar{\phi}_{jk} f_j}{\bar{a}_k(i\omega - \bar{\lambda}_k)} \quad (3)$$

For the damping of interest, i.e. with a damping under 10 per cent of critical damping, the first term in equation (3) is dominant as the real part μ_k of the eigenvalue λ_k is less than 10 per cent of the imaginary part σ_k, which means that:

$$| i\omega - \mu_k - i\sigma_k | << | i\omega - \mu_k + i\sigma_k | \quad (4)$$

at resonance speed $\omega = \Omega_k$

The maximum reading at unbalance excitation is obtained at rotation frequency Ω_k according to ref (1) for mode k

$$\Omega_k = \frac{\sigma_k^2 + \mu_k^2}{\sqrt{\sigma_k^2 - \mu_k^2}} \quad (5)$$

Now choose the direction of the unbalance so that the generalized force $\phi_{jk} f_j$ is as effective as possible. Suppose that there is a harmonic excitation of unbalance forces f_j along the rotor. The affected force in a rotor station can be described in complex formula by:

$$f_{xj} = m_j \underline{e}\, \Omega^2$$

$$f_{yj} = -i m_j \underline{e}\, \Omega^2 \quad (6)$$

where the displacement of the centre of gravity $\underline{e}$ is complex according to:

$$\underline{e} = e_c + i e_s \quad (7)$$

where

$$e_c = |e| \cos \phi_e \qquad e_s = |e| \sin \phi_e \quad (8)$$

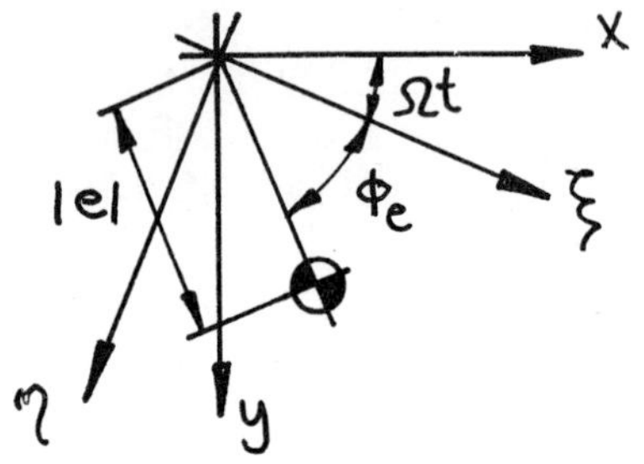

Fig 1 Mass eccentricity

The orbit of the left hand eigenvector ϕ is an ellipse which can be described as a prograde part and a retrograde part according to Figure 2.

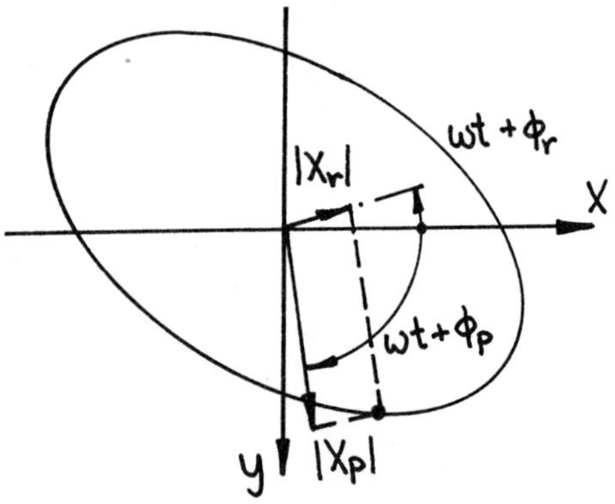

Fig 2 Elliptical orbit

The prograde and retrograde transformation can be described as follows:

$$\underline{x}_p = 1/2(\underline{x} + i\underline{y})$$

$$\underline{x}_r = 1/2(\underline{x} - i\underline{y}) \quad (9)$$

where

$$\underline{x} = x_c + i\, x_s \qquad \underline{y} = y_c + i\, y_s \quad (10)$$

$$x_c = |x| \cos \phi_x \qquad x_s = |x| \sin \phi_x$$

$$y_c = |y| \sin \phi_y \qquad y_s = -|y| \cos \phi_y \qquad (11)$$

$\underline{x}_r$ and $\underline{x}_p$ can be similarly described.

Equation (9) results in

$$\underline{x} = \underline{x}_p + \underline{x}_r$$

$$\underline{y} = -i\underline{x}_p + i\underline{x}_r \qquad (12)$$

The generalized force in rotor station j as a results of unbalance can be written:

$$\phi_j^T f_j = m\Omega^2 [\,\underline{x}_p + \underline{x}_r, \; -i\underline{x}_p + i\underline{x}_r\,] \begin{bmatrix} \underline{e} \\ -i\underline{e} \end{bmatrix}$$

$$\phi_j^T f_j = 2m\Omega^2 \underline{x}_r \,\underline{e} \qquad (13)$$

which shows, together with the equation (3), that only the retrograde part of the left-hand eigenvector can substantially contribute to the generalized force. In order to maximize the contribution from all rotor stations, all vector contributions must run in the same direction. For example the imaginary part in

$$\phi_j^T f_j = (x_{rc}\, e_c + x_{rs}\, e_s) + i\,(x_{rs}\, e_c + x_{rc}\, e_s) \qquad (14)$$

can be set to zero according to

$$0 = x_{rs} e_c + x_{rc} e_s = |x|\,|e| \sin(\phi_r + \phi_e) \qquad (15)$$

which results in ϕ_e being set to $-\phi_r$.
When it is impossible to find an unbalance distribution which is orthogonal to other left-hand vectors, two different models have been tested. The first has a mass eccentricity according to:

$$|\underline{e}|_j \sim |\underline{x}_r|_j \qquad (16)$$

which is only orthogonal to other left-hand eigenvectors for the particular case of a non-rotating shaft. The force can in this case be shown regarding equation (15):

$$f_j = \frac{\bar{\phi}_j \text{ retrograde}}{|\phi_j \text{ retrograde}|_{max}} \cdot m_j \Omega^2 \qquad (17)$$

The other model of mass eccentricity can be shown as follows:

$$f_j = \frac{\bar{\phi}_j \text{ retrograde}}{|\phi_j \text{ retrograde}|} \cdot m_j \Omega^2 \qquad (18)$$

i.e. the mass eccentricity is a unit measured over all rotor stations.

Forces due to the left-hand eigenvector's retrograde angular displacement has not been included in the model. This is partly because in most cases it reduces sensitivity, and partly because it is not physical at certain modes where the retrograde component of the angular displacement does not comply with the transverse displacement retrograde component.

3 CALCULATION OF UNBALANCE SENSITIVITY

Unbalance sensitivity for mode k is defined here as;

$$\left| \Psi_{ik} \sum_j \frac{\phi_{jk} f_j}{a_k(\lambda - \lambda_k)} + \bar{\Psi}_{ik} \sum_j \frac{\bar{\phi}_{jk} f_j}{\bar{a}_k(\lambda - \bar{\lambda}_k)} \right|_{max} \qquad (19)$$

where the unbalance distribution f_j has been chosen according to equation (17), 'modal eccentricity', and (18), 'unity eccentricity', and the excitation frequency Ω_k according to equation(5). Where eigenvalue λ_k is connected to the rotating speed of the rotor, the eigenvalue problem must be solved at different speeds.

The dependence of speed can be worked out with varying accuracy according to calculation method. In the case of discrete matrix representation, an approximate solution can be to divide the run-up into a number of areas with constant speed.

Along a rotor different parts have different unbalance potential. As an example, the area around a shaft coupling has high unbalance potential. These areas can be handled through a weightening of the generalized force calculated in equation (13). In the same way other areas can be of less interest to keep the vibration low as, for example, a smooth shaft. In the calculation of deflection, equation (19), these araes can be weightened down. This wiegthening process is highly coupled to the machine type, and has the drawback that the calculated unbalance sensitivity is not uniquely defined.

A comparison of international standards for permitted residual unbalance and permitted shaft displacement on site for a steam or gas turbine shows that the calculated sensitivity has to be lower than 5 in the operating area and 25 is acceptable outside the operating area.

4 SAMPLE CALCULATION

To demonstrate the usability of being able to easily estimate unbalance sensitivity in a rotor, a relatively complicated rotor system has been analysed. To minimize unbalance sensitivity in a twin-shaft gas turbin, a squeeze film damper has been introduced. Unbalance response calculations soon proved to be very time consuming to optimize the damper location and performance owing to difficulties in choosing the right unbalance distribution. In order to eliminate this problem, system damping was studied by solving the eigenvalue problem. These values are inadequate as the system damping could not be increased for all modes through squeeze film damping. The obvious question is which modes are more interesting than others? It is well known that retrograde whirl modes cannot be excited by unbalance. In a complicated rotor system it is difficult to distinguish prograde and retrograde whirl modes, so this criteria for unbalance sensitivity was not directly useful. The method of approximately determining unbalance sensitivity described in this paper has proved useful for the choice of damper.

4.1 Calculation model

A twin-shaft gas turbine is described with substructure component-mode synthesis, ref(2), where the substructures are described by modal parameters.The calculation model is described schematically in Figure 3.

The rotors are described with complex prograde and retrograde whirling modes calculated on a lumped mass model with free-free boundary conditions, ref(3). The high pressure rotor is described with 30 modes and the low pressure rotor with 22 modes. Unbalance tests with uncoupled rotors indicated that natural frequencies of up to twelve times the rotor`s angular speed must be included in the high pressure rotor model. The low pressure rotor was modelled with ten times the rotor`s angular speed.

The rotors are connected to each other dynamically through the stator which is described with calculated modes. Three dimensional shell elements were employed in the finite element modelling of the stator (11661 degrees of freedom), and an ADINA program was run yielding 21 useful modes. In order to simplify the calculations of the stator, it has been assumed to be symmetrical. The modes obtained were real with modal damping taken as one percent of the critical.

The bearings are of the tilting pad type where the 8 linearized stiffness and damping coefficients are calculated at the frequency ratio of one between rotation and vibration. The squeeze film damper is of the centering type and is described by a very simple linear non-cavitated model which presupposes a circular whirl orbit as shown in ref (4).

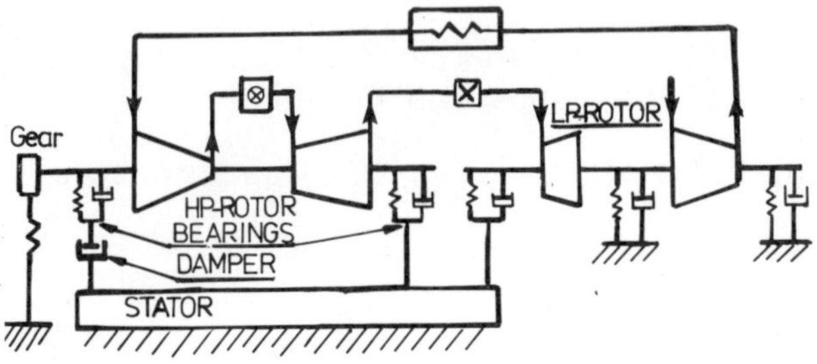

Fig 3 Calculation model

4.2 Numerical results

The orthogonality of the two suggested 'modal unbalance' distributions have been tested in the following way. For each eigenvalue an unbalance distribution according to equations (17) and (18) has been calculated. This unbalance has been used to excite two different calculation models. One is the calculation model described with the corresponding eigenvector (marked 'one mode' in the result) and the other is a calculation model for the rotor system where all eigenvectors are involved (marked 'summing up' in the result). If the unbalance distribution for one eigenvector is orthogonal to other eigenvectors, the calculated response for the two models should be identical.

The results from the two models are shown in Figures 4 and 5, when the squeeze film damper is not mounted. The results show that both models work fairly well (when comparing 'one mode' and 'summing up') . The 'modal eccentricity' model seems to be slightly better, and the 'unity eccentricity' has a higher calculated sensitivity to unbalance, as expected. In the area of high modal density, the conformity between `one mode` and the `summing up` for a particular mode is not satisfactory, but over the area the suggested model gives a good estimation of the sensitivity to unbalance for this particular machinery. In the area of sparse modal density, the conformity is very good.

The `unity eccentricity` model has been used in Figure 6 where both the eigenvalue (lefthand side in the diagram) and the sensitivity to unbalance (righthand side in the diagram) have been plotted for different damper value of the squeeze film damper. Only the most important modes have been plotted. The plot shows that the system damping ζ and the sensitivity to unbalance do not correspond to each other very well. As an example on curve A, when damping changes from 4% to 45% of critical damping the corresponding sensitivity to unbalance changes from 8 to 3. Curve B shows that even with increasing damping the sensitivity to unbalance can increase. It is obvious that the system damping is an insufficient basis for sensitivity to unbalance.

Figures 7 and 8 show that the right-hand and left-hand eigenvectors are very similar in this case because of the symmetry in this particular machinery. Figure 9 shows the calculated response which has the same shape as the right-hand eigenvector. In a machine with great asymmetry, such as journal bearings or gyroscopic effects, it can be impossible to find a good unbalance distribution just from the calculated unbalance response shape. This is because the response shape reflects only the right-hand eigenvector and not the left-hand eigenvectors which influence the generalized force. Figure 10 shows the response in the LP-rotor when the HP-rotor is excited.

5 CONCLUSION

The suggested method for calculation of sensitivity to unbalance is a practical tool for judgment in the analysis stage. The excitation model 'modal eccentricity' gives the best approximation of the sensitivity. The results from the sample calculation showed that the approach was very useful for optimazation of the squeeze film damper.

REFERENCES

(1) Choy, K.C. et al. Dynamic Analysis of Flexible Rotor-Bearing Systems using a Modal Approach. NASA Report No. UVA/528144 /MAE78/106.1978.

(2) Olausson, H.-L., and Torby, B.J. Complex Modal Analysis in Turbine Design. 4th International Modal Analysis Conference. 1986. page 189-195

(3) Olausson, H.-L. A Combined Sturm Sequence and Riccati Transfer Matrix Method for Eigenproblem Solution of a Flexible Rotor. ASME, 11th Biennial Conference on Mechanical Vibration and Noise. 1987. page 553-559.

(4) Li, D. F. Dynamic Analysis of Complex Multi-Level Flexible Rotor Systems. Ph.D. dissertation. University of Virginia.1978.

(5) Lund, J. W. Modal Response of a Flexible Rotor in Fluid-Film Bearing. ASME Paper no.73-DET-98, 1973.

(6) Lund, J. W. Stability and Damped Critical Speeds of a Flexible Rotor in Fluid-Film Bearings. ASME Paper no. 73-DET-103, 1973.

APPENDIX

The equations of motion are given by

$$\mathbf{M}\ddot{\mathbf{x}} + \mathbf{C}\dot{\mathbf{x}} + \mathbf{K}\mathbf{x} = \mathbf{f}(t) \tag{A1}$$

Using the identity

$$\mathbf{M}\dot{\mathbf{x}} = \mathbf{M}\mathbf{v} \tag{A2}$$

equation (A1) can be rewritten as

$$\begin{bmatrix} 0 & \mathbf{M} \\ \mathbf{M} & \mathbf{C} \end{bmatrix}\begin{bmatrix} \dot{\mathbf{v}} \\ \dot{\mathbf{x}} \end{bmatrix} + \begin{bmatrix} -\mathbf{M} & 0 \\ 0 & \mathbf{K} \end{bmatrix}\begin{bmatrix} \mathbf{v} \\ \mathbf{x} \end{bmatrix} = \begin{bmatrix} 0 \\ \mathbf{f}(t) \end{bmatrix} \tag{A3}$$

or

$$\mathbf{A}\dot{\mathbf{q}} + \mathbf{B}\mathbf{q} = \boldsymbol{\gamma} \tag{A4}$$

where

$$\mathbf{q}^T = \left[\mathbf{v}^T \; \mathbf{x}^T \right] \tag{A5}$$

having a dimension equal to twice the number of degrees of freedom.

For the homogeneous case $\gamma = 0$. Assuming $\mathbf{q} = \underline{\mathbf{q}}\, e^{\lambda t}$ to be a solution of the problem

$$\left[\lambda \mathbf{A} + \mathbf{B} \right] \underline{\mathbf{q}} = \mathbf{0} \tag{A6}$$

Since **A** and **B** may not be symmetric, left and right-hand eigenvectors must be introduced. For the j´th mode´s right-hand eigenvector

$$\left[\lambda_j \mathbf{A} + \mathbf{B} \right] \mathbf{h}_j = \mathbf{0} \tag{A7}$$

and for its left-hand eigenvector

$$\left[\lambda_j \mathbf{A}^T + \mathbf{B}^T\right] \mathbf{u}_j = 0$$

(A8)

where the eigenvalues are identical for both eigenvector sets. With $\mathbf{v} = \dot{\mathbf{x}}$,

$$\underline{\mathbf{q}}^T = \left[\lambda \underline{\mathbf{x}}^T \;\; \underline{\mathbf{x}}^T\right] e^{\lambda t}$$

(A9)

It can be shown that $\mathbf{h}_j$ and $\mathbf{u}_j$ can be written in terms of the complex mode forms as follows:

$$\mathbf{h}_j^T = \left[\lambda_j \boldsymbol{\Psi}_j^T \;\; \boldsymbol{\Psi}_j^T\right]$$

$$\mathbf{u}_j^T = \left[\lambda_j \boldsymbol{\Phi}_j^T \;\; \boldsymbol{\Phi}_j^T\right]$$

(A10)

or

$$\mathbf{H} = \begin{bmatrix} \boldsymbol{\Psi}\boldsymbol{\Lambda} \\ \boldsymbol{\Psi} \end{bmatrix}$$

$$\mathbf{U} = \begin{bmatrix} \boldsymbol{\Phi}\boldsymbol{\Lambda} \\ \boldsymbol{\Phi} \end{bmatrix}$$

(A11)

With distinct eigenvalues it can be shown that

$$\mathbf{U}^T\mathbf{A}\mathbf{H} = \text{diag}(a)$$

$$\mathbf{U}^T\mathbf{B}\mathbf{H} = \text{diag}(b)$$

(A12)

where diag(a) and diag(b) are diagonal matrices. These are the new orthogonality conditions.
Using complex normal coordinates ζ

$$\underline{\mathbf{q}} = \mathbf{H}\boldsymbol{\zeta}$$

(A13)

Then from equation (A4)

$$\mathbf{A}\mathbf{H}\dot{\boldsymbol{\zeta}} + \mathbf{B}\mathbf{H}\boldsymbol{\zeta} = \boldsymbol{\gamma}$$

(A14)

With $\dot{\underline{\mathbf{q}}} = \lambda \underline{\mathbf{q}}$ and employing the orthogonality conditions

$$[\,\lambda\text{diag}(a) + \text{diag}(b)\,]\,\zeta = \mathbf{U}^T\gamma \qquad \text{(A15)}$$

With complex conjugated eigenvalues, equation (A15) can be rewritten as

$$\underline{\mathbf{q}} = \mathbf{H}\begin{bmatrix} \text{diag}(\mathbf{d}) & \\ & \text{diag}(\bar{\mathbf{d}}) \end{bmatrix}\mathbf{U}^T\boldsymbol{\gamma}$$

(A16)

where the diagonal terms are

$$d_i = \frac{1}{a_i\lambda + b_i}$$

$$\bar{d}_i = \frac{1}{\bar{a}_i\lambda + \bar{b}_i}$$

(A17)

Using the complex mode forms (A10), the transfer function can be written from equation (A16)

$$r_{ij} = \sum_k \frac{\psi_{ik}\phi_{jk}}{a_k(\lambda - \lambda_k)} + \frac{\bar{\psi}_{ik}\bar{\phi}_{jk}}{\bar{a}_k(\lambda - \bar{\lambda}_k)}$$

(1)

Where the eigenvalue $\lambda_k = -a_k/b_k$ and the normalization constant

$$a_k = 2\lambda_k \boldsymbol{\Phi}_k^T \mathbf{M} \boldsymbol{\Psi}_k + \boldsymbol{\Phi}_k^T \mathbf{C} \boldsymbol{\Psi}_k$$

(A18)

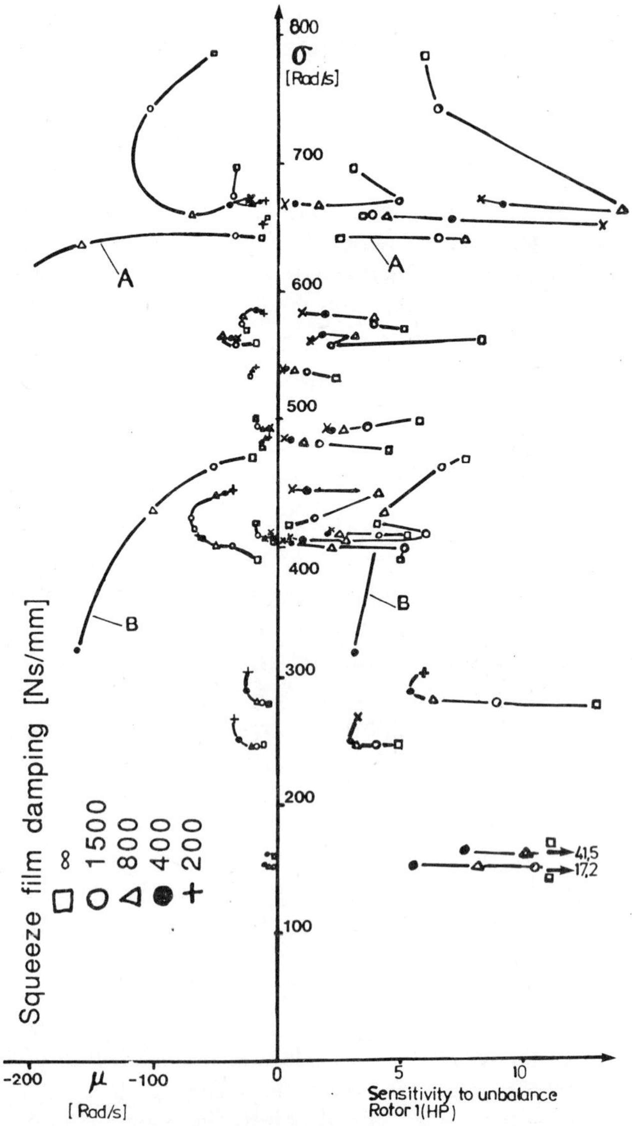

Fig 6 Calculated eigenvalues and sensitivities to unbalance for a twin-shaft gas turbine; variation in squeeze film damping

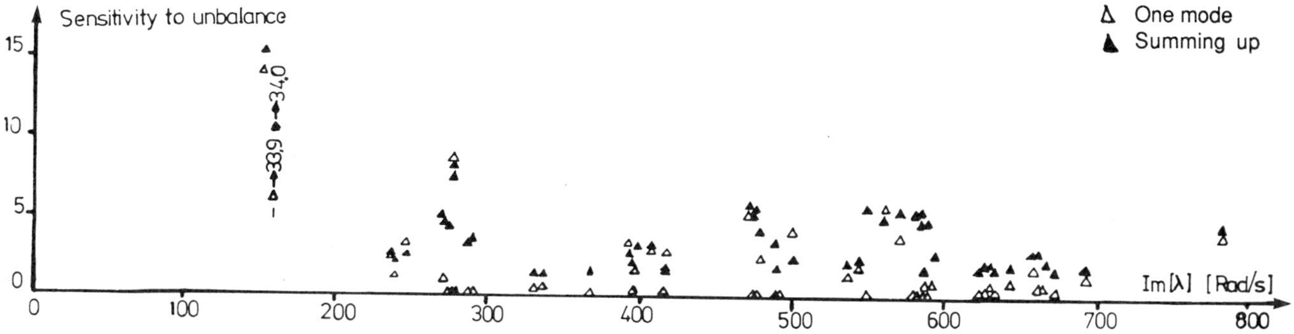

Fig 4 Calculated sensitivity to unbalance 'modal mass eccentricity'

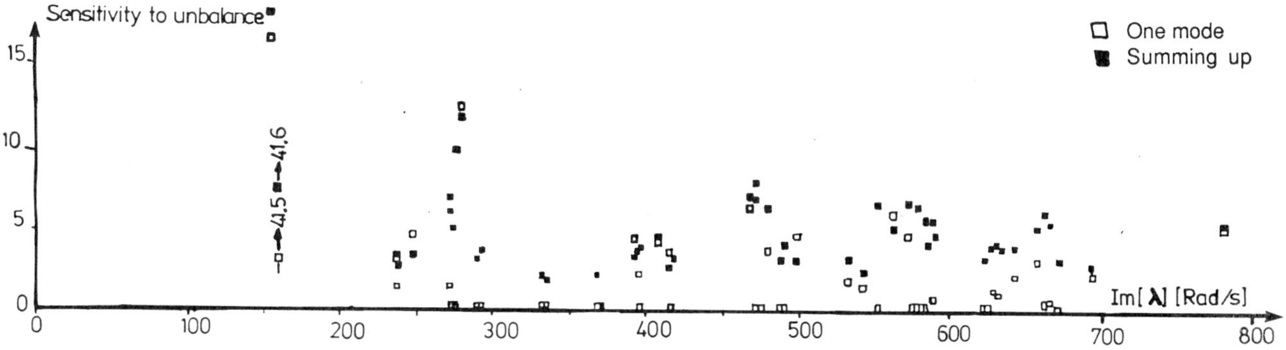

Fig 5 Calculated sensitivity to unbalance 'unity mass eccentricity'

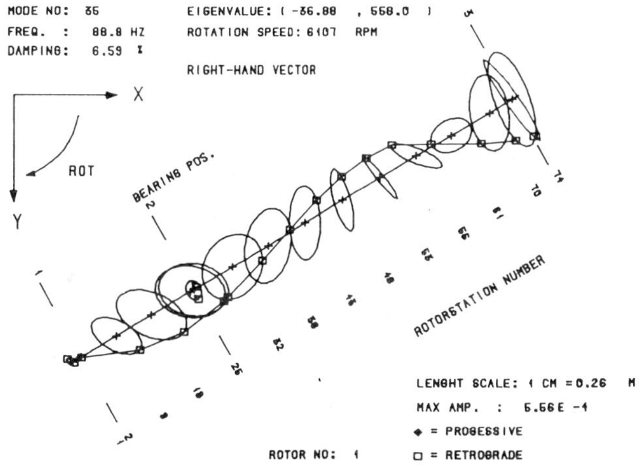

Fig 7 Right-hand eigenvector – HP rotor

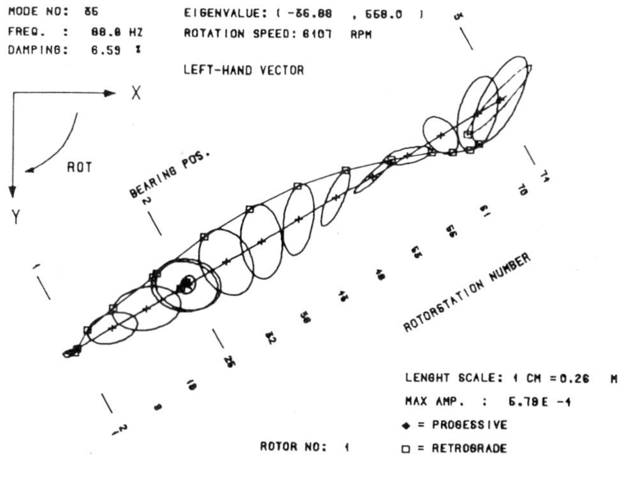

Fig 8 Left-hand eigenvector – HP rotor

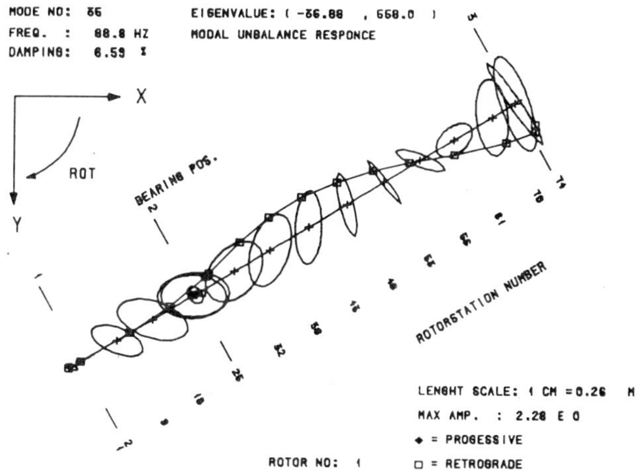

Fig 9 Modal response – HP rotor

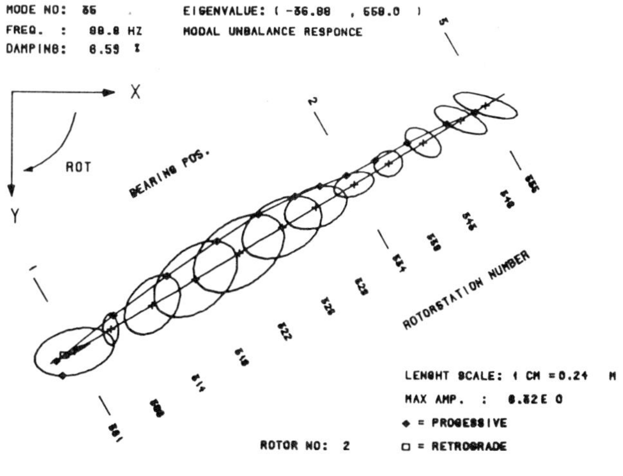

Fig 10 Modal response – LP rotor

C296/88

Unbalance response analysis of rotor bearing systems with spin speed dependent parameters

S-W HONG, BSc, MS and **C-W LEE**, BSc, MS, PhD
Department of Mechanical Engineering, Korea Institute of Science and Technology, Cheongryang, Seoul, Korea

SYNOPSIS : An efficient unbalance response analysis method for rotor bearing systems with spin speed dependent parameters is developed by utilizing a generalized modal analysis scheme. The spin speed dependent eigenvalue problem of the original system is transformed into the spin speed independent eigenvalue problem by introducing a lambda matrix, assuming the bearing dynamic coefficients are well approximated by polynomial functions of spin speed. This method features that it requires far less computational effort in unbalance response calculations and that the influence coefficients are readily available. In addition, the critical speeds and the corresponding logarithmic decrements can be readily identified from the resulting eigenvalues.

INTRODUCTION

The unbalance response analysis of flexible rotor bearing systems is often accomplished in order to investigate the structural dynamic characteristics [1,2] and/or provide the datum required for further rigorous applications, e.g., balancing [3,4]. When a rotor bearing system is represented as a discrete model with matrices of constant elements, its unbalance response calculation is straightforward with resort to either the direct calculation method or the modal analysis method. However, the spin speed dependency, mostly due to the gyroscopic effect and the bearing properties, has prohibited the direct use of the standard modal analysis to calculate the unbalance response of the system, since it is found to be inefficient to calculate the modal responses everytime when the spin speed is incremented [5].

On the other hand, the most commonly used technique to calculate unbalance responses has been the direct computational method utilizing FEM [6] and transfer matrix method [1,7]. Although the direct computational method avoids the difficulty in applying the modal analysis technique, in particular, to complex structures, the unbalance response calculation requires repetitive inversions of complex matrices, of which size tends to become large in many finite element models. Another common drawback in the use of the direct calculation method arises from the fact that the response sensitivity (influence coefficient) analysis is not easy.

In this paper, a new method is suggested, which transforms the spin speed dependent eigenvalue problem of the original system into the spin speed independent eigenvalue problem with complex matrices by introducing a lambda matrix, so that the standard modal analysis technique can be directly applied to obtain the unbalance response of the spin speed dependent rotor bearing system modeled by FEM. The essential feature of this method is to obtain a new generalized eigenvalue problem (or latent value problem) not containing any spin speed dependent parameters, assuming that the bearing dynamic properties are well approximated by polynomial functions of spin speed. In practice, most bearing dynamic parameters are well approximated by low order polynomials of spin speed, at least, over the operating speed range of interest.

The proposed method has significant advantages in computation and physical interpretation of the results compared with the conventional modal analysis technique. In addition, the proposed method produces accurate unbalance response calculations for gyroscopic systems with spin speed independent bearing properties, and, furthermore, the critical speeds and the corresponding logarithmic decrements can be readily identified from the resulting eigenvalues.

EQUATION OF MOTION [8]

The finite element equation of motion of a typical rotor bearing system can be written, neglecting the shear deformation and the internal/external damping of the shaft, as

$$\boldsymbol{M}^{s+d}\ddot{\boldsymbol{q}}+(\boldsymbol{C}^b(\Omega)+\Omega\boldsymbol{G}^{s+d})\dot{\boldsymbol{q}}+(\boldsymbol{K}^s+\boldsymbol{K}^b(\Omega))\boldsymbol{q}=\boldsymbol{f}(t) \qquad (1)$$

where the global coordinate vector $\boldsymbol{q}$ and the force vector $\boldsymbol{f}$ are represented, based on the nodal coordinates system as shown in Fig.1, by

$$\boldsymbol{q}=\begin{Bmatrix}\boldsymbol{y}\\ \boldsymbol{z}\end{Bmatrix} \;;\; \boldsymbol{f}=\begin{Bmatrix}\boldsymbol{f}_y\\ \boldsymbol{f}_z\end{Bmatrix} \qquad (2)$$

Here, $\boldsymbol{y}$ and $\boldsymbol{z}$ include the nodal coordinates such as q_1–q_4 and q_5–q_8, respectively, and $\boldsymbol{f}_y$ and $\boldsymbol{f}_z$ are the force components corresponding to $\boldsymbol{y}$ and $\boldsymbol{z}$. The spin speed (Ω) dependent global bearing stiffness and damping matrices, $\boldsymbol{K}^b(\Omega)$ and $\boldsymbol{C}^b(\Omega)$, respectively, are generally nonsymmetric and indefinite, the symmetric shaft plus disk mass matrix $\boldsymbol{M}^{s+d}$ and the shaft stiffness matrix $\boldsymbol{K}^s$ are positive definite and semidefinite, and the shaft plus disk gyroscopic matrix $\boldsymbol{G}^{s+d}$ is skew symmetric. The global matrices are of the order NxN, N being the dimension of the coordinate vector. It will prove convenient to introduce a coordinate vector, $\boldsymbol{q}_b$, which includes only the nodal coordinates of bearing elements, represented by

$$\underset{2n_b\text{x}1}{\boldsymbol{q}_b}=\begin{Bmatrix}\boldsymbol{q}_{by}\\ \boldsymbol{q}_{bz}\end{Bmatrix}=\underset{2n_b\text{x}N}{\boldsymbol{I}_b}\;\underset{N\text{x}1}{\boldsymbol{q}} \qquad (3)$$

where n_b denotes the number of bearings and $\boldsymbol{I}_b$, consisting of zeros and ones, is the transform matrix of the order $2n_b$xN.

GENERALIZED EIGENVALUE PROBLEM FOR UNBALANCE RESPONSE CALCULATION

Suppose the spin speed dependent properties of bearings are real and continuous over $0 \le \Omega \le \Omega_{max}$, then, given an ϵ, one can find the sufficiently high polynomial orders n_c and n_k such that

$$\left|\boldsymbol{C}^b(\Omega)-\sum_{i=0}^{n_c}\boldsymbol{C}_i^b\Omega^i\right| \le \epsilon$$
$$\left|\boldsymbol{K}^b(\Omega)-\sum_{i=0}^{n_k}\boldsymbol{K}_i^b\Omega^i\right| \le \epsilon \qquad (4)$$

where $\boldsymbol{C}_i^b$, $\boldsymbol{K}_i^b$ are the NxN real constant coefficient matrices, n_c and n_k are the orders of polynomials corresponding to $\boldsymbol{C}^b$ and $\boldsymbol{K}^b$, respectively, and Ω_{max} is the maximum spin speed of interest. The above statement is known to be Weierstrass approximation theorem [9]. Substituting eq.(4) into eq.(1) and letting $\boldsymbol{f}=\boldsymbol{f}_0e^{j\Omega t}$ and $\boldsymbol{q}=\boldsymbol{q}_0e^{j\Omega t}$, one obtains the response due to the excitation synchronous to the spin speed, i.e.,

$$\boldsymbol{q}_0 = \boldsymbol{H}(j\Omega)\,\boldsymbol{f}_0 \qquad (5)$$

where j denotes the unit imaginary number and

$$\boldsymbol{H}(j\Omega)=[-\Omega^2\boldsymbol{M}^{s+d}+j\Omega^2\boldsymbol{G}^{s+d}+j\sum_{i=0}^{n_c}\boldsymbol{C}_i^b\Omega^{i+1}+\sum_{i=0}^{n_k}\boldsymbol{K}_i^b\Omega^i+\boldsymbol{K}^s]^{-1}$$

Here, it will prove convenient to introduce a lambda matrix [10] of the form

$$\boldsymbol{D}_r(\lambda)=\boldsymbol{A}_r\lambda^r+\boldsymbol{A}_{r-1}\lambda^{r-1}+..+\boldsymbol{A}_1\lambda+\boldsymbol{A}_0 \qquad (6)$$

where $r=\max(n_c+1,\, n_k,\, 2)$ and

$$\boldsymbol{A}_0=\boldsymbol{K}_0^b+\boldsymbol{K}^s$$
$$\boldsymbol{A}_1=(\boldsymbol{K}_1^b+j\boldsymbol{C}_0^b)/j$$
$$\boldsymbol{A}_2=\boldsymbol{M}^{s+d}-j\boldsymbol{G}^{s+d}+(\boldsymbol{K}_2^b+j\boldsymbol{C}_1^b)/j^2$$
$$\boldsymbol{A}_i=(\boldsymbol{K}_i^b+j\boldsymbol{C}_{i-1}^b)/j^i, \quad i\ge 3$$

Here A_i, $i=0,1,..,r$, are the NxN complex matrices independent of spin speed. The order of the lambda matrix is always greater than or equal to 2, and A_i for $i \geq 3$ becomes a sparse matrix. Once the lambda matrix is constructed, then it holds

$$H(j\Omega) = [D_r(\lambda)]^{-1}_{\lambda=j\Omega} \tag{7}$$

Adopting the terminology and methodology in [10], the latent value problem and the adjoint associated with the lambda matrix become, respectively,

$$D_r(\lambda)\ \boldsymbol{u} = 0 \tag{8}$$

$$\overline{D}_r^T(\lambda)\ \boldsymbol{v} = 0 \tag{9}$$

where $\boldsymbol{u}$ and $\boldsymbol{v}$ denote the right and left latent vectors, respectively, and '—' denotes the complex conjugate. Since the sparse matrices A_i, $i \geq 3$, in general, are singular, $D_r(\lambda)$ inevitably becomes an irregular lambda matrix but with the rank N, so that the number of latent roots $s = 2N+2(r-2)n_b$. The solution method for eqs.(8) and (9) is based upon an equivalent eigenvalue problem and the adjoint represented by

$$(\lambda\ A + B)\ R = 0 \tag{10}$$

$$(\bar{\lambda}\ \overline{A}^T + \overline{B}^T)\ L = 0 \tag{11}$$

where

$$A = \begin{bmatrix} A'_r & A'_{r-1} & \cdots & A'_3 & A_2 & A_1 \\ & I_1 & \cdots & 0 & 0 & 0 \\ & & \cdots & . & . & . \\ & & & I_1 & 0 & 0 \\ & 0 & & & I_b & 0 \\ & & & & & I_2 \end{bmatrix}_{sxs}$$

$$B = \begin{bmatrix} 0 & 0 & \cdots & 0 & 0 & A_0 \\ -I_1 & 0 & \cdots & 0 & 0 & 0 \\ & -I_1 & \cdots & . & . & . \\ & & \cdots & . & . & . \\ & 0 & & -I_1 & 0 & 0 \\ & & & & -I_2 & 0 \end{bmatrix}_{sxs}$$

where $A'_k = A_k\ I_b^T$ for $k \geq 3$, I_1 and I_2 denote the identity matrices of the orders $2n_b$x$2n_b$ and NxN, respectively, and the right and left eigenvectors R and L are represented by

$$R = [\ U^T\ \ \boldsymbol{u}^T\ ]^T \tag{12}$$

$$L = [\ \boldsymbol{v}^T\ \ V^T\ ]^T \tag{13}$$

Here U and V are the vectors of the order $(s-N)$x1 determined by eqs.(10) and (11).

The eigenvalue problem and the adjoint problem, eqs.(10) and (11), yield the eigenvalues, λ_i, $i=1,2,..,s$, and the corresponding right and left eigenvectors, R_i and L_i, respectively, which are biorthonormalized so as to satisfy

$$\overline{L}_i^T\ A\ R_j = \delta_{ij} \tag{14}$$

$$\overline{L}_i^T\ B\ R_j = -\lambda_i\ \delta_{ij}$$

It should be noted that $\boldsymbol{u}_i$ and $\boldsymbol{v}_i$ generally form dependent vector sets while R_i and L_i form independent vector sets.

The inverse of the lambda matrix can be expressed as a finite series of partial fractions, using the latent values and the corresponding right and left latent vectors, i.e.,

$$[D_r(\lambda)]^{-1} = \sum_{i=1}^{s} \frac{\boldsymbol{u}_i\boldsymbol{v}_i^T}{\lambda - \lambda_i} \tag{15}$$

Using eqs.(5) and (15) together with eq.(7), one obtains the response due to the spin speed synchronous excitation, in the partial fraction form, as

$$q_0 = \sum_{i=1}^{s} \frac{\boldsymbol{u}_i\boldsymbol{v}_i^T f_0}{j\Omega - \lambda_i} = \alpha(\Omega)\ f_0 \tag{16}$$

where $\alpha(\Omega)$ is the influence matrix.

Notice that eq.(16) requires only algebraic summations of s partial fractions to obtain the system response for given speed Ω, once the latent value problem is solved. The significant improvement in computational efficiency with the use of this final fractional formula is expected when compared to the direct calculation method which requires the inversion of large complex matrix everytime whenever the spin speed is varied. Moreover, since the role of the eigenvalues in eq.(16) is consistent with that of the eigenvalues associated with the constant parameter system, the eigenvalues can be used to calculate the critical speeds and the corresponding logarithmic decrements for synchronous responses.

For unbalance response calculation, the force vector f_0 in eq.(16) is expressed in complex form, following the notation given in Fig.2, as

$$f_0 = \begin{Bmatrix} W \\ -j\,W \end{Bmatrix} \Omega^2 \tag{17}$$

The unbalance vector, W, can be represented by

$$W=\{w_{y1}+jw_{z1}\ 0\ w_{y2}+jw_{z2}\ 0 \ldots w_{yn}+jw_{zn}\ 0\}^T \tag{18}$$

where n is the number of nodal points, and w_{yi} and w_{zi}, $i=1,2,..,n$, are the y and z directional components, respectively, of the unbalance located at the nodal point i.

It is of use to investigate the response sensitivity with respect to the locations of sensors or balancing planes before balancing or identification is conducted. Consideration of such response sensitivity coefficients is straightforward by using the proposed formula (16). For example, the response sensitivity $s_{ij}(\Omega)$ may be defined as the magnitude of major whirl radius $R_{maj}(\Omega)$ measured at the nodal point i when a unit unbalance is attached to the nodal point j, i.e.,

$$s_{ij}(\Omega) = R_{maj}(\Omega) \tag{19}$$

i : sensor location ; j : unbalance location

where

$$R_{maj}(\Omega) = \frac{1}{2}\{\sqrt{(y_{ic}(\Omega)+z_{is}(\Omega))^2 + (z_{ic}(\Omega)-y_{is}(\Omega))^2} + \sqrt{(y_{ic}(\Omega)-z_{is}(\Omega))^2 + (z_{ic}(\Omega)+y_{is}(\Omega))^2}\}$$

and

$$y_{ic}(\Omega) = Re\{y_i(\Omega)\} \quad y_{is}(\Omega) = -Im\{y_i(\Omega)\}$$
$$z_{ic}(\Omega) = Re\{z_i(\Omega)\} \quad z_{is}(\Omega) = -Im\{z_i(\Omega)\}$$

Here, y_i and z_i are the complex synchronous responses at the i–th nodal point in the y and z directions, which may be easily obtained from eq.(16) without further computational efforts. This response sensitivity denotes a gain of the influence coefficient invariant to the sensor direction. Determination of the sensor or unbalance plane location for balancing is rather simple by using this response sensitivity.

NUMERICAL EXAMPLES

In this section, two illustrative examples are taken to demonstrate the underlying ideas of the proposed method.

Example 1 Overhung Rotor Bearing System

This example deals with computation of critical speeds by using the proposed method for an overhung rotor bearing system as shown in Fig.3.

In undamped and isotropic/orthotropic bearing cases, the computed eigenvalues are found to be the same as the forward and backward critical speeds obtained by conventional methods [6,11]. When critical speeds cannot be determined with resort to the conventional methods [6,11], i.e., dampings or nonorthotropic properties are present, the damped critical speed has been generally defined as the damped natural frequency with which the spin speed coincides [12]. As far as the synchronous response is concerned, such a definition may lead us to a confusion since the damped natural frequency changes when the spin speed is changed, necessitating whirl speed charts. Figure 4 shows the whirl speed chart which is generated by using the modal transform method [8] with 20 Ritz base vectors. In Table 1, the results obtained by the proposed method and the conventional method utilizing Fig.4 are compared. The logarithmic decrements obtained from the conventional method represent the relative stability of the poles while those obtained by the proposed method indicate the magnification factors at the poles (eigenvalues).

Example 2 Uniform Shaft Bearing System

Consider a uniform rotating shaft supported at both ends by two identical plain journal bearings [12]. The detailed specifications of the system are given in Table 2. The bearings are assumed as ideal short bearings over the operating speed range of interest 1000–12000 rpm. The spin speed dependent dynamic stiffness and damping coefficients, obtained according to the formula in [13], are shown in Figs.5–(a) and 5–(b), respectively. The shaft is modelled as an assemblage of four equal length finite elements and the value of r is taken such that $r = n_c+1 = n_k$ in the simulation.

The normalized unbalance responses at the mid span of the shaft are shown in Fig.6, which provides important information on balancing and identification. In unbalance response calculation, r=4 was found to be adequate. The response sensitivity matrix at 7000 rpm is given in Table 3. The response sensitivity matrix reveals that the system is non–self–adjoint as one would expect since the matrix does not appear to be symmetric. The matrix given in Table 3 suggests that the best location for sensor and balancing plane in the balancing at 7000 rpm would be the center of the shaft.

DISCUSSION AND CONCLUDING REMARKS

A new method for unbalance response analysis of rotor bearing systems with spin speed dependent properties is developed. The proposed method possesses two essential advantages : computational efficiency and physical insight into unbalance response analysis. The computational efficiency of the proposed method is significant when unbalance response plots over the operating speed range of interest is required. In addition, it provides the influence coefficient matrix invariant to unbalance distribution without further computational efforts. Since the method is based on the spin speed dependency approximatd by polynomial functions of spin speed, it yields exact unbalance response analysis for gyroscopic systems with no other spin speed dependent parameters. Furthermore, the damped critical speeds and the corresponding logarithmic decrements which are independent of spin speed are readily identified from the resulting eigenvalues.

Finally, it is worthwhile to mention that the proposed method can be easily extended to harmonic response analysis, one of the most important areas in the forced vibration analysis of rotor bearing systems. Moreover, the proposed generalized eigenvalue problem is of use also in the analysis of frequency dependent parameter systems [14].

REFERENCES

1. Lund, J.W. and Orcutt, F.R., Calculations and experiments on the unbalance response for a flexible rotor, *Trans. ASME, Journal of Engineering for Industry*, Series B, 1967, **89**(4), 785–796.
2. Gladwell, G.M.L. and Bishop, R.E.D., The vibration of rotating shafts in flexible bearings, *Journal of Mechanical Engineering Science*, 1959, 1(3), 195–206.
3. Saito, S. and Azuma, T., Balancing of flexible rotors by the complex modal method, *Trans.ASME, Journal of Vibrations, Acoustics, Stress, and Reliability in Design*, Jan. 1983, **105**, 94–100.
4. Parkinson, A.G., Darlow,M.S. and Smalley, A.J., A theoretical introduction to the development of a unified approach to flexible rotor balancing, *Journal of Sound and Vibration*, 1980, **68**(4), 489–506.
5. Nordmann, R., Modal analysis in rotor dynamics, *Dynamics of Rotors; Stability and System Identification* Chap.1, edited by Mahrenholtz, International Center for Mechanical Science, 1984.
6. Nelson, H.D. and McVaugh, J.H., The dynamics of rotor bearing system using finite elements, *Trans. ASME, Journal of Engineering for Industry*, May 1976, **99**, 593–600.
7. Rao, J.S., *Rotor Dynamics*, John Wiley & Sons, 1983.
8. Kim, Y.D. and Lee, C.W., Finite elements analysis of rotor bearing system using a modal transformation matrix, *Journal of Sound and Vibration*, 1986., **111**, 441–456.
9. Davis, P.J., *Interpolation and Approximation*, Dover Publication, Inc., New York, 1975
10. Lancaster, P., *Lambda Matrices and Vibrating Systems*, Pergamon Press, 1966.
11. Childs, D.W. and Graviss, K., A note on critical–speed solutions for finite–eLememt–based rotor models, *Trans. ASME, Journal of Mechanical Design*, April 1982, **104**, 412–416

12. Lund,J.W., Stability and damped critical speeds of a flexible rotor in fluid film bearings, *Trans.ASME, Journal of Engineering for Industry*, May 1974, **96**, 509–517.
13. Kirk, R.G. and Gunter, E.J., Stability and transient motion of a plain journal mounted in flexible damped supports, *Trans. ASME, Journal of Engineering for Industry*, May 1976, **99**, 576–592.
14. Hong, S.W. and Lee, C.W., Frequency and time domain analysis of frequency dependent parameter systems, Submitted to *Journal of Sound and Vibration*

Table 1 Damped critical speeds/logarithmic decrements of overhung rotor system determined by the proposed method and the conventional method

eigenvalue : $\lambda_k = \sigma_k + j\omega_k$
logarithmic decrement : $\delta_k = 2\pi\sigma_k/\omega_k$

ω_k(rpm)/δ_k

mode	present	conventional
1	2454.41/0.0384	2454.54/0.0385
2	2535.33/0.0201	2535.44/0.0201
3	13100.16/0.9428	13104.37/0.9467
4	15367.07/0.6987	15365.14/0.7021
5	18195.42/0.2354	18188.98/0.2528
6	20733.53/0.3871	20754.93/0.3644

Table 2 Specifications for uniform shaft bearing system

Shaft

.shaft length	1.27 m
.shaft diameter	10.16 cm
.shaft density	7833 kg/m^3
.Young's modulus	2.068x10^{11} N/m^2

Bearings (assume ideal short bearing ; L/D = 1/4)

.bearing length	2.54 cm
.journal diameter	10.16 cm
.oil viscosity	6.9 cp
.bearing clearance	0.051 mm
.static bearing load	395 N

Table 3 Response sensitivity matrix w.r.t. sensor and unbalance locations

	N_s				
N_u	1	2	3	4	5
1	.581E–2*	.134E–1	.171E–1	.136E–1	.551E–2
2	.134E–1	.610E–1	.831E–1	.591E–1	.137E–1
3	.173E–1	.831E–1	.116E–0	.831E–1	.173E–1
4	.137E–1	.591E–1	.831E–1	.610E–1	.134E–1
5	.551E–2	.136E–1	.171E–1	.134E–1	.581E–2

N_u : unbalance location (node) ;
N_s : sensor location (node)

* denotes the major whirl radius at the sensor location when a unit unbalance (kg.m) is attached to the selected unbalance location at the rotation speed of 7000 rpm. (unit = m/(kg.m))

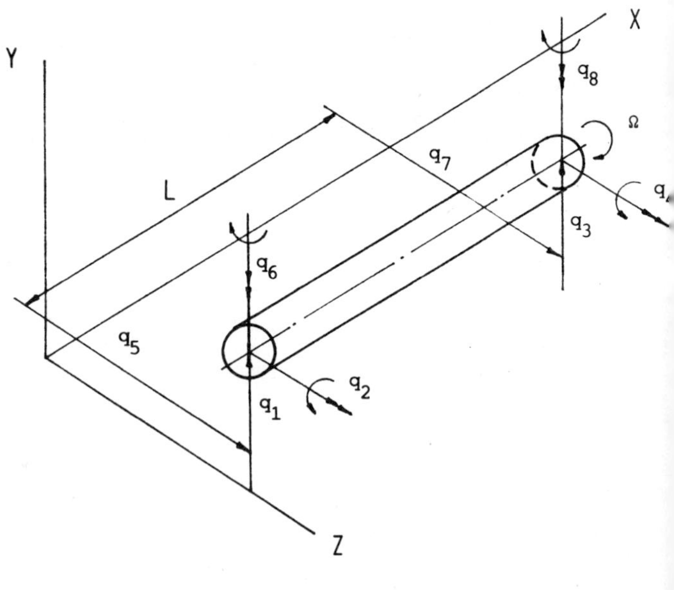

Fig 1 Typical nodal coordinates system [8]

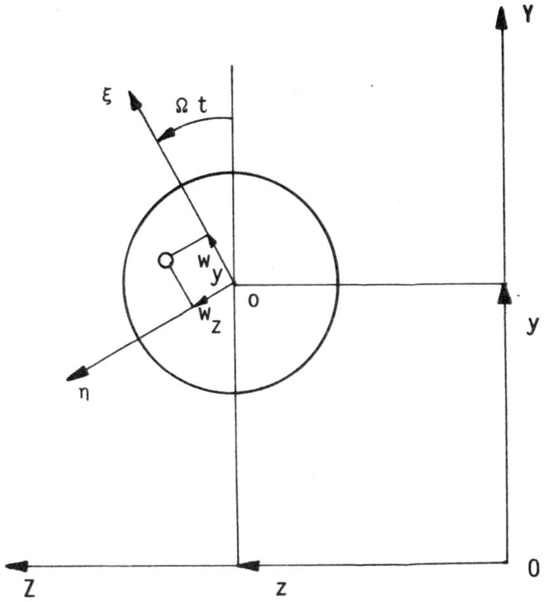

Fig 2 Coordinates for unbalance representation

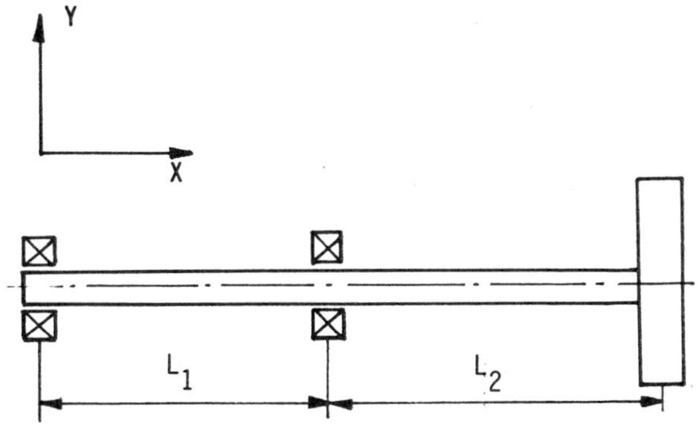

Shaft

. shaft length	1.2 m (L_1=L_2=0.6 m)
. shaft diameter	6 cm
. shaft density	7833 kg/m^3
. Young's Modulus	2.068×10^{11} N/m^2

Disk

. mass	7.5 kg
. polar mass moment of inertia	0.0368 kgm^2
. transverse mass moment of inertia	0.0190 kgm^2

Bearings

. stiffness and damping coefficients

$k_{yy} = 2.548 \times 10^7$ N/m $k_{zz} = 3.806 \times 10^7$ N/m

$c_{yy} = c_{zz} = 7000$ Ns/m the Others zero

Fig 3 Overhung rotor system

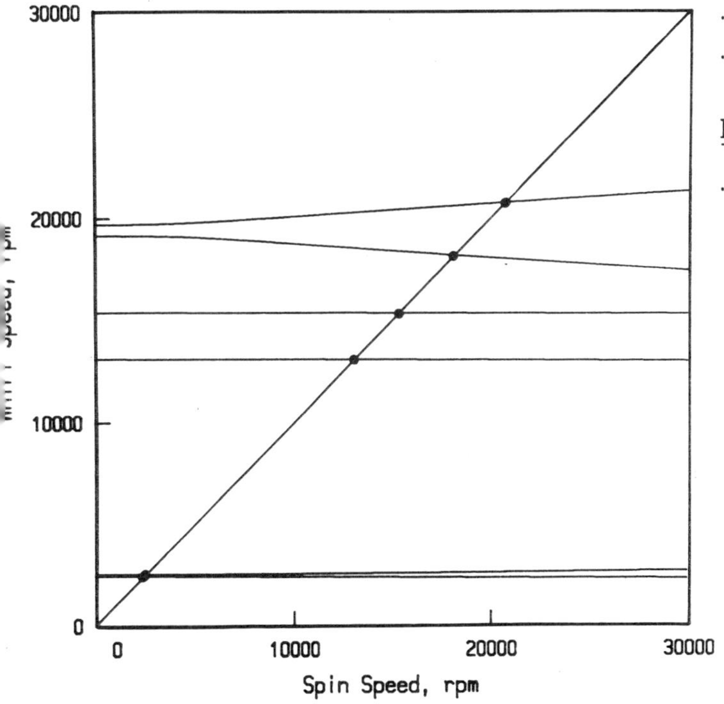

Fig 4 Whirl speed of overhung rotor system

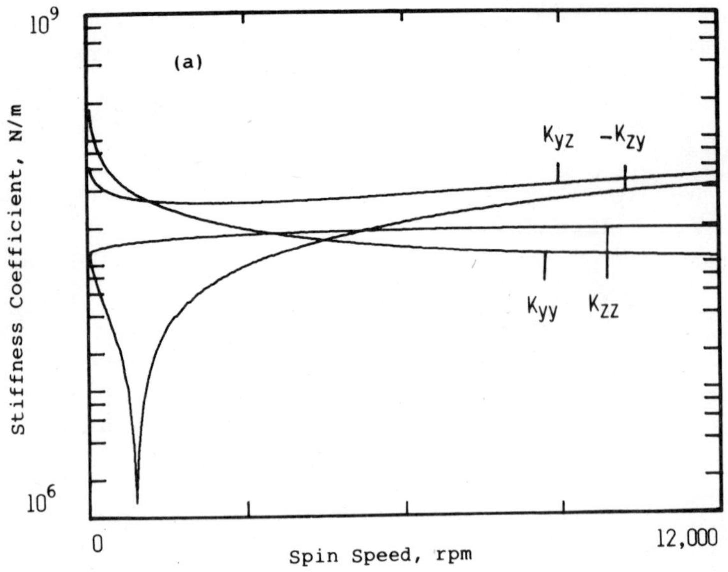

(a) Stiffness coefficients

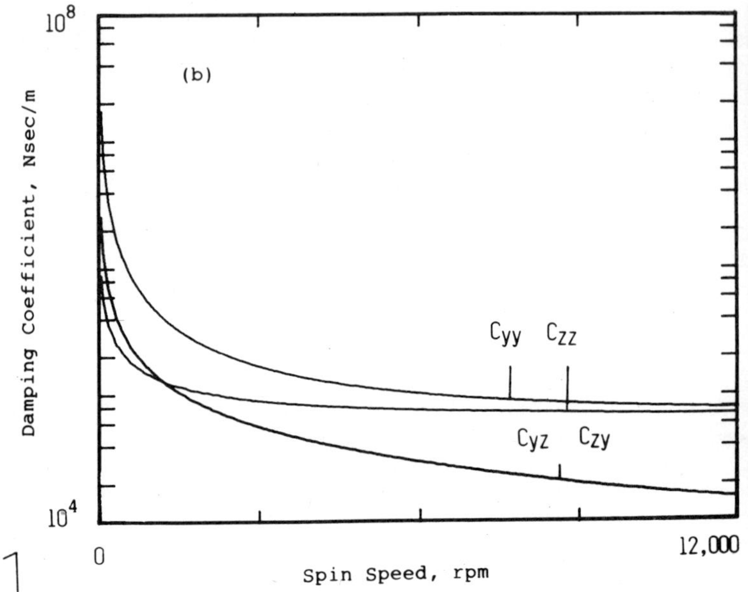

(b) Damping coefficients

Fig 5 Spin speed dependent bearing properti[illegible]

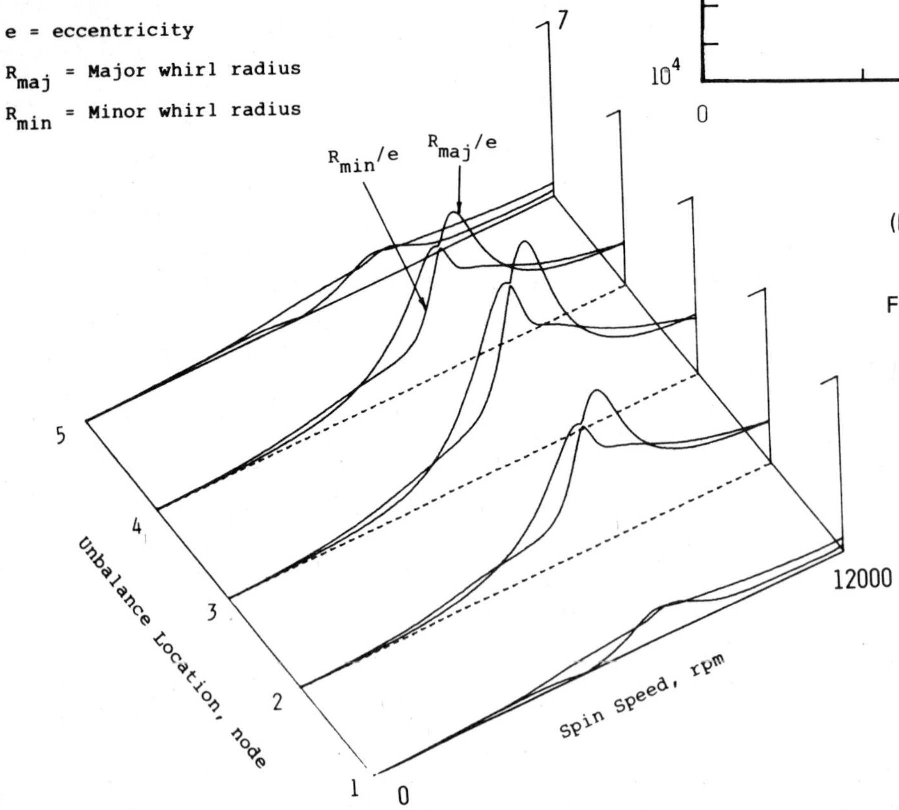

Fig 6 Normalized unbalance responses at the mid-span of uniform rotor bearing system

is linearly related to the difference in temperature rise of the conductors, such that:-

$$T_{gh}-T_{gn} = \lambda\ (\Theta_h-\Theta_n) \quad .....(A9)$$

then substituting from equation A9 into equation A8 gives;-

$$V = Z_e\ (\Theta_h-\Theta_n) \quad .....(A10)$$

where

$$Z_e = Z\ (1 + \lambda) \quad .....(A11)$$

Therefore, it follows from equation A10 that the vibration is related to the temperature rise difference between the 'hot' and 'normal' conductors, which from equation A4 is:-

$$\Theta_h-\Theta_n = k_h I^2\ [1-\alpha k_h I^2]^{-1} - k_n I^2 \ .....$$

$$.....[1-\alpha k n I^2]^{-1} \quad .....(A12)$$

To assist in understanding the form of equation A12 it is instructive to expand using the Bionomial Theorem giving:-

$$\Theta_h-\Theta_n = I^2(h_h-k_n) + \alpha I^4\ (k_h{}^2-k_n{}^2) \quad ....$$

$$+ \ ... + \sum_{p\,=\,3\text{ to }\infty} \alpha^{p-1} I^{2p}(k_h{}^p-k_n{}^p) \quad .....(A13)$$

Equation A13 shows that for finite values of temperature coefficient of resistance, α, the temperature difference has a dependence on rotor current which includes terms with powers greater than two.

The values of k_n and k_h can be defined by rearranging equation A4 such that:-

$$k_n = \Theta_n I^{-2}\ (\alpha\ \Theta_n+1)^{-1} \quad .....(A14)$$

and

$$k_h = \Theta_h I^{-2}\ (\alpha\ \Theta_h+1)^{-1} \quad .....(A15)$$

APPENDIX B

PREDICTED DUCT BLOCKAGE

The coefficient 'k' is the ratio of the conductor resistance at gas temperature to the convective heat transfer coefficient, equation A6. These are local values representing the difference between the 'hot' and 'normal' conductors. A change in the value of k may be interpreted as a change either in resistance or the heat transfer coefficient, the latter being considered to be the case for cooling duct restriction in the rotor. Therefore, assuming that the resistance remains constant by neglecting the difference in gas temperatures between 'hot' and 'normal' conductors, then it follows from equation A6 that the ratio of the local heat transfer coefficients is:-

$$h_h h_n{}^{-1} = k_n k_h{}^{-1} \quad .....(B2)$$

For turbulent flow of fluid through a pipe the dimensionless Nusselt number, Nu, is related to the Reynolds number, Re (2) by:-

$$Nu \propto Re^{0.75} \quad .....(B3)$$

Then to a first approximation, assuming the gas and material properties are unaffected by the temperature difference between the two sides of the rotor, the ratio of the mean gas velocities are:-

$$v_h v_n{}^{-1} = (h_h/h_n)^{1.333} \quad .....(B4)$$

Furthermore, if the restriction is the result of a localised reduction of the area of the holes at inlet or outlet to the cooling ducts, and this is represented by a simple sharp edged orifice the effective reduction in area is:-

$$A_h A_n{}^{-1} = v_h v_n{}^{-1} \quad .....(B5)$$

In the derivation of equation B5 it is assumed that the overall pressure remains unaffected by the local restrictions.

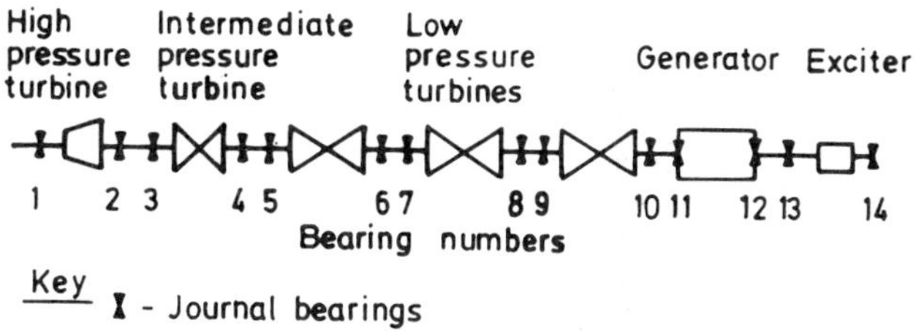

Fig 1 Typical layout of 500 MW turbine-generator

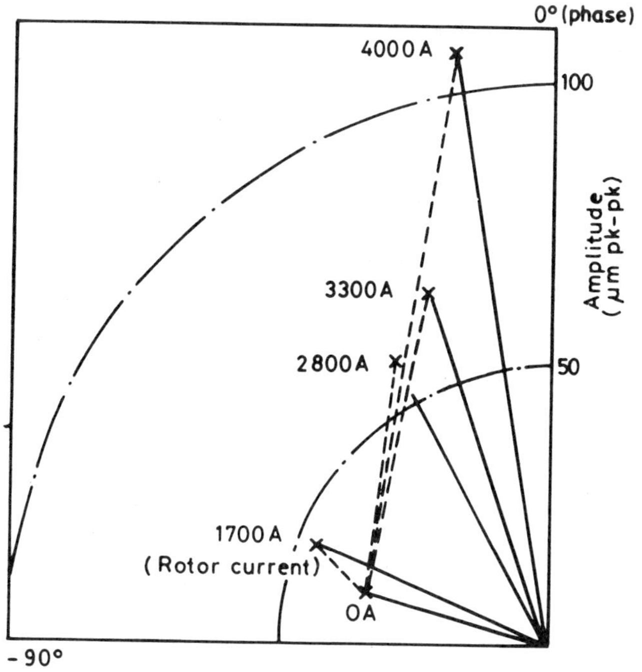

Fig 3 Variation of vibration at bearing 11 with rotor current

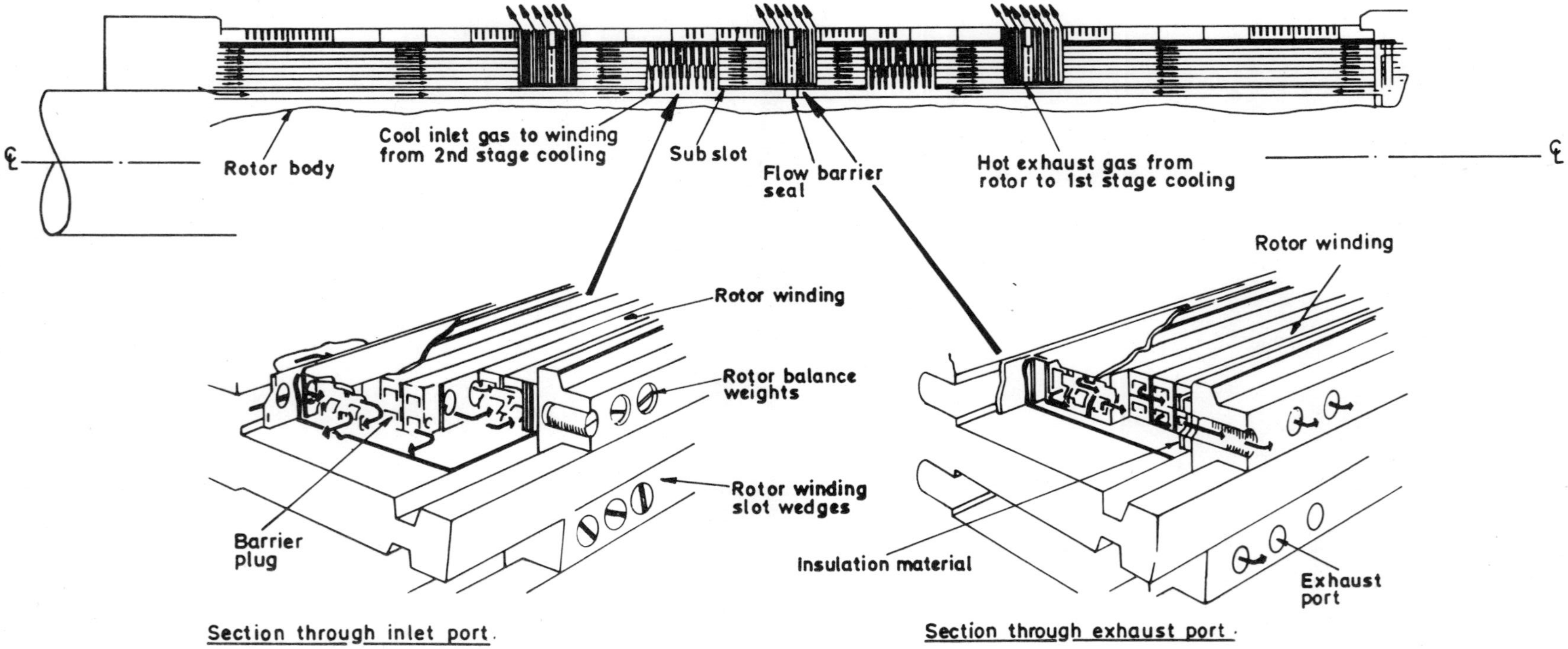

Fig 2 Generator rotor cooling arrangement

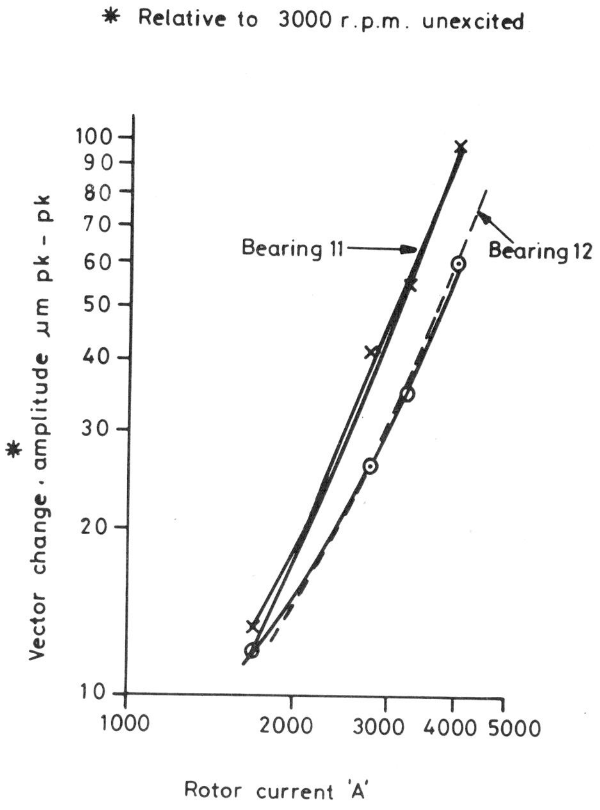

Fig 4 Variation of vibration vector changes with rotor current for bearings 11 and 12

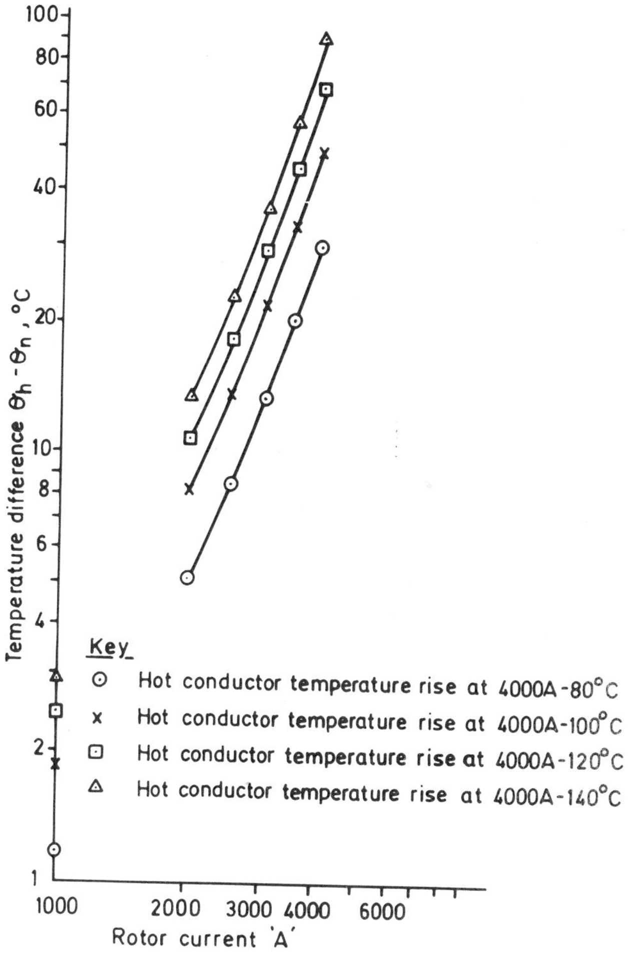

Fig 6 Predicted variation of temperature difference with rotor current

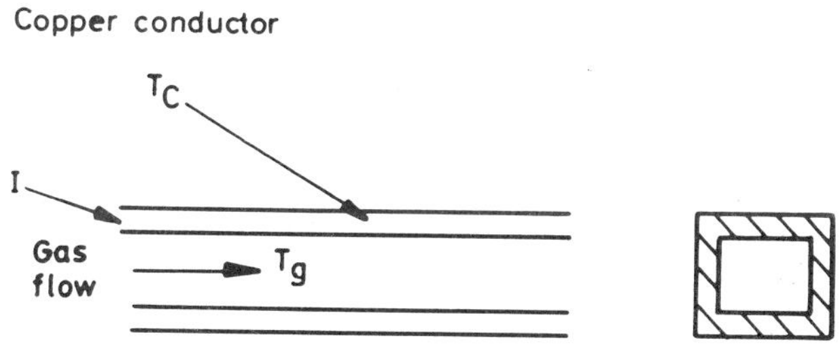

Fig 5 Simplified representation of gas-cooled rotor conductor

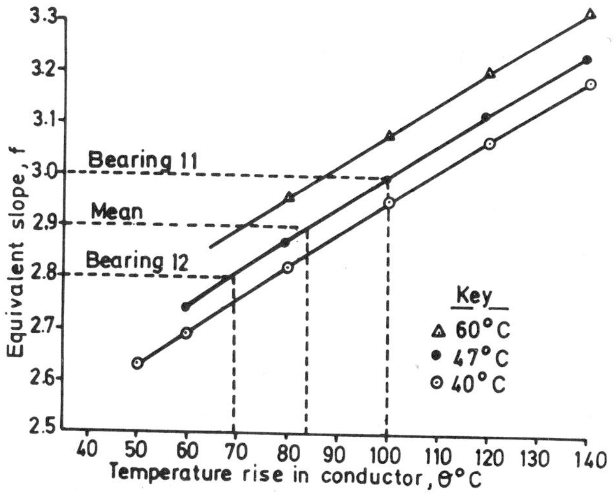

Fig 7 Variation of equivalent slope with temperature rise

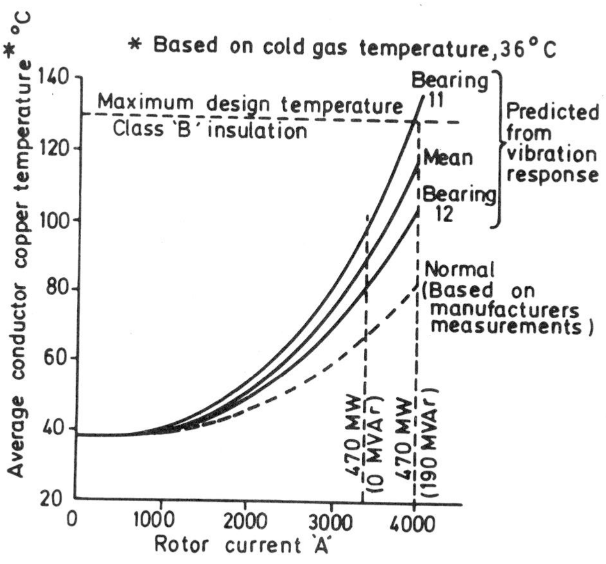

Fig 8 Variation of average conductor temperature with rotor current

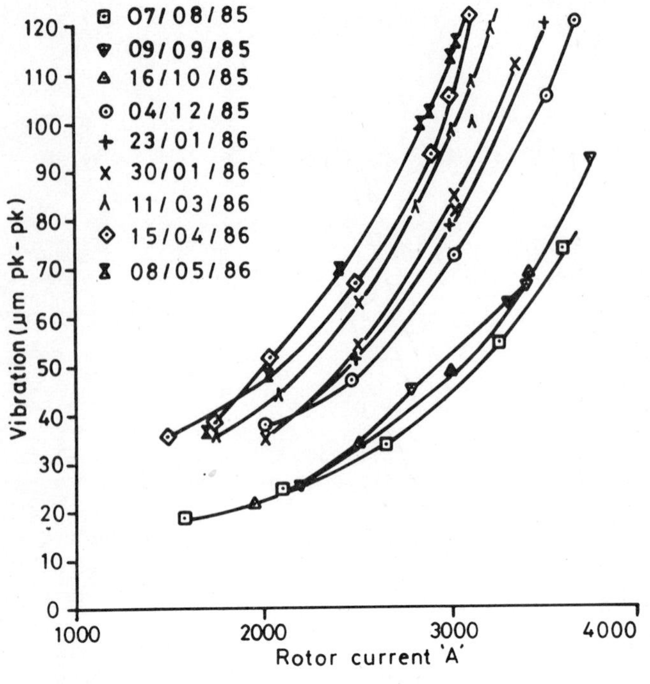

Fig 9 Variation of vibration with rotor current for selected times during nine months of operation

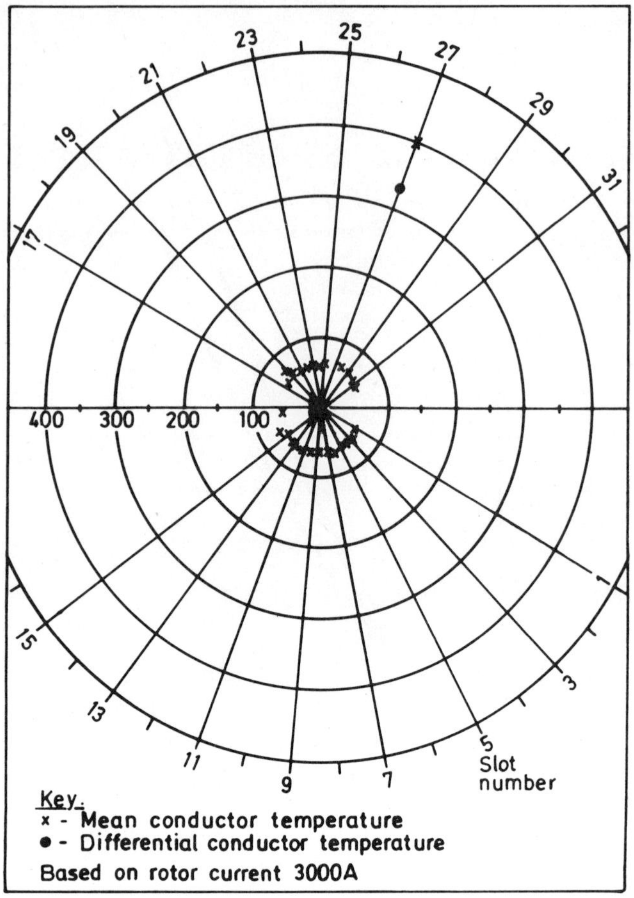

Fig 10 Predicted temperature rise for measured blockages

C297/88

Unbalance and vibration control in a large geotechnical centrifuge

A J A GIERS, Dipl-Ing and **D THELEN**, Dr-Ing
Carl Schenck AG, Darmstadt, West Germany

SYNOPSIS: A system is described where by the static and couple unbalances caused by the load or changes of the load during rotation are automatically counteracted while the centrifuge is rotating. In addition to unbalance excitation, other dynamic excitations are expected to occur such as shock due to the collapse of the model under test or from an exciter acting on the model. To overcome the problems involved, the centrifuge arms are provided with coulomb damping in two directions for many modes. Furthermore, by using remote control the radial stiffness of the main bearings can be changed by a ratio of 1 : 20 thus changing all natural frequencies of the system.

1 INTRODUCTION

As stated in (1), there is a large variation in the calculation methods available for solving geotechnical and foundation engineering problems. These methods must be checked by tests and experiments in order to ensure that they provide a reliable prediction of soil behaviour.

Frequently large structures like dams, offshore structures, multi story buildings etc. are the objects of interest. In this cases it is desirable to use scaled down models to save time and cost in testing. If during the test the stresses in the model can be maintained equal to that in the full scale structure, the translation of model results becomes much simpler.

The condition of equal stress can be met if the model is tested under high acceleration. The acceleration must be chosen as the same multiple of the gravity as the selected model scale. (For example: model scale 100 - acceleration needed is 100 g.)

For a centrifuge there are in general three important applications in geotechnics:

(a) Model investigations to predict prototype behaviour.
(b) Investigations aimed at checking calculation methods.
(c) Parametric studies.

2 THE INSTALLATION

Fig. 1 shows an artist's impression of the complete installation. The centrifuge has a rotor arm length of 6 m and is designed for a model mass including container of 5.5 ton at an acceleration of 300 g. With its capacity of 16.5 MN it is the largest capacity geotechnical centrifuge in the world. With full load the container reaches a speed of about 500 km/hour. With reduced load 500 g acceleration is permissible, which means more than 620 km/hour.

The centrifuge proper is rotating in a chamber 3.5 m high and 15 m diameter under reduced pressure (10 - 20 millibars.) In this way the windage losses are reduced to almost zero, thus also avoiding any excessive heat generation in the test cell. The hinged suspended container can be locked in position whilst in motion. The model can, therefore also be loaded from the side.

The container can be equipped with a hydraulic vibration exciter for frequencies up to 120 Hz. The axis of the container are hollow with a big glass window (diameter 600 mm) for television observation of the model during testing.

3 THE AUTOMATIC BALANCING SYSTEM

It was decided to install a balancing system with the following capacity:

Static unbalance in X direction (Fig 2)
0 - 35 000 kgm
Couple unbalance around Y-axis
0 - 4400 kgm^2

The couple unbalance is due to the fact that the container can be clamped in any angular position. This causes couples if the center of mass of the container including model does not coincide with the axis of rotation of the container.

Furthermore it must be possible to compensate the unbalances during rotation by means of a fast acting automatic system.

The dimensions of the machine made it possible to install two adjustable balancing systems 2.2 m apart in the Z-direction each having a capacity of ± 7200 kgm. In addition 6 inter-changeable plates (having a total capacity of 27 800 kgm) are provided, which can be bolted to the rotor.

With the mechanism shown in Fig 3, 2 pairs of masses (each 1600 kg with a radius 0.67 m) can be rotated in each plane in opposite direction in synchronism around the main axis of the centrifuge. The neutral position of the masses shown in Fig 4 is the Y-axis. With all masses rotated into the 0 or 180° position in this figure, the max. capacity is reached. The masses are driven by high pressure oil hydraulic motors. The oil flow is controlled by a proportional valve.

On the upper and lower main bearing of the centrifuge (Fig 5) a pick up senses the vibration caused by the unbalance in the system. The two signals are filtered in an unbalance measuring device and mixed in such a way that two output signals are generated which represent the unbalance in the two planes of the compensation masses (plane separation). Each output signal is used as the input to a PID controller which in turn controls the position of the above mentioned oil flow control valve. The control circuit is basically used for position control. The differential portion of the control circuit dominates the system. A typical result using PD characteristics is shown in Fig. 6 compared with P characteristic of the control. The input is a step function.

It should be mentioned that the control circuit operates in the speed range 30 to 217 rpm. For such low speeds the time constants of the filter system are rather long and therefore the system reacts slowly.

4 CONTROL OF DYNAMIC FORCES

Two sources of dynamic loads on the system are foreseen. The major one, which at the same time is the one most difficult to predict, is excitation due to the break up of models during test. This can result in very sudden changes of the position of masses. The assumptions made are that loads due to shock will result in forces in the X and Z direction of 0.5 MN and moments around the Y-axis of 2.4 MNm.

The second source stems from the hydraulic vibration exciter rotating with the container. This provides frequencies up to 120 Hz. Therefore the main resonance frequencies of the system were specified to be > 100 Hz in the X direction and 5 - 10 Hz in the Y and Z-direction.

Many different design studies where conducted but it was finally realized that a flexible structure with damping was safest even though the resonance frequencies in such a structure are numerous and cannot be kept within the desired ranges.

The radial support stiffness of the main bearings of the centrifuge can be changed, by remote control, by a factor 1 : 20. For this purpose the complete weight of the machine is carried by a vertical center bar beneath an axial bearing on the lower side (see Fig 2).

The outer ring of each of the radial bearings (Fig 7) is held by a membrane which on its outer circumference ends in a circular wedge. This wedge can be opened by (hydraulic means) so that the bearings are free to move in a radial direction. Only a small spring force connects the stator element with the membrane.

The four arms of the centrifuge (Fig 2) have a cross section as shown in Fig 8. It can be seen, that the solid cross section is divided over a length of 3360 mm into 6 elements. These elements are clamped together by hexagonal cross section bolts. On each side, therefore, a flat of the bolt connects via friction to the slots. Because the bolts are arranged side by side over the full length of the slots, all bending modes of the structure are damped. The damping characteristic is of the dry friction (coulomb damping) type.

The natural frequencies of the main structure have been calculated using different procedures. In the X direction, calculations where made with the main bearings clamped and unclamped. This was also done for rotation around the Y axis. In the Y and Z direction, the natural frequencies were calculated assuming the elements in the arms to be either slipping or solidly connected, with and without the effect of the centrifugal forces at full speed. The result for the most important modes are shown in the following table.

Table 1: Natural frequencies

	Resonance frequency (Hz)			
mode	arms solid		arms slipping	
	$\Omega = 0$*	$\Omega = \Omega_{max}$	$\Omega = 0$	$\Omega = \Omega_{max}$
Longitudinal X main bearing clamped	24.2/160	24.2/160	24.2/160	24.2/160
Longitudinal X main bearing free	9.5/155	9.5/155	9.5/155	9.5/155
Bending Y	5.6	7.1	2.8	5.18
Bending Z	7.1	8.3	2.9	5.2

* first/second

5 CATASTROPHIC FAILURE

In the event of the container, including the model breaking up and being lost, the arms of the centrifuge will also break because the link between the arms in the Y-direction via the container bearings is lost. The unbalance due to losing the arms will be so large, that the main bearing can (also) suffer, but may not be completely destroyed before the central rotor comes to a standstill.

If only the bottom of the container and the model are lost, no more damage would be expected because the remainder of the container will keep the arms together and the resulting unbalance forces can easily be handled by the main bearings. The same applies if the automatic balancing system malfunctions and generates unbalance forces up to its maximum capacity.

6 CONCLUSIONS

A centrifuge has been designed with the unique new feature of a high capacity automatic balancing system to counteract static and couple unbalances in the system.

The technique of using rigid structures employed in other centrifuges to control the dynamic forces, has been abandoned. Instead a flexible system is used including a series of damping elements for all relevant modes.

At the time of writing this paper, the machine has made the first successful runs with tests on system parts. The results completely confirm the theoretical predictions. The author will report on the final test runs, when presenting this paper.

7 REFERENCE

(1) The Delft Geotechnics Centrifuge
Leaflet published by "Delfts Geotechnics"
Delft, Netherlands

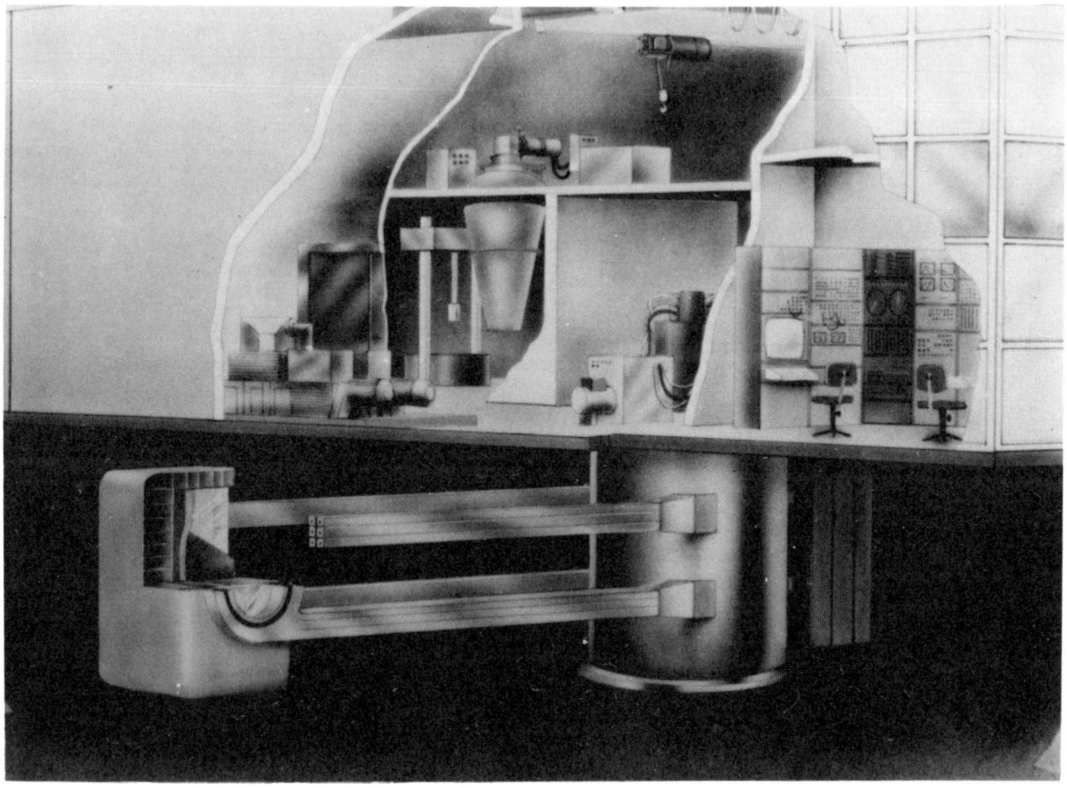

Fig 1 The geotechnics centrifuge at Delft

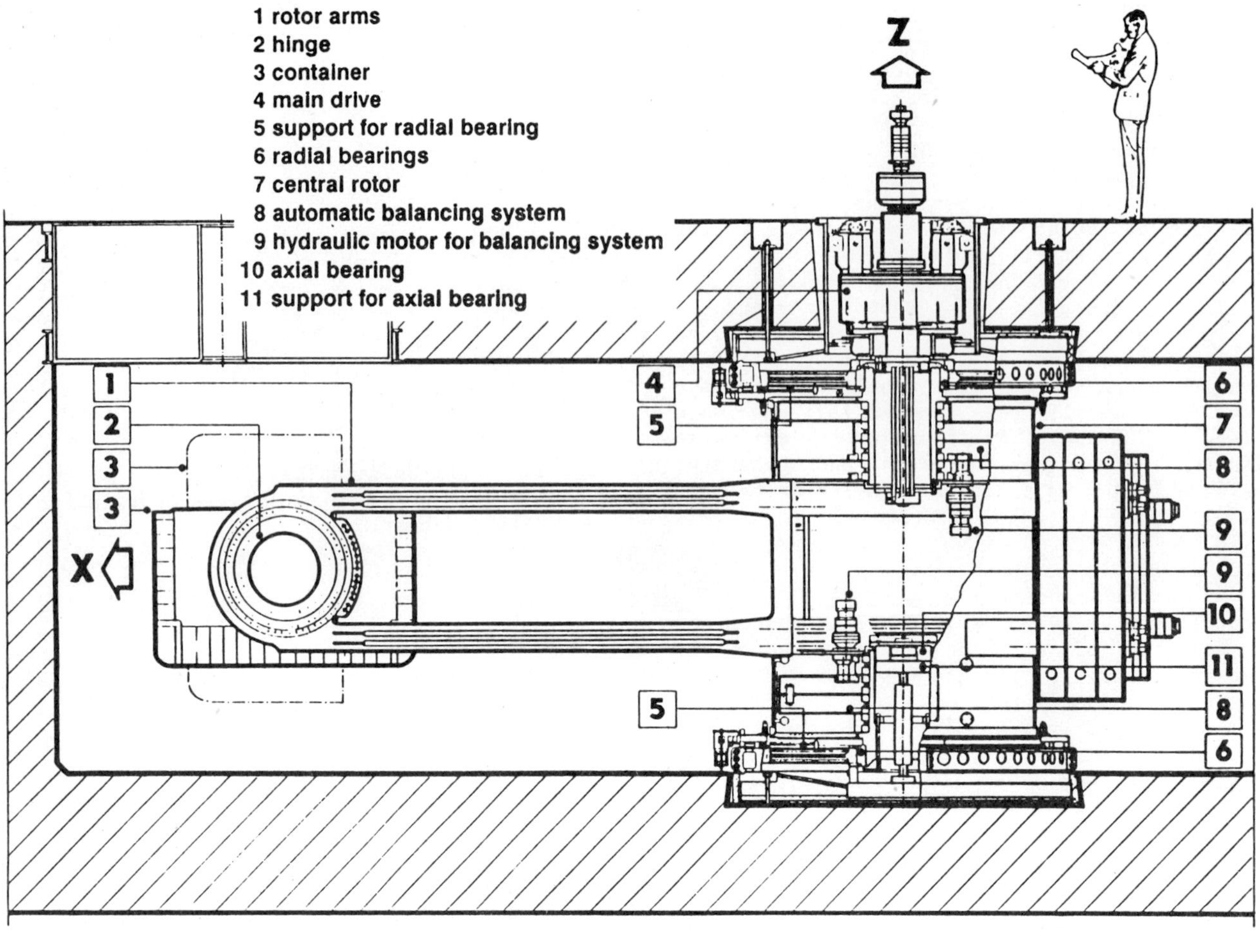

Fig 2 Longitudinal section of the rotor

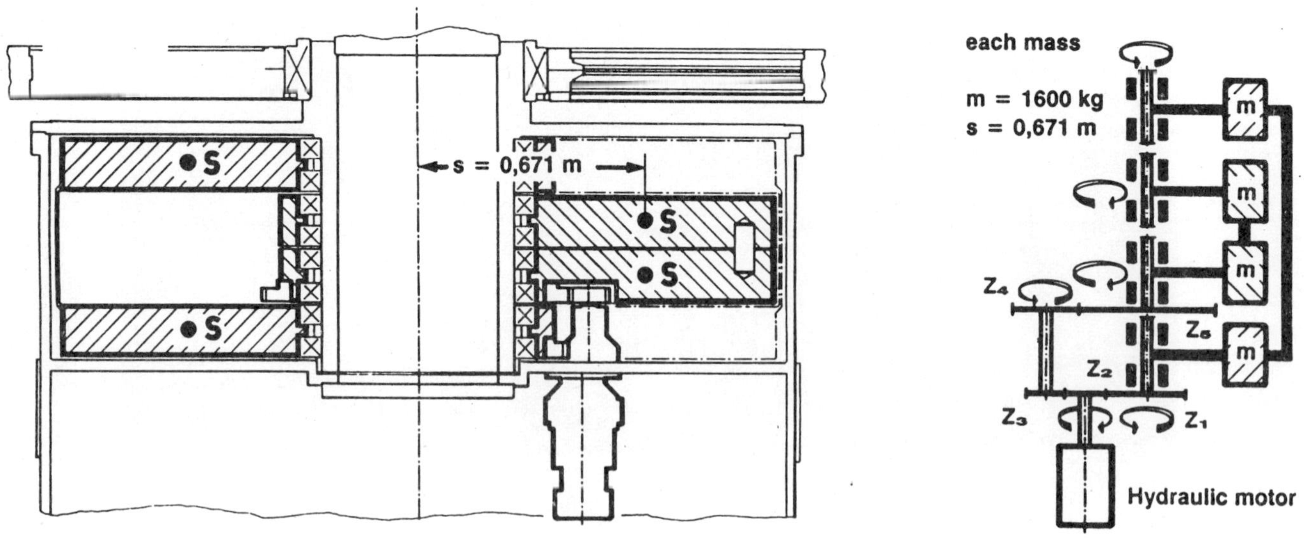

Fig 3 Balance masses and drive

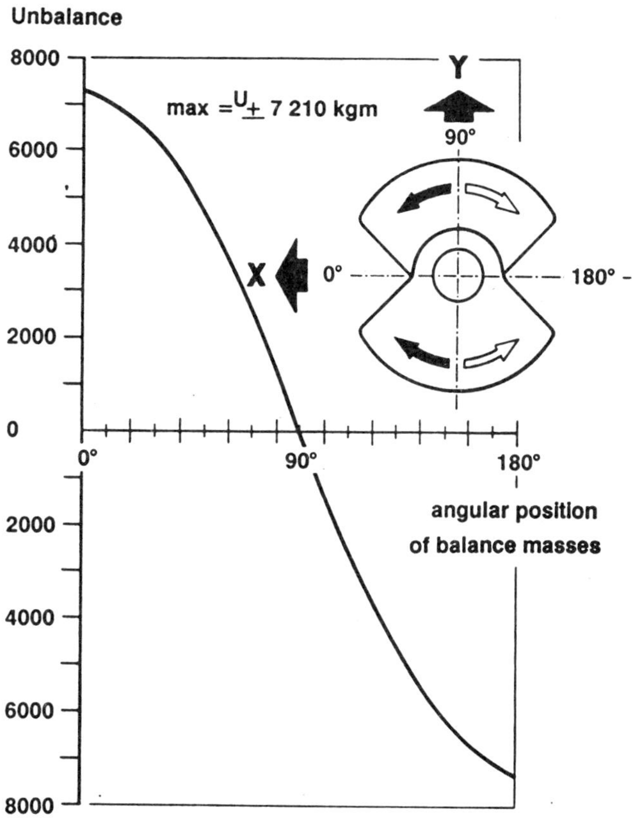

Fig 4 Balance capacity of the automatic system

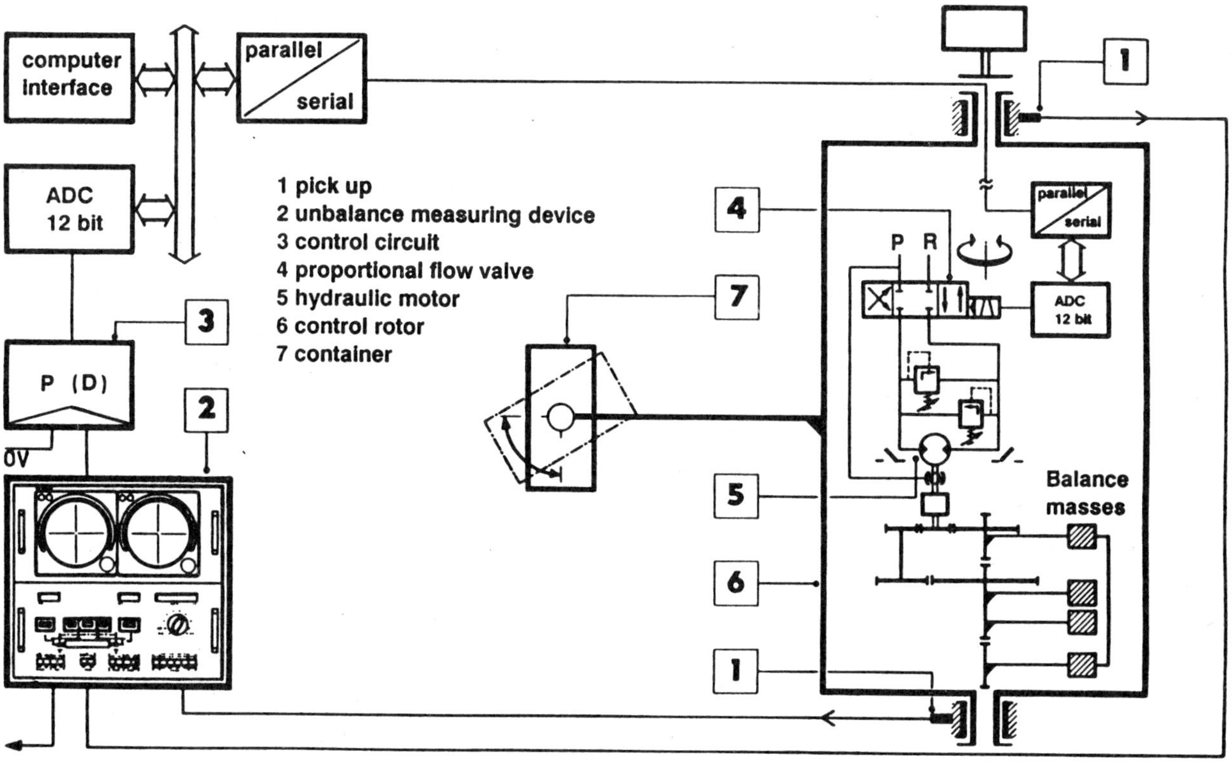

Fig 5 Complete automatic balancing system for one plane (schematic)

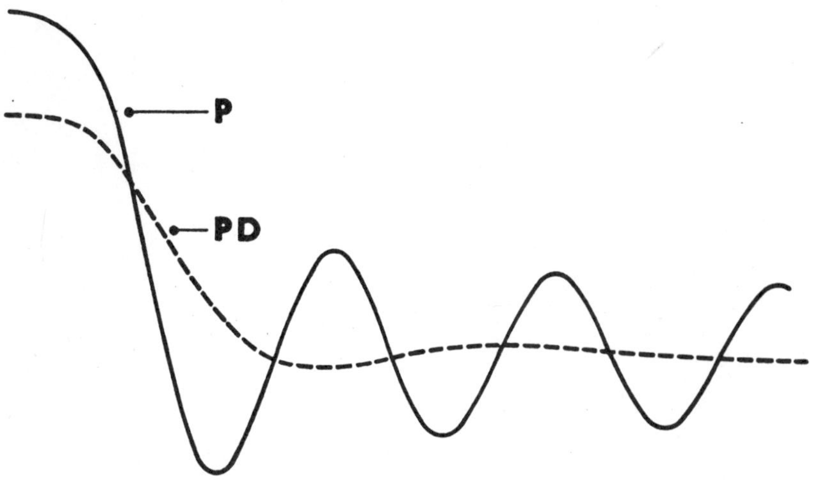

Fig 6 Reaction of control circuit to step excitation

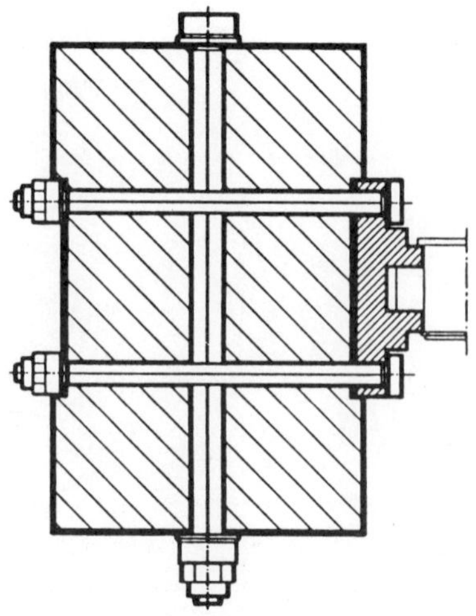

Fig 8 Cross-section through a centrifuge arm

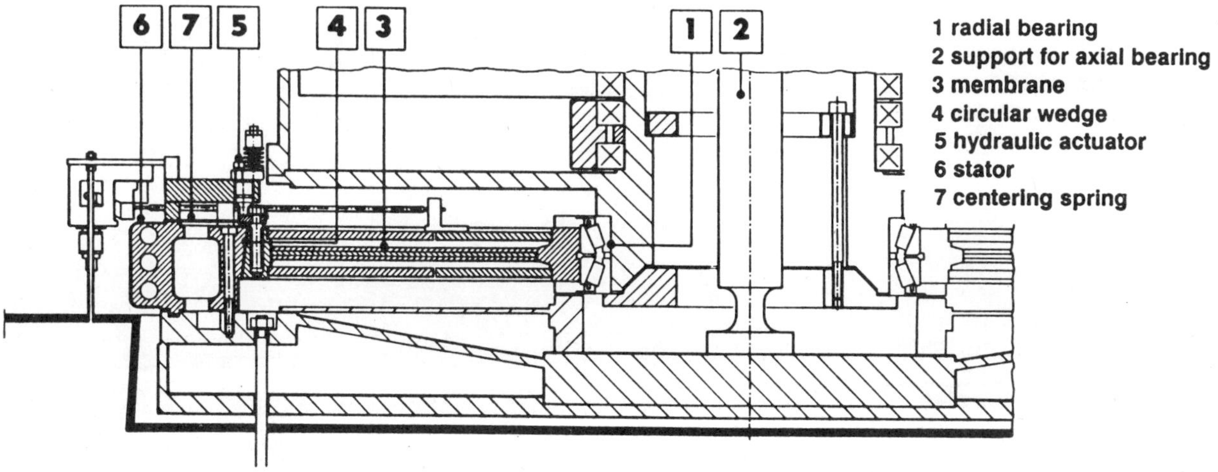

Fig 7 The lower radial bearing

C271/88

Problems associated with low frequency vibrations induced by rotating ride equipments

M S LEONG, BSc, PhD, PEng, MIEM
Research and Consultancy Unit, University of Technology Malaysia, Kuala Lumpur, Malaysia

SYNOPSIS The paper examines certain types of large rotating leisure ride equipments: centrifuge, gyroscopic and rollercoaster rides. The paper presents past experiences on vibration problems associated with such equipment whilst operating in buildings and elevated structures. Problems relate to low frequency vibrations generated by these equipments. High magnitude forcing frequencies were observed well within coincidence of building and structural natural frequencies. Equipment vibration characteristics and its influence on flexible foundations are examined. The behaviour of rotating elements in the ride equipment generating these vibrations are discussed. Field data from installed equipments are presented. This includes a case study of a particular unit causing concern to structural integrity from the vibrations.

1 INTRODUCTION

Carousels and other similar merry-go-round leisure equipments were once relatively simple innocuous low speed rotating ride equipments. The sight and sound certainly bring about reminiscence of many childhood nostalgia. Over the years variations of such carousels had evolved, tending towards those of a centrifuge. Such centrifuge rides are not unlike those used in flight training programmes. This paper examines several variation of such large rotating leisure ride equipments: basic centrifuge, spinning gyroscopic and rollercoaster rides. Whilst these equipments generally operate relatively safely and without much bother out of doors, potential problems however do arise when such equipments are located in buildings, suspended structures or flexible foundations. Vibration forcing characteristics, and interaction with the support foundation then become issues of concern. With commercial moves for amusement theme parks or arcades to be located in a single "all under one roof" complexes, there exists an increasing tendency for such ride equipments to be installed in buildings and similar structures.

2 RIDE EQUIPMENT

2.1 Equipment Consideration

Centrifuge rides generally consist of an electrical or hydraulic operated central drive unit, from which a spindle is driven. This central drive spindle assembly carries a series of extended boom-arms, much like a scaled up bicycle wheel. The near end of the boom is attached to the central drive spindle with a pivoted joint, and the far radial end of the boom extended out carring a passenger carriage bogie; Figure 1. The far end of the boom usually rests an a "fulcurm" point with guide or support wheels attached to the boom. These wheels then traverse along the periphery circular tracks; as in the case as illustrated in Figure 1. The passenger compartment is then left suspended off the free end of the boom. Variation exists where instead of wheels at a "fulcrum" location, the passenger compartment bogie has pairs of wheels located onto a rail track. This is similar to a simple railway coach undercarriage wheels and track arrangement. In the basic ride, as in a carousel, the track or guide rails are level and the ride spins like a large wheel with passenger cars at the rim on a level plane. Rides are now much more sophisticated, where the track or guide rails at the rim are with troughs and crests. Passengers thus experience a sensation of centripetal acceleration, as well as rise and fall motion in the vertical axis, as the centrifuge rotates.

An extension of such centrifuge is the gyroscopic ride, Figure 2. The passenger carriages are located on a platform, and this carriage platform during various period of its operation is tilted at an angle to the vertical axis of the supporting turntable. The carriage platform tilting motion is actuated by hydraulic drives. The supporting turntable is merely a ring beam which is driven by motors off a set of tyres or rollers. The tilted carriage platform rotates about its polar axis, and the whole platform precess about the vertical axis of the turntable with a different precession speed. The ride thus operate with a motion similar to that of a gyroscope.

Another common type of ride equipment is the rollercoaster. This is similar to a scaled down train moving within a closed-loop set of rail tracks. Passenger compartments are placed over a carriage bogie; and the undercarriage traction assembly comprises of pairs of wheels located onto parallel railway track. The drive mechanism consists of a driving wheel located on a driving car; this wheel moving along a toothed rack set in the track, Figure 3. The closed loop could be of any configuration but would generally include some tight spiral loops, as well as crest and troughs. Centripetal acceleration and vertical rise and fall motion are thus generated onto the passengers. Unlike the centrifuge ride, the forcing onto the foundation are generally transients as the carriage traverse over a particular track area.

2.2 Dynamic Consideration

Dynamic behaviour and vibrational forcing generated by properly manufactured and installed ride equipment are usually not of any great concern when it is operated outdoors. It is when such equipments are located within buildings, particularly on suspended or elevated slabs and structures, that the understanding of the dynamic bahaviour of the ride equipment becomes vital. Suspended structures as generally known are not infinitely rigid,and have varying degree of flexibility depending very much on structural mass, stiffness,and damping.

The installation of mechanical equipments on suspended slabs and structures are however by no means new. Industrial building services equipments for example, with proper vibration isolation are installed routinely in most buildings. This include large pumps, motors, industrial fans and blowers, centrifugal and reciprocating chillers, cooling towers and other rotating equipments for building air conditioning and associated services. Such equipments are usually electrically driven, and fundamental forcing frequencies for most machines are approximately 25 Hz and/or its harmonics (for 50 Hz electrical supply). Vibration isolation with elastomer mounts, and spring isolators of adequate transmissibility rates would usually satisfactorily resolve potential vibration transmission into the supporting structure.

Installation and operation of the above leisure ride equipments on suspended or elevated structures are however unique and unorthodox. Publication of technical literature on such equipments and its application are almost non-existent. Dynamic behaviour of such centrifuge and other ride equipments had not been previously considered. An understanding of the forcing and vibration characteristics generated by the centrifuge rotation is thus required.

Such equipments are relatively heavy, and forcing generated from its operation is significant in magnitude. Forcing in the radial direction would be obvious by virtue of centrifugal forces generated by the carriage rotation about a spin axis. Significant forcing in the vertical direction is also induced. What is of further concern is the frequency range which the forcings are generated from the equipment. Installation of any rotating equipment on a flexible foundation, and in this case the suspended floor slab, results in dynamic interaction between the floor slab and equipment. Vertical natural frequency of suspended floors in the fundamental mode is typically in the range of 5 to 30 Hz. There is the possibility of resonance, and increased transmission of vibration. This is further aggravated by additional difficulties in adequate attenuation of low frequency vibration forcing by conventional elastomer or spring isolators. Low frequency vibration forcing generated by the centrifuge ride equipments are thus of great concern.

Excitation of the longitudinal and transverse vibration modes are less likely, due to greater rigidity in these planes in most typical building structures. In vertically slender structures however, excitation of the structure in these planes are indeed possible.

3 CASE STUDY

Centrifuge Ride No.1

A case study is presented here where past experiences relating to a particular centrifuge ride equipment are discussed. Vibratory behaviour and its influence on the supporting slab foundation were particularly interesting.

3.1 Background

This leisure ride equipment was a basic centrifuge ride of the type supported by parallel wheels on tracks as described in Sec. 2.1. (The particular case examined here however is not that of the equipment as illustrated in the photograph of Figure 1). This equipment was installed on the third floor of a shopping complex building. Prior to the installation of the equipment, an additional 100 mm reinforced screeding was paved over the existing 150 mm reinforced concrete structural slab. The equipment was then installed directly anchored into the floor without any form of isolation.

Intense feelable vibrations were observed during the operation of the centrifuge ride. Within a month of operation, cracks were observed in the floor and were confirmed to be in the screeding up to 100 mm deep. Cracks propagation were emanating radially outwards from the ride equipment. Vibration levels off the floor slab at various location adjacent to the equipment, using seismic transducers, indicated levels of 0.260 m/s2 RMS acceleration, 0.120 mm peak-peak displacement. Typical frequency spectrum of structural slab vibration acceleration was as shown in Figure 4 (a). Significant low frequency vibration forcing from 9 to 80 Hz were observed.

Attempts were then made by the owners to install the ride equipment on resilient neoprene-in-compression pads, with nominal total static deflection of 3 to 5 mm. It would have been obvious at that stage in time that such isolation medium would not be adequate due to the presence of dominant low frequency vibration components. Not surprisingly then, the floor slab vibration levels with the neoprene isolation did not improve significantly. Measured vibration levels were typically 0.220 m/s2 RMS acceleration, 0.120 mm peak-peak displacement; Figure 4 (b).

Structural damage from the vibration was a primary concern. A generally accepted indicator of structural damage potential from vibration is the vibration velocity (Ref.1). Vibration velocity of the floor slab with neoprene isolation was typically from 1.0 to 1.5 mm/s RMS, 4.8 to 5.5 mm/s peak-peak velocity. Typical velocity frequency spectrum on the floor slab adjacent to an isolated footing is shown in Figure 4 (c). Note in particular the strong vibration component of 9 - 10 Hz. The floor slab vibration not surprisingly reflected forcing characteristics generated from the equipment. Figure 5 shows typical frequency spectrum of the equipment vibration motion on the rail footing: with strong forcing component at 9, 45 Hz and others, as shown. Frequency spectrum shown were for steady state constant ride rotating speed. Vibration levels at the equipment footing were in the order of 3.5 mm/s RMS, 22.0 mm/s peak-peak velocity. Vibration measurements using seismic tranducers at various equipment locations evidently showed predominant forcing generated off the track. As the carriages move along the track when the centrifuge rotated, vertical forcing was induced from the wheels-track rolling motion. Further discussions are presented Sec.4.

3.2 Structural Damage

Structural damage in buildings induced by vibration are still currently debated and universally accepted damage criteria quantified, (Ref.1). Early criteria had been proposed from several sources (Ref. 2,3), but were thought less applicable for continuous steady state vibration as such early criteria had been based on transient vibrations from blasting. An ISO Draft Proposal 4866 - 1975 (Ref. 4) exists, providing provisional guidance amongst other recommendations on vibration velocity limits for structural damage. "Threshold damage" is defined as visible cracking in non-structural members, whilst "minor damage" is defined as visible cracking in structural members.

Measured vibration data of the slab with the above particular centrifuge ride excitation was plotted against this damage criterion, Figure 6. In accordance to ISO DP 4866 - 1975 vibration levels were indeed approaching limits of structural concern. Site observations of the screeding cracks obviously confirmed such concern.

4 VIBRATION FORCING CHARACTERISTICS

It is apparent that in centrifuge, gyroscopic and rollercoaster rides, significant vibration forcing are generated in the vertical axis off the rolling motion of the carriage wheels on the tracks. Dominant vibration forcing frequencies, f, would be a function of ride rotational speed, N, and the number of wheels, n.

Thus $f = fn(n, N)$ (1)

N would vary from 0 to the maximum rotational speed of the ride.

Vibration forcing frequency would thus inevitably include a "wheel passing" frequency and its harmonics; which is the multiplication of the rotational speed with the number of wheels, and its multiples. This wheel passing frequency phenomenon was generally true, and were observed in the frequency spectra of most rides examined. In the previous case study of the ride equipment, the total number of carriages were twenty, with a pair of forward and rear wheels, and the ride rotated approximately at at one cycle per eight to ten seconds. The 9 Hz component observed would most probably be that of the second harmonics of the fundamental forcing frequency of 4.5 Hz.

Other mechanical forcing would also be present, which could include drive assembly and other traction forcing, and periodic/random excitation generated off imperfections in the tracks, coupled with mechanical noises.

The above discussions had so far been limited to the vertical plane, where as previously indicated, represents the most likely plane to be excited by the equipment when located on a suspended floor. Centrifugal forces are indeed induced in the radial direction, and by virtue of a large rotating assembly with mass-radius up to 9m, some degree of mass unbalance with unequal pasenger loading is inevitable. In the centrifuge ride, centrifugal forces are to some extent reduced by virtue of equipment symmetry. In the rollercoaster however centrifugal forces generated off the spiral loops are unbalanced, and significant radial forcing generated off the rail tracks.

An additional and extremely important observation was that during the course of a typical ride, the equipment would go through a run-up, steady and/or fluctuating continuous speed operation, and run-down per complete ride cycle. The ride operations would inevitably have a speed range varying from zero to a maximum operational speed. The period of dwell in each speed phase would very much depend on the speed-time profile selected or built-in to the ride controller. Maximum speeds do vary from 5 to 10 rpm. With most typical number of carriages and wheel pairs, major forcing frequencies thus occur in the range from 0 to 25 Hz or more. With such a varying speed range it is inevitable that the foundation fundamental mode natural frequency could be excited. Coincidence of steady state operating speeds with foundation natural frequencies is a real possibility. Further to this, vibration forcing components in the low frequency range are relatively intense.

Additional measured data on actual equipments, some of which were tested on ground foundation and/or suspended floors, are presented. Such data generally confirmed significant low frequency forcing components.

a. Centrifuge Ride No. 2

Vibration acceleration data were obtained on the centrifuge ride illustrated in Figure 1, on solid ground foundation. This was a different unit from the one examined in Sec. 3.1. The measurements confirmed significant vibration components particularly at 15, 20, 24 and 43 Hz; Figure 7 (a).

Field data were then obtained form the same equipment upon its installation on an elevated floor in a building. Figure 7 (b) shows the vibration velocity spectrum for the equipment at the footing of the guide rail. This equipment was installed on a pneumatically isolated steel base platform. The frequency spectrum distinctly showed concentration of vibration components in the lower frequencies.

b. Gyroscopic Ride

This ride was similar to that as previously illustrated in Figure 2. Measured data were obtained with the equipment operating on solid ground foundation. Velocity frequency spectrum confirmed significant forcing at 13, 16, 22 and 35 Hz; with other components as low as 3 Hz, Figure 8 (a). Measurements at the drive mechanism assembly showed much less significant forced vibration levels, Figure 8 (b). This illustrates for example the significant contribution of the wheels-rollers rolling motion in generating the low frequency forcing.

c. Rollercoaster

Vibration forcing at the support foundation of this equipment exhibited a time history, as explained earlier when the carriage traverse along the track over that area. A vibration time trace of the rollercoaster of Figure 3 is as shown in Figure 9 (a). Frequency spectrum of the vibration of the track footing on ground foundation confirmed significant vibration components as low as 4 and 8 Hz. Upon installation of the equipment on elevated floors in a building, frequency spectra of the equipment at the track footing reflected such frequency components; Figure 9 (b). Measurements taken on board the undercarriage, adjacent the wheel axle of a passenger car, showed dominant forced frequencies at such low frequencies of 1.2, 3.2 and 5.6 Hz; Figure 9 (c).

5 VIBRATION ISOLATION

The previous examples consistently illustrated low frequency forcing content generated from the ride equipments. Dynamic interaction with flexible foundation need to be carefully examined, especially in the design and selection of vibration isolation systems. While it is not within the scope of this paper to examine in detail the analysis and selection for vibration isolation of these equipments on suspended floors, a brief discussion is presented.

Several pertinent issues must be addressed in the design for isolation:

- Loading
- Transmissibility
- Safety

Properly engineered isolation solutions would adequately cater for the static and dynamic loading generated by the rides. Consideration of force and motion transmission into the supporting structure are however much more involved. Problems do exist when common spring or elastomer isolators with natural frequencies nominally from 2 to 12 Hz are used. The case seen in Sec. 3 for example demonstrated the futility in using isolators not capable of attenuating the forced vibrations with such low frequency content. Isolators static deflection is however not the sole controlloing variable in transmissibility performance of isolators used in equipment installed on flexible foundations. Analysis of the ride equipment-foundation model should ensure that the floor dynamic characteristics, isolator properties and equipment forcing characteristics be adequately represented.

Safety considerations of the ride equipments used by the general public are also of vital importance. With the equipment now installed on resilient mounts such consideration cannot be overlooked.

6 CONCLUSIONS

Some rotating leisure ride equipments had progressively developed into relatively large centrifuge rides, or similar variations. Commercial decision to locate such equipments on flexible floors or structures has dire consequences, if not adequately addressed. Past experiences and measured data had confirmed significant low frequency vibratory forcing, well within fundamental natural frequency of the supporting foundation. Structural damage arising from such installations is a real possibility. Vibration forcing were generally associated with the centrifuge rotational behaviour and the traction wheels passing frequency.

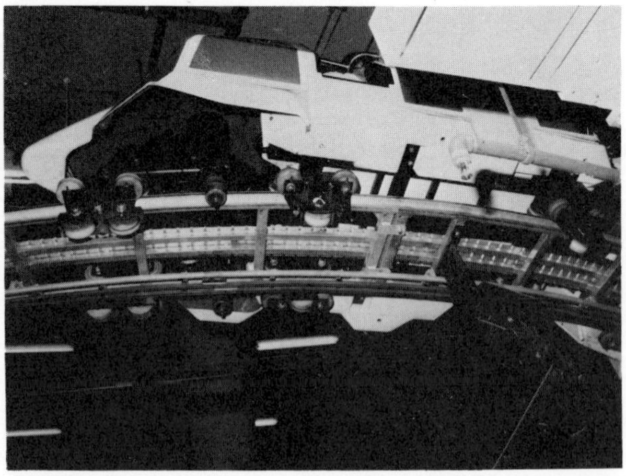

Fig 3 View of the undercarriage traction of a typical roller-coaster ride equipment

7 REFERENCES

(1) MACINANTE, J.A., Seismic Mountings for Vibration Isolation, 1984, Vibration Criteria, pp. 77- 80, John Wiley & Sons.

(2) STEFFENS, R.J., Some Aspects of Structural Vibration, Proceedings of the Symposium on Vibrations in Civil Engineering, 1966, pp. 1 - 30, Butterworths London.

(3) Great Britain Building Research Station, Vibrations in Buildings, Part 1, BRS Digest No. 117, 1970.

(4) International Organisation for Standardization, 1975, Evaluation and Measurement of Vibration in Buildings, Draft Proposal DP 4866 - 1975.

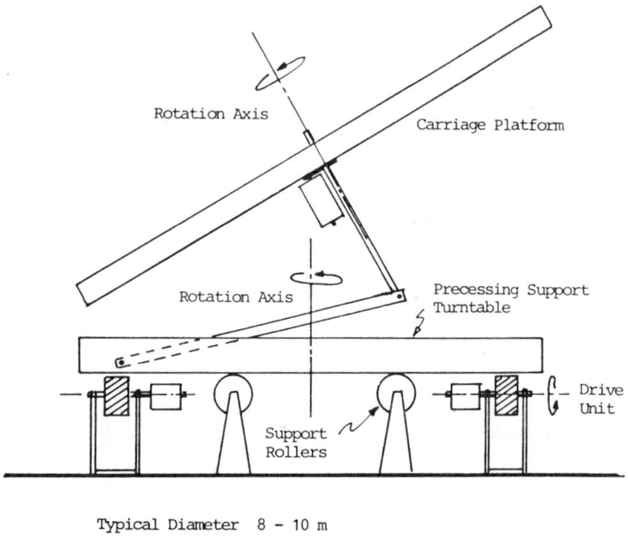

Fig 2 Schematic assembly of a gyroscopic ride equipment

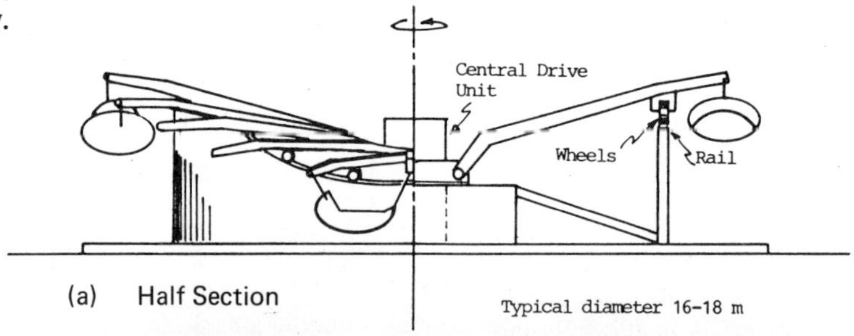

(a) Half Section

(b) Photograph showing top view

Fig 1 Illustration of a typical centrifuge ride equipment

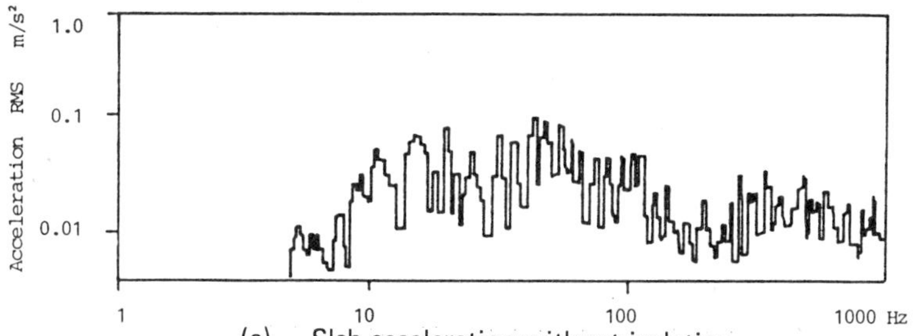

(a) Slab acceleration without isolation

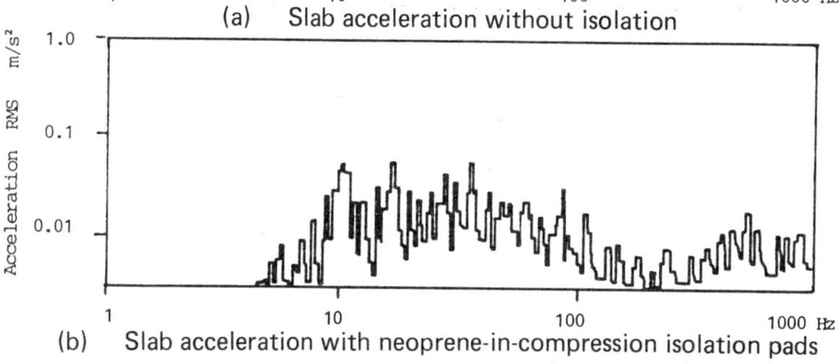

(b) Slab acceleration with neoprene-in-compression isolation pads

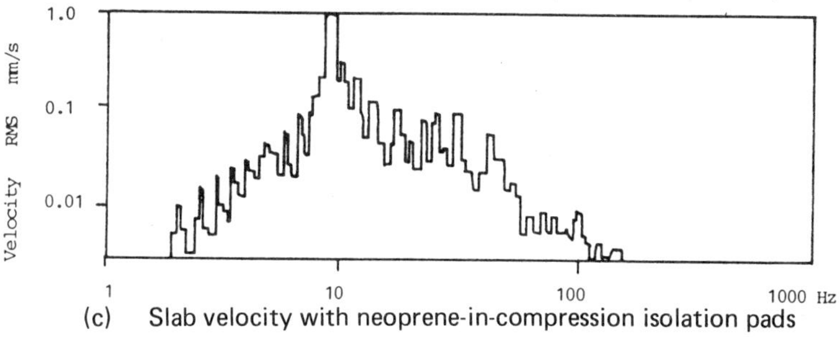

(c) Slab velocity with neoprene-in-compression isolation pads

Fig 4 Frequency spectra of elevated floor slab excited by rotating centrifuge ride equipment

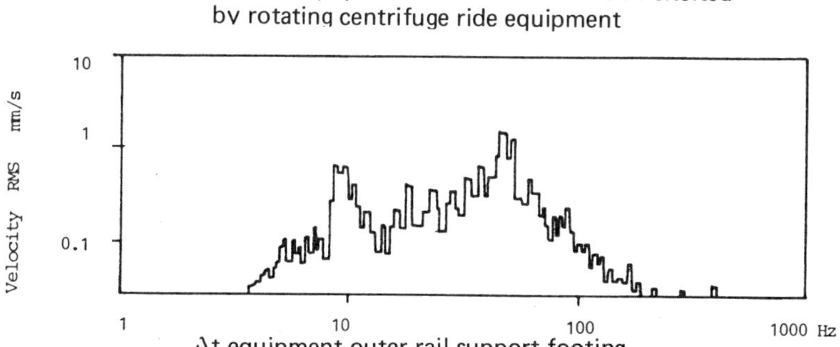

At equipment outer rail support footing

Fig 5 Frequency spectrum of rotating centrifuge equipment at track support footing on elevated slab

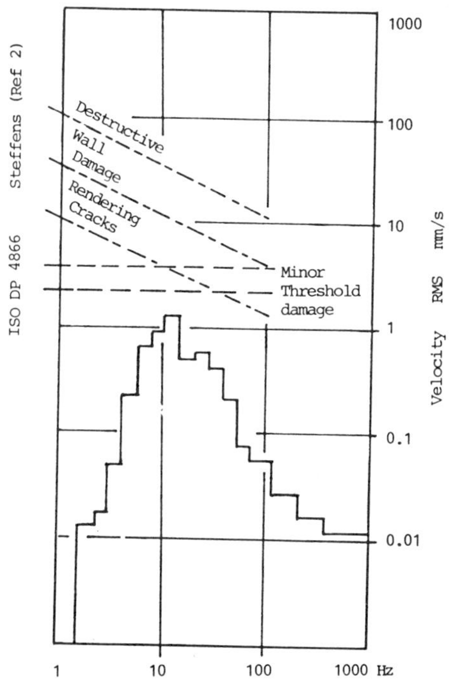

Fig 6 Potential structural damage caused by rotating centrifuge ride equipment on elevated slab

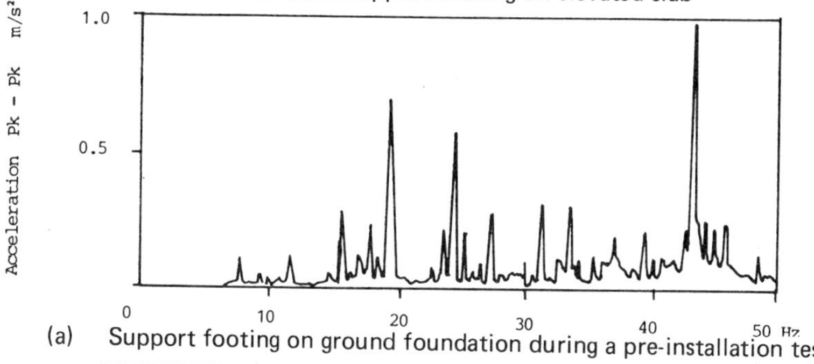

(a) Support footing on ground foundation during a pre-installation test

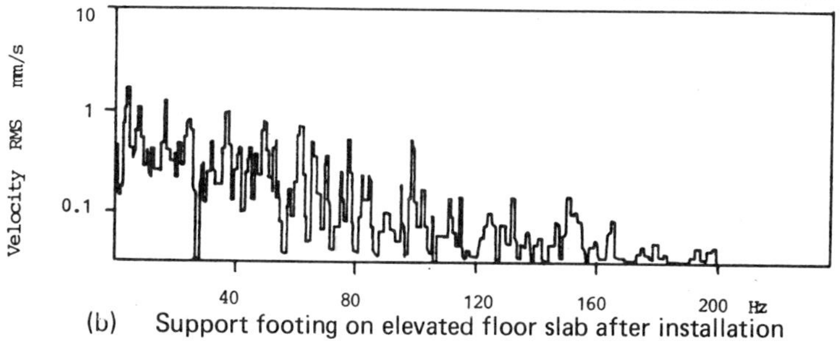

(b) Support footing on elevated floor slab after installation

Fig 7 Frequency spectra of a rotating centrifuge ride equipment

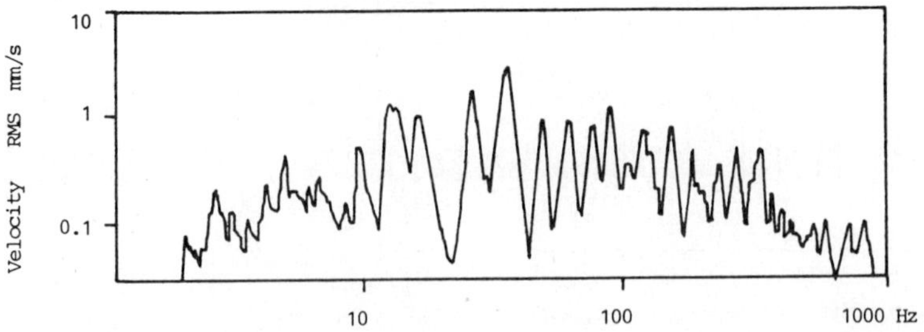

(a) Bearing at roller support footing with equipment on ground

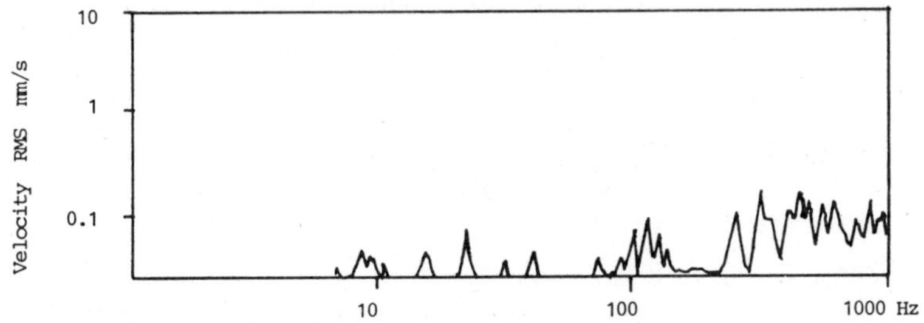

(b) Bearing at drive mechanism support with equipment on ground

Fig 8 Frequency spectra of a rotating gyroscopic ride equipment

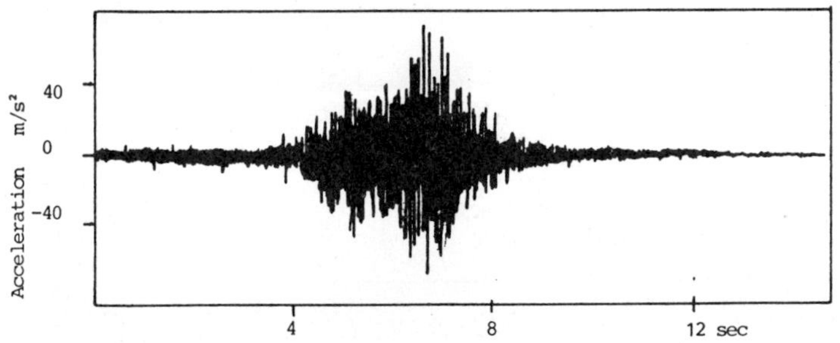

(a) Typical time history, equipment installed on ground

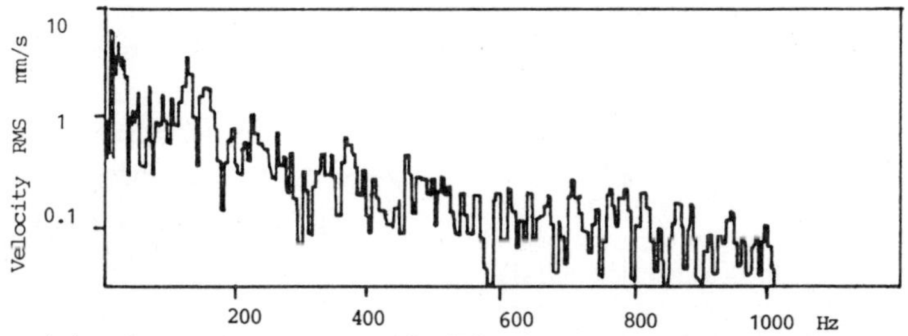

(b) Frequency spectrum of rail footing with equipment on elevated slab

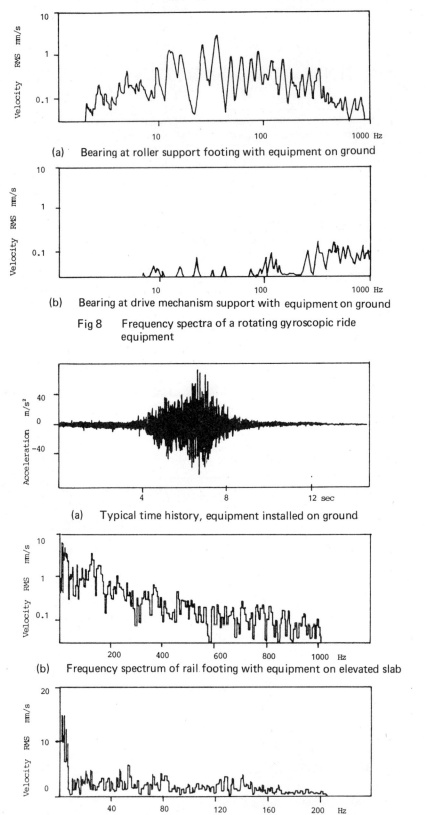

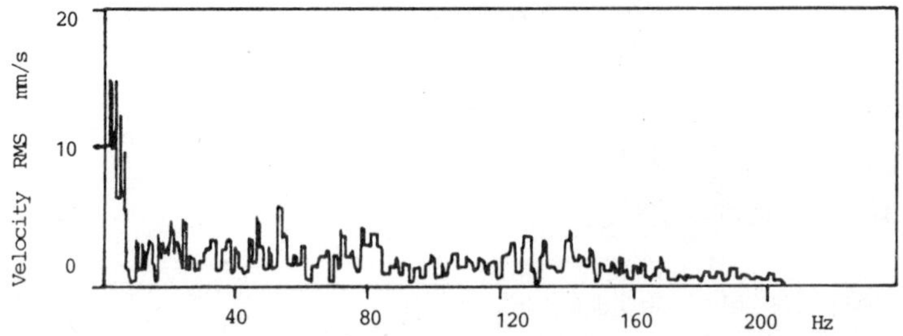

(c) Frequency spectrum of undercarriage of passenger car with equipment on elevated slab

Fig 9 Time history and frequency spectra of a roller-coaster in motion

C253/88

Conservative and nonconservative coupling in dynamic systems

S H CRANDALL, PhD, MASME
Department of Mechanical Engineering, Massachusetts Institute of Technology, Cambridge, Massachusetts, USA
J W MROSZCZYK, PhD
Atlantic Applied Research Corporation, Burlington, Massachusetts, USA

SYNOPSIS When undamped vibratory subsystems with a common resonance interact, the coupled system has a pair of undamped modes with frequencies above and below the subsystem resonance when the coupling is conservative. In the nonconservative case the two modes have the same frequency but the amplitude of one decays while the amplitude of the other grows exponentially. A canonical case where the subsystem resonances vary with a parameter is described. A simple heuristic procedure is developed for identifying whether the coupling is conservative or nonconservative and is applied to two rotordynamic instability problems: aircraft propeller-engine whirl and the whirling of rotors partially filled with liquid.

1 INTRODUCTION

Undamped vibratory systems with two or more degrees of freedom can sometimes be conveniently analysed as two subsystems which interact through a coupling mechanism. The uncoupled subsystems are analysed separately and then the behaviour of the coupled system is predicted by introducing the coupling. When the coupling is light there is little interaction when the uncoupled natural frequencies of the first subsystem are well separated from those of the second subsystem. However, when the uncoupled subsystems have a common natural frequency there is a strong interaction. The nature of the interaction depends on whether the coupling mechanism is conservative or nonconservative. With conservative coupling the two uncoupled modes with the same frequency are transformed into a pair of undamped natural modes of the combined system with *separate* natural frequencies, one higher and one lower than the common uncoupled frequency. With nonconservative coupling the two uncoupled modes with the same frequency are transformed into a pair of modes with *equal* frequencies and time dependent amplitudes, one which grows exponentially and one which decays exponentially. Because of the presence of a mode with exponential growth the combined system is said to be unstable. In certain systems the natural frequencies of the subsystems vary with a tuning parameter such as the rotational speed of a rotor or the flow velocity past an aeroelastic structure. In such systems the parameter values which cause the uncoupled subsystems to have a common natural frequency are of particular interest. If the coupling is conservative this parameter value marks a transition in the nature of two natural modes of the coupled system while if the coupling is nonconservative this parameter value marks the center of an instability range for the parameter. In this paper these points are illustrated by an idealized system with two single-degree-of-freedom subsystems coupled by simple conservative and nonconservative mechanisms. In more complicated systems it is often difficult to identify the nature of the coupling mechanisms. A heuristic procedure for deciding whether the coupling in the neighborhood of a particular cross-over of subsystem natural frequencies is conservative or not is developed. This procedure is illustrated by applying it to an aircraft propeller-engine system and to a rotor partially filled with liquid. In these systems the same physical coupling mechanism is conservative in one speed range and nonconservative in another speed range.

2 CANONICAL SYSTEM

Consider small motions of the mass particle m on the frictionless horizontal plane of Fig. 1(a). For most values of the tuning parameter α there are two distinct natural modes: vibration along the x-axis and vibration along the y-axis. The squares of the natural frequencies of these modes are displayed in Fig. 1(b) as functions of the tuning parameter. For small motions the x and y motions are uncoupled and proceed independently of one another even when $\alpha = 0$ and the two motions have the same natural frequency. We now consider two coupling mechanisms.

2.1 Conservative coupling

Imagine that the plane of Fig. 1(a) is replaced by a shallow frictionless saddle surface as indicated in Fig. 2(a) and that gravity acts downward pressing the mass particle against the surface. As long as the elevation z of the surface is of second order in the horizontal displacements x and y the horizontal components of the spring forces are unaffected to first order. The surface reaction on the mass introduces additional horizontal force components $-mg\,\partial z/\partial x$ and $-mg\,\partial z/\partial y$ in the x- and y-directions respectively. In Fig. 2(a) the dashed lines are contours of equal elevation and the solid lines represent the directions of the resultant horizontal force component applied to the mass particle by the surface reaction. If the surface is represented by

$$z = \epsilon\omega_0^2 xy/g \tag{1}$$

the equations of motion of the coupled system take the form

$$\begin{Bmatrix} \ddot{x} \\ \ddot{y} \end{Bmatrix} + \omega_0^2 \begin{bmatrix} 1-\alpha & \epsilon \\ \epsilon & 1+\alpha \end{bmatrix} \begin{Bmatrix} x \\ y \end{Bmatrix} = 0 \qquad (2)$$

For the coupled system ($\epsilon > 0$) the natural modes of vibration are back and forth oscillations along straight-line paths through the origin. The squares of the corresponding natural frequencies are displayed in Fig. 2(b). Note that at the uncoupled crossover point $\alpha = 0$, the resonance curves for the coupled system repel one another with the squares of the two natural frequencies differing by $2\epsilon\omega_0^2$. The mode corresponding to the higher frequency is a back and forth motion along the 45° line joining the high points of the saddle while the mode corresponding to the lower frequency is along the -45° line joining the low points of the saddle. Note also that the change in natural frequency due to the coupling diminishes rapidly as the tuning parameter is moved away from the crossover.

2.2 Nonconservative coupling

Imagine that the plane of Fig. 1(a) is replaced by the lubricated turntable of Fig. 3(a) which exerts a tangential drag force ϵkr on the mass particle whenever it has a radial displacement r. The coupled equations of motion for the mass particle then take the form

$$\begin{Bmatrix} \ddot{x} \\ \ddot{y} \end{Bmatrix} + \omega_0^2 \begin{bmatrix} 1-\alpha & \epsilon \\ -\epsilon & 1+\alpha \end{bmatrix} \begin{Bmatrix} x \\ y \end{Bmatrix} = 0 \qquad (3)$$

For most values of the tuning parameter α the natural modes of vibration are back and forth oscillations along straight-line paths passing through the origin. The squares of the corresponding natural frequencies are displayed in Fig. 3(b). In the unstable range $-\epsilon < \alpha < \epsilon$ the free vibrations of the coupled system are linear combinations of two elliptical spirals, one with increasing radius having the same sense as the turntable rotation, and one with decreasing radius in the opposite sense. The cyclic frequency remains ω_0 throughout this range. When $\alpha = 0$ the growth rate of the forward logarithmic spiral, $\epsilon\omega_0/2$, can be obtained by equating the rate at which the tangential drag force does work $\epsilon kr \cdot r\omega_0$, to the rate at which the stored energy, kr^2, increases.

The conservative coupling mechanism of Fig. 2 leads to a stiffness matrix in (2) which is symmetric while the nonconservative coupling mechanism of Fig. 3 leads to a stiffness matrix in (3) which is antisymmetric. A similar result occurs for the matrix multiplying the acceleration vector in the case of inertial coupling. In the case of coupling through a matrix multiplying the velocity vector, conservative coupling leads to an antisymmetric matrix [1]. If the matrix is symmetric, the coupled system is stable if the matrix is positive semi-definite and unstable if it is not.

3 IDENTIFICATION OF COUPLING TYPE

In more complicated systems the nature of the coupling is not always apparent and it may not be easy to transform the equations into the canonical forms discussed above. The following procedure permits identification of the coupling as conservative or nonconservative by simple physical arguments which don't require the equations of motion for the coupled system. It is necessary to know how the resonances of each subsystem vary with the tuning parameter so that all crossover points can be located. For each crossover point the type of coupling is determined by whether the coupled resonances repel each other as in Fig. 2(b) or attract one another as in Fig. 3(b). The argument proceeds as follows. Consider the tuning parameter set near to, but not right at, the crossover point. Imagine an uncoupled vibration of the subsystem mode in question for the first subsystem. Initially the corresponding mode of the second subsystem is at rest. Then consider the addition of infinitesimal coupling which excites an infinitesimal response from the second subsystem. The phasing or sense of this response must be determined. Finally determine whether the small motion of the second subsystem tends to raise or lower the frequency of the original vibration. If the frequency change is such as to represent repulsion of the coupled frequencies the coupling is conservative. In the opposite case the coupling is nonconservative.

3.1 Illustration

Consider the uncoupled system of Fig. 1 and suppose that the tuning parameter α has a fixed value in the range $-1 < \alpha < 0$. We begin with the mass particle vibrating in the x-direction with natural frequency ω given by $\omega^2 = \omega_0^2(1-\alpha)$. If coupling is introduced gradually in such a manner that steady-state motion is maintained the path of the particle will slowly deviate from the x-axis and simultaneously the frequency of oscillation will shift. By considering infinitesimal coupling we can consider these effects sequentially. We take first the case of the conservative coupling mechanism represented by Fig. 2(a). As the particle moves back and forth on the x-axis the surface reaction pushes it in the *negative* y-direction when x is positive, and in the *positive* y-direction when x is negative. This excites a small y-oscillation but since the natural frequency of y-motion is smaller than the forcing frequency, the y-displacement is 180° out of phase with the exciting force. This implies that the coupled vibration is along a line through the origin running from the *third* quadrant to the *first* quadrant. This is the first step of the argument. Next it can be seen from Fig. 2(a) that such a path involves moving uphill on the saddle as the particle moves away from the origin. This adds additional restoring force to that provided by the springs and thereby tends to *increase* the natural frequency. A shift of the resonance curve in this direction represents *repulsion* of the curves as verified by Fig. 2(b).

If the same argument is repeated with the nonconservative coupling mechanism represented by Fig. 3(a) the final result is reversed because the introduction of the coupling produces a drag in the *positive* y-direction when x is positive and a drag in the *negative* y-direction when x is negative. Since the y-motion is excited at a frequency above its resonance the y-displacement is 180° out of phase with the excitation and the introduction of small coupling moves the modal path from the

x-axis to a slightly tipped line running from the second quadrant to the fourth quadrant. From a vector diagram of the x-axis and y-axis spring forces plus the drag force it can be seen that the resultant force on the mass particle is a restoring force, acting along the tipped modal path, which is *smaller* than that when the path was along the x-axis. This results in a *decrease* in natural frequency and an *attraction* of resonance curves as verified by Fig. 3(b).

4 AIRCRAFT ENGINE-PROPELLER WHIRL

At the 2nd International Conference on Vibrations in Rotating Machinery forced whirling of engine-propeller systems was discussed [2] in terms of a two-dimensional model. Subsequently [3] a three-dimensional model was shown to have a range of unstable whirl analogous to self-excited ground resonance of helicopters [4]. Here we consider a simplified two-dimensional model which retains the basic instability mechanism of both the propeller-engine system and the helicopter ground resonance. In Fig. 4 each propeller blade has mass m, first moment of mass S, and second moment of mass I, where the moments are taken about the hub. Coupled whirling of the system involves a whirling motion of the engine block plus a pattern of individual blade vibration such that the center of mass of the propeller whirls about its hub. The analysis is considerably simplified when the system is treated as two subsystems which are then coupled. The first subsystem is the engine block plus the rotating but nonvibrating propeller. The natural frequency ω_M, in either the x_1 or x_2 direction, or in either a forward or backward whirl, is given by

$$\omega_M^2 = K/(M_e + 3m) \tag{4}$$

independently of the rotation rate Ω. The second subsystem involves the vibration of the propeller with respect to a frame which rotates at rate Ω about a fixed hub location. In our model we take the blades to be rigid and pivoted at the center so that small pendular oscillations in the plane of rotation are permitted. See Fig. 5. The oscillations of the separate blades are independent as long as the center O is motionless. Since the centrifugal and Coriolis forces on the blades are all collinear with O the only torque acting on a blade is the elastic restoring torque $k\phi_i$. The natural frequency of each blade is ω_0 where

$$\omega_0^2 = k/I \tag{5}$$

independently of the rotation rate Ω. In place of the three independent angles ϕ_i it is convenient to introduce the average rotation θ and the displacement components ξ, η of the mass center of the three blades, with respect to the rotating frame. Here ξ is measured along the undeflected direction of the first blade and η is measures at right angles to this direction. The relationships between these coordinates are

$$\begin{aligned} 3\theta &= \phi_1 + \phi_2 + \phi_3 \\ 3m\xi &= S(-\phi_2 + \phi_3)\sqrt{3}/2 \\ 3m\eta &= S(\phi_1 - \phi_2/2 - \phi_3/2) \end{aligned} \tag{6}$$

Thus a free vibration in which each blade vibrates at frequency ω_0 with its own amplitude ϕ_i may be viewed as a superposition of three modes of the propeller as a whole: a torsional vibration with amplitude θ, plus translational vibrations of the mass center along the ξ and η axes. Since the torsional vibration of the propeller does not couple to the translations of the engine block we omit further consideration of the θ-mode. Finally the translational modes may be replaced by forward and backward circular whirls of the mass center with frequency ω_0.

4.1 Nature of the coupling

At this stage we know that the uncoupled mount system can execute forward and backward whirls of frequency ω_M with respect to a nonrotating frame, and that the uncoupled propeller center of mass can execute forward and backward whirls of frequency ω_0 with respect to a frame rotating at rate Ω. Viewed from the nonrotating frame these uncoupled natural frequencies are displayed in Fig. 6. Coupling is introduced by including the relative displacements of the propeller mass center in writing the equations of motion of the engine plus propeller on the mounts and by including the inertia forces due to the hub motion in writing the equations of relative pendular motion of the propeller blades. Formal equations may be written [5] but are not required in order to identify the nature of the coupling at the crossover points in Fig. 6. For example, consider the crossover at $\Omega = \omega_0 + \omega_M$. Imagine that at a slightly lower speed the engine block is executing an uncoupled forward whirl at frequency ω_M while the propeller is undeflected. Now introduce small coupling. The whirling of the hub excites a small whirl of the propeller mass center at the same frequency. Viewed from the rotating frame this frequency represents a backward whirl which is somewhat slower than the natural whirl frequency ω_0. The response of the propeller mass center will thus be in phase with the excitation. In the coupled whirl the mass center of the propeller is displaced outward of the hub. Returning to the nonrotating frame we see that a small outward displacement of the propeller mass as the engine-propeller system whirls on the mount springs requires a slightly *smaller* whirl frequency. This dipping of the upper resonance curve to the left of the crossover at $\Omega = \omega_0 + \omega_M$ is an indication that the curves attract one another and that the coupling is *nonconservative.* A similar argument applied to the crossover at $\Omega = \omega_0 - \omega_M$ indicates that the curves repel each other there and that the coupling is *conservative.* These heuristic results are verified by the coupled resonance curves, obtained from a complete analysis [5], shown in Fig. 7.

5 WHIRLING OF A ROTOR PARTIALLY FILLED WITH LIQUID

In 1962 Kollman [6] reported experiments which indicated that a rotor with a short cylindrical chamber partially filled with liquid is unstable for a range of supercritical speeds. Theoretical explanations were given independently by Kuipers [7] and Wolf [8]. A flexibly mounted rotor with a cylindrical cavity of radius

a and axial length L rotates at rate Ω. When partially filled with liquid the centrifugal force plasters the liquid against the cylindrical wall leaving a cylindrical hole or void of radius b. The system can execute small vibrations in which the rotor displaces transversely and surface waves propagate around the free surface of the fluid. Of particular interest are the steady state modes of free vibration in which the rotor center executes a circular whirl of radius r synchronized with a surface wave whose wavelength is exactly one circumference. The effect of such a wave is to cause a circular void of radius b to whirl around within the rotor at some whirl radius ϵ. When considered as a coupled vibratory system this system is quite similar to the engine-propeller system. The first subsystem is the rotor of mass m_r plus the rotating liquid annulus of mass $m_l = \pi\rho L(a^2 - b^2)$ which is temporarily restrained from moving from its centered position with respect to the rotor. If the effective stiffness of the rotor shaft and bearings is k the natural whirling frequency of the the first subsystem ω_M is given by

$$\omega_M^2 = k/(m_r + m_l) \tag{7}$$

The second subsystem is the liquid in a rotor which rotates at rate Ω but does not move transversely. Small motions of the fluid with free surface in the centrifugal field are conveniently studied in a rotating frame [5]. The subsystem has two natural whirling modes for the circular void: a forward whirl and a backward whirl. With respect to the rotating frame the frequencies of these whirls are given by

$$n = \Omega(1 \pm \sqrt{\gamma + 1})/\gamma \tag{8}$$

where $\gamma = (a^2 + b^2)/(a^2 - b^2)$. Note that in contrast to the propeller-engine system these frequencies depend linearly on the rotation rate. A convenient way to describe the effects of these fluid modes [9] is to consider the actual fluid to be the superposition of a total fluid mass $m_t = \pi\rho a^2$ which would fill the rotor plus a *negative* hole mass $m_h = \pi\rho b^2$ located at the void.

5.1 Nature of the coupling

When the whirl frequencies (7) and (8) are viewed from the same reference frame it is seen that there are crossovers at the following speeds

$$\Omega = \omega_M \left[\gamma/(\gamma + 1 \pm \sqrt{\gamma + 1})\right] \tag{9}$$

and the complete solution [5] shows that the coupling is conservative for the smaller of (9) and nonconservative for the large. Here we shall show how the qualitative nature of the coupling can be determined without appealing to the detailed equations. Consider for example the crossover for the larger of (9). Imagine an uncoupled whirl of the first subsystem for a slightly lower rotation rate. The rotor whirls at at its uncoupled frequency ω_M but the liquid remains centered in the rotor. Now introduce small coupling. The motion of the rotor excites a small fluid response at frequency ω_M. Viewed from the rotating frame the exciting whirl is a backwards whirl with frequency lower than the natural backward whirl frequency given by (8). This means that the responses of a positive mass would be in phase with the excitation. By treating the hole as a negative mass we conclude that in the coupled mode the hole center moves inward with respect to the whirling center of the rotor. Returning to the nonrotating frame we see that the effect of the inward migration of the hub is to move the center of mass of the rotor and the actual fluid slightly outward which requires a whirl frequency somewhat smaller than ω_M. This is an indication that the resonances are attracting one another in the neighborhood of the crossover and thus that the coupling is nonconservative at the crossover. Similar arguments on either side of each crossover give predictions of the qualitative nature of the coupling which are in complete agreement with the complete solution [5]. It is interesting to note that in the two rotordynamic examples considered herein the crossovers which result when the subsystem whirls in the rotating and nonrotating frames have the same sense, either both backward or both forward, the coupling is conservative while the crossovers which result from subsystem whirls of opposite sense in their respective frames are nonconservative. If this could be generalized to a wider class of problems it would provide an even simpler method for identifying the nature of dynamic coupling.

REFERENCES

[1] Adams, M.L., and Padovan, J., Insights into linearized rotordynamics. *J. Sound and Vibn.*, 1981, 76, 129–142.

[2] Crandall, S.H., and Dugundji, J., Forced backward whirling of aircraft propeller-engine systems. Proc. 2nd Intl. Conf. Vibs. Rotating Machinery, Cambridge, England, 1980, 265–270, I.Mech.E.

[3] Crandall, S.H., and Dugundji, J., Resonant whirling of aircraft propeller-engine systems. *J. Appl. Mechs.*, 1981, 48, 929–935.

[4] Coleman, R.P., and Feingold, A.M., Theory of self-excited mechanical oscillations of helicopter rotors with hinged blades. NACA TR-1351, 1958.

[5] Mroszczyk, J.W., Instability of dynamic systems based on uncoupled sub-system responses with applications to rotor dynamics. Ph.D. Thesis, 1986, Mechanical Engineering Department, M.I.T., Cambridge, Mass.

[6] Kollman, F.G., Experimentelle und theoretische Untersuchen über die Kritischen Drehzahlen flüssigkeits-gefüllter Hohlkörper. *Forschung auf dem Gebiete des Ingenieurwesens*, 1962, (4) 115–123 and (5) 147–153.

[7] Kuipers, M., On the stability of a flexibly mounted rotating cylinder partially filled with liquid. *Appl. Scientific Research*, 1964, A13, 121–137.

[8] Wolf, J.A.Jr., Whirl dynamics of a rotor partially filled with liquid. *J. Appl. Mechs.*, 1968, 35, 676–682.

[9] Crandall, S.H., The physical nature of rotor instability mechanisms. Rotor Dynamic Instability (M.L. Adams, ed.), 1983, ASME Special Publication AMD Vol. 55, 1–18.

$$K_1 + K_3 = m\omega_0^2 (1-\alpha)$$

$$K_2 + K_4 = m\omega_0^2 (1+\alpha)$$

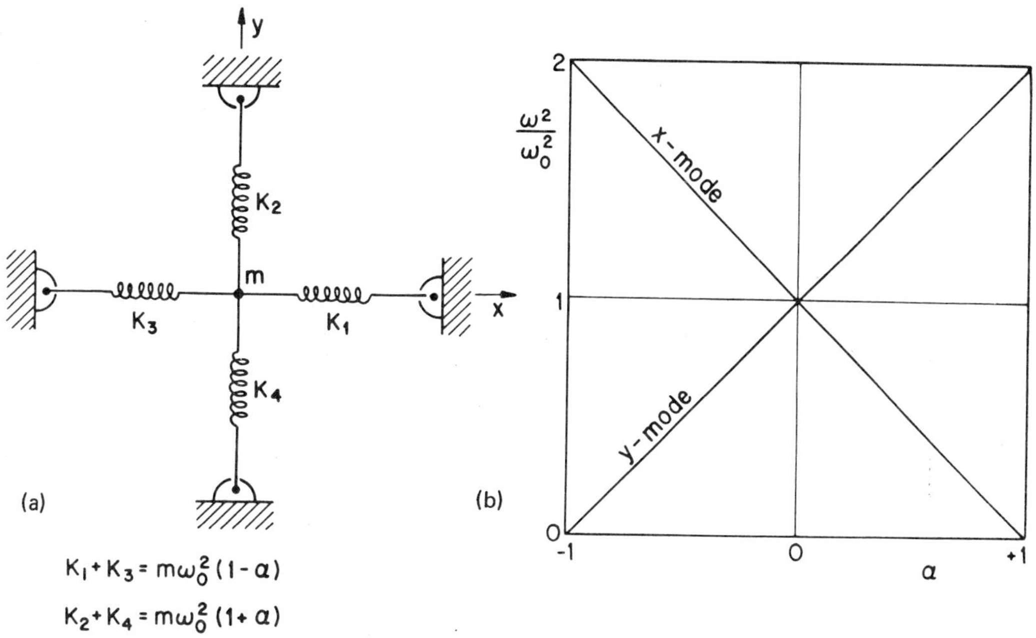

Fig 1   (a) Stiffness in χ- and γ-directions depend on tuning parameter α;
         (b) Natural frequencies ω of uncoupled modes

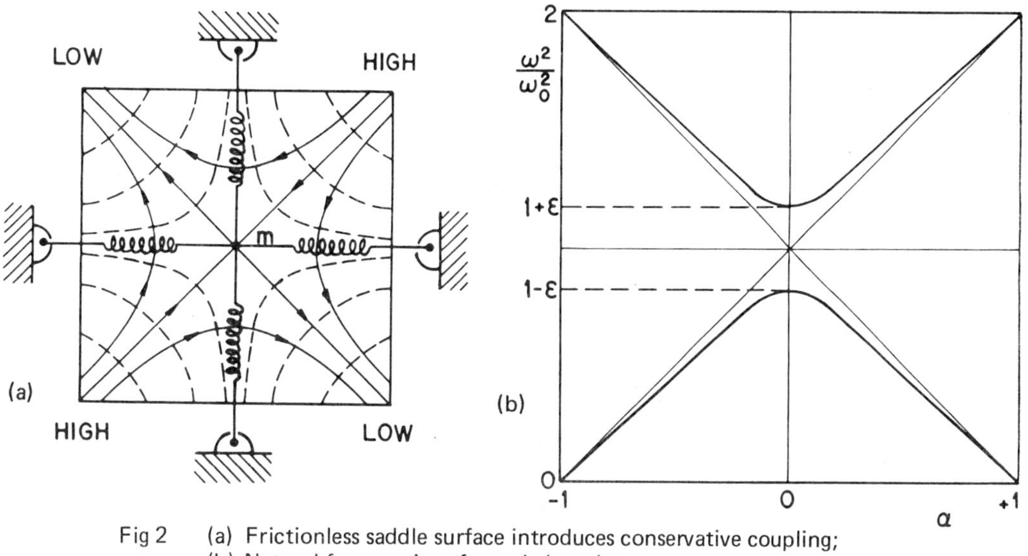

Fig 2   (a) Frictionless saddle surface introduces conservative coupling;
         (b) Natural frequencies of coupled modes

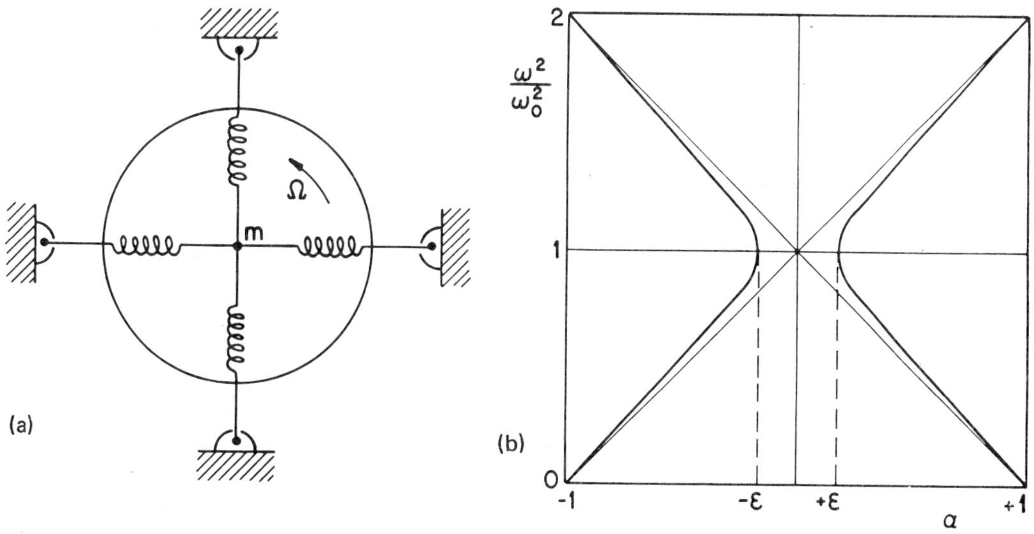

Fig 3   (a) Lubricated turntable introduces nonconservative coupling;
         (b) Natural frequencies of coupled modes

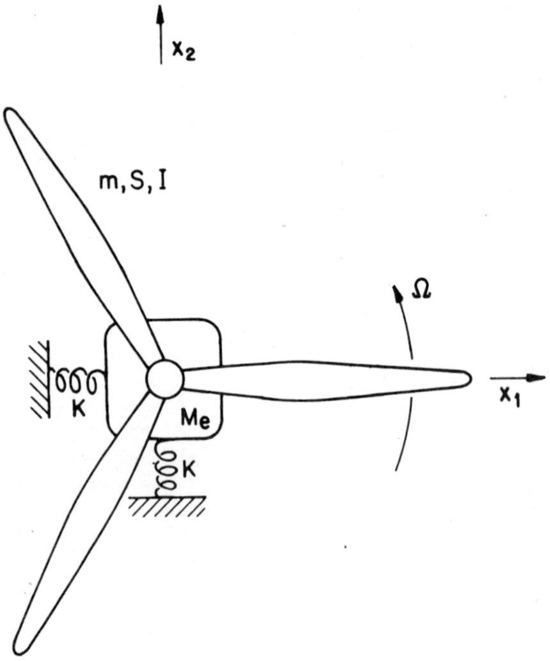

Fig 4 Engine with mass M_e supported on mounting springs drives propeller at steady rotation rate Ω

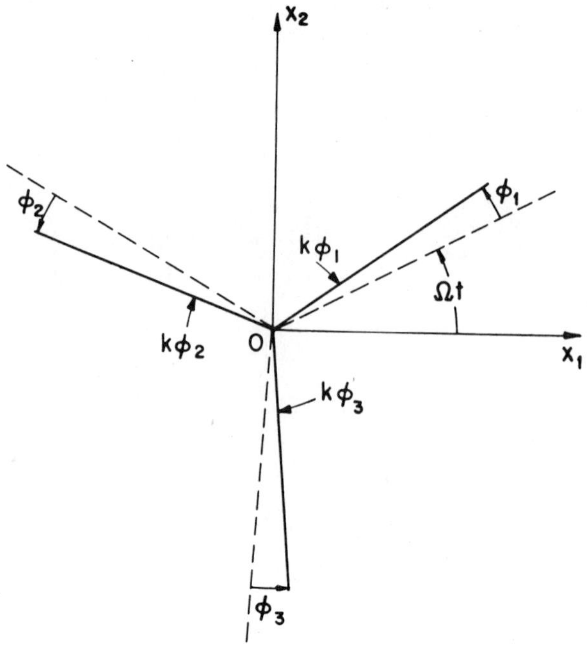

Fig 5 Propeller hub exerts elastic restoring torques $k\phi_i$ opposing the pendular displacements ϕ_i

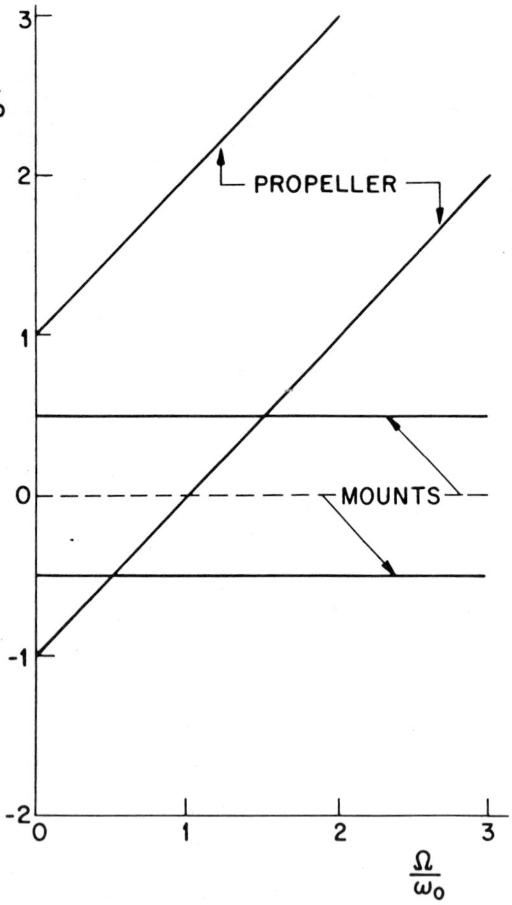

Fig 6 Uncoupled natural whirl frequencies with crossovers at $\Omega = w_o - w_M$ and $\Omega = w_o + w_M$

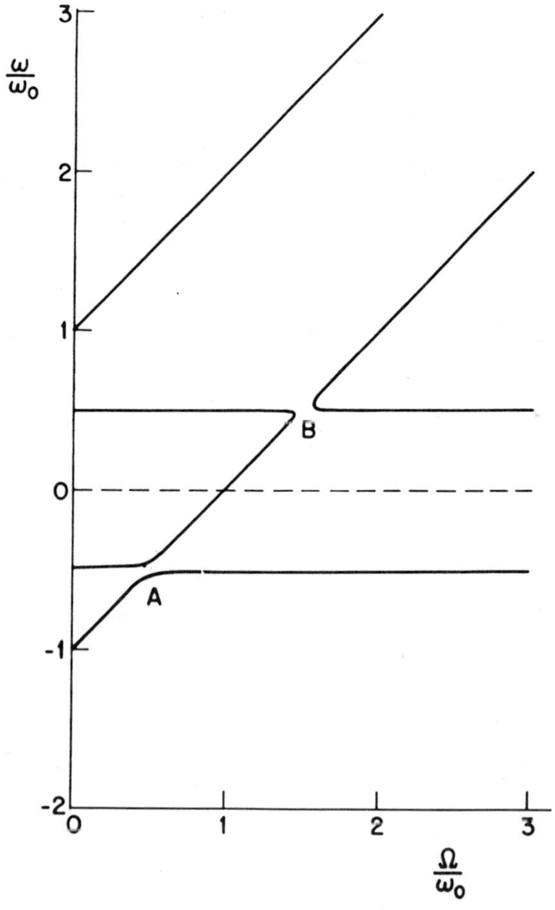

Fig 7 Coupled whirl frequencies with conservative coupling at A and nonconservative coupling at B

Effects of loose rotating parts on rotor dynamics

A MUSZYNSKA, PhD, MASME
Bently Rotor Dynamics Research Corporation, Minden, Nevada, United States of America

SYNOPSIS Dynamic response of a rotor with a loose rotating part (restricted, however, in axial direction) is discussed in this paper. Rotor first bending mode, shaft/loose part friction and external fluid drag are taken into consideration in the model. Results bring modified natural frequencies of the system, conditions of self-excited vibration inception and steady-state synchronous vibrations modulated in a 'beat' form. Experimental results illustrate analytical predictions.

NOTATION

a	Modal mass modification factor
A,B	Amplitudes of rotor responses
b	Rotor modal displacement factor
C_1,C_2	Constants of integration
D,K,M	Rotor generalized (modal) damping, stiffness and mass respectively
I	Polar moment of inertia of the loose element
$j=\sqrt{-1}$	
m	Modal unbalance mass
M_ℓ	Loose element mass
S	Normal force
r	Shaft radius
r_u	Modal unbalance radius
R	Loose element radius
t	Time
$u=(R-r)M_\ell/\nu$	
$z=x+jy$	Rotor lateral displacement
$z_1=x_1+jy_1$	Loose element lateral displacement
α,β	Phases of rotor responses
μ	Dry friction coefficient
ν	Fluid dynamic drag coefficient
ϑ	Rotation angle of loose element
$\psi=\psi(\omega-\dot{\vartheta})$	Friction function
ω	Rotor rotative speed
ω_ℓ	Loose element rotative speed
Ω	Complex eigenvalue
$\Omega_{1,2}$	System natural frequencies

1 INTRODUCTION

Looseness and eventual involvement in rotative motion of machine elements, such as disks or thrust collars mounted on rotating shafts, or such as bearings mounted in bearing pedestals, represent an important machine malfunction. A disconnected disk or a thrust collar will rotate at lower speed than that of the rotating shaft, and, may displace axially. A loose bearing may start rotating, dragged into rotative motion by rotating shaft.

The clearances, friction conditions between the shaft and the loose part, as well as loose part tangential drag (such as external aerodynamic drag), play an important role in the rotor dynamic response. While perturbing the normal operation of the machine the looseness-related dynamic phenomena can, however, be relatively easy to identify and eventually correct, as they cause very characteristic modifications of rotor responses. A loose rotating part usually carries an unbalance which changes the balance state of the machine. This results in a modification of synchronous vibrations. A loose rotating part generates additional subsynchronous self-excited vibrations of the rotor.

In spite of the fact that looseness appears as a fairly common malfunction of rotating machines [1] the literature on this subject is extremely scarce [2]. Some references are available on devices, which use loose parts for rotor automatic balancing purposes [3-6].

Dynamic responses of a rotor with a loose rotating part (fixed, however, in the axial direction) are investigated in the paper, both analytically and experimentally.

2 MATHEMATICAL MODEL

The physical model of a shaft rotating at a constant speed ω and carrying a loose element with the mass M_ℓ is illustrated in Fig. 1.

In the mathematical model of the system the first lateral mode of an isotropic rotor is taken into consideration:

$$aM\ddot{z}+D\dot{z}+Kz+S(1+j\psi)e^{j\vartheta} = mr_u\omega^2e^{j\omega t} \quad (1)$$

$$M_\ell\ddot{z}_1-S(1+j\psi)e^{j\vartheta} = 0, \quad (2)$$

$$I\ddot{\vartheta}+\nu\dot{\vartheta}^2r-rS\psi = 0, \quad (3)$$

$$z_1 = bz-(R-r)e^{j\vartheta}, \qquad aM+bM_\ell = M, \quad (4)$$

$$\psi = \psi(\omega-\dot{\vartheta}) \quad (5)$$

where

$$z = x+jy,\ z_1 = x_1+jy_1,\ ^{\bullet}=d/dt,\quad j = \sqrt{-1}.$$

Eqs. (1) and (2) describe lateral motion of the rotor and the loose element respectively. M, D, and K are generalized (modal) mass, viscous damping and stiffness of the correctly operating rotor (no looseness); mr_u is modal unbalance. I and M_ℓ are the polar moment of inertia and mass of the loose element. Nondimensional parameter 'a' indicates changes in modal mass due to the loose part involvement in separate rotation. Coefficient 'b' determines the amount of modal mass associated with the loose part, so that $aM+bM_\ell = M$. 'b' depends on the axial position of the loose part (Fig. 1). 'S' is the normal force at the shaft/loose element contacting surface, $\psi=\psi(\omega-\dot{\vartheta})$ is the fric tion function of the shaft/loose element relative angular velocity (for the simplest case ψ is constant, and equal to the dry friction coefficient, μ). It is assumed that the loose part motion occurs in a single plane (is restricted from axial displacement). '$\nu\dot{\vartheta}^2r$' is the external aerodynamic drag moment, proportional to the square of the loose part angular velocity $\dot{\vartheta}$. The coefficient 'ν' includes both pressure and surface drag effects, and it depends on the environment fluid density and loose part surface area exposed to the fluid dynamic drag.

Eq. (4) describes a geometric constraint: it is assumed that during rotation the shaft maintains contact with the loose element. Since only first lateral mode of the shaft is considered, 'bz' represents a fraction of displacement 'z', predicted from the shaft modal shape and axial location of the loose element (Fig. 1).

It is assumed that the loose part centre of gravity coincides with its geometric centre. It is easy, however, to alleviate this assumption and attach to the loose element a portion of the modal unbalance.

For the four unknowns z, z_1, ϑ, and S, Eqs. (1) to (4) can be transformed to the following form:

$$aM\ddot{z}+D\dot{z}+Kz+(I\ddot{\vartheta}/r+\nu\dot{\vartheta}^2)(1/\psi+j)e^{j\vartheta} = mr_u\omega^2e^{j\omega t} \qquad (5)$$

$$M_\ell b\ddot{z}-(R-r)M_\ell[\frac{d^2}{dt^2}(e^{j\vartheta})]-(I\ddot{\vartheta}/r+\nu\dot{\vartheta}^2)(1/\psi+j)e^{j\vartheta} = 0 \qquad (6)$$

$$S = (I\ddot{\vartheta}/r+\nu\dot{\vartheta}^2)/\psi \qquad (7)$$

Eqs. (5), (6), (7), and (4) represent the mathematical model of the system.

3 SOLUTION: LOOSE PART STEADY ROTATION AND FREE VIBRATIONS OF THE SYSTEM

Assume that the rotor is balanced, i.e., $mr_u=0$, and the solutions describing rotor free vibrations are of the form:

$$z = C_1e^{j\Omega t} \qquad \vartheta = \Omega t - j\, \ell n\, C_2 \qquad (8)$$

where C_1 and C_2 are constants of integration and Ω is the complex 'eigenvalue.' The solution (8) assumes that rotative frequency of the loose element is constant. Substituting (8) into Eqs (5) and (6) two algebraic equations are obtained:

$$\begin{aligned} &C_1(K-aM\Omega^2+j\Omega D) + C_2\nu\Omega^2(1/\psi+j) = 0 \\ &-C_1M_\ell b\Omega^2+C_2[(R-r)M_\ell\Omega^2-\nu\Omega^2(1/\psi+j)] = 0 \end{aligned} \qquad (9)$$

The necessary conditions for solutions (8) to exist for any initial condition is that the discriminant of (9) equals zero:

$$\begin{aligned} &(K-aM\Omega^2+j\Omega D)[(R-r)M_\ell\Omega^2-\nu\Omega^2(1/\psi+j)]+ \\ &+M_\ell b\Omega^2\nu\Omega^2(1/\psi+j)=0 \end{aligned} \qquad (10)$$

Eq. (10) resembles the characteristic equation for linear systems. Roots of this equation give complex eigenvalues Ω. Since external damping and restoring forces applied to the loose element were neglected in this study Eq. (10) has a double root Ω=0. Eq. (10) reduces, therefore, to:

$$(f_1-jf_2)\Omega^2-jD\Omega-K = 0 \qquad (11)$$

where

$$f_1 = bM_\ell \frac{\psi^2+1 - (R-r)M_\ell\psi/\nu}{\psi^2+[1-(R-r)M_\ell\psi/\nu]^2} + aM$$

$$f_2 = bM_\ell \frac{(R-r)M_\ell\psi^2/\nu}{\psi^2+[1-(R-r)M_\ell\psi/\nu]^2}$$

with two roots

$$\Omega = \frac{D(jf_1-f_2)}{2(f_1^2+f_2^2)} \pm \frac{1}{\sqrt{2}}\{\sqrt{f_3+\sqrt{f_3^2+f_4^2}} + j\sqrt{-f_3+\sqrt{f_3^2+f_4^2}}\} \qquad (12)$$

where

$$f_3 = \frac{D^2(f_2^2-f_1^2)}{4(f_1^2+f_2^2)^2} + \frac{Kf_1}{f_1^2+f_2^2}, \quad f_4 = \frac{Kf_2}{f_1^2+f_2^2} - \frac{D^2f_1f_2}{2(f_1^2+f_2^2)^2}$$

The functions f_1 and f_2 are illustrated in Fig. 2.

The complex roots (12) will have positive imaginary parts, i.e., they would assure stability of the corresponding solutions (8) if:

$$\frac{Df_1}{\sqrt{2}(f_1^2+f_2^2)} > \sqrt{-f_3+\sqrt{f_3^2+f_4^2}} \qquad (13)$$

which yields the condition

$$D/\sqrt{K} > f_2/\sqrt{f_1} \qquad (14)$$

Since both functions f_1 and f_2 depend on ψ which, in turn, is a function of the relative velocity $\omega-\Omega$, the condition (14) involves rotative speed. At the threshold of stability, i.e., when

$$D/\sqrt{K} = f_2/\sqrt{f_1} \qquad (15)$$

two roots (12) are real, and represent the system natural frequencies:

$$\Omega_{1,2} = -\frac{Df_2}{2(f_1^2+f_2^2)} \pm \sqrt{\left(\frac{Df_2}{2(f_1^2+f_2^2)}\right)^2 + \frac{Kf_1}{f_1^2+f_2^2}} \qquad (16)$$

or taking (15) into account

$$\Omega_{1,2} = -\frac{\sqrt{K/f_1}}{2(Kf_1+D^2)}[D^2 \mp \sqrt{D^4+4Kf_1(Kf_1+D^2)}] \qquad (17)$$

Note that this analysis yields only approximate results; the existence of zero roots of Eq. (10) requires consideration of nonlinear terms for the full analysis of the system stability.

The solution (8) describes free vibrations. After the stability threshold is exceeded (condition (14) violated) free vibration amplitudes start increasing, ending up in limit cycles of self-excited vibrations. The frequencies of these vibrations differ very little from the frequencies (17). Full analysis of the self-excited vibrations requires specific nonlinear functions to be included in the system model.

4 NATURAL FREQUENCIES FOR PARTICULAR CASES

4.1 Clearance negligibly small and/or low friction and/or high aerodynamic drag.

For R-r=0 or $\psi=0$ or $\nu=\infty$ functions f_1 and f_2 reduce to:

$$f_1 = bM_\ell + aM = M, \qquad f_2=0 \tag{18}$$

The natural frequencies (16) therefore, become $\Omega_{1,2} = \pm\sqrt{K/M}$ which is the natural frequency of the properly operating rotor (no looseness).

4.2 Rotor damping neglected.

If D=0 then the natural frequencies (16) reduce to:

$$\Omega_{1,2} = \pm\sqrt{K/f_1} \tag{19}$$

The positive value (19) versus ψ is illustrated in Fig. 3. For

$$\psi < \frac{1}{(R-r)M_\ell/\nu+1} \quad \text{and} \quad \psi > \frac{1}{(R-r)M_\ell/\nu-1}$$

the natural frequency increases. When

$$\frac{1}{(R-r)M_\ell/\nu+1} < \psi < \frac{1}{(R-r)M_\ell/\nu-1}$$

the natural frequency decreases. Note that ψ is a function of the relative angular velocity $\omega-\Omega$, thus the rotor natural frequency becomes a function of the rotative speed ω.

If ψ is constant (and equal to the dry friction coefficient μ), it can be seen in Fig. 3 that the natural frequency is a nonmonotonous function of the friction coefficient. In the realistic range of friction coefficient values ($\mu<0.2$), for high values of the parameter $u=(R-r)M_\ell/\nu$, at low μ the natural frequency Ω_1 increases. At higher μ it decreases. If 'u' is relatively low, only the increasing range of Ω_1 versus μ is meaningful.

5 CONSTANT NORMAL FORCE AND CONSTANT FRICTION FORCE

Eq. (3) can be easily solved assuming that S=const and $\psi(\omega-\dot{\vartheta})\equiv\mu$. The loose element angular velocity for an initial condition $\dot{\vartheta}(t_n)=\omega$ is as follows:

$$\dot{\vartheta}(t) = \sqrt{\frac{\mu S}{\nu}}\,\frac{\omega+\sqrt{\mu S/\nu} + (\omega-\sqrt{\mu S/\nu})e^{-\gamma(t-t_n)}}{\omega+\sqrt{\mu S/\nu} - (\omega-\sqrt{\mu S/\nu})e^{-\gamma(t-t_n)}} \tag{20}$$

where $\gamma = 2r\sqrt{\mu S\nu}/I$ and t_n is any time, $t_n \geq 0$.

The loose part angular velocity (20) as a function of time and rotative speed ω representing the initial condition is illustrated in Fig. 4. For any ω, ϑ asymptotically approaches the value $\sqrt{\mu S/\nu}$, resulting from a balance of friction and fluid dynamic drag forces.

6 EXPERIMENTAL RESULTS

Figs. 5 and 6 present spectrum cascade plots of the vertical vibration response of an experimental rotor illustrated in Fig. 7. Both spectra indicate existence of self-excited subsynchronous vibrations. Very similar responses of a centrifugal compressor rotor with a 'loose' oil ring seal was documented in [7].

Spectrum cascade in Fig. 6 shows that self-excited vibrations have lower frequency; the loose disk in the rotor rig was for this experiment equipped with light plastic blades which significantly increased the aerodynamic drag moment ('ν' higher).

As the analysis above indicated, the loose rotating part generates vibrations with rotative speed-dependent frequency. It is obvious that if this frequency coincides with the rotor natural frequency, vibrations are significantly amplified (Fig. 8).

Rotor subsynchronous vibrations due to loose rotating parts may be very easily confused with fluid-generated self-excited vibrations of 'oil whirl' and 'whip' type [8].

7 STEADY-STATE BEAT VIBRATIONS

For a constant angular speed of the loose part, Eqs. (1) to (4) yield the following equation:

$$M\ddot{z}+D\dot{z}+Kz=mr_u\omega^2e^{j\omega t}-M_\ell\omega_\ell^2(R-r)e^{j\omega_\ell t} \tag{21}$$

where, for each ω, the loose part angular velocity $\dot{\vartheta}=\omega_\ell$ can be calculated from (3) as a solution of equation:

$$\nu\dot{\vartheta}^2 = S\psi(\omega-\dot{\vartheta})$$

The loose part generates an additional unbalance exiting force.

The solution of Eq. (21) describing rotor forced vibrations due to combined unbalance is as follows:

$$z(t) = Ae^{j(\omega t+\alpha)} + Be^{j(\omega_\ell t+\beta)} \tag{22}$$

where the amplitudes and phases as functions of the corresponding frequencies are as follows:

$$A = \frac{mr_u\omega^2}{\sqrt{(K-M\omega^2)^2+D^2\omega^2}} \qquad \alpha = \arctan\frac{D\omega}{M\omega^2-K}$$

$$B = \frac{M_\ell\omega_\ell^2(R-r)}{\sqrt{(K-M\omega_\ell^2)^2+D^2\omega_\ell^2}} \qquad \beta = \arctan\frac{D\omega_\ell}{M\omega_\ell^2-K}$$

When the frequencies ω and ω_ℓ are close Eq. (22) can be transformed into a form:

$$z(t) \approx [A-B+2B\cos(\frac{(\omega-\omega_\ell)t+\alpha-\beta}{2})]e^{j(\omega t+\alpha)} \tag{23}$$

Eq. (23) indicates that the response amplitude is periodically modulated by a periodic function of frequency $(\omega-\omega_\ell)/2$. Since ω and ω_ℓ are close, their difference will be small, thus the period of modulation high. This type of vibrational response is known as 'beat.' Larger loose part unbalance produces higher modulations in the response amplitude. If B>A/3 then the response amplitude (22) periodically decreases to zero. Fig. 9 illustrates the beat vibration in the form of time base and orbital response. Note that modulation affects also the response phase. The Keyphasor (once-per-turn) mark, superimposed on the response, periodically oscillates.

In rotating machinery the beat vibrations at operating speed are often observed. They may be caused by various reasons (such as slip frequency in induction motors and generators or circulating cavitation bubbles in suction pumps). Most often, however, they are generated by loose rotating parts.

8 FINAL REMARKS

Dynamic phenomena occurring when the rotor of a rotating machine is carrying a loose part involved in rotational motion are discussed in this paper. A mathematical model of the rotor system is proposed. The model is limited to the isotropic shaft first bending mode. This model allows for determination of threshold of stability and the system natural frequencies for the loose part steady rotation. It also allows for steady-state solutions due to original and loose part-related unbalance. When the shaft and loose part rotative frequencies are close these solutions have a specific form of beat vibrations.

Experimental results confirm the analytical predictions. In particular, they show the existence of subsynchronous self-excited vibrations with frequencies depending on shaft rotative speed.

The presented model can certainly be extended. Further investigation should include specific nonlinearities, nonstationary rotation of the loose element, more shaft modes, effect of shaft angular acceleration and various unbalance distribution on the rotor response, effect of various external fluid drag forces, and effect of variable conditions of friction at the rotor/loose part surfaces.

REFERENCES

(1) Vibration Analysis on Sundstrand 5112 at NSMS, BNN Report 840409, 1984.

(2) MUSZYNSKA, A., Rotor Instability Due to Loose Rotating Part, Instability in Rotating Machinery, NASA Conf. Publ. 2409, Carson City, NV, 1985.

(3) DEN HARTOG, J. P., Mechanical Vibrations, McGraw-Hill Book Co. Inc., New York, 1956.

(4) THEARLE, E. L., Automatic Dynamic Balancers, Machine Design, v. 22, No. 9, 10, 11, 1950.

(5) PLAINEVAUX, J. E., Auto-équilibrage de pièces en Rotation, Revue Génerale de Mécanique, No. 76, 1955.

(6) DIMENTBERG, F. M., SHATALOV, K. T., GUSAROV, A. A., Kolebanya Mashin, (Machine Vibrations), Mashinostroyenye, Moskva 1964.

(7) KIRK, R. G., SIMPSON, M., Full Load Shop Testing of an 18000 HP Gas Turbine Driven Centrifugal Compressor for Offshore Platform Service: Evaluation of Rotor Dynamics Performance, Instability in Rotating Machinery, NASA Conf. Publ. 2409, 1985.

(8) MUSZYNSKA, A., Whirl and Whip -- Rotor/Bearing Stability Problems; Journal of Sound and Vibration, 110(3), 1986, 443-462.

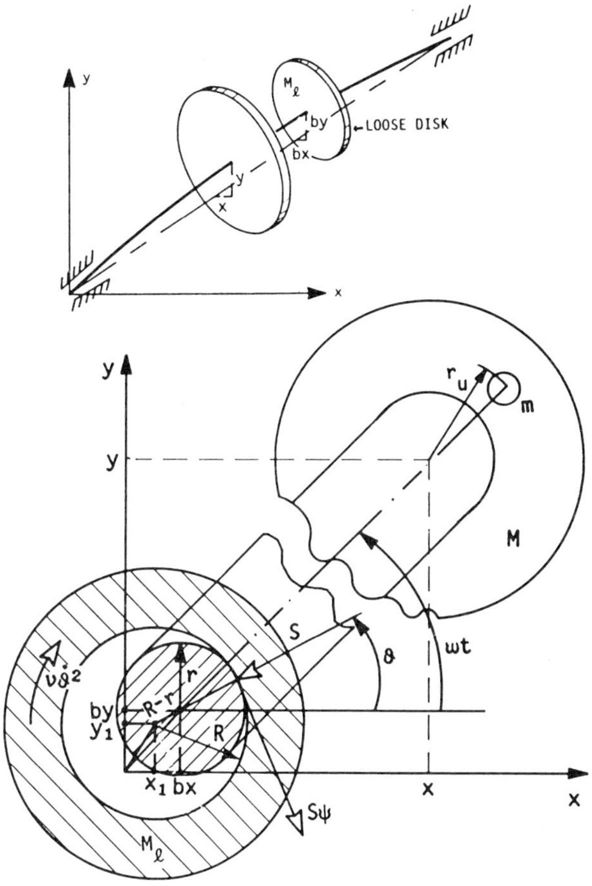

Fig 1 Physical model of the rotating shaft with a loose element of mass M_ℓ

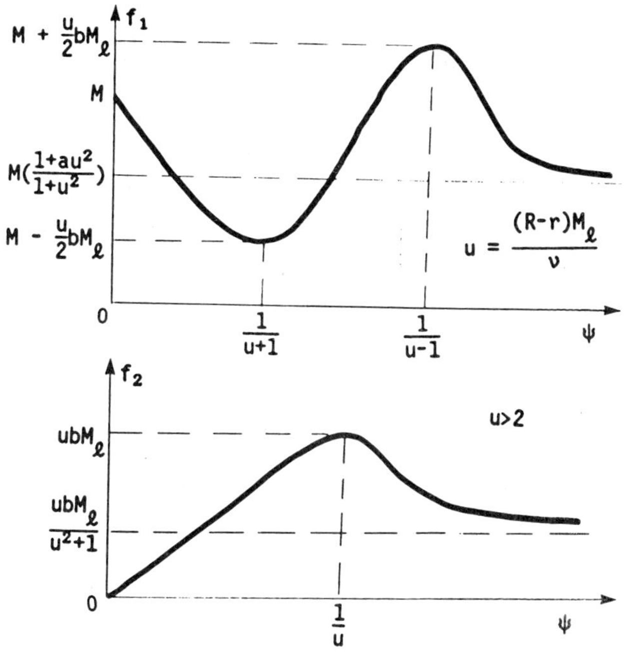

Fig 2 Functions of f_1 and f_2 versus friction function ψ

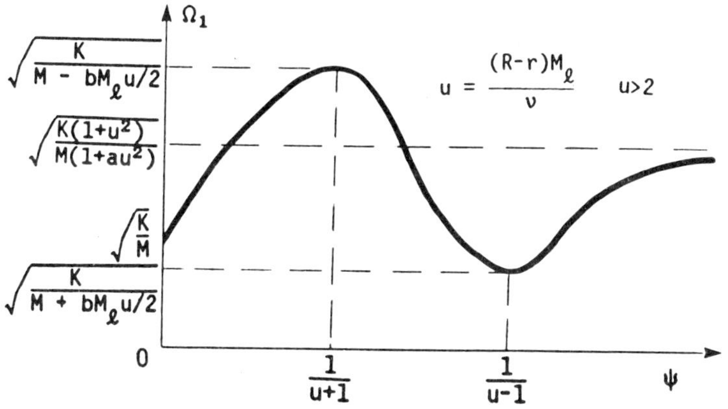

Fig 3 System natural frequency versus friction function ψ

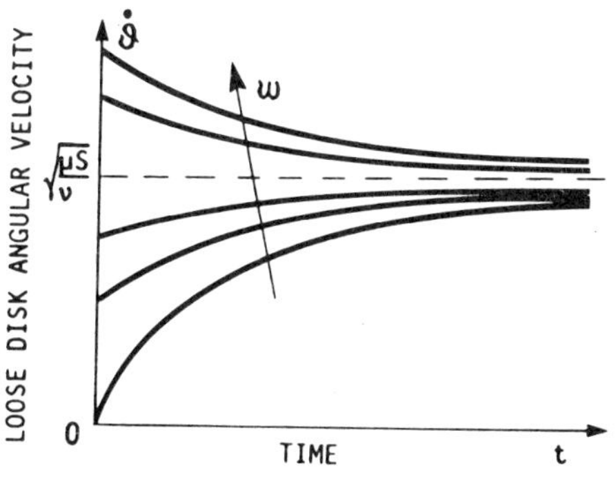

Fig 4 Loose part angular velocity versus time

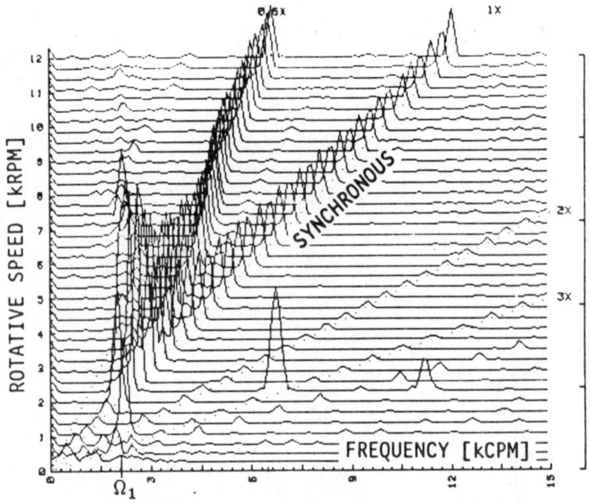

Fig 5 Spectrum cascade of vertical vibration response during start-up of an experimental rotor with a loose disc. Note the existence of subsynchronous self-excited vibrations with rotative speed-dependent frequency. Note also vibrations with frequency close to rotor natural frequency of the first bending mode. Higher harmonics appear due to non-linear effects in the system

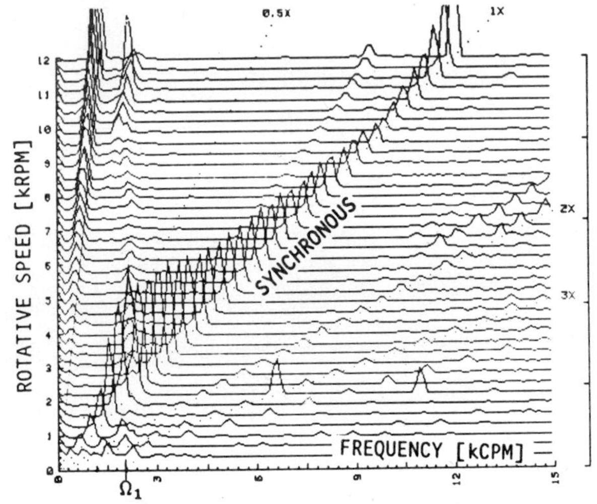

Fig 6 Spectrum cascade of vertical vibration response during start-up of the experimental rotor with a loose disc equipped with light plastic blades (increase of 'ν'). Note subsynchronous self-excited vibrations with the rotor natural frequency and with rotative speed-dependent frequencies, this time much lower than in the case illustrated in Fig 5

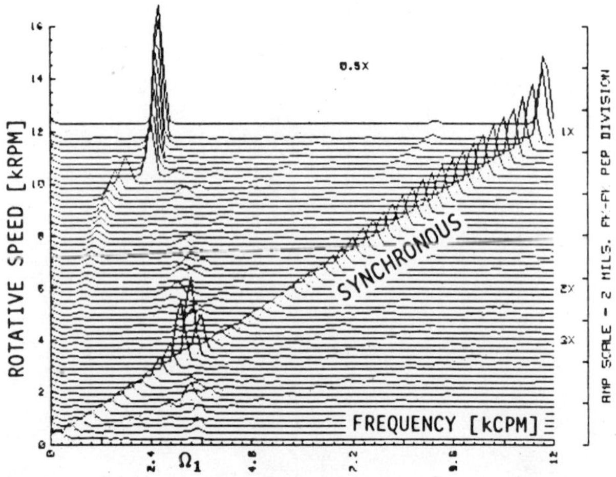

Fig 8 Spectrum cascade of vertical vibration response during start-up of the experimental rotor with a modified air drag effect

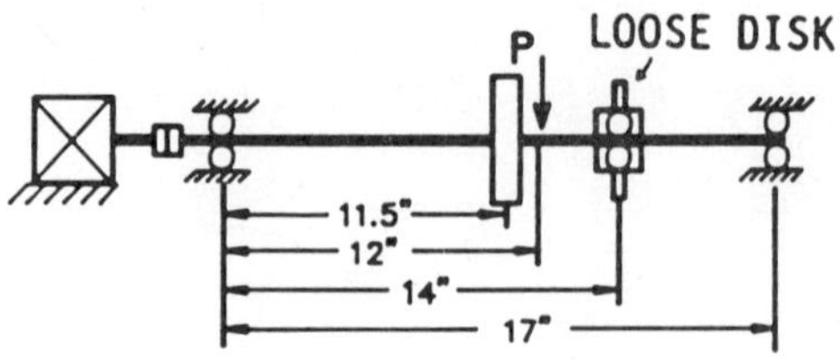

Fig 7 Experimental rotor rig. 'P' is the XY displacement probe location. The rig has been equipped with BRDRC data acquisition and processing system

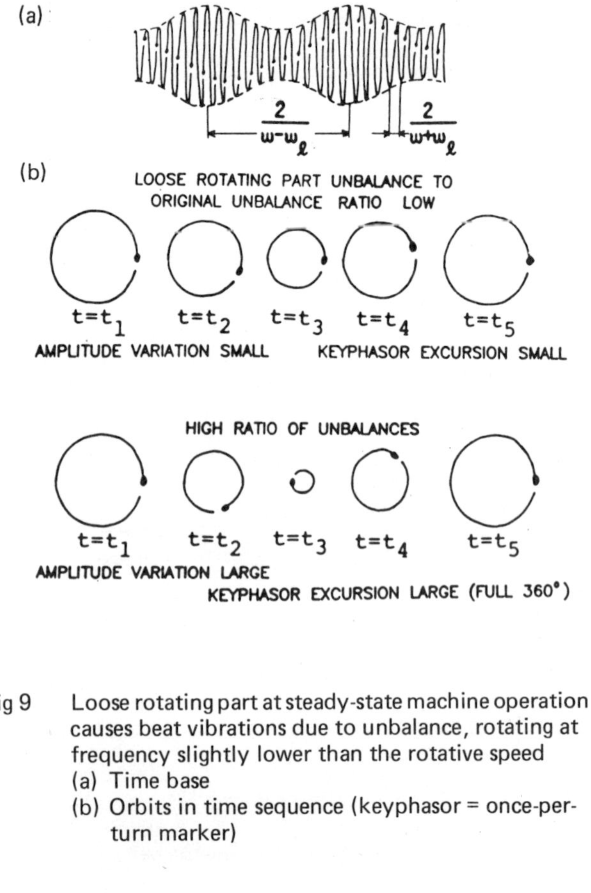

Fig 9 Loose rotating part at steady-state machine operation causes beat vibrations due to unbalance, rotating at frequency slightly lower than the rotative speed
(a) Time base
(b) Orbits in time sequence (keyphasor = once-per-turn marker)

C291/88

Vibration transmission characteristics of a marine diesel drive-shaft installation

R KINNS, MA, MASc, PhD, FIOA, CEng, FIMechE and **Z CHOWANIEC**, BSc, PhD, AMIMechE
Yard Limited, Glasgow

SYNOPSIS Diesel engines require a high degree of noise and vibration isolation for quiet ship installation. However, there are a multiplicity of different paths by which fluctuating forces can be transmitted to the ship structure. In the case of a geared propulsion diesel engine, these include the rotating drive shaft that transmits power from the engine to the gearbox. It is shown how two independent techniques for estimation of fluctuating force transmission via the drive shaft yield results that are in satisfactory agreement over an important frequency range.

1 INTRODUCTION

Modern ships often have to satisfy severe constraints on capital and operating costs while meeting demanding requirements for noise and vibration performance. This applies particularly to underwater noise, where the background levels due to a vessel's own machinery tend to limit the capability of that vessel to use its sonar systems to maximum effect, especially at low speeds.

Diesel engines are attractive in terms of capital cost and fuel economy. Possible installations include diesel generator sets that supply variable speed electric motors attached to the propeller shafts, and arrangements in which the engines drive propeller shafts via reduction gearing. However, diesel engines exhibit intrinsically high source vibration, which means that a high degree of isolation is required if they are to be used in quiet ships.

Fluctuating forces can be transmitted to the ship structure by a wide range of different transmission paths from a diesel engine. In the case of a geared propulsion engine, these paths include the rotating drive shaft between the engine and the reduction gearbox. The aim of this paper is to show how fluctuating forces transmitted via a typical drive shaft arrangement can be estimated in two independent ways that give satisfactory agreement. This agreement gives confidence that the effect of drive shaft transmission can be ranked correctly in the overall assessment of diesel isolation system performance, before a given layout is committed to detailed design and build.

1.1 Design of diesel isolation systems

In recent years, there has been considerable effort to improve understanding of the factors that govern the degree of vibration isolation that is possible using passive isolation systems (1). Such passive systems use at least one stage of soft mounts between the diesel engine and its foundation in the ship, in order to reduce levels of fluctuating forces applied to the ship structure over the frequency range of interest.

Two stage systems, using a set of soft mounts between the diesel engine and an intermediate raft and another set of soft mounts between this raft and the ship foundation, have also been developed for noise-critical installations. These are now common for fixed-speed diesel generator installations, where vibrational energy is concentrated at multiples of cycle repetition frequency (half the rotational frequency in the case of a four-stroke engine) and where resonances can be positioned in frequency away from significant disturbing forces. It has proved possible to reduce hull excitation via the mounting system to levels that are low enough to be negligible in many circumstances (1). The same applies to airborne excitation of the hull, once an acoustic enclosure has been specified for the diesel engine.

The source vibration characteristics of a fixed-speed diesel generator show only a weak dependence on power output. On the other hand, diesel source vibration shows a marked dependence on engine speed, which can be exploited in the case of propulsion diesel engines which drive propeller shafts via reduction gearing and whose speed can be varied in proportion to ship speed. The fall in vibration is typically 15 dB in most $^1/_3$-octave bands when the engine speed is reduced by a factor of 2.5. The same applies to airborne and fluid-borne noise. It is shown in (1) how inevitable coincidence between mounting system resonances with major disturbing frequencies in variable speed diesel installations can be made to be acceptable.

The fall in vibration as engine speed is reduced represents a key potential advantage for geared propulsion systems, which can only be realised in full if the degree of diesel vibration isolation that is possible with fixed speed diesel generators can be matched using a variable-speed geared propulsion arrangement. This depends critically on the isolation properties of the diesel drive shaft arrangement.

The gearbox itself is a source of hull excitation which is not present in diesel-electric systems and cannot be ignored in an overall assessment. Consideration of likely hull excitation due to the gearing and drive shaft between the diesel engine and gearbox led to the arrangement shown in Figure 1 for a recent ship design, which includes soft mounts for the gearbox.

1.2 Description of the diesel drive shaft installation

Figure 1 shows the connection of the double-mounted diesel engine to the single-mounted gearbox. The engine drive flange is attached directly to a torsional coupling which uses oil-damped metal springs to provide flexibility. The torsional coupling is connected in series with a flexible link coupling. These couplings are often used together in situations where the misalignment requirements for the installation exceed the allowable misalignment capability of the torsional coupling on its own. The combination is required here in order to cope with the relative movements that will arise in a seaway, as a result of using separate mounting systems for the diesel engine and gearbox.

The flexible link coupling is connected to a torque tube which drives the gearbox via a rubber block coupling. The combination of couplings was chosen to match the mechanical requirements for the installation while reducing the fluctuating forces that are transmitted to the gearbox to acceptable levels over the whole of the frequency range from 10 Hz upwards. The frequency range of greatest significance for an installation of this type, where the gearbox is itself isolated from the hull, extends up to about 100 Hz. This low frequency range therefore forms the central focus of the study.

1.3 Estimation of gearbox response from dynamometer measurements

It is only possible to obtain comprehensive data from a machinery installation of this complexity, when all components are available and assembled at a late stage in the build program. Nevertheless, it is vital to establish the gearbox vibration that is likely to occur in the ship due to excitation via the diesel drive shaft, in order to allow consideration of any changes from initial design that might be necessary to meet the overall ship requirements. The diesel installation and the drive shaft might be run on a test bed using a dynamometer to absorb the power of the engine. The problem then is to develop a method for estimation of gearbox vibration from feasible measurements.

Measured dynamometer vibration is not likely to correspond closely to the vibration that will be observed on the gearbox, partly because of the influence of self-induced gearbox and dynamometer vibration and partly because of differences in the forced response characteristics of the gearbox and dynamometer. However, fluctuating force levels transmitted at multiples of cycle repetition frequency might be estimated during dynamometer operation and used to predict gearbox vibration behaviour.

In order to determine the acoustical acceptability of the drive shaft arrangement shown in Figure 1, it was necessary to evaluate the dynamic properties of the gearbox and its rotating elements. First, the vibration response of the gearbox to excitation by given force levels in different directions was estimated. Secondly, permissible fluctuating force levels were derived, from consideration of overall ship acoustic requirements. Considerable work was then undertaken to explore the influence of the gearbox bearing oil films on these transmitted forces. Thus, it was possible to assess the extent to which forces transmitted via the roller bearings of a test dynamometer might be modified in changing to plain journal bearings in the ship. The emphasis in this paper is on the estimation of fluctuating forces transmitted to the dynamometer during factory tests, as a critical part of the prediction of overall ship acoustic behaviour.

2 DETERMINATION OF FLUCTUATING FORCE TRANSMISSION

The fluctuating forces transmitted via the drive shaft from the diesel engine to the test dynamometer were estimated in two independent ways. First, the forces were estimated entirely from measurements made on the test dynamometer. This required shaker tests on the dynamometer, in order to determine a number of transfer functions relating vibration levels at various points around the dynamometer to the forces acting at the dynamometer bearings. These transfer functions, used in conjunction with the measured vibration levels on the dynamometer with the engine on test, allowed an estimate of fluctuating forces transmitted to the dynamometer to be made. It is important to appreciate, however, that the forces derived in this way may not be due entirely to **transmission** via the drive shaft, but might include components due to unbalance of the rotating assembly, misalignment of the drive shaft and dynamic forces generated within the dynamometer.

Secondly, the fluctuating forces were estimated using a mathematical model of the drive shaft arrangement and measurements of diesel source vibration. It was assumed for this purpose that the vibration of the diesel drive flange was the same as the power average vibration above-mounts on the diesel engine in each direction. It was further assumed that the dynamometer acted as a high impedance termination for the drive shaft assembly. The dynamic properties of the drive shaft will be described in terms of its **apparent mass**; that is, the ratio of transmitted fluctuating force to input acceleration at the drive flange. Necessarily, force estimates derived in this way do not include any influence from misalignment, unbalance or dynamometer behaviour. They must therefore be expected to be approximate lower bounds on fluctuating forces that will arise in a real installation.

In order to allow a compact description of the two types of force estimate, estimates derived via shaker trials on the dynamometer will be called **measured**, while those derived via the mathematical model of the drive shaft will be called **predicted.** It is to be understood that measured forces are actually indirect estimates, obtained via measurements of vibration and transfer functions. The transfer functions themselves are specified in terms of **inertance**; that is, the ratio of induced acceleration to applied force.

3 EXPERIMENTAL DETERMINATION OF FORCES TRANSMITTED BY DIESEL DRIVE SHAFT

3.1 Description of test arrangement

In the test installation, the diesel engine was connected to a water brake dynamometer via the drive shaft assembly. The engine was double mounted in exactly the same way as intended for the shipboard installation, but the dynamometer was clamped to a low, steel framework which was bolted firmly to the concrete floor. The purpose of the framework was solely to raise the centreline of the dynamometer rotor in order to bring it into alignment with the engine crankshaft.

The dynamometer rotor was supported on one rolling element bearing at each end of the rotor. These bearings were expected to act like pinned joints at frequencies of interest, so that no significant moments would be transmitted to the dynamometer structure at either bearing. Figure 2 shows a schematic drawing of the test dynamometer and shows the excitation (shaker) and response (accelerometer) positions in the vicinity of the aft bearing. The connection between the rubber block coupling and the dynamometer necessitated the use of a stub shaft. Consequently, an additional (oil lubricated) pedestal bearing was installed as shown in the drawing, in order to avoid overloading the aft dynamometer bearing.

Access to the temporary pedestal bearing was very restricted, and as the bearing was not fitted with oil seals, any accelerometer located close to it would have been subject to serious oil contamination, thereby jeopardising the quality of the accelerometer signals. However, it was thought likely that fluctuating forces applied to the pedestal bearing would have a similar effect on dynamometer vibration to the same forces applied to the aft dynamometer bearing, because of their proximity and attachment to the same foundation structure. Vibration measurements and the necessary shaker tests were therefore conducted on the aft dynamometer bearing.

3.2 Measurement procedure

A technique for the determination of bearing forces within a rotating machine is described in (2), and forms the basis of the method by which fluctuating forces transmitted to the dynamometer via the diesel drive shaft were estimated. Essentially, the technique requires a number of shaker trials to be conducted, so that the vibration response at the bearing caps can be deduced for a known input force at several positions around the machine. Usually, the number of shaker trials should be at least 1.5 times the number of bearing forces to be determined, if the errors in the force estimates are to be kept within acceptable limits.

Shaker trial results are then interpreted reciprocally, i.e. as though the forces had been applied to the bearing caps and the responses measured around the machine (at what were the original shaker positions). These shaker trials should be conducted with the machine running, unless it can be shown that the effect of machine operation on transfer functions is negligible. This permits the bearing forces to be deduced from straightforward vibration measurements made at the original shaker positions, when the machine is running under the required operating conditions. A multiple regression technique, equivalent to an ordinary least squares procedure, is used to minimise errors in the magnitude and relative phase of bearing force estimates, wherever possible. In the event, the trials and analysis had to be simplified, but the basic analysis framework was exploited as far as possible.

It had been planned to conduct all shaker trials with the dynamometer running in strict accordance with the analysis techniques described in (2). However, the available time had to be cut dramatically at the time of measurement owing to circumstances outside the control of the trials team. Thus, it was decided to conduct most of the shaker trials with the dynamometer stopped in order to minimise background noise and allow shaker trials to be conducted rapidly by judicious use of fast sweep and random excitation techniques. A few check measurements were made with the dynamometer in operation using tonal excitation at selected frequencies. These indicated that although there were detailed changes in transfer functions when the dynamometer was rotating, they would not be expected to cause gross errors in fluctuating force estimates. The transfer inertance results reported in this paper were all obtained with the dynamometer stopped, using swept sine excitation techniques.

Figure 3 shows the magnitude and phase of the point inertance function in the athwartships direction, measured on the dynamometer base, adjacent to the aft bearing, and the transfer inertance in the same direction to the response position on the aft bearing (see Figure 2). The phase of the point inertance lies in the range 0-180°, as indeed it must, as energy is being supplied to the structure rather than extracted from it.

In this case, the phase of the transfer inertance differs uniformly from that of the point inertance by 180° simply because the response location was on the opposite side of the dynamometer to the excitation.

The magnitudes of the point and transfer inertances shown in Figure 3 are very similar up to about 140 Hz and distinct resonances at 20 and 28 Hz are clearly apparent in both functions. The similarity between the point and transfer inertances suggests that these resonances are in fact whole-body resonances of the dynamometer on its foundation. Figure 4 shows a similar pair of results for vertical excitation on the base, the vertical responses being measured at the point of application and at the bearing cap (Figure 2). Again, the point and transfer inertances are very similar in magnitude and phase. It is not surprising, in the light of these results, that the dynamometer vibration levels, up to about 140 Hz, are similar at the bearing and base locations when the dynamometer is operating.

The **transfer** functions shown in Figures 3 and 4 are just two from a whole array of functions that could be deduced from available recordings. Inspection of transfer function behaviour suggested that useful estimates of the net fluctuating forces that are transmitted to the dynamometer bearings via the drive shaft, when the engine is running, might be derived by assuming that all athwartships and vertical vibration on the dynamometer base is due to athwartships and vertical excitation respectively at the aft dynamometer bearing.

In reality, forces will be transmitted via the two dynamometer bearings and the additional pedestal bearing. Also, the excitation of the dynamometer will consist of forces in each of three directions and moments about each of the three axes (resulting from the **forces** acting at the bearings). In addition, all of these components will be cross-coupled to some extent. However, it is the net forces applied to the gearbox in the athwartships and vertical directions that are expected to have the greatest influence on ship acoustic performance. Thus, it is useful to think in terms of equivalent net forces in the athwartships and vertical directions that best account for the observed vibration.

3.3 Derivation of force estimates

In order to derive the required estimates, the transfer functions shown in Figures 3 and 4 must be interpreted reciprocally, i.e. as though the forces were applied to the bearing caps and the vibration levels were measured on the dynamometer base. It has to be recognised, however, that the detailed variations with frequency shown in Figures 3 and 4 are likely to be modified when the dynamometer is rotating. The measured transfer functions will also vary with precise accelerometer and force gauge locations. For these reasons, the transfer functions were smoothed by power averaging the detailed functions in $^{1}/3$-octave bands. The results of this smoothing are shown in Figure 5.

The narrowband vibration spectra measured in the athwartships and vertical directions on the dynamometer base when the diesel engine was running at a rotational frequency of about 15 Hz are shown in Figure 6. The observed tonals occur at multiples of cycle repetition frequency. The tonal at 192.5 Hz is known to be due primarily to the water brake, and not transmission via the drive shaft. For each direction, dividing each engine-related tonal acceleration by the value of the appropriate $^{1}/3$-octave band transfer inertance gave the corresponding bearing force component at that tonal frequency. Integrating these tonal forces in $^{1}/3$-octave bands gave the force estimates shown in Figure 7. Much lower force levels arise at lower diesel speeds, in accord with the vibration characteristics of the engine.

No estimate was made in the 25 Hz $^{1}/3$-octave band because this band did not contain any multiple of cycle repetition frequency at the given engine speed. No results are shown in $^{1}/3$-octave bands above 125 Hz because dynamometer vibration at multiples of cycle repetition frequency was either too low to be significant or could be attributed to operation of the dynamometer. It is evident from Figure 7 that the fluctuating force levels in the two directions appear to be similar. The most significant forces are in the 63 Hz and 80 Hz $^{1}/3$-octave bands. The high force levels in the 16 Hz $^{1}/3$-octave band are due to unbalance in the rotating assembly, which was subsequently reduced by trim balancing of the drive shaft arrangement.

4 NUMERICAL DETERMINATION OF FORCES TRANSMITTED BY DIESEL DRIVE SHAFT

4.1 Description of drive shaft numerical model

The transfer apparent mass functions for the diesel drive shaft in the athwartships and vertical directions were predicted from a numerical model of the drive shaft, the components of which are shown in Figure 8. The flexible elements of the torsional and flexible link couplings and of the rubber-block coupling were represented as light spring elements having specific values of radial and rotational stiffness and damping. The remaining components of the model were point mass elements having specific values of mass and transverse inertia. The model assumes a guided end constraint at the point of excitation and a fixed (rigid) end constraint at its termination, these being the conditions most closely resembling those believed to occur in practice.

The only difference between the athwartships and vertical models was in the value of damping associated with the radial stiffness of the torsional coupling. It arose because this coupling was believed to behave like a squeeze film damper when there was any relative radial movement between its inner and outer members, the extent of the damping being different in the athwartships and vertical directions.

Because the flexible elements were represented as light spring elements, the masses of these components being apportioned equally to adjacent point mass elements, wave effects occurring in rubber elements could not be reproduced by this model. Consequently, the calculated apparent mass functions were underestimated at frequencies in excess of about 100 Hz, at which the first wave effect was expected to occur. However, the main interest was in dynamic behaviour at low frequencies, because of the anticipated characteristics of the gearbox and its isolation system in the ship.

Figure 9 shows the expected transfer effective mass functions in the athwartships and vertical directions, relating the force that would be transmitted to the dynamometer to a prescribed vibration acceleration level at the diesel engine drive flange. The resonances occurring at 56 and 86 Hz represent translational and rotational modes respectively of the drive shaft torque tube and connected masses. Figure 10 shows the same transfer functions in $^1/3$-octave band format.

4.2 Derivation of force estimates

To calculate the fluctuating forces transmitted by the drive shaft, the diesel engine drive flange vibration was required. Unfortunately, this could not be measured directly, so it was decided to make the assumption that the drive flange vibration would be similar to the power average above-mount vibration, in the athwartships and vertical directions. This is believed to be plausible, because the engine crankshaft bearing oil films are thought likely to constrain the crankshaft vibration to be close to that of the crankcase at frequencies in excess of two or three times rotational frequency.

Figure 11 shows the $^1/3$-octave band power average vibration measured above mounts on the diesel engine in the athwartships and vertical directions, for the same engine operating conditions that were used for the dynamometer tests. (The engine vibration data and dynamometer vibration data were acquired at different times.)

Multiplying the $^1/3$-octave band engine acceleration by the appropriate $^1/3$-octave band transfer apparent mass gave the required force levels transmitted by the drive shaft. The results of this calculation are shown in Figures 12 and 13 for the athwartships and vertical directions respectively. Also plotted are the corresponding estimates derived from measurements made on the dynamometer. Note that, in addition to the assumptions made already, the fluctuating force estimates derived from the numerical model can only be precise if the dynamometer bearings act as a rigid constraint on the drive shaft. Otherwise, the assumption of a fixed end constraint at the termination point of the model is a convenient approximation.

5 CONCLUSIONS

Precise agreement between the predicted and measured force estimates in Figures 12 and 13 should not be expected. The main reason for this is the exclusion of any effects due to unbalance, misalignment or dynamometer excitation in the predicted force estimates. This applies particularly to the 16 Hz $^1/3$-octave band, which contains rotational frequency. It is known that the measured levels were governed primarily by shaft unbalance at the time of the trials, because substantial vibration reductions in the 16 Hz $^1/3$-octave band were achieved by subsequent trim balancing.

Despite the several assumptions and approximations that had to be made in deriving the two independent fluctuating force estimates at each frequency, there is an encouraging degree of agreement between the two sets of data in Figures 12 and 13, in the important frequency range covering $^1/3$-octave bands from 30 Hz to 100 Hz. It appears from this agreement that a relatively simple numerical model of the drive shaft, combined with diesel vibration data, can be used to estimate **transmission** of fluctuating forces to the gearbox. However, components due to shaft **excitation** appear to be dominant at lower frequencies. Both mechanisms of gearbox vibration excitation must be considered in consideration of any proposed drive shaft arrangement.

ACKNOWLEDGEMENT

The authors are grateful for the collaboration and support of their clients and colleagues, in carrying out extensive theoretical and experimental work on a short timescale.

REFERENCES

(1) KINNS, R. Some Observation on the Achievable Properties of Diesel Isolation Systems. Proceedings of ISSA'86 on Shipboard Acoustics, The Hague, 7-9 October 1986, pp265-278 (Martinus Nijhoff Publications, Dordrecht).

(2) KINNS, R. The Deduction of Bearing Forces in Rotating Machinery. Proceedings of Euromech Colloquium 122 on the Numerical Analysis of the Dynamics of Ship Structures, Ecole Polytechnique, Paris, 3-5 September 1979, pp345-361 (ATMA, Paris).

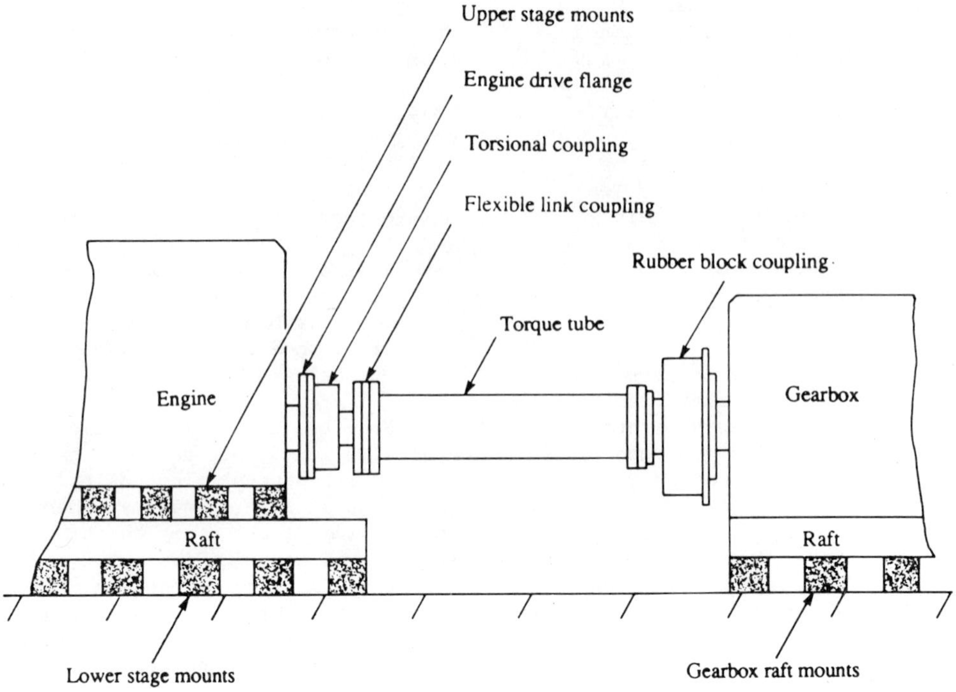

Fig 1 Diesel driveshaft installation

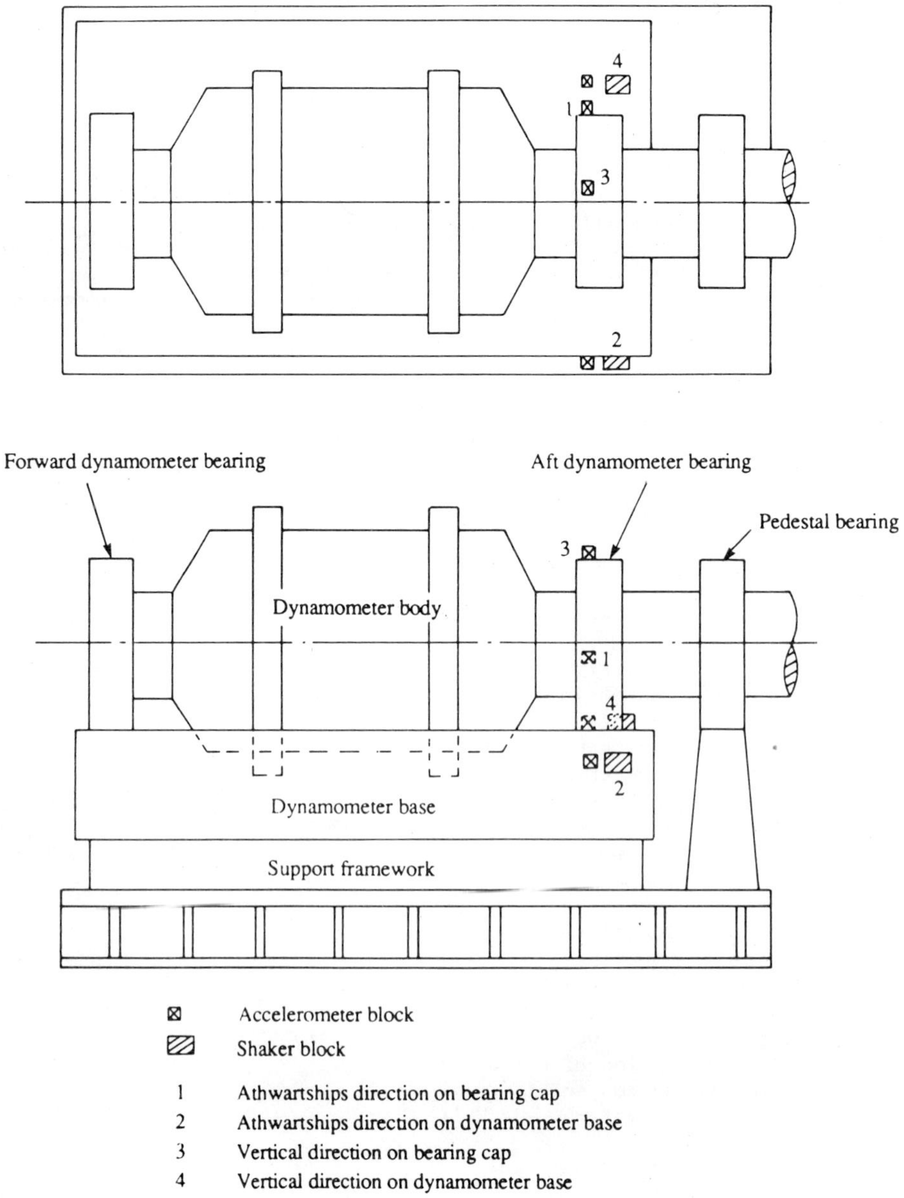

Fig 2 Dynamometer installation showing measurement and excitation positions

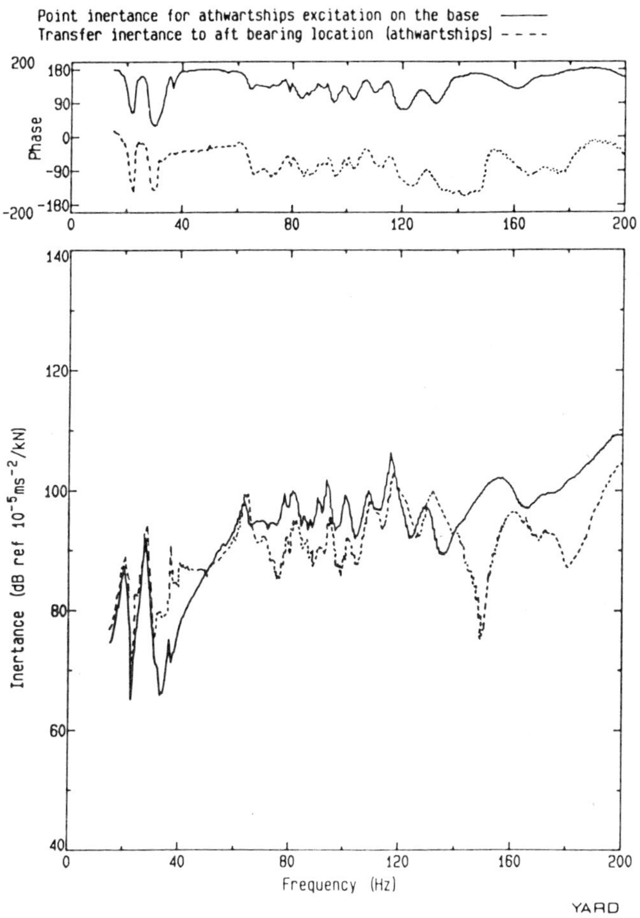

Fig 3 Point and transfer inertances measured on the dynamometer base in the athwartships direction

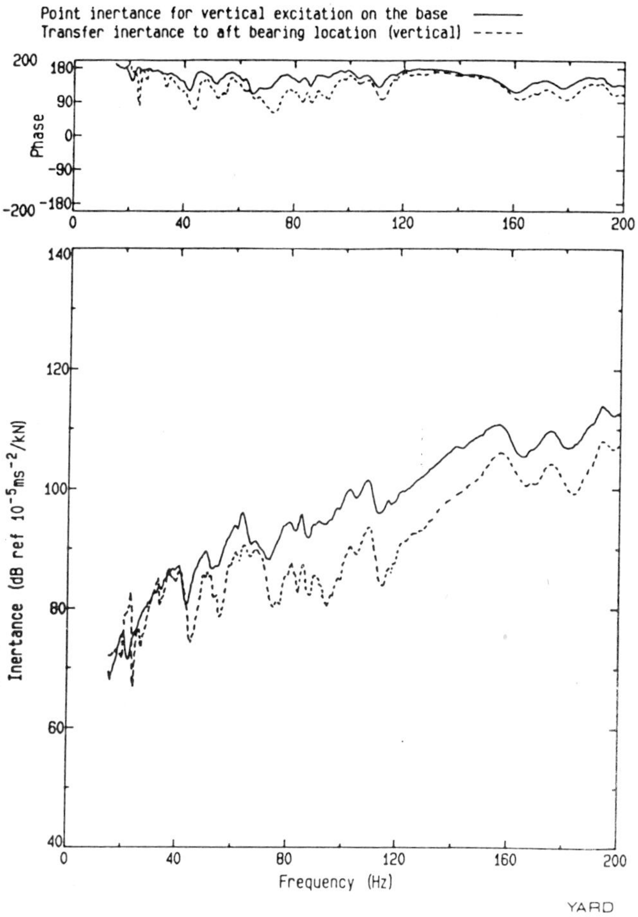

Fig 4 Point and transfer inertances measured on the dynamometer base in the vertical direction

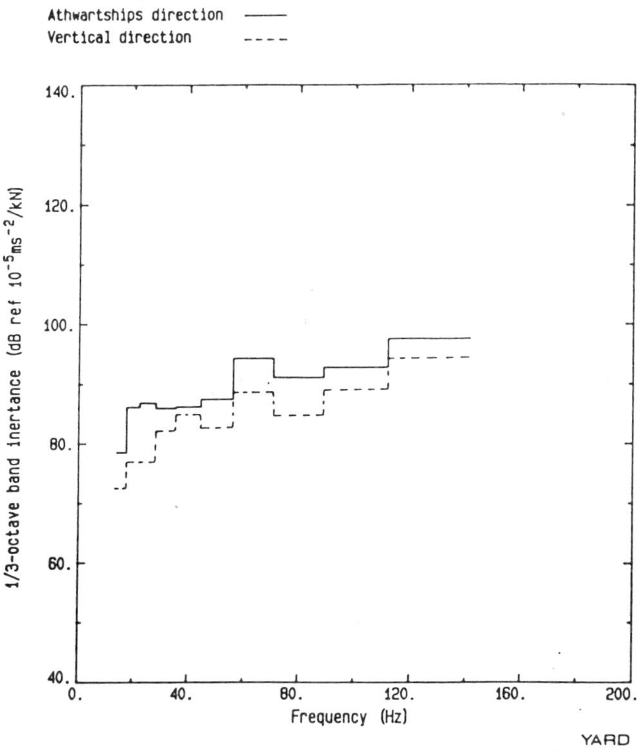

Fig 5 Measured transfer inertances in one third-octave bands showing response on dynamometer base resulting from excitation at dynamometer aft bearing cap

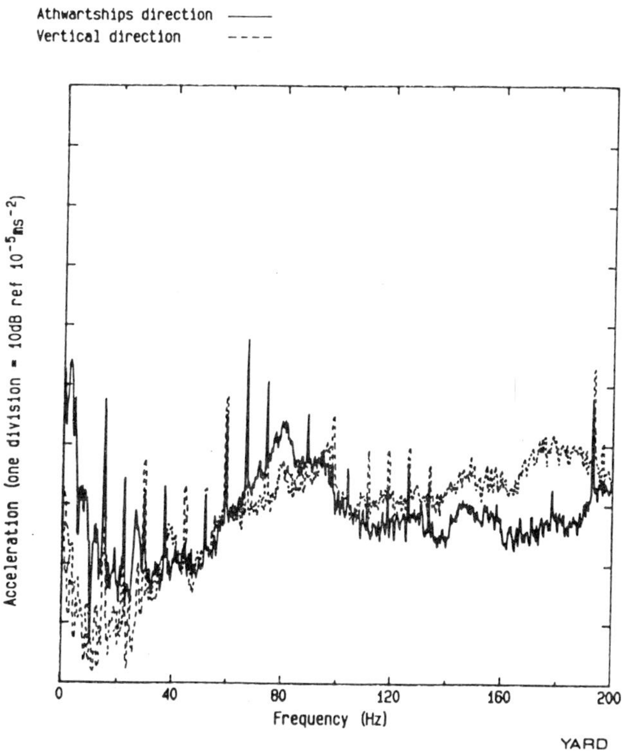

Fig 6 Typical vibration spectra measured on dynamometer base

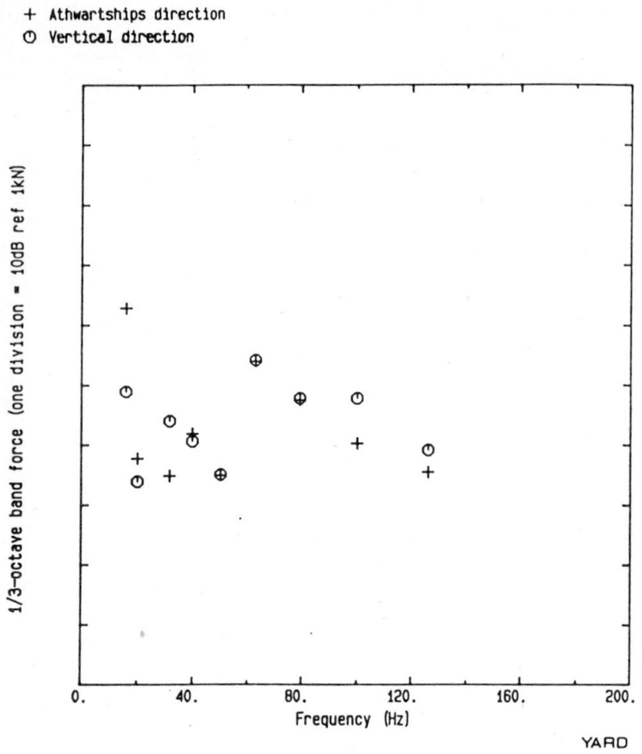

Fig 7 One third-octave band forces derived from shaker trials and vibration measurements on the test dynamometer

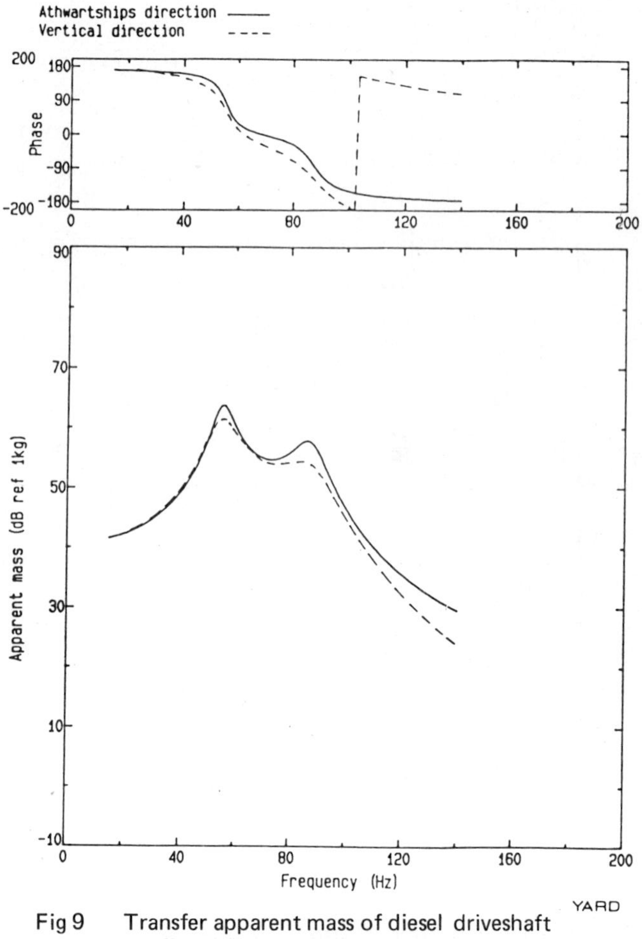

Fig 9 Transfer apparent mass of diesel driveshaft predicted from numerical models

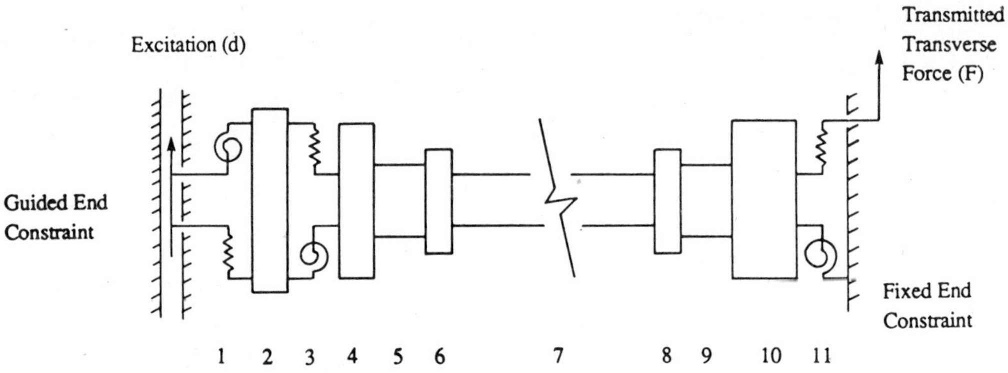

1 Torsional coupling flexible elements
2 Torsional/flexible link coupling inner members
3 Flexible link coupling flexible elements
4 Flexible link coupling outer member
5 Light, stiff beam element
6 Torque tube flange
7 Torque tube
8 Torque tube flange and watertight cover
9 Light, stiff beam element
10 Rubber block coupling inner member
11 Rubber block coupling flexible elements

Fig 8 Numerical model of diesel drive-shaft in athwartships and vertical directions

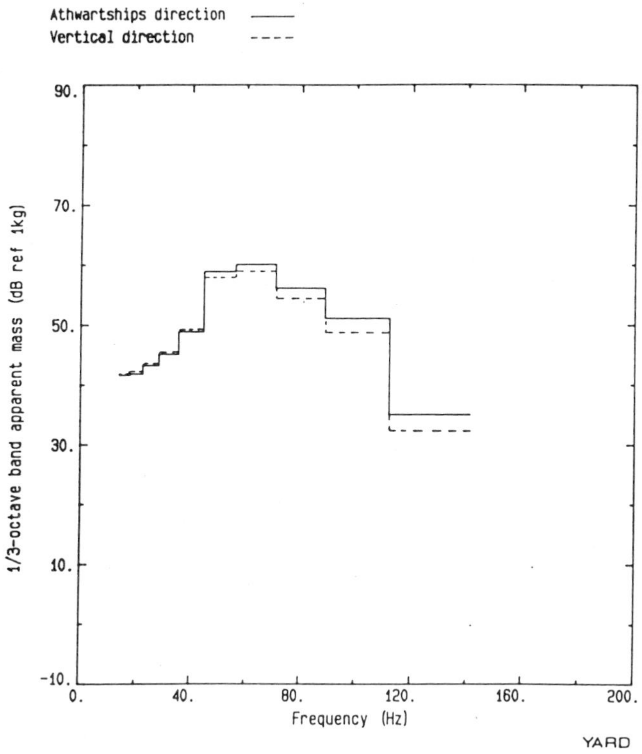

Fig 10 One third-octave band transfer apparent mass of diesel drive-shaft predicted from numerical models

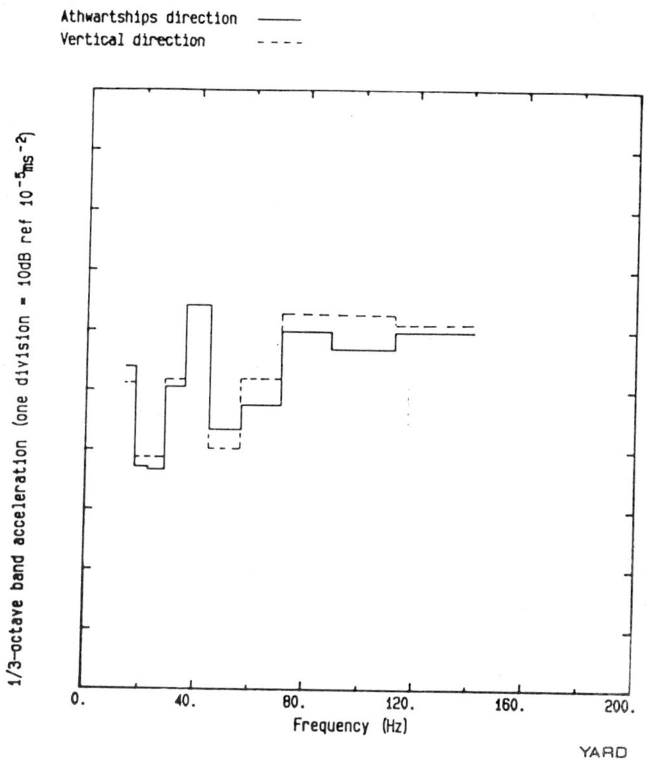

Fig 11 One third-octave band power average vibration measured above mounts on diesel engine

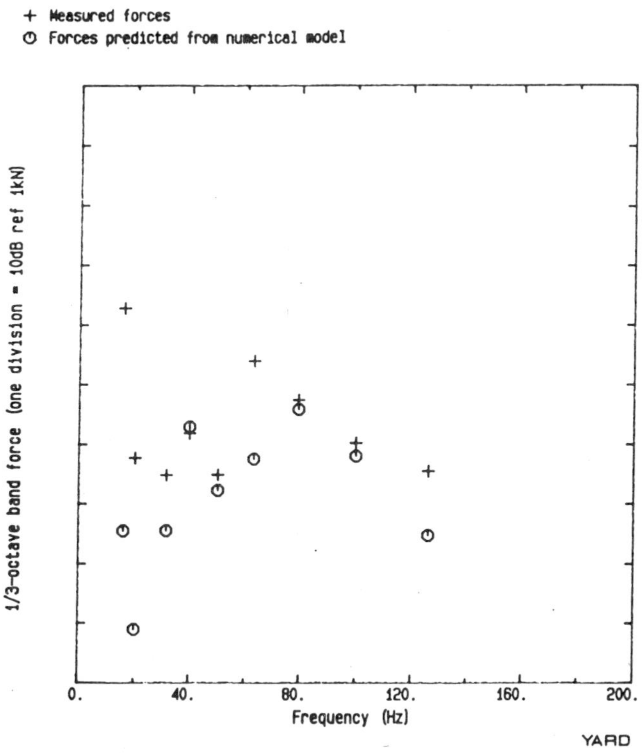

Fig 12 Measured and predicted one third-octave band forces transmitted to dynamometer by the diesel drive-shaft (athwartships direction)

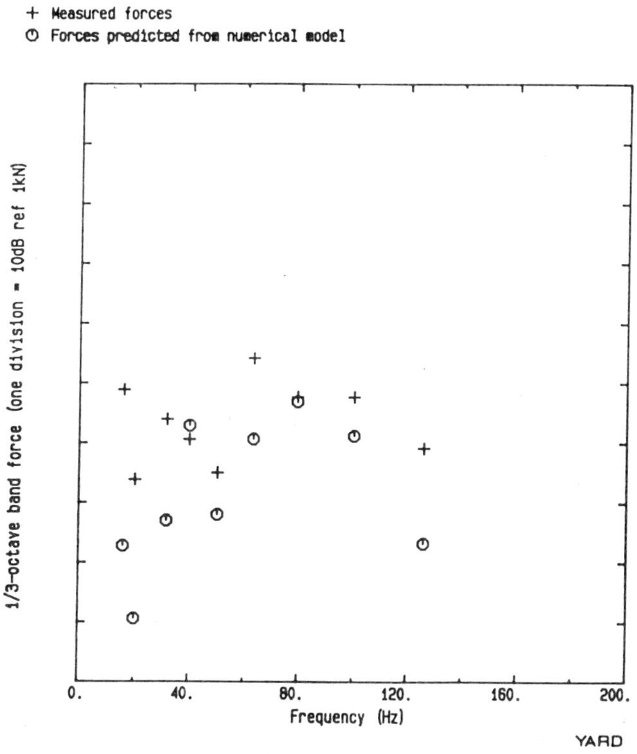

Fig 13 Measured and predicted one third-octave band forces transmitted to dynamometer by the diesel drive-shaft (vertical direction)

C288/88

Response of two elastically supported rigid rotors sharing a common discontinuously non-linear support

R D NEILSON, BSc, PhD and **A D S BARR**, BSc, PhD, CEng, FIMechE, FRSE
Department of Engineering, University of Aberdeen, Aberdeen

SYNOPSIS The vibration response of a two rotor system is considered. The system consists of two identical rigid rotors mounted collinearly on three elastic supports, the central support being discontinuously nonlinear in nature. The equations of motion of the system are derived and the responses computed from the model compared with those obtained from a corresponding experimental rig. Reasonably good agreement is found. The results reveal that over certain speed ranges, spectral sidebands, which diverge with shaft speed, or frequency components at combinations of the two shaft speeds may occur depending on the level of imbalance on the rotors. Similar responses have been found in full scale aeroengine vibration testing.

NOTATION

A, A_1, A_2	Principal moments of inertia.
C, C_1, C_2	Polar moments of inertia.
C_{p_i} (i=1,2,...6)	Damping coefficients of the coordinates p_i.
g_2	Radial clearance at the common support.
k_1, k_2, k_3	Linear support stiffnesses.
K_2	Secondary (snubber) stiffness at the common support.
ℓ, ℓ_1, ℓ_2	Lengths of the rotors.
ℓ_{T1}, ℓ_{T2}	Imbalance torque arms.
M, M_1, M_2	Masses of the rotors.
m_{R1}, m_{R2}	Equivalent imbalance masses for translational motion of the rotors.
m_{T1}, m_{T2}	Equivalent imbalance masses for rotational motion of the rotors.
p_i (i=1,2,...,6)	Modal coordinates of the nonrotating system.
T	Kinetic energy of the rotor system.
V	Potential energy of the rotor system.
x_i (i=1,2,3)	Generalised coordinates.
y_i (i=1,2,3)	Generalised coordinates.
ϕ_1, ϕ_2	Phase angles between the imbalance torques and the imbalance forces.
θ	Initial phase angle between the imbalance forces on the two rotors.
Ω_1, Ω_2	Angular velocities of the rotors.

1 INTRODUCTION

In the past a number of researchers have investigated the effects of a bearing or other clearance on the vibration response of a rotor system (1-9). The observed phenomena resulting from such effects include subharmonic vibration (1-5), asynchronous vibration (6) and the presence of spectral sidebands which diverge with increasing shaft speed (7-9). In none of these cases, however, is the influence of of the excitation from imbalance on a second rotor or shaft taken into account. In current multi-spool gas turbine aeroengine design such effects may be important, since in general, any element of the engine, and particularly inter-shaft bearings may be subject to multi-frequency excitation. Currently the presence of sidebands, subharmonics or other frequency components in the response spectrum of an engine is increasingly being used in the diagnosis of possible faults. Consequently the determination of typical response spectra for various fault conditions is important.

The work reported here considers the response of an elastically supported two rotor system having a bearing clearance effect at a common bearing support.

2 THEORY

To allow for multi-frequency excitation a model which consists of two elastically supported rigid rotors which can be run at different speeds was derived. These rotors are mounted on three elastic supports, the central support being common to both rotors. The remote supports are defined to be linear while the common support has a piece-wise linear, radially symmetric characteristic of the form shown in Fig 1. Fig 2 shows the model and the coordinate system used initially to describe the motion of the rotors.

Using this coordinate system the kinetic energy of the system may be readily obtained and is

$$T = \frac{1}{2}\left\{\frac{M_1}{4}[(\dot{x}_1+\dot{x}_2)^2+(\dot{y}_1+\dot{y}_2)^2]\right.$$

$$+\frac{M_2}{4}[(\dot{x}_2+\dot{x}_3)^2+(\dot{y}_2+\dot{y}_3)^2]$$

$$+\frac{A_1}{\ell_1^2}[(\dot{x}_1-\dot{x}_2)^2+(\dot{y}_2-\dot{y}_1)^2]$$

$$+\frac{A_2}{\ell_2^2}[(\dot{x}_2-\dot{x}_3)^2+(\dot{y}_3-\dot{y}_2)^2]$$

$$+C_1\Omega_1^2+\frac{2C_1\Omega_1}{\ell_1^2}(\dot{y}_2-\dot{y}_1)(x_1-x_2)$$

$$+C_2\Omega_2^2+\frac{2C_2\Omega_2}{\ell_2^2}(\dot{y}_3-\dot{y}_2)(x_2-x_3)\Bigg\} \quad (1)$$

Similarly the potential energy may be derived and for the case of no contact (i.e. $[x_2^2+y_2^2]^{1/2} < g_2$) is

$$V = \tfrac{1}{2}[k_1(x_1^2+y_1^2) + k_2(x_2^2+y_2^2) + k_3(x_3^2+y_3^2)] \quad (2)$$

while for the case of contact with the secondary spring at the common support (i.e. the case $[x_2^2+y_2^2]^{1/2} > g_2$) this becomes

$$V = \tfrac{1}{2}[k_1(x_1^2+y_1^2) + k_2(x_2^2+y_2^2) + K_2([x_2^2+y_2^2]^{\frac{1}{2}} - g_2)^2 + k_3(x_3^2+y_3^2)] \quad (3)$$

Application of Lagrange's equations to equations (1) to (3) and the inclusion of the effects of small arbitrary imbalance on the two rotors yields two sets of inertially coupled equations (see Appendix A). This inertial coupling may be removed and and the equations transformed into a form more amenable to numerical solution by use of an appropriate coordinate transformation. In this instance the transformation to the modal coordinates of the nonrotating system was chosen. This corresponds to the matrix transformation given in Appendix B. Application of this transformation and the inclusion of modal damping yields the following two sets of equations.

For $[x_2^2+y_2^2]^{1/2} < g_2$ the equations governing the motion of the system are

$$M\ddot{p}_1 + C_{P1}\dot{p}_1 + 2kp_1 = \frac{m_{R1}\Omega_1^2 r_1}{2}\cos\Omega_1 t + \frac{m_{R2}\Omega_2^2 r_2}{2}\cos(\Omega_2 t+\theta) \quad (4)$$

$$M\ddot{p}_2 + C_{P2}\dot{p}_2 + 2kp_2 = \frac{m_{R1}\Omega_1^2 r_1}{2}\sin\Omega_1 t + \frac{m_{R2}\Omega_2^2 r_2}{2}\sin(\Omega_2 t+\theta) \quad (5)$$

$$\frac{A}{\ell^2}\ddot{p}_3 + C_{P3}\dot{p}_3 + \frac{C}{4\ell^2}[2(\Omega_1+\Omega_2)\dot{p}_4 + (\Omega_1-\Omega_2)\dot{p}_6] + \frac{k}{2}p_3 = -\frac{m_{T1}\ell_{T1}\Omega_1^2 r_1}{4\ell}\cos(\Omega_1 t+\phi_1) + \frac{m_{T2}\ell_{T2}\Omega_2^2 r_2}{4\ell}\cos(\Omega_2 t+\theta+\phi_2) \quad (6)$$

$$\frac{A}{\ell^2}\ddot{p}_4 + C_{P4}\dot{p}_4 - \frac{C}{4\ell^2}[2(\Omega_1+\Omega_2)\dot{p}_3 + (\Omega_1-\Omega_2)\dot{p}_5] + \frac{k}{2}p_3 = -\frac{m_{T1}\ell_{T1}\Omega_1^2 r_1}{4\ell}\cos(\Omega_1 t+\phi_1) + \frac{m_{T2}\ell_{T2}\Omega_2^2 r_2}{4\ell}\cos(\Omega_2 t+\theta+\phi_2) \quad (7)$$

$$(\frac{M}{4}+\frac{A}{\ell^2})\ddot{p}_5 + C_{P5}\dot{p}_5 + \frac{C}{2\ell^2}[2(\Omega_1-\Omega_2)\dot{p}_4 + (\Omega_1+\Omega_2)\dot{p}_6] + kp_5 = -\frac{m_{R1}\Omega_1^2 r_1}{4}\cos\Omega_1 t - \frac{m_{T1}\ell_{T1}\Omega_1^2 r_1}{2\ell}\cos(\Omega_1 t+\phi_1) + \frac{m_{R2}\Omega_2^2 r_2}{4}\cos(\Omega_2 t+\theta) - \frac{m_{T2}\ell_{T2}\Omega_2^2 r_2}{2\ell}\cos(\Omega_2 t+\theta+\phi_2) \quad (8)$$

$$(\frac{M}{4}+\frac{A}{\ell^2})\ddot{p}_6 + C_{P6}\dot{p}_6 - \frac{C}{2\ell^2}[2(\Omega_1-\Omega_2)\dot{p}_3 + (\Omega_1+\Omega_2)\dot{p}_5] + kp_6 = -\frac{m_{R1}\Omega_1^2 r_1}{4}\sin\Omega_1 t - \frac{m_{T1}\ell_{T1}\Omega_1^2 r_1}{2\ell}\sin(\Omega_1 t+\phi_1) + \frac{m_{R2}\Omega_2^2 r_2}{4}\sin(\Omega_2 t+\theta) - \frac{m_{T2}\ell_{T2}\Omega_2^2 r_2}{2\ell}\sin(\Omega_2 t+\theta+\phi_2) \quad (9)$$

while for $[x_2^2+y_2^2]^{1/2} > g_2$ equations (8) and (9) remain the same and equations (4) to (7) become

$$M\ddot{p}_1 + C_{P1}\dot{p}_1 + 2kp_1 + \tfrac{1}{2}K_2(p_1+p_3) - \tfrac{1}{2}K_2 g_2\left\{\frac{(p_1+p_3)}{[(p_1+p_3)^2 + (p_2+p_4)^2]^{\frac{1}{2}}}\right\}$$

$$= \frac{m_{R1}\Omega_1^2 r_1}{2} \cos\Omega_1 t$$

$$+ \frac{m_{R2}\Omega_2^2 r_2}{2} \cos(\Omega_2 t+\theta) \qquad (10)$$

$$M\ddot{p}_2 + C_{P2}\dot{p}_2 + 2kp_2 + \tfrac{1}{2}K_2(p_2+p_4)$$

$$- \tfrac{1}{2}K_2 g_2 \left\{ \frac{(p_2+p_4)}{[(p_1+p_3)^2 + (p_2+p_4)^2]^{\frac{1}{2}}} \right\}$$

$$= \frac{m_{R1}\Omega_1^2 r_1}{2} \sin\Omega_1 t$$

$$+ \frac{m_{R2}\Omega_2^2 r_2}{2} \sin(\Omega_2 t+\theta) \qquad (11)$$

$$\frac{A}{\ell^2}\ddot{p}_3 + C_{P3}\dot{p}_3 + \frac{C}{4\ell^2}[2(\Omega_1+\Omega_2)\dot{p}_4$$

$$+(\Omega_1-\Omega_2)\dot{p}_6] + \frac{k}{2}p_3 + \tfrac{1}{2}K_2(p_1+p_3)$$

$$- \tfrac{1}{8}K_2 g_2 \left\{ \frac{(p_1+p_3)}{[(p_1+p_3)^2 + (p_2+p_4)^2]^{\frac{1}{2}}} \right\}$$

$$= \frac{m_{T1}\ell_{T1}\Omega_1^2 r_1}{4\ell} \cos(\Omega_1 t+\phi_1)$$

$$+ \frac{m_{T2}\ell_{T2}\Omega_2^2 r_2}{4\ell} \cos(\Omega_2 t+\theta+\phi_2) \qquad (12)$$

$$\frac{A}{\ell^2}\ddot{p}_4 + C_{P4}\dot{p}_4 - \frac{C}{4\ell^2}[2(\Omega_1+\Omega_2)\dot{p}_3$$

$$+(\Omega_1-\Omega_2)\dot{p}_5] + \frac{k}{2}p_3 + \tfrac{1}{2}K_2(p_1+p_3)$$

$$- \tfrac{1}{8}K_2 g_2 \left\{ \frac{(p_2+p_4)}{[(p_1+p_3)^2 + (p_2+p_4)^2]^{\frac{1}{2}}} \right\}$$

$$= - \frac{m_{T1}\ell_{T1}\Omega_1^2 r_1}{4\ell} \sin(\Omega_1 t+\phi_1)$$

$$+ \frac{m_{T2}\ell_{T2}\Omega_2^2 r_2}{4\ell} \sin(\Omega_2 t+\theta+\phi_2) \qquad (13)$$

The numerical solution of the equations of motion was undertaken using techniques developed by Borthwick (10,11) and later used and described by Neilson (7) and Neilson and Barr (8,9). These algorithms, based on the 4th order Runge-Kutta integration algorithm, solve the response in a piece-wise fashion using interpolation procedures to prevent loss of order of the solution at points of discontinuity.

The resulting computed responses were analysed for spectral content using standard Fourier techniques.

3 EXPERIMENTAL STUDIES

To give comparison with the results obtained from the model an experimental rig was designed and constructed.

The rig (Fig 3) consisted of two identical rigid rotors 900mm long and 90mm in diameter running in rolling element bearings which were mounted in three collinear elastically supported housings. The middle housing was common to both rotors and had a discontinuously nonlinear support stiffness. The elasticity of the bearing supports was effected by suspending each of the bearing housings on four flexible rods. The other ends of these rods were mounted into large support blocks which were in turn bolted to a massive cast iron bedplate. The discontinuous stiffness was provided by placing a similarly, but more stiffly supported ring around the common bearing housing. A radial gap between the two parts permitted some motion of the housing prior to contacting the secondary spring stiffness.

Two 1.5kW variable speed D.C. motors with thyristor controllers were used to drive the rotors through light shafts incorporating universal joints. No mechanical coupling existed between the rotors themselves, these being free to operate at different speeds and even in different directions if required. The only constraint applied to the system, therefore, was that the adjacent ends of the two rotors had the same displacement at all times. This was consistent with the model described earlier.

The response of the rig was monitored by six eddy current displacement transducers mounted in pairs at each of the bearing housings. Four l.e.d. opto-switches and two discs with holes drilled in them were also used to give indications of the shaft speeds and to provide once-per-rev phase markers to indicate the position of the imbalance masses.

In order to limit the amount of data generated by the model and the rig it was decided to investigated only particular shaft speed ratios between the two rotors. Given the origins of the research it was considered that the most appropriate speed ratio to investigate was one typical of a twin spool gas turbine aeroengine. Fig 4 shows the speed ratio which was chosen after consultation with industrial sources. For convenience a small op-amp circuit was built into one of the motor control units to give automatic speed setting of both rotors in the correct speed ratio from a single control.

The tests undertaken consisted of accelerating the rotors slowly over the range 700 to 2300r.p.m. of the low speed rotor. During this the shaft speed was held constant for a short period at increments of 80r.p.m. of the low speed rotor and a sample of the response obtained. The response for decreasing shaft speed was also obtained, in a similar manner, in order to account for nonlinear phenomena such as jumps.

During each test the two shaft speed signals, the phase markers and the outputs from one pair of displacement transducers were recorded using a Racal Store 7DS tape recorder. The vibration responses were subsequently analysed for spectral content using a Hewlett Packard 5423A analyser and 9826A microcomputer and plotted in the form of waterfall plots.

4 RESULTS AND DISCUSSION

The model was initially set up to simulate the rig and results obtained from both the rig and the model for a single configuration of discontinuous stiffness (K_2 = 5.86MN/m, g_2 = 0.216mm) but for a variety of configurations of imbalance on the two rotors. These were chosen to simulate possible arbitrary imbalance situations in a "real" system and to excite the three rigid rotor modes of the rig. The imbalance levels were also selected to cause amplitudes of vibration large enough to cause contact with the secondary spring stiffness at the common housing.

To prevent confusion in the discussion of the results, the term engine order will be used to denote the fundamental and harmonics of the shaft speeds, i.e. first engine order denotes the fundamental of the shaft speed, second engine order denotes response at twice the fundamental etc. Furthermore the low speed rotor will be referred to as rotor 1 and the high speed rotor as rotor 2.

4.1 Spectral Response

Figures 5(a-f) show the spectral response at the common support for three different imbalance configurations.

Figure 5(a) shows the theoretical response for the case where 750gm.mm was mounted on rotor 1 88.6mm inboard of the common support and 1500gm.mm was mounted on rotor 2 at a similar position. For increasing shaft speed the response remains linear until Ω_1 = 860r.p.m. when contact with the secondary spring stiffness occurs. This results in the generation of spectral sidebands about the first engine order of rotor 2. These sidebands are somewhat asymmetric in amplitude about the engine order and diverge with increasing shaft speed until they disappear at around Ω_1 = 1500r.p.m.. Around and above this speed a combination order or harmonic at $2\Omega_2$-Ω_1 is also present, although discrimination between particular components is somewhat difficult due to the broadband nature of some of the responses which is at times almost suggestive of randomness.

An interesting facet of the results is the fact that the resonance of Ω_2 with the third mode appears to override completely the resonances of Ω_1 with the first and second modes at 20.3Hz (1218r.p.m.) and 23.3Hz (1398r.p.m.) respectively.

Comparison with the experimental response of Fig 5(b) shows reasonably good correlation. If anything the sidebands and combination orders are more clearly defined on the spectra.

Fig 5 (c) depicts the theoretical response for 1500gm.mm on rotor 1 and 2250gm.mm on rotor 2. In this case the combination orders are more apparent than in the previous example but there are also frequency components at $(3\Omega_2$-$\Omega_1)/2$ and $(3\Omega_1$-$\Omega_2)/2$ present over part of the speed range. Comparison with the rig response (Fig 5(d)) shows a similar general response, although the spectra are less "noisy" and the more complex combination orders are not so apparent.

The response of the system for 2250gm.mm on rotor 1 and 1500gm.mm on rotor 2 (Figs 5(e-f)) show the combination orders well, although the amplitude of these components is somewhat larger in the theoretical result than in the experiment.

Having observed the phenomena present in the overall response of the system it is relevant to consider these types of response in more detail.

4.2 Sideband Response

More detailed examination of a particular occurrence of the sideband response indicated that the orbit was unsteady and underwent periodic collapse. Furthermore, transformation of the response from x,y coordinates into r,θ coordinates revealed the radial component of the response to be more or less periodic. These characteristics are consistent with the mechanism of sideband generation advanced by Neilson (7) and Neilson and Barr (9) for a single rotor system. Consequently it may be assumed that this response is analagous to that of the single rotor system.

This conjecture may possibly be further substantiated by the fact that the occurrence of sidebands was most noticeable in tests in which the imbalance on the high speed rotor (rotor 2) was larger than that on the low speed rotor (rotor 1). This coupled with the speed difference between the two rotors results in the imbalance effect on rotor 2 being dominant. It might therefore be expected that the response would be similar to that of the single rotor system in which only a single excitation exists.

4.3 Combination Orders

Examples of the combination response are depicted in Fig 6 for the case of 2250gm.mm mounted on rotor 1, 1500gm.mm mounted on rotor 2 and with rotor 1 running at 1820r.p.m. (30.3Hz). Examination of the time histories reveals that the response is of a regular beating form which is reflected in the orbits which collapse and reform.

By plotting the radial component of the theoretical response at the common housing (Fig 7) it may be seen that, as was noted for the sideband response, this component is virtually periodic. Consequently this type of response may, in some ways, be regarded as a type of sideband response in which the sidebands are spaced at the difference between the two shaft speeds.

This may be shown to be reasonable since the period of the oscillations in the radial direction corresponds to that of the difference between the two shaft speeds. This arises because the the largest amplitude of the forced response occurs at the the top of the beat cycle and therefore contact with the secondary spring is most likely to occur at this point. As a result contact is liable to happen regularly at each cycle of the beat frequency. The resulting frequency and amplitude modulation of the response will also therefore be at this frequency, generating "sidebands" at combinations of the shaft speeds.

The form of the time histories is somewhat akin to the truncated beating response proposed by Ehrich (12) in his consideration of sum and difference frequencies in a rotor system with an asymmetric clearance effect. In his model one of the sources of excitation was due to self-excited instability attributable to fluid trapped in the rotor rather than due to imbalance on a second rotor. Ehrich predicted that such truncated beating responses would give rise to sum and difference and other combination frequencies in the spectrum. The results described in this paper tend to add validity to this hypothesis, but also show that the clearance need not be asymmetric for similar phenomena to occur.

It should be noted that the existence of combination orders is not limited to dynamic systems with motion dependent discontinuities, but is instead a characteristic of nonlinear systems in general when subject to multi-frequency excitation. Examples of this type of response may be found in chapters 8 and 10 of (13) and (14) respectively.

4.4 Further Discussion

The reasonably good agreement between the theoretical and experimental responses suggests that the observed phenomena are genuine. This is further substantiated by comparison with test results from an actual aeroengine undergoing test prior to acceptance. The response, depicted in Fig 8, is a plot of the major frequency components against percentage of the maximum speed of rotor 2 and displays both of the characteristic types of response described earlier. Above 85 per cent of maximum shaft speed sidebands may be seen emanating from the first engine order of rotor 2. In addition, although perhaps less prominent, a component of the type $2\Omega_2-\Omega_1$ is present over much of the speed range. The engine, which incorporated an inter-shaft bearing, was thought to suffer from a clearance effect. This hypothesis tends to be confirmed by the preceding results.

As was noted earlier, the two types of nonlinear response appear to occur for fairly readily prescribed imbalance conditions and speed ranges. In cases where the imbalance on one of the rotors is dominant and the speed of that rotor is near resonance with the third mode, sidebands occur. For the shaft speed ratio and imbalance configurations studied, this can happen only for rotor 2 and the effect is seen most clearly when the imbalance on rotor 2 is as large or larger than that on rotor 1. Later numerical tests with 1500gm.mm on rotor 1 and zero imbalance on rotor 2 showed that this configuration would produce sidebands about Ω_1 after a slight reduction in the damping in the system.

The presence of broadband spectra in the responses of the system which were noted earlier, appeared after investigation to be be connected with the damping, or lack of it, in the secondary spring stiffness and the effect that this had on the effective damping ratios present in the the system during contact with the secondary stiffness. During the preliminary numerical tests it was found necessary to add damping to the secondary spring to prevent the generation of "noisy" spectra. This is probably reasonable since energy would undoubtedly be lost in the contact between the bearing housing and the snubber ring in the rig.

This problem, however, raises the whole question of the presence of seemingly wideband or chaotic responses in rotor systems with discontinuous nonlinearities. Such spectra are occasionally seen in industry during testing of aeroengines and a possible degeneration into chaos in an experimental single rotor system has been noted by one of the authors (7). Such phenomena are also reasonably well documented for single degree of freedom oscillators (see for example (15)), but to the best of the authors' knowledge little has been published on rotor systems except (16). This is obviously an area of importance for further research particularly in terms of fault diagnosis and condition monitoring.

5 CONCLUSIONS

The reasonable agreement between the response computed from the model and those obtained from the rig tends to indicate that the model of the system is adequate, at least qualitatively and in most respects quantitatively as well. Both model and rig display several nonlinear phenomena depending on the level of imbalance present on the rotors. For cases where the imbalance on one of the rotors is dominant, sidebands which diverge with shaft speed may be present about the engine order of that rotor. This phenomenon has been previously documented for a single rotor system. In most cases combination orders of the two shaft speeds are present and jumps into or out of this type of response may occur depending on whether the response for increasing or decreasing speed is being considered. The existence of broadband possibly chaotic spectra has also been noted.

The results correlate well with responses occasionally obtained from real aeroengines which may display any or all of these response characteristics (see Fig 8).

ACKNOWLEDGEMENTS

The authors wish to acknowledge both the financial support for this work given by the Ministry of Defence and the extensive support and technical liaison of the Bristol and Derby divisions of Rolls Royce Aeroengines Ltd..

REFERENCES

(1) Ehrich, F. F. Subharmonic vibration of rotors in bearing clearance. A.S.M.E. Design Conference and Show, 1966, Paper 66-MD-1, pp1-4.

(2) Ehrich, F. F. and O'Connor, J. J. Stator whirl with rotors in bearing clearance. J. Eng. Ind., 1967, 89, 381-389

(3) Bently, D. E. Subrotative speed dynamic action of rotating machinery. A.S.M.E. Petroleum Engineering Conference, Dallas, 1974, Paper 74-PET-16, pp1-8.

(4) Childs, D. W. Fractional frequency rotor motion due to nonsymmetric clearance effects. J. Eng. Power, 1982, 104, 533-541.

(5) Muszynska, A. Partial lateral rotor to stator rubs. I. Mech. E. 3rd. International Conference on Vibrations in Rotating Machinery, University of York, 1984, pp327-335.

(6) Day, W. B. Nonlinear rotordynamics analysis. N.A.S.A. Report CR 171425, 1985.

(7) Neilson, R. D. Dynamics of simple rotor systems having motion dependent discontinuities. Ph.D. Thesis, 1986, University of Dundee.

(8) Neilson, R. D. and Barr, A. D. S. Spectral features of the response of a rigid rotor mounted on discontinuously nonlinear supports. Seventh World Congress on the Theory of Machines and Mechanisms, Seville, 1987, pp1799-1803.

(9) Neilson, R. D. and Barr, A. D. S. Dynamics of a rigid rotor mounted on discontinuously nonlinear elastic supports. Accepted for publication in the Journal of Mechanical Engineering Sciences.

(10) Borthwick, W. K. D. The numerical solution of transient structural dynamics problems with stiffness discontinuities. Ph.D. Thesis, 1984, University of Dundee.

(11) Borthwick, W. K. D. The numerical solution of discontinuous structural systems. Second International Conference on Recent Advances in Structural Dynamics, University of Southampton, 1984, pp307-316.

(12) Ehrich, F. F. Sum and difference frequencies in vibration of high speed rotating machinery. J. Eng. Ind., 1972, 94, 181-184.

(13) Struble, R. A. Nonlinear Differential Equations. 1962, McGraw-Hill, New York.

(14) Nayfeh, A. H., Introduction to Perturbation Techniques., 1981, Wiley, New York.

(15) Thomson, J. M. T. and Ghaffari, R., Chaos after period doubling bifurcations in the resonance of an impact oscillator. Physics Letters, 1982, 91A, 1, 5-8

(16) Szczgielski, W. and Schweitzer, G., Dynamics of a high speed rotor touching a boundary. IUTAM/IFTOMM Symposium on Dynamics of Multibody Systems, CISM-Udine, Sept. 16-20, 1985.

APPENDIX A

The equations of motion in the original coordinate system are

$$\frac{M}{4}(\ddot{x}_1+\ddot{x}_2) + \frac{A}{l^2}(\ddot{x}_1-\ddot{x}_2) - \frac{C\Omega_1}{l^2}(\dot{y}_2-\dot{y}_1) + kx_1$$

$$= \frac{m_{R1}\Omega_1^2 r_1}{2}\cos\Omega_1 t + \frac{m_{T1}\ell_{T1}\Omega_1^2 r_1}{\ell}\cos(\Omega_1 t+\phi_1) \qquad \text{(A1)}$$

$$\frac{M}{4}(\ddot{y}_1+\ddot{y}_2) - \frac{A}{\ell^2}(\ddot{y}_2-\ddot{y}_1) - \frac{C\Omega_1}{\ell^2}(\dot{x}_1-\dot{x}_2) + ky_1 = \frac{m_{R1}\Omega_1^2 r_1}{2}\sin\Omega_1 t + \frac{m_{T1}\ell_{T1}\Omega_1^2 r_1}{\ell}\sin(\Omega_1 t+\phi_1) \qquad \text{(A2)}$$

$$\frac{M}{4}(\ddot{x}_1+2\ddot{x}_2+\ddot{x}_3) + \frac{A}{\ell^2}(-\ddot{x}_1+2\ddot{x}_2-\ddot{x}_3) + \frac{C\Omega_1}{\ell^2}(\dot{y}_2-\dot{y}_1) - \frac{C\Omega_2}{\ell^2}(\dot{y}_3-\dot{y}_2) + 2kx_2 = \frac{m_{R1}\Omega_1^2 r_1}{2}\cos\Omega_1 t - \frac{m_{T1}\ell_{T1}\Omega_1^2 r_1}{\ell}\cos(\Omega_1 t+\phi_1) + \frac{m_{R2}\Omega_2^2 r_2}{2}\cos(\Omega_2 t+\theta) + \frac{m_{T2}\ell_{T2}\Omega_2^2 r_2}{\ell}\cos(\Omega_2 t+\theta+\phi_2) \qquad \text{(A3)}$$

$$\frac{M}{4}(\ddot{y}_1+2\ddot{y}_2+\ddot{y}_3) + \frac{A}{\ell^2}(-\ddot{y}_1+2\ddot{y}_2-\ddot{y}_3) + \frac{C\Omega_1}{\ell^2}(\dot{x}_1-\dot{x}_2) - \frac{C\Omega_2}{\ell^2}(\dot{x}_2-\dot{x}_3) + 2ky_2 = \frac{m_{R1}\Omega_1^2 r_1}{2}\sin\Omega_1 t - \frac{m_{T1}\ell_{T1}\Omega_1^2 r_1}{\ell}\sin(\Omega_1 t+\phi_1) + \frac{m_{R2}\Omega_2^2 r_2}{2}\sin(\Omega_2 t+\theta) + \frac{m_{T2}\ell_{T2}\Omega_2^2 r_2}{\ell}\sin(\Omega_2 t+\theta+\phi_2) \qquad \text{(A4)}$$

$$\frac{M}{4}(\ddot{x}_2+\ddot{x}_3) - \frac{A}{\ell^2}(\ddot{x}_2-\ddot{x}_3) + \frac{C\Omega_1}{\ell^2}(\dot{y}_3-\dot{y}_2) + kx_3 = \frac{m_{R2}\Omega_2^2 r_2}{2}\cos(\Omega_2 t+\theta) - \frac{m_{T2}\ell_{T2}\Omega_2^2 r_2}{\ell}\cos(\Omega_2 t+\theta+\phi_2) \qquad \text{(A5)}$$

$$\frac{M}{4}(\ddot{y}_2+\ddot{y}_3) + \frac{A}{\ell^2}(\ddot{y}_3-\ddot{y}_2) + \frac{C\Omega_1}{\ell^2}(\dot{x}_2-\dot{x}_3) + ky_3 = \frac{m_{R2}\Omega_2^2 r_2}{2}\sin(\Omega_2 t+\theta) - \frac{m_{T2}\ell_{T2}\Omega_2^2 r_2}{\ell}\sin(\Omega_2 t+\theta+\phi_2) \qquad \text{(A6)}$$

for $[x_2^2+y_2^2]^{1/2} < g_2$ while for $[x_2^2+y_2^2]^{1/2} > g_2$ equations (A3) and (A4) become

$$\frac{M}{4}(\ddot{x}_1+2\ddot{x}_2+\ddot{x}_3) + \frac{A}{\ell^2}(-\ddot{x}_1+2\ddot{x}_2-\ddot{x}_3) + \frac{C\Omega_1}{\ell^2}(\dot{y}_2-\dot{y}_1) - \frac{C\Omega_2}{\ell^2}(\dot{y}_3-\dot{y}_2) + (2k+K_2)x_2 - K_2 g_2\left\{\frac{x_2}{[x_2^2+y_2^2]^{\frac{1}{2}}}\right\} = \frac{m_{R1}\Omega_1^2 r_1}{2}\cos\Omega_1 t - \frac{m_{T1}\ell_{T1}\Omega_1^2 r_1}{\ell}\cos(\Omega_1 t+\phi_1) + \frac{m_{R2}\Omega_2^2 r_2}{2}\cos(\Omega_2 t+\theta) + \frac{m_{T2}\ell_{T2}\Omega_2^2 r_2}{\ell}\cos(\Omega_2 t+\theta+\phi_2) \qquad \text{(A7)}$$

$$\frac{M}{4}(\ddot{y}_1+2\ddot{y}_2+\ddot{y}_3) + \frac{A}{\ell^2}(-\ddot{y}_1+2\ddot{y}_2-\ddot{y}_3) + \frac{C\Omega_1}{\ell^2}(\dot{x}_1-\dot{x}_2) - \frac{C\Omega_2}{\ell^2}(\dot{x}_2-\dot{x}_3) + (2k+K_2)y_2 - K_2 g_2\left\{\frac{y_2}{[x_2^2+y_2^2]^{\frac{1}{2}}}\right\} = \frac{m_{R1}\Omega_1^2 r_1}{2}\sin\Omega_1 t - \frac{m_{T1}\ell_{T1}\Omega_1^2 r_1}{\ell}\sin(\Omega_1 t+\phi_1) + \frac{m_{R2}\Omega_2^2 r_2}{2}\sin(\Omega_2 t+\theta) + \frac{m_{T2}\ell_{T2}\Omega_2^2 r_2}{\ell}\sin(\Omega_2 t+\theta+\phi_2) \qquad \text{(A8)}$$

APPENDIX B

The following is the coordinate transformation used to obtain the equations of motion in the modal coordinates of the nonrotating system.

$$
\begin{Bmatrix} x_1 \\ y_1 \\ x_2 \\ y_2 \\ x_3 \\ y_3 \end{Bmatrix} = \begin{bmatrix} 1 & 0 & -1 & 0 & -1 & 0 \\ 0 & 1 & 0 & -1 & 0 & -1 \\ 1 & 0 & 1 & 0 & 0 & 0 \\ 0 & 1 & 0 & 1 & 0 & 0 \\ 1 & 0 & -1 & 0 & 1 & 0 \\ 0 & 1 & 0 & -1 & 0 & 1 \end{bmatrix} \begin{Bmatrix} p_1 \\ p_2 \\ p_3 \\ p_4 \\ p_5 \\ p_6 \end{Bmatrix} \tag{B1}
$$

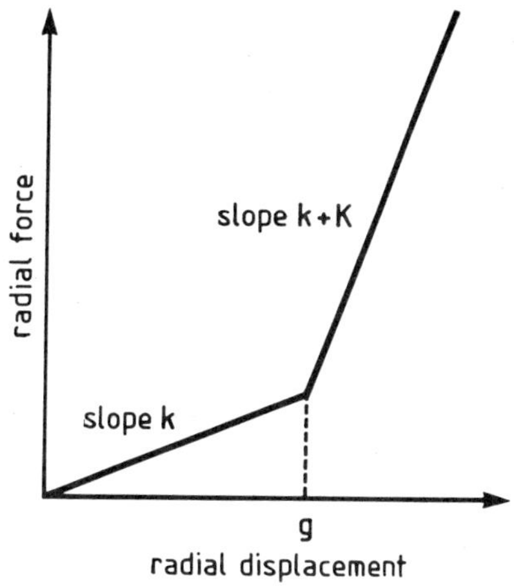

Fig 1 Radial stiffness characteristic

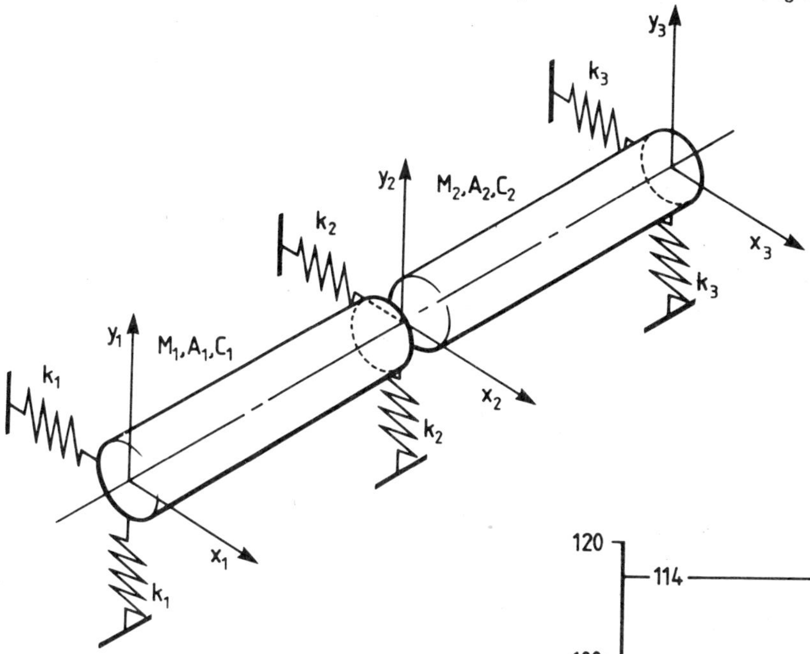

Fig 2 Rotor coordinate system

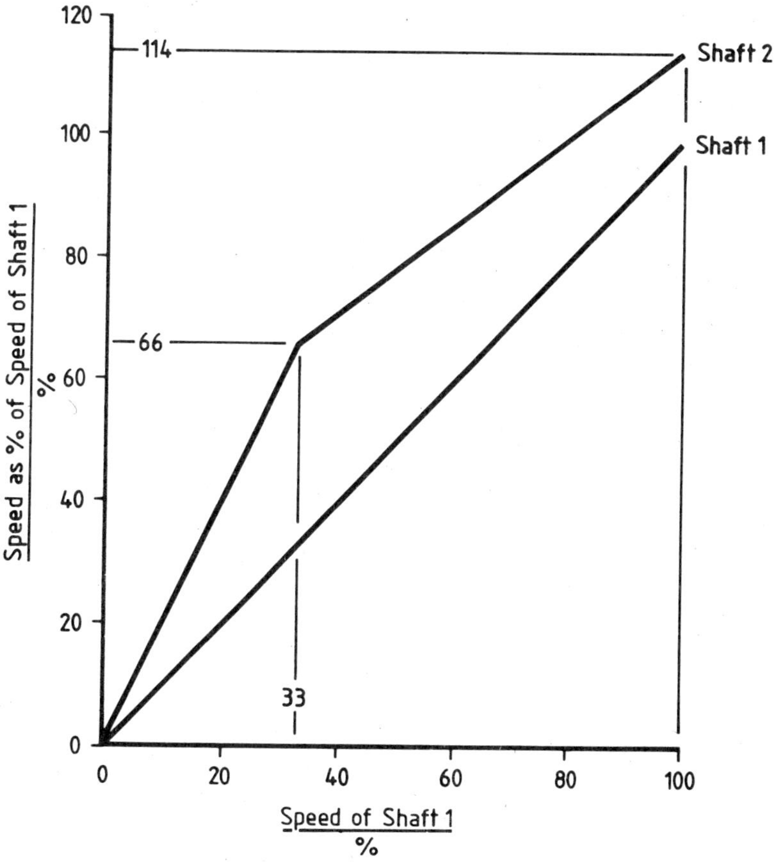

Fig 4 Rotor speed ratio

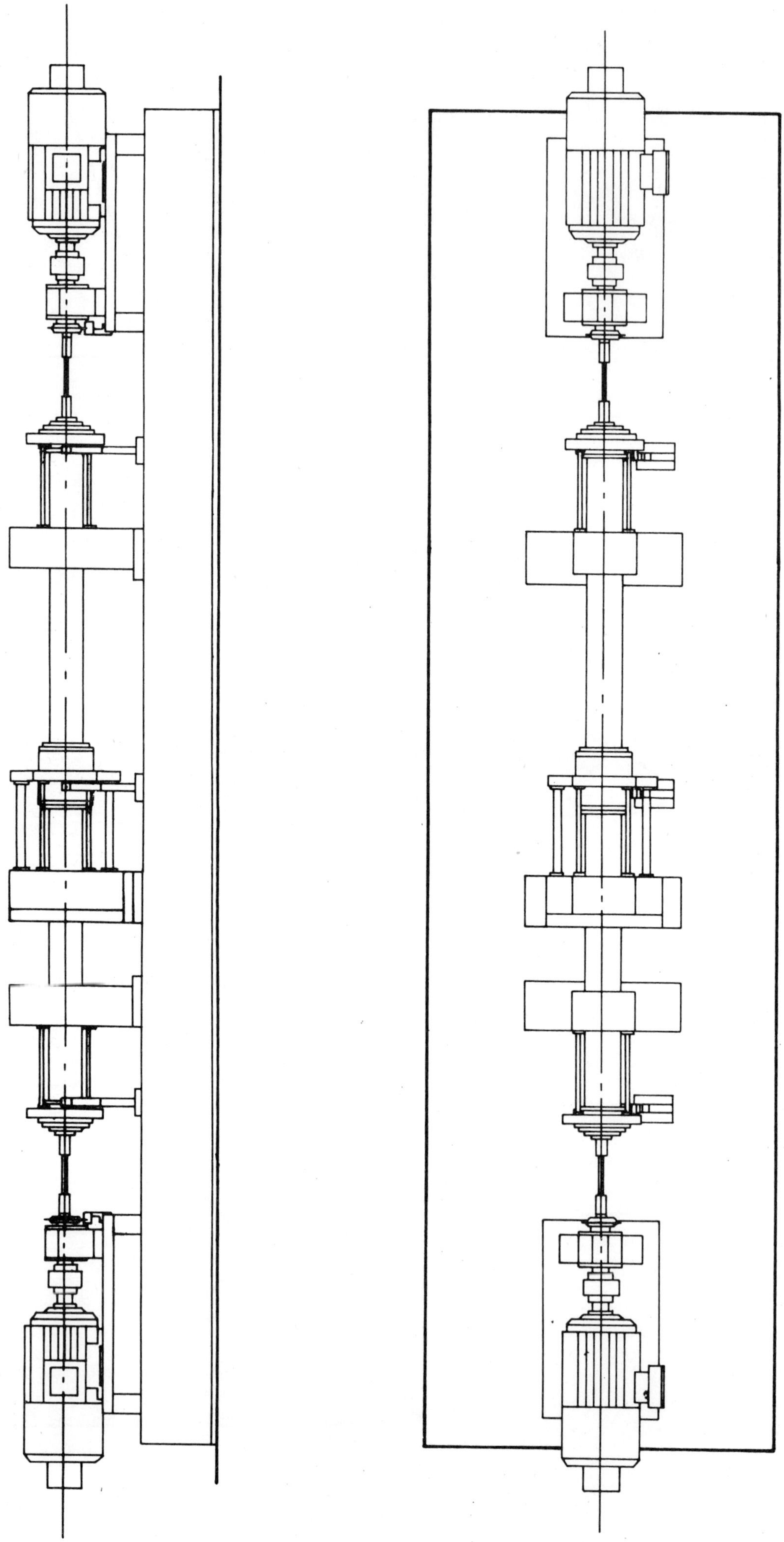

Fig 3 General arrangement of the two-rotor rig

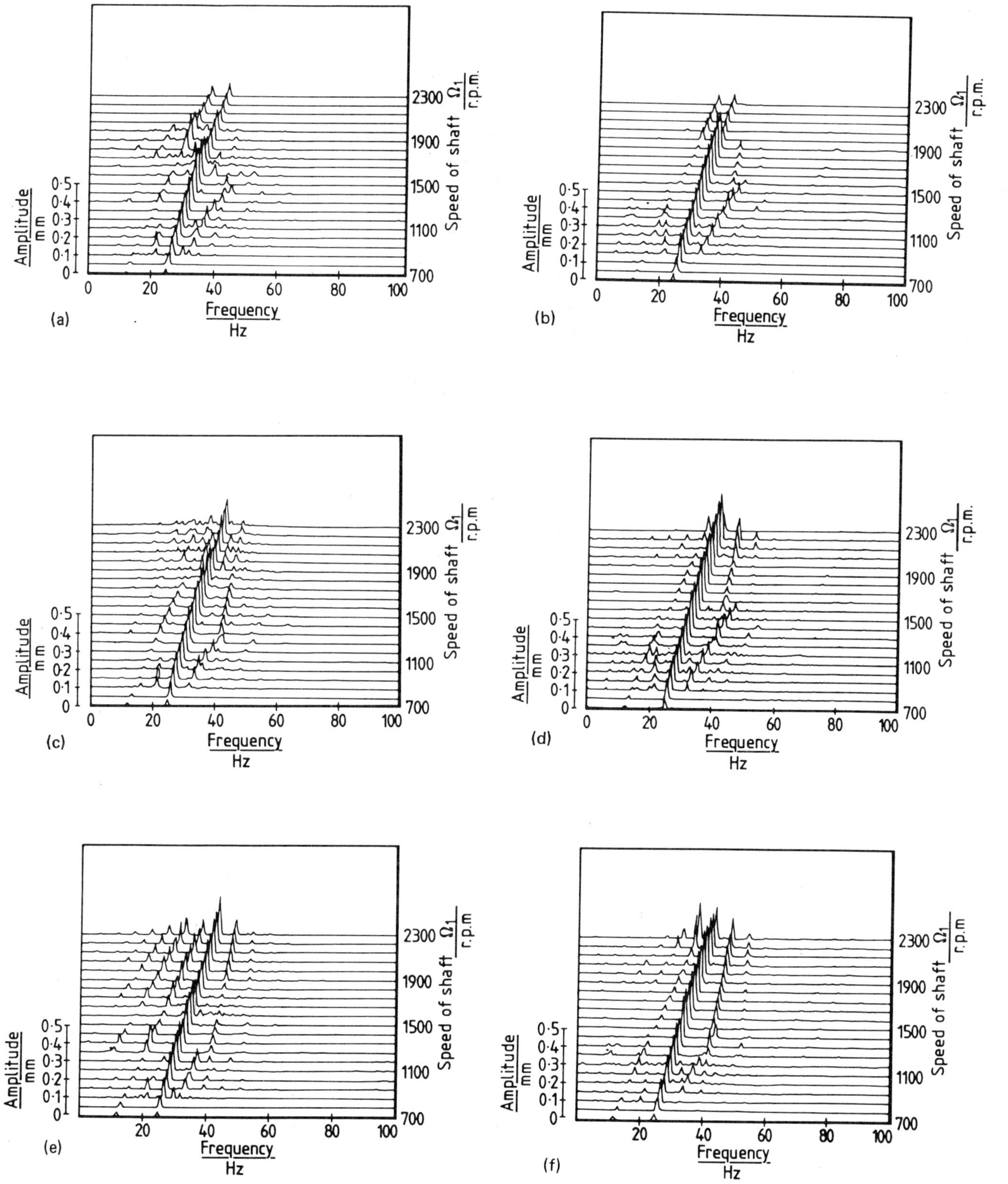

Fig 5 Examples of combination orders and sideband response in (a) the model and (b) the rig with 750 gm.mm on rotor one and 1500 gm.mm on rotor two;
(c) the model and (d) the rig with 1500 gm.mm on rotor one and 2250 gm.mm on rotor two;
(e) the model and (f) the rig with 2250 gm.mm on rotor one and 1500 gm.mm on rotor two.
In all cases the imbalances are mounted 88.6 mm from the common housing on both rotors

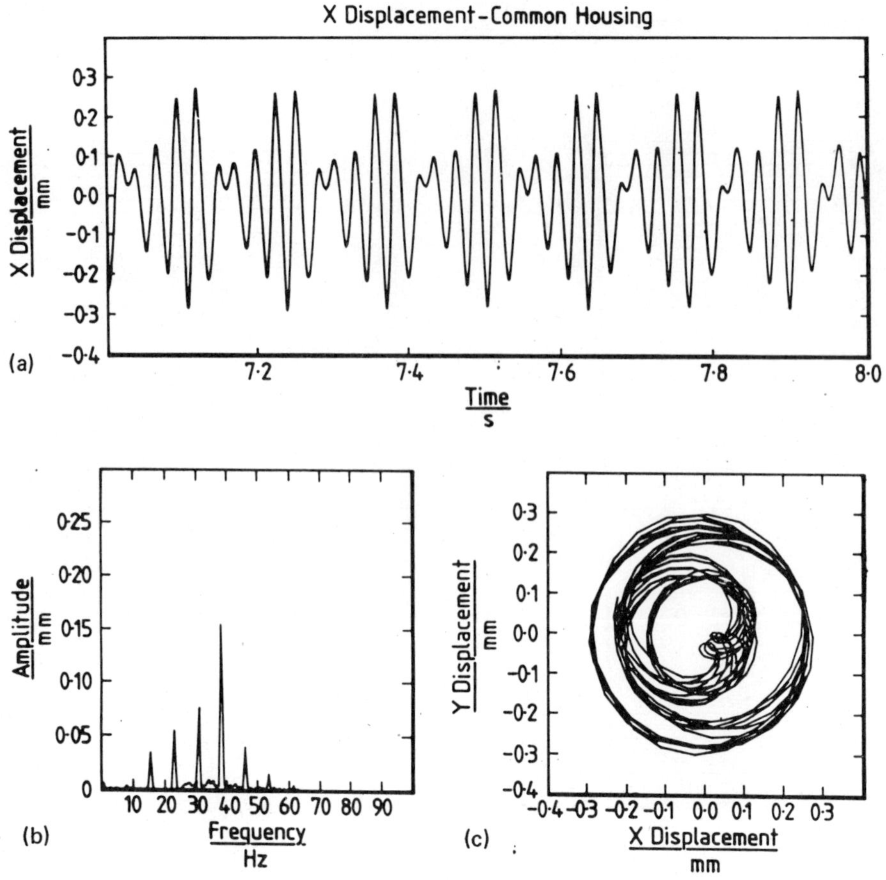

Fig 6 Time history, spectrum and orbit of the combination response at Ω_1 = 1820 r/min

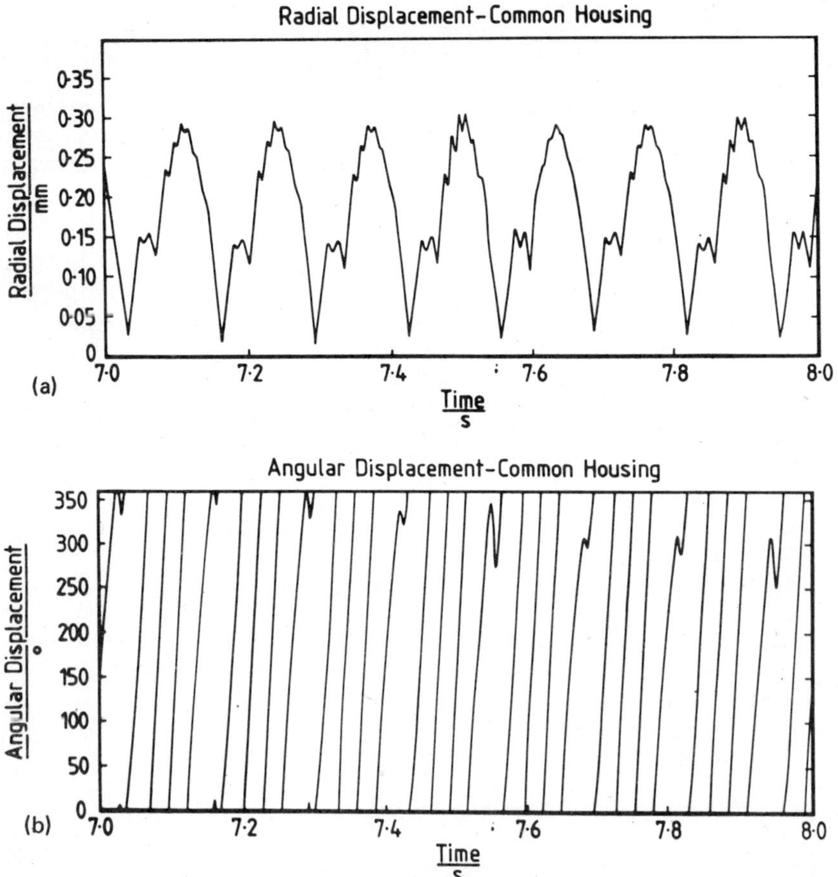

Fig 7 Radial and angular displacement corresponding to the response of Fig 6

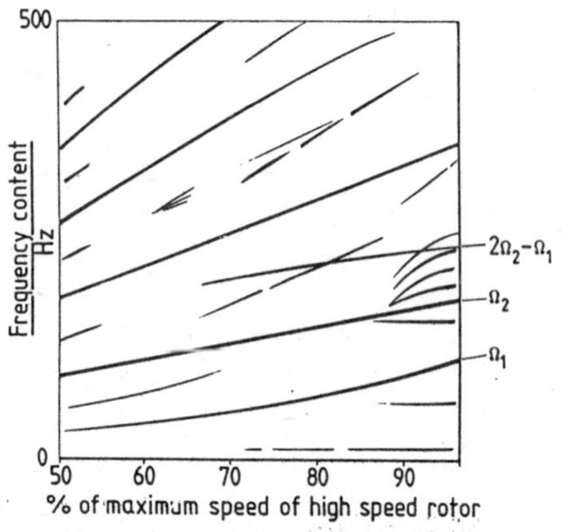

Fig 8 Aeroengine test response showing combination orders and sidebands

C276/88

Dynamic response of a dual-rotor system by extended transfer matrix method

K D GUPTA, MTech, **K GUPTA**, PhD and **J S RAO**, DSc
Department of Mechanical Engineering, Indian Institute of Technology, New Delhi, India
K V B SARMA, PhD
BHEL, Ramachandrapuram, India

SYNOPSIS This paper presents a general formulation for the problem of the steady -state unbalance response of a dual rotor system with a flexible intershaft bearing using an 'extended' transfer matrix method, where the transfer matrix assumed a dimension of (33 x 33). The validity of the formulation is established by comparing the results obtained through a computer program with closed form solutions available for some simple cases. Some interesting phenomena on the nature of steady - state whirl orbits of the dual rotor system are described.

1 INTRODUCTION

Because of definite certain advantages two spool system with intershaft bearing(s) is used in modern gas turbine engines. This method of rotor support minimizes shaft deflections caused by rotor unbalance and improves engine efficiency, performance and reliability by eliminating the static support structure in the aerodynamic flow path. It also facilitates easy mounting with engine casing and gives rise to compactness in the overall structure.

Cross - exciting vibration between the inner and outer shafts is effected through the intershaft bearing and thus the dynamic response of one rotor also depends upon the dynamic behaviour of the other. And this is what makes the mathematical analysis of dual rotor system different from that of a straight rotor.

Very few research papers are available on dual rotor system , and they mostly deal with transient response[(1),(2),(3)]. Li et al (4) investigated steady-state response of a dual rotor system in that a set of linear equations was generated by using forces and angular displacement as initial parameters for the rigid supports and the intershaft bearing ; and the solution is obtained by applying the connecting and boundary conditions. However, the connecting conditions at the intershaft bearing do not represent a general case.

The present paper is essentially an extension over classical transfer matrix method for a straight rotor - in that the transfer matrix formulated assumes a dimension of(33x33). Effects of one rotor over the other have been kept separately , thus doubling the number of elements of the state vector at any station on the rotor . This becomes imperative owing to the fact that the effects of one rotor will have frequency corresponding to its running speed. And effects of one rotor over the other are taken care of by setting proper junction conditions. The formulation is in a general manner so as to accommodate any type of support and intershaft bearing. The validity of the formulation is established by comparing the results obtained with those of some special cases for which closed form solutions are available.

The distinction of steady-state whirl orbits of a dual rotor from that of a straight rotor is brought out by typical case studies.

NOTATION

[A],[B] [C],[D]	represent transfer matrices, each of dimension (33 x33)
[B]	bearing matrix
[Bj]	junction matrix =[B]-[I]
Cyy,Czz	direct damping coefficients
Cyz,Czy	cross coupled damping coefficients
E	modulus of elasticity
e	eccentricity
[F]	field matrix
[I]	unit matrix
Kyy,Kzz	direct stiffnesses
Kyz,Kzy	cross coupled stiffnesses
l	sectional length
My,Mz	moments about y- and z- axes respectively
m	additional mass at section or bearing
[P]	point matrix
Uy,Uz	unbalance components (see fig. 2)
Vy,Vz	shear force in y- and z-directions respectively

v,w	deflection along y- and z- axes respectively
X,Y,Z	fixed co-ordinate system
$\bar{X},\bar{Y},\bar{Z}$	rotating co-ordinate system (see figure 2 and 3)
{Z}	state vector at any station
$\{Z\}_{OA}$, $\{Z\}_{OB}$, $\{Z\}_{OC}$, $\{Z\}_{OD}$	represent end conditions of the dual rotor system
ω_m, ω_n	inner and outer rotor speeds respectively
θ, ϕ	slope in x-z and y-z planes respectively

Subscripts

i,o	pertaining to inner and outer rotors respectively
m,n	pertaining to the effect of inner and outer rotor speeds respectively
c	cosine component
s	sine component
y	along y-axis
z	along z-axis
j	pertaining to a junction

Superscripts

R	right to a section
L	left to a section
I	pertaining to inner rotor
O	pertaining to outer rotor
T	transpose of a matrix

2 ANALYSIS

Figure 1 represents the mathematical model of a dual rotor system used for the present analysis. The system consists of two co-axial rotors -the outer rotor and the inner rotor, interlinked through an intershaft bearing.

2.1 State vector

The state vector at any station is based on :

(i) the sine and cosine components of each quantity constituting the state vector, i.e., deflection, slope, bending moment and shear force in both x-z and x-y planes ; and

(ii) each quantity of the state vector is assumed to have two components - one corresponding to the inner rotor speed and the other to the outer rotor speed.

Hence,

for x-z plane

$$w = w_{cm}\cos\omega_m t + w_{sm}\sin\omega_m t + w_{cn}\cos\omega_n t + w_{sn}\sin\omega_n t$$

$$\theta = \theta_{cm}\cos\omega_m t + \theta_{sm}\sin\omega_m t + \theta_{cn}\cos\omega_n t + \theta_{sn}\sin\omega_n t$$

$$My = Mycm\cos\omega_m t + Mysm\sin\omega_m t + Mycn\cos\omega_n t + Mysn\sin\omega_n t$$

$$Vz = Vzcm\cos\omega_m t + Vzsm\sin\omega_m t + Vzcn\cos\omega_n t + Vzsn\sin\omega_n t$$

for x-y plane

$$v = v_{cm}\cos\omega_m t + v_{sm}\sin\omega_m t + v_{cn}\cos\omega_n t + v_{sn}\sin\omega_n t$$

$$\phi = \phi_{cm}\cos\omega_m t + \phi_{sm}\sin\omega_m t + \phi_{cn}\cos\omega_n t + \phi_{sn}\sin\omega_n t\ .$$

$$Mz = Mzcm\cos\omega_m t + Mzsm\sin\omega_m t + Mzcn\cos\omega_n t + Mzsn\sin\omega_n t$$

$$Vy = Vycm\cos\omega_m t + Vysm\sin\omega_m t + Vycn\cos\omega_n t + Vysn\sin\omega_n t \qquad \ldots\ldots\ldots(1)$$

Thus, the state vector at any station can be written as

$$\{Z\}^T = [\{Szm\}^T\{Szn\}^T\{Sym\}^T\{Syn\}^T]$$

$$\{Szm\} = \begin{Bmatrix} -w_{cm} \\ \theta_{cm} \\ Mycm \\ Vzcm \\ -w_{sm} \\ \theta_{sm} \\ Mysm \\ Vzsm \end{Bmatrix} ; \quad \{Szn\} = \begin{Bmatrix} -w_{cn} \\ \theta_{cn} \\ Mycn \\ Vzcn \\ -w_{sn} \\ \theta_{sn} \\ Mysn \\ Vzsn \end{Bmatrix}$$

$$\{Sym\} = \begin{Bmatrix} v_{cm} \\ \phi_{cm} \\ Mzcm \\ -Vycm \\ v_{sm} \\ \phi_{sm} \\ Mzsm \\ -Vysm \end{Bmatrix} ; \quad \{Syn\} = \begin{Bmatrix} v_{cn} \\ \phi_{cn} \\ Mzcn \\ -Vycn \\ v_{sn} \\ \phi_{sn} \\ Mzsn \\ -Vysn \end{Bmatrix} \qquad \ldots\ldots\ldots\ldots(2)$$

2.2 Transfer matrices

Based on the above form of state vector, the field, point and bearing transfer matrices can be written as follows

Field matrix

$$\underset{32X32}{[\bar{F}]} = \begin{bmatrix} [F] & [\emptyset] & [\emptyset] & [\emptyset] \\ [\emptyset] & [F] & [\emptyset] & [\emptyset] \\ [\emptyset] & [\emptyset] & [F] & [\emptyset] \\ [\emptyset] & [\emptyset] & [\emptyset] & [F] \end{bmatrix} \qquad \ldots\ldots(3)$$

where,

$$\underset{8X8}{[F]} = \begin{bmatrix} [f] & [\emptyset] \\ [\emptyset] & [f] \end{bmatrix}$$

$$\underset{4x4}{[f]} = \begin{bmatrix} 1 & \ell & \ell^2/2EI & \ell^3/6EI \\ \emptyset & 1 & \ell/EI & \ell^2/2EI \\ \emptyset & \emptyset & 1 & \ell \\ \emptyset & \emptyset & \emptyset & 1 \end{bmatrix}$$

Point matrix

$$\underset{32x32}{[P]} = \begin{bmatrix} [Pm] & [\emptyset] & [\emptyset] & [\emptyset] \\ [\emptyset] & [Pn] & [\emptyset] & [\emptyset] \\ [\emptyset] & [\emptyset] & [Pm] & [\emptyset] \\ [\emptyset] & [\emptyset] & [\emptyset] & [Pn] \end{bmatrix} \qquad \ldots\ldots\ldots(4)$$

where,

$$[Pm]_{8\times8} = \begin{bmatrix} 1 & 0 & 0 & 0 & 0 & 0 & 0 & 0 \\ 0 & 1 & 0 & 0 & 0 & 0 & 0 & 0 \\ 0 & 0 & 1 & 0 & 0 & 0 & 0 & 0 \\ m\omega_m^2 & 0 & 0 & 1 & 0 & 0 & 0 & 0 \\ 0 & 0 & 0 & 0 & 1 & 0 & 0 & 0 \\ 0 & 0 & 0 & 0 & 0 & 1 & 0 & 0 \\ 0 & 0 & 0 & 0 & 0 & 0 & 1 & 0 \\ 0 & 0 & 0 & 0 & m\omega_m^2 & 0 & 0 & 1 \end{bmatrix}$$

[Pn] can be obtained by substituting ω_n in place of ω_m in the above matrix.

Bearing matrix

The bearing matrix for a general 8 coefficients bearing is given as,

$$[B]_{32\times32} = \begin{bmatrix} [Bmz_1] & [0] & [Bmz_2] & [0] \\ [0] & [Bnz_1] & [0] & [Bnz_2] \\ [Bmy_2] & [0] & [Bmy_1] & [0] \\ [0] & [Bny_1] & [0] & [Bny_2] \end{bmatrix} \qquad \ldots\ldots\ldots(5)$$

where,

$$[Bmz_1] = \begin{bmatrix} 1 & 0 & 0 & 0 & 0 & 0 & 0 & 0 \\ 0 & 1 & 0 & 0 & 0 & 0 & 0 & 0 \\ 0 & 0 & 1 & 0 & 0 & 0 & 0 & 0 \\ -Kzz+m\omega_m^2 & 0 & 0 & 1 & -Czz\,\omega_m & 0 & 0 & 0 \\ 0 & 0 & 0 & 0 & 1 & 0 & 0 & 0 \\ 0 & 0 & 0 & 0 & 0 & 1 & 0 & 0 \\ 0 & 0 & 0 & 0 & 0 & 0 & 1 & 0 \\ -Czz\,\omega_m & 0 & 0 & 0 & -Kzz+m\omega_m^2 & 0 & 0 & 1 \end{bmatrix}$$

$$[Bmz_2] = \begin{bmatrix} 0 & 0 & 0 & 0 & 0 & 0 & 0 & 0 \\ 0 & 0 & 0 & 0 & 0 & 0 & 0 & 0 \\ 0 & 0 & 0 & 0 & 0 & 0 & 0 & 0 \\ Kzy & 0 & 0 & 0 & Czy\,\omega_m & 0 & 0 & 0 \\ 0 & 0 & 0 & 0 & 0 & 0 & 0 & 0 \\ 0 & 0 & 0 & 0 & 0 & 0 & 0 & 0 \\ 0 & 0 & 0 & 0 & 0 & 0 & 0 & 0 \\ -Czy\,\omega_m & 0 & 0 & 0 & Kzy & 0 & 0 & 0 \end{bmatrix}$$

Matrices $[Bnz_1]$ and $[Bnz_2]$ can be obtained from $[Bmz_1]$ and $[Bmz_2]$ respectively by replacing ω_m with ω_n.

Matrices $[Bmy_1]$ and $[Bmy_2]$ are obtained from $[Bmz_1]$ and $[Bmz_2]$ respectively by replacing subscripts z with y, and y with z. Similarly, matrices $[Bny_1]$ and $[Bny_2]$ can be easily obtained.

Point matrix for unbalance mass

From figures 2 through 4, the equilibrium relations at an unbalance station on inner rotor can be written as

$$V_z^R = V_z^L + m\ddot{w} - \omega_m^2 U_z \cos\omega_m t + \omega_m^2 U_y \sin\omega_m t$$
$$V_y^R = V_y^L + m\ddot{v} - \omega_m^2 U_y \cos\omega_m t - \omega_m^2 U_z \sin\omega_m t \qquad \ldots\ldots\ldots\ldots(6)$$

Substituting w and v from equation (1) and equating the cosine and sine components, following relations are obtained,

$$V_{zcm}^R = V_{zcm}^L - m\omega_m^2 w_{cm} - \omega_m^2 U_z$$
$$V_{zsm}^R = V_{zsm}^L - m\omega_m^2 w_{sm} + \omega_m^2 U_y$$
$$V_{zcn}^R = V_{zcn}^L - m\omega_n^2 w_{cn}$$
$$V_{zsn}^R = V_{zsn}^L - m\omega_n^2 w_{sn}$$
$$V_{ycm}^R = V_{ycm}^L - m\omega_m^2 v_{cm} - \omega_m^2 U_y$$
$$V_{ysm}^R = V_{ysm}^L - m\omega_m^2 v_{sm} - \omega_m^2 U_z$$
$$V_{ycn}^R = V_{ycn}^L - m\omega_n^2 v_{cn}$$
$$V_{ysn}^R = V_{ysn}^L - m\omega_n^2 v_{sn} \qquad \ldots\ldots\ldots(7)$$

Hence, point matrix for unbalance mass on inner rotor can be written as

$$\begin{Bmatrix} \{Szm\} \\ \{Szn\} \\ \{Sym\} \\ \{Syn\} \\ 1 \end{Bmatrix}^R = \begin{bmatrix} [Pm] & [0] & [0] & [0] & \{M_{zm}^I\} \\ [0] & [Pn] & [0] & [0] & \{M_{zn}^I\} \\ [0] & [0] & [Pm] & [0] & \{M_{ym}^I\} \\ [0] & [0] & [0] & [Pn] & \{M_{yn}^I\} \\ \{0\}^T & \{0\}^T & \{0\}^T & \{0\}^T & 1 \end{bmatrix} \begin{Bmatrix} \{Szm\} \\ \{Szn\} \\ \{Sym\} \\ \{Syn\} \\ 1 \end{Bmatrix}^L \qquad \ldots\ldots\ldots(8)$$

where,

$$\{M_{zm}^I\}_{8\times1} = \begin{Bmatrix} 0 \\ 0 \\ 0 \\ -U_z\omega_m^2 \\ 0 \\ 0 \\ 0 \\ U_y\omega_m^2 \end{Bmatrix} ; \quad \{M_{ym}^I\}_{8\times1} = \begin{Bmatrix} 0 \\ 0 \\ 0 \\ U_y\omega_m^2 \\ 0 \\ 0 \\ 0 \\ U_z\omega_m^2 \end{Bmatrix}$$

$$\{M_{zn}^I\}_{8\times1} = \{M_{yn}^I\}_{8\times1} = \{0\}$$

Similarly, the point mass for unbalance on the outer rotor can be obtained by replacing the superscript I by O and putting

$$\{M_{zn}^O\}_{8\times1} = \begin{Bmatrix} 0 \\ 0 \\ 0 \\ -U_z\omega_n^2 \\ 0 \\ 0 \\ 0 \\ U_y\omega_n^2 \end{Bmatrix} ; \quad \{M_{yn}^O\}_{8\times1} = \begin{Bmatrix} 0 \\ 0 \\ 0 \\ U_y\omega_n^2 \\ 0 \\ 0 \\ 0 \\ U_z\omega_n^2 \end{Bmatrix}$$

$$\{M_{zm}^O\}_{8\times1} = \{M_{ym}^O\}_{8\times1} = \{0\}$$

2.3 Transfer matrix relations and junction conditions

In view of equation (8), for the unbalance response analysis, the transfer matrices will assume dimension of (33x33). From figure 5, following relations are obtained

Transfer matrix relations:-

$$\{Z\}^{L}_{ij} = [A]\{Z\}_{OA}$$

$$\{Z\}_{OB} = [B]\{Z\}^{R}_{ij}$$

$$\{Z\}^{L}_{oj} = [C]\{Z\}_{OC}$$

$$\{Z\}_{OD} = [D]\{Z\}^{R}_{oj} \qquad \ldots\ldots\ldots(9)$$

Junction conditions

$$\{Z\}^{R}_{ij} = [I]\{Z\}^{L}_{ij} + [Bj]\{Z\}^{L}_{ij} - [Bj]\{Z\}^{L}_{oj}$$

$$= [[I]+[Bj]]\{Z\}^{L}_{ij} - [Bj]\{Z\}^{L}_{oj}$$

Similarly,

$$\{Z\}^{R}_{oj} = [[I]+[Bj]]\{Z\}^{L}_{oj} - [Bj]\{Z\}^{L}_{oj} \qquad \ldots\ldots\ldots(10)$$

From equations (9)and(10), the system equations can be written in matrix form as

$$\begin{bmatrix} [AI] & [BI] & [CI] & [DI] \\ [AO] & [BO] & [CO] & [DO] \end{bmatrix}_{66\times132} \begin{Bmatrix} \{Z\}_{OA} \\ \{Z\}_{OB} \\ \{Z\}_{OC} \\ \{Z\}_{OD} \end{Bmatrix}_{132\times1} = \{0\} \qquad \ldots\ldots\ldots(11)$$

where,

$$[AI] = [[I]+[Bj]][A]$$

$$[BI] = -[B]^{-1}$$

$$[CI] = -[Bj][C]$$

$$[DI] = [0]$$

$$[AO] = -[Bj][A]$$

$$[BO] = [0]$$

$$[CO] = [[I]+[Bj]][C]$$

$$[DO] = -[D]^{-1} \qquad \ldots\ldots\ldots(12)$$

It can be shown that if [Bj] is set equal to null matrix, equation (11) reduces to that of two uncoupled straight rotors.

2.4 Solution of the system equations

Considering free-free ends on both rotors so that all components of moments and shear forces are zero and leaving out 33rd and 66th rows (as they simply give identity equations), and finally taking the constant terms to the right hand side, the system equations assume the form

$$[AA]_{64\times64}\ \{Zo\}_{64\times1} = \{BB\}_{64\times1} \qquad \ldots\ldots\ldots(13)$$

where, {Zo} contains 64 non-zero elements of the state vectors $\{Z\}_{OA}$, $\{Z\}_{OB}$, $\{Z\}_{OC}$ and $\{Z\}_{OD}$. By solving the above equations, unknown state vector components at the rotor ends are obtained; and with the help of which the state vector containing the unbalance response can be determined.

3 RESULTS AND DISCUSSION

Based on the present formulation, a general computer program has been developed and some preliminary results obtained are reported. A simple dual rotor system as shown in figure 6 is considered.Unbalance on each mass is taken equal to 0.001 kg-m(unbalances on each rotor are in phase).Shaft diameter is equal to 1.5 cm and shaft elasticity(E) is equal to 210×10^9 N/m .

3.1 Rotors supported on isotropic bearings

Case 1: Rigid supports with no intershaft bearings, i.e.,support stiffnesses are assumed to be very large(Ksupport/Kshaft $\sim 10^6$ in the present analysis) and intershaft bearing is assumed to have zero stiffnesses. In such a situation the dual rotor system gets degenerated into two straight rotors running at different speeds.Also closed form solution for each decoupled rotor can be obtained. The results obtained through dual rotor program are compared with the following exact solution of each rotor derived using influence co-efficients.

$$w_1 = \frac{\{\alpha_{11} + m_2\omega^2(\alpha_{12}^2 - \alpha_{11}\alpha_{22})\}P_1 + \alpha_{12}P_2}{(1-m_1\alpha_{11}\omega^2)(1-m_2\alpha_{22}\omega^2) - m_1 m_2 \alpha_{12}^2\omega^4} \qquad \ldots\ldots\ldots(14)$$

The response is checked at two speeds i.e., 2000 and 2500 RPM which is 0.254mm and 0.135mm respectively , and matches exactly with the closed form solution given by equation (14).

Case2: All supports including intershaft bearing are assumed to be rigid(Ksupport/Kshaft $\sim 10^6$ in the present analysis) so as to simulate rigid link between the two shafts.Results obtained at junction on inner and outer rotors are

$$w_{cij}(t) = w_{coj}(t) = 0.0$$

$$w_{sij}(t) = w_{soj}(t) = w_{sm}\sin\ t + w_{sn}\sin\ t$$
$$= -0.1883\times10^{-3}\ \sin 209.44t$$
$$-0.1557\times10^{-2}\ \sin 261.88t$$

As from physical consideration the components of deflection and slope at the junction point on the inner and outer rotors should be equal, which indeed is the case as shown above.

Case 3: A general case of support and intershaft bearing stiffnesses is considered here,i.e.,

Ksupport = 1 MN/m
Kintershaft = 1 MN/m
(in both z- and y- directions)

As is well known that the superposition of two perpendicular harmonic motions, with same angular frequency, give rise in general to an elliptical motion. Thus, in case of a straight rotor, the whirl orbits are either circular or elliptical depending upon the nature of the support stiffnesses. However, for a dual rotor the unbalance response will consist of two different response components belonging to the inner and outer rotors running at different speeds. The individual motion w(t) and v(t) given by equation(1) will be periodic (for $\omega_m \simeq \omega_n$, beatig phenomena will be exhibited) whose period is given by

$$T = 2\pi/[(\omega_n-\omega_m)/(p-q)] \quad ....(15)$$

where, p and q are the smallest integers such that

$$p/q = \omega_n/\omega_m$$

But the superposition of w(t) and v(t) will amount to combination of two distinct orbital motions of different angular frequencies. Hence, the resulting motion will not describe any single closed orbit ; but will exhibit periodicity whose period is given by equation (15). The number of loops in one period is given by

$$n = \{Max[\omega_n,\omega_m]\}\cdot\left\{\frac{(p-q)}{(\omega_n-\omega_m)}\right\} \quad ....(16)$$

For =1200 RPM , = 1000 RPM,
T= 0.6 sec ; n=6

Figures 7 and 8 depict the whirl orbits in one period of the respective cases at the 2nd mass from left on the inner rotor and figure 9 that at 2nd mass on the outer rotor. It is noted that the size of orbits varies between lower and upper limits(for $\omega_n \simeq \omega_m$, the lower limit would be an infinitesimally small orbit).

3.2 Dual rotor system with hydrodynamic bearings

Dual rotor system considered for analysis is shown in figure 10

Case 1 : Intershaft bearings

Kyy = Kyz = Kzy = Kzz = 0.0
Cyy = Cyz = Czy = Czz = 0.0

Support bearings

Kyy = 17.5 MN/m
Kyz = 10.0 MN/m
Kzy = 10.0 MN/m
Kzz = 17.5 MN/m
Cyy = 0.0
Cyz = 0.0
Czy = 0.0
Czz = 0.0

As is obvious, the dual rotor system gets decoupled and the results obtained through computer program are identical with the result available in literature (5).

Case 2: For general representation, support and intershaft bearings are considered as follows

Kyy = 17.5 MN/m
Kyz = 10.0 MN/m
Kzy = 10.0 MN/m
Kzz = 17.5 MN/m
Cyy = 700.0 KN-s/m
Cyz = 400.0 KN-s/m
Czy = 700.0 KN-s/m
Czz = 400.0 KN-s/m

The unbalance response is shown in figures 11 and 12. As expected, they exhibit elliptical loops; the time period T and number of loops n are given by equations (15) and (16) respectively.

CONCLUSION

The 'extended transfer matrix method' is an exact method and a conventional tool for determining the steady-state unbalance response, critical speeds of a dual rotor system . The method can also be extended for stability analysis of dual rotor system.

ACKNOWLEDGEMENT

This work was supported by a grant from Aeronautics Research and Development Board, Government of India and is gratefully acknowledged. Help received from BHEL , Ramachandrapuram is also appreciated.

REFERENCES

(1) CHILDS, D. W. Modal transient rotordynamic model for dual rotor jet engine. TRANS ASME , Aug 1976, PP 876-882.

(2) LI, D. F. and GUNTER, E. J. Component mode Synthesis of Large Rotor Systems. TRANS ASME, Vol 104, July 1982 , PP 552-559.

(3) LI, Q. and HAMILTON, J. F. Investigation of the Transient Response of a Dual-Rotor System with Intershaft Squeeze Film Damper. Journal of Engineering for Gas Turbines and Power, Vol.108, Oct 1986, PP 613-618.

(4) LI, Q. , YAN, L. and HAMILTON, J. F. Investigation of steady-state response of a dual rotor system with intershaft squeeze film damper , Journal of Engineering for Gas Turbines and Power, Vol.108, Oct 1986, PP 605-612.

(5) RAO, J. S., SARMA, K.V.B. and GUPTA, K. Transient Analysis of Rotors by Transfer Matrix Method, Proc. 11th Biennial Conference on Mechanical Vibration and Noise, Boston, Sept. 1987, pp 545-552.

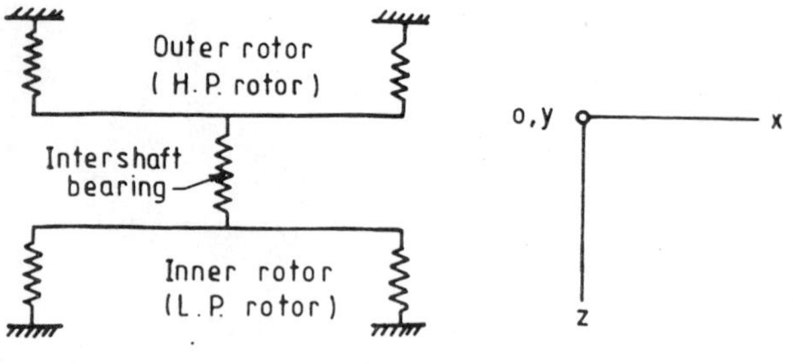

Fig 1 Theoretical model of dual-rotor system

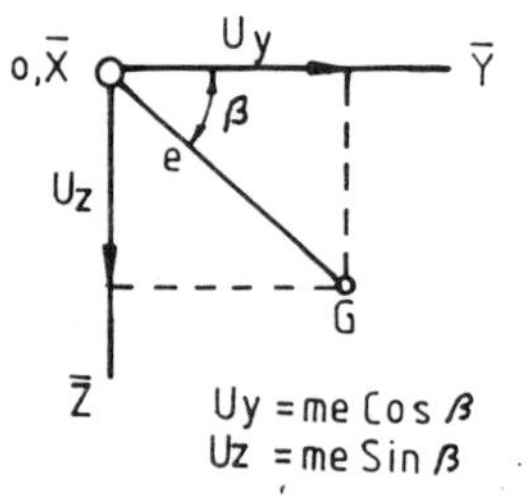

Fig 2 Unbalance in rotor coordinates

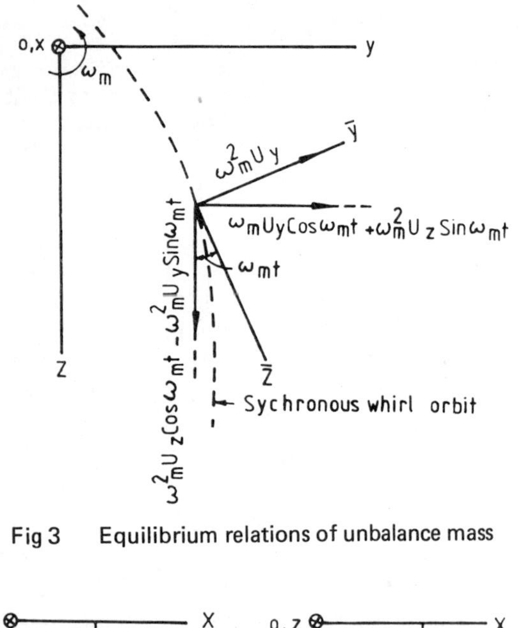

Fig 3 Equilibrium relations of unbalance mass

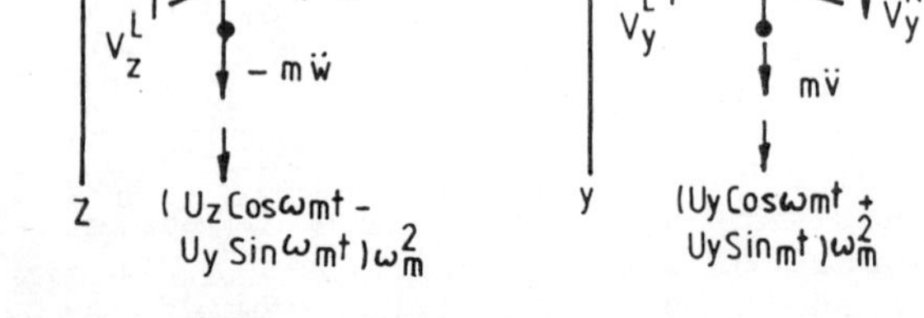

Fig 4 Equilibrium relations of unbalance mass in x–z and x–y planes

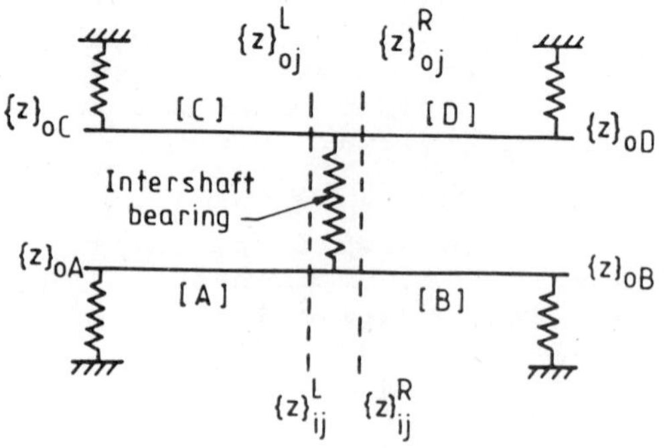

Fig 5 Mathematical model of dual-rotor system

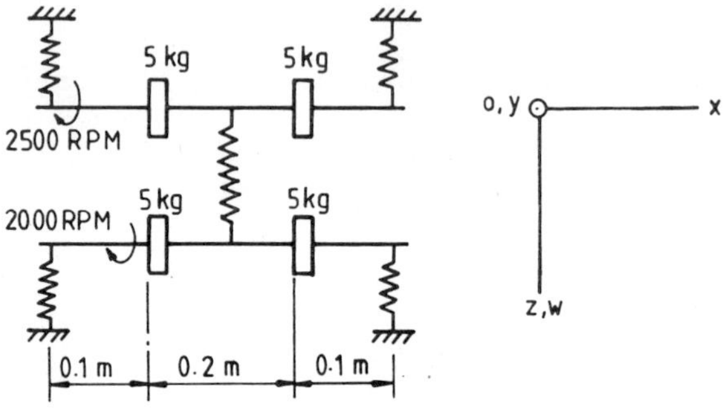

Fig 6 Test model of dual-rotor system

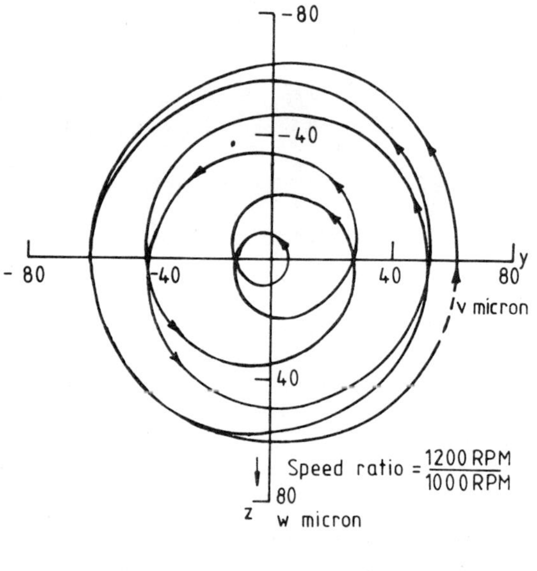

Fig 7 Whirl orbit at inner rotor in one full period

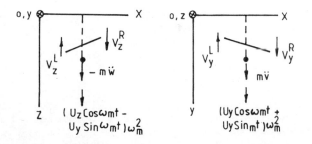

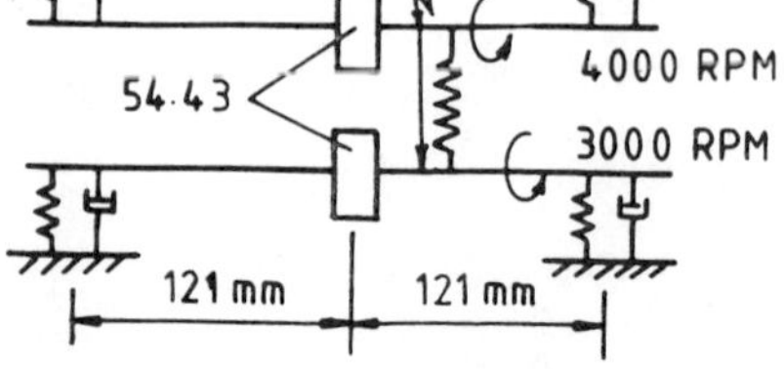

Fig 10 Simple dual-rotor on hydrodynamic bearings

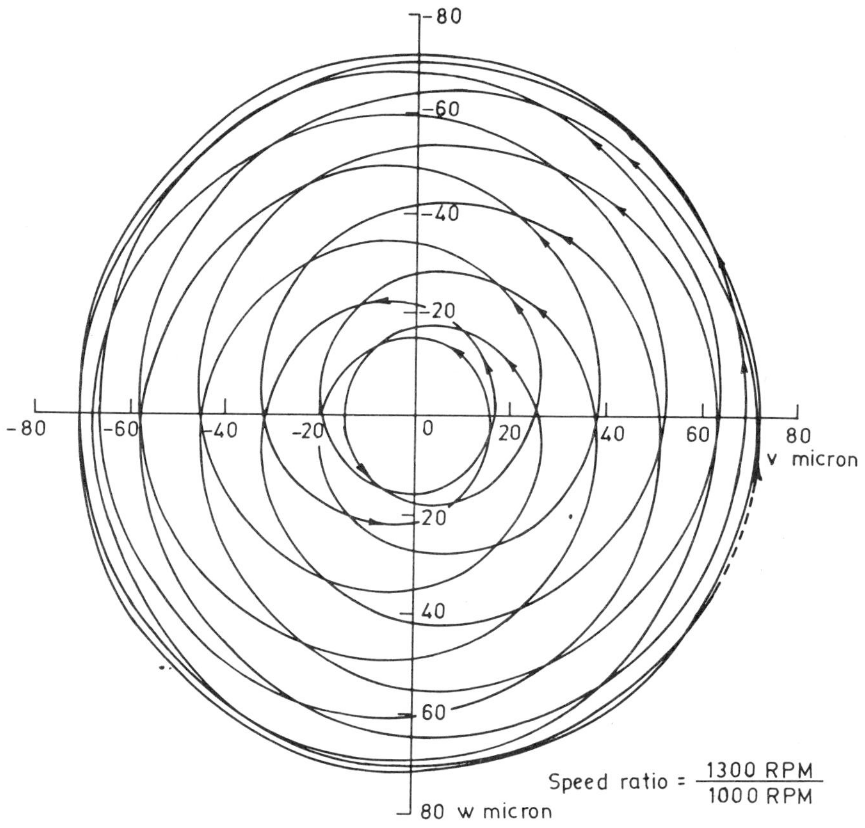

Fig 8 Whirl orbit at inner rotor in one full period

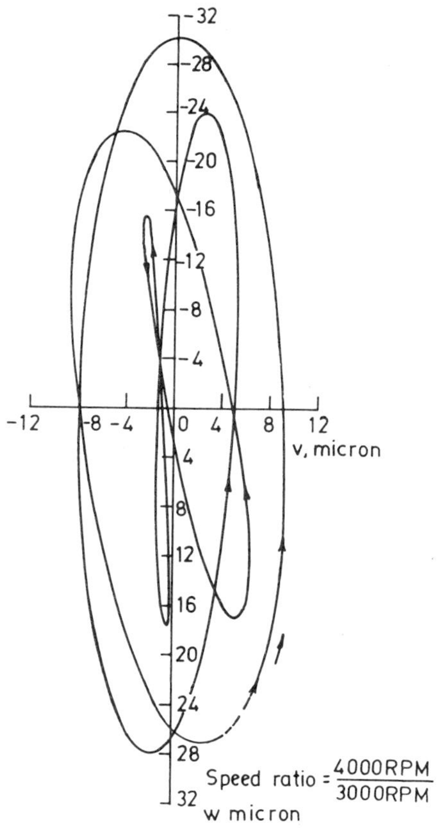

Fig 11 Whirl orbit on inner rotor in one full period (with hydrodynamic bearings)

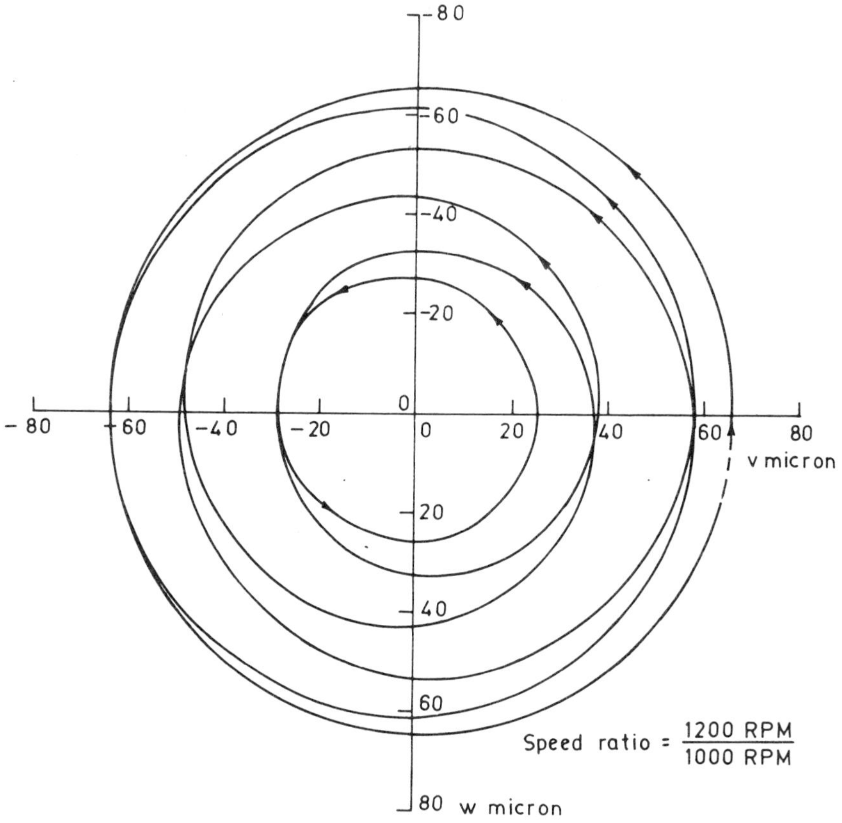

Fig 9 Whirl orbit at outer rotor in one full period

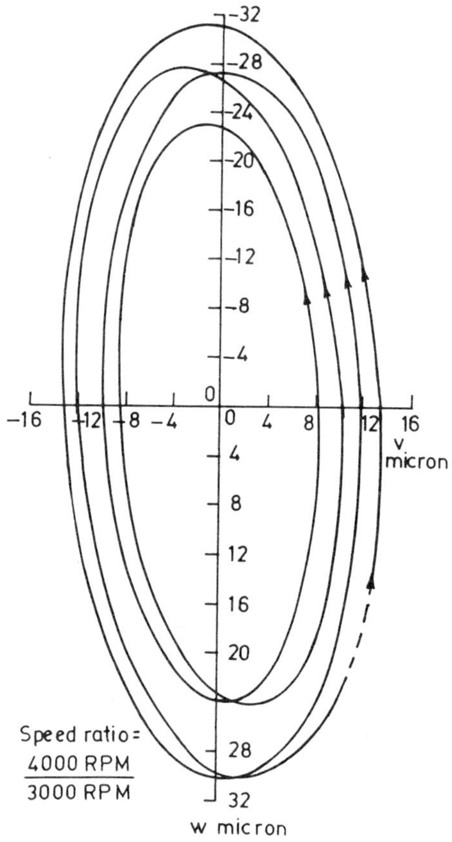

Fig 12 Whirl orbit on outer rotor in one full period (with hydrodynamic bearings)

C245/88

The dynamic analysis of a multi-shaft rotor- bearing-case system

Z-C ZHENG, MCMES, MCSVE and **Y HU**, MSc
Department of Engineering Mechanics, Tsinghua University, Beijing, People's Republic of China

SYNOPSIS The Gyroscopic Mode Synthesis has been further developed to be capable of dynamic analysing of multi-shaft rotor-bearing-case systems. As a numerical example, calculations have been performed for a dual rotor-bearing-case system.

NOTATION

A,B,C	component state vector matrices, modal coordinates
A_s, B_s	system state vector matrices, modal coordinates
f_b	force vector coupled on component boundary coordinates
f_{bn}	non-linear part of force vector coupled by bearings
f_u	external force vector on component
F_s	external force vector on system, modal coordinates
K_{blm}, C_{blm}	stiffness and damping matrices of a intermediary bearing
K_{bls}, C_{bls}	stiffness and damping matrices of a support bearing
M,G,D,K	component mass, gyroscopic damping, damping and stiffness matrices
N_s	non-linear force vector coupled of system
q_k	component modal coordinate vector
u	component generalized coordinate vector
v	system generalized coordinate vector
X	component displacement vector
Y,Z	component state vector, physical coordinates
α, β	transformation matrices
Λ	component precessional frequency matrix
$\underline{\Phi}$, $\overline{\Psi}$	constraint and gyroscopic mode matrices
Ω	spin speed
b,i,s	subscripts for boundary, interior and system
1 or 2	superscripts for shaft component 1 or 2
I or II	superscripts for subsystem I or II

The Project supported by National Natural Science Foundation of China

INTRODUCTION

During the design of an engine it is necessary to know the dynamic effect of case on the vibration behaviour of the entire engine with increased accuracy. Since the case of engines is thin wall shell structures with non-symmetric constraint, it is more reasonable to simulate it using thin shell element rather than beam element. For this situation, the traditional Transfer Matrix Method is not suitable for the dynamic analysis of rotor-case systems. Although Finite Element Method can be applied to such large order systems, it may be time consuming to set up and very costly to solve in terms of computer time and storage.

The method of Component Mode Synthesis allows for substantial reduction in the size of the overall system problem while still retaining the essential dynamic characteristics. Many researchers have presented a variety of approaches to analyse the vibration of rotor-bearing systems successfully [1] [2] [3] [4] [5], ZHENG et al developed a Gyroscopic Mode Synthesis[6]. By introducing the idea of connecting springs and dampers, and using gyroscopic modes instead of complex modes in mode synthesis, the method differs from complex mode synthesis as well as from classical component mode synthesis.

This paper extends the work of ZHENG et al[6], and allows the method for the dynamic analysis of multi-shaft rotor-bearing-case systems. The spoke-case is incorporated with the multi-shaft rotor-bearing subsystem as another subsystem, and the support bearings are considered as separate connection elements. After mode development for every subsystem and component with fixed boundary coordinates, we form the global system equations utilizing the concept of the equilibrium requirements of the boundary forces, in which various complex factors have been taken into account. The introduction of the idea of connecting springs and dampers, and the use of the gyroscopic modes instead of complex modes in the mode synthesis allow for considering various factors such as the non-linearity and asymmetry of bearings and spoke-case, gyroscopic effect of shafts and disks, and damping. Owing to the reduced sufficiently DOF of the system and the asymmetric matrix synthesis equation with real coefficients, the method is also very frugal in computer time and storage compared with the complex mode synthesis.

ANALYTICAL DEVELOPMENT

The system is divided into two main subsystems on the basis of its natural speciality: the multi-shaft rotor-bearing system and the spoke-case system. The support bearings by which the motion of rotor and case is coupled are considered as separate connecting elements.

The Multi-Shaft Rotor-Bearing Subsystem

Mechanical model. When analysing multiple shaft rotor-bearing systems, we can obtain the normal modes of fixed interface and the constraint modes for the components by using the fixed interface component mode synthesis method. The intermediary bearings among rotors are regarded as the coupling elements which include elastic and damping forces. Fig.1 (a) illustrates how to divide the system into components 1 and 2, and their connection by the intermediary and support bearings; (b) is the model for modal analysis of the gyroscopic and constraint modes with fixed boundary points: (c) indicates the coupling forces caused by the bearings and how to form the integrated system through the equilibrium requirements.

The motion equation of a shaft component descretized by FEM is of the form

$$M\ddot{X} + \Omega G\dot{X} + D\dot{X} + KX = \bar{f}_b + f_u \qquad (1)$$

The displacement vector of Eq.(1) is partitioned into boundary and interior coordinates as

$$X = \{X_i^T X_b^{(1)T} X_b^{(1,II)T}\}^T = \{X_i^T X_b^T\}^T \qquad (2)$$

where $X_b^{(1)}$ and $X_b^{(1,II)}$ represent the rotor-rotor and rotor-case boundary coordinates respectively. Correspondingly, the $\bar{f}_b$ consists of two parts. Eq.(1) is arranged in compatible block shape

$$\begin{bmatrix} M_{ii} & M_{ib} \\ M_{bi} & M_{bb} \end{bmatrix}\begin{Bmatrix} \ddot{X}_i \\ \ddot{X}_b \end{Bmatrix} + \Omega\begin{bmatrix} G_{ii} & G_{ib} \\ G_{bi} & G_{bb} \end{bmatrix}\begin{Bmatrix} \dot{X}_i \\ \dot{X}_b \end{Bmatrix} +$$

$$\begin{bmatrix} D_{ii} & D_{ib} \\ D_{bi} & D_{bb} \end{bmatrix}\begin{Bmatrix} \dot{X}_i \\ \dot{X}_b \end{Bmatrix} + \begin{bmatrix} K_{ii} & K_{ib} \\ K_{bi} & K_{bb} \end{bmatrix}\begin{Bmatrix} X_i \\ X_b \end{Bmatrix} = \begin{Bmatrix} 0 \\ \bar{f}_b \end{Bmatrix} + \begin{Bmatrix} f_{ui} \\ f_{ub} \end{Bmatrix} \qquad (3)$$

where M, K and D are symmetric matrices, G is a skew symmetric matrix.

The first term on the right side of Eq.(3) represents the boundary force which may be a non-linear function of spin speed and relative motion of the boundary points between the components. The second is external excitation such as unbalance force.

First order form. Introducing the state vector Z as follows:

$$Z = \{\dot{X}_i^T \dot{X}_b^T X_i^T X_b^T\}^T \qquad (4)$$

Eq.(1) may be rewritten in the first order form

$$\begin{bmatrix} M & 0 \\ 0 & K \end{bmatrix}\dot{Z} + \begin{bmatrix} \Omega G & K \\ -K & 0 \end{bmatrix}Z + \begin{bmatrix} D & 0 \\ 0 & 0 \end{bmatrix}Z = \begin{Bmatrix} \bar{f}_b \\ 0 \end{Bmatrix} + \begin{Bmatrix} f_u \\ 0 \end{Bmatrix} \qquad (5)$$

It is convenient to order the state variables into boundary and interior subsets by applying the transformation

$$Z = \alpha Y \qquad (6)$$

with

$$Y = \begin{Bmatrix} Y_i \\ Y_b \end{Bmatrix} = \{\dot{X}_i^T X_i^T \dot{X}_b^T X_b^T\}^T \qquad (7)$$

to Eq.(5) and premultiplying α^T to yield

$$\begin{bmatrix} A_{ii} & A_{ib} \\ A_{bi} & A_{bb} \end{bmatrix}\begin{Bmatrix} \dot{Y}_i \\ \dot{Y}_b \end{Bmatrix} + \begin{bmatrix} B_{ii} & B_{ib} \\ B_{bi} & B_{bb} \end{bmatrix}\begin{Bmatrix} Y_i \\ Y_b \end{Bmatrix}$$

$$+ \begin{bmatrix} D_{ii} & D_{ib} \\ D_{bi} & D_{bb} \end{bmatrix}\begin{Bmatrix} Y_i \\ Y_b \end{Bmatrix} = \bar{F}_b + F_u \qquad (8)$$

where $\bar{F}_b = \{0 \quad 0 \quad \bar{f}_b^T \quad 0\}^T$

$$F_u = \{f_{ui}^T \quad 0 \quad f_{ub}^T \quad 0\}^T$$

Component mode development. The undamped free vibration equation for the interior coordinates is obtained by setting the boundary coordinate state vector in Eq.(8) to zero

$$A_{ii}\dot{Y}_i + B_{ii}Y_i = 0 \qquad (9)$$

where

$$A_{ii} = \begin{bmatrix} M_{ii} & 0 \\ 0 & K_{ii} \end{bmatrix}, \qquad B_{ii} = \begin{bmatrix} \Omega G_{ii} & K_{ii} \\ -K_{ii} & 0 \end{bmatrix}$$

A_{ii} is symmetric and positive definite. Eq.(9) represents a gyroscopic eigenproblem. According to the method of solving a gyroscopic eigenproblem[7], every precessional frequency ω_l and its assocciated gyroscopic modes Φ_l and Ψ_l $(l=1, 2, \ldots, k)$ can be obtained and arranged to form a tri-diagonal skew symmetric frequency matrix Λ_k and a gyroscopic modal matrix Ψ_k

$$\Lambda_k = Block-diag\begin{pmatrix} 0 & \omega_l \\ -\omega_l & 0 \end{pmatrix} \qquad (10)$$

$$\Psi_k = \left[\Phi_1 \Psi_1 \cdots\cdots \Phi_k \Psi_k\right] = \begin{bmatrix} \Psi_{kv} \\ \Psi_{kx} \end{bmatrix} \qquad (11)$$

where Ψ_k has been normalized, satisfying the generalized orthogonal relations

$$\Psi_k^T A_{ii} \Psi_k = I, \qquad \Psi_k^T B_{ii} \Psi_k = \Lambda_k \qquad (12)$$

The static constraint modes are determined as follows:

$$\Phi = -K_{tt}^{-1}K_{tb} \quad , \quad \underline{\Phi} = \begin{bmatrix} \Phi & 0 \\ 0 & \Phi \end{bmatrix} \qquad (13)$$

The modal transformation may then be given

$$Y = \begin{Bmatrix} Y_t \\ Y_b \end{Bmatrix} = \begin{bmatrix} \Psi_k & \underline{\Phi} \\ 0 & I \end{bmatrix} \begin{Bmatrix} q_k \\ Y_b \end{Bmatrix} = \beta u \qquad (14)$$

In order to reduce the DOF, matrix Ψ_k includes only the first k order (pairs) kept modes of the component.

Substituting Eq.(14) into the state vector Eq.(8) and premultiplying by β^T, we obtain

$$A\dot{u} + (B + C)u = F + F_b \qquad (15)$$

$$A = \begin{bmatrix} I & A_{12} & 0 \\ A_{21} & A_{22} & 0 \\ 0 & 0 & A_{33} \end{bmatrix} , \quad B = \begin{bmatrix} A_k & B_{12} & 0 \\ B_{21} & B_{22} & B_{23} \\ 0 & B_{32} & 0 \end{bmatrix}$$

$$C = \begin{bmatrix} C_{11} & C_{12} & 0 \\ C_{21} & C_{22} & 0 \\ 0 & 0 & 0 \end{bmatrix} , \quad F = \begin{Bmatrix} \Psi_{kv}^T f_{ut} \\ \Phi^T f_{ut} + f_{ub} \\ 0 \end{Bmatrix}$$

$$F_b = \begin{Bmatrix} 0 \\ f_b \\ 0 \end{Bmatrix} , \quad u = \begin{Bmatrix} q_k \\ \dot{X}_b \\ X_b \end{Bmatrix}$$

where A is symmetric, and B is skew symmetric. The sub-matrices are expressed from the sub-elements of M, D, G and K as follows:

$$A_{12} = \Psi_{kv}^T M_{tt}\Phi + \Psi_{kv}^T M_{tb}$$

$$A_{22} = \Phi^T M_{tt}\Phi + \Phi^T M_{tb} + M_{bt}\Phi + M_{bb}$$

$$A_{33} = B_{23} = \Phi^T K_{tb} + K_{bb}$$

$$C_{11} = \Psi_{kv}^T D_{tt}\Psi_{kv}$$

B12, C12, and B22, C22 have the same form as A12 and A22 but using $\Omega \cdot G$ or D instead of M.

The boundary forces by bearings. The linear characteristics of stiffness and damping of a bearing between component 1 and 2 has the expression

$$K_{bml} = \begin{bmatrix} K_{yy} & K_{yz} \\ K_{zy} & K_{zz} \end{bmatrix} , \quad C_{bml} = \begin{bmatrix} C_{yy} & C_{yz} \\ C_{zy} & C_{zz} \end{bmatrix}$$

For systems with several bearings, the matrices are augmented to form the following matrices

$$K_{bm} = Block - diag(K_{bml})$$

$$C_{bm} = Block - diag(K_{bml})$$

Thus, the boundary forces between two components may be expressed by

$$\begin{Bmatrix} f_b^{(1,2)} \\ f_b^{(2,1)} \end{Bmatrix} = -\begin{bmatrix} C_{bm} & K_{bm} & -C_{bm} & -K_{bm} \\ -C_{bm} & -K_{bm} & C_{bm} & K_{bm} \end{bmatrix} \begin{Bmatrix} \dot{X}_b^{(1)} \\ X_b^{(1)} \\ \dot{X}_b^{(2)} \\ X_b^{(2)} \end{Bmatrix} + \begin{Bmatrix} f_{bn}^{(1,2)} \\ f_{bn}^{(2,1)} \end{Bmatrix} \qquad (16)$$

where the second term on the right side of the equation is non-linear term.

Subsystem equation of motion. Through the equilibrium requirements of the coupling forces caused by the intermediary bearings, we obtain the truncated subsystem equation

$$A_I\dot{v}_I + B_I v_I = F_b^{(I,II)} + F_I + N_I \qquad (17)$$

where $F_b^{(I,II)}$ is the coupling forces caused by the support bearings and F_I is the external force vector, N_I is the nonlinear part of intermediary bearing forces.

The Spoke-Case Subsystem

Fig.2 (a) illustrates the realistic spoke-case structure, the support bearings are fixed to the spokes: (b) is the model for modal analysis with fixed boundary; (c) indicates the coupling forces caused by the support bearings.

Because of the thin shell configuration and the non-symmetric constraints, the case can be modelled into thin shell elements and the spoke into truss elements. The motion equation is of the following form

$$M\ddot{X} + KX = \bar{f}_b + f_u \qquad (18)$$

Similarly, we develop the modal analysis for the spoke-case system in the same steps as for the rotor system. Finally, we also obtain the modal transformation which gives a truncated subsystem equation of sufficient reduced DOF

$$A_{II}\dot{v}_{II} + B_{II}v_{II} = F_b^{(II,I)} + F_{II} + N_{II} \qquad (19)$$

$$A_{II} = \begin{bmatrix} I & A_{12} & 0 \\ A_{21} & A_{22} & 0 \\ 0 & 0 & A_{33} \end{bmatrix} , \quad B_{II} = \begin{bmatrix} A_k & 0 & 0 \\ 0 & 0 & B_{23} \\ 0 & B_{32} & 0 \end{bmatrix}$$

$$v_{II} = \begin{Bmatrix} q_k^{(II)} \\ \dot{X}_b^{(II)} \\ X_b^{(II)} \end{Bmatrix} , \quad F_b^{(II,I)} = \begin{Bmatrix} 0 \\ f_b^{(II,I)} \\ 0 \end{Bmatrix}$$

$$F_{II} = \begin{Bmatrix} \Psi_{kv}^T \quad f_{ut}^{(II)} \\ \Phi^T \quad f_{ut}^{(II)} + f_{ub}^{(II)} \\ 0 \end{Bmatrix}$$

where $F_b^{(II,I)}$ is the coupling forces of the support bearings and F_{II} is the external force vector on the spoke-case, A_{II} is symmetric and B_{II} is skew symmetric. The sub-matrices have the same expression form as in the rotor system

System Equation Assembly and its Solution

As for the support bearings which couples the motion of rotor and case, we have similar boundary force expression to Eq.(16)

$$\begin{Bmatrix} f_{bs}^{(I,III)} \\ f_{bs}^{(III,I)} \end{Bmatrix} = -\begin{bmatrix} C_{bs} & K_{bs} & -C_{bs} & -K_{bs} \\ -C_{bs} & -K_{bs} & C_{bs} & K_{bs} \end{bmatrix} \begin{Bmatrix} \dot{X}_b^{(I)} \\ X_b^{(I)} \\ \dot{X}_b^{(III)} \\ X_b^{(III)} \end{Bmatrix} + \begin{Bmatrix} f_{bn}^{(I,III)} \\ f_{bn}^{(III,I)} \end{Bmatrix} \qquad (20)$$

where $K_{bs} = Block-diag(K_{bsl})$

$C_{bs} = Block-diag(C_{bsl})$

$$K_{bsl} = \begin{bmatrix} K_{yy} & K_{yz} \\ K_{zy} & K_{zz} \end{bmatrix}, \qquad C_{bsl} = \begin{bmatrix} C_{yy} & C_{yz} \\ C_{zy} & C_{zz} \end{bmatrix}$$

System equation of motion. Collection of all the modal coordinates of each subsystem and boundary coordinates, and arrangement of them in sequence, yield the overall set of independent coordinates of system

$$v_s = (q_I^{(1)T}, q_I^{(2)T}, \cdots, q_I^{(s)T}, q_{II}^T, Y_{bI}^{(1,2)T}, Y_{bI}^{(2,1)T}, \cdots,$$

$$Y_{bI}^{(s-1,s)T}, Y_{bI}^{(s,s-1)T}, Y_{bII}^{(1)T}, Y_{bII}^{(2)T}, \cdots, Y_{bII}^{(l)T})^T, \quad (l \le s) \qquad (21)$$

We sum over all subsystems utilizing the equilibrium condition of the boundary force (caused by the support bearings: expression (20)), to give

$$A_s \dot{v}_s + B_s v_s = F_s + N_s \qquad (22)$$

where A_s and B_s is the assemblage of two subsystem matrices A_I, A_{II} and B_I, B_{II}, respectively. The matrices of linear stiffness and damping of the bearings are arranged in the lower right area corresponding to the boundary coordinates. F_s is the collection of the external force vectors with modal coordinates such as unbalance force and inertia force caused by base motion. Whereas N_s is the nonlinear part of bearing coupling force. The nonlinearities in the component can also be contained in N_s, and the details can be consulted in author's work[8][9].

We have obtained the truncated overall system equation in which various complex factors have been considered. Owing to the separation of the system into subsystems, every subsystem and component can be analyzed in detail and the asymmetry and nonlinearity of the bearings can be fully taken into account. Using the gyroscopic modal synthesis, the order of equation can be reduced sufficiently and calculation can be simplified.

Eigenvalue and response analysis. In the homogeneous form of Eq.(22), let

$$v_s = V e^{\lambda t}$$

This yields

$$(B_s + \lambda A_s) V = 0 \qquad (23)$$

Using a complex eigenvalue program of real coefficient matrices, the complex eigenvalues λ_r and the associated complex modes V_r can be obtained

$$\lambda_r = n_r + j\omega_r \quad , \quad V_r$$

The ω_r is the damped natural frequencies of whirl and the n_r represents the damping exponent which is negative for stable modes. The critical speeds are determined using a whirl speed map.

It is easy to obtain the linear response for the known external force vector $\mathbf{F_s}$. For example, in the case of unbalance response, let

$$F_s = F_y \cos\Omega t + F_z \sin\Omega t = F_1 e^{j\Omega t} + F_2 e^{-j\Omega t} \qquad (24)$$

and substitute Eq.(24) into Eq.(22), to yield

$$(j\Omega A_s + B_s) X_1 = F_1$$

$$(-j\Omega A_s + B_s) X_2 = F_2$$

Then, X_1 and X_2 solved from above equations provide conjugate vector pair.

The nonlinear response can be calculated by applying numerical integration.

NUMERICAL EXAMPLE

In a dual rotor-case system shown in Fig.3 and 4, inner shaft (component 1) and outer shaft (component 2) are modelled as an assemblage of 8 and 6 elements respectively. The case is fixed by three simple supports. Three support bearings are mounted on three sets of spokes separately, and the eight spokes of each set are distributed on the case. The case is modelled with 55 thin shell elements and each spoke with a truss element. All axial freedom is neglected. In order to determine the first 5 forward frequencies of the system, we retain the 12, 12 and 10 modes of inner shaft, outer shaft and spoke-case in synthesis.

In realistic working process, the spin speeds of both shafts are varying, we only chose the following case for calculation: Ω_1 =6000 rpm, Ω_2=5000-20000 rpm, there is an unbalance amount at node 1 on the outer shaft.

The critical speeds determined by whirl map (Fig.5) are

n_{cr1}=6800rpm, n_{cr2}=7000rpm

n_{cr3}=11200rpm, n_{cr4}=13250rpm

n_{cr5}=18000rpm,

The approximate modes for the first 3 critical speeds are shown in Fig.6, Fig.7 and Fig.8 display the curves of unbalance response of node 4 on the inner shaft and node 3 on the case. It can be seen the response crests well correspond with critical speeds.

CONCLUSION

The method of Gyroscopic Mode Synthesis has been developed to analyse the dynamic characteristics of multi-shaft rotor-bearing-case systems, and a general computer program, DAPOR-2 is written based on the method. The foundation system is incorporated with rotor system by introducing the connecting elements and the overall equation with order having been significantly reduced is given after mode development for both rotors and foundation.

The following advantages in the application of Gyroscopic Mode Synthesis are further verified in this paper: (1) It can use real mode programs for calculation of the gyroscopic modes in component mode development; (2) The synthesis equation of the system is an asymmetric matrix equation with real coefficients. Both of the advantages offer promise to Gyroscopic Mode Synthesis as a method of decreasing analysis time and cost in comparison with the complex mode synthesis.

ACKNOWLEDGEMENT

The project was supported by the National Natural Science Foundation of China. The authors wish to thank T.Y. Liu, C.X. Yue and L.X. Zhang of Shenyang Aero Engine Research Institute for their assistance.

REFERENCES

(1) HASSELMAN, T. K. and KAPLAN, A. Dynamic Analysis of Large Systems by Complex Mode Synthesis. ASME Journal of Dynamic Systems, Measurement and Control, 1974, Vol.96, No.3, pp.327-333.

(2) GLASGOW, D. A. and NELSON, H. D. Stability Analysis of Rotor-Bearing Systems Using Component Mode Synthesis. ASME Journal of Mechanical Design, 1980, Vol.102, No.2, pp.352-359.

(3) NELSON, H. D. and MEACHAM, W. L. Transient Analysis of Rotor-Bearing Systems Using Component Mode Synthesis. ASME paper No.81-GT-110, 1981 Gas Turbine Conference, Houston, Texas, Mar. 1981.

(4) NELSON, H. D., MEACHAM, W. L., FLEMING, D. P. and KASCAK, A. F. Nonlinear Analysis of Rotor-Bearing System Using Component Mode Synthesis. ASME Journal of Engineering for Power, 1983, Vol.105, pp.606-614.

(5) LI, D. F. and GUNTER, E. J. Component Mode Synthesis of Large Rotor Systems. ASME Journal of Engineering for Power, 1982, Vol.104, No.3, pp.552-560.

(6) ZHENG, Z. C. et al, Gyroscopic Mode Synthesis in the Dynamic Analysis of a Multi-Shaft Rotor-Bearing System. ASME paper No.85-IGT-73, 1985 Beijing International Gas Turbine Symposium and Exposition, Beijing, P.R.C., Sept.1985.

(7) MEIROVITCH, L. A New Method of Solution of the Eigenvalue Problem for Gyroscopic Systems. AIAA Journal, 1974, Vol.12, No.10, pp.1337-1342.

(8) ZHENG, Z. C. and TAN, M. Y. Numerical Method in Dynamic Response of Nonlinear Systems. Applied Mathematics and Mechanics, 1983, Vol.4, No.4, pp.93-101.

(9) ZHENG, Z. C. and TAN, M. Y. The Extension of Modal Synthesis Techniques to Nonlinear Systems. ICNM, Shanghai, China, 1985.

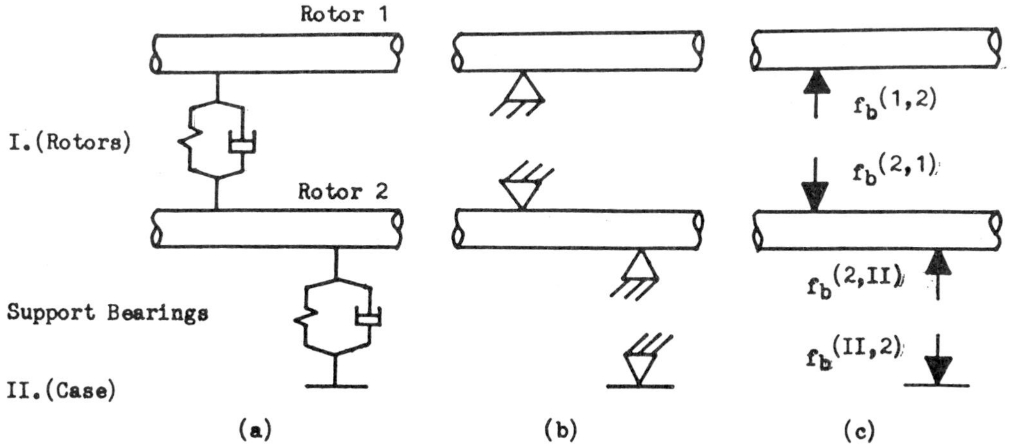

Fig 1 Rotor system divided into components

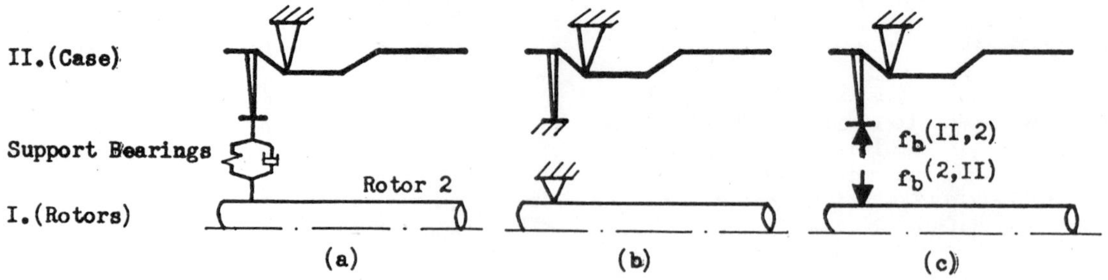

Fig 2 Case system

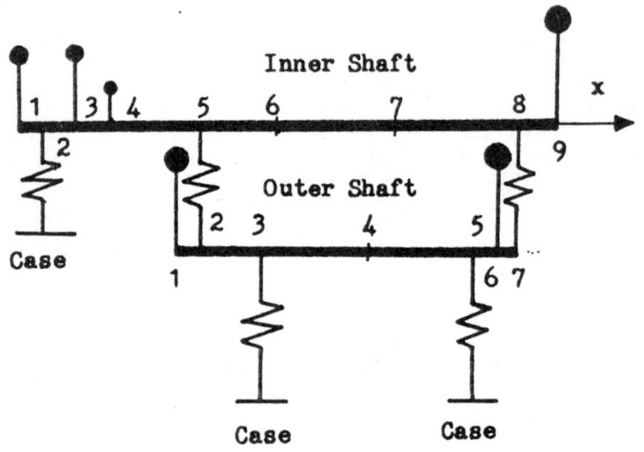

Fig 3 Dual rotors — schematic model

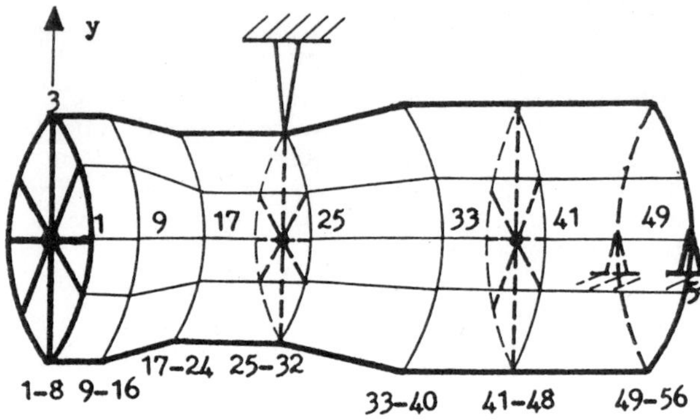

Fig 4 Case — schematic model

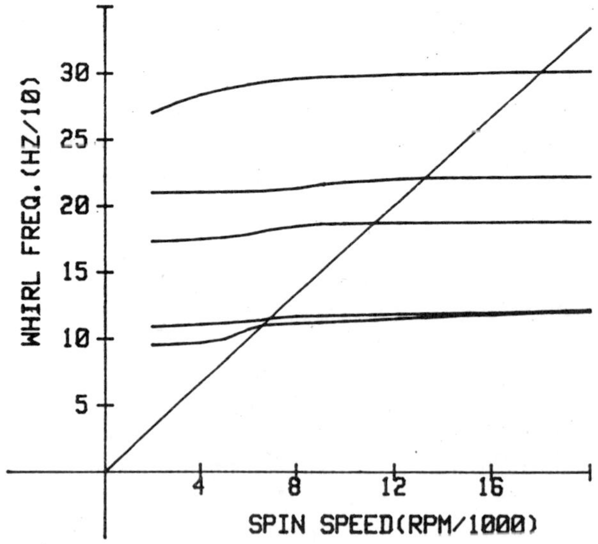

Fig 5 Whirl speed map

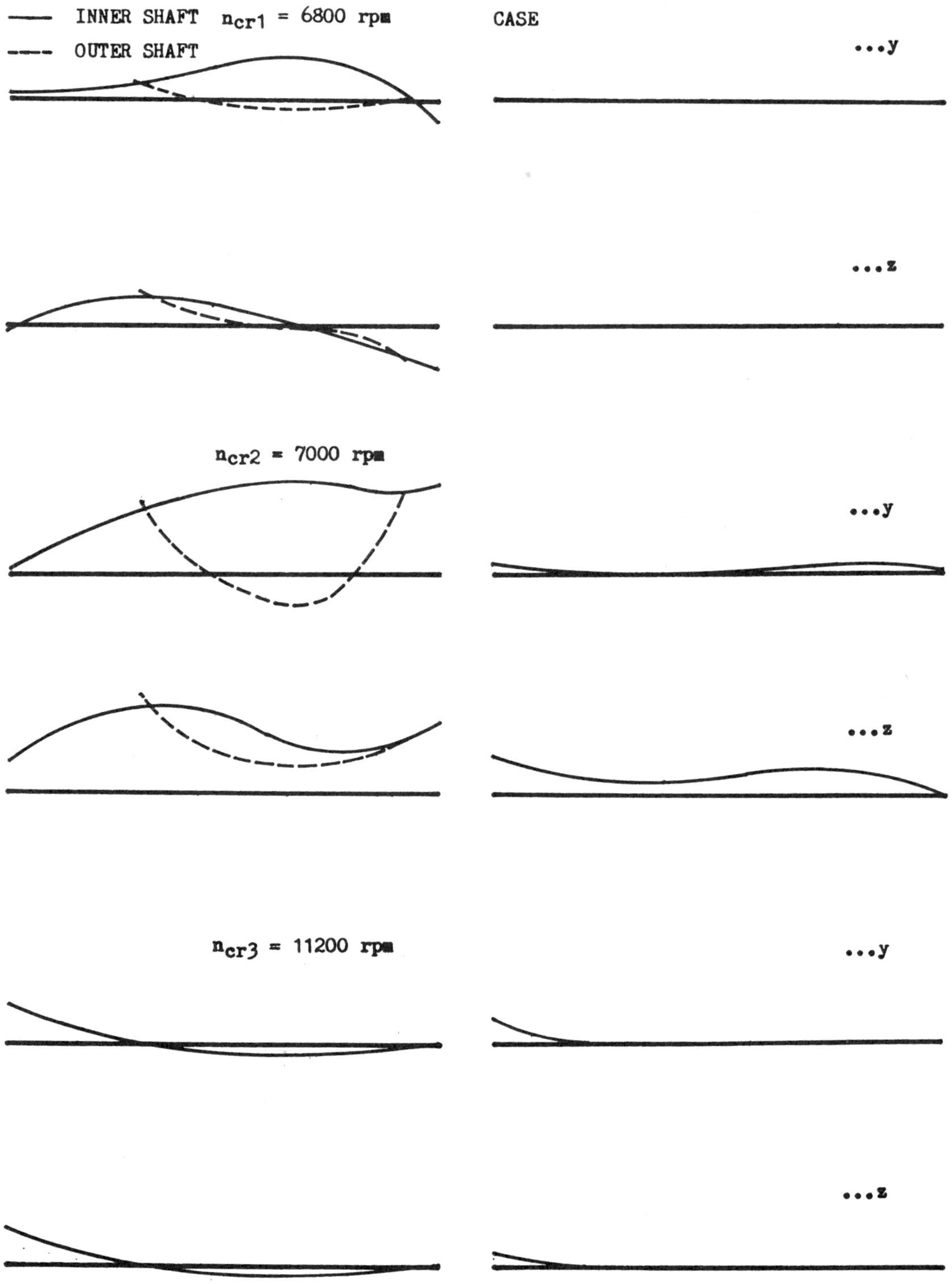

Fig 6 Mode shapes

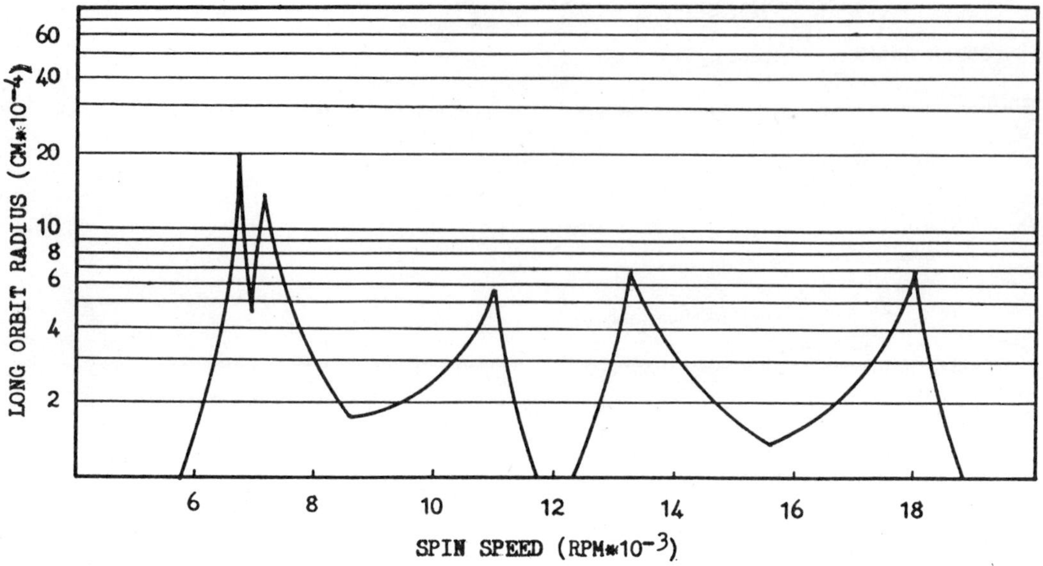

Fig 7 Unbalance response, fourth node on inner shaft

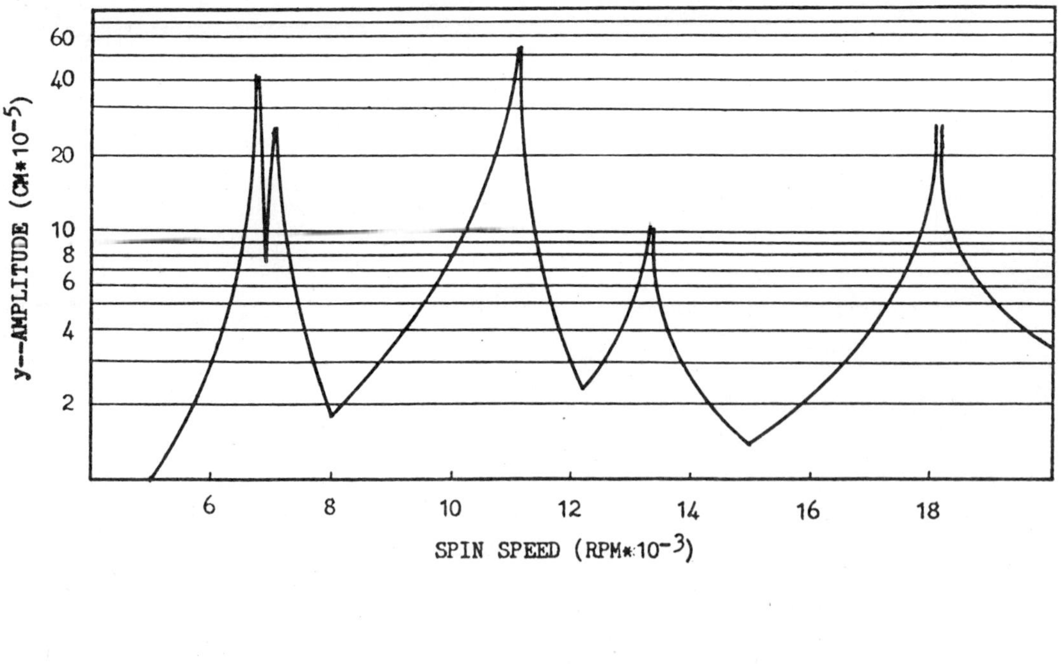

Fig 8 Unbalance response, third node on case

C298/88

A finite element approach to the aeroelastic analysis of flexible rotors attached to flexible supports

A D GARRAD, BA, PhD, CEng, MIMechE, MRAeS, FIMA and **M H PATEL**, BSc, PhD
Garrad Hassan and Partners, London

SYNOPSIS The various means of performing the dynamic analysis of flexible rotors on flexible supports are described and, in particular, the difficulties of using finite element techniques are discussed. A mathematical model which overcomes those difficulties and thus includes the rotation of the rotor relative to the support and aeroelastic effects is derived and some results obtained from it are presented.

1 INTRODUCTION

Any discussion of rotors attached to flexible supports naturally starts with consideration of helicopter dynamics. The helicopter methods predate the recent rapid increase in computing power and tend to use what can be termed the "classical modal" approach. The advantage of such a method is that the final model has relatively few degrees of freedom. The algebra involved in attaining that reduced model may well be substantial, but methods do exist for automating the generation of these equations both algebraically, see for example Garrad and Quarton [1] and numerically, see Done, Juggins and Patel [2].

The peculiarity which separates rotors with flexible supports from other dynamically active structures is the presence of two distinct components: the rotor and the support which are rotating with respect to one another. This relative rotation gives rise to complicated coupling which produces very large algebraic expression during the development of the equation of motion using a Lagrangian approach. It also produces periodic terms in the final equations of motion. Again [1] gives a detailed description of this problem. The advent of fast computers as every day tools and the almost universal use of finite element methods for structural analysis suggests an alternative approach to the modal method. One alternative method, which will be described here, was pioneered by Lobitz [2], for use on wind turbines. It consists of the formation of the mass and stiffness matrices of the rotor and the tower as separate entities which are then coupled together in a suitable way that accounts for their relative rotations. The aeroelastic effects are also included. The result is a very complete final model with the disadvantage that it has a large number of degrees of freedom. The advantage is that the algebra involved in deriving the model is very limited and thus error free and a direct interface is possible between the standard finite element idealisation and the dynamic analysis. It is also very easy to change the structural model for example by adding another blade or changing the nature of the hub.

The purpose of the aerodynamic part of this analysis is twofold: to provide an input excitation to the structure and to provide the aerodynamic damping and hence the aerodynamic feedback. If the load cases are suitably defined the aerodynamic program can compute the variation of distributed loads with azimuth. In this context azimuth is taken to mean the angle through which the rotor has turned about its axis.

The distinction between the two approaches is then quite clear. Whichever method is adopted there is a considerable amount of analysis to be performed. Using the modal method it is done during the derivation of the equation of motion either by hand or using computer algebra [1,2] whereas using the present method it is done numerically by the computer. The advantages and disadvantages of each method can only be properly evaluated by use.

The most obvious applications of this type of approach are wind turbines and helicopters, since for these structures the degree of periodicity in the equations of motion is very high. Other applications include gas turbines and aircraft propellers.

2 BASIC MATHEMATICAL MODEL

The basic mathematical model used in the present analysis will be described in this section. It is conveniently split into two parts: aerodynamics and structural dynamics.

2.1 Aerodynamics

The aerodynamics model used for this project is based on standard strip theory. The blade is divided radially into chordwise strips. A momentum balance is performed for each strip and interference velocities are calculated using an iterative procedure. Once the momentum balance has been obtained the flow may be perturbed to determine aerodynamic damping derivatives and applied loads.

The purpose of the aerodynamic part of this analysis is twofold: to provide an input excitation to the structure and to provide the aerodynamic damping and hence the aeroelastic feedback. If the load cases are suitably defined the aerodynamic program can compute the variation of distributed loads with azimuth. For both the damping and the loads the quantities naturally appear expressed per unit length. Since they are to be used in a finite element analysis they must be integrated to form point values.

The aerodynamic damping matrices and the nodal forces are calculated before entering the dynamic part of the analysis. The nodal forces are conveniently expressed as Fourier series which are then read by the dynamics section, at present the aerodynamic damping matrices are considered constant although a Fourier representation of them could also be used.

2.2 Dynamics

2.2.1 Model Structure

The complete system consists of two major components: the rotor and the tower. The rotor may have any number of blades although to date only one, two or three have been used. The main complication for the structural model is the relative rotation between the rotor and the tower. Structures commonly deform about some time invariant mean position so that if the load were removed at any particular instant the structure would return to the same position independent of the time at which the event occurs. For a flexible rotor on a flexible structure there is no mean position invariant with time since, even in the absence of a load, the relative position of the tower and rotor will change. The distinction may be summarised in terms of the mass and stiffness matrices of the complete coupled system. For a "normal" structure they are time independent. For a structure which has some fixed and some rotating parts they are periodic. Standard finite element packages do not allow for such structures and so it is this feature that precludes their direct use.

The structural components are modelled as beam elements and because of the relative simplicity which this limitation introduces the mass and stiffness matrices are generated internally by the computer program described in the text below. The equation of motion for the tower and blade i can be written as:

$$[M_T]\ddot{\underline{u}}_T + [C_T]\dot{\underline{u}}_T + [K_T]\underline{u}_T = \underline{F}_T + \underline{F}_{BiT} \qquad (2.2.1)$$

$$[M_{Bi}]\ddot{\underline{u}}_{Bi} + [C_{Bi}+C_{\Omega i}]\dot{\underline{u}}_{Bi} + [K_{Bi}-S_{\Omega i}]\underline{u}_{Bi} = \underline{F}_{Bi} + \underline{F}_{TBi} \qquad (2.2.2)$$

where the subscript T refers to the tower or support and B_i refers to the blade i. [M], [C] and [K] have their usual meanings. $\underline{F}_T$ and $\underline{F}_{Bi}$ are externally applied forces. $\underline{F}_{BiT}$ is the force that blade i applies to the tower head and $\underline{F}_{TBi}$ is the force that the tower head applies to blade i. Their inclusion in the analysis helps to understand the underlying principles. Equation 2.2.1 is straightforward since it is written in the stationary frame of reference. For the rotor, which is idealised in the rotating frame, some additional terms associated with the rotation also appear. They are denoted by the subscript Ω. The applied force F_{Bi} has aerodynamic, centrifugal and gravitational terms. At present it is assumed that all the external loads applied to the structure come from the rotor but it is a simple matter to include directly applied tower forces in the vector $\underline{F}_T$. This facility will prove very useful for systems where the support structure is subject to appreciable loads.

Equations 2.2.1 and 2.2.2 can be combined into one larger equation of motion that includes the tower and all the blades:

$$[M']\ddot{\underline{u}}' + [C']\dot{\underline{u}}' + [K']\underline{u}' = \underline{F}' \qquad (2.2.3)$$

Hitherto the structures have been considered separately. In fact they are of course connected. Only the tower head and the rotor hub are involved in the connection. The connecting matrix is quite simple and very versatile since it determines the nature of the hub. The details of the matrix itself will be discussed below in the next section; for the present we shall simply denote it by [L].

It is convenient to use a single vector $\underline{u}$ to describe the state of the system as a whole. $\underline{u}$ is related to the individual displacement vectors $\underline{u}_T$ and $\underline{u}_{Bi}$ by:

$$\begin{Bmatrix} \underline{u}_T \\ \underline{u}_{B1} \\ \vdots \\ \underline{u}_{Bn} \end{Bmatrix} = [L]\,\underline{u} \qquad (2.2.4)$$

Equivalent relationships can be derived for the velocities and accelerations. Substitution of equations (2.2.4) into (2.2.3) and premultiplication by $[L]^T$ gives rise to the complete system equation of motion:

$$[M]\ddot{\underline{u}} + [C]\dot{\underline{u}} + [K]\underline{u} = \underline{F} \qquad (2.2.5)$$

where

$$[M] = [L]^T[M'][L]$$

$$[C] = [L]^T[C'][L] + 2[L]^T[M][\dot{L}]$$

$$[K] = [L]^T[K'][L] + [L]^T[C'][\dot{L}] + [L]^T[M'][\ddot{L}]$$

and

$$\underline{F} = [L]^T\underline{F}' \qquad (2.2.6)$$

Equation 2.2.5 is the equation of motion used in the computer implementation.

2.2.2 The Connecting Matrix

The connecting matrix [L] is rather a special feature of this analysis since it provides a means of coupling the finite element models of the individual components. There is also a close analogy between the physical hub and the mathematical connecting matrix which is useful. [L] has already been defined in equation 2.2.4 but deserves some more detailed discussion in its own right.

Consider a single blade i connected to the support. Recall that the support equations are written in stationary coordinates and the blade equation is written in rotating coordinates. Let the azimuthal rotation about z-axis be ψ. The two reaction vectors F_{TBi} and F_{BiT} of equations 2.2.1 and 2.2.2 must be equal and opposite. They are applied at single points of connection and are therefore (6x1) vectors. F_{TBi} is written in the rotating axes and F_{BiT} in the stationary set. For equilibrium, then

$$\underline{F}_{TBi} = -[T_\psi]\underline{F}_{BiT} \qquad (2.2.7)$$

where $[T_\psi]$ is a (6x6) rotation matrix which transforms a vector from the stationary set into the rotating set.

The use of the connecting matrix removes some of the degrees of freedom associated with the tower head and the blade roots. The choice of which degree to retain is arbitrary. In the present analysis, it is required that any rotor and hub connection can be used. Under this rather onerous requirement it is easier to retain all the support degrees of freedom and express the blade root displacement in terms of them via the transformation matrix $[T_\psi]$. The connecting matrix affects only the support head and blade root so most of it is in fact an identity matrix. It is convenient to denote the non-identity part of [L] by the symbol $[L^*]$. Using this notation

$$[L] = \begin{bmatrix} [I] & [O] & [O] \\ [O] & [L^*] & [O] \\ [O] & [O] & [I] \end{bmatrix} \qquad (2.2.8)$$

where $[L^*]$ is a (12x12) matrix. Furthermore, since all the support degrees of freedom are to be retained, the upper left hand (6x6) matrix of $[L^*]$ is also the identity. For the present it will be assumed that the blade is rigidly connected to the support. This assumption is realistic for a rigid hub rotor with an infinitely stiff power train. The blade root displacements and rotations can therefore be easily expressed in the stationary axes using the matrix $[T_\psi]$

$$\underline{u}(\text{ blade in rotating axes}) = [T_\psi]\ \underline{u}(\text{tower})$$

The form of the connecting matrix is now clear:

$$[L^*] = \begin{bmatrix} [I\ (6,6)] & [O(6,6)] \\ [T\ (6,6)] & [O(6,6)] \end{bmatrix} \qquad (2.2.9)$$

where [O] is the zero matrix.

The force vector $\underline{F}'$ in equation (2.2.3) will have the form

$$\underline{F}' = \{\ \underline{F}_1, \ldots \underline{F}_{th} + \underline{F}_{BiT}, \underline{F}_{br} + \underline{F}_{TBi}, \ldots\}^T$$

where $\underline{F}$ denotes an applied force vector and the subscripts th and r denote "tower head", or support, and "blade root" respectively. Since the present discussion is addressed to the connection details the externally applied forces can conveniently be put to zero. When premultiplied by the transpose of the connecting matrix $[L]^T$ to obtain the force vector in (2.2.5) we obtain:

$$\underline{F} = [L]^T\ \underline{F}' =$$

$$\{\ \underline{O}, \ldots\ \underline{F}_{BiT} + [T_\psi]^T\ \underline{F}_{TBi},\ O, \ldots\}^T \qquad (2.2.10)$$

but from (2.2.7)

$$\underline{F}_{TBi} = -[T_\psi]\underline{F}_{BiT}$$

so the connecting load in equation (2.2.10) becomes

$$\underline{F}_{BiT} - [T_\psi]^T\ [T_\psi]\underline{F}_{BiT} = \underline{O} \qquad (2.2.11)$$

The connecting loads are therefore removed by the premultiplication by $[L]^T$.

Precisely the same argument applies for the other blades. The null result of equation (2.2.12) shows that the connecting loads can be completely ignored in the analysis which is a very useful simplification.

To summarise, the blade is connected to the support using the connecting matrix $[L^*]$ defined in equation (2.2.10), use of this matrix automatically takes care of the internal reactions between the tower and the blade. Different types of hub connection can easily be accommodated by suitable modification of $[L^*]$. The transformation of the complete system of equations is shown in equation (2.2.5) where the full expression for the connecting matrix [L] defined in equation (2.2.8) must be used:

2.2.3 A Simple example of the Use of a Connecting Matrix

The effect of the connecting matrix manipulations on the equation of motion of the system has quite important implications on the numerical aspects of its solution. The numerical problems will be discussed below but the problems associated directly with the connecting matrix will be introduced here.

The complete structure is idealised as an encastre system of beam elements to represent the support and a free-free system for the blade. For illustrative purposes the beam elements can be replaced by springs and masses as shown in Figure 1.

The equations of motion of the three masses can be written as a single matrix equation:

$$\begin{bmatrix} m_1 & 0 & 0 \\ 0 & m_2 & 0 \\ 0 & 0 & m_3 \end{bmatrix} \ddot{\underline{x}} + \begin{bmatrix} k_1 & 0 & 0 \\ 0 & k_2 & -k_2 \\ 0 & -k_2 & k_2 \end{bmatrix} \underline{x} = \underline{f}' = \begin{matrix} f_1 \\ f_2 \\ f_3 \end{matrix} \qquad (2.2.12)$$

where x_i and f_i are the displacement and applied force at node i. The task of the connecting matrix is to link masses 1 and 2 so that they share the same displacements. For the real problem the support degrees of freedom have been made masters and the blade freedoms have been made slaves , so here the displacement of mass 1 will be retained and that of mass 2 will be removed. Mathematically:

$$\begin{Bmatrix} x_1 \\ x_2 \\ x_3 \end{Bmatrix}_{New} = \begin{Bmatrix} x_1 \\ x_1 \\ x_3 \end{Bmatrix}_{Old} \qquad (2.2.13)$$

or

$$\underline{x} = [L]\,\underline{x}'$$

$$\text{where} \quad [L] = \begin{bmatrix} 1 & 0 & 0 \\ 1 & 0 & 0 \\ 0 & 0 & 1 \end{bmatrix}$$

In this simple case [L] is not a function of time so all its time derivitives are zero. Referring to equation (2.2.6) the new coefficient matrices in the equation of motion can be obtained:

$$[M] = [L]^T[M'][L] = \begin{bmatrix} (m_1+m_2) & 0 & 0 \\ 0 & 0 & 0 \\ 0 & 0 & m_3 \end{bmatrix}$$

$$[K] = [L]^T[K'][L] = \begin{bmatrix} (k_1+k_2) & 0 & -k_2 \\ 0 & 0 & 0 \\ -k_2 & 0 & k_2 \end{bmatrix}$$

and

$$\underline{f} = [L]^T\underline{f}' = \{ (\underline{f}_1 + \underline{f}_2), 0 , f_3 \}^T$$

The new equations of motion are then:

$$\begin{aligned} (m_1+m_2)\ddot{x}_1 + k_1\dot{x}_1 + k_2(x_1-x_2) &= (f_1+f_2) \\ 0 \quad + \quad 0 \quad + \quad 0 &= 0 \\ m_3\ddot{x}_3 \quad - k_2x_1 + k_2x_3 &= f_3 \end{aligned} \qquad (2.2.14)$$

The first and last equations are exactly as expected and could be derived easily by inspection. The middle redundant equation is of no consequence physically, its presence merely demonstrates that one of the original degrees of freedom has been removed. Numerically its retention would obviously give rise to difficulties. In the real problem the connecting matrix is periodic but is used in exactly the same way as in the simple example above.

2.3 SOLUTION TECHNIQUES

The mathematical model described here is a very complete representation of the system and as such could be used to investigate both aeroelastic stability and forced response. Some refinements are required to the model to make it suitable for stability analysis and the solution techniques for the two approaches are quite different. The present work has concentrated exclusively on forced response.

2.3.1 Newmark Beta

It is common practice when dealing with structural models with potentially very high natural frequencies to use the "Newmark Beta" integration method. Other standard methods such as Runge Kutta require the use of prohibitively small time steps to retain numerical stability. The Newmark Beta method is unconditionally stable for linear differential equations with constant coefficients. It may not be unconditionally stable for this application since the system contains periodic coefficients however no problems of stability have yet been encountered.

The Newmark Beta implementation used gives a recurrence relation between values at successive time points which reduces to the solution of a set of simultaneous algebraic equations.

The connecting matrix reduces the number of independent degrees of freedom in the system and introduces singularities as shown in equation (2.2.13). As a result of these singularities, the solution has to be performed with a reduced set of degrees of freedom. The reduced form of the equation is solved using the Crout factorisation method. For every other step in the computation, complete specification of the structure's motion is required.

3 PROGRAM IMPLEMENTATION

The mathematical model described above in section 2 has been implemented as a Fortran program. The aerodynamic loads are computed and stored together with the aerodynamic damping matrices for use by the dynamics code. The mass and stiffness matrices of the support and a single blade are computed using a standard finite element package and are also stored for later use. These data are then supplied to the dynamics code (known as Finite Element Analysis of Rotating Structures or FEARS) which links the structural models together and computes the response of the turbine to the applied loads including the aeroelastic effects introduced by the aerodynamic damping. The output from the program is a time history of deflections and loads.

4 SAMPLE RESULTS

Although the computer program has wide applicability it was developed specifically for wind turbine applications. The results obtained with it are therefore for wind turbine response. The test case used is the Danish Nibe B turbine shown in Figure 2. It is rated at 630kW and is 40m in diameter. This machine was chosen because its structural and aerodynamic details are publically available and a DEFU report [5] gives a large amount of high quality experimental data.

The structural model used to describe the machine is shown in Figure 3. Six nodes were used to describe each blade and three were used for the tower giving a total of 102 unconstrained degrees of freedom.

The applied aerodynamic loads were limited to an upper frequency of 5P. There is, in fact, considerable energy at frequencies higher than this and future results will contain the fundamental and the first eight harmonics. The code can accommodate frequencies up to 10P but can be extended if required.

Figure 4 shows the out-of-plane, or flatwise, and in-plane, or edgewise, root bending moments for a variety of load cases. The main features which describe the load case are the wind speed and yaw angle. The yaw angle is the angle between the wind vector and the main shaft of the turbine, i.e. the degree of misalignment. The response in Figures 4(a) and (b) is near the blade root. It is evident that both loads contain a large element of once-per-rev (1P) loading. In 4(a) this component is due to the 1P aerodynamic load introduced by the yawed flow in conjunction with some gravity loading introduced by the blade coning and shift tilt. The higher harmonic content is near the blade resonant frequency and is excited by the passage of the blade past the tower. The in-plane response shown in Figure 4(b) is dominated by the 1P gravity loading with a very small amount of dynamic response. The predicted results are compared with the measured Nibe data. The general shape of the two sets of curves seem to agree quite well.

Figures 4(c) and (d) show the response of the tower head from the beginning of the computation of the response until steady state is reached. The nature of the response clearly changes from the transient at the tower resonant frequency to the steady state response at 3P because of the three-bladed rotor. It is clear from Figures 4(c) and (d) that the tower motion is predominantly in the fore-aft direction with a little side-to-side motion.

One of the strengths of this type of analysis is that deflections and loads at any nodal point in the structure may be obtained very easily. Using the modal approach it is sometimes quite difficult to obtain the internal loads unless large numbers of modes are used.

5 CONCLUDING REMARKS

A complete finite element analysis of a support-rotor system has been derived and implemented as a computer program. The limited results that have been obtained to date appear promising but considerable further validation is considered important. Despite the large number of degrees of freedom that are involved the computing resources are fairly modest. The relative merits of this approach and the more conventional modal approach remains to be established by use of both of the systems, but initial reactions indicate that the approach described here is very attractive.

References

1 Garrad A D and Quarton D C The Use of Symbolic Computing as a Tool in Wind Turbine Dynamics. Journal of Sound and Vibration 1986 109(1),65-78.

2 Done G T S, Juggins P T W and Patel M H Further Experience with a New Approach to Helicopter Aeroelasticity. Proceedings of Thirteenth European Rotorcraft Forum, Arles - France, September 1987, Paper No. 6.11.

3 Lobitz D W A NASTRAN Based Computer Program for Structural Dynamic Analysis of HAWT's. Proceedings of EWEA Conference Hamburg 1984.

4 Lobitz D W, Carne T, Nord A R and Watson R A Finite Element Analysis and Modal Testing of a Rotating Wind Turbine. Sandia Report SAND82-0345.

5 Madsen P H, Hansen K and Frandsen S Nibe Maleprogram Rotorbelastninger DEFU Rapport EEV 85-04 (In Danish).

Acknowledgements

The work described in this paper was supported by the U K Department of Energy. Permission to publish the results is gratefully acknowledged.

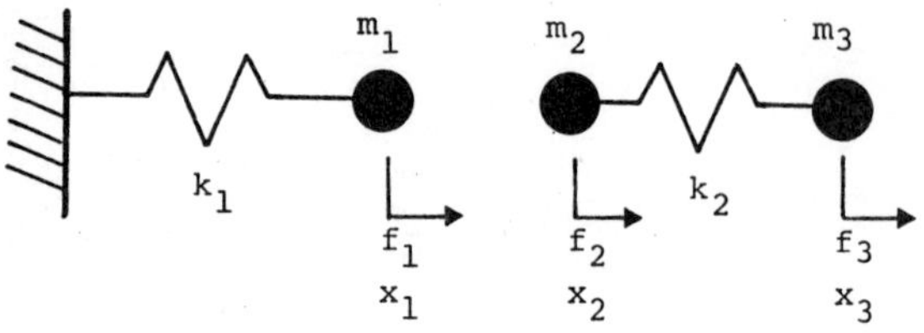

Fig 1 Simple illustration of coupling using a connecting matrix

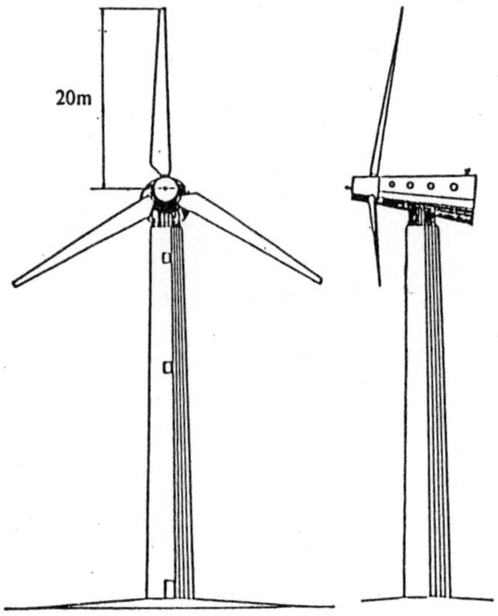

Fig 2 Nibe B wind turbine; 630kW, 40m diameter

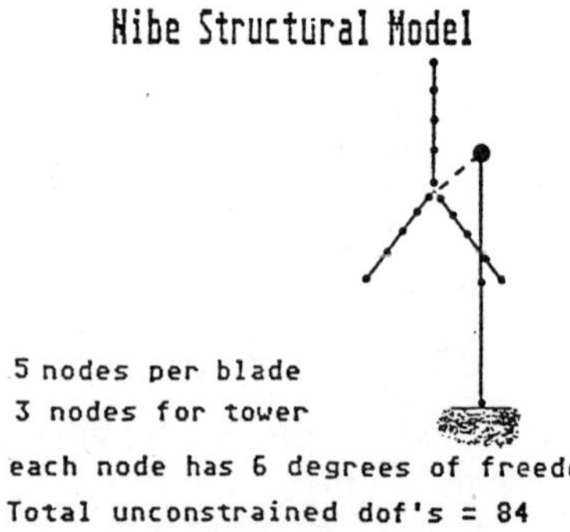

Fig 3 Basic finite element representation of the turbine

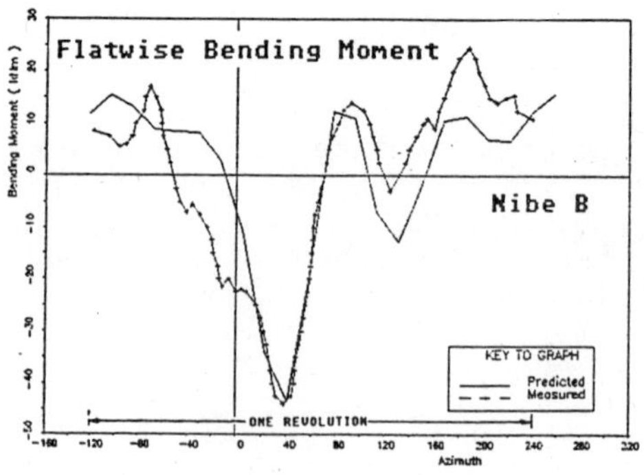

Fig 4a Comparison of measured and predicted bending moment

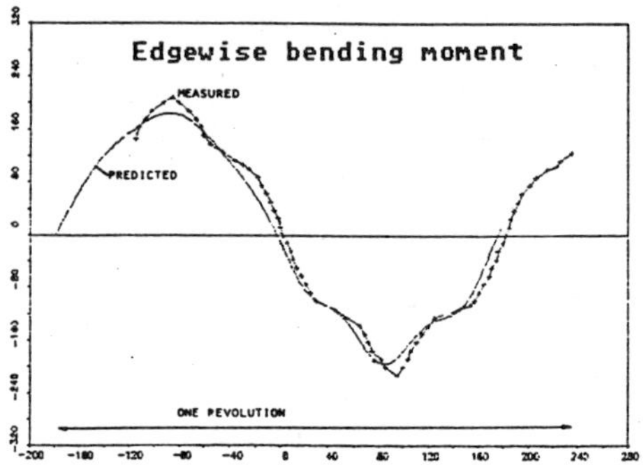

Fig 4b Comparison of measured and predicted bending moment

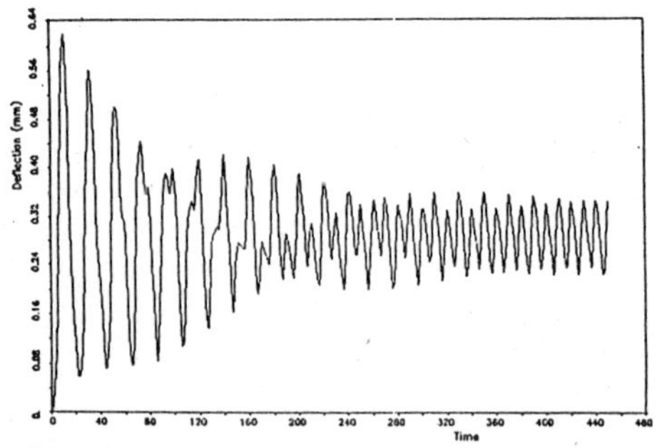

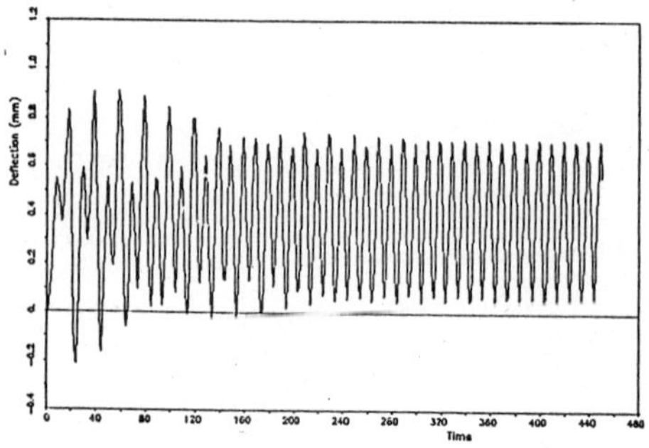

Fig 4c Tower head motion – side-to-side

Fig 4d Tower head motion – fore-aft

C269/88

An expert system interface to a suite of rotordynamics programs

G B THOMAS, BA, CEng, MIMechE
Central Electricity Generating Board, Harrogate, Yorkshire
R C THOMAS, MSc, MPhil and **C C LAI**, BSc
Department of Computer Studies, University of Leeds, Leeds

SYNOPSIS An expert system interface to a suite of rotor dynamics programs is described. It performs an elementary task of matching model output to actual turbogenerator vibration peaks. The system appears to work faster and more accurately than the human counterpart. The paper is presented in the hope that it will introduce the subject to practising engineers.

NOTATION

a	Structural stiffness coefficient
a_{xx} to b_{yy}	Oil film coefficients.
C	Bearing radial clearance
W	Bearing static load
x	Horizontal displacement of journal relative to bearing
y	Vertical displacement of journal relative to bearing
Ω	Rotational speed

1 INTRODUCTION

Rotordynamics methods which were originally developed and used for design calculations have for the last 15 years been of considerable practical use in assisting the solution of a wide variety of vibration problems on high merit turbogenerators. Defects such as unbalances, bends, cracks and instability have been encountered. These are shown by the vibration monitoring equipment as changes in the pattern of on-load or run-down bearing pedestal vibration. We aim to simulate these vibration patterns by using computer programs to find the response to likely excitations. This process demands reasonably accurate models, consisting of three parts - coupled shafts, oil films and structural components. The programs, collectively named SPAR, were developed from those (1) originally written by one of us for use in the design of steam turbines. An expert system interfaces SPAR and performs the process of matching, so that the amplitude and damping of vibration peaks of the model correspond to that of the actual turbogenerator; this is done by adjustment of the data parameters as described below.

There are a number of extant expert system interfaces, see for example Uschol et al (2), Bundy (3), and Bennett & Englemore (4). These programs control a software package by feeding task specifications to it, and then taking the output from the package for analysis by the interface. The present work follows the same pattern, and is more fully reported by Lai (5). The interface sets up the data for SPAR, schedules the appropriate programs, assesses the success of the run and continues to modify the data and re-run the programs until an acceptable match is obtained.

2 THE SPAR PROGRAMS

SPAR operates as a self-contained system installed on various minicomputers together with the latest available model of each relevant turbogenerator. It is available for interactive calculations using these standard models or with subsidiary data included to trim the model to fit monitored data. Input files are set up by a control program from dialogue with the user and the appropriate rotordynamics programs are scheduled by that program to run in sequence. Fig. 1 shows this modular system of programs linked through binary files which transfer processed data and remain as input for later calculations to avoid repetition.

2.1 The Shaft Model and the program SHAFT

The input to SHAFT consists of two files. The larger Standard Machine Data file contains the unchanging shaft sectional data, positions and dimensions of the bearings and suggested positions for the master freedoms, mass-stations and journals. SHAFT converts the shaft sectional data into stiffness and mass matrices and also calculates the smoothly aligned bearing reactions. A small additional Machine Job data file allows the position of the master freedoms to be changed to suit particular conditions. SHAFT calculates the mass matrix in relation to the master freedoms by a condensation process indirectly generating a Timoshenko Beam model.

Shaft sectional data are initially compiled from drawings, idealised according to empirical rules. They have been extensively checked by shaking individual shafts to compare their calculated and measured natural frequencies, and corrected where necessary either empirically or by a technique described in (6) to an accuracy of about two per cent.

2.2 Pinned natural frequencies

The program PINNED calculates the pinned natural frequencies and modal shapes of the coupled shafts from the matrices generated by SHAFT without any further data. Though this information is sometimes useful in its own right, the main purpose of this program is to assist the solution method in RESPONSE.

2.3 Alignment calculations

The initial configuration of the coupled shafts is "smoothly aligned" i.e. without shear force or bending moment at the couplings. For this condition the bearing reactions have already been calculated by SHAFT. The program ALIGN allows one or more bearings to be displaced by a specified distance from the aligned positions, calculating the new bearing reactions. For this calculation the bearing characteristic is represented by a relationship due to Ettles (7). ALIGN also calculates the shaft catenary and self-weight bending stress for a given alignment.

2.4 The programs FLEX and STAB

These two programs calculate natural frequencies and modal shapes, of the coupled shafts alone supported on simple springs in the case of FLEX and of the complete system by STAB, which also calculates the damping of each mode. This is of particular use in comparing sensitivity to oil and steam whirl. FLEX also supplies the free-free eigenvectors needed by KITE, a time stepping response program (8) used chiefly for "breathing" crack calculations.

2.5 The program RESPONSE

The components of the model are brought together in the program RESPONSE. The coupled shafts are defined by the matrix/modal model from SHAFT and PINNED.

The bearing oil film forces are defined by the usual equations:-

$$P_x = \frac{W}{C}\left[\, a_{xx}x + a_{xy}y + b_{xx}\frac{\dot{x}}{\Omega} + b_{xy}\frac{\dot{y}}{\Omega}\,\right]$$

$$P_y = \frac{W}{C}\left[\, a_{yx}x + a_{yy}y + b_{yx}\frac{\dot{x}}{\Omega} + b_{yy}\frac{\dot{y}}{\Omega}\,\right]$$

in which the eight oil film coefficients (a_{xx} - b_{yy}) are tabulated for a series of Sommerfeld numbers and interpolated within the program for each bearing according to its own Sommerfeld number which depends on rotational speed and the bearing load W. The best available oil film coefficients are used although they are not known with the same degree of confidence as the shaft model. The factor W/C, W being the static load of each bearing in turn, dimensionalises the coefficients.

In the smoothly aligned condition the bearing loads W are as calculated by SHAFT, otherwise the bearing loads modified by ALIGN are input to RESPONSE through the file F4 (Fig. 1) and are used in interpolating and dimensionalising the coefficients.

In some rotordynamics calculations it is necessary to use a detailed structural model which may be obtained by calculation or by direct measurement of receptances in shaking tests, but for the methods described in this paper two simpler models are used. The "Single Coefficient" model expresses the pedestal stiffnesses, both horizontal and vertical, in terms of an overall dimensionless stiffness coefficient "a" which is dimensionalised at each bearing by the same factor W/C that is applied to the oil film coefficients. The second "Damped Spring" model consists of a table of flexibility and damping, given separately for the horizontal and vertical directions. During a response run using the single coefficient model the program prepares a table of equivalent flexibilities (= C/(aW)) for use as the second type of model for subsequent calculations, to be modified as the calculation proceeds. Damping and pedestal coupling can also be included in this second model.

RESPONSE calculates the shaft and pedestal vibration due to unbalances and/or bends. This output is given in tabulated or graphical form, plotting run-down curves or shaft shapes as required.

3. MODELLING A LARGE TURBOGENERATOR

The measured response of the LP and Generator shafts of a particular design is shown by the run-down curves on Fig. 2. Associating peaks with a particular shaft can be a difficult problem and the figures given are the result of experience of the particular design of turbogenerator. This is a problem that will be tackled by future Expert Systems. In this section we consider the modelling process carried out by hand in terms of the Single Coefficient model outlined above.

First consider the Generator Shaft. Fig. 2a shows the vertical peaks at 750 r/min and 2000 r/min. After running RESPONSE with several Single Coefficient values the coefficient of a = 1.5 was selected, giving peaks at 730 r/min and 2020 r/min.

Now consider the response of LP1 Shaft shown in Figs. 2b and 2c. Fig. 1350 and 1550 shows peaks at 1350 r/min and 1550 r/min for the first critical. These were identified as horizontal and vertical respectively by comparison with the calculated version shown in Fig. 3. These were matched by coefficients of 1.3 horizontal and 1.3 vertical. The second vertical critical of LP1 is above the running speed of 3000 r/min, but the second horizontal critical is shown clearly by Fig. 2c at 2500 r/min. It was matched by a coefficient of 1.0.

4. THE TECHNIQUE OF MATCHING

The first rule for matching simulated and observed criticals is to run response calculations with unbalances on each shaft using the Single Coefficient model with several values of coefficient in succession. By running the calculation over a series of speeds the peak speeds of each shaft can be compared with the measured values, so showing the coefficient value appropriate to each shaft. First criticals are calculated with a single unbalance in phase, and seconds using a pair of unbalances in anti-phase.

From these results a table of flexibilities can be built up for the Damped Spring model. The damping of each spring can be set up to match the shape of the resonance curves. Further information may be available from site balancing or from the measured bearing vibration response to known defects such as bends or blade loss. Since pedestals are, in effect, part of the measuring equipment the calculated vibration due to a given defect is proportional to the assumed flexibility. This value should check with that derived from frequency comparison. Coupling can be included at this stage to propagate vibration down the shafts. A Tuned Substructure is used when a resonance is clearly structural.

5. THE EXPERT SYSTEM INTERFACE

The way the interface works is heavily influenced by the way the expert engineer uses the package; accordingly such expert-influenced information is called a "user-heuristic". The interface is implemented in POP11, a general purpose Artificial Intelligence programming language which is implemented in an environment called POPLOG, see Barrett & Sloman (9). Two features make it particularly attractive for this work: first it can interface with Fortran fairly easily, and second it comes with a large library of modules which provide a good toolkit for getting started. It also happens to be available at Leeds University.

The input to SPAR is normally through an interactive dialogue which in Fortran terms is taken from a given channel corresponding to the terminal. The channel has been altered to become a file, generated by the interface, which consists of a head, body and tail defined as POP11 procedures. Thus make_head creates the head of the input file, and so on:

```
define make_head(speed_unit,pedestal_coeff)->
head;

define make_body(number_of_unbalances,
  first_mass station,
    first_unbalance_weight,
    first_unbalance_phase,
  second_mass_station,
    second_unbalance_weight,
    second_unbalance_phase)-> body;

define make_tail(number_of_speeds_in_run,
lowest speed,interval,
    number_of_bearings_to_plot,
    first_bearing,second_bearing)-> tail;

define make_input_file(file,head,body,tail);
```

These declarations show the procedures for the three components and one to build the whole file. During a run, only the head and tail alter. (The reader unfamiliar with POP11 should note that the symbols -> mean assign).

Two of RESPONSE's output files, shown in Fig. 1, are important: the one showing the flexibilities of all bearings (recall section 2.5) and the one with run-down plots for the bearings in question. RESPONSE has been altered so that an output file is generated instead of a terminal display, and to cut out unnecessary reporting of summary information.

The first task before any simulation can begin is to run RESPONSE with the output switched so that all the mass station positions are reported. These depend upon the way Response is set-up for a given turbogenerator, therefore this data has to be collected by a special run rather than hard-wired into the interface.

Clearly the main requirement of the interface is to be able to take an observed first or second critical and match it up with the calculated critical on the run-down plots; this is achieved by adjustment of the single coefficient which varies the pedestal flexibilities. This requirement can be coded straight into POP11 as a procedure definition:

```
define match_critical(first_or_second, bearing,
    observed_critical_speed) -> results;
```

The procedure body contains code to carry out this instruction. From the user's point of view an important effect of POP11's interactive nature is that the above, less the word define, becomes a command. Thus the interface can be asked to do a match and put the output into a structure called results or on the terminal.

In order to match a critical the interface has to have a starting point from which to iterate towards the solution. There is a user-heuristic to perform this which is to look in the middle and then move away to the extreme values. Accordingly match_critical has a list of suitable values which are tried in turn. The sequence of initial values is given in a list which is created using the POP11 statement:-

```
[3.5 2.25 4.25 1 6] -> init_coefficient_limit;
```

Failure to find any critical for any value stops the matching. Obviously a procedure is required to find a critical given a fixed single coefficient:

```
define find_critical(first_second, bearing,
  pedestal_coeff) -> results;
```

Now procedure match_critical can first call find_critical and upon successful completion move on to iterate towards a match given an initial critical for a corresponding pedestal coefficient. The procedure to do this is defined below:

```
define match_critical_given(first_or_second,
  bearing, observed_critical_speed,
  given_starting_point) -> results;
```

This is the heart of the interface. However it is not all that is required because once the pedestal coefficient has been found for either the first or second critical, the expert would normally check that a reasonable result is obtained for the other critical. Often, however, a value which suits one critical does not suit the other as well, so further iterations are needed. The process can be thought of as finding the flexibility of a given bearing by coefficient adjustment (see section 3 above). The top-level module in the system does this:

```
define find_flex(bearing,
  observed_first_critical_or_nil,
  observed_second_critical_or_nil) -> results;
```

The overall interaction between the procedures is shown in Fig. 4.

The major details of the software are coded as a Production System and implemented with the "prodsys" library in POPLOG. There are two reasons for adopting the production system approach as opposed to conventional procedural programming: first it provides an easy way to code up user-heuristics which should show returns when the project is extended; second the matching process to control the firing of rules is far easier in a production system than in a procedural system using "goto" or "case" statements.

The basic components of a production system are a rule-base, a database and an interpreter or control system. The rules consist of a set of pre-conditions followed by the action to take if all the pre-conditions are evaluated as true; for example, one of the rules in find_critical is:

```
IF the current_state is 'ready_to_simulate'
THEN make_head...
  make_tail...
  make_input_file...
  run response_package
  set current_state to 'simulation_finished'
```

The database contains all the data that the rules need to use. Pre-conditions are evaluated with respect to the relevant database items and the action parts update database items. One such item is the current_state, which the above rule sets to "simulation_finished". There is also a set of database states which correspond to goal or "problem solved" states; when these are reached processing terminates. The interpreter checks to see which rules could fire, and chooses one according to some strategy so that a virtual search space is covered for example depth first.

5.1 find_critical

The RESPONSE package can consume a large quantity of resources, so it is important to run it efficiently. Accordingly a number of user-heuristics are adopted in this module.

There is a table of upper and lower bounds for first and second critical speeds for each shaft. Each range is divided into three sub-ranges for which corresponding subranges of coefficient also exist. When given a coefficient find critical determines on which sub-range it lies, and then runs the simulation on its corresponding subrange of speeds. If the critical is not found, the next subrange is searched.

The accuracy is altered during the search. Initially intervals of 100 rpm are used and the results are inspected for amplitude as well as phase so that it can be certain to have identified a possible first or second critical. This process requires simulation results for the pair of bearings at either end of the shaft. Once the range is down to within 100 rpm, the interval is reduced to 25 rpm; in this case only one bearing is used in the simulation as no phase check is performed. Thus computation is saved.

The procedure find_critical employs eight rules one of which is shown above. Some of the rules are also used in match_critical_given.

5.2 match_critical_given

This procedure matches a critical given an initial coefficient and corresponding critical. The basic approach is iteratively to adjust the coefficient until the calculated critical matches the observed data. In general this is a monotonic process in which the range of speeds simulated can be reduced successively so as to cut the computational load. A simple bisection method is adopted to compute the next coefficient. The upper and lower bounds on the coefficient are held in up_coeff and lw_coeff. If the observed critical is less than the calculated critical then:

```
pedestal_coeff ->coeff;
(pedestal_coeff +lw_coeff)2 -> pedestal_coeff;
```

similarly for the converse. The method is just about adequate although it can overshoot occasionally, but faster convergence could be obtained if the rate of change of critical with coefficient were known. In fact this data is available elsewhere in the package as part of FLEX; extensions to the interface which will use this are in hand.

match_critical_given can terminate in one of three ways: either a match is achieved, or the coefficient has gone out of the legitimate range, or a fixed maximum number of iterations has been exceeded. There are 11 major rules.

6. KNOWLEDGE ELICITATION AND DECISION SUPPORT

It is well known that the acquisition of rules from an expert is a very difficult process, see for example Waterman (10). The simulation models themselves are in a continuous state of evoluntionary improvement, so any expert system will have to be built against a changing background. In addition, the technology for "deep knowledge modelling" of rotating machinery happens to be in its infancy. Other work is reported for example in (11).

Accordingly the approach adopted here is not to build a system which attempts to automate the judgemental parts of the engineer's job, rather provide support. In particular it is recognised that the initial matching task is laborious and repetitive, not often requiring ad-hoc decisions. It is thus an ideal candidate for mechanisation, as any success here will liberate the experts for more demanding work elsewhere. In fact this approach is well documented in Decision Support literature, see for example Keen & Scott-Morton (12). The methodology concentrates on identification of the major task for support, followed by the provision of software tools which automate those parts of the task which can reliably be predefined while leaving the judgemental parts to the human decision-maker. A side-effect of the introduction of such Decision Support Systems is always a learning process by the decision-maker: this is just what is required in the current context and has happened to some extent.

7. RESULTS

As far as can be ascertained the program works well; it certainly does do the matching task in a manner which imitates the engineer's approach. The current software only matches verticals and ignores amplitude data, but tests show that it performs a match in all cases which the expert deems possible, but does not match those cases he expects to be unmatchable. It is also quite clear that the program works faster than the human couterpart: typically a job which would take a half day or a day can be computed within a few minutes. The system checked the results in this paper, performing 29 SPAR runs in 66 elapsed minutes on an averagely loaded VAX780 computer. About 20 CPU minutes were consumed. Manually this took about 15 minutes per SPAR run or even longer depending upon thinking time, i.e. more than one working day. The user was free to do other work during most of the 66 minute period, which would not have been the case in the manual method. There is implicit in these results a warning that this expert system approach may well boost the engineers' productivity, but at a potentially very high computing cost. It is unfortunate that in spite of this success, it is not yet in regular use because of access problems; it is hoped this will be overcome shortly.

A major result is that in some senses the system is more accurate than the expert. The computer goes down to 25 rpm checks and the phase, whereas the expert did not always. As a result of this some doubt was cast on certain speeds which had always been assumed to be criticals, and more importantly the whole problem of the identification of peaks as criticals has been highlighted; in fact this has happened in parallel with other work elsewhere. This last point is expected to become the focus for another sub-set of the eventual suite of expert systems.

ACKNOWLEDGEMENT

The work undertaken by Mr. K.J. Manning prior to the present work is gratefully acknowledged. He was supported by SERC. This paper has been published with the permission of the Director of Plant Engineering, Electrical/C & I, Operational Engineering Division Northern Area, CEGB.

REFERENCES

(1) THOMAS, G.B. Rotordynamics. Conference on Vibrations in Rotating Machinery, February 1972, (Inst. Mech. Eng.).

(2) USCHOLD, M., HARDING, N., MUETZELFELDT, R., & BUNDY, A. An Intelligent Front End for Ecological Modelling. In O'Shea, T. (ed) Advances in Artificial Intelligence, 1985, 13-22 (North-Holland).

(3) BUNDY, A. Intelligent Front Ends. Pergamon Infotech State of the Art Report on Expert Systems, 1984, 194-203 (Pergamon Press).

(4) BENNETT, J.S. & ENGLEMORE, R. SACON: A Knowledge-based Consultant for Structural Analysis. Proc 6th Int Joint Conf on Artificial Intelligenc, Tokyo, 1979- 47-9 (IJCAI).

(5) LAI, C.C. Development of the Front End Interface for SPAR-2. 1987, Dept of Computer Studies, University of Leeds, England.

(6) THOMAS, G.B. & Littlewood, P. A Technique for Modelling Rotors from Measured Vibration Characterisitics. Conference on Vibrations in Rotating Machinery, 1982, C341/80 (Inst. Mech. Eng.).

(7) ETTLES, C. An Investigation of Bearing Misalignment Problems in 500MW Turbo-generator Sets at West Burton Power Station. Sept 1971, Imperial College, London.

(8) THOMAS, G.B. The Application of Non-linear Methods to Turbogenerator Rotordynamics. Symposium on Steam Turbines for the 1980's, 1979, C186/79 (Inst. Mech. Eng.).

(9) BARRETT, R. & SLOMAN, A. POP11: A Practical Language for Artificial Intelligence, 1986 (Ellis Horwood).

(10) WATERMAN, D.A. A Guide to Expert Systems, 1986 (Addison-Wesley).

(11) MILNE, R. & ELDRIDGE, R. Reasoning About Relationships, 1986, Intelligent Applications Ltd., Livingstone Village, Scotland, EH54 7AY

(12) KEEN, P.W.G. & SCOTT-MORTON, S.M. Decision Systems: An Organisational Perspective, 1978 (Addison-Wesley).

GBT/TEXT/B7
AES/AB/2
18.4.88

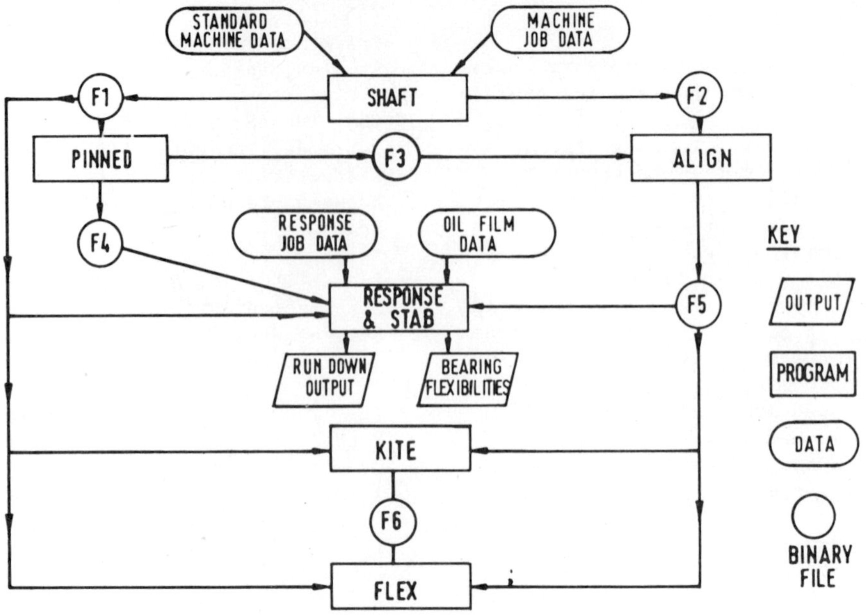

Fig 1 Flowchart of SPAR programs

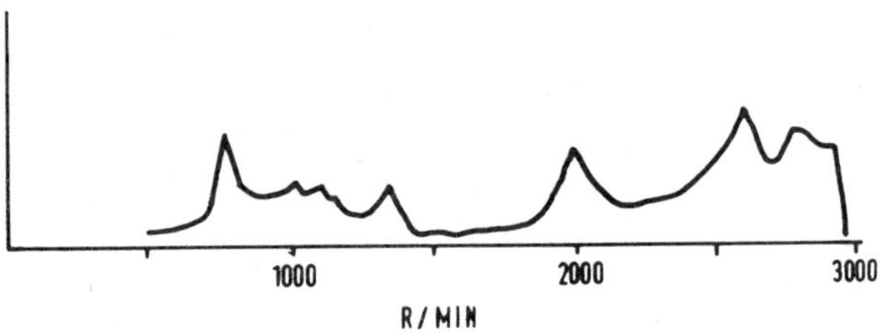

Fig 2a Large turbogenerator measured response — vertical generator bearing

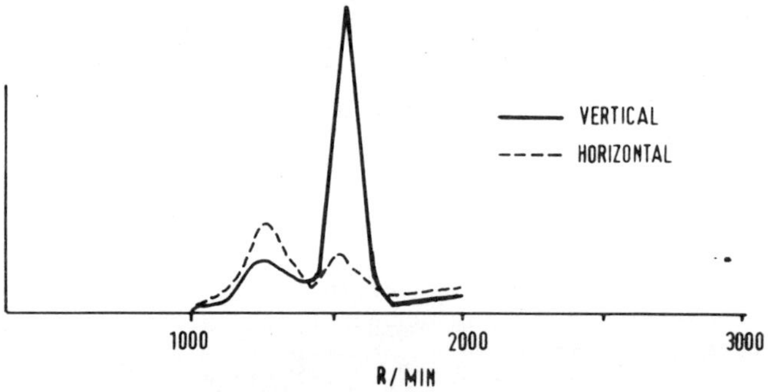

Fig 3 Large turbogenerator calculated response — LP turbine bearing

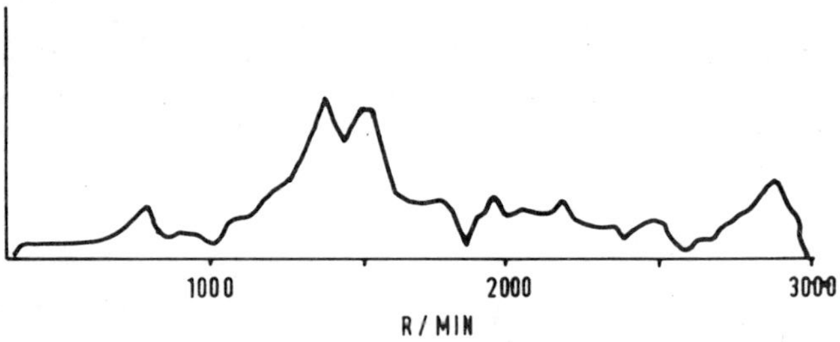

Fig 2b Large turbogenerator measured response — vertical LP turbine bearing

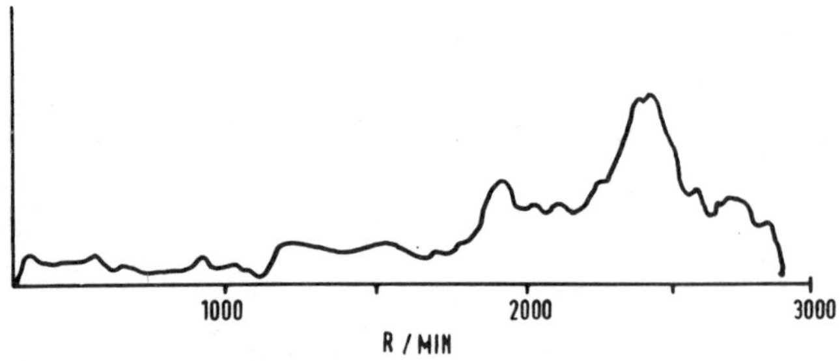

Fig 2c Large turbogenerator measured response — horizontal LP turbine bearing

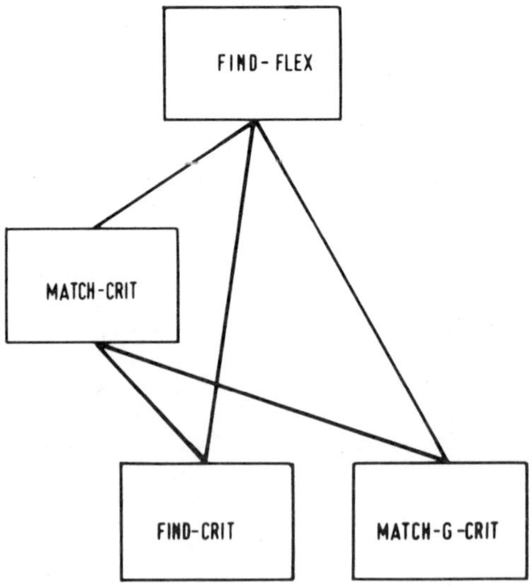

Fig 4 Expert system — overall interaction between procedures

C283/88

Non-linear dynamic analysis of flexible rotors

M GÉRADIN, PhD and **N KILL**
LTAS, University of Liège, Liège, Belgium

SYNOPSIS The nonlinear dynamics of flexible rotors is investigated using either approximate techniques such as equivalent linearization methods or nonlinear transient analysis. Common nonlinear interaction forces (rubbing, dead-band clearances) are considered. The proposed methods are applied to the dynamic analysis of the turbopump rotor of a cryogenic engine.

1. Introduction

The dynamic behavior of jet engines, turbogenerators or turbopumps under special operations (hard landing for the jet engines, blade losses at running speeds, seismic events for the turbogenerators,...) where vibration amplitudes become of the order of the clearances in the bearings, seals or squeeze film dampers has to be studied by a nonlinear approach. Highly nonlinear forces may then occur in the interaction devices (bearings, seals, squeeze film dampers) which can induce nonlinear vibrating phenomena such as subharmonic resonance and instability limit cycles due to the stiffening of the interaction devices [1,2,3,4]. Other nonlinear forces are related either to rotor-stator rubs or to the presence of radial clearance at the rolling bearings. These nonlinearities induce subharmonic resonances as well as amplitude jumps in some cases (large excitations) [4,5,6].

The nonlinear dynamic behavior of such rotating machinery has then to be modelled accordingly. A first approach consists of using equivalent linearization methods in order to perform nonlinear steady state analysis. This approach is feasible under the condition that the non-linearities remain small, and it provides only approximate solutions. It gives no information on the stability of the system but allows to locate the resonances and to determine their sensitivity with respect to changes in load intensities and in nonlinear characteristics (such as clearances, for example).

A second approach consists to perform a nonlinear transient analysis, in which case it is necessary to adopt economical solution methods. Computation cost depends mainly on the size of the numerical model and on its characteristic period. The size of the model can be reduced first by the use of component mode synthesis [7] to represent the linear parts of the structure. Secondly, powerful time integration schemes have to be adopted. Explicit integration schemes are most often used to perform the time stepping. However, our experience has revealed that either implicit or semi-implicit methods are preferable since they provide stability with much larger time steps.

The nature of the dynamic system of equations is briefly presented in section 2. Equivalent linearization methods are described in section 3 and applied to common nonlinear interaction forces. The nonlinear transient analysis is then considered in section 4.

Finally, the methods of analysis described are applied in section 5 to the nonlinear dynamic analysis of the turbopump rotor of a cryogenic engine. The rotor is supported by rolling bearings and the damping is introduced by a series of seals with dynamic characteristics function of rotation speed. Non-linearities correspond to clearances between bearing outer races and bearing carriers. The rotor is modelled by beam finite elements and component mode synthesis is performed in order to reduce the size of the system. This application is used to compare the efficiency of the different algorithms to predict the nonlinear response.

2. System of equations

When a discretization using for example the finite element method is performed, the system of equations describing the rotor dynamics has the following form

$$\mathbf{M}\ddot{\mathbf{q}}(t) + \mathbf{B}(\Omega)\dot{\mathbf{q}}(t) + \mathbf{K}(\Omega)\mathbf{q}(t) + \mathbf{f}\big(\mathbf{q}(t), \dot{\mathbf{q}}(t), \Omega(t)\big) = \mathbf{g}(t) \tag{2.1}$$

where $\mathbf{M}$ is the symmetric mass matrix, $\mathbf{B}(\Omega)$ and $\mathbf{K}(\Omega)$ are respectively the generalized damping and stiffness matrices, $\Omega(t)$ is the rotation speed, $\mathbf{q}(t)$ is the vector that collects the degrees of freedom of the discrete system, $\mathbf{f}\big(\mathbf{q}(t), \dot{\mathbf{q}}(t), \Omega(t)\big)$ corresponds to the nonlinear interaction forces and $\mathbf{g}(t)$ is the vector related to the excitation. The dynamics of the rotating parts is described in an inertial frame under the assumption of circumferential geometric and material isotropy of the rotor. The dependence of the generalized stiffness and damping matrices with respect to the rotation speed results from gyroscopic effects, circulatory forces related to structural damping of the rotors and from the presence of interaction elements with damping and stiffening characteristics function of rotation speed (seals, hydrodynamic bearings,...).

Different kinds of nonlinear interaction forces may be considered. In the case of contact between a rotor and a stator, the force may be described as follows. Assuming a beam model, the rotor and the stator are described as two cylinders of non deformable section. If the transversal displacements of the rotor correspond to $u_r(t), v_r(t)$ and those of the stator to $u_s(t), v_s(t)$, the gap between the rotor and the stator is given by

$$\delta(t) = \sqrt{\left(u_r(t) - u_s(t)\right)^2 + \left(v_r(t) - v_s(t)\right)^2} \quad (2.2)$$

The nonlinear force corresponds to a radial force $f_\rho(\delta(t))$ and to a tangential force $f_\theta(\delta(t))$ which induce reactions in the the transverse plane. The radial force may be described by a large number of nonlinear functions. The most widely used corresponds to dead-band clearance characterized by stiffness k and dead-band $[-\delta_a, \delta_a]$

$$f_\rho(\delta) = \begin{cases} k\delta - k\delta_a & \text{if } \delta \geq \delta_a; \\ 0 & \text{if } |\delta| \leq \delta_a; \\ k\delta + k\delta_a & \text{if } \delta \leq -\delta_a \end{cases} \quad (2.3)$$

Other kinds of asymmetric functions such as bi-linear function or non-linear function described analytically may be adopted.

When rubbing occurs, it induces a tangential force proportional to the radial force and opposed to the precession of the rotor

$$\mathbf{f}_\theta = \mu|\mathbf{f}_\rho|\mathbf{e}_\theta \quad (2.4)$$

where μ is the dry friction coefficient and $\mathbf{e}_\theta$ is the unit vector related to the direction of the tangential force.

3. Equivalent linearization methods

If the excitating force is harmonic

$$\mathbf{g}(t) = \Re\{\mathbf{g}e^{i\omega t}\} \quad (3.1)$$

the solution of (2.1) can be considered as quasi-harmonic when the non linear character of the system is small and when the system is excited in the neighborhood of a resonance. Thus, it is possible to perform a linearization of the system of equations (2.1) and to approximate its solution by solving the modified system using a non-linear steady-state analysis. Different ways exist to linearize the system of equations (2.1). The first one corresponds to the harmonic balance method. It consists to equalize the first terms of the expansion of the non linear forces in Fourier and Taylor series. Let us develop the non-linear force vector as a Taylor series in the neighborhood of the equilibrium position and conserve only the terms of the development proportional to the displacements and the velocities

$$\mathbf{f}(\mathbf{q}, \dot{\mathbf{q}}) \simeq \mathbf{f}(0,0) + \frac{\partial \mathbf{f}}{\partial \mathbf{q}}\mathbf{q} + \frac{\partial \mathbf{f}}{\partial \dot{\mathbf{q}}}\dot{\mathbf{q}} \quad (3.2)$$

If the solution is supposed harmonic,

$$\mathbf{q}(t) = \Re\{\mathbf{q}e^{i\omega t}\} \quad (3.3)$$

the displacement vector corresponds to

$$\mathbf{q}^T = \{q_1 \cos\alpha_1, q_2 \cos\alpha_2, \cdots, q_N \cos\alpha_N\} \quad (3.4)$$

where $\alpha_j = \omega t + \theta_j$ and θ_j is the phase related to the displacement $q_j(t)$.

Let us suppose next that each function $f_j(\mathbf{q}, \dot{\mathbf{q}})$ can be expanded in a Fourier series. These functions are also separable with respect to the displacements

$$\mathbf{f}(\mathbf{q}, \dot{\mathbf{q}}) = \sum_l \mathbf{f}_l(q_l, \dot{q}_l) \quad (3.5)$$

The Fourier series expansion yields to

$$\mathbf{f}_l(q_l, \dot{q}_l) = \sum_{\nu=0}^{\infty}\{\mathbf{a}_{\nu_l}\cos(\nu_l\omega t) + \mathbf{b}_{\nu_l}\sin(\nu\omega t)\} \quad (3.6)$$

where the components of the vectors $\mathbf{a}_{\nu_l}$ and $\mathbf{b}_{\nu_l}$ are given by

$$a_{\nu_{jl}} = \frac{1}{\pi}\int_0^{2\pi} f_{jl}(q_l\cos\alpha_l, -\omega ql\sin\alpha_l)\cos(\nu\alpha_l)\,d\alpha_l \quad (3.7)$$

$$b_{\nu_{jl}} = \frac{1}{\pi}\int_0^{2\pi} f_{jl}(q_l\cos\alpha_l, -\omega ql\sin\alpha_l)\sin(\nu\alpha_l)\,d\alpha_l \quad (3.8)$$

The comparison between the Taylor and Fourier series expansions leads to the following modified expression of the system of equations

$$\mathbf{M}\ddot{\mathbf{q}}(t) + \mathbf{B}_0(\Omega)\dot{\mathbf{q}}(t) + \mathbf{B}_L(q_1, q_2, \cdots)\dot{\mathbf{q}}(t) + \mathbf{K}_0(\Omega)\mathbf{q}(t) + \mathbf{K}_L(q_1, q_2, \cdots)\mathbf{q}(t) = \mathbf{g}(t) \quad (3.9)$$

where the terms of the matrices $\mathbf{K}_L$ and $\mathbf{B}_L$ depend on displacement amplitudes as follows

$$\{\mathbf{K}_L\}_{jl} = \frac{a_{1_{jl}}}{q_l} \quad \text{and} \quad \{\mathbf{B}_L\}_{jl} = \frac{-b_{1_{jl}}}{\omega q_l} \quad (3.10)$$

As displacements and forces are considered harmonic with pulsation ω, the system of equations (3.9) can be rewritten in the form

$$\{-\omega^2\mathbf{M} + i\omega\mathbf{B}_0(\Omega) + i\omega\mathbf{B}_L(q_1, q_2, \cdots, q_N) + \mathbf{K}_0(\Omega) + \mathbf{K}_L(q_1, q_2, \cdots, q_N)\}\mathbf{q} = \mathbf{g} \quad (3.11)$$

where N is the number of degrees of freedom.

Let us note that, in the least-square sense, the adopted method is equivalent to minimize over one cycle the error related to the approximation on the displacement field. This error is defined by

$$e = \sum_{l=1}^{N} e_l \tag{3.12}$$

with

$$e_l = \sum_{j=1}^{N} \int_0^T |f_{jl}(q_l, \dot{q}_l) - \{\mathbf{K}_L\}_{jl} q_l - \{\mathbf{B}_L\}_{jl} \dot{q}_l|^2 \, d\tau \tag{3.13}$$

Its minimization yields to the conditions

$$\frac{\partial e_l}{\partial \{\mathbf{K}_L\}_{jl}} = 0 \quad \text{and} \quad \frac{\partial e_l}{\partial \{\mathbf{B}_L\}_{jl}} = 0 \quad \text{for all } j \text{ and } l \tag{3.14}$$

These conditions provide the explicit relationships

$$-\int_0^T f_{jl}(q_l, \dot{q}_l) q_l \, d\tau + \{\mathbf{K}_L\}_{jl} \int_0^T q_l^2 \, d\tau + \{\mathbf{B}_L\}_{jl} \int_0^T q_l \dot{q}_l \, d\tau = 0 \tag{3.15}$$

$$-\int_0^T f_{jl}(q_l, \dot{q}_l) \dot{q}_l \, d\tau + \{\mathbf{K}_L\}_{jl} \int_0^T q_l \dot{q}_l \, d\tau + \{\mathbf{B}_L\}_{jl} \int_0^T \dot{q}_l^2 \, d\tau = 0 \tag{3.16}$$

If we consider that the displacement $q_l(t)$ has the form

$$q_l(t) = q_l \cos(\omega t + \theta_l) \tag{3.17}$$

the resolution of the system of equations (3.15-16) yields to the equivalent stiffness and damping coefficients given by (3.10). The harmonic balance method and the minimization of the error in the least-square sense provide thus the same approximations as long as the functions $f_j(\mathbf{q}, \dot{\mathbf{q}})$ are separable. When the non-linear functions are no longer separable, the error due to the approximation is given by

$$e = \sum_{j=1}^{N} \int_0^T |f_{jl}(\mathbf{q}, \dot{\mathbf{q}}) - \sum_{l=1}^{N} \{\{\mathbf{K}_L\}_{jl} q_l - \{\mathbf{B}_L\}_{jl} \dot{q}_l\}|^2 \, d\tau \tag{3.18}$$

and the minimization of the error yields then to the $2N^2$ conditions

$$-\int_0^T f_{jl}(q_1, q_2, \cdots, \dot{q}_1, \dot{q}_2, \cdots) q_l \, d\tau + \sum_{k=1}^{N} \{\mathbf{K}_L\}_{jk} \int_0^T q_k q_l \, d\tau + \sum_{k=1}^{N} \{\mathbf{B}_L\}_{jk} \int_0^T q_l \dot{q}_k \, d\tau = 0 \tag{3.19}$$

$$-\int_0^T f_{jl}(q_1, q_2, \cdots, \dot{q}_1, \dot{q}_2, \cdots)) \dot{q}_l \, d\tau + \sum_{k=1}^{N} \{\mathbf{K}_L\}_{jk} \int_0^T q_k \dot{q}_l \, d\tau + \sum_{k=1}^{N} \{\mathbf{B}_L\}_{jk} \int_0^T \dot{q}_l \dot{q}_k \, d\tau = 0 \tag{3.20}$$

If we consider that the displacements have the form (3.4), the equivalent coefficients are given by the solution of a linear system of $2N^2$ equations. This approximate expression yields to fully equivalent stiffness and damping matrices. However, the modified system keeps always the form (3.11). Let us note that the application of the harmonic balance method in the case of non separable functions would yield to a different solution since the linearized coefficients are calculated according to a privileged degree of freedom [12].

A sweeping procedure is used to locate the resonances of system (3.11) which is then solved for different rotation speeds by application of the Newton-Raphson process. When dealing with systems having high number of degrees of freedom, their solution cannot be performed directly. The frontal method with substructuring technique [8] can then be used. The process can be applied at very low cost if the non linear interaction elements are placed in the last substructure. The computation is organized as follows

(i) First, the load vector is condensed and a first contribution of the load to the displacement vector is calculated.

(ii) The non-linear system limited to the last substructure is solved using the Newton-Raphson method.

(iii) The displacements are finally restored in the different substructures associated with the linear part of the structure.

Linearization of some non-linear interaction forces

Non linear stiffness connecting two DOF

Let us consider a non linear stiffness force $f(\delta)$ connecting two DOF q_1 and q_2. Since the displacements are assumed harmonic, the difference between both displacements is given by

$$\delta(t) = q_1(t) - q_2(t) = \Re\{\delta e^{i\omega t}\} \tag{3.21}$$

This kind of non-linearity may occur, for example, in supporting structures.

The equivalent linearization process can be applied to $f(\delta)$ since it is always possible to define the degrees of freedom $(q_1 + q_2)$ and $(q_1 - q_2)$ instead of q_1 and q_2. The linearization leads to the approximation

$$f(\delta) \simeq f_L(\delta) = k_e(\delta)\delta \tag{3.22}$$

with the equivalent stiffness $k_e(\delta)$. The equivalent stiffness for usual kinds of non-linear forces is given hereafter.

Thus, the application of the equivalent linearization process to the dead-band clearance case of equation (2.3) yields to the following equivalent stiffness

$$k_e = k\left[1 - \frac{2}{\pi} \arcsin\left(\frac{\delta_a}{\delta}\right) - \frac{2}{\pi}\left(\frac{\delta_a}{\delta}\right)\sqrt{1 - \left(\frac{\delta_a}{\delta}\right)^2}\,\right] \tag{3.23}$$

when the motion amplitude δ exceeds δ_a. Let us note that the equivalent stiffness is function of both parameters δ_a and k as well as amplitude δ.
In the case of the bi-linear function with dead-band clearance $[-\delta_a, \delta_a]$, stiffness k_1 in intervals $[\delta_a, \delta_b]$ and $[-\delta_b, -\delta_a]$ and stiffness k_2 in intervals $[\delta_b, \infty]$ and $[-\infty, -\delta_b]$, one obtains

$$\begin{aligned} k_e = k_2\{1 & \\ & - \frac{2}{\pi}(1 - \frac{k_1}{k_2})\left[\arcsin\left(\frac{\delta_b}{\delta}\right) + \left(\frac{\delta_b}{\delta}\right)\sqrt{1 - \left(\frac{\delta_b}{\delta}\right)^2}\right] \\ & - \frac{2}{\pi}\frac{k_1}{k_2}\left[\arcsin\left(\frac{\delta_a}{\delta}\right) + \left(\frac{\delta_a}{\delta}\right)\sqrt{1 - \left(\frac{\delta_a}{\delta}\right)^2}\right]\} \end{aligned} \quad (3.24)$$

when the motion amplitude δ is larger than δ_b. When the motion amplitude is contained in the interval $[\delta_a, \delta_b]$, one reobtains the expression (3.23).

The linearized stiffness matrix related to the degrees of freedom q_1 and q_2 is then easily generated.

Contact between rotor and stator
The nature of the interaction forces was described in section 2. The forces acting on the rotor are given by

$$\begin{aligned} f_{u_r}(t) &= f\big(\delta(t)\big)\frac{u_r(t) - u_s(t)}{\delta(t)} \\ f_{v_r}(t) &= f\big(\delta(t)\big)\frac{v_r(t) - v_s(t)}{\delta(t)} \end{aligned} \quad (3.25)$$

where $f\big(\delta(t)\big)$ is the nonlinear force resulting from the combination of the radial and the tangential components. The forces acting on the stator have opposite signs.

In general $\delta(t)$ varies in both time and direction as the precession of the rotor describes an ellipse. The equivalent linearization is then performed according to (3.19-20) where the coefficients of the linear system of $2N^2$ equations are evaluated using numerical integration.

When the problem is circumferentially isotropic, the gap $\delta(t)$ varies in direction but is constant as the precession of the rotor is a circumference. It is then possible to obtain the exact nonlinear steady-state response by application of the Newton-Raphson process.

4. Nonlinear transient analysis

The temporal integration of the system of equations describing the non-linear dynamics of the system may be performed using direct integration. Either pure implicit algorithms or implicit-explicit multi-corrector schemes may be used. The latter algorithms allow to adopt large time steps even when the excitation frequencies are high.

By comparison with linear transient analysis it is now necessary to solve at each time step a non-linear system by application of the Newton-Raphson method. Just as for the non-linear steady state analysis, it is possible to organize the solution scheme in order to place the non-linear interaction elements in the last substructure when the frontal method with substructuring technique is adopted.

The integration schemes and the associated solution strategies as well as some of the non-linear interaction elements are described hereafter.

The time interval is split in variable time steps of length h. The state of the system at time t_{n+1} is obtained by integrating velocities and displacements using Newmark schemes with two parameters β and γ which define the approximation of the acceleration in the considered interval. If we consider that $\beta h^2 \ddot{\mathbf{q}}_{n+1}$ corresponds to the implicit displacement correction $\triangle \mathbf{q}_{n+1}$, these schemes yield to

$$\dot{\mathbf{q}}_{n+1} = \dot{\mathbf{q}}_n + (1-\gamma)h\ddot{\mathbf{q}}_n + \frac{\gamma}{\beta h}\triangle \mathbf{q}_{n+1} \quad (4.1)$$

$$\mathbf{q}_{n+1} = \mathbf{q}_n + h\dot{\mathbf{q}}_n + (\frac{1}{2} - \beta)h^2\ddot{\mathbf{q}}_n + \triangle \mathbf{q}_{n+1} \quad (4.2)$$

Thus, the state of the system at time t_{n+1} is calculated in three steps as it is usual with implicit integration schemes

- prediction of velocities and displacements
- determination of corrections to displacements by application of the Newton-Raphson process to the non-linear system
- correction of velocities and displacements

It is also possible to overcome the inconvenience due to the temporal dependence of the iteration matrix by using implicit-explicit algorithms. The generalized stiffness and damping matrices are split into their constant parts $\mathbf{K}_{0_E}$ and $\mathbf{B}_E$ and their time dependent parts, respectively. Only the constant part of the structural matrices is kept to build the iteration matrix. The determination of the contribution of the linear part of the iteration matrix to the inverse is only needed when the time interval is modified. The implicit-explicit multi-corrector scheme adapted to the non-linear case is composed of two iterative procedures and is organized as follows

- (i) prediction of velocities and displacements

$$\begin{aligned} \dot{\mathbf{q}}_{n+1}^{(1)} &= \dot{\mathbf{q}}_n + (1-\gamma)h\ddot{\mathbf{q}}_n \\ \mathbf{q}_{n+1}^{(1)} &= \mathbf{q}_n + h\dot{\mathbf{q}}_n + (\frac{1}{2} - \beta)h^2\ddot{\mathbf{q}}_n \\ \ddot{\mathbf{q}}_{n+1}^{(1)} &= 0 \\ i &= 1 \quad i \text{ is the iteration counter} \end{aligned} \quad (4.3)$$

- (ii) evaluation of the residual vector at step i

$$\begin{aligned} \mathbf{r}_{n+1}^{(i)} = \mathbf{g}_{n+1} - \mathbf{M}\ddot{\mathbf{q}}_{n+1}^{(i)} - \mathbf{B}\big(\dot{\phi}(t_{n+1})\big)\dot{\mathbf{q}}_{n+1}^{(i)} \\ - \mathbf{K}\big(\dot{\phi}(t_{n+1}), \ddot{\phi}(t_{n+1})\big)\mathbf{q}_{n+1}^{(i)} - \mathbf{f}(\mathbf{q}_{n+1}^{(i)}, \dot{\mathbf{q}}_{n+1}^{(i)}) \end{aligned} \quad (4.4)$$

– (iii) check of convergence of the residual vector

$$\| \mathbf{r}_{n+1}^{(i)} \| < \epsilon \quad (4.5)$$

If the test is satisfied, then proceed to next time step else go to (iv).

– (iv) evaluation of displacement corrections by application of the iterative Newton-Raphson process to the system of equations:

$$\{\frac{1}{\beta h^2}\mathbf{M} + \frac{\gamma}{\beta h}\mathbf{B}_E + \mathbf{K}_{0E}\}\Delta\mathbf{q}_{n+1}^{(i+1)} + \mathbf{f}(\mathbf{q}_{n+1}, \dot{\mathbf{q}}_{n+1}) = \mathbf{r}_{n+1}^{(i)} \quad (4.6)$$

– (v) correction phase

$$\begin{aligned} \ddot{\mathbf{q}}_{n+1}^{(i+1)} &= \ddot{\mathbf{q}}_{n+1}^{(i)} + \frac{1}{\beta h^2}\Delta\mathbf{q}_{n+1}^{(i+1)} \\ \dot{\mathbf{q}}_{n+1}^{(i+1)} &= \dot{\mathbf{q}}_{n+1}^{(i)} + \frac{\gamma}{\beta h}\Delta\mathbf{q}_{n+1}^{(i+1)} \\ \mathbf{q}_{n+1}^{(i+1)} &= \mathbf{q}_{n+1}^{(i)} + \Delta\mathbf{q}_{n+1}^{(i+1)} \\ i &= i+1 \quad \text{and go to (ii)} \end{aligned} \quad (4.7)$$

Note that numerical damping can be introduced by application of the HHT algorithm to the non-linear system of equations. The dynamic equilibrium equation to be solved at time t_{n+1} is then modified as follows

$$\begin{aligned} &\mathbf{M}\ddot{\mathbf{q}}_{n+1} + (1-\alpha)\mathbf{B}(\dot{\phi}(t_{n+1}))\dot{\mathbf{q}}_{n+1} + \alpha\mathbf{B}(\dot{\phi}(t_n))\dot{\mathbf{q}}_n \\ &+ (1-\alpha)\mathbf{K}(\dot{\phi}(t_{n+1}), \ddot{\phi}(t_{n+1}))\mathbf{q}_{n+1} \\ &+ \alpha\mathbf{K}(\dot{\phi}(t_n), \ddot{\phi}(t_n))\mathbf{q}_n + (1-\alpha)\mathbf{f}(\mathbf{q}_{n+1}, \dot{\mathbf{q}}_{n+1}) \\ &+ \alpha\mathbf{f}(\mathbf{q}_n, \dot{\mathbf{q}}_n) = (1-\alpha)\mathbf{g}_{n+1} + \alpha\mathbf{g}_n \end{aligned} \quad (4.8)$$

where α is chosen in the interval $[0, 1/3]$. In specific cases such as contact problems the introduction of numerical damping is essential for acceptable convergence behavior of the iteration procedure.

Solution of the non-linear system

The incremental equations (4.6) are solved by application of the Newton-Raphson method. At step $k+1$, the contribution of the non-linear forces to the tangent iteration matrix is given by

$$\frac{\gamma}{\beta h}\frac{\partial \mathbf{f}}{\partial \dot{\mathbf{q}}}\bigg]_{(\mathbf{q}_{n+1}^{(k)}, \dot{\mathbf{q}}_{n+1}^{(k)})} + \frac{\partial \mathbf{f}}{\partial \mathbf{q}}\bigg]_{(\mathbf{q}_{n+1}^{(k)}, \dot{\mathbf{q}}_{n+1}^{(k)})} \quad (4.9)$$

and results from a Taylor series expansion of the non-linear forces in the neighborhood of the current state.

Again, just as with the steady state analysis, the iteration matrix is inverted using the frontal method with substructuring technique where the non-linear interaction elements are placed in the last substructure.

Application to rubbing between a rotor and a stator

Let us consider the simplified case of the dead-band clearance with the radial force described as in (2.3).

When rubbing between rotor and stator occurs, a tangential force proportional to the radial contact force and opposed to the precession of the rotor is induced. With the transient analysis it is necessary to determine the direction of rubbing. It is not the case with equivalent linearization techniques since it is then supposed that the rotor performs forward precessions. In order to determine the direction of rubbing, let us consider the evolution of the point of contact. During the time interval Δt the motion of this point is the combination of

– a displacement due to the prescribed rotation of the shaft

$$\Omega r \Delta t \quad (4.10)$$

where Ω is the rotation speed and r the rotor radius;

– a displacement due to the whirl of the shaft

$$\omega \delta \Delta t \quad (4.11)$$

where ω is the whirling angular speed and δ the relative radial displacement of the shaft with respect to the stator.

The whirling angular speed corresponds to

$$\omega = \frac{u_r(t)\dot{v}_r(t) - \dot{u}_r(t)v_r(t)}{u_r^2(t) + v_r^2(t)} \quad (4.12)$$

The direction of the frictional force depends on the sign of the expression

$$\dot{x} = \omega\delta + \Omega r \quad (4.13)$$

When the radial force due to the stiffness of the stator is defined by (2.3), it is then possible to describe the interaction force as follows

$$\begin{Bmatrix} f_{u_r}(t) \\ f_{v_r}(t) \\ f_{u_s}(t) \\ f_{v_s}(t) \end{Bmatrix} = \begin{bmatrix} k & -\mu e k & -k & \mu e k \\ \mu e k & k & -\mu e k & -k \\ -k & \mu e k & k & -\mu e k \\ -\mu e k & -k & \mu e k & k \end{bmatrix} \begin{Bmatrix} u_r(t) \\ v_r(t) \\ u_s(t) \\ v_s(t) \end{Bmatrix} - \frac{k\delta_a}{\delta(t)} \begin{Bmatrix} \Delta u(t) - \mu e \Delta v(t) \\ \Delta v(t) + \mu e \Delta u(t) \\ -\Delta u(t) + \mu e \Delta v(t) \\ -\Delta v(t) - \mu e \Delta u(t) \end{Bmatrix} \quad (4.14)$$

where $e = \text{sign}(\dot{x})$ and

$$\begin{aligned} \Delta u(t) &= u_r(t) - u_s(t) \\ \Delta v(t) &= v_r(t) - v_s(t) \end{aligned} \quad (4.15)$$

The tangent matrix related to these non-linear forces is generated according to (4.9). However, as the derivative of e with respect to $\Delta u(t), \Delta v(t), \Delta\dot{u}(t), \Delta\dot{v}(t)$ is always equal to zero except when the sign of $\dot{x}$ changes, particular attention has to be given to the numerical treatment of this discontinuity. We have chosen to modify the function

$$e = \text{sign}(\dot{x}) = \begin{cases} 1 & \text{if } \dot{x} > 0; \\ 0 & \text{if } \dot{x} = 0; \\ -1 & \text{if } \dot{x} < 0. \end{cases} \quad (4.16)$$

into

$$e = \begin{cases} 1 & \text{if } \dot{x} > \epsilon; \\ \frac{\dot{x}}{\epsilon} & \text{if } |\dot{x}| < \epsilon; \\ -1 & \text{if } \dot{x} < -\epsilon. \end{cases} \quad (17)$$

The choice of ϵ is somewhat arbitrary and is problem related since it has the meaning of an absolute velocity. In our case ϵ was chosen equal to 1.0 10^{-8}.

5. Application

In order to compare the different proposed algorithms the nonlinear dynamic analysis of the turbopump rotor of a cryogenic engine is performed. The computations were performed using the finite element code SAMCEF [9] developed by the Aerospace Laboratory of the University of Liège.

Description of the rotor

The turbopump consists of a 70 cm long, 35 kg rotor carrying two impellers in series and two turbine wheels. The rotor is positioned radially by two sets of duplex ball bearings located in front of the first impeller and behind the turbine. Non-linearities correspond to clearances between bearing outer races and bearing carriers. The clearances are 19.25 μm in width. The nominal speed corresponds approximately to 34000 r/min and a power of 11000 kw is transmitted by the shaft. There are several seals between the rotor and the stator introducing damping and circulatory forces. They correspond to labyrinth seals located in front and behind each impeller as well as to a smooth interstage seal between the impellers. These seals together with the aerodynamic cross-coupling forces at the turbine and at the first stage of the pump have important effects on the dynamic behavior of the rotor.

Finite element model

The classical beam approach is used to model the shaft. The disks attached to the shaft are assumed to be rigid. The ball bearings, seals and fluid forces are represented by interaction elements with eight coefficients which are function of the rotation speed. The nonlinearities related to the flexible bearing carriers at both pump and turbine bearing locations are modelled by nonlinear interaction element with stiffness defined by (2.3). The stiffness k corresponds to 1.25 10^8 N/m.

The finite element model consists of 316 DOF and includes 115 beam elements and 15 interaction elements. The inertial frame approach is adopted owing to the dynamic characteristics of the seals and the fluid forces. Component mode synthesis is used to reduce the size of the model. The rotor is represented by a super-element with 52 boundary DOF and 20 normal modes. The boundary DOF are located at the bearings and seals as well as on the discs (impellers, turbines).

Results

The excitation of the rotor corresponds to five unbalances of 50 gr-mm. The unbalances are located at both turbine discs, at both impellers and at the first stage of the pump.

A nonlinear steady-state analysis is performed in the speed range of 30 000 to 40 000 r/min. As the model and the excitation are circumferentially isotropic, the exact nonlinear steady-state response is obtained by application of the Newton-Raphson process.

Plots of displacement amplitudes of both pump and turbine bearings as functions of rotation speed are represented in figures 1 a,b. It can be seen that the displacement of the turbine bearing is larger than at the pump bearing level.

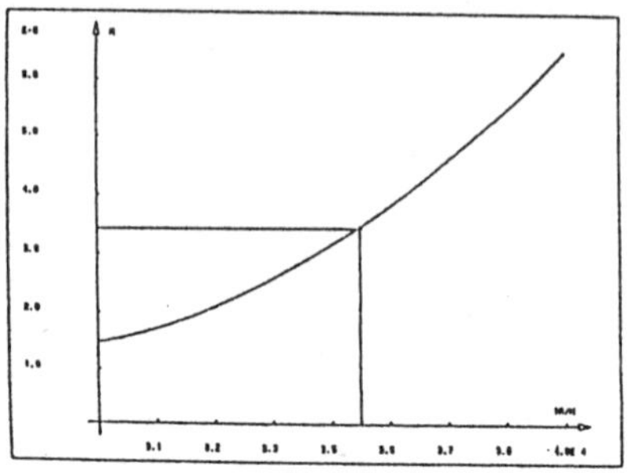

Fig 1a Pump bearing displacement amplitude

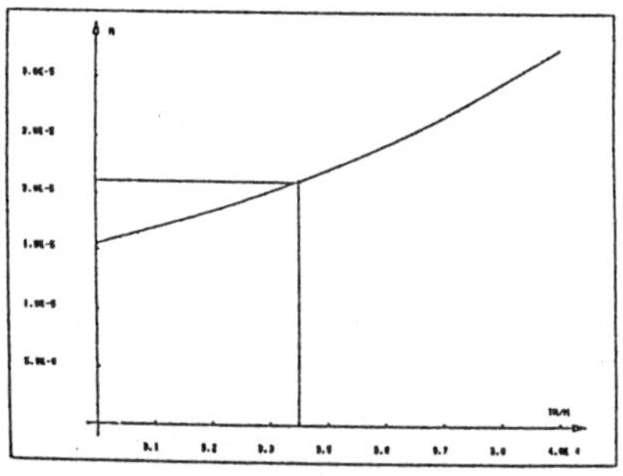

Fig 1b Turbine bearing displacement amplitude

A critical speed analysis is performed in the same speed range using linearized stiffness at the bearing carriers. The eigensolutions are obtained using either bi-iteration or Lanczos algorithm [11]. A plot of eigenvalues as functions of rotation speed is represented in figures 2 a,b. The rotor is stable and a stiffening effect appears at 30 400 r/min as contact occurs at the turbine bearing carrier.

A nonlinear transient analysis is also performed when the rotor is running at 34 000 r/min. The excitation corresponds to sudden unbalances occuring at time $t = 0$. The value and the location of these unbalances are the same as with the steady-state analysis. The transient analysis is performed during 0.01 sec. As the rotation speed is constant, an implicit-explicit multi-corrector scheme is used.

Plots of displacement of both pump and turbine bearings as functions of time are represented in figures 3 a,b. Orbits of the rotor at the same locations are represented in figures 4 a,b.

It can be seen that the nonlinear steady-state analysis yields mean values of rotor transient displacements. Owing to the sudden character of the unbalances, transient effects are always present in the time range of 0.01 sec and displacements thus perform oscillations around these mean values.

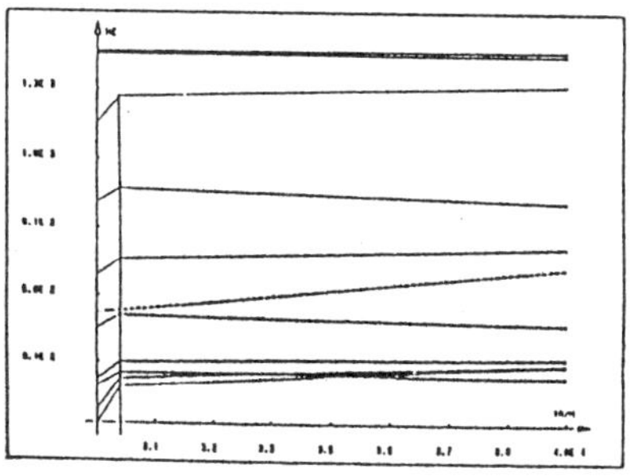

Fig 2a Eigenfrequencies versus rotation speed

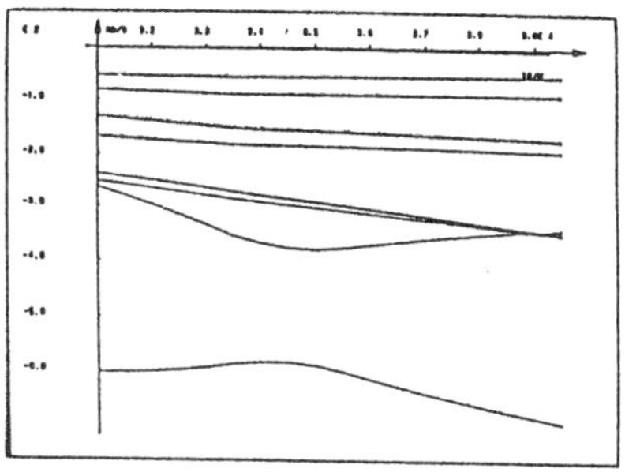

Fig 2b Real parts of the eigenvalues versus rotation speed

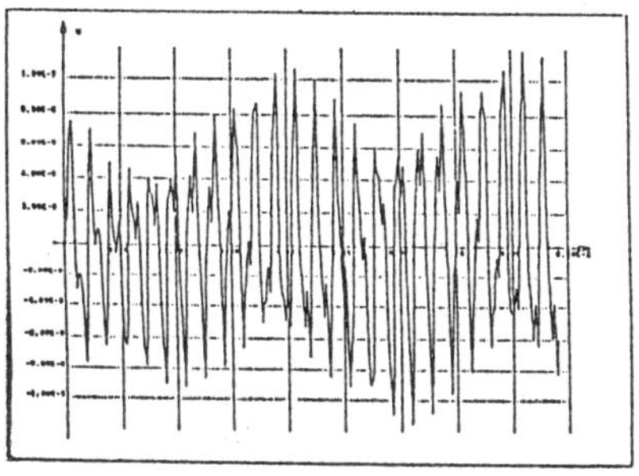

Fig 3a Transient pump bearing displacement

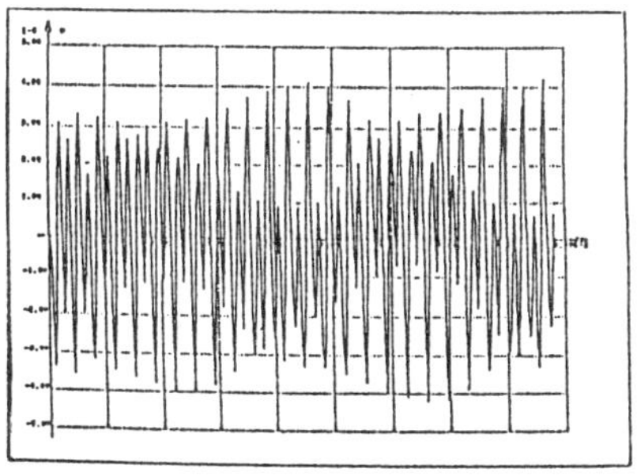

Fig 3b Transient turbine bearing displacement

Both analyses yield to similar results in a qualitative manner.

Conclusions

Different algorithms for nonlinear dynamic analysis of rotating systems have been presented and applied to the dynamic analysis of the turbopump rotor of a cryogenic engine. The application shows the abil-

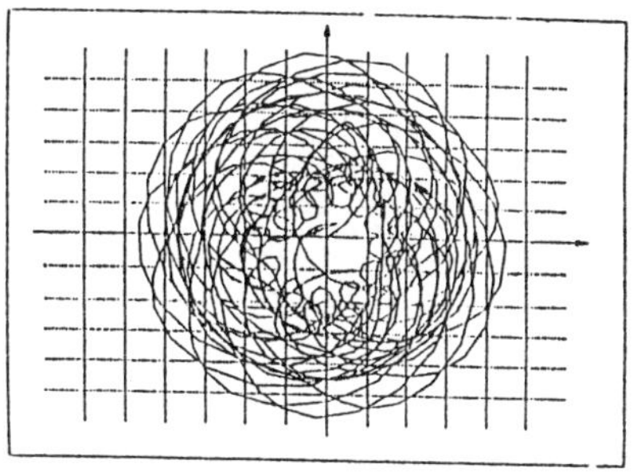

Fig 4a Pump bearing orbit

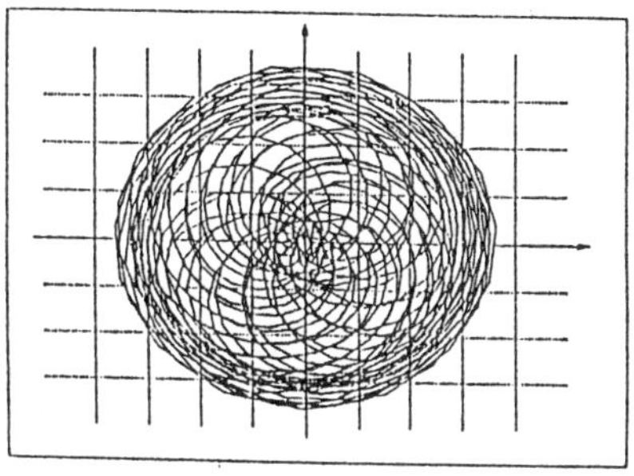

Fig 4b Turbine bearing orbit

ity of the nonlinear steady-state analysis to predict mean values of the results.

In order to obtain finer and more accurate results, the different algorithms could be applied to rotors modelled by the 3-dimensional formalism proposed in [10].

References

[1] Adams M.L. , *Non-linear Dynamics of flexible multi-bearing rotors*, Jl. of Sound and Vibr., Vol. 71 (1), 1980, pp 129-144

[2] Adams M.L. ,Padovan J.,Fertis D.g., *Engine Dynamic Analysis With General Non-linear Finite-Element Codes, Part 1: Overall Approach and Development of Bearing Damper Element*,ASME, Jl. of Eng. for Power, Vol. 104, July 1982, pp 586-593

[3] Tondl A., Some Problems of Rotor Dynamics, Chapman & Hall, London, 1965

[4] Gunter E.J., Barrett L.E., Allaire P.E., *Design of Nonlinear Squeeze-Film Dampers for Aircraft Engines*, ASME, Jl. of Lubrication Techn., Jan. 1977, pp 57-64

[5] Muszinska A., *Partial lateral rotor to stator rubs*, I. Mech. E., 1984, pp 327-335

[6] Childs D.W., *Fractional-Frequency Rotor Motion Due to Nonsymmetric Clearance Effects*, ASME, Jl. of Eng. for Power, Vol. 104, July 1982, pp 533-541

[7] Craig R.R., Bampton M.C., *Coupling of Substructures for Dynamic Analyses*, AIAA Jl., 6(7), 1968, pp 1313-1319

[8] Beckers P., Sander G., *Improvement of the Frontal Solution Technique*, LTAS Report SA-72, University of Liège, 1979

[9] SAMCEF manuals, Samtech, Liège, 1987 edition

[10] Géradin M., Kill N., *Dynamic Analysis of Struc-*

tures with Flexible Rotors, 2nd International Symposium on Aeroelasticity and Structural Dynamics, Aachen, April 1-3, 1985
[11] Géradin M., Kill N., *Eigenvalue Algorithms for Stability and Critical Speeds Analysis of Rotating Systems*, 6th IMAC Conference, Orlando Fl., February 1-4, 1988
[12] Flügge W., *Handbook of engineering mechanics*, Mac Graw Hill Book Company, 1962, pp 65-33, 65-35

C280/88

The transient response of a three-level rotor system subjected to a seismic environment

R SUBBIAH, PhD, MASME, **S ZHOU**, MS, MASME and **N F RIEGER**, PhD, MASME
Stress Technology Incorporated, Rochester, New York, United States of America

SYNOPSIS

Transient transfer matrix method has been proposed to model a three-level rotor-bearing support system and to analyse the dynamic response of rotor system subjected to support excitations. The forces impressed by the secondary support equipment (pedestal, foundation, and casings, etc.) on the primary system (rotor-bearing) at a time instant had been evaluated using a time-marching scheme. With the knowledge of the known support forces, the response of rotor-bearing (primary system) structure has been obtained at the same time instant using transient transfer matrix approach. This formulation provides greater flexibility to include both linear and non-linear sub-models without difficulty. Support seismic excitations of moderate nature have been simulated to study their influence on the dynamic behavior of simple rotor systems using the transient support model. The formulation establishes constant and compact system matrices.

1) NOTATION

a acceleration

c damping coefficient

k stiffness coefficient

m mass

q generalized displacement

t time

x,y displacements in x, y directions

A,D time constants

B,E time variables

F force component

ΔT time interval

Subscripts:

f representing foundation

i time instant

n representing the station number

p representing pedestal

r representing rotor

2) INTRODUCTION

Rotors are found wide usage in industry. They are either installed on the ground or on floating bases such as moving vehicles, ships, etc. Depending on the size, usage and the complexity of operation, such rotors are supported on one or more levels consisting of casings, pedestals, and foundation supports. These multi-level rotor systems may be subjected to external excitations due to mass unbalances or the support excitations or both. In order to achieve a safe design operating condition of the rotor support systems, it is highly essential to obtain the dynamic characteristics of rotors subjected to various types of external forces.

In rotordynamic analyses, few researchers had developed support models. Of particular note are the works of Ruhl et. al. [1], Subbiah et. al. [2], Zhou et. al. [3], and Armentrout et. al. [4]. While Ruhl and Subbiah used finite element model to study the behavior of rotors subjected to support motions, Zhou used the transfer matrix approach. Later, Lund [5] studied the response of rotors subjected to base excitations that are random in nature with simple one-level support model. Tessarzik [6] et. al. conducted experimental studies to obtain the dynamic response of rotating machinery to external random vibration. Quite recently, Inagaki [7] developed numerical method to study the three-level rotor support systems based on Myklestad-Prohl transfer matrix method. All the above authors used frequency domain approach to solve the multi-level rotordynamic problem. While this approach is good enough to obtain the dynamic characteristics of the rotor systems for steady state and off-resonance conditions, this may not provide the dynamic information of the system for initial transients, close-resonance conditions, threshold instability speeds, large amplitude motions, and time-dependent external excitations of various kinds. Under these circumstances, a more realistic approach would be the construction of a transient dynamic model to study the response of the rotor support system.

In the present work, a three-level rotor support transient model has been developed to study the dynamic characteristics of the rotor system. This work is the extension of the previous work performed on the transient analysis of one-level rotor support system by Subbiah et. al. [8]. The main advantage of this formulation [9] is the use of a small (9 x 9) constant and manageable form of matrices as against the finite element model wherein the proposed number of dynamic d.o.f. is directly proportional to the number of elements implicit in the problem and also against the transfer matrix formulations in [3] and [7] where the size of the matrices are at least (17 x 17) for a one-level support system. Moreover, the system matrices are formulated in a more flexible manner to adapt any time-marching schemes without difficulty. Therefore, this transient model has been used to study the rotordynamic system subjected to the seismic support motions.

Realistic excitations such as due to earthquakes ([10], [11]) may be non-stationary and non-Gaussian and better results may be obtained using (i) time history specification of support excitations, or (ii) by response spectra specification of support. The proposed rotor support model can be used either with a time-history analysis or a response spectra method in order to study the system response due to support motions.

3) SUPPORT MODEL

The three-level (bearing, pedestal, and foundation) support model of a single disk rotor system is shown in Figure 1. The rotor shaft section has been formulated as point and field matrices representing the mass and stiffness properties of the shaft respectively [8]. The point and field matrix expressions derived from the model show that the point matrix properties are time-dependent, whereas the field matrix properties are time invariant for a symmetric shaft section. At the supports, only point matrices have been developed corresponding to the mass points. The equations of motion of a three-level support system can be written as:

$$m_r \ddot{x}_r + c_{xx} (\dot{x}_r-\dot{x}_p) + k_{xx} (x_r-x_p) + c_{xy} (\dot{y}_r-\dot{y}_p) + k_{xy} (y_r-y_p) = F_X \quad (1)$$

$$m_r \ddot{y}_r + c_{yy} (\dot{y}_r-\dot{y}_p) + k_{yy} (y_r-y_p) + c_{yx} (\dot{x}_r-\dot{x}_p) + k_{yx} (x_r-x_p) = F_Y \quad (2)$$

$$m_p \ddot{x}_p + c_{px} (\dot{x}_p-\dot{x}_f) + k_{px} (x_p-x_f) + c_{xx} (\dot{x}_p-\dot{x}_r) + k_{xx} (x_p-x_r) + c_{xy} (\dot{y}_p-\dot{y}_r) + k_{xy} (y_p-y_r) = 0 \quad (3)$$

$$m_p \ddot{y}_p + c_{py} (\dot{y}_p- \dot{y}_f) + k_{py} (y_p-y_f) + c_{yy} (\dot{y}_p-\dot{y}_r) + k_{yy} (y_p-y_r) + c_{xy} (\dot{x}_p-\dot{x}_r) + k_{xy} (x_p-x_r) = 0 \quad (4)$$

$$m_f \ddot{x}_f + c_{xf} \dot{x}_f + k_{xf} x_f + c_{px} (\dot{x}_f-\dot{x}_p) + k_{px} (x_f-x_p) = 0 \quad (5)$$

$$m_f \ddot{y}_f + c_{yf} \dot{y}_f + k_{yf} y_f + c_{py} (\dot{y}_f-\dot{y}_p) + k_{py} (y_f-y_p) = 0 \quad (6)$$

Equations (1) through (6) may be solved using a suitable numerical time-marching scheme. In this paper, the transient model developed by Kumar et. al. [9] for general structural dynamic systems will be extended to formulate the time-marching scheme. As such, the algorithm is briefly described as follows:

4) TIME-MARCHING ALGORITHM

Most of the numerical integration schemes used in structural analysis are truncated Taylor's series as shown:

$$q_n (t_i) = q_n (t_{i-1}) + \Delta T \dot{q}_n (t_{i-1}) + \frac{\Delta T^2}{2} \ddot{q}_n (t_{i-1}) + \ldots \quad (7)$$

where

$$\Delta T = (t_i - t_{i-1})$$

Different integration schemes may be derived by replacing the derivatives in equation (7) by finite differences. Here a simplest finite difference scheme is chosen to explain the methodology. In addition, it is assumed that the acceleration is constant during the time internal (t_i-t_{i-1}) and is equal to the average of the acceleration values at t_i and t_{i-1}, and hence:

$$\ddot{q}_n = a = \frac{\ddot{q}_n (t_i) + \ddot{q}_n (t_{i-1})}{2} \quad (8)$$

and

$$\dot{q}_n = a \, \Delta T = \frac{\ddot{q}_n (t_i) + \ddot{q}_n (t_{i-1})}{2} \Delta T \quad (9)$$

Substitution of relationships (8) and (9) into equation (7) results in:

$$q_n(t_i) = q_n(t_{i-1}) + \Delta T\, \dot{q}_n(t_{i-1}) + \Delta T^2 \frac{\ddot{q}_n(t_i) + \ddot{q}_n(t_{i-1})}{4} \qquad (10)$$

Equation (10) can be rewritten as:

$$\ddot{q}_n(t_i) = A_n(t_i)\, q_n(t_i) + B_n(t_i) \qquad (11)$$

where

$$A_n(t_i) = \frac{4}{\Delta T^2}$$

and

$$B_n(t_i) = -A_n(t_i)\left[q_n(t_i-1) + \Delta T\, \dot{q}_n(t_i-1) + \frac{\Delta T^2}{4}\ddot{q}_n(t_i-1)\right] \qquad (12)$$

Similarly, by substituting equation (11) into equation (9) results:

$$\dot{q}_n(t_i) = D_n(t_i)\, q_n(t_i) + E_n(t_i) \qquad (13)$$

where

$$D_n(t_i) = \frac{2}{\Delta T}$$

and

$$E_n(t_i) = -D_n(t_i)\left[q_n(t_{i-1}) + \frac{\Delta T}{2}\dot{q}_n(t_{i-1})\right] \qquad (14)$$

From equations (11) and (13), it can be seen that the time constants $A_n(t_i)$, $D_n(t_i)$ and the corresponding time variables $B_n(t_i)$ and $E_n(t_i)$ are all functions of the system properties at time t_i, and the response quantities $q_n(t_{i-1})$, $\dot{q}_n(t_{i-1})$ and $\ddot{q}_n(t_{i-1})$ at the previous time instant which are all known at time instant t_i. It should be noted that the time constants and the corresponding time variables can, in general, be obtainable and used for any integration schemes suitably.

Using the time-marching procedure discussed above, equations (1) through (6) can be solved for any initial acceleration, velocity or displacement conditions at the foundation mass point. This results in the evaluation of forces at the support levels of the secondary equipment at a time instant. Then, following the transient transfer matrix procedures [8] for the rotor-bearing system (primary equipment), the overall transfer matrix can be obtained in terms of end-to-end state vectors. Since the forces at the support points are already known (using the free-free boundary condition), the responses at various points along the rotor can be evaluated at the same time instant t_i. This procedure is repeated as time marches from one instant to the other.

The matrices that have been handled in the present approach are small and manageable in size and hence the procedure ensures considerable savings in computational effort.

5) RESULTS AND DISCUSSIONS

A transient three-level rotor-support model has been developed and used to obtain the response of the rotor system subjected to support seismic excitations. In general, the seismic excitations are random in nature, i.e., non-stationary and non-Gaussian and hence a rigorous time-history response [12] calculation may result in a realistic approach for the evaluation of the response of the structure. The main advantage of this time-history method is that once the structure model has been developed, the rest of the procedure results in the exact solution of the governing equations of motion for various excitation characteristics. In order to demonstrate the validity of the model in the present work, simple rotor systems have been studied for initial transients only.

Depending on the severity of the base excitations and the secondary support properties, the initial transient motion may develop large amplitudes at the vital points of the structure. Hence, in this study, more emphasis is made to obtain the initial transient motions at the disk locations of the rotor system for the balanced and unbalanced rotor operating conditions.

Zhou [3] et. al. have studied a simple rotor supported on bearings of equal stiffness and damping values on both x and y directions. This rotor has been used to verify the response of the structure for mild seismic base excitations of 0.1g along x and y directions. They considered the fully balanced rotor and found that the secondary support equipment introduced critical frequencies at rotor speeds of about 155 rpm and 1550 rpm respectively.

The steady state response plot is shown in Figure 2. The initial transient plot for this rotor (Figure 3) shows that the base excitations of the order of 0.1g do not cause serious damage to the primary equipment.

However, in real rotor structures, the primary rotor equipment may have some unbalances at the heavy mass points. Consequently, these unbalance forces in conjunction with the forces impressed by the secondary equipment may develop large response levels which may be detrimental to the operation of the structure. As such, another rotor system ([8] ROTOR-1) for which the secondary equipment properties are available, has been studied. An eccentricity value of 1.084×10^{-4} kgm has been used at the disk location and severe seismic excitations of the order 1g (x-direction) and 2g (y-direction) have been introduced at the base. Initial transient response plots (Figures 4 and 5) confirm that the unbalanced rotor develops considerably larger response levels (Figure 5) at the disk location than the perfectly balanced rotor responses at the corresponding disk location (Figure 4).

6) TIME MARCHING SCHEME

It is important to note that the time step selection is crucial in this approach. In the present approach, Houbolt algorithm is used because it provides a stable solution [8] for the non-symmetric system such as the rotor-bearing system. However, the validity of the algorithm is not verified for larger systems. In the present work, a time step of 1/40 of first natural period has been used.

7) CONCLUSIONS

A transient three-level support model has been developed. Mild and large seismic base excitation levels were used to study their influence on the response of a single disk rotor system and it has been concluded that:

- o Unbalances present in the primary rotor equipment adds to the severity of damage impressed by the excitation of the secondary input equipment considerably.
- o Since the system matrices are compact, the transient support model can be extended for rigorous time-history response analysis of support synthesized seismic qualifications.
- o The transient model provides greater flexibility to analyse the influence of non-linear support models.

8) REFERENCES

1) Ruhl, R. L., Conry, T. F., and Steger, R. L., 'Unbalanced Response of a Large Rotor-Pedestal-Foundation System Using an Elastic Half-Space Soil Model,' ASME Journal of Mechanical Design, Vol. 102, pp. 311-319, April 1980.

2) Subbiah, R., Bhat, R. B., and Sankar, T. S., 'Response of Rotors Subjected to Random Support Excitations,' ASME Journal of Vibration, Acoustics, Stress, and Reliability in Design, Vol. 107, pp. 453-459, October 1985.

3) Zhou, S., Rieger, N. F., 'Development and Verification of an Unbalance Response Analysis Procedure for Three-Level Multi-Bearing Rotor-Foundation Systems,' ASME Paper No. 85-DET-113.

4) Armentrout, R. W., Gunter, J. E., and Humphris, R. R., 'Dynamic Characteristics of a Jeffcott Rotor Including Foundation Effects,' ASME Paper No. 83-DET-98, Presented at Design Engineering Technical Conference, Dearborn, Michigan, September 1983.

5) Lund, J. W., 'Response Characteristics of a Rotor With Flexible Damped Supports,' Symposium of International Union of Theoretical and Applied Mechanics, Lyngby, Denmark, pp. 319-349, August 12-16, 1974.

6) Tessarzik, J. M., Chiang, T., and Badgely, R. H., 'The Response of Rotating Machinery to External Random Vibration,' ASME Journal of Engineering for Industry, Vol. 96, No. 2, pp. 477-489, May 1974.

7) Inagaki, T., 'Numerical Computation Method for Rotor-Foundation (or casing) Dynamics: (Based on the Myklestad-Prohl Transfer Matrix Method),' Proceedings of Rotating Machinery Dynamics, Vol. 2, ASME 11th Biennial Conference on Mechanical Vibration and Noise, Boston, Massachusetts, pp. 383-390, September 27-30, 1987.

8) Subbiah, R., Rieger, N. F., 'On the Transient Analysis of Rotor-Bearing Systems,' Proceedings of Rotating Machinery Dynamics, 1987, Vol. II, pp. 525-536. Also accepted for publication in the ASME Journal of Vibration, Acoustics, Stress, and Reliability in Design.

9) Kumar A. Selva, Sankar, T. S., 'A New Transfer Matrix Method for Response Analysis of Large Dynamic Systems,' Computers and Structures, Vol. 23, No. 4, pp. 545-552, 1986.

10) Asmis, G. J. K., 'Response of Rotating Machinery Subjected to Seismic Excitation,' IMechE Conference Publications, Engineering Design for Earthquake Environments,' C192/78, pp. 215-226, 1978.

11) Hadjian, A. H., 'Support Motions for Mechanical Components During Earthquakes,' IMechE Conference Publications, 'Engineering Design for Earthquake Environments,' C173/78, pp. 27-46, 1978.

12) Masri, S. F., Richardson, J. F., and Young, G. A., 'Evaluation of Seismic Analysis Techniques for Nuclear Power Plant Piping and Equipment,' IMechE Conference Publications, Engineering Design for Earthquake Environments, C178/78, pp. 75-90, 1978.

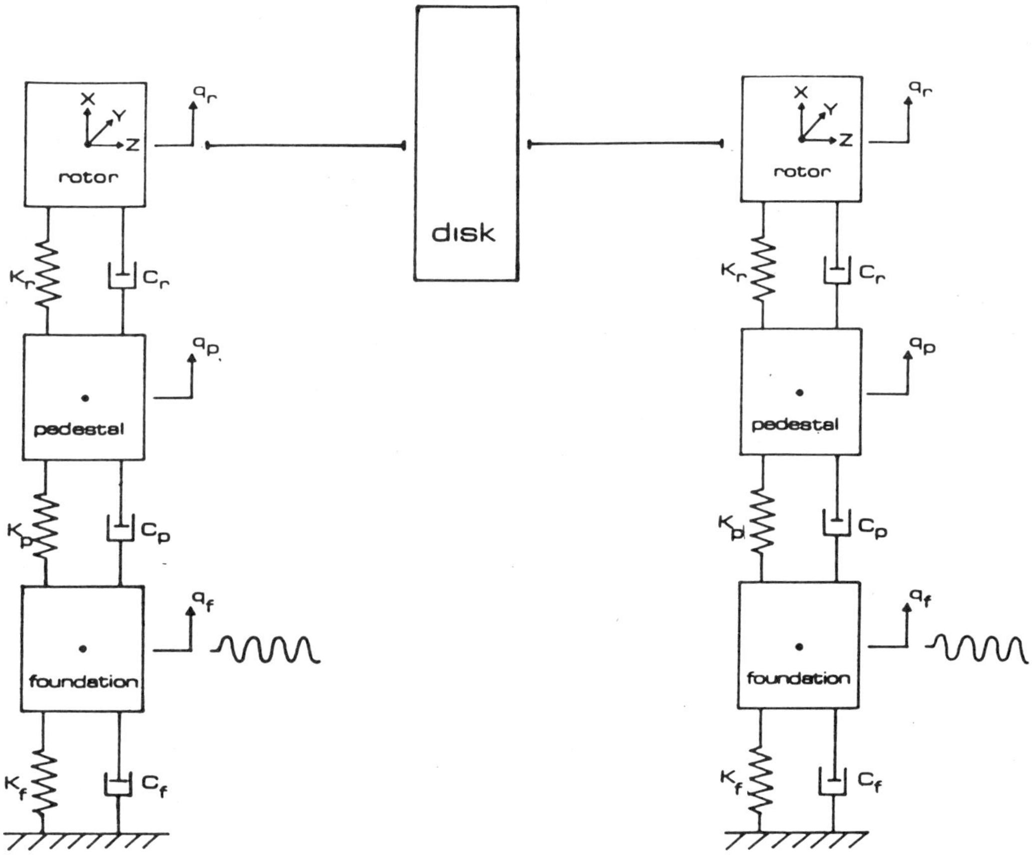

Fig 1   Rotor system support model

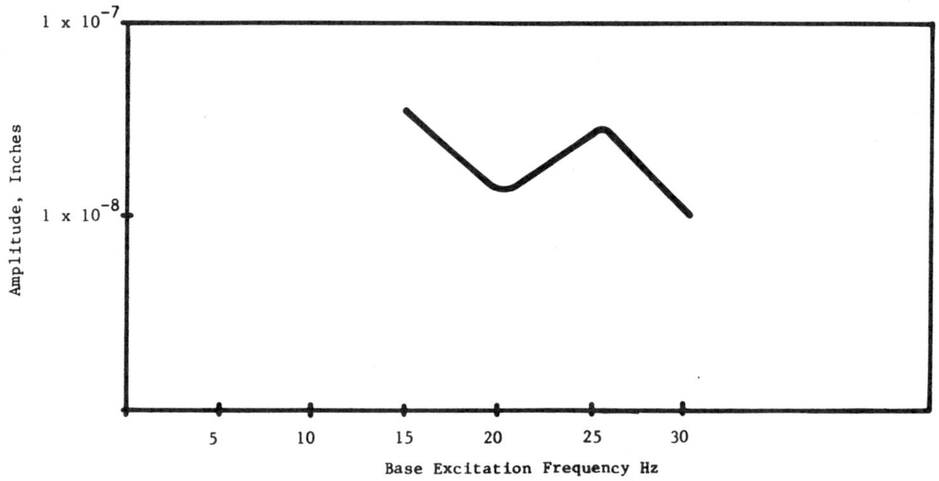

Fig 2   Frequency response at disk location (rotor speed = 1000 r/min)

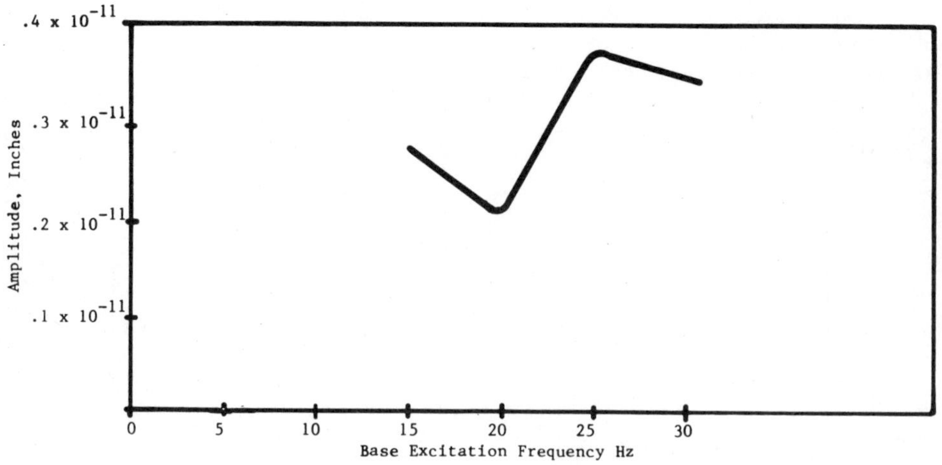

Fig 3 Initial transients at disk location (rotor speed = 1000 r/min)

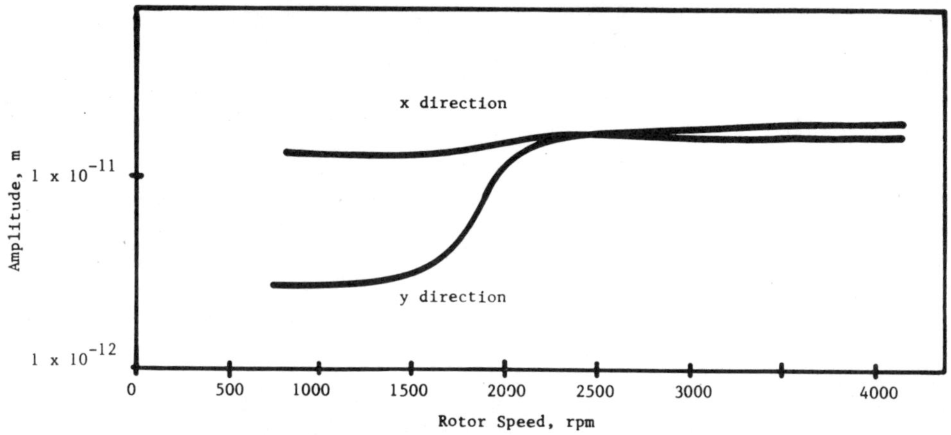

Fig 4 Initial transient response at disk location (perfectly balanced rotor)

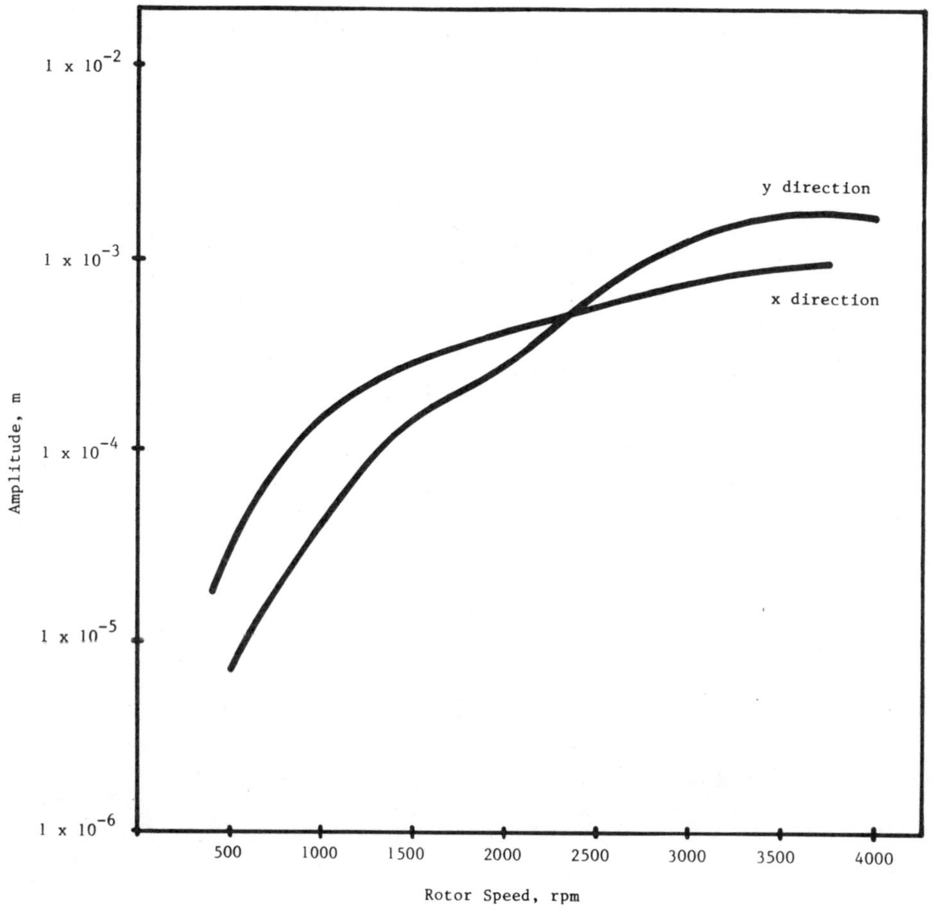

Fig 5 Initial transient response at disk location (unbalanced rotor condition)

C248/88

Transient response of a rotor beam excited by impulsive loading at the fixed end

J F BRATT, MS, MNSCE
Division of Machine Design, Norwegian Institute of Technology, University of Trondheim, Trondheim, Norway

SYNOPSIS Transient response to impulsive excitation of a rotating beam is studied using the central finite difference method and a recurrence algorithm. The influence of design parameters on shock spectra is discussed.

NOTATION

L	length of beam
w(i,j)	deflection of beam neutral axis at section i along the x-axis at instant j along time axis
w_m	maximum amplitude of simple harmonic deflection
a_m	maximum acceleration
x	distance along beam neutral axis
t	time
ω	excitation frequency, natural bending frequency of rotating beam
τ	natural bending period of rotating beam

1 INTRODUCTION

Several papers on impulse response for rotating beams deal with fracture and fatigue resulting from foreign-object damage to e.g. jet engines. The present analysis deals with responses to impact in the direction of and along the axis of rotation. This results in vibration perpendicular to he plane of rotation. Such vibration constitutes a problem where rotor foundations in linear motion (e.g. rotors on vehicles) abruptly stop as a result of accidents or when there is a sudden change in motion of the rotor foundation.

To find the rotor beam response steady-state solutions to transfer matrix equations, experimental impact results and transient solutions of the governing differential equation expressed in terms of differences are examined. The primary objectives of the investigation are:

(a) To establish a method of solution for the transient response of a rotating beam analysis, and thereby observe the influnce of variation of the system parameters and of the nature of the excitation.

(b) To verify the suggested trend of deflection transmissibility versus rotor r.p.m. in a shock spectrum for the sets of parameters investigated.

Minor flow-induced vibration and rotating unbalance are not considered in this analysis.

The deformation of the beam, which is a simplified model of a rotor blade, is assumed to obey the Bernoulli-Euler theory [1]. It is further assumed that the beam design permits some variation in the choice of material density, equivalent flexural rigidity and internal damping. Thus, in addition to accomplishing the technical function the designer may control dynamic behaviour, such as shock response [2].

2 EQUATION OF MOTION

The geometry and the stiffness of the beam are given by the length L, the cross-sectional area A and the equivalent bending stiffness EI [3]. In general the beam consists of a stiff filament compatible with a suitable matrix material.

It is assumed that the extension of the hub is small; hence the beam length is measured from the axis of rotation. The equation of motion is stated as follows:

$$EI \frac{\partial^4 w}{\partial x^4} + m \frac{\partial^2 w}{\partial t^2} - m\Omega^2 \frac{\partial}{\partial x}\left(\frac{L^2-w^2}{2} \frac{\partial w}{\partial x}\right) + cI \frac{\partial^5 w}{\partial x^4 \partial t} = 0 \qquad (1)$$

w is the deflection of the neutral axis, m is the mass per unit length, Ω is the rotational speed and c is the internal damping coefficient of the viscous type (Stokes law). The first two terms are the familiar expressions for bending deflection of a continuous beam. The component of the centrifugal force normal to the deflected beam neutral axis is regarded as a shear force resulting in the 3rd partial differential term. The last term is theoretically the most difficult to form. Observed data from experiments are therefore used. Equation (1) is the same as that for damped, free, transverse vibrations. External damping terms are considered small; generally they are beyond the beam designer's control.

By varying the properties of the reinforcement, (e.g. Boron reinforced Aluminium vanes with Titanium core or Graphite/Epoxy blades), one may control the 1st term in equation (1). The second and third terms are primarily controlled by the density of the matrix material; the last term is influenced by the damping of the matrix material.

3 BOUNDARY CONDITIONS

The excitation acts at the fixed end of the beam. In compliance with standard practice the controlled acceleration here is assumed to be a half-cycle sine pulse with zero initial and final displacements. Another "reference" pulse is the versed sine pulse. However, in an actual, uncontrolled collision situation the most likely pulse may be an acceleration sine pulse with zero initial displacement and velocity. A comparison of the beam tip response for these 3 cases is shown in figure 1. The built-in end displacements are given by:

$$w(0,t) = \frac{a_m}{2}\left(\frac{t^2}{2} - \frac{1}{\omega^2}(1-\cos\omega t) - \frac{\pi}{\omega}t\right) \quad (2)$$

$$w(0,t) = w_m \sin\omega t \quad (3)$$

$$w(0,t) = \frac{a_m}{2}\left(\frac{t^2}{2} - \frac{1}{\omega^2}(1-\cos\omega t)\right) \quad (4)$$

respectively. In these expressions the displacements correspond to a shock acceleration of 30 g_m/s^2, and the pulse duration is $15 \cdot 10^{-3}$ s. At the built-in end and at the free end the following additional conditions prevail for the above 3 excitation impulse cases.

$$\left.\frac{\partial w}{\partial x}\right| (0,t) = 0 \quad (5)$$

$$\left.\frac{\partial^2 w}{\partial x^2}\right| (L,t) = 0 \quad (6)$$

$$\left.\frac{\partial^3 w}{\partial x^3}\right| (L,t) = 0 \quad (7)$$

4 SPIN TEST

To evaluate the damping term in equation (1) the logarithmic decrement is measured experimentally. The spin test stand is described in reference [2]. Two identical, symmetrically mounted, instrumented cantilever beams with a total length of 0.3 m, are rotated in air at room temperature and pressure. Damping decrements versus r.p.m. are obtained. Only small variation in the measured decrement with r.p.m. is observed. The decrement is increased by about 40% when a beam of epoxy matrix material is substituted by one of rubber-like matrix material, the glass-fibre reinforcement remaining the same in both cases.

A temporary, small reduction in r.p.m. during impact is observed in the experiments. The effect on the transient response is likely small; the integration of equation (1) is carried out with an assumed constant r.p.m. during the entire observed period.

5 NUMERICAL SOLUTION

The integration of the equation of motion (1) with the given end conditions is accomplished by the method of finite differences. One difference equation, corresponding to equation (1), is written for each one of selected sections along the beam, thus establishing a set of linear simultaneous equations of the form

$$[A]\{w(i,j+1)\} = \{B\} \quad (8)$$

The coefficient matrix [A] and the right hand side vector {B} contain terms which are functions of the system mechanical properties, geometry, rotational speed, length of subdivisions of the beam Δx, time increments Δt, deflections $w(i,j)$ and $w(i,j-1)$. $\{w(i,j+1)\}$ is the unknown deflection vector.

Ten subdivisions are used along the beam axis. This number is sufficient for the estimation of the lower modes of vibration. Due to damping and finite strain energy in the system the higher modes become less pronounced and play a minor role in the assessment of the impulse spectrum.

The time divisions are somewhat less obvious. Small Δts are not sufficient for the solution to be stable. The number of divisions must be great enough so that Δt becomes considerably less than the shortest vibration period to be investigated [4]. Accordingly 2000 time increments were used in the calculation.

A recurrence algorithm is applied using a Fortran program for solving equation (8). The deflection gradients along the time axis at t = 0 are in accordance with expressions (2), (3) and (4) above. Numerically unstable situations occur at angular velocities above 2000 r/s (19098 r.p.m.). Still, the region from 0 up to this speed is regarded relevant and sufficient for practical purposes, e.g. fast rotating vanes and blades. Certain combinations of equivalent bending stiffness and density distribution also result in too high second natural frequency for numerical solution to be performed. However, the results for the region investigated clearly bring about the effect of varying the system parameters on the shock spectrum. If the number of increments were increased by e.g. a factor of 10 more detailed information might be gathered.

Steady state solutions for the first mode using the transfer matrix method is accomplished by modified text book programs, e.g. [5]. The results indicate variations in the natural, undamped frequency in the r.p.m.-region mentioned above.

6 RESULTS

The results of the integration of the equation of motion (1) are most conveniently shown in diagrams 1 to 5. Due to small time steps ($\Delta t = 1.5 \cdot 10^{-5}$ s) the curves are representative to a high degree of accuracy. Only the beam tip displacements are shown. Following the termination of the impulse period the amplitudes of the lowest and second lowest vibration modes are recognized. In figure 6 are shown the deflected shapes of the beam centre line at selected instants during the impulse period. Also in diagrammatic form is indicated the dependence of the system natural frequency on the r.p.m., see figure 7.

The trends indicated in figures 2 to 6 are based on the "reference" half-sine impulse excitation, maximum deflection = $6.7 \cdot 10^{-3}$ m or $0.822 \cdot 10^{-3}$ m. The system variables are the equivalent bending stiffness, mass per unit length, structural damping and rotational speed. Unless otherwise stated the parameters in figures 1-6 are EI = 37.11 Nm^2, m = 0.67 kg/m, cI = $0.1 \cdot 10^{-3}$ Nsm^2 and Ω = 1000 r/s.

7 DISCUSSION

The results prove that the analytical method is representative and accurate; the system vibrational behaviour during and following an impulse is clearly demonstrated.

An improvement is possible as to impulsive response. The design quantities most likely to be varied are EI, m and to a less extent cI. Figure 2 indicates that increasing EI from 22 to 48 Nm^2, while the other parameters are kept constant, reduces the deflection transmissibility from 1.14 to 1.10. This change corresponds to a reduction in natural period from $4.655 \cdot 10^{-3}$ to $3.793 \cdot 10^{-3}$ s.

The residual spectrum depends upon the displacement and velocity at the tip of the beam at the termination instant of the impulse period. This is in accordance with the spin test results. The classical oscillator diagram, e.g. [6], is different.

An increase in mass per unit length from 0.5 to 1.0 kg/m, while the other parameters are kept fixed, increases the tip deflection transmissibility by about 6%, as shown in figure 3; the change in natural period is 5.1%.

The effect of altering the internal damping is indicated in figure 4. A beam tip transmissibility reduction of 7.6% results from increasing the damping coefficient cI from 0 to 0.01. This is a marked improvement for the set of data considered. As expected only minor changes in the damped natural period is noted.

Although in many cases not subject to variation for design reasons variation in rotational speed is nevertheless considered in figure 5. As shown, increasing the rotational speed increases ω. In figure 5 the pulse period is $5.25 \cdot 10^{-3}$ s with a maximum base deflection $0.822 \cdot 10^{-3}$ m. This corresponds to a maximum acceleration of 30 g. Allowing the beam r.p.m. to rise, as indicated, alters the ratio of pulse duration to τ from 0.952 to 1.178 (beyond the peak of the response spectrum, [2]) and reduces the transmissibility by 9.6%.

Conventional steady state analysis cannot predict the impulse responses dealt with above. The difference method with recurrence loops mentioned appears to be a superior approach. However, a transfer matrix analysis of the steady-state dependence of ω on r.p.m. is shown in figure 7. The stiffening effect of the centrifugal forces is clearly recognized for all material combinations, S, C, A and G, [7].

The computed beam tip deflections, figures 2 to 5, confirm that a substantial reduction in transmissibility and accordingly diminished stress levels result when the operating range of a rotating beam lies beyond the impulsive range in the response spectrum.

ACKNOWLEDGEMENT

This research is supported, in part, by NTNF through SINTEF, The Foundation for Scientific and Industrial Research at the Norwegian Inst. of Technology.

REFERENCES

[1] AMADA, S., Dynamic response of a beam subjected to impulsive like rotations. JSME, vol. 29, No. 252, June 1986.

[2] BRATT, J.F., Response of rotating reinforced beams to shock excitation. Transactions CSME, Vol, 11, No. 1, 1987.

[3] TSAI, S.W. and HAHN, H.T., Introduction to composite materials. Technomic Publishing Co. Pa, USA, 1980.

[4] RAO, S.S., Mechanical Vibrations. Addison-Wesley. Reading, Ma, USA, 1986.

[5] LALANNE, M. et al, Mechanical vibrations for engineers, John Wiley, Chichester, 1983.

[6] HARRIS, M.H. and CREDE, C.E., Shock and vibration handbook. McGraw-Hill Book Co., 2nd ed. New York, 1976.

[7] BRATT, J.F., Response of a reinforced rotating beam excited by impulsive loading at the fixed end. Proc. 11th CANCAM, Edmonton, Canada, June 1987.

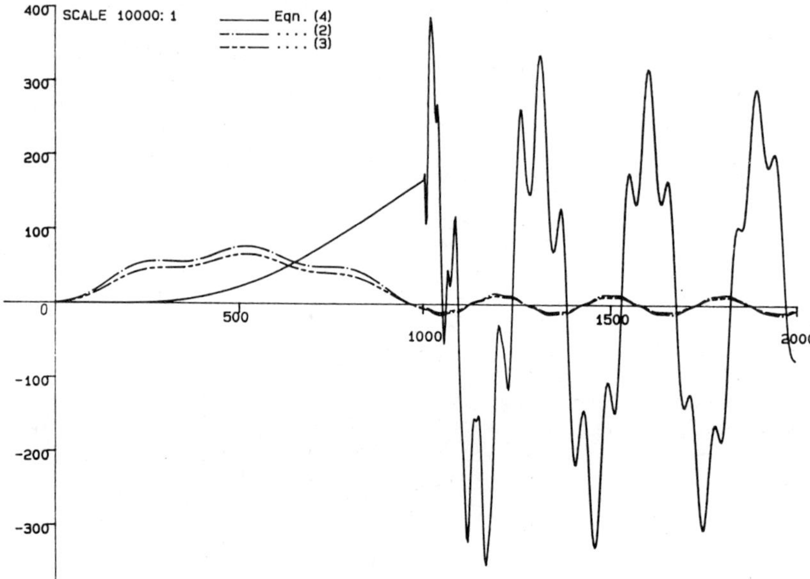

Fig 1 Responses to excitations given by equations (2), (3) and (4); beam tip deflection in m versus number of time increments ($\Delta t = 1.5 \cdot 10^{-5}$ s)

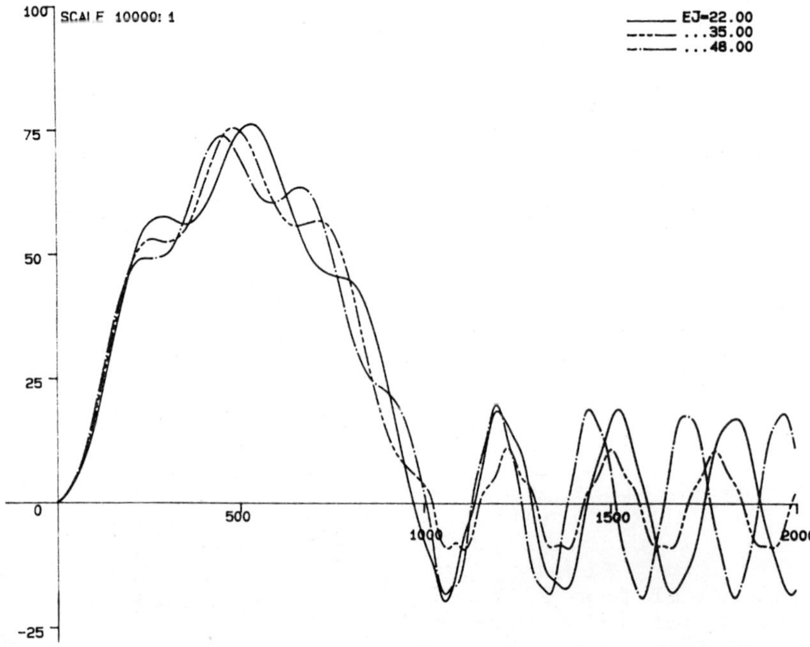

Fig 2 Influence of varying the equivalent bending stiffness EI ; beam tip deflection in m versus number of time increments ($\Delta t = 1.5 \cdot 10^{-5}$ s)

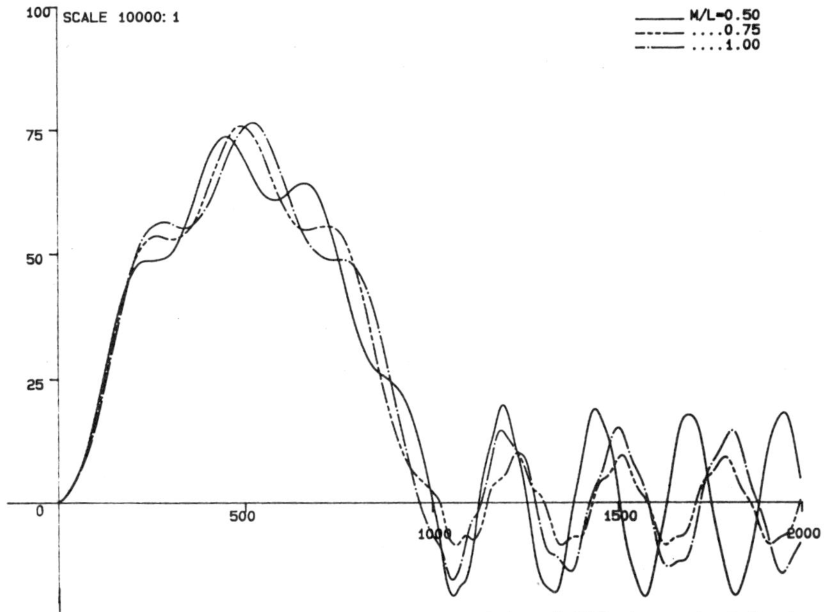

Fig 3 Variation of the mass per unit length M/L; beam tip deflection in m versus number of time increments ($\Delta t = 1.5 \cdot 10^{-5}$ s)

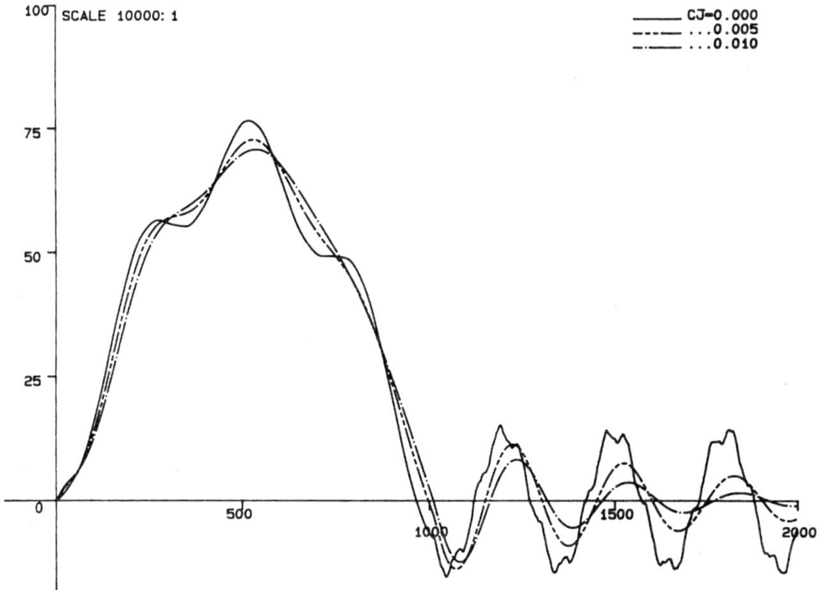

Fig 4 Variation of structural damping cl; beam tip deflection in m versus number of time increments ($\Delta t = 1.5 \cdot 10^{-5}$ s)

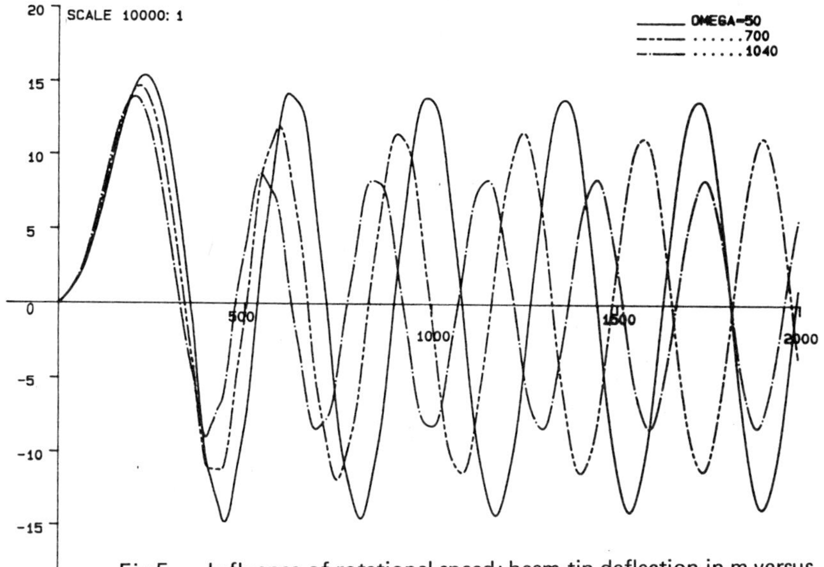

Fig 5 Influence of rotational speed; beam tip deflection in m versus number of time increments ($\Delta t = 1.5 \cdot 10^{-5}$ s)

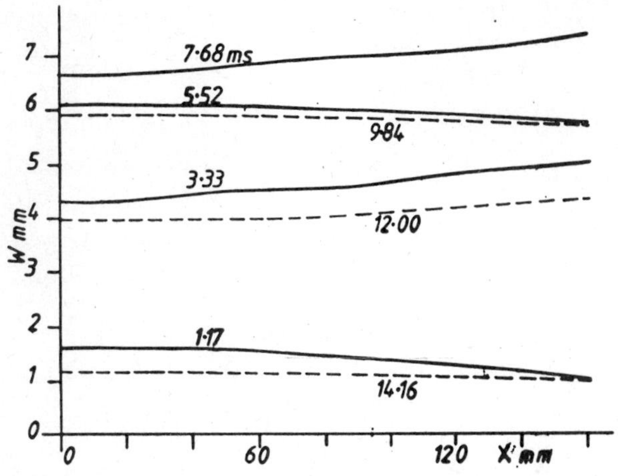

Fig 6 Deflected beam at various instants during the impulse

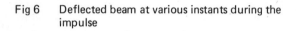

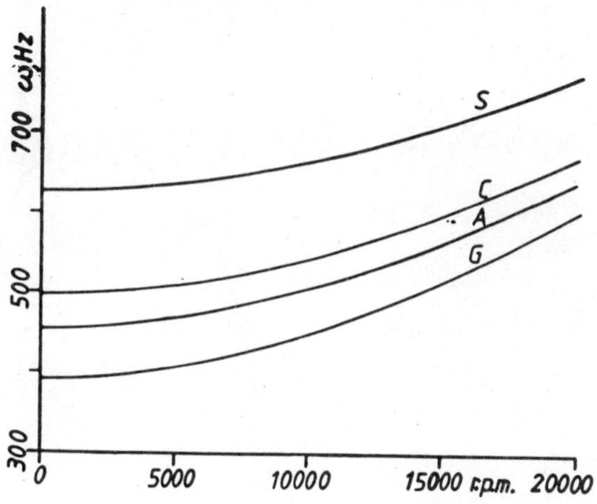

Fig 7 Variation of the equivalent bending stiffness EI versus r/min

C322/88

Unstable vibration of a rotor with a transverse crack

A TAMURA
Faculty of Engineering, Takushoku University, Tokyo, Japan
Y IWATA and H SATO
Faculty of Technology, Kanazawa University, Kanazawa, Japan

1 INTRODUCTION

A transverse crack which results from a fatigue or a shock to rotors of rotating machines may cause a catastrophic accident, furthermore may cause a system down. In recent years, a technique for early detection of transverse cracks has been tried using changes of vibrational characteristics during the operation of rotors. There are many papers (1 - 8) consisting of analyses and/or experiments for cracked rotors. It has been shown in these papers that the crack mainly influenced components of once, twice and three times of the rotational speed, and that the characteristic of response curves near the critical speed is dependent on the unbalance location because of the influence of the opening and closing of the transverse crack, but the experiments of rotors with a large transverse crack may not have been carried out because of difficulties in making the crack and for safety of the experiments.

An imitation crack, which simulates the characteristic of the practical transverse crack, is devised and a rotor system having such a crack was made up in this experiment. It is possible to vary the crack depth up to a large value and the rotor can be safely operated in the experiment because the crack is not propagated. After the validity of the imitation crack is confirmed, the vibrational characteristic of the rotor with the transverse crack has been qualitatively investigated from the measured result of the amplitude response curve. It was newly observed that an unstable vibration occurs in the region near the rotational speed at 2/3 of the critical speed in the large crack case. Such an unstable vibration was only expected by an analogue computer analysis in the other Ref (6). In this paper, the unstable vibration of rotors with the transverse crack is analytically investigated and the influences of the crack depth and of the damping on the unstable region, are discussed.

2 EXPERIMENTAL EQUIPMENT

A transverse crack in a rotor opens and closes during rotation in the case of horizontal rotors, because the rotors are deflected by the gravity. For example, if the crack direction agrees with the gravity direction, the crack is open and if the crack direction agrees with the anti-gravity direction, the crack is closed. If the crack direction exists between both directions, there are both open and closed sections in the crack. Since the area ratio of the open and the closed sections continuously varies according to the rotation, it is considered that the flexural rigidity of the rotor also continuously varies. Because it is difficult to make a large crack in the experiment, the imitation crack to satisfy the above-mentioned characteristic of the crack has been made up and the experiment of the cracked rotor has been carried out.

In the case where the operation of rotors with the transverse crack is done in laboratories, the crack is small because the shafts of rotors are generally slender. Then it is considered difficult to make the crack and that the errors due to making the crack has an influence on the crack characteristic. In this experiment, the large crack was imitated by the rotor shown in Fig 1. This rotor consists of a cylinder (A), two thin disks (B) and two shafts (C) and is symmetric. The two disks are fixed at both of the sides of the cylinder by 24 bolts respectively and each shaft is set up in the centre of each disk. If some of the bolts fixing the disks are removed, the removed disk section opens and closes with the rotation of the rotor similar to the behaviour of a crack. Since the disk deformation becomes large because of the thinness, and the degree of opening at the crack section becomes large, the errors of making the crack and of the assembly do not considerably influence the opening and closing of the crack section. The rotor was balanced by adding weights equal to the eccentricity of removed bolts, and the coupled vibration of a parallel mode and a conical mode of the cylinder is eliminated because the bolts are symmetrically removed. The rotor assembled was supported in both ends by deep groove ball bearings and was coupled to a motor by a rubber joint. The assembled rotor is shown in Fig 2. The horizontal and the vertical vibrations of the cylinder were measured by two non-contact displacement sensors which were set near the middle of the two bearings. 0. 7. 9 and 11 bolts were removed from every side of the cylinder in the experiment. The crack depth is expressed by a crack parameter corresponding to the number of bolts removed. The crack parameter is a non-dimensional parameter, the crack depth due to removing bolts is divided by an outside diameter of the cylinder; the relation between the crack parameter and the number of the removed bolts is shown in Table 1.

In the case of the cracked rotor, the static deflection due to the self-weight of the rotor is dependent on the rotational angle because the flexural rigidity continuously varies with the rotation of the rotor. The static deflection of the rotor used in this experiment was measured for the horizontal and the vertical directions respectively and the result is shown in Fig 3. The measured value denotes the relative static deflection of the non-cracked rotor. It is found from Fig 3 that the static deflection of the cracked rotor has a once per revolution periodic variation and that the proportion of the variation becomes large with increasing the crack parameter. In the case of the vertical deflection curves, the deflection at 180° is largest because of the fully opened crack section. But the deflection at 0° is not smallest in spite of the fully closed crack section and the deflection at 90° is smaller than that at 0°. The reason for this is that the large deflection of the disks is mainly caused at 0° by a non-restraint of the thin disk at the crack section. But the influence of the static deflection near 0° becomes small in comparison with the maximum deflection with increasing values of the crack parameter. If Fig 3 is compared to the result of paper (1) in which the horizontal and vertical static deflections were measured for the rotor with a practical transverse crack, the qualitative agreement of both results is seen in the case of the large crack, therefore, it is considered that the characteristic of the practical crack is sufficiently simulated by the imitation crack used in this experiment and the rotor shown in Fig 1 can be regarded as a rotor with a transverse crack.

3 EXPERIMENTAL RESULT

Amplitude-rotational speed response curves in the horizontal and the vertical directions were measured for every crack parameter shown in Table 1. Since the response curves of both directions are similar, the horizontal response curves only are shown in Fig 4. The amplitude of y-axis is a half of the difference between the maximum and the minimum displacements measured. Arrows on the figure are directions for a slow sweep of the rotational speed and the rotational speed for every peak is given in the figure. From Fig 4(a), where the crack parameter is zero, the critical speed is 2290 rpm, and a small peak exists near 1/2 of the critical speed because the flexural rigidity of the experimental rotor has a very small asymmetry. In the Fig 4(b) case, when the crack parameter is 0.24, it was impossible to pass 1/2 peak owing to its magnitude. The 1/2 peak was passed by means of restraining the rotor deflection and the response curve was recorded from a neighbourhood of the critical speed by slowly reducing the rotational speed. Furthermore, it was entirely impossible to pass the critical speed; thus the peaks near the critical speed and near 1/2 of that, are sensitive to existence of the crack, and a new peak appeared near 1/3 of the critical speed. In the Fig 4(c) case, where the crack parameter is 0.35, the height of the 1/3 peak and a region of the 1/2 peak were larger than those of the Fig 4(b) case and a new peak appeared near 1/4 of the critical speed. From reports that the components of integer multiples of the rotational speed are included in the vibration wave (2), (4-6) and (8), it follows that the resonance occurs at sub-multiples, that is 1/2. 1/3. 1/4 etc, of the critical speed. If the response curve was measured after adding an unbalance, the peak near the critical speed only became large and the other peaks did not change at all. Therefore it is considered that the additional resonances near 1/2, 1/3 and 1/4 of the critical speed are caused by a gravity because there is no force to act on the rotor except an unbalance force and a gravity force.

In the Fig 4(d) case, where the crack parameter is 0.46, heights of the 1/3 and the 1/4 peaks and a region of the 1/2 peak, became larger and a 1/5 peak appeared. Also the amplitude suddenly diverged near 2/3 of the critical speed, and such a phenomenon did not exist in the response curves for the crack parameter less than 0.35. Because the peak of this phenomenon completely differs from the forms of the other peaks and the amplitude increases with time, the 2/3 peak is not the peak due to a resonance. Therefore it is considered that the vibration becomes unstable near 2/3 of the critical speed and it is necessary to give attention to the operation of large cracked rotors near this rotational speed. Such an unstable vibration at 2/3 of the critical speed was only expected by an analogue computer analysis (6). The existence of this unstable vibration region is confirmed theoretically afterward in this paper.

4 UNSTABLE VIBRATIONAL ANALYSIS

From the experiment, it became clear that the unstable vibration occurred near 2/3 of the critical speed. To prove theoretically the occurrence of the unstable vibration for a rotor with transverse crack, the unstable region for a simple cracked rotor is analytically investigated by using the method of H. Ota and K. Mizutani (9).

As shown in Fig 5, x and y axes are set to the horizontal and the vertical directions respectively, and ξ and η axes are fixed on the cross section of the rotor to agree with principal axes of the flexural rigidity during the full open crack. If the rotor is rotating with angular speed ω, ξ and η are rotational coordinates with angular speed ω for x-y coordinates. If the influence of gravity for the rotor supported horizontally are considered, spring constants $k\xi$ and $k\eta$ to ξ and η respectively are assumed as follows:

$$\begin{aligned} k\xi &= \frac{1}{2}\{(k_0+k_1)+(k_0-k_1)\cos\Omega t\} \\ k\eta &= \frac{1}{2}\{(k_0+k_2)+(k_0-k_2)\cos\Omega t\} \end{aligned} \qquad (1)$$

where the deflection of the rotor is very small. Such simple expressions were also assumed in the paper (7). k_0 is a spring constant in the case of no crack (that is, the crack is close), k_1 and k_2 are spring constants for ξ and η directions during the full open crack respectively.

Supposing a simple rotor consisting of a disk and a cracked massless shaft, the equations of motion of the free vibration can be expressed on the rotational co-ordinates as follows:

$$M(\frac{d^2\xi}{dt^2} - 2\Omega\frac{d\eta}{dt} - \Omega^2\xi)+C(\frac{d\xi}{dt} - \Omega\eta)+k_\xi\xi = Mg\ sin\Omega t$$
$$M(\frac{d^2\eta}{dt^2} + 2\Omega\frac{d\xi}{dt} - \Omega^2\eta)+C(\frac{d\eta}{dt} + \Omega\xi)+k_\eta\eta = Mg\ cos\Omega t \qquad (2)$$

where M is a mass of the disk and g is a gravitational acceleration. An external viscous damping coefficient C is considered in Equation (2) to investigate the influence of the damping. A gyroscopic effect and a slope motion of the disk are neglected in Equation (2) to simplify the analysis. Transforming Equation (2) into the expressions on x-y fixed co-ordinates and substituting Equation (1), the equations of motion of the simple cracked rotor on the fixed co-ordinates are obtained. Then, using a complex form z=x+jy, the equation of motion becomes as follows:

$$M\frac{d^2z}{dt^2} + C\frac{dz}{dt} + Kz + \Delta K(e^{j\Omega t}+e^{-j\Omega t})z - \Delta k(e^{j\Omega t} - 2e^{j2\Omega t} + e^{j3\Omega t})\bar{z} = 0 \qquad (3)$$

where $\bar{z}$ is a conjugate complex form to z and K, ΔK and Δk are the following expressions:

$$K = (2k_0+k_1+k_2)/4$$
$$\Delta K = (2k_0-k_1-k_2)/8 \qquad (4)$$
$$\Delta k = (k_1-k_2)/8$$

The static deflection of the rotor can be calculated by exchanging a right side of Equation (3) with the self-weight jMg. k_0=33.0 kN/m, k_1=30.8 kN/m and k_2=21.9 kN/m were measured in the case of crack parameter 0.46 and M is 5.63 kg neglecting the mass of shaft parts in Fig 1. These values are substituted into Equation (3) and the horizontal and the vertical static deflections are calculated. The result is shown in Fig 6. Since Fig 6 qualitatively agrees with Fig 3, it is shown that the assumption of the spring constants of Equation (1) is valid.

Equation (3) is transformed into the following non-dimensional form:

$$\frac{d^2u}{d\tau^2} + 2\xi\frac{du}{d\tau} + u + \kappa_1(e^{j\omega\tau}+e^{-j\omega\tau})u - \kappa_2(e^{j\omega\tau}-2e^{j2\omega\tau}+e^{j3\omega\tau})\bar{u} = 0 \qquad (5)$$

where

$$\omega=\Omega/\Omega_n,\quad \Omega_n= \sqrt{K/M},\quad \xi=C/(2\sqrt{MK}),\quad \tau=\Omega_n t$$
$$\kappa_1=\Delta K/K,\quad \kappa_2=\Delta k/K,\quad u=z/s,\quad s=Mg/K \qquad (6)$$

κ_1 is a parameter with respect to the down of the rotor stiffness due to the crack, κ_2 is a parameter with respect to an asymmetry of the rotor stiffness and those parameters have very small values. The measurement values of κ_1 and κ_2 for the crack parameters of the experiment are shown in Table 1. It is found from Table 1 that κ_1 and κ_2 correspond to the crack depth. In the case of the undamped system (ζ=0) for Equation (5), using the method of successive approximation of the paper (9), the approximate solution becomes as follows:

$$u_0 = Ae^{jp\tau}+ae^{j(\omega+p)\tau}+be^{j(\omega-p)\tau}+ce^{j(-\omega+p)\tau}$$
$$+de^{j(2\omega-p)\tau}+fe^{j(3\omega-p)\tau}+\alpha e^{j(-p)\tau} \qquad (7)$$
$$+\beta e^{j(2\omega+p)\tau}+\gamma e^{j(-2\omega+p)\tau}+\delta e^{j(4\omega-p)\tau}$$

where p denotes an angular speed of the whirling motion. A is the amplitude of the whirling motion. a, b, c, d and f denote amplitudes with respect to a very small value. α, β, γ and δ denote amplitudes with respect to square order of a very small value. Amplitudes of small values more than cube order are neglected. It is assumed that the damped solution is also approximated by Equation (7), and the approximate solution of Equation (5) is expressed by the following form to judge a stability of the solution:

$$u = e^{-q\tau}u_0 \qquad (8)$$

Then every amplitude of the whirling motion in Equation (7) has a complex value because a phase lag is caused by the damping. Substituting Equation (8), to which λ=p+jq is applied, into Equation (5) and comparing the coefficient of each harmonic component, the following frequency equation is obtained:

$$(M\lambda^2 + C\lambda + K)u = 0 \qquad (9)$$

where u, M, C and K are shown in Appendix.

κ_1 = 0.05 and κ_2 = 0.04 close to κ_1 = 0.056 and κ_2 = 0.038 in the case of the crack parameter 0.46 are selected and the undamped (ζ=0) natural frequency λ for any ω is calculated. The result becomes as Fig 7. The shadowed portion shows an unstable region in which an imaginary part of λ becomes negative and in which the vibration expressed by Equation (8) diverges. The unstable vibrations occur near twice, once, 2/3 and 1/2 of the critical speed and the unstable region becomes narrow with such an order. In particular, the unstable region near 1/2 is very narrow. The effect of κ_1 and κ_2 on the unstable region is investigated from Fig 8. Fig 8(a) shows the effect of κ_2 for κ_1=0.05 constant and Fig 8(b) the effect of κ_1 for κ_2=0.05 constant. It is found that κ_1 does not at all influence the unstable region and that κ_2 mainly has the influence on the unstable region. The influence of the external viscous damping factor ζ in the case of Fig 8(a) is calculated as Fig 9. The unstable regions become narrow due to damping, and, in particular, the unstable region near 1/2 of the critical speed disappears with only a very small damping. The unstable region near 2/3 of the critical speed was observed in the experiment and confirms this analytical result, but it was difficult to confirm the existence of unstable regions except 2/3 in the experiment because it was impossible to distinguish the unstable vibration from the resonance in the response curve.

5 CONCLUSIONS

The following conclusions resulted from the experiment and unstable vibrational analysis of the cracked rotor.

(1) The rotor with the imitation crack devised in this experiment adequately simulates the characteristic of flexural rigidity of practical cracked rotors.

(2) From the response curves obtained by the experiments, it was found that additional resonances at 1/2, 1/3, 1/4 and 1/5 of the critical speed were caused by the crack and gravity.

(3) It was observed that the unstable vibration occurred in the region near the rotational speed at 2/3 of the critical speed in the large crack case.

(4) From the unstable vibration analysis, it is confirmed that the unstable regions exist at twice, once, 2/3 and 1/2 of the critical speed, and the influences of the crack depth parameters and of the damping on the unstable regions is shown in Fig 8 and Fig 9.

REFERENCES

(1) ZIEBARTH, H., SCHWERDTFEGER, H. and MUHLE, E.-E., 'Auswirkung von Querrissen auf das Schwingungsverhalten von Rotorn'. VDI-Berichte, 320 (1978), p.37.

(2) INAGAKI, T., KANKI, H. and SHIRAKI, K., 'Transverse Vibrations of a General Cracked-Rotor Bearing System'. ASME J. Mechanical Design, 104 (1982), p.345.

(3) DIMAROGONAS, A.D. and PAPADOPOULOS, C.A., 'Vibration of Cracked Shafts in Bending'. J. Sound and Vibration, 91-4 (1983), p.583.

(4) MAYES, I.W. and DAVIES, W.G.R., 'Analysis of the Response of a Multi-Rotor-Bearing System Containing a Transverse Crack in a Rotor'. ASME J. Vibration, Acoustics, Stress and Reliability in Design, 106 (1984), p.139.

(5) DAVIES, W.G. and MAYES, I.W., 'The Vibrational Behaviour of a Multi-Shaft, Multi-Bearing System in the Presence of a Propagating Transverse Crack'. ASME J. Vibration, Acoustics, Stress and Reliability in Design, 106 (1984), p.146.

(6) GROBOWSKI, B., 'Dynamics of Rotors Stability and System Identification' (Edited by Mahrenholtz). SPRINGER-VERLAG (1984), p.423.

(7) SCHMIED, J. and KRAMER, E., 'Vibrational Behaviour of a Rotor with a Cross-Sectional Crack'. IMechE. Conf. 1984, p.183.

(8) NELSON, H.D. and NATARAJ, C., 'The Dynamics of a Rotor System with a Cracked Shaft'. ASME J. Vibration, Acoustics, Stress and Reliability in Design, 108 (1986), p.189.

(9) OTA, H. and MIZUTANI, K., 'Influence of Unequal Pedestal Stiffness on the Instability Regions of a Rotating Asymmetric Shaft'. ASME J. Applied Mech., 45 (1978), p.400.

Table 1 Crack parameter

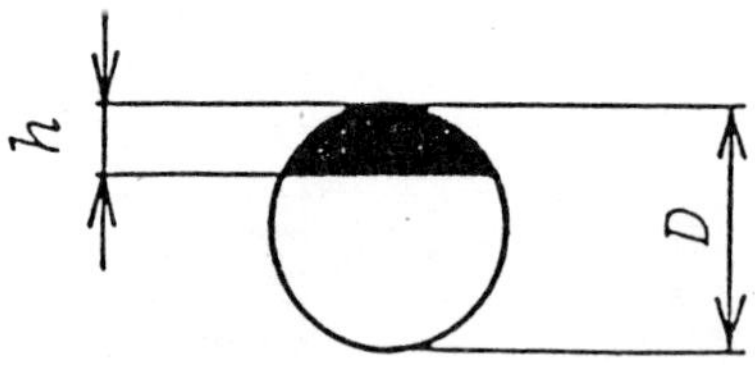

Removed bolts	0	7	9	11
Crack parameter(h/D)	0.0	0.24	0.35	0.46
κ_1	0.0	0.018	0.035	0.056
κ_2	0.0	0.011	0.030	0.038

APPENDIX

$$u = \{A,\ a,\ b,\ c,\ d,\ f,\ \alpha,\ \beta,\ \gamma,\ \delta\}^{\tau} \tag{10a}$$

$$M = -I \ (I \text{ is a unit matrix}) \tag{10b}$$

$$C = \begin{bmatrix}
j2\zeta & 0 & 0 & 0 & 0 & 0 & 0 & 0 & 0 & 0 \\
0 & -2\omega+j2\zeta & 0 & 0 & 0 & 0 & 0 & 0 & 0 & 0 \\
0 & 0 & 2\omega+j2\zeta & 0 & 0 & 0 & 0 & 0 & 0 & 0 \\
0 & 0 & 0 & 2\omega+j2\zeta & 0 & 0 & 0 & 0 & 0 & 0 \\
0 & 0 & 0 & 0 & 4\omega+j2\zeta & 0 & 0 & 0 & 0 & 0 \\
0 & 0 & 0 & 0 & 0 & 6\omega+j2\zeta & 0 & 0 & 0 & 0 \\
0 & 0 & 0 & 0 & 0 & 0 & j2\zeta & 0 & 0 & 0 \\
0 & 0 & 0 & 0 & 0 & 0 & 0 & -4\omega+j2\zeta & 0 & 0 \\
0 & 0 & 0 & 0 & 0 & 0 & 0 & 0 & 4\omega+j2\zeta & 0 \\
0 & 0 & 0 & 0 & 0 & 0 & 0 & 0 & 0 & 8\omega+j2\zeta
\end{bmatrix} \tag{10c}$$

$$K = \begin{bmatrix}
1 & \kappa_1 & -\kappa_2 & \kappa_1 & 2\kappa_2 & -\kappa_2 & 0 & 0 & 0 & 0 \\
\kappa_1 & 1-\omega^2+j2\omega\zeta & 2\kappa_2 & 0 & -\kappa_2 & 0 & 0 & 0 & 0 & 0 \\
-\kappa_2 & 2\kappa_2 & 1-\omega^2-j2\omega\zeta & 0 & \kappa_1 & 0 & 0 & 0 & 0 & 0 \\
\kappa_1 & 0 & 0 & 1-\omega^2-j2\omega\zeta & -\kappa_2 & 2\kappa_2 & 0 & 0 & 0 & 0 \\
2\kappa_2 & -\kappa_2 & \kappa_1 & -\kappa_2 & 1-4\omega^2-j4\omega\zeta & \kappa_1 & 0 & 0 & 0 & 0 \\
-\kappa_2 & 0 & 0 & 2\kappa_2 & \kappa_1 & 1-9\omega^2-j6\omega\zeta & 0 & 0 & 0 & 0 \\
0 & -\kappa_2 & \kappa_1 & 0 & 0 & 0 & 1 & 0 & 0 & 0 \\
0 & \kappa_1 & -\kappa_2 & 0 & 0 & 0 & 0 & 1-4\omega^2+j4\omega\zeta & 0 & 0 \\
0 & 0 & 0 & \kappa_1 & 0 & -\kappa_2 & 0 & 0 & 1-4\omega^2-j4\omega\zeta & 0 \\
0 & 0 & 0 & -\kappa_2 & 0 & \kappa_1 & 0 & 0 & 0 & 1-16\omega^2-j8\omega\zeta
\end{bmatrix} \tag{10d}$$

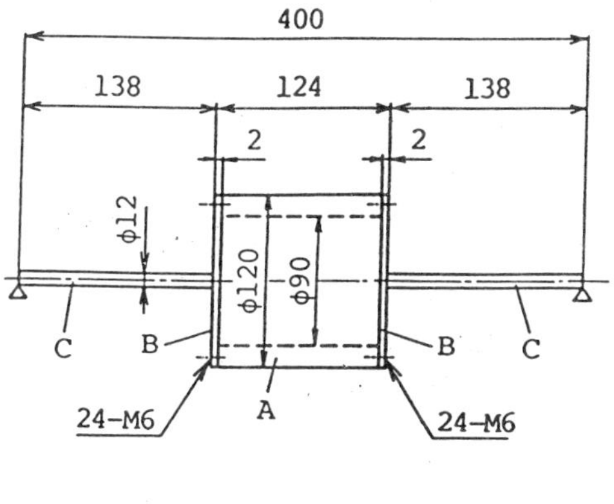

Fig 1 Rotor with imitation crack (A: cylinder, B: thin discs, C: shafts)

Fig 2 Assembled rotor

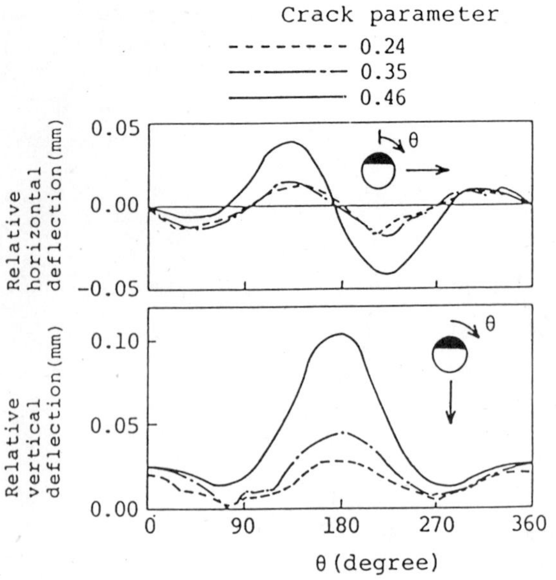

Fig 3 Variation of static deflection of rotor with imitation crack by self-weight

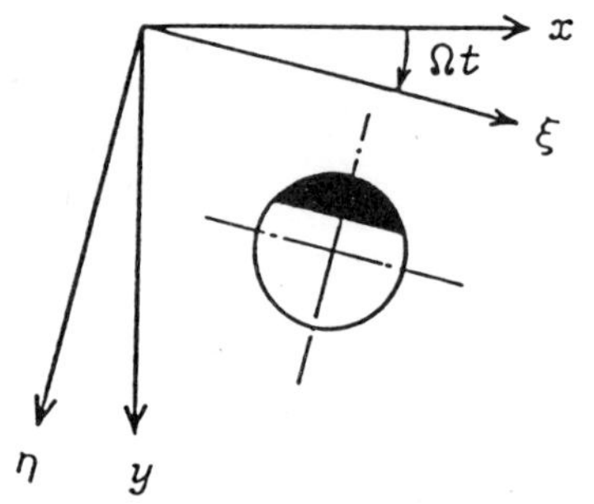

Fig 5 Coordinate system

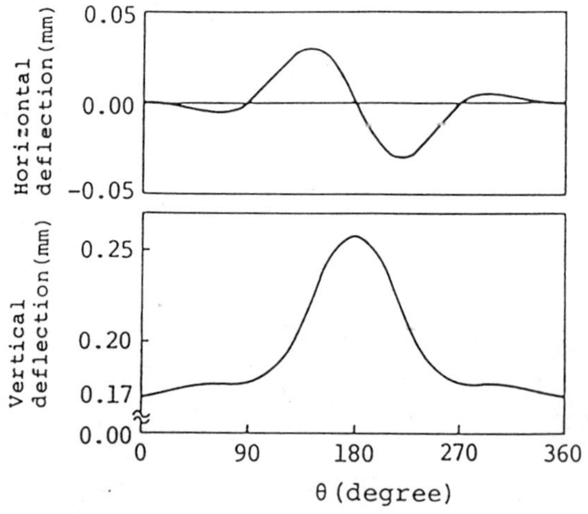

Fig 6 Calculation result of static deflection

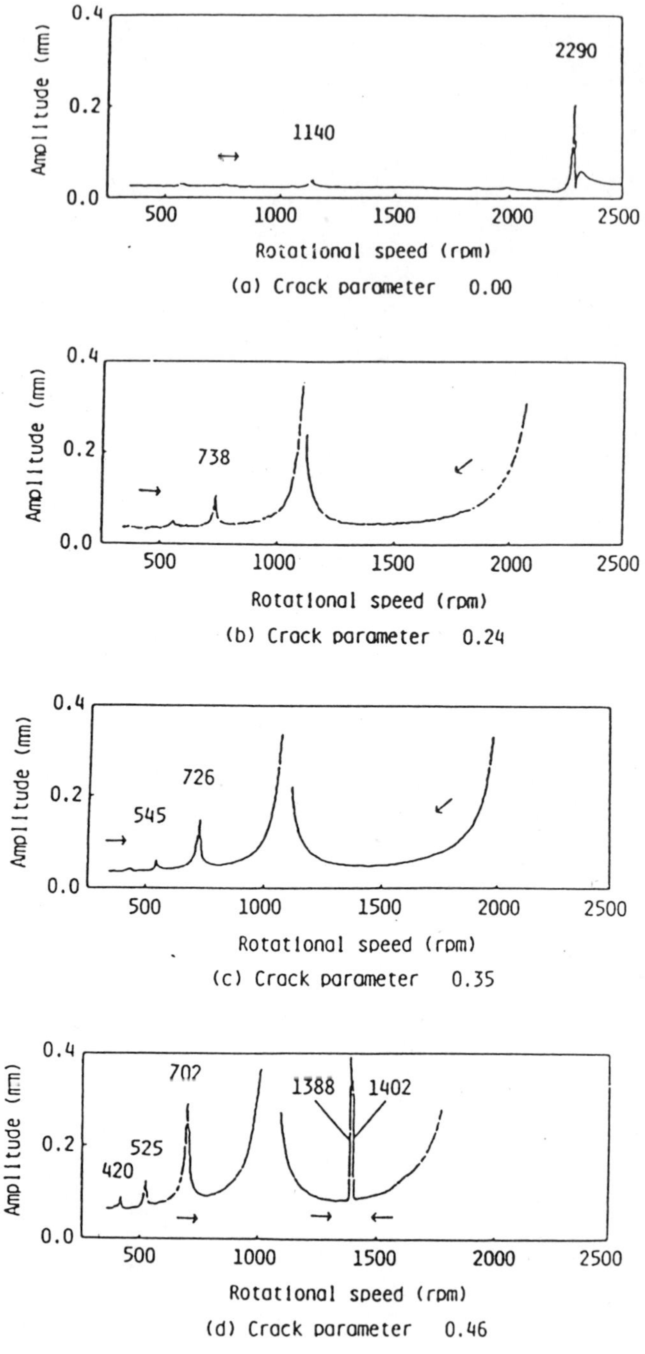

Fig 4 Horizontal amplitude–response curves

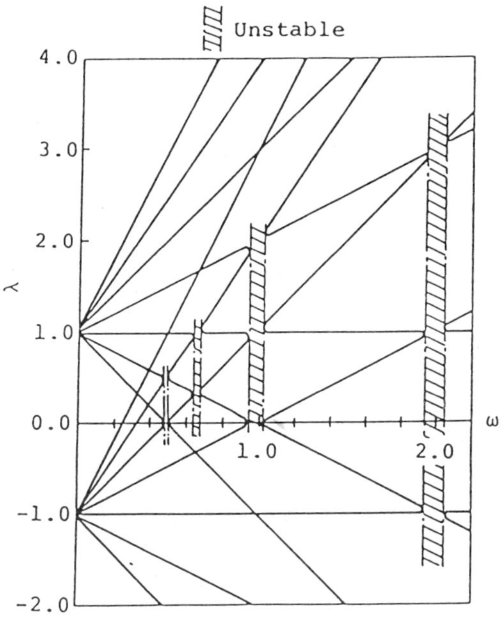

Fig 7 λ–ω graph ($\kappa_1 = 0.05, \kappa_2 = 0.04$)

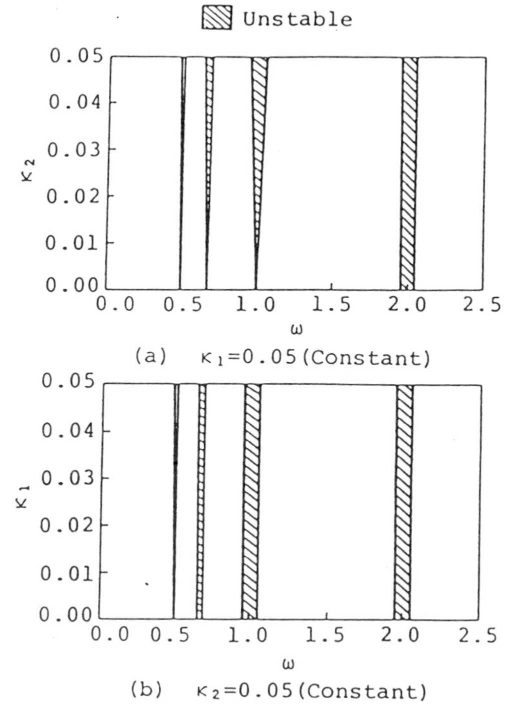

Fig 8 Influence of κ_1 and κ_2 on unstable region ($\zeta = 0$)

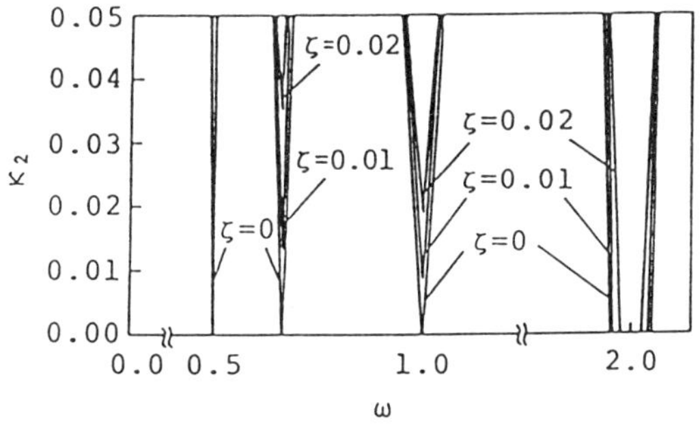

Fig 9 Influence of damping

C323/88

The development and practical application of rotordynamic analysis techniques for large steam turbine generators

P M McGUIRE, BSc and **D W PRICE**, MA
GEC Turbine Generators Limited, Rugby, Warwickshire

SYNOPSIS Theoretical rotor dynamic analysis methods have undergone continuous refinement in order to provide designers with more accurate predictions of rotor behaviour. Details of the distinctive features of the method employed by GEC Turbine Generators are described. The way in which these methods complement existing diagnostic techniques to identify various rotor related vibration phenomena is discussed and two typical case histories are described.

1 INTRODUCTION

Large steam turbine generator units for power generation, such as the one shown in Fig. 1, consist of a number of coupled rotors supported by a series of fluid film journal bearings and a complex steel or concrete foundation. Typically the overall rotor length may be in the region of 50 metres and the total rotor mass is several hundred tonnes.

The rotor dynamic behaviour is determined by the characteristics of the rotors, bearings and foundation. Knowledge of the effects of these is therefore necessary in order to ensure that satisfactory designs are produced, which will operate trouble free for the many years of continuous service that are required.

This paper describes how the well established basic theory of rotordynamics is used to predict the performance of new designs and to aid the identification of various vibration phenomena on existing plant.

The application of the theory is described by examining which terms in the basic equations are included and which can be omitted, together with analytical and experimental evidence to justify the decisions. No detailed description is given of the methods of calculating, for example, mass and stiffness matrices or bearing and foundation characteristics, since these are adequately dealt with elsewhere (1) and (2). The paper does, however, highlight some of the distinctive features of the calculation of rotordynamic behaviour as routinely carried out by GEC Turbine Generators which have not been dealt with previously.

In-service vibration phenomena encountered on large steam turbine generators are summarised and guidelines are provided for their analysis and identification. Finally two examples are given to illustrate how theory and practice are used to solve site problems.

2 THEORY

The calculation method follows the usual procedure of representing the rotor by a number of discrete reference points distributed along its length. The basic equation of motion of the rotating shaft in its fullest form is

$$\mathbf{M}.\ddot{\mathbf{x}} + \boldsymbol{\Omega}\mathbf{G}.\dot{\mathbf{x}} + \mathbf{K}.(\mathbf{x} - \mathbf{x}_b) =$$
$$\mathbf{P} + (\mathbf{I}-(\mathbf{I} + \mathbf{B}.\mathbf{F}^{-1})^{-1}).\mathbf{Q} +$$
$$\mathbf{S}.\mathbf{x} - (\mathbf{I} + \mathbf{B}.\mathbf{F}^{-1})^{-1}.\mathbf{B}.\mathbf{x} \qquad (1)$$

The various terms in this equation are discussed below.

x is a column vector containing horizontal and vertical, linear and angular displacements of each reference point. It is possible by assuming that the effect of the internal inertial moments on the stiffness terms is negligible, to eliminate the angular displacement freedoms. Calculation shows that this introduces an error in calculated critical speeds of less than 0.5 per cent. The saving in computational effort is so great that this reduction is always done.

M and **K** are the rotor mass and stiffness matrices. They are real matrices; damping in the rotor is ignored. The stiffness matrix is calculated by the usual transfer matrix beam theory, and the mass matrices by the Lagrangian method, assuming for this purpose that the dynamic deflected shape is the same as the static deflected shape. It was common practice in the past to ignore shear deflection terms in the stiffness matrix and rotary inertia of cross sections in the mass matrix. However, calculations show that this can introduce a significant error in calculated critical speeds. Consequently, it is current practice to include shear and rotary inertia in all our calculations.

ΩG is the gyroscopic matrix, Ω being the rotational speed of the shaft. It is calculated in similar manner to the mass matrix, by the Lagrangian method. Its value is twice the rotary inertia component of the mass matrix, but it acts at right angles to the mass term. This term has a negligible effect on the results of a forced response calculation for rotors of the type with which the authors are commonly concerned. It is therefore not included in the standard response calculation, but it does have a noticeable effect on stability and is retained for that calculation. It should be noted that for rotors with significantly different geometry (e.g. overhung rotors) it may be necessary to include this term in both forced response and stability calculations.

P is a column vector representing the forces applied to the rotor at each reference point. Because rotational freedoms have been eliminated it is not possible for moments to be applied. For a normal forced response calculation this vector represents unbalance distribution, increasing as Ω^2, to excite the significant modes and simulate in service effects. For a stability calculation it is set to **0**.

Q is a column vector of forces applied to the stationary part of the system at the bearing housings. This facility is incorporated to analyse the effects of external excitations (e.g. earthquakes) and also provides a means for analysing the results of impedance tests at site (see for example Section 5.1). For a normal response or stability calculation it is set to **0**.

B is the complex bearing matrix incorporating stiffness and damping terms, and cross-coupling between horizontal and vertical directions. The terms of the matrix can be calculated by solving the hydrodynamic equations for the oil films. Historically the calculations had to be backed up by careful experimental determination of bearing stiffness and damping coefficients. However recent tests at GEC on a simplified bearing have improved the understanding of bearings, and a solution of the full thermohydrodynamic equations, with account taken of the thermal distortion of the bearing surface can now give accurate calculation of the bearing matrix. (3)

F is the complex foundation matrix with inertia, stiffness, damping and cross-coupling terms. It is calculated by representing the foundation by a finite element model. For the purpose of a rotor-dynamics calculation a full dynamic analysis of the foundation is reduced to a receptance matrix for forces applied at the rotor bearing supports.

S is a matrix representing the self excitation forces which may be produced in a large turbine generator set. They are principally steam forces generated in the leakage passages of the turbine and recent work has identified the blade tips and diaphragm seals as the main sources of excitation. The forces can be calculated by means of detailed analysis of the flow in these passages, (4), (5), (6). These steam excitation forces are important in stability calculations.

$\mathbf{x_b}$ is a column vector representing a rotor bend. More precisely it contains for each mass point the difference between the elastic neutral axis and the centre of mass. If in a forced response calculation we put $\mathbf{G} = \mathbf{Q} = \mathbf{S} = \mathbf{0}$ and $\mathbf{P} = \Omega^2\ \mathbf{p}$, equation 1 can be written

$$(-\Omega^2\ \mathbf{M} + \mathbf{K} + (\mathbf{I} + \mathbf{B}.\mathbf{F}^{-1})^{-1}.\mathbf{B}).\mathbf{x} = \Omega^2\ \mathbf{p} + \mathbf{K}.\mathbf{x_b} \qquad (2)$$

This demonstrates that the effect of a bend is dependent only on the rotor stiffness matrix, and cannot therefore be defined uniquely at all speeds by a single equivalent unbalance distribution.

Equation 2 can be solved easily for a continuous bend, but a bend can be discontinuous, for example as a result of eccentric couplings. This leads to severe numerical problems, as the right hand side is now trying to represent a system of forces that will put steps in a rotor. Although formally more complicated, a better representation is

$$(-\Omega^2\mathbf{M} + \mathbf{K} + (\mathbf{I} + \mathbf{B}.\mathbf{F}^{-1})^{-1}.\mathbf{B}).\ (\mathbf{x} - \mathbf{x_b})$$
$$= \Omega^2\ \mathbf{p} + (\Omega^2\ \mathbf{M} - (\mathbf{I} + \mathbf{B}.\mathbf{F}^{-1})^{-1}.\mathbf{B}\ ) .\mathbf{x_b}$$

and to solve for $(\mathbf{x} - \mathbf{x_b})$. This remains continuous, the right hand side remains finite, and **x** is finally recovered, complete with steps by adding back $\mathbf{x_b}$.

With all the facilities described above, forced response and stability calculations can be carried out for a full rotor, bearing, foundation system, due to any system of applied forces, including rotor bends, and eccentric couplings. The actual manipulation of the equations is fully optimised to take account of the particular form of each matrix, and the sophisticated techniques employed permit full analysis over a wide range of operating speeds to be carried out at moderate cost.

3 EXPERIMENTAL JUSTIFICATION

Great care has been taken to verify all stages of the calculation wherever possible.

3.1 Rotor Characteristics - M, K and X

By suspending a rotor in flexible slings from a crane (Fig. 2) and exciting it impulsively, or with a swept sine wave the free-free natural frequencies and modes can be obtained. These give a direct check on the combined assumptions in the calculations of **M** and **K**, and of the reduction of rotational degrees of freedom in **x**.

Table 1 lists the results of such tests on HP, IP and LP rotors of a modern series of turbines. The calculated results with and without shear deflection and rotary inertia are presented, to demonstrate the importance of their inclusion. It is seen that the agreement between test and the full calculation is excellent.

		1st Mode	2nd Mode	3rd Mode
HP Turbine	Calculated (a)	9682 (1.061)	15668 (1.116)	26647 (1.207)
	Calculated (b)	9042 (0.991)	13993 (0.997)	22559 (1.022)
	Measured	9120 (1)	14040 (1)	22080 (1)
IP Turbine	Calculated (a)	6256 (1.075)	11909 (1.203)	20400 (1.273)
	Calculated (b)	5853 (1.006)	10331 (1.044)	16946 (1.058)
	Measured	5820 (1)	9900 (1)	16020 (1)
LP Turbine	Calculated (a)	5267 (1.064)	7934 (1.087)	18204 (1.141)
	Calculated (b)	4960 (1.002)	7231 (0.991)	14595 (0.914)
	Measured	4950 (1)	7296 (1)	15960 (1)

Table 1. Natural frequencies of free-free modes of turbine rotors - rev/min. Comparison of measured frequencies with frequencies calculated (a) not including and (b) including shear deflection and rotary inertia terms.

3.2 Bearing Matrix - B

Bearing coefficients are measured on a large bearing rig, capable of testing full size bearings at a wide range of operating conditions (7). Thus effects of scale do not invalidate the results. Recent tests on a full-size, but simplified single arc bearing have shown the importance of using a full thermohydrodynamic analysis for bearing analysis, and also led to an understanding of the effect of bearing thermal distortion on calculated coefficients, particularly at low Sommerfeld No. (high load). Fig. 3 shows the measured coefficients compared with calculated values using successively a normal hydrodynamic solution without distortion, a full thermohydrodynamic solution without distortion, and then with distortion. It can be seen that the full calculation gives very much improved agreement between test and theory. When the groups in which the coefficients are used in calculations are considered, the agreement between measurement and calculation is even better than for the individual coefficients.

3.3 Self-Excitation Matrix - S

The ability to calculate steam excitation forces generated in leakage paths is relatively new and laboratory verification of the methods in all cases is not yet available. Measurement of these forces at site is only possible by observing their effects on rotor stability, and steam excited instability is very rare on modern machines. Supporting evidence for our calculations comes from a successful diagnosis and cure of an instability problem on an older unit. This case is discussed more fully in section 5.2.

3.4 Overall Accuracy

A check on foundation calculations and the combination of rotor, bearing and foundation models is the comparison of site response to calculated response. This is somewhat complicated by the fact that the site response is to an unknown residual out-of-balance, whereas the calculated response has to be to a prescribed out-of-balance distribution. However, the principle features of peak responses should occur at the same speeds in each.

Figs. 4 and 5 show the calculated response for the coupled HP rotor of a 660 MW unit. Fig. 4 assumes a rigid foundation, and thus indicates the critical speeds of the rotor itself; Fig. 5 shows the effect of the foundation, with its large number of natural frequencies. Superimposed on this calculated response curve is a run-down trace from site. The rundown was measured at 45^0 to the horizontal and should reflect features from both horizontal and vertical responses. It can be seen that the agreement between the peaks on the curves is good, giving high confidence in the overall methods of calculation.

4 ANALYSIS & IDENTIFICATION OF IN-SERVICE VIBRATION PHENOMENA

The analysis techniques described above can be used to simulate the response of rotor systems to various excitation sources. They are therefore of assistance in diagnosing the causes of particular phenomena which may produce unacceptable rotor vibration, and in assessing the effectiveness of possible remedial actions. Such methods complement but do not replace established diagnostic techniques which are used to analyse vibration data obtained from operational plant.

This section briefly describes how some of these phenomena can be identified and possible remedial actions.

4.1 Vibration Frequency

The single most important parameter when analysing vibration data is the predominant frequency (or frequencies). There are three main frequency regions which categorise rotor vibration, namely:

a) vibration frequency corresponding to shaft rotational speed (one per rev.) This covers the majority of cases encountered. Vibrations in this group are invariably related to some form of rotor unbalance and therefore in most but not all cases they can be counteracted or eliminated by improvements to rotor balance.

b) vibration frequency corresponding to integer multiples of shaft rotational speed. This is often related to asymmetry of the rotor, for instance a poorly compensated generator rotor.

c) vibration frequency less than rotational speed (sub-synchronous) and generally not simply related to it. Vibrations which occur in this group are generally of an unstable nature and may increase to damaging levels in very short periods of time.

Identification of the frequency characteristics is the first step in any investigation into rotor vibration.

Thereafter it may be necessary to examine the vibration vectors (amplitude and phase) at various axial positions along the rotor and establish whether these remain steady or vary with time or changes in particular operating conditions (e.g. temperature, load). If there is some variation it is important to establish whether this follows a consistent pattern or is random in nature.

4.2 Inherent Unbalance

It is inevitable that there will be some inherent unbalance present in a coupled rotor system even though the individual rotors have been finely balanced by the manufacturer. The unbalance may be due to the cumulative effects of the residual unbalance in each rotor, permanent bends, lack of concentricity between individual rotors due to assembly errors or inadequate tolerances on coupling faces and bolt holes, or a combination of all three.

In this case, the unbalance will produce steady one per rev vibration vectors which are repeatable and do not vary with time and changes in operating conditions at a given speed. The most common remedial action is to reduce the excitation by improving the rotor balance. However, if the major cause of the unbalance is lack of concentricity between rotors the magnitude of the induced unbalance may be in excess of that which can be reasonably corrected by in situ balancing alone and the correct solution in that case is to improve the concentricity.

4.3 Change of Unbalance

Changes in unbalance can occur for a variety of reasons, but normally fall into the following three basic categories:

a) component loss, e.g. turbine blades

b) induced rotor bends

c) component movement

4.3.1 Component Loss

If a component such as a turbine blade becomes detached from the rotor there will be a step change in rotor unbalance. The effect on the one per rev. component of vibration will be influenced by the magnitude, axial location and angular position of the initial residual unbalance relative to the induced unbalance. Dependent on these three factors the amplitude of the measured vibration may increase, decrease or remain unchanged and monitoring of vibration amplitude alone will not always show that such a change has taken place. It is therefore necessary to examine both the amplitude and phase of the one per rev. vibration.

It is not possible to identify precisely the axial position along the rotor where the change of unbalance has occurred because the same modal unbalance effect can be obtained by means of a number of different unbalance combinations. However, in practice, a good assessment can be made by taking measurements at a number of positions, and at speeds where the rotor line adopts different modal shapes. Further confirmation can then be obtained by using the analysis techniques to simulate the vibration behaviour.

If a change of unbalance has been diagnosed it is important that steps be taken to ascertain the reason for the change, since although the resultant vibration levels may not be excessive, they may indicate that there is potential for further unacceptable changes.

4.3.2 Rotor Bends

a. Permanent bends

Permanent bends in rotors provide a steady one per rev. excitation of the rotor system and cannot be separated by vibration measurements from the effects of inherent unbalance (see section 4.2). The presence of a permanent bend would normally be detected by slow roll shaft eccentricity measurements.

In those cases where the excitation produced by the bend is not excessive a satisfactory reduction in vibration can be achieved by an in situ trim balance. As shown in section 2 excitation produced by a bend is independent of rotational speed and therefore there is only one speed at which the effect of the bend can be completely compensated by balancing. However for most practical cases the errors involved at other speeds are not normally significant.

It may not be practical to trim balance rotors with larger bends and in these cases the necessary remedial action involves heat treating the rotor to minimise the bend followed by light machining.

Some high temperature rotors can develop a progressively increasing bend due to differential creep of the rotor material. It is only possible to discriminate between this and a permanently bent rotor by monitoring over a long period of time. However, as the creep bend develops at a slow rate a satisfactory solution can be achieved by occasional trim balance adjustments. Heat treatment straightening of creep bends is a less satisfactory solution because experience has shown that there is a period of accelerated creep immediately following this process which results in the original creep bend returning.

b. Temporary bends

From time to time transient bending of the rotors may take place and this can be identified by examination of the one per rev. vibration vectors. The two main reasons for transient bending are either loss of clearance leading to contact between the rotor and the stator at critical points such as glands; or changes in temperature.

Loss of clearance combined with the orbit of the shaft will eventually result in intermittent contact between the rotor and stator. If this contact is severe, frequency components at harmonics of shaft rotational speed will be observed in the vibration spectrum. Generally, the construction of steam turbine glands is such that this is of secondary importance and the most significant effect is due to the localised frictional heating of the shaft surface which causes the rotor to bend. The consequent unbalance effect vectorially adds to the inherent unbalance already present in the rotor system to produce a new resultant unbalance vector.

The dynamics of the shaft orbit are such that for a particular rotor system at a given speed there is a constant relationship between the angular position of the unbalance vector and the point of contact (and hence the plane of the transient bend). Therefore, the angular position of the resultant unbalance vector will be different from that for the initial unbalance vector and there will be a corresponding new point of contact lagging the first. As long as contact is maintained this process continues and therefore over a period of time the resultant unbalance vector will progressively move around the rotor against the direction of rotation. If the vibration amplitude and phase are plotted over a period of time on a vector diagram, this effect will appear as a characteristic rotating vector which, provided there are no other significant effects present, will follow a cyclic variation of increasing and decreasing vibration.

Two special cases exist when the point of contact is either in phase or 180° out of phase with the initial unbalance. The former case normally occurs when the rotor is running below its first critical speed and represents the most dangerous condition, since the initial unbalanceand induced unbalance are exactly in phase. Therefore no rotation of the unbalance vector will occur and the deflection will progressively increase until the shaft yields locally and on cooling the rotor develops a permanent bend. For the second case, when the initial unbalance is 180° out of phase with the point of contact, the effects of the induced bend will oppose the initial unbalance and will therefore provide a self balancing effect which will reduce the size of the shaft orbit and temporarily eliminate the contact between the rotor and stator.

Bending induced in shafts due to heating effects (other than rubbing) generally occurs at a fixed angular position and therefore, although the magnitude of the induced unbalance changes, the angular position will remain constant. In most cases this change is transient and varies in a repeatable way with temperature. It may be due, for example, in turbines, to differences in coefficient of expansion or heat transfer on each side of the rotor; and in generators to uneven distribution of cooling gas in rotor windings or variation in friction coefficient between the windings and the rotor body. The main characteristic of thermally induced rotor bends is that the variation of vibration amplitude and phase as the bend develops follows a straight line relationship when plotted on a vector diagram instead of the rotating vector produced by a rub.

In those cases where the rotor bends due to loss of clearance, the only satisfactory remedial action is to ensure that the unit is built with adequate clearance, or that the reasons for loss of clearance are identified and rectified. Also the gland design should be reviewed and where necessary modified to reduce the sensitivity of the rotor to a gland rub. If transient bends occur due to thermal effects these should be minimised as far as possible. However, if it can be shown that the thermally induced vibration follows a consistent and repeatable pattern, then a satisfactory compromise may be achieved by carrying out a compensating balance which will ensure that at all times the vibration remains within acceptable limits.

4.3.3 Component Movement

Rotors having components which may move in service are often characterised by inconsistent one per rev. vibration which may vary significantly from run to run. The often quoted classical example is entrapped fluid within a rotor bore. Alternatively, small movements of shrunk-on components may occur as the interface stresses are modified by rotational effects or transient heating during start-up or shutdown. The main identification feature of this is either inconsistent behaviour of the vibration vectors or temperature dependency, although in this case the symptoms may be similar to those for bearing misalignment (see section 4.4).

In some cases the movement may not itself be a problem as far as the mechanical integrity of the turbine is concerned, provided that it can be shown by carrying out repeat runs that the worst combined effects of the initial inherent unbalance and the largest transient unbalance due to the movement do not exceed a level which is regarded as satisfactory for long term operation of that unit. In this respect, the use of in-situ balancing to produce a better than normal state of balance may be used to enable the unit to continue in service until the next suitable overhaul date when the reasons for the movement may be found and rectified.

4.4 Bearing Misalignment

Alignment changes can occur on large steam turbine generator sets for a variety of reasons, e.g. push-pull forces acting on bearing support pedestals/beams as cylinders expand and contract, external pipe thrust forces, temperature differentials, vacuum loads. The primary effect of such alignment changes is to modify the static bearing load and hence the dynamic characteristics of the oil film. This may change the sensitivity of the rotor system to unbalance and other external excitation forces, and if excessive, may cause rotor natural frequencies to fall within the rated speed range.

It is necessary to differentiate between the effects of changing sensitivity and changing unbalance. This can be done by correlating changes in vibration with various operating parameters such as unit load, temperatures, time, vacuum etc, by measurements of shaft position within the bearing, jacking oil pressures and bearing height movements. Furthermore, analysis of run down data can assist by showing that critical speeds have changed.

As misalignment effects become more severe, erratic sub-synchronous vibration may be encountered. This is dealt with in greater detail below, (section 4.5).

In those cases where the machine vibration is adversely affected by misalignment, efforts should be made to limit the alignment changes. On new machines this can be achieved by ensuring that support beams are designed with adequate stiffness to resist the applied forces and that the magnitude and distribution of these forces is controlled. For existing machines, care should be taken to ensure that all sliding surfaces are in good condition, keys have adequate clearance and that there are no abnormal temperature differentials due to steam leaks or poor lagging.

Some degree of misalignment may be inevitable as the machine operating conditions change. In such cases, provided that the changes can be shown to be consistent, a satisfactory solution may be achieved by deliberately misaligning the bearings to compensate for the changes.

Alternatively, it may be necessary to instal different bearing types which are less susceptible to changes in bearing load. Such bearings are generally of the parasitic load type which are capable of maintaining an active load within the bearing even though the gravity load from the rotor has been removed.

4.5 Rotor Stability

Sub-synchronous, unstable vibrations of turbine rotor systems may be encountered due to inherent instability of the hydrodynamic journal bearings or because the overall system damping provided by the rotor / bearing systems becomes insufficient to contain the effect of external destabilising forces produced by steam flow variations within the turbine. The former case is the classic oil whirl situation and the latter is commonly referred to as steam whirl. In some cases the non-linear nature of the bearings may prevent total instability and introduce some limit cycle behaviour.

For new designs the most effective way of improving the stability margin is by the use of stiff rotor designs, particularly for high pressure rotors where the potential for steam excitation is the greatest. Increases in the system damping for both new and existing designs can be obtained by using alternative bearings, such as the parasitic load types referred to above. These include, fixed and tilting pad, multi-wedge bearings.

In those cases where steam excitation effects are important, significant reductions of the excitation forces can be achieved by careful attention to the detailed design of internal clearance passages and to the setting of the rotor position within the cylinder. One such example of this is discussed below.

4.6 Rotor Asymmetry

Asymmetry of the rotor system relative to its axis of rotation will cause changes in the self weight bending deflection of the rotor during each revolution. This results in a steady twice per rev. excitation of the rotor system. The most common example of this is in two-pole generators where the effect is minimised by machining compensating slots in the pole faces.

Transverse cracks in a rotor will produce asymmetry and therefore regular monitoring of the vibration spectrum for components at harmonics of running speed provides a means of identifying the growth of such cracks. These measurements are also complemented by run down analysis, which takes advantage of the dynamic magnification at resonance peaks which will be excited, not only by the fundamental unbalance excitation, but also by the harmonic components at corresponding sub-multiples of shaft speed. However, the changes obtained, even for large crack depths, are relatively small and sophisticated trend analysis techniques are often necessary in order to make meaningful predictions.

Theoretical assessments of the effects of rotor asymmetry indicate that parametric instability of the rotor system can occur. However the authors have no evidence of this happening in practice on large steam turbine generator rotor systems.

5 CASE STUDIES

5.1 Rotor Critical Speed

On an overseas 1200 MW, 1800 RPM machine, a larger than usual vibration of the LP / Generator beam occurred during the initial commissioning. This vibration appeared to reach a peak at running speed, and it was suspected that this was due to a flexural rotor critical speed. Unfortunately, for reasons unconnected with the turbine generator, it was not possible to run the machine for a considerable time.

During this downtime, a shaker test on the stationary unit was commissioned, with the rotor supported by jacking oil. These tests showed a major resonance at 1600 RPM which it was feared might move to 1800 RPM when the unit was at full speed.

The stationary system was analysed using the standard models for the rotor and foundation mass, stiffness and damping. For the bearings, on jacking oil, the vertical support was assumed to be a stiff spring and all other oil film stiffness and damping coefficients were taken as zero. The calculation reproduced the major features of the measured response closely, thus confirming the accuracy of the basic model. This is shown in Fig. 6 where for the sake of clarity the test curve has been arbitrarily scaled to lie above the calculated curve.

Having validated the model in this way the rotating system was analysed, using a full bearing coefficients representation of the oil films, and unbalance forces applied directly to the rotor. This showed that the full rotating system was clear of any major flexural critical speeds near to the running speed (Fig. 7) and that the suspected resonance was a minor structural resonance responding to inherent unbalance in the rotor system. It was judged that a reduction in vibration levels could therefore be obtained by balancing to reduce the excitation strength. On recommissioning, the unit was successfully balanced at site and a satisfactory reduction in vibration obtained.

5.2 HP Rotor Instability

Load dependent HP rotor instability was experienced on a CEGB 500 MW unit for which,unlike current practice, the HP rotor was a flexible design with its first flexural critical speed well below rated speed. Experience with other similar units had shown that the instability was sensitive to relatively small changes and in this particular case it could be partially controlled by closing one of the two governing valves feeding the lower half of the cylinder inlet, simulating partial admission.

Development of methods for calculating steam forces in leakage passages enabled an estimate of destabilising forces to be made. This showed that the largest forces were generated in the blade tip seals and that these were proportional to steam flow (and therefore load) and at right angles to the rotor displacement. In order to overcome the immediate problem with the operating unit it was decided to take advantage of these steam forces. The dynamic component of the force at right angles to rotor displacement is a destabilising force, but the static component modifies the bearing load. Consequently a static displacement of the HP rotor relative to the cylinder was made in a horizontal direction. This resulted in a static downward force of approximately 20 000 N on the HP rotor at full load. This extra bearing load served to increase the oil film damping and stabilise the rotor without the need to close a governing valve.

To provide greater security on a second similar unit fitted with new blading, the tip seal arrangement was analysed in greater detail using the analysis method described earlier. The gap behind the tip seals was opened out, and calculation showed that this should be sufficient to prevent an instability problem. In addition another modification developed by the CEGB to reduce diaphragm shaft seal forces by modification of gland inlets was also fitted, and the rotor was built displaced in the cylinder as described above. On return to service the machine ran satisfactorily up to full load without instability.

6 CONCLUSIONS

Theoretical analysis techniques for the prediction of rotordynamic behaviour have been developed and improved so that they conform with the results of experimental verification tests. These methods are used during the design of new rotor systems and complement established methods for diagnosing and solving specific vibration problems which may occur on operational plant.

7 ACKNOWLEDGEMENT

The authors are grateful to GEC Turbine Generators Ltd for permission to publish this paper.

REFERENCES

1. THOMAS, G.B. Rotordynamics
Conference on vibration in rotating systems
I.Mech.E. February 1972.

2. McGUIRE, P.M. The Dynamic Design of Large Turbine Generator Rotor Systems
Conference Steam Turbines for the 1980's
I.Mech.E. October 1979.

3. COWKING, E.W. Thermohydrodynamic Analysis of Multi Arc Journal Bearings.
Tribology International August 1981.

4. WYSSMAN, H.R. PHAM, T.C. and JENNY, R.J. Prediction of Stiffness and Damping Coefficients for Centrifugal Compressor Labyrinth Seals.
Trans. ASME, October 1984, Vol. 106, p.920

5. THOMAS, H.J. Unstable Oscillations of Turbine Rotors due to Steam Leakage in the Clearance of Sealing Glands and the Buckets.
Bulletin de l'AIM, 1958, Vol. 71 No. 11/12 p.1039.

6. BENCKERT, H. and WACHTER, J. Flow Induced Spring Constants of Labyrinth Seals.
Conference-Vibrations in Rotating Machinery.
I.Mech.E. September 1980.

7. MORTON P.G. Dynamic Characteristics of Bearings, Measurement Under Operating Conditions.
GEC Jnl. Science and Technology 1975 42 (1)

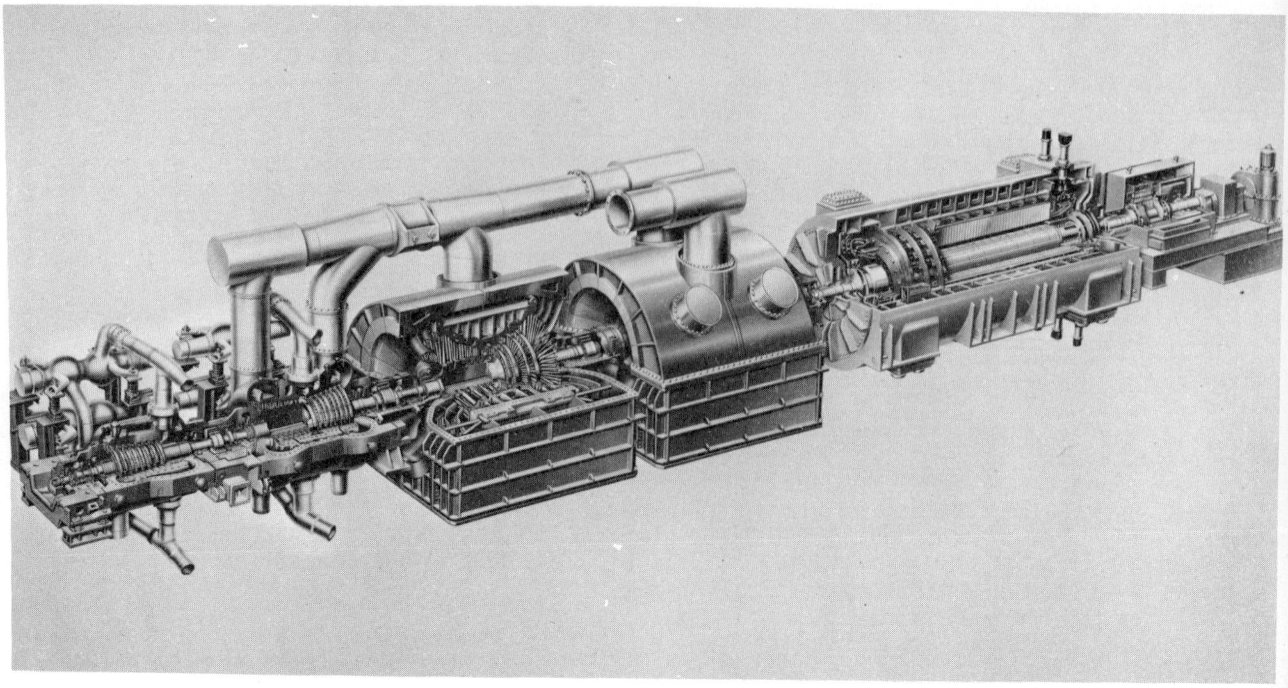

Fig 1 Sectional diagram of a modern 500–700 MW turbine generator unit

Fig 2 IP rotor supported on flexible slings for free–free vibration test

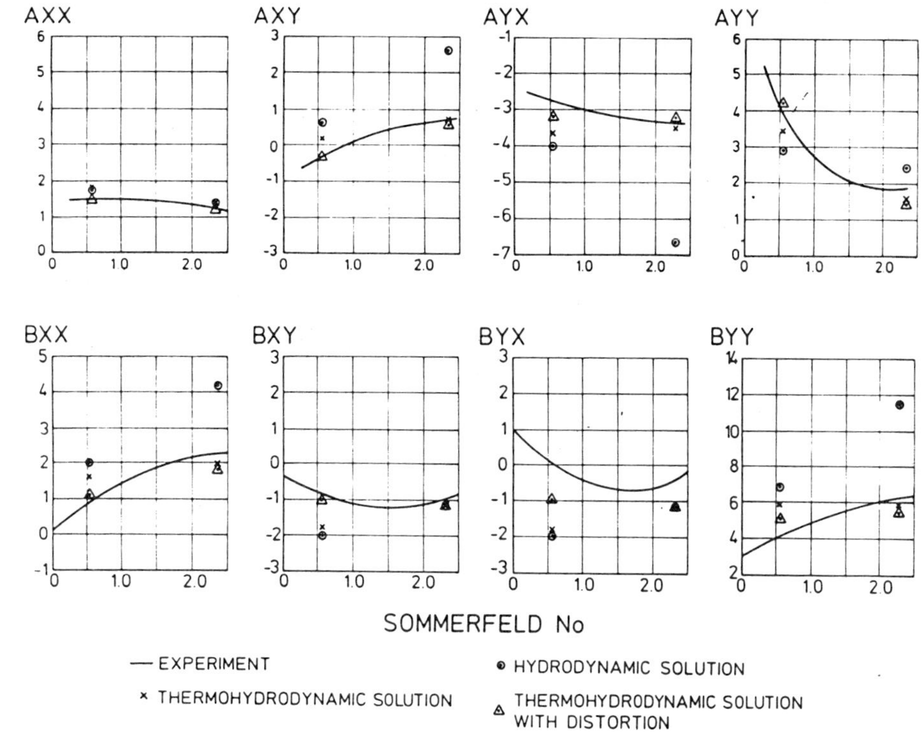

Fig 3 Comparison of calculated and measured bearing stiffness (A) and damping (B) coefficients

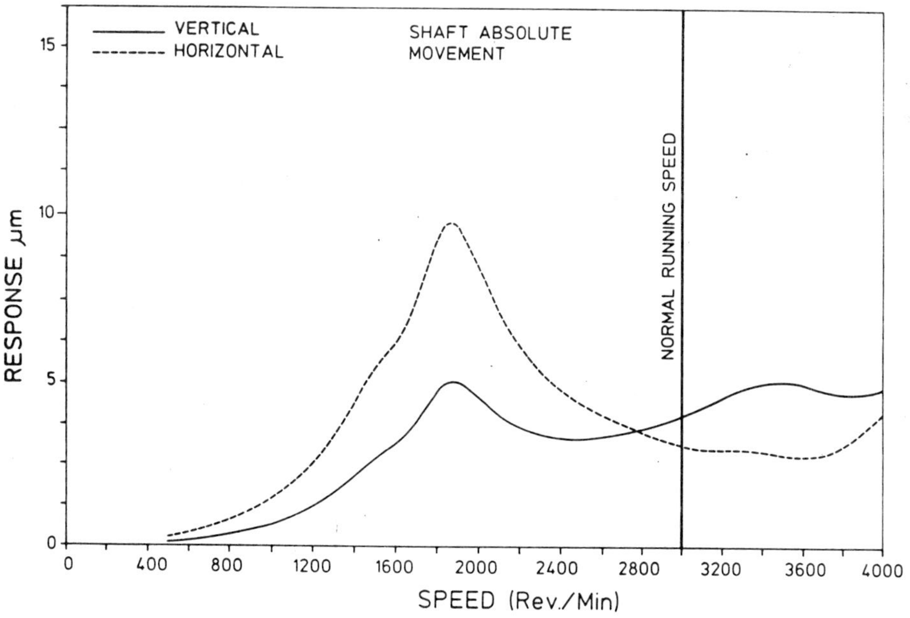

Fig 4 Typical response calculation for a coupled HP turbine rotor on rigid foundations

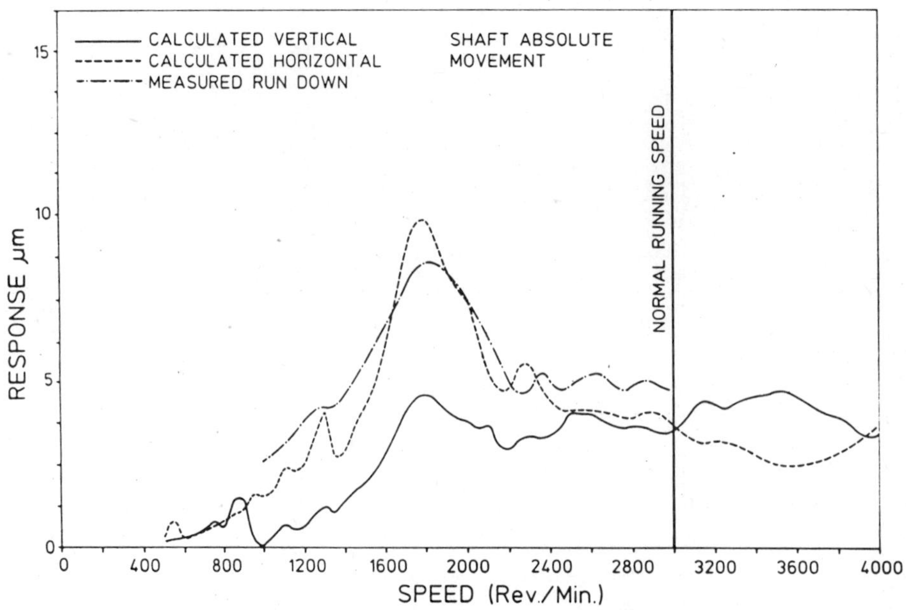

Fig 5 Typical response calculation for a coupled HP turbine rotor on its foundation compared with site data

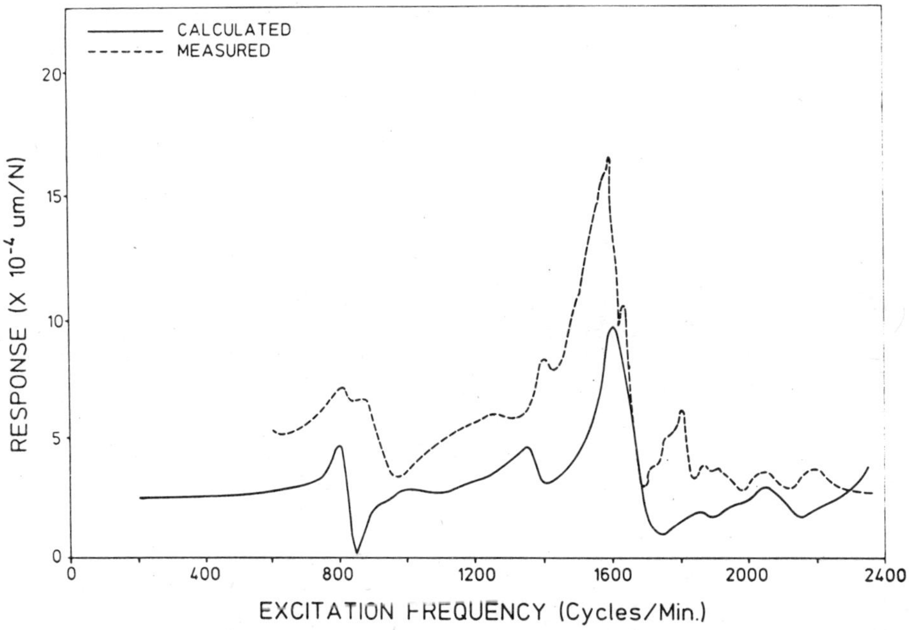

Fig 6 Response of the LP/generator beam of a 1200 MW unit to excitation of the bearing housing — rotor stationary

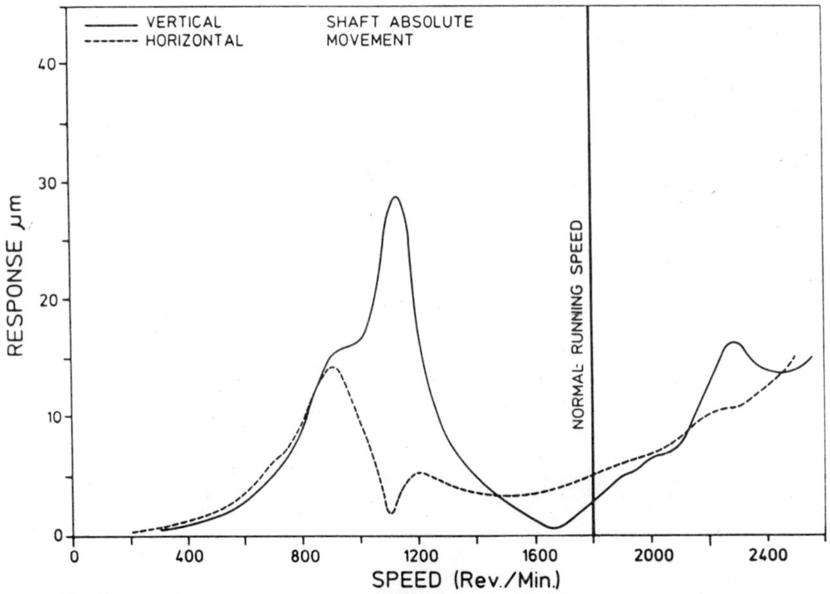

Fig 7 Calculated response of LP rotor to out-of-balance for the operating 1200 MW unit — foundation fully represented